PLANT AND INSECT NEMATODES

PLANT AND INSECT NEMATODES

edited by

William R. Nickle

United States Department of Agriculture
Agricultural Research Service
Beltsville Agricultural Research Center
Beltsville, Maryland

MARCEL DEKKER, INC.   New York and Basel

Library of Congress Cataloging in Publication Data
Main entry under title:

Plant and insect nematodes.

Includes index.
1. Plant nematodes. 2. Nematode diseases of plants.
3. Nematoda. 4. Insects--Parasites. 5. Insect control--
Biological control. 6. Plant nematodes--Control. 7.
Plant nematodes--United States. 8. Nematode diseases
of plants--United States. 9. Nematoda--United States.
10. Insects--United States--Parasites. 11. Insect con-
trol--Biological control--United States. 12. Plant nem-
atodes--Control--United States. I. Nickle, William R.,
[date]
II. Title: Insect nematodes.
SB998.N45P63 1984 632'.65182 84-4937
ISBN 0-8247-7079-X

MARCEL DEKKER, INC.
270 Madison Avenue, New York, New York 10016

Current printing (last digit):
10 9 8 7 6 5 4 3 2 1

PRINTED IN THE UNITED STATES OF AMERICA

Foreword

During this period of accelerated research in agriculture in which new technology and concepts are being developed, it is appropriate that a comprehensive text on plant- and insect-parasitic nematodes be made available.

Nematodes have long been recognized as deterrents to crop production and parasites of insects which in turn affect crop yields and are detrimental to man and animals. The contributors to the various chapters have been selected for their expertise and active research experience in the selected topic areas. Their knowledge of the subject matter is reflected in the in-depth coverage of selected subject areas that have been categorized under the major crops and insect orders. This book places strong emphasis on crops and insect problems of the North American continent. However, many of the problems and projected solutions cross continental borders and provide information that is readily adaptable for use by a wide range of research workers, advanced students, and technical experts. Since many of the contributors are from North America, the text addresses current agricultural problems in the United States and provides information on ways to control specific plant-parasitic nematodes or to use nematodes as biological agents for the control of insects. In most cases, the nematodes discussed in the book extend geographical borders and their proposed control measures have broad applications.

The history, development, and importance of plant and insect nematology are discussed in separate chapters. Photographs of some of the pioneers in these two fields are presented.

An example of the wide distribution of nematode problems is seen in the presence of the soybean cyst nematode that was originally described in Japan, but found locally in the United States (North Carolina) and later discovered in the mid-South region of the United States. Currently, the soybean cyst nematode infests soils throughout the Southeastern United States, the Mississippi Valley, and with localized infestations in Iowa, Minnesota, Wisconsin, Delaware, and Maryland. Thus far, the development of resistant cultivars appears to be the best solution for raising soybeans in infested soils. Considerable progress is evident in the twofold to threefold increase in yields when resistant cultivars are used instead of susceptible cultivars.

The potato cyst nematode stands high on the list of important plant-parasitic nematodes in the United States, not because of its wide distribution, but because of its confinement through effective federal quarantine which has kept the infestations isolated in one region of the United States. Since potatoes are grown throughout the country, the potential for spread and econo-

mic damage is significant, especially when compared to the extensive infestations that occur in Europe and the expenses incurred for its economic control.

The chapter on citrus describes an old but persistent disease of citrus
termed *spreading decline* which is caused by the burrowing nematode. The
drastic method of "pull and treat" has been effective, but has involved considerable resources. Infested plants were removed and burned, the original
field site cleaned of root debris, and the soil treated with nematicides prior to
placing the land in the fallow state for up to two years. Currently, these procedures have been modified to emphasize nematode exclusion through regulatory programs in Florida. The authors provide a balanced view of the problems associated with other serious nematodes, such as the citrus nematode and
the lesion nematode which cause serious crop losses and reduction in root vig-
or. The challenges of seeking new nematode control measures are implied in
the loss of certain nematicides and the search for possible biological control
measures.

Nematode parasitism of corn and rice, two of the world's most important food crops, are covered very well by experienced field nematologists
actively working on these problems. Experts from Texas deal with the nematodes, such as root-knot nematodes, affecting cotton production, and workers in North Carolina discuss nematode problems in tobacco production. Nematode parasites of alfalfa and other forages, wheat and other cereals, and
grasses, including turf, are major crops affected by plant-parasitic nematodes.
These associations are discussed in detail with the aid of excellent photographs. The most serious nematode parasites of the normally difficult area
of vegetable and truck crops are skillfully brought out by two USDA experts.
In the southern part of the United States the cultivation of peanuts is important, and a short chapter is devoted to the nematode parasites of peanuts.
Nematodes accumulate in large numbers on perennials such as peaches and
other tree crops, and their treatment is discussed at length. In the western
part of the United States, sugarbeets are widely grown and nematode parasites of this crop are discussed. Sugarcane nematode problems are taken up
by a Louisiana expert.

Although this book is structured to emphasize the major crop plants
and insects affected by nematodes, specific information is available on the
taxonomy and morphology of the nematode, host symptomology under nematode
stress, life cycle of the nematode, economic loss in crop production, distribution of the nematodes, nationally and internationally, and various biological
or chemical control measures. Within the scope of nematode damage to crops,
there are specific examples of host-parasite interactions that involve other
disease-causing agents, such as fungi, bacteria, and viruses. It is recognized that several plant-pathogenic fungi and bacteria are more destructive
in the presence of plant-parasitic nematodes than as single causal agents of
plant diseases.

Although several dagger nematode species cause extensive root damage
by restricting root elongation by inhibiting root tip tissues, the dagger nematode of grape was the first nematode proven to be a vector of a plant virus,
the grape fanleaf virus. These observations were the precursor to a number
of studies of other crops and of two families of nematodes that have since
been implicated in transmitting rod-shaped, rigid particles (Tobraviruses)
and nepo viruses that are polyhedral shaped and isodiametric.

Still another disease complex is illustrated by the pine wilt disease that
involves the pinewood nematode which transfers from a wood-boring beetle
to the growing tips of pine trees. During insect feeding, the nematodes en-

Preface

This book is an attempt to bring together the knowledge of many of the leading workers on plant and insect nematodes into a comprehensive volume for students, researchers, and administrators. It deals with the major nematode pests of our North American food crops and the main nematode parasites of some of the major pest insect groups. No single author can handle such a wide field of nematology. An edited book is required, written by carefully selected internationally eminent contributors. Several of the authors have spent their entire professional careers doing research combating the nematode parasites of a single crop or pest insect, and therefore have the practical knowledge necessary for such an undertaking.

Plant pathologists, entomologists, nematologists, and other agricultural scientists will have a need for the content of this book. The few available books are out of date and most are discipline rather than crop or pest oriented. An enormous number of published papers have been sifted through to bring out the essence of that which is important to the scientists of today, who do not have the luxury of unlimited time and funds for a general literature search or the knowledge needed for the weeding-out process.

Each major crop is represented with its major nematode parasites, associated symptomatology, life cycle, economic loss (which averages 10%), distribution, and up-to-date control measures. In an effort to minimize overlap, the insect nematode chapters are separated into major insect groupings such as nematode parasites of Lepidopterans, Orthopterans, etc.

I am indebeted to the 35 nematologists who have contributed to this book, for their knowledge, cooperation,and patience and to Ms. Shari Blohm for helping to keep the chapters organized and to keep them moving toward completion. I also want to thank my wife Cathy for her assistance with some of the typing and for helping to proofread the galley.

William R. Nickle

Contributors

Kenneth R. Barker, B.S., M.S., Ph.D. Professor, Department of Plant
Pathology, North Carolina State University, Raleigh, North Carolina

Robin A. Bedding, B.Sc., A.R.C.S., Ph.D., D.I.C. Principal Research
Scientist, Division of Entomology, Commonwealth Scientific Industrial Organisation, Tasmanian Regional Laboratory, Hobart, Tasmania, Australia

Wray Birchfield, Ph.D. Research Plant Pathologist, United States Department
of Agriculture, Agricultural Research Service, Department of Plant Pathology,
Louisiana State University, Baton Rouge, Louisiana

B. B. Brodie, Ph.D. Research Plant Pathologist, United States Department
of Agriculture, Agricultural Research Service, Department of Plant Pathology,
Cornell University, Ithaca, New York

W. M. Dowler, B.S., M.S., Ph.D.* Staff Scientist, National Program Staff,
United States Department of Agriculture, Agricultural Research Service,
Beltsville, Maryland

George Fassuliotis, B.S., M.S., Ph.D. Zoologist, United States Vegetable
Laboratory, United States Department of Agriculture, Agricultural Research
Service, Charleston, South Carolina

Christopher J. Geden Department of Entomology, University of Massachusetts,
Amherst, Massachusetts

Roger Gordon, B.Sc., Ph.D. Associate Professor, Department of Biology,
Memorial University of Newfoundland, St. John's, Newfoundland, Canada

---

Present affiliation
*Laboratory Director, Plant Disease Research Laboratory, United States
Department of Agriculture, Agricultural Research Service, Frederick,
Maryland

G. D. Griffin, Ph.D. Nematologist, Crops Research Laboratory, United States Department of Agriculture, Agricultural Research Service, Utah State University, Logan, Utah

Charles M. Heald, Ph.D. Research Nematologist, United States Department of Agriculture, Agricultural Research Service, Weslaco, Texas

John P. Hollis, Jr., B.S., M.S., Ph.D.* Professor, Department of Plant Pathology and Crop Physiology, Louisiana State University, Baton Rouge, Louisiana

A. W. Johnson, B.S., M.S., Ph.D. Supervisory Research Plant Pathologist, United States Department of Agriculture, Agricultural Research Service, Coastal Plain Experiment Station, Tifton, Georgia

Harry K. Kaya, Ph.D. Associate Professor of Entomology, Associate Nematologist, Division of Nematology, University of California, Davis, California

Sman Keoboonrueng Ministry of Agriculture, Bangkok, Thailand

Lorin R. Krusberg, B.S., M.S., Ph.D. Professor of Plant Pathology, Department of Botany, University of Maryland, College Park, Maryland

George Blanchard Lucas, B.Sc., M.Sc., Ph.D. Professor, Department of Plant Pathology, North Carolina State University, Raleigh, North Carolina

Yasuharu Mamiya, Ph.D. Chief, Laboratory of Nematology, Division of Forest Protection, Forestry and Forest Products Research Institute, Ibaraki, Japan

Norman A. Minton, B.S., M.S., Ph.D. Nematologist, United States Department of Agriculture, Agricultural Research Service, Coastal Plain Experiment Station, Tifton, Georgia

William R. Nickle, B.S., M.S., Ph.D. Zoologist, Nematology Laboratory, Plant Protection Institute, United States Department of Agriculture, Agricultural Research Service, Beltsville, Maryland

Gregory R. Noel, B.S., M.S., Ph.D. Research Plant Pathologist, Department of Plant Pathology, United States Department of Agriculture, Agricultural Research Service, University of Illinois, Urbana, Illinois

Don C. Norton, B.S., M.S., Ph.D. Professor, Department of Plant Pathology, Seed and Weed Sciences, Iowa State University, Ames, Iowa

John H. O'Bannon, B.S., M.S., Ph.D. Research Nematologist, United States Department of Agriculture, Agricultural Research Service, Irrigated Agriculture Research and Extension Center, Prosser, Washington

----

Present affiliation
*Biotron, Inc., Baton Rouge, Louisiana, and Biotron Tropical Agricultural Experiment Station, Burrell Boom, Belize

Calvin C. Orr, Ph.D. Zoologist, United States Department of Agriculture, Agricultural Research Service, Texas Agricultural Experiment Station, Lubbock, Texas

James J. Petersen, B.S., M.S., Ph.D.* Research Entomologist, Gulf Coast Mosquito Research Laboratory, United States Department of Agriculture, Agricultural Research Service, Lake Charles, Louisiana

Dewey J. Raski, B.Sc., Ph.D. Professor, Division of Nematology, University of California, Davis, California

Donald P. Schmitt, B.S., M.S., Ph.D. Associate Professor, Department of Plant Pathology, North Carolina State University, Raleigh, North Carolina

Arnold E. Steele, B.A., M.S. Zoologist, United States Department of Agriculture, Agricultural Research Service, Salinas, California

John G. Stoffolano, Jr., B.S., M.S., Ph.D. Professor, Department of Entomology, University of Massachusetts, Amherst, Massachusetts

Armen C. Tarjan, B.Sc., M.Sc., Ph.D. Professor of Nematology, Department of Entomology and Nematology, University of Florida, Gainesville, Florida

Cyril H. S. Thong, B.Sc., Ph.D. Senior Laboratory Instructor, Department of Biological Sciences, Centre for Pest Management, Simon Fraser University, Burnaby, Vancouver, British Columbia, Canada

Seymour D. Van Gundy, B.A., Ph.D. Professor of Nematology and Plant Pathology, Chairman, Department of Nematology, University of California, Riverside, California

John M. Webster, B.Sc., Ph.D., D.I.C. Professor, Department of Biological Sciences, Centre for Pest Management, Simon Fraser University, Burnaby, Vancouver, British Columbia, Canada

E. J. Wehunt, B.S.A., M.S., Ph.D. Southeastern Fruit and Tree Nut Laboratory, United States Department of Agriculture, Agricultural Research Service, Booneville, Arkansas

Harold E. Welch, B.A., M.A., Ph.D. Professor, Department of Zoology, University of Manitoba, Winnipeg, Manitoba, Canada

Wilhelmus M. Wouts, Ph.D. Nematologist, Department of Scientific and Industrial Research, Entomology Division, Mt. Albert Research Center, Auckland, New Zealand

---

Present affiliation
*Research Entomolgist, Department of Entomology, United States Department of Agriculture, Agricultural Research Service, University of Nebraska, Lincoln, Nebraska

Contents

PLANT AND INSECT NEMATODES

Chapter 1

Importance of Agricultural Plant Nematology

W. M. Dowler* *USDA, Agricultural Research Service, Beltsville, Maryland*

Seymour D. Van Gundy *University of California, Riverside, California*

I. INTRODUCTION

The study of nematodes is generally concentrated in four areas, based on feeding habits of nematodes: animal and human parasites, insect parasites, plant parasites, and the fungal-bacterial feeders. Historically, the study of animal and human parasites came first and is concentrated in the veterinary and medical sciences and in parasitology. In the past decade, the use of fungal-bacterial feeders has become an important biological model system for geneticists, biologists and medical scientists. These "free-living" nematodes may well become the *Escherichia coli* of the animal kingdom (Zuckerman, 1980). The study of plant-parasitic nematodes and their role in crop protection is a relatively small and young science in comparison with the other crop pest disciplines; entomology, plant pathology, and weed science. They are one of the least understood pest problems currently confronting agriculture throughout the world. These interesting organisms have been observed on crops for

---

**Present affiliation:* Plant Disease Research Laboratory, USDA, Agricultural Research Service, Frederick, Maryland

hundreds of years, but it was not until the 1940s and the discovery of soil fumigants that we began to recognize their importance to agriculture.

Plant-parasitic nematodes are found in all soils, in irrigation water, and even in blowing dust in many of our agricultural regions of the world. Almost any crop grown suffers economic damage at one time or another from these pests. Nematodes, because of their secluded environment and inaccessibility, are one of the most difficult soil pest problems to identify, demonstrate, and control.

The importance of this volume lies in succeeding chapters, which discuss specific nematode-crop interactions and their importance to agricultural production. Therefore, the purpose of this chapter is merely to point out some historical highlights and the general problem areas of nematology research and to discuss some general aspects concerning the role of nematodes in agriculture today. There are also a series of chapters dealing with entomophilic nematology. Within the last ten years there has been a resurgence of interest in the nematode parasites of insects and their potential for managing insect pests. The high cost of pesticides and the environmental contamination problems associated with pesticides have economically forced a new and expanded emphasis on biological control of all crop, animal, and human pests. It is sufficient to say that there are a wide variety of nematode species that kill insects and have potential as a control tactic in an integrated pest management program.

II. HISTORICAL HIGHLIGHTS

A. Development of Plant Nematology in Europe

The first plant-parasitic nematodes were observed by Needham in 1743 in blighted wheat kernels. But it was not until 1775-1776 that Roffredi and Fontana independently demonstrated beyond any doubt the animal nature of Needham's discovery. No further significant developments occurred until the latter half of the nineteenth century, when Berkeley observed nematode galls (*Meloidogyne* spp.) on greenhouse-grown cucumbers. Kühn described a nematode (*Ditylenchus dipsaci*) from malformed floral heads of fuller's teasel, and Schacht reported the occurrence of a nematode (*Heterodera schachtii*) on sugar beet roots. Bastian's *Monograph of the Anguillulidae* in 1865 became the first cornerstone of the science of nematology. This work was quickly followed by the well-known morphological and taxonomic work of Bütschli (1873) and de Man (1876). Raski (1959) and Thorne (1961) present detailed accounts of the early history and development of nematology.

At the turn of the century, a number of scientists in many countries became active nematologists and developed the foundation of twentieth century nematology. Some of the more important ones were Micoletzky and Fuchs in Austria, Filipjev in Russia, deConinck in Belgium, Goodey in England, Goffart in Germany, and Cobb in the United States. Schuurmans Stekhoven's translation of I. N. Filipjev's work led to the publication of *A Manual of Agricultural Helminthology* in 1941, which is still a valuable research tool. Tom Goodey's books *Plant Parasitic Nematodes and the Diseases They Cause* in 1933 and *Soil and Freshwater Nematodes* in 1951 are still considered classic and valuable tools in nematology libraries. The Society of European Nematologists was founded in 1955, and the first issue of *Nematologica* was published in 1956. Finally, the works of Wallace on *The Biology of Plant Parasitic Nematodes* in 1963 and of deConinck published in the *Traite' de Zoologie* in 1965 complete the foundation series of important nematology works in Europe. Key leaders in the development of twentieth century nematology research and

Fig. 1 Fifty years of nematology leadership of the Nematology Section of the USDA Plant Industry Station. (A) N. A. Cobb, 1907-1932; (B) G. Steiner, 1932-1956; (C) A. L. Taylor, 1956-1964; (D) J. M. Good, 1964-1973.

training centers in Europe include Jones at Rothamsted, Peters in Imperial College, Seinhorst and Oostenbrink at Wageningen, deConinck at Ghent, Goffart and Weischer at Munich, Ritter at Antibes, Braun in Paris, and Luc in Africa.

B. Development of Plant Nematology in the United States

Although observations of nematode problems in agriculture had been made by Atkinson, Neal, Halstead, Stone, and Smith before the turn of the century, the major turning point in the development of nematology was the hiring of N. A. Cobb (Fig. 1A) by the U.S. Department of Agriculture in 1907. Not only were his scientific contributions outstanding, but he was also instrumental in promoting and developing the first center for nematology research and training, which achieved section status in 1929. His students and colleagues formed the foundation and leadership for nematology as we know it today and he is often referred to as the "father of nematology" in the United States.

Cobb's teachings imprinted on nematology in many ways. He was a stickler for detail and accurate illustrations. His employment of Chambers brought quality to taxonomic illustrations that has never been equaled. He made major discoveries in the areas of nematode taxonomy, morphology, and methodology. Many of his techniques carry his name and are still used today. He was also instrumental in removing the study of plant parasitic and "free-living" nematodes from helminthology and establishing a separate branch of science called *nematology*, which has remained active and grown into a society with over 750 members.

Some of the important architects who came principally from an apprentice association with Cobb were E. Buhrer, B. G. Chitwood, J. R. Christie, W. D. Courtney, G. Steiner, G. Thorne, and Jocelyn Tyler. Chitwood, although a very difficult man to work with and study under, was a brilliant scientist and observer of biology. His book, *An Introduction to Nematology*, in 1950 was an outstanding compilation of his knowledge of nematodes and still serves as a guide for any beginning student of nematology. Chitwood (Fig. 2A) started his career with Cobb in 1928 then transferred to the Bureau of Animal Industry from 1931-1937 and then back to the Bureau of Plant Industry until 1949, when he joined the staff at Catholic University for two years. Chitwood then worked a series of short-term nematology assignments in Kentucky, Michigan, Florida, California, and Hawaii from 1942 to 1962. His variety of research activities and job locations had a major impact on many nematologists throughout the United States. In the early fifties while Chitwood gave a series of lectures at Catholic University in Washington on nematodes, his students desiring to receive formal graduate work in nematology were enrolled at the University of Maryland, took their nematology course work at Catholic University, and carried out their dissertation research under the direction of personnel at the Division of Nematology of the USDA Plant Industry Station. Some of the early trainees who entered nematology by this route and who have made major contributions to nematology include E. J. Cairns, A. M. Golden, W. R. Jenkins, J. N. Sasser, and A. C. Tarjan.

Christie spent most of his nematology career at the USDA Research Station in Florida promoting nematology research and training in the southeastern United States. Christie's book in 1959 was the first compilation of the state of the knowledge directed solely at plant parasitic nematodes. His *Plant Nematodes, Their Bionomics and Control* contains a wealth of information for agricultural workers. He also pioneered research that demonstrated the importance of ectoparasitic nematodes in agriculture.

Fig. 2 Some deceased leaders in American nematology who contributed strongly to development of the Western branch of nematology and the first department of nematology in the United States. (A) B. G. Chitwood, 1907-1972; (B) G. Thorne, 1890-1975; (C) M. W. Allen, 1912-1974 and, (D) S. A. Sher, 1923-1976.

G. Steiner (Fig. 1B) followed Cobb in directing the nematology section of the USDA. In addition, he worked on plant parasitic nematodes and insect parasitic nematodes and made contributions to our morphological and taxonomic knowledge in these groups. Steiner was also influential in tutoring and developing the training of individuals at the Plant Industry Station.

One of his proteges was Al Taylor (Fig. 1C), who developed the first nematology program at the USDA Research Station in Tifton, Georgia and then returned to Washington, D.C. to follow Steiner as the leader of the Nematology Section. He brought together an outstanding cadre of current day nematologists into the USDA which include V. H. Dropkin, B. Y. Endo, G. Fassuliotis, J. Feldmesser, A. M. Golden, and J. M. Good. It was under Taylor's leadership that Beltsville became a leading nematology research center with a strong staff of research workers at various USDA Field Research Stations. One of his protégés, J. M. Good (Fig. 1D), started his career, as did Taylor, at Tifton, and then became the fourth leader of the nematology program at Beltsville.

Gerald Thorne (Fig. 2B) developed the first permanent nematology field station in the western United States, which was located in Salt Lake City, Utah. Thorne's major emphasis was on assembling the first major taxonomic collection of nematodes. He wrote monographs on over 60 genera and the first major textbook, *Principles of Nematology* in 1961. He was influential in developing the Western branch of nematology research and training in this part of the country. Although many nematologists were trained by Thorne, one of the most significant was Merlin Allen (Fig. 2C). The first formal graduate university course in nematology was given by Allen in 1948 at the University of California at Berkeley. Students who took this course and later became responsible for the development of the first and only department of nematology in the United States were such dedicated nematologists as W. H. Hart, D. J. Raski at the University of California, Davis, and S. A. Sher (Fig. 2D) at the University of California, Riverside. Another student of Allen was Jensen, who started a nematology research and training program at Oregon State University. W. R. Nickle, the editor of this book, was also one of Allen's students.

From 1950 to 1960, there was a phenomenal growth in nematology in the United States. The development of 1,3-D in 1943 by Walter Carter at the Pineapple Research Institute in Hawaii and the introduction of EDB by the Dow Chemical Company in 1945 as nematicides for plant-parasitic nematodes in agricultural production gave impetus for a new young cadre of research workers to enter nematology. This enthusiasm and interest in nematology was strengthened by the organization of a Southern Regional Graduate Summer Session in Nematology at North Carolina State College by J. N. Sasser in 1959. This course brought together a central and national focus on the importance of nematodes and produced a textbook on the state of the art at that time (Sasser and Jenkins, 1960). It also provided a nucleus of scientists who formed a Society of Nematologists in 1960. The Society grew by about 100 members per year through 1965 and now numbers about 750 members worldwide. Also in 1960, Tarjan drew together a *Checklist of Plant and Soil Nematodes*.

In 1967, Jenkins and Taylor published the second textbook, *Plant Nematology*. At the same time, there was an upsurge in tropical nematology with the formation of an Organization of Tropical American Nematologists in 1967 and the publication of *Tropical Nematology* by Smart and Perry in 1968.

Other important early nematology research and training centers include Cornell University under the direction of W. F. Mai, who produced the first pictorial key to plant-parasitic nematodes (Mai and Lyon, 1960) and a manual on nematode control (Mai et al., 1968), Virginia Polytechnic Institute under the direction of L. I. Miller, and University of Illinois under the guidance of M. B. Linford. In addition, Canada had several celebrated nematologists, who include A. D. Baker (1962), J. E. Bosher, R. J. Hastings, R. H. Mulvey, and W. B. Mountain.

In the 1970s, a rapid series of publications on plant-parasitic nematodes appeared in the literature. They include two volumes on *Plant Parasitic Nematodes*, edited by Zuckerman et al. in 1971; Webster's *Economic Nematology* in 1972; Tarjan and Hopper's *Nomenclatorial Compilation of Plant and Soil Nematodes* in 1974; Norton's *Ecology of Plant-Parasitic Nematodes* in 1978, Dropkin's textbook *Introduction to Plant Nematology* in 1980, and finally, Maggenti's textbook *General Nematology* in 1981.

Centers of nematology research and training now flourish in many countries of Europe and Africa, and in India, Australia, and Japan.

Soil-inhabiting nematodes, including the plant-parasitic forms, are now being recognized as important members of the soil fauna as well as one of the most numerous. They are receiving considerable attention in ecological studies of disturbed agricultural soils and undisturbed native soils. This interest in soil nematodes has been largely responsible for increased collection and identification of new plant-feeding nematodes. The number of plant-parasitic genera and species has grown worldwide from 55 and 1239, respectively, in 1967, to 134 and 2622, respectively, in 1980. These numbers will continue to increase rapidly as interest in nematodes grows with increasing numbers of trained nematologists. Fortunately, the appearance of new economic nematode pest problems in agriculture is much slower. In 1979, the pine wood nematode was found in the United States (Dropkin and Foudin, 1979), and in 1980 a new root-knot nematode was discovered (Santo et al., 1980). More recently, in 1981, the corn cyst nematode was found for the first time in Maryland (Sardanelli et al., 1981).

Past nematode problems, as well as new ones, continue to plague agricultural production throughout the world. Meanwhile, the number of chemicals used to manage pest problems continues to shrink and their use becomes more regulated. Thus, the control of nematode pest problems is slowly moving from a strong chemical base to an integrated, multifaceted control strategy, combining where possible the use of biological, chemical, cultural, and genetic methods.

III. ROLE OF NEMATODES IN CROP LOSSES

Having discovered the ubiquitous presence of nematodes, one might have presumed that their role in crop damage would be readily understood by now. However, damage caused by nematodes is often subtle and difficult to distinguish from damage due to other factors. In addition, their microscopic size makes them difficult to detect even though they may be present in large numbers.

The symptoms of nematode damage most often involve reduced plant size and yield, and sometimes yellowing of the top part of the plant, all of which can be due to any number of problems affecting the roots. Improper nutrition, lack of water, too much water, root-rotting fungi, or viruses may all cause similar symptoms that are not readily distinguishable. In fact, usually only after these and other more readily identifiable causes have been probed and eliminated does the plant health practitioner think of the possibility of nematodes being the problem and then proceed to search for evidence. The numbers of nematodes present also affect the severity of the symptoms, and they may be additive to problems caused by other stress factors. They may also be vectors of bacterial and virus diseases. Thus, small numbers of nematodes, which may be involved in a problem, may often go undetected as the major plant pest. In addition, by the time the plant shows serious symptoms of damage and begins to wilt, or perhaps dies, the nematode may leave the roots and not be present when attempts are made to discover the cause of the problem. All these factors tend to make diagnosis of nematodes and their role in plant diseases difficult.

Nematodes cause diverse types of damage to plants, depending on feeding habit. One of the most obvious symptoms that is readily discernible is the galling caused by the root-knot nematode, an endoparasite. On legumes, these galls are sometimes confused with the lateral galls formed by the nitrogen-fixing bacteria. They are sometimes confused with galls formed on root tips caused by feeding of ectoparasitic nematodes, such as the sheath nematode. Galls produced by the nematode are used as a measure of numbers of nematodes present in the soil, as well as a tool for screening varieties to determine possible resistance or susceptibility. Another general root symptom is the discoloration and formation of necrotic lesions on roots. The lesion nematode, a migratory endoparasite, produces somewhat readily identifiable lesions in certain host plants, or marked differences in the length and appearance of roots. The ectoparasites, quite often during root feeding, pierce the plant root, then move on to other sites, leaving little evidence of their presence. If their feeding is restricted to root tips, they may cause excessive branching and reduced root elongation. These symptoms may result in markedly less healthy plants, but the problems of identification are much more difficult to determine without soil analysis. Some nematodes are vectors for viruses, and in this case the virus may be the major problem, although it cannot attack the plant without the benefit of the nematode as a vector. There is considerable evidence that the combination of the two pests creates a worse problem than the sum of the damage caused by the two organisms alone.

Our present estimates and determinations of crop loss from any given factor leave much to be desired. Experiments to measure crop loss are difficult to design because of the many overlapping interactions involved. Nonetheless, to determine the importance of various factors in the role of crop losses, it is necessary to do the best job we can in making such estimates and determinations. In 1971, the Society of Nematologists undertook a major crop loss estimation project. These results (Society of Nematologists, 1971) indicated that average losses caused by plant-parasitic nematodes on all our major crops were on the order of 10%. Losses in individual situations may run as high as 50%. Other scientists have estimated losses in the range of 5-15%, and it is likely that 10% is a good average figure. It takes very little figuring to determine that the magnitude of such a loss has great importance to our food supply and that of the world. Based on current crop reporting estimates, this could amount to annual losses in the neighborhood of $500,000,000.

As an example, consider the damage nematodes cause to soybeans. Soybeans are rapidly becoming the leading crop in the United States. Nematodes are causing serious problems in soybean production, particularly the soybean cyst nematode, which has spread to almost all the soybean-growing regions. Nematodes other than the soybean cyst are also important pests to contend with in soybean production. As our demands for soybeans increase, the crop expands to new acreages, and we tend toward continuous cropping, we can expect additional nematode pest pressures on this crop.

Nematodes are serious pests of other field crops, vegetables, and ornamentals. Some of the most serious nematode problems are those of long-term perennial tree and vine crops. In annual crops, the problem can be solved for the next season by use of resistant varieties, rotation to nonhost crops, or preplant chemical fumigation. Perennial crops must be managed to live with the problem from 2 to 70 years. We are hampered in fighting this pest because of (1) the shortage of nematologists, (2) the difficulty in identifying the nematode as the pest involved, and (3) lack of good integrated control strategies.

IV. ROLE OF NEMATODE CONTROL IN INTEGRATED PEST MANAGEMENT

Although integrated pest management has been touted recently as a new method of pest control, it is really something farmers have been doing for many years. We are simply referring to the best combination of chemical, cultural, biological, and genetic methods for the most effective control of pests (diseases, insects, nematodes, and weeds) that yields an economic return to the grower. Obviously, nematodes must be considered in any pest management program, since they are known to be problems in almost any soil or cropping situation. We must continue to add to our knowledge of nematodes in order to better advise growers how to manage this pest in combination with all the other factors they must consider in crop production. An excellent review of the status of nematode control and associated research needs has been presented by the National Academy of Science (Mai et al., 1968). Some important components of an integrated pest management system that are discussed in succeeding chapters include proper identification, quarantine, host-plant resistance, cultural practices, biological agents, and chemicals.

A. Identification

Assuming we are able to identify nematodes as the problem, the next question is, what species, biotype, or race do we have in this particular field. Before any sound recommendation can be made, an accurate identification of the nematode species is needed. For example, the common root-knot nematode occurs worldwide as 40 different species, of which six species with at least six specialized races are most prevalent in agricultural fields. Recommendations on crop rotations and resistant varieties are only helpful if accurate nematode identifications are made in the laboratory. Consequently, nematode-identification laboratories are becoming common in both the public and private sector of agriculture. In the research laboratory, the scanning electron microscope has become a helpful tool in resolving morphological differences that were difficult to evaluate under the light microscope. There is a continuing and important need for taxonomists in nematology.

B. Chemicals

After we have clearly and accurately identified the nematode, how do we go about treating to prevent damage from this pest? The complexity of organisms in the soil and the difference in the makeup of soils make it difficult to determine the best mode of attack to use in alleviating the problem. Since the 1940s, chemical soil fumigants have been the first line of defense against nematodes. These pesticides are effective in destroying nematodes in the soil environment, but they must usually be applied when certain conditions are

present. This may involve soil temperature, soil moisture, presence or absence of host plants, organic matter in the soil, and soil type. More recently,
nonfumigant insecticide-nematicides have been developed that are effective
for a relatively short time, but may allow a buildup of nematodes, which can
cause problems at a later time in another crop. Some of these chemicals may
affect beneficial organisms in the soil, so that again the choice of chemicals
becomes critical. Currently, about 65% of the nematicide usage in the United
States occurs on four crops: corn, soybeans, cotton, and tobacco. Although
chemicals will continue to be an important tactic in integrated pest management,
other tactics are becoming just as important.

C. Quarantine

Since nematodes move relatively short distances under their own power, this
gives us the possibility of preventing the spread of nematodes through quarantine or other restrictions whenever possible. This method has been used
successfully to restrict the golden nematode thus far to New York and the
burrowing nematode of citrus to Florida. It is hoped that the corn cyst nematode can be prevented from spreading, if it is found to be localized in a
given area. Again, the microscopic nature of the nematode makes it difficult
to detect and thus it may be transmitted in soil, plant tissue, various produce,
or even in dust carried by the wind, or in mud on the feet of birds. Nevertheless, sanitation is an extremely important part of the program to reduce
problems due to nematodes in individual fields. It is important to carefully
remove soil from machinery when moving from a nematode-infested field to a
clean or chemically treated one. It is equally important to be sure all seed
and planting stock are free of nematode pests before planting.

D. Host-Plant Resistance

Ideally, the best method of controlling the nematode would be to develop and
use resistant crop hosts. This method has worked well in some crops, such
as soybeans, where we now have several resistant varieties that perform extremely well in areas infested with the soybean cyst nematode and its five
races. We now have many varieties of various crops resistant to nematodes,
but the proliferation of new races, the development of multiple nematode resistance, and the lack of new gene pools make this a continuing battle. Nematode resistance is quite often available in wild plant species, and it then becomes a problem for the plant breeder to incorporate this resistance into commercially acceptable varieties. Obviously, we face an even greater problem
in development of desired resistance in perennial crops, such as fruit and nut
trees.

E. Cultural Practices

We have had some success using specific land management and cultural practices to control nematodes. Such practices as flooding, fallow, growing selected cover crops, crop rotation, use of trap crops, and use of organic amendments have all been used to alleviate the nematode problem. A major problem
with this method of control is the current demand for the farmers to keep their
land occupied with profitable crops, so that there is little opportunity or desire for fallow or even crop rotation to a less desirable crop to assist with ne-

matode control. Nonetheless, good growers will take these practices into account as they plan their farming operations with an eye to effective pest control.

F. Biological Control

A relatively new procedure for alleviating the nematode problem is that of biological control. Although we can agree that this method has great potential for assisting with this problem, we also know we have a long way to go before this method of control becomes practical. We know that nematodes are attacked by a large variety of organisms, but the problem of getting this organism into the biological complexity of the soil mass in the proper ways and at the proper time to attack the nematode is formidable. In spite of these difficulties, we know that biological control is an existing phenomenon in England in the situation involving the oat cyst nematode (Kerry, 1981). After several years in small grain, a beneficial organism presumably is built up to the proper number to aid significantly in controlling the oat cyst nematode. Undoubtedly, this method will become more important as we learn more about how to manage the organism involved.

V. CONCLUSIONS

It is obvious that nematodes are important pests to consider in agriculture. If we accept the 10% estimate of crop losses as being anywhere near accurate, it is apparent that this is a serious problem for consideration. It is also apparent that we need to know much more about nematodes and the role they play in the complex array of pests. It would be nice to assume that numbers of nematologists would double within the next few years and we would be able to probe more readily the areas where we lack information. However, to be realistic we know that this is not probable, and it will be necessary for us to do a better job with what we have in putting information together to most effectively manage and control this important pest.

REFERENCES

Baker, A. D. (1962). *Check Lists of the Nematode Superfamilies Dorylaimoidea, Rhabditoidea, Tylenchoidea and Aphelenchoidea.* E. J. Brill, Leiden.
Bastian, H. C. (1865). Monograph on the Anguillulidae. *Linn. Soc. Lond. Trans. 25*: 73-184.
Bütschli, O. (1873). Beiträge zur Kenntnis der freilebenden Nematoden. *Nova Acta Acad. Leop. Carol. 36*: 1-124, plates 17-27.
Chitwood, B. G., and Chitwood, M. G. (1950). *An Introduction to Nematology.* Monumental Printing, Baltimore.
Christie, J. R. (1959). *Plant Nematodes, Their Bionomics and Control.* Univ. Florida, Gainesville.
de Coninck, L. (1965). Class of nematodes. In *Treatise on Zoology*, vol. 4, part 2 (Pierre-P. Grassé, ed.). Masson and Co., Paris, pp. 3-217; 387-432.
Dropkin, V. H. (1980). *Introduction to Plant Nematology.* Wiley, New York.
Dropkin, V. H., and Foudin, A. S. (1979). Report of the occurrence of *Bur-*

saphelenchus lignicolus-induced pine wilt disease in Missouri. *Plant Dis. Rep. 63*: 904-905.

Goodey, T. (1933). *Plant Parasitic Nematodes and the Diseases They Cause.* Methuen, London.

Goodey, T. (1951). *Soil and Freshwater Nematodes.* Methuen, London.

Jenkins, W. R., and Taylor, D. P. (1967). *Plant Nematology.* Reinhold, New York.

Kerry, B. R. (1981). Progress in the use of biological agents for control of nematodes. In *Beltsville Symposium in Agricultural Research*, vol. 5, *Biological Control in Crop Production* (G. C. Papavizas, ed.). Allenheld, Osmun, N.J.

Maggenti, A. R. (1981). *General Nematology.* Springer-Verlag, New York.

Mai, W. F., and Lyon, H. H. (1960). *Pictorial Key to Genera of Plant Parasitic Nematodes.* Cornell Univ. Press, Ithaca, New York.

Mai, W. F., Cairns, E. J., Krusberg, L. R., Lownsbery, B. F., McBeth, C. W., Raski, D. J., Sasser, J. N., and Thomason, I. J. (1968). Control of plant-parasitic nematodes. In *Principles of Plant and Animal Pest Control*, vol. 4. National Academy of Sciences, Pub. 1966, Washington, D.C., pp. 1-172.

de Man, J. G. (1876). Onderzoekingen over vrij in de aarde levende Nematoden. *Tijds. Chr. Nederland, Dierk. Vereen. 2*: 78-196.

Norton, D. C. (1978). *Ecology of Plant-Parasitic Nematodes.* Wiley, New York.

Raski, D. J. (1959). Historical Highlights of Nematology, Chapter 34. In *Plant Pathology—Problems and Progress 1908-1958* C. S. Holton, G. F. Fisher, R. W. Fulton, H. Hart, and S. E. A. McCallam, (eds.). Univ. Wisc. Press, Madison.

Santo, G. S., O'Bannon, J. H., Finley, A. M., and Golden, A. M. (1980). Occurrence and host range of a new root-knot nematode (*Meloidogyne chitwoodi*) in the Pacific Northwest. *Plant Dis. 64*: 951-952.

Sardanelli, S., Krusberg, L. R., and Golden, A. M. (1981). Corn cyst nematode, *Heterodera zeae*, in the United States. *Plant Dis. 65*: 622.

Sasser, J. N., and Jenkins, W. R. (1960). *Nematology Fundamentals and Recent Advances with Emphasis on Plant Parasitic and Soil Forms.* Univ. North Carolina Press, Chapel Hill.

Smart, G. C., Jr., and Perry, V. G. (1968). *Tropical Nematology.* Univ. Fla. Press, Gainesville.

Society of Nematologists. (1971). Estimated crop losses due to plant-parasitic nematodes in the United States. Special publication No. 1 (supplement to the *Journal of Nematology*).

Tarjan, A. C. (1960). *Check List of Plant and Soil Nematodes.* Univ. Fla. Press, Gainesville.

Tarjan, A. C. and Hopper, B. E. (1974). *Nomenclatorial Compilation of Plant and Soil Nematodes.* Society of Nematologists. DeLeon Springs, Florida.

Thorne, G. (1961). *Principles of Nematology.* McGraw-Hill, New York.

Wallace, H. R. (1963). *The Biology of Plant Parasitic Nematodes.* Edward Arnold, London.

Webster, J. M. (1972). *Economic Nematology.* Academic Press, New York.

Zuckerman, B. M. (Ed.). (1980). *Nematodes as Biological Models. Behaviorial and Developmental Models,* vol. 1; *Aging and Other Model Systems,* vol. 2., Academic Press, New York.

Zuckerman, B. M., Mai, W. F. and Rohde, R. A. (eds.) (1971). *Plant Parasitic Nematodes.* Academic Press, New York.

Chapter 2

Nematode Parasites of Soybeans

Donald P. Schmitt *North Carolina State University, Raleigh, North Carolina*

Gregory R. Noel *USDA, Agricultural Research Service, University of Illinois, Urbana, Illinois*

I. INTRODUCTION

A. History of the Crop

There is conflicting evidence about the origin and early history of the soybean
[*Glycine max* (L.) Merr.]. Reports indicate an origin somewhere between 2338
and 2383 B.C. (Probst and Judd, 1973). Soybean is considered to be native
to Asia and probably originated in North and Central China. *Glycine soja* Sieb.
and Zucc. is probably the progenitor of *G. max*. Soybean seeds were used for
centuries in the Orient for food, and were considered to be one of the five
essential foods of the Chinese civilization.

The first domestication of the soybean probably occurred in the eastern
half of North China around the eleventh century B.C. (Hymowitz, 1970). The
cultivated form was introduced into Korea from North China and disseminated
to Japan between 200 B.C. and the third century A.D. (Nagata, 1959). There
has been some speculation that the soybean may have been introduced from
Central China to southern Japan. The first soybean shipments were made to
Europe about 1908 and attracted worldwide attention (Probst and Judd, 1973).

There is evidence that soybeans were present in the United States in the
early 1800s (Probst and Judd, 1973), but the first reference in recorded scientific literature of the soybean being tested in the United States was in an 1879
report from the Rutgers Agricultural College Farm. Later, more tests were
conducted in several locations throughout the United States, first as a forage
crop and then as a seed crop. There were about 50,000 acres of soybean in
1907, which was used primarily for forage. Some oil was extracted from seed
in 1915, and by 1929, large quantities of seed were being crushed for oil. By
the mid-1920s, the acreage had increased to 1.8 million. The importance of
soybean as a seed crop exceeded its value as a forage crop by 1941. Several
reasons for the early increase in acreage include the increased use of machinery, acreage restrictions placed on corn, wheat, and cotton, and the price
support programs on soybeans during World War II. The trend of growing
more soybeans has continued to the present. For example, 12.5 million acres
were planted in 1949, 43.2 million acres in 1971, and 68.1 million acres in 1981.
This rapid expansion in the United States has been followed recently in Brazil,

Colombia, Mexico, and the USSR. Moderate increases have occurred in Canada, Indonesia, and South Korea. In contrast, there were marked declines in production in the Peoples Republic of China and in Japan.

B. Importance of Nematodes

Over 100 species of plant-parasitic nematodes are associated with soybeans (Good, 1973; Rebois and Golden, 1978; Robbins, 1982a). Undoubtedly, many more species would be found if all areas where soybeans are grown were surveyed extensively and the nematodes identified to species. These associations, however, are not sufficient evidence of pathogenic host-parasite relationships. For example, the ring nematode, *Criconemella ornata* (Raski) Luc and Raski, often found in large numbers in soybean fields, reproduces on weeds (Barker et al., 1982). Weed-free plots contained 54 nematodes per 500 cm^3 of soil compared to over 5000 in crabgrass-infested plots.

Soybean yield losses due to nematodes range from negligible to 100% on a field-by-field basis. Attempts have been made to establish losses on statewide and national bases (Feldmesser, 1970; Whitney, 1978). Many of these estimates are subject to much skepticism because most are based on best-guess estimates, nematicide tests, or comparative yields of resistant and susceptible cultivars. These estimates, however, do provide a reasonable measure of the relative importance of nematodes.

The sensitivity of soybean cultivars to a given nematode species varies considerably. Substantial efforts are being made to determine the relationship of these nematodes to crop damage. The soybean cyst nematode, *Heterodera glycines* Ichinohe, and root-knot nematodes, *Meloidogyne* spp., are receiving the greatest attention because of their damage potential and widespread occurrence. Other soybean-parasitic nematodes are being studied less intensively to determine their role in soybean production.. Action thresholds or some other measure of nematode damage potential are useful guidelines for nematologists, IPM specialists, and soybean producers in making wiser control decisions. Current action thresholds usually are conservative for several reasons: (1) it is impossible to predict the rainfall patterns for the growing season, and thus the amount of damage; (2) most threshold numbers are based on fall sampling from which planting time population densities are predicted; (3) species and races are often undetermined; and (4) the variation in distribution of nematodes within a typical field results in sampling errors.

C. General Control Considerations

The persistence and dynamic changes of nematode populations in the soil require constant vigilance and up dated management decisions in order to avoid major yield losses. The development of cultivars with resistance to the major nematode species was an important step in crop protection. Crop rotation can be employed to control a number of soybean-parasitic nematodes. It is especially useful when nonhosts, resistant, and susceptible cultivars are used in combination. Nematicidal treatments may be necessary for certain nematode problems or may be used as a complementary control.

Control of soybean-parasitic nematodes must be aimed at reducing the initial inoculum or preventing infection because the damage done to the soybean during its early growth has the greatest effect on reducing yield potential. Crop rotation and nematicides reduce the initial population density, whereas resistant cultivars prevent or inhibit nematode feeding and development.

Control of soybean-parasitic nematodes by nematicides can be altered by chemical and physical factors in the soil environment. Rate (Rodriguez-

Kabana et al., 1979), method of application (Minton et al., 1979; Rodriguez-Kabana et al., 1981), soil physical characteristics, including hard-pans (Minton and Parker, 1975), other pesticides (Schmitt and Corbin, 1981), and non-target organisms (Kinloch and Schenck, 1978) can all affect nematicide activity. The best method of application and optimum rate must be determined for each specific nematode community, set of soils, and environments. For example, a band application of systemic nematicides was more effective in controlling nematodes and increasing yields in Alabama (Rodriguez-Kabana et al., 1981), although the in-furrow application was slightly better than the band application in Georgia (Minton and Parker, 1980).

Thus, the major control tactics for nematodes on soybeans include crop rotation, resistance, and nematicides. Acceptable yields can be achieved in fields infested with pathogenic nematodes by the inclusion of sound control practices (Kinloch, 1980).

II. THE SOYBEAN CYST NEMATODE

A. History and Distribution

"Yellow dwarf" disease was first reported in 1915, although it was noticed as early as 1881 (Riggs, 1977). The disease was initially thought to be caused by *Heterodera schachtii* Schmidt. The soybean cyst nematode was found in Korea in 1936 and in the Peoples Republic of China in 1938. It was identified for the first time in the United States in 1954 (Winstead et al., 1955). The nematode was named *Heterodera glycines* in 1952 by Ichinohe, 71 years after it was first noted causing disease in the Orient.

This serious pest of soybeans may have been introduced or may be indigenous to the United States. The yellow dwarf symptoms observed on Japanese soybean were identical in North Carolina, where *H. glycines* was first identified in the United States. However, the Japanese and North Carolina populations differed morphologically (Golden and Epps, 1965). *H. glycines* can survive in bags shipped from Japan (Epps, 1968), thus providing evidence that bags brought into the Mississippi Delta could have been a source of introduction.

H. glycines is currently found in Japan, Korea, the Peoples Republic of China, and the United States. This pest has been found in 22 of the 30 soybean-producing states. The nematode can spread and increase rapidly once it is introduced into a new area (Riggs, 1977). It is effectively disseminated by a wide range of means such as wind, water, machinery, birds, and soil beds (Riggs, 1977).

B. Species and Races

The cysts of *H. glycines* are lemon shaped (Thorne, 1961). They are 560-850 μm long and 350-590 μm wide. Bullae are present. The females pass through a yellow stage before becoming a dark-brown color. The cyst wall has a rugose pattern of irregular short zigzag lines. A gelatinous matrix is produced at the vulval cone, usually containing some eggs. Males are approximately 1.3 mm long. Second-stage juveniles average 450 μm long. About half of the tail is hyaline.

Four races of the soybean cyst nematode were described in 1970 (Golden et al., 1970), and a fifth race was proposed in 1979 (Inagaki, 1979) that is very similar to race 2 but cannot reproduce on Peking soybean (Table 1). This race scheme has been useful, largely because cultivars could be bred for resistance to designated races.

Table 1 Characterization of Races of *Heterodera glycines*

Race	Reproduction[a]				
	Pickett	Peking	PI88788	PI90763	Lee
1	-	-	+	-	S
2	+	+	+	-	S
3	-	-	-	-	S
4	+	+	+	+	S
5	+	-	+	-	S

[a]Reproduction on a cultivar or line less than 10% of that on Lee is given a negative sign (-) and an amount of 10% or more is given a positive sign (+); S = susceptible host.
Source: From Golden et al. (1970); Inagaki (1979).

The occurrence of biotypes in most fields lead Riggs et al. (1981) to suggest that another system of classification be studied. They recommend that "the best procedure for determining soybean cyst nematode races should be decided by a group of 5-7 nematologists and plant breeders who have been involved in soybean cyst nematode work." *H. glycines* has developed races rapidly under field conditions (Price et al., 1978). Even different populations of the same race had different indices of parasitism, indicating that qualitative and quantitative differences exist among populations of the same race (Triantaphyllou, 1975). This nematode apparently possesses several genes for parasitism. Thomas (1974) has suggested that the nematode may contain as many as 10 genes for parasitism.

In addition to race shifts in field populations, the ability of the races to interbreed and successfully produce progeny further enhances the chances for a population to successfully parasitize a resistant cultivar. For example, crossing of races 1 or 3 with 2 or 4 resulted in the ability of some progeny to parasitize a resistant host (Price 1978).

Unfortunately, the establishment of any race classification system will probably break down over time because of this pest's genetic diversity.

The present race classification is largely qualitative. A quantitative scheme giving information on the percentage of control by resistant cultivars would enable decision makers to determine if resistance was adequate or if additional control tactics would be required.

C. Life Cycle and Biology

H. glycines has four juvenile stages and an adult stage (Fig. 1). Embryonic development within the eggs results in a first-stage juvenile. The first molt occurs within the egg. The second-stage juvenile hatches and moves out of the cyst or gelatinous matrix and into the soil. This hatching occurs spontaneously when the egg is not in diapause (Ross, 1963), although there is some evidence of a hatching stimulant (Okada, 1972; Masamure et al., 1982). Glycinoeclepin A is a hatching stimulant extracted from kidney bean (Masamure et al., 1982). The hatched second-stage juvenile moves through the soil and penetrates the root, generally well behind the zone of differentiation. After

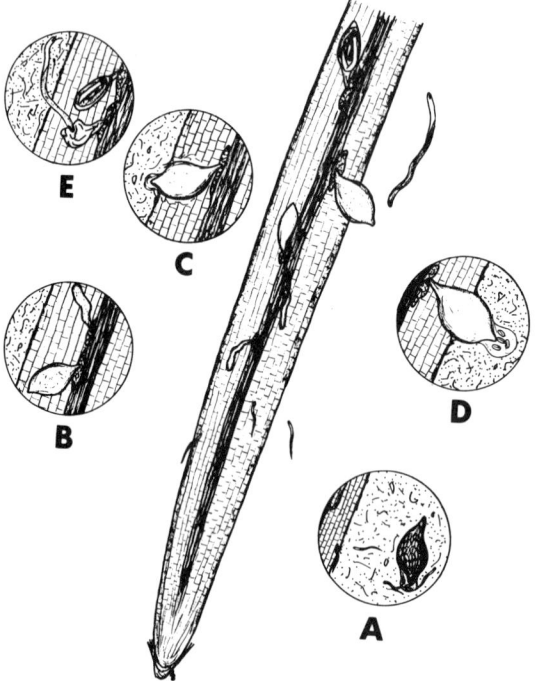

Fig. 1 Stages in the life cycle of *Heterodera glycines*. (A) Second-stage juveniles penetrating the root. (B) Swollen juveniles in the root. (C-D)
Adult female. (E) Adult males. (Illustrated by A. L. Bostian.)

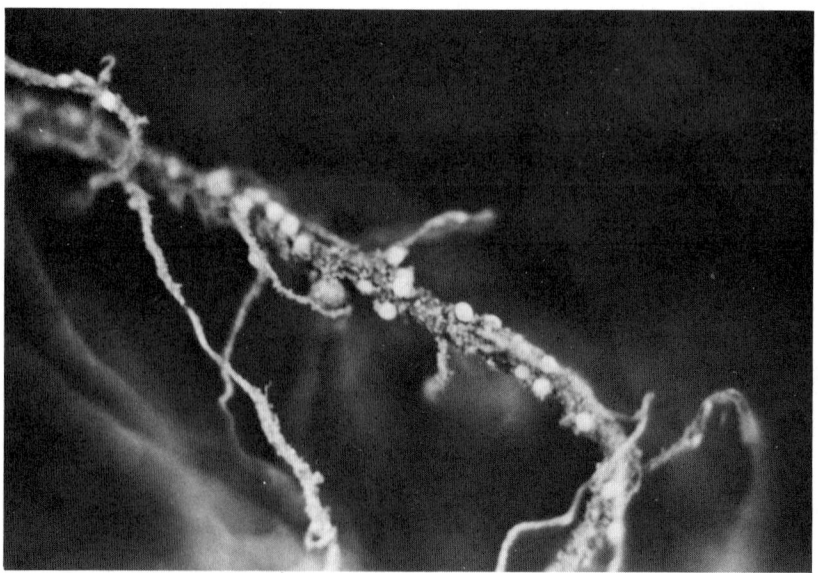

Fig. 2 *H. glycines* females on roots of soybean.

penetration, the second-stage juvenile migrates through the cortex to the vascular tissue and places its head adjacent to the stele. The nematode then becomes sedentary and begins to feed and swell. The nematode passes through the third- and fourth-juvenile stages and becomes an adult. The adult male matures faster than the female and emerges from the juvenile cuticles, leaves the root, and begins its search for females. The developing female ruptures the epidermis, and the adult protrudes from the root (Fig. 2). The male moves through the soil to the female and inseminates her. The fertilized eggs develop and are largely retained within the female, though some eggs may be deposited into a gelatinous matrix surrounding the posterior end of the female. The expected ratio of males to females is 1:1 (Koliopanos and Triantaphyllou, 1972). The ratio recovered from a sample is rarely 1:1, partially due to a differential death rate of males and females (Koliopanos and Triantaphyllou, 1972). In addition, the cysts persist in soil, whereas males are ephemeral. Temperature affects the nematode's rate of development (Hamblen et al., 1972; Ross, 1964b). The largest number of penetrations occurs at 28°C. The most rapid development occurs at 28-31°C, with little or no development at or below 15°C and at or greater than 33°C. *H. glycines* required 7512 heat units (degrees above 10°C multiplied by hours) to complete its life cycle (Ichinohe, 1955).

The eggs within the cysts are able to survive for long periods of time; some have survived for 11 years (Inagaki and Tsutsumi, 1971). This ability to survive long periods may largely be due to the contraction of the cyst wall and dense deposition of phenolic substances in the cyst wall (Kondo and Ishibashi, 1975). A jellylike matrix within the cyst, the egg shell, and the nematode cuticle could be additional protection for the nematode (Kondo and Ishibashi, 1975).

Fig. 3 High infestation levels of the soybean cyst nematode resulted in these severely stunted and dead soybeans in Massac County, Illinois. (Courtesy of B. J. Jacobsen.)

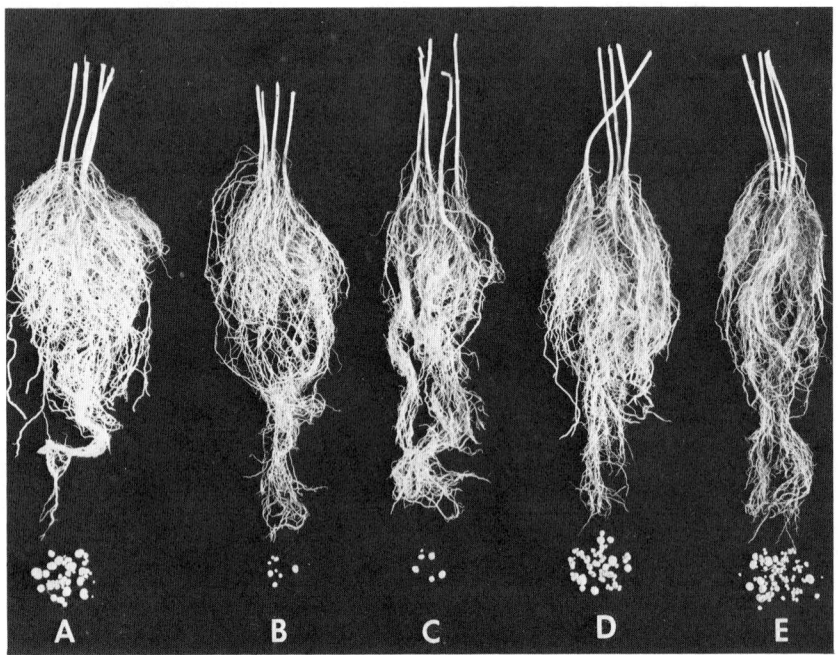

Fig. 4 Influence of *H. glycines* on nodule development. (A) *R. japonicum.*
(B-C) *R. japonicum* and *H. glycines* race 1. (D-E) *R. japonicum* and *H. glycines* race 4. (Courtesy of K. R. Barker.)

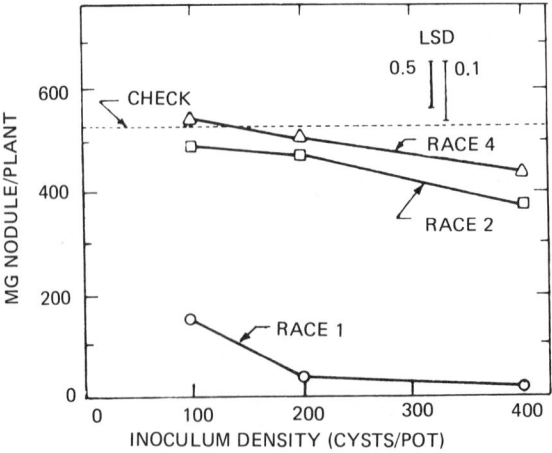

Fig. 5 Comparative effects of *H. glycines* races 1, 2, and 4 on nodule weight at 55 days after inoculation. (From Lehman et al., 1971.)

D. Associated Symptoms and Damage

Symptoms of injury caused by *H. glycines* vary from slight stunting to severe stunting, chlorosis, and death (Fig. 3). Plants showing the greatest degree of injury are those that have the greatest amount of inhibition of root nodulation (Lehman et al., 1971) (Figs. 4, 5). Race 1 had a greater effect on the inhibition of nodulation, nitrogen fixation, and nodule efficiency than races 2 or 4.

In addition to its effect on nitrogen fixation, *H. glycines* causes other physiological and morphological changes in the roots. The resulting damage involves primarily cell wall breakdown leading to the development of syncytia (Endo, 1964; Gipson et al., 1971). More tissue distortion usually occurs with females than with males. Syncytial development can continue into the secondary cambium region, which then inhibits secondary growth of phloem and xylem (Endo, 1964), restricting the growth of the plant. The syncytium becomes necrotic after the nematode has completed its life cycle and feeding ceases (Fig. 6) (Endo, 1964).

E. Relationship of Population to Yield

Systems used throughout the United States to predict crop damage are based on cyst, egg, and/or second-stage juvenile counts. More research will be required to determine which parameter(s) will be best for predicting crop loss. Cysts (J. N. Sasser, personal communication), eggs (Bonner, 1981), second-stage juveniles (Schmitt, unpublished), and eggs plus second-stage juveniles (Noel et al., 1980) have given fair to good relationships between nematode numbers at planting and yield. Some of the relationships were linear (Bonner, 1981; Schmitt, unpublished) and others are quadratic (Noel et al., 1980; Schmitt, unpublished). The soil texture may be important in these relationships because there is a tendency toward linear slopes in sandy soils and quadratic curves in other soil types (Schmitt, unpublished).

F. Control

1. Crop Rotation

Crop rotation is an effective and practical means of controlling *H. glycines* because few crops are susceptible (Fig. 7). Generally, growing a nonhost for two years is required before a susceptible soybean cultivar can be grown with the expectation of a full yield potential (Fig. 7). Soybean yields in fields infested with *H. glycines* were two to three times greater with a two year rotation and seven to eight times greater with three and four year rotations with nonhosts than with monocultured soybeans (Ross, 1962). Growing a nonhost for one year may be adequate if there are just a few nematodes and overseasoning conditions are unfavorable for survival. Overwinter survival was nearly 100% in some winters in North Carolina, but less than 10% during other winters (Schmitt, unpublished; Bostian and Schmitt, unpublished).

2. Resistance

Resistance to *H. glycines* is a type of hypersensitive reaction in which the tissue affected by the nematode deteriorates and the nematode fails to develop (Figs. 8, 9). Second-stage juveniles penetrate the root tissue of resistant cultivars as readily as susceptible cultivars (Endo, 1965). Within a few days

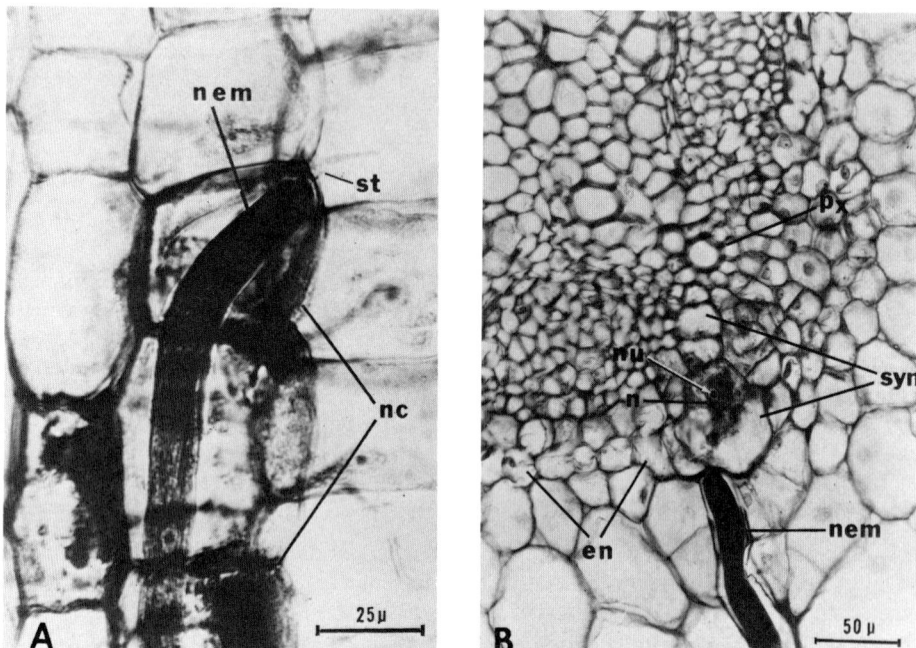

Fig. 6 Penetration and infection of the soybean cyst nematode, *H. glycines*, in Lee soybean roots, a susceptible reaction: (A) Intracellular migration of juvenile (nem) through cortex causing necrosis of cells (nc); note stylet (st) penetration of cortical cell wall (1 day after inoculation). (B) Nematode (nem) in contact with early stage of syncytium (syn) extending from the endodermis (en) into the protoxylem region (px). Enlarged nucleus (n) surrounds hypertrophied nucleolus (nu). Other nuclei are present but not readily distinguishable (2 days after inoculation). (C) Adult female nematodes (nem) adjacent to syncytia (syn) in vascular region. After the nematode matures and stops feeding, the syncytium deteriorates (ds) (24 days after inoculation). (D) Vascular damage from syncytial development; syncytia inhibited wide regions of secondary xylem and phloem tissue. Note the cluster of nuclei in a syncytium (27 days after inoculation). (Courtesy of B. Y. Endo.)

after infection of the resistant cultivar, the syncytia become necrotic and collapse (Fig. 10). A deposition of secondary wall materials surrounds the necrotic tissue of the diseased area (Riggs et al., 1973). Parenchyma tissue often invades the degenerated and collapsed syncytial site (Endo, 1965).

Continuous or frequent use of resistant cultivars results in a race shift that eventually renders the resistant cultivar useless. Most field populations apparently are a mixture of genotypes. Selection forces imposed by resistant cultivars will change the gene frequency (Triantaphyllou, 1975; McCann et al., 1982). Populations grown under greenhouse conditions retained the same gene frequency for parasitism if they were cultured on susceptible soybeans, but resistance genes induced a change in the frequency of genes for parasitism (Triantaphyllou, 1975; Riggs et al., 1977). For example, the index of parasitism (reproduction on resistant line/reproduction on susceptible line

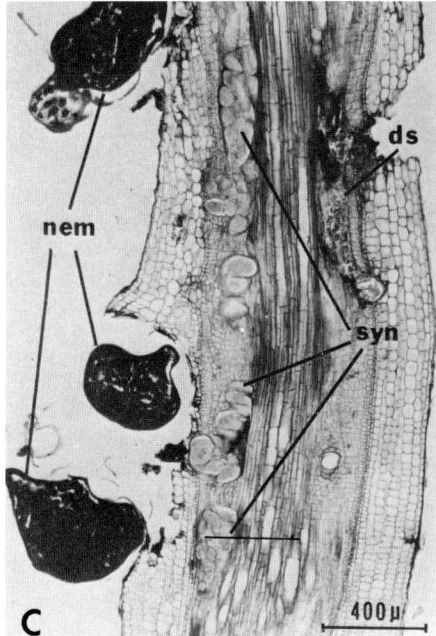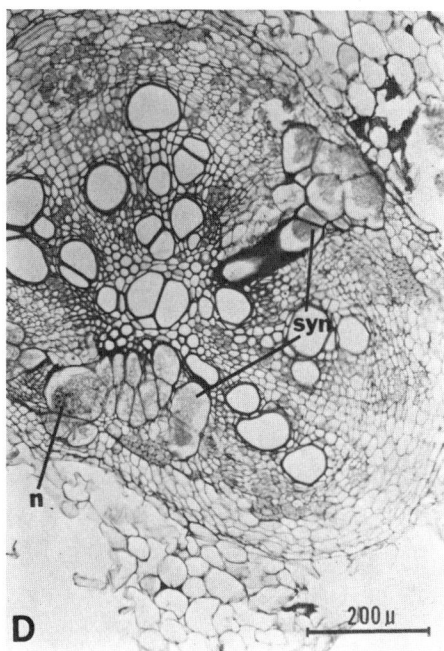

Fig. 6 (Continued)

×100) of a population of *H. glycines* from Johnston County, North Carolina, following continuous propagation for seven generations on susceptible soybeans remained the same as the original population (Triantaphyllou, 1975). The index of parasitism increased from 3 to 76 following propagation for seven generations on the resistant cultivar Peking (Table 2).

3. Nematicides

Certain nematicides are effective in controlling *H. glycines*. Their effectiveness is dependent upon soil type, soil moisture, and organic matter. Fumigants are effective in killing this nematode if the proper rate and method of application are used. The systemic nematicide aldicarb is effective over a wide range of soil types. Phenamiphos and carbofuran are effective in some soils with a low organic matter content. The decision to use a nematicide should be based on the results of soil assays, cost versus expected return, and available equipment.

4. Integrated Control

Long-term control of *H. glycines* requires the integration of crop rotation, resistant cultivars, nematicides, and good land management. In fields where *H. glycines* is the primary problem, a rotation of nonhosts, resistant, and susceptible cultivars can be effectively used without the need for a nematicide if the susceptible cultivar is grown no more than every third year. Infrequent use of resistant cultivars should minimize the rate of shift in the nematode

Fig. 7 Influence of crop rotation on control of *H. glycines* on soybean.

Fig. 8 Resistant and susceptible soybeans growing in a *H. glycines*-infested field. (Courtesy of J. P. Ross.)

Fig. 9 The use of resistant cultivars is one method used to control the soybean cyst nematode. In this Fayette County, Illinois, field, the susceptible cultivar Union is injured severely while the resistant cultivar, Franklin, is quite vigorous. (Courtesy of B. J. Jacobsen.)

Table 2 Changes in the Degree of Parasitism of a Johnston County, North Carolina, Population of *H. glycines*[a]

| | Field population originally checked on | | Field population propagated for 7 generations on | | | |
| | | | Lee and checked on | | Peking and checked on | |
	Lee	Peking	Lee	Peking	Lee	Peking
No. cysts per plant	108	2	120	4	103	76
Index of parasitism		2.8		3.3		74

[a]Following continuous propagation for seven generations on a susceptible cultivar (Lee) and a resistant cultivar (Peking).
Source: From Triantaphyllou (1975).

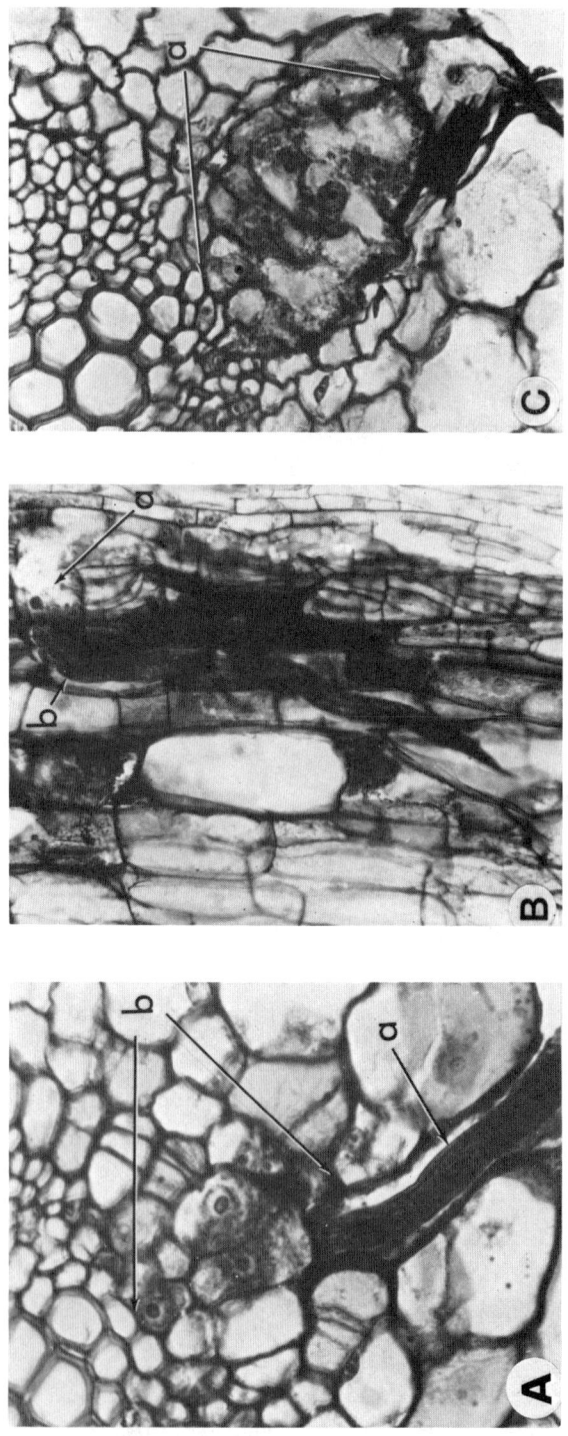

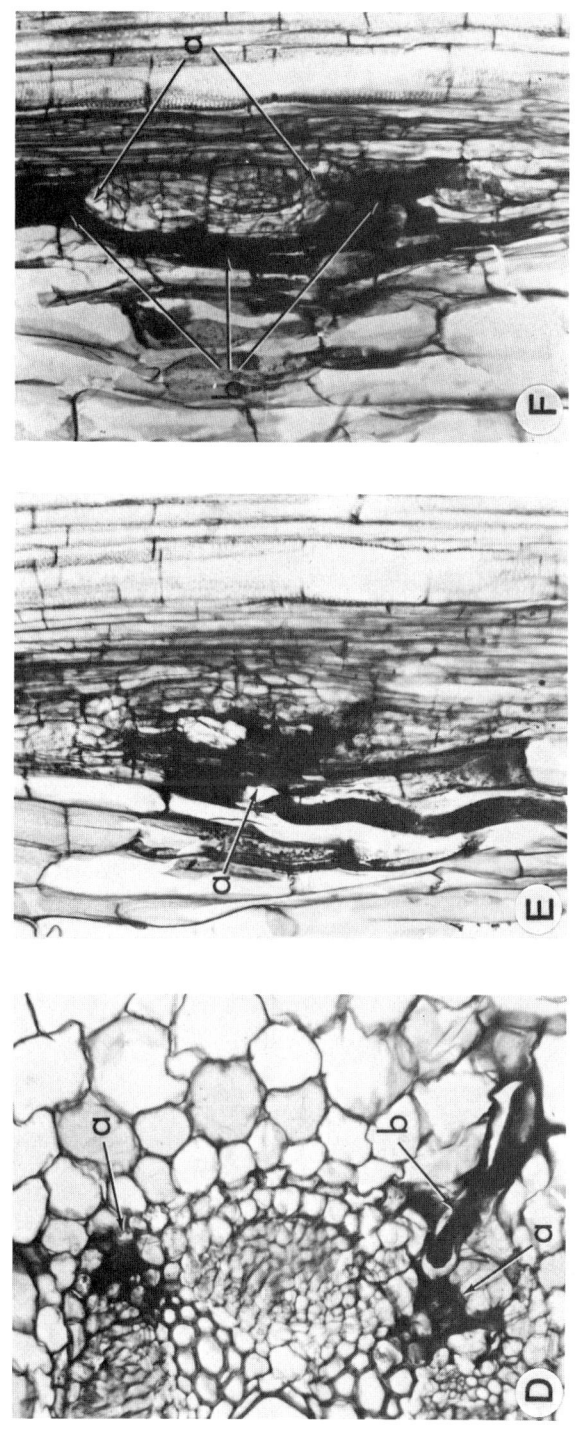

Fig. 10 Resistant host-parasite reaction of Peking soybean root invaded by *H. glycines*. (A) Second-stage juvenile (a) at feeding site where tissue stimulation (b) has occurred (three days). (B) Enlarged second-stage juvenile (b) in contact with a vacuolate syncytium (a) (three days). (C) Moderately developed syncytium (a) showing signs of deterioration (four days). (D) Necrotic regions (a) of former syncytial development site and deteriorated second-stage juvenile (b) (eight days). (E) Longitudinal section of nematode feeding site (a) (10 days). (F) Thin-walled parenchyma cells (a) developed in region of the collapsed syncytial unit (b). (Courtesy of B. Y. Endo.)

gene frequency. Nematicides may be complementary if mixed biotypes or other nematode species are present.

III. THE RENIFORM NEMATODE

A. History and Distribution

The reniform nematode, *Rotylenchulus reniformis* Linford and Oliveira, was described in 1940 from cowpea roots in Hawaii (Linford and Oliveira, 1940). Soybean is one of the many hosts of *R. reniformis* (Carvalho, 1958; Fassuliotis and Rau, 1967; Peacock, 1956; Timm, 1964; Linford and Yap, 1940; Ayala and Ramirez, 1964). Infestations of *R. reniformis* have been reported from many countries in the tropics and warmer areas of the temperate zone (Ayala and Ramirez, 1964; Carvalho, 1958; Dasgupta et al., 1968; Siddiqi, 1972). In the United States, infestations have been reported from Alabama, Arkansas, California, Florida, Georgia, Louisiana, South Carolina, Texas, and the Commonwealth of Puerto Rico (Ayala and Ramirez, 1964; Fassuliotis and Rau, 1967; Riggs, personal communication).

B. Species

R. reniformis is sexually dimorphic. The males and immature females are vermiform. The immature female assumes a "C" shape. Mature females are arcuate ventrally and become saccate in the shape of a kidney (Fig. 11). The female stylet is somewhat delicate and approximately 16-18 µm long. The male has a poorly developed stylet and esophagus with a reduced metacorpus and indistinct valve. In the female, the median esophageal bulb is strongly

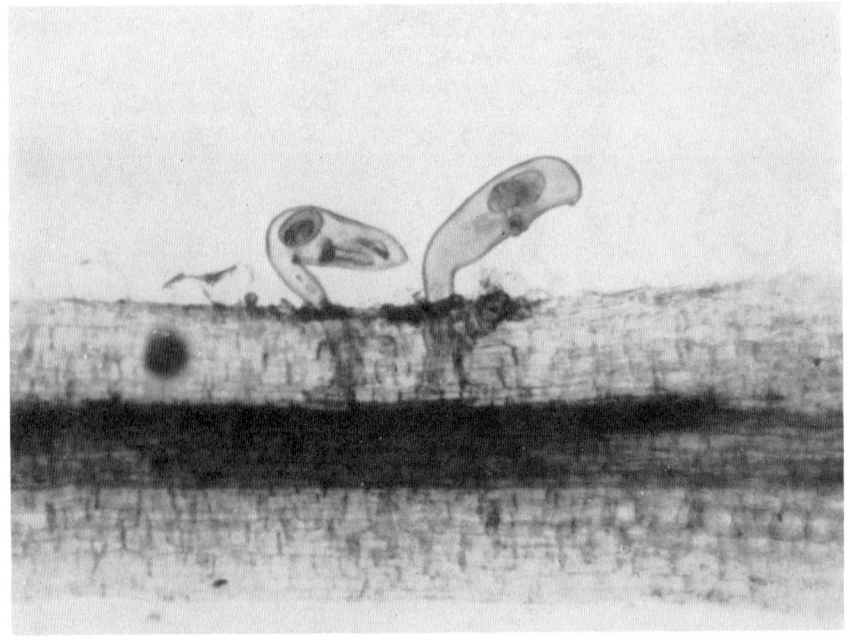

Fig. 11 *R. reniformis* females on soybean roots. (Courtesy of R. V. Rebois.)

developed and the basal bulb extends over the intestine, forming a large flattened lobe. Females have two opposed ovaries and the vulva is located 60-75% of the distance from the lip region to the terminus. Females produce a gelatinous matrix from six unicellular pear-shaped glands that empty into the vulva (Sivakumar and Seshadri, 1971). Most of the eggs are deposited externally in the matrix (Sivukumar and Seshadri, 1971).

C. Life Cycle

The life history of *R. reniformis* on soybean has not been determined completely, but is probably similar to that on other hosts, such as bhendi (*Abelmoschus esculentus* L.) (Sivakumar and Seshadri, 1971). Eggs are normally deposited into the protective gelatinous matrix. A mobile vermiform first-stage juvenile forms in about four days. The first molt occurs within the next 24 hours inside the egg and then the second-stage juvenile hatches. The third- and fourth-stage juveniles are inactive and a stylet is not formed. The fourth-stage female juveniles become the infective immature females. The immature females feed semiendoparasitically with the anterior one-third of their bodies inside the root. The area of the body immediately surrounding the vulva bulges, and the female attains the characteristic reniform shape. An average of 12 eggs per day were produced, but only 45 eggs were found per egg mass. The life cycle from egg to egg required 24-29 days on bhendi (Sivakumar and Seshadri, 1971); 17 to 23 days on cotton (Birchfield and Brister, 1962), and 19 days on soybean at 29.5°C (Rebois, 1973). *R. reniformis* produced 39 eggs per egg mass on soybean (Peacock, 1956).

   R. reniformis survived in air-dried (3.3% moisture) soil maintained at 20-25°C for seven months. Numbers of immature females and males were approximately 270 per 250 cm^3 of air-dried soil and 2160 per 250 cm^3 nondried soil (Birchfield and Martin, 1967).

   Two races (A and B) of this nematode have been characterized by the differential reaction on cowpea, castor, and cotton (Dasgupta and Seshadri, 1971). Race A completes its life cycle on all three differential hosts, but race B would not complete its life cycle on castor or cotton.

   Temperature is an important edaphic factor that will affect the completion of the life cycle as well as the amount of damage caused by *R. reniformis*. At 27 days after inoculation, no egg masses were found on roots of plants grown at 15 or 36°C. Population development and the effect of the nematode on root growth were greater at 29.5°C than at 36.0, 21.5, or 15.0°C (Rebois, 1973).

D. Associated Symptoms and Damage

The foliar symptoms for *R. reniformis* infection include stunting and chlorosis. The severely stunted root system may become necrotic. The most common sign of the nematode is the soil-covered egg masses on the root.

   Immature females penetrate the root intercellularly and become oriented perpendicularly to the longitudinal axis of the root with the posterior portion of the body outside the root (Rebois et al., 1970). The posterior portion of of the female swells, causing the epidermis to rupture. Cortical cells through which the nematode passes become disorganized and devoid of cytoplasm. Ruptured cortical cells with torn edges are folded centripetally and form a partial stricture around the female (Fig. 12) (Rebois et al., 1975). The nematode presses its lips against one endodermal cell, which becomes the first syncytial cell (Fig. 13). It is the only endodermal cell involved in the early development of the syncytium since the syncytium is composed primarily of parenchymal cells. During the first two days after infection, the cell walls of the pericycle begin to dissolve and break down (Rebois et al., 1975). The syncytia are formed by the ninth day after infection. Endodermal cells near the infection site break down and collapse and the outermost primary xylem vessels and innermost cortical cells are crushed by the enlarging syncytium. Approximately 100-200 cells may eventually be associated with syncytial formation and feeding of one female.

Feeding from pericyclic tissue adjacent to the outermost vessel of the protoxylem pole resulted in maximum female development, egg mass production, and syncytial formation (Rebois et al., 1970). Feeding at the protoxylem poles resulted in the inability of the nematode to complete the life cycle.

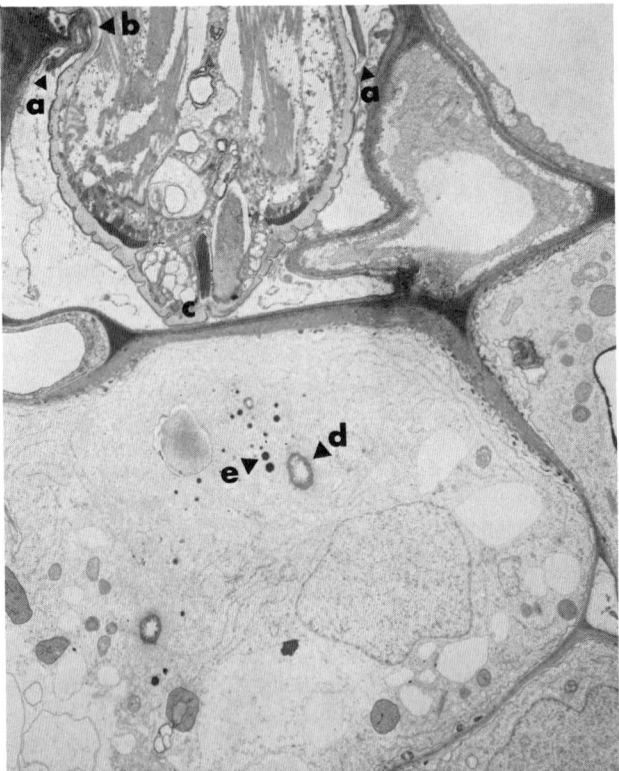

Fig. 12 Head of *R. reniformis* embedded in cell showing torn, penetrated cell walls (a) and constricted nematode body (b) in Lee soybean two days after inoculation. Nematode lips (c) are pressed against cell wall in juxtaposition with an endodermal cell, which shows some wall lysing. Nematode secretion (d) is evident and associated with dense inclusions (e) that resemble secretion granules or possibly hydrolytic reaction products. (Courtesy of R. V. Rebois.)

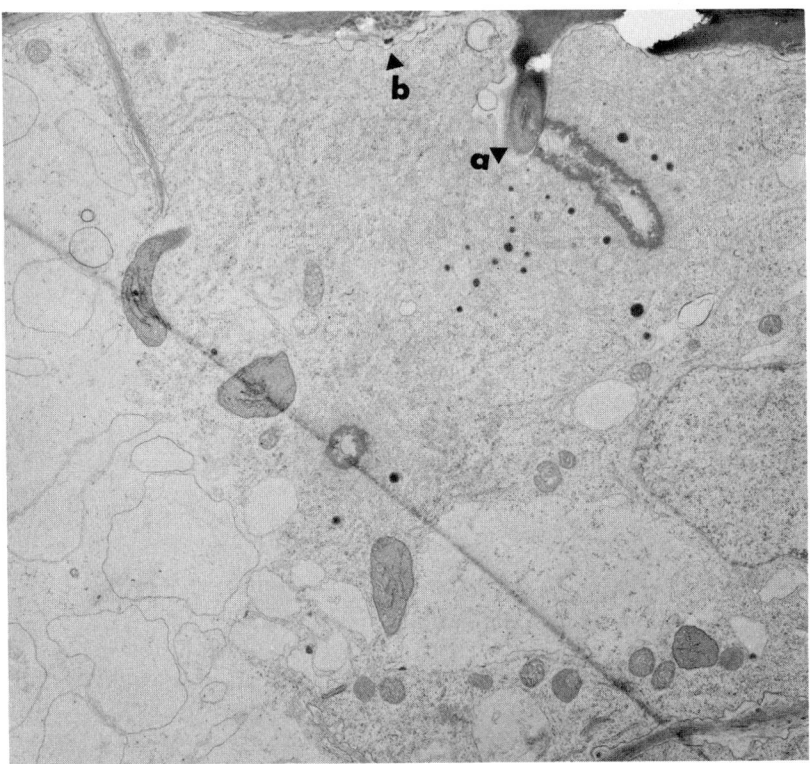

Fig. 13 Stylet (a) of *R. reniformis* inserted into an endodermal cell identified by Casparian strip. Wall near stylet is lysing (b) in Lee soybean two days after inoculation, and separated vestigial wall in advanced stage of lysis separates original pericycle from endodermal cell. The plastid suggests that cytoplasmic flow is away from the stylet. (Courtesy of R. V. Rebois.)

E. Control

1. Rotation

The production of grass crops for two or more years reduces the population of the reniform nematode sufficiently to allow production of a crop without the use of other control tactics (Birchfield, personal communication). Many plants are hosts, limiting the numbers of rotation options available.

2. Resistance

Resistant cultivars are the best means of controlling the reniform nematode. Several cultivars are available (Birchfield et al., 1971; Lim and Castillo, 1979). Many of these also have resistance to *H. glycines*. The genes for resistance to these two nematodes may be separate, but linked (Birchfield et al., 1971).

3. Nematicides

Nematicides, primarily phenamiphos and aldicarb, are effective in reducing the populations of the reniform nematode but have not given adequate yield increases to justify their use (Birchfield, personal communication).

IV. ROOT-KNOT NEMATODES

A. History and Distribution

Root-knot nematodes (*Meloidogyne* spp.) are more widely distributed throughout the world than any other major group of plant-parasitic nematodes (Sasser,
1977). Since *Meloidogyne* spp. are so widely distributed, they probably have
parasitized soybeans wherever the crop is grown. The most common and well
known species, *M. incognita* (Kofoid and White) Chitwood, *M. javanica* (Treub)
Chitwood, *M. arenaria* (Neal) Chitwood, and *M. hapla* Chitwood, are the major
species recovered from soybean fields. *M. incognita* has been the most important species on soybean. *M. javanica* and *M. arenaria* are becoming increasingly important in the warmer climates of the world. *M. hapla* occurs in cold-
er climates and can cause yield loss of soybeans.

M. hapla, M. incognita, M. javanica, and M. arenaria, are worldwide in
distribution (Sasser, 1977). This wide distribution is attributed to a number
of factors (Sasser, 1977): a wide host range, production of large amounts of
eggs that are subject to frequent transport, parthenogenetic reproduction,
and survival under a wide range of soil, moisture, and temperature conditions.

B. Species and Races

The four common species of *Meloidogyne* are known to parasitize soybean: the
peanut root-knot nematode, *M. arenaria*; the southern root-knot nematode, *M.
incognita*; the northern root-knot nematode, *M. hapla*; and the javanese root-
knot nematode, *M. javanica*. *M. bauruensis* (Lordello) Esser, Perry, and Taylor and *M. inornata* Lordello are reported on soybean in Brazil (Lordello,
1956a, b; Esser et al., 1976).

M. hapla and M. javanica usually can be distinguished from M. incognita
and *M. arenaria* on the basis of perineal patterns (Chitwood, 1949). The
separation of *M. arenaria* and *M. incognita* using perineal patterns is difficult
because their morphology overlaps (Eisenback et al., 1981) (Fig. 14). Coup-
ling head shapes and stylet morphology of males (Fig. 15) with the perineal
patterns of females, all four species can usually be distinguished from one
another. *M. bauruensis* is similar to *M. javanica* and was initially described
as a subspecies of the latter species (Lordello, 1956a). *M. inornata* is morphologically similar to *M. incognita*.

Several races of root-knot nematodes are characterized cytologically and
by host differential tests (Eisenback, et al., 1981) (Table 3). *M. incognita
wartellei* Golden and Birchfield (Golden and Birchfield, 1978) was initially
designated as the Wartell race of *M. incognita*. Five populations of *M. incognita* from western Tennessee varied considerably in their reaction to 12 different soybean cultivars (Bernard, 1980). Resistance appeared to be more dependent on the nematode population than on the cultivar.

C. Life Cycle and Biology

The life cycle of *Meloidogyne* spp. is similar on soybean and most other hosts.
The one-celled egg passes through embryogenesis, resulting in a first-stage
juvenile within the egg. This juvenile molts to the second stage within the
egg. The second-stage juvenile emerges from the egg, moves through the
soil, and penetrates the root near the root tip. Giant cells are initiated in the
root and the second-stage juvenile begins to swell.

Soon after the nematode starts to swell, it undergoes three more molts.
The third- and fourth-stage juveniles do not possess a stylet, but the stylet

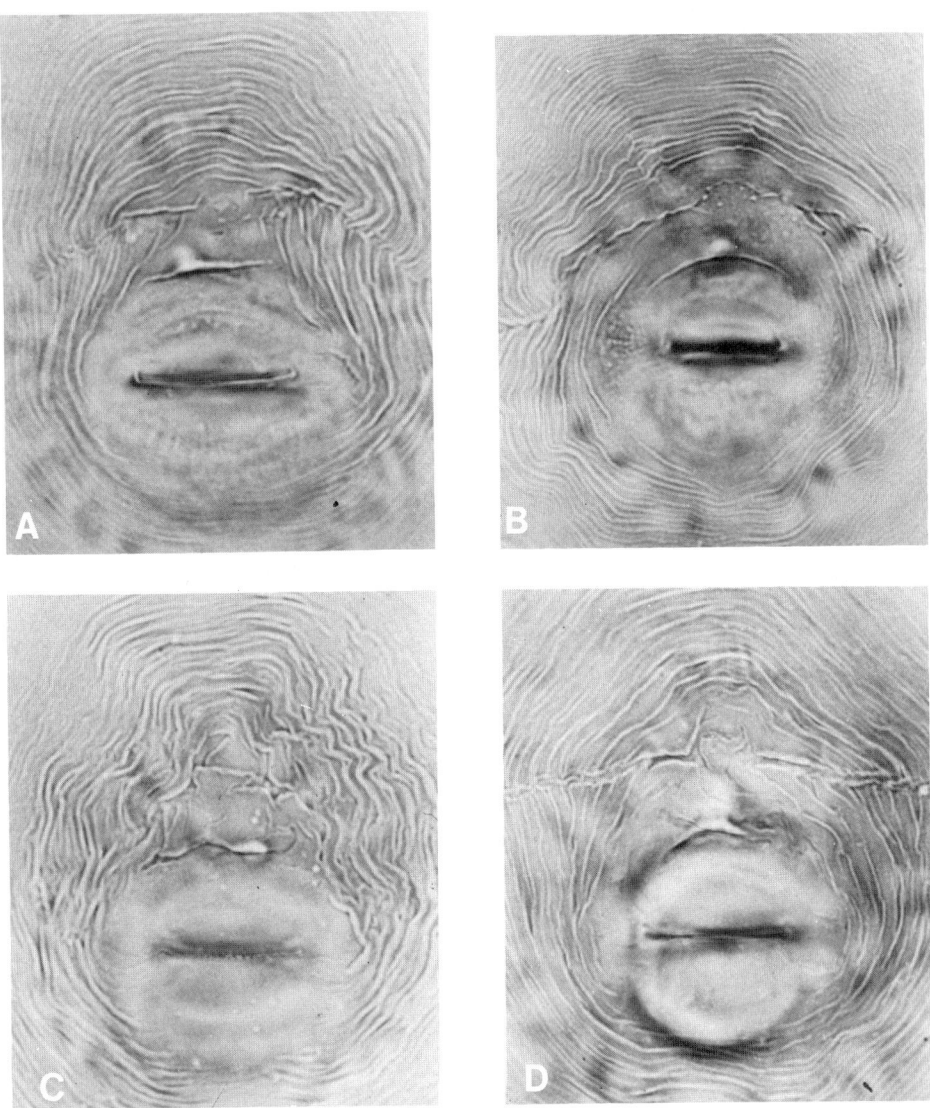

Fig. 14 Perineal patterns of the four common species of *Meloidogyne*: (A) *M. arenaria*, (B) *M. hapla*, (C) *M. incognita*, (D) *M. javanica*. (Courtesy of J. D. Eisenback.)

reappears when the nematode undergoes the final molt and becomes an adult. The females extrude a gelatinous matrix into which eggs are deposited. In the four common species, the males that emerge from the roots after the fourth molt are not normally involved in reproduction.

The rate of development of *Meloidogyne* spp., particularly *M. incognita* is slower on soybean than on good hosts, such as tomato or tobacco. About 12,000 heat units are required for *M. incognita* to complete development and begin laying eggs on soybean (Dropkin, 1963). Approximately 6500-8000 heat units were required for root-knot females to reach maturity on tomato (Tyler,

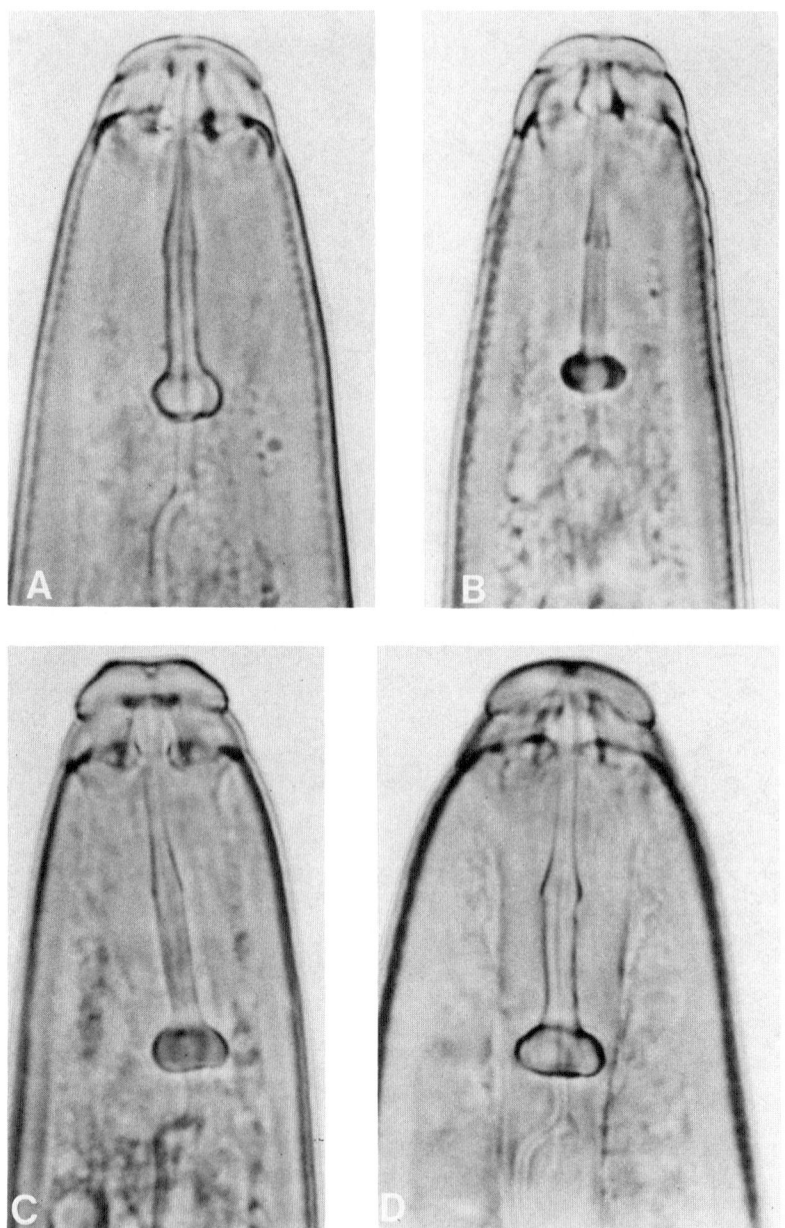

Fig. 15 Head morphology of the four common species of *Meloidogyne*: (A) *M. arenaria*, (B) *M. hapla*, (C) *M. incognita*, (D) *M. javanica*. (Courtesy of J. D. Eisenback.)

Table 3 Differential Host and Cytological Races of *M. incognita*, *M.* and *M. hapla*.

Nematode species	Host reaction		Race
	Host races		
M. incognita	Cotton "Deltapine 16"	Tobacco "NC 95"	
	-	-	1
	-	+	2
	+	-	3
	+	+	4
M. arenaria	Peanut		
	+		1
	-		2
	Cytological races	Chromosome no.	
M. hapla		n = 14-17	A
		2n = 30-31	
		3n = 43-48	B
M. arenaria		2n = 50-56	A
		2n = 34-37	B
M. incognita		2n = 32-36	B
		3n = 40-46	A

Source: From Eisenback et al. (1981).

1933). The life cycle extends beyond 39 days in soybean (Gommers and Drop-kin, 1977). The optimum temperature range for reproduction of *M. incognita* is between 24 and 35°C (Dropkin, 1963; Nardacci and Barker, 1979).

D. Associated Symptoms and Damage

The root gall is the most characteristic symptom of root knot (Fig. 16). Galls vary in size and shape, depending on the species and population density of nematodes in the gall. Generally, the galls caused by *M. hapla* are small with numerous fibrous roots emerging from the gall, and those induced by *M. incognita*, *M. javanica*, and *M. arenaria* are larger.

The above-ground symptoms of root-knot nematode infection vary from slight stunting to severely suppressed growth and chlorosis of the plant. Early foliage senescence may be associated with root knot. The four common *Meloidogyne* spp. suppress plant growth and yield. Yield losses associated with *M. incognita* on soybean were suppressed up to 90% in Florida (Kinloch, 1974a). The severity of damage they cause is dependent upon initial popula-tion density and the environmental conditions during the growing season. For example, *M. incognita* causes much greater damage in Alabama and Florida than it does farther north in North Carolina. The nematode-induced morpho-logical and physiological changes can directly affect growth and yield. Addi-tional effects on the plant can result from interactions with soil microflora and microfauna. Nitrogen fixation is probably not altered sufficiently to be in-

Fig. 16 Root galling of soybean caused by *M. incognita* (A) and *M. javanica* (B). (Photo of *M. javanica* galling is courtesy of K. R. Barker.)

Table 4 Influence of *M. incognita* (Mi) on Nodule Development and Nitrogen-Fixation Capacity (Measured in µmoles Ethylene) by *Rhizobium japonicum* on Forrest and Lee 68 Soybeans at Different Time Intervals

Treatment	Time after inoculation (days)			
	50	75	100	135
Number of nodules				
Forrest + Mi	130.0	217.4	163.1	131.7
Forrest	107.0	177.4	171.6	98.4
Lee 68 + Mi	96.9	155.7	125.9	30.4**
Lee 68	99.3	124.7	135.1	97.4
Dry weight of nodules (mg)				
Forrest + Mi	409.9	423.0	477.7*	386.9
Forrest	420.4	541.7	692.9	347.7
Lee 68 + Mi	525.9	548.4	639.2	105.2**
Lee 68	401.0	697.3	639.3	534.4
µmoles Ethylene				
Forrest + Mi	7.8**	36.6**	3.7*	2.4
Forrest	14.3	12.0	5.2	2.1
Lee 68 + Mi	10.0**	9.0*	2.2	0.6*
Lee 68	5.3	14.1	3.9	1.2

Asterisks (*, **) indicate a significant difference from respective nematode-free controls at P = 0.05 and 0.01. Significance was determined from LSDs for a combined analysis including all time intervals. (All data were transformed to natural logs for statistical analyses.)
Source: From Baldwin et al. (1979).

volved in this effect (Baldwin et al., 1979) (Table 4). A resistant cultivar (Forrest) fixed more N_2 in inoculated plants than controls at 75 days after inoculation but less at 50 and 100 days after inoculation. On a susceptible cultivar (Lee 68), N_2 fixation was greater in inoculated plants than controls at 50 days after inoculation, but decreased from then until the termination of the experiment at 135 days (Baldwin et al., 1979).

The morphological changes associated with root-knot nematode infection occur very quickly after infection (Kaplan et al., 1979). Within one day after infection most second-stage juveniles have migrated to the protostele. At two days after inoculation, some nuclei in the phloem cells were binucleate or in the process of mitosis. Multinucleate cells were associated with the head region of many of the juveniles present in the stele on the third day. Giant cells were evident around second-stage juveniles at the sixth day after inoculation, and extensive giant cell formation was associated with third- and fourth-stage juveniles by the eighth day. Twelve days after inoculation, large giant cells with granular cytoplasm were associated with fourth-stage juveniles with enlarged ovaries.

There is a lack of connective tissue between the normal xylem and phloem of the primary root and the lateral root, in the area of the giant cell (Byrne et al., 1977). The root-stele tissue is triarch instead of tetrarch in the area of the infected sites. The orientation was altered from the normal 120° orientation for a triarch root to ones that were 90°-90°-180°.

Major metabolic changes occur in the plant at the nematode feeding site
(Gommers and Dropkin, 1977). Protein in the giant cells of soybean infected
with *M. incognita* was increased 3.3-fold. The free amino acids increased sixfold, and glucose was much higher in the nematode-induced transfer cells
(Gommers and Dropkin, 1977).

E. Relationship of Numbers of *Meloidogyne* spp. to Crop Response

Initial population density is inversely proportional to plant growth and yield
(Fig. 17). Standard regression models or the Seinhorst regression model fit
the data (Kinloch, 1982; Nardacci and Barker, 1979). The numbers of nematodes required to cause crop injury vary and are modified by environmental
factors. For example, there was a high tolerance to *M. incognita* at low temperatures (18 and 22°C) and a low tolerance at high soil temperatures (30°C)
(Nardacci and Barker, 1979).

Moisture, particularly at the time of pod filling, has a major impact on
the yield of soybeans. Soil moisture changed the magnitude of the relationship
of nematode numbers and crop response, but not the slope of the curve (Barker, 1982). Yields were greater as the soil moisture was increased.

F. Effects of Microflora and Microfauna on the Symptoms and Damage Caused by Root–Knot Nematodes

Some interactions of microorganisms with *Meloidogyne* species may be adverse,
whereas others may be beneficial. The yields of soybeans were less in treatments involving *Fusarium oxysporum* or *F. solani* and *M. incognita* than either
organism alone (Goswami and Agarwal, 1978). A similar relationship existed
with *H. glycines* and *M. incognita* (Ross, 1964a).

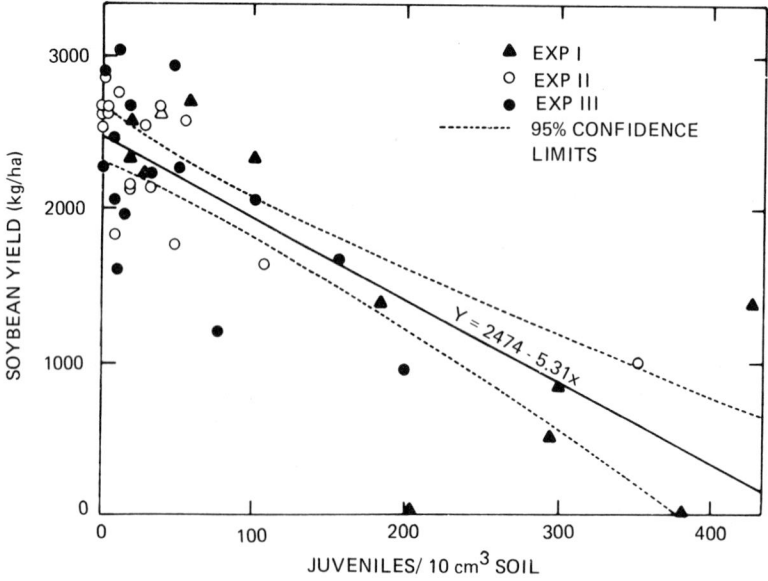

Fig. 17 Relationship between soybean yield and numbers of *M. incognita*
juveniles. (From Kinloch, 1982.)

The presence of some microorganisms may result in less damage from a nematode than if that nematode alone were present. *Glomus macrocarpus* limited the yield loss caused by *M. incognita* (Kellam and Schenck, 1980), but such interactions differ with various cultivars (Schenck et al., 1975), thereby making it impossible to predict without further experimentation which combinations will be beneficial and which will be harmful.

G. Control

Mixed populations of two or more species of *Meloidogyne* plus the wide host range of this group of nematodes complicates development of a satisfactory control strategy. The differential control of these species by nematicides further confounds the situation. It is possible to develop a solution, but diagnosis, including accurate identification of species and sometimes races, is important.

1. Crop Rotation

Crop rotation schemes are relatively difficult to develop for root-knot nematode control in most areas of the world because of their wide host range and the possible presence of several species in a field. With the proper identification of species, rotation systems are possible and practical. Corn and cotton are good hosts of *M. incognita*, whereas peanut is not a host (Johnson et al., 1975). A corn and soybean rotation or a corn or soybean monoculture will maintain moderate to high numbers of this nematode (Kinloch, 1974b). Peanut, soybean, clover, and potato are good hosts for *M. hapla* (Nishizawa, 1978). Many grass crops are nonhosts of this nematode.

The effect of rotation in controlling root-knot nematodes is favored by increasing the time that a nonhost is grown in relation to a susceptible host. Such combinations should effectively keep root-knot nematodes at population levels that will not damage the crop.

2. Resistance

Breeding efforts have been directed primarily at *M. incognita*, resulting in the release of many resistant cultivars. Some cultivars are resistant to *M. arenaria* and/or *M. javanica*. There are varying degrees of resistance in soybeans to root-knot nematodes, particularly to *M. incognita*. In addition, the behavior of geographical isolates (physiological races) is modified by cultivars, and conversely the cultivars respond differently when exposed to the isolates (Boquet et al., 1975).

The environment may modify the amount of damage caused by the nematodes. In North Carolina, resistant cultivars perform as well or better with high population densities as with no nematodes in microplot tests (Barker, 1982). Resistant cultivars frequently outperform a susceptible cultivar treated with a nematicide in *M. incognita*-infested fields in Alabama and Florida (Kinloch and Hinson, 1973; Rodriguez-Kabana and Thurlow, 1980). The resistant cultivars yielded still more by the addition of some nematicides. Figure 18 shows dramatically the effect of planting a root-knot resistant soybean variety. The background area was planted with the resistant variety Jackson. This grower from Florida then ran out of this resistant variety and planted the susceptible variety Hood (in the foreground) to finish off his field.

Juveniles of *M. incognita* penetrated roots of susceptible and resistant cultivars equally (Veech and Endo, 1970). Cortical and vascular cells became necrotic in resistant soybean in response to nematode feeding (Veech and

Fig. 18 Results showing the effect of root-knot nematode (*M. incognita*) on the susceptible soybean variety Hood in Escambia County, Florida. (Courtesy of R. V. Rebois.)

and Endo, 1970). The incompatible response of resistant soybean was dependent upon glyceollin accumulation (Kaplan et al., 1979). Glyceollin in the susceptible cultivar Pickett was less than 15 μg/g of fresh root, but in the resistant cultivar Centennial, a high level of accumulation occurred (Kaplan et al., 1980a) (Fig. 19). The highest concentration of the glyceollin was in the stele.

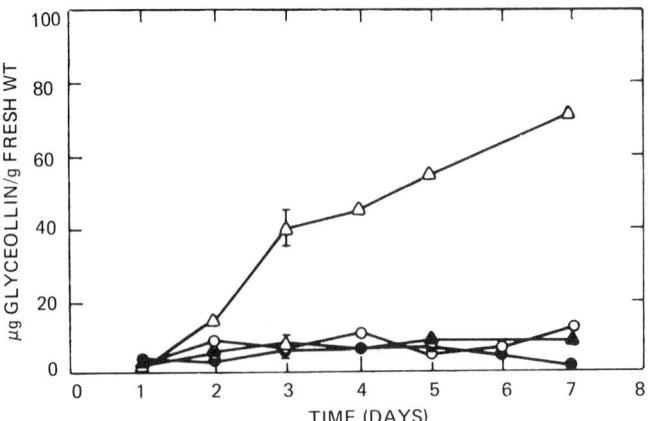

Fig. 19 Accumulation of glyceollin in soybean roots infected with root-knot juveniles. △ = Centennial × *M. incognita,* o = Pickett 71 × *M. incognita,* ▲ = Centennial × *M. javanica,* ● = Pickett 71 × *M. javanica.* (From Kaplan et al, 1980a.)

Glyceollin was nematistatic to *M. incognita* but not *M. javanica* (Kaplan et al., 1980b). This material inhibited respiration in soybean mitochondria as well as the electron transport system. Glyceollin may have a similar effect upon the mitochondria of *M. incognita*.

Continuous use of resistant cultivars is an unwise practice because heavy selection pressure could lead to the buildup of dense populations able to attack the resistant soybean. For example, a few *M. incognita* juveniles were able to penetrate resistant soybean roots and complete their life cycles (Veech and Endo, 1970). A population of sufficient density could build up in a few generations to damage a resistant soybean.

3. Chemical Control

Several nematicides are effective for controlling root-knot nematodes. The fumigants are generally more effective in controlling the four common species than the nonfumigants (Kinloch, 1980). Some chemicals are more effective against one species of *Meloidogyne* than another species. For example, ethoprop is ineffective for controlling *M. hapla* on soybeans in North Carolina but does provide some control of *M. incognita* (Schmitt, unpublished).

The method of application must be that which achieves optimal diffusion of the chemical. The fumigant DBCP was most effective when it was injected 20 cm deep in controlling *M. incognita* in Georgia (Minton et al., 1979). Subsoiling to a depth of approximately 40 cm deep to break a hardpan was equivalent to the nematicide treatment in increasing yields and decreasing root-knot index (Minton and Parker, 1975).

V. THE LANCE NEMATODES

A. History and Distribution

Several species of *Hoplolaimus* have been associated with soybeans. *Hoplolaimus columbus* Sher, the most pathogenic lance nematode, was found in the 1950s and described in 1963 (Sher, 1963). At that time *H. columbus* was known only from three counties in South Carolina, but was reported from seven additional South Carolina counties in 1968 (Fassuliotis et al., 1968). By 1982, *H. columbus* infestations were found in the coastal plain areas of Georgia, North Carolina, South Carolina, and portions of Louisiana (Astudillo and Birchfield, 1980; Barker, personal communication; Bird et al., 1974). *H. columbus* appears to be indigenous to the southeastern United States. This nematode parasitizes many weeds and with the advent of intensified crop production it became an agricultural pest (Bird and Hogger, 1973).

H. galeatus (Cobb) Thorne has been associated with soybeans, but not given much significance since it does not appear to cause severe damage. More recently, *H. magnistylus* Robbins has been found on soybeans from Arkansas, Mississippi, and Tennessee (Robbins, 1982b).

B. Species

Lance nematodes are relatively long vermiform nematodes (length = 0.8-1.8 mm) and possess a massive well-developed stylet. Stylet length ranges from 30-61 μm with anteriorly projected basal knobs. Females have two outstretched ovaries, and the vulva is located 50-60% of the body length from the anterior end. The tail is rounded. *Hoplolaimus* spp. also usually have one phasmid anterior to the vulva and one posterior to the vulva. The lateral field has four or fewer incisures. *H. columbus* has one incisure in the lateral field.

The gross morphology of males is similar to that of females. The caudal alae extend around the terminus to just anterior to the spicules. Males are common for *H. galeatus* and *H. magnistylus*, but rare for *H. columbus*.

C. Life Cycle and Biology

The life cycle of *H. columbus* on soybean required 45-49 days at temperatures between 21 and 27°C (Smith, 1969). The nematode reproduces parthenogenetically, although males have been found in some field populations (Fassuliotis, 1974). Each female produces at least 15 eggs, but the fecundity has been difficult to establish due to nematode migration (Fassuliotis, 1975). Embryogenesis is typical for plant-parasitic nematodes, with the eggs hatching 9-15 days after oviposition (Fassuliotis, 1974).

The nematode is predominantly endoparasitic (Lewis et al., 1976). Greater root populations of adults and juveniles occurred at 30°C than at 25 or 20°C (Nyczepir and Lewis, 1979). They are oriented parallel to the root axis; however, a few partially encircle the vascular region. Viable *H. columbus* was recovered five years after placement in soil with a moisture content of 0.2-0.3% (Fassuliotis, 1971). A unique basal granular layer in the cuticle may give *H. columbus* the ability to survive osmotic stress and desiccation (Lewis and Huff, 1976).

D. Associated Symptoms and Damage

H. columbus, the Columbia lance nematode, feeds as an ectoparasite and as an endoparasite. Hosts of *H. columbus* which are of primary economic importance are cotton and soybean. Chlorosis, stunting, and reduced pod production are the primary above-ground symptoms of *H. columbus* infection of soybean under field conditions (Fig. 20).

Fig. 20 Field damage to soybean caused by *H. columbus* in South Carolina. (Courtesy of S. A. Lewis.)

Nematode migration results in ruptured cortical cells and abnormally dark, enlarged nuclei in the cortex (Fig. 21). Cells containing brown, granular cytoplasm were also observed in tissues where *H. columbus* fed. The damage to cortical tissues results in the sloughing of these tissues and the lack of secondary roots. *H. columbus* will also congregate at an entry point or wound (Lewis et al., 1976).

E. Interactions with Other Organisms

H. columbus is aggressive and can compete well with other nematodes as the primary nematode pest (Bird et al., 1974; Fassuliotis et al., 1968; Fortnum and Lewis, 1978a). For example, *M. incognita* gall indices were less when Davis soybean was also infected with *H. columbus* (Fortnum and Lewis, 1978a). Numbers of juveniles and adults of *H. columbus* in the soil were more numerous in the presence of *Cylindrocladium crotalariae*, but numbers were unaffected in the root (Fortnum and Lewis, 1978a). *Pratylenchus scribneri* population densities and *M. incognita* gall indices were lower in the presence of *H. columbus* or *C. crotalariae*. Necrosis caused by *C. crotalariae* was not affected by the nematodes. Nodulation of soybean roots by *Rhizobium japonicum* was lower with the concomitant infection of *P. scribneri*, *H. columbus*, and *C. crotalariae*. There was a positive correlation between root populations of *Rhizoctonia* spp. and *H. columbus* and *M. incognita* at the time soybeans were flowering (Fortnum and Lewis, 1978b).

F. Control

The Columbia lance nematode is most effectively controlled by breaking a hard pan, if one exists, and applying EDB (Steve Lewis, personal communication). Resistant cultivars and some nonfumigant nematicides (aldicarb and phenamiphos) give fair control of this nematode (Steve Lewis, personal communication).

VI. STING NEMATODES

A. History and Distribution

Belonolaimus longicaudatus Rau may be indigenous to the southeastern United States and is the sting nematode of greatest economic importance in soybean production. The first record of sting nematode on soybeans is from the Holland, Virginia, area (Owens, 1951). This nematode was probably *B. longicaudatus* but was identified as *B. gracilis* (Steiner, 1949). *B. longicaudatus* may be a composite taxon involving two or more species (Robbins, personal communication).

The nematode is distributed primarily in the southern Atlantic Seaboard states from Virginia to Florida. Infestations also occur in Alabama, Arkansas, Louisiana, Oklahoma, Texas, Kansas, and New Jersey (Christie, 1959; Dickerson et al., 1972; Holdeman, 1955; Myers, 1979; Riggs, 1961; Russell and Sturgeon, 1969).

Many plants are parasitized by *B. longicaudatus*. The widely cultivated and important field crops, soybean, cotton, corn, and peanut, are excellent hosts for *B. longicaudatus*. Strawberry, small grains, forage crops, vegetables, ornamentals, trees, and turfgrass are also good hosts. A large number of weeds are excellent or good hosts (Robbins and Barker, 1973; Holdeman and Graham, 1953).

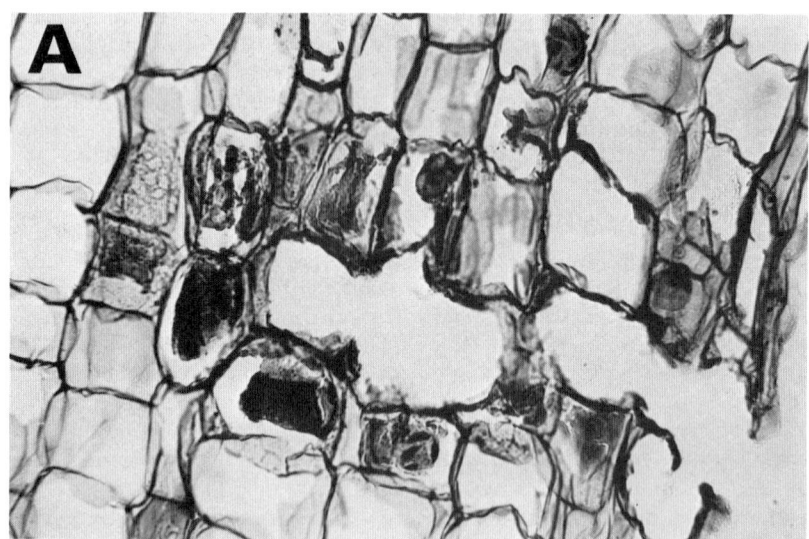

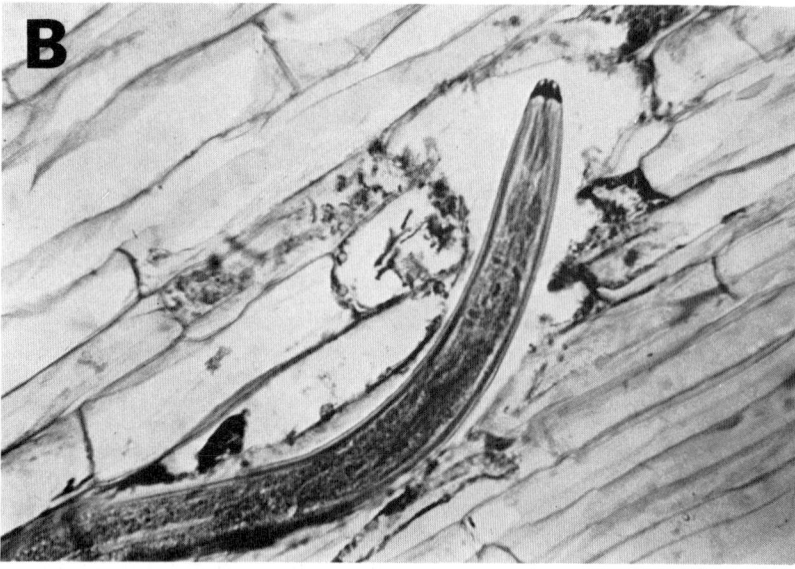

Fig. 21 *H. columbus* in roots of soybean. (A) Epidermal-cortical region infected and vacated by the nematode. (B) Adult in cortex with broken cells. (C) Nematode pathway through the cortex. (D) Nematode penetrating endodermis. (Courtesy of S. A. Lewis.)

B. Species

The sting nematodes are among the largest plant-parasitic nematodes. Adults of *B. longicaudatus* range from 2.0 to 3.0 mm long. The lip region is hemispheric and offset by a deep constriction. The lateral field consists of a single incisure extending from the base of the lip region to near the tail terminus. The stylet is long (100-140 μm) and thin. The stylet knobs are rounded. The vulva is located 46-54% of the body length from the anterior end and the tail is rounded.

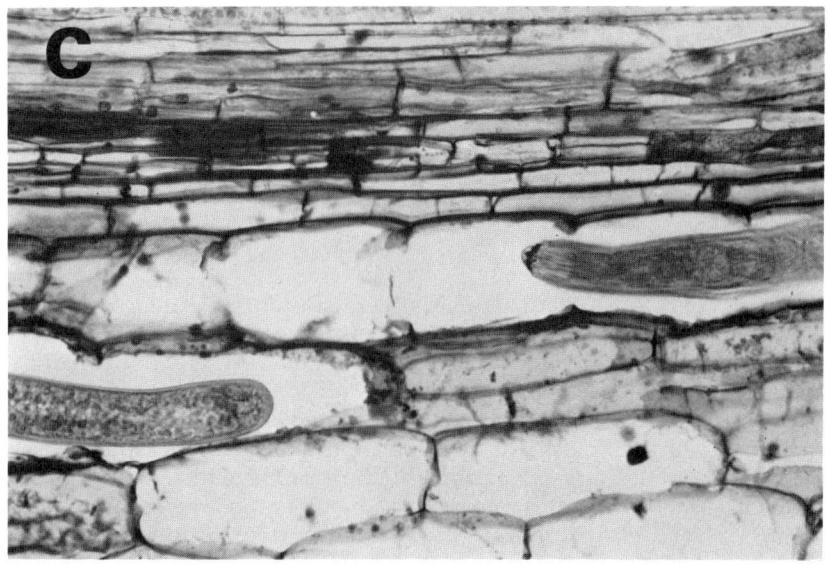

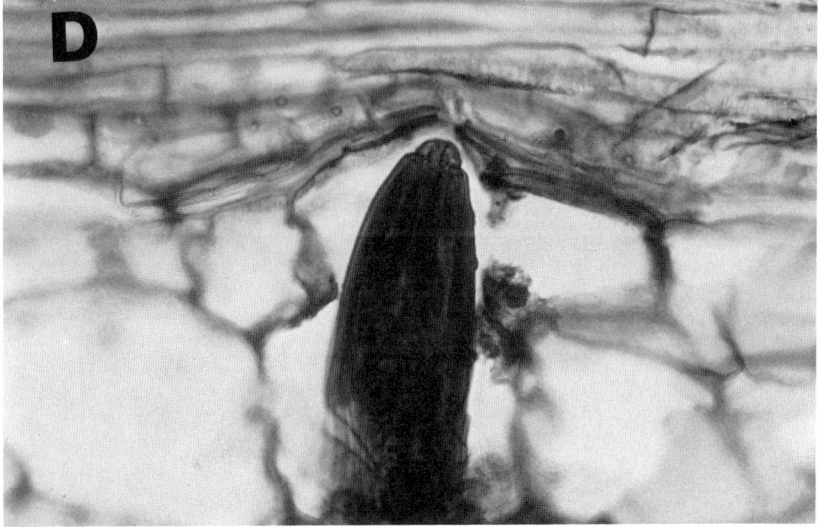

Fig. 21 (Continued)

The gross morphology of the male is similar to the female. The male tail is narrower and more pointed than the female tail. The caudal alae extend around the terminus to approximately 20 or 25 μm anterior to the spicules.

C. Life Cycle and Biology

The life cycle is presumed to have four juvenile stages and an adult stage. *B. longicaudatus* normally occurs in sands or loamy sands with a sand content of 84-94% (Miller, 1972; Robbins and Barker, 1971) but was recovered from muck soils in Florida (Rhoades, 1980). The optimum temperature for reproduction is approximately 30°C (Boyd and Perry, 1970; Robbins and Barker,

1971). Populations of this nematode decline rapidly during the winter, and they are barely detectable in the spring.

There are a number of biotypes of the sting nematode. Three Florida populations of *B. longicaudatus* were differentiated as races on tomato, rough lemon, peanut, and strawberry (Abu-Gharbieh and Perry, 1970). Of three North Carolina populations and one Georgia population of *B. longicaudatus*, only the Georgia population increased on cucumber and dandelion (Robbins and Barker, 1973).

D. Associated Symptoms and Damage

Above-ground symptoms of *B. longicaudatus* include stunting, chlorosis, and wilting and are similar to those of many other causes. Infested areas in a field vary in shape and size. Root symptoms include necrosis, stubby roots, and lack of feeder roots (Fig. 22).

Nematode feeding sites on bean (*Phaseolus vulgaris* L.) were at root apices, just posterior to the apices, or at the base of lateral roots (Standifer and Perry, 1960). A yellow area developed at the feeding site. A few days after the onset of feeding near the root apices, root tips became swollen; the roots curved in the form of a "J", and growth apparently continued. Lesions on bean roots formed at the root apices and zone of maturation (Standifer, 1959). Lesions often extended from the cortex into the stele and consisted of a cavity encircled by injured cells. Some biochemical process in the nematode or host is the most probable cause of the severe damage caused by the sting nematode (Standifer, 1959).

E. Control

Crop rotation and chemical control are effective means of controlling *B. longicaudatus*. Selection of rotation crops may be difficult because several nonhost crops are not feasible for many growers to produce. Tobacco and watermelon,

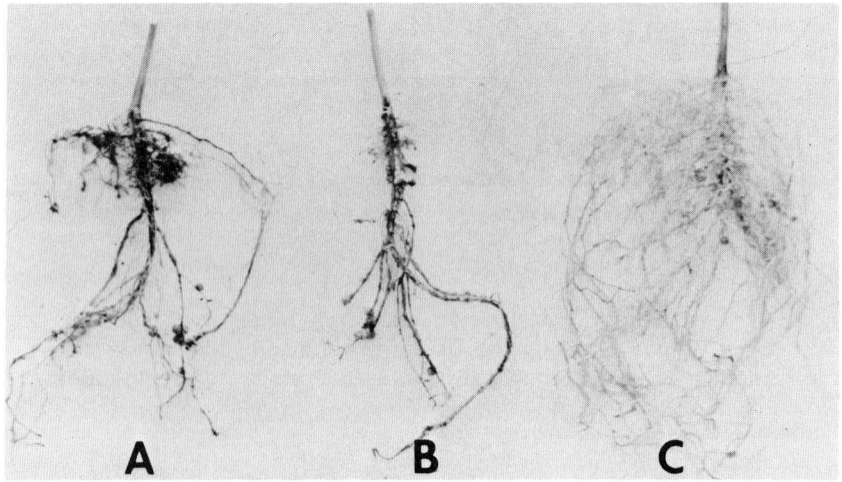

Fig. 22 Damage caused to soybean roots by *B. longicaudatus*: (A) Georgia population, (B) North Carolina population, (C) Control. (Courtesy of K. R. Barker.)

for instance, are excellent rotational crops, but do not fit into certain growers' production systems.

Nematicides may be most practical for control of this devastating pest. Phenamiphos has given excellent control of *B. longicaudatus* and significant yield increases when compared to several nematicides tested in North Carolina (Schmitt, unpublished). Aldicarb has also provided significant control and yield increases.

VII. THE LESION NEMATODES

A. History and Distribution

Pratylenchus brachyurus (Godfrey) Filipjev and Schuurmans Stekhoven attacked soybeans used as an indicator for root-knot nematode infestations in pineapple fields (Godfrey, 1929). This nematode, identified as *P. pratensis* (deMan) Filipjev, was reported on soybeans from the Hawaiian Islands in 1940 (Parris, 1940). The cultivars Abura and Rio Grande were severely damaged in Sao Paulo, Brazil (Lordello, 1955) by lesion nematodes, probably *P. brachyurus* (Lordello et al., 1958) and *P. pratensis* (Kloss, 1960).

Other species of *Pratylenchus* associated with soybeans include *P. penetrans* (Cobb) Filipjev and Schuurmans Stekhoven (Gotoh and Ohshima, 1963; Rebois and Cairns, 1968); *P. coffeae* (Zimmermann) Filipjev and Schuurmans Stekhoven (Gotoh and Ohshima, 1963); *P. hexincisus* Taylor and Jenkins (Ferris et al., 1971; Rebois and Cairns, 1968); *P. neglectus* (Rensch) Filipjev and Schuurmans Stekhoven (Gotoh and Ohshima, 1963; Rebois and Cairns, 1968; Ferris et al., 1971); *P. crenatus* Loof (Ferris et al., 1971); *P. alleni* Ferris (Burns, 1971; Ferris et al., 1971); *P. scribneri* Steiner (Ferris et al., 1971); *P. agilis* Thorne and Malek (Golden and Rebois, 1978); and *P. zeae* Graham (Endo, 1959; Rebois and Cairns, 1968; Golden and Rebois, 1978). Although *P. vulnus* Allen and Jensen has not been associated with soybeans in the field, this plant is a host (Acosta and Malek, 1979).

P. scribneri may be the most widely distributed species in the United States. It is associated with soybeans in both subtemperate and temperate regions (Dickerson, 1979; Rebois and Golden, 1978). *P. penetrans* occurs in soybean-producing areas of the midwestern, southern, and eastern United States (Ferris et al., 1971; Golden and Rebois, 1978; Taylor et al., 1958; Jenkins, et al., 1957). In addition, *P. penetrans* has been associated with soybean in Japan (Gotoh and Ohshima, 1963). Although it was found in Maryland (Jenkins et al., 1957), *P. brachyurus* is limited primarily to subtemperate, subtropical, and tropical regions (Acosta, 1977; Fielding and Hollis, 1956; Lordello et al., 1958; Parris, 1940; Rebois and Cairns, 1968).

The wide overlapping geographical distribution of lesion nematodes that parasitize soybean may be due, in part, to the large number of common hosts including corn, cotton, tobacco, wheat, alfalfa, apple, peach, and citrus (Goodey et al., 1965). In addition, large numbers of weeds serve as hosts and reservoirs of many species of *Pratylenchus* (Manuel et al., 1980).

B. Species

Lesion nematodes, initially called meadow nematodes, were assigned to the genera *Tylenchus* and *Anguillulina* until 1936, when *P. pratensis* was described. *Pratylenchus* spp. are migratory endoparasites of subterranean plant parts (Fig. 23). The nematodes are vermiform ranging from 0.3 to 0.9 mm long. The anterior end of the nematode is broadly flattened with the off-

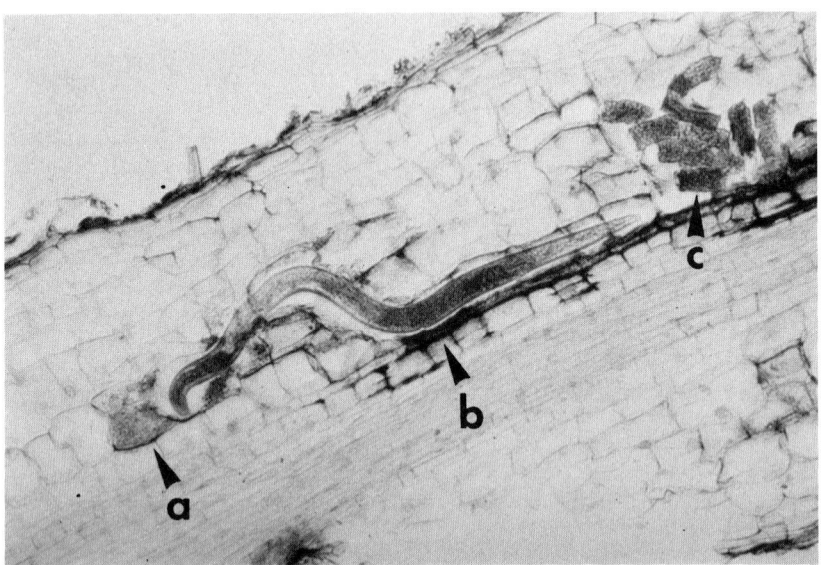

Fig. 23 Longitudinal section of a soybean root 18 days after infection with granular cytoplasm (a) around the anterior end of a root-lesion nematode, necrosis around its body (b), necrosis of endodermal cells, and sections of nematodes (c) in a multilayer cavity forming in the cortex. (Courtesy of N. Acosta and R. B. Malek.)

set lip region bearing two to four annules. The lateral fields contain four to six incisures. The terminus is bluntly rounded. *Pratylenchus* spp. have a well-developed but rather short stylet, ranging from 15 to 20 μm long, with conspicuous basal knobs. In most *Pratylenchus* spp. females, the vulva is approximately 70-80% of the distance from the head to the terminus. Males are common in some species and either rare or unknown in others.

Species of lesion nematodes vary in their temperature optimum, and the optimum may be different within species, depending on the soybean cultivar (Acosta and Malek, 1979; Dickerson, 1979; Lindsey and Cairns, 1971). The optimum temperature for *P. penetrans* and *P. vulnus* is 25°C (Acosta and Malek, 1979). These two species were able to reproduce at 15°C. *P. brachyurus*, *P. coffeae*, *P. neglectus*, and *P. zeae* reproduce best at 30°C (Acosta and Malek, 1979). Temperatures from 30 to 35°C resulted in the most reproduction for *P. scribneri* and 30 to 37.5°C for *P. alleni* (Acosta and Malek, 1979; Dickerson, 1979).

The greatest amount of root invasion by *P. alleni* occurred at pH 6.0 and the least at 4.0 (Burns, 1971). The highest concentration of potassium and thickest outer epidermal walls occurred at pH 4.0. More *P. alleni* occurred in roots grown in soil maintained from field capacity to temporary wilting than above field capacity (Norton and Burns, 1971). There were also more males in the drier regime.

C. Life Cycle

The lesion nematodes have four juvenile stages and an adult stage. Eggs, usually deposited in roots, undergo embryogenesis resulting in a first-stage juvenile. One molt occurs inside the egg, and the second-stage juvenile hatches and begins to feed. The juveniles will molt three more times before they become adults. The second-, third-, and fourth-stage juveniles and adults are infective and feed on root tissue.

D. Associated Symptoms and Damage

Chlorosis and stunting are the foliar symptoms most commonly associated with lesion nematode infection. Plants in some areas of a field may be killed. These nematodes can severely suppress root growth (Ferris and Bernard, 1962; Zirakparvar, 1981). Lesions are typically associated with *Pratylenchus* infection and may be observed on primary and secondary roots (Fig. 24) (Acosta and Malek, 1981). Some thickening of intact cell walls adjacent to the nematode-affected cells occurs (Lindsey and Cairns, 1971). The alteration of cells may result from a toxic reaction (Brooks and Perry, 1967). Suitability of soybean lines as hosts differs among species of *Pratylenchus* (Acosta et al., 1979; Endo, 1967; Zirakparvar, 1981).

E. Relationship of Numbers of Lesion Nematodes to Crop Loss

The relationship of numbers of lesion nematodes and yield of soybean is modified by cultivar and soil type. Yield losses of Forrest soybeans were linearly related to population densities of *P. brachyurus* in sandy soils but curvilinearly related in a sandy-clay loam (Schmitt and Barker, 1981). Lee 68 soybeans grown in microplots of a loamy sand were very sensitive to *P. penetrans*. These results corroborated field observations in which large numbers of *Pratylenchus brachyurus* were associated with damage in sandy to sandy-loam soils.

F. Interactions with Other Organisms

Some lesion nematodes can increase the severity of diseases caused by other agents. Concomitant infection of *Rhizoctonia solani* Kuehn and *P. brachyurus* increased postemergence damping-off of Pickett soybean from 37 to 79% when compared to *R. solani* alone 21 days after planting (Lindsey and Cairns, 1971). Plant weight was suppressed by *P. penetrans* and air pollutant combinations (Weber et al., 1979).

G. Control

1. Rotation

Crop rotation can be an effective tactic for reducing populations of lesion nematodes, but can be complicated by the occurrence of two or more species in a single field. Numbers of *P. hexincisus, P. scribneri,* and *P. penetrans* increase on corn and soybeans (Ferris and Bernard, 1967). *P. neglectus* increases on soybean, oats, and wheat, and *P. hexincisus* increases on oats (Ferris and Bernard, 1967). Forage mixtures and cotton are also good hosts of *P. scribneri*, but winter-grown wheat resulted in small population densities (Rodriguez-Kabana and Collins, 1980). Population densities of *Pratylenchus* spp. (mixture of *P. zeae* and *P. brachyurus*) were generally higher on monocultured corn or soybean than on cotton or peanut (Johnson et al., 1975). In soil infested with *P. zeae* and *P. brachyurus*, *P. zeae* becomes dominant when corn is planted and *P. brachyurus* becomes dominant when soybean is planted (Endo, 1967).

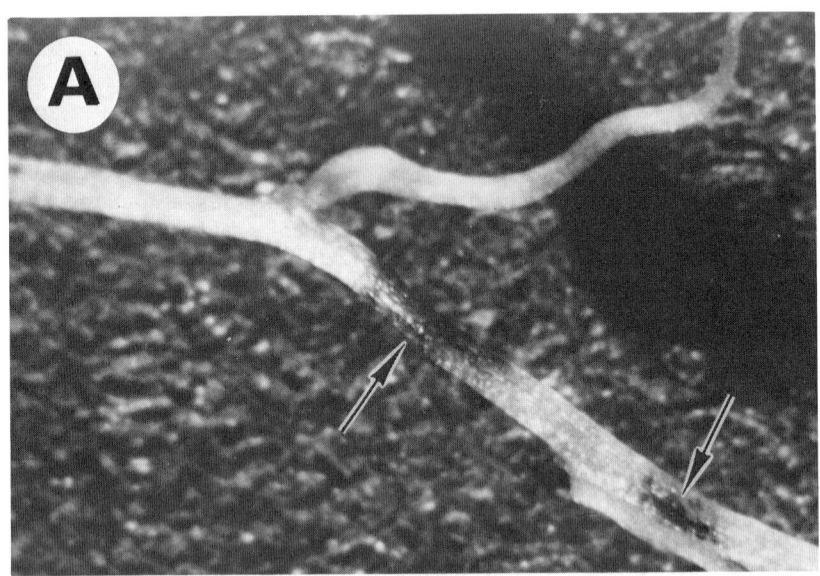

Fig. 24 Stages of lesion development in soybean 45 days after planting in *P. scribneri*-infested soil. (A) Young roots with small lesions with 1-50 nematodes. (B) Older roots with necrotic areas of coalesced lesions containing many nematodes and eggs. (Courtesy of N. Acosta and R. B. Malek.)

2. Resistance

Little is known about the resistance of soybean to lesion nematode. There is some evidence that either resistance or tolerance to these nematodes does exist. Forrest soybean is resistant to *P. scribneri* (Acosta et al., 1979), although it was one of the most susceptible to *P. brachyurus* (Schmitt and Barker, 1981). The cultivar Essex is tolerant to *P. brachyurus* (Schmitt and Barker, 1981).

3. Nematicides

There are few published data on the effect of nematicides on increased soybean yields resulting from control of lesion nematodes. Of several nematicides tested, only DBCP increased soybean yields in a field with a mixture of *P. brachyurus* and *P. zeae* (Schmitt, unpublished).

VIII. OTHER NEMATODES

Several genera of nematodes that cause little to no damage, or at least not practical to control, are associated with soybean. A possible exception is the dagger nematode, *Xiphinema americanum* Cobb. This nematode caused significant loss of several cultivars in a sandy soil in North Carolina (Schmitt, unpublished). Other nematodes frequently encountered in soybean fields are *Tylenchorhynchus, Helicotylenchus, Scutellonema,* and *Rotylenchus* (Fielding and Hollis, 1956; Jenkins et al., 1957; Rebois and Cairns, 1968), *Criconemella*, and *Paratrichodorus* (Coursen et al., 1958; Jenkins et al., 1957; Mai et al., 1960; Taylor et al., 1958). *Tylenchorhynchus claytoni* Steiner suppressed yields of soybean by 21% in microplots (Ross, et al., 1967), but control of this nematode with nematicides did not give significant yield increases (Schmitt, unpublished). *Criconemella ornata* (Barker et al., 1982), *Scutellonema* spp., and *Paratrichodorus minor* (Colbran) Siddiqi did not affect yield (Schmitt, unpublished). Numbers of the latter nematode are negatively correlated with yield in experiments with nematicides. In these experiments, however, *H. glycines* or *M. hapla* were the primary pathogens and population numbers of *P. minor* followed the same trend as numbers of these nematodes (Schmitt, unpublished).

IX. OUTLOOK

Effective tactics have been developed for the control of many soybean-parasitic nematodes. Optimum utilization of these tactics is rarely achieved because of conservative approaches to control. Before a high degree of precision can be achieved in predicting crop loss from nematodes, reliable pathogenicity indices must be developed for the major pathogenic nematodes as well as damage threshold numbers under expected environmental conditions. Scientists must strive to understand the nematode's biology and population dynamics as well as their interactions with soybeans under various crop management systems. When we understand the nematode's behavior and the impact and mechanisms of control tactics, economical control of soybean-parasitic nematodes can be achieved without adverse side effects.

REFERENCES

Abu-Gharbieh, W. I., and Perry, V. G. (1970). Host differences among Florida populations of *Belonolaimus longicaudatus* Rau. *J. Nematol.* 2: 209-216.

Acosta, N. (1977). Host-parasite relationships of lesion nematodes, *Pratylenchus* spp. Filipjev, and soybean *Glycine max* (L.) Merr. Ph.D. Dissertation, Univ. of Illinois.

Acosta, N., and Malek, R. B. (1979). Influence of temperature on population development of eight species of *Pratylenchus* on soybean. *J. Nematol.* 11: 229-232.

Acosta, N., and Malek, R. B. (1981). Symptomatology and histopathology of soybean roots infected by *Pratylenchus scribneri* and *P. alleni*. *J. Nematol. 13*: 6-12.

Acosta, N., Malek, R. B., and Edwards, D. I. (1979). Susceptibility of soybean cultivars to *Pratylenchus scribneri*. *J. Agric. Univ. Puerto Rico 63*: 103-110.

Astudillo, G. E., and Birchfield, W. (1980). Pathology of *Hoplolaimus columbus* on sugarcane. *Phytopathology 70*: 565.

Ayala, A., and Ramirez, C. T. (1964). Host-range, distribution, and bibliography of the reniform nematode, *Rotylenchulus reniformis*, with special reference to Puerto Rico. *J. Agric. Univ. Puerto Rico. 48*: 140-161.

Baldwin, J. G., Barker, K. R., and Nelson, L. A. (1979). Effects of *Meloidogyne incognita* on nitrogen fixation in soybean. *J. Nematol. 11*: 156-161.

Barker, K. R. (1982). Influence of soil moisture, cultivar, and population density of *Meloidogyne incognita* on soybean yield in microplots (abstract) *J. Nematol. 14*: 429.

Barker, K. R., Schmitt, D. P., and Campos, V. P. (1982). Response of peanut, corn, tobacco and soybean to *Criconemella ornata*. *J. Nematol. 14*: 576-581.

Bernard, E. C. (1980). Reassessment of resistance among soybeans to *Meloidogyne incognita* (abstract). *J. Nematol. 12*: 215.

Birchfield, W., and Brister, L. R. (1962). New hosts and nonhosts of reniform nematode. *Plant Dis. Rep. 46*: 683-685.

Birchfield, W., and Martin, W. J. (1967). Reniform nematode survival in airdried soil (abstract). *Phytopathology 57*: 804.

Birchfield, W., Williams, C., Hartwig, E. E., and Brister, L. R. (1971). Reniform nematode resistance in soybeans. *Plant Dis. Rep. 55*: 1043-1045.

Bird, G. W., Brooks, O. L., and Perry, C. E. (1974). Dynamics of concomitant field populations of *Hoplolaimus columbus* and *Meloidogyne incognita*. *J. Nematol. 6*: 190-194.

Bird, G. W., and Hogger, C. (1973). Nutsedges as hosts of plant-parasitic nematodes in Georgia. *Plant Dis. Rep. 57*: 402.

Bonner, M. J. (1981). The relationship of field population densities of *Heterodera glycines* to soybean growth and yield. M. S. Thesis, North Carolina State University.

Boquet, D. J., Williams, C., and Birchfield, W. (1975). Resistance in soybeans to five Louisiana populations of the root-knot nematode. *Plant Dis. Rep. 59*: 197-200.

Boyd, F. T., and Perry, V. G. (1970). Effects of seasonal temperatures and certain cultural treatments on sting nematodes in forage grass. *Proc. Soil Crop Sci. Soc. Fla. 30*: 360-365.

Brooks, T. L., and Perry, V. G. (1967). Pathogenicity of *Pratylenchus brachyurus* to citrus. *Plant Dis. Rep. 51*: 569-573.

Burns, N. C. (1971). Soil pH effects on nematode populations associated with soybeans. *J. Nematol. 3*: 238-245.

Byrne, J. M., Pesacreta, T. C., and Fox, J. A. (1977). Vascular pattern change caused by a nematode, *Meloidogyne incognita*, in the lateral roots of *Glycine max* (L.) Merr. *Am. J. Bot. 64*: 960-965.

Carvalho, J. C. (1958). *Rotylenchus elisensis* nova especie associada com raizes de soja. *Rev. Inst. Adolfo Lutz 17*: 43-46.

Chitwood, B. G. (1949). "Root-knot nematodes"–Part I. A revision of the genus, *Meloidogyne* Goeldi 1887. *Proc. Helm. Soc. Wash, 16*: 90-104.

Christie, J. R. (1959). The sting and awl nematodes. In *Plant Nematodes, Their Bionomics and Control*. Agricultural Exp. Sta., Gainesville, Fla., pp. 126-136.

Coursen, B. W., Rohde, R. A., and Jenkins, W. R. (1958). Additions to the host lists of the nematodes *Paratylenchus projectus* and *Trichodorus christiei*. *Plant Dis. Rep. 42*: 456-460.

Dasgupta, D. R., Raski, D. J., and Sher, S. A. (1968). A revision of the genus *Rotylenchulus* Linford and Oliveira, 1940 (Nematoda: Tylenchidae). *Proc. Helm. Soc. Wash. 35*: 169-192.

Dasgupta, D. R., and Seshadri, A. R. (1971). Races of the reniform nematode *Rotylenchulus reniformis* Linford and Oliveira, 1940. *Ind. J. Nematol. 1*: 21-24.

Dickerson, O. J. (1979). The effects of temperature on *Pratylenchus scribneri* and *P. alleni* populations on soybeans and tomatoes. *J. Nematol. 11*: 23-26.

Dickerson, O. J., Willis, W. G., Daniello, F. J., and Pair, J. C. (1972). The sting nematode, *Belonolaimus longicaudatus*, in Kansas. *Plant Dis. Rep. 56*: 957.

Dropkin, V. H. (1963). Effect of temperature on growth of root-knot nematodes in soybeans and tobacco. *Phytopathology 53*: 663-666.

Eisenback, J. D., Hirschmann, H., Sasser, J. N., and Triantaphyllou, A. C. (1981). A guide to the four common species of root-knot nematodes (*Meloidogyne* spp.) with a pictorial key. International *Meloidogyne* Project, North Carolina State University, Raleigh.

Endo, B. Y. (1959). Responses of root-lesion nematodes, *Pratylenchus brachyurus* and *P. zeae*, to various plants and soil types. *Phytopathology 49*: 417-421.

Endo, B. Y. (1964). Penetration and development of *Heterodera glycines* in soybean roots and related anatomical changes. *Phytopathology 54*: 79-88.

Endo, B. Y. (1965). Histological responses of resistant and susceptible varieties, and backcross progeny to entry and development of *Heterodera glycines*. *Phytopathology 55*: 375-381.

Endo, B. Y. (1967). Comparative population increase of *Pratylenchus brachyurus* and *P. zeae* in corn and in soybean varieties Lee and Peking. *Phytopathology 57*: 118-120.

Epps, J. M. (1968). Survival of soybean cyst nematodes in seed bags. *Plant Dis. Rep. 52*: 45.

Esser, R. P., Perry, V. G., and Taylor, A. L. (1976). A diagnostic compendium of the genus *Meloidogyne* (Nematoda: Heteroderidae). *Proc. Helm. Soc. Wash. 43*: 138-150.

Fassuliotis, G. (1971). Tolerance of *Hoplolaimus columbus* to high osmotic pressures, desiccation, and high soil temperatures. *J. Nematol. 3*: 309-310.

Fassuliotis, G. (1974). Host range of the columbia lance nematode *Hoplolaimus columbus*. *Plant Dis. Rep. 58*: 1000-1002.

Fassuliotis, G. (1975). Feeding, egg laying, and embryology of the columbia lance nematode, *Hoplolaimus columbus*. *J. Nematol. 7*: 152-158.

Fassuliotis, G., and Rau, G. J. (1967). The reniform nematode in South Carolina. *Plant Dis. Rep. 51*: 557.

Fassuliotis, G., Rau, G. J., and Smith, F. H. (1968). *Hoplolaimus columbus*, a nematode parasite associated with cotton and soybeans in South Carolina. *Plant Dis. Rep. 52*: 571-572.

Feldmesser, J. (Chairman). (1970). Estimated crop losses from plant-parasitic nematodes in the United States. Soc. Nematol., Special Publ. No. 1.

Ferris, V. R., and Bernard, R. L. (1962). Injury to soybeans caused by *Pratylenchus alleni*. *Plant Dis. Rep. 46*: 181-184.

Ferris, V. R., and Bernard, R. L. (1967). Population dynamics of nematodes in fields planted to soybeans and crops grown in rotation with soybeans. I. The genus *Pratylenchus* (Nemata: Tylenchida). *J. Econ. Ent. 60*: 405-410.

Ferris, V. R., Ferris, J. M., Bernard, R. L., and Probst, A. H. (1971). Community structure of plant-parasitic nematodes related to soil types in Illinois and Indiana soybean fields. *J. Nematol. 3*: 399-408.

Fielding, M. J., and Hollis, J. P. (1956). Occurrence of plant-parasitic nematodes in Louisiana soils. *Plant Dis. Rep. 40*: 403-405.

Fortnum, B. A., and Lewis, S. A. (1978a). Interaction between *Cylindrocladium* root rot and polyspecific nematode populations on soybean. *J. Nematol. 10*: 287.

Fortnum, B. A., and Lewis, S. A. (1978b). Populations of certain fungi in soybean roots and rhizosphere related to infection by *Meloidogyne incognita* and *Hoplolaimus columbus*. *Proc. Am. Phytopathol. Soc. 4*: 222-223.

Gipson, I., Kim, K. S., and Riggs, R. D. (1971). An ultrastructural study of syncytium development in soybean roots infected with *Heterodera glycines*. *Phytopathology 61*: 347-353.

Godfrey, G. H. (1929). A destructive root disease of pineapple and other plants due to *Tylenchus brachyurus* n. sp. *Phytopathology 19*: 611-629.

Golden, A. M., and Birchfield, W. (1978). *Meloidogyne incognita wartellei* n. subsp. (Meloidogynidae), a root-knot nematode on resistant soybeans in Louisiana. *J. Nematol. 10*: 269-277.

Golden, A. M., and Epps, J. M. (1965). Morphological variations in the soybean-cyst nematode (abstract). *Nematologica 11*: 38.

Golden, A. M., Epps, J. M., Riggs, R. D., Duclos, L. A., Fox, J. A., and Bernard, R. L. (1970). Terminology and identity of infraspecific forms of the soybean cyst nematode (*Heterodera glycines*). *Plant Dis. Rep. 54*: 544-546.

Golden, A. M., and Rebois, R. V. (1978). Nematodes on soybeans in Maryland. *Plant Dis. Rep. 62*: 430-432.

Gommers, F. J., and Dropkin, V. H. (1977). Quantitative histochemistry of nematode-induced transfer cells. *Phytopathology 67*: 869-873.

Good. J. M. (1973). Nematodes. In *Soybeans: Improvement, Production, and Uses* (B. E. Caldwell, ed.). Agronomy No. 16, American Society of Agronomy, pp. 527-543.

Goodey, J. B., Franklin, M. T., and Hooper, D. J. (1965). *T. Goodey's The Nematode Parasites of Plants Catalogued Under Their Hosts*, 3rd edition. Comm. Agric. Bur.

Goswami, B. K., and Agarwal, D. K. (1978). Interrelationships between species of *Fusarium* and root-knot nematode, *Meloidogyne incognita* in soybean. *Nematol. Medit. 6*: 125-128.

Gotoh, A., and Ohshima, Y. (1963). *Pratylenchus*-arten and ihre geographische verbreitung in Japan (Nematoda: Tylenchida). *Jpn. J. Appl. Ent. Zool. 7*: 187-199.

Hamblen, M. L., Slack, D. A., and Riggs, R. D. (1972). Temperature effects on penetration and reproduction of soybean-cyst nematode (abstract). *Phytopathology 62*: 762.

Holdeman, Q. L. (1955). The present known distribution of the sting nematode, *Belonolaimus gracilis*, in the coastal plain of the southeastern United States. *Plant Dis. Rep. 39*: 5-7.

Holdeman, Q. L., and Graham, T. W. (1953). The effect of different plant species on the population trends of the sting nematode. *Plant Dis. Rep. 37*: 497-500.

Hymowitz, T. (1970). On the domestication of the soybean. *Econ. Bot. 24*: 408-421.

Ichinohe, M. (1955). Studies on the morphology and ecology of the soybean nematode, *Heterodera glycines*, in Japan. Report No. 48, Hokkaido Natl. Agric. Exp. Sta., Kotoni, Sapporo, Japan.

Inagaki, H. (1979). Race status of five Japanese populations of *Heterodera glycines*. *Jpn. J. Nematol. 9*: 1-4.

Inagaki, H., and Tsutsumi, M. (1971). Survival of the soybean cyst nematode *Heterodera glycines* Ichinohe. (Tylenchida: Heteroderidae) under certain storing conditions. *Appl. Ent. Zool. 6*: 156-162.

Jenkins, W. R., Taylor, D. P., Rhode, R. A., and Coursen, B. W. (1957). Nematodes associated with crop plants in Maryland. Univ. Maryland Bull. A-89.

Johnson, A. W., Dowler, C. C., and Hauser, E. W. (1975). Crop rotation and herbicide effects on population densities of plant-parasitic nematodes. *J. Nematol. 7*: 158-168.

Kaplan, D. T., Keen, N. T., and Thomason, I. J. (1980a). Association of glyceollin with the incompatible response of soybean roots to *Meloidogyne incognita*. *Physiol. Plant Pathol. 16*: 309-318.

Kaplan, D. T., Keen, N. T., and Thomason, I. J. (1980b). Studies on the mode of action of glyceollin in soybean incompatibility to the root knot nematode, *Meloidogyne incognita*. *Physiol. Plant Pathol. 16*: 319-325.

Kaplan, D. T., Thomason, I. J., and Van Gundy, S. D. (1979). Histological study of the compatible and incompatible interaction of soybeans and *Meloidogyne incognita*. *J. Nematol. 11*: 338-343.

Kellam, M. K., and Schenck, N. C. (1980). Interactions between a vesicular-arbuscular mycorrhizal fungus and root-knot nematode on soybean. *Phytopathology 70*: 293-296.

Kinloch, R. A. (1974a). Response of soybean cultivars to nematicidal treatments of soil infested with *Meloidogyne incognita*. *J. Nematol. 6*: 7-11.

Kinloch, R. A. (1974b). Nematode and crop response to short-term rotations of corn and soybean. *Proc. Soil Crop Sci. Soc., Fla. 33*: 86-88.

Kinloch, R. A. (1980). The control of nematodes injurious to soybean. *Nematropica 10*: 141-153.

Kinloch, R. A. (1982). The relationship between soil populations of *Meloidogyne incognita* and yield reduction of soybean in the coastal plain. *J. Nematol. 14*: 162-167.

Kinloch, R. A., and Hinson, K. (1973). The Florida program for evaluating soybean (*Glycine max* (L.) Merr.) genotypes for susceptibility to root-knot nematode disease. *Proc. Soil Crop Sci. Soc. Fla. 32*: 173-176.

Kinloch, R. A., and Schenck, N. C. (1978). Nematodes and fungi associated with soybeans grown under various pesticide regimes. *Soil Crop. Sci. Soc. Fla. 37*: 224-227.

Kloss, G. R. (1960). Catalogo de nematoides fitofagos do Brasil. Boleti Fitossanitario III, No. 182.

Koliopanos, C. N., and Triantaplyllou, A. C. (1972). Effect of infection density on sex ratio of *Heterodera glycines*. *Nematologica 18*: 131-137.

Kondo, E., and Ishibashi, N. (1975). Ultrastructural changes associated with the tanning process in the cyst wall of the soybean cyst nematode, *Heterodera glycines* Ichinohe. *Appl. Ent. Zool.* *10*: 247-253.

Lehman, P. S., Huisingh, D., and Barker, K. R. (1971). The influence of races of *Heterodera glycines* on nodulation and nitrogen-fixing capacity of soybean. *Phytopathology 61*: 1239-1244.

Lewis, S. A., and Huff, T. F. (1976). Cuticle anatomy of *Hoplolaimus columbus*. *J. Nematol. 8*: 293.

Lewis, S. A., Smith, F. H., and Powell, W. M. (1976). Host-parasite relationships of *Hoplolaimus columbus* on cotton and soybean. *J. Nematol. 8*: 141-145.

Lim, B. K., and Castillo, M. B. (1979). Screening soybeans for resistance to reniform nematode disease in the Philippines. *J. Nematol. 11*: 275-282.

Lindsey, D. W., and Cairns, E. J. (1971). Pathogenicity of the lesion nematode, *Pratylenchus brachyurus*, on six soybean cultivars. *J. Nematol. 3*: 220-226.

Linford, M. B., and Oliveira, J. M. (1940). *Rotylenchulus reniformis*, nov. gen., n. sp., a nematode parasite of roots. *Proc. Helm. Soc. Wash. 7*: 35-42.

Linford, M. B., and Yap, F. (1940). Some host plants of the reniform nematode in Hawaii. *Proc. Helm. Soc. Wash. 7*: 42-44.

Lordello, L. G. E. (1955). Nematodes attacking soybean in Brazil. *Plant Dis. Rep. 39*: 310-311.

Lordello, L. G. E. (1956a). Nematoides que parasitam a soja na regiao de Bauru. *Bragantia 15*: 55-64.

Lordello, L. G. E. (1956b). *Meloidogyne inornata* sp. n., a serious pest of soybean in the state of Sao Paulo, Brazil. Nematoda (Heteroderidae). *Rev. Brasil Biol. 16*: 65-70.

Lordello, L. G. E., Zamith, A. P. L., and vaz de Arruda, H. (1958). Nematodeos que prejudicam as culturas da soja e do algodoeiro no estado de S. Paulo e sua interferencia nos planos de rotacao. *Rev. Agric. 33*: 161-167.

Mai, W. F., Crittenden, W. H., and Jenkins, W. R. (1960). Distribution of stylet-bearing nematodes in the northeastern United States. New Jersey Agric. Exp. Sta. Bull. 795.

Manuel, J. S., Reynolds, D. A., Bendixen, L. E., and Riedel, R. M. (1980). Weeds as hosts of *Pratylenchus*. Ohio Agric. Res. Devel. Cent. Res. Bull. 1123.

Masamure, T., Anetai, M., Takasugi, M., and Katsui, N. (1982). Isolation of a natural hatching stimulus, glycinoeclepin A, for the soybean cyst nematode. *Nature 297*: 495-496.

McCann, J., Luedders, V. D., and Dropkin, V. H. (1982). Selection and reproduction of soybean cyst nematodes on resistant soybeans. *Crop Sci. 22*: 78-80.

Miller, L. I. (1972). The influence of soil texture on the survival of *Belonolaimus longicaudatus*. *Phytopathology 62*: 670-671.

Minton, N. A., and Parker, M. B. (1975). Interaction of four soybean cultivars with subsoiling and a nematicide. *J. Nematol. 7*: 60-64.

Minton, N. A., and Parker, M. B. (1980). Response of soybeans and nematodes to at-planting applications of fumigant and nonfumigant nematicides. *Nematropica 10*: 31-37.

Minton, N. A., Parker, M. B., Brooks, O. L., and Perry, C. E. (1979). Effects of nematicide placement on nematode populations and soybean yields. *J. Nematol. 11*: 150-155.

Myers, R. F. (1979). The sting nematode. *Belonolaimus longicaudatus* from New Jersey. *Plant Dis. Rep. 63:* 756-757.

Nagata, T. (1959). Studies on the differentiation of soybeans in the world with special regard to that of Southeast Asia. *Proc. Crop Sci. Soc. Jpn. 28:* 79-82.

Nardacci, J. F., and Barker, K. R. (1979). The influence of temperature on *Meloidogyne incognita* on soybean. *J. Nematol. 11:* 62-70.

Nishizawa, T. (1978). Annual population changes of soil nematodes in the field with continuous cropping or rotation. *Kasetsart J. 12:* 3-33.

Noel, G. R., Bloor, P. V., Posdal, R. F., and Edwards, D. I. (1980). Influence of *Heterodera glycines* on soybean yield components and observations on economic injury levels (abstract). *J. Nematol. 12:* 232-233.

Norton, D. C., and Burns, N. (1971). Colonization and sex ratios of *Pratylenchus alleni* in soybean roots under two soil moisture regimes. *J. Nematol. 3:* 374-377.

Nyczepir, A. P., and Lewis, S. A. (1979). Relative tolerance of selected soybean cultivars to *Hoplolaimus columbus* and possible effects of soil temperature. *J. Nematol. 11:* 27-31.

Okada, T. (1972). Hatching stimulant in the egg of the soybean cyst nematode,*Heterodera glycines* Ichinohe. *Appl. Ent. Zool. 7:* 234-237.

Owens, J. V. (1951). The pathological effects of *Belonolaimus gracilis* on peanuts in Virginia. *Phytopathology 41:* 29.

Parris, G. K. (1940). A checklist of fungi, bacteria, nematodes and viruses occurring in Hawaii and their hosts. *Plant Dis. Rep. Suppl. 121:* 77-82.

Peacock, F. C. (1956). The reniform nematode in the Gold Coast. *Nematologica 1:* 307-310.

Price, M., Caveness, C. E., and Riggs, R. D. (1978). Hybridization of races of *Heterodera glycines*. *J. Nematol. 10:* 114-118.

Probst, A. H., and Judd, R. W. (1973). Origin, U.S. history and development, and world distribution. In *Soybeans: Improvement, Production and Uses* (B. E. Caldwell, ed.). Agronomy No. 16. American Society of Agronomy, pp. 1-15.

Rebois, R. V. (1973). Effect of soil temperature on infectivity and development of *Rotylenchulus reniformis* on resistant and susceptible soybeans, *Glycine max*. *J. Nematol. 5:* 10-13.

Rebois, R. V., and Cairns, E. J. (1968). Nematodes associated with soybeans in Alabama, Florida, and Georgia. *Plant Dis. Rep. 52:* 40-44.

Rebois, R. V., Epps, J. M., and Hartwig, E. E. (1970). Correlation of resistance in soybeans to *Heterodera glycines* and *Rotylenchulus reniformis*. *Phytopathology 60:* 695-700.

Rebois, R. V., and Golden, A. M. (1978). Nematode occurrences in soybean fields in Mississippi and Louisiana. *Plant Dis. Rep. 62:* 433-437.

Rebois, R. V., Madden, P. A., and Eldridge, B. J. (1975). Some ultrastructural changes induced in resistant and susceptible soybean roots following infection by *Rotylenchulus reniformis*. *J. Nematol. 7:* 122-139.

Rhoades, H. L. (1980). Reproduction of *Belonolaimus longicaudatus* in treated and untreated muck soil. *Nematropica 10:* 139-140.

Riggs, R. D. (1961). Sting nematode in Arkansas. *Plant Dis. Rep. 45:* 392.

Riggs, R. D. (1977). Worldwide distribution of soybean-cyst nematode and its economic importance. *J. Nematol. 9:* 34-39.

Riggs, R. D., Hamblen, M. L., and Rakes, L. (1977). Development of *Heterodera glycines* pathotypes as affected by soybean cultivars. *J. Nematol. 9:* 312-318.

Riggs, R. D., Hamblen, M. L., and Rakes, L. (1981). Infra-species variation in reactions to hosts in *Heterodera glycines* populations. *J. Nematol. 13*: 171-179.

Riggs, R. D., Kim, K. S., and Gipson, I. (1973). Ultrastructural changes in Peking soybeans infected with *Heterodera glycines. Phytopathology 63*: 76-84.

Robbins, R. T. (1982a). Phytoparasitic nematodes associated with soybean in Arkansas. *J. Nematol. 14*: 466.

Robbins, R. T. (1982b). Description of *Hoplolaimus magnistylus* n. sp. (Nematoda: Hoplolaimidae). *J. Nematol. 14*: 500-506.

Robbins, R. T., and Barker, K. R. (1971). Reproductive responses of *Belonolaimus longicaudatus* to soil type and temperature. *J. Nematol. 3*: 328.

Robbins, R. T., and Barker, K. R. (1973). Comparisons of host range and reproduction among populations of *Belonolaimus longicaudatus* from North Carolina and Georgia. *Plant Dis. Rep. 57*: 750-754.

Rodriguez-Kabana, R., and Collins, R. J. (1980). Relation of fertilizer treatments and cropping sequence to populations of *Pratylenchus scribneri. Nematropica 10*: 121-129.

Rodriguez-Kabana, R., King, P. S., and Pope, M. H. (1981). Comparison of in-furrow applications and banded treatments for control of *Meloidogyne arenaria* on peanuts and soybeans. *Nematropica 11*: 53-67.

Rodriguez-Kabana, R., Penick, H. W., and King, P. S. (1979). Control of nematodes on soybeans with planting-time applications of ethylene dibromide. *Nematropica 9*: 61-66.

Rodriguez-Kabana, R., and Thurlow, D. L. (1980). Evaluation of selected soybean cultivars in a field infested with *Meloidogyne arenaria* and *Heterodera glycines. Nematropica 10*: 50-55.

Ross, J. P. (1962). Crop rotation effects on the soybean cyst nematode population and soybean yields. *Phytopathology 52*: 815-818.

Ross, J. P. (1963). Seasonal variation of larval emergence from cysts of the soybean cyst nematode, *Heterodera glycines. Phytopathology 53*: 608-609.

Ross, J. P. (1964a). Interaction of *Heterodera glycines* and *Meloidogyne incognita* on soybeans. *Phytopathology 54*: 304-307.

Ross, J. P. (1964b). Effect of soil temperature on development of *Heterodera glycines* in soybean roots. *Phytopathology 54*: 1228-1231.

Ross, J. P., Nusbaum, C. J., and Hirschmann, H. (1967). Soybean yield reduction by lesion, stunt, and spiral nematodes (abstract). *Phytopathology 57*: 463-464.

Russell, C. C., and Sturgeon, R. V. (1969). Occurrence of *Belonolaimus longicaudatus* and *Ditylenchus dipsaci* in Oklahoma. *Phytopathology 59*: 118.

Sasser, J. N. (1977). Worldwide dissemination and importance of the root-knot nematodes *Meloidogyne* spp. *J. Nematol. 9*: 26-29.

Schenck, N. C., Kinloch, R. A., and Dickson, D. W. (1975). Interaction of endomycorrhizal fungi and root-knot nematode on soybean. In *Endomycorrhizas* (F. E. Sanders, B. Mosse, and P. B. Tinker, eds.). Academic Press, New York, pp. 607-617.

Schmitt, D. P., and Barker, K. R. (1981). Damage and reproductive potentials of *Pratylenchus brachyurus* and *P. penetrans* on soybean. *J. Nematol. 13*: 327-332.

Schmitt, D. P., and Corbin, F. T. (1981). Interaction of fensulfothion and phorate with preemergence herbicides on soybean parasitic nematodes. *J. Nematol. 13*: 37-41.

Sher, S. A. (1963). Revision of the Hoplolaiminae (Nematoda). II. *Hoplolaimus* Daday, 1905 and *Aorolaimus* n. gen. *Nematologica 9*: 267-295.

Siddiqi, M. R. (1972). *Rotylenchulus reniformis*. C. I. H. Descriptions of Plant Parasitic Nematodes, Set 1, No. 5. Comm. Inst. Helm., England.

Sivakumar, C. V., and Seshadri, A. R. (1971). Life history of the reniform nematode, *Rotylenchulus reniformis* Linford and Oliviera, 1940. *Ind. J. Nematol. 1*: 7-20.

Smith, F. H. (1969). Host-parasite and life history studies of the lance nematode (*Hoplolaimus columbus*, Sher, 1963) on soybeans. Ph.D. Dissertation, Univ. Ga., Athens.

Standifer, M. S. (1959). The pathologic histology of bean roots injured by sting nematodes. *Plant Dis. Rep. 43*: 983-986.

Standifer, M. S., and Perry, V. G. (1960). Some effects of sting and stubby-root nematodes on grapefruit roots. *Phytopathology 50*: 152-156.

Steiner, G. (1949). Plant nematodes the growers should know. *Proc. Soil Sci. Soc. Fla. IV-B*: 72-117.

Taylor, D. P., Anderson, R. V., and Haglund, W. A. (1958). Nematodes associated with Minnesota crops. I. Preliminary survey of nematodes associated with alfalfa, flax, peas, and soybeans. *Plant Dis. Rep. 42*: 195-198.

Thomas, J. D. (1974). Genetics of resistance to races of the soybean cyst nematode. M.S. Thesis, Univ. Arkansas, Fayetteville.

Thorne, G. (1961). *Principles of Nematology*. McGraw-Hill, New York.

Timm, B. W. (1964). A preliminary survey of the plant-parasitic nematodes of Thailand and the Philippines. South-East Asia Treaty Org. Secretariat—General, Bangkok.

Triantaphyllou, A. C. (1975). Genetic structure of races of *Heterodera glycines* and inheritance of ability to reproduce on resistant soybeans. *J. Nematol. 7*: 356-364.

Tyler, J. (1933). Development of the root-knot nematode as affected by temperature. *Hilgardia 7*: 391-415.

Veech, J. A., and Endo, B. Y. (1970). Comparative morphology and enzyme histochemistry of root-knot resistant and susceptible soybean. *Phytopathology 60*: 896-902.

Weber, D. E., Reinert, R. A., and Barker, K. R. (1979). Ozone and sulfur dioxide effects on reproduction and host-parasite relationships of selected plant-parasitic nematodes. *Phytopathology 69*: 624-628.

Whitney, G. (Chairman). (1978). Southern states soybean disease loss estimate—1977. *Plant Dis. Rep. 62*: 1078-1079.

Winstead, N. N., Skotland, C. B., and Sasser, J. N. (1955). Soybean-cyst nematode in North Carolina. *Plant Dis. Rep. 39*: 9-11.

Zirakparvar, M. E. (1981). Susceptibility of soybean cultivars and lines to *Pratylenchus hexincisus*. *J. Nematol. 14*: 217-220.

Chapter 3
Nematode Parasites of Corn

Don C. Norton *Iowa State University, Ames, Iowa*

I. INTRODUCTION

More than 100 million hectares are devoted to corn production worldwide. Considering all direct and indirect uses, corn is the most important crop in the United States, and, with wheat and rice, they constitute the three most important crops in the world.

Pathogenicity by nematodes on corn is a concept documented only relatively recently. Papers by Graham (1951) and Christie (1953) are important because they were some of the early works that implicated nematodes as pathogens of corn.

Because mixtures of populations of plant-parasitic nematodes usually exist in a corn field, evaluation of loss by any given species is difficult. Damage cannot always be correlated with numbers of nematodes because few nematodes are often associated with a "declining" root system and many with a "healthy" root system.

In some instances, injury caused by such species as *Belonolaimus longicaudatus*, *Longidorus breviannulatus*, and *Paratrichodorus minor* have rather distinctive root symptoms and, unless they are mixed, one can with some degree of certainty identify damage by these species. Some nematodes, such as species of *Pratylenchus*, can cause a brushy effect of the roots. But the only safe procedure is to isolate and identify the nematodes. Whatever the cause, control should be aimed at greater nematode mortality than natality until population densities are well below the injury threshold, not only for the current crop but also for a following susceptible crop. The injury or economic threshold numbers are difficult to determine. The nematode species, tolerance by the host, climatic and edaphic factors, and their interrelationships, govern the amount of damage done to the crop and the feasibility of economic control.

Discussion of nematode damage to corn by individual species may be misleading because it may imply a one nematode-one disease type of situation. While this might be essentially true in some instances, damage by polyspecific communities is probably more common. It is unnatural to discuss nematodes separately, but such a system is convenient if for no other reason than insufficient work has been done with polyspecific nematode communities.

In the treatment that follows, the genera are arranged by the importance of their species as pathogens of corn. This arrangement necessarily is subjective. The species listed at the heading of each section are those for which there is relatively good documentation concerning parasitism and pathogenicity. Because corn nematology is a relatively new development, species merely touched on, or not mentioned at all, may assume importance with further research. New discoveries are being made daily, and the present treatment is not to be considered definitive. Distributions are given only for areas where the trouble occurs or where work has been done on corn, and not for the overall distribution of the nematode.

A final word on terminology. In the United States "corn" refers to "maize" as used in nearly all other countries. "Corn" in other countries can be used for all cereals and sometimes means any hard edible seed, grain, or kernel. The term *corn* for *Zea mays* L. and closely related species will be used in this treatment.

II. ROOT-LESION NEMATODES

A. Species

The major root lesion organisms are: *Pratylenchus brachyurus* (Godfrey 1929) Filipjev and Schuurmans Stekhoven 1941, *P. delattrei* Luc 1958, *P. hexincisus*

Taylor and Jenkins 1957, *P. penetrans* (Cobb 1917) Filipjev and Schuurmans Stekhoven 1941, and *P. zeae* Graham 1951. Many other species have been isolated from corn roots, and although several probably are pathogenic, little documentation has been made.

Species of *Pratylenchus* are small nematodes ranging from 0.30 to 0.75 mm. They possess a tylenchoid esophagus with basal glands that overlap the anterior part of the intestine. The lips are frequently low. The vulva is about 68-89% from the anterior end of the nematode. The functional female gonad is monodelphic and anterior. Males are common in some species and unknown in others.

B. Associated Symptoms

Gross symptoms of damage caused by species of *Pratylenchus* vary with the degree of nematode infestation and the environmental conditions. Aboveground symptoms range from severe stunting with no yield to losses demonstrated only by carefully controlled experiments. Stunting frequently occurs in patches that often coincide with patches of soil conditions that allow increase of the nematode to enormous numbers. Chlorosis or other discoloration often is evident in severe instances, but frequently is absent in mild infestations. Because mixed populations of parasitic nematodes make diagnosis difficult in the field, studies under controlled conditions are necessary to provide information on the pathogenic capabilities of a species in the absence of other pathogenic organisms.

Reduction of plant height, stalk diameter, and stalk and root weights of inoculated plants compared with noninoculated ones has been demonstrated for some species (Dickerson et al., 1964; Zirakparvar, 1980). These symptoms are common in the field when large populations of *Pratylenchus* spp. occur and agree with the negative correlations of yield with nematode numbers (Bergeson, 1978; Egunjobi, 1974; Miller et al., 1963).

Graham (1951) reported that early water-soaked root lesions containing mostly *P. zeae* could be easily overlooked. Later the lesions become distinctly discolored, as has been found with other species (Fig. 1) (Zirakparvar, 1980). Graham found that single lesions contained up to 80 eggs and 80-100 nematodes in lesions 5 mm long. Feeding in the fibrous roots can result in the destruction of the cortical parenchyma, resulting in sloughing off of this tissue (Ogiga and Estey, 1975; Zirakparvar, 1980). Severe pruning of the roots can occur (Fig. 2). That the environment can affect symptoms is exemplified by the work of Dickerson et al. (1964), who found that weights of corn roots infested with P. *penetrans* were significantly different at 20 and 24°C compared with noninoculated plants, but differences in top weights were significant only at 20°C.

C. Life Cycle of the Nematodes

Species of *Pratylenchus* are basically migratory endoparasites. They have the common developmental stages for plant nematodes, i.e., the egg, four juvenile stages, and the adult. After the undifferentiated egg, all stages remain vermiform. Numbers generally increase during the growing season, with the number of generations depending on the environmental conditions. Usually the nematodes live only in the cortical parenchyma (Graham, 1951; Zirakparvar, 1980). Olowe and Corbett (1976), however, found that with excised roots, *P. brachyurus* and *P. zeae* invaded all parts of corn roots including the stele. Cavities were formed in the cortex where there was little necrosis, but much necrosis occurred in the stele. A dense-staining substance formed that

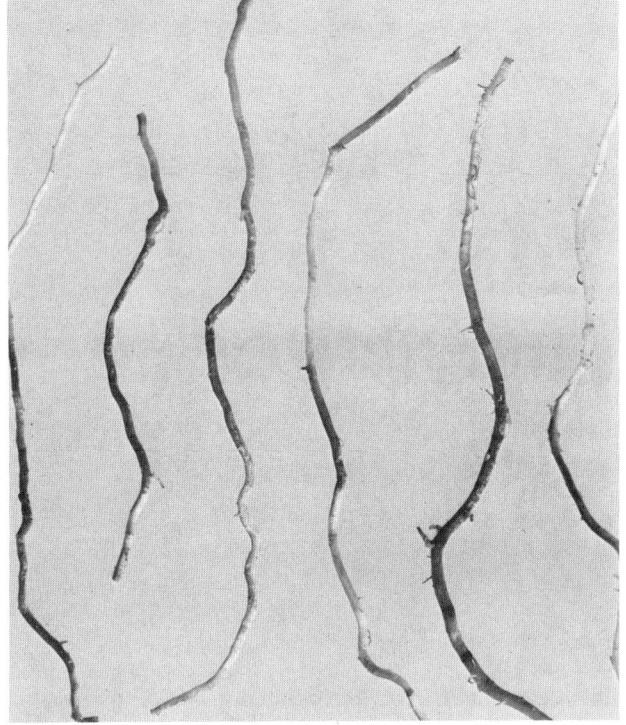

Fig. 1 Lesions on corn roots invaded by *P. hexincisus*. (From Zirakparvar, 1980.)

occluded the xylem and phloem elements. *P. zeae* caused more mechanical damage than did *P. brachyurus*, but there was less necrosis in the excised roots. Both species penetrated the root hair zone, and many *P. brachyurus* penetrated the root tips but *P. zeae* did not. Numbers of *P. hexincisus* per gram of dry root were greater in fibrous roots than in coarse roots (Zirakparvar, 1979). In work with *P. neglectus* and corn tissue cultures, Mountain (1954) found that adult and late juvenile stages of the nematode were most frequently attracted to the root. The root tip region attracted more nematodes than did other portions of the root, and the nematodes fed on the meristematic cells of the root tip and the larger cells of the root cap. Because of the compactness of the root, it was not possible to see if nematodes were in the root tip tissue. It is probable that penetration rarely occurred at the growing point because nematodes were seldom found in this region in stained tissue. Penetration by large numbers of nematodes resulted in the formation of many lateral roots, as also found in axenic cultures with *P. penetrans* (Ogiga and Estey, 1975). Occasionally a slight host-cell hypertrophy occurred when many nematodes invaded an area. Living nematodes were observed in the smaller lateral roots and appeared to have migrated from tissues of the main root into newly developed lateral ones.

Little research has been done on the duration of the life cycle of *Pratylenchus* spp. in corn. In what was probably a mixture of mostly *P. zeae* and *P. brachyurus*, 16-32% of the eggs hatched in 15-20 days at 24-27°C. The period from egg to maturity was 35-40 days (Graham, 1951). Using tempera-

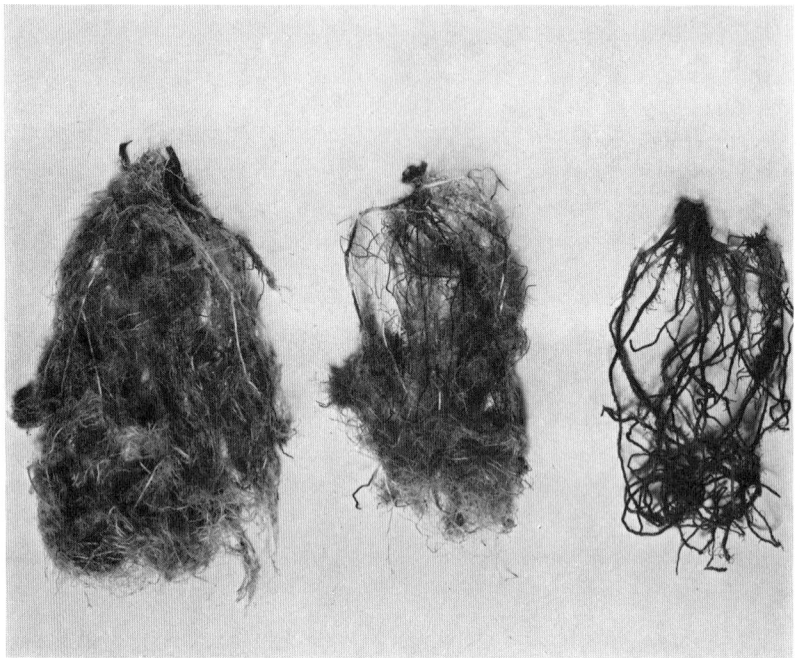

Fig. 2 Root damage caused by *P. hexincisus*. Left to right: noninfested, plants inoculated with 5000 nematodes per pot, and plants inoculated with 20,000 nematodes per pot. After three months. (From Zirakparvar, 1980.)

tures from 15 to 35°C, the optimum generation time in excised roots was three and four weeks for *P. zeae* and *P. brachyurus*, respectively (Olowe and Corbett, 1976). Reproduction of both nematodes was greatest at 30°C.

In corn, *P. brachyurus, P. hexincisus*, and *P. zeae* develop best at 30°C, but 25°C is frequently favorable (Olowe and Corbett, 1976; Zirakparvar et al., 1980). On the other hand, *P. penetrans* reproduced best at 24°C (Dickerson et al., 1964) but invaded corn best at 20°C (Townshend, 1972). Similarly, although population development of *P. brachyurus* in excised roots was greatest at 30°C, invasion was greatest at 20°C, which was also the best temperature for root growth (Olowe and Corbett, 1976). Frequently there is a range of conditions for which development is favorable. For example, although *P. hexincisus* does best in sandy soils, the nematode is widespread and can obtain large populations in the finer soils of the midwestern U.S. corn belt. Zirakparvar et al. (1980) found that the number of heat units necessary for *P. hexincisus* to reach a level to cause significant reduction in corn biomass varied with the soil.

There are many factors besides temperature, soil type, and moisture that can regulate populations of *Pratylenchus* spp. Among these in the agroecosystems are the tillage systems employed. Because control of nematodes in corn is aimed at maintaining populations below injury thresholds and not eradication, any measure, however subtle, should be employed if economical and compatible with other management systems.

With greater interest in conservation tillage, we should be aware of the nematode changes that occur under different tillage systems employing corn. Although All et al. (1977) found no differences in populations of *P. zeae* between a no-till and a conventional till in Georgia tests, Thomas (1978) found that *Pratylenchus* spp. were greater in no-till ridge plots than six other tillage regimes. Numbers per gram of dry root were lowest in the fall-plowed plots.

D. Percentage and Dollar Loss

Percentage and dollar losses are difficult to estimate for lesion and other nematodes attacking corn. Degrees of infestations vary annually. Experiments usually are not randomly located, and the percentage of yield increases with nematode control may not be representative of large areas. In addition, a target species may not be solely responsible for the yield differences. Because of the mixed populations of lesion nematodes with those of other genera, and possibly other pathogenic organisms, a degree of subjectiveness usually is involved in loss judgments. Although one might criticize this subjectiveness, it probably applies equally to other disciplines involving pests living in a polyspecific community. Therefore, several approaches should be used to "establish" pathogenicity.

Although at least 15 species of *Pratylenchus* have been associated with corn, not all have been implicated as parasites or pathogens. The species best documented as being important pathogens are *P. brachyurus*, *P. hexincisus*, and *P. zeae*. The insidious nature of the lesion nematodes may result in more injury to corn than we realize. Damage probably is more widespread but less intense or dramatic than some of the more locally distributed nematodes. Nevertheless, the many tests where increased yields are obtained with control of the lesion nematodes, where these are the most visible parasitic species present, and where other factors cannot be demonstrated as being seriously involved (Bergeson, 1978; Egunjobi, 1974; Norton et al., 1978), point to these nematodes as being important. Yield increases of 13-14% in Indiana (Bergeson, 1978), 31% in Georgia (Johnson and Chalfant, 1973), 10% in Iowa (Norton et al., 1978), 54% in South Dakota (Smolik, 1978), and 33 to over 100% in South Africa (Walters, 1978) have been attributed largely to control of nematodes over small or wide areas and in which lesion nematodes were the primary target species or were common in the plots.

E. Distribution

Members of *Pratylenchus* are ubiquitous, and it is usual to find at least one species in every corn field. They seem to be favored by cultivation. Although worldwide, certain species are more conspicuous in some areas than in others. For example, *P. brachyurus* and *P. zeae* are a problem in the warmer climates, such as the southern United States, Nigeria, and South Africa, but are conspicuous by their absence in the northern corn belt in the United States. *P. brachyurus* also is associated with corn in Brazil and the Transvaal Highland in Africa. *P. neglectus* occurs around corn in Germany, Yugoslavia, and North America. *P. penetrans* is associated with corn in North America, Canada, and South Africa. *P. delattrei* is a pathogen of corn in India.

F. Control

After an assessment of losses, the need for and methods of control should be indicated. The philosophy of control essentially is one of management of the nematode population. Consideration of future crops must also be made, as well as those of monetary and environmental aspects. Although economical chemical treatments have given impetus to corn nematology, and have a place in control, all approaches to control should be considered.

1. *Chemical Control*

Because soil fumigants were usually monetarily prohibitive for dryland corn, nematode problems on corn were slow to be investigated. With the advent of organophosphates and carbamates, many of which were first used as insecticides, a tool was provided for practical use in many corn fields.

Their use often controlled *Pratylenchus* spp. and other nematodes. Reduction of nematodes by fumigants and granular nematicides are frequently correlated with increased yields (Bergeson, 1978; Dickson and Waites, 1978; Johnson and Chalfant, 1973; Norton and Hinz, 1976; Norton et al., 1978; Zirakparvar, 1979).

2. *Sanitation*

Because *Pratylenchus* spp. can survive the dry season in corn roots and many weeds in Nigeria, removal of these roots after harvest is important (Egunjobi and Bolaji, 1979).

3. *Rotation*

Several studies indicate that some crops may effectively reduce populations of lesion nematodes (Ferris, 1967; Johnson et al., 1975; Singh, 1967). For example, populations of *P. brachyurus* and *P. zeae* were generally reduced by cotton and peanuts in Georgia tests (Johnson et al., 1975). Rotations need to be formulated regionally for individual nematode species. Indications are that a given corn cultivar is not equally susceptible to all species of *Pratylenchus*. There are also conflicting reports concerning susceptibilities of corn to the same nematode species. These may be due to cultivar differences or differences in methodology among various investigators.

4. *Resistance*

Resistance to *Pratylenchus* spp. within *Zea mays* has been investigated little, but sufficient information has been accumulated to show that resistance has promise as a useful control measure. For example, Thomas (1980) found that commercial hybrids differed in degree of susceptibility to lesion nematodes and that the relative degree of colonization per gram of dry root was fairly consistent in seven different soils.

5. *Soil Amendments*

Because of the high cost of manufactured nematicides, soil amendments using natural products have been examined as an alternative. In Nigeria, farm yard manure and several plant products significantly reduced populations of *P. brachyurus* and increased corn growth. Although some of the yield increase was believed to be a fertilizer effect, part of the decrease in nematode numbers was considered to be due to nematicidal or nematostatic effects of the amendments (Egunjobi and Larinde, 1975).

III. STING NEMATODES

A. Species

Females of *Belonolaimus* spp. (*B. gracilis* Steiner 1949, *B. longicaudatus* Rau 1958, and *B. nortoni* Rau 1963) are slender and can be as long as 3 mm. The cuticle is coarsely striated with a single lateral line on each side. The lips

are often off-set and contain 7-10 or more annules. Stylets are long and slender with well-developed knobs. The basal esophageal bulb overlaps the anterior part of the intestine. Female tails are usually bluntly rounded.

After Rau described *B. longicaudatus*, it was realized that many of the
earlier reports of *B. gracilis* were actually of *B. longicaudatus*. The latter
species is the one reported most commonly on corn in the southeastern United
States.

B. Associated Symptoms

Probably the main field symptom is severe stunting of the plants where dense
infestations occur (Fig. 3). Stunted plants often are in well-defined areas
but may also be spaced erratically throughout the field. Chlorosis may be
present, but the leaf veins often remain green. Wilting of the foliage during
the heat of the day and recovering at night occurs with severe infestations.
Stalks of infested plants are thinner than healthy ones, and the plant may
tassle when small. Roots are often severely pruned, resulting in a coarse,
stubby appearance (Fig. 4). Necrosis of the root occurs in the later stages
of the disease. In the sand hills of Nebraska, root damage was most noticeable
about 25 cm below the soil surface (Kerr and Wysong, 1979).

C. Life Cycle of the Nematode

The nematode usually is severely damaging only in highly sandy soils. It most
often feeds as an ectoparasite on the root tips and along the sides of the roots.
Occasionally juveniles are found within the roots (Christie et al., 1952).

Fig. 3 Effect of chemicals on corn growth in soil heavily infested with *B.
longicaudatus*. The two rows of stunted corn were nontreated. Florida.
(From Rhoades, 1979.)

Fig. 4 Injury caused by *B. longicaudatus*. Note the reduced root system and root pruning. (Courtesy of H. L. Rhoades.)

D. Percentage and Dollar Loss

Under favorable conditions the sting nematodes are some of the most destructive pests of corn (Fig. 3). In tests in Orlando fine sand in Florida, grain yields averaged 75%, with a maximum of 121%, higher in plots treated to control sting nematodes than in the nontreated plots (Johnson and Dickson, 1973).

E. Distribution

As far as is known, *Belonolaimus* spp. occur only in the United States. It used to be thought that they occurred primarily in the southern states and along the relatively mild Atlantic coastal plains. Occurrences in southern Kansas around wheat and sorghum and with stunted corn in North Central Nebraska has drastically changed our thinking. The species reported most commonly in the southern states is *B. longicaudatus*. *B. nortoni* is associated with corn in Texas, and the occurrences in Nebraska are very close to this species, if not identical. The species in Nebraska has also been found around native grasses in unplowed prairies, indicating that it is probably native to the area.

F. Control

In the absence of known resistance, most control has been with chemicals. Excellent control has been obtained in sandy soils with fumigants (Rhoades, 1979), but good to excellent control has also been obtained with granular carbamate and organophosphate nematicides (Brodie, 1968; Johnson and Dickson, 1973; Rhoades, 1978; 1979).

IV. NEEDLE NEMATODES

A. Species

Females of *Longidorus breviannulatus* Norton and Hoffman 1975 measure about 4-5 mm long, V = 43-50%, odontostyle 81-88 μm, guiding ring from anterior of nematode = 21-26 μm. Males are rare.

B. Associated Symptoms

Damage may be noticed within two weeks after seedling emergence. Oval to oblong patches of stunted chlorotic plants occur (Fig. 5). A purple discoloration characteristic of phosphorus deficiency may be noticed in heavily parasitized plants. Stunted plants appear to be under drought stress.

On the roots, the radicle is abbreviated and the terminals of the seminal and crown roots are frequently slightly swollen (Fig. 6). Pruning of the laterals may be severe. Scarcity of small feeder roots may give the root system a coarse appearance. The roots may have a pronounced yellow discoloration. Root symptoms frequently decrease with depth. Under irrigation or good rainfall, there may be a brushlike proliferation of the roots in the upper layer of the soil. Although there are no distinct root lesions, older portions of the crown roots become tan to brown, and the seminal roots may be destroyed. Seedlings die if parasitism is heavy. Surviving plants may become as tall as noninfected ones, but the stalks are more slender and ears, if formed at all, are small (Malek et al., 1980).

Fig. 5 Field heavily infested with *L. breviannulatus*. Note stunted and sparser stand in area where infestation was greatest. (Courtesy Iowa Agriculture and Home Economics Experiment Station of Ames, Iowa.)

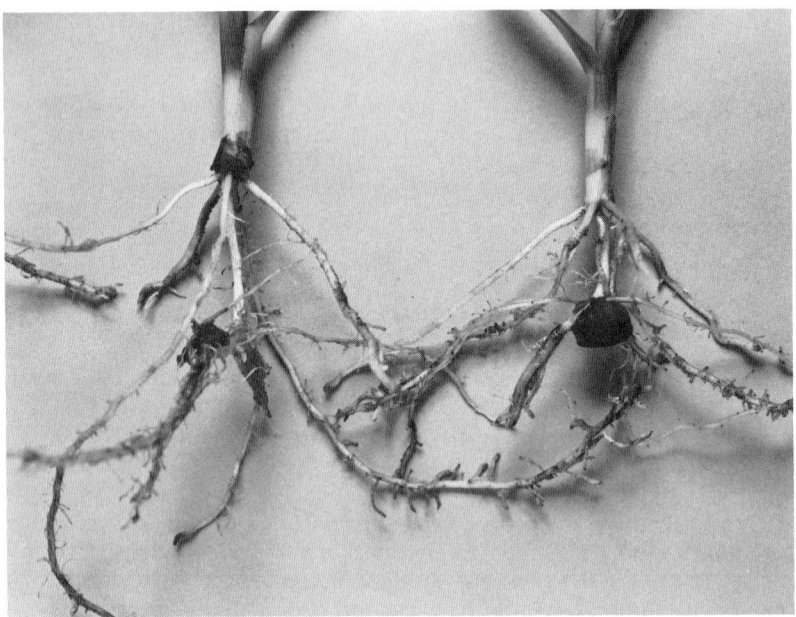

Fig. 6 Young corn roots that were heavily parasitized by *L. breviannulatus* in the field. Note the stubby root effect of lateral roots and the near absence of fibrous roots. (Courtesy of Iowa Agriculture and Home Economics Experiment Station, Ames, Iowa).

C. Life Cycle of the Nematode

Although the nematode can occur in soils with as low as 49% sand, largest populations usually occur in soil with a sand content of 90% or more. Most nematodes are usually found early in the growing season. They may be difficult to find at harvest. Feeding habits and generation time of the nematodes are unknown.

D. Percentage and Dollar Loss

This can be one of the most devastating diseases of corn where the nematode occurs. Because large populations of the nematode evidently are restricted to highly sandy soils, severe damage is likewise restricted to these soils. In Illinois, yields were reduced up to 62% in severely affected areas compared with those in lightly or noninfested areas. Pathogenicity has been demonstrated in the greenhouse (Malek et al., 1980).

E. Distribution

This nematode is known around corn in Delaware, Illinois, Indiana, and Iowa (Malek et al., 1980) but probably is damaging corn in other areas.

F. Control

No control is known at present, but crop rotation may help. There is no evidence that soybeans are a host (Malek et al., 1980). Populations have declined after this crop in both Iowa and Illinois.

V. STUBBY-ROOT NEMATODE

A. Species

The female of *Paratrichodorus minor* (Colbran 1956) Siddiqi 1973 is usually less than 1 mm long with a curved stylet, didelphic gonad, and a near-terminal anus.

B. Associated Symptoms

Christie (1953) was perhaps the first to call attention to this nematode as a major parasite of corn. The field symptoms include stunting, chlorosis, and reduced yields (Fig. 7). Affected roots are smaller, with fewer and shorter rootlets, than healthy plant roots. The roots may be coarse and often with devitalized tips giving a "stubby-root" appearance (Fig. 8). Discoloration and lesions usually are lacking. Severity of the symptoms varies with the nematode population and the age of the plant when most feeding occurs.

C. Life Cycle of the Nematode

P. minor is an ectoparasite feeding largely on the root tips, which results in cessation of terminal growth of the rootlets, contributing to the stubby-root effect. Corn is a favored host, with nematode populations generally increasing around the plant (Johnson et al., 1975; Brodie and Murphy, 1975), although six years in continuous corn suppressed populations in a sandy loam in Georgia (Brodie et al., 1969). Johnson and Nusbaum (1968) reported that five corn varieties were suitable hosts, but populations were influenced by the cultivar

Fig. 7 Poor stand of corn due to stubby-root nematodes in a dry season. (Courtesy of K. R. Barker.)

Fig. 8 Root damage and stunting caused by *P. minor* and *B. longicaudatus*. The stubby roots on these plants are typical of *P. minor* damage. (Courtesy of K. R. Barker.)

and the presence of other parasitic nematodes. In combination with *Pratylenchus zeae*, populations of *P. minor* were higher on all cultivars than when *P. minor* was used alone. Populations of *P. minor* were reduced on one cultivar in combination with *Tylenchorhynchus claytoni*.

D. Percentage and Dollar Loss

Mixed populations of parasitic nematodes occurring in a field and actual losses due to an individual species are difficult to ascertain, but *P. minor* is considered a major pest on corn where populations are large. Yield increases are usually obtained where nematode populations are controlled by chemicals. Johnson and Chalfant (1973) found that nonvolatile chemicals reduced populations of *Belonolaimus longicaudatus* and *P. minor* more than did soil fumigants. Yield increases were attributed to control of these nematodes as well as *Pratylenchus zeae*. Brodie (1968) also found that control of *P. minor*, among others, resulted in yield increases.

E. Distribution

Common in sandy soils in the Atlantic and Gulf Coastal Plain states from Mass-achusetts to Texas, the nematode is also known in the Midwest, Oregon, and California.

F. Control

Chemicals (see above) often give effective control. The nematode has a wide host range, making rotation difficult in many areas. However, Johnson et al. (1975) found that populations were suppressed by peanuts and soybeans in Georgia.

VI. ROOT-KNOT NEMATODES

A. Species

The causal organisms are *Meloidogyne arenaria* (Neal 1889) Chitwood 1949, *M. arenaria thamesi* Chitwood in Chitwood, Specht, and Havis 1952, *M. graminis* (Sledge and Golden 1964) Whitehead 1968, *M. incognita* (Kofoid and White 1919) Chitwood 1949, *M. incognita acrita* Chitwood 1949, *M. javanica* (Treub 1885) Chitwood 1949, and *M. naasi* Franklin 1965. The female is generally pear shaped, with a fragile white cuticle, and may be up to 2 mm in diameter. She possesses a tylenchoid esophagus and a small stylet. The eggs are deposited in a gelatinous matrix from the near-terminal vulva.

B. Associated Symptoms

Symptoms of corn infection by *Meloidogyne* spp. vary considerably with the nematode species and the host cultivar involved. Depending on the degree of infestation, field symptoms vary from being barely visible to that of severe stunting with consequent yield reduction. Damage may be erratic throughout a field or it may occur in definite patches. Chlorosis or other off-color of the foilage may be present.

   Although galling or other swelling on roots can be caused by several biological agents, the occurrence of root knot should always be considered. Galls (Fig. 9) may be large or small, terminal, or further back on the root. Proliferation of fibrous roots may be present. The swelling may be slight on some cultivars.

C. Life Cycle of the Nematodes

Under favorable conditions, the life cycle can be completed in about 30 days. After invasion of the host by the second-stage juvenile, the second, third, and fourth molts occur before the formation of the sexually mature adults. The nematodes enlarge during each stage, resulting in a saccate or pear-shaped adult female. The adult male is vermiform. Most juveniles penetrate root apices, but the zone of elongation and more mature tissues also are in-vaded. Giant cells develop in the meristematic and stelar tissue within six days after inoculation. Adult females are found in 20-24 days after inocula-tion, and the life cycle is completed in about 30-35 days.

   Development of *M. incognita* in a susceptible cultivar of corn is similar to responses in other hosts (Baldwin and Barker, 1970a) with respect to giant cell formation (Fig. 10). Egg-laying females developed within 25 days after

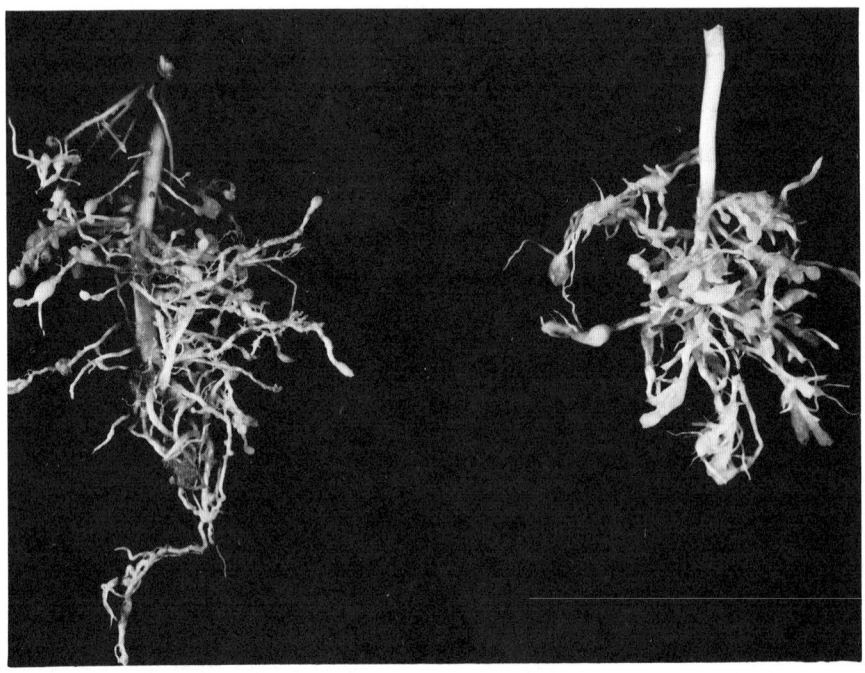

Fig. 9 Corn roots infected with *M. incognita*. Note the numerous galls. (From Baldwin and Barker, 1970.)

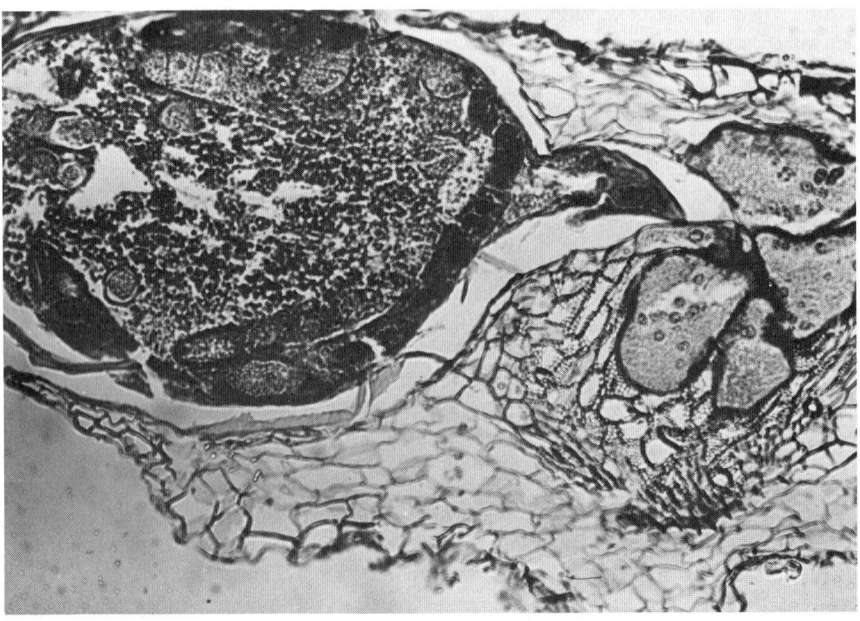

Fig. 10 Female of *M. incognita* and giant cells in susceptible corn. (Courtesy of J. G. Baldwin.)

inoculation. In a poor host, however, giant cells collapsed and were associated with apparently dead juveniles. More juveniles penetrated the susceptible than the poor cultivar by eight days after inoculation.

D. Percentage and Dollar Loss

Because of the worldwide distribution of the root-knot nematodes and the
damage they are known to cause, these nematodes must be ranked as important. Many cultivars are resistant to some species of *Meloidogyne*, and corn
is sometimes used in rotation to control the nematode (Baldwin and Barker,
1970b). But there are frequent reports indicating that *M. arenaria* (Baldwin and Barker, 1970b) and *M. incognita* (Baldwin and Barker, 1970b; Johnson et al., 1975) increase after cropping to corn.

E. Distribution

Although individual species are restricted geographically, the genus is worldwide in distribution.

F. Control

From the many conflicting reports of susceptibilities to a given nematode species, it seems that corn is variable in its reaction to *Meloidogyne* species.
Some information is available concerning resistance in corn against the root-
knot nematodes (Baldwin and Barker, 1970b; Davide, 1979; Johnson, 1975;
Miller, 1973; Nelson, 1957), but as far as I know there have been no mass
screenings for resistance. Sasser (1954) and Dropkin (1959) found susceptibility to different root-knot species varied with the host.

Corn is generally thought to be a nonhost for *M. hapla*, but two biotypes
have been reported to reproduce on corn (Ogbuji and Jensen, 1974). Because
corn is generally resistant to *M. hapla* and is a principal crop in the North-
central United States, it probably acts as a natural control for the nematode.

Good control of *M. incognita* has been obtained with chemicals, but the
degree of control among chemicals has been variable. Fumigants have generally been better than granular nematicides.

Other types of control for root knot have been attempted or observed in
conjunction with other experiments. Brodie and Murphy (1975) found that six
weeks fallow after tomato, a good host, reduced the population of *M. incognita*
more than when corn was planted within one week after tomato harvest.

VII. CYST NEMATODES

A. Species

The causal organisms are *Heterodera avenae* Filipjev 1934, *H. oryzae* Luc and
Berdon Brizuela 1961, *H. zeae* Koshy, Swarup, and Sethi 1971, and *Punctodera
chalcoensis* Stone, Moss, and Mulvey 1976. Adult females are pear or lemon
shaped and measure about 0.4-0.7 mm long and 0.3-0.7 mm wide. The stylet
is about 25 μm long. The thick-walled adult females are pearly white, turning
light to dark brown as the cyst matures. Sometimes there is a yellow phase
in the transition from white to brown. The cyst wall may have zigzag patterns
that are sometimes rugose, depending on the species. Eggs are generally retained in the cyst until they hatch.

B. Associated Symptoms

Stunting frequently occurs in irregular patches of a few to several meters in diameter. In Canada, Fushtey (1965) noted that plants infected with *H. avenae* were two feet tall in mid-August with no signs of flower development, whereas healthy plants were six feet tall with tassels fully formed. Two weeks later the infected plants were as tall as healthy ones but were thinner and had just started to tassel. In India, plants affected by *H. zeae* were described as stunted, pale in color, and having narrow leaves (Koshy et al., 1970). Aerial parts of plants affected with *P. chalcoensis* were stunted and chlorotic. The many short laterals of the stunted root system gave a bottle brush effect (Stone et al., 1976). Fertilizing may partially mask the symptoms (Sosa Moss and Gonzáles, 1973).

C. Life Cycle of the Nematodes

Although the life cycle has not been worked out for all species, it seems that it usually follows the developmental pattern typical of other cyst nematodes. After invasion of the plant the juveniles go through a series of molts, enlarging during each stage. At the last molt the pear- to lemon-shaped white female breaks through the root surface and turns into the brown cyst. However, Fushtey (1965) failed to observed mature cysts or white females of *H. avenae* on the surface of roots in the field. Histological observations revealed that apparently mature females developed within the roots to a limited extent, but egg formation did not occur in the cultivars used (Johnson and Fushtey, 1966). In the cultivar used by Saefkow and Lücke (1979) the nematode developed normally, breaking through the root surface in the usual manner (Fig. 11). Development in corn took about one week longer than in oats. The life cycle of *H. avenae* was completed in two months on corn in the greenhouse (Swarup et al., 1964).

D. Percentage and Dollar Loss

Loss figures generally are not available, but damage may be considerable in some instances. Several reports indicate that corn is not a good host for *H. avenae* (Gill and Swarup, 1971; Rivoal, 1973), but Yadav and Verma (1971) reported corn to be a good host. There are indications, however, that *H. avenae* can cause damage to corn in heavily infested soil even though corn may be a poor host (Hughes, 1975). Although oats and wheat are generally regarded as better hosts for the nematode than is corn, Vallotton and Perrier (1976) considered corn a moderate host because the rate of reproduction was slight. Corn had a depressive effect on the nematode, but the authors stated that corn is sensitive to *H. avenae* because small numbers of the nematode caused serious damage. In Swiss romande, damage occurs in a variety of soils but especially in light ones. A cold, wet period in spring facilitates hatching of the nematode and invasion of the root (Vallotton and Perrier, 1976). In a test in Germany where corn was used for ensilage, losses varied from 5.5 to 39.2% among 10 cultivars (Lücke and Saefkow, 1978).

In India, *H. zeae* often occurs on corn along with *H. avenae* and both are considered of economic importance (Koshy and Swarup, 1971). The loss by *P. chalcoensis* is unknown but is regarded as potentially serious (Stone et al., 1976).

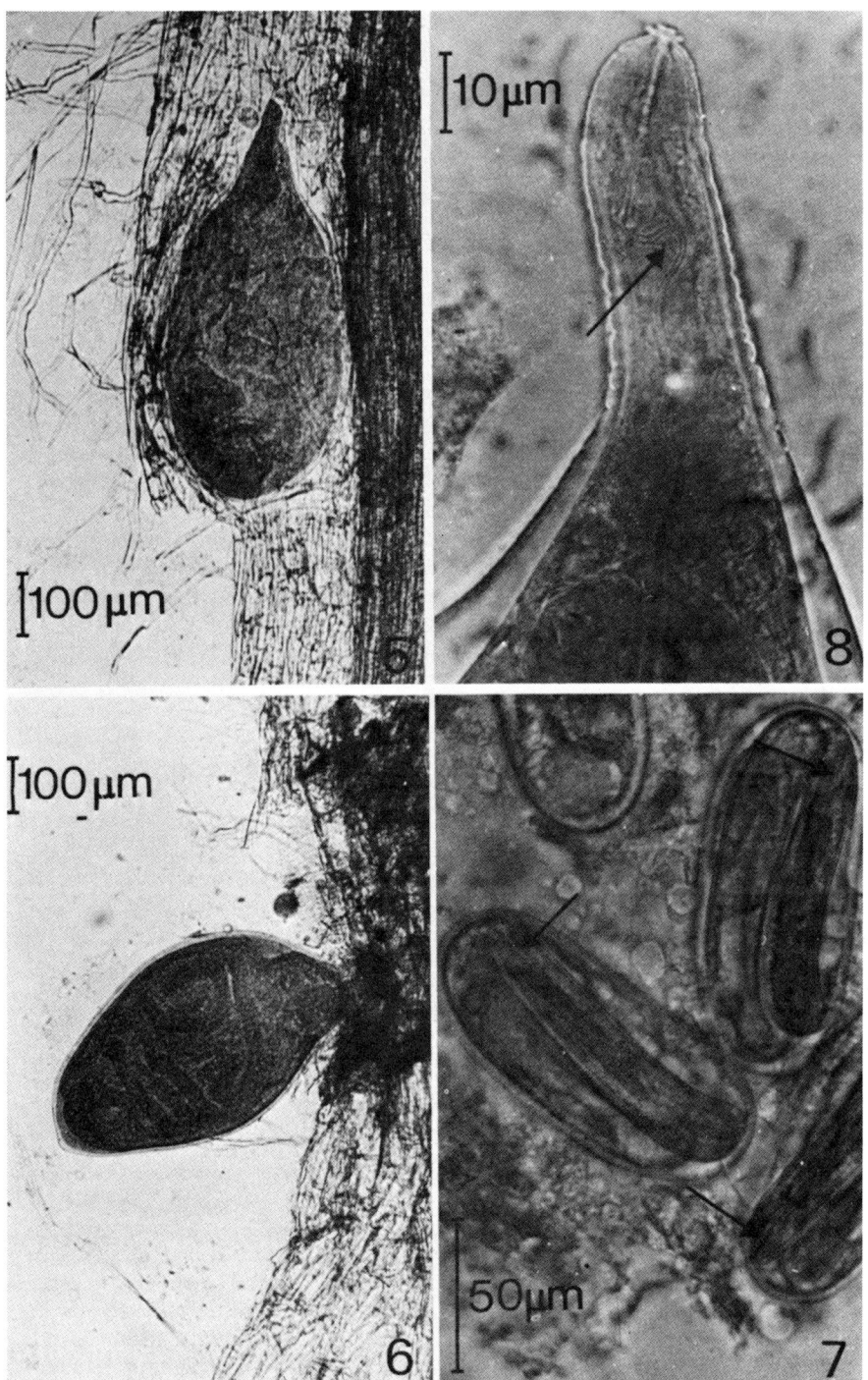

Fig. 11 (5) The swollen female breaking through the root tissue. (6) About
30-40 days after infection. Mature female with distinct ovaries and vulva
cone. (7) 84 days after infection. First L2 with distinct stylet (arrow) in
the eggs. (8) At this time the anterior portion of the digestive system of the
female is entire an functioning. Arrow: opening of the dorsal esophageal
gland. (From Saefkow and Lücke, 1979.)

E. Distribution

Corn can potentially be a host of these nematodes wherever the nematodes occur. *H. avenae* is known to attack corn in Canada, Europe, and India. *H. zeae* was reported first from India (Koshy and Swarup, 1971), but it is now also known in Egypt, Pakistan, and Maryland in the United States (Sardanelli et al., 1981). In Maryland, about 325 hectares are known to be infested (L. R. Krusberg, personal communication, September, 1981). *P. chalcoensis* is known only from Mexico, where it is found above 2000 m (Stone et al., 1976).

F. Control

Crop rotation seems to be the favored method of control, but there is still much to learn concerning host ranges. As with many of the cyst nematodes, suscep- tibility in the field seems to be rather narrow, but maintenance at low levels on other crops and weeds must be kept in mind. If *H. avenae* is a problem on corn, noncereals should be used in rotation. Merny and Cadet (1978) rated several plants for susceptibility to *H. oryzae*, as did Stone et al. (1976) for *P. chalcoensis*, and Gill and Swarup (1971) for *H. avenae*. Susceptibility to *H. zeae* varied greatly among corn cultivars (Srivastava and Swarup, 1975). Wheat, oats, barley, and sorghum were listed as nonhosts or poor hosts.

VIII. STEM AND BULB NEMATODES

A. Species

Ditylenchus dipsaci (Kühn 1857) Filipjev 1936 are slender nematodes that range from 1.0 to 1.4 mm in length. The stylet is small, with strongly developed knobs. The esophagus is tylenchoid with a slender isthmus. The vulva is about 80% from the anterior of the nematode.

B. Associated Symptoms

Ditylenchus dipsaci causes a condition known as a toppling disease because the plants may lodge. Circular to elongated patches 5-60 m in diameter of lodged plants appear in late June to early July (Fig. 12). The internodes are shortened. The base of the stalk is swollen and becomes blackened and ne- crotic. Secondary roots become nonexistent, which explains the weakness of the plant. In France, damage is more severe in plants sown in April than in May (Caubel, 1974).

C. Life Cycle of the Nematode

Under favorable conditions the life cycle can be completed in about three weeks. Cool, wet weather after planting is necessary for infection and development of the disease. Nematodes invade the stalk, causing the swelling and necrosis (Caubel and Rivoal, 1972). Populations from 10 different plants, including corn, alfalfa, beet, narcissus, and oats reproduced on corn (Caubel, 1973).

D. Percentage and Dollar Loss

In 1971 and 1972 in France, 114 of 788 infested hectares of corn were destroyed (Caubel, 1973).

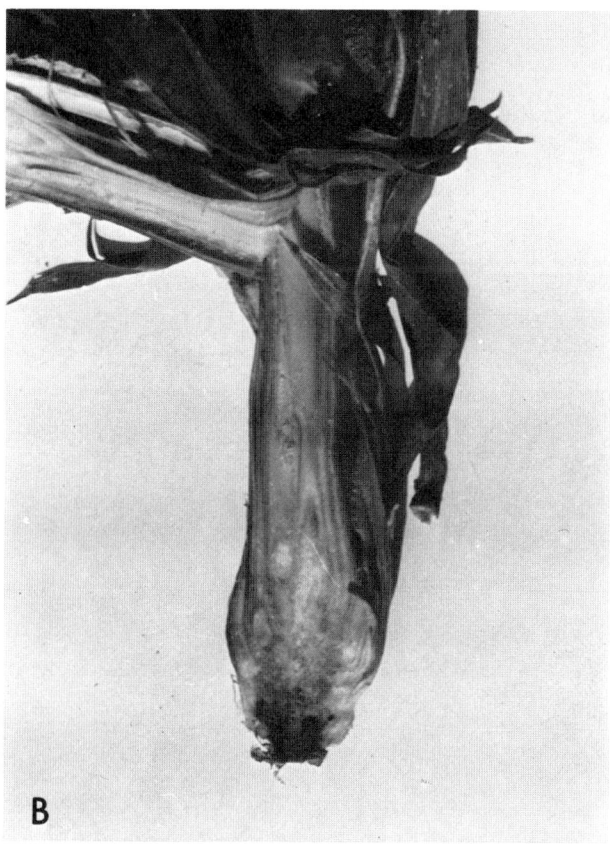

Fig. 12 (A) Corn attacked by *D. dipsaci*. (Courtesy of G. Caubel.) (B) Necrosis caused by *D. dipsaci* on corn stalk. (Courtesy of G. Caubel, Institut National de la Recherche Agronomique, Le Rheu, France.)

E. Distribution

Stem and bulb nematodes are reported from France, Germany, and Yugoslavia.

F. Control

In western France, control was best achieved by careful crop rotation and thorough weeding (Caubel, 1973). Chemical control was possible but not economical in tests in Germany (Hirling, 1974).

IX. LANCE NEMATODES

A. Species

Hoplolaimus aegypti Shafiee and Koura 1968, *H. columbus* Sher 1963, *H. galeatus* (Cobb 1913) Filipjev and Schuurmans Stekhoven 1941, and *H. indicus* Sher 1963 are rather stout, vermiform nematodes with a tylenchoid esophagus overlapping the anterior portion of the intestine. Stylet knobs project forward. The length of the adult female can be over 1 mm. Female gonads are didelphic. One phasmid is anterior to midbody and the other is posterior.

B. Associated Symptoms

Stunting by *H. galeatus* can occur early in the season (Fig. 13). Although affected plants can reach the same height as noninfected ones by the end of the season, they remain more spindly with greatly reduced yields (Norton and Hinz, 1976).

Fig. 13 Area of field heavily infested with *H. galeatus* with a smaller infestation of *Pratylenchus* spp. June 19, 1980. (Courtesy of Iowa Agriculture and Home Economics Experiment Station.)

In South Carolina corn appears to be tolerant to *H. columbus* but can maintain populations that pose a threat to other susceptible crops (Lewis and Smith, 1976). The nematode evidently has a wide host range, and both sweet and field corn are good to excellent hosts (Fassuliotis, 1974; Lewis and Smith, 1976).

H. indicus is associated with stunted plants and patchy growth in corn fields in India (Haider et al., 1978). In a sandy loam soil, an initial population of 100 nematodes around 10-day-old corn seedlings caused symptoms. A gradual reduction in plant growth occurred as inoculum was increased logarithmically to 100,000 nematodes. Affected plants were stunted and weighed less than noninoculated ones. The leaves became pale yellow and died at the tip. At the highest inoculum level, most of the plant was dead by the middle of the third month. Roots become deep brown, are devoid of fibrous roots, and are smaller compared with noninoculated ones.

Movement in the roots by *H. indicus* was both intercellular and intracellular (Haider et al., 1978). The cortical parenchyma around the feeding site became granular, thick walled, and finally disintegrated, leaving cavities. There was no sloughing off of the epidermal and cortical tissues. In some cases, cells of the endodermis, pericycle, and vascular parenchyma stained darkly.

Moderate numbers of *H. aegypti* were associated with stunted corn in the United Arab Republic (Shafiee and Koura, 1968).

C. Life Cycle of the Nematodes

These nematodes are generally migratory endoparasites but at times can be semiendoparasitic. Large numbers of *H. galeatus* are found only in sandy or well-drained soils.

D. Percentage and Dollar Loss

Tests in Iowa resulted in increased yields of 26% in heavily infested soil where most of the parasitic biomass was *H. galeatus* (Norton and Hinz, 1976). Neither soil insects nor other nematodes were believed to be a major factor in the damage. With many situations, as with *H. indicus*, loss figures are not given but stunting in the field is obvious in association with large numbers of nematodes.

E. Distribution

H. galeatus is widely distributed in the United States, but large populations occur mostly in sandy soils. *H. columbus* is known from South Carolina and Georgia. *H. aegypti* occurs in the United Arab Republic and *H. indicus* in India.

F. Control

H. galeatus has been collected around many crops and has been shown to parasitize many types of plants. Evidently its host range is extensive (Ahmad and Chen, 1980). Crop rotation might be useful if resistant crops fit into the agronomic practices. On corn, the fumigant 1,3-D provided better control than carbofuran and other granular nematicides (Norton and Hinz, 1976; Norton et al., 1978).

X. SPIRAL NEMATODES

A. Species

The *Helicotylenchus* spp. *H. digonicus* Perry in Perry, Darling, and Thorne 1959, *H. dihystera* (Cobb 1893) Sher 1961, and *H. pseudorobustus* (Steiner 1914) Golden 1956, are didelphic nematodes with a tylenchoid esophagus that overlaps the anterior part of the intestine. Lips are often conical and continuous with the body contour. The body is arcuate to spiral when at rest. Stylets range from 22 to 30 μm in the species discussed.

B. Associated Symptoms

Mild stunting and reduced yields occur where populations are large, perhaps in conjunction with adverse environmental conditions. Small, light to dark brown lesions occurred on corn inoculated with *H. pseudorobustus* (Taylor, 1961).

C. Life Cycle of the Nematode

These nematodes are generally considered to be semiendoparasites on corn with feeding restricted mostly to the cortical parenchyma. However, Taylor (1961) found that *H. pseudorobustus* can invade roots entirely after artificial inoculation. These nematodes can occur in large numbers in a wide variety of soils.

D. Percentage and Dollar Loss

Damage in the field has been difficult to evaluate because of mixtures of these nematodes with those of other genera. Thus, most work on losses has been done in the greenhouse using monospecific cultures.

Perry et al. (1959) found that *H. digonicus* has no effect on sweet corn, but the nematode caused significant stunting to W-464 field corn in greenhouse tests (Griffin, 1964).

The status of *H. dihystera* as a pathogen or even a parasite on corn is uncertain. Several reports indicate that corn is not a good host and that field populations are suppressed by the plant (Brodie et al., 1969; Johnson et al., 1975) or at least do not increase appreciably (Singh, 1976). Field populations around 15 sweet corn cultivars varied from 0 to 137 nematodes per 150 cm^3 of soil in Georgia (Johnson, 1975). However, Sharma and Loof (1977) found 80,000 per 100 g of soil associated with popcorn in Brazil.

In contrast to *H. dihystera*, *H. pseudorobustus* is a common and well-known parasite of corn, especially in the midwestern United States (Ferris and Bernard, 1971; Norton et al., 1978). In spite of its commonness, good data on damage by this nematode have been difficult to obtain because of populations of mixed species of nematodes in the field. Norton (1977) considered the nematode only a mild pathogen under greenhouse conditions. Although significant differences in top and root weights were obtained in inoculated plants compared with noninoculated ones, numbers were over 200,000 per 1500 cm^3 of soil after one year with no severe damage occurring. Sometimes such high numbers are found in a field that it makes one speculate that this nematode might be causing damage.

E. Distribution

Spiral nematodes are seemingly ubiquitous with corn.

F. Control

Where control is needed, chemicals might be of value. Granular nematicides were only moderately effective against *H. pseudorobustus* in Iowa tests (Norton et al., 1978), and Singh (1976) found fumigants more effective against *H. dihystera* than some granular nematicides. In tests using an Indian population of *H. dihystera*, six corn cultivars varied greatly in susceptibility (Rao and Swarup, 1974). Several crops were nonhosts. Significant differences in increase of *H. pseudorobustus* occurred in 10 corn lines in greenhouse tests, and no lines were considered to be resistant (Norton, 1977). Numbers of *H. pseudorobustus* were generally negatively correlated with the amount of nitrogen applied (Castaner, 1966). Seed treatment of corn with oxamyl conferred short-term control against *H. dihystera* (Truelove et al., 1977). If this technique can be perfected, it could have considerable merit.

XI. OTHER NEMATODES

A. Dagger Nematode

Xiphinema americanum Cobb 1913 is a slender, didelphic nematode up to 2 mm long. The esophagus is dorylaimoid; the spear is 130 μm long and is flanged at the base.

Although *X. americanum* is a common associate of corn in the United States, there is surprisingly little information concerning its pathogenicity considering that the nematode is important on many other crops. There are many reports that populations of the nematode remain low to moderate around corn, or that populations do not increase as rapidly around corn as with other plants, especially perennials (Johnson et al., 1975; Miller, 1980; Christie, 1953). However, Ferris and Bernard (1971) reported corn to be as good a host as other crops. There is speculation that the life cycle takes a year or more to complete. In annual crops the life cycle may be broken annually by cultivation, resulting in essentially a dilution effect. Also, the nematode is sensitive to sudden changes in edaphic conditions resulting in greater perturbations and seasonal fluctuations than with some nematodes. Large populations of the nematode found in sandy soils early in the season makes this species a good suspected pathogen. Pathogenicity data are difficult to obtain in the field because the nematode usually occurs with other known pathogenic species.

B. Awl Nematode

Females of the nematode *Dolichodorus heterocephalus* Cobb 1914 are similar to those of *Belonolaimus* except that the tail is abruptly acute and the basal esophageal bulb does not overlap the anterior part of the intestine.

This nematode feeds at the surface of the root, usually at the root tip, causing growth to cease. Seed germination may be poor. New roots are often attacked as soon as they are formed, resulting in few secondary roots. Lesions, probably due partly to secondary organisms, may occur along the roots. The above-ground portions of the plant can be stunted along with other symptoms that occur as a result of an impaired root system. The nematode usually is found only in wet areas, including irrigated ones, where it can increase under agricultural conditions. The nematode has been found around corn in Florida and at Tifton, Georgia (Perry, 1953).

C. Ring Nematodes

The markedly annulate, stubby nematodes of *Criconemella* spp. range from 0.2 to 1.0 mm in length. The stylet is robust for the size of the nematode. The esophagus has a fused metacorpus and procorpus and the isthmus is followed by a narrow basal bulb.

Criconemella spp. associated with corn usually occur with mixed populations of other species. Large populations around corn usually occur only in highly sandy soils. In a three year study on nematicide-treated and untreated plots, *C. ornata* (Raski 1958) Luc and Raski 1981) together with *Helicotylenchus dihystera* and *Paratrichodorus minor*, were considered mainly responsible for reduction in yield of sweet corn (Johnson, 1975).

Johnson (1975) found that populations of the nematode varied considerably among 15 sweet corn cultivars. In the field, the nematode increased rapidly on Coker 71 corn and peanuts but was suppressed by cotton and soybeans in rotation studies (Johnson et al., 1975). Seneca Chief was a good host for the nematode (Johnson and Chalfant, 1973).

Other species of *Criconemella* and related genera have been associated with corn, but biological data were not provided. Inbread WF 9 was listed as a "congenial" host for *C. onoensis* (Luc) (Alhassan and Hollis, 1969).

D. Burrowing Nematode

The burrowing nematode *Radopholus similis* (Cobb 1893) Thorne 1949 is mostly a tropical and subtropical species. Work on corn seems to be limited, although corn is a good host (Edwards and Wehunt, 1971; Keetch, 1972; Martin et al., 1969).

Symptoms include brown to reddish-black lesions along the roots (Fig. 14). Damage is usually limited to cortical areas.

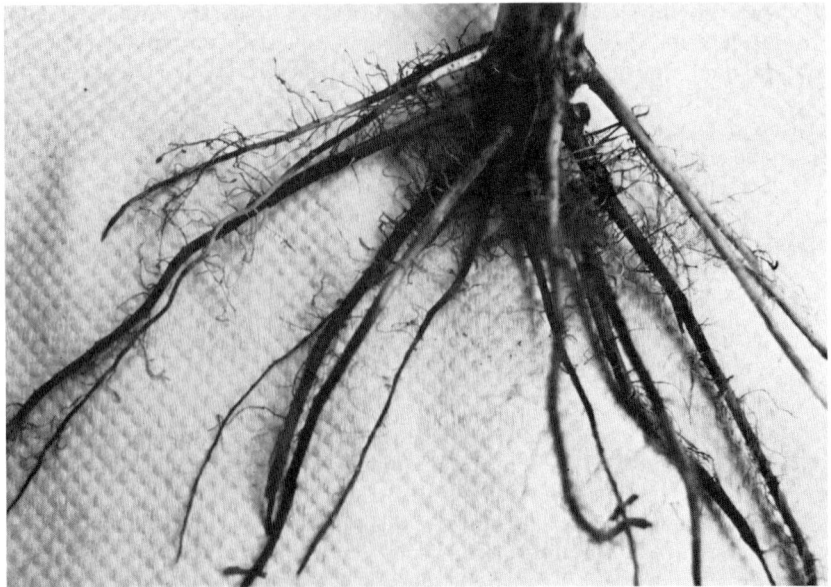

Fig. 14 Injury to corn roots by *R. similis*. (Courtesy of D. P. Keetch.)

E. Reniform Nematodes

Three species of *Rotylenchulus* have been associated with corn but relatively little information is available.

Rotylenchulus borealis Loof and Oostenbrink 1962 is known on corn in Yugoslavia and Bulgaria. Corn is a good host for *R. parvus* Williams 1960 (Sher, 1961; Dasgupta and Raski, 1968; Furstenburg and Heyns, 1978) and is the dominant species of the genus in corn in South Africa (Van den Berg, 1978). The mean number of eggs per egg mass was 16, 24, and 13 around corn, barley, and red kidney bean, respectively (Dasgupta and Raski, 1968). On barley, the nematode reproduced best at 30°C, 20-35°C being the range tested. Survival varied with temperature and storage time.

Plant reactions to *R. reniformis* Linford and Oliveira 1940 have varied. Field corn, *Zea mays* var *rugosa* and *Zea mays* var *indentata* were resistant (Birchfield and Brister, 1962; Singh, 1975) but corn was a good host for one race of the nematode (Dasgupta and Seshadri, 1971).

F. Stunt Nematodes

Tylenchorhynchus and *Quinisulcius* are didelphic, medium-sized nematodes with a tylenchoid esophagus that does not overlap the intestine. The *Tylenchorhynchus* species discussed have four lines in the lateral field, but *Quinisulcius* has five.

Members of these genera are common associates of corn, but in most instances pathogenicity has not been demonstrated. They are generally considered mild pathogens at best, but more work is necessary. The pertinent information available is summarized in Table 1.

G. Pin Nematodes

Some of the smallest plant-parasitic nematodes known, species of *Paratylenchus* and the related genus *Gracilacus* are easily overlooked. Some species may be less than 0.3 mm. Many have been associated with corn but their importance is not known.

Paratylenchus minutus Linford in Linford, Oliveira, and Ishii 1949 was seen to feed on corn, but nothing is known of its pathogenicity to the crop (Linford et al., 1949).

H. Sheath Nematode

Species of *Hemicycliophora* are occasionally reported associated with corn, but documentation as parasites seems only to be with *H. parvana* Tarjan 1952 in Florida.

The nematode reproduced well on sweet corn, increasing 85 times in five months in the greenhouse (Ruehle and Christie, 1958). There was no noticeable injury to the host under these conditions.

I. Miscellaneous Occurrences

1. *Aglenchus agricola (De Man 1884) Meyl 1961*

This nematode parasitized corn under gnotobiotic conditions. No gross pathogenic effects were noticed by Deubert et al. (1967). In dixenic cultures, *A. agricola* increased penetration of the vascular stele of corn by *Fusarium roseum*, but the presence of the nematodes generally did not increase invasions by *Pythium ultimum* or *F. roseum* (Kisiel et al., 1969).

Table 1 *Tylenchorhynchus* and *Quinisulcius* spp. Associated with Corn

Species	Remarks	Reference
Tylenchorhynchus agri Ferris 1963	Dent corn and Golden Cross Bantam sweet corn were good hosts	Coates-Beckford and Malek (1978)
	Populations not related to nitrogen in the field	Castaner (1966)
T. clarus Allen 1955	Sweet corn Golden Blend a good host	Edongali and Lownsbery (1980)
T. claytoni Steiner 1937	Parasitism but no gross pathological effects under gnotobiotic conditions	Deubert et al. (1967)
	Inbreds varied in suscepti-bility; susceptible lines had greatly reduced root systems and nematode popu-lations increased	Nelson (1957)
	Suppressed by *Paratricho-dorus minor* in some but not all corn cultivars	Johnson and Nusbaum (1968)
T. maximus Allen 1955	Caused stunting of corn in greenhouse	Griffin (1964)
T. nudus Allen 1955	Common associate in mid-western United States	
T. vulgaris Upadhyay, Swarup, and Sethi 1972	Corn a good host as indicated by buildup in field; on good hosts plant growth is stunted; pH range of 5.5-7.7 and sand loam or loam soil favored reproduction; wide host range; India	Upadhyay and Swarup (1972)
	Considerable differential in response among 17 cultivars	Upadhyay and Swarup (1976)
Quinisulcius acutus (Allen 1955) Siddiqi 1971	Increased around corn in crop rotation studies in Kansas	Dickerson et al. (1978)

2. *Anguina tritici (Steinbuch 1799) Chitwood 1935*

Sweet corn was a host by inoculation (Limber, 1976).

3. *Aphelenchoides arachidis Bos*

Large numbers of this nematode were found in the roots of corn in Nigeria (Bos, 1977).

4. *Scutellonema brachyurum (Steiner, 1938) Andrássy 1958*

S. brachyurum was associated with corn in South Africa (Walters, 1979). Field and sweet corn were not good hosts in the greenhouse (Kraus and Lewis, 1979).

5. *Subanguina radicicola (Greef 1872) Paramonov 1967*

Corn was slightly infected by material from American beachgrass (Halisky et al., 1977).

REFERENCES

Ahmad, M., and Chen, T. A. (1980). Effect of certain environmental factors and host plants on reproduction of *Hoplolaimus galeatus. Plant Dis. 64:* 479-480.

Alhassan, S. A., and Hollis, J. P. (1969). Ring nematode ratings of rice area rotation crops (abstract). *Phytopathology 59:* 111.

All, J. N., Kuhn, C. W., Gallaher, R. N., Jellum, M. D., and Hussey, R. S. (1977). Influence of no-tillage-cropping, carbofuran, and hybrid resistance on dynamics of maize chlorotic dwarf and maize dwarf mosaic diseases of corn. *J. Econ. Entomol. 70:* 221-225.

Baldwin, J. G., and Barker, K. R. (1970a). Histopathology of corn hybrids infected with root knot nematode, *Meloidogyne incognita. Phytopathology 60:* 1195-1198.

Baldwin, J. C., and Barker, K. R. (1970b). Host suitability of selected hybrids, varieties and inbreds of corn to populations of *Meloidogyne* spp. *J. Nematol. 2:* 345-350.

Bergeson, G. B. (1978). Control of the lesion nematode (*Pratylenchus* spp.) in corn with carbofuran. *Plant Dis. Rep. 62:* 295-297.

Birchfield, W., and Brister, L. R. (1962). New hosts and nonhosts of reniform nematode. *Plant Dis. Rep. 46:* 683-685.

Bos, W. S. (1977). A preliminary report on the distribution and host-range of the nematode *Aphelenchoides arachidis* Bos in the north of Nigeria. *Samaru Agric. Newsletter 19:* 21-23.

Brodie, B. B. (1968). Systemic pesticides for control of sting and stubby-root nematodes on vegetables. *Plant Dis. Rep. 52:* 19-23.

Brodie, B. B., Good, J. M., and Adams, W. E. (1969). Population dynamics of plant nematodes in cultivated soil: Effect of sod-based rotations in Cecil sandy loam. *J. Nematol. 1:* 309-312.

Brodie, B. B., and Murphy, W. S. (1975). Population dynamics of plant nematodes as affected by combinations of fallow and cropping sequence. *J. Nematol. 7*: 91-92.

Castaner, D. (1966). The relationship of numbers of *Helicotylenchus microlobus* to nitrogen soil amendments. *Iowa State J. Sci. 41*: 125-135.

Caubel, G. (1973). Problemes posés par le németode des tiges (*Ditylenchus dipsaci* Kühn) en culture de mais. *C. R. Seances Acad. Agric. (France) 59*: 627-640.

Caubel, G. (1974). Etude de la pénétration de *Ditylenchus dipsaci* dans les plantules de mais. (Abstract). In *Simposia Internacional XII Nematologia Sociedad Europea Nematologos. 1974*: 19-20.

Caubel, G., and Rivoal, R. (1972). Observations sur les attaques de deux németodes nuisibles au mais en 1971. *Phytoma 239*: 15-18.

Christie, J. R. (1953). Ectoparasitic nematodes of plants. *Phytopathology 43*: 295-297.

Christie, J. R., Brooks, A. N., and Perry, V. G. (1952). The sting nematode, *Belonolaimus gracilis*, a parasite of major importance on strawberries, celery, and sweet corn in Florida. *Phytopathology 42*: 173-176.

Coates-Beckford, P. P., and Malek, R. B. (1978). Host preferences of the stunt nematode *Tylenchorhynchus agri*. *Plant Dis. Rep. 62*: 793-796.

Dasgupta, D. R., and Raski, D. J. (1968). The biology of *Rotylenchulus parvus*. *Nematologica 14*: 429-440.

Dasgupta, D. R., and Seshadri, A. R. (1971). Reproduction, hybridization, and host adaption in physiological races of the reniform nematode, *Rotylenchulus reniformis*. *Indian J. Nematol. 1*: 128-144.

Davide, R. G. (1979). Reactions of different crops to infection by *Meloidogyne arenaria* isolated from fig and the influence of temperature on development of the nematode. *Plant Dis. Rep. 63*: 207-211.

Deubert, K. H., Norgren, R. L., Paracer, S. M., and Zuckerman, B. M. (1967). The influence of *Tylenchus agricola* and *Tylenchorhynchus claytoni* on corn roots under gnotobiotic conditions. *Nematologica 13*: 56-62.

Dickerson, O. J., Darling, H. M., and Griffin, G. D. (1964). Pathogenicity and population trends of *Pratylenchus penetrans* on potato and corn. *Phytopathology 54*: 317-322.

Dickerson, O. J., Franz, T. J., and Lash, L. D. (1978). Influence of crop rotation on nematode populations in Kansas (abstract). *J. Nematol. 10*: 284.

Dickson, D. W., and Waites, R. E. (1978). Chemical control of *Trichodorus christiei*, *Pratylenchus zeae*, *P. brachyurus*, and *Criconemoides* sp. on field corn in Florida (abstract). *J. Nematol. 10*: 285.

Dropkin, V. H. (1959). Varietal responses of soybeans to *Meloidogyne*—a bioassay system for separating races of root-knot nematodes. *Phytopathology 49*: 18-23.

Edongali, E. A., and Lownsbery, B. F. (1980). Reproduction of mixed populations of *Tylenchorhynchus clarus* and *Pratylenchus* spp. on 10 host plants. *Plant Dis. 64*: 458-459.

Edwards, D. I., and Wehunt, E. J. (1971). Host range of *Radopholus similis* from banana areas of Central America with indications of additional races. *Plant Dis. Rep. 55*: 415-418.

Egunjobi, O. A. (1974). Nematodes and maize growth in Nigeria. I. Population dynamics of *Pratylenchus brachyurus* in and about the roots of maize and its effects on maize production at Ibadan. *Nematologica 20*: 181-186.

Egunjobi, O. A., and Bolaji, E. I. (1979). Dry season survival of *Pratylenchus* spp. in maize fields in western Nigeria. *Nematol. Medit. 7*: 129-135.

Egunjobi, O. A., and Larinde, M. A. (1975). Nematodes and maize growth in Nigeria. II. Effects of some amendments on populations of *Pratylenchus brachyurus* and on the growth and production of maize (*Zea mays*) in Nigeria. *Nematol. Medit. 3:* 65-73.

Fassuliotis, G. (1974). Host range of the Columbia lance nematode, *Hoplolaimus columbus*. *Plant Dis. Rep. 58:* 1000-1002.

Ferris, J. M. (1967). Factors influencing the population fluctuation of *Pratylenchus penetrans* in soils of high organic content. I. Effect of soil fumigants and different crop plants. *J. Econ. Entomol. 60:* 1708-1714.

Ferris, V. R., and Bernard, R. L. (1971). Effect of soil type on population densities of nematodes in soybean rotation fields. *J. Nematol. 3:* 123-128.

Furstenberg, J. P., and Heyns, J. (1978). The effect of cultivation on nematodes. Part I. *Rotylenchulus parvus. Phytophylactica 10:* 77-80.

Fushtey, S. G. (1965). The oat cyst nematode, *Heterodera avenae* Wollenweber, on corn, *Zea mays*, in Ontario. *Can. Plant Dis. Sur. 45:* 105-106.

Gill, J. S., and Swarup, G. (1971). On the host range of the cereal cyst nematode, *Heterodera avenae* Woll. 1924, the causal organism of "Molya" disease of wheat and barley in Rajasthan, India. *Indian J. Nematol. 1:* 63-67.

Graham, T. W. (1951). Nematode root rot of tobacco and other plants. *S. Car. Agr. Expt. Sta. Bull. 390.* 25pp.

Griffin, G. D. (1964). Association of nematodes with corn in Wisconsin. *Plant Dis. Rep. 48:* 458-459.

Haider, M. G., Nath, R. P., and Prasad, S. S. (1978). Studies on the lance nematode *Hoplolaimus indicus* I—Pathogenicity and histopathogenesis on maize. *Indian J. Nematol. 8:* 9-12.

Halisky, P. M., Ahmad, M., and Glover, L. B. (1977). Observations on the root gall nematode parasitizing American beachgrass in New Jersey. *Plant Dis. Rep. 61:* 48-49.

Hirling, V. W. (1974). Schädliche Nematoden an Mais in Baden-Wurtemberg. I. Die Umfallkrankheit beim Mais durch *Ditylenchus dipsaci*—Befall. *Anzeiger Schadlingskunde, Pflanzen Umweltschutz 47:* 33-39.

Hughes, R. C. (1975). The scope of efficient pesticide use on oil rape and maize. In British Insecticide and Fungicide Conference (8th), Brighton. *Proc. 3:* 1019-1024.

Johnson, A. W. (1975). Resistance of sweet corn cultivars to plant-parasitic nematodes. *Plant Dis. Rep. 59:* 373-376.

Johnson, A. W., and Chalfant, R. B. (1973). Influence of organic pesticides on nematode and corn earworm damage and on yield of sweet corn. *J. Nematol. 5:* 177-180.

Johnson, A. W., Dowler, C. C., and Hauser, E. W. (1975). Crop rotation and herbicide effects on population densities of plant-parasitic nematodes. *J. Nematol. 7:* 158-168.

Johnson, A. W., and Nusbaum, C. J. (1968). The activity of *Tylenchorhynchus claytoni, Trichodorus christiei, Pratylenchus brachyurus, P. zeae* and *Helicotylenchus dihystera* in single and multiple inoculations on corn and soybean (abstract). *Nematologica 14:* 9.

Johnson, J. T., and Dickson, D. W. (1973). Evaluation of methods and rates of application of three nematicide-insecticides for control of the sting nematode on corn. *Proc. Soil Crop Sci. Soc. Fla. 32:* 171-173.

Johnson, P. W., and Fushtey, S. G. (1966). The biology of the oat cyst nematode *Heterodera avenae* in Canada. II. Nematode development and related anatomical changes in roots of oats and corn. *Nematologica* 12: 630-636.

Keetch, D. P. (1972). Some host plants of the burrowing eelworm, *Radopholus similis* (Cobb) in Natal. *Phytophylactica* 4: 51-57.

Kerr, E. D., and Wysong, D. S. (1979). Sting nematode, *Belonolaimus* sp., in Nebraska. *Plant Dis. Rep.* 63: 506-507.

Kisiel, M., Deubert, K., and Zuckerman, B. M. (1969). The effect of *Tylenchus agricola* and *Tylenchorhynchus claytoni* on root rot of corn caused by *Fusarium roseum* and *pythium ultimum*. *Phytopathology* 59: 1387-1390.

Koshy, P. K., and Swarup, G. (1971). Distribution of *Heterodera avenae*, *H. zeae*, *H. cajani* and *Anguina tritici* in India. *Indian J. Nematol.* 1: 106-111.

Koshy, P. K., Swarup, G., and Sethi, C. L. (1970). *Heterodera zeae* n. sp. (Nematoda: Heteroderidae), a cyst-forming nematode on *Zea mays*. *Nematologica* 16: 511-516.

Kraus, S. H., and Lewis, S. A. (1979). *Scutellonema brachyurum*: host plants and pathogenicity on cotton. *Plant Dis. Rep.* 63: 688-691.

Lewis, S. A., and Smith, F. H. (1976). Host plants, distribution, and ecological associations of *Hoplolaimus columbus*. *J. Nematol.* 8: 264-270.

Limber, D. P. (1976). Artificial infection of sweet corn seedlings with *Anguina tritici* Steinbuch (1799) Chitwood, 1935. *Proc. Helm. Soc. Wash.* 43: 201-203.

Linford, M. B., Oliveira, J. M., and Ishii, M. (1949). *Paratylenchus minutus*, n. sp., a nematode parasitic on roots. *Pac. Sci.* 3: 111-119.

Lücke, E., and Saefkow, M. (1978). Untersuchungen über Befall und Zystenbildung durch das Getreidezystenälchen am Mais. *Z. Pflanzenkrankheiten Pflanzenschutz* 85: 385-392.

Malek, R. B., Norton, D. C., Jacobsen, B. J., and Acosta, N. (1980). A new corn disease caused by *Longidorus breviannulatus* in the Midwest. *Plant Dis.* 64: 1110-1113.

Martin, G. C., James, G. L., Bissett, J. L., and Way, J. I. (1969). Trials with field crops and *Radopholus similis* with observations on *Pratylenchus* sp., *Meloidogyne* sp., and other plant parasitic nematodes. *Rhod. J. Agric. Res.* 7: 149-157.

Merny, G., and Cadet, P. (1978). Penetration of juveniles and development of adults of *Heterodera oryzae* on different plants. *Rev. Nematol.* 1: 251-255.

Miller, L. I. (1973). Development of a Virginia isolate of *Meloidogyne arenaria* on eighteen inbred lines of *Zea mays* (abstract). *Vir. J. Sci.* 24: 110.

Miller, P. M. (1980). Reproduction and survival of *Xiphinema americanum* on selected woody plants, crops, and weeds. *Plant Dis.* 64: 174-175.

Miller, R. E., Boothroyd, C. W., and Mai, W. F. (1963). Relationship of *Pratylenchus penetrans* to roots of corn in New York. *Phytopathology* 53: 313-315.

Mountain, W. B. (1954). Studies of nematodes in relation to brown root rot of tobacco in Ontario. *Can. J. Bot.* 32: 737-759.

Nelson, R. R. (1957). Resistance in corn to *Meloidogyne incognita* (abstract). *Phytopathology* 47: 25-26.

Norton, D. C. (1977). *Helicotylenchus pseudorobustus* as a pathogen on corn, and its densities on corn and soybean. *Iowa State J. Res.* 51: 279-285.

Norton, D. C., and Hinz, P. (1976). Relationship of *Hoplolaimus galeatus* and *Pratylenchus hexincisus* to reduction of corn yields in sandy soils in Iowa. *Plant Dis. Rep.* 60: 197-200.

Norton, D. C., Tollefson, J., Hinz, P., and Thomas, S. H. (1978). Corn yield increases relative to nonfumigant chemical control of nematodes. *J. Nematol.* 10: 160-166.

Ogbuji, R. O., and Jensen, H. J. (1974). Two Pacific Northwest biotypes of *Meloidogyne hapla* reproduce on corn and oat. *Plant Dis. Rep.* 58: 128-129.

Ogiga, I. R., and Estey, R. H. (1975). Penetration and colonization of *Brassica rapa* and *Zea mays* root tissues by *Pratylenchus penetrans*. *Phytoprotection* 56: 23-30.

Olowe, T., and Corbett, D. C. M. (1976). Aspects of the biology of *Pratylenchus brachyurus* and *P. zeae*. *Nematologica* 22: 202-211.

Perry, V. G. (1953). The awl nematode, *Dolichodorus heterocephalus*, a devastating plant parasite. *Proc. Helm. Soc. Wash.* 20: 21-27.

Perry, V. G., Darling, H. M., and Thorne, G. (1959). Anatomy, taxonomy, and control of certain spiral nematodes attacking blue grass in Wisconsin. *Univ. Wis. Agr. Exp. Sta. Res. Bull.* 207: 24pp.

Rao, V. R., and Swarup, G. (1974). Susceptibility of plants to the spiral nematode *Helicotylenchus dihystera*. *Indian J. Nematol.* 4: 228-230.

Rhoades, H. L. (1978). Influence of nonfumigant nematicides and DBCP on *Belonolaimus longicaudatus* and yield of field corn in central Florida. *Plant Dis. Rep.* 62: 91-94.

Rhoades, H. L. (1979). Evaluation of nematicides and methods of their application for control on nematodes on field corn. *Nematropica* 9: 43-47.

Rivoal, R. (1973). Accidents végétatifs et parasitaires au début de la croissance du mäis dans l'ouest de la France. La maladie vermiculaire du mäis cuasée par le nématode à kyste des céréales, *Heterodera avenae*. Sci. Agron. Rennes 1973: 223-224.

Ruehle, J. L, and Christie, J. R. (1958). Feeding and reproduction of the nematode *Hemicycliophora parvana*. *Proc. Helm. Soc. Wash.* 25: 57-60.

Saefkow, M., and Lücke, E. (1979). Die Entwicklung von *Heterodera avenae* in maiswurzelm. *Nematologica* 25: 309-313.

Sardanelli, S., Krusberg, L. R., and Golden, A. M. (1981). Corn cyst nematode, *Heterodera zeae*, in the United States. *Plant Dis.* 65: 622.

Sasser, J. N. (1954). Identification and host-parasite relationships of certain root-knot nematodes (*Meloidogyne* spp.). *Md. Agr. Exp. Sta. Bull.* A-77 (*Tech*): 30pp.

Shafiee, M. F., and Koura, F. (1968). *Hoplolaimus aegypti* n. sp. (Hoplolaimidae: Tylenchida; Nematoda) from U.R.A. *Bull. Zool. Soc. Egypt* 22: 117-120.

Sharma, R. D., and Loof, P. A. A. (1977). Nematodes of the cocoa region of Bahia, Brazil. VII. Nematodes associated with vegetables. *Soc. Brasil Nematol. II. Pub.* 2: 125-133.

Sher, S. A. (1961). Revision of Hoplolaiminae (Nematoda) I. Classification of nominal genera and nominal species. *Nematologica* 6: 155-169.

Singh, N. D. (1975). Studies on selected hosts of *Rotylenchulus reniformis* and its pathogenicity to soybean (*Glycine max*). *Nematropica* 5: 46-51.

Singh, N. D. (1976). Studies on the population dynamics of selected plant nematodes on three crops. *Plant Dis. Rep.* 60: 783-786.

Singh, S. D. (1967). On two new species of the genus *Aphelenchoides* Fischer, 1894 (Nematoda: Aphelenchoididae) from north India. *J. Helminthol.* 41: 63-70.

Smolik, J. D. (1978). Influence of previous insecticidal use on ability of carbofuran to control nematode populations in corn and effect on corn yield. *Plant Dis. Rep.* 62: 95-99.

Sosa Moss, C., and González, P. C. (1973). Respuesta de maiz chalqueño fertilizado y no fertilizado a 4 diferentes niveles de *Heterodera punctata Raza* Mexicana (Nematoda: Heteroderidae) (abstract). *Nematropica 3:* 13-14.

Srivastava, A. N., and Swarup, G. (1975). Preliminary studies on some graminaceous plants for their susceptibility to the maize cyst nematode, *Heterodera zeae* Koshy et al., 1970. *Indian J. Nematol. 5:* 257-259.

Stone, A. R., Sosa Moss, C., and Mulvey, R. H. (1976). *Punctodera chalcoensis* n. sp. (Nematoda: Heteroderidae) a cyst nematode from Mexico parasitizing *Zea mays. Nematologica 22:* 381-389.

Swarup, G., Prasad, S. K., and Raski, D. J. (1964). Some *Heterodera* species from India. *Plant Dis. Rep. 48:* 235.

Taylor, D. P. (1961). Biology and host-parasite relationships of the spiral nematode, *Helicotylenchus microlobus. Proc. Helm. Soc. Wash. 28:* 60-66.

Thomas, S. H. (1978). Population densities of nematodes under seven tillage regimes. *J. Nematol. 10:* 24-27.

Thomas, S. H. (1980). Response of plant-parasitic nematodes to corn hybrids and edaphic factors *(Abstract). J. Nematol. 12:* 239.

Townshend, J. L. (1972). Influence of edaphic factors on penetration of corn roots by *Pratylenchus penetrans* and *P. minyus* in three Ontario soils. *Nematologica 18:* 201-212.

Truelove, B., Rodriguez-Kabana, R., and King, P. S. (1977). Seed treatment as a means of preventing nematode damage to crop plants. *J. Nematol. 9:* 326-330.

Upadhyay, K. D., and Swarup, G. (1972). Culturing, host range and factors affecting multiplication of *Tylenchorhynchus vulgaris* on maize. *Indian J. Nematol. 2:* 139-145.

Upadhyay, K. D., and Swarup, G. (1976). Reaction of some maize varieties against *Tylenchorhynchus vulgaris. Indian J. Nematol. 6:* 105-106.

Vallotton, R., and Perrier, J. J. (1976). *Heterodera avenae*, le nématode à kyste des céréales, un parasite du mais peu connu en Suisse romande. *Rev. Suisse Agric. 8:* 160-174.

Van den Berg, E. (1978). The genus *Rotylenchulus* Linford & Oliveira, 1940 (Rotylenchulinae: Nematoda) in South Africa. *Phytophylactica 10:* 57-64.

Walters, M. C. (1978). Present status of knowledge of nematode damage and control in South Africa. Proc. Third S. African Maize Breeding Symposium. *Comm. Dept. Agric. Tech. Serv. Repub. S. Africa 152:* 62-66.

Walters, M. C. (1979). The possible status of parasitic nematodes as limiting factors in maize production in South Africa. Proc. Second S. African Maize Breeding Symp. 1976. *Tech. Commun. Dept. Agric. Tech. Serv. Repub. S. Africa. 142:* 112-118.

Yadav, B. S., and Verma, A. C. (1971). Cereal cyst eelworm and other nematodes associated with maize in Rajasthan. *Indian J. Nematol. 1:* 97-98.

Zirakparvar, M. E. (1979). Population changes of *Pratylenchus hexincisus* as influenced by chemicals in fibrous and coarse roots of corn. *Plant Dis. Rep. 63:* 55-58.

Zirakparvar, M. E. (1980). Host range of *Pratylenchus hexincisus* and its pathogenicity on corn, soybean, and tomato. *Phytopathology 70:* 749-753.

Zirakparvar, M. E., Norton, D. C., and Cox, C. P. (1980). Population increase of *Pratylenchus hexincisus* on corn as related to soil temperature and type. *J. Nematol. 12:* 313-318.

Chapter 4
Nematode Parasites of Rice

John P. Hollis, Jr.* *Louisiana State University, Baton Rouge, Louisiana*

Sman Keoboonrueng *Ministry of Agriculture, Bangkok, Thailand*

I. INTRODUCTION

There is a critical shortage of basic research information in the soil microbiology of rice culture and natural aquatic plant ecosystems over most of the globe. Two reasons for this shortage emerge when relevant comparisons are made with corresponding knowledge of wheat, corn, soybean, cotton, and natural upland prairie ecosystems. There have been a lack of facilities and of

Present affiliation: Biotron, Inc., Baton Rouge, Louisiana, and Biotron Tropical Agricultural Experiment Station, Burrell Boom, Belize

scientific personnel in the underdeveloped countries where rice is grown, in spite of countervailing efforts by the Rockefeller and Ford Foundations. Japan, a country preeminent in rice research, has encountered soil problems in rice primarily on iron-deficient soils, and extensive research data have not been applicable elsewhere.

Research in the Louisiana State University Department of Plant Pathology on submerged soil ecosystems began approximately three decades ago, a short time later plant-parasitic nematodes were first investigated and subsequently identified as possible components of such ecosystems. Submerged soil ecosystems—including rice areas—constitute approximately one-half of our Louisiana environment and play major roles in both the ecology and economy of Louisiana and related areas of the U.S. Gulf Coast.

The broad aims of this chapter are to place soil and plant nematodes in scientific and technological perspective and to assemble information on plant parasitic nematodes in rice culture on a worldwide basis. The Librarian, International Rice Research Institute (Manila, Philippines) has kindly made statistics available on rice acreage in the major producing countries (Table 1), and on types of rice culture (Table 2); this enables evaluation of market potential for chemicals and resistant rice cultivars for control of nematodes distributed both worldwide and regionally (Table 3).

The data presented in Tables 1, 2, and 3 show 346 million total rice acres; 263 million are estimated to be infested with damaging levels of plant parasitic nematodes; 160 million acres are estimated economically treatable for nematode control at sixteen 1975 U.S. dollars per acre for a total market potential of 2 billion 575 million dollars.

It should be emphasized there is no obvious quantitative relation between the market for rice nematode control measures and the economic value increase in the worldwide rice crop resulting from successful nematode control; we can be sure that benefits of output would greatly exceed those of input.

From the standpoint of nematode distribution, the several aspects of rice culture can be treated under four categories; these are modified from the hydrology of rice lands presented by Moormann and Van Breemen (1978): (1) Upland rice = pluvial rice, mostly nonbounded by ridges, nonleveled, seeded dry or transplanted, depending on rainfall or surface water for moisture, soil not saturated, generally well drained; (2) Upland-lowland = phreatic rice lands = fluxial rice lands, nonbounded, nonleveled, water supply from rain water or surface flow, and also ground water, soil inundated or saturated with water only part of season; (3) well-drained paddy fields, bounded by ridges and leveled, flooded all season but drainable and of unsaturated soil profiles during early part of rice-growing season and between rice crops, or when water is removed from paddy; (4) poorly drained paddy with bulk of rice roots in a soil profile saturated with water throughout growing season and generally throughout the year.

In simplest terms we may refer to the three types of rice culture in terms of soil moisture levels: low-intermediate, intermediate-saturated, and saturated. The saturated category can be divided into oxidized-reduced and reduced with reference to biochemical reduction of the upper soil profile layers containing the bulk of organic matter and rice roots. The classification is much more simplistic from the nematological than from the diverse ethnic-agronomic standpoint and can be further described in terms of soil moisture and soil reduction characteristics at different key points during the rice-growing season; namely, planting, six weeks after planting, midseason, end of season before drainage, and two weeks after drainage of (paddy fields) before harvest. This scheme is hardly inclusive of many unusual rice-growing practices,

Table 1 Rice Acreage Statistics by Country (Millions of Acres)[a] for 1979
(Generally Refers to Harvested Area)

Country number	Country name	Rice acreage
1 Africa		11.172
2	Egypt	1.142
3	Guinea	1.112
4	Ivory Coast	0.939
5	Liberia	0.531
6	Malagasy Rep.	2.594
7	Nigeria	1.384
8	Sierra Leone	0.988
9	Zaire	0.827
10	Others[b]	1.736
11 North America		5.031
12	USA	3.016
13	Others[b]	2.016
14 South America		18.107
15	Brazil	15.067
16	Colombia	0.988
17	Others[b]	2.053
18 Asia		306.382
19	Afghanistan	0.519
20	Bangladesh	24.700
21	Burma	12.721
22	Cambodia	1.359
23	People's Republic of China	86.944
24	Taiwan	1.734
25	India	91.390
26	Indonesia	21.489
27	Iran	0.741
28	Japan	6.113
29	North Korea	1.814
30	South Korea	2.914
31	Laos	1.717
32	Malaysia	2.001

Table 1 (Continued)

Country number	Country name	Rice acreage
33	Sabah	0.155
34	Sarawak	0.292
35	Coast Malaysia	1.870
36	Nepal	3.075
37	Pakistan	4.915
38	Philippines	16.179
39	Sri Lanka	1.662
40	Thailand	19.816
41	North Vietnam	12.597
42	South Vietnam	7.077
43	Others[b]	0.343
44 Europe		0.961
45 USSR		1.482
46 World (total acres)		346.289

[a]All data conversions from hectares to acres are from IRRI (Palacpac, 1980) World Rice Statistics.
[b]Countries with less than 0.494 million acres in 1972.

such as deep-water paddy and harvest from boats in Bangladesh, but suffices with understandable and easily stated modification for most cultural areas around the globe.

When we consider rice nematodes on a worldwide basis, we are concerned first with species of *Hirschmanniella* and second with *Criconemella* species (Table 3). Root-knot, rice stylet, and stem nematodes round out those species that attack paddy (flooded) rice root systems.

The stem nematode *Ditylenchus angustus* (Filipjev) is of regional distribution, confined to some 10 countries.

White tip nematode, *Aphelenchoides besseyi* Christie, is capable of attacking developing florets and seed initials of rice plants on a worldwide basis.

Miscellaneous nematodes include those species confined to partially unsaturated soils (soils not saturated throughout the growing season) or soils known as upland soils, including those supplied water by irrigation, such that water-saturated soil conditions are not maintained during the entire growing season. For the sake of brevity and simplicity, *miscellaneous nematodes will be considered to attack rice only under upland growing conditions, or in ecological situations such that upland soil (unsaturated with water) is used for growing rice during a sufficient period of time to stimulate natural population behavior to a point detected by the usual soil extraction procedures.*

Miscellaneous nematodes are considered probable damage agents only on upland soils defined in Table 2 and include the genera *Heterodera, Pratylenchus, Helicotylenchus, Hoplolaimus, Hemicycliophora,* and *Rotylenchulus,* in

Table 2 Acreage of Rough Rice, Lowland and Upland, Specified Countries[a]

Country number	Country name	Area (millions of acres)		
		Lowland	Upland[b]	Total in upland (%)
18 Asia				
25	India[c]	64.687	27.195	29.6
23	People's Republic of China	94.800	17.400	15.5
20	Bangladesh	23.163	2.470	9.6
26	Indonesia	17.342	2.808	13.9
28	Japan	5.459	0.079	.01
30	South Korea	2.954	0.047	.02
38	Philippines	7.647	1.020	11.8
24	Taiwan	1.773	0.015	.01
1 Africa				
4	Ivory Coast	0.049	0.667	93.2
8	Sierra Leone	0.193	0.615	76.1
5	Liberia	0.032	0.316	90.8
—	Gambia	0.040	0.025	38.5
14 South America				
15	Brazil	2.225	10.073	81.9
16	Colombia	0.284	0.383	54.7
Totals		220.648	63.113	
Average % upland of total acreage (weighted for total acreages × % upland)				22.2

[a]All data conversions from hectares to acres are from IRRI (Palacpac, 1980), World Rice Statistics. Data were compiled by IRRI for individual years from 1971 to 1978 for the different countries, except Colombia (1963 only).
[b]Upland rice refers to rice grown on both flat and sloping fields, not ridged; prepared and seeded under dry conditions and depending upon rainfall moisture. In Asia, upland refers to both high and lowland summer crop grown without irrigation.
[c]The statistics for upland rice in India and the People's Republic of China (because of the importance of these countries in rice culture) *include* rice irrigated intermittently; for example 60% of this category in India does not have an assured water supply and 40% in China is assumed to fall below water-saturated levels during part of the growing season because of partial irrigation.

Table 3   Plant-Parasitic Nematodes in World Rice Production

Nematode species	Country of occurrence (numbers from Table 1)	Acres infested (millions)	Estimated from 346 million total acres worldwide		Cultivars/ chemical market value[a] 1975 (U.S.) millions dollars
			Loss (%)	Acres to be treated (millions)	
Rice-root nematodes[b] (*Hirschmanniella* spp.)	1-46	200	25	100	1600
Ring nematodes[b] (*Criconemella* spp.)	1-46	60	40	30	480
Root-knot nematodes (*Meloidogyne* spp.)	1-46	50	<1	None	None
Rice stylet nematodes (*Tylenchorhynchus* spp.)	1-46	200	<1	None	None
Stem nematode[b] (*Ditylenchus angustus*)	1, 2, 6, 18, 20, 21, 25, 32, 38, 40	10	30	10	160
White tip nematode[b] (*Aphelenchoides besseyi*)	1-46	3	2	Seed treatment	15

Miscellaneous nematodes in aggregate (upland rice):					
ring nematodes[b] (*Criconemella* spp.),					
root-knot nematodes[b] (*Meloidogyne* spp.),					
cyst nematodes[b] (*Heterodera* spp.),					
lesion nematodes[b] (*Pratylenchus* spp.),	1, 4, 5, 8, 14, 15, 16, 18, 20, 23, 24, 25, 26, 28, 38	63	30	20	320
spiral nematodes (*Helicotylenchus* spp.),					
lance nematodes[b] (*Hoplolaimus* spp.),					
sheath nematodes[b] (*Hemicycliophora* spp.),					
reniform nematodes (*Rotylenchulus* spp.),					
Total	46	263		160	2575

[a]Nematode control costs per acre are estimated at sixteen 1975 U.S. dollars.
[b]Nematodes that cause damage to rice sufficient to justify attempts to control them.

addition to some genera normal to paddy rice. It is necessary to emphasize that although these "upland genera" occur frequently in paddy rice, such occurrences almost always appear in the growing season in association with weed hosts and/or carryover from a previous crop. Their parasitism on rice, if occurring, will be considered on evidence for the individual genera and species.

Likewise, *Meloidogyne* spp. are not adapted for successful parasitism on rice, although paddy rice is often found with small populations of a particular species. Evidence will be presented to suggest that root-knot nematodes are of very limited importance in rice and that damage-causing capabilities are confined almost entirely to their occurrence in upland rice.

II. RICE-ROOT NEMATODES (*Hirschmanniella* spp.)

Hirschmanniella spp. attacking rice are probably at least seven in number, distributed as one or more species in all rice-growing countries and infesting some 200 million acres (Table 3). Sher (1968, in his revision of the genus *Hirschmanniella*, lists 15 species; *H. oryzae* (Soltwedel), *H. spinacaudata* (Sch. Stek.), *H. mucronata* (Das), *H. caudacrena* Sher, and *H. belli* Sher are five species that have economic significance because their effects on rice have been investigated in different parts of the world.

Extensive studies of nematode bionomics, damage to the rice plant, feeding on rice roots, nematode-host reactions, and population dynamics have been made with *H. oryzae* in Japan, Indonesia, and India. Work in the United States has focused on both *H. oryzae* and *H. caudacrena* (originally thought to be *H. oryzae*), in Africa on *H. spinacaudata*, and in India on *H. mucronata*; current practice is simply to refer to the rice-root nematodes under consideration as *Hirschmanniella* species.

There is considerable evidence in the original papers (Van der Vecht and Bergman, 1952; Kawashima, 1964a, b; Kawashima and Fujinuma, 1965; Panda and Rao, 1969; Rao et al. 1969; Sivakumar and Seshadri, 1969; Rao and Panda, 1970; Venkitesan et al., 1980), and in reviews (Ichinohe, 1964; 1966; Taylor, 1965; Taylor et al., 1966; Ou, 1972) supporting the 25% yield loss estimate in rice made for *Hirschmanniella* species (Table 3).

Hirschmaniella species have an extensive host range on grasses and sedges (Van der Vecht and Bergman, 1952; Whitlock, 1957; Kawashima, 1963; Venkitesan and Charles, 1979). Extensive surveys of rice-root nematodes in flooded fields in Louisiana from 1954 to 1962 revealed a common incidence of high populations; however, the introduction of an herbicide for grassy weed control in 1961 (3,4-dichloro-propionanilide), commonly known as propanil, stam or rogue, and its increasing widespread use in rice culture during the 1960s was correlated with a marked reduction in prevalence and populations of *Hirschmanniella* species in the late 1960s. It is the author's view that partial control of both grasses and sedges in Louisiana rice fields, resulting from the common use of this herbicide, reduced populations of rice-root and other nematode species because many of the weed species eliminated by the chemical were more congenial, tolerant, or better for reproduction of rice-root and other rice parasitic nematodes than rice itself. This situation may occur also in other rice-growing areas practicing successful grassy weed control.

Extensive host range studies of *Criconemella onoensis* Luc in the 1960s revealed patterns of grass and sedge weed host involvement equal to that for rice-root nematodes; however, data on populations of *C. onoensis* in rice fields for the 1950s are lacking because of extraction methods used. Population data before and after herbicide introduction are available only for rice-root nematodes.

It has been our privilege at Louisiana State University to conduct research on the biology, pathology, and epidemiology of a new species of *Hirschmanniella*: *H. caudacrena* Sher. The earlier work of Fielding and Hollis (1956) and Whitlock (1957) dealt with *H. oryzae* (Van Breda de Haan); the later work was with *H. caudacrena* Sher (1968).

The relevant research on *H. caudacrena* rice relations was conducted by my student Sman Keoboonrueng, and presented in his doctoral dissertation entitled "Effects of Rice-Root Nematode, *Hirschmanniella oryzae* (Van Breda de Haan 1902) Luc and Goodey 1963 on Rice Seedlings" (Keoboonrueng, 1971). The justification for estimates presented in Table 3 rests upon the assumption, or at least the anticipation, that the several species of *Hirschmanniella*, collectively of worldwide distribution in rice, are roughly equivalent in their damage capabilities on rice. We have abundant evidence for *H. oryzae*. Extensive evidence has not been presented heretofore for another species; therefore, such evidence is presented in this chapter by summarizing Keoboonrueng's unpublished results with *H. caudacrena*.

In our present state of knowledge the anticipation of equivalence among the five economic species of *Hirschmanniella* may claim such support as the great argument from analogy is capable of supplying. The detailed data presented below for *H. caudacrena* add considerable weight to this analogy.

A. General Observations on the Feeding of *H. caudacrena* on Rice Roots

Feeding of *H. caudacrena* on rice roots was observed in the laboratory with the use of special chambers constructed by Hollis et al. (1959). The method for the culture and inoculation of rice seedlings was the same as that used for determination of phenolic compounds and enzymatic activities, except that the seedlings were grown in chambers instead of beakers. The chambers contained 1% water agar in a layer 1 cm thick. Observations were made by inversion of the chamber on the microscope stage.

Within 24 hours after inoculation, several nematodes were found moving along the primary seminal root and occasionally stopping and pressing the root with their anterior ends. The nematodes invaded the piliferous region of the root and the region in which secondary roots emerged; they were never found feeding on the root tip or on secondary roots.

Within 48 hours after inoculation, brown lesions occurred on roots at a distance of 1-2 cm from the seeds (root length at this stage was about 4 cm). Examination of brown lesions under a microscope revealed brown patches on the surface of the root. Usually one lesion developed on a root, but two lesions were sometimes found 3-5 mm apart. The lesions usually occurred first on the underside of a root (the side that faces the bottom of the chamber). The brown necrotic lesions were sometimes only small spots, but mostly they were elongated up to 1 cm in length. The length of brown lesions increased with time after inoculation.

Secondary roots were found to emerge from the primary root 3-4 days after inoculation of the rice seedling chambers with *H. caudacrena*. In some cases, secondary roots in the necrotic area turned brown and never emerged from the primary root. Sometimes necrosis occurred on one side of the root and secondary roots were produced on the other side.

The seedlings were pulled out from the agar and the roots were fixed in formalin-acetic acid-alcohol solution (FAA). Root tissues were embedded by the paraffin method, and sections were cut 16 μm thick with a rotary microtome and stained with safranin and fast green.

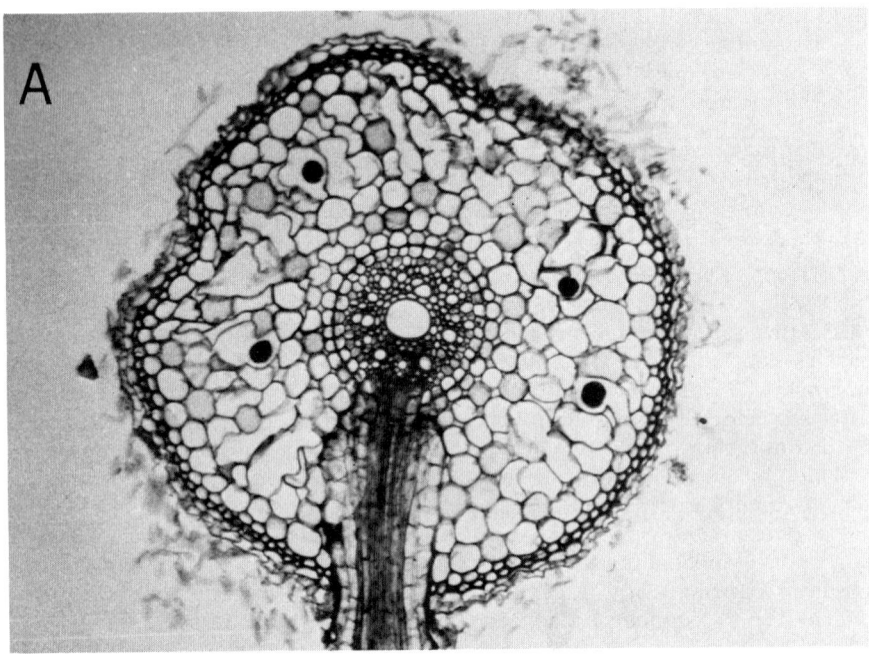

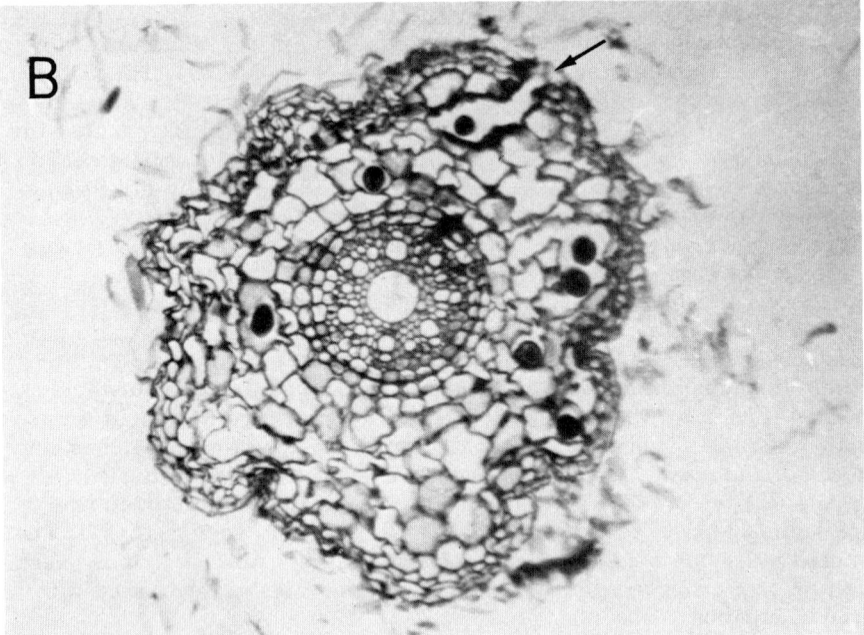

Fig. 1 Transverse sections of rice roots showing the rice root nematode (*H. caudacrena*) in cortical cells. (A) Four cortical cells, each containing a transverse segment of a single nematode. (B) Tunnel made by a nematode entering the cortex.

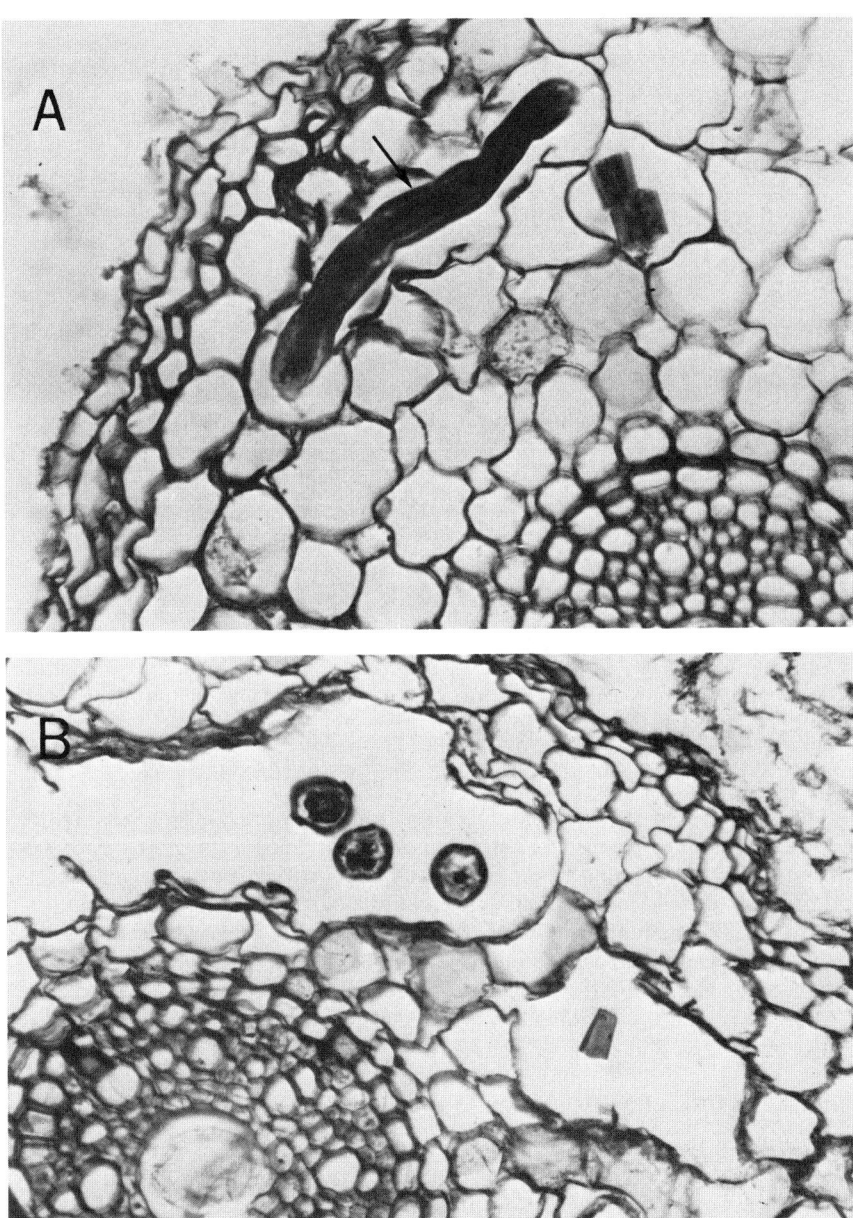

Fig. 2 Transverse sections of rice roots showing the rice-root nematode (*H. caudacrena*) in cortical cells. (A) One nematode occupying several cortical cells. (B) Cavity in cortex occupied by three nematodes seven days after inoculation of seedlings.

Stained transverse sections of roots cut in the vicinity of infection sites showed that *H. caudacrena* took up feeding positions at random in parenchyma cells of the cortex. The feeding was always intracellular (Fig. 1A). On penetration, the nematode made a tunnel that was almost perpendicular to the root

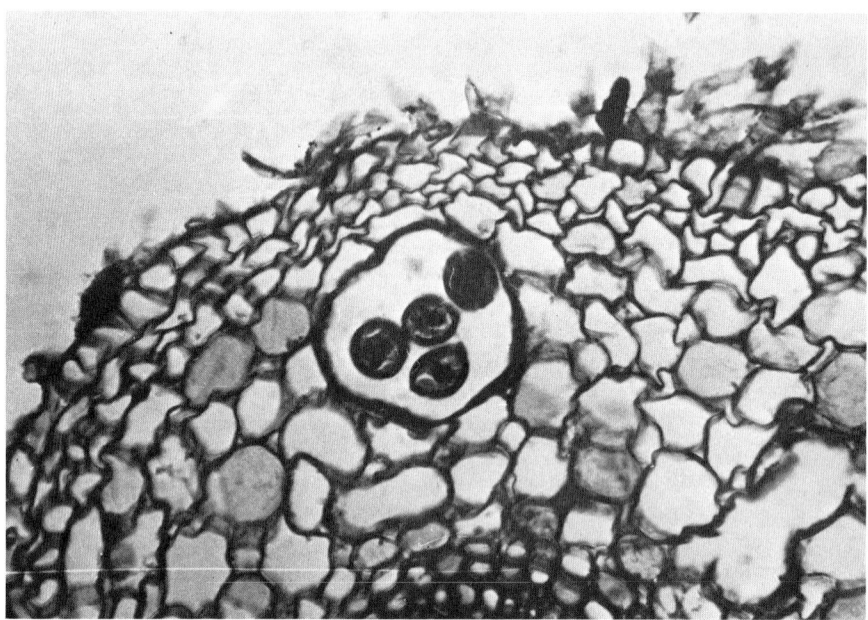

Fig. 3 Transverse section of root cortex showing a rounded large cell containing four specimens of rice-root nematode (*H. caudacrena*) in transverse section
seven days after inoculation of seedlings.

surface (Fig. 1B). The depth of penetration was variable. After penetration,
the nematode usually became oriented parallel to the long axis of the root and
moved either up or down; nematode movement occurred in both directions. In
some instances, the nematode moved transversely across the cortex and
damaged five or six parenchyma cells (Fig. 2A). Single nematodes feeding in
the root did not always disrupt adjacent cells (Fig. 1A), but when several
nematodes were associated with an infection, a cavity was always formed that
included them (Fig. 2B). In some cases a rounded large cell was developed
(Fig. 3).

B. Nematode–Host Reaction Studies

The effect of *H. caudacrena* on rice seedlings in the greenhouse was as follows.
Roots of rice seedlings grown in steam-sterilized field soil were healthy and
white. In infested plants, roots showed brown lesions at a distance of a few
centimeters from the culms. No knots or other abnormalities were seen on the
infected roots. The primary or solitary seminal root was attacked first; then
adventitious roots were attacked. The brown lesion expanded on an individual
root and at last the whole root rotted. Young, newborn roots, which had not
been attacked by the nematode, were white.
      Tables 4, 5, and 6 show the means of six measurements of roots and
shoots. Table 7 shows the statistical significance of the results presented in
Tables 4-6. When the data from two-week-old seedlings were analyzed statistically, there were some significant differences between the control and the
treatment with nematodes, except for the observation on the number of roots
in which the reduction was not statistically significant. The nematode caused

Table 4 Mean[a] of Length, Number, Dry Weight of Roots and Length, and Fresh and Dry Weight of Shoots[b]

Treatment		Total root length (cm)	Total no. roots	Root dry weight (g)	Shoot length (cm)	Shoot fresh weight (g)	Shoot dry weight (g)
Nematode infested	Mean	94.14	137.00	0.0018	28.25	0.1293	0.0211
	Standard deviation	31.34	36.57	0.0006	4.83	0.0173	0.0027
	Standard error	12.79	14.93	0.0002	1.97	0.0071	0.0011
Nematode free	Mean	157.21	226.33	0.0036	33.33	0.1728	0.0271
	Standard deviation	57.48	91.93	0.0007	1.82	0.0265	0.0030
	Standard error	23.46	37.52	0.0003	0.74	0.0108	0.0012

[a]Average of six seedlings.
[b]Two-week-old rice seedlings in greenhouse.

Table 5 Mean[a] of Length, Number, Dry Weight of Roots and Length, and Fresh and Dry Weight of Shoots[b]

Treatment		Total root length (cm)	Total no. roots	Root dry weight (g)	Shoot weight (cm)	Shoot fresh weight (g)	Shoot dry weight (g)
Nematode infested	Mean	942.34	1252.00	0.0189	46.05	0.8178	0.1227
	Standard deviation	231.07	231.69	0.0046	3.43	0.2123	0.0304
	Standard error	94.31	94.57	0.0019	1.40	0.0867	0.0124
Nematode free	Mean	1317.99	1587.17	0.0258	49.73	1.0439	0.1567
	Standard deviation	397.83	411.07	0.0051	2.58	0.1460	0.0189
	Standard error	162.38	167.78	0.0021	1.05	0.0596	0.0077

[a]Average of six seedlings.
[b]Two-week-old rice seedlings in greenhouse.

Table 6 Mean[a] of Length, Number, and Dry Weight of Roots and Shoot Length, Fresh Weight, and Dry Weight[b]

Treatment		Total root length (cm)	Total no. roots	Root dry weight (g)	Shoot length (cm)	Shoot fresh weight (g)	Shoot dry weight (g)
Nematode infested	Mean	1478.17	1490.33	0.0649	56.98	1.6643	0.3284
	Standard deviation	496.33	515.51	0.0146	2.88	0.2176	0.0258
	Standard error	202.58	210.41	0.0060	1.18	0.0888	0.0105
Nematode free	Mean	2523.94	2372.83	0.1949	58.85	1.8753	0.3373
	Standard deviation	355.22	330.94	0.0330	1.13	0.1864	0.0468
	Standard error	144.99	135.08	0.0135	0.46	0.0761	0.0191

[a]Average of six seedlings.
[b]Six-week-old rice seedings in greenhouse.

Table 7 Effect of *H. caudacrena* Sher on Rice Seedlings Growing in Greenhouse[a]

Seedling age (weeks)	Measurement	Treatment mean difference (nematode free) - (nematode infested)	Standard error of mean difference	t Value[b]
2	Total root length	63.07	26.72	2.360*
	Total root number	89.33	40.38	2.212
	Root dry weight	0.0018	0.0004	4.500**
	Shoot length	5.08	2.10	2.419*
	Shoot fresh weight	0.0445	0.0129	3.450**
	Shoot dry weight	0.0060	0.0013	4.615**
4	Total root length	375.65	187.78	2.000
	Total root number	335.17	192.60	1.740
	Root dry weight	0.0069	0.0028	2.464*
	Shoot length	3.68	1.75	2.103
	Shoot fresh weight	0.2261	0.1052	2.149
	Shoot dry weight	0.0340	0.0146	2.329*
6	Total root length	1045.77	249.12	4.197**
	Total root number	882.50	250.04	3.529**
	Root dry weight	0.1300	0.0467	2.784*
	Shoot length	1.8800	1.2700	1.480
	Shoot fresh weight	0.2110	0.1169	1.805
	Shoot dry weight	0.0089	0.0218	0.4083

[a]Statistical significance of the data presented in Tables 1 and 3.
[b]*Significant (at 5% level of probability). **Highly significant (at 1% level of probability).

Table 8 Diameters[a] of Primary and Secondary Roots of Rice Seedlings Grown in the Greenhouse in Nematode-Free and Nematode-Infested Soil

Seedling age (weeks)	Primary roots (μm)			Secondary roots (μm)		
	Nematode infested	Nematode free	t Value	Nematode infested	Nematode free	t Value
2	480.99	511.05	0.929	89.56	95.49	1.699
4	629.22	648.71	0.692	87.91	90.32	1.840
6	715.71	736.69	0.872	86.40	93.71	3.973[b]

[a]Mean diameters of roots from six seedlings.
[b]Highly significant (at 1% level of probability).

highly significant reductions in root dry weight, shoot fresh weight, and dry weight. Significant reductions were found in total root length and length of shoot.

Four-week-old seedling data showed significant reductions in root dry weight and shoot dry weight. For measurements of total root length, total number of roots, length of shoot, and shoot fresh weight, the data showed reductions but they were not statistically significant.

In six-week-old seedlings, the nematode caused highly significant reductions in total root length and total root number, and a significant reduction in root dry weight. Reductions of shoot length, shoot fresh weight, and dry weight were not statistically significant.

The means of diameters of roots at three different ages and their statistical analyses are shown in Table 8. The nematode caused reductions in diameters of both primary and secondary roots of all ages, but these reductions were not statistically significant except for the secondary roots in six-week-old seedlings.

Root-shoot ratios of rice seedlings are shown in Fig. 4. The ratios were slightly increased with increase in the age of the plant and were highest in six-week-old seedlings.

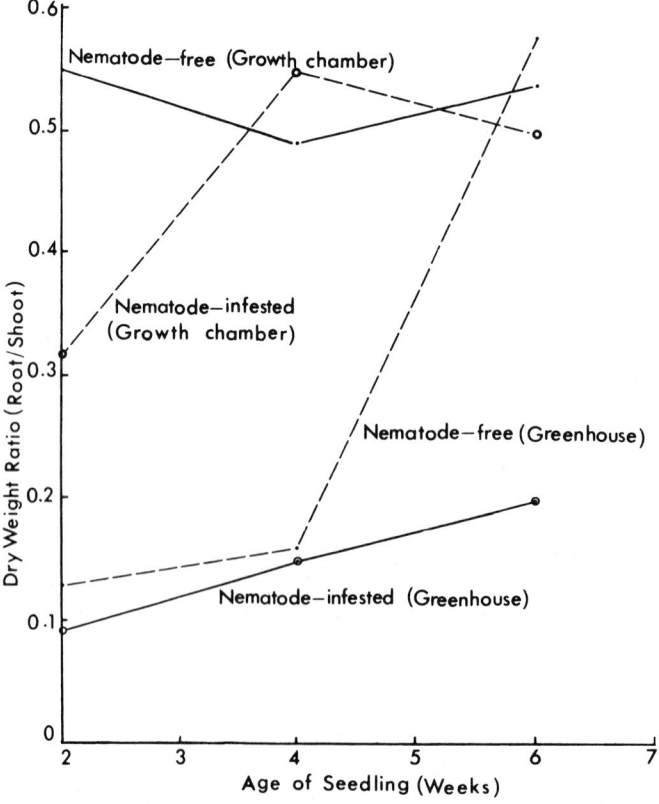

Fig. 4 Root-shoot ratios of rice seedlings growing in the greenhouse and in a growth chamber.

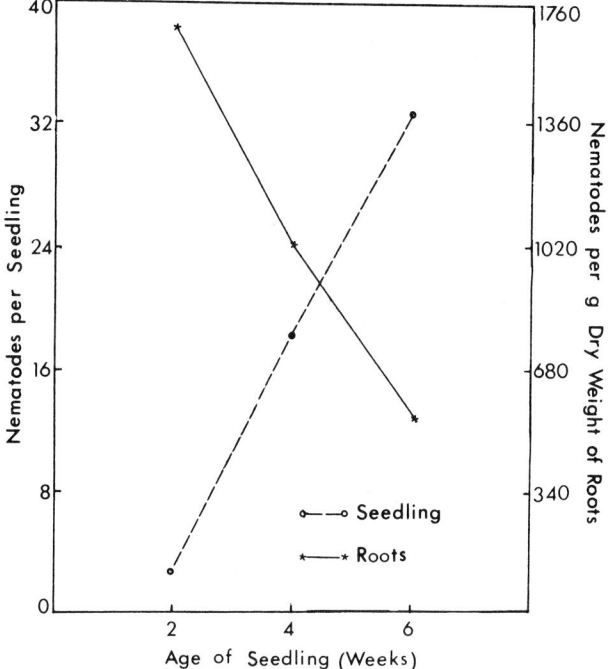

Fig. 5 Number of nematodes per rice seedling and calculated number of nematodes per gram dry weight of roots.

Figure 5 shows the average number of the nematodes per seedling and calculated number per gram dry weight of roots. The number of nematodes per seedling increased from 2.8 in two-week-old seedlings to 18.3 and 32.8 in four- and six-week-old seedlings, respectively. By contrast, the calculated number of nematodes per gram dry weight of roots decreased from 1629.7 in two-week old seedlings to 1024.7 in four-week-old and 545.0 in six-week-old seedlings.

The effects of *H. caudacrena* on rice seedlings in a growth chamber were as follows. When rice seedlings were grown in a growth chamber under controlled conditions, the seedlings produced more roots and the root-shoot ratio was higher than in the greenhouse (Fig. 4). The root-shoot ratio was lowest in two-week-old seedlings, intermediate in six-week-old seedlings, and highest in four-week-old seedlings.

Tables 9, 10, and 11 show the means obtained from measurements made on rice seedlings at two, four, and six weeks of age, respectively. Table 12 shows the statistical significance of the results presented in Tables 9, 10, and 11. When the data from two-week-old seedlings were analyzed statistically, there was a significant difference in all cases between the control treatment and the treatment with nematodes. *H. caudacrena* caused significant reductions in total length of roots, total root number, length of shoot, shoot fresh weight, and dry weight, and it caused a highly significant reduction in root dry weight.

Four-week-old seedling data showed a significant difference between the control treatment and the treatment with the nematode in all measurements except that of the root dry weight. The nematode caused highly significant re-

Table 9 Mean[a] of Length, Number, and Dry Weight of Roots, and Shoot Length, Fresh Weight, and Dry Weight[b]

Treatment		Total root length (cm)	Total no. roots	Root dry weight (g)	Shoot length (cm)	Shoot fresh weight (g)	Shoot dry weight (g)
Nematode infested	Mean	1894.08	1183.08	0.0351	33.05	0.5288	0.1113
	Standard deviation	1101.02	474.94	0.0159	2.00	0.0937	0.0240
	Standard error	317.84	137.10	0.0046	0.58	0.0270	0.0069
Nematode free	Mean	2752.80	1622.17	0.0808	34.98	0.6598	0.1469
	Standard deviation	659.49	326.05	0.0297	2.49	0.1624	0.0401
	Standard error	190.38	94.12	0.0086	0.72	0.0469	0.0116

aAverage of 12 seedlings.
bTwo-week-old rice seedlings in growth chamber.

Table 10 Mean[a] of Length, Number, and Dry Weight of Roots, and Shoot Length, Fresh Weight, and Dry Weight[b]

Treatment		Total root length (cm)	Total no. roots	Root dry weight (g)	Shoot length (cm)	Shoot fresh weight (g)	Shoot dry weight (g)
Nematode infested	Mean	1874.78	2873.00	0.1184	37.29	0.7886	0.2153
	Standard deviation	742.37	440.35	0.0229	1.63	0.1396	0.0394
	Standard error	214.30	127.12	0.0066	0.47	0.0403	0.0114
Nematode free	Mean	4008.63	2256.75	0.1310	39.68	0.9907	0.2684
	Standard deviation	902.47	404.86	0.0301	1.51	0.2116	0.0570
	Standard error	260.52	116.87	0.0087	0.44	0.0611	0.0165

[a]Average of 12 seedlings.
[b]Four-week-old rice seedlings in growth chamber.

Table 11 Mean[a] of Length, Number, and Dry Weight of Roots, and Shoot Length, Fresh Weight, and Dry Weight[b]

Treatment		Total root length (cm)	Total no. roots	Root dry weight (g)	Shoot length (cm)	Shoot fresh weight (g)	Shoot dry weight (g)
Nematode infested	Mean	2499.11	1651.42	0.1335	38.81	0.9275	0.2657
	Standard deviation	574.25	370.06	0.0732	2.32	0.2166	0.0633
	Standard error	165.77	106.83	0.0211	0.67	0.0625	0.0183
Nematode free	Mean	2963.26	1899.00	0.1749	39.98	1.0688	0.3254
	Standard deviation	679.99	363.50	0.0445	2.67	0.2402	0.0844
	Standard error	196.30	104.93	0.0128	0.77	0.0693	0.0244

[a]Average of 12 seedlings.
[b]Six-week-old rice seedlings in growth chamber.

Table 12 Effect of *H. caudacrena* Sher on Rice Seedlings Growing in Growth Chamber[a]

Seedling age (weeks)	Measurement	Treatment mean difference (nematode free) - (nematode infested)	Standard error of mean difference	t Value[b]
2	Total root length	858.72	370.49	2.318*
	Total root number	439.09	166.30	2.640*
	Root dry weight	0.0457	0.0098	4.663**
	Shoot length	1.93	0.92	2.098*
	Shoot fresh weight	0.1310	0.0541	2.421*
	Shoot dry weight	0.0356	0.0135	2.637*
4	Total root length	1133.85	337.34	3.361**
	Total root number	362.17	172.68	2.097*
	Root dry weight	0.0126	0.0109	1.156
	Shoot length	2.39	0.64	3.734**
	Shoot fresh weight	0.2021	0.0732	2.761*
	Shoot dry weight	0.0531	0.0201	2.642*
6	Total root length	464.15	256.93	1.807
	Total root number	247.58	149.74	1.653
	Root dry weight	0.0414	0.0247	1.676
	Shoot length	1.80	1.02	1.765
	Shoot fresh weight	0.1413	0.0933	1.515
	Shoot dry weight	0.0597	0.0305	1.957

[a]Statistical significance of the data presented in Tables 9 to 11.
[b]*Significant (at 5% level of probability). **Highly significant (at 1% level of probability).

Table 13 Diameters[a] of Primary and Secondary Roots of Rice Seedlings Grown in a Growth Chamber in Nematode-Free and Nematode-Infested Soil

Seedling age (weeks)	Primary roots (μm)			Secondary roots (μm)		
	Nematode infested	Nematode free	t Value[b]	Nematode infested	Nematode free	t-Value[b]
2	693.97	788.23	3.918**	95.75	106.05	9.196**
4	789.38	813.82	1.168	94.01	103.17	11.171**
6	748.03	793.84	2.144*	90.92	96.57	3.844**

[a]Mean diameters of roots from 12 seedlings.
[b]*Significant (at 5% level of probability), **Highly significant (at 1% level of probability).

ductions in total root length and length of shoot. Significant reductions were found in total root number, shoot fresh weight, and dry weight.

The data of six-week-old seedlings showed no significant differences between the control and the nematode-infested treatments.

Table 13 shows the mean differences in diameters of roots and their statistical significance. The nematode caused significant or highly significant reductions in diameters of primary and secondary roots at all ages of seedlings, with the exception of the diameters of the primary roots in four-week-old seedlings.

C. Nematode–Soil Amendments

The results in Table 14 indicate that neither the water from the nematode suspension nor the culture of bacteria and fungi from the water of the nematode suspension caused a statistically significant reduction in total root length; however in other treatments, a reduction in root length was found. Noninfested rice roots when added to soil caused reductions in the length of roots, but greater reductions were caused by treatments with nematode suspensions and noninfested rice roots plus the nematode.

Table 15 shows the statistical significance of different treatments on rice seedling root numbers. Water from the nematode suspensions or the culture of the microorganisms in that water did not cause a significant reduction of root numbers. Noninfested and infested rice roots, added to the soil, caused a reduction in numbers of roots similar to that caused by the water from the nematode suspension and the culture of bacteria and fungi. The nematode and noninfested rice roots plus the nematode caused a statistically significant reduction in the number of roots, and this reduction was greater than that of other treatments.

Statistically significant differences in root dry weight of rice seedlings are shown in Table 16. Noninfested rice roots plus the nematode caused the highest reduction in root dry weight, but equivalent reductions were caused

Table 14 Statistical Significance of Root Length of Rice Seedlings

Treatment	Mean[a]	Indication of significance[b]
Nematode free	496.44	1
Culture of bacteria and fungi[c]	471.31	1 2
Water from nematode suspension	470.99	1 2
Noninfested rice roots added	381.17	2 3
Infested rice roots added	320.93	3 4
Nematode suspension added	270.18	4
Noninfested rice roots plus nematode suspension	215.75	4

[a]Average of 10 seedlings (cm).
[b]Values in the column followed by numbers in common are not significantly different at the 5% level.
[c]Culture of the bacteria and fungi from water of nematode suspension.

Table 15 Statistical Significance of Number of Roots and Rice Seedlings

Treatment	Mean[a]	Indication of significance[b]
Nematode free	529.60	1
Water from nematode suspension	465.20	1 2
Culture of bacteria and fungi[c]	458.10	1 2
Noninfested rice roots added	414.60	2
Infested rice roots added	361.70	2
Nematode suspension added	257.10	3
Noninfested rice roots plus nematode suspension	236.80	3

[a]Average of 10 seedlings.
[b]Values in the column followed by numbers in common are not significantly different at the 5% level.
[c]Culture of the bacteria and fungi from water of nematode suspension.

by noninfested rice roots, infested roots, and *H. caudacrena* added to soil alone. Microorganisms or cultures of those microorganisms from the nematode suspension water did not cause significant reductions in root dry weight.

Table 17 shows the statistical significance of differences in the shoot length of rice seedlings. Water from the nematode suspension or the culture of bacteria and fungi from that water did not cause statistically significant reductions in the length of the shoot. Nematodes or noninfested rice roots, when added to the soil, caused a reduction in shoot length. Infested rice

Table 16 Statistical Significance of Root Dry Weight of Rice Seedlings

Treatment	Mean[a]	Indication of significance[b]
Water from nematode suspension	0.0164	1
Culture of bacteria and fungi[c]	0.0163	1
Nematode free	0.0153	1
Noninfested rice roots added	0.0138	2
Infested rice roots added	0.0132	2
Nematode suspension added	0.0132	2
Noninfested rice roots plus nematode suspension	0.0088	3

[a]Average of 10 seedlings (g).
[b]Values in the column followed by numbers in common are not significantly different at the 5% level.
[c]Culture of the bacteria and fungi from water of nematode suspension.

Table 17 Statistical Significance of Shoot Length of Rice Seedlings

Treatment	Mean[a]	Indication of significance[b]
Water from nematode suspension	47.73	1
Culture of bacteria and fungi[c]	47.10	1
Nematode free	46.64	1
Nematode suspension added	43.01	2
Noninfested rice roots added	42.22	2
Noninfested rice roots plus nematode suspension	39.37	3
Infested rice roots added	38.60	3

[a]Average of 10 seedlings (cm).
[b]Values in the column by numbers in common are not significantly different at the 5% level.
[c]Culture of the bacteria and fungi from water of nematode suspension.

roots or noninfested rice roots plus the nematode caused the highest reductions.

The results in Table 18 indicate that the water from the nematode suspension or the culture of the bacteria and fungi from that water did not cause reductions in fresh weight of shoots. Noninfested rice roots or the nematode, when added to the soil, caused reductions of shoot fresh weight. Reductions in shoot fresh weight were highest with additions of infested rice roots or noninfested rice roots plus the nematode.

Table 18 Statistical Significance of Fresh Weight of Shoots of Rice Seedlings

Treatment	Mean[a]	Indication of significance[b]
Water from nematode suspension	0.8173	1
Nematode free	0.7635	1
Culture of bacteria and fungi[c]	0.7493	1
Noninfested rice roots added	0.6284	2
Nematode suspension added	0.6152	2
Infested rice roots added	0.4552	3
Noninfested rice roots plus nematode suspension	0.4499	3

[a]Average of 10 seedlings (g).
[b]Values in the column followed by numbers in common are not significantly different at the 5% level.
[c]Culture of the bacteria and fungi from water of nematode suspension.

Table 19 Statistical Significance of Shoot Dry Weight of Rice Seedlings

Treatment	Mean[a]	Indication of significance[b]
Water from nematode suspension	0.1417	1
Nematode free	0.1372	1 2
Culture of bacteria and fungi[c]	0.1359	1 2
Nematode suspension added	0.1167	2 3
Noninfested rice roots added	0.1063	3
Infested rice roots added	0.0817	4
Noninfested rice roots plus nematode suspension	0.0799	4

[a]Average of 10 seedlings (g).
[b]Values in the column followed by numbers in common are not significantly different at the 5% level.
[c]Culture of the bacteria and fungi from water of nematode suspension.

In Table 19, the results show that the water from the nematode suspension or the culture of the microorganisms in that water did not cause statistically significant reductions in dry weight of shoots. However, other treatments caused reductions in shoot dry weight. The nematode and noninfested rice roots, when added to the soil, caused reductions in the dry weight of shoots. The highest reductions in shoot dry weight resulted from additions of infested rice roots and noninfested rice roots plus the nematode.

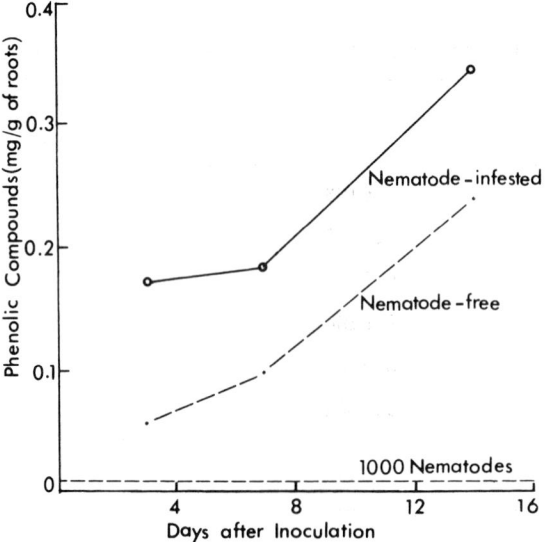

Fig. 6 Total phenolic compounds found in nematode-infested and nematode-free rice roots and in 1000 specimens of *H. caudacrena*.

D. Total Phenolic Compounds in Rice Roots and Nematodes

The results on determinations of total phenolic compounds (Fig. 6) indicate that the infestation of nematodes caused an increase in total phenols. At three and seven days after inoculation the amount of total phenolic substances in nematode-infested roots was about two times greater than in nematode-free roots; this difference became less at 14 days after inoculation. It was found also that *H. caudacrena* contained trace amounts of phenolic compounds (7.5 µg per 1000 nematodes).

E. Enzymatic Activity

1. Polyphenol Oxidase

Polyphenol oxidase activities in nematode-infested and nematode-free rice roots at three, seven, and fourteen days after inoculation are shown in Figs. 7, 8, and 9, respectively. In all cases, the activity of the enzyme in nematode-infested roots was higher than in nematode-free roots. In another experiment, rice roots were punctured with a sterilized needle and the enzyme activity was determined at three and seven days after root injury. It was found that three days after root injury, the polyphenol oxidase activity was greater than in healthy roots, but it was not as high as that in nematode-infested roots (Fig. 10). Seven days after the roots were punctured, the enzyme activity was at the same level as in healthy roots (results not shown).

2. Catalase

The results (Fig. 11) show that three days after inoculation of rice roots with the nematode the catalase activity was about two times greater than that in nematode-free roots. At seven days after inoculation the difference in enzyme activity was less than on the third day. At 10 and 14 days after inoculation there was almost no difference in the activity of the enzyme between the two treatments.

3. β-Glucosidase

The activities of β-glucosidase enzyme in rice roots at seven days after inoculation and in *H. caudacrena* are shown in Table 20. Activity of the enzyme in nematode-infested roots was almost two times greater than that in nematode-free roots. It was found also that *H. caudacrena* secreted or excreted β-glucosidase into the nematode suspension water.

Table 20 β-Glucosidase Activity of Rice Root and Nematode Homogenate Extracts and Nematode Wash Water

Treatment	g P-nitrophenol released per g root fresh weight
Nematode-infested roots[a]	3100
Nematode-free roots	1750
1000 *H. oryzae*	226
H. oryzae wash water	20

[a]Rice roots extracted 7 days after inoculation with *H. caudacrena*.

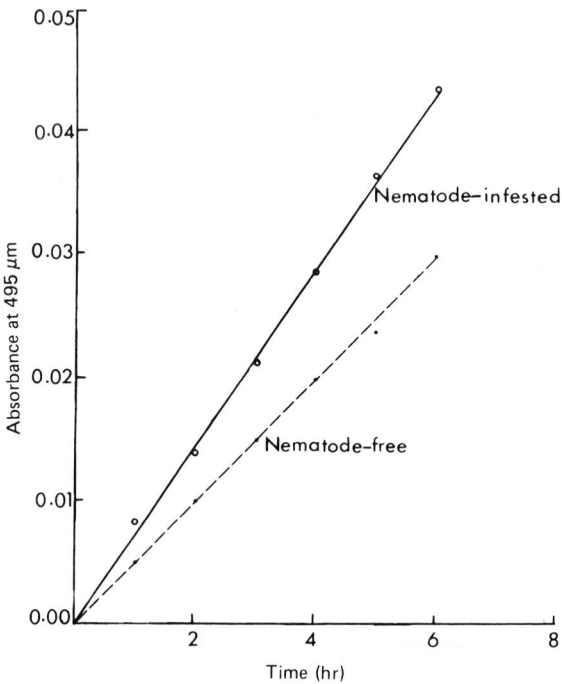

Fig. 7 Polyphenol oxidase activity in nematode-infested and nematode-free rice roots at three days after inoculation of rice seedlings with *H. caudacrena*.

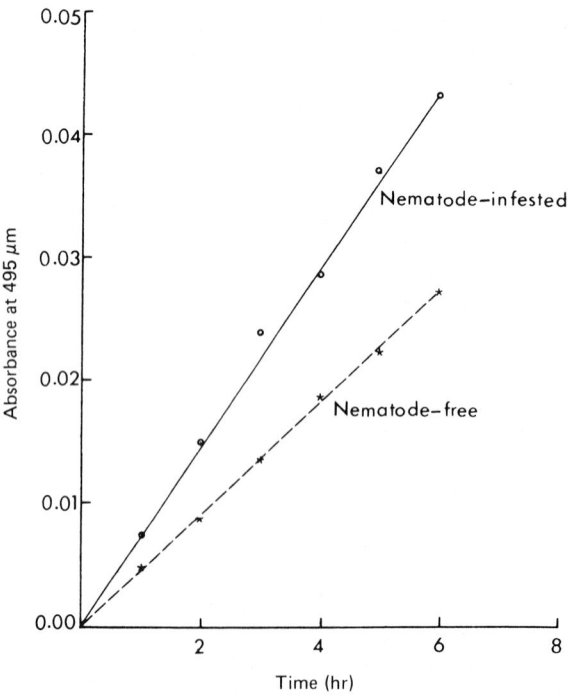

Fig. 8 Polyphenol oxidase activity in nematode-infested and nematode-free rice roots at seven days after inoculation of rice seedlings with *H. caudacrena*.

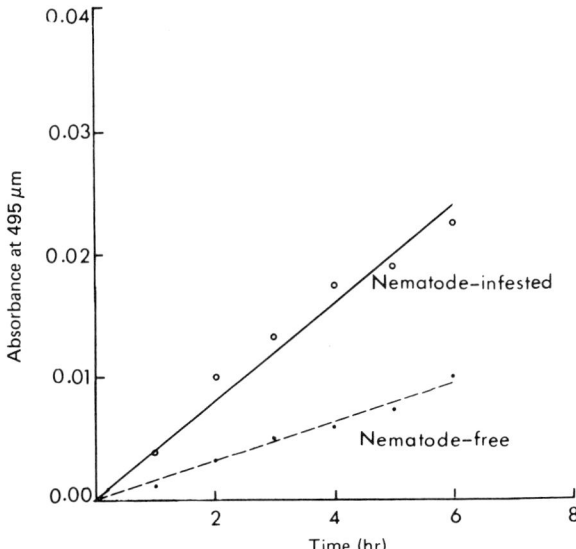

Fig. 9 Polyphenol oxidase activity in nematode-infested and nematode-free rice roots at 14 days after inoculation of rice seedlings with *H. caudacrena*.

The following points emerge in summary. Feeding of *H. caudacrena*, as observed in the laboratory, was intracellular on the primary root in the region of emergence of secondary roots and in the root hair region but not at the root tip. Feeding caused disruption of cell walls, formation of cavities, and necrosis of roots.

The infestation by *H. caudacrena* of rice root seedlings growing in the green house caused significant or highly significant reductions in total root length, root dry weight, shoot length, shoot fresh weight, and dry weight in two-week-old seedlings. The retardation of growth of rice seedlings declined in four- and six-week-old seedlings. In growth chamber tests, the nematode caused significant or highly significant reductions in root and shoot measurements at two and four weeks after inoculation; then the plants tended to recover and the retardation of growth disappeared in six-week-old seedlings.

Bacteria and fungi from nematode wash or suspension water did not cause significant reductions in growth of rice seedlings. Rice seedlings grown in soil amended with noninfested or infested rice roots showed an inhibition in growth that became more severe when specimens of *H. caudacrena* were added also to rice seedlings growing under these conditions.

Infestation of rice seedling roots by *H. caudacrena* caused an increase in phenolic compounds in the roots. The nematode itself contained a trace amount of total phenols, 7.5 µg per 1000 nematodes.

Polyphenol oxidase activity in nematode-infested rice roots was higher than in nematode-free roots. A slight increase in the activity of this enzyme was found also in roots three days after mechanical injury was made. Catalase activity increased in the roots only at an early stage of nematode infestation. The activity of this enzyme declined to a normal level in roots at about 10 days after inoculation of seedlings. The enzyme was found also in nematode homogenates. The activity of β-glucosidase enzyme increased in nematode-infested rice seedling roots. The nematode itself contained this enzyme

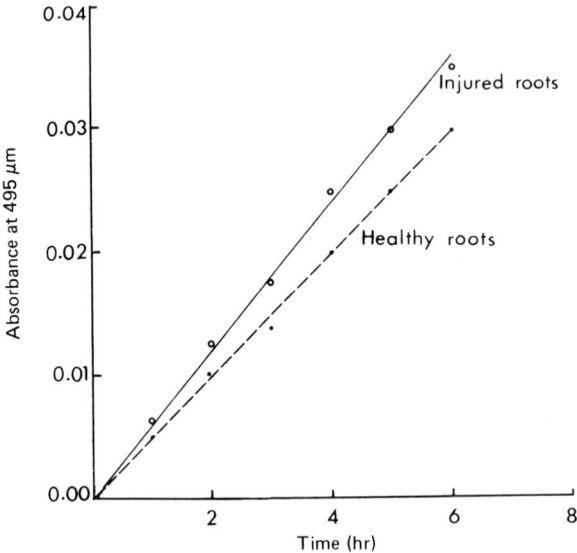

Fig. 10 Polyphenol oxidase activity in healthy and injured rice roots three days after roots were punctured with a sterilized needle.

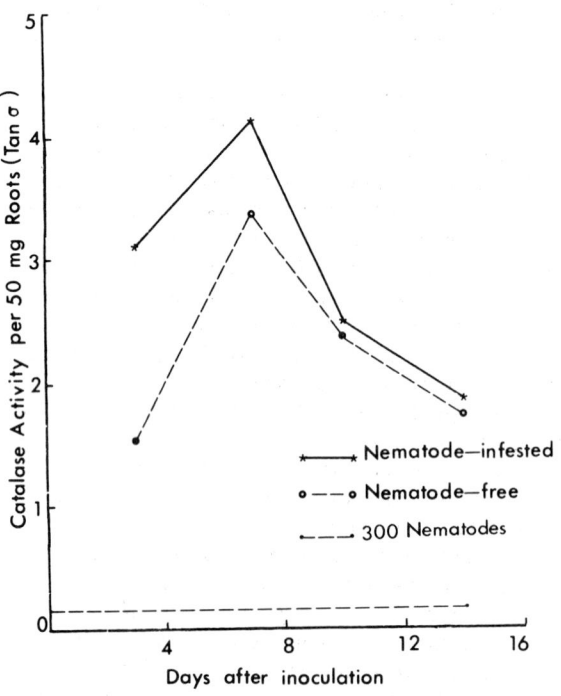

Fig. 11 Catalase activity in nematode-infested and nematode-free rice roots and in 300 specimens of *H. caudacrena*.

and excreted the enzyme into the suspension water. Increase in the activity of this enzyme in the host came from both the host response to nematode infestation and from the nematode excretion of the enzyme.

III. RING NEMATODES (*Criconemella* spp.)

Ring nematodes are uniquely damaging to rice, causing apical knots on secondary roots (compare Fig. 12, normal seedling roots, with Fig. 13, infected seedling roots). There are an estimated 5-10 species on a worldwide basis in rice; all of our current knowledge of ring nematode-rice pathology rests upon work performed in Louisiana with *C. onoensis* Luc and an undescribed species, which probably is *C. reedi* Diab and Jenkins. Ring nematodes (species unspecified) have been reported in rice field surveys by Timm and Ameen (1960), Taylor (1965), and Taylor et al. (1966). Imamura (1931) found *C. komabaensis* (Imamura) in Japan. Timm (1956) reported *C. rustica* (Micoletzky) on rice in Pakistan.

    C. onoensis has been reported on rice from Guinea and Ivory Coast, West Africa, along with *C. curvata* Raski, *C. palustris* Luc, and *C. sphaerocephala* Taylor (Luc, 1959, 1970), from Surinam (Maas, 1970), and from Louisiana and Texas (Hollis, 1969a, b). Recently, *C. onoensis* has been found on rice in Belize, Central America (Hollis, unpublished data, 1978). Identifications of *C. onoensis* populations from Louisiana and Texas were made by D. Raski and M. Golden.

    Discovery and elucidation of the ring nematode problem in rice required more than 22 years of continuous effort from 1954 to 1977 (Embabi, 1965; Hollis, 1977, 1980).

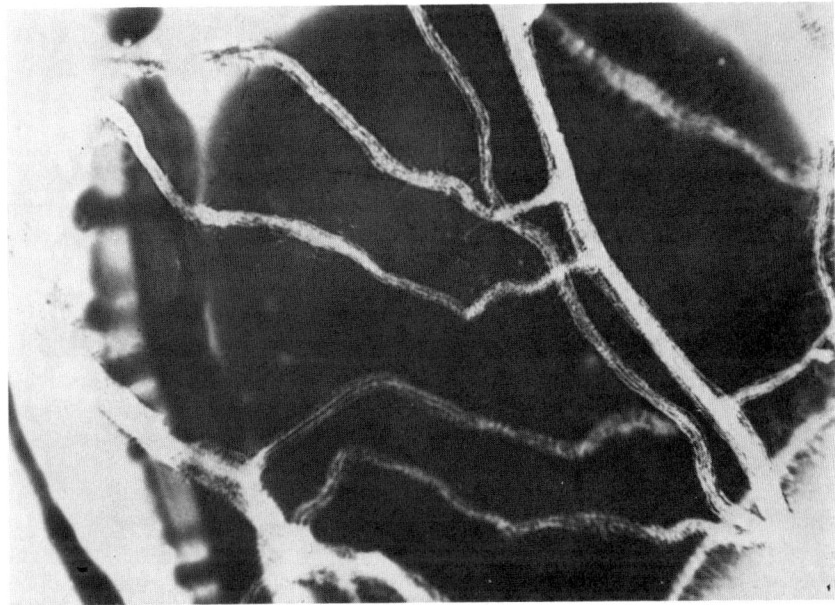

Fig. 12 Normal seedling rice roots.

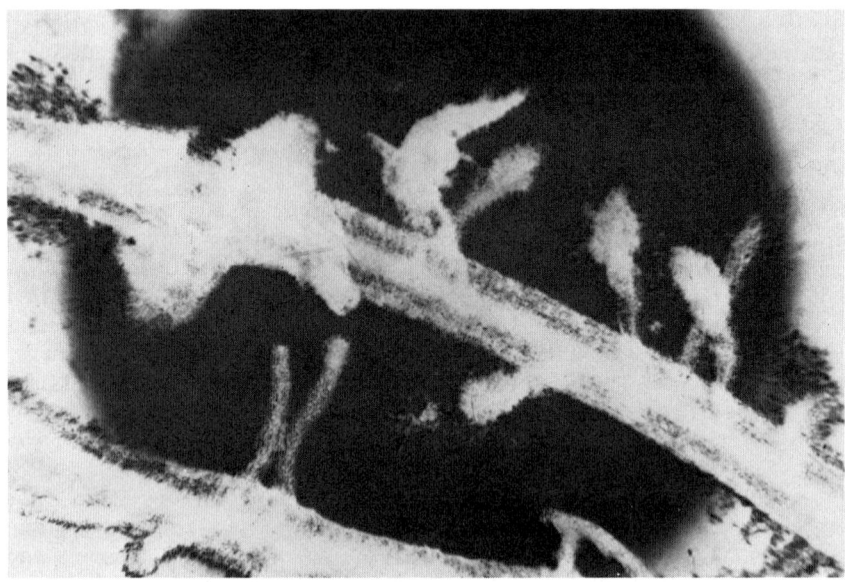

Fig. 13 Seedling rice roots attacked by *C. onoensis* (Luc).

Fig. 14 Rice plants growing in moist soil in greenhouse: (A) Seedlings attacked by small numbers (100 per pint soil) ring nematodes, *C. onoensis* (Luc). (B) Control plants (no nematodes).

The uniqueness of ring nematode disease of rice, probably typical of many nematode diseases of plants that are still unknown, can only be comprehended from a summary statement:

Nematicidal control of ring nematode lowered rice yields in plots infested with yellow nutsedge (*Cyperus esculentus*). Control of nematodes stimulated weed growth more than rice growth. Nematode control plus weed removal produced large, significant increases in rice yield. This is the first report of a migratory root-parasitic nematode, and of a ring nematode, (*Criconemella onoensis*), of uniform distribution across many soil series, in a large region, producing significant and symptomless yield reductions in a major crop. Common association in paddy rice of nematodes and weed hosts suggests similar interactions may be widespread.

Relevant points in the research effort (1) through (7) that required attack and solution in serial order before the next one emerged and became known, are as follows.

1. Ring nematodes in small numbers were occasionally detected in rice field soil samples. A change in extraction procedure in 1965, involving a shorter settling time (of the short, heavy-bodied ring nematode specimens) followed by extensive survey in Louisiana and Texas, showed that *C. onoensis* was indeed the most prevalent plant-parasitic nematode, infesting one million acres of rice in these states. The survey data showed a strong inverse relation between ring nematode prevalence and rice yields (Hollis 1967, 1969b; Hollis et al., 1968).

2. Populations of *C. onoensis* at the level of 100 per pint moist soil in greenhouse pots produced severe stunting, yellowing, and galling of roots (Figs. 12, 13, and 14). This was the first and remains the only instance of a nematode common and widespread in paddy rice causing severe effects on rice seedlings.

3. Microplot tests in 1968 with systemic insecticide-nematicides demonstrated that 50-100 × entomological dosage levels were required for nematode control (Hollis, 1969a, b). Subsequently, phenamiphos at 25 lb/acre was found the most efficient nematicide and dosage relation (Hollis, 1973).

4. Successful nematode control tests in the field could be conducted only by eliminating the yield-reducing effects of the rice water weevil, *Lissorhoptrus oryzophilus* Kuschel, as distinct from those of nematodes. This was accomplished in the successful tests by treatment at flooding time of the entire field, including the experimental area, with 0.5 lb/acre (a.i.) of carbofuran (Furadan R) 2,3-dihydro-2,2-dimethyl-7-benzofuranyl methylcarbamate in the form of 3% granules applied by airplane. The extensive background of work and cooperation of J. Gifford, USDA entomologist, is acknowledged in this connection.

5. Organic phosphates or carbamates for both insect and nematode control in rice can be applied only after application for grassy weed control of (propanil, Stam R, or Rogue) 3,4-dichloro-propionanilide (DCPA). This is a serious handicap for nematode control because plants are 10-12 in. high before protection arrives and have already suffered severe ring nematode damage. Application of insecticide-nematicides before DCPA causes severe yellowing and stunting of rice plants resulting from interaction of the chemicals in plant tissues. Fortunately, the use of carbofuran after application of DCPA under experimental conditions in Louisiana results in satisfactory control and prevents damage by rice water weevil.

6. Host studies with *C. onoensis* showed that *Cynodon dactylon* (L.) Pers., *Paspalum hydrophilum* Henr., and *Cyperus iria* L., among others, were susceptible-tolerant hosts that produced and supported high nematode populations, whereas rice cultivars tested, yellow nut sedge (*Cyperus esculentus* L.), *Fimbristylis milacea* (L.) Vahl., and *Eleocharis* spp. R. Br., among others, were resistant-intolerant hosts damaged by the nematodes but limiting their population buildup because of injury to root tissue (Alhassan and Hollis 1968, 1969; Hollis, 1972a, b; Hollis and Joshi, 1975).

7. Hand pulling of *C. esculentus* from rice-nematode control plots (Fig. 15) emerged as a necessary practice for two reasons: (a) lack of a chemical control for grassy weeds that would permit (be compatible with) preplant, time of planting, or postplant applications on very young rice seedlings (not more than 2-3 in. high) and (b) weed competition from resistant-intolerant hosts, notably *C. esculentus*, which depressed the yield of rice in nematicide-treated plots (Fig. 16).

At the time of this writing, some 27 years from the beginning of work and through the satisfaction of the above requirements, it has been possible to present detailed evidence for the first time from field experiments of decreased production of rice by *C. onoensis* sufficient to justify strenuous efforts to develop control measures. One million acres in Louisiana and Texas are infested with economically damaging levels of ring nematodes, both *C. onoensis* and *C. reedi*. An estimate of 30 million acres of ring nematode damage worldwide is conservative. A detailed analysis of all survey and

Fig. 15 Rice field plots treated with phenamiphos for control of ring nematodes [*C. onoensis* (Luc)] showing alleyways used for movement during hand weeding for removal of *Cyperus esculentus* L. (yellow nut sedge).

Fig. 16 (A) Stimulation of *Cyperus esculentus* L. by phenamiphos at 25 lb/acre
(a.i.) in absence of hand weeding. (B) Control (no treatment). (Courtesy
of E. J. Brill Leiden, The Netherlands.)

yield data (Fig. 17) show the low numbers per pint of soil causing high yield
reductions, with a midpoint of 450 nematodes and a reduction in yield of 20%.
Approximately one-tenth this number causes a loss of 5% in yield of rice.

The effects of phenamiphos (Table 21) are shown in control of total plant
parasitic and ring nematodes at two locations. Total leaf nitrogen of rice was
reduced significantly in the southwestern Louisiana test by numbers of ring
nematodes averaging 110 per pint of soil (5.5 per 1/20 pint).

Rice production in the United States is mainly in four states: Arkansas,
California, Louisiana, and Texas. Ring nematodes are a production problem

Table 21 Effects in Paired Comparison (3 × 3 latin square) of Phenamiphos at 25 lb (a.i.) per Acre on Numbers of Plant-Parasitic Nematodes per 1/20 Pint Soil from La Belle Rice Plots in Belize Mangrove Peat (Black Sand) and Louisiana Coastal Marsh (Sandy Loam) Soils in 1979[a]

| Paired comparison | Total plant-parasitic nematodes[b] in mangrove peat soil after treatment | | | |
| | 53 days | | 100 days | |
	Phenamiphos	Control	Phenamiphos	Control
1	1	167	5	56
2	0	173	5	11
3	0	87	4	21
4	0	85	0	47
5	0	103	0	3
6	0	90	2	16
7	45	115	5	63
8	0	91	1	0
9	0	158	0	8
Mean	5.11	118.78	2.44	30.97
Significance level	$P < 0.01$		$P < 0.05$	

[a]Thanks are due Mel Kallal Dow Chemical Co. (USA) for financial support, Fred Donaldson Mobay Chemical Corp. for generous supplies of Phenamiphos (15 g), and E. A. Epps, Leonard Devold, and Pedro V. Carrasco for determining total nitrogen of rice leaf samples.
[b]M. arenaria (Neal) root-knot nematode; C. onoensis (Luc) rice ring nematode; R. reniformis Linford and Oliveira, reniform nematode; X. americanum (Cobb), dagger nematode; Hoplolaimus spp. lance nematode.
[c]C. onoensis (Luc), rice ring nematode; C. reedi type ring nematode.

only in the coastal prairie regions of Louisiana and Texas. The Northeast Mississippi river bottom area of Louisiana is free of these nematodes, as is all of Arkansas and California; yields in these ring-nematode-free areas are very high when compared with infested areas.

IV. ROOT-KNOT NEMATODES IN RICE (Meloidogyne spp.)

Root-knot nematodes have been known to attack rice since "a spot in the field occurrence" was noticed by Tullis (1934) in Arkansas and the organism referred to as Heterodera marioni (Cornu). Ichinohe (1955) found M. incognita var acrita Chitwood on upland rice in Japan. M. javanica (Treub), M. arenaria subsp. thamesi Chitwood, M. exigua, and a new species, M. graminicola

| Coastal marsh sandy loam soil after treatment | | | | |
| Ring nematode (21 days) | | Total leaf nitrogen (87 days) | | |
Phenamiphos	Control	Phenamiphos	Control	Difference
2	7	3.87	3.58	0.29
2	5	3.89	3.51	0.38
0	7	3.75	3.38	0.37
0	1	3.80	3.16	0.64
3	5	3.98	3.73	0.20
1	5	3.93	3.68	0.25
1	9	3.93	3.73	0.20
0	8	4.00	3.72	0.28
1	3	3.99	3.85	0.14
1.11	5.55	3.90	3.59	0.31
$P < 0.01$			$P < 0.05$	

Golden and Birchfield have been reported in various parts of the world (Hollis in Louisiana and Texas ricefield surveys 1964-1976) (Kanjanasoon, 1964; Hashioka, 1963; Israel et al., 1964; Golden and Birchfield, 1968).

Root-knot nematodes in rice are of academic-scientific, specifically of ecological and physiological interest; these nematodes, although a problem in nurseries of rice seedlings used for transplanting, cannot survive, except very marginally, in paddy rice. Future work will indicate their economic importance in upland rice. Extremely small numbers of *M. graminicola* are found in 20-30% of rice fields (paddy) in Louisiana and Texas during midseason.

Root knotting, although an efficient mode of parasitism for *Meloidogyne* spp. under moist soil conditions on numerous hosts, is not detectable on a recurring basis in flooded soils, and the nematodes are held at very low population levels. This is observationally correlated by a failure of normal female

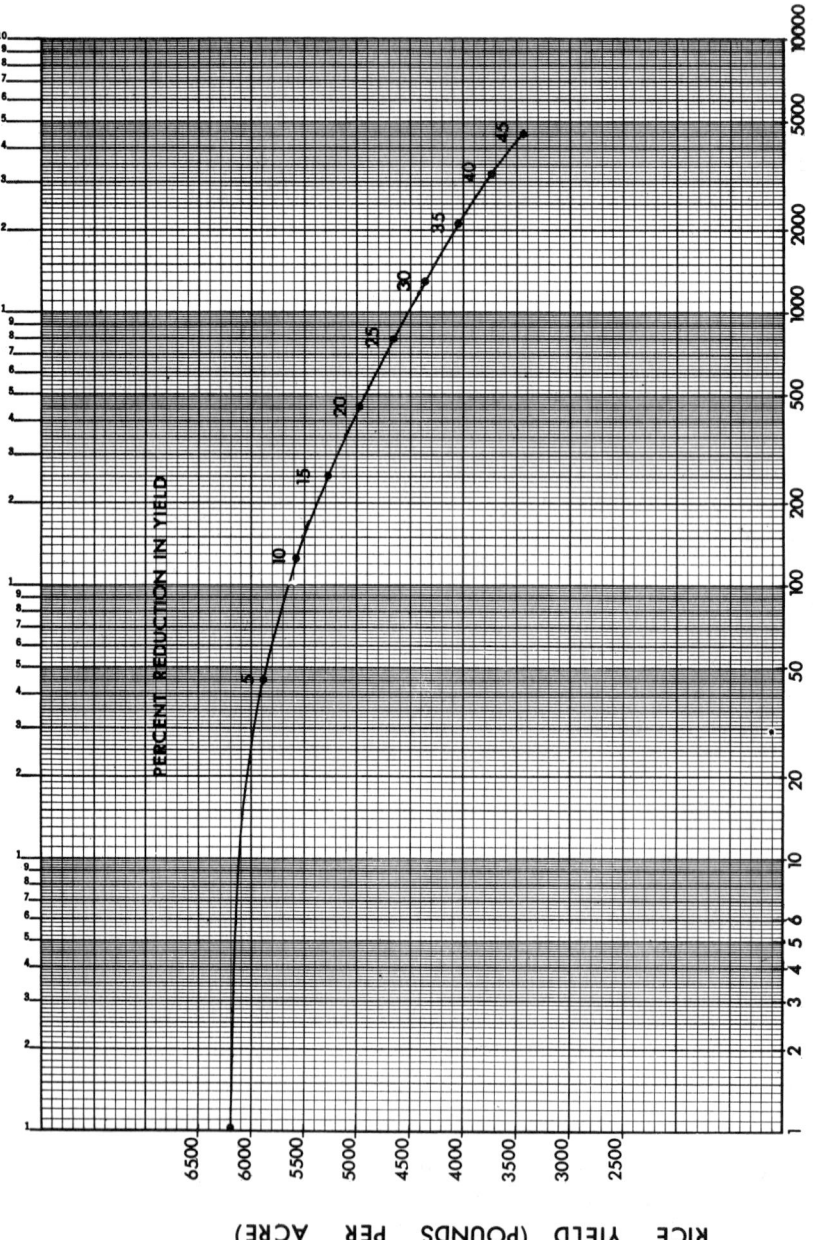

RING NEMATODES PER PINT OF SOIL (BEFORE FLOODING)

Fig. 17 Correlation between ring nematode numbers and rice yields in Louisiana (the curve is based on data from four rice regions—northeastern Louisiana, southwestern Louisiana, Beaumont, Texas, and Eagle Lake, Texas; and particularly on data from northeast of Rayne, northeast of Crowley, southeast of Gueydan, and southwest of Gueydan). Additionally, the curve is in agreement with data from chemical and cultural control experiments and grower practices and with experience relative to the severity of nematode attacks on crop plants.

maturation, gall development, and an absence of egg laying and formation of egg masses in greenhouse pots of rice seedlings.

We have assembled evidence in Louisiana State University Laboratories that provides a reasonable construct of events relating to natural control of root-knot nematodes in paddy rice. Measurement of respiration of small nematodes with a YSI oxygen monitor (YSI Model 55, Yellow Springs Instrument Co., Inc., Yellow Springs, Ohio 45387) included second-stage larvae of *M. incognita*. The average respiration rate was 0.13×10^{-4} µl O_2 per nematode per minute for *M. incognita*, and this level was comparable for *Rotylenchulus reniformis* Linford and Oliveira 0.18×10^{-4} and *Tylenchorhynchus martini* Fielding 0.39×10^{-4} (Ibrahim et al., 1975). Thus oxygen release from rice roots (Armstrong, 1969) should be a factor in the development of the host-parasite relation between nematodes and rice plants. We have researched the development of anaerobiosis in rice soils for many years and have found this to be the case. This work has been summarized in a paper on the ecology of *Beggiatoa*, and a literature list has been provided in this summary paper to meet the needs of explanation in the context of nematode-rice relations (Hollis, 1979).

A direct attack on the problem of root-knot-oxygen-rice plant relations was made by Joshi et al. (1975). *M. incognita* infections (galls) caused a significant reduction in the amount of oxygen released from three-week-old seedlings of the rice cultivars Bluebelle and Saturn. Reductions of oxygen release from seedlings, calculated as microliters per plant per minute, were inversely proportional signficantly to the number of nematodes per pot and the number of galls per plant.

We can interpret the above data to merge with events of: (1) nematode oxygen requirements and (2) reduction of oxygen released from galled plants. The conclusion follows: nematode development, reproduction, and second-stage larvae (infective larvae) survival in flooded soil outside the plant are incompatible with the limiting oxygen supplies caused by reduction of oxygen release from seedlings. This reduction of oxygen supply to root tips is brought about by the mode of parasitism (gall formation) and disruption of cortical tissue through which oxygen must diffuse.

Additional evidence on root-knot nematodes indicates some differences of experience worth citing. Page and Bridge (1979) speculate on the possibility of significant yield reductions in deep-water rice in Bangladesh by *M. graminicola*. They state that the nematode probably causes the most damage to the aman (winter) crop sown between March and May, and harvested November to December.

> Typical field symptoms are yellowing and stunting, associated with the the presence of *M. graminicola* females within roots and with some galling. *M. graminicola* can obviously tolerate flooding—large numbers of it were found in roots submerged in water as deep as 1.5m, which was expected to reach more than 4-m depth later. The soil temperatures at the sites ranged from 31 to 36°C. The main *M. graminicola* damage appeared to be reduction in the ability of the deepwater rice plant to elongate so that it could not keep pace with the rising floodwater. In some situations, that could result in the complete drowning out of the crop.

A critical first question is the condition of the soil - whether biochemically oxidized or reduced, and whether deep water is in effect a supplier of oxygen to the soil—enabling *M. graminicola* to develop in deep-water paddy to a degree not observed in paddy rice under shallow flood.

V. RICE STYLET NEMATODE (*Tylenchorhynchus martini* Fielding)

This stylet nematode found widely in rice and sugarcane throughout the world was described by Fielding (1956) and named *T. martini* in honor of Dr. W. J. Martin of Louisiana State University. The name *stunt nematode*, while applicable to *T. claytoni* Steiner, the tobacco stunt nematode, which does indeed damage tobacco (Graham, 1954), is incorrect for *T. martini* and practically all species of *Tylenchorhynchus*, which although efficient parasites building up to large populations on many host plants, cause no damage: for example, extremely high rice yields commonly occur in the presence of very high populations of *T. martini* in Louisiana and Texas.

T. *martini* has been an extremely useful test animal for relating nematology to other disciplines. We have discovered a swarming phenomenon, determined it to be caused by an oncogenic virus transmitted through the egg stage as the unstable provirus, and then converted into the stable transformed state (swarming virus) in the second generation (approximately six weeks) after egg hatch (Ibrahim et al., 1973, 1975, 1978; Hollis et al., 1981). Swarming populations in greenhouse pots have been found extremely pathogenic to rice (Joshi and Hollis, 1976); since swarming incidence in the field is very low and inconsistent, this evidence has not been linked to nematode-rice relations.

There have been recent claims of observed damage to rice by *Tylenchorhynchus* (Israel et al., 1966), including *T. martini*, *T. elegans* Siddiqi, and *T. indicus* Siddiqi. Page and Bridge (1979) stated *T. martini* probably caused significant yield reductions in deep-water rice in Bangladesh. These observations, although undoubtedly accurate, run directly counter to additional Louisiana experience in the late 1950s where numerous, extensive *T. martini* pathogenicity trials involving high populations in the greenhouse gave consistently negative results.

VI. STEM NEMATODE (*Ditylenchus angustus* Filipjev)

This typical host-specific stem nematode, closely related to *D. dipsaci* Kühn, was discovered and described by Butler (1913) in eastern Bengal (Bangladesh) along with its incident disease *ufra*, as it is known locally.

The short review of the disease by Ou (1972) is excellent and needs supplementing only with recent reports published in *International Rice Research Newsletter* (Manila, Philippines).

Symptoms include incomplete emergence, chlorosis, malformation, and masked. The nematode, a surface feeder on growing-point tissues of seedlings, moves in relative humidity levels of 85-100% while feeding on epidermal cells; desiccation results in the anabiotic (dormant) state characteristic of stem and bud nematodes. The most severe infestations occur during the rainy season crop (July to December) aman; less effects result during tropical autumn, aus (September to January), which includes a little dry season as the sun returns south. The least severe infestations are in winter (January to April) boro, dry season characterized by lower temperatures.

A variant of *D. dipsaci* and hence of *D. angustus*, *D. destructor* Thorne, has been implicated as a fungus feeder (some strains), but there is no evidence this is common among stem nematodes.

Ecologically, *D. angustus* is a problem in deep-water rice, principally in Bangladesh and Thailand, although reports of ufra disease occurrence in shallow paddy are known (Ou, 1972; Bakr, 1978).

The reports of Cox and Rahman (1979a-c, 1980), Cox (1980), and Cox et al. (1980) provide a current summary of ufra disease. The authors successfully developed and used minature deep-water tanks for screening chemicals against ufra by transplanting diseased tillers from the field during August. Latin square treatments of Furadan 3 G levels were compared; the results showed many types of chemicals could be screened effectively.

The nematode primary infestation comes from residues of the preceding crop; secondary infestation is water borne during the growing season. Obviously, for a crop sown during the first rains, residues are patterned over a field and primary infestation is spotty. The secondary spread by flood water results in continuous distribution of inoculum and also long-distance spread, but there is often a period of up to two months between planting and flood (primary and secondary infestations). This is a prime opportunity gap for application of chemical control measures prior to flood.

Laboratory incubation of infested panicles (incomplete emergence) collected from fields in November showed a rapid population decay during a three-month period suggesting that control of ufra in deep-water rice can be improved by prolonging the decay phase. Techniques suggested were: (1) planting early-flowering varieties, (2) late planting, or (3) transplanting during initial flood period.

Ufra disease is a prime candidate for successful control by the use of systemic nematicides; indeed, the forefront of rice nematode control efforts should be concentrated against ufra because of its high visibility. Successful candidate chemicals could then be tested against hidden nematode diseases in rice.

VII. WHITE TIP NEMATODE (*Aphelenchoides besseyi* Christie)

The white tip (foliar) nematode is borne in rice seed beneath the glumes in a dormant (anabiotic) condition for periods up to at least two years, and upon seed germination and/or other water absorption phenomena, changes again to an active condition, infesting and feeding ectoparasitically upon external surfaces of leaf sheaths, glumes, paleas and lemmas, and hulls of the rice plant.

Ou (1972) has provided in his book a thorough review of all aspects of the white tip disease—a name that adequately describes the leaf symptoms on rice.

In the Gulf Coast region of the United States, white tip is a curiosity on Calusa and Calora rice varieties imported from California, used in breeding programs and found only in nurseries. Extensive field surveys by Atkins in Louisiana and Texas during the 1960s showed absence of the nematode in field-collected seed lots of medium- and long-grain varieties. Earlier, in pathogenicity trials, Atkins and Todd (1959) had found the medium-grain varieties Zenith, Arkrose, and Calrose susceptible, but the long-grain varieties tested were resistant.

It was hypothesized by the author and M. C. Rush that Phostoxin (aluminum phosphide),* used extensively to control insects in rice storage bins in the Gulf Coast region, could control the nematode in seed lots and prevent its appearance and spread in the medium-grain varieties. Recently, Dr. E. C. McGawley conducted surveys that revealed small percentages of white

---

*Phostoxin produced by Degesch, GmbH (West Germany).

tip nematode infestations among seed lots of both medium- and long-grain cultivars collected from rice fields in Louisiana and Texas (McGawley et al., 1982).

White tip disease is controlled on small-holdings-grown rice throughout the world by seed selection practices and by hot-water seed treatment (52-54°C, 15 minutes).

Many of the newer organic phosphate or carbamate insecticide-nematicides as seed treatments will control white tip; however, since there is an adverse interaction among rice plant tissues, the use of these chemicals and of 3.4-dichloropropionanilide (DCPA), a chemical used widely for grassy weed control, is not possible. In a real sense, therefore, white tip is a DCPA-dependent disease since this chemical effectively blocks control of white tip by chemical seed treatment. Situations requiring investigation with reference to white tip disease can be anticipated by the following changes in rice culture: (1) elimination of Phostoxin, (2) elimination of DCPA, and (3) introduction of short-grain varieties.

In summary, white tip is now a minor disease of little or no interest except in extenuating circumstances, but certain cultural changes could make it a troublesome pest.

Along with ufra disease, white tip is an excellent, highly visible, and easily manipulated disease for use in screening systemic chemicals as a prelude to successful control of hidden nematode diseases in rice.

VIII. MISCELLANEOUS NEMATODES IN UPLAND RICE

The following genera and species are reported to occur on a worldwide basis in nonsaturated rice soils and are listed in descending order of estimated importance (see Table 3 for a partial list); *Criconemella* spp., *Meloidogyne graminicola* Golden and Birchfield, *Heterodera oryzae* Luc and Brizuela, *H. oryzicola* Jayaprakash, *Pratylenchus indicus* Rao and Prasad, *Hoplolaimus indicus* Das, *Hemicriconemoides* spp., *Hemicycliophora* spp. In spite of repeated efforts there is little or no evidence to support views for economic importance in rice of *Tylenchorhynchus martini* Fielding, *T. indicus* Siddiqi, *T. elegans* Siddiqi, *T. claytoni* Steiner, *Helicotylenchus* spp., *Heterodera graminophila* Golden and Birchfield, and *Meloidogyne* spp. *Rotylenchulus reniformis* Linford and Oliveira is not a rice nematode although often found in rice fields in association with weeds.

Upland rice nematodes have long been recognized in Japan (Okada, 1955); more recently workers in India, Philippines, Pakistan, and West Africa have taken interest in the evident problems they cause (Das and Rao, 1969; Fofie and Raymundo, 1979; Rao and Prasad, 1977; Castillo et al., 1978; Malik and Yasmeen, 1978; Timm, 1956, 1965; Brizuela and Merny, 1964; Rao and Jayaprakash, 1977; Raveendran and Mohandas, 1978). The recitation of nematode species, plant symptoms, host range, and control tests is a replay of the litany of events noted in numerous publications in Europe and in America in the 1950s for nematodes and damage to the major crops. This survey phase of work in rice provides a base for training nematologists and for strategic planning.

Both the recognition and control of upland and paddy rice nematode problems is in a lamentable, primitive state, localized and reduced to cultural practices aimed at preventing visible damage or total loss of the crop.

If our aim is to utilize the knowledge of invisible, hidden nematode damage revealed here and elsewhere, then we are impelled to outline an overall strategy for future development by the community of interests in rice research and development. This strategy should be aimed at useful levels of nematode control designed to bring about significant increases in rice yields wherever rice is grown. We would hope that such efforts have an impact on rice equal to or greater than the celebrated success of dwarf varieties in wheat production.

IX. SOIL CLASSIFICATION AND NEMATODES

There is to an unknown extent a relation between soil characteristics in global rice culture and the occurrence of nematodes that cause damage to rice. This relation rests upon both the work conducted in Asia, Africa, and at Louisiana State University in both the Louisiana and Texas rice areas and in the classification and descriptions of rice soils on the (USDA 1975) Soil Taxonomy Scheme by Moormann and van Breemen (1978).

The four most important soil orders described by these authors, Alfisols, Entisols, Inceptisols, and Ultisols, all have substantial portions of the following natural/cultural regimes in the vast rice-growing areas they occupy around the world: (1) pluvial (rainfed) upland, (2) phreatic (rainfed) upland-lowland [both (1) and (2) may be subject to brief and irregular or uncontrolled flooding], (3) irrigated lowland with flooding and (soil saturation) during growing season, and (4) fluxial lowland (swamp) flooded during growing season and with soil saturated throughout year. All of the important and widespread rice soil orders embrace variations ranging over most of these four natural/cultural regimes.

Alfisols are soils with clay deposits representing downward translocation without excessive depletion of bases and with a predominance of montmorillinite clays. These soils are found in the U.S. coastal prairie of Louisiana and Texas, where the ring nematode problem in rice has been elucidated and they include also the Udalfs suborder, which occurs in Arkansas, where nematode damage is unknown, and in the west part of the Texas rice area, where yields are very high and nematode damage is minimal.

Nematodes are of estimated importance as damage agents on approximately one-half the area of Alfisols, including substantially all of the pluvial-phreatic (upland) areas.

The order of Entisols, which include soils in which soil-forming processes are absent, are without a profile—only a thin A horizon may be developed. These soils are found in intermittently flooded coastal areas, tidal marshes, and back swamps. This order includes problem soils exhibiting toxicant diseases (Hollis, 1967; Hollis et al., 1975) and nutritional disorders, including zinc deficiency and iron toxicity. Nematode problems may occur but are not known. These soils are not found in southern United States rice-producing areas.

Inceptisols comprise the most important order in rice culture and make up the rice-growing alluvial plains (flood plains) and deltaic areas of the world. These soils have immature, weakly developed profiles and include the relatively small rice areas in Northeast Louisiana and in Arkansas, which are free of ring nematodes and produce high rice yields. However, pluvial-phreatic (upland) areas of these soils are of major extent, and the great variability of subgroups of Inceptisols (Aquepts, Tropepts, and Octhrepts suggests many soil conditions are favorable to the development of nematode problems.

Ultisols resemble Alfisols in the downward translocation of clay; this movement has resulted in intensive leaching with depletion of bases. Ultisols are characterized by lower base saturation of subsoils, a low subsoil pH, and dominance in the clay complex of kaolinite and sesquioxide clays. Ultisols exhibit great variability in physiographic position and parent materials, rank in importance for rice growing with Alfisols, and are exceeded in geographical range around the world only by Inceptisols. Ultisols are foremost in pluvic-phreatic, rainfed, and temporary flood rice culture (upland), and consequently rice grown on these soils is subject to severe nematode damage. A postulated high level of nematode damage on Ultisols resulting from natural/cultural upland regimes is correlated with low rice yields on these soils (suborder aquults) in Southeast Asia (Thailand, Cambodia, and Vietnam), on the east coast of the Malay peninsula, and in the humid parts of West Africa.

Histosols (peat soils) of equatorial regions cover an estimated 75 million acres in the rain forest belts of Asia, Africa, and America. These soils have great potential value for extensive rice culture. The author operates The Biotron Tropical Agricultural Experiment Station in Belize* on a mineral Histosol that contains all nematode genera known to attack rice and some that do not attack rice. Several nematode species, including *M. arenaria* race 2 (Neal), *R. reniformis* (Linford and Oliveira), and *C. onoensis* (Luc), build to high numbers during crop development, ranking mangrove peat soil in the highest category of nematode damage capability. Rice nematicide trials on microplots have yielded outstanding results.

Several additional orders of rice soils of either local or minor importance in world rice culture are Vertisols, Mollisols, Oxisols, Aridisols, and Spodosols. Vertisols are fine clay soils difficult to work and not much used because of the extremely narrow moisture range between their plastic and solid states. These soils of the Uderts suborder are important in the Lake Charles, Louisiana, and Beaumont, Texas areas and are heavily infested with ring nematodes *C. onoensis* and *C. reedi*. Spodosols are coastal or riverine sands (sand piles). Attempts to grow rice invariably fail. Oxisols, old volcanic formations containing iron and aluminum oxides, are highly fertile but not important in rice culture because of their uses for other crops and occurrence of water shortages. Aridisols are fertile soils of arid areas containing high salinity; rice can be grown only where good water is available; generally, rice cannot compete with other crops for this high-quality water. Mollisols are the fertile brown soils developing under grasslands or forest, better known as chernozems of the old Russian classification schemes. These soils are more common in the tropics than previously known. Nematode problems are anticipated on Mollisols, the worldwide geographical extent of which is not known. These soils, of minor occurrence in Louisiana and Texas rice areas, are heavily infested with ring nematodes.

X. NEMATODE CONTROL STRATEGY

Rice is the prime international crop linking temperate and tropical reaches of the world in common agricultural problems. It is necessary to attack these problems, whether economic, agronomic, pathologic, or of other specific discipline, on two levels: (1) locally and (2) internationally. The paddy and upland rice-growing provinces and districts find useful the common fund of knowledge in numerous publications and the special adaptations distilled for their local use; these methods (primarily cultural and rotational) are known and will not be treated here. We will be concerned with international attempts

---

*Operated by Biotron, Inc., 833 Dentation Dr., Baton Rouge, LA 70808.

to assess nematode damage and to improve nematode control in rice based on an assessment of international market potentials for technological innovation.

A broad approach to the study of nematodes is essential to encompass some possibilities for rational development; this was outlined in the first part of the chapter. The possibilities are vast: it is likely that nematodes are universal animals to a much greater extent than are humans (distributed much more abundantly throughout the universe). Limited knowledge of nematode activities and significance on one planet tells us already that these animals will make great space travelers and colonizers; however, in order to fully understand nematodes and to bring a large measure of their potentialities under some kind of control, we will need much more information that can be acquired from a monoplanetary base. Large generalizations about universal animals are irrational; however, there is a greater danger: as a ghettoized animal, man must continue to look beyond his lowly state; nematodes provide another kind of universal entity, more concrete than Carl Sagan's stars, that will help us to avoid ghettoization of the human spirit.

International nematode disease control research demands the broadest possible approach:

1. Selection, breeding, and genetics (including genetic engineering) of rice for resistance/immunity to *Hirschmanniella* and *Criconemella* (30% of total effort).
2. Chemical control research with systemic-action materials that render rice plants resistant/immune by any mechanism. Test animals *A. besseyi* (white tip nematode) and *D. angustus* (stem nematode) with results considered applicable to all nematodes (30% of total effort).
3. Compatibility studies with nematicide–insecticides, herbicides, other pesticides, rice varieties, rice field weeds (30% of total effort).
4. Ecological, geographical, and host range studies on nematodes known to damage rice (10 percent of total effort).

XI. BASIS FOR RESEARCH AND DEVELOPMENT

The strategy outlined above comprehends innovation and support of research and development from proven resource areas:

1. Multinational corporations following survey and recognition of market potentials.
2. High-technology companies with expertise in the fields of biochemistry and genetic engineering in cooperation with other resource areas.
3. Foundations with distinguished records in support of work toward solutions of visible problems; when nematode problems become "visible," resource agencies support work toward their control.
4. Individual government agencies such as the United States Department of Agriculture and the Department of State for International Development (AID) support visible problem research.
5. The World Bank and affiliated financial institutions support visible problem research.
6. International government-sponsored agencies, such as the Food and Agriculture Organization of the United Nations, support visible problem research.

Generally, all resource agencies have been short-term and "belly -oriented" toward visible rice problems demanding immediate solution or presenting built-in, population-intensive market potential; the tenor of efforts has been for short-term gains on visible problems with the rice farmers themselves unaware of political factors involved in the decision-making processes.

There has been a notable lack of long-term "profit-oriented" approaches to rice problems. The private resource sector, involving multinational corporations and multinational high-technology companies, affords the only means of breaching the sociopolitical barriers that have stultified rice research for the past quarter century. In the past, even our most progressive multinational corporations have not been able to find a basis for long-range planning toward rational goals and have confined their efforts to sales development in visible, demonstrated, but relatively insignificant areas of rice research, such as seed treatment chemicals, solely for the purpose of establishing a thick stand of rice seedlings, all in the face of irrefutable scientific evidence that there is no relation whatever within normal limits between stand and yield!

This summary chapter presents evidence for the first time making it possible to plan long-range rational solutions to important, limiting and invisible, hidden problems of rice production; namely, rice root and ring nematodes, in terms of both plant modification and chemical treatment approaches.

The total market potential for nematode control worldwide is estimated as 2 billion 575 million dollars on a treatment cost level involving chemicals/cultivars set at sixteen 1975 U.S. dollars per acre. On a usual 1:4 ratio of treatment cost to response value in 1975 U.S. dollars, the annual loss in rice worldwide due to plant parasitic nematodes is 10.3 billion dollars.

ACKNOWLEDGMENTS

The rice nematode studies at Louisiana State University over the past quarter century owe whatever success may be claimed or assigned to a relatively small number of cooperators whose names have been listed in previous publications. Foremost has been the role of the Louisiana Agricultural Experiment Station in sponsorship; the five individual cooperators listed in alphabetical order are: Alphonse Habetz, Jr., Crowley, LA; H. G. Hardee, Jr., Gueydan, LA.; W. O. Miller, Dow Chemical Co., Wayside, MS; Elmo E. Shipp, Mobay Chemical Co., Memphis, TN.; L. J. Theriot, Jr., Gueydan, LA.

REFERENCES

Alhassan, S. A., and Hollis, J. P. (1968). Removal of Phosphorus-32 from rice roots by nematodes (abstract). *Phytopathology 58*: 725.

Alhassan, S. A., and Hollis, J. P. (1969). Ring nematode ratings of rice area rotation crops (abstract). *Phytopathology 59*: 111.

Armstrong, W. (1969). Rhizosphere oxidation in rice: An analysis of intervarietal differences in oxygen flux from the roots. *Physiol. Plant. 22*: 293-303.

Atkins, J. G., and Todd, E. H. (1959). White tip diseases of rice. III. Yield tests and varietal resistance. *Phytopathology 48*: 189-191.

Bakr, M. A. (1978). Occurrence of ufra disease in transplanted rice. *Int. Rice Res. Newsletter 3*: 16.

Brizuela, R., and Merny, G. (1964). Biologie d'*Heterodera oryzae* Luc and Berndon 1961. I. Cycle du parasite et reactions histologiques de l'hôte. *Rev. Path. Veg. Ent. Agric. Fr. 43*: 43-53.

Butler, E. J. (1913). Ufra disease of rice. *Agric. J. India 8*: 205-220.

Castillo, M. B., Arceo, M. B., and Litsinger, J. A. (1978). Populations of parasitic nematodes of grain legumes in rice-based cropping in the Philippines. *Int. Rice Res. Newsletter 3*: 19.

Cox, P. G. (1980). Symptoms of ufra disease of deepwater rice in Bangladesh. *Int. Rice Res. Newsletter 5*: 18.

Cox, P. G., and Rahman, L. (1979a). The overwinter decay of *Ditylenchus angustus*. *Int. Rice Res. Newsletter 4*: 14.

Cox, P. G., and Rahman, L. (1979b). Synergy between benomyl and carbofuran in the control of ufra. *Int. Rice Res. Newsletter 4*: 11.

Cox, P. G., and Rahman, L. (1979c). The ufra nematode population in deepwater rice in Bangladesh. *Int. Rice Res. Newsletter 4*: 10-11.

Cox, P. G., and Rahman, L. (1980). Components of yield loss in ufra. *Int. Rice Res. Newsletter 5*: 18.

Cox, P. G., Rahman, L., and Hannan, M. A. (1980). Simulation of an ufra attack. *Int. Rice. Res. Newsletter 5*: 19-20.

Das, P. K., and Rao, S. (1969). Life history and pathogenicity of *Hoplolaimus indicus* Sher incidence in rice (abstract). *All India Nematology Symposium*, Aug. 21-22, New Delhi, India, p. 13.

Embabi, M. S. (1965). Quantitative nematode-root relations in rice and cotton. Ph.D. Thesis, Louisiana State University, Baton Rouge.

Fielding, M. J. (1956). *Tylenchorhynchus martini*, a new nematode species found in the sugar cane and rice fields of Louisiana and Texas. *Proc. Helm. Soc. Wash. 23*: 47-48.

Fielding, M. J. and Hollis, J. P. (1956). Occurrence of plant-parasitic nematodes in Louisiana soils. *Plant Dis. Rep. 40*: 403-405.

Fofie, A. S., and Raymundo, S. A. (1979). Parasitic nematodes in continuously cropped uplands. *Int. Rice Res. Newsletter 4*: 17-18.

Golden, A. M., and Birchfield, W. (1968). Rice root-knot nematode (*Meloidogyne graminicola*) as a new pest of rice. *Plant Dis. Rep. 52*: 423.

Graham, T. W. (1954). The tobacco stunt nematode in South Carolina (abstract). *Phytopathology 44*: 332.

Hashioka, Y. (1963). The rice stem nematode *Ditylenchus angustus* in Thailand. *Plant Prot. Bull. FAO 11*: 97-102.

Hollis, J. P. (1967). Concentration of toxicants in the rice soil profile (abstract). *Phytopathology 57*: 460.

Hollis, J. P. (1969a). Chemical control of soil nematodes in rice fields. *Phytopathology 59*: 1031.

Hollis, J. P. (1969b). Genesis of a soil nematode problem in Louisiana rice. *Int. Rice Comm. Newsletter 18*: 19-28.

Hollis, J. P. (1972a). Competition between rice and weeds in nematode control tests. *Phytopathology 62*: 764.

Hollis, J. P. (1972b). Nematicide-weeds interaction in rice fields. *Plant Dis. Rep. 56*: 420-424.

Hollis, J. P. (1973). A summary of ring nematode studies. *65th Annual Report,* Experimental Station, Crowley, Louisiana, pp. 153-174.

Hollis, J. P. (1977). Loss in yield of rice caused by the ring nematode *Criconemoides onoensis* revealed by the elimination of yellow nutsedge, *Cyperus esculentus*. *Nematologica 23*: 71-78.

Hollis, J. P. (1979). Ecology of *Beggiatoa*. *Acta Phytopathol. Acad. Sci. Hung. 14*: 419-439.

Hollis, J. P. (1980). The ring nematode problem in rice. *La. Agric. 23:* 6-7.

Hollis, J. P., Embabi, M. S., and Alhassan, S. A. (1968). Ring nematode disease of rice in Louisiana. *Phytopathology 58:* 728-729.

Hollis, J. P., Ibrahim, I. K. A., and Joshi, M. M. (1981). Standardized transmission procedure for nematode swarming virus (abstract). *Phytopathology 71:* 226.

Hollis, J. P., and Joshi, M. M. (1975). Weed effect on chemical control of ring nematodes in rice. *Ann. Proc. Am. Phytopathol. Soc.* 44-45.

Hollis, J. P., Whitlock, L. S., Atkins, J. G., and Fielding, M. J. (1959). Relations between nematodes, fumigation and fertilization in rice culture. *Plant Dis. Rep. 43:* 33-40.

Ibrahim, I. K. A., Joshi, M. M., and Hollis, J. P. (1973). The swarming virus disease of *Tylenchorhynchus martini* (abstr. No. 555). *Proc. 2nd. Intern. Cong. Plant Pathol.*, Minneapolis, Sept. 5-13.

Ibrahim, I. K. A., Joshi, M. M., and Hollis, J. P. (1975). Respiration measurements of four nematode species with the biological oxygen monitor. *Alex. J. Agr. Res. 23:* 161-165.

Ibrahim, I. K. A., Joshi, M. M., and Hollis, J. P. (1978). Swarming disease of nematodes: Host range and evidence for a cytoplasmic polyhedral virus in *Tylenchorhynchus martini*. *Proc. Helm. Soc. Wash. 45:* 233-238.

Ichinohe, M. (1955). Two species of the root-knot nematodes in Japan. *Jpn. J. Appl. Zool. Nogyo-Gijutsu (Tokyo) 18:* 356-359.

Ichinohe, M. (1964). A review of studies on nematodes attacking rice. Paper presented at the 10th session working party on rice productions and protection. *Int. Rice Comm. FAO*, Manila, Philippines.

Ichinohe, M. (1966). Present status of nematological research on rice in Japan. Paper presented at the 11th session working party on rice production and protection. *Int. Rice. Comm.* FAO, Lake Charles, Louisiana.

Imamura, S. (1931). Nematodes in the paddy field, with notes on their population before and after irrigation. *J. Coll. Agr. Imperial Univ. (Tokyo) 11:* 193-240.

Israel, P., Rao, Y. S., and Rao, V. N. (1964). Rice nematode-host and parasite relationship. Paper presented at the 10th session working party on rice production and protection. *Int. Rice Comm. FAO*, Manila, Philippines.

Israel, P., Rao, Y. S., and Rao, V. N. (1966). Rice parasitic nematodes. Paper presented at the 11th session working party on rice production and protection. *Int. Rice Comm. FAO*, Lake Charles, Louisiana.

Joshi, M. M., and Hollis, J. P. (1976). Pathogenicity of *Tylenchorhynchus martini* swarmers to rice. *Nematologica 22:* 123-124.

Joshi, M. M., Ibrahim, I. K. A., and Hollis, J. P. (1975). Oxygen release from rice seedlings: Effect of *Meloidogyne incognita* and *Helminthosporium oryzae* infections. *Acta Phytopathol. Acad. Sci. Hung. 10:* 51-53.

Kanjanasoon, P. (1964). Rice root-knot nematodes and their host plants. Paper presented at the 10th session working party on rice production and protection. *Int. Rice Comm. FAO*, Manila, Philippines.

Kawashima, K. (1963). Investigations on *Hirschmanniella oryzae*. I. Varietal susceptibility to the nematode. II. Susceptibilities of weeds to the nematode. III. Durable effect of fungicides. *Ann. Rep. Soc. Plant Prot. N. Jn. 14:* 111; 112-113; 158-159.

Kawashima, K. (1964a), Damage by nematode and its control (translation). *Nogyo Gitjutsu 19:* 75-78.

Kawashima, K. (1964b). Studies on *Hirschmanniella oryzae*. IV. On the soil reduction and nematode injury (translation) *Ann. Rep. Soc. Plant Prot. N. Jn. 15*: 131-132.

Kawashima, K., and Fujinuma, T. (1965). On the injury to the rice plant caused by rice-root nematode (*Hirschmanniella oryzae*) injury to the rice seedlings. *Bull. Fukushima Pref. Agric. Exp. Sta. 1*: 57-64.

Keoboonrueng, S. (1971). Effects of rice-root nematode, *Hirschmanniella oryzae* (Van Breda de Haan 1902) Luc and Goodey 1963 on rice seedlings. Ph.D. dissertation, Louisiana State University, Baton Rouge.

Luc, M. (1959). Nouveaux Criconematidae de la zone tropical (Nematoda: Tylenchida). *Nematologica 4*: 16-22.

Luc, M. (1970). Contribution a l'etude de genre *Criconemoides* Taylor, 1936 (Nematoda: Criconematidae). *Cah. Orstom, Ser. Biol. 11*: 69-131.

Maas, P. W. (1970). Tentative list of plant parasitic nematodes in Surinam, with descriptions of two new species of Hemicycliophorinae. *Bull. Van Het Landbouwptoefstation*, Suriname, South America.

Malik, R., and Yasmeen, Z. (1978). Nematodes in paddy fields of Pakistan. *Int. Rice Res. Newsletter 3*: 16-17.

Moormann, F. R., and van Breemen N. (1978). Rice: Soil, water, land. *Int. Rice Res. Inst.*, Los Banos, Philippines.

Okada, T. (1955). On morphological characters of the upland rice nematode, *Heterodera* sp. (abstract, paper Ann. Meet). (Japanese). *Appl. Zool. Entomol. Jpn.* 1955: 11.

Ou, S. H. (1972). *Rice Diseases*. Commonwealth Mycological Institute. Kew, England.

Page, S. L., and Bridge, J. (1979). Root and soil parasitic nematodes of deepwater rice areas in Bangladesh. *Int. Rice Res. Newsletter 4*: 10.

Palapac, A. C. (1980). World rice statistics. *IRRI* (Dept. Agric. Econ.) 130 p.

Panda, M., and Rao, Y. S. (1969). Evaluation of losses caused by incidence of *Hirschmanniella mucronata* Das in rice (abstract). *All India Nematology Symposium*, Aug. 21-22, New Delhi, India, p. 1.

Rao, Y. S., Biswas, H., Panda, M., Rao, P. R., and Rao, V. N. (1969). Screening rice varieties for their reaction to root and root-knot nematodes (abstract). *All India Nematology Symposium*, Aug. 21-22, New Delhi, India, p. 59.

Rao, Y. S., and Jayaprakash, A. (1977). Leaf chlorosis due to infestation by a new cyst nematode. *Int. Rice Res. Newsletter 2*: 5.

Rao, Y. S., and Panda, M. (1970). Study of plant parasitic nematodes affecting rice production in the vicinity of Cuttack (Orissa) India. *Final Tech. Rep. (USPL 480 Project)*. Central Rice, Memphis, Tennessee.

Rao, Y. S., and Prasad, J. S. (1977). Root-lesion nematode damage in upland rice. *Int. Rice Res. Newsletter 2* : 6-7.

Raveendran, V. J., and Mohandas, C. (1978). Cyst nematode infestation of rice in Kerala State, India. *Int. Rice Res. Newsletter 3*: 15.

Sher, S. A. (1968). Revision of the genus *Hirschmanniella* Luc and Goodey, 1963 (Nematoda: Tylenchoidea). *Nematologica 14*: 243-275.

Sivakumar, C. V., and Seshadri, A. R. (1969). Histopathology of the rice root infested by *Hirschmanniella oryzae* (van Breda de Haan, 1902) Luc and Goodey, 1964 (Abstr.) *All India Nematology Symposium*, Aug. 21-22, New Delhi, India, p. 1.

Taylor, A. L. (1965). Final report of a nematode survey in Thailand. *FAO*.

Taylor, A. L., Kaosiri, T., Sittichai, T., and Guangwuwon, D. (1966). Experiments on the effects of nematodes on the growth and yield of rice in Thailand. *FAO Plant Prot. Bull. 14*: 17-23.

Timm, R. W. (1956). Nematode parasites of rice in East Pakistan (abstract). *Proc. Pak. Sci. Conf.*, 12th (1960), Sec. B, 25-26.

Timm, R. W. (1965). A preliminary survey of the plant parasitic nematodes of Thailand and the Philippines. Director of Cultural Affairs, SEATO, SEATO Headquarters, Bangkok, Thailand.

Timm, R. W., and Ameen, M. (1960). Nematodes associated with commercial crops in East Pakistan. *Agric. Pakistan 3*: 1-9.

Tullis, E. C. (1934). The root-knot nematode on rice. *Phytopathology 24*: 938-942.

Van der Vecht, J., and Bergmann, H. H. (1952). Studies on the nematode "*Radopholus oryzae*" (Van Breda de Haan) Thorne and its influence on the growth of the rice plant. *Contrib. Cen. Agric. Res. Sta.* (Bogor, Indonesia), 82 p.

Venkitesan, T. S., and Charles, J. S. (1979). The rice root nematode in lowland paddies in Derala, India. *IRRI Newsletter 4*: 21.

Venkitesan, T., Charles, J. S., and Nair, V. R. (1980). Checking infection of rice root dips to increase yield. *Int. Rice Res. Newsletter 5*: 9-10.

Whitlock, L. S. (1957). Notes on *Radopholus oryzae* (Nematoda, Phasmidia) with a key to the genus and description of a new species *R. paludosus*. M. S. Thesis, Louisiana State University, Baton Rouge.

Chapter 5

Nematode Parasites of Cotton

Charles M. Heald *USDA, Agricultural Research Service, Weslaco, Texas*

Calvin C. Orr *USDA, Agricultural Research Service, Texas Agricultural Experiment Station, Lubbock, Texas*

I. INTRODUCTION AND ECONOMIC IMPORTANCE

Cotton is a major crop in the warmer parts of every continent of the world.
It has been cultivated by humans for thousands of years, becoming a major
crop with the development of the cotton gin. The enormous expansion of cotton acreage, with much of this acreage continuously cropped to cotton, caused
numerous problems including the increased incidence of nematodes and diseases.

Plant-parasitic nematodes present some of the most difficult pest problems
encountered in our agricultural economy. Each year these minute organisms
extract an ever-increasing toll from almost every cultivated acre in the world.
The amount may be relatively small in most instances, but the aggregate represents a staggering total which the farmers and consumers of the world can ill
afford to lose (Thorne, 1961). In the United States the annual loss of cotton
to nematodes is 3-6% of the crop; however, losses exceed 50% in many fields.
In numerous tests conducted on root-knot nematode-infested soils on the
High Plains of Texas, cotton yields were increased an average of 20% from the
use of a soil fumigant (Orr and Brashears, 1977).

Underestimating or lack of knowledge of the damage caused by parasitic
nematodes adversely affects scientific investigation in many disciplines of
agricultural research. For example, the presence of a damaging nematode
population can lead to serious misinterpretation in plant-breeding nurseries,
as well as in field studies in soils, efficacy tests of fungicide and insecticides,
and fertility studies.

Nematodes are microscopic animals that cause injury to cotton by feeding
on the roots of their host. They are transparent, wormlike in shape, and
virtually impossible to see with the unaided eye. Nematodes derive food from
roots by puncturing cell walls and extracting the contents through a hollow
stylet or spear-shaped structure located in the head of the animal. Some nematode species feed on surface tissue while remaining vermiform and migratory
throughout the life cycle. In other species sexual dimorphism is pronounced,
where females become sedentary with the posterior portion swelling to an oval,
pear, or sausage shape upon entering the adult stage.

II. ROOT-KNOT NEMATODE

A. History

The first investigations of *Meloidogyne* species as an economically important
nematode was by Jobert in 1878 on coffee in Brazil. Ten years later, Goeldi
investigated the same problem and named the root-knot nematode *Meloidogyne
exigua*. In 1889, Atkinson published a classic work on root-knot nematodes
of cotton. During this period root-knot nematodes were known as *Heterodera
radicicola*, and later as *H. marioni*. Root-knot nematodes were considered a
single species until Chitwood (1949) described or redescribed the four most
common and widely distributed species of the genus *Meloidogyne: M. incognita*,

M. javanica, M. arenaria, and *M. hapla.* By 1976, at least 36 species of the genus *Meloidogyne* had been named and sufficiently well described to fulfill the requirements of the International Rules of Zoological Nomenclature. Taylor and Sasser (1978) reported that 180 *Meloidogyne* populations collected in countries from five continents of the world showed a remarkable uniformity of host preference, perennial pattern, mode of reproduction, chromosome number, and chromosome behavior during maturation of oocytes, to one or another of the four major species groups listed above. They confirmed the existence of only four widespread races of *M. incognita,* two races of *M. arenaria,* one race of *M. javanica,* and one race of *M. hapla.* Only *M. incognita* among these species is known to be parasitic on cotton, and only races 3 and 4 of this species are known to attack cotton.

B. Distribution

M. incognita, the species that infests cotton, is found predominately in the warmer climates of the world between 35° S and 35° N latitudes. In the United States it occurs in all cotton-producing states. Cotton root-knot nematodes thrive in sandy soils but are rarely a problem in clay soils. The precise factors for soil type preference are not known, although aeration and moisture stress are undoubtedly important. Infestations of the nematode are abundant in high rainfall as well as in semiarid regions of the cotton belt. In addition to cotton, *M. incognita* is known to infest more than 700 other plant species.

C. Associated Symptoms

Root-knot nematode infestations are characterized by distinctive spindle-shaped galls or knots on roots that are readily recognizable when lifted from the soil with a spade. Galls on the tap root cause extensive injury to cotton, but galls on lateral roots are less debilitating to the plant. Gall size varies depending on the susceptibility of the host, the number of nematodes present, and the coalescence of adjacent galls. Usually the galls on cotton are not as large or as numerous as those found on more susceptible plants, i.e., tomato, okra, or green bean.

Root-knot galling causes a disruption and disorganization of the vascular system that impedes the absorption and upward translocation of water and nutrients. The reduced efficiency of water and nutrient transport by infested plants causes stunting of above-ground plant parts and occasionally yellowing of the foliage. Plant symptoms of root-knot nematode attack are easily confused with mineral deficiencies, drought, soil-borne diseases, or other root system disorders.

The severity of nematode damage to cotton is related to population density of the nematode, environmental conditions, and stage of growth of the plant. Cotton seedlings may be killed during the first weeks of growth, resulting in thin stands. Heavily infested plants in the field may exhibit temporary wilting on hot afternoons even in the presence of adequate soil moisture and, under drought conditions, older plants may be killed. Because root-knot nematodes are rarely uniformly distributed throughout a field, areas of short plants with thin stands are surrounded by healthy normal plants resulting in an irregular, uneven appearance of the crop.

D. Mode of Infection

Parasitic larvae have been shown to migrate in the soil up to 12 cm to locate a root. Probably root exudates are responsible for attracting larvae to their surfaces (Bird, 1974; Peacock, 1956). Penetration occurs mostly within the first 2 cm of the root tip. Once in the root, larvae move intracellularly and intercellularly through the cortex to the stele, where they become sedentary. Provascular cells, where feeding begins, respond to the injected nematode salivary secretions by undergoing marked changes. The nuclei and nucleoli enlarge and divide and the cytoplasm becomes dense and granular, indicating increased metabolic activity. The larva feeds on four to eight cells in the provascular tissue, causing them to develop into giant cells. These giant cells become permanent feeding sites for the remainder of the nematode's life. In addition to the transformation of immature cells into giant cells, other cells adjacent to the nematode are affected. Pericycle cells divide repeatedly, and cortical cells enlarge to give rise to the distinctive root gall around the developing larva. Abnormal xylem surrounding the giant cells is a distinctive feature.

Histochemical observations of giant cells reveal that they are metabolically hyperactive and serve as metabolic sinks. How giant cells are formed is the subject of controversy. One hypothesis is that giant cells form as a result of karyokinesis (nuclear division) and hypertrophy (cell enlargement) without concomitant cytokinesis (cytoplasmic division). Alternatively, it is suggested that the walls between adjacent cells at the feeding site of the nematode dissolve, probably in response to metabolites produced by the nematode, thus forming the large multinuclear structure (Bird, 1974). In either case, the nematode is an active partner in the process. Nematode peroxidase has been suggested as a critical factor in syncytium initiation and development. This could be the case as this enzyme has indole acetic acid (IAA) oxidase activity.

E. Nematode Disease Complex

The soil is a complex system inhabited by many organisms. It is reasonable to assume that many physical and physiological interactions occur among the many pathogens found about the roots of a single plant. Nematodes are important in the susceptibility of plants to attack by other organisms. This has been confirmed in studies where the effects of an important nematode and fungal or bacterial pathogen of plant roots have been examined simultaneously (Norton, 1978; Powell, 1971).

F. *Fusarium* Wilt of Cotton

The classic nematode-fungus complex of cotton is that produced by the root-knot nematode *M. incognita* and the fungus *Fusarium oxysporum* Schlect. f. sp. *vasinfectum* (Atk.) Snyd. and Hans. By itself, *F. oxysporum* causes a severe vascular wilt from which plants usually die. The synergistic effect of root-knot nematodes on *Fusarium* wilt development in cotton was first noted almost 100 years ago, and for many years nematodes have complicated cotton breeding programs for *Fusarium* wilt resistance (Smith and Taylor, 1941). *Fusarium* wilt-resistant cotton varieties developed in sandy root-knot infested coastal plains soils always possessed root-knot resistance, but wilt-resistant varieties developed in other areas were frequently root-knot susceptible. *Fusarium* wilt resistance has now been incorporated into many cotton varieties, but where root-knot nematodes are present even the best wilt-resistant cottons succumb unless they are root-knot nematode resistant as well.

The synergistic effect of root-knot nematodes on *Fusarium* wilt of cotton has been demonstrated with certainty in many laboratories and greenhouse experiments conducted under controlled conditions. A study compared various combinations of five logarithmically expanded inoculum levels of *Fusarium* and root-knot larvae that showed a very strong correlation between root-knot nematode population level and *Fusarium* wilt incidence. As few as 50 nematodes were sufficient to consistently induce wilt at a very low level of *Fusarium* inoculum (650 propagules per plant) (Garber et al., 1979).

G. *Verticillium* Wilt

Verticillium wilt is a vascular disease of cotton and other plants that is somewhat similar to *Fusarium* wilt, although symptoms usually appear later in the growing season. The causal organism, *Verticillium dahliae* Kleb., is one of the most widely distributed and destructive pathogens in agricultural soils. It is found throughout the world and on a national basis causes annual losses to cotton yields of about 3%. Although nematodes, especially root-knot and *Pratylenchus* spp., have been shown to be quite important in the development and severity of *Verticillium* wilt on some crop plants, very little information is available on cotton. However, cotton varieties developed for resistance to the root-knot *Fusarium* complex frequently possess some degree of *Verticillium* resistance (Sappenfield, 1963).

H. Cotton Seedling Disease

Cotton seedling diseases cause losses in cotton yields each year. Symptoms include pregermination seed decay, seedling root rot, preemergence damping off, and postemergence damping off caused by partial or complete girdling of the seedling at or near the soil surface (also referred to as soreshin). The causes of cotton seedling disease include a diversity of pathogenic nematodes and fungi and the dominant species involved vary with geographic locality. Some nematodes play a minor role in disease development, and others, such as root-knot, play a very important role.

Where *Rhizoctonia solani* Kuehn is the predominant fungal pathogen causing soreshin, the presence of the cotton root-knot nematode significantly increases disease incidence. Preplant fumigation of *R. solani* root-knot-infested cotton fields with ethylene dibromide (EDB) markedly increased seeding survival and cotton yields (Reynolds and Hanson, 1957). Greenhouse and field experiments further examined the effects of differential treatment for seedling disease in root-knot-infested soil by using the fungicide pentachloronitrobenzene (PCNB) and the nematicide, 1,2-dibromo-3-chloropropane (DBCP). Fumigation for nematodes with DBCP reduced seedling disease caused by *R. solani* and *Thielaviopsis basicola* (Berk. and Br.) Ferr. to a level comparable to that achieved with fungicide (White, 1962). Root-knot nematodes increase the duration of susceptibility of cotton plants to soil-borne pathogens.

For many years it was observed that cotton seedling disease was more common in lighter soils, especially during cool, wet springs. Disease severity varied with the inoculum level of each pathogen of *R. solani* and *M. incognita* (Arndt and Christie, 1937). The optimum temperature for disease development was found to lie between 18 and 21°C, which confirmed observations in the field. Where only *R. solani* and *M. incognita* were present, increased soil particle size reduced seedling weight. Moreover, hypocotyl lesions were increased by an increase in soil particle size when *R. solani* was present and root galling was increased by particle size when *M. incognita* was present. When both organisms were present, increased particle size synergistically increased lesions and significantly decreased seedling survival (Carter, 1975a, b).

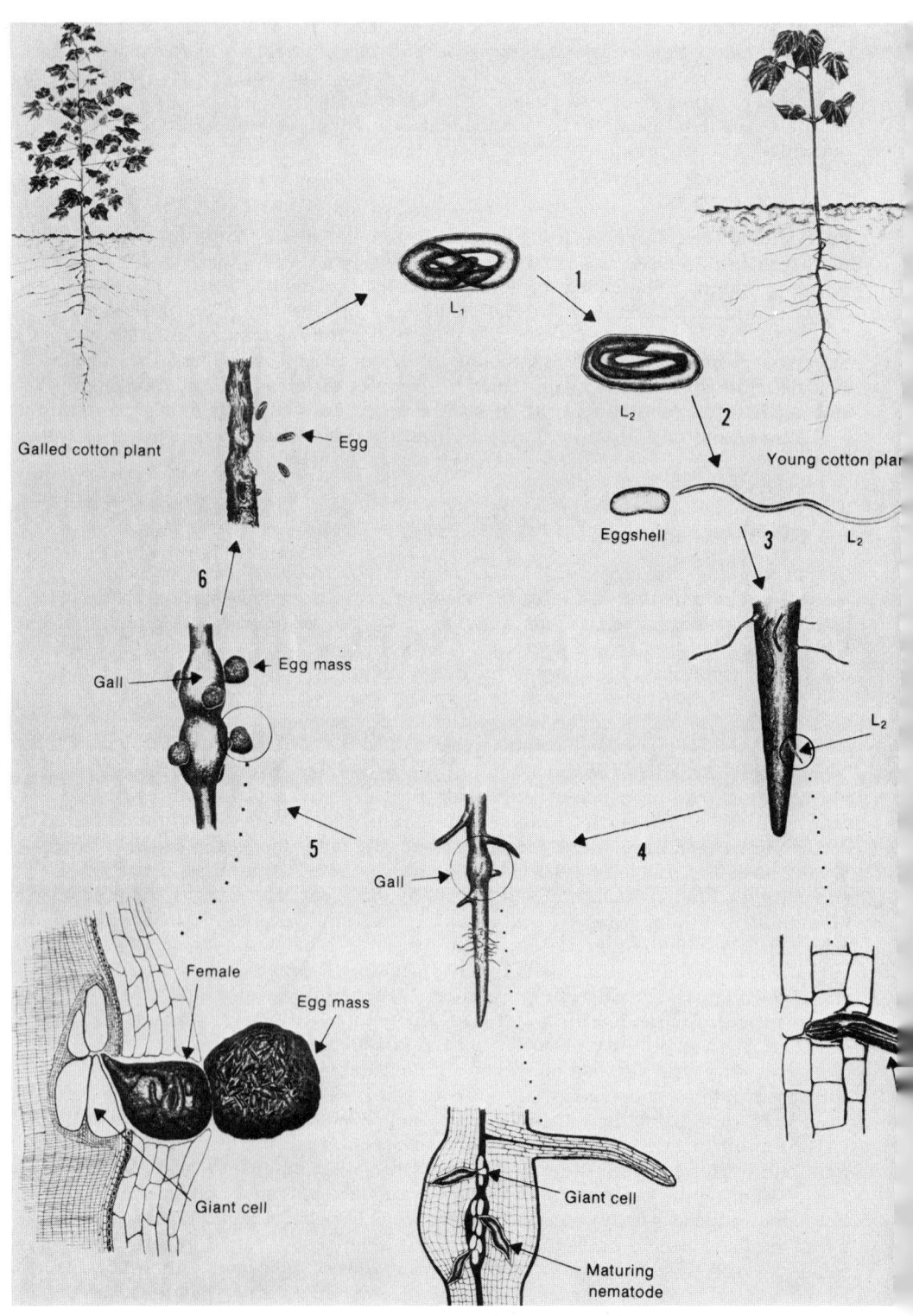

Galled cotton plant

Young cotton plant

Egg

L₁

L₂

Eggshell

L₂

Egg mass

Gall

L₂

Gall

Female

Egg mass

Giant cell

Giant cell

Maturing nematode

I. Mechanisms

Mechanisms involved in the synergistic injury to plants from nematode-fungal interaction have been investigated. Mechanical injury to roots by nematodes may provide fungi with a portal of entry. This is very likely to occur with endomigratory nematodes, such as lesion, which move into, through, and out of roots and destroy many cortical cells while feeding. Increased concentrations of plant exudates in the rhizoplane as a result of mechanical injury may provide increased nutrients for soil-borne fungi and may provide stimuli for the growth of fungal mycelia toward roots via chemotrophisms. The pronounced effects of root-knot nematodes, reniform nematodes, and other sedentary endoparasitic species are somewhat more complex.

There is evidence that cotton roots are physiologically predisposed by root-knot nematodes to fungal attack. Root-knot nematodes cause considerable hypertrophy and hyperplasia in the feeding area and induce the formation of syncytial nurse or giant cells, which are morphologically and physiologically distinct from all cells of normal host tissue. In examining the histopathology of cotton roots concomitantly infected by root-knot and *Fusarium*, abundant fungal growth in the root-knot feeding site region was observed. *Fusarium* incidence was much greater when cotton plants were inoculated with root-knot two or four weeks before inoculating with *Fusarium*. This effect is explained in terms of physiological predisposition of the plants by the nematodes. The information available to date strongly underlines the importance of nematode-fungus interactions in the development of plant disease.

J. Life Cycle

Figure 1 shows the general life cycle of the root-knot nematode. Eggs of *Meloidogyne* are elongate ovate bodies about 77×32 μm, usually contained in a gelatinous matrix binding them together in an egg or egg mass sac. As many as 1000 eggs in an egg mass may be produced by a single female. An individual egg passes through five stages in its development from the unsegmented egg to the adult female or male. Molting occurs four times during this process. When deposited an egg is filled with undifferentiated granular material that often contains a visible nucleus. Embryonic development progresses through repeated cell divisions until a larva is formed. Larvae move about in the flexible egg shell, emerging through a hole or slit created by repeated thrusting with the stylet. The first molt has occurred before emergence from the egg shell.

Second-stage or preparasitic larvae can be found during their migration through the soil in search of a host plant. Eventually they find and penetrate a root of a suitable host. Within 48 hours after penetration they assume a final feeding position in or near the vascular cylinder, and giant cells are developed. Second-stage larvae remain viable in the soil for a few weeks at summer temperatures. When hatched in the fall the second-stage larvae can overwinter and complete the life cycle the next spring. Another

Fig. 1 Life cycle of the root-knot nematode (*M. incognita*). (1) Larva (L_1) within the egg molts to parasitic larvae (L_2). (2) Larva hatches from the egg, and (3) seeks out root and becomes parasitic. (4) Gall formation of the root induced by nematode penetration. (5) Mature females with egg masses deposited in gelatinous matrix. (6) Eggs released into soil from the decomposing egg mass.

molt resulting in the third-stage occurs about one week after entering a root. As the nematode develops further and approaches sexual maturity, the size and metabolic activity of the giant cell also increases. By day 12 after penetration, the giant cell is typically thick walled, metabolically very active, and supporting a nearly mature female nematode. With constant feeding the female continues to grow into a pearly-white pear-shaped adult oriented in the gall with its posterior near the periphery of the gall. The female nematode secretes a gelatinous matrix as eggs are deposited to protect the eggs from dehydration as well as from predators. The time between entrance and egg deposition is 19-22 days.

During this third stage, the genital primordium develops to an elongated mass of cells forming a "clear" area in the posterior third of the body. At this point sexes can be differentiated. With further development, females undergo the final molt and become flask shaped (Triantaphyllou and Hirschmann, 1960). Females inside roots live and reproduce for extended periods of three months or more.

In males, the fourth-stage larvae complete the final molt but remain coiled within the second-stage cuticle. After emerging from the second- and third-stage cuticles, the elongate male, complete with stylet, esophagus with median bulb, spicules, and sperm in the testes, may migrate toward a female. Mating may occur but it not essential to reproduction since parthenogenesis is normal in this genus (Triantaphyllou, 1970).

Meloidogyne eggs and larvae in the soil are sensitive to environmental extremes, the optimum temperature being 25-30°C for hatching, mobility, growth, and reproduction (Wallace, 1964). Activity is limited above 40°C, or below 5°C. At optimum temperature, a life cycle from egg to egg is completed in about 30 days.

Moisture is another important factor in hatching and survival of nematodes. As soil dries at the beginning of drought seasons, *Meloidogyne* eggs are subject to osmotic stresses. Hatching ceases, but development in the egg continues, so all live eggs soon contain second-stage larvae. If eggs become too dry the larvae die, but if they survive until moisture is available, they hatch and infect plants (Dropkin et al., 1958).

III. RENIFORM NEMATODE

A. History

The genus *Rotylenchulus* was established by Linford and Oliveira (1940), with the description of *R. reniformis* from cowpea roots grown in soil from a pineapple field on the Island of Oahu, Hawaii. In their original description, they state that the nematode is probably the same one observed in 1931, also from cowpea roots. Smith (1940) found cotton in Georgia heavily infested with the new species of nematode that was being described from Hawaii by Linford. The next year, Smith and Taylor (1941) reported *R. reniformis* on roots of both cotton and cowpea in Baton Rouge, Louisiana. Since that time the reniform nematode has been reported on numerous crops causing economic crop losses in many parts of the world.

B. Distribution

The reniform nematode is found primarily in tropical and subtropical areas of the world. Investigators (Holdeman et al., 1977) reported that more than 38 countries have found the nematode infesting a wide range of crops. The most concentrated occurrence outside the tropics appears to be in the United States, where the nematode is found in all coastal states from South Carolina to Texas and California. Smith and Taylor (1941) found the nematode in Baton Rouge, Louisiana, about a year after Smith (1940) first reported it from Georgia. Steiner (1949) found the reniform nematode at Quincy, Florida, on tomatoes and coffee weed. In Texas, Norton (1959) reported the nematode present in the Lower Rio Grande Valley, and the same year Minton and Hopper (1959) found cotton in eastern Alabama infested with *R. reniformis*. Fassuliotis and Rau (1967) examined soil samples from soybean fields near Sycanne, South Carolina, and found them to be infested with the reniform nematode. In San Bernardino County, California, researchers (Holdeman et al., 1977) found the nematode confined to ornamental plants. In 1981, Fox (personal communication) reported the reniform nematode in Mississippi from fields previously cropped in cotton. The wide distribution of the nematode in tropical and subtropical areas of the world make it a potential threat to cotton wherever it is grown.

C. Associated Symptoms

The reniform nematode injury to cotton has been described by Jones et al. (1959); Birchfield and Jones (1961); Brodie (1963); Minton and Hopper (1959); and Heald and Heilman (1971). In summary, these workers found that injury to cotton by the reniform nematode is evident very early in seedling cotton. By the third or fourth leaf stage, the seedling ceases rapid growth and takes on a light-green chlorotic appearance. With heavy infestation the margin of the leaf may take on a purple cast. Many plants are lost at this stage of injury. Other plants continue to grow slowly to maturity, maintaining their off color. These symptoms are often confused with fertilizer deficiencies, diseases, salt injury, and other soil problems. Roots of infested plants are smaller than noninfested plants with some loss of small secondary roots. The depleted root system results in fewer blooms, smaller leaves, reduction in boll size and yield reduction up to 60%. In fields that have been infested for several years, damage to cotton will be uniform, whereas new field infestations will have depressed or uneven growth in areas of high nematode populations.

D. Mode of Infection

Birchfield (1962) studied the mode of infection of the reniform nematode on cotton. He found that epidermal cells were destroyed as the fourth-stage female entered the root, resulting in a slight browning and necrosis of surrounding cells. Cortical parenchyma cells were destroyed as the nematode extended its head through the endodermis and pericycle and began to feed in the phloem. Cells to either side of the feeding site stained a darker color, indicating a toxin had been produced by the nematode and translocated in the cells. The nematode showed no specificity for age of most tissue, although young succulent roots were frequently infested near root tips. Necrosis of phloem and parenchyma collapse ultimately resulted in severe root pruning of seedling roots and consequent dwarfing. The reniform nematode has also been found associated with diseases of cotton. Neal (1953, 1954) probably was among the first to recognize the association of the reniform nematode and *Fusarium* wilt of cotton. He reported on the high incidence of *Fusarium* wilt in a susceptible variety of cotton that was dependent upon the presence of a heavy population of the reniform nematode. In greenhouse studies he concluded that the reniform nematode enhanced the wilt index of susceptible cotton varieties but was not able to alter resistance in wilt-resistant cotton varieties as did the root-knot nematode. Brodie and Cooper (1964) found that when reniform nematode populations were increased to 20,000 per 500 g of soil in combination with *Rhizoctonia solani*, susceptibility to the seedling disease was greater than in the absence of the nematode.

E. Life Cycle

The life cycle of the reniform nematode was investigated by Sivakumar and Seshadri (1971). Their work showed that it took approximately four days from the time the egg was laid to formation of the first stage larva in the egg. Twenty-four hours later the larvae molted within the egg shell. Six to seven days after the egg was laid the second-stage larva forces its way out of the egg. The second-, third-, and fourth-stage larva do not feed, and sex can be determined at the third-stage. After the fourth molt (Fig. 2), the immature female becomes infective and penetrates the root by inserting the anterior one-third of the body into the tissue. Within 24 hours the female begins to enlarge on the ventral side around the vulva region. The female continues to swell and becomes reniform in shape by the fourth or fifth day after infection. The males develop to maturity, never feeding, but are usually found in the proximity of the female feeding site or coiled around her body (Fig. 3). The female secretes a gelatinous matrix that envelopes the body, providing protection and a medium in which she can deposit eggs (Fig. 4).

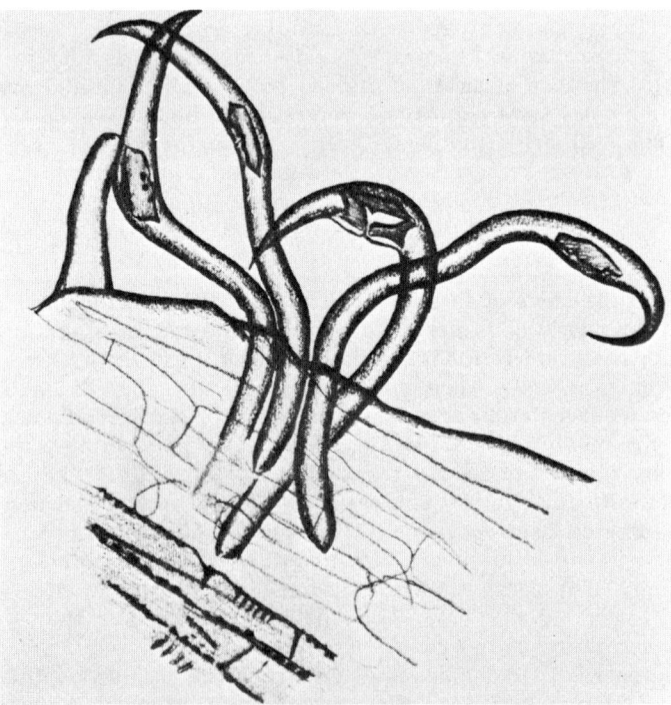

Fig. 2 Infective immature female reniform nematodes (*R. reniformis*) penetrate root system and feed in vascular system.

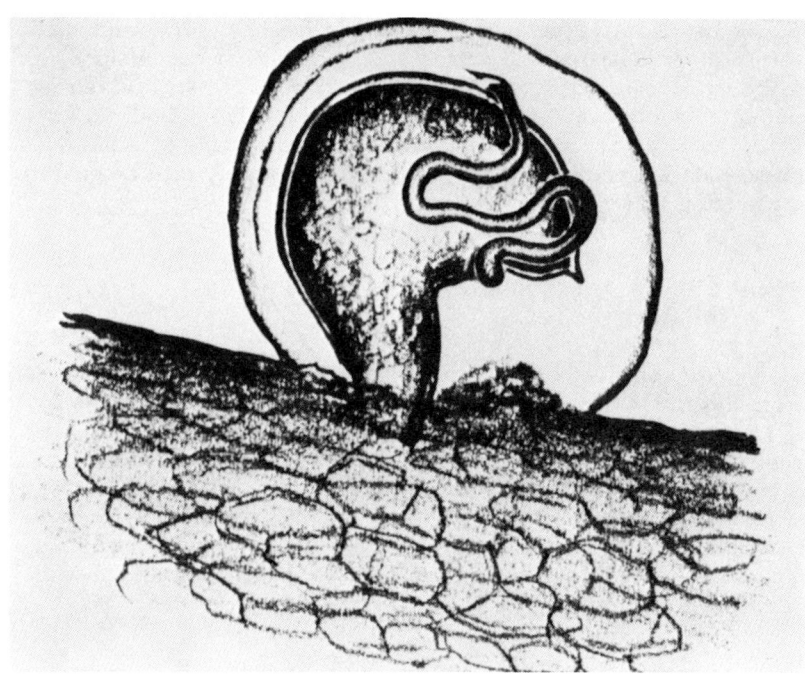

Fig. 3 Maturing female reniform nematode (*R. reniformis*) with reniform male coiled around her body.

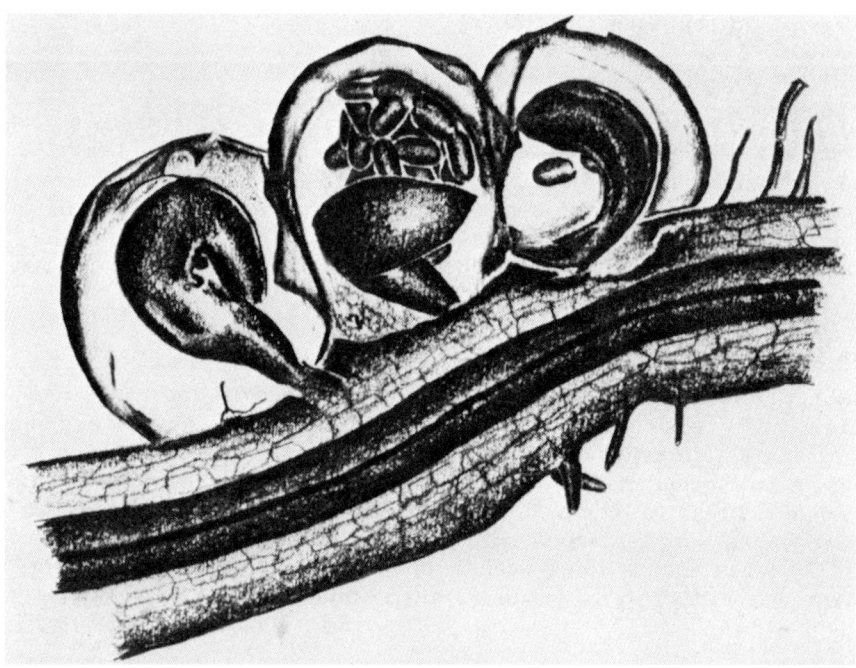

Fig. 4 Mature female reniform nematode (*R. reniformis*) with eggs deposited in gelatinous matrix.

The life cycle took 24-29 days from egg to egg. Birchfield (1962) studied the life cycle of the reniform nematode on cotton and made similar conclusions, except he found the cycle was completed in 17-23 days. Linford and Oliveira (1940) also conducted a life cycle study with egg to egg taking about 25 days. Temperatures and host can account for differences in total days of a life cycle. The reniform nematode is generally considered to be bisexual, with a sex ratio of 1:1 and reproducing by cross fertilization.

IV. STING NEMATODE

A. History

The genus *Belonolaimus* was established by Steiner (1949) with the description of *B. gracilis* on slash pine at Ocala, Florida, and longleaf pine seedlings in other parts of Florida. Rau (1958) described *B. longicaudatus*, which is believed to be the primary species observed by Steiner (1949) and not *B. gracilis*, which is now known to be limited to a very small area of Florida. Therefore, the first nematode reported on cotton from Virginia by Owens (1951) was most likely *B. longicaudatus*. Holdeman and Graham (1954) reported the presence of the sting nematode in several cotton fields in South Carolina associated with *Fusarium* wilt of cotton.

B. Distribution

The nematode has been reported only in the United States and appears to be limited by soil types. Primarily the sting nematode is found in the Southeastern Coastal Plain, typically in light sandy soil (Holdeman, 1955). However, it has also been reported as far north as Connecticut and New Jersey and as far west as Texas and Louisiana.

C. Associated Symptoms

The sting nematode is one of the most devastating nematodes that feeds on cotton. Plants are severely stunted and chlorotic and often die due to heavy population pressure. Infested areas in the field may vary in size, but it is not difficult to define the margin because of stunted plants. Injury on the roots of cotton plants (Fig. 5) appear as discolored and shrunken minute lesions (Graham and Holdeman, 1953). When the new root tip is attacked it may often die or appear to break off.

D. Mode of Infection

The sting nematode is an ectoparasite and feeds on epidermal and cortical cells by puncturing the cells and removing the cell contents. This nematode is one of the largest plant parasites and is capable of causing extreme damage as it feeds along a root. Holdeman and Graham (1954) found that the sting nematode was able to break resistance of cotton to *Fusarium* wilt by its feeding action; therefore, in the presence of both pests a total crop failure is almost certain. The importance of this nematode in cotton has decreased as the cotton industry has shifted to the western United States.

E. Life Cycle

A detailed study of the life cycle has not been reported. Larvae are not greatly different from adults, and all stages appear to be able to feed. Males are present but much shorter than the adult female.

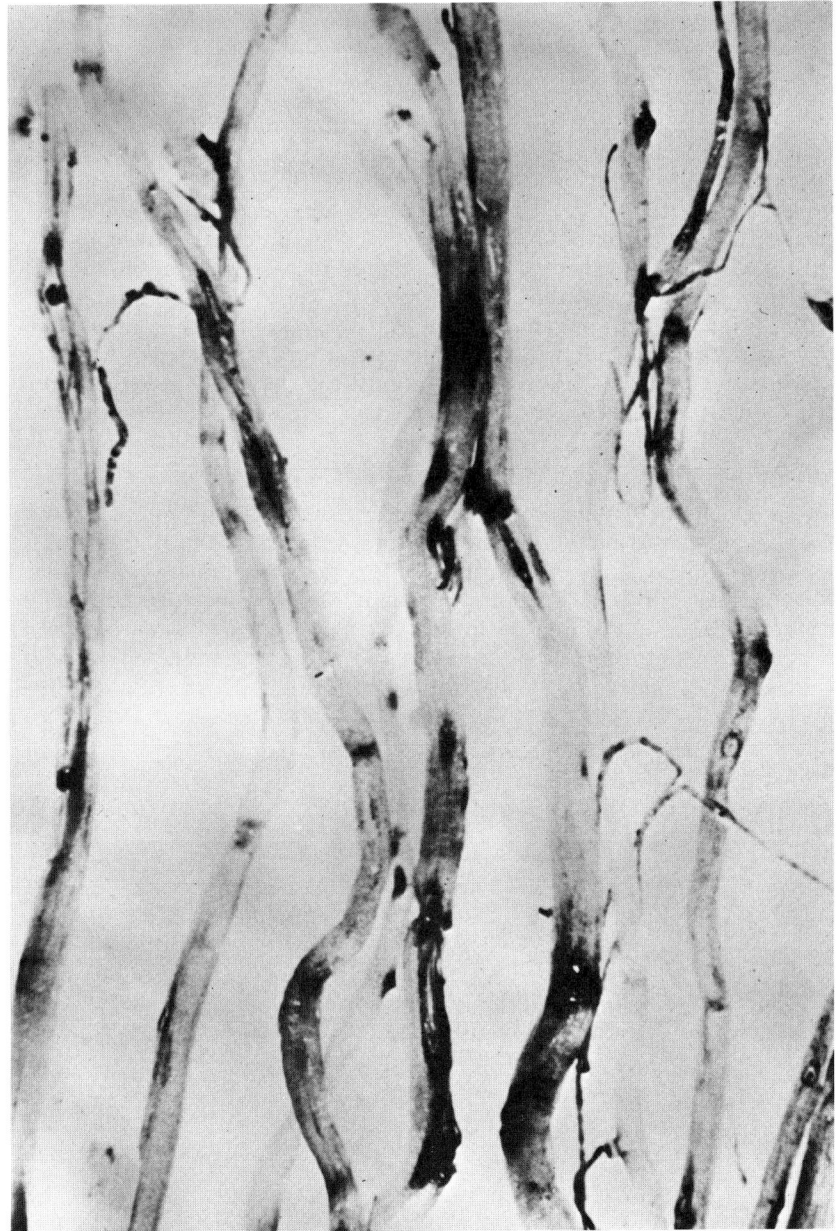

Fig. 5 Discolored and shrunken lesions on cotton root caused by feeding of the sting nematode (*B. longicaudatus*).

V. LANCE NEMATODE

A. History

Two species of the lance nematode, *Hoplolaimus columbus* Sher and *H. galeatus* (Thorne) Sher, are known to cause economic loss to cotton. *H. columbus* (the Columbia lance nematode) was first reported from Richland County, South Carolina, in the early 1950s (Lewis et al., 1976) and since then has been detected in more than 30,400 ha of prime agricultural land in that state. Krusberg and Sasser (1956) reported *H. galeatus* (then *H. coronatus*) to cause severe stunting of cotton in North Carolina. Later it was reported in several locations in the southeastern United States Coastal Plains. *Hoplolaimus* spp. are competitive in soils. Other genera, such as *Meloidogyne* spp. and *Pratylenchus* spp., were present only in small numbers where *Hoplolaimus* was abundant, indicating that when *Hoplolaimus* was the dominant species, other plant-parasitic nematodes were unable to compete (Krusberg and Sasser, 1956).

B. Distribution

The lance nematode *H. columbus* was described in 1963. Since that time it has been reported to be economically important in the Atlantic Coast states from North Carolina to Georgia. The distribution of *H. galeatus* is more extensive. This nematode is also found causing crop losses in the southeastern United States. Furthermore, it is indigenous in native prairie soils of Kansas (Orr and Dickerson, 1967), in other western states, as well as in several countries abroad.

Fig. 6 Injury to cotton caused by heavy infestation of the lance nematode (*H. columbus*).

C. Associated Symptoms

Heavy infestations of *H. columbus* in soils cause extensive stunting of cotton
(Fig. 6) followed by chlorosis. Brownish yellow discoloration is observed in
the epidermal cells at the point of nematode entry and to a depth of two to
five cells along the path of nematode movement. Similar symptoms have been
observed for *H. galeatus*. Almost complete defoliation of cotton has been reported in fields heavily infested with lance nematode following drought conditions (Bird et al., 1974; Fassuliotis et al., 1968).

D. Mode of Infection

Lance nematodes are both ecto- and endoparasitic. They migrate freely from
the root while feeding ectoparasitically. When feeding endoparasitically, they
are found embedded two or three cell layers inside the epidermis, lying parallel with the longitudinal root axis, and with the anterior end extended into the
conductive tissue. Cavities are formed, and cell walls appear thicker where
nematodes fed. Endodermal injury is similar to that of the cortex but less extensive. In some cells of the phloem parenchyma, abnormal cell division
occurs, forming cork tissue. Occasionally xylem elements are punctured and
tyloses are formed. Tyloses cause plugging of xylem elements, which is partially responsible for the stunting of plants (Krusberg and Sasser, 1956).

Cortical root tissue infested by *H. columbus* contains cells that are abnormally dark and contain enlarged nuclei and dense granular cytoplasm.
Brownish-yellow discoloration continues along the path of the nematode through
the cortex, and eggs are deposited as nematodes move through the tissue.

E. Life Cycle

No detailed study has been made of the life cycle of lance nematodes. Since
eggs are frequently laid within root tissue, egg hatch very likely occurs within the root. Eggs are also deposited in the soil as females migrate between
feeding sites. Otherwise the life cycle is probably similar to that of most
plant parasitic genera of the Tylenchida. The sex ratio is approximately equal
for males and females.

VI. CONTROL

A. Chemical

Successful chemical control for nematodes began in the mid-1940s, when a
group of chemicals known as halogenated hydrocarbons were shown to be excellent nematicides. Included among these chemicals were 1,3-D (1,3-dichloro-
propene) and related hydrocarbons and EDB (ethylenedibromide) (Fig. 7).

Fumigants are usually injected 6-10 in. deep into the soil, and nematodes
are killed as the chemical diffuses through the soil. Several factors are important in soil fumigation: soil temperature, soil moisture, soil type, compaction, and sorption. Soil temperature and moisture suitable for planting is optimum for soil fumigation. Coarse-textured soils low in organic matter are
more favorable for fumigation because gas diffuses more readily through the
soil profile with less sorption to soil particles and organic matter.

The objective of soil fumigation in cotton production is to reduce the nematode population so that seedlings may become established. Fumigants have
a relatively short residual activity, and their ovicidal properties are moderate

Fig. 7 Chemical control of the root-knot nematode (*M. incognita*). Small plants are stunted from nematode infection. Larger surrounding plants are growing in soil fumigated for nematode control.

to weak. However, if the seedling tap root is protected from nematode infestation for a few weeks, plants produce satisfactorily in spite of later nematode infestation on secondary roots.

Nonfumigant and systemic compounds are the newest types of nematicides. Although these chemicals are usually less effective than soil fumigants for nematode control (Orr and Brashears, 1977), they have some advantage in ease of application and handling. Systemic nematicides are marketed as granular formulations that require soil incorporation and moisture to release the chemical. Nematodes are killed by contact with the chemical in the soil or by feeding on root tissue above the point of toxicant absorption by the plant. Other postulated modes of action are repellancy, which prevents feeding, or by interference with reproduction.

B. Crop Rotation

Crop rotation has long been an accepted practice among agriculturalists. Rotations are practiced for many reasons, including improved fertility, nitrogen fixation, soil tilth, added organic matter, and control of nematodes and disease. The management decisions for rotations are based on economics. Effective rotation decisions must include crops that will accomplish the objectives for rotation and contribute to the financial well-being of the farming enterprise. Because plant-parasitic nematodes are obligate parasites with limited mobility, rotating with nonhost crops reduces nematode populations through starvation of larvae and by limiting nematode reproduction. The selection of the rotational crop becomes important because of the wide host range of most nematodes and the necessity for the alternate crop to be resistant or immune.

Examples of crops used in rotation with cotton for control of the cotton root-knot nematode include small grains, peanuts, or grasses. Small grains, such as wheat or barley, are resistant to *M. incognita*. Similarly, peanuts are immune to *M. incognita*, and cotton is resistant to all root-knot species except races 3 and 4 of *M. incognita*. Small grains and other grass crops have the additional advantage of reducing populations of fungal pathogens often associated with root-knot nematodes.

An important cultural requirement is weed control, in both the cotton crop and the rotation crop. Many weeds are good hosts for nematodes. Their presence in the field can maintain populations, thus reducing or preventing the success of the rotation.

C. Host Plant Resistance

If any characteristic of the plant or any interaction between host and parasite retards entry or reproduction of the parasite and injury to the host is slight or does not occur, the plant is termed resistant. All gradations between susceptible and resistant hosts are found. Three general resistance responses in cotton to nematode parasitism have been described: (1) necrosis, (2) lack of galling; and (3) reduced larval penetration (Dropkin and Webb, 1967). Far fewer giant cells are initiated in resistant roots. Larvae apparently "probe" cells, comparable to those that would become giant cells in a susceptible plant, but for some reason they find them unacceptable. These larvae remain transitory until they perish by starvation, or possibly by antibiosis.

A few of the larvae that enter the root of a resistant cotton initiate giant cells. Nuclear and nucleolar hypertrophy, and perhaps some karyokinesis occur. However, only slight cell hypertrophy takes place and the giant cell usually collapses and disintegrates prior to thickening of the giant cell wall. This leaves an "empty gall" with no discernible giant cell or nematode or with only fragments of them indicating their presence. Only a few of the larvae that manage to initiate giant cells live sufficiently long to complete their life cycle. When they do complete the life cycle, the giant cells and the nematode develop similarly to those in susceptible cotton (McClure et al., 1974).

The factor responsible for preventing or inhibiting giant cell development in resistant plants has been elusive. Quantitative and qualitative studies of preinfectional levels of free sugars, total lipids, phenols, fatty acids, gossypol, and so on, would indicate these metabolites are not responsible for resistance. Postinfection changes in the concentrations of secondary metabolites, however, have been implicated in the resistance of cotton to root-knot nematode. Preinfection concentrations of terpenoid aldehydes (gossypol and gossypol-related compounds) vary among cotton varieties but do not correlate with the relative resistance of the varieties to the root-knot nematode. However, the ability to synthesize additional terpenoid aldehydes in response to infection does correlate with the relative resistance of certain cotton varieties to the nematode. These terpenoids are toxic to root-knot nematodes in concentrations below that found in host cells of such cultivars (Veech, 1979). The first superior root-knot nematode *Fusarium* wilt-tolerant variety was Auburn 56, developed from the cross of Cook 307 × Coker 100 wilt. Later studies have shown that Bayou strains are superior to Auburn 56 in preventing the buildup of nematode populations during the growing season (Jones and Birchfield, 1967). The recent release of Auburn 623 from Alabama represents a high level of tolerance of *Fusarium* wilt root-knot disease and was developed from a cross of Clevewilt-6 × Louisiana Mexico wild (Birchfield and Jones, 1961).

Genes conditioning tolerance to wilt and nematodes are identified by strain, variety, or species in which they are found (Hare, 1965). In Egyptian cotton (*Gossypium barbadense*) tolerance is conditioned by one dominant and one or more minor genes. A dominant gene imparting resistance to the wilt-nematode complex was found in *G. herbaceum*. Two complementary dominant genes and a third gene expressing an inhibitory effect were found in *G. arboreum* (Kelker et al., 1947). The cultivar Cook 307 carries one major dominant and several modifying genes for wilt tolerance and additional genes for root-knot nematode tolerance. Seabrook, a variety of Sea Island, has two dominant genes that are additive in conferring a high level of wilt tolerance (Smith and Dick, 1960). Root-knot tolerance in Clevewilt-6 is a quantitative character with relatively few genes involved. A high level of root-knot tolerance is found in *G. barbadense* var *darwinnii*. At least two recessive genes are present in this source tolerance (Wiles, 1957). Although there has been an appreciable amount of research toward developing root-knot nematode resistance in cotton, there has been very little work to develop resistance to other nematodes that cause injury to cotton.

REFERENCES

Arndt, C. H., and Christie, J. R. (1937). The comparative role of certain nematodes and fungi in the etiology of damping-off, or soreshin, of cotton. *Phytopathology 27*: 569-572.

Birchfield, W. (1962). Host-parasite relations of *Rotylenchulus reniformis* on *Gossypium hirsutum*. *Phytopathology 52*: 862-865.

Birchfield, W., and Jones, J. E. (1961). Distribution of the reniform nematode in relation to crop failure of cotton in Louisiana. *Plant Dis. Rep. 45*: 671-673.

Bird, A. F. (1974). Plant response to root-knot nematodes. *Annu. Rev. Phytopathol. 12*: 69-85.

Bird, G. W., Brooks, O. L., Perry, C. E., Futral, J. G., Canerday, T. D., and Boswell, F. C. (1974). Influence of subsoiling and soil fumigation on the cotton stunt disease complex, *Hoplolaimus columbus* and *Meloidogyne incognita*. *Plant Dis. Rep. 58*: 541-544.

Brodie, B. B. (1963). Pathogenicity of certain parasitic nematodes on cotton seedlings and their relationship to post-emergence damping-off by *Rhizoctonia solani* Kuhn and *Phythium debaryanum* Hesse (abstract). *Dissertation 23*: 4491-4492.

Brodie, B. B., and Cooper, W. E. (1964). Relation of plant parasitic nematodes to post-emergence damping-off of cotton. *Phytopathology 54*: 1023-1027.

Carter, W. W. (1975a). Effects of soil temperatures and inoculum levels of *Meloidogyne incognita* and *Rhizoctonia solani* on seedling disease of cotton. *J. Nematol. 7*: 229-233.

Carter, W. W. (1975b). Effects of soil texture on the interaction between *Rhizoctonia solani* and *Meloidogyne incognita* on cotton seedlings. *J. Nematol. 7*: 234-236.

Chitwood, B. G. (1949). Root-knot nematodes. I. A revision of the genus *Meloidogyne*, Goeldi, 1887. *Proc. Helm. Soc. Wash. 16*: 90-104.

Dropkin, V. H., Martin, G. C., and Johnson, R. W. (1958). Effect of osmotic concentrations on hatching of some plant parasitic nematodes. *Nematologica 3*: 115-126.

Dropkin, V. H., and Webb, R. E. (1967). Resistance of axenic tomato seedlings to *Meloidogyne incognita acrita* and to *M. hapla. Phytopathology* 57: 584-587.

Fassuliotis, G., and Rau, G. J. (1967). The reniform nematode in South Carolina. *Plant Dis. Rep. 51:* 557.

Fassuliotis, G., Rau, G. J., and Smith, F. H. (1968). *Hoplolaimus columbus,* a nematode parasite associated with cotton and soybeans in South Carolina. *Plant Dis. Rep. 52:* 571-572.

Garber, R. H., Jorgenson, E. C., Smith, S., and Hyer, A. H. (1979). Interaction of population levels of *Fusarium oxysporum* f. sp. *vasinfectum* and *Meloidogyne incognita* on cotton. *J. Nematol. 11:* 133-137.

Graham, T. W., and Holdeman, Q. L. (1953). The sting nematode, *Belonolaimus gracilis* Steiner: A parasite of cotton and other crops in South Carolina. *Phytopathology 43:* 434-439.

Hare, W. W. (1965). The inheritance of resistance of plants to nematodes. *Phytopathology 55:* 1162-1167.

Heald, C. M., and Heilman, M. D. (1971). Interaction of *Rotylenchulus reniformis,* soil salinity and cotton. *J. Nematol. 3:* 179-182.

Holdeman, Q. L. (1955). The present known distribution of the sting nematode, *Belonolaimus gracilis* in the coastal plains of the southeastern United States. *Plant Dis. Rep. 39:* 5-8.

Holdeman, Q. L., Cordas, D., Watson, T., Matsumoto, R., and Siddiqui, I. (1977). Fact finding study on the reniform nematode *Rotylenchulus reniformis. State of Calif. Dept. Food and Agric. Div. Plant Ind. Study Team 1973-1974.*

Holdeman, Q. L., and Graham. T. W. (1954). Effects of the sting nematode on expression of *Fusarium* wilt of cotton. *Phytopathology 44:* 683-685.

Jones, J. E., Newsom, L. D., and Finley, E. L. (1959). Effect of the reniform nematode on yield, plant characters, and fiber properties of upland cotton. *Agron. J. 51:* 353-356.

Jones, J. E., and Birchfield, W. (1967). Resistance of the experimental cotton variety, Bayou and related strains to root-knot nematodes and *Fusarium* wilt. *Phytopathology 57:* 1327-1331.

Kelker, S. G., Chawdhari, G. S., and Hiremath, N. B. (1947). Inheritance of *Fusarium* resistance in Indian cotton. *Proc. Third Conf. Cotton. Growing Publ. India ICCC* Bombay, pp. 125-162.

Krusberg, L. R., and Sasser, J. N. (1956). Host-parasite relationships of the lance nematode in cotton roots. *Phytopathology 46:* 505-510.

Lewis, S. A., Smith, F. H., and Powell, W. M. (1976). Host-parasite relationships of *Hoplolaimus columbus* on cotton and soybeans. *J. Nematol. 8:* 141-145.

Linford, M. B., and Oliveira, J. M. (1940). *Rotylenchulus reniformis,* Nov. gen. N. sp., a nematode parasite of roots. *Proc. Helm. Soc. Wash. 7:* 35-42.

McClure, M. A., Ellis, K. C., and Nigh, E. L. (1974). Post-infection development and histopathology of *Meloidogyne incognita* in resistant cotton. *J. Nematol. 6:* 21-26.

Minton, N. A., and Hopper, B. E. (1959). The reniform and sting nematodes in Alabama. *Plant Dis. Rep. 43:* 47.

Neal, D. C. (1953). *Fusarium* wilt resistance and fiber properties of some new cotton hybrids and selections (abstract). *Phytopathology 43:* 292.

Neal, D. C. (1954). The reniform nematode and its relationship to the incidence of *Fusarium* wilt of cotton at Baton Rouge, Louisiana. *Phytopathology 44:* 447-450.

Norton, D. C. (1959). Plant parasitic nematodes in Texas. *Tex. Agric. Exp.*

Sta. Misc. Pub. 321.

Norton, D. C. (1978). *Ecology of Plant-Parasitic Nematodes.* John Wiley and Sons, New York, 268 pp.

Orr, C. C., and Brashears, A. D. (1977). Aldicarb and DBCP for root-knot nematode control of cotton. *Plant Dis. Rep. 62:* 623-624.

Orr, C. C., and Dickerson, O. J. (1967). Nematodes in true prairie soils of Kansas. *Trans. Kan. Acad. Sci. 69:* 317-334.

Owens, J. V. (1951). The pathological effects of *Belonolaimus gracilis* on peanuts in Virginia (abstract). *Phytopathology 41:* 29.

Peacock, F. C. (1956). The reniform nematode in the Gold Coast. *Nematologica 1:* 307-310.

Powell, N. T. (1971). Interactions between nematodes and fungi in disease complexes. *Ann. Rev. Phytopathol. 9:* 253-274.

Rau, G. J. (1958). A new species of sting nematode. *Proc. Helm. Soc. Wash. 25:* 95-98.

Reynolds, H. W., and Hanson, R. G. (1957). *Rhizoctonia* disease of cotton in presence or absence of the cotton root-knot nematode in Arizona. *Phytopathology 47:* 256-261.

Sappenfield, W. P. (1963). *Fusarium* wilt root-knot nematode and *Verticillium* wilt resistance in cotton: Possible relationships and influence on cotton breeding methods. *Crop Sci. 3:* 133-135.

Sivakumar, C. V., and Seshadri, A. R. (1971). Life history of the reniform nematode, *Rotylenchulus reniformis* Linford and Oliveira, 1940. *Indian J. Nematol. 1:* 7-20.

Smith, A. L. (1940). Distribution and relation of meadow nematode, *Pratylenchus pratensis*, to *Fusarium* wilt of cotton in Georgia (abstract). *Phytopathology 30:* 710.

Smith, A. L., and Dick, J. B. (1960). Inheritance of resistance to *Fusarium* wilt in upland and Sea Island cottons as complicated by nematodes under field conditions. *Phytopathology 50:* 44-48.

Smith, A. L., and Taylor, A. L. (1941). Nematode distribution in the 1940 regional cotton-wilt plots (abstract). *Phytopathology 31:* 771.

Steiner, G. (1949). Plant nematodes the grower should know. Proc. Soil Sci. Soc. Flor., 1942, 4-B: 72-117. Reprinted: *Fla. Dept. Agric. Bull. 131.*

Taylor, A. L., and Sasser, J. N. (1978). Biology, identification, and control of root-knot nematodes. *N. Carolina State Univ. Graphics*, 111 pp.

Thorne, G. (1961). *Principles of Nematology.* McGraw-Hill, New York.

Triantaphyllou, A. C. (1970). Cytogenetic aspects of evolution of the family Heteroderidae. *J. Nematology 2:* 26-32.

Triantaphyllou, A. C., and Hirschmann, H. (1960). Post-infestion development of *Meloidogyne incognita* Chitwood, 1949. *Ann. Inst. Phytopathol. Benaki 3:* 1-11.

Veech, J. A. (1979). Histochemical localization and nematoxicity of Terpenoid aldehydes in cotton. *J. Nematology 11:* 240-246.

Wallace, H. R. (1964). *The Biology of Plant Parasitic Nematodes.* St. Martins, New York, pp. 163-191.

White, L. V. (1962). Root-knot and seedling disease complex of cotton. *Plant Dis. Rep. 46:* 501-504.

Wiles, A. B. (1957). Resistance to root-knot nematode in cotton (abstract). *Phytopathology 47:* 37.

Chapter 6
Nematode Parasites of Potato

B. B. Brodie *USDA, Agricultural Research Service, Cornell University, Ithaca, New York*

I. INTRODUCTION

A. The Crop

The potato of commerce, *Solanum tuberosum* L., has its origins in the South American Andes where it was an ancient cultivated crop when the Spaniards arrived in the early sixteenth century (Hawkes, 1978). The Andean potato belongs to the subspecies *S. tuberosum* subsp. *andigena* (Juz. et Buk.) Hawkes, from which subspecies *tuberosum* was derived after it was brought to Europe following the Spanish conquest. The potato first appeared in Europe during the last quarter of the sixteenth century. There is was regarded as a botanical curiosity and was not taken into widespread cultivation until some 200 years later (Hawkes, 1978). Evidence indicates two separate introductions of potato into Europe; the first being about 1570 (Spain) and the second about 1590 (England). From these two introductions, the potato has spread throughout the world.

The exact date potatoes first arrived in North America is questionable. Some suggest that they arrived in 1621 from England via Bermuda and were grown in the colony of Virginia where they were erroneously assumed to be native (Hawkes, 1978). Others believe that they did not arrive until 1719, when they were recorded being grown in New Hampshire by the Scotch Irish who settled there (Plaisted, 1971). The first introduction of South American origin was in 1848, when Goodrich obtained a variety from Bogota, Colombia, which failed to survive (Plaisted, 1971). In 1851, Goodrich obtained some varieties from the Panamanian Counsel that were believed to have come from Chile (Hawkes, 1978). Garnet Chili, a variety arising from this material, made an outstanding contribution to potato breeding in the United States and is in the pedigree of many present day cultivars.

From meager introductions of often-times unadapted plants, the potato has evolved into one of the world's major food crops, occupying over 15 million

hectares that produce in excess of 250 million metric tons annually wi
annual worth of over 1.2 billion dollars in the United States alone (Ur.
States Department of Agriculture, 1980). After first being esteemed a
delicacy and eaten by royalty or the rich, potatoes now figure in the d
diets of millions of people. In the United States, about 20 million metri
are produced annually, of which in excess of 10 million metric tons are c
sumed in human diets.

B. The Nematodes

During its evolution as a basic food crop, the potato obviously gained acceptance by nematode and human alike. A recent bibliography of nematode pests
of potato lists 67 nematode species representing 24 genera that have been reported associated with potato culture (Jensen et al., 1979). Undoubtedly,
many of these species are of little or no importance in potato production, and
others are known to have a major impact on potato yields. The most feared
and obviously the most damaging nematodes of potato, the potato cyst nematodes, are the subject of quarantines in most countries where they occur.
Other important nematodes of potato, such as potato rot, root lesion, root-
knot, and stubby-root nematodes, are distributed worldwide and cause tremendous losses to the potato crop when host and parasite meet under conditions suitable for disease development. Still other nematodes, such as false
root-knot, pseudo stem, and sting nematodes, are less widely distributed but
where they do occur cause losses that are nonetheless real to the potato grower.

In the United States, estimates hold that 10% of the potato crop is lost
annually to nematodes (Committee on Crop Losses, 1971). Because nematodes
parasitize roots and tubers, and there are no concise diagnostic symptoms on
above-ground portions of the plants that signify their attack, much of this
loss goes unrecognized. Consequently, nematode damage to potatoes is often
attributed to other causes. For example, attack by large numbers of nematodes causes unthrifty top growth occasionally accompanied with yellowing of
the foliage resembling symptoms associated with poor root growth that could
be caused by many biotic or abiotic factors. Attack by low numbers of nematodes results in no above-ground symptoms but often reduces tuber yields
and quality.

As the world population increases, there will be a greater demand for
potatoes as a basic food, and land suitable for growing potatoes will become
more scarce. Consequently, potatoes will be grown more frequently on the
best potato land and, because monoculture encourages nematode density increases, damage to potatoes caused by nematodes will increase dramatically.

II. POTATO CYST NEMATODES

A. History

It is generally agreed that potato cyst nematodes originated in the Andean region of South America where they coevolved with their preferred host, the
potato (Mai, 1977). Some 250 years after its introduction into Europe, the
potato had increased in popularity until many peasant farmers had come to
depend on the crop. Then potato blight suddenly appeared in the mid nineteenth century, bringing famine to Ireland. After the potato blight disaster,
many collections of potato tubers were brought to Europe from South America
to breed for resistance to the disease. Apparently, potato cyst nematodes

were brought along with the tuber collections to Europe. They were first
found in 1881 in Germany, about 30 years after the introduction of breeding
material began (Evans and Brodie, 1980). When first discovered, potato cyst
nematodes were thought to be a strain of the sugar beet cyst nematode. In
1923, Wollenweber determined that the potato strain of cyst nematodes was a
different species from the sugar beet strain (Jones, 1970). Potato cyst nematodes became commonly referred to in Europe as the "potato root eelworm."

Potato cyst nematodes were first found in the United States in 1941 on
Long Island, New York, where they apparently had caused field symptoms
since the late 1930s (Mai, 1977). Because of its yellow or golden phase during
development, the nematode became known in the United States as the "golden
nematode" (Chitwood, 1951). It is generally believed that the golden nematode was introduced to Long Island as early as 1920 via military equipment returning from Europe (Evans and Brodie, 1980).

B. The Nematodes

The causal organisms are *Globodera rostochiensis* (Wollenweber 1923) Behrens 1975 and *G. pallida* (Stone 1973) Behrens 1975. For many years,
potato cyst nematodes were thought to consist of a single species, *Heterodera rostochiensis* Wollenweber 1923, denoting the place where they were
first found, Rostock, Germany. When potato cultivars resistant to potato cyst nematodes came into usage, populations of the nematodes were found
that could reproduce on them. Such populations were distinguished by their
ability to reproduce on plants with resistant genes derived from either *S.
tuberosum* subsp. *andigena* or from *S. multidissectum* Hawkes. It was later
noted that some of the populations had a prolonged white or cream color instead of a golden female phase before becoming a brown cyst. This difference
and additional ones in general morphology led to the separation of potato cyst
nematodes into two species (Stone, 1973). The populations with white or
cream-colored females were designated a new species, *H. pallida* Stone 1972
while those with golden females remained *H. rostochiensis*. Later, the subgenus *Globodera* was elevated to generic rank in which the round cyst nematodes were placed (Behrens, 1975). The potato cyst nematodes thus assumed
their present designation of *G. rostochiensis* and *G. pallida*.

Females of the two species can be distinguished by their color, golden
yellow in *G. rostochiensis* and white or cream in *G. pallida*. Certain morphological differences in cysts and second-stage juveniles also distinguish the
two species. Cysts of *G. rostochiensis* have a greater average anal-vulval
distance, 60 μm compared with 44 μm for *G. pallida*. A greater number of
cuticular ridges between the anus and vulva occur on *G. rostochiensis* cysts,
12.6 compared with 12.2 on *G. pallida*. The stylets of *G. rostochiensis* juveniles are typically shorter (±21 μm) than those of *G. pallida* (±23 μm) and
have more rounded basal knobs (Evans and Stone, 1977).

Mature females (cysts) of potato cyst nematodes are difficult to distinguish from other members of the genus *Globodera*. They are subspherical
with a protruding neck and are about 500-800 μm in length (Fig. 1). This
great variation in size is probably due to host nutrition during development.
Mature cysts are light to dark brown, and an irregular pattern of subsurface
punctuations occur over most of the body area (Mulvey and Stone, 1976).

Both species of potato cyst nematodes contain pathogenic variants, designated pathotypes. The first variant noted among potato cyst nematodes
was designated pathotype B, with the original population becoming pathotype

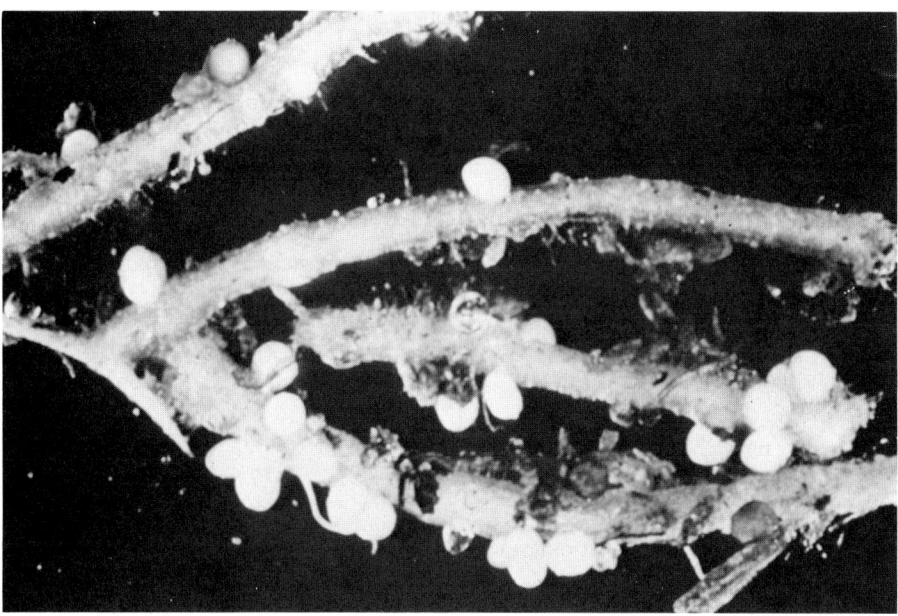

Fig. 1 Adult females and cysts of *G. rostochiensis* on potato roots. (Courtesy of W. F. Mai.)

A. With concurrent work on variation in potato cyst nematodes in several countries, the designation of pathotypes became confusing as identical pathotypes were designated with different letters or names. In 1977, representatives from England, Germany, and The Netherlands presented an "international scheme" for identifying and classifying pathotypes of potato cyst nematodes (Kort et al., 1977). Concurrently, investigations of variation in potato cyst nematodes in the Andean region of South America resulted in a new system of "race" (pathotype) classification (Canto-Saenz and deScurrah, 1977). Table 1 gives these two classification systems together, showing their similarity. Although both systems are improvements over the previous ones of classifying variants of potato cyst nematodes, they need to be merged into a single system to prevent further confusion.

Potato cyst nematodes are among the most highly specialized and successful plant-parasitic nematodes. They are sedentary endoparasites that exhibit marked sexual dimorphism. Females enlarge to become rounded, flask-shaped bodies when mature, but males regain their original, though enlarged, vermiform shape. At maturity, females die and their body walls tan to become tough protective cysts containing up to 500 eggs, each with second-stage juveniles in a quiescent state. It is this stage of development that provides the nematode with a highly specialized survival mechanism. In many ways, the cysts mimic tubers produced by its host. Once formed, cysts lie dormant in the soil until suitable conditions return for their revival. In addition to suitable temperature and moisture, the presence of host roots is required to stimulate encysted juveniles to emerge. In the absence of a host, encysted juveniles will remain viable in a quiescent state for many years with only a few emerging each year.

Table 1 Current Systems for Identifying and Classifying "Races" or Pathotypes of Potato Cyst Nematodes

Differential host	Designation of *Globodera rostochiensis* pathotypes				
	European: Ro1	Ro4	Ro2	Ro3	Ro5
	South American: R_1A	R_1B	R_2A	R_3A	$-^a$
Solanum tuberosum ssp. *tuberosum*	+	+	+	+	+
S. tuberosum ssp. *andigena* (H_1)	-	-	+	+	+
S. kurtzianum KTT/60.21.19	-	+	-	+	+
S. vernei GLKS.58.1642.4	-	+	-	-	+
S. vernei $(VT^n)_2 62.33.3$	-	-	-	-	+

Differential host	Designation of *Globodera pallida* pathotypes					
	European: Pa1 $-^a$	$-^a$	$-^a$	Pa2	Pa3	
	South American: P_1A	P_1B	P_2A	P_3A	P_4A	P_5A
Solanum tuberosum ssp. *tuberosum*	+	+	+	+	+	+
S. multidissectum (H_2)	-	-	+	+	+	+
S. kurtzianum KTT/60.21.19	+	+	-	+	+	+
S. vernei GLKS.58.1642.4	+	+	+	-	+	+
S. vernei $(VT^n)_2 62.33.3$	-	+	-	-	-	+

[a]No comparable pathotype reported.

C. Life Cycle

The active part of the life cycle (Fig. 2) begins in the spring, when second-stage juveniles emerge from the eggs after stimulation by substances emanating from host plant roots. The second-stage juveniles enter the host roots near the tip, and cut through the cell walls, leaving a trail of ruptured cells. Eventually they come to rest with their heads toward the stele and begin feeding either on pericycle, cortex, or endodermis cells. Their hollow stylet pierces the cells, injects saliva, and later withdraws cell contents. The juveniles, by injecting saliva, induce cell enlargement and breakdown of the cell wall, which in turn causes formation of a large syncytial transfer cell with dense, granular cytoplasm (Jones and Northcote, 1972). The nematode continues feeding from the transfer cell until its development is complete, a period that takes two or three months depending upon temperature. Once the juvenile is sedentary, it undergoes a series of three molts through the third and fourth juvenile stages to the adult. Sex is distinguishable at the third juvenile stage, and once sex is determined, it is irreversible.

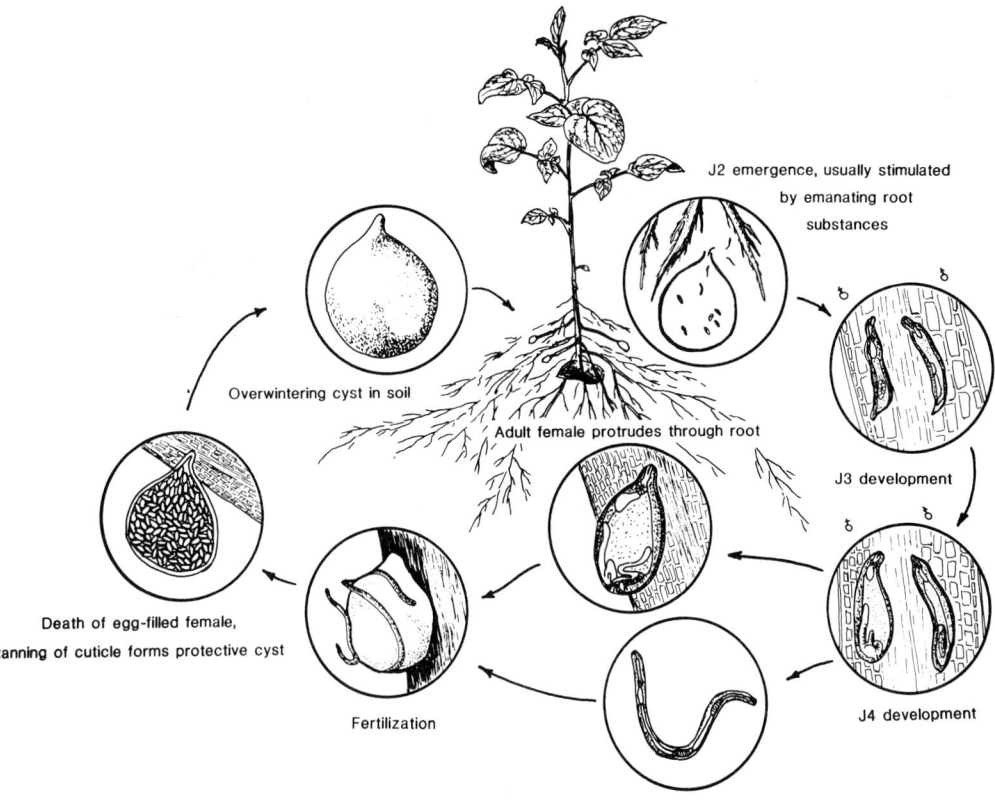

Fig. 2 Life cycle of *G. rostochiensis*. (Illustrated by M. Brucato.)

Fourth-stage males remain coiled within the saclike third-stage cuticle and emerge from the root after the final molt. When the adult males emerge, they are vermiform, about 1 mm long. They live for about 10 days in the soil and apparently do not feed (Evans and Stone, 1977). The fourth-stage females are saccate and are enclosed in the third-stage cuticle. Adult females enlarge as their gonads increase in size, eventually rupturing the root cortex so that their spherical body is exposed outside the root with only the head and "neck" embedded in the root. The females then release a substance that attracts many males. Fertilization, which is essential in *Globodera* spp., is accomplished when the males coil around the vulval area of the female. Each female may undergo multiple matings with many different males (Evans and Stone, 1977).

The embryos develop within the egg up to the formation of second-stage juveniles while still within the female's body. The female dies and the cuticle tans to form a tough, leathery cyst (Fig. 3) that contains up to 500 embryonated eggs (Fig. 4). When potatoes are harvested, the cysts are detached from the roots and become free in the soil, where they overwinter. When potatoes are planted, exudates from their roots stimulate juveniles to emerge from eggs and the life cycle is again initiated. Generally, only one generation is produced each year but there is some evidence for a partial second (Evans and Stone, 1977). If no host crop is planted, about one-third of the juveniles

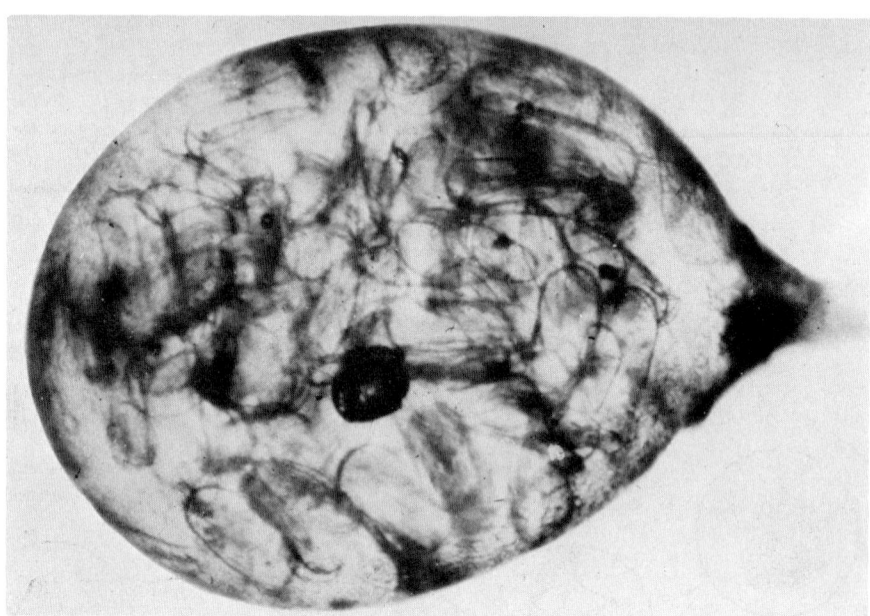

Fig. 3 Adult female of *G. rostochiensis* with eggs. (Courtesy of W. F. Mai.)

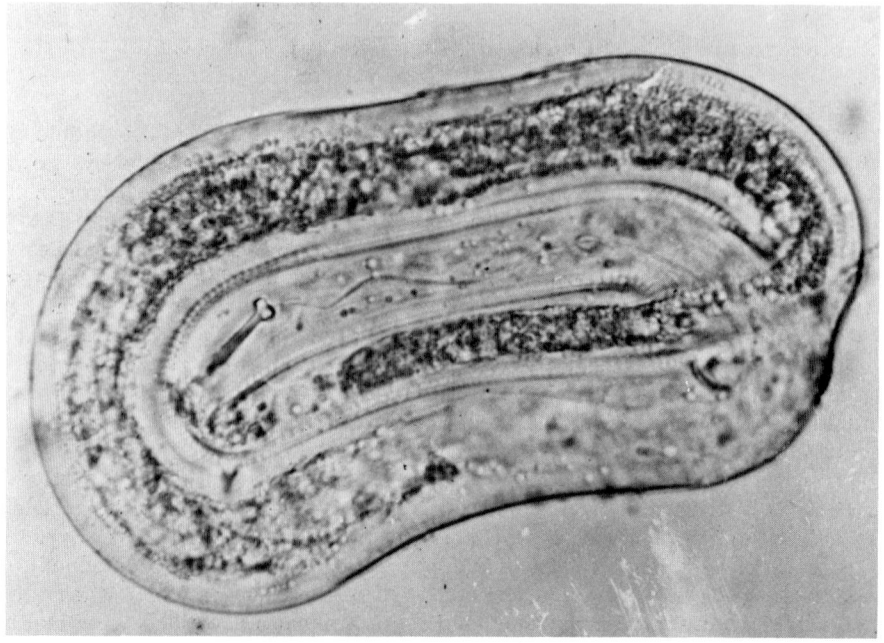

Fig. 4 Egg of *G. rostochiensis* with second-stage juvenile. (Courtesy of W. F. Mai.)

within the eggs will hatch each year, but some may survive for as long as 20 years (Jones, 1970).

The life cycles of the two species are essentially the same except for differences in temperature adaptation, a fact that may have influenced their geographic distribution. *G. pallida* is better adapted at temperatures between 10 and 18°C than is *G. rostochiensis* (Franco, 1979). At 20°C there appears to be no measurable differences in activity of the two species, but at 25°C, *G. rostochiensis* is better adapted than is *G. pallida*. When soil temperatures rise above 30°C for a lengthy period, neither species will establish (Mai and Harrison, 1959; Evans and Trudgill, 1978).

D. Associated Symptoms and Damage

Like most plant-parasitic nematodes, potato cyst nematodes do not cause distinctive above-ground symptoms that are of diagnostic value. Because the nematodes attack roots, infected plants exhibit typical symptoms of water or mineral deficiency stress. The foliage of plants infected with potato cyst nematodes has a sickly, yellow appearance and the plants may die prematurely (Fig. 5). High nematode densities cause severe stunting and in some cases stop growth completely. Close examination of the roots of infected plants at the time of flowering reveals minute pearly white bodies, which are immature females that have erupted through the root epidermis. When the females mature, they turn dark brown and most of them become dislodged from the roots when the plants are lifted for examination.

The first signs of potato cyst nematode infestation in a potato field are small patchy areas of poor growth. These areas are at first roughly circular

Fig. 5 Damage to potato plant by *G. rostochiensis* showing typical midday wilting. (Courtesy of U.S. Department of Agriculture.)

in shape and represent the initial foci of infestation. In subsequent years, patches of poor growth may become oblong due to spread by cultivation. Repeated cropping of the field to potatoes results in spread of the nematode over the entire field and to other fields. Current sampling techniques will not detect the first cysts that reach a field. By the time their progeny are detected or cause poor plant growth, they have spread to other parts of the field or to distant fields.

Because potato cyst nematodes are primarily debilitating parasites, the damage they cause to potatoes is reflected in reduced size and number of tubers produced. The damage they cause is manifest through disruption of the conducting tissues of the root. This disruption is brought about by formation of feeding sites (syncytia). The syncytia are formed by incorporation of adjacent cells, mainly phloem parenchyma, through cell wall dissolution (Evans and Trudgill, 1978). Protuberances from the syncytia form next to xylem vessels. Boundary formation and microtubules are associated with ends of the protuberances and serve to increase the surface area of the syncytial cell wall relative to its volume and allows for increased flow of solutes across the plasma membrane (Jones and Northcote, 1972). Thus, the syncytia become metabolic sinks for excess nutrients rather than the nutrients being stored in the form of tubers. The invading juveniles are considered to cause little or no damage to the plants other than slight necrosis of cortical cells (Evans and Stone, 1977).

E. Distribution

Prior to 1900, potato cyst nematodes were known to occur only in Germany. In the first quarter of the twentieth century they were found in many other parts of Europe and were thought to be endemic, although there was no evidence to support a theory of European origin. It was known that their main host crop, the potato, originated in South America (Hawkes, 1978). In 1951, evidence was presented indicating a South American origin of potato cyst nematodes when cysts were discovered in soil from potatoes in the stores of a ship from Peru arriving in the United States (Mai, 1977). From that time onward it was generally agreed that potato cyst nematodes originated in the Andean region of South America. From South America they have spread throughout the world to almost any area suitable for producing potatoes. Most likely they were transported in soil adhering to potato tubers, but they also can be transported by any means that can transport soil. Their routes of distribution are a matter of speculation, but apparently Europe became a secondary center of distribution and the nematodes were spread with seed potatoes of improved cultivars developed in Europe and exported throughout the world (Evans and Trudgill, 1978). Most of the infestations found throughout the world, including some in Central and South America, appear to have emanated from Europe (Evans and Stone, 1977). An exception may be that cysts appear to have been transported from Peru to Japan in contaminated guano sacks (Inagaki and Kegasawa, 1973).

Evans and Stone (1977) list 48 countries from which one or both species of potato cyst nematodes have been found. *G. rostochiensis* appears to be more widely spread than does *G. pallida*, but this may be because the latter was more recently described and is less well known.

Much information is available on the distribution of the two species. Both species occur in central and western European countries, but in the south and east of Europe only *G. rostochiensis* is found (Evans and Stone, 1977). In South America, only *G. pallida* occurs in Colombia, Ecuador, and most of Peru. In the south of Peru both species occur and both often can be found infecting potato in the same field. Both species also have been reported from Bolivia and Venezuela, but in Chile only *G. rostochiensis* has been found (Evans et al., 1975). Both species have been reported on the North American continent, but only *G. rostochiensis* has been found in the United States (Mai, 1977). In Canada, *G. rostochiensis* occurs on Vancouver Island (Spears, 1968), and both species are reported from Newfoundland (Stone et al., 1977).

It is speculated that *G. rostochiensis* was introduced to the United States on Long Island, New York, by 1920 or even earlier on military equipment returning from Europe. Strict quarantine measures were imposed on movement of any potentially contaminated material from Long Island by 1944, but in 1967 an infestation was discovered in Steuben County in western New York. Since 1967, infestations of *G. rostochiensis* have been discovered in three additional counties of western New York (1974, 1976, and 1983) and additional infestations were found in Steuben County in 1968, 1976, 1979, 1980, and 1981. Infestations discovered in Delaware and New Jersey appear to be isolated occurrences and were contained so that no further spread occurred (Mai, 1977).

It is interesting to note that the time from its likely introduction until the discovery of *G. rostochiensis* in the United States was approximately 30 years, a period corresponding to the time between the introduction of potato cyst nematodes into Europe and their discovery. Furthermore, the sporadic discovery of new infestations in western New York from 1967 onward corresponds to sporadic recording of new infestations from 1881 onward in Europe. Although at a slower rate of spread, it appears that the pattern of potato cyst nematode infestations in Europe is being duplicated in the United States some 100 years later.

F. Estimated Losses

If left uncontrolled, potato cyst nematodes are capable of causing a 100% loss in potato yields. Reliable estimates of actual yield loss caused by potato cyst nematodes are not available for many countries where the nematodes occur. Scientists in the United Kingdom, where losses on infested land have been limited by crop rotation, estimate losses caused by potato cyst nematodes to be about 9% of the annual potato yield (Evans and Stone, 1977). The percentage loss is greater in some countries, and the figures given usually do not reflect indirect losses, such as the cost of enforced rotations where alternate crops are of less value, chemical application, and legislative control (quarantine and regulation) (Evans and Brodie, 1980). The value of the world crop loss caused by potato cyst nematodes is, therefore, very great, and the potential for loss is greatest where average yields are highest. Jones (1972) places the nematode density at which significant yield losses occur at 15 eggs per gram of soil. The damage threshold (i.e., nematode density at which significant crop losses occur) is influenced by climate, soil type, and price of the crop, all of which vary with country and season.

The losses caused by potato cyst nematodes are particularly important in such areas of the world as the South American Andes, where the climate is generally suitable for only a few crops other than potatoes (Hawkes, 1978). Frequently, in such areas nematode densities reach a level where potato yields are not much greater than the amount of seed planted (Mai, 1977).

Because of strict quarantine and regulatory procedures, potato cyst nematodes do not cause yield losses in the United States. The systematic survey (sponsored by the Federal Government) of U.S. potato lands reveals the

Table 2 Costs When Potato Cyst Nematodes Are Either Widespread or Virtually Absent

Nematodes widespread		Nematodes virtually absent	
G[a]	Yield loss (10% of total?)	FS	Quarantine of imports
G	Hidden loss of enforced rotation (potatoes not grown on best potato land)	FS	Quarantine of exports
		FS	Surveillance for spread and appearance of new races
G	Chemicals for economic control	FSG	Elimination procedures
		FS	Resistance breeding
FS	Quarantine of imports	FS	Nematology research
FS	Quarantine of exports		
SG	Extension service		
FS	Resistance breeding		
FS	Nematology research		

[a]G indicates cost borne by grower. S indicates cost borne by state agencies. F indicates cost borne by federal agencies.

presence of potato cyst nematode infestations before densities reach damaging levels and such infested fields are fumigated. Consequently, only indirect losses are realized from potato cyst nematode infestation. Such losses are manifest in the cost of maintaining a quarantine and in restriction of activities of other agricultural industries, such as nurseries, which lowers the total agricultural income for a region. When taking into consideration the cost of such items as border inspections, field surveys, treatment of soil, decontamination of equipment and containers, and constant surveillance of the quarantined area, indirect losses are considerable.

On a global basis, potato cyst nematodes present two extreme situations, one in which the pest is endemic and one in which the pest is virtually absent but constantly threatens to invade. Both situations have their associated costs, one the costs associated with yield losses and control measures, the other the cost of quarantine procedures. Table 2 (Evans and Brodie, 1980) lists these costs and who bears them under the two situations. Estimating dollar amounts for costs involved in the two situations is difficult, but as a tenth of the present U.S. potato crop is worth 150 million dollars, losses would escalate should the nematode become more widespread. Also, more land would be required to produce the same overall potato yield.

G. Control

When an important crop pest such as the potato cyst nematode becomes endemic in a country, it is dealt with differently than a pest that is limited in distribution but has the potential of becoming endemic given the chance. As indicated above, the United States operates a quarantine and survey system to confine the potato cyst nematode to its known areas of infestation, whereas South American and European countries use control measures to minimize crop losses due to the nematode, which is considered endemic to South America and over most of Europe.

Because potato cyst nematode eggs are protected inside cysts and most of them remain unhatched until a host crop is grown, they are the most difficult of all nematodes to control. Once potato cyst nematode infestations are established, three primary control tactics can be employed to reduce their numbers and minimize crop losses. These tactics include growing nonhost crops (crop rotation), growing resistant potato cultivars, and the use of nematicidal chemicals.

1. Crop Rotation

Because of the limited host range of potato cyst nematodes (potato, tomato, eggplant,and wild *Solanum* spp.), nonhosts play an important role in reducing population densities and minimizing crop losses. Depending upon climate, nematode densities in the soil decline 30-50% annually when a nonhost crop is grown (Jones, 1970; Brodie, 1976a). Crop rotation is used both to suppress the increase of low densities and to reduce high densities. The length of time required to reduce damaging densities to below damaging levels varies with the initial density. When the nematode has reached equilibrium density (a density that would establish in a monoculture system), six to seven years of nonhost are required before potatoes can again be grown profitably (Jones, 1972). Although the relative unimportance of choice of nonhost crop has been emphasized, some evidence suggests that some nonhosts are more effective than others in reducing nematode numbers in the soil (Hesling et al., 1961).

2. Resistance

Several commercial cultivars of potatoes with resistance to *G. rostochiensis* are available. This resistance is conditioned by a single dominant gene first discovered in a clone of *S. tuberosum* subsp. *andigena* (Ellenby, 1954) and bred into *S. tuberosum*. Resistance conferred by this gene (H_1) is almost complete and appears to be stable, but it is effective against only two pathotypes of *G. rostochiensis* and not effective at all against pathotypes of *G. pallida* (see Table 1). Several countries report one or more high-yielding, commercial cultivars containing the gene H_1 that are resistant to *G. rostochiensis*.

Resistance to other pathotypes of *G. rostochiensis* and to those of *G. pallida* has been less striking and less exploited. Additional genes from *S. tuberosum* subsp. *andigena* have been reported to offer some resistance to *G. rostochiensis* and one pathotype of *G. pallida*. These genes have been of some value in potato-breeding programs in Europe (Howard and Fuller, 1971).

Although recent evidence indicates that resistance to *G. pallida* is available in *S. tuberosum* subsp. *andigena* (deScurrah, personal communication), wild species of *Solanum* have been most often utilized in breeding for resistance to pathotypes of *G. pallida*. Notable successes have been achieved with *S. vernei, S. multidissectum, S. chacoense*, and *S. spegazzinii* (Ross, 1972). Resistance to potato cyst nematodes in wild species of *Solanum* appears to be polygenic and is much more difficult to breed because of differences in ploidy levels. However, considerable efforts are being made in several countries to transfer resistance from diploid to tetraploid potatoes. Dutch workers report success in transfer of resistance to *G. pallida* from *S. vernei* to *S. tuberosum* (Evans and Stone, 1977).

Monoculture of cultivars that are resistant to only one pathotype in fields with mixed populations will select pathotypes that can reproduce upon them. This selection of aggressive pathotypes within species is likely to be slow, but when *G. pallida* replaces *G. rostochiensis* as a result of growing cultivars resistant to *G. rostochiensis*, the shift is relatively rapid (Evans and Stone, 1977). Shift in pathotypes or species also occurs to some extent when susceptible cultivars are grown (Kort and Jaspers, 1973), suggesting competition between species.

When suitable resistant cultivars are grown, the numbers of potato cyst nematodes are reduced by 80-95% each year (Brodie, 1976a). The value of resistant cultivars lies in their ability to stimulate hatch but prevent development of females. When population densities are extremely high, resistant cultivars may suffer some damage from massive invasion of juveniles into the roots (Evans and Stone, 1977).

3. *Chemical Control*

Much research has been done on chemically treating soil to control potato cyst nematodes. The early research on chemical control concentrated on soil fumigation. Soil fumigants such as 1,3-dichloropropane (Telone), methyl bromide, and methyl isothiocynate liberators (Vapam, Vorlex, Dazomet) have been used successfully in some countries (Chitwood, 1951; Jones, 1970). Dosage of these soil fumigants depends upon the degree of control desired. A double treatment is required in those situations where population densities are reduced to levels where the nematode can not be detected (Spears, 1968). Much less fumigant is needed to protect the crop for a single growing season.

The best chemicals developed so far to provide control for a single growing season are the oximecarbamates, aldicarb and oxamyl. These compounds are effective at low dosages (3-5 kg/ha) and their effectiveness appears to be independent of soil type (Whitehead, 1973; Brodie, 1980). Some organophosphates, such as phenamiphos and ethoprop, are also effective, but their effectiveness is limited somewhat by soil type, being less effective in organic soils (Whitehead, 1975). These nonvolatile compounds, particularly if they are systemic, present a problem of residue in plant products if applied too close to harvest. Also, aldicarb has been recently implicated in the contamination of ground water on Long Island, New York. All nematicides are harmful to humans and require special handling during application. Also, because they are expensive, large yield increases are required to justify their usage.

4. *Integrated Control*

None of the aforementioned control tactics when used singly has successfully dealt with the potato cyst nematode problem. The most promising approach to successful control of potato cyst nematodes is by integration of control tactics into a management system or control strategy. Such an approach is popularly referred to as integrated pest management (IPM). Certain characteristics of potato cyst nematodes, particularly their narrow host range and response to host resistance, make them particularly amenable to management strategies. The key to successful management of potato cyst nematode densities is to keep the nematodes at a disadvantage by always exerting a negative influence on population growth. Table 3 gives results of several years' study of managing densities of *G. rostochiensis* (Evans and Brodie, 1980). Such management systems are more effective in suppressing low nematode densities than they are in reducing high densities (Brodie, 1976a).

Table 3 Population Densities of *Globodera rostochiensis* After Various Management Practices

Practice	Eggs per gram of soil after no. year(s)						
	1	2	3	4	5	6	7
Monoculture of susceptible cultivar	0.8	2.9	6.7	43	96	104	117
Alternating susceptible cultivar and nonhost	1.8	0.7	3.1	1.8	11.1	6.1	8
Alternating susceptible and resistant cultivars	0.4	0.5	1.1	1.2	4.9	0.5	1.6
Rotating nonhost, resistant, and susceptible cultivars	1.2	0.3	0.1	2.4	0.6	0.4	5
Rotating nonhost, resistant, and susceptible cultivar with aldicarb soil treatment	0.5	0.7	1.0	0.3	0.3	0.1	0.6
Monoculture of resistant cultivar	0.1	0.1	0.1	0.2	0.1	0.2	0.2

The control program for *G. rostochiensis* in the United States is unique in that its objective is to manage densities of *G. rostochiensis* below levels at which spread occurs, while in most countries, the objective is to manage densities below plant-damaging levels. Consequently, the U.S. program represents one of the most extensive pest management systems ever attempted (Mai, 1977). It has as its main component extensive surveys to determine occurrence and distribution of the pest. Once detection is positive, the infested areas are regulated and several policies go into effect. These regulations involve: (1) prevention of seed potato production in regulated areas, (2) prevention of host crop production on land with detectable nematode densities, (3) prevention of use of reusable containers in potato production, (4) regulation of movement of such items as farm machinery, top soil, and plant material; and (5) massive soil fumigation to reduce nematode numbers below detectable levels. Once fumigation is successful, potato production is resumed, utilizing approved nematode management systems involving resistant cultivars, nonhost crops, and nematicides. The success of this approach hinges on the fact that only one pathotype of *G. rostochiensis* is known to exist in the United States. If another pathotype were discovered, the system would break down until resistance to the new pathotype was available.

III. POTATO ROT AND STEM NEMATODES

A. History

A progressive dry rot of potato tubers, which was not associated with any characteristic symptom on the stems and leaves, was reported by Kühn in 1888 as being caused by a nematode (Thorne, 1961). In the same year Ritzema Bos reported an entirely different type of nematode infestation of potatoes in which

not only the tubers were attacked but also the stems and leaves were stunted
and malformed (Thorne, 1961). In the ensuing years either one or both types
of damage was reported from many parts of Europe.

In North America, Atkinson (1889) illustrated a nematode he found in
potato tubers that, according to Thorne (1961), appears to be very similar,
if not identical to the potato rot nematode. The first confirmed infestation of
the potato rot nematode on the North American continent was when Blodgett
(1943) discovered an infestation near Aberdeen, Idaho. In 1945, its presence
was confirmed in six potato fields on Prince Edward Island, Canada, and the
following year it was intercepted on Long Island, New York, in seed potatoes
shipped from Prince Edward Island (Thorne, 1961). No additional infestations
were reported until 1953 when it was found in potatoes from several fields in
Wisconsin. That same year, the potato rot nematode was identified from collections of infected tubers in Western Canada in a potato field that had followed
bulbous iris. Circumstantial evidence indicated that the pest had been present
in these localities for several years (Thorne, 1961).

In recent years, the seriousness of the potato rot nematode in North
America has declined, possibly owing to successful soil fumigation or removal
of infested land from production. It continues to be a very serious pest of
potatoes in the European portion of the USSR, Central Asia, and parts of
Germany and The Netherlands where the stem nematode is also a serious
problem (Kirjanova and Krall, 1971; Winslow and Willis, 1972).

B. The Nematodes

These are *Ditylenchus destructor* Thorne 1945 and *D. dipsaci* (Kühn 1857)
Filipjev 1936. When the tuber rot disease of potato was first described, the
causal organism was identified as *Anguillula dipsaci* Kühn 1857. Almost 100
years passed before it became known as *D. dipsaci*. Afterward, the complexity
of *D. dipsaci* was recognized and *D. destructor* was separated out as the potato rot nematode leaving *D. dipsaci* as the potato stem nematode (Thorne, 1961).

Morphologically the two species are very similar, but biologically, they
differ in some important ways. Although *D. dipsaci* may infect tubers, usually
its infection of potatoes is confined to the above-ground portions of the plant.
On the other hand, *D. destructor* infection is confined to the underground
stolons and tubers. In addition, *D. dipsaci* is an obligate parasite of higher
plants, whereas *D. destructor* thrives both on higher plants and on fungi.
Moreover, *D. destructor* differs from *D. dipsaci* in that it does not produce
the characteristic "nematode wool" on infected plant parts, is unable to withstand desiccation, and does not carry over from one season to the next in the
preadult stage (Thorne, 1961).

All stages of *D. destructor* may be found in the host tissue or in surrounding soil. The nematodes move into, out of, or within the plant tissues
at will, thus being classified as migratory endoparasites.

C. Life Cycle

For a convenient starting point, the life cycle (Fig. 6) of *D. destructor* begins
when the nematodes enter small potato tubers through lenticels or eyes of the
tuber. The nematodes may have been surviving in the soil, on fungi or weed
hosts, or may have been introduced by planting infected seed pieces. Once
inside the tuber, the nematodes feed in the tissues just beneath the skin
(Thorne, 1961). They first exist singly or in small numbers, and small white
lesions mark their location. Nearby healthy tubers may become infested by
nematode migration from their diseased neighbors (Winslow and Willis, 1972).

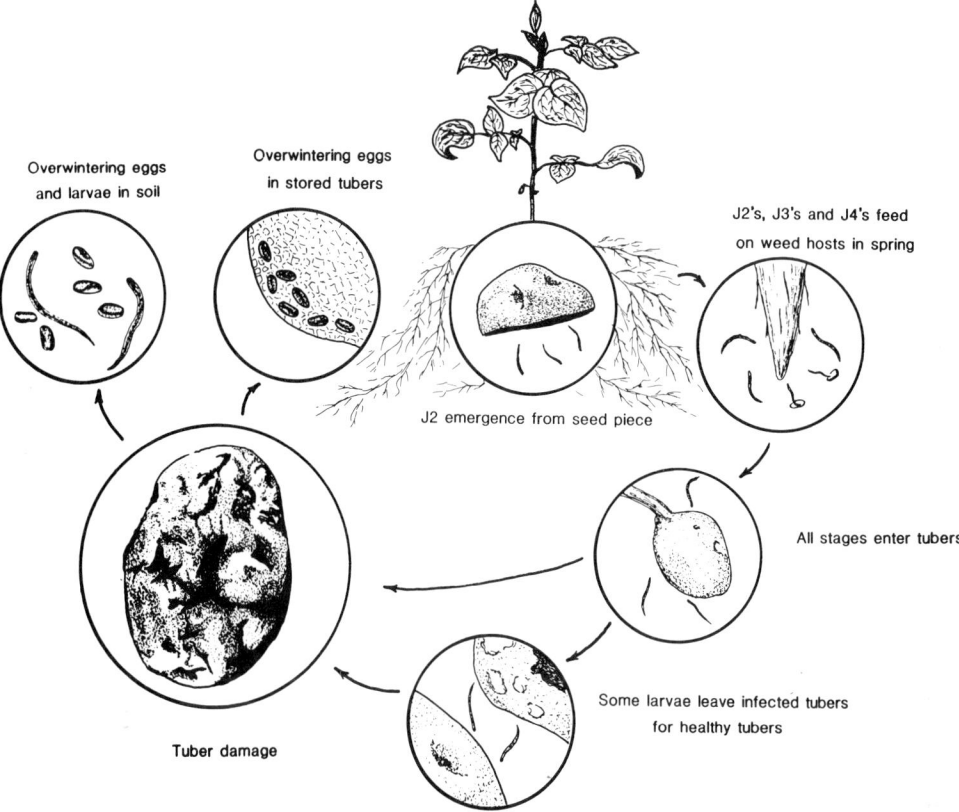

Overwintering eggs
and larvae in soil

Overwintering eggs
in stored tubers

J2's, J3's and J4's feed
on weed hosts in spring

J2 emergence from seed piece

All stages enter tubers

Some larvae leave infected tubers
for healthy tubers

Tuber damage

Fig. 6 Life cycle of *D. destructor*. (Illustrated by M. Brucato.)

The nematodes continue to live and reproduce in harvested tubers in storage or they may be left behind in the soil where they survive on fungi and weeds until another potato crop is planted. Depending upon the types of secondary invaders, tubers in storage may have various kinds of wet or dry rot. Under conditions of wet rots, the nematodes are killed and thus fail to repeat the life cycle. If dry rot ensues in storage, the nematodes survive in the tuber, usually as eggs, and, if such tubers are planted, the eggs hatch and the life cycle is again initiated (Thorne, 1961).

Although *D. dipsaci* primarily attacks stems and leaves, its life cycle differs only slightly from *D. destructor*. The life cycle begins when *D. dipsaci* invades the aerial parts of the plant during wet weather. Penetration may take place anytime from plant emergence up to the time tubers are formed. The nematode is thought to be well adapted to invading stem tissues by virtue of pectolytic enzymes it produces that dissolve the middle lamella between cells allowing the nematode to move through the tissue (Riedel and Mai, 1971). Once the nematode is inside the protected environment of stem or leaf tissues, multiplication is rapid and the plant is severely stunted and distorted. The nematode may occasionally move through infected stem (stolon) tissues and infect tubers, or tubers may become infected via the soil (Kirjanova and Krall, 1971). Heavy stem infection is confined to the lower regions of the stem.

At harvest, many of the nematodes return to the soil and overwinter in preadult stages. Some of the nematodes may be trapped inside dried plant tissues, where they undergo anabiosis and lie in a quiescent state until suitable growing conditions return and the life cycle is again initiated.

D. Associated Symptoms and Damage

The above-ground portions of the plant do not reveal attack by *D. destructor*. Symptoms of *D. destructor* attack are manifest in the tubers. Small white, chalky or light-colored spots that can be seen just below the surface of the tuber when it is peeled are the earliest symptoms of infection (Thorne, 1961). The infected tissues are dry and granular and infected spots may coalesce, the affected tissue darkening gradually through greyish to dark brown or black, as secondary organisms, such as bacteria, fungi, or saprophytic nematodes, invade. The nematode is seldom found in the dark tissues, being confined to the white, mealy tissue at the advancing edge of the lesion. The skin of the tuber is not attacked but becomes paper thin and cracks as underlying infected tissues dry and shrink (Fig. 7).

Unlike the lesions caused by *D. destructor*, which tend to be superficial, those caused by *D. dipsaci* may extend throughout the tuber. Tuber tissues infected with *D. dipsaci* become spongy and are yellowish to brown in color. Although as noted earlier, distinct differences between the two species may occur in tuber symptoms, diagnosis is often difficult and species identification is necessary. In addition, *D. dipsaci* attacks above-ground parts of the plant, causing typical stunting and thickening with a distortion of stems and leaf petioles, which may occur without tuber symptoms.

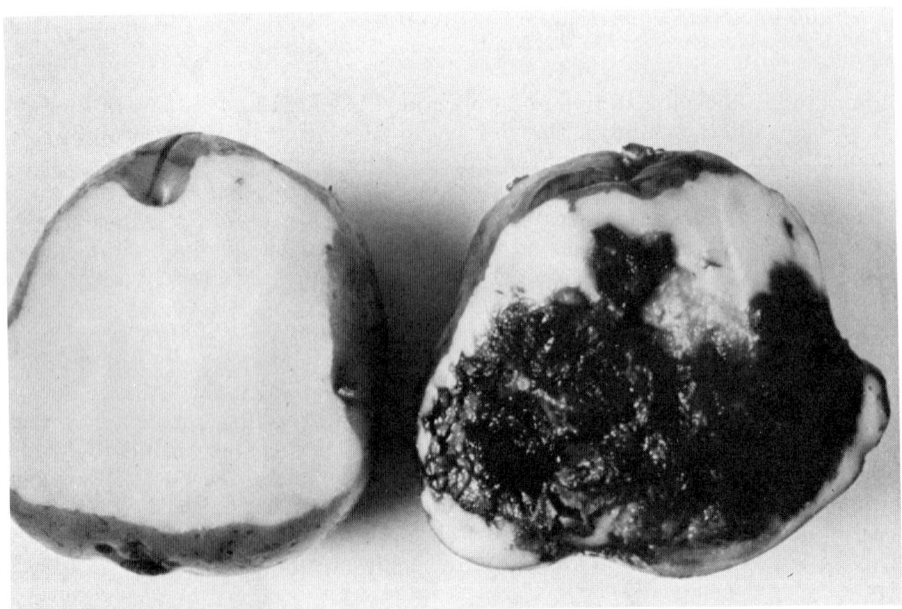

Fig. 7 Tuber damage caused by *D. destructor* (rt) compared with healthy tuber (lt). (Courtesy of W. F. Mai.)

Both species are primary pathogens, invading tubers via eyes, lenticels, or stolons. Under suitable environmental conditions in storage or in the field, a bacterial soft rot may ensue following tuber infection by either species, causing complete destruction of the tubers.

E. Distribution

Both species are found throughout the world, but have been reduced to rather minor importance on potatoes in some countries. *D. dipsaci* is a major pest of several crops throughout the world but is regarded as a serious pest of potatoes only in Germany, The Netherlands, and parts of the USSR (Kirjanova and Krall, 1971). It appears that the race of *D. dipsaci* that attacks potatoes has a very limited distribution.

Generally, *D. destructor* is a much more important pest of potatoes. *D. destructor* reaches its greatest importance in temperate regions, probably owing to its inability to withstand drying rather than a direct temperature relation. The nematode has been reported from many areas of Europe and is especially widespread in the USSR (Kirjanova and Krall, 1971; Evans and Trudgill, 1978). It is also known to occur in South Africa, some areas of the Mediterranean region, South America, and isolated areas of North America.

F. Estimated Losses

Estimates of actual losses caused by *D. dipsaci* are apparently lacking in the literature. Although its seriousness as a potato pest cannot be denied, its occurrence is so localized that losses are of extreme local concern.

Much, if not all, of the losses attributed to *D. dipsaci* prior to 1945 were no doubt caused by *D. destructor*. This nematode is considered by many to be the second ranking nematode pest of potato, with potato cyst nematodes ranking first. In the field of its inital discovery in the United States, Thorne (1961) noted that "tubers were so severely damaged that most of the crop was left lying in the field." However, in recent years losses from *D. destructor* in the United States and Canada have declined, perhaps owing to avoiding infested land and to soil fumigation with ethylene dibromide.

Judging from the literature, the greatest losses from *D. destructor* occur in the USSR. In Estonia, 80-90% infestation of tubers is not uncommon and annual losses in the USSR are reported to be in excess of 150,000 metric tons (Kirjanova and Krall, 1971). Most severe losses occur when infested tubers are stored and rots ensue, which often results in loss of the entire crop to *D. destructor*.

G. Control

The use of healthy "noninfected" seed is an essential step in the control of *D. destructor*. Other phytosanitary measures that can be employed to reduce *D. destructor* attack include destruction of infected plant parts left in the field and control of weed hosts. In addition, late planting when soil moisture is inadequate for optimum invasion reduces losses (Kirjanova and Krall, 1971). Early harvest is helpful in providing healthy seed stock.

1. Chemical

Excellent control of *D. destructor* in the United States has been achieved with soil fumigation using ethylene dibromide (EDB) (Thorne, 1961). The efficacy of EDB was enhanced by the fact that *D. destructor* inhabits the upper portions of the soil. Unfortunately, fumigation with EDB proved uneconomical in the USSR, where the nematode is a particular menace (Kirjanova and Krall, 1971). Also, EDB proved ineffective in the heavy clay soils of The Netherlands, where *D. dipsaci* is a problem (Evans and Trudgill, 1978).

2. Resistance

Some success has been realized in identifying resistance to *D. destructor*. Some commercial varieties have been observed to suffer less damage than do others. In addition, some wild species of *Solanum* have been reported to be resistant to *D. dipsaci* (Olefir, 1972).

3. Crop Rotation

Because of the wide host range of *D. destructor*, crop rotation was initially considered inadvisable for its control. However, Russian scientists report satisfactory control by crop rotation with small grains, vetch, and lupine provided potatoes are grown only once in three or four years (Kirjanova and Krall, 1971). Crop rotation is more effective when used in combination with the aforementioned sanitary practices.

IV. ROOT-KNOT NEMATODES

A. History

The history of root-knot nematodes on potatoes is difficult to trace because of the early confusion of nomenclature. Neal (1889) described a root-knot disease of potato in Florida and named the responsible organism *Anguillula arenaria* Neal 1889. Later all the root-knot nematodes were placed in the species *Heterodera marioni* (Cornu 1879). Up to 1949, all the literature pertaining to root knot on potatoes refer to the causal organism as *H. marioni*. Chitwood (1949) placed the root-knot nematodes in the genus *Meloidogyne* Goeldi 1887 and recognized five species and one variety. Since that time several additional species of *Meloidogyne* have been described. Franklin (1979) recognized 35 species, five of which are reported to attack potatoes. In 1980, a new species of root knot was described from potatoes in the northwestern United States (Santo et al., 1980).

Greatest reproduction and survival of root-knot nematodes occur in sandy soils at temperatures of 25°C and above. They are not presently a major, worldwide problem on potatoes, but certain species have become established and cause severe damage to potatoes in localized areas. These nematodes reach their greatest importance in tropical and warm temperature climates and are less of an economic problem in northern latitudes and high elevations of southern latitudes where soil temperatures are cool. Because of the extreme susceptibility of potato to root-knot nematodes, successful extension of the range of potato culture into warmer climates would bring with it a serious root-knot problem. This problem would be confounded by interaction of root-knot with other disease-causing organisms, such as bacterial wilt caused by *Pseudomonas solanacearum*, which is also prevalent in warmer soils.

B. The Nematodes

At least six species of *Meloidogyne* are known to parasitize potatoes. These species differ in distribution and amount of damage they cause. As in the case of most nematodes, each species has a common or descriptive name that relates to the climate in which they are found or to their preferred host.

Species known to parasitize potatoes are *M. arenaria* (Neal 1889) Chitwood 1949 (peanut root-knot nematode); *M. incognita* (Kofoid and White 1919) Chitwood, 1949 (southern root-knot nematode); *M. hapla* Chitwood 1949 (northern root-knot nematode); *M. javanica* (Treub 1885) Chitwood 1949 (Javanese root-knot nematode); *M. thamesi* Chitwood in Chitwood, Specht, and Havis 1952 (Thames' root-knot nematode); and *M. chitwoodi* Golden et al. 1980 (Columbia root-knot nematode).

Root-knot nematodes are sedentary endoparasites that exhibit sexual dimorphism. Second-stage juveniles emerge from the egg and are vermiform. Upon entering the plant they undergo a series of molts in which the third and fourth stages are flask shaped and the fifth-stage females are pear shaped. The fifth-stage males emerge from the fourth-stage flask-shaped cuticle as vermiform adult males.

Reproduction in *Meloidogyne* spp. commonly occurs in the absence of males through a process called mitotic parthenogenesis (Bird, 1978). Sexes can be distinguished in second-stage juveniles when the genital primordium in potential males is rod shaped and in potential females is V shaped. Under environmental or physiological stress, sex reversal may occur, causing many potential females to become males.

Although there are some exceptions, *Meloidogyne* spp. usually overwinter in the egg stage. When conditions are favorable, they hatch and infect potatoes. The number of generations produced depends on temperature. In tropical climates up to 12 generations are produced in one year. However, in the temperate zone, where potatoes are grown, only two to three generations are produced a year. Each female produces on the average 300-400 eggs.

C. Life Cycle

The life cycle of all species of *Meloidogyne* is essentially the same. Growth rates of different species may differ slightly. For example, the thermal optima for *M. hapla* for embryogenesis and growth are about 5°C less than for *M. javanica* or *M. incognita* (Bird, 1978).

The life cycle of root-knot nematodes shall be considered to have started when second-stage juveniles emerge from the egg and move through the soil in search of roots (Fig. 8). They enter the roots at or near the meristematic region just behind the root cap. Entry is gained by means of thrusting the stylet into the root until an opening is made. Upon entering the root, the juvenile comes to rest close to the developing vascular system and becomes sedentary.

Once in position, the juveniles begin to feed, increase in size, and induce alteration in the cells on which they feed. The interaction of the nematode and host results in the development of syncytia (giant cells) from which the nematode derives its food. Syncytia are characterized by their size, dense cytoplasm, numerous large amoeboid-shaped nuclei, and thick cell walls with numerous projections jutting into the cytoplasm (Bird, 1978). The syncytia act as transfer cells, with the nematode, becoming the nutrient sink.

Shortly after syncytial formation the nematode undergoes three molts after which the female grows rapidly and becomes pear shaped. The males emerge from the root after the fourth molt and are vermiform. Depending upon species, males may or may not be functional in reproduction.

The females begin to lay eggs that are extruded in a gelatinous matrix which remain attached to the female. A single female is capable of producing 500-1000 eggs. The life cycle is completed with embryogenesis in the egg leading to the formation of first-stage juveniles. There juveniles molt once

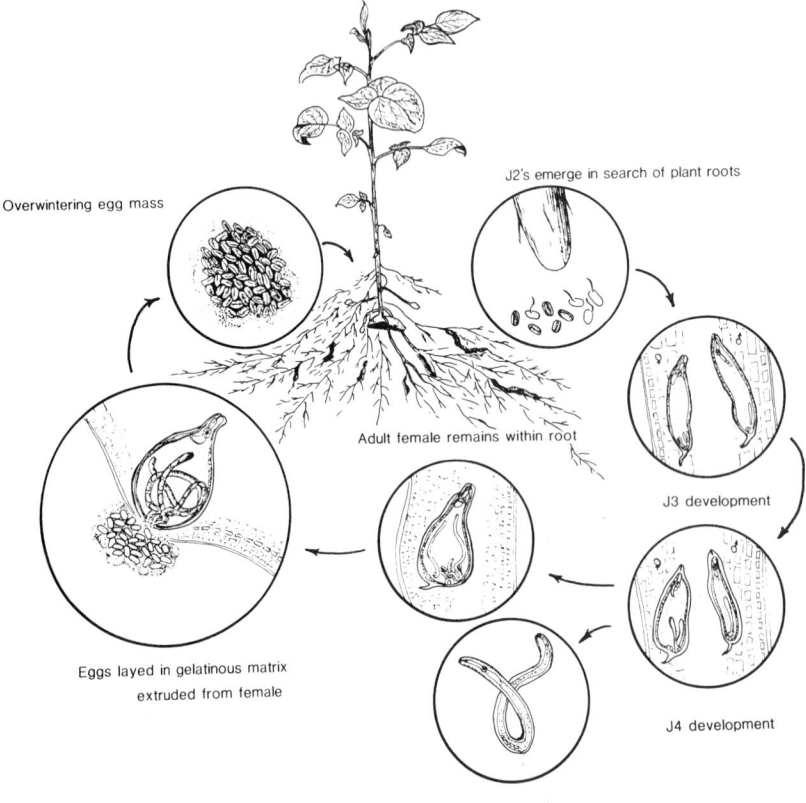

Fig. 8 Life cycle of *Meloidogyne* spp. (Illustrated by M. Brucato.)

inside the egg to give rise to second-stage juveniles which, under optimum conditions, emerge from the egg about 10 days after being laid. The second-stage juveniles are the infective stage and may enter roots to initiate a second generation. The number of generations is dependent upon favorable temperature and moisture. In temperate zones where potatoes are grown, generation time is about four to six weeks and the number of generations per growing season is limited to three or four (Winslow and Willis, 1972). During the winter, juveniles disappear from the soil and the nematode overwinters in the egg stage (Bird, 1978).

D. Associated Symptoms and Damage

Above-ground symptoms of potatoes infected with root-knot nematodes are of no diagnostic value because other biotic and abiotic stresses may cause similar symptoms. Depending upon nematode density and environmental conditions, infected plants may show varying degrees of stunting, chlorosis, and wilting. Root-knot disease of potatoes is identified by the presence of knots or galls of varying size and shapes on the roots (Fig. 9). Gall shape ranges from almost spherical (*M. arenaria*) to a very rough and irregular appearance (*M. hapla*). In addition to galling, *M. hapla* causes initiation of extensive lateral root formation, giving the entire root system a "wiry" appearance. Individual

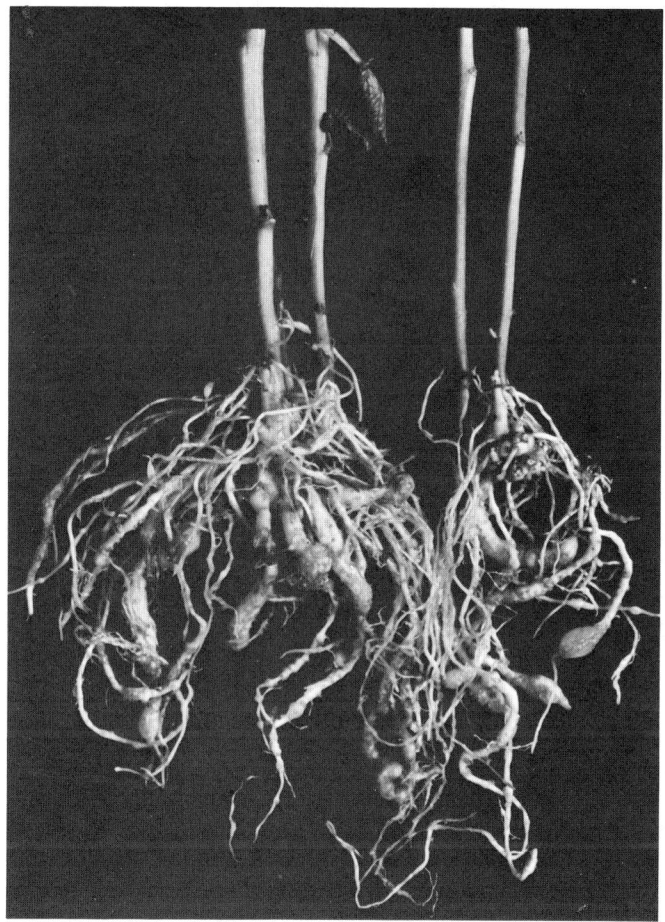

Fig. 9 Galls on potato roots caused by *Meloidogyne incognita*. (Courtesy of P. Jatala.)

gall size depends upon nematode density, species of *Meloidogyne*, size of root, temperature, and possibly other environmental factors. When nematode densities are high and favorable environmental conditions prevail, tubers are infected and display galling that gives the tubers a "warty" appearance (Fig. 10). Tubers may become infected and not show galls. White, pear-shaped females can be found inside the galls and usually attached to the females, cream-colored to brownish egg masses can be seen outside the root.

It is not uncommon for *Meloidogyne* spp. to infect potato roots and not produce galls. This situation is particularly true with *M. chitwoodi* (Santo et al., 1980). Also, certain wild tuber-bearing species of *Solanum* when infected with *M. incognita* do not produce galls (Jatala, 1975).

Root-knot nematode damage to potatoes results in a reduction in the size and number of tubers produced as well as a reduction in tuber quality. Indirect damage also occurs when root-knot nematodes predispose the potato plant to other pathogens, such as *Pseudomonas solanacearum*, the causal agent of bacterial wilt of potatoes.

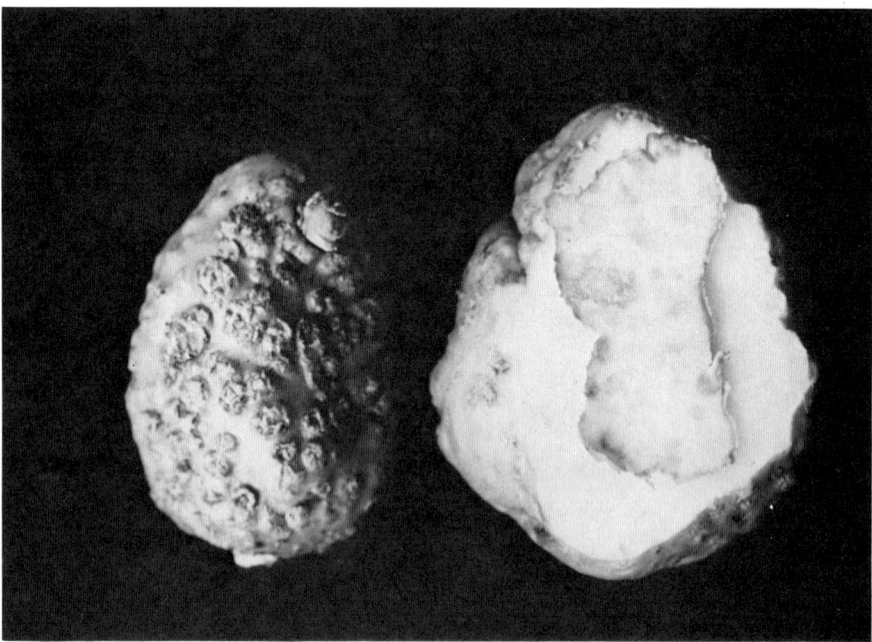

Fig. 10 Potato tuber infected with *M. incognita*. (Courtesy of P. Jatala.)

E. Role in Disease Complexes

Root-knot nematodes are noted for their role in predisposing plants to other
organisms, resulting in disease complexes. Disease complexes involving root-
knot nematodes on potatoes have been recorded for *M. incognita* and *P. solanacearum* (Jatala and Martin, 1977), *M. hapla* and *Verticillium albo-atrum* (Jacobsen et al., 1979), and possibly *M. incognita* and *Rhizoctonia solani* (Joubert
and Dalmasso, 1972). The most notable interaction on potato involves *M. incognita* and *P. solanacearum*. The nematode is thought to provide entry sites
for the bacterium as well as alter the host physiology so that the host is more
quickly colonized by the bacterium (Jatala and Martin, 1977). Potato plants
infected with both the nematode and bacterium show early and more severe
symptoms of wilt and die sooner than those infected with the bacterium alone.
Similar responses of the plant are noted when both *M. hapla* and *V. albo-atrum*
infect potatoes.

F. Distribution

Root-knot nematodes are distributed worldwide, but are often limited to specific areas by temperature and soil type. The so-called northern root-knot nematode, *M. hapla*, has an optimum temperature of 25°C. Other species, except *M. chitwoodi*, have even higher temperature optima and cannot survive
extremely cold temperatures, which obviously limit their geographic range.
Consequently, root-knot nematodes reach their greatest economic importance
in tropical and warmer areas of temperate climates and of lesser importance
in northern latitudes and high elevations of southern latitudes.

Because potatoes are generally grown in the cooler climates of the world,
root knot is not presently a worldwide economic problem on potatoes. How-

ever, certain species have become established and cause severe losses to potatoes in relatively localized areas. The *M. incognita* group (including *M.*
incognita acrita) is perhaps the most widely distributed species of root-knot
nematodes and occurs throughout the tropical, subtropical, and Mediterranean areas of the world as well as the warm or sheltered areas of the temperate
zone (Franklin, 1979; Sasser, 1979). Although cultivated potatoes are extremely susceptible to *M. incognita*, they usually escape severe damage from
this species because they are not grown in its presence or grown during the
part of the year when *M. incognita* is inactive. However, potatoes planted
when temperatures are conducive to *M. incognita* activity suffer severe
damage. For example, potatoes grown in the irrigated valleys of Peru suffer
from both root and tuber infection. *M. arenaria* has been reported from potatoes on most continents but damage is usually slight (Winslow and Willis, 1972).
In Australia and Africa, *M. javanica* is the dominant root-knot nematode species attacking potatoes and under proper conditions of soil temperature
causes severe damage (Bird, 1978).

Because of its lower temperature optima and ability to survive severe
winter temperatures, *M. hapla* is the species most frequently encountered in
potato culture. It has been reported attacking potatoes throughout the
northern areas of the temperate zone. *M. hapla* is the dominant species on
potato in Europe, North America, Japan, and high elevations of South America.
The temperature requirements for *M. hapla* more closely match those of the
potato, thus being a menace wherever potatoes are grown.

A new species of root-knot nematode, *M. chitwoodi*, has recently been
found attacking potatoes in the United States. It is apparently endemic to
the United States, as it has been found in a relatively large area of the Pacific Northwest in the states of Washington, Oregon, and Idaho. Taxonomically,
this species is closely related to *M. hapla* but differs from *M. hapla* in host
range and temperature requirements (Golden et al., 1980). It was probably
confused with *M. hapla* for many years. *M. chitwoodi* reproduces well on
grain crops and at lower temperatures (15-25°C) than does *M. hapla*. It may
eventually become the predominant root-knot species attacking potatoes.

G. Estimated Losses

Root-knot nematodes cause losses to the potato crop by a reduction in the number and size of tubers produced as well as by lowering the marketability of
infected tubers. In tropical areas where root-knot nematodes reach their
greatest potential for damage, losses to the potato crop is presently estimated
to be 24% (Sasser, 1979). Because of the extreme susceptibility of potato to
root knot, this figure will, of course, be much higher if potatoes are grown
in the warmer months or the range of potato culture expands into the lower
tropics as proposed by the International Potato Center. Presently, potatoes
escape much of the damage that could occur in the tropics because they are
grown at high elevations or during cooler months of the year.

Compared with the situation in the tropics, potato losses due to root knot
in cool temperature climates are insignificant, but nonetheless real to the farmer. In localized areas where conditions favor the nematode, such as sandy
soil and rather sheltered areas, serious economic losses occasionally do occur.
In the United States, losses of 25% in potential potato yield were recorded
when conditions favored root-knot infection of potatoes (Sitterly and Fassuliotis, 1965).

H. Control

1. *Chemical*

The most popular and perhaps the most consistent means of controlling root-knot nematodes on potatoes is with chemicals. Soil fumigation with such halogenated hydrocarbons as D-D or ethylene dibromide (EDB) has been the most successful. The effectiveness of such treatments is independent of the species involved. When nematode densities are high, fumigation usually results in a striking reduction in nematode population densities (99%) and increase in potato yields (55%) (Winslow and Willis, 1972).

More recently nonvolatile nematicides, such as aldicarb, phenamiphos, and oxamyl, have shown promise of excellent control of root-knot nematodes on potatoes (Abdel-Rahman et al., 1974). These compounds have also been used effectively as a seed-piece dip to reduce nematode infection of seed (Rodriguez-Kabana, 1977). Such compounds have the added feature of being insecticidal and provide excellent control of certain insects. They are also extremely toxic to humans and must be handled with extreme caution.

2. *Crop Rotation*

One of the oldest means of reducing damage caused by nematodes is crop rotation. On first sight, rotations appear to be very effective in controlling root-knot nematodes as these nematodes have no mechanism of long-term survival without a host plant. Up to 99% reduction in *M. hapla* population densities occurred in one year when no host was present (Kirjanova and Krall, 1971). Although root-knot species differ somewhat in their host range, the root-knot nematodes as a whole record over 2000 plant species as host (Bird, 1978), making choices of alternate crops in rotations very difficult. One case of apparent success with rotations is in Rhodesia, where certain grasses are grown in rotation with potatoes to control *M. javanica* (Jensen et al., 1979). It is, therefore, important to correctly identify the species of *Meloidogyne* involved before embarking upon a scheme of control by crop rotation. In the northwestern United States, recommended control of *M. hapla* on potatoes is rotation with grain crops. Recently, the predominant species of root knot on potatoes in that area was identified as *M. chitwoodi*, to which grains are susceptible (Santo et al., 1980).

3. *Resistance*

The use of resistance is regarded as the most biologically sound and environmentally safe method to control plant-parasitic nematodes. Although sources of resistance to several species of *Meloidogyne* in wild and cultivated species of *Solanum* have been identified, they have been little exploited. This lack of enthusiasm to develop root-knot resistant potato cultivars perhaps relates to the priority of the problem in relation to the priority of the potato cyst nematode problem. Nevertheless, good resistance to *M. incognita*, *M. javanica*, and *M. arenaria* has been identified in *S. sparsipilum* (Jatala and Rowe, 1976). Resistance to *M. hapla* exists in *S. tuberosum* subsp. *andigena* (Brodie and Plaisted, 1977). Efforts are underway to combine these two sources of resistance into a single tetraploid population suitable for transferring resistance to cultivated potatoes.

4. Biological Control

An exciting means of controlling root-knot nematodes on potatoes has recently emerged. This method involves the use of a fungus, *Paecilomyces lilacinus*, which is parasitic on eggs of *M. incognita*. Recent reports indicate that introduction of this fungus into soil infested with *M. incognita* reduced production of viable eggs by 70% (Jatala et al., 1979). The use of biological control agents will undoubtedly gain impetus as constraints on the use of pesticides increase.

V. ROOT LESION NEMATODES

A. History

A disease of potato tubers in Tennessee associated with large numbers of nematodes was described in 1889 (Scribner, 1889). In 1941, severe nematode injury to potatoes was noted on Prairie Island, Platt River, Nebraska. In 1943, the responsible nematode was described as a species of *Pratylenchus* (Thorne, 1961). Because *Pratylenchus* spp. frequently cause lesions on roots colonized by other organisms and the combined action of these associated organisms produce conspicuous lesions, members of this genus became known as "root lesion" nematodes. They are also referred to in some literature as "meadow" nematodes.

Since their first description as parasites of potatoes, root lesion nematodes have been reported associated with poor potato growth in practically every country in which potatoes are grown. It appears that several species of *Pratylenchus* are either indigenous to different countries or have been distributed throughout the world with infested plant material. Although many of them appear to be isolated occurrences, at least 15 species of *Pratylenchus* are reported associated with potato culture. Maximum damage to potatoes from root lesion nematodes usually occurs on coarse-textured soils; partly because the nematodes prefer this soil type and partly because it is the preferred soil type for potato culture.

Although the common name "root-lesion" nematodes infer that *Pratylenchus* spp. are primarily root parasites, some, but not all, species are also known to damage potato tubers, causing a severe reduction in tuber quality.

B. The Nematodes

At least 15 species of *Pratylenchus* have been reported associated with potato culture. Four species, *P. penetrans* Cobb 1917, *P. pratensis* de Man 1880, *P. brachyurus* Godfrey 1929, and *P. scribneri* Steiner 1943 have been reported most frequently as causing major damage to potatoes. Other species reported associated with potatoes include *P. alleni* Ferris 1961, *P. andinus* Lordello et al. 1961, *P. cerealis* Haque 1966, *P. coffeae* (Zimmermann 1898) Filipjev and Stekhoven 1941, *P. crenatus* Loof 1960, *P. hexincisus* Taylor and Jenkins 1957, *P. loosi* Loof 1960, *P. neglectus* (Rench 1924) Filipjev and Stekhoven 1941, *P. teres* Khan and Singh 1975, *P. thornei* Sher and Allen 1953, and *P. vulnus* Allen and Jensen 1951. For the most part, these latter species are isolated occurrences and apparently cause little damage, except *P. coffeae*, which has been reported to cause severe damage to potatoes in Japan (Winslow and Willis, 1972).

Lesion nematodes are migratory endoparasites, and all stages of the nematode are vermiform and active. Although second-stage juveniles emerge from the egg, all stages, including the second, third, fourth, and fifth stages, are infective (Thorne, 1961). Entry is usually behind the root cap, but they may also enter other unsuberized surfaces of roots, rhizomes, and tubers. Entry is accomplished primarily by mechanical pressure and cutting action of the stylet rather than enzymatic action. Once inside the roots, the nematode usually, but not always, excretes substances that cause necrosis of root cells. This necrosis usually occurs ahead of the area penetrated and serves as infection courts for secondary organisms to invade the root. Entry and movement through the roots may be either intercellular or intracellular (Mai et al., 1981).

Males are common in some species of *Pratylenchus* but not in others (Winslow and Willis, 1972). Bisexual reproduction is known to occur in species in which males are abundant. Females lay eggs either in the roots or in soil. Eggs may be deposited singly or in small groups. Females have been observed to lay one egg per day, but because of their migratory nature, total egg production per female is not known.

C. Life Cycle

The nematodes overwinter in soil or roots in all stages of development except egg-laying adults (Mai et al., 1981). More survive the winter in roots than in soil (Kable and Mai, 1968). When suitable temperature and moisture are

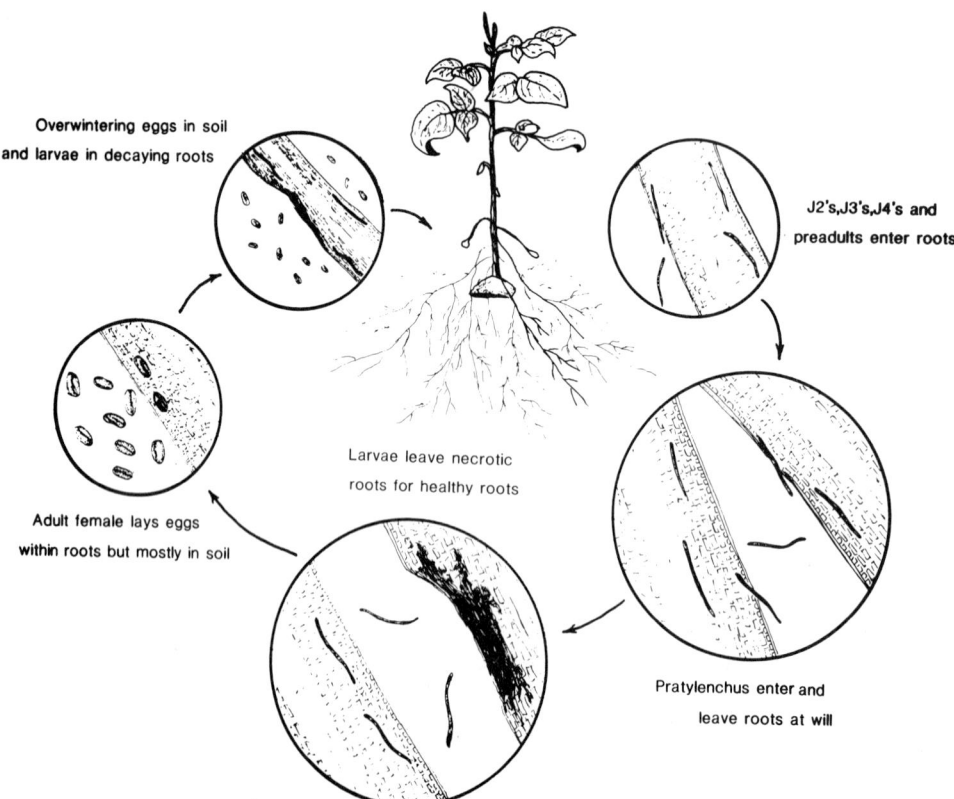

Overwintering eggs in soil and larvae in decaying roots

J2's, J3's, J4's and preadults enter roots

Adult female lays eggs within roots but mostly in soil

Larvae leave necrotic roots for healthy roots

Pratylenchus enter and leave roots at will

Fig. 11 Life cycle of *Pratylenchus* spp. (Illustrated by M. Brucato.)

achieved, the nematodes become active and the life cycle (Fig. 11) commences when a source of food is available. Second-stage juveniles emerge from eggs, and these juveniles and other stages up to and including the fifth stage force their way into the roots and migrate between and through cortical cells. The nematodes feed from the cortical cells as they migrate through the root. Cells in the path of the nematode may become necrotic. The extent of necrosis depends on the action of secondary invaders, which are primarily bacteria and fungi. When necrosis of the roots becomes severe, the nematodes tend to leave in search of nonnecrotic roots. Development through the various stages is accomplished as the nematodes feed, moving in and out of the roots at will.

Upon maturity, which takes anywhere from 28-65 days depending upon host, nematode species, and temperature, the female nematodes deposit eggs either in the roots or in the soil (Winslow and Willis, 1972). If time permits, the eggs will hatch, liberating second-stage juveniles that invade new roots or tubers and produce a second generation. When the plants senesce, the nematodes become inactive and overwinter in the stage to which they have developed. They become active when favorable conditions return and the life cycle is again initiated.

D. Associated Symptoms and Damage

Above-ground symptoms caused by high population densities of lesion nematodes resemble typical plant stress. In the field, areas of poor growth appear, plants are less vigorous, turn yellow, and cease to grow during the latter part of the growing season (Fig. 12). Infected plants will mature somewhat earlier than healthy plants.

Fig. 12 Potato field showing damage caused by *P. penetrans*. (Courtesy of W. F. Mai.)

Fig. 13 Potato tuber infected with *P. penetrans* (rt) compared with healthy tuber (lt). (Courtesy of W. F. Mai.)

Pratylenchus spp. attack both roots and tubers. The only species that have been reported to attack tubers are *P. penetrans*, *P. brachyurus*, *P. scribneri*, and *P. pratensis*. Symptoms on tubers vary with species and may appear as sunken surface lesions, scabs, pimples (*P. brachyurus*, *P. scribneri*) or wart-like protuberances (*P. penetrans*) (Cunningham and Mai, 1947), varying from brown to black in the early season to purple in storage (Fig. 13). Lesions are shallow, remaining in the cortical tissues and rarely penetrating the tuber surface in excess of 0.5 mm (Winslow and Willis, 1972). Although tubers are invaded as soon as they are formed, symptoms usually do not appear until three to four months later (Cunningham and Mai, 1947).

Symptoms on potato roots appear as typical sunken lesions ranging from dark brown to black (Fig. 14). High nematode densities often cause extensive root necrosis resulting from coalescing of lesions and destruction of unsuberized feeder roots. Although lesion nematodes alone are fully capable of destroying plant cells and causing lesions, such lesions are not of the magnitude of those observed in the field where other organisms are present. Lesion formation and root death usually occur ahead of the invading nematode, and the nematodes often migrate out of severely necrotic areas to enter healthy tissues.

E. Interaction with Other Organisms

The nature of injury caused by *Pratylenchus* spp. make the roots particularly suitable for invasion by other organisms. Workers for some time have noted an increase in *Verticillium* wilt and *Rhizoctonia* disease of potatoes in the presence of high densities of *Pratylenchus* spp. (Cetas and Harrison, 1963). Infection of potatoes by *P. penetrans* increases symptom expression and reduces the

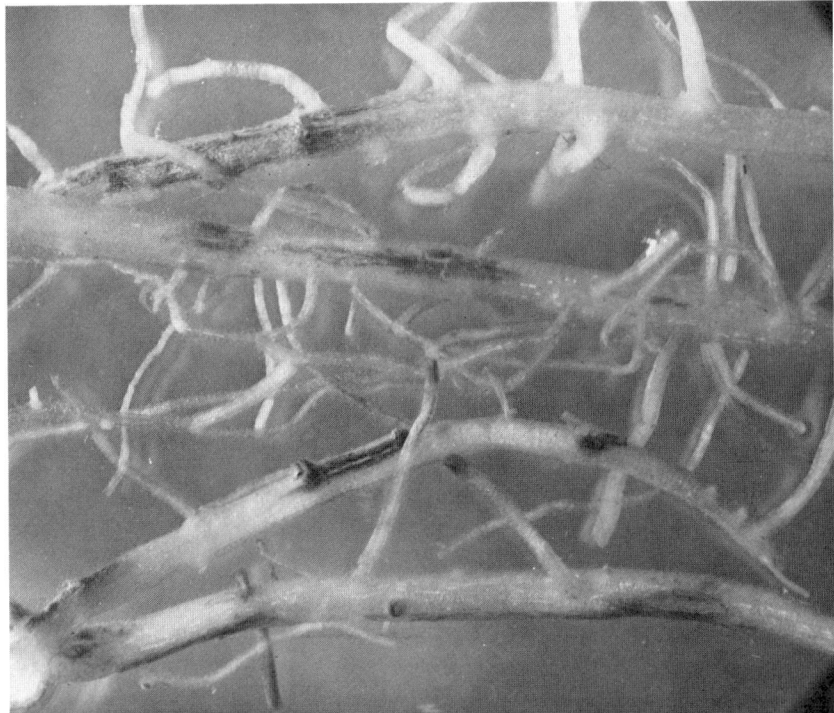

Fig. 14 Lesions on potato roots caused by *P. penetrans*. (Courtesy of W. F. Mai.)

incubation period of *Verticillium albo-atrum* and *V. dahliae*, but, in some cases *V. albo-atrum* infection suppresses numbers of *P. penetrans* in potato roots (Burpee and Bloom, 1978). In addition to increasing symptom expression in *Verticillium*-susceptible varieties, *P. thornei* increases the severity of wilt in *Verticillium*-tolerant varieties to the extent that such varieties often suffer 30-40% loss of yield (Krikun and Orion, 1977).

Most of the proof of *Pratylenchus*-fungus interaction has been gained through experiments on nematode control that lessened the fungus disease. In most cases the *Pratylenchus-Verticillium* complex of potatoes has been successfully controlled by soil fumigation or application of aldicarb to control *Pratylenchus* spp. (Mai et al., 1981).

F. Distribution

Temperature requirements vary with different species, causing different species to predominate in different climates. For example, *P. penetrans* is an important pest of potatoes in Europe and northern potato-producing areas of the United States, but is much less important or even absent in warmer climates. Likewise, *P. brachyurus* predominates on potatoes in the southern United States and parts of Africa where temperatures are higher. *P. coffeae* appears to be the dominant species of *Pratylenchus* attacking potatoes in Japan, but *P. scribneri* and *P. pratensis* appear to predominate in tropical and subtropical climates. It is not uncommon to find two or more species of *Pratylenchus* on potatoes in the same field but one species tends to predominate over the other.

G. Estimated Losses

Lesion nematodes inflict losses to the potato crop either by lowering plant vigor with a subsequent reduction in tuber number and size or by infecting of tubers, which lowers tuber quality. Although the common name root lesion nematode indicates that *Pratylenchus* spp. are root parasites, many species also attack tubers. The species most notable for parasitizing potato tubers are *P. penetrans*, *P. scribneri*, *P. brachyurus*, and *P. pratensis*. The necrotic depressions or protuberances formed on potato tubers by *Pratylenchus* can lower tuber quality by 10-20% (Kirjanova and Krall, 1971). The percentage reduction in tuber quality is dependent upon nematode population density.

The most widespread and important species attacking potatoes is *P. penetrans*. In Europe and parts of North America, this species is capable of causing a growth decline of 50-70% and a yield decline of 10-50% (Jensen et al., 1979). Severe damage and yield losses occur most commonly in coarse-textured soils, although the nematode can survive in heavier soils.

H. Control

1. Chemical

Soil fumigants such as D-D, EDB, Telone II, and Vorlex offer good control of *Pratylenchus* spp. on potatoes but are practical only when yields and prices of potatoes are high. Also, such fumigants are less effective in fine-textured soils than they are in coarse-textured soils (Winslow and Willis, 1972). Vorlex has a more dramatic effect in soil infected with both root lesion nematodes and the *Verticillium* wilt fungus. Under such conditions, fumigation with Vorlex increased yields by 10-14% (Hawkins and Miller, 1971). Fall application of soil fumigants is more feasible in potato culture and appears to be equally as effective in controlling *Pratylenchus* spp. as is spring application.

Nonvolatile nematicides in the organophosphate and oximecarbamate groups have proven to be more practical than soil fumigants in controlling *Pratylenchus* spp. on potatoes. Such compounds as aldicarb, oxamyl, and phenamiphos are less phytotoxic than are soil fumigants. Such compounds can be applied directly in the row during the planting operation and are effective on a broad range of soil types. Oxamyl has also been used successfully as a seed-piece dip to control *P. brachyurus* (Rodriguez-Kabana and Ingram, 1977). The organophosphate and oximecarbamate compounds are extremely poisonous to humans and must be used with extreme caution.

2. Resistance

Although resistance to *Pratylenchus* spp. has been reported in some commercial cultivars and in the wild diploid species *Solanum sparsipilum*, little or no effort has been made to exploit this resistance. The widely used commercial cultivar Russet Burbank suffers less damage from *P. penetrans* attack than do other cultivars but fails to reduce population densities, suggesting tolerance to *P. penetrans* (Bernard and Laughlin, 1976). The cultivar Peconic, which is resistant to *G. rostochiensis*, appears to be less susceptible to *P. penetrans* than are similar cultivars that are susceptible to *G. rostochiensis* (Fawole and Mai, 1976). Resistance in Peconic is expressed as a lower increase in density of *P. penetrans* resulting from fewer eggs laid by each female. Other *G. rostochiensis*-resistant selections have a higher degree of resistance to *P. penetrans* than does Peconic. Some resistance to *P. pratensis* has been found in native cultivars of *S. tuberosum* subsp. *andigena* in South America, but the degree of resistance was not substantial enough to incorporate into a breeding program. Resistance to *P. pratensis* in the wild diploid *S. sparsipilum* approaches immunity (Jatala and Kaltenbach, 1978), but no effort has been made to exploit this resistance.

3. Crop Rotation

Crop rotation offers little in the way of controlling root lesion nematodes of potatoes because of the large number of species involved and the relatively wide host range of most species, particularly the predominant species, *P. penetrans* and *P. brachyurus*. Some success has been realized in controlling species of limited host range and geographic distribution such as *P. pratensis* by crop rotation (Thorne, 1961). Marigold (*Tagetes* spp.) very effectively reduces population densities of most species of *Pratylenchus*. Growing of *Tagetes* spp. to control *Pratylenchus* spp. may be feasible for small garden plots but would be impractical on a field-scale basis, except where *Tagetes* spp. are normally grown for seed.

The choice of winter cover crop can have a marked effect on population densities of *Pratylenchus* spp. Rye, the most commonly used cover crop in the potato producing area of the northeastern United States, suffers no visible damage from *P. penetrans* but is an excellent host for this nematode. Potatoes grown after rye suffer considerable damage from *P. penetrans*. Even the density-reducing effects of growing a resistant cultivar such as Peconic can be negated if a rye cover crop is used (Fawole and Mai, 1976).

VI. FALSE ROOT-KNOT NEMATODE

A. History

Cobb was apparently the first to record specimens of *Nacobbus* when he illustrated a male and juvenile from sugar beet in 1918 which he considered to be *Heterodera schachtii* Schmidt 1881. According to Thorne (1961), these illustrations clearly indicate that the nematodes belong to the genus *Nacobbus* Thorne and Allen 1944. Thus, when the genus *Nacobbus* was erected, its members were referred to as "Cobb's root-gall nematodes." Because the galls they produce are similar in appearance to those produced by some species of *Meloidogyne*, in more recent literature members of *Nacobbus* are referred to as "false root-knot nematodes."

The earliest record of natural infection of solanaceous plants by false root-knot nematodes is that of Franklin (1959) when describing a new species, *Nacobbus serendipiticus* Franklin 1959 from tomatoes. False root-knot was not recognized as an important disease of potatoes until Lordello et al. (1961) described the subspecies *N. serendipiticus bolivianus* Lordello et al. 1961 attacking potatoes in Cochabamba, Bolivia. This subspecies was later found to be an important pest of potatoes in Peru, Argentina, Chile, and Ecuador et al., 1981). Because the galls produced on potato roots appear in beadlike fashion, the disease in Latin America is often referred to as "Rosario," making reference to rosary beads.

Sher (1970) revised the genus *Nacobbus* and recognized only two species. He synonymized *N. batatiformis* Thorne and Schuster 1956, originally described from sugar beets, *N. serendipiticus* from tomato, and *N. serendipiticus bolivianus* from potato with *N. aberrans* (Thorne 1935) Thorne and Allen 1944. The fact that the three species synonymized under *N. aberrans* are reported to have quite different host ranges suggests that they may be different races.

B. The Nematodes

This is *Nacobbus aberrans* (Thorne 1935) Thorne and Allen 1944. The nematodes exhibit pronounced sexual dimorphism, with the adult female transformed into a swollen, irregularly shaped body, but the male remains a typical filiform nematode with a short tail. Infection of plant roots causes formation of conspicuous galls that may or may not contain living females. Females discharge a portion of their eggs in a gelatinous matrix outside their body but a portion of the eggs remain inside the female body (Sher, 1970). They appear to survive and reproduce best at temperatures of 20-26°C but are often found attacking potatoes grown at 15-18°C (Mai et al., 1981). *N. aberrans* seems to not be affected by soil type. Thus, considering its temperature adaptability and ability to parasitize potatoes grown in an array of soil types, it is puzzling why *N. aberrans* has not been recorded attacking potatoes outside South America. There is evidence that *N. aberrans* occurring on potatoes in the South American Andes consists of two or more races differing in pathogenicity (Jatala and Golden, 1977). *N. aberrans* (syn. *N. batatiformis*) that occurs on sugar beet in the western United States is apparently yet another race as it has never been reported attacking potatoes.

C. Life Cycle

Although *N. aberrans* mostly overwinters in the egg stage, some quiescent preadult nematodes may survive and infect plant roots (Thorne, 1961). Presuming that winter survival is in the egg stage, the first molt occurs within the egg and second-stage juveniles emerge and invade small roots, beginning the life cycle (Fig. 15). Inside the roots the juveniles move intracellularly to find a favorable location for feeding. Once feeding begins cells in the feeding site (vascular tissue) increased in size followed by a necrosis of the cortical cells. The juveniles undergo two molts and increase in size. At this point in the life cycle, the nematodes may either leave the roots or continue feeding on the already established sites. Those that continue feeding initiate gall development and produce eggs, completing their life cycle. Depending upon temperature and the "race" of nematode involved, generation time takes 25-50 days (Mai et al., 1981).

A portion of the nematodes that leave the roots complete the final molt and become males or active females. The young females enter larger roots and establish themselves with their heads near the stele. As they once again feed, the surrounding cells enlarge and galls develop. The posteriors of the females extend toward the cortex, and at the surface of the root an opening is formed through which eggs are discharged into a gelatinous matrix exuded by the nematode.

Any infective stage of *N. aberrans* may invade tubers but they seldom develop to maturity. Usually they lie in a semiquiescent state inside the tubers with which they may be disseminated over long distances and become active when the tuber is planted (Mai et al., 1981).

D. Associated Symptoms and Damage

There are no specific above-ground symptoms of false root-knot attack that are of diagnostic value. Infected plants usually show typical symptoms of poor root growth that could be attributed to a variety of factors. Such symptoms appear as irregular patches in a field and increase with age of infestation. Individual infected plants are stunted, show signs of chlorosis, and tend to wilt at mid day.

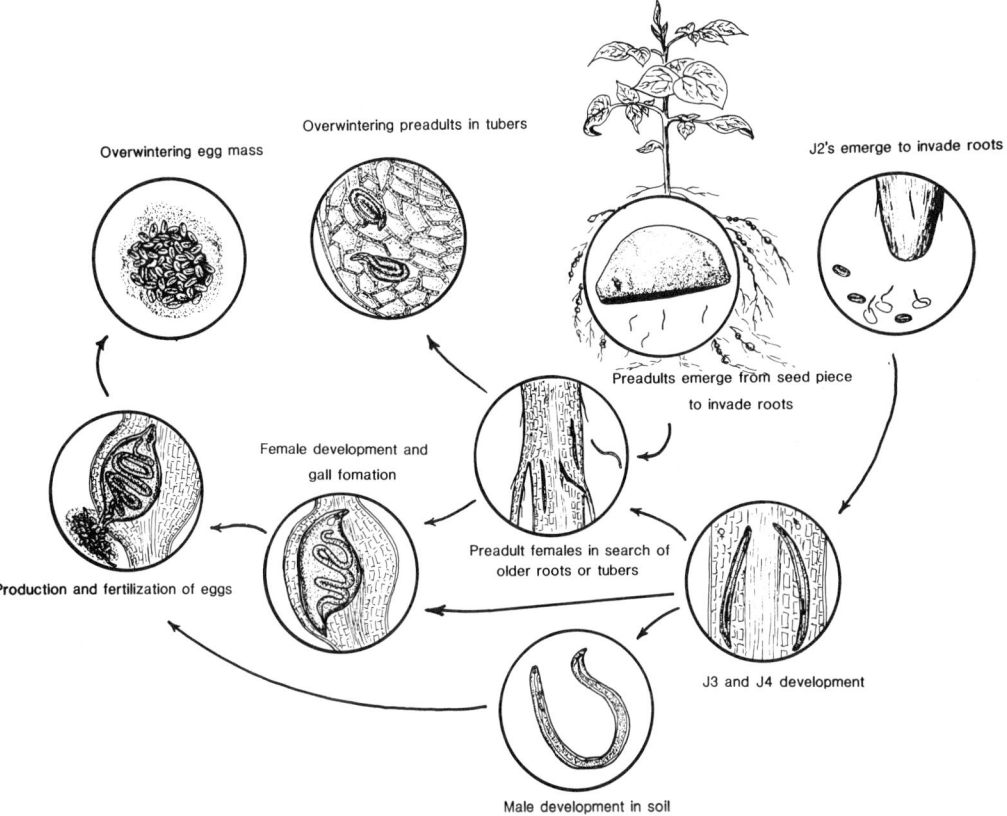

Fig. 15 Life cycle of *N. aberrans*. (Illustrated by M. Brucato.)

Obvious signs of false root-knot attack can be found on the roots, where galls similar to those produced by *Meloidogyne* spp. are evident. The galls are usually spherical in shape and similar to those caused by *M. arenaria* but more obviously occur in a beadlike fashion (Fig. 16). Gall shape and size may vary with nematode density, "race" of *Nacobbus* involved, and root size. Normally the galls lack extension of lateral roots resulting in abnormal fibrous root growth. Massive invasion by juveniles may cause death and deterioration of small roots. The nematode is also capable of attacking potato tubers, but no obvious symptoms or damage result as they penetrate only 1-2 mm deep (Mai et al., 1981).

Damage to potatoes from *N. aberrans* is caused primarily from a weakened and functionally reduced root system that reduces tuber numbers and size.

E. Distribution

Information on the geographic distribution of false root-knot nematodes on potatoes is far from complete. Although *N. aberrans* has been reported from several countries, it has been found damaging potatoes only in South America, where it is apparently endemic to the Andean region of Peru and Bolivia. Elsewhere in South America, *N. aberrans* is reported attacking potatoes in

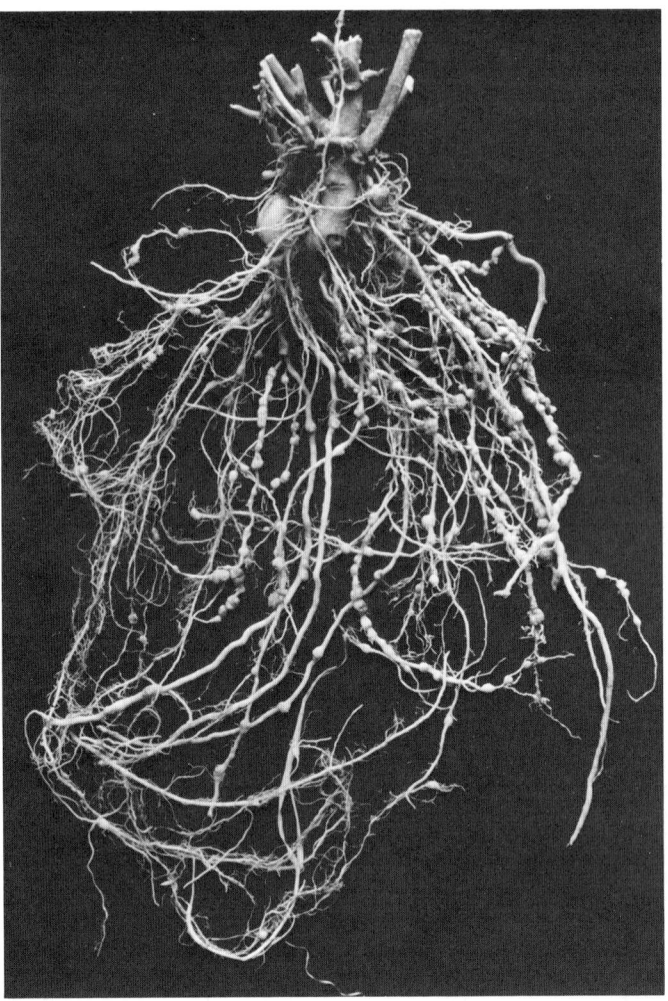

Fig. 16 Galls on potato roots caused by *N. aberrans*. (Courtesy of P. Jata-la.)

Ecuador, Argentina, and Chile. Outside South America, *N. aberrans* has been reported on other crops in the United States (sugar beets), England (tomatoes), and in Mexico, Holland, India, and the USSR on various crops.

Because of its rather wide distribution, ability to survive and reproduce under a wide range of temperatures (15-26°C) and soil types, and extensive host range, it is surprising that *N. aberrans* has not been encountered more frequently on potatoes. Undoubtedly host-specific races exist and the "potato race" has a rather limited geographic distribution, being confined primarily to the South American Andes. Although *N. aberrans* can survive limited periods of desiccation (Mai et al., 1981), its survival mechanisms apparently are not sufficient for worldwide distribution such as is the case with *Globodera* spp. on potatoes.

F. Estimated Losses

Very little information is available on losses to potato caused by *N. aberrans*. The nematode is reported to damage potatoes grown at elevations of 2000-4200 m in South America, where the growers basically rely on potatoes as a major source of income. Where high infestations of *N. aberrans* occur in these areas, yield losses of 55% are not uncommon and losses up to 90% have been reported (Mai et al., 1981). Because *N. aberrans* often occurs together with root-knot and potato cyst nematodes and the galls it produces are often infected with *Spongospora subterranea*, the combined losses to potatoes can be devastating.

G. Control

1. Chemical

Only a few efforts have been made toward the control of *N. aberrans* on potatoes. In South America, acceptable control has been achieved with organophosphate and oximecarbamate nematicides (Mai et al., 1981). Considering the similarity between *N. aberrans* and *Meloidogyne* spp., they would be expected to respond similarily to nematicide treatment.

2. Rotation

Because *N. aberrans* is reported to have a rather extensive host range, selection of nonhost crops to use in rotation schemes would be difficult. There is, however, some evidence that certain members of the *Gramineae* and most of the *Leguminosae* are resistant to *N. aberrans* (Mai et al., 1981). Rotations with such crops could be relatively short because *N. aberrans* populations decline very rapidly in the absence of a host crop.

3. Resistance

A number of *Solanum* spp., including commercial cultivars of *S. tuberosum* subsp. *andigena*, are reported to possess resistance to *N. aberrans* (Alarcon, and Jatala, 1977). The diploid species *S. sparsipilum*, which also possesses resistance to certain *Meloidogyne* spp., appears to presently be the best source of resistance to *N. aberrans* (Mai et al., 1981). In breeding programs in both Peru and Bolivia attempts are being made to transfer resistance to *N. aberrans* into commercially acceptable cultivars. Although there is evidence of the existence of several races of *N. aberrans* and resistance is temperature dependent (Alarcon and Jatala, 1977), resistance appears to offer the best and most economical means of controlling this nematode.

VII. ECTOPARASITIC AND OTHER NEMATODES OF LOCALIZED IMPORTANCE

A. History

Several ectoparasitic nematodes and other species of localized importance to potato production have been reported throughout the world. Certain of these nematodes, such as the stubby-root and dagger nematodes, are important in potato production not only because of the direct damage they may cause but also because they are known to transmit viruses. Other nematodes such as sting and pseudo potato rot nematodes, cause severe damage in localized areas of some countries. Still others, such as spiral, stunt, ring, burrowing, and foliar nematodes, although associated with damaged potatoes, lack positive proof of pathogenicity to potatoes and are cited as potato pests on circumstantial evidence. In greenhouse tests, the reniform nematode has been demonstrated to be pathogenic to potatoes.

It would be space and time consuming to trace the history of the above-mentioned nematodes on potatoes. Many of them have been reported only once and the extent of their involvement in damage to potato and yield losses is unclear. Perhaps the most important event in the history of ectoparasitic nematodes on potato was the discovery that certain species of stubby-root nematodes transmit tobacco rattle virus to potatoes (Sol and Seinhorst, 1961).

B. The Nematodes

1. *Stubby-Root Nematodes*

These include *Trichodorus* and *Paratrichodorus* spp. Until 1974, all stubby-root nematodes were classified under the genus *Trichodorus* Cobb 1913. Ten species of *Trichodorus* were cited as pathogens of potatoes. With the revision of the genus *Trichodorus*, five of these species were placed in a new genus, *Paratrichodorus* Siddiqi 1974 and the remaining five were left in the genus *Trichodorus*. Most of the literature pertaining to stubby root nematodes on potatoes refers to the genus *Trichodorus*. The generic name *Paratrichodorus* is just now emerging in the literature.

The two species of stubby-root nematodes that were first shown to transmit tobacco rattle virus (TRV) to potatoes, *T. christiei* Allen 1957 and *T. pachydermis* Seinhorst 1954, were placed in the genus *Paratrichodorus* (Siddiqi 1974). Later *P. christiei* (Allen 1957) Siddiqi 1974 and *P. minor* (Colbran 1956) Siddiqi 1974 were synonymized, becoming a single species, *P. minor*.

Other species of stubby-root nematodes reported to transmit TRV to potatoes include *P. allius* (Jensen 1963) Siddiqi 1974, *P. nanus* (Allen 1957) Siddiqi 1974, *P. teres* (Hooper 1962) Siddiqi 1974, *T. cylindricus* Hooper 1962, *T. primitivus* (de Man 1880) Micoletzky 1922, *T. proximus* Allen 1957, *T. similis* Seinhorst 1963, and *T. sparsus* Szczygiel 1968. Tobacco rattle virus is also reported to be transmitted to potatoes by *Xiphinema coxi* Tarjan 1964 (Jensen et al., 1979).

2. *Sting Nematode*

This is *Belonolaimus longicaudatus* Rau 1958. Sting nematodes cause serious reduction in potato yield in a rather isolated area of Northeast Florida (Weingartner et al., 1975). Apparently because of their size (females often over 2 mm long), these nematodes are limited in distribution by soil type, requiring soils with at least 80% sand for survival (Brodie, 1976b). They feed ectoparasitically on root tips and along the sides of small roots. Feeding results in dark, sunken lesions along the root axis and necrotic root tips (Thorne, 1961).

3. *Reniform Nematode*

This is *Rotylenchulus reniformis* Linford and Oliveira 1940. When *R. reniformis* was first described from Hawaii, potato was reported as a host crop (Linford and Yap, 1940). It was later reported on potatoes in India and Egypt and recently shown to be pathogenic to potato in greenhouse trials in the mainland United States (Rebois et al., 1978). The reniform nematode is closely related to *Nacobbus* spp. It is a sedentary parasite but is usually not completely embedded in the root. Like *Nacobbus* spp., eggs are deposited in a gelatinous matrix outside the female body. *R. reniformis* exhibits sexual dimorphorism in that adult males are vermiform and adult females swell to become kidney shaped or reniform, thus the species name *reniformis*.

4. Nematodes of Uncertain Parasitism on Potato

There are several species of plant-parasitic nematodes that have been found associated in some way with potatoes. The pathogenicity of most of these species has been established solely on circumstantial evidence, although a few species have been proven pathogenic in controlled experiments or observed in the roots of damaged potato plants.

The species *Neotylenchus vigissi* (Skarbilovich 1952) Goodey 1963 is reported by Russian workers as a parasite of potato leaves, stems, and tubers. This nematode is referred to in Russian literature as "pseudo stem nematode." Because high infestations of *N. vigissi* reduce potato yields up to 50%, Russian scientists believe it should be added to the list of nematode pathogens of potato (Kirjanova and Krall, 1971).

Certain species of nematodes have been cited as potato pathogens on the basis of their presence in diseased plant parts. Such is the case with *Paraphelenchus pseudoparietinus* (Micoletzky 1922) Micoletzky 1925, *Radopholus similis* (Cobb 1893) Thorne 1949, *Aphelenchoides fragariae* (Ritzema Bos 1891) Christie 1937, and an unidentified species of *Meloinema*.

Several species of nematodes are cited as potato pests solely on circumstantial evidence. This evidence was gained either via chemical control experiments in which potato yields were increased or via discovery of new nematode species in soil about potato roots, thus listing potato as the type host. Those species where their number were reduced and chemical treatment resulted in increased potato yields are *Tylenchorhynchus claytoni* Steiner 1937, *T. dubius* (Bütschli 1873) Filipjev 1936, *T. martini* Fielding 1956, *Criconemella onoensis* Luc, *Helicotylenchus dihystera* (Cobb 1893) Sher 1961, *Hoplolaimus galeatus* (Cobb, 1913) Sher 1961, *Longidorus elongatus* (de Man 1876) Thorne and Swanger 1936, *L. leptocephalus* Hooper 1961, and *Xiphinema index* Thorne and Allen 1950.

A number of new species of nematodes have been described from soil around the roots of potatoes. These species include *Aphelenchoides curiolis* Gritsenko 1971, *Ditylenchus solani* Husain and Khan 1975, *D. tenuidens* Gritsenko 1971, *Longidorus nirulai* Siddiqi 1965, *Criconemella peruensiformis* de Grisse 1967, *Nothotylenchus geraerti* Kheiri 1971, *N. taylori* Husain and Khan 1974, *N. tuberosis* Kheiri 1971, *Nygolaimus solaniphilus* Sabava 1974, *Rotylenchulus stakmani* Husain and Khan 1965, and *Scutylenchus mamillatus* Tobar 1966.

A detailed assessment of the pathogenicity of these species and the damage they cause to potatoes is needed before they can be considered with certainty to be potato parasites.

C. Typical Life Cycle

Because eggs are not deposited in masses or retained in their bodies and because of their migratory nature, the life cycles of ectoparasitic nematodes are difficult to study. Consequently, in most cases all that is known is their feeding habits, temperature relations, and time required for egg production.

Because of the general absence of these nematodes from soil in the winter months, it is assumed that for the most part they overwinter in the egg stage.

Once suitable temperatures return and sufficient moisture and host roots are present, juveniles emerge from the eggs and begin to feed. Some species feed only by injecting their stylets into root cells (*Belonolaimus* spp.), and others may force the anterior portions of their bodies into the root (*Helicotylenchus* spp.). Apparently all stages of the nematodes feed on the plant, but in some cases certain stages of development can proceed in the absence of feeding. Eggs are deposited singly in the soil and hatch within a few days given proper moisture and temperature. Newly hatched juveniles may feed on plant roots or develop freely in the soil. At the onset of unfavorable environmental conditions, juvenile and adult stages disappear, leaving the eggs to over-winter. Some species complete their life cycle rather rapidly (*Trichodorus* spp. and *Paratrichodorus* spp. in 16 days) and others take longer.

The life cycle of *R. reniformis* is somewhat unique among plant-parasitic nematodes in that only the adult female requires nutrition (Thorne, 1961). Apparently there is no molt in the egg. First-stage juveniles emerge from the egg and undergo three molts without feeding, producing adult males and females no larger than the juveniles themselves. For further development, the females find a host and embed themselves partly, or occasionally entirely, in the root. The anterior portion soon enlarges to the familar kidney shape, and eggs are deposited in a gelatinous matrix extruded from the nematode. The life cycle is usually completed in 15-25 days (Thorne, 1961).

D. Associated Symptoms and Damage

1. Stubby-Root Nematodes

These nematodes are ectoparasitic, feeding on root cells from outside the root with occasional entrance of their heads into the cells. Feeding occurs in the meristematic region of the root tissues causing cessation of root growth that results in numerous stunted roots, giving the root system a "stubby-root" appearance. Stunted roots show little or no necrosis, discoloration, or other symptoms of injury. Infected plants show unthrifty growth, stunting, and wilting.

Stubby-root nematodes are important pests of potatoes not so much for the direct damage they cause to the crop but more importantly because they transmit a virus to potatoes. This virus, tobacco rattle virus, causes a disease of potato tubers called "spraing" or "corky ring spot." The virus also causes stem mottle consisting of yellowish rings and line patterns or mottling, together with malformation of the leaves.

Tubers infected with TRV become irregularly shaped during early stages of growth. The skin tissue cracks within arc-shaped lesions, and brown concentric rings develop on the surface of many of the tubers. Rusty-brown, irregularly shaped lesions that have a corky texture may appear in the flesh of the tuber. In late season, tubers exhibit deep cracks and shallow corky depressions on their surface which renders them unmarketable (O'Brien and Rich, 1976).

2. Sting Nematodes

There are no published descriptions of symptoms of sting nematode infection of potatoes. Descriptions of symptoms of sting nematode attack on other crops indicate that roots exhibit dark, shrunken lesions along the root axis and at the tip with lesions sometimes girdling the root causing it to break (Thorne, 1961). Field symptoms consist of the usual decline and dying of small plants with stunting and chlorisis of those remaining. In potatoes, such symptoms would be expected to be accompanied by a reduction in the number and size of tubers (Weingartner et al., 1975).

3. Reniform Nematode

There is no clearly published record of symptoms of *R. reniformis* on potatoes. On other crops, it is known that only females are parasitic and under conditions of high infestations, the tap root may be destroyed resulting in severe damage to the plant (Thorne, 1961). This nematode can severely reduce tuber yield and quality but it does not infect tubers (Rebois et al., 1978).

4. Nematodes of Uncertain Parasitism on Potato

Of the several species of nematodes for which only circumstantial evidence of their parasitism of potatoes exist, only three species have actually been observed infecting potatoes, ten species have been implicated as potato pathogens via chemical control experiments, and 13 species described from soil around potato roots. Except for those observed infecting potatoes, the literature offers no indication of the type of symptoms or the extent of damage caused by these species.

The two species of *Neotylenchus* reported associated with potatoes in Russia apparently are severe pathogens of potatoes in localized areas. They attack both stems and tubers. The symptoms and damage produced are similar to those caused by *Ditylenchus* spp. Apparently in the European part of the USSR, *Neotylenchus* spp. inflict as much damage to potato as do *Ditylenchus* spp. (Kirjanova and Krall, 1971).

Meloinema sp. has been observed infecting potato roots and *Radopholus similis* infecting potato tubers, but no mention was made of the type of symptoms or damage they produced (Jatala, 1977; Martin, 1971).

Much detailed experimental work is needed to elucidate the damage to potato caused by those nematodes implicated as potato pathogens via chemical control experiments or those described from soil around potato roots.

E. Distribution

The stubby-root nematodes (*Trichodorus* and *Paratrichodorus*) are distributed worldwide with certain species being limited by soil type and temperature requirements. Certain species predominate in certain parts of the world. In the United States, *P. minor* (*P. christiei*) and *P. allius* are the most common species associated with potatoes, while *P. pachydermus* and *T. primitivus* are most frequently found associated with potato in Europe (Winslow and Willis, 1972).

Sting nematodes are limited in distribution primarily because of their soil and temperature requirements. These nematodes have been reported mainly from tropical and subtropical areas. They have been found associated with potatoes in a relatively small area of northeastern Florida. Their lack of importance as potato pests is probably due to the fact that potatoes are not grown to a great extent in areas where sting nematodes exist or at the time of year they are active.

The reniform nematode, *R. reniformis*, is also primarily a parasite in tropical and subtropical areas. It has been reported as a parasite of potatoes in Egypt, Hawaii, and India (Rebois et al., 1978). The nematode occurs in the southern United States, but its parasitism on potatoes has been limited to greenhouse studies. Again, *R. reniformis* is probably not an important pest of potatoes because potatoes are not grown where high populations of the nematode exist.

Members of the other genera of nematodes associated with potatoes are apparently widespread. Most of them are reported from tropical or subtropical areas of the world. Thus, their occurrence on potatoes is infrequent because potatoes are not grown to any extent in those areas.

F. Estimated Losses

Limited information is available on estimated losses caused by these nematodes. Tubers infected with corky ring spot transmitted by *Trichodorus* or *Paratrichodorus* are rendered useless. The incidence of corky ring spot in northeast Florida was decreased 32-54% when stubby root nematodes were controlled. Yields of tubers were increased 50-150% when sting nematodes were controlled (Weingartner et al., 1975).

Scientists in the USSR report that *Neotylenchus* spp. cause as much damage as *Ditylenchus* spp. in the European part of the USSR (Kirjanova and Krall, 1971).

There is no mention of damage or losses caused by the several nematode species described from potato soils.

G. Control

In general, these nematodes are controlled by soil fumigation. Some species are controlled better than others. For example, stubby-root nematodes have been more difficult to control by soil fumigation than have other species, apparently because they reside deeper in the soil.

The nonvolatile nematicides, particularly aldicarb and oxamyl, have been used with good success to control stubby-root, sting, and reniform nematodes (Weingartner et al., 1975).

Little or no effort has been made to breed potato cultivars that are resistant to these nematodes. Apparently some resistance to reniform nematodes is present in some of the commercial potato cultivars (Rebois et al., 1978).

Because most of these nematodes have relatively wide host ranges among crop species, rotation as a means of control is not generally suggested or practiced.

REFERENCES

Abdel-Rahman, T. B. Elgindi, D. M., and Oteifa, B. A. (1974). Efficacy of certain systemic pesticides in the control of root-knot and reniform nematodes of potato. *Plant Dis. Rep. 58*: 517-520.

Alarcon, C., and Jatala, P. (1977). Efecto de la temperatura en la resistencia de *Solanum andigena* a *Nacobbus aberrans*. *Nematropica 7*: 2-3.

Atkinson, G. F. (1889). A preliminary report upon the life history and metamorphosis of a root gall nematode *Heterodera radicicola* (Greef) Müller, and the injuries caused by it upon the roots of various plants. *Alabama Agric. Exp. Sta. Bull. No. 9.*

Behrens, E. (1975). *Globodera* Skarbilovich, 1959, eine selbständige Gattung in der Unterfamilie Heteroderinae Skarbilovich, 1947 (Nematoda: Heteroderidae). "1. Vortragstagung zu aktuellen Problemen der Phytonematologie am 29.5.1975 in Rostock" pp. 12-26.

Bernard, E. C., and Laughlin, C. W. (1976). Relative susceptibility of selected cultivars of potato to *Pratylenchus penetrans*. *J. Nematol. 8*: 239-242.

Bird, A. F. (1978). Root-knot nematodes in Australia. *CSIRO Aust. Div. Hort Res. Tech. Paper No. 2.*

Blodgett, E. C. (1943). Stem nema on potato: A new potato disease in Idaho. *Plant Dis. Rep. 27:* 658-659.

Brodie, B. B (1976a). Managing population densities of *Heterodera rostochiensis. J. Nematol. 8:* 280.

Brodie, B. B. (1976b). Vertical distribution of three nematode species in relation to certain soil properties. *J. Nematol. 8:* 243-247.

Brodie, B. B. (1980). Control of *Globodera rostochiensis* in relation to method and time of application of nematicides. *J. Nematol. 12:* 215-216.

Brodie, B. B., and Plaisted, R. L. (1977). Breeding for resistance to root-knot nematodes in potatoes. *Nematropica 7:* 2.

Burpee, L. L., and Bloom, J. R. (1978). The influence of *Pratylenchus penetrans* on the incidence and severity of *Verticillium* wilt of potato. *J. Nematol. 10:* 95-99.

Canto Saenz, M. A., and de Scurrah, M. M. (1977). Races of the potato cyst nematode in the Andean region and a new system of classification. *Nematologica 23:* 340-349.

Cetas, R. C., and Harrison, M. B. (1963). Evaluation of fumigants for control of early maturity wilt of potatoes on Long Island. *Phytopathology 53:* 347-348.

Chitwood, B. G. (1949). Root-knot nematodes. Part I. A revision of the genus *Meloidogyne* Goeldi, 1887. *Proc. Helm. Soc. Wash. 16:* 90-104.

Chitwood, B. G. (1951). The golden nematode of potatoes. *USDA Circular 875.*

Committee on Crop Losses. (1971). Estimated crop losses due to plant-parasitic nematodes in the United States. *Society of Nematologists* (USA) Special Publication, No. 1.

Cunningham, H. S., and Mai, W. F. (1947). Nematodes parasitic on the Irish potato. *Cornell Extension Bull. 712.*

Ellenby, C. (1954). Tuber forming species and varieties of the genus *Solanum* tested for resistance to the potato root eelworm *Heterodera rostochiensis* Wollenweber. *Euphytica 3:* 195-202.

Evans, K., and Brodie, B. B. (1980). The origin and distribution of the golden nematode and its potential in the U.S.A. *Am. Potato J. 57:* 79-89.

Evans, K., Franco, J., and de Scurrah, M. M. (1975). Distribution of species of potato cyst nematodes in South America. *Nematologica 21:* 365-369.

Evans, K., and Stone, A. R. (1977). A review of the distribution and biology of the potato cyst-nematodes *Globodera rostochiensis* and *G. pallida. PANS 23:* 178-189.

Evans, K., and Trudgill, D. L. (1978). Pest aspects of potato production. Part I. Nematode pests of potatoes. In *The Potato Crop* (P. M. Harris, ed.). Chapman and Hall, London, pp. 440-469.

Fawole, B., and Mai, W. F. (1976). Population dynamics of *Pratylenchus penetrans* in a potato-rye crop rotation. *Proc. Am. Phytopathol. Soc. 2:* 97.

Franco, J. (1979). Effect of temperature on hatching and multiplication of potato-cyst nematodes. *Nematologica 25:* 237-244.

Franklin, M. T. (1959). *Nacobbus serendipiticus* n. sp., a root-galling nematode from tomatoes in England. *Nematologica 4:* 286-293.

Franklin, M. T. (1979). Economic importance of Meloidogyne in temperate climates. In *Root-Knot Nematodes (Meloidogyne Species) Systematics,*

Biology and Control (F. Lamberti and C. E. Taylor, eds.), Academic, New York, pp. 331-339.

Golden, A. N., O'Bannon, J. H., Santo, G. S., and Finley, A. M. (1980). Description and SEM observations of *Meloidogyne chitwoodi* n. sp. (Meloidogynidae), a root-knot nematode on potato in the Pacific Northwest. *J. Nematol.* 12: 319-327.

Hawkes, J. G. (1978). History of the potato. In *The Potato Crop* (P. M. Harris, ed.). Chapman and Hall, London, pp. 1-14.

Hawkins, A., and Miller, P. M. (1971). Row treatments of potatoes with systemics for meadow nematode (*P. penetrans*) control. *Am. Potato J.* 48: 21-25.

Hesling, J. J., Pawelska, K., and Shepherd, A. M. (1961). The response of potato root eelworm, *Heterodera rostochiensis* Wollenweber and beet eelworm, *H. schachtii* Schmidt to root diffusates of some grasses, cereals and of *Tagetes minuta*. *Nematologica 6*: 207-213.

Howard, H. W., and Fuller, J. M. (1971). Resistance to the cream and white potato cyst nematodes. *Plant Pathol.* 20: 32-35.

Inagaki, H., and Kegasawa, K. (1973). Discovery of the potato cyst nematode, *Heterodera rostochiensis* Wollenweber, 1923. (Tylenchida: Heteroderidae) from Peru guano. *Appl. Entomol. Zool.* 8: 97-102.

Jacobsen, B. J., MacDonald, D. H., and Bissonette, H. L. (1979). Interaction between *Meloidogyne hapla* and Verticillium wilt disease of potato. *Phytopathology 69*: 288-292.

Jatala, P. (1975). Root-knot nematodes (*Meloidogyne* species) and their effects on potato quality. *6th Triennial Conference of the EAPR, 15-19.* Wageningen, Netherlands, p. 194.

Jatala, P. (1977). A new nematode attacking potatoes in Peru. *Nematropica* 7: 10.

Jatala, P., and Golden, A. M. (1977). Taxonomic status of *Nacobbus* species attacking potatoes in South America. *Nematropica 7*: 9-10.

Jatala, P. and Kaltenbach, R. (1978). Reaction of some Peruvian potato cultivars to *Pratylenchus pratensis* (abstract). *J. Nematol.* 10: 290.

Jatala, P., Kaltenbach, R., and Bocangel, M. (1979). Biological control of *Meloidogyne incognita acrita* and *Globodera pallida* on potatoes. *J. Nematol.* 11: 303.

Jatala, P., and Martin, C. (1977). Interactions of *Meloidogyne incognita acrita* and *Pseudomonas solanacearum* on field grown potatoes. *Proc. Am. Phytopathol. Soc.* 4: 177-178.

Jatala, P., and Rowe, P. R. (1976). Reaction of 62 tuber-bearing *Solanum* species to the root-knot nematode, *Meloidogyne incognita acrita*. *J. Nematol.* 8: 290.

Jensen, H. J., Armstrong, J., and Jatala, P. (1979). Annotated bibliography of nematode pest of potato. International Potato Center, Lima, Peru.

Jones, F. G. W. (1970). The control of the potato cyst-nematode. *R. Soc. Arts. J.* 117: 179-199.

Jones, F. G. W. (1972). Management of nematode populations in Great Britain. *Proc. Tall Timbers Conf. February 24-25, 1972,* Tall Timbers Res. Sta., Tallahassee, Fl., pp. 81-107.

Jones, M. G. K., and Northcote, D. H. (1972). Nematode-induced syncytium: A multinucleate transfer cell. *J. Cell Sci.* 10: 789-809.

Joubert, J., and Dalmasso, A. (1972). Possible interaction between *Rhizoctonia solani* and *Meloidogyne incognita* on potato. *Ann. Phytopathol.* 4: 409.

Kable, P. F., and Mai, W. F. (1968). Overwintering of *Pratylenchus penetrans* in a sandy loam and a clay soil at Ithaca, New York. *Nematologica* 14: 150-152.

Kirjanova, E. S., and Krall, E. L. (1971). *Plant Parasitic Nematodes and Their Control*, Vol. II. Nauka Publishers, Leningrad Section, Leningrad (English translation, Amerind Publishing Co. Pvt. Ltd. New Delhi).

Kort, J., and Jaspers, C. P. (1973). Shift of pathotypes of *Heterodera rostochiensis* under susceptible potato cultivars. *Nematologica* 19: 538-545.

Kort, J., Ross, H., Rumpenhorst, H. H., and Stone, A. R. (1977). An international scheme for identifying and classifying pathotypes of potato cyst-nematodes *Globodera rostochiensis* and *G. pallida*. *Nematologica* 23: 333-339.

Krikun, J., and Orion, D. (1977). Studies of the interaction of *Verticillium dahliae* and *Pratylenchus thornei* on potato. *Phytoparasitica* 5: 67.

Linford, M. B., and Yap, F. (1940). Some host plants of the reniform nematode in Hawaii. *Proc. Helm. Soc. Wash.* 7: 42-44.

Lordello, L. G. E., Zamith, A. P. L., and Boock, O. J. (1961). Two nematodes found attacking potato in Cochabamba, Bolivia. *Anais Acad. Brasil. Ciencias 33*: 209-215.

Mai, W. F. (1977). Worldwide distribution of potato-cyst nematodes *Heterodera rostochiensis*, and *H. pallida* and their importance in crop production. *J. Nematol. 9*: 30-34.

Mai, W. F., Brodie, B. B., Harrison, M. B., and Jatala, P. (1981). Nematode diseases of potatoes. In *Compendium of Potato Diseases* (W. J. Hooker, ed.). American Phytopathology Society, St. Paul, MN., pp. 93-101.

Mai, W. F., and Harrison, M. B. (1959). The golden nematode. *Cornell Extension Bull. No. 870.*

Martin, G. C. (1971). Infection and reproduction of the burrowing nematode in potato. *FAO Plant Protection Bull. 19*: 91-92.

Mulvey, R. H., and Stone, A. R. (1976). Description of *Punctodera matadorensis* n. gen., n. sp. (Nematoda: Heteroderidae) from Saskatchewan with lists of species and generic diagnoses of *Globodera, Heterodera* and *Sarisodera. Can. J. Zool. 54*: 772-785.

Neal, J. C. (1889). The root-knot disease of the peach, orange and other plants in Florida, due to the work of *Anguillula. USDA Bull. No. 20.*

O'Brien, M. J., and Rich, A. E. (1976). Potato diseases. *USDA Agriculture Handbook No. 474.*

Olefir, V. V. (1972). Nematodnye bolezni sel'skokhozyaistvennykh kul-tur i mery bor'by s nimi. *Tezisy soveshchaniya. Moskva dekabr'* 1972. Moscow, USSR VASHNIL 101-103.

Plaisted, R. L. (1971). 400 years of potato evolution. *N.Y. Food Life Sci. Q. 4*: 24-26.

Rebois, R. V., Eldridge, B. J., and Webb, R. E. (1978). *Rotylenchulus reniformis* parasitism of potatoes and its effect on yields. *Plant Dis. Rep. 62*: 520-523.

Riedel, R. M., and Mai, W. F. (1971). Pectinases in aqueous extracts of *Ditylenchus dipsaci. J. Nematol. 3*: 28-38.

Rodriguez-Kabana, R., and Ingram, E. G. (1977). Treatment of potato seed-pieces with oxamyl for control of plant-parasitic nematodes. *Plant Dis. Rep. 61*: 29-31.

Ross, H. (1972). Mejoramiento de la papa en Alemania para resistencia a las dos especies de nematodo del quiste. In *Prospect for the Potato in the Developing World* (E. R. French, ed.). International Potato Center,

Lima, Peru, pp. 181-190.

Santo, G. S., O'Bannon, J. H., Finley, A. M., and Golden, A. M. (1980). Occurrence and host range of a new root-knot nematode (*Meloidogyne chitwoodi*) in the Pacific Northwest. *Plant Dis. 64:* 951-952.

Sasser, J. N. (1979). Economic importance of *Meloidogyne* in tropical countries, in *Root-Knot Nematodes (Meloidogyne species), Systematics, Biology and Control* (F. Lamberti and C. E. Taylor, eds.). Academic, New York, pp. 359-374.

Scribner, F. L. (1889). Diseases of the Irish potato. *Tenn. Agr. Exp. Sta. Bull. 2:* 27-43.

Sher, S. A. (1970). Revision of the genus *Nacobbus* Thorne and Allen, 1944 (Nematoda: Tylenchoidea). *J. Nematol. 2:* 228-235.

Sitterly, W. R., and Fassuliotis, G. (1965). Potato losses in South Carolina due to the cotton root-knot nematode, *Meloidogyne incognita acrita. Plant Dis. Rep. 49:* 723.

Sol, H. H., and Seinhorst, J. W. (1961). The transmission of rattle virus by *Trichodorus pachydermus. Tijdschr. Piziekt. 67:* 307-309.

Spears, J. F. (1968). The golden nematode handbook, survey, laboratory, control and quarantine procedures. *USDA Agriculture Handbook No. 353.*

Stone, A. R. (1972). *Heterodera pallida* n. sp. (Nematoda: Heteroderidae), a second species of potato cyst-nematode. *Nematologica 18:* 591-606.

Stone, A. R., Thompson, P. R., and Hopper, B. E. (1977). *Globodera pallida* present in Newfoundland. *Plant Dis. Rep. 61:* 590-591.

Thorne, G. (1961). *Principles of Nematology.* McGraw-Hill, New York.

United States Department of Agriculture. (1980). *Agricultural Statistics.* U.S. Government Printing Office, Washington, D.C., pp. 181-183.

Weingartner, D. P., Shumaker, J. R., Smart, G. C., and Dickson, D. W. (1975). A new nematode control program for potatoes grown in Northeast Florida. *Proc. Fla. State Hort. Soc. 88:* 175-182.

Whitehead, A. G. (1973). Control of cyst nematodes *Heterodera* spp. by organophosphates, oximecarbamates and soil fumigants. *Ann. Appl. Biol. 75:* 439-453.

Whitehead, A. G. (1975). Chemical control of potato cyst-nematode. *Res. Rev. 1:* 17-23.

Winslow, R. D., and Willis, R. J. (1972). Nematode diseases of potato. In *Economic Nematology* (J. M. Webster, ed.). Academic, New York.

Chapter 7

Nematode Parasites of Tobacco

Kenneth R. Barker and George Blanchard Lucas *North Carolina State University, Raleigh, North Carolina*

I. INTRODUCTION

Nematodes pose serious hazards to tobacco production throughout the world. Fortunately, the short growing season required for tobacco often favors escape from the severest effects of nematode injury, particularly, when favorable seasons allow continuous lush growth (Graham et al., 1964). The striking effects of environmental factors and cultural practices on tobacco growth and quality have long been recognized (Chapman, 1920). Nevertheless, species of *Meloidogyne* or *Pratylenchus* often cause important yield losses of this crop as well as impairment of quality. Several control tactics usually are necessary for profitable production of tobacco, especially in warm geographic regions.

Although individual genera or species of nematodes will be treated separately in this chapter, it should be emphasized that interactions between various species of nematodes and/or associated microflora occurring in the polyspecific communities are often the norm (Lucas, 1975; Milne, 1972; Powell, 1971, 1979). In addition to competition, synergistic interactions from the viewpoint of crop damage or nematode population increase are also likely (Seinhorst, 1970). Experimental combinations of various nematode species frequently result in suppression of population increases (Johnson and Nusbaum, 1970; Miller and Wihrheim, 1968), but the effects of such interactions on yield under field conditions have received little attention. *Globodera tabacum* often occurs by itself in tobacco fields and may be inhibitory to *Pratylenchus penetrans* or other nematode species when mixed under experimental conditions (Miller and Wihrheim, 1968). In contrast, a nematode root-disease complex involving *Meloidogyne incognita* and *Pratylenchus* species has been described (Graham et al., 1964). Precise determinations of the etiology of such diseases is difficult. More importantly, much of the damage, often attributed to nematodes such as *Meloidogyne* and *Pratylenchus* species, is the result of interactions of nematodes with associated fungi or bacteria (Powell, 1971, 1979; Lucas et al., 1955; Sasser et al., 1955).

Nematode damage often resembles or, in fact, causes nutrient deficiencies on many crops. Leaves of infected plants typically exhibit nitrogen deficiency, and potassium deficiency is often seen. Deficiencies of such minor elements as boron may be more severe in burley tobacco parasitized by *M. incognita* or *M. javanica* than in healthy plants (Arcia et al., 1976).

By 1952, methods of studying the pathogenicity of nematodes and control measures were open to serious question (Chitwood and Oteifa, 1952). The refinements of concepts for investigating nematodes by Steiner (1953), Mountain (1960), and others set the stage for more precise characterization of nematode diseases on tobacco and other crops. Because of the high value per hectare of tobacco, much of the research on species of nematodes attacking this crop has focused on the related nematode development and crop responses. As a result, most of the research on the host-parasite relationships, ecology, and control has been accomplished on tobacco. Numerous cultivars of tobacco resistant to *M. incognita* are available, and sources of resistance or tolerance are available for certain other nematodes attacking this crop (Lucas, 1975; Milne, 1972; Clayton et al., 1958; Schweppenhauser, 1968; Slana and Stavely, 1979).

II. ROOT-KNOT NEMATODES

Although tobacco is a host for numerous species of plant-parasitic nematodes, species of *Meloidogyne* are the most widespread and important. These pests are especially damaging in sandy or sandy loam soils in warm climates. As indicated previously, much of the total yield loss associated with these nematodes is the result of their concerted interactions with numerous fungi and bacteria (Powell, 1971, 1979).

A. History

Roots galled by root-knot nematodes were first described by Berkeley (1855). The different specific names that have been applied to root-knot nematodes were reviewed by Chitwood (1949), who described most of the species that parasitize tobacco. That *Meloidogyne* spp. enhance the development of black shank of tobacco has been known for 50 years (Tisdale, 1931; Tyler, 1933). Nevertheless, recognition of the magnitude of crop damage caused by these nematodes and the development of effective control tactics have occurred only in the last 30 years. Root knot probably has been present in tobacco-growing areas of the southeastern United States for centuries, but it received only limited attention until approximately 1950. At that time, striking evidence became available that root knot and disease complexes associated with root knot were causing excessive losses, as indicated by a combination of factors: (1) the spread of black shank into root-knot nematode-infested fields; (2) occurrence of several dry, hot years in succession; (3) the use of newly developed cultivars, which were especially susceptible to *Meloidogyne* spp.; and (4) the increased use of chemical soil treatments. The research following these demonstrations of the magnitude of the disease problem brought about a reduction of the estimated disease losses from 5 to 10% in the 1950s to less than 1% in the late 1970s for most tobacco-producing areas (Todd, 1976). This reduction in losses to nematodes in tobacco production reflects a new era in nematology that began in the early 1940s. Major findings that occurred during this 35-year period include: (1) the discovery of DD as a soil fumigant (Carter, 1943); (2) the discovery of races of the root-knot nematode (Christie and Albin, 1944); (3) the delineation and description of the major species of *Meloidogyne* (Chitwood, 1949); (4) the demonstration of the role of these nematodes in disease complexes on tobacco and other crops (Powell, 1971); (5) the development of multi-disease resistant cultivars; and (6) the emergence of modern concepts of integrated pest management, nematode population dynamics, and other improved tactics for nematode control (Lucas, 1975; Rabb and Guthrie, 1970; Todd, 1979).

B. Species

Four species of root-knot nematodes are frequently associated with tobacco. These include *M. incognita*, *M. javanica*, *M. arenaria*, and *M. hapla*. The relative importance of the major *Meloidogyne* species on tobacco is difficult to assess because of variations in ecological adaptation and host-parasite relationships between different populations within the various species (Daulton and Nusbaum, 1961; Graham, 1968; Nusbaum, 1969). Overall, *M. incognita* is the most important species that parasitizes tobacco, closely followed by *M. javanica* (Nusbaum, 1969). *M. arenaria*, although highly virulent, is encountered less frequently than these two species. Large numbers of *M. hapla* usually are necessary to cause a serious crop loss. *M. incognita* occurs most commonly in temperate zones, whereas *M. javanica* is often the more important nematode in subtropical and tropical soils. Nevertheless, these species may occur together in the same field and may be associated with *M. hapla* or *M. arenaria*. A fifth species, *M. grahami*, has been described (Golden and Slana, 1978), but this nematode may be a variant of *M. incognita*

as initially indicated (Graham, 1968). In any case, this race or species of
nematode has the capacity to attack the cultivars that have been developed
for high resistance to *M. incognita* (Graham, 1968).

C. Associated Symptoms

The effects of root-knot nematodes on tobacco are so distinct, that Clayton
et al. (1944) suggested that a description was unnecessary. Severe stunting,
nitrogen deficiency, and wilting on hot days characterize plants affected by
root knot. The characteristic wilting is caused by increased transpiration
during the first two months after infection with *M. incognita* (Odihirin, 1971).
Root galls and decay, however, undoubtedly suppress water uptake as the
disease progresses, resulting in a decline in transpiration. Severely infected
plants may die, especially during dry weather. A key diagnostic characteristic to root-knot damage is the uneven distribution of affected plants within
the same field. In some areas plants may be severely affected, whereas
plants in other parts of the field may be free of symptoms. In addition to
nitrogen deficiency, other mineral deficiencies, such as potassium, or drought
injury, may occur even when moisture and fertilizer are adequate. The
leaves of the lower portion of the plant may show yellow and/or necrotic
margins and tips, which may result in destruction of as much as one-third or
more of the leaf area. Striking flagging of such leaves is commonly followed
by premature yellowing and rim firing of the older leaves, which sometimes
must be harvested green or immature if harvested at all. The magnitude of
these types of symptoms depends on the initial densities of the nematodes.
In fact, low numbers may have only a slight effect on plant quality or yield.
The key characteristic symptoms of root knot on tobacco are galls on the
roots (Fig. 1). The size of the galls may vary from that of a pinhead (2 mm)
for single infections of *M. hapla* to many times the thickness of the root. The
galls are irregular and spindle shaped or spherical, and the size and shape
of the galls varies with different species of nematodes. Usually, *M. javanica*
will cause the most extensive, very large galls of 2-5 cm in diameter, whereas
M. hapla induces the formation of many very small galls throughout a normal-
size root system. *M. arenaria* often causes more of beaded types of galls on
tobacco as compared to those induced by *M. incognita*, which often are similar
to the larger galls caused by *M. javanica*.

An equally important symptom associated with *M. javanica*, *M. arenaria*,
and *M. incognita* is the extensive root decay and necrosis (Fig. 1) (Clayton
et al., 1944, 1958). Interactions of these nematodes with numerous fungi and
bacteria are responsible for this damage (Powell, 1971, 1979; Lucas et al.,
1955; Sasser et al., 1955). Leaves of root-knot nematode infected plants also
may be more susceptible to attack by the brown-spot fungus *Alternaria
alternata* and several other foliage pathogens (Milne, 1972; Powell, 1971, 1979;
Tsamagari and Tanaka, 1954). Because of the available information in several
detailed treatments of the interactions of nematodes and other pathogens on
tobacco (Lucas, 1975; Milne, 1972; Powell, 1971, 1979) the topic will not be
discussed extensively here. Nevertheless, the importance of the changes in
host physiology induced by *Meloidogyne* spp. (Powell, 1971; Van Gundy et al.,
1977) in enhancing plant diseases cannot be overemphasized.

Secondary characteristics of severe infestation of *Meloidogyne* in tobacco
fields include uneven stunting of plants (Fig. 2) and extensive weed cover as
the weeds become more competitive for water and nutrients. As the crop

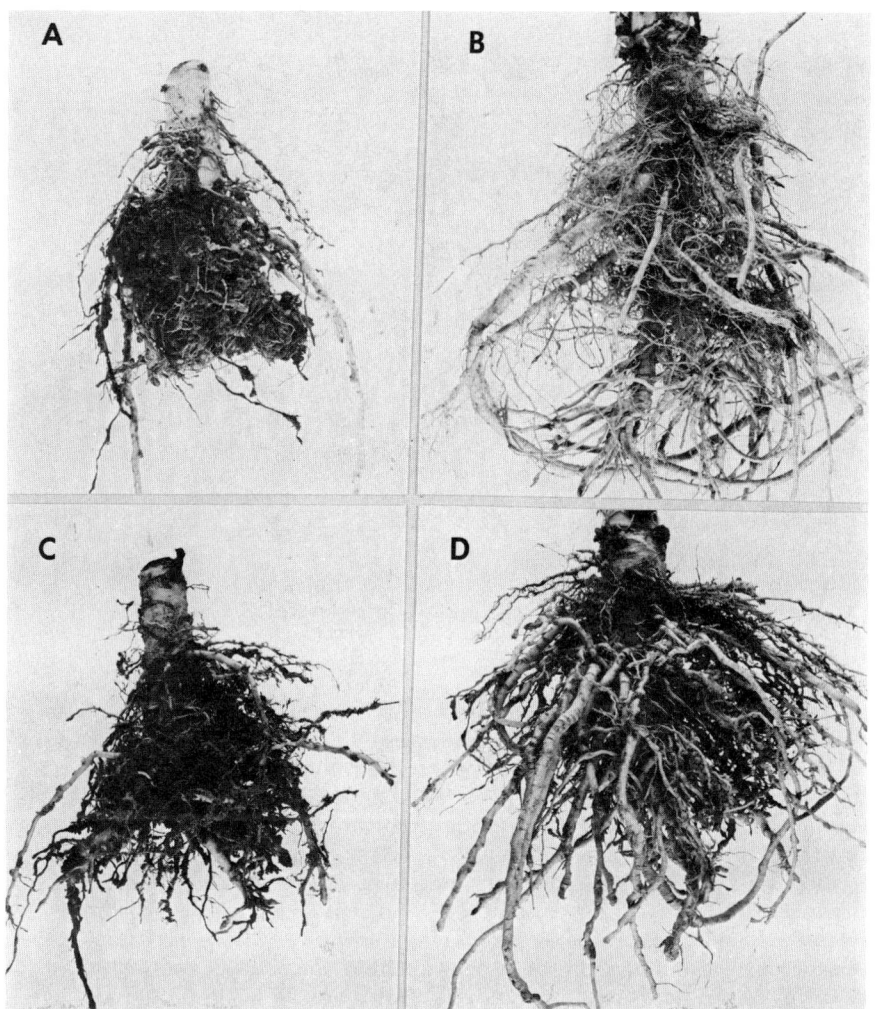

Fig. 1 Tobacco root galls induced by *Meloidogyne* species in microplots. (A) Severe galls and necrosis induced by *M. arenaria* on cv. McNair 944. (B) Limited galls (with no necrosis) caused by *M. arenaria* on cv. Speight G-28. (C) Severe galls and necrosis induced by *M. javanica* on cv. McNair 944. (D) Moderate galls and slight necrosis caused by *M. javanica* on cv. Speight G-28. (From Barker, unpublished data.)

approaches the final harvest period, heavily parasitized plants usually produce no suckers, whereas extensive growth of new shoots occurs on noninfected plants.

D. Life Cycle

The life cycles of *Meloidogyne* spp. on tobacco are typical of the genus (Milne, 1972). Reproduction of *M. incognita*, *M. javanica*, and *M. arenaria* is exclusively parthenogenetic; *M. hapla* is facultatively parthenogenetic

Fig. 2 Uneven growth and stunting of tobacco plants growing in a field
heavily infested with root-knot nematodes. (Courtesy of F. A. Todd.)

(Triantaphyllou and Hirschmann, 1980). Millions of *Meloidogyne* eggs are
found in a given hectare of soil. The first molt occurs while the juvenile is
developing inside the egg, and the second-stage or infective juvenile emerges
from the egg. These juveniles may move considerable distances up to 50-75
cm, even in short periods of time (Prot, 1976) and thereby may find the roots
of tobacco. A large number of compounds associated with plant roots are
attractive to nematodes (Viglierchio, 1979). After penetrating a susceptible
or resistant plant, a juvenile enters and migrates into the axial cylinder of
the root where the life cycle is completed. The late second-stage juveniles
continue to feed and become somewhat sausage shaped. The nematodes' salivary secretions injected into the host cells include compounds that stimulate
the plant to develop giant cells or syncytia (Bird, 1979). The nematode continues to feed only on these giant cells except for a short period as the third
and fourth juvenile stages. As a female reaches maturity and undergoes the
final molt, it commences feeding again, and may produce from 300 to more
than 2000 eggs. Normally, many thousands of eggs are deposited in a gelatinous matrix in masses along the surface of the roots, whereas others may be
formed inside the roots. As these newly developed eggs are produced, they
hatch and secondary infections occur. After emergence, males often move
to the eggs outside the roots and tend to aggregate in the egg masses of
females even though most do not participate in the reproductive process.

The time required for completion of the life cycle varies with species
and environmental conditions. For example, *M. javanica* may require only 21
days in the summer but 56 or more days during cool weather (Milne and
du Plessis, 1964). Furthermore, tropical populations of this species tend to
be more tolerant of high temperatures than those from temperate climates,
but more sensitive to cold (Daulton and Nusbaum, 1961).

The life cycle of *Meloidogyne* spp. may be affected by initial population
densities and environmental conditions (Triantaphyllou, 1973). When food is
scarce, or with large numbers of infective juveniles, greater numbers of

males often develop. For certain species, such as *M. incognita*, sex reversal may be responsible for this observed difference in sex ratios. The use of some growth regulators (maleic hydrazide) also may influence the sex ratios of root-knot nematodes on tobacco (Triantaphyllou, 1973; Nusbaum, 1958).

E. Distribution

Because *Meloidogyne* spp. are able to parasitize more than 3000 plant species, they are distributed throughout much of the world (Sasser, 1980). Still, species of *Meloidogyne* have limitations in their geographic distribution. *M. hapla*, the northern root-knot nematode, generally occurs in the cooler regions of the world, yet it may be found in tropical countries, such as Brazil (Lordello and Monteiro, 1974). *M. incognita*, the southern root-knot nematode, typically is found in temperate zones as well as in the tropics, whereas *M. javanica* and *M. arenaria* occur in the tropics as well as in limited distribution in temperate zones. *Meloidogyne* spp. have great temporal as well as spatial distribution variation. Being obligate parasites the vertical distribution is largely dependent on root growth (Wallace, 1973). The extensive host damage characteristic of *M. javanica*, *M. arenaria*, and *M. incognita* and the deposition of hundreds to thousands of eggs at given foci are partially responsible for the uneven horizontal distribution of these nematodes. The temporal changes in distribution of this nematode are especially striking (Barker et al., 1969). Greatest numbers tend to occur at midseason to shortly after final harvest of tobacco, depending on initial densities. Where high initial densities occur, the peak populations will occur at midseason, followed by a rapid decline due to damage to the root system of the host. In contrast, the peak populations for foci with low initial numbers will occur at the time of final harvest or late summer to early fall (Fig. 3).

The practice of moving tobacco transplants from various geographic localities and within grower communities undoubtedly has added to the wide distribution of these nematodes. Frequently, growers who have had no nematodes have borrowed or purchased plants that were infected with *Meloidogyne* spp., thereby facilitating the wider distribution of this most imporant pathogen of tobacco.

F. Estimated Crop Losses

Meloidogyne spp. cause two types of direct losses on tobacco. These are (1) altered physiological processes and stunted plants, and (2) the unusually large number of disease complexes of the roots and the tops involving these nematodes as described by Powell (1971), which are of equal or greater importance. For example, the current estimates of approximately 1% loss to root-knot nematodes in North Carolina (Todd, 1976) probably are too low as they do not include the effects of these disease complexes.

The secondary economic losses resulting from nematodes are of much greater significance than those from direct losses. Because taxes on tobacco are greater than the value of the crop, there is at least a dollar's worth of indirect loss in taxes for each dollar loss in yield (Milne, 1972). Furthermore, the added costs of production via chemical soil treatments may exceed the value of direct losses. For example, the cost of chemical soil treatments, often used by farmers as "insurance" against nematode and related soilborne diseases in North Carolina, amounts to about $20 million annually, whereas the estimated direct yield losses caused by nematodes are estimated at $5 million per year (Todd, 1979). During the last decade, quantification of

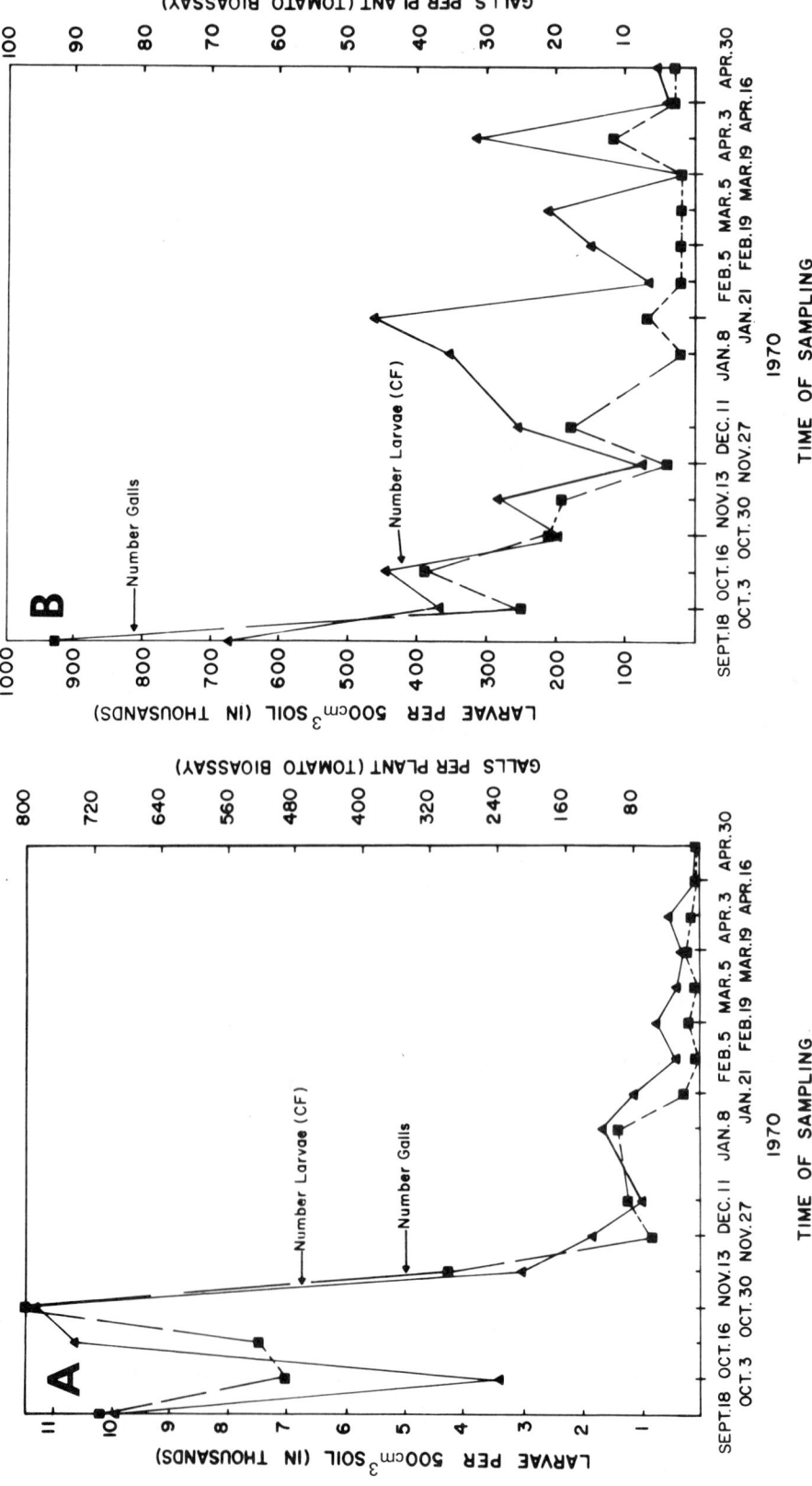

Fig. 3 Seasonal population fluctuations of *M. incognita* following tobacco as determined by centrifugal flotation (CF) and bioassay on tomato (number galls). (A) Severely infested field. (B) Lightly infested field. (From Barker, unpublished data.)

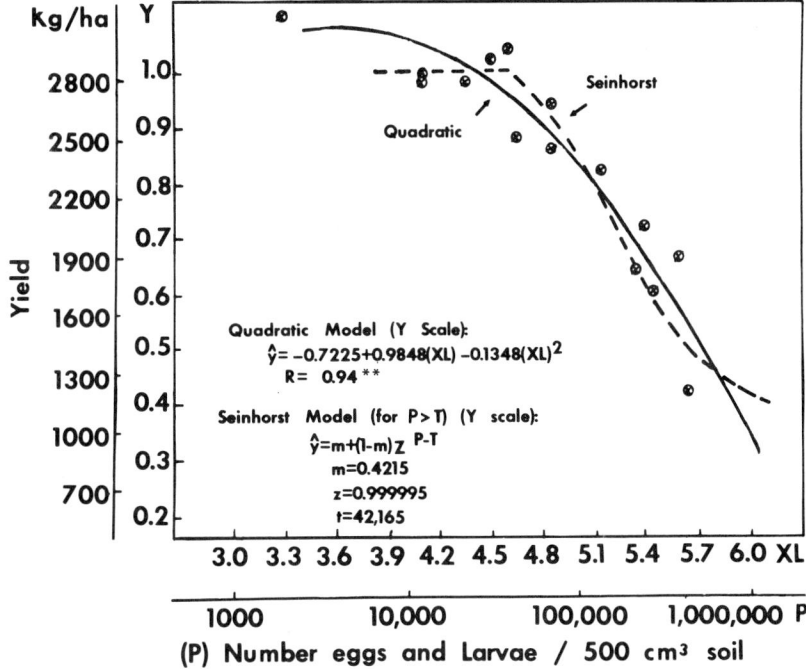

Fig. 4 Relationship of midseason numbers of eggs and juveniles of *M. incognita* to yield of tobacco (initial nematode numbers have a similar but linear relationship to yield). (After Barker et al., 1981.)

epidemics, including crop losses has received increasing study (Ferris, 1976; Zadoks, 1972; Zadoks and Schein, 1979). With respect to nematode diseases, initial numbers of nematodes, root-gall and/or root-necrosis indices may be used as a basis for predicting or estimating losses to these nematodes on tobacco (Figs. 4 and 5) (Barker et al., 1981). The ideal timing for recording meaningful gall indices, however, may vary with geographic regions (Barker et al., 1981; Brodie and Dukes, 1972).

Generally, losses to *Meloidogyne* spp. will be much less in heavy soils as compared to sandy soils (Fig. 5C, D). Based on microplot experiments, *M. arenaria* and *M. javanica* may induce approximately 13-19% yield loss per 10-fold increase in initial numbers, whereas *M. incognita* may cause a 5-10% loss in yield compared to about 3% loss in yield for *M. hapla* for each 10-fold increase in initial numbers (Barker et al., 1981).

Meloidogyne spp. induce many physiological changes in infected tobacco that result in altered quality. An important effect is a greater amount of nicotine in infected plants (Zacheo et al., 1974), but this influence depends on many factors (Weybrew et al., 1953).

G. Control

There is no practical way known at present to completely eradicate nematodes from tobacco soils. Therefore, the goal of nematode control is to keep the populations at a level where they cause minimal damage. The object is

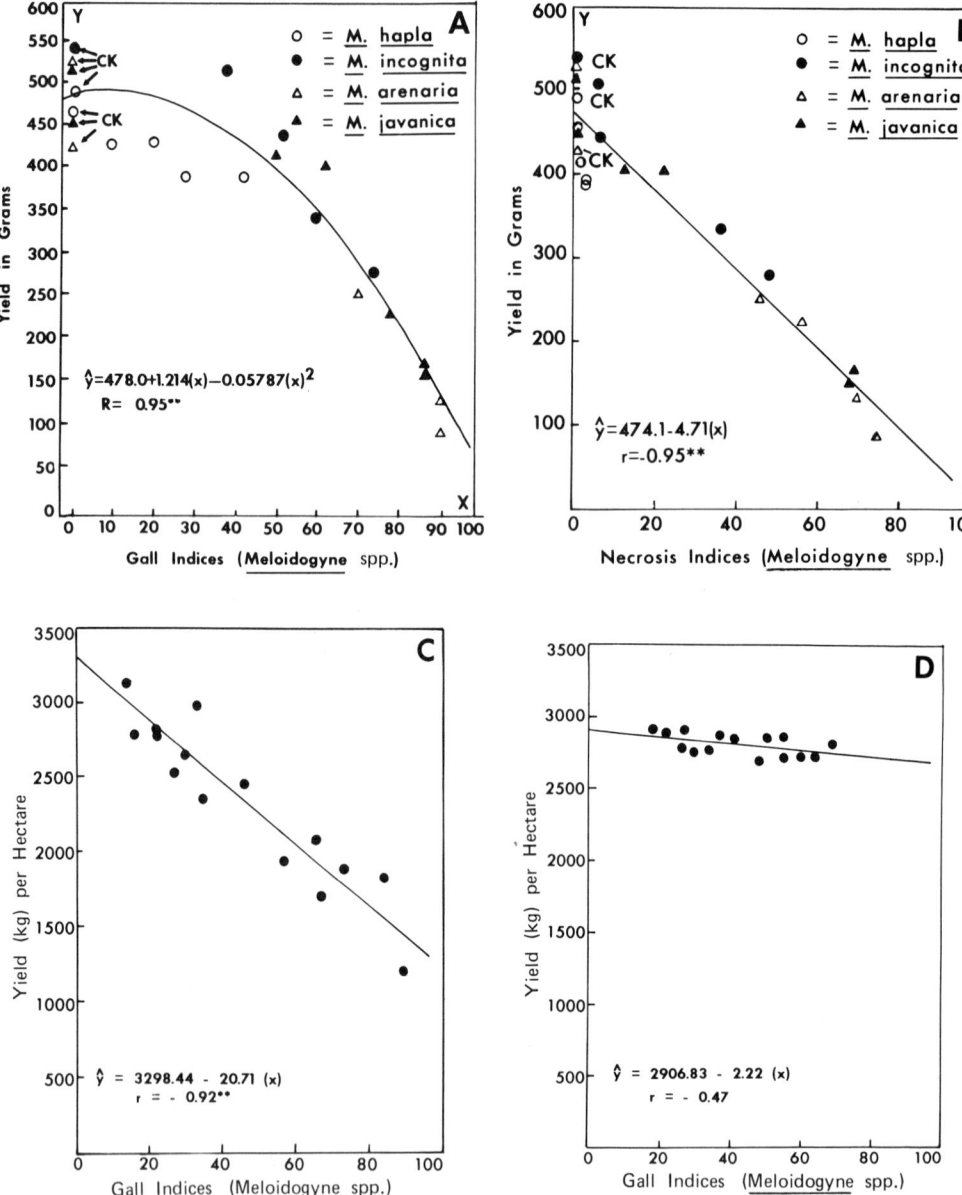

Fig. 5 Relative tobacco yield losses caused by four *Meloidogyne* spp. Relationship of (A) root galls and (B) root necrosis associated with four *Meloidogyne* species to yield in microplots. Regressions of (C) root galls in a coarse sandy soil and (D) a fine-textured soil to tobacco yields in the field. (After Barker et al., 1981.)

to provide the tobacco crop with every economically and ecologically reasonable advantage against nematode attack. This goal requires the use of a continuous integrated control program carried out from one year to the next, including: (1) the destruction of the roots and debris from previous crops; (2) the nematode-free transplants: (3) cultivars resistant to *M. incognita*; (4) the invaluable effects of desirable cropping sequences or rotation; and (5) the availability of effective chemical soil treatments.

Although some of the concepts of "integrated pest management" are of recent origin (Rabb and Guthrie, 1970), Tyler (1933) suggested that we must learn to live with nematodes. She further indicated that a well-planned combination of practices will go much further toward control of nematodes than any of the recommended treatments alone.

Control of nematodes on tobacco should commence with the seed bed. Prior to the 1950s, many growers in North Carolina cut the trees and burned portions of the wood on small areas of forest land for seedbeds, thereby obtaining pest control for tobacco seedlings. Similar approaches have been used in other regions (Milne, 1972). The advent of methyl bromide as a soil treatment and the availability of low-cost plastic covers resolved the problems of nematode as well as weed and other pest control in tobacco seedbeds (Nusbaum, 1969). This practice became almost universal among tobacco growers in the southeastern United States (Todd and Clayton, 1956) and in many other countries. Less expensive chemical soil treatment [EDB(ethylene dibromide), D-D (mixture of dichloropropene and dichloropropane, 1,3-dichloropropene)] and various nonfumigants are most commonly recommended in other countries, especially where only a nematicide is required (Milne, 1972).

In the last 20 years, considerable progress has been made in nematode diagnostic and advisory programs. Fields to be planted to tobacco can be sampled (Barker and Nusbaum, 1971) and relative population densities of *Meloidogyne* spp. determined. Based on this information, cropping history, and the history of diseases associated with nematode disease complexes, advice can be offered for developing an integrated control program.

Tactics that can be utilized in such a program have been well developed for root knot on tobacco. The extensive research by Clayton and co-workers (1958) and many others in tobacco research resulted in the development of a wide range of cultivars with vertical resistance to *M. incognita*. Based on evaluations with NC-95, races 2 and 4 of the four designated races of *M. incognita* may reproduce on these resistant cultivars (Sasser, 1980). Fortunately, about 75% of the populations of this nematode fail to induce galls on the resistant NC-95. These root knot-resistant cultivars, as well as resistant breeding lines, also have moderate resistance to root rot (Clayton et al., 1958) and brown-root rot caused by *Pratylenchus* spp. (Graham, 1965). High populations of *M. incognita*, however, will cause considerable stunting and often suppress yields of these resistant cultivars. This response is the result of a hypersensitive reaction of the root tissues induced by the nematodes (C. J. Nusbaum, unpublished). As indicated earlier, biotypes of *M. incognita* (Graham, 1968) or a new species *M. grahami* (Golden and Slana, 1978) have appeared that attack these resistant cultivars. As yet, no commercial cultivars resistant to *M. javanica* or *M. arenaria* are available, but sources of resistance have been identified (Lucas, 1975; Schweppenhauser, 1968; Slana and Stavely, 1979; Slana, 1978). The *M. incognita*-resistant cultivars also have some resistance (probably horizontal) or tolerance to *M. arenaria* and *M. javanica* (Fig. 1) (Barker, 1978).

Nematode damage has been reduced greatly through widespread use of *M. incognita*-resistant cultivars. Their extreme sensitivity to certain strains of the vein-banding virus (potato virus y, PVY) poses a threat to their continued usefulness (Lucas, 1975). Should PVY become widespread, cultivars with nematode resistance from other sources (Slana, 1978) could become available.

Although the development of effective chemical soil fumigants (Fig. 6) is of recent origin, Bessey as early as 1911 tested CS_2 as a fumigant for root knot (Bessey, 1911). Matthews reported nematicidal properties of chloropicrin in 1919, but DD was not discovered until 1943 by Carter (Lucas, 1975). In 1949, it is estimated that only 100 acres of tobacco were fumigated in North Carolina. By 1952, the fumigant usage had jumped to 10,000 and in 1956 more than 100,000 hectares were treated with nematicides. This was increased to more than 70% of the acreage by 1973. Up until the 1960s the primary fumigants were DD and EDB. At that time attention was shifted to nonfumigant nematicides that could be applied at planting. Currently, most of the chemical soil treatments, except where tobacco is grown continuously, are of the nonfumigant types, such as ethoprop and related materials (Lucas, 1975; Milne, 1972; Todd, 1979; Brodie and Dukes, 1972; Brodie and Good, 1973).

The most effective chemical soil treatments for specific geographic regions and soil types must be identified (Todd, 1976; Barker et al., 1981; Brodie and Good, 1973). For example, the involvement of *Meloidogyne* spp. in disease complexes often determines the type of cultivar rotation scheme, as well as the chemical soil treatment needed in given problem fields (Todd,

Fig. 6 Field control of root knot by soil fumigation. Left, untreated (yield = 3100 Kg/ha). Right, treated with DD (row treatment with 80 liters/ha; yield = 4940 Kg/ha). (Courtesy of F. A. Todd.)

1979; Nusbaum and Todd, 1970). In addition, consideration of weed as well as other disease problems is also essential in selection of pesticides (Brodie, 1970).

The use of fumigants has its hazards. Shallow placement of fumigants in hot, dry sandy soils results in a rapid loss from the soil surface (Milne, 1972). Under these conditions, the fumigant should be placed to depths of 23-28 cm to obtain adequate control (Daulton, 1967). In contrast, when fumigants are applied to cold wet soils, the fumigants do not volatilize readily, but remain in the soil. If the grower does not wait at least two to four weeks before transplanting, injury, indicated by stunted plants with swollen stems, may occur. In fact, losses in yield from such stunting from fumigants may occur in a high percentage of fields, especially in those that have low numbers of nematodes (Nusbaum, 1960).

Utilizing halogenated hydrocarbons, including DD, may result in detrimental effects, such as suppressed yields, increases in total nitrogen and alkaloids, and a decrease in reducing sugars in the cured leaf (Nusbaum, 1960; Elliot et al., 1972; McCants et al., 1959). Fall fumigation may also interfere with normal nitrification processes (Elliot and Mountain, 1963). The organic phosphate and carbamate nematicides are being used more widely for all nematode species (Todd, 1979).

We also have major problems in selecting the most efficient nematicide. For example, certain of the increasingly popular nonfumigants provide adequate control of moderate initial numbers of *M. incognita*, but may fail to prevent damage by *M. arenaria* (Barker et al., 1981). This species, as well as *M. javanica*, is highly virulent and also appears to be more difficult to kill. The fact that most nonfumigants are very water soluble increases the chances that heavy rains may leach them through the soil before sufficient control is attained.

Rotation, the oldest tactic for controlling plant diseases, has great potential for minimizing tobacco losses to *Meloidogyne* spp. Where land is available, the use of nonhost crops, such as fescue, may greatly enhance yields (Milne, 1972; Nusbaum and Ferris, 1973). Numerous crop sequences, which suppress populations of a given species of nematodes, have been developed (Lucas, 1975; Milne, 1972; Nusbaum and Ferris, 1973; Olthof, 1979; Sasser and Nusbaum, 1955). A root knot-resistant cultivar of corn, as apparently detected (for *M. javanica*) by Norse (1972), would greatly increase the utility of rotation in control of this disease on tobacco. Rotation schemes may improve soil structure and provide benefits in addition to nematode control (Nusbaum and Ferris, 1973). The cropping system has a major influence on the composition of nematode communities as well as the seasonal fluctuations in population densities of individual species. The desired length of rotation or frequency of tobacco crops over time is very important and is influenced by many factors. Generally, a longer rotation is necessary in areas with long growing seasons and mild or frost-free winters than in those with short growing seasons and severe winters (Nusbaum, 1969). Two-year rotation schemes are fairly effective in North Carolina, but three-year rotations are highly desirable. Unfortunately, as tobacco production has become more mechanized, many growers cannot or choose not to use long-term rotations.

The application of cultural practices as a control for root knot on tobacco warrants special consideration. The use of an early planting date may greatly minimize damage to *Meloidogyne* spp. as compared to late planting (Milne, 1972). Nusbaum (1969) and Todd (1979) have expanded the concept of post-harvest root or host destruction as described by Atkinson in 1889 into a

system called "R-9-P" (reduce 9 pests). This approach results in cessation of
pest reproduction and provides for the control of nematodes, several diseases,
insects, and weeds. Many nematode eggs and juveniles are killed when exposed to direct sunlight; the eggs and juveniles also are killed as they are
air dried. Thus, repeated plowing or disking after harvest allows the nematodes as well as the crop roots to be exposed to the sun and the wind.
Where desirable, such fields so treated may be planted to cover crops to
minimize soil erosion.

Nematode damage on heavy tobacco types (dark air-cured, cigar) may
be limited by providing greater rates of fertilizer (Milne, 1972). The detrimental effects of excessive nitrogen on the quality of flue-cured tobacco
makes this practice inadvisable.

Biocontrol of *Meloidogyne* spp. has promise, but practical use still must
be developed. Certain species of *Tagetes* (marigolds) and their a-terthienyl
exudates give significant reductions of *M. javanica* populations (Daulton and
Curtis, 1963). The recently described parasitism of *Meloidogyne* eggs by
fungi has much potential in biocontrol of these nematodes (Stirling and
Mankau, 1978).

III. ROOT-LESION NEMATODES

Species of *Pratylenchus*, the lesion nematodes, have received less attention
on tobacco than have the root-knot nematodes, but they may cause significant
yield losses. Lesion nematodes are a primary factor involved in brown root
rot of tobacco and general stunting of many other plants. These migratory
endoparasites cause no distinct symptoms except for root necrosis; losses
caused by them, therefore, are often underestimated. Further, the indirect
losses resulting from the interactions of these nematodes with various fungi
and/or bacteria may be overlooked.

A. History

The history of the brown root rot disease is given in detail by Lucas (1975).
It was first described in 1919 by Johnson, who considered the causal agent
to be a species of *Fusarium*. Lehman (1931) associated *Pratylenchus pratensis*
with roots exhibiting this disease. Steiner (1945) indicated that species of
Pratylenchus were causing serious crop damage and should receive intensive
study. *Pratylenchus* species were associated with brown root rot in South
Carolina (Graham and Heggestad, 1959) and other regions (Mountain, 1954;
Valleau and Johnson, 1947). Monoxenic cultures of *Pratylenchus neglectus*
(*minyus*) were used to demonstrate that the nematode alone induced brown
root rot of tobacco (Mountain, 1954). Although lesion nematodes apparently
are the principal cause of brown root rot of tobacco (Lucas, 1975; Mountain,
1954; Jenkins, 1948), the interactions of these nematodes with various fungi
undoubtedly contribute significantly to the total loss caused by this disease
(Inagaki and Powell, 1969). In recent years, little brown root rot has been
observed on tobacco in the United States.

B. Species

In addition to *Pratylenchus pratensis*, *P. neglectus*, and *P. brachyurus*,
other species that have been reported on tobacco include *P. zeae*, *P. hexincisus*, *P. thornei*, *P. vulnus*, and *P. penetrans*. *P. zeae*, however, reproduces so poorly on tobacco, populations usually are not maintained (Southards,
1965).

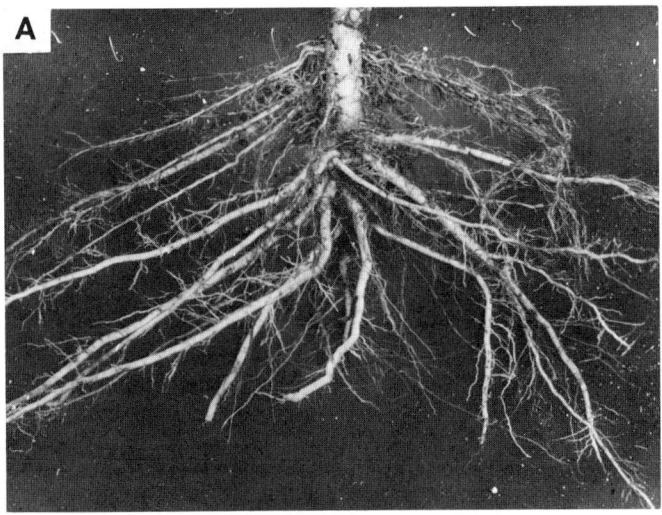

Fig. 7 Tobacco root damage by *P. brachyurus*. (A) Damaged root with characteristic necrosis. (B) Luxuriant root system of plant from methyl-bromide treated plot. (Courtesy of C. J. Nusbaum.)

C. Associated Symptoms

Brown root rot usually occurs in well-defined patches, but sometimes occurs throughout a field (Lucas, 1975). Shoot symptoms are limited largely to the stunting and severe wilting in the afternoon during dry periods (Jenkins, 1948). The earliest root symptoms involve a colorless, water-soaked appearance just behind the growing point. Most lesions occur in the region of elongation, but some may be initiated in the younger roots, including the growing point. Parasitized roots exhibit varying degrees of damage, with cortical lesions varying from pale yellow to almost black (Fig. 7). These lesions usually girdle the entire feeder root; lesions often break open and the cortex tissues slough off in a sleevelike fashion leaving only the vascular cylinder (giving a common diagnostic symptom), with subsequent pruning and death of the roots. Large numbers of nematodes often cause excessive root pruning with the result that the roots have a stubby appearance.

If roots are stained with a suitable dye (Thorne, 1961), large numbers of various stages of the nematode as well as eggs and frequently large numbers of microbivorous nematodes will be detected throughout the root system. Staining of suspected roots or extraction from such roots may be necessary to confirm the presence of these nematodes.

D. Life Cycle

In contrast to *Meloidogyne* spp., the eggs of *Pratylenchus* are deposited singly, each gravid female averaging one egg a day for about 30 days (Lucas, 1975). Eggs deposited in root tissue hatch in 6-17 days, depending on the species and environmental conditions, especially temperature and moisture. The length of the life cycle may vary from 28 days for *P. neglectus* to 35 days for *P. brachyurus*, and other species may require as much as 65 days under optimum conditions. *P. brachyurus*, *P. neglectus*, and *P. zeae* reproduce parthenogenetically, whereas *P. penetrans* and several other species reproduce amphimictically (Triantaphyllou and Hirschmann, 1980; Roman and Triantaphyllou, 1969). Eggs, juveniles, and adults may overwinter in soil and in living or dead roots.

E. Distribution

Species of *Pratylenchus* are generally worldwide in distribution. They are spread by transplants, drainage water, and contaminated wind-blown soil, and by various instruments. *Pratylenchus* spp. thus are generally worldwide in distribution, although *P. penetrans*, *P. thornei*, and *P. neglectus* are more common in temperate regions (Milne, 1972; Corbett, 1969). *P. penetrans*, *P. neglectus*, and *P. crenatus* have been associated with tobacco in Canada (Olthof and Hopper, 1973). *P. penetrans* also occurs on tobacco in New Zealand (Canter-Visscher, 1969). *P. brachyurus* and *P. zeae* also tend to be more commonly found in tropical areas than *P. penetrans*, *P. thornei*, and *P. neglectus*. The general use of *M. incognita*-resistant cultivars, which may support only limited reproduction of *Pratylenchus* spp. (Graham, 1965), and less frequent rotations with other crops, may be partially responsible for these nematodes declining on tobacco in North Carolina and other areas.

F. Estimated Crop Losses

In the 1950s, losses to *Pratylenchus* were estimated to be as great as 1% or more (Lucas, 1975). However, during the last 30 years, losses have declined in the southeastern United States and Canada, probably as a result of wide-scale use of R-9-P, preplant fumigation, and use of cultivars that are poor hosts. Losses caused by *Pratylenchus* spp. also may be limited by the nematode's inability to migrate through various soil types. In contrast to the movement of *M. javanica* of 50 cm or more in three days (Prot, 1976), *P. zeae* migrated only 5 cm in a clay soil during a four-month period (Endo, 1959). Even in a favorable Norfolk sandy loam, relatively few *P. zeae* migrated as much as 12.5 cm. More importantly, tobacco is generally a poor host for *Pratylenchus* spp.

G. Control

The basic control measures used for root knot also apply for brown root rot. These tactics, as discussed earlier, should be fitted into an overall pest management program. Methyl bromide gives excellent control of root-knot and lesion nematodes in the seedbed. Field fumigation with EDB or DD gives good control in sandy soils, but DD tends to be more effective (Nusbaum, 1960; Nusbaum and Sasser, 1955; Owens and Ellis, 1951).

The selection of crops in rotation schemes poses a problem because of the wide host range of *Pratylenchus* spp. In fact, some authors (Valleau and Johnson, 1947) have indicated that brown root rot is less severe on continuous burley tobacco than with numerous rotations. Growth of bluegrass, corn, rye, and many legumes allows the buildup of large populations that may injure a subsequent crop of tobacco. However, oats, sweet potato, and several other crops support little reproduction of given species such as *P. brachyurus* (Endo, 1959). The relative low winter mortality of such species as *P. penetrans* (ranges from 40 to 65%) may compound the problems in their control (Olthof, 1971).

No cultivars highly resistant to *Pratylenchus* spp. have been developed. Limited research does show that certain cultivars have more tolerance to *P. brachyurus* than others (Southards and Nusbaum, 1967). As indicated earlier, some *M. incognita*-resistant cultivars (NC-95) support only limited reproduction of *Pratylenchus*, but may be severely stunted by this nematode (Graham, 1965). Certain burley cultivars are somewhat resistant to *P. penetrans* (Olthof, 1968). Reliable methods for evaluating resistance or tolerance of tobacco to *Pratylenchus* are available (Southards, 1965; Southards and Nusbaum, 1967; Graham and Ford, 1968).

Biocontrol and cultural practices may have potential in managing this nematode. Extracts from decomposing rye and timothy are toxic to *P. penetrans* as well as *M. incognita* (Sayre et al., 1965).

IV. CYST NEMATODES

Until recently, *Globodera tabacum* was considered to be the only important cyst nematode on tobacco. All species of this genus have relatively narrow host ranges, but a rather wide range of plants (largely in Solanaceae) support some reproduction of *G. tabacum* (Harrison and Miller, 1969). Currently, a new threat is posed by the recently described species *G. solanacearum* (Fig. 8A).

A. History

The tobacco cyst nematode, *G. tabacum*, was first found to be parasitizing the roots of shade tobacco in Connecticut in 1951 (Lownsbery and Lownsbery, 1954). A second tobacco cyst nematode was found on flue-cured tobacco in Virginia in 1960 (Nusbaum, 1969), and was named *G. solanacearum* (Miller and Gray, 1972). Another very closely related species from Virginia and North Carolina, the horsenettle nematode, was described as *G. virginiae* (Miller and Gray, 1968).

B. Species

Until 1976, the tobacco cyst nematodes were members of *Heterodera*. The group of round cyst nematodes to which these belong was distinguished as the new genus *Globodera* (Mulvey and Stone, 1976). The tobacco cyst

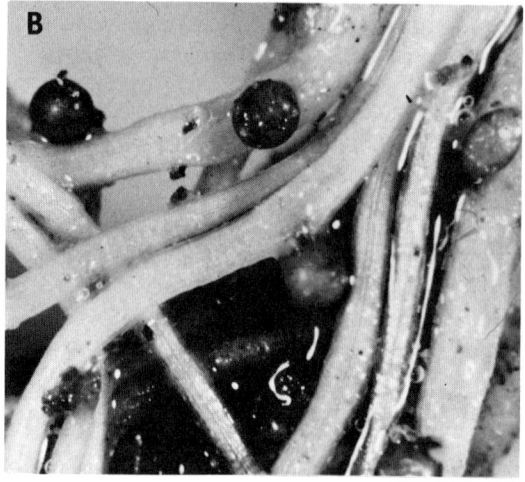

Fig. 8 *G. solanacearum* on flue-cured tobacco. (A) Patchy stunting in field.
(B) Cysts on roots. (Courtesy of J. A. Fox.)

nematode, *G. tabacum*, which resembles the golden nematode *G. rostochiensis*,
attacks shade and flue-cured tobacco (Milne, 1972). Brownish cysts are
easily observed on infected roots (Fig. 8B). A second tobacco cyst nematode,
G. solanacearum, reproduces on tobacco, tomato, and eggplant, but not on
potato (Nusbaum, 1969). Most flue-cured cultivars grown in Virginia and
North Carolina are susceptible to this nematode. The horsenettle nematode,
G. virginiae, has a similar host range but reproduces poorly on several cultivars of flue-cured tobacco (Nusbaum, 1969). *G. solanacearum* and *G. virginiae* can be identified on suitable differential hosts (L. I. Miller, unpublished
data). All three species are very closely related, but they do have some
important differences in host preferences.

C. Associated Symptoms

Foliage symptoms for the tobacco cyst nematodes usually include severe stunting and dark green plants in scattered parts of the field (Fig. 8A), wilting in midafternoon, and a greatly suppressed root system. The presence of the dark brown oval cysts (Fig. 8B) of about 0.5 mm in diameter on the roots aids greatly in identification of this pest.

D. Life Cycle

Most of the eggs of *Globodera* spp. are retained within the female body, which as the female dies, develops into a brown, leatherlike cyst. The eggs and these cysts are resistant to breakdown by soil organisms, to drying, and to action of chemical soil treatments. Eggs and juveniles contained in the soil may survive for years in the absence of the host plant. With the introduction of a host plant, the eggs (probably stimulated by root exudates) hatch and emerge from the cyst, with the second-stage juveniles infecting the seedlings. The life cycle is similar to that of *Meloidogyne*, except all stages parasitize the enlarged syncytial cells near the xylem tissue. Mature males emerge from the roots and fertilize the females. About 20 days are required for *G. tabacum* to complete its life cycle and produce eggs. This nematode usually has about four to five generations on a single tobacco crop. The generation time of *G. solanacearum* is 34 days on plants grown at 24-29°C (Miller and Gray, 1972). In the field, the life cycle varies from 32 to 82 days, depending upon the time of the year. In Virginia, two or more generations usually occur each season.

E. Distribution and Estimated Crop Losses

Where tobacco cyst nematodes occur in large numbers, yield losses of 25-50% or greater may occur. Although species of *Globodera* usually have limited geographic distribution, some of them occur in fine as well as coarse-textured soils (Jones et al., 1969). Fortunately, these nematodes are of economic importance in only a few limited geographic regions. Nevertheless, they pose a possible major problem to tobacco production. *G. virginiae*, although it reproduces slowly, may cause considerable stunting of burley tobacco (Miller, 1977). *G. solanacearum*, in addition to affecting tobacco growth directly, suppresses the reproduction of mycorrhizal fungi (Fox and Spasoff, 1972). On shade tobacco, *G. tabacum* is so competitive that it inhibits the reproduction of other nematodes, such as *P. penetrans* (Miller and Wihrheim, 1968).

F. Control

Certain nonfumigant and fumigant nematicides appear to give excellent control of these nematodes. The presence of resistant cysts, however, make efficient control difficult, especially with nonfumigant nematicides. Such materials as aldicarb kill emerged juveniles of *G. tabacum*, but fail to kill many eggs within the cysts (Miller, 1970). Interestingly, applications of benomyl or thiabendazole tend to suppress the invasion of *G. tabacum* on shade tobacco (Miller, 1969). These fungicides are used for control of various fungi. Unfortunately, the suppression effects of these materials are short-lived.

Although species of *Globodera* may reproduce on a number of plants, their general host range is relatively limited. Thus, the development of effective rotation systems should be utilized in the integrated pest control programs. Major additions of organic matter retard the invasion of tobacco by *G. tabacum* (Miller et al., 1968). Considerable effort has been initiated in the development of breeding lines resistant to these nematodes. *Nicotiana glutinosa*, *N. paniculata*, *N. plumbaginifolia*, and *N. longiflora*, and certain crosses, supported very little reproduction of *G. solanacearum* (Baalawy and Fox, 1971), although there was extensive juvenile invasion of these species and breeding lines. Resistance to this nematode in burley and dark-fired tobacco apparently is polygenic (Miller et al., 1972). As yet, no commercial cultivars are available to any of these cyst nematodes.

V. BULB AND STEM NEMATODE

A. History, Species and Races, Distribution, and Estimated Crop Losses

The bulb and stem nematode, *Ditylenchus dipsaci*, has a long history of races that attack many crops (Sturhan, 1971). Despite its wide distribution as a foliage pathogen on alfalfa and a pathogen of many bulb crops, the stem nematode has been reported to parasitize and cause "stem break" on tobacco only in Germany, France, Holland, and Switzerland (Lucas, 1975; Vallotton and Corbaz, 1976). The absence of this disease in major flue-cured tobacco-growing areas, such as the United States, leaves a question as to whether the races occurring in such regions are able to parasitize tobacco.

This nematode, first described by Kühn in 1857, has been studied extensively and has numerous host-specific races (Sturhan, 1971; Goodey, 1952). The conditions conducive for the development of this disease within only a few countries are not fully understood, but the complex genetics of the races of this nematode (Triantaphyllou and Hirschmann, 1980) could be responsible.

B. Associated Symptoms

During wet weather, the stem nematodes enter the leaves or stem of tobacco seedlings and induce the development of small, yellow swellings or galls that may extend up to 40 cm above the soil level (Fig. 9A, B). These galls rot (Fig. 9C), the plant stops growing, and eventually the stem breaks and the plant falls over. The swellings increase in number and the tissue dies prematurely. The lower leaves fall off and the upper leaves turn yellow. Severe wilting occurs only in advanced stages, since the vascular tissue remains functional for a considerable period of time after onset of the disease. Younger plants may be more severely stunted than older plants (Fig. 9).

C. Life Cycle

The fourth-stage juveniles of *D. dipsaci* have the unusual capacity to become cryptobiotic and may remain in this inactive metabolic state for 20 or more years. Juveniles (J-4) and eggs survive freezing. Optimum conditions for survival are low moisture and soil temperature near or below $0^{\circ}C$. The life cycle usually is completed in 19-23 days (Lucas, 1975). Reproduction is of the amphimictic type. Only a few of the more than 20 races of *D. dipsaci* (Sturhan, 1971) reproduce on tobacco. The race question clouds the issue of the distribution of the nematode species and strains that attack tobacco (Vallotton and Corbaz, 1976). The race that reproduces on alfalfa in the United States has not been detected on tobacco. Although a good host, the

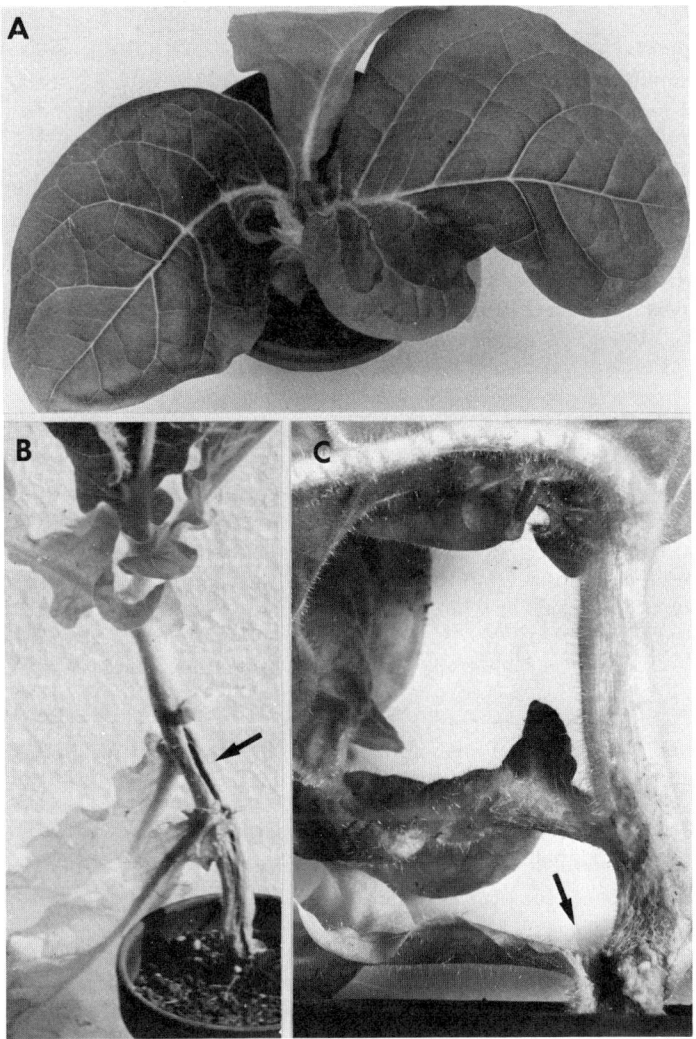

Fig. 9 Symptoms of *D. dipsaci* on tobacco. (A) Misshapen foliage of infected plant. (B) Severe necrosis and stem splitting (arrow) sometimes associated with *D. dipsaci*. (C) Swellings and necrosis caused by this nematode (arrow). (After Vallotton and Corbaz, 1976.)

vigorous growth habit of tobacco generally limits the damage caused by this nematode.

D. Control

Removal of diseased plants from the field is advisable where stem break occurs. Rotations of nonhost crops for long duration may be effective in some instances. The fumigant nematicides give good control, and the systemic nonfumigants, such as aldicarb, also have promise (Vallotton and Corbaz, 1976).

VI. OTHER NEMATODES

Ectoparasitic nematodes are of limited importance on tobacco. Several reports, however, include descriptions of direct yield losses from this group of nematodes. Ectoparasitic nematodes, especially migratory forms, are difficult to characterize as pathogenic or parasitic. In fact, Croll and Matthews (1977) suggested that the externally feeding nematodes should be referred to as microbrowsers rather than parasites. Certain of these nematodes (*Trichodorus* spp., *Paratrichodorus* spp., and *Xiphinema* spp.) may be more important as virus vectors than as direct suppressants of tobacco growth.

Although little information is available, the sedentary endoparasite *Rotylenchulus reniformis* apparently causes considerable damage to tobacco in some countries (Reddy, 1978). The role of this nematode and most ectoparasites as disease agents on tobacco need further evaluation.

A. History, Specific Taxa, Distribution, and Estimated Crop Losses

Ectoparasitic nematodes associated with tobacco received considerable attention in the 1950s. For example, *Tylenchorhynchus claytoni* at that time was thought to cause severe damage on tobacco because of its reproduction and association with this crop (Krusberg, 1959; Patel et al., 1962). Nevertheless, subsequent research by Nusbaum (unpublished data) and others have shown this nematode to have little effect on the yield of tobacco. Numerous other ectoparasitic nematodes have been associated with this crop. *Xiphinema* spp. and trichodorid species could be important in that they serve as vectors for certain viruses.

Many species of ectoparasitic nematodes may cause limited damage on tobacco. Taxa frequently associated with tobacco include *T. claytoni*, *Paratylenchus* spp., *Xiphinema americanum*, *Helicotylenchus* spp., *Criconemella* spp., *Rotylenchus* spp., *Tetylenchus nicotianae*, *Paratrichodorus* spp., and *R. reniformis* (Lucas, 1975; Milne, 1972; Reddy, 1978; Barker, 1974). Severe stunting may be caused by *Longidorus elongatus* (Marks and Elliot, 1973). *Trichodorus* spp. may be important on tobacco in limited geographic regions (Meagher, 1969; Wyss, 1973).

Because of the subtle damage caused by ectoparasitic nematodes and their interactions as virus vectors, no accurate disease loss data are available. Microplot experiments with *T. claytoni* and *Criconemella ornata* showed these two nematodes to have no effect on yield (Barker, unpublished). *Paratrichodorus lobatus*, as reported in Australia (Meagher, 1969), may cause severe losses in yield.

B. Associated Symptoms

Although the literature has conflicts concerning the stunt nematode *Tylenchorhynchus* spp., there are reports in the United States and other countries that show these nematodes are associated with stunting and a retarded root system (Lucas, 1975; Patel et al., 1962). Affected roots have no distinct symptoms, but may be sparsely developed, and fail to elongate normally. Some ectoparasites, especially *Tylenchorhynchus*, *Helicotylenchus*, and *Hoplolaimus* may be found in the root cortical tissue.

Where stubby-root nematodes occur, more severe damage can be encountered, particularly on younger plants. Growth is retarded, the foliage wilts readily, and the plants have limited ability to undergo drought stress. These nematodes feed largely at the root tips, killing root hairs and rhizodermis cells, and thereby debilitate roots resulting in a stubby-root effect (Meagher,

1969; Wyss, 1973). When large numbers of nematodes are present, these stubby roots may become discolored with necrotic lesions. Another migratory ectoparasite, *Xiphinema americanum*, is known to feed on tobacco and transmit tobacco ringspot virus (McGuire, 1964), but this nematode causes no distinct symptoms on tobacco roots.

C. Life Cycle

The type of reproduction and life cycle of the ectoparasitic nematodes associated with tobacco vary greatly. Species such as *X. americanum*, *C. ornata*, *Helicotylenchus dihystera*, and *Paratrichodorus minor* reproduce by parthenogenesis. *T. claytoni* is amphimictic. This species requires about one month to complete its life cycle at temperatures of 30-34°C (Krusberg, 1959). In contrast, *X. americanum* apparently completes only one or two life cycles per year.

D. Control

Control programs for ectoparasitic nematodes on tobacco can generally be limited to those regions in which they have been definitely demonstrated to cause yield losses. Undoubtedly, the ability of *X. americanum* to serve as the reservoir of the tobacco ringspot virus needs to be considered where this virus occurs, as in the case of *Trichodorus* spp. for the tobacco rattle virus. Where these nematodes are of economic importance, chemical soil treatments should be effective. The soil fumigants often fail to give good control of some trichodorid nematodes, which may occur at lower soil depths and apparently escape fumigant action (Weingartner et al., 1980). In contrast, many of the nonfumigants are effective against these pests. The effective use of rotation systems is more difficult because of the wide host range of this group of nematodes. As yet, damage caused by these nematodes has not been sufficient to warrant the development of resistant cultivars.

VII. SUMMARY

Many species of nematodes attack tobacco, but *Meloidogyne* spp. are responsible for most of the associated yield losses. Estimated losses to nematodes undoubtedly are too low, as they usually fail to include the damage caused by fungi and bacteria that parasitize the galled roots. Furthermore, the importance of various nematodes in many countries remains to be determined. Much of the basic research and the empirical integrated control systems for nematodes on tobacco should serve as useful models for other crops. The systems approach in North Carolina includes the determination of the type and magnitude of the problems present and the selection of suitable control tactics, including disease-free plants, resistant cultivars, cultural practices (especially R-9-P), rotation, and chemical soil treatments. The widespread use of combined tactics, especially healthy plants, resistant cultivars, some crop sequences, and R-9-P has greatly reduced the occurrences and densities of *Meloidogyne* and *Pratylenchus* species in many regions. The availability of highly resistant cultivars alone has increased the returns from tobacco production by billions of dollars over the last 20 years. Nevertheless, there is opportunity to enhance production efficiency by developing more comprehensive integrated pest management systems, including the reduction of pesticide usage as insurance. This approach and increased emphasis on additional sources of resistance are essential as new nematode problems pose additional hazards to this crop. *Meloidogyne arenaria* and resistance-breaking strains of *M. incognita* are being detected more frequently than in the past. Highly virulent species of *Globodera* are appearing in flue-cured tobacco fields. Therefore, although much progress has been made in understanding nematode problems on tobacco and reducing related yield losses, these problems are not static, but require continuous monitoring and research.

REFERENCES

Arcia, M. A., Vargas, M., Casanova, E., and Meredith, J. A. (1976). Efecto de los nematodos *Meloidogyne incognita* y *Meloidogyne javanica* sobre la deficiencia de boro en plantas de tabaco burley. *Nematropica* 6: 63-67.

Atkinson, G. F. (1889). Nematode root-galls. *Ala. Agric. Exp. Sta. Bull.* 9: 1-54.

Baalawy, H., and Fox, J. A. (1971). Resistance to Osborne's cyst nematode in selected *Nicotiana* species. *J. Nematol.* 3: 395-398.

Barker, K. R. (1974). Influence of geographic area and previous crop on occurrence and densities of plant-parasitic nematodes in North Carolina. *Plant Dis. Rep.* 58: 991-995.

Barker, K. R. (1978). Relative sensitivity of flue-cured tobacco cultivars to four species of *Meloidogyne* (abstract). *J. Nematol.* 10: 281.

Barker, K. R., and Nusbaum, C. J. (1971). Diagnostic and advisory programs. In *Plant Parasitic Nematodes*, vol. I (B. M. Zuckerman, W. F. Mai, and R. A. Rohde, eds.). Academic, New York, pp. 281-301.

Barker, K. R., Nusbaum, C. J., and Nelson, L. A. (1969). Seasonal population dynamics of selected plant-parasitic nematodes as measured by three extraction procedures. *J. Nematol.* 1: 232-239.

Barker, K. R., Todd, F. A., Shane, W. W., and Nelson, L. A. (1981). Interrelationships of *Meloidogyne* species with flue-cured tobacco. *J. Nematol.* 13: 67-78.

Berkeley, M. J. (1855). The roots of a variety of plants . . . (*Vibrio* forming cysts on cucumber). *Gardeners Chron.* 14: 220.

Bessey, E. A. (1911). Root-knot and its control. *USDA Bur. Plant Ind. Bull. 217.*

Bird, A. F. (1979). Morphology and ultrastructure. In *Root-knot Nematodes* (Meloidogyne) *species* (F. Lamberti and C. E. Taylor, eds.). Academic, New York, pp. 59-84.

Brodie, B. B. (1970). Use of non-selective and mixtures of selective pesticides for multiple pest control. *Phytopathology* 60: 12-15.

Brodie, B. B., and Dukes, P. D. (1972). The relationship between tobacco yield and time of infection with *Meloidogyne javanica*. *J. Nematol.* 4: 80-83.

Brodie, B. B., and Good, J. M. (1973). Relative efficacy of selected volatile and non-volatile nematicides for control of *Meloidogyne incognita* on tobacco. *J. Nematol.* 5: 14-18.

Canter-Visscher, T. W. (1969). The association of root-lesion nematodes with reduced growth of tobacco in the Nelson district. *N. Zeal. J. Agric. Res.* 12: 423-426.

Carter, W. (1943). A promising new soil amendment and disinfectant. *Science* 97: 383-384.

Chapman, G. H. (1920). Tobacco investigations. Progress report including miscellaneous observations on tobacco. *Mass. Agric. Exp. Sta. Bull.* 195: 38 p.

Chitwood, B. G. (1949). Root-knot nematodes — Part I. A revision of the genus *Meloidogyne* Goeldi, 1887. *Proc. Helm. Soc. Wash. 16*: 90-104.

Chitwood, B. G., and Oteifa, B. A. (1952). Nematodes parasitic on plants. *Annu. Rev. Microbiol. 6*: 151-184.

Christie, J. R., and Albin, F. E. (1944). Host-parasitic relationships of the root-knot nematode, *Heterodera marioni*. I. The question of races. *Proc. Helm. Soc. Wash. 11*: 31-37.

Clayton, E. E., Gaines, J. G., Smith, T. E., Shaw, K. J., and Graham, T. W. (1944). Control of flue-cured tobacco root diseases by crop rotation. *USDA, Farmer's Bull. 1952*: 1-12.

Clayton, E. E., Graham, T. W., Todd, F. A., Gaines, J. G., and Clark, F. A. (1958). Resistance to the root-knot disease of tobacco (Part II). *Tob. Sci. 2*: 58-63.

Corbett, D. C. M. (1969). *Pratylenchus pinguicaudatus* n. sp. (Pratylenchinae: Nematoda) with a key to the genus *Pratylenchus*. *Nematologica 15*: 550-556.

Croll, N. A., and Matthews, B. E. (1977). *Biology of Nematodes*. Blackie, Glasgow.

Daulton, R. A. C. (1967). Injection depth important in soil fumigation in Rhodesia. *Down to Earth 22*: 20-21.

Daulton, R. A. C., and Curtis, R. F. (1963). The effects of *Tagetes* spp. on *Meloidogyne javanica* in Southern Rhodesia. *Nematologica 9*: 357-362.

Daulton, R. A. C., and Nusbaum, C. J. (1961). The effect of soil temperature on the survival of the root-knot nematodes *Meloidogyne javanica* and *M. hapla*. *Nematologica 6*: 280-294.

Elliot, J. M., Marks, C. F., and Tu, C. M. (1972). Effects of nematicides on *Pratylenchus penetrans*, soil microflora, and flue-cured tobacco. *Can. J. Plant Sci. 52*: 1-11.

Elliot, J. M., and Mountain, W. B. (1963). Influence of spring and fall application of nematicides on *Pratylenchus penetrans* and quality of flue-cured tobacco grown with various forms of nitrogen. *Can. J. Soil Sci. 43*: 18-26.

Endo, B. Y. (1959). Response of root-lesion nematodes, *Pratylenchus brachyurus* and *P. zeae*, to various plants and soil types. *Phytopathology 49*: 417-421.

Ferris, H. (1976). Development of a computer-simulation model for a plant-nematode system. *J. Nematol. 8*: 255-263.

Fox, J. A., and Spasoff, L. (1972). Interaction of *Heterodera solanacearum* and *Endogone gigantea* on tobacco (abstract). *J. Nematol. 4*: 224-225.

Golden, A. M., and Slana, L. J. (1978). *Meloidogyne grahami* n. sp. (Meloidogynidae), a root-knot nematode on resistant tobacco in South Carolina. *J. Nematol. 10*: 355-361.

Goodey, J. B. (1952). Investigations into the host ranges of *Ditylenchus destructor* and *D. dipsaci*. *Ann. Appl. Biol. 39*: 221-228.

Graham, T. W. (1965). Tobacco varieties used to show differential response of root-lesion and root-knot nematodes. *Plant Dis. Rep. 49*: 822-826.

Graham, T. W. (1968). A new pathogenic race of *Meloidogyne incognita* on flue-cured tobacco. *Tob. Sci. 13*: 43-44.

Graham, T. W., and Ford, Z. T. (1968). Inoculation methods with the root-lesion nematode *Pratylenchus brachyurus* and symptom expression on tobacco. *Tob. Sci. 12*: 16-19.

Graham, T. W., Ford, Z. T., and Currin, R. E. (1964). Response of root-knot-resistant tobaccos to the nematode root disease complex caused by *Pratylenchus* spp. and *Meloidogyne incognita acrita*. *Phytopathology 54*: 205-210.

Graham, T. W., and Heggestad, H. E. (1959). Growth response and root decay development in certain tobacco varieties and breeding lines infected with root-lesion nematodes. *Tob. Sci. 3*: 172-178.

Harrison, M. B., and Miller, L. I. (1969). Additional hosts of the tobacco cyst nematode. *Plant Dis. Rep. 53*: 949-951.

Inagaki, H., and Powell, N. T. (1969). Influence of the root-lesion nematode on black shank symptom development in flue-cured tobacco. *Phytopathology 59*: 1350-1355.

Jenkins, W. A. (1948). Root rot disease complexes of tobacco in Virginia. I. Brown root rot. *Phytopathology 38*: 528-541.

Johnson, A. W., and Nusbaum, C. J. (1970). Interactions between *Meloidogyne incognita, M. hapla,* and *Pratylenchus brachyurus* in tobacco. *J. Nematol. 2*: 334-340.

Johnson, J. (1919). Fusarium root-rot of tobacco (abstract). *Phytopathology 9*: 49.

Jones, F. G. W., Larbey, D. W., and Parrott, D. M. (1969). The influence of soil structure and moisture on nematodes especially *Xiphinema, Longidorus, Trichodorus* and *Heterodera* spp. *Soil Biol. Biochem. 1*: 153-165.

Krusberg, L. R. (1959). Investigations on the life cycle, reproduction, feeding habits and host range of *Tylenchorhynchus claytoni* Steiner. *Nematologica 4*: 187-197.

Lehman, S. G. (1931). Research in botany. A new tobacco root disease. *N.C. Agric. Exp. Sta. Ann. Rep. 54*: 73-76.

Lordello, L. G. E., and Monteiro, A. R. (1974). Preliminary data on a nematode damaging to the coffee tree. In Trabalhos apresentados a reuniao de nematologia, *Soc. Bras. Nematol. 1*: 13-15.

Lownsbery, B. F., and Lownsbery, J. W. (1954). *Heterodera tabacum* n. sp., a parasite of solanaceous plants in Connecticut. *Proc. Helm. Soc. Wash. 21*: 42-47.

Lucas, G. B. (1975). *Diseases of Tobacco* (3rd ed.). Biological Consulting Association, Raleigh, North Carolina.

Lucas, G. B., Sasser, J. N., and Kelman, A. (1955). The relationship of root-knot nematodes to Granville wilt resistance in tobacco. *Phytopathology 45*: 537-540.

Marks, C. F., and Elliot, J. M. (1973). Damage to flue-cured tobacco by the needle nematode *Longidorus elongatus. Can. J. Plant Sci. 53*: 689-692.

McCants, C. B., Skogley, E. O., and Woltz, W. B. (1959). Influence of certain soil fumigation treatments on the response of tobacco to ammonium and nitrate forms of nitrogen. *Soil Sci. Soc. Am. Proc. 23*: 466-469.

McGuire, J. M. (1964). Efficiency of *Xiphinema americanum* as a vector of tobacco ringspot virus. *Phytopathology 54*: 799-801.

Meagher, J. W. (1969). Nematodes as a factor in citrus production in Australia. *Proc. First Int. Citrus Symp. 2*: 999-1006.

Miller, L. I. (1977). Pathogenicity of *Globodera virginiae* to Kentucky 16 burley tobacco (abstract). *Proc. Am. Phytopathol. Soc. 4*: 217.

Miller, L. I., Fox, J. A., and Spasoff, L. (1972). Genetic relationship of resistance to *Heterodera solanacearum* in dark-fired and burley tobacco (abstract). *Phytopathology 62*: 778.

Miller, L. I., and Gray, B. J. (1968). Horsenettle cyst nematode, *Heterodera virginiae* n. sp., a parasite of solanaceous plants. *Nematologica 14*: 535-543.

Miller, L. I., and Gray, B. J. (1972). *Heterodera solanacearum* n. sp., a parasite of solanaceous plants. *Nematologica 18*: 404-413.

Miller, P. M. (1969). Suppression by benomyl and thiabendazole of root invasion by *Heterodera tabacum*. *Plant Dis. Rep. 53*: 963-966.

Miller, P. M. (1970). Failure of several non-volatile and contact nematicides to kill eggs in cysts of *Heterodera tabacum*. *Plant Dis. Rep. 54*: 781-783.

Miller, P. M., Taylor, G. S., and Wihrheim, S. E. (1968). Effects of cellulosic soil amendments and fertilizers on *Heterodera tabacum*. *Plant Dis. Rep. 52*: 441-445.

Miller, P. M., and Wihrheim, S. E. (1968). Mutual antagonism between *Heterodera tabacum* and some other parasitic nematodes. *Plant Dis. Rep. 52*: 57-58.

Milne, D. L. (1972). Nematodes of tobacco. In *Economic Nematology* (J. M. Webster, ed.). Academic, London, pp. 159-186.

Milne, D. L., and du Plessis, D. P. (1964). Development of *Meloidogyne javanica* (Treub.) Chit., on tobacco under fluctuating soil temperatures. *S. Afr. J. Agric. Sci. 7*: 673-680.

Mountain, W. B. (1954). Studies of nematodes in relation to brown root-rot of tobacco in Ontario. *Can. J. Bot. 32*: 737-759.

Mountain, W. B. (1960). Theoretical considerations of plant-nematode relationships (J. N. Sasser and W. R. Jenkins, eds). University of North Carolina Press, Chapel Hill, pp. 419-431.

Mulvey, R. H., and Stone, A. R. (1976). Description of *Punctodera matadorensis* n. gen., n. sp. (Nematoda: Heteroderidae) from Saskatchewan with lists of species and generic diagnoses of *Globodera* (n. rank), *Heterodera*, and *Sarisodera*. *Can. J. Zool. 54*: 772-785.

Norse, D. (1972). Nematode populations in a maize-groundnut tobacco rotation and the resistance of maize varieties to *Meloidogyne javanica*, *Pratylenchus*, and *Helicotylenchus*. *Trop. Agric. 49*: 355-360.

Nusbaum, C. J. (1958). The response of root-knot-infected tobacco plants to foliar applications of maleic hydrazide (abstract). *Phytopathology 48*: 344.

Nusbaum, C. J. (1960). Soil fumigation for nematode control in flue-cured tobacco. *Down to Earth Summer*: 1-4.

Nusbaum, C. J. (1969). In *Proceedings of the Symposium on Tropical Nematology* (J. A. Ramos, ed.). University of Puerto Rico, Mayaguez Campus, Agric. Exp. Sta. Rio Piedras, pp. 58-67.

Nusbaum, C. J., and Ferris, H. (1973). The role of cropping systems in nematode population management. *Ann. Rev. Phytopathol. 11*: 423-440.

Nusbaum, C. J., and Sasser, J. N. (1955). Comparison of Dowfume W-85 and D-D mixture applied as liquids and impregnated on vermiculite for nematode control (abstract). *Phytopathology 45*: 349-350.

Nusbaum, C. J., and Todd, F. A. (1970). The role of chemical soil treatments in the control of nematode disease complexes of tobacco. *Phytopathology 60*: 7-12.

Odihirin, R. A. (1971). Effects of root-knot and lesion nematodes on transpiration and water utilization by tobacco plants. *J. Nematol. 3*: 321-322.

Olthof, T. H. A. (1968). Races of *Pratylenchus penetrans*, and their effect on black root rot resistance of tobacco. *Nematologica 14*: 482-488.

Olthof, T. H. A. (1971). Seasonal fluctuations in population densities of *Pratylenchus penetrans* under a rye-tobacco rotation in Ontario. *Nematologica 17*: 453-459.

Olthof, T. H. A. (1979). Effects of *Pratylenchus penetrans* and *Meloidogyne hapla* on potential crops for the tobacco *Nicotiana tabacum* growing areas of southwestern Ontario Canada. *Can. J. Plant Sci. 59*: 1117-1122.

Olthof, T. H. A., and Hopper, B. E. (1973). Distribution of *Pratylenchus* spp. and other stylet-bearing nematode genera in soils in the flue-cured tobacco area of southern Ontario. *Can. Plant Dis. Surv. 53*: 31-33.

Owens, R. G., and Ellis, D. E. (1951). The efficacy of certain chemical soil treatments against meadow nematodes. *Phytopathology 41*: 123-126.

Patel, G. J., Desai, M. V., and Shah, H. M. (1962). Stunt nematodes and tobacco in Gujarat, India. *Plant Dis. Rep. 46*: 173-174.

Powell, N. T. (1971). Interactions between nematodes and fungi in disease complexes. *Annu. Rev. Phytopathol. 9*: 253-274.

Powell, N. T. (1979). Internal synergisms among organisms inducing disease. In *Plant Disease — An Advanced Treatise*, vol. 4 (J. G. Horsfall and E. B. Cowling, eds.). Academic, New York, pp. 113-133.

Prot, J. C. (1976). Amplitude et cinetique des migrations du nematode *Meloidogyne javanica* sous l'influence d'un plant de tomate. *Cah. OSTROM (Ser. Biol.) 11*: 157-166.

Rabb, R. L., and Guthrie, F. E. (eds). (1970). *Concepts of Pest Management*, Conf. Proc. North Carolina State University, Raleigh.

Reddy, D. B. (Compiler). (1978). Tech. Document, FAO P1. Prot. Committee Southeast Asia and Pacific Region, No. 117.

Roman, J., and Triantaphyllou, A. C. (1969). Gametogenesis and reproduction of seven species of *Pratylenchus*. *J. Nematol. 1*: 357-362.

Sasser, J. N. (1980). Root-knot nematodes: A global menace to crop production. *Plant Dis. 64*: 36-41.

Sasser, J. N., Lucas, G. B., and Powers, H. R., Jr. (1955). The relationship of root-knot nematodes to black shank resistance in tobacco. *Phytopathology 45*: 459-461.

Sasser, J. N., and Nusbaum, C. J. (1955). Seasonal fluctuations and host specificity of root-knot nematode populations in two-year tobacco rotation plots. *Phytopathology 45*: 540-545.

Sayre, R. M., Patrick, Z. A., and Thorpe, H. J. (1965). Identification of a selective nematicidal component in extracts of plant residues decomposing in soil. *Nematologica 11*: 263-268.

Schweppenhauser, M. A. (1968). Recent advances in breeding tobacco resistant to *Meloidogyne javanica*. *CORESTA Inf. Bull. 1*: 9-20.

Seinhorst, J. W. (1970). Dynamics of populations of plant-parasitic nematodes. *Ann. Rev. Phytopathol. 8*: 131-156.

Slana, L. J. (1978). Studies on resistance to *Meloidogyne* in *Nicotiana* species. Ph.D. thesis, Unversity of Maryland, College Park.

Slana, L. J., and Stavely, J. R. (1979). Reaction of *Nicotiana* species to *Meloidogyne javanica*. *Phytopathology 69*: 537.

Southards, C. J. (1965). Host-parasite relations of the lesion nematodes, *Pratylenchus brachyurus*, *P. zeae*, and *P. scribneri*, and flue-cured tobacco. Ph.D. thesis, North Carolina State University, Raleigh.

Southards, C. J., and Nusbaum, C. J. (1967). Genetic variability of tobacco response to *Pratylenchus brachyurus*. *Phytopathology 57*: 18-21.

Steiner, G. (1945). Meadow nematodes as the cause of root destruction. *Phytopathology 35*: 935-937.

Steiner, G. (1953). Changes in basic concepts in plant nematology. *Plant Dis. Rep. 37*: 203-205.

Stirling, G. R., and Mankau, R. (1978). Parasitism of *Meloidogyne* eggs by a new fungal parasite. *J. Nematol. 10*: 236-240.

Sturhan, D. (1971). Biological races. In *Plant Parasitic Nematodes*, vol. II (B. M. Zuckerman, W. R. Mai, and R. A. Rohde, eds.). Academic, New York, pp. 51-71.

Thorne, G. (1961). *Principles of Nematology*. McGraw-Hill, New York.

Tisdale, W. B. (1931). Development of strains of cigar wrapper tobacco resistant to black shank (*Phytophthora nicotianae* Breda de Haan). *Fla. Agric. Exp. Sta. Bull. 226.*

Todd, F. A. (1976). *1977 Tobacco Information*, North Carolina Agricultural Extension Service AG-46, p. 59.

Todd, F. A. (1979). In *1980 Tobacco Information*, North Carolina Agricultural Extension Service AG-187, pp. 32-53.

Todd, F. A. and Clayton, E. E. (1956). *N.C. Agric. Exp. Sta. Tech. Bull. 119.*

Triantaphyllou, A. C. (1973). Environmental sex differentiation of nematodes in relation to sex management. *Annu. Rev. Phytopathol. 11*: 441-462.

Triantaphyllou, A. C., and Hirschmann, H. (1980). Cytogenetics and morphology in relation to evolution and speciation of plant-parasitic nematodes. *Annu. Rev. Phytopathol. 18*: 333-359.

Tsamagari, H., and Tanaka, I. (1954). Root-knot formation (*Heterodera marioni* (Cornu) Goodey) and its symptoms of tobacco plant. *Kagoshima Tob. Exp. Sta. Bull. 9*: 26-29.

Tyler, J. (1933). The root-knot nematode. *Calif. Agric. Exp. Sta. Circ. 330*: 1-34.

Valleau, W. D., and Johnson, E. M. (1947). The relation of meadow nematodes to brown root-rot of tobacco. *Phytopathology 37*: 838-841.

Vallotton, R., and Corbaz, R. (1976). A study of lodging in tobacco caused by *Ditylenchus dipsaci* in the French part of Switzerland; means of chemical control. *Ann. Tabac 13*: 49-68.

Van Gundy, S. D., Kirpatrick, J. D., and Golden, J. (1977). The nature and role of metabolic leakage from root-knot nematode galls and infection by *Rhizoctonia solani*. *J. Nematol. 9*: 113-121.

Viglierchio, D. R. (1979). Selected aspects of nematode physiology. In *Root-knot Nematodes (Meloidogyne species)* (F. Lamberti and C. E. Taylor, eds.). Academic, New York, pp. 115-153.

Wallace, H. R. (1973). *Nematode Ecology and Plant Disease*, Crane, Russak, New York.

Weingartner, D. P., Smart, G. C., Jr., and Shumaker, J. R. (1980). Population dynamics of trichodorid nematodes in Florida Irish potato soils following soil fumigation (abstract). *J. Nematol. 12*: 241.

Weybrew, J. A., Jones, G. L., Mann, T. J., Woltz, W. G., Hutchenson, T. B., Nusbaum, C. J., and Van Bavel, C. H. M. (1953). Factors affecting the nicotine content of flue-cured tobacco. *N.C. Agric. Exp. Sta. Rep. No. 8.*

Wyss, U. (1973). Reaktion von cytoplasma und zellkern in von *Trichodorus similis* besaygten Wurzelhaaren. Wirt: *Nicotiana tabacum* - *Mitt. Biol. Bundersanst. Land Forst., Berl. -Dahlem, 151*: 305-306.

Zacheo, G., Lamberti, F., and Durbin, R. D. (1974). Effect of *Meloidogyne incognita* (Kofoid and White) Chitwood on the nicotine content of tobacco (*Nicotiana tabacum* L.). *Nematologica Med. 2*: 165-170.

Zadoks, J. C. (1972). Methodology of epidemiological research. *Annu. Rev. Phytopathol. 10*: 253-276.

Zadoks, J. C., and Schein, R. D. (1979). *Epidemiology and Plant Disease Management*, Oxford University Press, New York.

Chapter 8

Nematode Parasites of Alfalfa, Cereals, and Grasses

G. D. Griffin *USDA, Agricultural Research Service, Utah State University, Logan, Utah*

I. ALFALFA

A. Introduction

Alfalfa (*Medicago sativa* L.) is the only known forage that was cultivated before recorded history, and its exact origin is not accurately defined. The most likely center of origin, however, is in southwestern Asia, probably Iran (Bolton, 1962), but alfalfa can be traced back to antiquity, as far back as recorded records go. It was apparently the principal fodder of cavalry and chariot horses of the ancient Persians, Greeks, and Romans, and was probably carried by caravans throughout the world, as far as China and India.

Alfalfa has had many common names throughout its cultivated history, but is known primarily as alfalfa or lucerne; the two names seem to be closely associated with the advance of the crop across North Africa and into Europe.

Alfalfa was introduced into the new world by the early European explorers and conquerors. Stewart (1926) stated that when the conquest of Mexico and Peru was over, the natives had alfalfa in lieu of their gold and their monarchs.

Alfalfa was probably introduced into the southwestern United States from Central America. It eventually found its way into California around 1850, and then eastward into the intermountain region and the Great Plains area of the United States. It was also introduced into the New England colonies from Europe in the early 1700s (Bolton, 1962).

The history of alfalfa until the 1900s has been one of introduction and distribution. However, the twentieth century has seen emphasis placed on increasing alfalfa yields and adaptability. Disease control has also played an important role in the world status of alfalfa today, and only through genetic breeding and agronomic practices in controlling disease organisms, including nematodes, has the status of alfalfa grown to the important stature it has today in the world food supply.

Of the plant-parasitic nematodes on alfalfa, the alfalfa stem nematode, *Ditylenchus dipsaci*, the root-knot nematodes, *Meloidogyne* spp., and the root lesion nematodes, *Pratylenchus* spp., are the most important economically. Their importance is measured not only in their parasitism of alfalfa, but by the important role(s) they play in the disease complexes with other plant pathogens.

B. Alfalfa Stem Nematode [*Ditylenchus dipsaci* (Kühn) Filipjev]

1. History and Distribution

Ditylenchus dipsaci is the most important nematode pathogen on alfalfa, and has accompanied alfalfa seed to practically every corner of the globe. It was first reported on alfalfa in Germany in 1881 and later in the United States in 1923 (Thorne, 1961). It is most frequently reported as a serious pest in areas of heavy soil, high rainfall, heavy spring rains, or in irrigated fields.

One of the peculiarities of *D. dipsaci* is that it contains several distinct biological races. These, however, vary from location to location and several reports have shown race variability. Green (1981) found several weeds and wild plants to be hosts of *D. dipsaci* in England, which may be a source of different biological races. Edwards and Taylor (1963) found a race in Illinois that was able to reproduce on several plant species. Sturhan (1971) stated that more than 20 biological races have been described; these attack more than 300 plant species, and are differentiated only by hosts. Although a plant species may not be a host for a given race, *D. dipsaci* is able to infect and parasitize nonhost plant tissue. An alfalfa race from the United States caused characteristic symptoms on nonhost seedlings of sweet clover, onion, tomato, sugarbeet, and wheat. Although unable to reproduce on any of the cultivars, except sainfoin, *D. dipsaci* caused plant mortality of seedlings ranging from 20% on sugarbeet and tomato to 100% on onion (Griffin, 1975; Griffin et al., 1975; Griffin and Waite, 1971).

Seed was once considered the major avenue for dissemination of *D. dipsaci*, and up to 17,000 nematodes have been found in a pound of fine screenings from uncleaned seed (Thorne, 1961). Brown (1957) and Bingefors (1960) reported similar epidemiological patterns; however, anything that moves nematode-infested soil and alfalfa tissue will move the nematodes. They are easily spread by machinery during harvest season and by rain and irrigation water from infected alfalfa debris left in the fields. The reuse of waste irrigation water is probably the most common method of nematode dissemination (Faulkner and Bolander, 1970).

2. Biology and Symptomology

All stages of the nematode, except the first stage, which molts in the egg, are able to attack the alfalfa plant. The nematode enters primordial bud tissue and migrates into the developing buds. Infected stems become enlarged and are usually discolored, the nodes swell, and the internodes become shortened (Figs. 1 and 2). The disease symptoms are due to enzymes, such

Fig. 1 *D. dipsaci* on alfalfa. Left, uninoculated control. Center and right, nematode infected. Note stunting and swelling. (Courtesy of the Department of Nematology, University of California, Riverside.)

Fig. 2 *D. dipsaci*, susceptible Ranger alfalfa infected (left) and noninfected (right). Note swollen crown buds.

as pectinase, propectinase, amylase, and invertase, that are secreted by the nematode. Pathological changes, however, may not only be due to enzymes secreted by the nematode, but also to a physiological imbalance of auxins produced by the plant. Growing stems may succumb to the infection or over- come the swelling and grow normally. Stem necrosis results as the nematode multiplies, and after long periods of parasitism accompanied by moderate temperatures and high humidity. Blackening of the stem can be observed up to a foot or more above the ground (Fig. 3). Great numbers of nematodes (several thousand per gram of fresh tissue) can be found in the matrix inside the blackened stem. The numbers of stems per crown become fewer as the alfalfa crown is destroyed, and eventually the entire plant dies (Fig. 4). The stand of alfalfa is thinned, leaving space for weeds and grasses (Fig. 5).

   Soil temperature is very important in the host-parasite relationships of *D. dipsaci* on alfalfa (Griffin, 1968; Perret, 1971). In some cases, when heavy nematode infection is combined with warm humid weather, the nematode may migrate into the leaf tissue, causing a curling and distortion of the leaves. Plants may show "white flagging" (Fig. 6), a symptom attributed to nematode infection of the leaf tissue and destruction of the chloroplasts (Campbell and Griffin, 1973); infected alfalfa leaves exhibit a gradient of leaf discoloration from pale green to completely white. There is a reduction in the lamellar structure of the chloroplasts and a reduction in the starch gran- ules and an accompanying loss of chlorophyll and protein.

   D. dipsaci may also infect the alfalfa seed if the influorescences are in- fected (Brown, 1957). Roots may occasionally be infected, causing internal cavities (Krusberg, 1961) or there may be gall-like outgrowths that may girdle the root crown. Large numbers of the nematodes may survive periods

Fig. 3 Blackening of susceptible Ranger alfalfa stem tissue parasitized by *D. dipsaci*.

Fig. 4 Susceptible Moapa alfalfa succumbing to parasitism by *D. dipsaci* in southern Utah.

Fig. 5 Field infested with weeds after death of Ranger alfalfa due to *D. dipsaci*.

Fig. 6 White flagging of second cutting of susceptible Ranger alfalfa caused by *D. dipsaci* infection.

of environmental stress in the alfalfa crown tissue. Only small numbers of
nematodes have, however, been recovered from alfalfa field soil at any time of
the year. Tseng et al. (1968) found a maximum number of 50 nematodes in
400 cm^3 of soil at any time during the year. Similar findings were found by
Lewis and Mai (1960) on onion and by Wallace (1962) on oats. There appear
to be differences, however, in the ability of *D. dipsaci* to survive in the
absence of a host plant. Chitwood et al. (1940) reported that *D. dipsaci* was
unable to survive overwintering in an onion field, and Edwards (1951) stated
that this same species was able to survive a five year rotation in an onion
field in Australia. This may be due to differences in soil moisture regimes.
We have found that the alfalfa race of *D. dipsaci* is unable to survive more
than two years under a nonhost cropping of grain under heavy rainfall, or
when standard irrigation practices are followed. However, we have found
fourth-stage juveniles surviving in an anhydrobiotic state for over six years
in fallow soil in Nevada where there was little or no rainfall. This agrees
with the findings of Fielding (1951), who found that *D. dipsaci* survived more
than 20 years in a dormant stage. It appears that the parasite will survive
for long extended periods, unless activated by moisture in the absence of a
host, where the nematode soon succumbs. Wallace (1962) stated that over
90% of *D. dipsaci* became active when nematode-infested dry soil was immersed
in water. Therefore, differences in longevity of this nematode appear to depend on availability of host tissue in the soil, soil moisture relationships, and
possible physiological differences within races and nematode populations.

D. *dipsaci* is unique in its ability to parasitize plants over a large temperature range. *D. dipsaci* was able to acclimatize to the temperature at which
reproduction occurred, and there is a direct correlation between the acclimation temperature and the temperature at which maximum penetration of plant
tissue occurs (Griffin, 1974).

D. *dipsaci* is also unique in that individuals congregate together and
form what is commonly known as "nema wool." This phenomenon has been
observed in infected alfalfa tissue and in water extractions from plant tissue
and in host tissue. This may result from some pheromone or sexually attractive secretion exuded by the nematode, since we have also observed several
males congregating and entangling around a single female.

One of the important parasitic relationships of *D. dipsaci* is its relationship with other plant pathogenic organisms. This nematode plays an important
role in development of bacterial wilt *Corynebacterium insidiosum* (McCull)
H. L. Jens of alfalfa. Hawn and Hanna (1966) found that stem nematode was
able to break wilt resistance of alfalfa to the bacterial wilt.

3. Control

Resistance: Without resistant alfalfa cultivars, the alfalfa stem nematode
would be much more devastating to alfalfa than it is now. Resistance to the
alfalfa race of *D. dipsaci* was first observed in a Turkistan alfalfa selection
planted in nematode-infested fields in Northern Utah. From this Turkistan
selection, the alfalfa cultivar Nemastan was selected (Thorne, 1961), which
is also resistant to bacterial wilt (Fig. 7). Lahontan was later selected from
Nemastan (Fig. 8). Resistance to the stem nematode has also been found in
the Turkish alfalfa introduction, Kayseri. Deseret, a selection from Kayseri,
has resistance to downy mildew as well as to stem nematode (Pederson and
Griffin, 1977). Additionally, several alfalfa cultivars contain some degree of
stem nematode resistance; included are Washoe, Resistador, Caliverde 65, and
Appalachee. Nematode-resistant germ plasms, released from incorporation

Fig. 7 Alfalfa parasitized by *D. dipsaci*: Left, susceptible Utah common.
Right, resistant Nemastan. (Photo by G. Thorne.)

Fig. 8 Reaction of alfalfa to *D. dipsaci*. Lahontan, resistant; Ranger,
susceptible. (Courtesy O. J. Hunt.)

into existing or new alfalfa cultivars (Hartman et al., 1979; Peaden et al., 1976) have also been developed.

There are similar ultrastructural changes that occur within both resistant and susceptible alfalfa tissue upon nematode penetration. Although electron micrographs of tissue of nematode-infected plants at 15, 20, and $25^{\circ}C$ showed more lipid bodies than did noninfected plants after one day, swollen resistant (Lahontan) and susceptible (Ranger) plants showed similar anatomical changes three and seven days after inoculation. Plant damage, however, was more greatly magnified in susceptible than in resistant plants. Only the rate of infection and the degree of damage differed between resistant and susceptible alfalfa, with susceptible plants showing the greater degree of swelling and disorganization of cellular structures. The greatest injury occurred in both cultivars at the higher temperature. As plants increased in age, however, Lahontan plants were able to overcome the damage and make a normal growth (Chang et al., 1973, 1975).

Cultural: Reed et al. (1977) found *D. dipsaci* caused cavities in galled tissue of both susceptible (Buffalo) and resistant (Washoe) alfalfa, the cavities being smaller in Washoe. They also found that mechanical injury caused lignification of cavity walls. A two-to-three-year rotation, using nonhost crop such as grain, beans, or sugarbeet, will usually reduce the nematode population to below the detection level. However, recontamination from machinery, animals, or waste irrigation water can quickly negate the beneficial affects of crop rotation and the field can become reinfested quickly.

Certain agronomic practices can partially alleviate the parasitism of the *D. dipsaci* on alfalfa. In irrigated areas of the arid and semiarid western United States, the alfalfa stem nematode is usually a problem only in the first cutting of alfalfa. Inasmuch as each cutting required reinfection mainly from the soil, infection can be reduced if the alfalfa is cut when the top 5-8 cm of the soil is dry. This results in little or no nematode infection of subsequent plant growth. This effect can, however, be negated by rainfall or irrigation immediately before or after cutting, which raises the soil moisture to about 50% field capacity or above. Hence, the nematode can reduce quick tillering and plant regrowth, depending on the nematode population density. There is some infection from secondary growth of alfalfa stubble, but this is minimal.

We have found significant decreases in nematode infection in spring plant growth when fall burning is used for weed control. However, spring burning has the opposite effect, and is not recommended; this initiates growth of new plant tissue resulting in increased infection. Two- to three-year-old alfalfa fields have been completely devastated with this nematode because of spring burning.

Normally, the decline in yield varies with the source of nematode infestation. If the nematode source is from irrigation water or if the number of nematodes in the soil at the time of plant establishment is high, decline occurs in two to three years. If the source of infestation, however, is from seed or machinery or from a reduced population in the soil, the decline is more gradual and more time may be required to reduce the alfalfa stand by 50%. There is also a more rapid decline in the nondormant alfalfas than in the dormant alfalfa, since nondormant types do not survive as long under normal conditions, and there is a continual association between the plant and the parasite's active stages.

Chemicals: Many of the systemic nematicides have been used successfully in experimental trials, including oxamyl, phorate, thionazin, and fenamiphos (Griffin, 1967; Nigh et al., 1969). Timing is of major importance, and nematicides should be applied soon after spring growth is initiated. However, in most areas, the high cost of chemicals and their application make this type of control impractical.

C. Root-Knot Nematodes (*Meloidogyne* spp.)

Root-knot nematodes were originally considered to be a single species [*Heterodera marioni* (Cornu) Goodey] until Chitwood (1949) redescribed five species and assigned them to the genus *Meloidogyne* (Goeldi) Chitwood. Since then, about 50 species have been sufficiently differentiated to merit species status (Taylor and Sasser, 1978; Golden et al., 1980). Sasser (1977) stated that the most widespread and economically important of all root-knot species are *M. incognita*, *M. hapla*, *M. javanica*, and *M. arenaria*. These also contain the most important root-knot species parasitizing alfalfa.

1. Northern Root-Knot Nematode (Meloidogyne hapla Chitwood 1949)

History and Distribution. Although referred to as the northern root-knot nematode, *M. hapla* is widely distributed. It has been known as the most commonly found root-knot species in the northern hemisphere, where rhizosphere soil temperatures drop to $0°C$ or below and summer soil temperatures seldom exceed $25-30°C$. The predominance of *M. hapla*, however, has been questioned in the northern United States with the discovery of the Columbia root-knot nematode, *M. chitwoodi* Golden, O'Bannon, Santo, and Finley, 1980 (Golden et al., 1980).

M. hapla is found throughout the northern United States, southern Canada, northern Europe, and northern Asia, $40°$ S latitude and in the mountainous areas at lower latitudes. It has also been reportedly found in Africa, Australia, and South America (Taylor and Sasser, 1978).

Biology and Symptomology. *M. hapla* is considered a mild parasite on alfalfa. However, severe invasion of alfalfa seedlings from highly infested soil is capable of causing a high mortality in young alfalfa plants in newly seeded ground with an ultimate decrease in hay yields. A tolerance level of less than one nematode per cubic centimeter soil is reported for susceptible Washoe seedlings under greenhouse conditions (Inserra et al., 1980). Under light soil infestations, however, alfalfa plants may appear normal, resulting in satisfactory hay yields.

M. hapla is greatly reduced in numbers where the frost line extends below the rhizosphere, and it is practically nonpathogenic at soil temperatures below $15°C$. It is often difficult to find spring populations in cultivated soil after a severe winter when the frostline extends below the rhizosphere unless a bioassay method is used.

The greatest importance of *M. hapla* on alfalfa is that: (1) it creates an economic problem for highly susceptible crops that are grown in rotation with alfalfa, such as vegetable truck garden crops and susceptible field crops, and (2) it plays a role in the host-parasite relationship between alfalfa and other alfalfa pathogens. *M. hapla* in association with *Corynebacterium insidiosum* increased the incidence of bacterial wilt on bacterial wilt susceptible and resistant alfalfa cultivars (Hunt et al., 1971; Griffin and Hunt, 1972a). Griffin and Thyr (1978) showed an increase in the incidence of *Fusarium* wilt with *M. hapla*. They found that the combination of *M. hapla* and *Fusarium oxysporum* reduced plant growth, and sequential inoculations of the two pathogens significantly reduced plant growth at 20 and $25°C$, but not at 15 and $30°C$. There is also a synergistic interaction between *M. hapla* and *D. dipsaci* (Griffin, 1980). *D. dipsaci* is able to predispose root knot-resistant alfalfa to *M. hapla*. There were 26% of a root knot-resistant alfalfa selection galled when *D. dipsaci* preceded inoculation of eight-week-old plants by *M. hapla*. Single inoculations of eight-week-old plants by *M. hapla* failed to gall resistant plants.

The life cycle for *M. hapla* is similar to that of other root-knot nematode species as outlined by Taylor and Sasser (1978).

Newly hatched second-stage juveniles are attracted to metabolically active alfalfa tissue, and attraction is increased slightly with an increase in temperature from 15 to $30^{o}C$ (Griffin and Waite, 1971). The greatest juvenile orientation is to the root apical meristem and region of elongation. There is, however, some juvenile congregation in the upper hypocotyl and shoot meristem tissue, although penetration usually occurs only in the area of the root apical meristem.

Soil temperature plays an important role in penetration of alfalfa roots. Griffin (1969) found that 65% of susceptible alfalfa plants were infected after 21 days, but only 20% after 28 days at $5^{o}C$ due to the inability of the nematode to establish feeding sites at this low temperature. This, however, increased to 100% infected plants after 29 and 28 days at soil temperatures of $10-30^{o}C$. The percentage of penetration density increased with an increase in soil temperature from 5 to $25^{o}C$, and then decreased slightly at $30^{o}C$. Nematode development increased as temperature increased, and there was little or no nematode development at 5 and $10^{o}C$; maximum development occurred at 25 and $30^{o}C$. There was a significant difference, however, in the male-female ratio at the higher soil temperatures, this being 0.22 at $25^{o}C$ and 0.84 at $30^{o}C$. Nematode reproduction parallels nematode development; there is little or no reproduction at soil temperatures below $20^{o}C$, and maximum reproduction occurred at $25^{o}C$. The life cycle of *M. hapla* on alfalfa is approximately 30 days at $25^{o}C$, and each female is capable of producing about 100-350 eggs, depending on developmental and reproduction conditions.

A unique characteristic of *M. hapla* is its symptomology on host plant tissue. *M. hapla* galls are usually readily identified by the lateral root growth from galls that is characteristic of this nematode species. Gall size will differ, however, depending on the physiological response of the plant (Fig. 9A).

Control Resistance: Alfalfa growers and scientists have little latitude in determining a methodology than can be used to control *M. hapla* on alfalfa. Crop rotation, where a susceptible alfalfa variety is used, is not practical since alfalfa is capable of producing large populations of nematodes over the life of the alfalfa plant, depending on climatic zone. *M. hapla* also has a wide host range and will reproduce on most plants other than those of the grain and grass families. The most practical control methodology is that of resistant cultivars. Resistance to *M. hapla* was first recorded from Vernal alfalfa by Stanford et al. (1958). Since then, resistant germ plasm, Nevada Synthetic XX (Peaden et al., 1976), has been released to commercial alfalfa breeders (Fig. 9B).

The ultimate source of resistance to nematodes would be immunity. Goplen and Stanford (1959) stated that a Vernal selection (M-4) was also immune to nematode penetration by *M. hapla*. Griffin and Elgin (1977), however, found that M-4 was penetrated by *M. hapla* juveniles, but being unable to establish feeding sites the nematodes either died or migrated from the root tissue. This is similar to results found by Reynolds et al. (1970) with *M. incognita acrita* on African alfalfa.

A

B

Fig. 9 (A) Differential host response of root tissue to *M. hapla*. Top, sainfoin. Bottom, alfalfa. (B) Resistance of alfalfa to *M. hapla*. Left and right, Susceptible Ranger; center, Resistant Nevada Synthetic XX. (Courtesy of O. J. Hunt.)

Several factors may affect the resistance of alfalfa to *M. hapla*. Griffin and Hunt (1972b) found that *M. hapla* was able to penetrate and gall resistant alfalfa plants when inoculated in the pregerminated seed stage. However, *M. hapla* failed to gall the same selection when plants were not inoculated until eight weeks of age. Talboys (1964) suggests the possibility of low resistant zones occurring in roots of high resistant plants in young seedlings that are not found as the plants become older.

Chemicals are known to affect the resistance of alfalfa to *M. hapla*. Griffin and Anderson (1979) found that the herbicide EPTC (S-ethyl dipropylthiocarbamate) reduced the resistance of alfalfa to *M. hapla* galling.

2. *Southern Root-Knot Nematode* [Meloidogyne incognita *(Kofoid and White 1919) Chitwood 1949]*

History and Distribution. *M. incognita* is recognized as the most important root-knot species in relation to damage and economic loss to cultivated plants; this species consists of four races of nematodes, distinguished by host specificity. *M. incognita* is, however, unable to survive the harsh temperate climates as well as *M. hapla* and is, therefore, found only in areas with optimum temperature ranges of 25-30°C. *M. incognita* eggs are able to survive temperatures at or below freezing for only a very short time, and juveniles fail to survive at temperatures around 5°C. In the United States, *M. incognita* is not found above geographical zones where the average January temperature is below -1°C (Taylor and Sasser, 1978). Since less than 10% of the alfalfa and alfalfa mixtures acreage is found in *M. incognita* climatically adapted areas, it does not present as great a problem as does *M. hapla*.

Biology and Symptomology. *M. incognita* is also less pathogenic on alfalfa than is *M. hapla*. Chapman (1960) found that *M. hapla* was more pathogenic on Atlantic alfalfa than *M. incognita*, while *M. incognita* was more pathogenic on Kenland red clover than was *M. hapla*, showing a host preference of both root-knot nematodes. Chapman (1963) also compared the pathogenicity and development of *M. incognita* and *M. hapla* on Atlantic alfalfa, and what effect cutting would have on plant growth and nematode reproduction; he found that although *M. hapla* significantly reduced top growth below that of *M. incognita*, reproduction of *M. incognita* was greater than that of *M. hapla*. This would tend to support the fact that *M. incognita* does not affect root growth as much as does *M. hapla*. Reynolds (1955) found a similar relationship with *M. incognita* and *M. javanica* (Treub) Chitwood; *M. javanica* was more damaging to hardy and nonhardy alfalfa than was *M. incognita acrita*.

M. incognita has also been associated with disease complexes in alfalfa. McGuire et al. (1958) found that *Fusarium* wilt (*F. oxysporum*) was more severe on alfalfa in the presence of *M. incognita* than when associated with *M. arenaria* or *M. javanica*.

Chapman and Turner (1975) found that there was a positive relationship between growth suppression of "Buffalo" alfalfa roots and the degree of juvenile penetration of alfalfa tissue; the greater the penetration the less the root growth. Christie (1936) stated that roots begin to respond to the presence of juveniles even before they are invaded.

Control Resistance: The primary method of nematode control is with resistant cultivars. Reynolds (1955) found resistance to *M. incognita acrita* in nondormant African and Chilean alfalfa selections. Goplen et al. (1959) also found resistance to *M. incognita acrita* in an African alfalfa selection.

O'Bannon and Reynolds (1962) found resistance to *M. incognitu acrita* and *M. javanica* in African and Sirsa alfalfa. Nigh (1972) stated that another nondormant alfalfa, Mesa Sirsa, was resistant to *M. incognita*.

Reynolds et al. (1970) reported that although *M. incognita* juveniles penetrated resistant African, Moapa, and Sonora cultivars, and susceptible Lahontan alfalfa in similar numbers, all nematodes soon migrated from the resistant selections and the alfalfa tissue showed no morphological changes that would indicate nematode development or feeding. This agrees with the findings of Griffin and Elgin (1977) within *M. hapla* on alfalfa. In Nevada, Synthetic YY alfalfa germ plasm (a nondormant alfalfa type) has been released that contains resistance to *M. incognita* as well as resistance to some insect pests (Hartman et al., 1979).

3. Javanese Root Knot [Meloidogyne javanica *(Treub. 1885) Chitwood 1949]*

History and Distribution. The javanese root-knot nematode, *M. javanica*, is found under less severe climatic temperature conditions than *M. hapla*, and the northern limit is near the 7.2°C isotherm for average January temperatures (Taylor and Sasser, 1978).

M. javanica is commonly found parasitizing alfalfa in the arid Southwest of the United States. Infection and virulence is dependent on the alfalfa variety, crop rotation history, and soil type. Like most root-knot species, the biggest problem with *M. javanica* is found in sandy to sandyloam soils and like *M. hapla* it is mainly a problem on alfalfa because alfalfa is included in rotational programs with nematode-susceptible crops, such as cotton and melons.

Biology and Symptomology. Bird (1972) found that the optimum temperature for embryogenesis of *M. javanica* is between 25 and 30°C, although the mortality of nematode embryos was twice as great at 30°C as at 25°C. When comparing *M. javanica* development with that of *M. hapla*, Bird and Wallace (1965) found that the optimum temperatures for hatching, mobility, invasion, and growth was 25, 20, 15-20, and 20-25°C, respectively, for *M. hapla*, and 30, 25, 15-30, and 25-30°C, respectively, for *M. javanica*.

Reynolds and O'Bannon (1960) found that when comparing *M. javanica* to *M. incognita* on susceptible varieties, such as DuPuit, Caliverde, Lahontan, Zea, and Ranger, *M. javanica* was more virulent than *M. incognita*.

Control Resistance: O'Bannon and Reynolds (1962) found that 13 breeding lines of African and one of Sirsa selected for resistance to the spotted alfalfa aphid, *Therioaphis maculata* Buckton, were also resistant to *M. javanica*.

Another root-knot nematode species, *M. arenaria* (Neal 1889) Chitwood 1949, has approximately the same northern limitation, geographically, as *M. incognita* (Taylor and Sasser, 1978) and is less tolerant to cold climate conditions than *M. hapla*. It is of minor importance on alfalfa and is a concern to growers only on rare occasions.

D. Root Lesion Nematodes (*Pratylenchus* Filipjev 1934)

1. History and Distribution

The group of nematodes we know as "lesion nematodes" was first placed in the genus *Tylenchus*. Goffart (1929) and Goodey (1932) later placed them in the genus *Anquillulina*. The genus *Pratylenchus* was later established by Filipjev (1934), and revised by Sher and Allen (1953).

Root lesion nematodes are found throughout the world from the tropics to the most northern temperate areas. They attack a wide range of plants and play an important role in damaging both cultivated and noncultivated plants, including alfalfa (Chapman, 1958, 1959).

Pratylenchus species are among the most important plant-parasitic nematodes. Plant growth retardation is usually associated with this nematode, and plant stunting is usually spotty in a field. Damage caused by root lesion nematodes is difficult to evaluate accurately because other soil microorganisms usually invade the damaged nematode-infected roots. Above-ground symptoms are also difficult to assess, since other pathogens may produce similar symptoms. No symptoms develop when nematode numbers are low, but when numbers are high and environmental conditions are ideal, infected plants become stunted.

2. Biology and Symptomology

Pratylenchus is a migratory nematode genus that invades plant roots in both juveniles and adult stages.

Kable and Mai (1968) and Sontirat and Chapman (1970) suggested that only fourth-stage juveniles and adults penetrate the roots of alfalfa. Townshend (1978), however, found that roots of DuPuit alfalfa were also penetrated by third-stage juveniles. Females were able to penetrate alfalfa roots over a greater temperature range, and in greater numbers than males or juveniles. Females penetrated roots from 5 to 35°C with maximum penetration occurring at 10-30°C. Males and third-stage juveniles penetrated alfalfa roots at 10-30°C, with maximum penetration accuracy at 20°C. Townshend attributes the greater female penetration to the greater size of the posterior subventral digestive gland lobe secreting the enzyme utilized by the nematode in root penetration. Townshend further states "The superior infective capacity of female penetration aids the nematode in escaping adverse soil environments (temperature and moisture) and assures its survival in northern climates."

The life cycle of root lesion nematodes is simple. Females deposit eggs in root tissue or in the soil. Both juvenile and adult forms enter roots by forcing their way between or through the epidermal and cortical cells and then feed on the cell contents as they migrate through the root tissue. The points of entrance to the roots also provide access for other soil microorganisms.

Townshend and Stobbs (1981) found that *P. penetrans* penetrated the entire length of Saranac alfalfa except for the root tips. Lesions first appeared as water-soaked areas, but later became discolored, coalesced, and the coloration then intensified until the entire root may become blackened (Fig. 10); *P. penetrans* fed on the cortical tissue and never entered the stele. Polyderm developed beneath the endoderma. Root discoloration resulted in part from phenol oxidation from the formation of a ligninlike substance. Lesion formation could be negated with ascorbic acid, at time of feeding, that held the phenols in a reduced state.

Several species of *Pratylenchus* have been found associated with alfalfa. Chapman (1954) found that *Pratylenchus* spp. severely damaged alfalfa that had been spring sown in winter wheat fields. He found over 100,000 nematodes per gram of severely damaged alfalfa roots. He hypothesized that a high nematode population density plus competition with wheat under moist conditions resulted in the severe infection and destruction of alfalfa root tissue in the alfalfa-wheat planting. Chapman, however, further states that

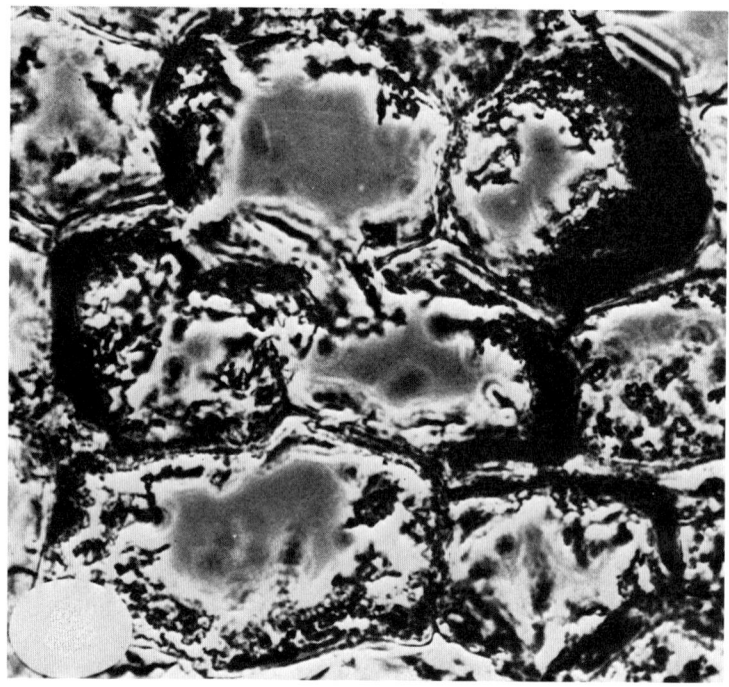

Fig. 10 Collapse of alfalfa root tissue caused by *P. penetrans*. (Courtesy J. L. Townshend and L. Stobbs. Reprinted by permission of the Canadian Phytopathological Society.)

after a summer fallow, fall-sown alfalfa made good growth and the roots were only slightly damaged. Stessel (1960) found *Pratylenchus* associated with poor stands and low yields of Narragansett alfalfa in Rhode Island. Santo et al. (1980) found *P. penetrans* reduced the growth of Washoe and Vernal alfalfa. *P. penetrans*, however, is apparently the most important lesion nematode on alfalfa. It is found throughout the United States, Canada, and Europe and has a wide host range. Like most plant-parasitic nematodes, however, *P. penetrans* is affected by certain edaphic factors that in turn affect the virulence of the nematode. Kimpinski and Willis (1981) found that the number of *P. penetrans* penetrating alfalfa roots increased as soil temperature increased from 10 to 30°C, and Suzuki and Willis (1974) found that root lesion nematodes affect the cold tolerance of alfalfa. They also stated that a pH of 5.2-6.9 was optimum for *P. penetrans* reproduction. This agrees with Dunn (1973), who found that the life cycle of *P. penetrans* was shortest at 30°C. Dickerson (1979), however, stated that the temperature limitations depend more on the host plant than on the nematode. Willis (1972) further states that *P. penetrans* significantly decreased alfalfa yields at pH 5.2 and 6.4, but not at pH 4.4 and 7.3. Willis (1976) also found that increased potassium usage resulted in lower rootlet yields in *P. penetrans*-infested soil, and yield increases were smaller from increased potassium. Endo (1959) found that infection and reproduction of *Pratylenchus* was also affected by soil types. *P. penetrans* has been found to also interact with other plant pathogens on alfalfa. Mauza and Webster (1982) found that a combination of *P. penetrans* and two species of *Fusarium* significantly reduced Vernal alfalfa growth below that of the pathogen. Chapman and Turner (1975) found that a combination of *P. penetrans* and *Meloidogyne incognita* significantly reduced alfalfa root growth below that of single inoculations of either nematode. They, however, found that *M. incognita* inhibited *P. penetrans* reproduction, especially when inoculation of *M. incognita* preceded that of *P. penetrans*. Edmunds and Mai (1966), however, found that *Trichoderma virde* and *F. oxysporum* enhanced penetration of alfalfa roots by *P. penetrans*.

3. Control

Root lesion nematodes have an extensive range of hosts, and more than one species may infect a given host. Crop rotation, therefore, is of less value as a control measure for lesion nematodes than for other nematode-host combinations. Crops differ in their susceptibility as hosts for root lesion nematodes, and alfalfa should not be rotated with crops that are considered good to excellent hosts for the lesion nematodes if it can be avoided.

Alfalfa cultivars resistant to root lesion nematodes are not yet available, and no useful level of resistance has been located among *Medicago* species. Recent studies, however, indicate that it may be possible to develop adapted cultivars with satisfactory resistance to one or more *Pratylenchus* species. Certain nematicides will control root lesion on alfalfa, but the cost of the chemical and its application is generally prohibitive (Thompson and Willis, 1970).

E. Ectoparasitic Nematodes

Several genera of nematodes have been reported associated with alfalfa. Thomason and Sher (1957) found that stubby-root nematodes affected the growth of alfalfa. Taylor et al. (1958) found *Criconemella, Helicotylenchus, Hoplolaimus, Paratylenchus, Paratrichodorus*, and *Tylenchorhynchus* associated with alfalfa in Minnesota. McGlohon et al. (1961) found similar genera in North Carolina; they found *Trichodorus, Tylenchorhynchus, Helicotylenchus*, and *Xiphinema* associated with alfalfa. Noel and Lownsbery (1978) found that *Tylenchorhynchus clarus* suppressed the growth of Moapa alfalfa at 21-27°C; nematodes caused a reduction in feeder root growth, but no lesions or other abnormal root growth were observed. Townshend and Potter (1973) found *Helicotylenchus digonicus* associated with the growth of alfalfa, which increased significantly on alfalfa during the growing season.

Norton (1963) found an inverse relationship between alfalfa yields and *Xiphinema americanum* in Iowa, and stunted plants were associated with high populations of nematodes. He also found that plant growth was reduced when soil was artificially inoculated with *X. americanum*.

Norton et al. (1968) also showed that *X. americanum* causes serious damage to alfalfa and red clover fields in the North Central region of the United States, and is the most common nematode species found. This agrees with Taylor et al. (1958), who found *X. americanum* in 75% of all samples collected from alfalfa fields in Minnesota.

Several viruses found on alfalfa are vectored by nematodes; alfalfa is a host for the pea early-browning virus, which is transmitted by a *Paratrichodorus* spp.

II. CEREALS

A. Introduction

Of all the plants cultivated, the cereals play the most important role in producing essential nutrition for humans. (Since corn and rice are being covered in other chapters, this section will be limited mainly to barley, wheat, oats, and rye.)

Cereal production began about 20,000 years ago, and barley was apparently the first cultivated cereal (Bland, 1971). Vavilov (1926) indicates two centers of origin for barley, i.e., North Africa and East Asia. Barley was probably the first plant cultivated, and has one of the widest cultivations of all crops known, from inside the Arctic Circle to the tropical regions of the globe where it is grown at elevations over 15,000 feet.

Wheat originated from certain wild grasses and was introduced to cultivation sometime after barley. It became important to the ancient civilizations of Egypt, Greece, and Persia (Bland, 1971). It has spread to all parts of the globe and is important in the food supply of many countries. It is suggested that wheat originated in Southwest Asia. The most commonly cultivated wheat belongs to the genus *Triticum*, and the most economically important are the bread wheats, *T. aestivum*.

Coffman (1946) suggests that the development of oats paralleled that of barley. Rye has been cultivated for about 2000 years, and O'Brien (1925) stated that its origin was about the time of the beginning of Christianity. Vavilov (1926) suggests that the origin of rye was in Southeast Asia, Asia Minor, or Persia.

Several genera of nematodes parasitize cereals. However, the most important nematodes belong to the genera *Heterodera*, *Meloidogyne*, *Ditylenchus*, and *Pratylenchus*.

B. Cereal Cyst Nematode (*Heterodera avenae* Wollenweber 1924)

1. History and Distribution

The oat cyst nematode, *Heterodera avenae*, was first seen on roots of wheat, oats, and barley in Germany (Kühn, 1874) (Fig. 11). It was originally classified as the "oat race" of *Heterodera schachtii*, then later classified as *H. schachtii* var. *avenae*. Schmidt (Thorne, 1961) raised the nematode to subspecies (*H. schachtii* sub. sp. *major*). Franklin reclassified it as *H. major* in 1940, but it was later renamed *H. avenae* Wollenweber because of taxonomic rank (Franklin et al., 1959).

It had originally been supposed that *H. avenae* occurred in only the temperate areas of the world, but it is now found to be cosmopolitan in distribution and is found throughout the world. Brown (1982) states that there are two million ha of wheatland infested in Australia. This parasite has been reported from 31 countries around the world, since its first detection in Germany (Kühn, 1874), to the last recorded incidences from New Zealand (Grandison and Halliwell, 1975) and the United States (Jensen et al., 1975) (Table 1).

It is evident from the areas of nematode infestations that the nematode has been disseminated throughout the world through the practice of cereal cultivation. Humans, in their cultivation of cereals, have undoubtedly been primarily responsible for the dissemination of *H. avenae* through the movement of soil with machinery, animals, water, etc. (Fig. 12A). It has been shown, however, that viable cysts can be rapidly disseminated from dust storms and strong winds (Meagher, 1977) (Fig. 12B).

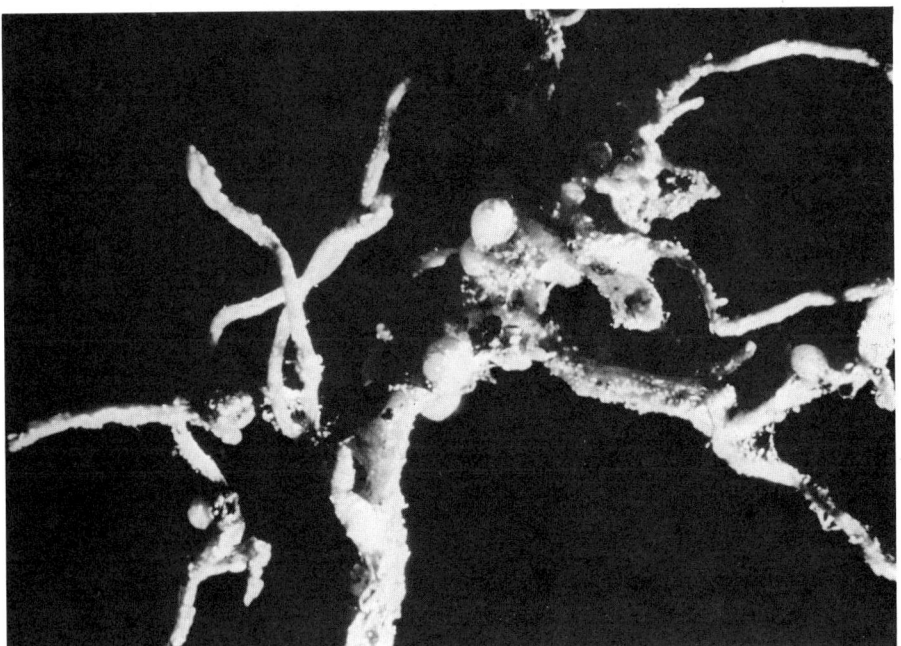

Fig. 11 *H. avenae* females on wheat root. (Courtesy R. H. Brown.)

2. Biology and Symptomology

H. avenae is a sexually dimorphic species, and the life cycle is similar to that of all cyst nematodes. Upon emerging from the cysts, the second-stage juveniles invade the cereal roots and develop through the normal cyst nematode cycle. The nematode undergoes morphological development, causing the formation of specialized cells, called *syncytia*, on which it feeds. This interferes with the ability of the plant to obtain adequate nutrients and water (Figs. 13 and 14). The time of penetration and development is dependent on edaphic and agronomic factors for a given geographical area and plant cultivar.

In Europe and Canada, autumn-sown cereals are penetrated by second-stage juveniles soon after seed germination, but development is terminated during the winter and is initiated again in the spring; females may be seen on the roots as early as April. In spring-sown fields, however, the nematode overwinters in the egg stage and invades newly planted spring grain at temperatures as low as 4^{0}C in mid-March (Decker, 1972). The juvenile soil population usually peaks in April or May and declines sharply during June. Females may be seen on the roots within 90 days after the penetration under optimum nematode development temperatures.

In Australia, eggs remain in a quiescent stage during the hot days and early fall months, and the time of planting and subsequent infection is dependent on the period of rainfall; seed is sown and eggs hatch soon after the spring rains. Although soil temperatures are declining in Australia and soil temperatures are rising in Europe and Canada, root penetration occurs at approximately the same time during the same month in the European and Canadian spring grain plantings and Australian fall grain plantings (Fig. 15A).

Table 1 World Distribution of Cereal Cyst Nematode, *H. avenae*

Country	Year
East Germany	1874
Holland	1891
Denmark	1897
Sweden	1897
England	1908
West Germany	1923
USSR	1925
Norway	1926
Australia	1930
Canada (Ontario)	1935
Scotland	1946
Tunisia	1953
Italy	1953
Japan	1954
Israel	1956
Belgium	1957
Peru	1958
India	1959
Poland	1960
France	1961
Eire	1962
Spain	1963
Portugal	1963
Northern Ireland	1964
Switzerland	1965
Greece	1966
Yugoslavia	1966
Bulgaria	1967
Czechoslovakia	1967
United States	1975
New Zealand	1975

Source: After Meagher (1977).

Although different periods of development are observed from different geographical areas, the life cycle is similar. There is only one generation per year, and the life cycle is between three and four months under optimum temperature development conditions. However, due to the different geographical conditions, females are usually observed in the roots of fall-planted grain in March or April, and on spring-planted grain, females may be seen from the middle of June. Cysts may be seen in February in India, but not until August to October, depending on the time of planting, in Australia.

H. avenae juvenile emergence is not affected by root leachates as are several *Heterodera* species. Several edaphic factors, however, affect juvenile hatching, and subsequently the degree of virulence of *H. avenae* to cereals. Gair (1965) found the nematode in all soil types, but Meagher (1972a) found that *H. avenae* was found most predominantly in soils with good physical structure and not in heavy poorly structured soils. He further stated that *H. avenae* is more virulent in the lighter soils. Fidler and Bevan (1963) also

Fig. 12 (A) Spotty effect of *H. avenae* on wheat in Australia due to initial infestation of field. (Courtesy of J. W. Meagher.) (B) Dissemination of *H. avenae* by dust storm in Australia. (Courtesy R. H. Brown.)

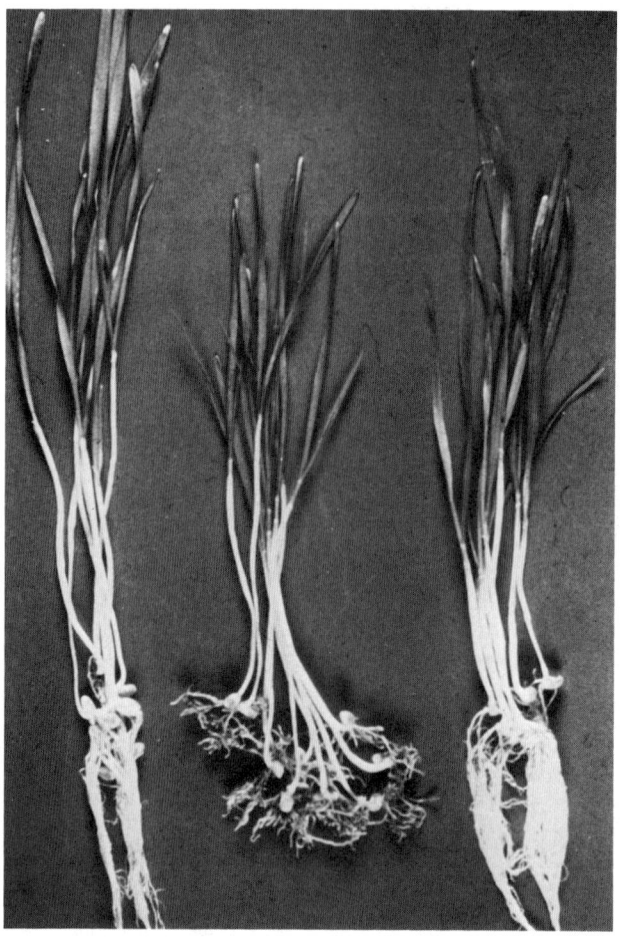

Fig. 13 Effect of *H. avenae* on the growth of wheat. Left and right, healthy growing plants. Center, nematode parasitized plants. Note shortened and increased lateral root growth. (After Rovira, 1982.)

showed the importance of soil types on the host-parasite relationship between oats and *H. avenae* (Fig. 15B).

Cotten (1962) showed that low temperature stimulates hatching of *H. avenae*, and Fushtey and Johnson (1966) stated that a minimum dormancy period of eight weeks at low temperatures was necessary before substantial hatching could be obtained. Therefore, soil temperature has a marked effect on the virulence of the cereal cyst nematode on grain. As previously stated, it has been shown (Meagher, 1972b; Dixon, 1963) that seed germination and the degree of nematode parasitism as affected by soil population density is dependent on the time of rainfall. When early planting is made in May at about 15°C, there is low nematode penetration of wheat roots and minimal damage. Late rainfalls, however, result in seed germination at soil temperatures of 10°C or less, resulting in an increased nematode hatch, greater nematode penetration, and greater plant damage (Figs. 16, 17, 18A, B).

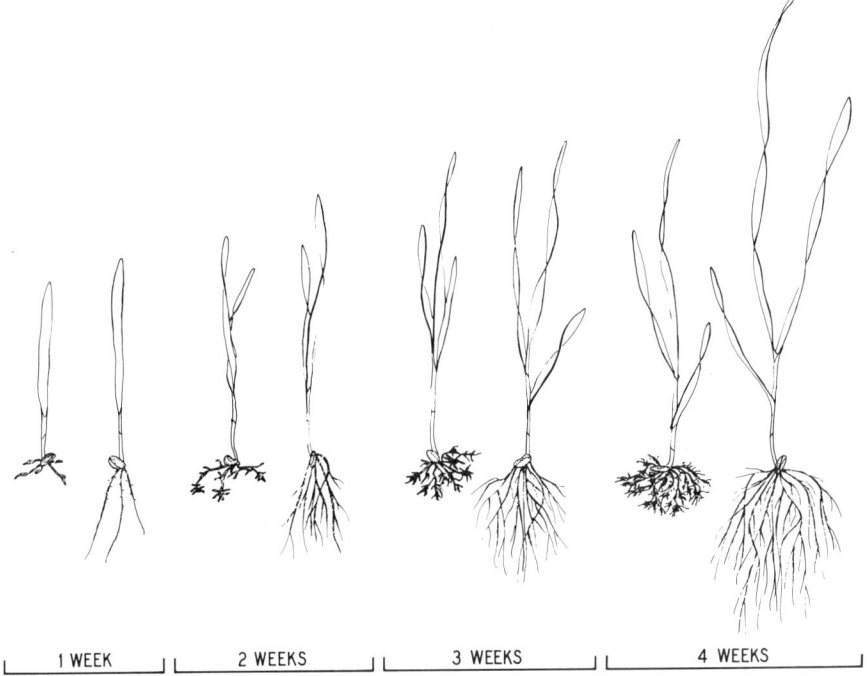

| 1 WEEK | 2 WEEKS | 3 WEEKS | 4 WEEKS |

Fig. 14 Diagramatic effect of *H. avenae* on grain over a four week period. Left, inoculated. Right, control. (Courtesy of J. W. Meagher.)

Females are initially white in appearance and cysts turn cream colored to brown and almost black. Nematode symptoms can readily be detected by a matted root resulting in a root growth proliferation (Fig. 19).

Heavily infected plants show a general overall stunted appearance, and a reduction in tillering. Cereal heads produced by stunted plants are usually small and are poorly filled. There may be a burning of the leaves, similar to that caused by a drought condition, due to the nematode interfering in the metabolic balance of the plant physiology and inhibiting hydrostatic water pressure that results in wilting. Decker (1972) stated that *H. avenae* retards plant development so that heads in nematode-infested soil remain upright, but heads in noninfested soil drop due to earlier ripening. Meagher (1972a) shows losses of 73-89% in wheat yield from *H. avenae* in Australia (Fig. 20). Spring oats is the most susceptible of the cereals to *H. avenae*. This is followed, in rank of susceptibility, by winter oats, spring wheat, winter wheat, spring barley, and winter barley and rye (Fig. 21). Dixon (1965) found that ten eggs or larvae per gram of soil resulted in losses of 381, 190.5, and 76.2 kg/ha for oats, spring wheat, and winter barley, respectively. Oat and wheat losses have been estimated at 47-64 and 23-50%, respectively, but barley losses have never exceeded 20% in Australia (Meagher, 1972a). Andersen (1960) showed that oat and spring barley yields fluctuated with soil population density variations. Oat yields were reduced by 21, 31, 49, 68, and 85% at initial soil population densities of 1.0, 2.5, 5.0, 10.0, and 20.0 larvae per gram of soil. This compared to spring barley losses of 16, 17, 21, 40, and 55% for similar nematode populations.

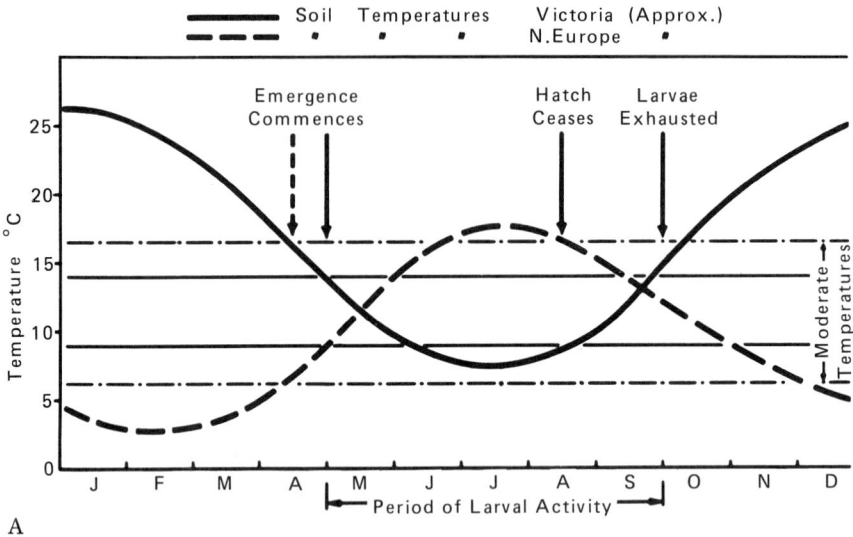

Comparison of Soil Temperatures in Victoria and N.Europe and its
Effect on Seasonal Emergence of Heterodera Avenae. Woll.

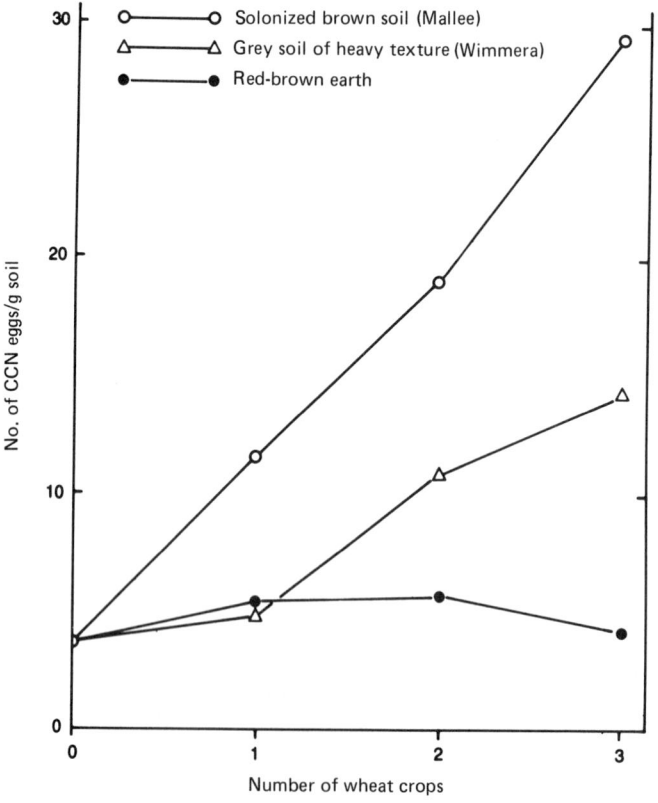

Fig. 15 (A) Diagramatic comparisons of *H. avenae* larval emergence and hatch
from two different geographical areas of the world. Note extreme differences
in soil temperatures. (Courtesy of J. W. Meagher.) (B) Effect of soil type
on population dynamics of *H. avenae*. (Courtesy of J. W. Meagher.)

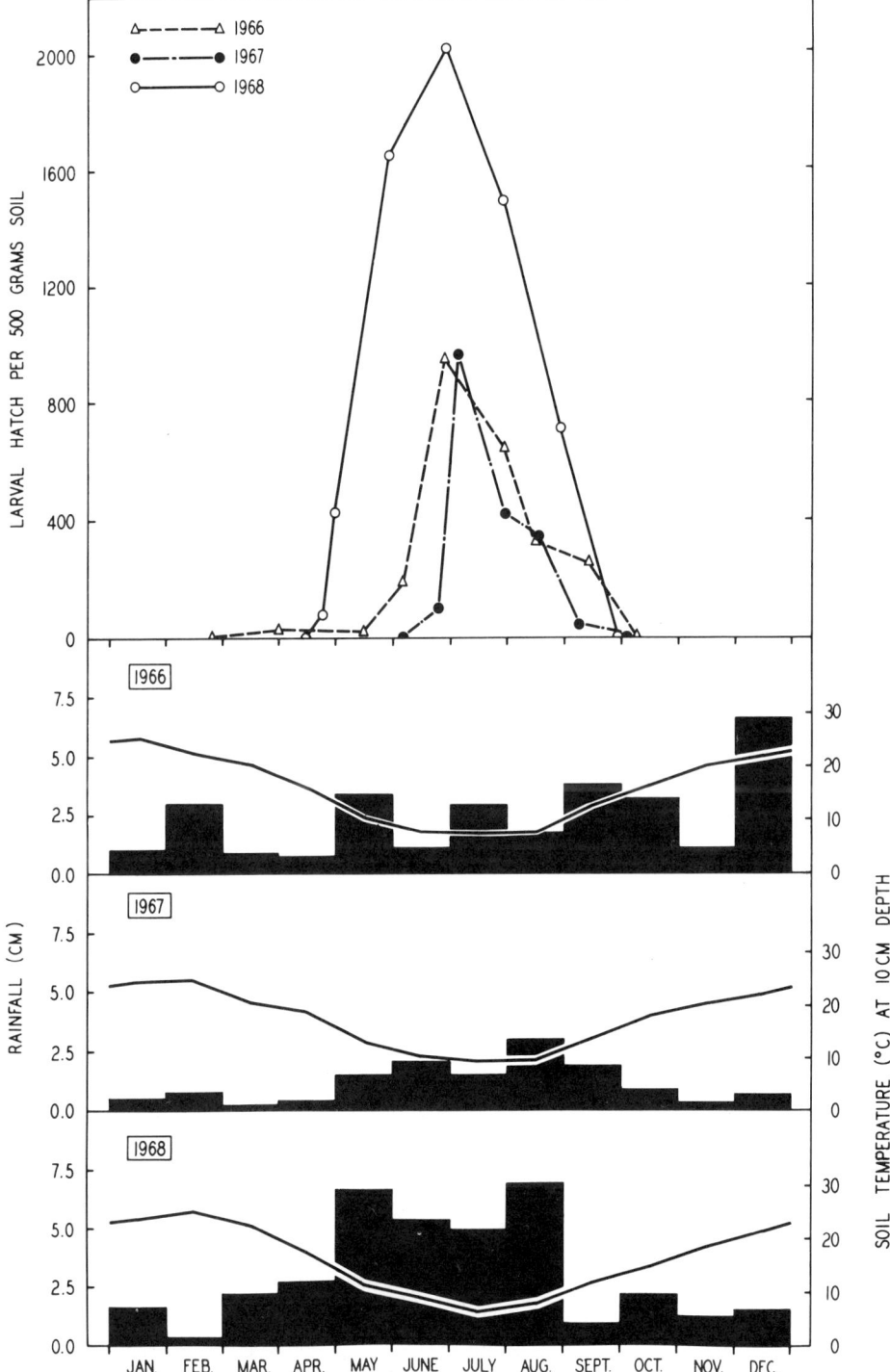

Fig. 16 Relationship of time and amount of precipitation to larval hatch of *H. avenae.* (Reprinted by permission of E. J. Brill.)

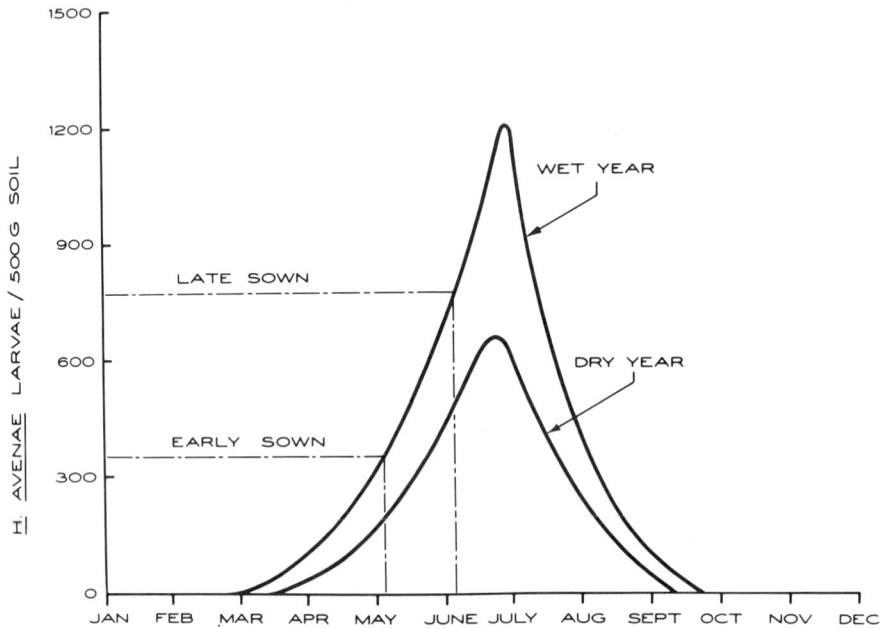

Fig. 17 The relationship of soil moisture to *H. avenae* hatch. Earlier sowing results in a lower rate of root penetration and subsequently better plant growth. (Courtesy of R. H. Brown.)

The damage threshold level of *H. avenae* differs between plant cultivars and geographical areas. Klejburg (Decker, 1972) found that one to three juveniles per cubic centimeter of soil is a moderate population, three to five juveniles per cubic centimeter of soil is severe, and over five is very severe. Andersen (1961), however, stated that wheat losses occur only when the nematode soil populations are greater than ten juveniles per cubic centimeter of soil. Meagher and Chambers (1971) and Meagher et al. (1978), however, found that economic threshold levels were affected by the association of *H. avenae* in *Rhizoctonia solani* Kühn. A combination of *H. avenae* and *R. solani* reduced wheat yields below that of single inoculations of either pathogen (Fig. 22).

Edaphic factors, such as soil pH (Duggan, 1963) and fertility also affect the degree of plant damage. This, however, is contingent more on the effect of environmental factors on plant growth as it enhances the virulence of the nematode.

Although *H. avenae* is considered economically important on cereals, it has been found on other cultivars, including fescue, ryegrass, brome, and Kentucky bluegrass.

3. Control

Cultural: Meagher (1972a) found that disease severity was greatest when a fallow-wheat or fallow-wheat-oats cropping was used. Meagher and Rooney (1966), however, found that improved wheat yields occurred when a wheat-legume rotation was used. Meagher and Brown (1974) found that an

A

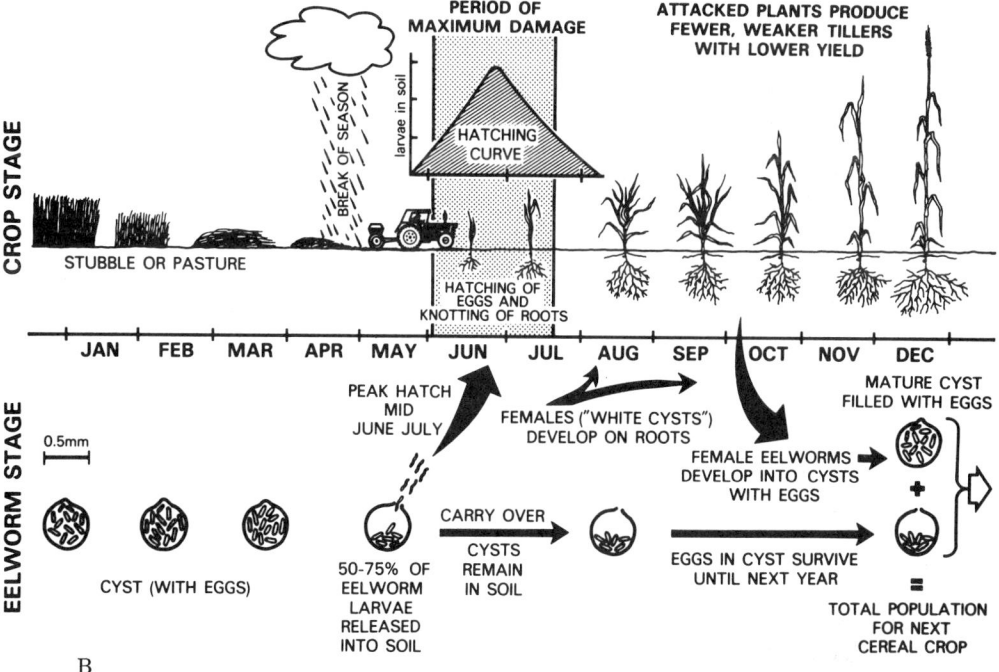

B

Fig. 18 (A) The effect of time of sowing of grain on the virulence of *H. avenae* on oats. Left, sown in early May. Right, sown in late June. (Courtesy of J. W. Meagher.) (B) Life cycle of *H. avenae* in Australia. (Courtesy of A. Rovira. Reprinted by permission CSIRO, Division of Soils, Adelaide, South Australia.)

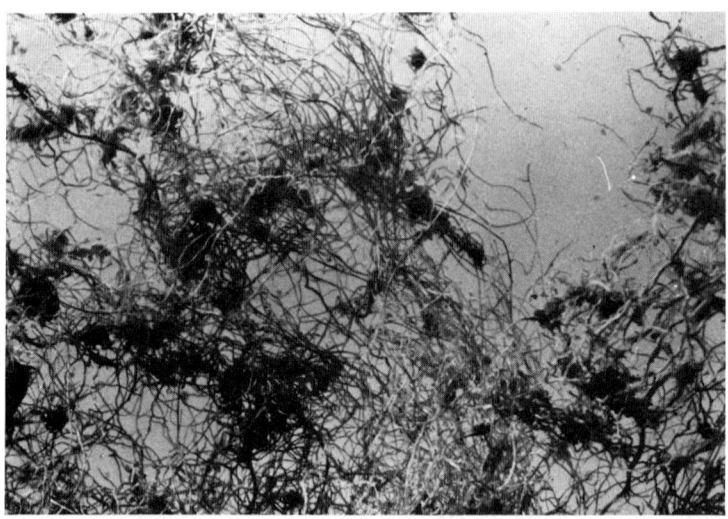

Fig. 19 Wheat roots parasitized by *H. avenae*. Note matting due to proliferation of root growth. (Courtesy of J. W. Meagher.)

Fig. 20 *H. avenae* on wheat. Left, sheath from noninfested area. Right, sheath from nematode-infested area. (Courtesy of J. W. Meagher.)

Fig. 21 Difference in tolerance and susceptibility of grain to *H. avenae*. Left, barley. Right, oats. (Courtesy of J. W. Meagher.)

Fig. 22 Relationship between *H. avenae* and *Rhizoctonia solani* on the growth of wheat. Left to right, nematode plus fungus, nematode alone, fungus alone, uninoculated control. (After Meagher et al., 1978.)

Fig. 23 The importance of resistance to *H. avenae* on the growth of cereal. Left, susceptible Olympic wheat. Right, resistant South Australian rye. (Courtesy of R. H. Brown.)

extended fallow reduced nematode soil populations to the lowest levels and produced the greatest subsequent wheat yields.

Barry et al. (1974) found that resowing of damaged cereal crops with wheat or barley was not effective in producing higher yields, and increased nitrogen failed to be economical.

A bioassay has been developed in Australia that is used before planting and gives an indication of the degree of nematode infestation and what the potential damage to the subsequently planted cereal crop may be (Rovira, 1982).

Resistance: Resistance is the major source of nematode control, and excellent plant growth has been obtained with susceptible cultivars used in rotation with resistant cultivars (Figs. 23 and 24). The search for nematode resistance was apparently initiated by Nilsson-Ehle (1920), and this was later picked up by Andersen (1961) in Denmark. The barley variety Dorst was released in 1951 as being resistant to *H. avenae*. Nielsen (1966) later

Fig. 24 The effect of crop rotation on subsequent growth of *H. avenae* susceptible wheat, cv. Olympic. (Top) Olympic wheat following five year rotation of resistant oats, cv. Avon. (Bottom) Olympic wheat following five year rotation of susceptible wild oats. (Courtesy of R. H. Brown.)

reported "Loros," a spring wheat variety, as being resistant. Cotten (1963) found resistance in oats and barley, and Andersen and Andersen (1970) found resistance in *Avena sterilis* and *A. strigosa* to *H. avenae*.

Cereal breeders have been successful in incorporating resistance into certain cereal varieties. This has been most successful with barley; in Europe, resistance can be traced to Dorst, LP-191, or Morocco selections. In Australia, resistance is found in Martin 303-2 and Morocco, and in India resistant germ plasm is found in Bajo-Aragon 1-1, Dalmatishe, and Martin 303-2.

A single dominant gene is also responsible for resistance in Loros spring wheat (Nielsen, 1966). Selections of Loros identified as Australian 10894 and 15577 are used in that country (Brown, 1974), and the world's first resistant wheat variety was recently released (Brown, 1982). India also has a wheat-screening program that has identified three USDA selections (PI 183868 from Turkey, PI 185205, and PI 185207 from Portugal) that did not support cyst formation (Brown, 1974).

Although complete resistance has not been found for cultivated oats, excellent progress has been reported and varieties should be available soon (Kort, 1972). As with other cereals, resistance appears to involve a single dominant gene. European plant breeders have developed several lines for resistance, such as CI 2094, 2154, and 3444; PI 175022, 175024, 185775, and "Silva". Such lines are resistant to most biotypes occurring in Europe.

Biotypes of *H. avenae* were first reported from Denmark by Andersen (1959). Since then, biotypes have been found in Holland by Kort et al. (1964), in Britain by Duggan (1958), Cotten (1963), and Fiddian and Kimber (1965), Neubert in Germany (1967), Brown in Australia (1969), and Mathur et al. (1974) in India.

Meagher and Brown (1974) found that the final populations of *H. avenae* differed from host to host and indicated what the nematode potential would be for subsequent plantings. Wild oat, *Avena fatua*, was the most efficient host, and moderately resistant oats cv. Avon, and rye cv. South Australia were the most inefficient hosts.

There was a direct correlation between subsequent wheat growth and the final nematode populations. A peak nematode population of 42.2 eggs per gram of soil was reached after three consecutive years of wild oat; plant growth was greatly reduced at this population ceiling and agrees with the findings of Duggan (1958).

Chemical: Chemicals are usually not considered in a nematode control program on cereals because of economic considerations. Recent studies, however, have shown that significant increases in wheat can be obtained with nematicides (Brown et al., 1970; Brown, 1972, 1973) (Fig. 25). Meagher et al. (1978) found that a methyl bromide-chloropicrin mixture (450 kg/ha) increased grain yields up to 323%. This compared to 170% increase for aldicarb (9 kg a.i. ha) and 130% for methyl bromide-chloropicrin (45 kg/ha). Brown et al. (1982) found that aldicarb (4 kg/ha), DBCP (7.4 1/ha), and EDB (at very low rates from 1.85 to 11.1 liters/ha) significantly increased wheat yields over those of controls. Increased yields have also been obtained with the use of fertilizers (Fig. 26), but it is not recommended since it is not economically feasible, and it aids in a nematode buildup in the soil (Brown, 1982).

Biological: Graham (1980), in a rainfall study, found that penetration of cereals by second-stage juveniles was greatest, but nematode reproduction was smallest on plants receiving the most water. He concluded that fungal parasites played an important role in population changes. Kerry et al.

Fig. 25 Chemical control of *H. avenae* on wheat. Left and right, chemical-treated plots. Center, untreated plot. (Courtesy of R. H. Brown.)

Fig. 26 Effect of application of ammonium sulfate (1 cwt/acre) on growth of wheat in *H. avenae*-infested soil. (Courtesy of J. W. Meagher.)

(1982a, b) found that *H. avenae* soil populations failed to increase on susceptible host plants after two years. They found that females were able to reproduce on plant roots, but few females produced eggs and their fecundity
was reduced. When the soil was treated with formalin (38% formaldehyde),
fecundity increased and nematode reproduction was increased by 30-1460%.
They found that the poor nematode reproduction was apparently due to
females being parasitized by fungi, *Verticillium chlamydosporium* Goddard,
or *Nematophthora gynophila* Kerry and Crump. They found that about 60%
of the females that failed to form egg-containing cysts were parasitized by a
fungus, especially *N. gynophila*. Gill and Swarup (1977) also found reduced
nematode production with a combination of *H. avenae*, *Fusarium moniliforme*,
and *Helminthosporium gramineum* on barley.

4. Heterodera latipons *(Franklin 1969)*

Hesling (1965) found morphological differences from a cyst nematode, collected from the roots of chlorotic wheat plants in Tripoli, and *Heterodera
avenae*. This nematode was recognized as a new species and designated *H.
latipons* by Franklin (1969). *H. latipons* has been reported from several
Mediterranean countries associated with the growth of wheat, including Israel,
Libya, Tunisia, and Italy, and in Bulgaria (Kort, 1972). Since *H. latipons*
cysts are similar in size and shape to *H. avenae*, it is suggested that this
species may be responsible for some of the previously described wheat losses
attributed to *H. avenae*.

5. Punctodera punctata *(Thorne 1928)*

Thorne (1928) described *P. punctata* from the roots of poorly growing wheat
in Canada. The cysts were lemon shaped with a rounded posterior end.
This nematode has since been reported from England, The Netherlands,
Mexico, Poland, Hungary, Germany, and Russia (Decker, 1972). *P. punctata*
is commonly found in pasture lands, and besides cereals is found to parasitize several weed hosts and tomato (Thorne, 1961; Spears, 1956).
      Another cyst nematode, *Heterodera hordecalis* (Andersson, 1974), has
been found attacking cereals and grasses in Sweden.

C. Cereal Root-Knot Nematode (*Meloidogyne naasi* Franklin 1965)

1. History and Distribution

The cereal root-knot nematode, *M. naasi*, is probably the most important
root-knot nematode affecting grain in most of the European countries, as well
as in the United States. It has been found in Belgium (D'Herde, 1965),
Britain (Franklin, 1965), The Netherlands (Kuiper, 1966), France (Schneider,
1967), Germany (Sturhan, 1973), Yugoslavia (Grujicic, 1967), Italy and
Malta (Inserra et al., 1975), the United States (Golden and Taylor, 1967),
and the USSR (Gooris and D'Herde, 1977). *M. naasi* has a wide host range
and attacks many economically important cultivars, such as sugarbeet and
onion, as well as barley, wheat, and rye, and several grass species. Gooris
and D'Herde (1977) list over 100 species that are hosts of *M. naasi*. Kuiper
(1966) and Schneider (1967) attributed losses of entire crops of spring barley to *M. naasi* in Holland and France.

2. Biology and Symptomology

Like all plant-parasitic nematodes, temperature is very important in the host-
parasite relationship of *M. naasi* on cereals. Gooris and D'Herde (1977)

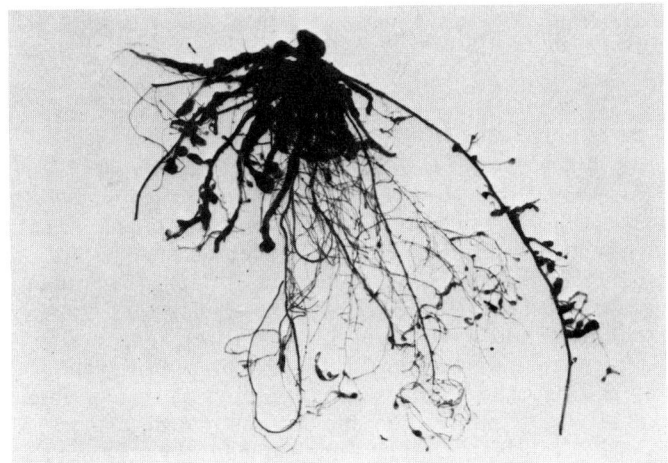

Fig. 27 Wheat roots galled by *M. naasi*. Note atypical galling of roots.
(Photo by K. Kuiper; courtesy Am. Phyto. Society.)

showed that several oat cultivars, including Nemaha, Garland, Jaycee, and
Clinton, are resistant to *M. naasi* at 15°C, but susceptible at 25°C. The
number of generations reportedly produced by *M. naasi* differs between geographical areas on different host plants. Franklin et al. (1971) found only
one generation per year in England on barley, while Schneider (1967) reported up to three generations per year in France on ryegrass. Kort (1972)
stated that there were two generations on barley in The Netherlands. Songul
and Dickerson (1976) found that the optimum temperature for development of
races was 26°C, and a life cycle was completed in 34 days.

Root galls caused by *M. naasi*, are atypical root-knot galls and do not
resemble those associated with any other *Meloidogyne* species (Fig. 27). They
are curved and spiral, or horseshoe in shape.

Irregular growth of stunted plants is usually associated with infested
areas, and symptoms are similar to those caused by *H. avenae*; areas of different size and shape of poorly growing yellowish plants are characteristic,
and mixed populations have been found (Fig. 28). The biological characteristics of *M. naasi* in relation to egg hatching are more like that of *Heterodera*
species than *Meloidogyne* spp. Watson and Lownsbery (1970) found that
hatching of *M. naasi* was stimulated by the cold, and temperature treatments
that consisted of 6-9°C incubation for a period of seven weeks followed by an
incubation period of 21-24°C resulted in maximum nematode hatch. Hatching
was also stimulated by a weak sodium hypochlorite solution. The hatch of
eggs in water was not changed by incubation at temperatures from -28°C to
39°C.

3. Control

There are supposedly five physiological races of *M. naasi*, as determined by
host differentiation, and the rate of nematode development (Michell et al.,
1973) differs among pathotypes. Hence, an understanding of the nematode
race involved is necessary, and only then can an accurate judgment be made
as to what cultivars are nonhosts and which can be used in a crop rotation
practice. It is, therefore, impossible to delineate any definite control

methodology in relation to crop rotation. Kort (1972) states that the planting of such crops as beet or onion following cereals should be avoided, and heavily infested land should be planted to a nonhost plant. The use of non-host plants will give optimum control, since the nematode is unable to survive for more than two years in the absence of a host plant.

4. *Other* Meloidogyne *Species*

The southern root-knot nematode, *M. incognita*, and the Javanese root-knot nematode, *M. javanica*, are able to parasitize and reproduce on several *T. aestivum* cultivars (Martin, 1958; Sasser, 1954; Thomason, 1962). Thomason found that nematode reproduction (light to moderate) was similar for *M. incognita* and *M. javanica* on barley and wheat, but *M. incognita* reproduction was greater than that of *M. javanica* on oats and rye. Under winter plantings of wheat, *M. javanica* soil populations decreased under both susceptible and resistant cereals, and by May the nematode populations had decreased 800%, the winter soil temperatures being too low for nematode development. However, Roberts and Van Gundy (1981) found that both *M. incognita* and *M. javanica* were able to reproduce on *T. aestivum* cv. Anza at 14-30°C. The rate of reproduction was directly proportional to soil temperature, at 14-30°C for *M. incognita* and 18-26°C for *M. javanica*. Anza was, however, tolerant to both nematode species and plant growth effect from either nematode was maximal. Roberts et al. (1981) found that when susceptible winter wheat was planted at a fall soil temperature of 21°C, *M. incognita* juveniles were able to penetrate the roots, mature, and reproduce by the time of harvest six months later. When wheat was planted at 16°C, however, the nematode failed to penetrate the wheat roots.

The Columbia root-knot nematode, *Meloidogyne chitwoodi*, is able to attack (Fig. 29) and reproduce on grain (Santo and O'Bannon, 1981). Wheat, barley, and oat roots parasitized by *M. chitwoodi* were significantly reduced. There were, however, no differences in top growth between parasitized and nematode-free plants. This nematode, however, has only recently been found and identified from the Pacific Northwest of the United States, and its potential importance on the growth of grain is not yet completely understood (Santo et al., 1980).

The British gall nematode, *M. artiellia* Franklin, is found in Britain and Greece. Although it attacks and reproduces on cereals, it appears to be relatively unimportant.

D. Stem Nematode [*Ditylenchus dipsaci* (Kühn) Filipjev]

1. *History and Distribution*

The stem nematode is found throughout the world, and attacks a great variety of plant cultivars. It is, however, a problem on cereals in heavy soils in areas of high rainfall and cool growing seasons. It is more important in areas where fall and winter grain is planted than in areas cultivated only by summer grains.

2. *Biology and Symptomology*

D. dipsaci attacks and invades the plant foliage and stem base. The nematode migrates intra- and intercellularly, causing a breakdown of the middle lamella, the cell tissue feeding on the plant tissue. This results in hypertrophy and hyperplasia of the plant cells, causing abnormal plant growth and distortion. Shortened plants and swelling and bursting of leaves and stems are usual symptoms attributed to nematode parasitism (Fig. 30).

Fig. 28 Mixed population of *H. avenae* and *M. naasi* on wheat in Southern Italy. Notice poor growth at top of field. (Courtesy of R. N. Inserra.)

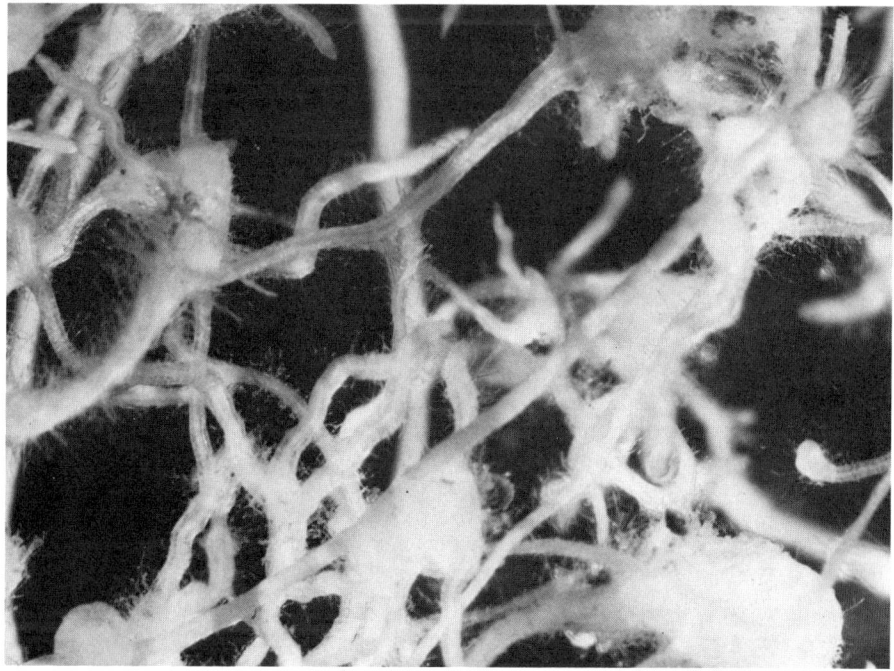

Fig. 29 *M. chitwoodi* galling the roots of Nugaines' wheat. (Courtesy of R. N. Inserra.)

Fig. 30 Parasitism of oats (center) by *D. dipsaci*. (Courtesy of S. D. Van Gundy.)

The occurrence of *D. dipsaci*-infected plants usually appears spotty in a field, and this may be seen by only an occasional area of stunting or it may be general depending on the degree of nematode infection. The primordial growth tissue may be destroyed in heavily infested soil, resulting in the death of the plant. Panicle and spike growth may be inhibited by nematode parasitism, resulting in a poor crop stand and reduced grain yields.

The degree of severity of *D. dipsaci* on cereals is dependent on environmental conditions as well as the degree of nematode soil infestation. Extended periods of cool moist weather enhance the parasitism of the nematode. It is important that soil moisture be optimum for nematode migration to infective loci before penetration occurs. As previously stated, heavy soils favor nematode persistence in the absence of a host plant.

3. *Control*

A number of races of *D. dipsaci* are able to parasitize cereals. Kort (1972) stated that the rye strain of *D. dipsaci* is more common in Europe, and the oat strain is more common in Britain. Kort also states that the rye strain attacks rye and oats as well as several other plant cultivars, including bean, corn, onion, tobacco, and clover. This strain also attacks a number of weed species found associated with the growth of cereals, including chickweed, bindweed, shepherd's purse, and plantain. The oat strain attacks oats, onion, pea, and bean but not rye. It also has as hosts several weed species, including chickweed, bindweed, clovers, and wild oats.

The occurrence of different biological races or strains makes *D. dipsaci* difficult to control. Eriksson (1965) and Sturhan (1966) were able to cross different strains resulting in different host ranges. However, proper crop rotation can minimize the importance of this nematode on cereals. Kort (1972) suggested that cereals be grown only after a nonhost, such as alfalfa.

The cultivation of spring rather than fall or winter wheat will usually reduce the optimum environmental conditions for the nematode, resulting in better plant growth.

E. Root-Gall Nematode [*Subanguina radicicola* (Grf.) Param.]

This nematode has been associated with plants for approximately 200 years, and is found in Britain, The Netherlands, Germany, the USSR, the United States, Canada, and the Scandinavian countries (Stessel and Golden, 1961; Jatala et al., 1973).

Subanguina radicicola causes characteristically shaped galls on the roots and may be confused with those caused by *M. naasi*, since mixed populations of these two nematodes are known to occur (Fig. 31). There are also apparent biological differences in populations of *S. radicicola*; in Scandinavian countries poor growth of barley "Krok" is known to result from nematode

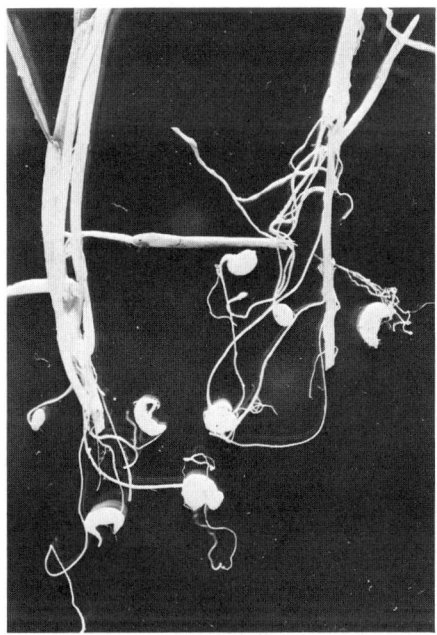

Fig. 31 *S. radicicola* galling grass roots. Note crescent-shaped galls. (Photo by K. Kuiper; courtesy Am. Phyto. Society.)

parasitism. This nematode also attacks rye in Sweden, but only wheat in Canada. The Scandinavian population heavily galls barley roots, but a Dutch population does not.

Infected roots show galling in about two weeks, and a life cycle is completed in approximately 67 days. This is, however, dependent on climatic conditions. *S. radicicola* can be controlled by rotating cereals with nonhost crops, such as legumes.

F. Wheat Cockle Nematode [*Anguina tritici* (Steinbuch 1799) Chitwood 1935]

1. History and Distribution

The genus *Anguina* is composed of many species of nematodes, of which *A. tritici* is important on the growth of wheat. This nematode has been reported from most of the European countries, Russia, Israel, Syria, Pakistan, India, China, Australia, New Zealand, Egypt, Brazil, and several areas in the United States.

2. Biology and Symptomology

A. tritici, like other *Anguina* species, is characterized by obese females that are usually curled. The life cycle is fairly simple. Second-stage juveniles invade the host plant and are carried up by the growing point. They penetrate the plant shoots, causing leaf curling (Fig. 32). Nematodes invade the inflorescence and mature to adult males and females inside galls or cockles that replace the kernels. Each female may produce several hundred eggs

Fig. 32 Curling of wheat shoots and leaves due to parasitism of *A. tritici*. (Photo by K. Kuiper; courtesy of American Phytopathological Society.)

Fig. 33 Parasitism of wheat heads (left) by *A. tritici*. Note growth with that of uninfected head on the right. (Courtesy of S. D. Van Gundy.)

inside the galls. The juveniles hatch, develop into the second stage, which becomes the dormant stage, in which they can survive adverse weather conditions in the galls or cockles. Upon finding favorable climatic conditions, the nematode becomes active, leaves the gall, and attacks new wheat seedlings, repeating the life cycle.

A. *tritici* is characterized by its ability to survive in a anhydrobiotic stage through several environmental conditions (Bloom, 1964). Bird (1981) states that it can remain viable over a temperature range of 300°C, and survive for a short period at 105°C. The ability of the nematode to lose water, accumulate lipid bodies, and show a negligible rate of respiration is characteristic of this and other *Anguina* species.

A. *tritici* obtained its fame from Turbevill Needham in 1743, who found this nematode in wheat cockles. This was the first recorded incident of a plant-parasitic nematode parasitizing plant tissue (Thorne, 1961).

A. *tritici* mainly attacks wheat and rye, and has been found wherever wheat is grown, apparently disseminated by cockles, which may contain several thousand nematodes (Fig. 33). However, with strict laws regulating the dissemination of seed and seed certification programs, this nematode has almost disappeared from most of the major wheat-growing areas of the world. It is, however, still common in the USSR and many Mediterranean countries and some undeveloped countries.

As stated, the nematode is disseminated by cockle-contaminated seed at the time of sowing. The nematodes are carried up with the growing plant tissue. Marcinowski (1909) found that galls were produced by undifferentiated flower buds infected by the nematode, and up to five galls may replace a single kernel.

Fig. 34 *A. tritici* on wheat. Black, infected cockles. White, noninfected
wheat kernels. (Courtesy of S. D. Van Gundy.)

    A. tritici infection may be indicated by small and dying plants, leaves
generally twisted due to nematode infection, heads irregularly stunted and
blackened by the appearance of the cockles (Fig. 34).
    The major method of control is by planting nematode-free seed.

G. Lesion Nematodes (*Pratylenchus* spp.)

Several species of lesion nematodes are found associated with cereals. Those
most commonly found are *P. crenatus*, *P. penetrans*, *P. thornei*, and *P.
neglectus*.
    Inserra et al. (1978) reported the association of *P. neglectus*, *P.
thornei*, and *P. penetrans* with cereals in Italy. *P. neglectus* was found in
53, 67, and 55% of samples collected from fields planted to wheat, barley, and
oats. *P. penetrans* was found only on wheat and oats, and *P. thornei* was
found only on wheat.
    P. thornei was shown to be a serious parasite of wheat in Mexico, where
yields were significantly increased with the use of a nematicide (Fig. 35)
(Perez, 1971; Perez et al., 1970; Van Gundy et al., 1974). They also found,
however, that the combination of crop rotation, proper soil fertility, and
winter planting when the soil temperature was 15°C gave adequate nematode
control and increased wheat yields.
    Orion et al. (1982) found that *P. thornei* was the dominant plant-para-
sitic nematode species associated with the cultivation of wheat in northern
Neger, a semiarid region in Israel. They found that under dry-land farming
nematode populations were high (300 per gram fresh plant tissue), but under
irrigation nematode populations were low and probably did not affect the

Fig. 35 The pathogenicity of *P. thornei* to wheat. Left, parasitized plant. Note stunting, decreased tillering, and dwarfed heads. Right, healthy growing plant. (Courtesy of S. D. Van Gundy et al., 1974.)

growth of wheat. This is similar to conditions reported by Graham (1980) about *Heterodera avenae* populations, where fungal parasites reduced the nematode populations. Orion et al. (1982) also stated that the nematode population could be reduced by 90% with a biannual fallowing, which increases grain yields by 50%. Chemical applications of metam-sodium resulted in good nematode control and increased grain yields by 100%.

Baxter and Blake (1968) found that *P. thornei* caused lysis of cells, cavities, and eventually destruction of the root cortex of wheat plants. They found that an increase in soil temperature resulted in a more rapid decline in the nematode soil population in the absence of a host plant.

Baxter and Blake (1967) found that *P. thornei* does not randomly invade host tissue, but is attracted to previously invaded root tissue. It is assumed that exudates from wounded tissue are the attraction agent causing a positive nematode response. Once inside the root tissue, nematode reproduction begins and destruction of the cortical tissue occurs.

Corbett (1970) found *P. fallax* associated with barley in England. He later (1972) found that *P. fallax* penetrates the root tip region of root hair development, and the junction of the main and lateral roots of wheat and barley, and that several nematodes may penetrate the same site. He found that internal cell tissues were severely damaged and collapsed before the appearance of external symptoms that would indicate early internal growth retardation. Choppin de Janvry (1971) found that soil type affects distribution of lesion nematodes. *P. pratensis* and *P. fallax* were found in light soils, and *P. neglectus* was found in heavy soils.

H. Other Plant-Parasitic Nematodes

Many other species of plant-parasitic nematodes have been found associated with the growth and decline of cereals. *Radopholus ritteri* (syn. *Pratylenchoides ritteri*) and *Zygotylenchus guevarai* were found associated with cereals in France, Italy, and Spain. Prasad and Gaur (1975) found that combined populations of *Tylenchorhynchus*, *Pratylenchus*, *Hoplolaimus*, and *Helicotylenchus* significantly reduced yields of wheat, *Tylenchorhynchus* being the most virulent of all these genera.

Krusberg (1959) found that wheat, oats, barley, and rye were excellent hosts of *Tylenchorhynchus claytoni*. Optimum reproduction occurred on wheat at soil temperatures of 21-27°C. There was a low rate of reproduction at soil temperatures below 18°C, which indicates that this nematode may be of minor importance on winter grains. Krusberg found, however, that the nematode is able to survive in fallowed soil at temperatures of 2-24°C.

Anderson (1979) found *Tylenchorhynchus dubius*, *Merlinius brevidens*, *M. microdorus*, *Pratylenchus neglectus*, and *P. crenatus* associated with barley in Denmark. Chhabra and Bindra (1971) found *Tylenchorhynchus eremicolus* attacking wheat in India, and plants were stunted and chlorotic; a DBCP application resulted in economic control. Potter and Townshend (1973) found several genera of plant-parasitic nematodes, including *Tylenchorhynchus* and *Helicotylenchus*, associated with cereals in Canada.

Brown and Sykes (1975) stated that the loss of barley yield to *Longidorus elongatus* in Britain was about 1 ton/ha per two nematodes per gram of soil.

Spaull (1980) found that *Paratrichodorus anemones* stunted the growth of spring wheat and barley. Some recovery was made in plant growth if nematode exposure was terminated after six weeks. Additional exposure decreased the plant's ability to recover. Grain yields decreased significantly with increased exposure up to eight weeks. *P. anemones* populations increased 200% and 300% under wheat and barley, respectively.

Jones (1979) showed a correlation between plant-parasitic nematodes and the yield of cereals. He stated that several nematode species, including *Helicotylenchus pseudorobustus* and *H. digonicus*, did not decrease the yields of cereal grain, and *Tylenchorhynchus dubius* and *Longidorus* spp. were associated with leaf chlorosis and reductions in barley yields.

Trichodorus species have been found on cereals (Figs. 36 and 37). Inserra et al. (1983) found that the tolerance level of *Paratrichodorus* spp. (close to *P. tunisrensis*) to wheat is 1.4 nematodes per cubic centimeter of soil (Figs. 38 and 39). This, however, is found to be between 0.1 and 0.25 nematodes per milliliter of soil in the absence of nematode mortality. The nematode increased in numbers 9- to 20-fold in six weeks. It was found that the nematode caused root disorganization and a proliferation of lateral roots. The nematode feeding on the root cap and apical meristem caused a

Fig. 36 Pathogenicity of *P. minor* on wheat cv. Triumph. (Courtesy of
C. C. Russell.)

Fig. 37 *Paratrichodorus* spp. on durum wheat cv. "Creso." Notice spotty
infestation. (Courtesy of R. N. Inserra.)

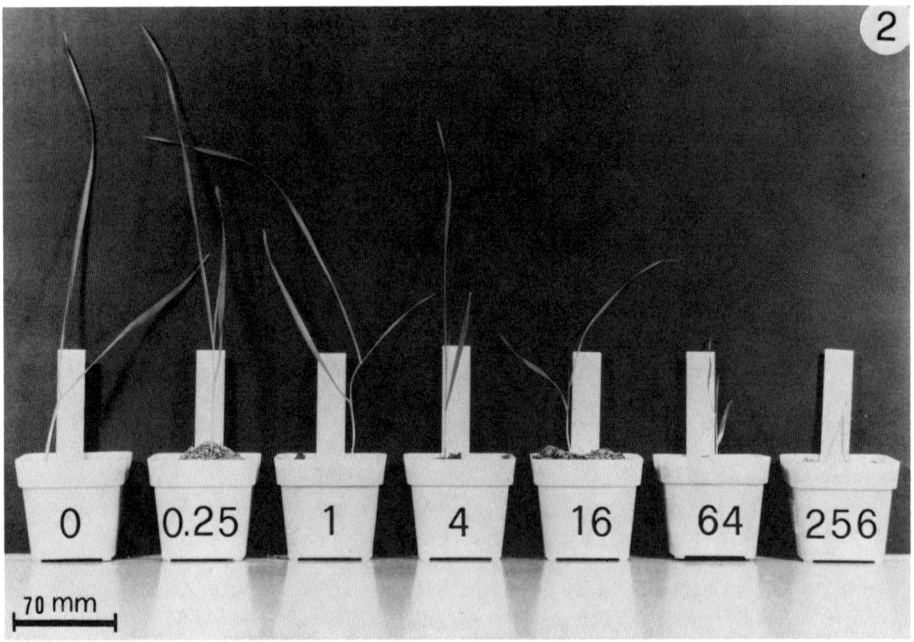

Fig. 38 Effect of increased densities (left to right) of *Paratrichodorus* spp. on growth of durum wheat cv. Creso, 30 days after sowing. (Courtesy of R. N. Inserra et al., 1983.)

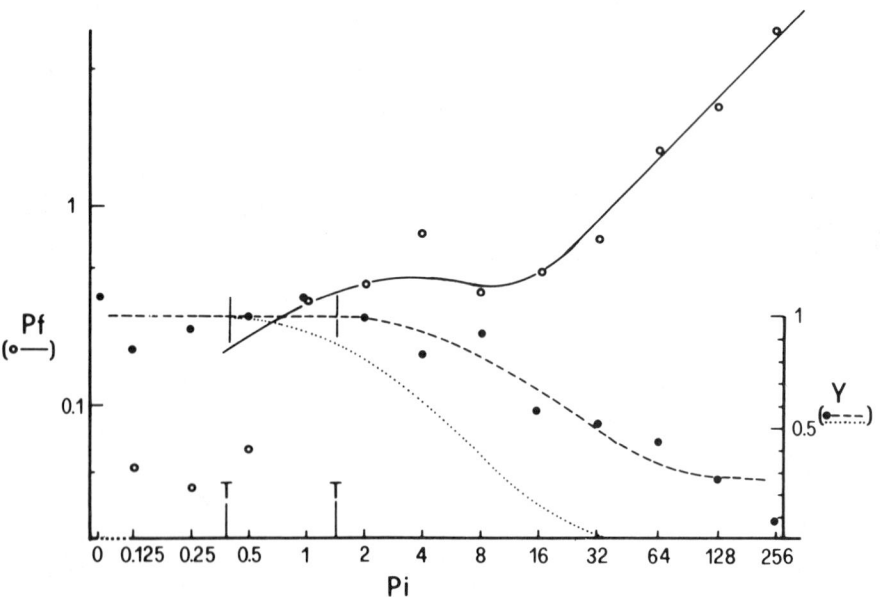

Fig. 39 Relation between initial nematode population density of *Paratrichodorus* spp., P_i, and relative weight of tops of wheat plant, y (o), and between P_i and final nematode density P_f (o). The dashed curve is according the Seinhorst's nematode-plant growth model (T = tolerance limit). The dotted curve indicates the same relationship as dashed curve, but in presence of small nematode mortality. The solid curve indicates the relation between P_i and P_f according to model with continuous redistribution of nematodes on root system. (Courtesy of R. N. Inserra et al., 1983.)

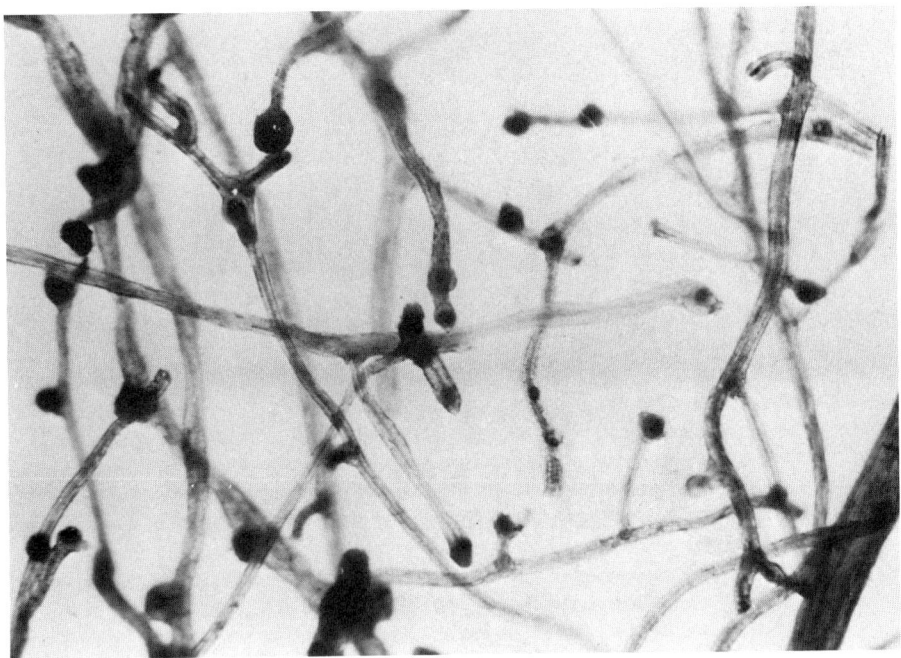

Fig. 40 Stubby adventitious roots of durum wheat, cv. Creso, due to the feeding activity of *Paratrichodorus* spp. (Courtesy of R. N. Inserra.)

cessation in root growth and a stubby-root formation. Irregular plant growth and stunting were associated with nematode infestations under field conditions (Fig. 40).

Russell and Perry (1966) found that *Paratrichodorus minor* (syn. *Trichodorus christiei*) fed most frequently in the region of root elongation and just behind the root cap of wheat and adults were able to penetrate root hairs with a single thrust of the stylet. Nematode entry into the roots occurred by forcing the loosely adhering root cap cells apart and extended for about half a body length into the root.

Coursen and Jenkins (1958) found pin nematode, *Paratylenchus projectus*, associated with tall fescue, and found that the nematode reduced plant growth after three months, and increased rapidly on root tissue. Similar results were found by Townshend and Potter (1976).

Langdon et al. (1961) reported parasitism of wheat in Oklahoma by *Merlinius brevidens* (syn. *Tylenchorhynchus brevidens*). Plants were stunted and associated with chlorosis and reduced tillering; roots were stunted and blackened. *M. brevidens* was found associated with a species of *Olpidium*, and a combination of the two pathogens reduced plant growth below that of single inoculation of either organism.

Hoplolaimus is known to parasitize wheat (Fig. 41). Richardson et al. (1978) found that *Hoplolaimus* spp. caused an occlusion of the phloem conductive cells, and there were many feeding sites near or in the branch root origin.

Townshend and Potter (1976) studied the association of plant-parasitic nematodes with grains and found that oats, rye, and barley were good hosts and wheat a fair host of *Helicotylenchus digonicus*.

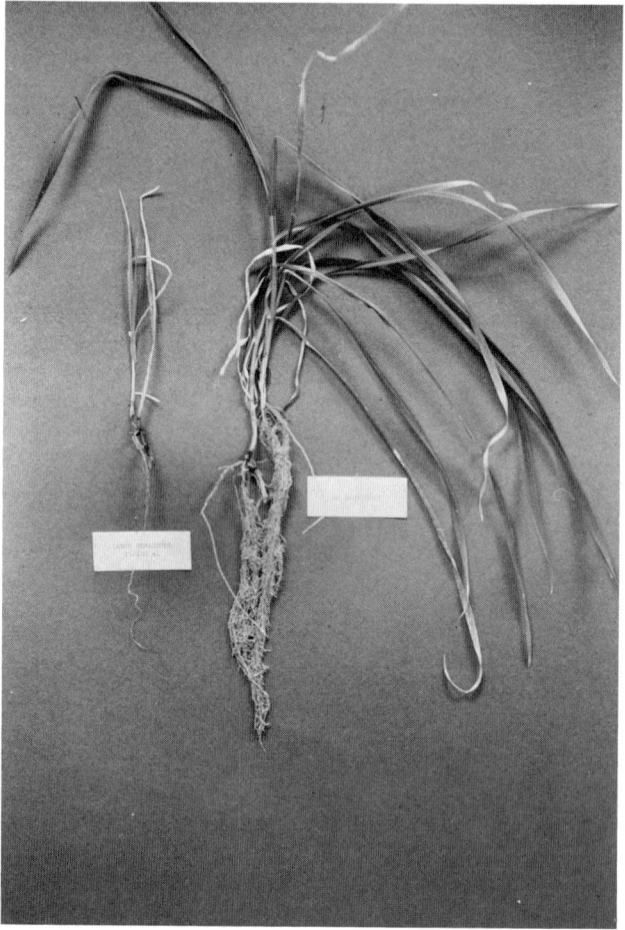

Fig. 41 Parasitism and stunting of wheat by *Hoplolaimus* spp. Left, inoculated. Right, uninoculated. (Courtesy of C. C. Russell.)

    Pepper (1963) found eight genera, including *Tylenchorhynchus, Xiphinema*, and *Aphelenchoides*, associated with the growth of barley and wheat in North Dakota.
    Norton (1959) found *Tylenchorhynchus, Helicotylenchus, Paratylenchus, Rotylenchus*, and *Xiphinema* associated with root rot of wheat in North and Central Texas.

III. GRASSES

A. Introduction

Pastures and rangelands play an important role in the production of livestock throughout the world. This was true in ancient times, when humans depended upon grazing land for feed for domestic as well as nondomestic animals. This is also true today in grassland agriculture that emphasizes the importance of grasses and legumes in livestock and land management.

.Proper grassland management not only provides food for increased beef production and other animal products, but it plays an important role in renewal of organic matter, erosion prevention, improved soil tilth, and the aesthetic beauty of the land.

As the world population increases, there is going to be a greater demand for increased yield from arable land including the world's grasslands. It is therefore important that a knowledge of the factors affecting grass production be understood.

Plant pathogens hamper the establishment of pastures and rangelands, decrease plant growth, and impair the plant quality. Although less attention has been directed to nematodes in the host-parasite relationships, studies have shown that nematodes are important in the ecosystem of turf, pasture, and rangeland diseases.

The importance of plant-parasitic nematodes on grasses has not been recognized until recently. Several species of nematodes, however, are limiting factors on the growth and production of grass species. The most important genera of plant-parasitic nematodes that attack grasses are *Meloidogyne*, *Heterodera*, *Anguina*, *Tylenchorhynchus*, *Pratylenchus*, *Ditylenchus*, *Hoplolaimus*, and *Xiphinema*.

B. Floral and Leaf Gall Nematodes (*Anguina* spp.)

There are more than 20 species of *Anguina*, and many of these are known to parasitize one or more species of grass. They form galls on the inflorescences, stems, and leaves of host plants.

1. Bent Grass Nematode [Anguina agrostis (Steinbuch) Filipjev 1936]

History and Distribution. This is the most important species of *Anguina* associated with grasses. It is similar to *A. tritici* in its life cycle, and was first described on colonial bent grass, *Agrostis tenuis*, and has since been found on many grass species (Fig. 42). Marcinowski (1909) made an extensive study of this nematode. She listed several grass species as being hosts of this nematode. These include sheep fescue, *Festuca ovina*, annual bluegrass, *Poa annua*, Kentucky bluegrass, *P. pratensis*, and alpine bluegrass, *P. alpina*. For more than 100 years, however, there was conjecture whether Steinbuch, who discussed this nematode, had been involved in one or two nematode species. It was not until Goodey (1930) studied the nematode from creeping bentgrass, *A. palustris*, and redtop, *A. alba*, that an adequate description of the nematode was made. Courtney and Howell (1952) studied the life history of *A. agrostis* in the Pacific Northwest of the United States and found that this nematode caused a serious reduction in bentgrass seed production. They also found that *A. tenuis*, *A. canina*, *A. exarata*, and *A. alba* were hosts of this nematode. They found that eight grass species, including *Festuca phleum*, *Poa holcus*, and *Anthoxanthum*, were nonhosts. This differed from the population described by Marcinowski, which had a wider host range.

Biology and Symptomology. Second-stage juveniles invade the growing point of the young plant and eventually enter the inflorescence of the plant. They enter the developing ovules and produce spindle-shaped purple-colored galls. Equal numbers of males and females may be found in each gall.

Each female (one to three per gall) is capable of producing up to 1000 eggs over a two week period, and the life cycle is completed in about three to four weeks. This, however, depends on the plant host and the rate of

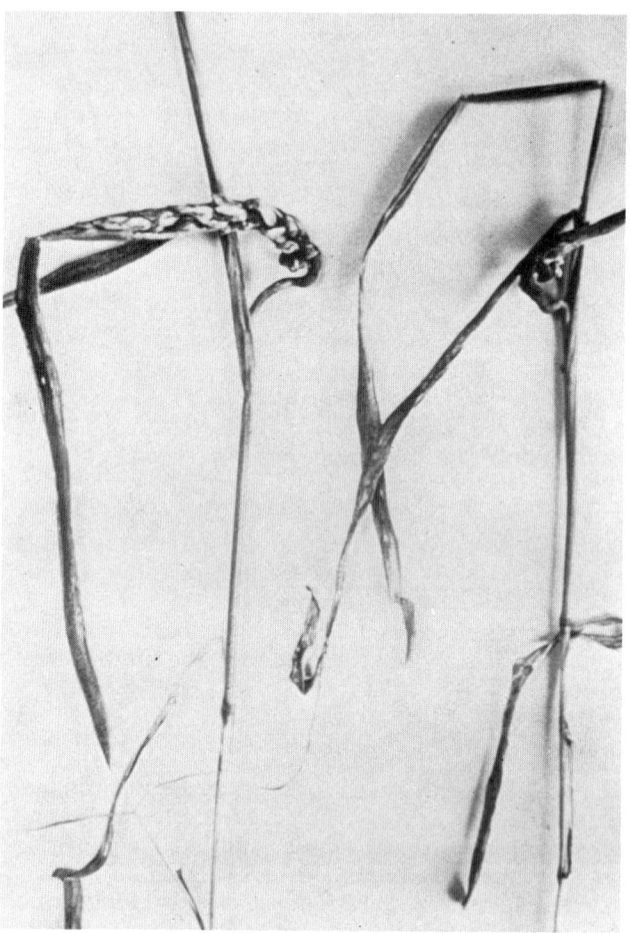

Fig. 42 *Anguina agrostis* parasitizing bluegrass. (Courtesy of the Department of Nematology, University of California, Riverside.)

plant development; there is but one nematode generation per year. *A. agrostis* is disseminated by infested seed and plant debris, threshing machines, water, etc. The nematode is capable of greatly reducing the yield of grass seed. Jensen (1961) states that a reduction of 50-75% may be observed in nematode-infested stands of Astoria bentgrass.

Ingestion of grass seed galls caused by *Anguina agrostis* may result in toxicity to animals. Like *A. tritici*, there is an association with the bacterium, *Corynebacterium* and *A. agrostis*. A nervous condition is observed in cattle and sheep after feeding on rye grass, *Lolium perenne*, and chewing fescue, *Festuca rubra commutata*, that were infected with *A. agrostis*. It was, however, found that *Corynebacterium rathayi* was associated with the nematode in the seed galls. Bird and Stynes (1977) found that the nematode acts as a vector for the bacterium by surface contamination. The galls are 20-30 times more toxic than the bacterial component (Stynes et al., 1979). It is assumed that the toxin restricts the flow of blood, which results in the toxic symptoms that include staggering, collapse, and sometimes death of the animal.

Control. Courtney and Howell (1952) found that plowing plant debris to a depth of 15-30 cm or planting a nonhost crop for one or more years was successful in controlling the nematode. Volunteer grass, however, would soon negate the control program. The most effective way, however, to control *A. agrostis* is by clean seed. A simple method is to soak the seed in brine, allowing the galls to float to the surface and then be removed and destroyed. The clean seed can then be washed with clean water. Crop rotation, fallow, burning of stubble, and mowing will result in good control (Jensen et al., 1958). The use of herbicides shows promise (Apt et al., 1960).

2. Anguina graminis *(Hardy) Filipjev 1936*

Anguina graminis (Hardy) Filipjev 1936 was first found in galls or the leaves of fescue. These galls resemble knots or swollen nodes. The nematode causes cell enlargement and hypertrophy. Hosts include sheep fescue, *Festuca ovina*, red fescue, *F. rubra*, and hard fescue, *F. ovina* var. *duriuscula*.

3. Anguina graminophila *(Goodey) Thorne 1961*

This species produces galls on the leaves of fine bentgrass, similar to those of *A. graminis*, and produce greenish-yellow galls, 1-15 mm long, due to cell hypertrophy and hyperplasia of epidermal, mesophyll, and vascular tissue.

Anguina tumefaciens, (Cobb) Filipjev and Schuurmans Stekhoven was found on Bradley grass, *Cynodon transvaalensis* in South Africa (Thorne, 1961). Nematode galls are found on stems and leaves and occur singly or in tandem groups.

4. Anguina agropyronifloris *Norton 1965*

Seed galls on *Agropyron* caused by *Anguina* were reported by Molliard (1904), and since then there have been several reports of a gall nematode on this genus (Bessey, 1905; Kirjanova, 1955). Collins (1966) and Norton and Everson (1963) found *Anguina* on western *Agropyron smithii* that was named as *Anguina agropyronifloris* by Norton (1965).

This species is host specific and is not parasitic on grass species that are commonly parasitized by *A. agrostis*. One phenomenon of *A. agropyronifloris* is that emerging juveniles from galls broken in water were predominantly outstretched but those of *A. agrostis* were coiled.

5. Anguina amsinckiae *(Steiner and Scott 1935)*

Anguina amsinckiae parasitizes common fiddleneck, *Amsinckia intermedia*, and is capable of causing pathological disorders in livestock. One would assume that this is due to the association of the nematode with a species of *Corynebacterium*.

This nematode induces "blinding" by galling the terminal meristem of the plant host. Leaf galls are formed by a curbing of the leaf (Nagamine and Maggenti, 1980).

C. Root-Knot Nematodes (*Meloidogyne* spp.)

1. Meloidogyne graminicola *Golden and Birchfield 1965*

Meloidogyne graminicola was described from the roots of barnyard grass, *Echinchloa colonum* L., by Golden and Birchfield (1965). Birchfield (1965) stated that the nematode caused galls that were initially white and spongy and became dark and necrotic with age. One of the interesting characteristics of this nematode is that in barnyard grass the female remains embedded within the galled root and eggs are deposited in the cortex. Most juveniles hatch within the galls and reinfect the same root. Besides barnyard grass, barley, *Poa annua*, *Alopecurus carolinianus*, and crowsfoot grass are hosts.

2. Meloidogyne graminis *(Sledge and Golden) Whitehead 1968*

Van Weerdt et al. (1960) reported an unidentified species causing a chlorosis and decline of St. Augustine grass. This nematode was later described by Sledge and Golden (1964) as *Hypsoperine graminis* and later synonomyzed by Whitehead (1968) as *Meloidogyne graminis*, and is called the pseudo root-knot nematode. There have been several reports of *M. graminis* reducing growth of different grasses, such as bermuda grass and zoysia grass (Dickerson, 1966; Bell and Krusberg, 1964).

Heald (1969) showed a reduction in root and top weights of 28.4% after eight months and a 68.9% reduction in clipping weights of tifdwarf bermudagrass as a result of *M. graminis* infection. *M. graminis* caused extensive damage to the vascular and cortical tissue. Giant cell formation resulted mostly in the xylem with fewer in the phloem. Typical egg masses were formed, and the hatched juveniles were often seen in the gelatinous matrix.

3. Meloidogyne naasi *Franklin 1965*

M. naasi has been reported parasitizing creeping bentgrass (*Agrostis palustris*) in the United States. It has been associated with wilting of creeping bentgrass in California and stunted plant growth in Illinois.

Sikora et al. (1972) found that *M. naasi* was highly pathogenic to creeping bentgrass. Michell et al. (1973) found that *M. naasi* was able to parasitize and reproduce on quackgrass, *Agropyron repens*; redtop, *Agrostis alba*; creeping bentgrass, *A. palustris*; colonial bentgrass; orchardgrass *Dactylis glomerata*; hairy crabgrass, *Digitaria sanquinalis*; meadow fescue, *Festuca pratensis*; Italian ryegrass, *Lolium multiflorum*; perennial ryegrass, *L. perenne*; annual bluegrass, *Poa annua*; Kentucky bluegrass, *P. pratensis*; and drought bluegrass, *P. trivialis*. Some differences were observed in relation to host specificity of different geographical nematode populations. Radewald et al. (1970) found that a population of *M. naasi* from seaside bentgrass (*A. palustris*) was able to reproduce on 23 different plant species, showing the wide host range of this nematode.

M. naasi was found to be highly pathogenic on creeping bentgrass in combination with *P. penetrans* and *Tylenchorhynchus agri* (Sikora et al., 1972).

4. Other Meloidogyne *Species*

Golden and Birchfield (1965) described a new root-knot species *M. graminicola* attacking grasses. Vargas and Pajuelo (1973) found that *Coix lachryma jobi* and gordua grass were susceptible and showed considerable field damage from parasitism by *M. hapla* and *M. incognita* in Peru.

Riggs et al. (1962) studied the effect of *M. arenaria*, *M. hapla*, *M. incognita*, and *M. javanica* on bermudagrass, *Cynodon* spp. Ten pasture and six lawn types of bermudagrass were tested and *M. incognita* was the most damaging and *M. hapla* caused the least damage (coastal and midland grasses were the most resistant of the pasture grasses, and Uganda was the most resistant lawn grass.

McGlohon et al. (1961) studied the effect of five species of *Meloidogyne* on the growth of 20 grass species. They found that tall oat grass, *Arrhenatherum elatius*, smooth bromegrass, *Bromus inermis*, orchardgrass, *Dactylis glomerata*, Italian ryegrass, *Lolium multiflorum*, Harding grass, *Phalaris tuberosa*, and Kentucky bluegrass, *Poa pratensis*, were moderate to extremely susceptible to *M. javanica*, *M. incognita*, *M. incognita acrita*, and *M. arenaria*. Only tall oat grass and smooth bromegrass showed any response to *M. hapla* where an occasional gall, but no egg mass, was observed.

Crabgrass, *Digitaria sanguinalis*, was susceptible only to *M. javanica*. Among those grasses showing resistance to the *Meloidogyne* spp. were fescue, coastal bermudagrass, carpet grass, Pensacola bahiagrass, lovegrass, switchgrass, Wilmington bahiagrass, Kentucky 31 tall fescue, Starr Millet, and sudangrass. Although not galled, egg masses of *M. javanica* and *M. incognita* were formed on the roots of Pensacola bahiagrass.

D. Root Lesion Nematodes (*Pratylenchus* spp.)

Pratylenchus spp. are typically vagrant parasites moving in and out of plant root tissue. Both juveniles and adults interrupt roots by forcing their way through or between cortical cells. There is a breakdown of root tissue by the progressive migration of the nematode through the root. There is, however, no notable hyperplasia or cell hypertrophy. Necrotic areas on roots are associated with nematode parasitism. These are initially small, but may increase in darkness and size, affecting a large portion of the root. Lesion nematodes create a favorable, environment for secondary invaders, and are often associated with disease complexes on plants. Several species of *Pratylenchus* have been found associated with or parasitizing grasses.

Troll and Rohde (1966) found *P. penetrans* in all root areas, except the root cap of annual ryegrass, creeping red fescue, and Kentucky bluegrass. There were, however, no lesions or discoloration of the root tissue observed, and a histological study showed only tearing of cortical parenchyma cells. *P. penetrans* significantly decreased the root weight of annual ryegrass. Shoot growth of any of these three grasses, however, was not adversely affected by the nematodes.

Parris (1957) found *Pratylenchus* spp. associated with St. Augustine grass, Zoysia grass, and centipede grass in Mississippi. Wetzel (1969) found high populations of *P. neglectus* around the roots of plants grown for seed, including *Arrhenatherum elatius*, *Lolium multiflorum*, *L. perenne*, *Phleum pratense*, *Dactylis glomerata*, and *Trisetum flavescens*. Good control was affected with metham-sodium. Good et al. (1959) found *P. brachyurus* associated with bermuda, centipede, St. Augustine, and zoysia grass in Georgia turf nurseries, and Taylor et al. (1963) found a species of *Pratylenchus* associated with bentgrass on golf courses in Illinois.

Lordello and Filho (1969) found *P. brachyurus* on gordua grass, *Melinus minutiflora*, and jarajua grass in Brazil. Lordello and Filho (1970) also found *P. zeae* on *Panicum maximum*, *P. purpurascens*, and *Brachara mutica*. *P. brachyurus* was extremely virulent to *B. mutica*, and inflicted severe crop losses.

Cole et al. (1973) found *Pratylenchus* associated with *Fusarium* spp. in turfgrass sod fields in Pennsylvania, and Townshend et al. (1973) found that 10 different grass cultivars were hosts of *P. neglectus*.

E. Ectoparasitic Nematodes

1. *Stunt Nematode* (Tylenchorhynchus *spp.*)

Tylenchorhynchus spp. are ectoparasitic nematodes that are widely distributed throughout the world. Some species, such as *T. claytoni*, are important plant parasites and parasitize a number of plants. Several species have been found associated with grasses.

Laughlin and Vargas (1972) found that inoculation of *Agrostis palustris* with 500 or 1000 *Tylenchorhynchus dubius* resulted in significant reduction of dry weights of leaves and roots after 90 days. Gross symptoms were most obvious at 10°C. *T. dubius* also reduced leaf and root weights of *Poa pratensis*, and the nematodes multiplied from 300 to 1290 per pot in 90 days. The nematodes fed primarily on the root hairs and epidermal cells immediately behind the meristematic region, only the stylet penetrating the cells and no necrosis appearing at feeding sites.

Vargas and Laughlin (1972) also found that *T. dubius* interacted with *Fusarium rosem* on *P. pratensis* var. Merion; the nematode was the dominant pathogen, and only in its presence did severe reduction in plant top and root growth occur. It is thought that *T. dubius* may predispose the grass to attack by the fungus.

Lukens and Miller (1973) showed that *T. dubius* was pathogenic to *P. pratensis* and *Agrostis* spp., but not to *Lolium perenne*. Jakobsen (1975) found that high numbers of *T. dubius*, up to 20 per milliliter soil, were associated with decreasing yields of *A. tenuis*.

Bridge and Hague (1974) studied the feeding habits of three species of *Tylenchorhynchus*. They found that *T. dubius, T. lamelliferus*, and *T. maximus* fed as browsing ectoparasites on the roots of ryegrass; *T. maximus*, however, also fed in aggregations on epidermal root cells, causing mechanical breakdown of epidermal, cortical, and undifferentiated vascular tissue. *T. maximus* also caused a reduction in main root growth, but this was compensated for by increased lateral root growth.

Smolik and Malek (1973) found that *T. nudus* reduced the growth of *P. pratensis* roots by 36% and crowns by 26%, and nematodes were more virulent in sandy loam than loam soil, and in soil with low moisture. This may explain the decline of bluegrass turf during hot and dry summers. The nematode population increased 800% over a four month period. Western wheatgrass growth was severely reduced by *Tylenchorhynchus robustus* in a greenhouse study (Fig. 43).

Taylor et al. (1963) found that *Tylenchorhynchus* was the most common of seven genera of nematode associated with the growth of bentgrass. Smolik and Malek (1972) found that *P. pratensis* was a favored host of *T. nudus*. Lucas et al. (1974) found that *T. claytoni* was associated with golf greens in North Carolina, and Perry et al. (1959) found *T. maximus* associated with Kentucky bluegrass in Wisconsin.

2. *Lance Nematode* (Hoplolaimus *spp.*)

Kelsheimer and Overman (1953) found *Hoplolaimus galeatus* (*coronatus*) associated with diseased patches of St. Augustine grass. Parris (1957) also found *Hoplolaimus* spp. associated with St. Augustine, zoysia, and centipede grasses.

Fig. 43 Western wheatgrass parasitized by *Tylenchorhynchus robustus*. (Courtesy of J. D. Smolik.)

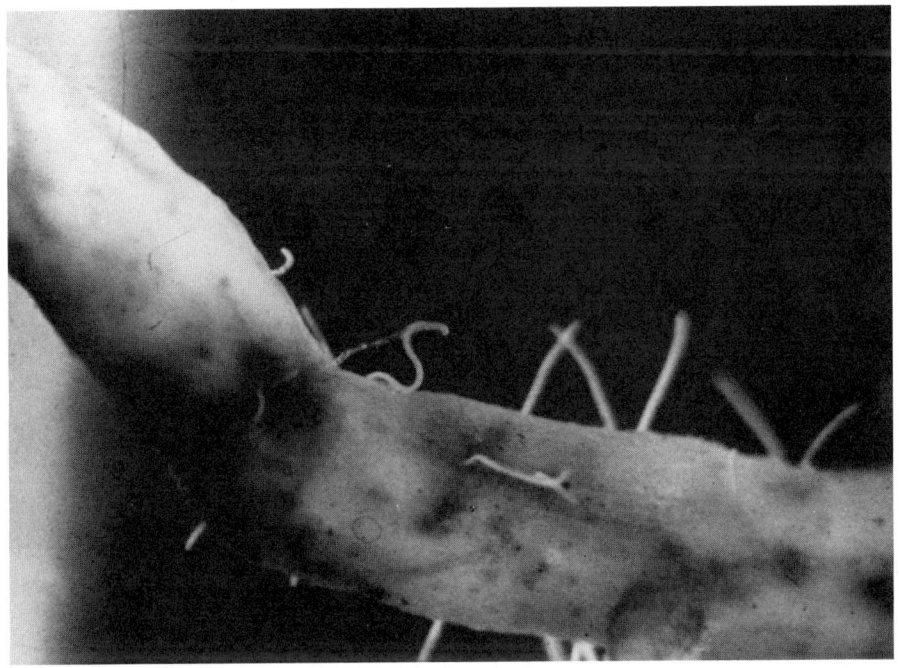

Fig. 44 *Hoplolaimus galeatus* parasitizing St. Augustine grass root. (Courtesy of D. W. Dickson.)

H. galateus parasitizes St. Augustine grass (Fig. 44), and Good et al. (1956) stated that *H. galeatus* was the most important plant-parasitic nematode on St. Augustine grass in peninsular Florida and caused extensive destruction of the root tissue as internal parasites.

Another species of *Hoplolaimus* found associated with grass is *H. concaudajuvencus* from around the roots of *Lolium multiflorum* and *Cynodon dactylon* (Golden and Minton, 1970).

Hoplolaimus was also found associated with bermudagrass golf greens in North Carolina (Lucas et al., 1974) and in golf greens in New York (Murdock et al., 1978).

3. *Spiral Nematodes* (Helicotylenchus *spp.*, Rotylenchus *spp.*)

Spiral nematodes are some of the most numerous plant-parasitic nematodes associated with the growth of plants. They are especially prevalent in pastures, lawns, golf greens, and in rangelands. It is not unusual to find two or more species occurring together.

Minton and Cairns (1957) found that Dallis, Bermuda, nut, Johnson, fescue, sweet Sudan, and orchard grasses were hosts of *Helicotylenchus nannus*.

Anderson (1974) described three new species, *H. cornurus*, *H. phalerus*, and *H. teleductus*, from the rhizosphere of grasses. Lucas et al. (1974) found *Helicotylenchus* associated with golf greens in North Carolina.

Perry et al. (1959) found high populations of *H. digonicus*, *H. microlobus*, and *Rotylenchus pumilus* associated with Kentucky bluegrass in Wisconsin, *H. digonicus* being the most numerous. *H. digonicus* reduced root weight and total plant weight by 55% (Figs. 45 and 46). Krupinsky et al. (1983) and Krupinsky et al. (1981) found *Helicotylenchus* spp. associated with blue grama and western wheatgrass in North and South Dakota.

Alby and Russell (1975) stated that there was a preference in attraction of *H. digonicus* to good hosts over poor hosts. Nematodes were observed partially submerged inside the root and feeding on cortical cell tissue. *Rotylenchus* spp. have also been found associated with the decline of grasses. Good et al. (1956) found *Rotylenchus* spp. associated with St. Augustine grass in Florida. Golden and Taylor (1956) described *R. christiei* from bermuda grass in Florida, and Cole et al. (1973) found *Rotylenchus* parasitizing turf in Pennsylvania.

4. *Ring Nematodes* (Criconemella *spp.*)

Pastures and grasslands are favorable habitats for this genus of nematodes, and it is not unusual to find ring nematodes, *Criconemella* spp., in turf, pasture, and rangeland grasses. Kelsheimer and Overman (1953), Good et al. (1956), and Parris (1957) found ring nematode associated with chlorotic growth of St. Augustine grass.

Johnson and Powell (1968) found that bermuda, St. Augustine, centipede, and emerald Zoysia grasses are good hosts of *C. lobatum* and there was a reduction in growth by the nematode. *C. ornata* caused observable damage to *P. pratensis* and *C. dactylon* (Feldmesser and Golden, 1972; Johnson, 1970). Cole et al. (1973) found ring nematodes associated with turfgrass in Pennsylvania, Lucas et al. (1974) found ring nematode associated with bermudagrass golf greens in North Carolina, and Safford and Riedel (1976) found *C. rustica* and *C. xenoplax* occurring in soil samples from golf green turf in Ohio.

Fig. 45 Kentucky bluegrass infected with *Helicotylenchus digonicus*. Left, uninfected. Right, infected. (After Perry et al., 1959.)

5. Other Plant-Parasitic Nematodes

Several other genera of plant-parasitic nematodes have been found associated with grasses. Winchester and Burt (1964) reported that *Belonolaimus longicaudatus* was responsible for leaf yellowing and stunting of bermudagrass. Johnson (1970) found that *B. longicaudatus* was pathogenic on grass (Fig. 47). Greco (1976) found a race of *D. dipsaci* that was pathogenic on *Poa trivialis*. Parris (1957) found *Trichodorus* spp. and *Pratylenchus* spp. associated with the decline of St. Augustine grass, zoysia grass, and centipede grass. Kelsheimer and Overman (1953) found *Belonolaimus gracilis*, *Radopholus similis*, and *Ditylenchus* spp. associated with diseased St. Augustine grass in Florida. Griffin (unpublished) found *X. americanum* associated with poor growth of smooth bromegrass in Wisconsin (Fig. 48). Good et al. (1956) stated that *Trichodorus* spp. and *Belonolaimus gracilis* were the major nematode pests of bermudagrass in Florida. Good et al. (1959) found *P. minor* (*T. christiei*), *X. americanum*, *Helicotylenchus nannus*, *P. brachyurus*, and

Fig. 46 Parasitism of Kentucky bluegrass by *Helicotylenchus digonicus*.
Left, uninfected. Right, parasitized grass. (After Perry et al., 1959.)

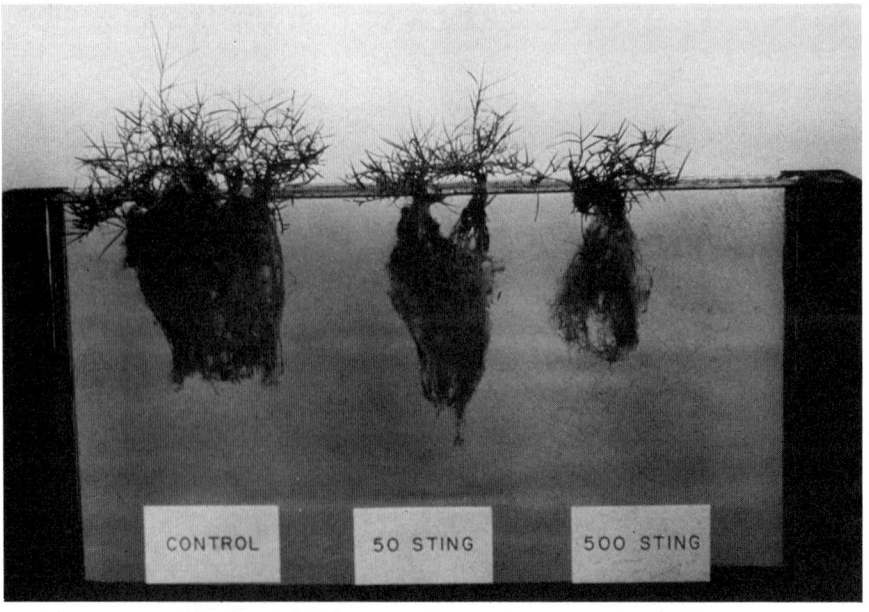

Fig. 47 Parasitism of tifdwarf bentgrass by *B. longicaudatus*. (Courtesy of
A. W. Johnson.)

Fig. 48 Parasitism of *Xiphinema americanum* on smooth bromegrass. Top, methyl bromide treated. Bottom, untreated.

Belonolaimus spp. associated with bermuda, centipede, St. Augustine, and zoysia grasses in Georgia. *Paratrichodorus minor* was found associated with bentgrass and bermudagrass in North Carolina (Lucas et al., 1974). Ratana-worabham and Smart (1970) found that *Criconemoides ornatus* at inocula rates of 1000 or greater reduced the growth of centipede grass.

Smolik (1973; 1974) found several plant-parasitic nematodes, including *Xiphinema* spp., *Tylenchorhynchus* spp., *Pratylenchus* spp., and *Helicoty-lenchus* spp. associated with *Buchloe dactylis*, *Bouteloua gracilis*, and *Agro-pyron smithii* in South Dakota. He found that *Tylenchorhynchus* spp. were dominant in grazed grassland areas, but *Xiphinema* spp., particularly *X. americanum*, showed a preference for ungrazed grasslands.

Marlatt and Perry (1971) found that *Aphelenchoides besseyi* infection increased the inflorescence and top growth production of wild grass, *Sporobolus poirettii*. This is the first report of a foliar nematode causing a plant growth stimulus that increased plant growth.

Control Chemical: Since it has been shown that nematodes can and do extract tremendous losses from grasses, it has been found that chemical control can produce increased plant growth (Figs. 49 to 54).

Murdock et al. (1977) tested the efficacy of using a nematicide to control *M. incognita* on Tifdwarf bermudagrass *Cynodon* spp. Phenamiphos and DBCP gave excellent control, but diazinonxylene and oxamyl were ineffective.

Manzelli (1955) found that 0-2,4-dichlorophenyl-0,0-diethyl phosphorothioate was effective in controlling nematodes on St. Augustine, annual rye, colonial bent, and bermudagrass. Tarjan (1965) controlled *Pratylenchus scribneri* on centipede turfgrass with chemical applications. *P. goodeyi*, *Paratrichodorus minor*, and *Hemicycliophora parvana* control, however, was effective after 12 weeks when dasanit was used. Wetzel (1969) found that metham-sodium significantly controlled *P. neglectus* on *Arrhenatherum elatius*, *Lolium multiflorum*, *Phleum pratense*, *Dactylis glomerata*, *Trisetum flavescens*, and *L. perenne* in Germany.

Perry and Dickson (1974) were able to control *Belonolaimus longicaudatus*, *Hoplolaimus galeatus*, and *Dolichodorus heterocephalus* on *Cynodon dactylon* with granular applications of carbofuran, aldicarb, fensulfothion, and phenamiphos, and injected DBCP. The grass responded faster to granular applications, especially aldicarb and phenamiphos, but DBCP-treated plots showed the best growth after eight weeks. Murdock et al. (1977) also obtained good control of *M. incognita* on tifdwarf bermudagrass with phenamiphos and DBCP. Perry et al. (1959) also obtained good control of *Helicotylenchus digonicus*, *H. microlobus*, *Rotylenchus pumilus*, and *H. platyurus* with DBCP.

Brodie and Burton (1967), using DBCP, obtained good control of *Belonolaimus longicaudatus* but not *Meloidogyne graminis* with a DBCP drench, and Alexander (1964) obtained effective control of *Helicotylenchus nannus*, *Pratylenchus* spp., *Criconemella* spp., and *Pratylenchus* spp. with a DBCP drench. Injection of DBCP reduced *B. longicaudatus* below detectable levels under Ormond bermudagrass turf (Smart and Perry, 1966). They also reported good control of *B. longicaudatus* with fensulfothion. However, fensulfothion failed to control *M. graminis*, as control of sting nematode resulted in a significant increase in the *M. graminis* population, showing a population competition between the two nematodes. Heald and Burton (1968) found that inorganic nitrogen increased reproduction of *B. longicaudatus* and *M. graminis* on Tifgreen bermudagrass, while nematode numbers were reduced with organic sludge.

Nutter and Christie (1958) obtained control of *B. longicaudatus*, *Criconemella* spp., *Hoplolaimus* spp., and *Dolichodorus* spp. with DBCP. They found that soil aeration, close mowing, soil moisture, and fertility were necessary for successful chemical treatments. Good control of *Trichodorus* spp. was obtained initially, but after eight weeks the nematode numbers increased considerably above those in untreated plots. Apt et al. (1960) found that the use of herbicides, including Dalaphon and maleic hydrazide, prevented seed heading of *Agrostis tenuis*, thus resulting in control of *Anguina agrostis*. The seed loss affected by the herbicides was compensated for by increased seed yields the following year. Winchester and Hayslip (1961)

Fig. 49 Chemical control of *B. longicaudatus* on bermudagrass. Left, treated. Right, untreated. (Courtesy of G. C. Smart.)

Fig. 50 Effect of experimental nematicide on growth of Bayshore bermudagrass parasitized by *B. longicaudatus*. (Courtesy of G. C. Smart.)

Fig. 51 Chemical control of *B. longicaudatus* with fensulfothion on bermuda-grass. (Courtesy of G. C. Smart.)

Fig. 52 Chemical control of *B. longicaudatus* on golf course with carbofuran. (Courtesy of G. C. Smart.)

Fig. 53 Chemical control (DBCP) of *B. longicaudatus* on bermudagrass.
Left, treated. Right, untreated. (Courtesy of G. C. Smart.)

found that *M. incognita* decreased rapidly under pangola grass, and the final
populations were similar to that found in clean fallowed soil.

Smolik (1973) found that grass clipping weights were increased from 28
to 59% with applications of oxamyl, there being a 45% increase in regrowth
weight obtained on the final clippings. Similar results had been obtained with
soil fumigation where yield increases of 35-67% were obtained with soil fumigation. Smolik also found that oxamyl reduced nematode populations of *Helicotylenchus*, *Paratylenchus*, *Pratylenchus*, *Tylenchorhynchus*, and *Xiphinema* by 87-96% in field studies and 84-96% in greenhouse studies.

Resistance: Since it is improbable that chemical control of plant-parasitic nematodes will be considered on other than turf, preferably golf greens,
it is necessary to develop other means of control, mainly resistance or tolerance. The area of breeding nematode-resistant grasses is one that has received little or no attention. Eriksson (1972) stated that breeding for resistant grass selections to plant-parasitic nematodes is practically a virgin area
of research.

McBeth (1945) found that several grasses, including bermuda, sudan,
and bahiagrasses, were resistant to *Meloidogyne* spp. (*Heterodera marioni*).
Riggs et al. (1962) found resistance to *M. incognita acrita* in bermudagrass.
Gaskin (1965) found resistance to *M. incognita acrita* in strains of *P. pratensis*, *P. compressa*, *Festuca cula*, *Agrostis alba*, *A. tenuis*, and Highland bentgrass. Boyd and Perry (1970) found resistance to *B. longicaudatus* in *Digitaria grazensis* and *D. procumbens*.

Fig. 54 Effect of oxamyl on growth of sod removed from native rangeland in South Dakota. Left, untreated. Right, treated. (Courtesy of J. D. Smolik.)

REFERENCES

Alby, T., and Russell, C. C. (1975). Attraction of certain grasses and ornamentals to *Helicotylenchus digonicus. J. Nematol.* 7: 319.

Alexander, P. M. (1964). Nematode and fairy ring control on Tifgreen Bermuda. *Golf Course Rep.* 5: 12-14.

Andersen, K., and Andersen, S. (1970). Changes in population of the cereal eelworm under grass species and resistant barley varieties. *Tijdschr. Planteavl.* 74: 559-565.

Andersen, S. (1959). Resistance of barley to various populations of the cereal root eelworm (*Heterodera major*). *Nematologica* 4: 91-98.

Andersen, S. (1960). Havreal problemer. *Sonderdr. Ans. Tolumands Bladet* 11: 45.

Andersen, S. (1961). Resistens mod havreal *Heterodera avenae. Meddelelse Nr. 68 fraden KG1.* Veterinaer-og Landboho, Skoles, Kopenhangen.

Anderson, H. J. (1979). Migratory nematodes in Danish barley fields. I. The qualitative and quantitative composition of the fauna. *Tijdschr. Planteval* 83: 1-8.

Anderson, R. V. (1974). Canadian species of the genus *Helicotylenchus* Steiner, 1945 (Nematoda: Hoplolaimidae), their identifying characteristics and descriptions of three new species. *Can. J. Zool.* 52: 1365-1381.

Andersson, S. (1974). *Heterodera hordecalis* n. sp. (Nematoda: Heteroderidae) a cyst nematode of cereals and grasses in southern Sweden. *Nematologica* 20: 445-454.

Apt, W. J., Austenson, H. M., and Courtney, W. D. (1960). Use of herbicides to break the life cycle of the bentgrass nematode, *Anguina agrostis* (Steinbuch 1799) Filipjev 1936. *Plant Dis. Rep.* 44: 524-526.

Barry, E. R., Brown, R. H., and Elliott, B. R. (1974). Cereal cyst nematode (*Heterodera avenae*) in Victoria: Influence of cultural practices on grain yields and nematode populations. *Aust. J. Exp. Agric. Anim. Husb.* 14: 566-571.

Baxter, R. I., and Blake, C. D. (1967). Invasion of wheat roots by *Pratylenchus thornei. Nature (Lond.)* 215: 1168-1169.

Baxter, R. I., and Blake, C. D. (1968). *Pratylenchus thornei* — a cause of root necrosis in wheat. *Nematologica* 14: 351-361.

Bell, A. A., and Krusberg, L. R. (1964). Occurrence and control of a nematode of the genus *Hypsoperine* on Zoysia and Bermuda grasses in Maryland. *Plant Dis. Rep.* 48: 721-722.

Bessey, E. A. (1905). A nematode disease of grasses. *Sci. N.S.* 21: 391-392.

Bingefors, S. (1960). Stem nematode in lucerne in Sweden. *I. Uppsala. Lantbrhogsk Ann.* 26: 217-233.

Birchfield, W. (1965). Host-parasite relations and host range studies of a new *Meloidogyne* species in southern U.S.A. *Phytopathology* 55: 1359-1361.

Bird, A. F. (1972). Influence of temperature on embryogenesis in *Meloidogyne javanica. J. Nematol.* 4: 206-212.

Bird, A. F., (1981). The *Anguina-Corynebacterium* association. In *Plant Parasitic Nematodes*, vol. III (B. M. Zuckerman and R. A. Rohde, eds.). Academic, New York, pp. 303-323.

Bird, A. F., and Stynes, B. A. (1977). The morphology of a *Corynebacterium* sp. parasitic on annual rye grass. *Phytopathology* 67: 828-830.

Bird, A. F., and Wallace, H. R. (1965). The influence of temperature of *Meloidogyne hapla* and *M. javanica*. *Nematologica 11*: 581-589.

Bland, B. F. (1971). *Crop Production: Cereals and Legumes*. Academic, New York.

Bloom, J. R. (1964). Effect of temperature extremes on the wheat seed gall nematode, *Anguina tritici*. *Plant Dis. Rep. 47*: 938-940.

Bolton, J. L. (1962). *Alfalfa, Botany, Cultivation, and Utilization*. Interscience, New York.

Boyd, F. T., and Perry, V. G. (1970). The effect of sting nematode on establishment, yields, and growth of forage grasses on Florida sandy soils. *Proc. Soil Crop Sci. Soc. Fla. 29*: 288-300.

Bridge, J., and Hague, N. G. M. (1974). The feeding behavior of *Tylenchorhynchus* and *Merlinius* species and their effect on growth of perennial ryegrass. *Nematologica 20*: 119-130.

Brodie, B. B., and Burton, G. W. (1967). Nematode population reduction and growth response of bermuda turf as influenced by organic pesticide applications. *Plant Dis. Rep. 51*: 562-566.

Brown, E. B. (1957). Lucerne stem eelworm in Great Britain. *Nematologica (Suppl.) 2*: 369-375.

Brown, E. B., and Sykes, G. B. (1975). Studies on the relation between density of *Longidorus elongatus* and yield of barley and potatoes. *Plant Pathol. 24*: 221-223.

Brown, R. L. (1982). The ecology and control of cereal cyst nematode (*Heterodera avenae*) in southern Australia. *J. Nematol.* (In Press.)

Brown, R. H. (1969). The occurrence of biotypes of the cereal cyst nematode (*Heterodera avenae*) Woll. in Victoria. *Aust. J. Exp. Agric. Anim. Husb. 9*: 453.

Brown, R. H., Meagher, J. W., and McSwain, N. K. (1970). Chemical control of the cereal cyst nematode (*Heterodera avenae*) in the Victoria Mallej. *Aust. J. Exp. Agric. Anim. Husb. 10*: 172-173.

Brown, R. H. (1972). Chemical control of the cereal cyst nematode (*Heterodera avenae*) in Victoria. A comparison of systemic and contact nematicides. *Aust. J. Exp. Agric. Anim. Husb. 12*: 662-666.

Brown, R. H. (1973). Chemical control of the cereal cyst nematode (*Heterodera avenae*) — a comparison of methods and rates of application of two systemic nematicides. *Aust. J. Exp. Agric. Anim. Husb. 13*: 587-591.

Brown, R. H. (1974). Further studies on the Victorian biotype of the cereal cyst nematode (*Heterodera avenae*). *Aust. J. Exp. Agric. Anim. Husb. 14*: 394-398.

Brown, R. H., and Pye, D. L. (1981). The effect of nematicide application and time of sowing on the cereal cyst nematode, *Heterodera avenae*, and the subsequent yield of wheat. *Aust. J. Plant Pathol. 10*: 17-18.

Brown, R. H., Pye, D. L., and Stratford, G. T. (1982). A comparison of low volume, in-row applications of nematicides at seeding, for control of the cereal cyst nematode (*Heterodera avenae*) in wheat. *Nematol. Medit. 10*: 9-19.

Campbell, W. F., and Griffin, G. D. (1973). Fine structure analyses of stem-nematode induced white flagging in *Medicago sativa*. *J. Nematol. 5*: 123-126.

Chang, D. C. N., Campbell, W. F., and Griffin, G. D. (1973). Ultrastructure changes induced by stem nematodes in hypocotyl tissue of alfalfa. *J. Nematol. 5*: 165-173.

Chang, D. C. N., Campbell, W. F., and Griffin, G. D. (1975). Effects of temperature on the fine structural responses in the hypocotyl region of alfalfa lines to *Ditylenchus dipsaci*. *J. Nematol.* 7: 239-246.

Chapman, R. A. (1954). Meadow nematodes associated with a failure of spring-sown alfalfa. *Phytopathology* 44: 542-545.

Chapman, R. A. (1958). The effect of root-lesion nematodes on the growth of red clover and alfalfa under greenhouse conditions. *Phytopathology* 48: 525-530.

Chapman, R. A. (1959). Development of *Pratylenchus penetrans* and *Tylenchorhynchus martini* on red clover and alfalfa. *Phytopathology* 49: 357-359.

Chapman, R. A. (1960). The effect of *Meloidogyne incognita* and *M. hapla* on the growth of Kenland red clover and Atlantic alfalfa. *Phytopathology* 50: 181-182.

Chapman, R. A. (1963). Development of *Meloidogyne hapla* and *M. incognita* in alfalfa. *Phytopathology* 53: 1003-1005.

Chapman, R. A., and Turner, D. R. (1975). Effect of *Meloidogyne incognita* on reproduction of *Pratylenchus penetrans* in red clover and alfalfa. *J. Nematol.* 7: 6-10.

Chhabra, H. K., and Bindra, O. S. (1971). Preliminary studies with DBCP on the control of *Tylenchorhynchus eremicolus* Allen infesting wheat in Ludhiana. *Indian J. Entomol.* 32: 275-276.

Chitwood, B. G. (1949). Root-knot nematodes. Part I. A revision of the genus *Meloidogyne* Goeldi 1887. *Proc. Helm. Soc. Wash.* 16: 90-104.

Chitwood, B. G., Newhall, A. C., and Clement, R. L. (1940). Onion bloat or eelworm rot, caused by the bulb or stem nematode, *Ditylenchus dipsaci* (Kühn) Filipjev. *Proc. Helm. Soc. Wash.* 7: 44-51.

Choppin de Janvry, E. (1971). Les nematodes des cereals. In *Les nemates des cultures*. Acta, Paris, pp. 273-292.

Christie, J. R. (1936). The development of root-knot nematode galls. *Phytopathology* 26: 1-22.

Coffman, F. A. (1946). Origin of cultivated oats. *J. Am. Soc. Agron.* 38: 983-1002.

Cole, H., Jr., Forer, L. B., Nelson, P. E., Bloom, J. R., and Jodon, M. H. (1973). Stylet nematode genera and *Fusarium* species isolated from Pennsylvania turfgrass sod production fields. *Plant Dis. Rep.* 57: 891-895.

Collins, D. D. (1966). The occurrence of seed-gall nematodes in a native population of western wheat grass (*Agropyron smithii*) in Montana. *Plant Dis. Rep. 50*: 45.

Corbett, D. C. M. (1970). Cereal *Pratylenchus* spp. in England and Wales and their recognition. *Plant Pathol.* 19: 6-10.

Corbett, D. C. M. (1972). The effect of *Pratylenchus fallax* on wheat, barley, and sugarbeet roots. *Nematologica* 18: 303-308.

Cotten, J. (1962). The effect of temperature on hatching in the cereal root eelworm. *Nature (Lond.) 195*: 308.

Cotten, J. (1963). Resistance in barley and oats to the cereal root eelworm *Heterodera avenae* Wollenweber. *Nematologica* 9: 81-94.

Coursen, B. W., and Jenkins, W. R. (1958). Host parasite relationship of the pin nematode, *Paratylenchus projectus* on tobacco and tall fescue. *Plant Dis. Rep. 42*: 865-872.

Courtney, W. D., and Howell, H. B. (1952). Investigation of the bent grass nematode, *Anguina agrostis* (Steinbuch 1799) Filipjev 1936. *Plant Dis. Rep. 36*: 75-83.

Decker, H. (1972). In *Plant Nematodes and Their Control (Phytonematology)* (N. M. Sveshnikova, ed.). Translated from Russian, USDA and NSF, Washington, D.C. Amerind Publ. Co. Put. Ltd., New Delhi.

D'Herde (1965). Een nieus wortelhnobbeloatze parasiet van de suikerbiete-teelt. *Landbouutijdschrift 18*: 879-884.

Dickerson, O. J. (1966). Some observations on *Hypsoperine graminis* in Kansas. *Plant Dis. Rep. 50*: 396-398.

Dickerson, O. J. (1979). The effect of temperature on *Pratylenchus scribneri* and *P. alleni* populations on soybean and tomatoes. *J. Nematol. 11*: 23-26.

Dixon, G. M. (1963). The effect of spring rainfall on the host-parasite relationship between the cereal root eelworm (*Heterodera avenae* Woll.) and the oat plant (*Avena sativa* L.). *Nematologica 9*: 521-526.

Dixon, G. M. (1965). The effect of cereal root eelworm (*Heterodera avenae* Woll.) on spring sown cereals. *VIII Int. Nematol. Symp. Antibes*, p. 26.

Duggan, J. J. (1958). Population studies on cereal root eelworm, *Heterodera major* (O. Schmidt 1930). *Econ. Proc. R. Dublin Soc. 4*: 103.

Duggan, J. J. (1963). Relationship between intensity of cereal root eelworm (*Heterodera avenae* Wollenweber 1924) infestation and pH value of soil. *Irish J. Agric. Res. 2*: 105-109.

Dunn, R. A. (1973). Extraction of eggs of *Pratylenchus penetrans* from alfalfa callus and relationship between age of culture and yield of eggs. *J. Nematol. 5*: 73-74.

Edmunds, J. E., and Mai, W. F. (1966). Effect of *Trichoderma viride*, *Fusarium oxysporum*, and fungal enzymes upon the penetration of alfalfa roots by *Pratylenchus penetrans*. *Phytopathology 56*: 1132-1135.

Edwards, D. I., and Taylor, D. P. (1963). Host range of an Illinois population of the stem nematode (*Ditylenchus dipsaci*) isolated from onion. *Nematologica 9*: 305-312.

Edwards, G. R. (1951). Insect pests of onions. *J. Dept. Agric. S. Austral. 54*: 558-561.

Endo, B. Y. (1959). Responses of root lesion nematodes, *Pratylenchus brachyurus* and *P. zeae* to various plant and soil types. *Phytopathology 49*: 417-421.

Eriksson, K. B. (1965). Crossing experiments with races of *Ditylenchus dipsaci* on callus tissue cultures. *Nematologica 11*: 244-248.

Eriksson, K. B. (1972). Nematode diseases of pasture legumes and turf grasses. In *Economic Nematology*, (J. M. Webster, ed.). Academic, New York, pp. 66-96.

Faulkner, L. R., and Bolander, O. J. (1970). Acquisition and distribution of nematodes in waterways of the Columbia Basin in Eastern Washington. *J. Nematol. 2*: 362-367.

Feldmesser, J., and Golden, A. M. (1972). Control of nematodes damaging home lawngrasses in two counties in Maryland. *Plant Dis. Rep. 56*: 476-480.

Fiddian, W. E. H., and Kimber, D. S. (1965). A study of biotypes of the cereal cyst nematode (*Heterodera avenae* Woll.) in England and Wales. *Nematologica 10*: 631.

Fidler, J. H., and Bevan, W. T. (1963). Some soil factors influencing the density of cereal root eelworm (*Heterodera avenae* Woll.) populations and their damage to the oat crop. *Nematologica 9*: 412-420.

Fielding, M. J. (1951). Observations on the length of dormancy in certain plant infecting nematodes. *Proc. Helm. Soc. Wash. 18*: 110-112.

Filipjev, I. N. (1934). The classification of the free living nematodes and their relation to the parasitic nematodes. *Smithsonian Misc. Coll.* *3216 89*: 1-63.

Franklin, M. T. (1965). A root-knot nematode *Meloidogyne naasi* n. sp. on field crops in England and Wales. *Nematologica 11*: 79-86.

Franklin, M. T. (1969). *Heterodera latipons* n. sp., a cereal cyst nematode from the Mediterranean region. *Nematologica 15*: 535-542.

Franklin, M. T., Clark, S. A., and Course, J. A. (1971). Population changes and development of *Meloidogyne naasi* in the field. *Nematologica 17*: 575-590.

Franklin, M. T., Thorne, G., and Oostenbrink, M. (1959). Proposal to stabilize the scientific name of the cereal eelworm (Class Nematoda). *Bull. Zool. Nomencl. 17*: 76-85.

Fushtey, S. G., and Johnson, P. W. (1966). The biology of the oat cyst nematode, *Heterodera avenae* in Canada. I. The effect of temperature on the hatchability of cysts and emergence of larvae. *Nematologica 12*: 313-320.

Gair, R. (1965). Cereal root eelworm. In *Plant Nematology* (J. F. Southey, ed.). Tech. Bull., 2nd ed., Ministry Agriculture Fish. pp. 199-211.

Gaskin, T. A. (1965). Susceptibility of bluegrass to root-knot nematodes. *Plant Dis. Rep. 49*: 89-90.

Gill, J. S., and Swarup, G. (1977). Effect of interaction between *Heterodera avenae* Woll. 1924, *Fusarium moniliforme* and *Helminthosporium gramineum* on barley plants and nematode reproduction. *Indian J. Nematol.* *7*: 42-45.

Goffart, H. (1929). Beobachtungen uber *Anguillulina pratensis* deMan. *Z. Parasitenkunde 2*: 97-120.

Golden, A. M., and Birchfield, W. (1965). *Meloidogyne graminicola* (Heteroderidae) a new species of root-knot nematode from grass. *Proc. Helm. Soc. Wash. 32*: 228-231.

Golden, A. M., and Minton, N. A. (1970). Description and larval heteromorphism of *Hoplolaimus concaudajuvencus* n. sp. (Nematoda: Hoplolaimidae). *J. Nematol. 2*: 161-166.

Golden, A. M., O'Bannon, J. H., Santo, G. S., and Finley, A. M. (1980). Description and SEM observation of *Meloidogyne chitwoodi* n. sp. (Meloidogynidae) a root-knot nematode on potato in the Pacific Northwest. *J. Nematol. 12*: 319-327.

Golden, A. M., and Taylor, A. L. (1956). *Rotylenchus christiei* n. sp., a new spiral nematode species associated with roots of turf. *Proc. Helm. Soc. Wash. 23*: 109-112.

Golden, A. M., and Taylor, D. P. (1967). The barley root-knot nematode in Illinois. *Plant Dis. Rep. 51*: 974-975.

Good, J. M., Christie, J. R., and Nutter, J. C. (1956). Identification and distribution of plant parasitic nematodes in Florida and Georgia. *Phytopathology 46*: 13.

Good, J. M., Steele, A. E., and Radcliffe, T. J. (1959). Occurrence of plant-parasitic nematodes in Georgia turf nurseries. *Plant Dis. Rep. 43*: 236-238.

Goodey, T. (1930). On *Tylenchus agrostis* (Steinbuch 1799). *J. Helm. 8*: 197-210.

Goodey, T. (1932). The genus *Anguillulina* Gerv. and v. Ben. 1859. vel *Tylenchus* Bastian 1865. *J. Helminth. 10*: 75-180.

Gooris, J. and D'Herde, C. J. (1977). *Studies on the Biology of* Meloidogyne naasi *Franklin 1965.* Ministry of Agriculture, Agricultural Research Administration State Agricultural Research Center, Ghent.

Goplen, B. P., and Stanford, E. H. (1959). Studies on the nature of resistance in alfalfa to two species of root-knot nematodes. *Agron. J.* 51: 486-488.

Goplen, B. P., Stanford, E. H., and Allen, M. W. (1959). Demonstration of physiological races within three root-knot nematode species attacking alfalfa. *Phytopathology 49*: 653-656.

Graham, C. W. (1980). The effect of rainfall and soil type on the population dynamics of cereal cyst nematode (*Heterodera avenae*) on spring barley (*Hordeum vulgare*) and spring oats (*Avena sativa*). *Ann. Appl. Biol.* 94: 243-253.

Grandison, G. S., and Halliwell, H. G. (1975). New pest of cereals. *N.Z. J. Agric. 130*: 64-65.

Greco, N. (1976). Weed host of *Ditylenchus dipsaci* in Puglia. *Nematol. Medit. 4*: 99-102.

Green, C. D. (1981). The effect of weeds and wild plants on the reinfestation of land by *Ditylenchus dipsaci*. In *Pests, Pathogens, and Vegetation* (J. M. Thresh, ed.). Pitman, London, pp. 217-224.

Griffin, G. D. (1967). Chemical control of the stem nematode, *Ditylenchus dipsaci*. *Plant Dis. Rep. 51*: 973-974.

Griffin, G. D. (1968). The pathogenicity of *Ditylenchus dipsaci* to alfalfa and the relationship of temperature to plant infection and susceptibility. *Phytopathology 58*: 929-932.

Griffin, G. D. (1969). Effects of temperature on *Meloidogyne hapla* in alfalfa. *Phytopathology 59*: 599-602.

Griffin, G. D. (1974). Effect of acclimation, temperature or infection of alfalfa by *Ditylenchus dipsaci*. *J. Nematol. 6*: 57-59.

Griffin, G. D. (1975). Parasitism of nonhost cultivars by *Ditylenchus dipsaci*. *J. Nematol. 7*: 236-238.

Griffin, G. D. (1980). The interrelationship of *Meloidogyne hapla* and *Ditylenchus dipsaci* on resistant and susceptible alfalfa. *J. Nematol. 12*: 287-293.

Griffin, G. D., and Anderson, J. L. (1979). Effects of DCPA, EPTC, and chlorpropham on pathogenicity of *Meloidogyne hapla* to alfalfa. *J. Nematol. 11*: 32-36.

Griffin, G. D., and Elgin, J. H., Jr. (1977). Penetration and development of *Meloidogyne hapla* in resistant and susceptible alfalfa under differing temperatures. *J. Nematol. 9*: 51-56.

Griffin, G. D., and Hunt, O. J. (1972a). Effect of temperature and inoculation timing on the *Meloidogyne hapla/Corynebacterium insidiosum* complex in alfalfa. *J. Nematol. 4*: 70-71.

Griffin, G. D., and Hunt, O. J. (1972b). Effect of plant age on resistance of alfalfa of *Meloidogyne hapla*. *J. Nematol. 4*: 87-90.

Griffin, G. D., Hunt, O. J., and Murray, J. J. (1975). Pathogenicity of *Ditylenchus dipsaci* to sainfoin (*Onobrychis viciaefolia* Scop.). *J. Nematol. 7*: 93-94.

Griffin, G. D., and Thyr, B. D. (1978). Interaction of *Meloidogyne hapla* and *Fusarium oxysporum* on alfalfa. *J. Nematol. 10*: 289.

Griffin, G. D., and Waite, W. W. (1971). Attraction of *Ditylenchus dipsaci* and *Meloidogyne hapla* by resistant and susceptible alfalfa seedlings. *J. Nematol. 3*: 215-219.

Grujicic, G. (1967). *Korenova nematoda (*Meloidogyne naasi *Franklin)*. *u Srbiji. Zastita Bilja 18, Beograd*, pp. 93-95.

Hartman, B. J., Hunt, O. J., Peaden, R. N., Jensen, H. J., Thyr, B. D., Faulkner, L. R., and Griffin, G. D. (1979). Registration of Nevada Synthetic YY Alfalfa. (Reg. No. GP 99). *Crop Sci. 19*: 416-417.

Hawn, E. J., and Hanna, M. R. (1966). Influence of stem nematode infestation on bacterial wilt reaction and forage yield of alfalfa varieties. *Can. J. Plant Sci. 47*: 203-208.

Heald, C. M. (1969). Pathogenicity and histopathology of *Meloidogyne graminis* infecting 'Tifdwarf' bermuda grass roots. *J. Nematol. 1*: 31-35.

Heald, C. M., and Burton, G. W. (1968). Effect of organic and inorganic nitrogen on nematode populations in turf. *Plant Dis. Rep. 52*: 46-48.

Hesling, J. J. (1965). *Heterodera* morphology and identification. In *Plant Nematology* (J. F. Southey, ed.). Tech. Bull. Minist. Agric. Fish. Fd. No. 7, pp. 103-130.

Hunt, O. J., Griffin, G. D., Murray, J. J., Pederson, M. W., and Peaden, R. N. (1971). The effect of root-knot nematodes on bacterial wilt in alfalfa. *Phytopathology 61*: 256-259.

Inserra, R. N., Di Vito, M., Vovlas, N., and Seinhorst, J. W. (1983). Relationship between *Paratrichodorus* sp. density and growth of wheat in pots. *J. Nematol. 15*: 79-87.

Inserra, R. N., Lamberti, F., Volvas, N., and Dandria, D. (1975). *Meloidogyne naasi* nell'Italia Meridionale ea Malta. *Nematol. Medit. 3*: 163-166.

Inserra, R. N., O'Bannon, J. H., and Santo, G. S. (1980). The effect of *Meloidogyne hapla* on growth of alfalfa seedlings. *Nematologica 26*: 379-381.

Inserra, R. N., Volvas, N., and Bradonisio, A. (1978). Nematodi endoparassiti associati a colture di cereali deperimento nell'Italia meridionale. *Nematol. Medit. 6*: 163-174.

Jakobsen, J. (1975). Nematoder pa graes. *Nord. Jordbrugsforsk. 57*: 514-515.

Jatala, P., Jensen, H. J., and Shimabukuro, R. A. (1973). Host range of the "grass root-gall nematode," *Ditylenchus radicicola*, and its distribution in Williamette Valley, Oregon. *Plant Dis. Rep. 57*: 1021-1023.

Jensen, H. J. (1961). Nematodes affecting Oregon agriculture. *Ore. Agric. Exp. Sta. Bull. 579*.

Jensen, H. J., Eshtiaghi, H., Koepsell, P. A., and Goetze, N. (1975). The oat cyst nematode, *Heterodera avenae* occurs in oats in Oregon. *Plant Dis. Rep. 59*: 1-3.

Jensen, H. J., Howell, H. B., and Courtney, W. D. (1958). Grass seed nematode and production of bentgrass seed. *Agric. Exp. Sta. Bull. No. 565*.

Johnson, A. W. (1970). Influence of organic pesticides on nematode populations and seed production of centipedegrass. *J. Nematol. 2*: 252-254.

Johnson, A. W., and Powell, W. M. (1968). Pathogenic capabilities of a ring nematode, *Criconemoides lobatum*, on various turf grasses. *Plant Dis. Rep. 52*: 109-113.

Jones, R. K. (1979). Migratory plant parasitic nematode on pests of cereals. *Ann. Appl. Biol. 92*: 257-262.

Kable, P. F., and Mai, W. F. (1968). Influence of soil moisture on *Pratylenchus penetrans*. *Nematologica 14*: 101-122.

Kelsheimer, E. G., and Overman, A. J. (1953). Notes on some parasitic nematodes found attacking lawns in the Tampa Bay area. *Proc. Fla. State Hort. Soc. 66*: 301-303.

Kerry, B. R., Crump, D. H., and Mullen, L. A. (1982a). Studies of the cereal cyst nematode, *Heterodera avenae* under continuous cereals, 1974-78. I. Plant growth and nematode multiplication. *Ann. Appl. Biol. 100*: 477-487.

Kerry, B. R., Crump, D. H., and Mullen, L. A. (1982b). Studies of the cereal cyst nematode, *Heterodera avenae* under continuous cereals, 1974-78. II. Fungal parasitism of nematode females and eggs. *Ann. Appl. Biol. 100*: 489-499.

Kimpinski, J. M., and Willis, C. B. (1981). Influence of soil temperature and pH on *Pratylenchus penetrans* and *P. crenatus* in alfalfa and timothy. *J. Nematol. 13*: 333-338.

Kirjanova, E. S. (1955). Quackgrass nematode-*Paranguina agropyri* Kiryanova n.g. n. sp. (Nematoda). *Trudi Zool. Inst. Akad. Nauk SSSR 18*: 42-52.

Kort, J. (1972). Nematode diseases of cereals of temperate climates. In *Economic Nematology* (J. M. Webster, ed.). Academic, New York, pp. 97-126.

Kort, J., Dantuma, G., and Van Essen, A. (1964). On biotypes of the cereal root eelworm (*Heterodera avenae*) and resistance in oats and barley. *Neth. J. Plant Pathol. 70*: 9-17.

Krupinsky, J. M., Barker, R. E., and Donald, P. A. (1983). Nematodes associated with vegetative collections of blue grama and western wheatgrass in the western Dakotas. *Plant Dis.* (In press).

Krupinsky, J. M., Donald, P. A., and Barker, R. E. (1981). *Helicotylenchus* spp. and *Tylenchorhynchus* spp. associated with grasses in the western Dakotas. *J. Nematol. 13*: 446-447.

Krusberg, L. R. (1959). Investigations on the life cycle, reproduction, feeding habits, and host range of *Tylenchorhynchus claytoni* Steiner. *Nematologica 4*: 187-197.

Krusberg, L. R. (1961). Studies on the culturing and parasitism of plant parasitic nematodes in particular *Ditylenchus dipsaci* and *Aphelenchoides ritzemabosi* on alfalfa tissues. *Nematologica 6*: 181-200.

Kühn, J. (1874). Über das VorKommen von Rübennematoden an den Wurzeln der Halmfrüchte. *Z. Wiss. Landw-Arch. Kgl. -Preuss. Landes-Okon. Kolleg. 3*: 47-50.

Kuiper, K. (1966). Einige bijzondere aaltjesaantastingen in 1965. *Tijdschr. Plant. 72*: 210.

Langdon, K. R., Struble, F. B., and Young, H. C., Jr. (1961). Stunt of small grains, a new disease caused by the nematode *Tylenchorhynchus brevidens*. *Plant Dis. Rep. 45*: 248-252.

Laughlin, C. W., and Vargas, J. M., Jr. (1972). Pathogenic potential of *Tylenchorhynchus dubius* on selected turfgrass. *J. Nematol. 4*: 277-280.

Lewis, G. D., and Mai, W. F. (1960). Overwintering and migration of *Ditylenchus dipsaci* in organic soils of southern New York. *Phytopathology 50*: 341-343.

Lordello, L. G. E., and Filho, T. M. (1969). Capins gordua e jaragua, hospedeiros novos de um nematoide migrador. *Solo 61*: 27.

Lordello, L. G. E., and Filho, T. M. (1970). Mais tres capins hospedeiros de nematoides migraderes. *Rev. Agric. Piracicaba 45*: 78.

Lucas, L. T., Blake, C. T., and Barker, K. R. (1974). Nematodes associated with bentgrass and bermudagrass golf greens in North Carolina. *Plant Dis. Rep. 58*: 822-824.

Lukens, R. J., and Miller, P. M. (1973). Injury to turfgrass by *Tylenchorhynchus dubius* and *Hoplolaimus* spp. *Phytopathology 63*: 204.

Manzelli, M. A. (1955). Progress report on the use of V-C 13 for the control of nematodes infesting turfgrasses. *Virginia J. Sci. 6*: 231-232.

Marcinowski, K. (1909). Parasitische und semiparasitische an Pflanzen lebenden Nematoden. *Arb. Kaisenlichen Biol. Anstalt Land-u. Fonst. Berlin 7*: 1-192.

Marlatt, R. B., and Perry, V. G. (1971). Growth stimulation of *Sporobolus poiretti* by *Aphelenchoides besseyi*. *Phytopathology 61*: 740.

Martin, G. C. (1958). Root-knot nematodes (*Meloidogyne* spp.) in the federation of Rhodesia and Nyasaland. *Nematologica 3*: 332-349.

Mathur, B. N., Arya, H. C., Mathur, R. L., and Handa, D. K. (1974). The occurrence of biotypes of the cereal cyst nematode (*Heterodera avenae*) in the light soils of Rajasthan and Haryana, India. *Nematologica 20*: 19-26.

Mauza, B. E., and Webster, J. M. (1982). Suppression of alfalfa growth by concommitant populations of *Pratylenchus penetrans* and two *Fusarium* species. *J. Nematol. 14*: 364-367.

McBeth, C. W. (1945). Tests on the susceptibility and resistance of several southern grasses to the root-knot nematode, *Heterodera marioni*. *Proc. Helm. Soc. Wash. 12*: 41-44.

McGlohon, N. E., Sasser, J. N., and Sherwood, R. T. (1961). Investigations of plant-parasitic nematodes associated with forage crops in North Carolina. *N.C. Agric. Exp. Sta. Tech. Bull. 148*.

McGuire, J. H., Walters, H. J., and Slack, D. A. (1958). The relationship of root-knot nematodes on the development of fusarium wilt in alfalfa (abstract). *Phytopathology 48*: 344.

Meagher, J. W. (1972a). Cereal cyst nematode (*Heterodera avenae* Woll.). Studies on ecology and content in Victoria. *Tech. Bull. 24 Dept. Agric. Vict.*

Meagher, J. W. (1972b). Cryptobiosis of the cereal cyst nematode (*Heterodera avenae*) and effects of temperature and relative humidity on survival of eggs in storage. *Nematologica 20*: 323-336.

Meagher, J. W. (1977). World dissemination of the cereal-cyst nematode (*Heterodera avenae*) and its potential as a pathogen of wheat. *J. Nematol. 9*: 9-15.

Meagher, J. W., and Brown, R. H. (1974). Microplot experiments on the effect of plant hosts on populations of cereal cyst nematode (*Heterodera avenae*) and on the subsequent yield of wheat. *Nematologica 20*: 337-346.

Meagher, J. W., Brown, R. H., and Rovira, A. D. (1978). The effect of cereal cyst nematode (*Heterodera avenae*) and *Rhizoctonia solani* on the growth and yield of wheat. *Aust. J. Agric. Res. 29*: 1127-1137.

Meagher, J. W., and Chambers, S. C. (1971). Pathogenic effects of *Heterodera avenae* and *Rhizoctonia solani* and their interaction on wheat. *Aust. J. Agric. Res. 22*: 189-194.

Meagher, J. W., and Rooney, D. R. (1966). The effect of crop rotations in the Victorian Wimmera on the cereal cyst nematode (*Heterodera avenae*), nitrogen fertilizer and wheat yield. *Aust. J. Exp. Agric. Anim. Husb. 6*: 425-431.

Michell, R. E., Malek, R. B., Taylor, D. P., and Edwards, D. T. (1973). Races of the barley root-knot nematode, *Meloidogyne naasi*. I. Characterization by host preference. *J. Nematol.* 5: 41-43.

Minton, N. A., and Cairns, E. J. (1957). Suitability of soybeans var. Ogden and twelve other plants as hosts of the spiral nematode. *Phytopathology* 47: 313.

Molliard, M. M. (1904). Structure de quelques Tylenchocecidies foliares. Bull. de Societe Botanique de France. *Session extraordinaire a Paris en aout 1904* 51: 101-112.

Murdock, C. L., Apt, W. J., and Tashiro, H. (1977). Effects of nematicides on root-knot nematodes in bermudagrass putting greens in Hawaii. *Plant Dis. Rep.* 61: 978-981.

Murdock, C. L., Tashiro, H., and Harrison, M. B. (1978). Plant-parasitic nematodes associated with golf putting-green turf in New York. *Plant Dis. Rep.* 62: 85-87.

Nagamine, C., and Maggenti, A. R. (1980). "Blinding" of shoots of a leaf gall in *Amsinckia intermedia* induced by *Anguina amsinckia* (Steiner and Scott 1934) (Nemata, Tylenchida), with a note on the absence of a rachis in *A. amsinckia*. *J. Nematol.* 12: 129-132.

Neal, J. C. (1889). The root-knot disease of peach, orange, and other plants in Florida due to *Anguillula*. *U.S. Dept. Agric. Div. Ent. Bull.* 20.

Needham, T. (1743). A letter concerning certain chalky tubulous concretions called malm; with some microscopical observations on the farina of the red lily, and of worms discovered in smutty corn. *Phil. Trans. R. Soc.* 42: 173; 174; 634-641.

Neubert, E. (1967). Über das Vorkommen von Biotypen des Haferzystenaichens (*Heterodera avenae* Wollenweber, 1924) im Norden der DDR. *Nbl. Dtsch. Pflanzenschutzd (Berlin)* 21: 66-68.

Nielsen, C. H. (1966). Investigations on the inheritance of resistance to cereal root eelworm, *Heterodera avenae*, in wheat. *Nematologica 12*: 575-578.

Nigh, E. L., Jr. (1972). Resistance of selected alfalfa clones to the root-knot nematode, *Meloidogyne incognita*. *Phytopathology 63*: 780.

Nigh, E. L., Jr., Ellis, K. C., and Hine, R. B. (1969). Systemics control alfalfa stem nematode. *Prog. Agric. Ariz.* 21: 10-11.

Nilsson-Ehle, H. (1920). Über die Resistenz gegen *Heterodera schachtii*, bei gewissen Gerstensorten, ihre Vererbungsweise un Bedeutung fur die Praxis. *Hereditas 1*: 1-34.

Noel, G. R., and Lownsbery, B. F. (1978). Effect of temperature on the pathogenicity of *Tylenchorhynchus clarus* to alfalfa and observations on feeding. *J. Nematol.* 10: 195-198.

Norton, D. C. (1959). Relationship of nematodes to small grains and native grasses in North and Central Texas. *Plant Dis. Rep.* 43: 227-235.

Norton, D. C. (1963). Population fluctuations of *Xiphinema americanum* in Iowa. *Phytopathology 53*: 66-68.

Norton, D. C. (1965). *Anguina agropyronifloris* n. sp. infecting florets of *Agropyron smithii*. *Proc. Helm. Soc. Wash.* 32: 118-122.

Norton, D. C., Dickerson, O. J., and Ferris, J. M. (1968). Nematology in the North Central Region. 1956-1966. *Agric. Home Econ. Exp. Sta. Iowa State Univ. Speciol Rep. 58*.

Norton, D. C., and Everson, L. E. (1963). *Anguina* on western wheatgrass, *Agropyron smithii*. *Proc. Assoc. Off. Seed Analysis 53*: 208-209.

Nutter, G. C., and Christie, J. R. (1958). Nematode investigations of putting green turf. *Proc. Fla. Sta. Hort. Soc. 71*: 445-449.

O'Bannon, J. H., and Reynolds, H. W. (1962). Resistance of alfalfa to two species of root-knot nematodes. *Plant Dis. Rep. 46*: 558-559.

O'Brien, D. G. (1925). *The Rye Crop.* (*Z. ges. Getreidew.* 5: 41). Gresham, London, p. 253.

Orion, D., Kirkun, J., and Amir, J. (1982). Population dynamics of *Pratylenchus thornei* and its effect on wheat in a semiarid region. *Proceedings of the XVIth International Symposium European Society Nematology* St. Andrews, Scotland. p. 48.

Parris, G. K. (1957). Screening Mississippi soils for plant parasitic nematodes. *Plant Dis. Rep. 41*: 705-706.

Peaden, R. N., Hunt, O. J., Faulkner, L. R., Griffin, G. D., Jensen, H. J., and Stanford, E. H. (1976). Registration of a multiple pest resistant alfalfa germplasm (Reg. No. GD 51). *Crop Sci. 16*: 125-126.

Pederson, M. W., and Griffin, G. D. (1977). Registration of Deseret alfalfa. (Reg. No. 78). *Crop Sci. 17*: 671.

Pepper, E. H. (1963). Nematodes in North Dakota. *Plant Dis. Rep. 47*: 102-106.

Perez, B. J. G. (1971). Evaluacion de la susceptibilidad del cultivo del trigo al ataque del nematado *Pratylenchus thornei* Sher y Allen, 1953. Tesis. Instituto Technologico y de Estudios Superiores de Monterrey, Mexico.

Perez, B. J. G., Van Gundy, S. D., Stolzy, C. H., Thomason, I. J., and Laird, R. J. (1970). *Pratylenchus thornei*, a nematode pest of wheat in Sonora, Mexico (abstract). *Phytopathology 60*: 1307.

Perret, J. (1971). La maladies vermiculaice des tiges de l'avione due a *Ditylenchus dipsaci.* In *Les nematodes des cultures.* Acta, Paris, pp. 293-326.

Perry, V. G., Darling, H. M., and Thorne, G. (1959). Anatomy, taxonomy, and control of certain special nematodes attacking bluegrass in Wisconsin. *Univ. Wisc. Res. Bull. 207.*

Perry, V. G., and Dickson, D. W. (1974). Nematode control on turfgrasses. *Nematropica 4*: 4.

Potter, J. W., and Townshend, J. L. (1973). Distribution of plant-parasitic nematodes in field crop soils of southwestern and central Ontario. *Can. J. Plant Dis. Surv. 53*: 39-48.

Prasad, S. K., and Gaur, H. S. (1975). Relation between nematode populations and yield of wheat. *Indian J. Nematol. 4*: 152-159.

Radewald, J. D., Pyeatt, L., Shybuya, F., and Humphrey, W. (1970). *Meloidogyne naasi*, a parasite of turfgrass in southern California. *Plant Dis. Rep. 54*: 940-942.

Ratanaworabham, S., and Smart, G. C., Jr. (1970). The ring nematode *Criconemoides ornatus* on peach and centipede grass. *J. Nematol. 2*: 204-208.

Reed, B. M., Richardson, P. E., and Russell, C. C. (1977). Histological and cytological effects of an alfalfa-nematode interaction (abstract). *Proc. Am. Phytopathol. Soc. 4*: 123-124.

Reynolds, H. W. (1955). Varietal susceptibility of alfalfa to two species of root-knot nematodes. *Phytopathology 45*: 70-72.

Reynolds, H. W., Carter, W. W., and O'Bannon, J. H. (1970). Symptomless resistance of alfalfa to *Meloidogyne incognita acrita. J. Nematol. 2*: 131-134.

Reynolds, H. W., and O'Bannon, J. H. (1960). Reaction of sixteen varieties of alfalfa to two species of root-knot nematodes. *Plant Dis. Rep. 44:* 441-443.

Richardson, P. E., Russell, C. C., and Reed, B. M. (1978). Histological responses by wheat plants to infection by *Hoplolaimus* sp. *Phytopathology 4:* 123.

Riggs, R. D., Dale, J. L., and Hamblen, M. L. (1962). Reaction of bermuda grass varieties and lines to root-knot nematodes. *Phytopathology 52:* 587-588.

Roberts, P. A. and Van Gundy, S. D. (1981). The development and influence of *Meloidogyne incognita* and *M. javanica* on wheat. *J. Nematol. 13:* 345-352.

Roberts, P. A., Van Gundy, S. D., and McKinney, H. E. (1981). The effect of soil temperature and planting date of wheat of *Meloidogyne incognita* reproduction, soil populations, and grain yield. *J. Nematol. 13:* 338-345.

Rovira, A. (1982). *Management Strategies for Controlling Cereal Cyst Nematode*. CSIRO, Div. of Soils, Adelaide, S. Aust. A. E. Keating, Port Melbourne, Australia.

Russell, C. C., and Perry, V. G. (1966). Parasitic habits of *Trichodorus christiei* on wheat. *Phytopathology 56:* 357-358.

Safford, J., and Riedel, R. M. (1976). *Criconemoides* species associated with golf course greens in Ohio. *Plant Dis. Rep. 60:* 405-408.

Santo, G. S., Evans, D. W., and Bower, D. B. (1980). Reaction of three alfalfa cultivars to several species of plant-parasitic nematodes. *Plant Dis. 64:* 404-405.

Santo, G. S., and O'Bannon, J. H. (1981). Pathogenicity of the Columbia root-knot nematode (*Meloidogyne chitwoodi*) on wheat, corn, oats, and barley. *J. Nematol. 13:* 548-550.

Santo, G. S., O'Bannon, J. H., Finley, A. M., and Golden, A. M. (1980). Occurrence and host range of a new root-knot nematode (*Meloidogyne chitwoodi*) in the Pacific Northwest. *Plant Dis. 64:* 951-952.

Sasser, J. N. (1954). Identification and host-parasite relationships of certain root-knot nematodes (*Meloidogyne* spp.). *Univ. Maryland Agric. Exp. Sta. Tech. Bull. A-77.*

Sasser, J. N. (1977). Worldwide dissemination and importance of the root-knot nematodes, *Meloidogyne* spp. *J. Nematol. 9:* 26-29.

Schneider, J. (1967). Un nouveau nematode du genre *Meloidogyne* parasite des cereales en France. *Phytoma 185:* 21-25.

Sher, S. A., and Allen, M. W. (1953). Revision of the genus *Pratylenchus* (Nematoda: Tylenchidae). *Univ. Calif. Pub. Zool. 57:* 441-470.

Sikora, R. A., Koshy, P. A., and Malek, R. B. (1972). Evaluations of wheat selections for resistance to the cereal cyst nematode. *Indian J. Nematol. 2:* 81-82.

Sikora, R. A., Taylor, D. P., Malek, R. B., and Edwards, D. I. (1972). Interaction of *Meloidogyne naasi*, *Pratylenchus penetrans*, and *Tylenchorhynchus agri* on creeping bentgrass. *J. Nematol. 4:* 162-165.

Sledge, E. B., and Golden, A. M. (1964). *Hypsoperine graminis* (Nematoda: Heteroderidae), a new genus and species of plant-parasitic nematode. *Proc. Helm. Soc. Wash. 31:* 83-88.

Smart, G. C., and Perry, V. G. (1966). Pathogenicity and host-parasite relationships of nematodes on turf in Florida. *Fla. Agric. Exp. Sta.* pp. 108-109.

Smolik, J. D. (1973). Effect of soil fumigation and clipping intensity on growth of blue grass and western wheatgrass. *Proc. S.D. Acad. Sci. 52*: 267-268.

Smolik, J. D. (1974). Nematode studies at the Cottonwood site. *US/IBP Grassland Biome Tech. Rep. No. 251*, Colorado State University.

Smolik, J. D., and Malek, R. D. (1972). *Tylenchorhynchus nudus* and other nematodes associated with Kentucky bluegrass turf in South Dakota. *Plant Dis. Rep. 56*: 898-900.

Smolik, J. D., and Malek, R. B. (1973). Effect of *Tylenchorhynchus nudus* on growth of Kentucky bluegrass. *J. Nematol. 5*: 272-274.

Songul, A. E., and Dickerson, O. J. (1976). Life cycle, pathogenicity, histopathology, and host range of race 5 of the barley root-knot nematode. *J. Nematol. 8*: 228-230.

Sontirat, S., and Chapman, R. A. (1970). Penetration of alfalfa roots by different stages of *Pratylenchus penetrans* (Cobb). *J. Nematol. 2*: 270-271.

Spaull, A. M. (1980). Effect of *Paratrichodorus anemones* on growth of spring wheat and barley. *Nematologica 26*: 163-169.

Spears, J. F. (1956). Occurrence of the grass cyst nematode, *Heterodera punctata*, and *Heterodera cacti* group cysts in North Dakota and Minnesota. *Plant Dis. Rep. 40*: 583-584.

Stanford, E. H., Goplen, B. P., and Allen, M. W. (1958). Sources of resistance in alfalfa to the northern root-knot nematode, *Meloidogyne hapla*. *Phytopathology 48*: 347-349.

Stessel, G. J. (1960). Effect of nematocides on *Pratylenchus pratensis* and alfalfa yields. *Phytopathology 50*: 656.

Stessel, G. J., and Golden, A. M. (1961). Occurrence of *Ditylenchus radicicola* (Nematoda: Tylenchida) in the United States on a new host. *Plant Dis. Rep. 45*: 26-28.

Stewart, G. (1926). *Alfalfa Growing in the United States and Canada.* MacMillan, New York.

Sturhan, D. (1966). Witrspflanzenuntersuchungen an Bastardpopulationen von *Ditylenchus dipsaci*-Rassen. *Z. Pflanzenkrankh. 73*: 168-174.

Sturhan, D. (1971). Biological races. In *Plant Parasitic Nematodes* (B. M. Zuckerman, W. F. Mai, and R. A. Rohde, eds.). Academic, New York, pp. 51-71.

Sturhan, D. (1973). *Meloidogyne naasi* — ein fur Deutschland neuer Getreide parasit. Nachrichtenblatt Deutsch. *Pflanzenschutzd (Braunschweiz) 25*: 102-103.

Stynes, B. A., Peterson, D. S., Lloyd, J., Payne, A. L., and Lanigan, G. W. (1979). The production of toxin in annual ryegrass, *Lolium rigidum*, infected with a nematode, *Anguina* sp. and *Corynebacterium rathayi*. *Aust. J. Agric. Res. 30*: 201-209.

Suzuki, M., and Willis, C. B. (1974). Root lesion nematodes affect cold tolerance of alfalfa. *Can. J. Plant Sci. 50*: 577-581.

Talboys, P. W. (1964). A concept of the host-parasite relationship in Verticillium wilt diseases. *Nature (Lond.) 202*: 361-364.

Tarjan, A. C. (1965). Rejuvination of nematized centipede grass turf with chemical drenches. *Proc. Fla. Sta. Hort. Soc. 77*: 456-461.

Taylor, A. L., and Sasser, J. N. (1978). Biology, identification, and control of root-knot nematodes (*Meloidogyne* species). *International Meloidogyne Project*. North Carolina State University Graphics.

Taylor, D. P., Anderson, R. V., and Haglund, W. A. (1958). Nematodes associated with Minnesota crops. I. Preliminary survey of nematodes associated with alfalfa, flax, peas, and soybeans. *Plant Dis. Rep. 42:* 195-198.

Taylor, D. P., Burton, M. P., and Hechler, H. C. (1963). Occurrence of plant parasitic nematodes in Illinois golf greens. *Plant Dis. Rep. 47:* 134-135.

Thomason, I. J. (1962). Reaction of cereals and sudan grass of *Meloidogyne* spp. and the relation of soil temperature of *M. javanica* populations. *Phytopathology 52:* 787-791.

Thomason, I. J. and Sher, S. A. (1957). Influence of the stubby root nematode of the growth of alfalfa. *Phytopathology 47:* 159-161.

Thompson, L. S., and Willis, C. B. (1970). Effect of nematicides on root-lesion nematodes and forage legume yields. *Can. J. Plant Sci. 50:* 577-581.

Thorne, G. (1928). *Heterodera punctata* n. sp., a nematode parasitic on wheat roots from Saskatchewan. *Sci. Agric. 8:* 707-710.

Thorne, G. (1961). *Principles of Nematology.* McGraw Hill, New York.

Townshend, J. L. (1978). Infectivity of *Pratylenchus penetrans* on alfalfa. *J. Nematol. 10:* 318-322.

Townshend, J. L., Eggens, J. L., and McCollum, N. K. (1973). Turfgrass hosts of three species of nematodes associated with forage crops. *Can. Plant Dis. Surv. 53:* 137-141.

Townshend, J. L., and Potter, J. W. (1973). Nematode numbers under cultivars of forage legumes and grasses. *Can. Plant Dis. Surv. 53:* 194-195.

Townshend, J. L., and Potter, J. W. (1976). Evaluation of forage legumes, grasses, and cereals as hosts of forage nematodes. *Nematologica 22:* 196-201.

Townshend, J. L, and Stobbs, L. (1981). Histopathology and histochemistry of lesions caused by *Pratylenchus penetrans* in roots of forage legumes. *Can. J. Plant Pathol. 3:* 123-128.

Townshend, J. L., Willis, C. G., Potter, J. W., and Santerre, J. (1973). Occurrence and population density of nematodes associated with forage crops in eastern Canada. *Can. Plant Dis. Surv. 53:* 131-136.

Troll, J., and Rohde, J. A. (1966). Pathogenicity of *Pratylenchus penetrans* and *Tylenchorhynchus claytoni* on turf grass. *Phytopathology 56:* 995-998.

Tseng, S. T., Allred, K. R., and Griffin, G. D. (1968). A soil population study of *Ditylenchus dipsaci* (Kühn) Filipjev in an alfalfa field. *Proc. Helm. Soc. Wash 35:* 57-62.

Van Gundy, S. D., Perez, B. J. G., Stolzy, L. H., and Thomason, I. J. (1974). A pest management approach to the control of *Pratylenchus thornei* on wheat in Mexico. *J. Nematol. 6:* 107-116.

Van Weerdt, L. G., Birchfield, W., and Esser, R. P. (1960). Observation on some subtropical plant-parasitic nematodes in Florida. *Proc. Soil Crop Sci. Soc. Fla. 19:* 443-451.

Vargas, F. O., and Pajuelo, C. (1973). Effecto del *Meloidogyne* spp. (Nematoda: Heteroderidae) en algunas especies forrajeras gramineas. *Anales Cietificos,* Lima, Peru *11:* 205-218.

Vargas, J. M., and Laughlin, C. W. (1972). The role of *Tylenchorhynchus dubius* in the development of fusarium blight of Merion Kentucky bluegrass. *Phytopathology 62:* 1311-1314.

Vavilov, N. I. 1926). Studies on the origin of cultivated plants. *Bull. Appl. Bot. Plant Breed Leningrad 16*: 170.

Wallace, H. R. (1962). Observations on the behavior of *Ditylenchus dipsaci* in soil. *Nematologica 7*: 91-101.

Watson, T. R., and Lownsbery, B. F. (1970). Factors influencing the hatching of *Meloidogyne naasi* and a comparison with *M. hapla*. *Phytopathology 60*: 457-460.

Wetzel, T. (1969). Untersuchungen über den wandern den Wurzelnematoden *Pratylenchus neglectus* an Futtergräsern. *Nematologica 15*: 193-200.

Whitehead, A. G. (1968). Taxonomy of *Meloidogyne* (Nematoda: Heteroderidae) with descriptions of four new species. *Trans. Zool. Soc. Lond. 31*: 263-401.

Willis, C. B. (1972). Effect of soil pH on reproduction of *Pratylenchus penetrans* and forage yield of alfalfa. *J. Nematol. 4*: 291-295.

Willis, C. B. (1976). Effect of potassium fertilization and *Pratylenchus penetrans* on yield and potassium content of red clover and alfalfa. *J. Nematol. 8*: 116-121.

Winchester, J. A., and Burt, E. O. (1964). The effect and control of sting nematodes on Ormond Bermudagrass. *Plant Dis. Rep. 48*: 625-628.

Winchester, J. A., and Hayslip, N. C. (1961). The effect of land management practices on the root knot nematode, *Meloidogyne incognita* in Florida. *Proc. Fla. Sta. Hort. Soc. 73*: 100-104.

Chapter 9

Nematode Parasites of Vegetable Crops

A. W. Johnson *USDA, Agricultural Research Service, Coastal Plain Experiment Station, Tifton, Georgia*

George Fassuliotis *USDA, Agricultural Research Service, Charleston, South Carolina*

I. INTRODUCTION

Most vegetable crops are attacked by one or more species of nematodes. In fields heavily infested with nematodes, the above-ground symptoms are similar to those caused by any condition that deprives a plant of an adequate and properly functioning root system. Chlorosis, premature leaf drop, and stunting usually are the first obvious evidence of the presence of large populations of nematodes. These symptoms alone are not sufficient to identify nematode damage, since it is difficult to distinguish between nematode injury and damage caused by other pathogens, pests, herbicides, or inadequate nutrition. With favorable growing conditions, especially ample moisture and fertility, most vegetables may produce an acceptable crop yield while harboring enormous numbers of nematodes.

The economic impact of plant-parasitic nematodes is based largely on their ability to incite damage to plant tissues, or to act in complexes with other organisms to cause losses in quantity and often quality of crop yields. The rhizosphere often contains several species of plant-parasitic nematodes. Some nematodes may be highly pathogenic, whereas others may be weak or nonpathogens. To prove pathogenicity, pure cultures of nematodes must be studied in association with the crop.

Crop damage estimates are notoriously difficult to formulate. A report of crop loss estimates that helped pinpoint the economic impact of plant-parasitic nematodes and the need for continuing research has been published (Crop Losses Committee, 1971). The factors considered in making the estimates included low yields of market quality resulting primarily from nematode infections, increased yield and/or quality after chemical treatments, and cost of chemical control measures. Damage caused by plant-parasitic nematodes in the United States on 24 vegetable crops is estimated to be approximately 11%, or 3.4 million metric tons (3.8 million tons), with average damage per hectare of $247 ($100 per acre) in 1967-1968.

During the past 50 years, research on nematicides, screening for resistance, and cultural practices have yielded a number of means of suppressing nematode damage to vegetable crops. The limited number of available nematicides offer the most immediate solutions for nematode control, but their use is usually justified only where total costs [chemical plus application costs, approximately $24-$100 per hectare ($10-$40 per acre)] can be returned three- to fivefold in value from increased crop yield. This approach is not feasible for the management of nematode populations in many low cash value vegetable crops where losses may be $24 or less per hectare; the loss is significant in total reduction of food supply, but too low to warrant the expense of chemical treatments.

This chapter discusses the major nematode species associated with vegetable crops and for which pathogenicity has been proven.

II. NEMATODE PARASITES OF SOLANACEOUS VEGETABLES (TOMATO, EGGPLANT, PEPPER)

Over 60 species representing 19 genera of plant-parasitic nematodes have been associated with tomato cultivation (Valdez, 1978). Of these, the root-knot (*Meloidogyne* spp.), root lesion (*Pratylenchus* spp.), and reniform (*Rotylenchulus reniformis* Linford and Oliveira) nematodes have been associated with major damage to vegetable crops.

A. Root-Knot Nematodes (*Meloidogyne* spp.)

1. Historical

Root knot has been known as a disease of vegetable crops since 1855, when Berkeley in England first described the disease as "vibrios forming excrescences on cucumber roots." The causal organism was later described as *Heterodera radicicola* by Müller in 1884. For 65 years root-knot nematodes were considered a single species and referred to by a number of designations: *Anguillula marioni* Cornu 1879; *A. arenaria* Neal 1889; *A. vialae* Lavergne 1901; *H. javanica* Treub 1885; *Tylenchus arenarius* Cobb 1890; *Meloidogyne exigua* Goeldi 1887; *Oxyurus incognita* Kofoid and White 1919; *Caconema radicicola* Cobb 1924; and *Heterodera marioni* (Cornu 1879) Marcinowski 1909 (Thorne, 1961). Prior to 1949, evidence was accumulating that the root-knot nematodes consisted of races that differed in their host preference and pathogenic expression of infection (Christie and Albin, 1944; Christie, 1946). Morphological differences among the populations were noted and the root-knot nematodes were reassigned to the genus *Meloidogyne* (Chitwood, 1949). *Meloidogyne exigua*, *M. javanica*, *M. arenaria*, *M. hapla*, *M. incognita*, and *M. incognita acrita* were recognized based on the perineal pattern. *M. incognita* was later synonymized with *M. incognita* by Triantaphyllou and Sasser (1960) after studies of the perineal patterns from single egg mass isolates showed that patterns ranged from the *incognita* type to the *acrita* type. Since 1949, 53 species have been described (personal communication, A. L. Taylor and J. N. Sasser).

Root-knot nematodes are worldwide in distribution. They are essentially hot weather organisms and are most important in regions where summers are long and winters, if any, are short and mild. Nevertheless, these parasites are not confined to the tropics and subtropics. In cool climates, where the average temperature of the coldest month of the year is near or below 0°C and the average temperature of the warmest month is about 15°C or above, the most common *Meloidogyne* species is *M. hapla* (Taylor and Sasser, 1978). Current information indicates that *M. hapla* is adapted to long time existence in the northern United States (Fig. 1) and southern Canada.

In the tropic zone, the most common *Meloidogyne* species are *M. incognita* and *M. javanica*. In North America, *M. javanica* is seldom found above 30° N and 35° S latitude and becomes more common nearer the equator. *M. incognita* and *M. arenaria* are common and widespread in the same regions. In the United States, the northern limit for continued existence of *M. incognita* is a few hundred miles farther north of the limits of *M. javanica* (Figs. 2 and 3).

Fig. 1 Distribution of *M. hapla* in the United States (Courtesy of Committee on Nematode Distribution, Society of Nematologists.)

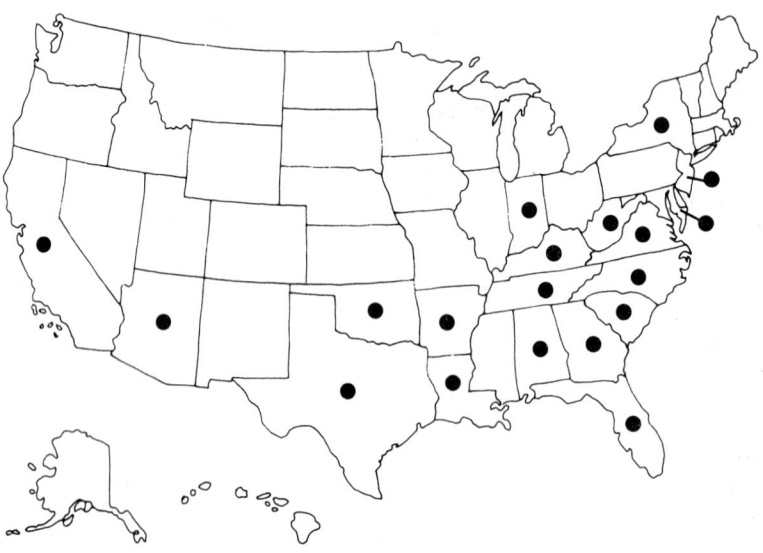

Fig. 2 Distribution of *M. incognita* in the United States (Courtesy of Committee on Nematode Distribution, Society of Nematologists.)

Fig. 3 Distribution of *M. javanica* in the United States (Courtesy of Committee on Nematode Distribution, Society of Nematologists.)

Thus, the area of the United States south of 35° S latitude is widely infested by *M. incognita*, *M. javanica*, and *M. arenaria* adapted to continuous existence in warm areas. North of 35° latitude the most common *Meloidogyne* species is *M. hapla*. These four species, as currently identified by taxonomists, are the most widespread and common *Meloidogyne* species in agricultural soils of the United States and of much of the rest of the world that has been surveyed. They very probably cause more damage to vegetable crops than all other *Meloidogyne* species combined (Sasser, 1977; Sasser et al., 1980).

More than 1000 living populations of *Meloidogyne* species have been studied from more than 70 cooperators in 75 countries of the world involved in the International Meloidogyne Project (IMP) (personal communication, J. N. Sasser). All populations have been given the differential host test (Taylor and Sasser, 1978) and have been identified by morphological study and by chromosome counts. This is not only the most extensive, but also the most accurate series of identifications ever made.

Over 95% of the samples were collected from crops growing in farm fields. They are the largest number of samples of *Meloidogyne* species from agricultural soils ever collected. When the samples were sorted by species, 52% were *M. incognita*, 30% *M. javanica*, 8% *M. arenaria*, and 8% *M. hapla* for a total of 98% (Table 1). The remaining 2% were *M. exigua* and other species.

By the differential host test, *M. incognita* is composed of four widely distributed biotypes referred to as races. Of these, race 1 is the most numerous. In the host test, it infects neither NC 95 tobacco nor Deltapine 16 cotton. Race 2 infects NC 95 tobacco, but not cotton. Race 3 infects cotton, but not tobacco. Race 4 infects both cotton and tobacco.

Table 1 Percentage Distribution of *Meloidogyne* Species
and Races in 1036 Samples from 75 Countries

Meloidogyne incognita	Race 1	= 71%
	Race 2	= 12
	Race 3	= 15
	Race 4	= 2
All samples of *M. incognita*		52%
Meloidogyne javanica		30%
Meloidogyne arenaria	Race 1	= 18%
Meloidogyne arenaria	Race 2	= 82%
Two races of *M. arenaria*		= 8%
Meloidogyne hapla		8%
Miscellaneous species		2%

Source: Courtesy of Dr. J. N. Sasser.

M. javanica has no clearly defined races, but aberrant populations exist;
in Egypt, peanuts are infected by the nematodes. *M. arenaria* has two races:
race 1 infects peanuts and race 2 does not. *M. hapla* shows no races in the
host test, but two races with differing chromosome numbers exist.

This distribution of species and races brings us to a new view of *Meloidogyne* species in agricultural soils. This new concept may be as important
and revolutionary for nematode control research as Chitwood's 1949 paper on
Meloidogyne. No work on the common species will be complete without identification to race. Much of the present literature on resistance must be revised to specify which races of *Meloidogyne* were tested.

The nematode samples often contained single or combinations of *Meloidogyne* species, usually *M. incognita*, *M. javanica*, and *M. arenaria*. Mixed
races are more difficult to detect by the host test, but they do occur, and
possibly very often. Populations that did not fit neatly into one or another
of the standard races have not been detected.

The uniformity of world species of *Meloidogyne* implies that plant breeding can be simplified. We can hope that a test of breeding lines against races
1, 2, 3, and 4 of *M. incognita* in the United States will be sufficient also for
Africa, Asia, Australia, and South America. If so, breeding is simplified,
and there is a worldwide market for seed of resistant cultivars, regardless of
where they are developed.

2. Biology and Life History

M. incognita, *M. javanica*, *M. arenaria*, and *M. hapla* are the most important
root-knot nematode species associated with tomato (Fig. 4), eggplant, and
pepper production.

All species of root-knot nematodes have substantially the same life history, but can be influenced greatly by soil temperature and host plant.

Fig. 4 Tomato roots heavily infected with a root-knot nematode, *M. incognita*, in a grower's field in Westmoreland County, Virginia, collected late July (30), 1975. (Courtesy of A. M. Golden.)

The general biology and pathogenicity of the root-knot nematodes on these crops are similar.

The pioneering work of Christie (1936) presented an excellent account of the life history and pathology of tomato roots infected with root-knot nematodes. Detailed studies on the growth and development of *M. javanica* and *M. incognita*, respectively, on tomato have been reported (Bird, 1959; Triantaphyllou and Hirschmann, 1960).

Growth of *M. javanica* takes the form of a sigmoid curve (Bird, 1959). The ovate eggs are usually found in the gelatinous egg sac surrounding the posterior end of the female. Upon hatching, the slender infective-stage juveniles are in the second stage of development; the first molt takes place within the egg. Juveniles are attracted toward roots from as far as 75 cm (Prot and Netscher, 1979). After searching for a suitable host root, the second-stage juveniles usually enter just behind the root cap, where there is intense meristematic activity. The nematodes migrate through undifferentiated root cells until they reach the developing vascular system where they begin to feed. Cell walls are pierced with the stylet, and secretions from the esophageal glands are extruded through the stylet which cause an alteration to the plant cells on which the nematodes feed. These feeding cells are commonly called giant cells that grow to various sizes, depending on the host plant and temperature. The chemical forces responsible for initiating these cells are not completely understood (Bird, 1972).

The nematode remains sedentary during feeding and enlarges in cross section but not in length. Cells of the genital primordium divide, developing into a two-pronged ovary in the female or an elongated structure in the male. About the 14th day, the juvenile molts and the stylet and esophageal bulb disappear. No further feeding takes place until after the final molt. Three molts take place between days 14 and 18. Shortly after the fourth molt, the stylet and median bulb are regenerated, the uterus and vagina are formed, and a perineal pattern is visible (Taylor and Sasser, 1978). The female assumes a flask shape during development, and after the last molt the female grows rapidly, becoming pear shaped.

In the developing male juvenile there is rapid metamorphosis; a vermiform body develops inside the cuticle, possessing a stylet, esophagus with median bulb, spicules, and sperm in the testis. The male develops faster than the female and bursts out of the cuticle in search of a female.

In cross-fertilizing (amphimictic) species, males are often found within the gelatinous matrix secreted by the female. The matrix is produced from six rectal glands into which 500-1000 eggs are extruded. Eggs are deposited in the one-cell stage, and mitosis begins a few hours later. The rate of embryonic development is temperature dependent. Embryos of *M. javanica* developed in 46-48 days at 15°C, 16-48 days at 20°C, 11-13 days at 25°C, and 9-10 days at 30°C (Bird, 1974).

Unlike some of the *Globodera* and *Heterodera* species, there is no apparent hatching factor necessary to induce juveniles to hatch from the eggs. Hatching is dependent on soil temperature and soil moisture.

The usual phenotypic response to parasitism by root-knot nematodes is the galling response, which is a separate phenomenon from the giant cell response. The size and character of the galls are influenced by host and parasite (Dropkin, 1969). Galls develop by cell hypertrophy, which starts within a few hours after penetration. However, galling can be induced by juveniles that are browsing on the surface of the tomato root. Galling may be induced by growth regulators introduced from the subventral glands of the infective-stage juvenile or when tryptophane is released upon hydrolysis of plant proteins by the juvenile, which react with endogenous phenolic acids to yield auxins (Bird, 1978).

Galls induced by *M. hapla* differ from those produced by the other species of *Meloidogyne*. Roots harboring *M. hapla* develop small spherical galls with extensive lateral root proliferation, often forming a dense mat when infection is severe. Terminal galls are more prevalent in infections by *M. hapla* than by *M. javanica* or *M. incognita*, since the former species more effectively suppresses mitotic activity. Lateral root initiation apparently is promoted due to the large amount of indoleacetonitrile (IAN) present in the eggs, juveniles, and tomato galls of *M. hapla* (Yu and Viglierchio, 1964).

The first effect the nematode induces upon root penetration is to retard differentiation among cells that would have given rise to the vascular elements. These and other cells in the vascular cylinder are induced to form giant cells. The exact manner by which the giant cells are induced by root-knot nematodes has yet to be clearly resolved. There are three schools of thought: (1) giant cells are formed by cell wall breakdown and incorporation of adjacent cells, (2) the cells enlarge and divide without incorporating adjacent cells, and (3) the cells form from a combination of the two (Dropkin, 1969).

M. hapla is a serious pest of tomatoes in the northern part of the United States and in Canada. Damage to tomatoes has been related to population density of *M. hapla* (Olthof and Potter, 1977). Low numbers of *M. hapla* (260 and 1840 juveniles per kilogram of soil) stimulated vegetative and root growth and resulted in higher yields than those with initial densities of 6120 and 27,950 juveniles. At the highest nematode density, fruit yield was suppressed by 40%.

3. Interactions with Other Organisms

Root-knot nematode infections generally increase the severity or predispose tomato plants to a more rapid and severe expression of other diseases. Wilting in *Fusarium* wilt-resistant tomato Chesapeake was induced only when the nematodes were present (Jenkins and Coursen, 1957). *M. incognita* appeared to be more capable in affecting wilt than *M. hapla*. However, wilt in resistant tomatoes was more severe with *M. hapla* infections than with *M. incognita* (Cohn and Minz, 1960). The nematodes may affect the hosts in different physiological manners. Although the nematodes may provide the *Fusarium* with infection courts through which the fungus could penetrate, in split-root experiments the physiological effects of the nematode were translocated to the entire plant, making it more susceptible to *Fusarium* wilt (Bowman and Bloom, 1966).

A synergistic growth reduction of the tomato Bonny Best was obtained when a combined infection of *M. javanica* and *Fusarium* occurred (Bergeson et al., 1970).

Tomato plants inoculated with *M. incognita* and *Rhizoctonia solani* subjected to continuous leaching to remove the exudates did not develop root decay. In contrast, when leachates were collected from *M. incognita*-infected roots and applied to roots of tomato inoculated with *R. solani* alone, a severe rot developed on tomato roots receiving the leachates. When roots were exposed to *M. incognita* and *R. solani* and not subjected to leaching, severe rot developed (Van Gundy et al., 1977).

When *M. incognita* and *Sclerotium rolfsii* Sacc. occurred together on eggplant, 25% of the plants wilted, whereas when only the fungus alone was present, only 6% wilted (Goswami et al., 1970).

Bacterial wilt, caused by *Pseudomonas solanacearum* E. F. Smith, is more severe in tomato and eggplant in the presence of *M. incognita*, indicating that the nematode is responsible for enhancing bacterial wilt in resistant cultivars. In tomato inoculated with *M. incognita* plus the bacterium, plants developed wilt one to three weeks earlier than those with the bacterium alone. The combination of the two organisms also suppressed the survival rate of wilt-resistant tomato plants 33-36% (Valdez, 1978).

Resistance of the *Verticillium* wilt-resistant tomato cultivar Gilat 38 was not affected by *M. javanica* and *Verticillium dahliae*, but in the susceptible cultivar, Rehobot 13, wilt symptoms and vascular discoloration were increased after inoculation with the nematode-fungus combination (Valdez, 1978).

When soils contained *M. incognita* and the fungus, *Ozonium texanum* var. *parasiticum*, together, germination of eggplant seed was severely depressed over soils containing either the nematode or fungus alone (Nath et al., 1979).

In another disease complex interrelationship, the combined effect of *M. incognita* and mycoplasmalike organisms that cause little leaf ("yellows") disease of eggplant was greater in reducing total plant growth from dual inoculations than from individual organisms (Dhawan and Sethi, 1977).

A positive interaction of *M. incognita* and tobacco mosaic virus on eggplant has been reported (Goswami and Chenulu, 1974).

4. Resistance

Resistance has been an important factor for economical control of root-knot nematodes. In tomatoes, resistance was developed through hybridization of *Lycopersicon esculentum* with the wild species *L. peruvianum*. In a series of crosses, selections, and backcrosses that spanned a 30 year period, plant breeders and nematologists have developed tomato cultivars with desirable horticultural traits containing the gene *Mi* for resistance to *M. incognita*, *M. javanica*, and *M. arenaria*. The gene, however, did not confer resistance to *M. hapla*. Resistance is inherited by one or more genes that act as a unit (Harrison, 1960). Over 50 cultivars and hybrids are resistant to root-knot nematodes, many of which are readily available from seed catalogs (Sasser and Kirby, 1979).

All cultivars of eggplant are susceptible to root-knot nematodes, but a high level of resistance is contained in a related wild species, *Solanum sisym-briifolium* (Fassuliotis, 1975). The wild species is resistant to both *M. incognita* and *M. javanica*, but highly susceptible to *M. arenaria*. However, in India, the wild species is resistant to *M. arenaria* (Verma et al., 1979). Hybridization of the two plant species has not been obtained with standard plant-breeding techniques, but through the fusion of somatically derived protoplasts, it may soon be proven technically feasible to transfer resistance from the wild species to the eggplant (Fassuliotis et al., 1981).

Peppers express a complete range of galling response from highly resistant to very susceptible. The small fruited hot peppers are resistant to *M. incognita*, *M. javanica*, and *M. arenaria*, but not to *M. hapla*. Bell and pimento types lack resistance. A root knot-resistant pimento pepper, Nemaheart, and a bell pepper, Mississippi 68, were developed over 10 years ago, but have not been used extensively. A pepper-breeding program recently renewed at the U.S. Vegetable Laboratory, Charleston, South Carolina, includes root-knot nematode resistance in the breeding objectives.

B. Root Lesion Nematodes (*Pratylenchus* spp.)

Pratylenchus spp. are probably the second-most important nematodes of economic importance to tomato cultivation. This crop is a host of at least five species, i.e., *P. brachyurus*, *P. coffeae*, *P. penetrans*, *P. scribneri*, and *P. vulnus* (Jensen, 1972). *Pratylenchus* spp. have a cosmopolitan distribution with a wide host range on other vegetables: Wando peas, pepper, spinach, lettuce, beets, radish, celery, onion, eggplant, and beans.

P. penetrans Sher and Allen 1953 is the most economically important plant-pathogenic nematode in the Northeast states (Mai et al., 1977), but is also found in other states (Fig. 5), in Canada (Potter and Olthof, 1974), and in western and central Europe (Loof, 1978).

1. Historical

The first species of *Pratylenchus* was described by de Man in 1880 under the name of *Tylenchus pratensis*. The name *Pratylenchus* was proposed by Filip-jev in 1934 to include nematodes that had been placed in the genera *Anguillu-lina* and *Tylenchus*. The genus was revised and 10 species were recognized, three of which were new (Sher and Allen, 1953). Another comprehensive taxonomic study supplemented this work in which 20 species were recognized (Loof, 1960). Currently, there are over 66 described species of *Pratylenchus*. The common name, root lesion nematode, has been designated to this genus because of the severe necrotic lesions that are produced in the infection sites, usually in the cortex.

2. Symptoms

The above-ground symptoms of injury by *P. penetrans* to plants in the cool northern states often is a gradual decline or lack of plant vigor rather than any rapid, striking change. Tops of infected plants are stunted and chlorotic and wilt readily.

The below-ground symptoms include root lesions and root necrosis. Lesions are ideal infection sites for a host of fungi and bacteria. Under field conditions, root necrosis is caused by a complex of organisms rather than *P. penetrans* alone. The extent of host injury is variable, depending in part on the phenolic content of the roots and the subsequent amount of browning.

Fig. 5 Distribution of *P. penetrans* in the United States (Courtesy of Committee on Nematode Distribution, Society of Nematologists.)

The diminished root system of infected plants is a poor absorber of water and nutrients.

The effect of various nutrient solutions on pathogenicity of *P. penetrans* to pepper has been studied (Shafiee and Jenkins, 1962). Except for phosphorus-deficient plants, growth of all inoculated plants was retarded as a result of nematode infection. Plants deficient in nitrogen showed most marked reduced growth. There is strong evidence for nematode involvement in what has been known as cold or winter injury.

Low populations of *P. penetrans* (360 nematodes/per kilogram of soil) stimulated tomato fruit production, whereas populations of 2000 or more nematodes per kilogram of soil suppressed fruit production (Potter and Olthof, 1977).

3. Life History and Biology

The life cycle is simple and reproduction is sexual, taking 30-90 days for completion, depending on soil temperature. After fertilization, the female deposits eggs singly in roots or in the soil. The first molt occurs in the egg, and the second-stage juvenile, which hatches from the egg, molts three more times between intervals of feeding. The nematodes overwinter in soil primarily as adults and fourth-stage juveniles or as eggs in roots.

P. penetrans is most often found in sandy soils, and the nematodes migrate further in coarse than in fine-textured soil.

There is extensive morphologic variation in *P. penetrans* which is affected by food and environment (Mai et al., 1977). The stylet length was the only characteristic in the genus or species that was stable which could be used for taxonomic purposes (Tarjan and Frederick, 1978).

Populations from different geographical locations and from axenic cultures displayed variations in the shape of the stylet knobs and tail tip. Host plants, nitrogen levels, and light intensity also affected stylet length and gonad development (Mai et al., 1977).

4. Interactions with Other Disease Organisms and Nematodes

A synergism between *Verticillium albo-atrum* and *P. penetrans* in the etiology of wilt has been reported (McKeen and Mountain, 1960). *P. penetrans* is not pathogenic to eggplant at 4000 nematodes per pot; however, in the presence of *Verticillium* wilt, severe crop damage can occur. At low and intermediate levels of the wilt organism the incidence of wilt was increased in the presence of the nematode. In the presence of the fungus, nematode reproduction increased. On tomato and pepper, the final population of *P. penetrans* was higher in the presence of *Verticillium* than in its absence (Mountain and McKeen, 1962; Olthof and Reyes, 1969). In pepper, the damage caused by *P. penetrans* and *Verticillium* was additive (Olthof and Reyes, 1969).

The role of *P. penetrans* in increasing susceptibility of tomato to *V. albo-atrum* was studied using a split-plot technique. No increase in susceptibility of tomato to the fungus occurred when the nematode was separated from the fungus (Conroy et al., 1972). *P. penetrans* did not alter the resistence of tomato or pepper to *Fusarium* wilt. Histological studies have shown that the wilt-susceptible tomato harbored much mycelium in the stele, but the nematode avoided this area by restricting its feeding sites to the outer cortex (Mai et al., 1977).

When *Verticillium dahliae* was present with *P. penetrans* in soil, reproduction of the nematode increased in the roots of tomato and eggplant, but not in pepper (Mountain and McKeen, 1962). *P. penetrans* caused extensive cortical necrosis, with the lesions extending from the epidermis to the vascular tissues (Mountain and McKeen, 1965). The mycelia have a distinct affinity for the necrotic areas, and microsclerotia germinated rapidly. When *P. penetrans* and *M. incognita* coinhabited tomato Campbell 146 roots, development and reproduction of the root-lesion nematode was inhibited. The presence of *P. penetrans* caused a reduction of galling in tomato roots (Estores and Chen, 1972).

P. penetrans without associated organisms suppressed tomato yields by 73% (Hastings and Bosher, 1938). Differences in suitability for *P. penetrans* reproduction and in susceptibility to damage was found in various cultivars of tomato. Eighteen tomato cultivars supported *P. penetrans* populations, but on the most suitable cultivars there were up to four times as many nematodes in final populations. Valiant tomato, a poor host, showed greater susceptibility to damage, producing more and larger lesions than Rutgers, a suitable host (Mai et al., 1977).

C. Reniform Nematode (*Rotylenchulus reniformis* Linford and Oliveira)

Rotylenchulus reniformis parasitizes many dicotyledonous plants, but few monocotyledonous plants in tropical and semitropical areas. *R. reniformis* infects roots of many vegetable crops (Linford and Yap, 1940). Besides tomato and eggplant, such common vegetables as artichoke, beet, kale, chard, cauliflower, cabbage, Chinese cabbage, squash, cucumber, carrot, okra, lettuce, bean, pea, radish, potato, cowpea, and sweet corn are also attacked.

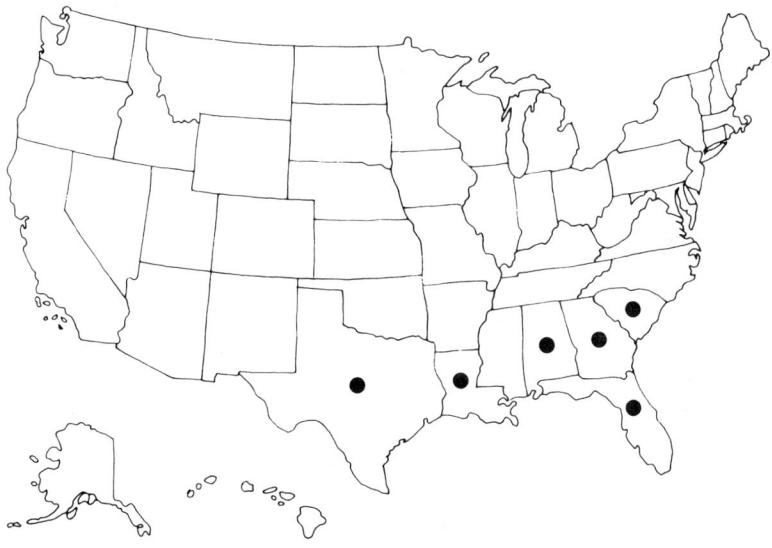

Fig. 6 Distribution of *R. reniformis* in the United States (Courtesy of Committee on Nematode Distribution, Society of Nematologists.)

1. Historical

The reniform nematode, *R. reniformis*, was observed for the first time in the roots of cowpea plants growing in pineapple fields of the Hawaiian Islands, and much of our information has come from this original work (Linford and Oliveira, 1940). The generic name was assigned because in some respects it was similar to spiral nematodes of the genus *Rotylenchus*. There are nine described species of *Rotylenchulus*. The only species that feeds on vegetables is *R. reniformis*.

The distribution of the reniform nematode in the continental United States is apparently limited to the southern and southeastern states, where it attacks many vegetable crops (Fig. 6).

2. Life History and Biology

An unusual characteristic of this nematode is that the juvenile can develop to preadult stages in water through a series of three superimposed molts without feeding. The first molt occurs in the egg. The young female penetrates the root cortex until partly or wholly embedded. After three days of feeding, the female begins to swell, assuming the characteristic kidney shape, and six days later begins to lay eggs in a gelatinous matrix attached to her body. Each female lays an average of 120 eggs, which hatch in about eight days. Males do not possess a stylet or esophagus and are therefore nonparasitic. Copulation takes place only after the female has penetrated a root and the body has begun to enlarge. Males remain vermiform and are found in the rhizosphere or in the gelatinous matrix. On eggplant, females complete development in 20-25 days and males in 12-16 days at 24-26°C (Singh and Khera, 1978).

Little is known about the biology of the reniform nematode on vegetable crops and its interrelationship with other organisms. Varied reports have been made on the feeding site of *R. reniformis*. Feeding takes place in the cortical region of the cowpea root (Linford and Oliveira, 1940). In tomato roots, the nematode fed near the stele of the root (Nath et al., 1969; Sivakumur and Seshadri, 1972). The cell adjacent to the nematode lips contained a feeding peg enclosing the stylet (Razak and Evans, 1976). Opposite the stylet tip a helical hyaline feeding tube was observed coiled in the feeding cell. This tube is thought to act as a filter through which the nematode obtains cell solutes.

Tomato is an excellent host for the reniform nematode, but differences in pathogenicity have been reported. *R. reniformis* is considered one of the major nematode pests of tomatoes in Puerto Rico, and significant reductions in tomato yield were produced in India with only 100 nematodes per plant (Rebois et al., 1973). Severe damage was reported to field-grown tomatoes in Colombia, South America, but in greenhouse tests no damage was observed when plants were inoculated with 1000 nematodes. Severe damage was observed only when plants were overwhelmed with 73,000 nematodes. Reniform nematode alone was more pathogenic to tomato than either *M. incognita* alone or both nematodes together (Román, 1978).

Control of *R. reniformis* by using resistant crops has been suggested. There is a high degree of resistance to *R. reniformis* in *L. esculentum*, PI 375937 (Rebois et al., 1973).

D. False Root-Knot Nematode (*Nacobbus aberrans* Thorne and Allen)

1. Historical

The false root-knot nematode, *N. aberrans*, was first found in England infesting tomato roots in a greenhouse, causing galls and associated with many lateral roots similar to those caused by *M. hapla*. *Nacobbus* often occurs together with *Meloidogyne* spp. and is therefore easily overlooked. Although generally not widely distributed in England, it appears to be an important nematode pathogen on tomatoes in Mexico (Marbán, personal communication).

2. Life History

There are four molts, the first occurring in the egg (Clark, 1967). Eggs hatch in 12-17 days in water at 15-18°C. Juveniles feed on the surface of the root or within the root. Entrance into the root is gained with repeated thrusts of the stylet against the cell wall until it is pierced. The nematode moves intercellularly, causing much cellular damage. The second molt may occur within the root or in the soil. The third-stage juvenile could enter and leave roots at will and molts in the root or in the soil. Fourth-stage juveniles in the cortex usually remain coiled and molt into adult males or females. The young females are long and slender with a well-developed vulva and move from the cortex to a position near the vascular cylinder. As eggs develop the bodies of the females swell into variable shapes, which are dependent on pressures applied by surrounding cells. Females produce an egg sac, which extends along the small channel formed by the entering nematode to the outside of the root surface.

Galls occur only in association with adult females. The galls contain a spindle of small cells in which starch grains are found. The life cycle on excised tomato roots was completed in 36 days at 25°C and 43 days at 20 or 30°C (Prasad and Webster, 1967). The sex ratio (males to females) was greater at 15 and 30°C than at 20 or 25°C.

N. aberrans was pathogenic at low population levels on pepper and tomato, causing significant losses in yield (Román, 1978).

E. Stubby-Root Nematode [*Paratrichodorus minor* (Colbran) Siddiqi]

1. Historical

The economic significance of members of the genus *Paratrichodorus* was first
demonstrated by Christie and Perry (1951), who showed that a species, later
named *P. minor*, was responsible for damage to roots of beets, celery, and
sweet corn in Sanford, Florida. The name *stubby root* was given to the
nematode because of the symptoms it produced on host plants. As a result of
recent taxonomic studies, *Trichodorus christiei* has been synonymized under
Paratrichodorus minor (Colbran) Siddiqi. There are presently about 45
species belonging to the genera *Trichodorus* and *Paratrichodorus* (Siddiqi,
1973).
P. minor is primarily found widespread in the southeastern United States
and has been reported in other states (Fig. 7).

2. Life History and Biology

P. minor is an ectoparasite that feeds mostly at root tips, but also along the
sides of young succulent roots. The primary effect of the feeding is malformation of lateral roots near the base of the stem and devitalization of the root
tip, which stops growth. Little is known about the life history. There is
evidence that populations build up rapidly. The nematode completes its life
cycle in 16-17 days at 30°C and in 21-22 days at 22°C (Rhode and Jenkins,
1957). At 20°C there was no increase in gravid females, and at 35°C no ju-

Fig. 7 Distribution of *P. minor* in the United States (Courtesy of Committee
on Nematode Distribution, Society of Nematologists.)

Fig. 8 Sweet corn roots damaged by *P. minor* (Courtesy of Dr. H. L. Rhoades, University of Florida, Sanford.)

veniles were recovered. Specimens that developed at the lower temperature were larger than those recovered at the higher temperature.

The effect of temperature and *P. minor* density on the growth and development of tomato has been studied (Högger, 1973). The tolerance limit, as determined by lateral root growth, was between 500 and 750 nematodes per plant. The optimal temperature for nematode reproduction was between 20 and 25°C.

P. minor is extremely pathogenic to eggplant and feeds vigorously on epidermal cells in the elongation and meristematic zones, causing browning and collapse of the epidermis and cessation of growth. Growth was drastically inhibited when eggplant was inoculated with 200 nematodes per plant (Schilt and Cohn, 1975).

Symptoms of injury by stubby-root nematodes on vegetable crops consist of general stunting of the entire plant. Yellowing in plants attacked by *P. minor* may be caused by a mineral deficiency (Christie and Perry, 1951). One month after inoculating sweet corn plants with 300 *P. minor* per pot, a yellow striping between leaf veins developed similar to that attributed to magnesium deficiency (Johnson, 1967). This symptom was associated with the presence of *P. minor* and *M. incognita* alone or combined. The influence of *P. minor* on the nutrient status of tomato has been studied by Maung and Jenkins (1959).

Feeding was restricted to the epidermal cells and outermost cortical cells (Rhode and Jenkins, 1957). The stylet extended 3-4 μm past the lip region, and the cell was punctured by a rasping motion rather than by direct thrust. During feeding the protoplast of the attacked cell shrank from the cell wall and a loss of meristematic tissue resulted. No definite root cap or region of elongation remained, and the region of mitosis was much smaller than usual. Protoxylem thickening developed almost to the apex of the root. Damage to roots was caused by decreased cell multiplication in root tips rather than by mechanical destruction of cells. The effect of damaged root tips on the development of the root system varies widely, depending on the kind of plant, the abundance of the nematodes, and the age of the plant when the feeding occurs. In most instances the root system is obviously and conspicuously abnormal (Fig. 8). Affected plants have a smaller root system and fewer and shorter rootlets than nonaffected plants. Usually, this is the "stubby-root" type of abnormality, although the result may sometimes be "coarse root." The root injury caused by stubby-root nematodes is deceptive because it often lacks discoloration, necrotic lesions, and other conditions usually associated with the disease. Sweet corn and certain other vegetable crops rarely show necrosis at any time.

F. Awl Nematode (*Dolichodorus heterocephalus* Cobb)

Tomatoes can be damaged by *D. heterocephalus*. Roots of young tomato plants were almost completely destroyed when grown in heavily infested soil (Perry, 1953).

III. NEMATODE PARASITES OF LEGUMINOUS VEGETABLES (BEANS, PEAS, COWPEAS)

The leguminous vegetables are important food crops from which a significant part of the world population obtains much of its dietary protein. They are cultivated over a wide range of environmental conditions and are eaten as tender green shoots and leaves, immature pods, or green or dry seeds. They are grown commerically to some extent in almost every state of the United States and in the majority of home gardens.

Twelve nematode genera consisting of 24 species have been encountered in association with the roots of beans. But the most frequent nematodes encountered of economic importance are species of *Meloidogyne* and *Pratylenchus* (Schwartz and Galvez, 1980).

In addition, stubby-root (*Paratrichodorus*), reniform (*R. reniformis*), sting (*Belonolaimus gracilis*), sugar beet (*Heterodera schachtii*), soybean cyst (*Heterodera glycines*), awl (*D. heterocephalus*), and stem and bulb (*Ditylenchus dipsaci*) nematodes can be damaging to these crops.

A. Root-Knot Nematodes (*Meloidogyne* spp.)

In severe infestations, yield losses of 50-90% with root-knot and 10-80% with root lesion nematodes have been reported (Schwartz and Galvez, 1980).

In the warmer climates the legumes are subject to attack by *M. incognita*, *M. javanica*, and *M. arenaria*. Less important species, such as *M. acronea*, *M. artiella*, *M. graminicola*, and *M. thamesi*, have been reported from beans and *M. hapla*, *M. ethiopica*, *M. africana*, and *M. kikuyensis* from cowpeas (also known as southern peas, blackeyed peas, blackeyed beans, field beans, or crowder peas) (Jensen, 1972).

The life cycle of the root-knot nematodes on beans is similar to that reported on tomato except that it proceeds at a faster rate.

In the southeastern United States, beans and cowpeas under commercial cultivation are usually planted in the spring of the year while temperatures are cool enough to suppress high nematode activity. A crop planted in late summer or early fall is subject to severe infection by root-knot nematodes. The general practice of many home gardeners is to plant beans throughout the season for a continuous supply of fresh green beans. Inevitably, bean losses are experienced as nematode reproduction increases. In southern Florida,

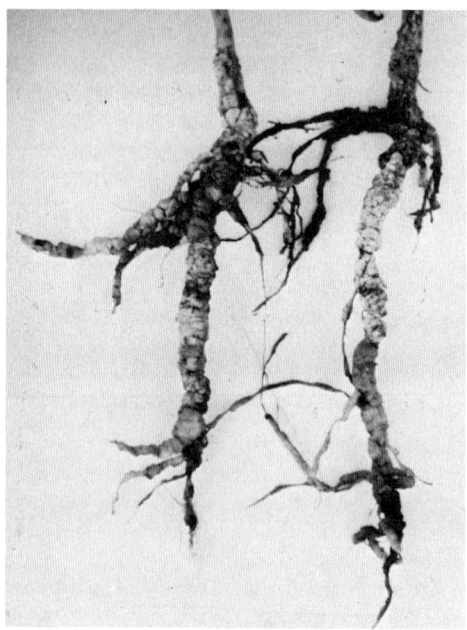

Fig. 9 Roots of beans, *Phaseolus vulgaris* cv. Provider, heavily infected with *M. incognita*.

Fig. 10 Field test of lima bean, *Phaseolus lunatus*, showing differential susceptibility to *M. incognita*. Allgreen (left) was extremely susceptible and stand was very poor, Jackson Wonder showed reduced growth, early maturation, and poor yield, L4116, a resistant breeding line selected from a cross Jackson Wonder × Nemagreen, showing vigorous plants with no chlorosis.

10-12 generations per year are possible depending on soil temperatures. This thermal relationship was demonstrated by Townshend (1937). The mean minimum temperature for complete development on beans is about 15°C, 3°C higher than found for root-knot nematodes on tomatoes. On beans, 22 days were required to complete a generation at a mean soil temperature of 27°C, whereas on tomato the rate of development was most rapid at 28°C, taking 23-26 days (Townshend, 1937).

   M. incognita is responsible for losses in beans (Fig. 9) and lima beans (Fig. 10) as high as 30% in Georgia, South Carolina, North Carolina, and Tennessee (Toler and Wester, 1966). In Kenya, as much as 60% yield loss was attributed to *M. incognita* and *M. javanica* infestations (Ngundo and Taylor, 1974).

   The reaction of bean roots to infection by root-knot nematodes is extremely variable, ranging from no galling to severe galling responses. On nongalled roots, females appear as strings of white pearls. Gall characteristics by *M. incognita*, *M. javanica*, and *M. arenaria* are similar. As discussed in Section II, galls produced by *M. hapla* can often be distinguished from those of the other root-knot nematode species by the matted appearance of the roots formed by a combination of relatively small galls and several lateral roots emanating from each gall.

Resistance

Resistance in beans has been reported only for *M. incognita*. Both pole-type and bush-type snap beans have been developed with resistance to this root-knot nematode species. Among the pole types are Alabama No. 1, Alabama No. 2, Alabama No. 8, Alabama No. 19, and Manoa Wonder. A nongalling bush type snap bean line, NemaSnap, was developed recently with a higher level of level of resistance to *M. incognita* than the pole types (Wyatt et al., 1983). Resistance was acquired through a wild type, PI 165426 (Fassuliotis et al., 1970). NemaSnap has performed successfully in heavily infested home gardens in South Carolina throughout the summer months. Yield in nontreated soil was twofold greater than the yield of a susceptible cultivar grown following soil fumigation (Wyatt et al., 1980).

   Lima bean cultivars with resistance are Hopi, Nemagreen, Western, and White Ventura. A breeding line, L-40215, is being developed at the U.S. Vegetable Laboratory primarily for the southern market. In tests conducted in California, this line showed resistance to *M. arenaria* (I. J. Thomason, personal communication).

   The resistance in the common bean is controlled by double-recessive genes (Blazey et al., 1964). Galling and nematode development and reproduction are independent phenomena, probably under separate genetic control. However, resistance expressed in the roots is not expressed in the stems. The juveniles invaded the hypocotyl during germination when bean seeds were planted too deeply in infested soil (Fassuliotis and Deakin, 1973). The resistance to root-knot nematode in lima beans is dominant and governed by a few genes (McGuire et al., 1961).

   "Iron" cowpea is resistant to root-knot nematodes (Webber and Orton, 1902). Some of the pioneering work on breeding for root-knot resistance demonstrated the feasibility of breeding for resistance (Orton, 1913). Using Iron as a resistant parent, a cultivar, Mississippi Silver, with resistance to *M. incognita*, *M. javanica*, and *M. arenaria*, was developed (Fig. 11) (Hare, 1967). A crowder-type cowpea, Worthmore, developed by crossing Mississippi Silver with Pink Eye Purple Hull, is resistant to *M. incognita*, *M. arenaria*,

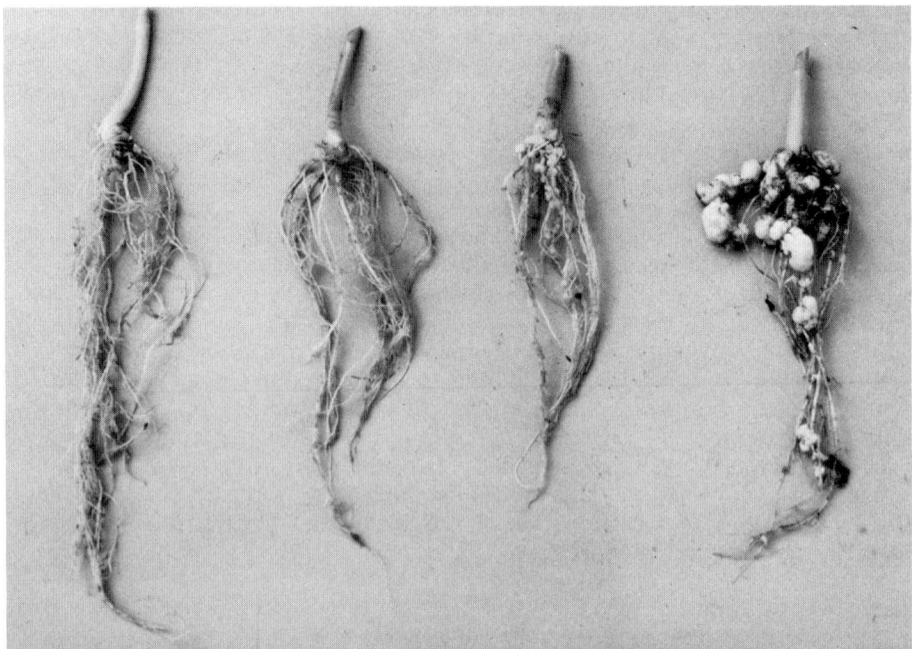

Fig. 11 Examples of effects on southern pea of the southern root-knot nematode (*M. incognita*) on plants of the susceptible breeding line CR 18-13-1 (right), the resistant Mississippi Silver (left), and the resistant F_1 of CR 18-13-1 × Mississippi Silver (center plants). The root of the susceptible plant is severely galled and only minimal disease and some bacterial nodulation are evident on the roots of the resistant plants. (After R. L. Fery and P. D. Dukes, 1980.)

M. hapla, and *M. javanica* (Gay, 1976; R. E. Motsinger, personal communication). The same single dominant gene, *Rk*, governs resistance to *M. incognita* and *M. javanica* (Fery and Dukes, 1980).

B. Root Lesion Nematodes (*Pratylenchus* spp.)

Beans are highly susceptible to *P. scribneri*. On Kentucky Wonder, pathogenicity tests in pot-grown plants indicated that the nematode increased nearly 75-fold over a 50 day period and affected the top growth. Lima bean Fordhook 52 is a poor host to *P. scribneri*. Attacked roots were able to overcome the nematode infection by eliciting a hypersensitive response that allowed only a minimal reproduction of the nematode (Thomason et al., 1976).

In greenhouse experiments bean Top Crop was an excellent host of *P. neglectus* (Olthof, 1979). After growing plants for four months inoculated with 50 nematodes per pot, 30,000 nematodes were recovered.

Life Cycle and Biology

A comprehensive review on the biology and ecology of *P. penetrans* has been published (Mai et al., 1977). In axenic studies on pea roots, the nematode penetrated the roots of plants behind the zone of elongation. The nematodes

probed the root epidermis within six hours after inoculation; however, most
of the nematodes penetrated the roots 12 hours after inoculation (Oyekan et
al., 1972). Invaded regions of the cortex showed orange discoloration. During feeding and reproduction activities, there was extensive breakdown of
the cortex. Gravid females probed the root endodermis with their stylet and
collapsed some endodermal cells. Tissue darkening was associated with phenolic content of the cells attacked or their ability to synthesize phenols after injury. The response in carrots and celery is similar to pea roots.

The life cycle is simple and reproduction is by bisexual cross fertilization.
In pea roots the male-female ratio was about 5:1. The first molt occurs in the
egg; the second-stage juvenile hatches from the egg and molts three more
times between intervals of feeding to become an adult. The complete life cycle
requires 30-92 days, depending upon temperature and host (Mai et al., 1977).

All juvenile stages and adults can invade roots, but more fourth-stage
juveniles and adults of both sexes penetrate than do second and third stages.
All stages of the nematode overwinter in the soil, and many eggs survive low
soil temperatures in roots. All stages can be found as deep as 60 cm (Dunn,
1972).

C. Reniform Nematode (*Rotylenchulus reniformis*)

The reniform nematode was discovered on the roots of cowpea plants (Linford
and Oliveira, 1940) (see Section II).

D. Spiral Nematode [*Rotylenchus robustus* (de Man) Filipjev]

Rotylenchus robustus is an ectoparasite that feeds on the roots of various
plants, although little is known of its actual host range. This nematode species has been recognized as an economic pest and causes severe injury to
peas (Seinhorst, 1954). The injury was characterized by reduced and decayed roots, accompanied by extreme stunting of top growth.

E. Awl Nematode (*Dolichodorus heterocephalus*)

The nematode has been found in large numbers in sunken lesions on hypocotyls of bean seedlings, and the seed coats of bean that failed to germinate
apparently because the nematodes had fed on the embryos. When seedlings
or young plants of bean were grown in soil that was heavily infested with *D.
heterocephalus*, the roots of the plants were almost completely destroyed
(Perry, 1953).

IV. NEMATODE PARASITES OF CUCURBITS

The Cucurbitaceae is a large botanical family supplying humans with food and
fiber. The major edible crops include cantaloupe, squash, pumpkin, cucumber, and watermelon. The major parasitic nematodes attacking these crops
are root-knot nematodes (*Meloidogyne* spp.) and reniform nematodes (*R. reniformis*).

A. Root-Knot Nematodes (*Meloidogyne* spp.)

Root knot is a severe disease of cucurbits caused by several species of *Meloidogyne. M. incognita* is the most common species, but *M. javanica, M. arenaria*, and *M. hapla* can be troublesome also. Infected plants may be chlorotic
and stunted. Heavily infected plants are weakened and eventually die before
producing marketable fruit.

Root knot has been severe frequently in southern states in late spring and fall cucumber crops. An average cantaloupe yield increase of 89% resulted from controlling *M. incognita* by soil fumigation (Bergeson, 1971).

Several screening programs have failed to identify any desirable amounts of resistance to the root-knot nematodes in the cucurbits (Fassuliotis, 1967, 1971).

Watermelon and cucumber react differently to various isolates of *M. hapla* (Thomason and McKinney, 1959; Winstead and Riggs, 1963). A virulent population of *M. hapla* appeared in Manitoba, Canada, attacking field-grown cucumbers (Zimmer and Walkof, 1968).

Despite considerable progress being made in developing root-knot resistance in other vegetable crops, none has been developed in the Cucurbitaceae because germ plasm with resistance to root-knot nematodes in the commercial-type cucurbits has not been found after extensive search.

The only known sources of resistance that may be potentially useful in a breeding program are found in noncultivated species of *Cucumis*, i.e., *C. anguria* (West India gherkin), *C. ficifolius*, *C. longipes*, *C. metuliferus*, and *C. heptadactylus* (Fassuliotis, 1970). Sexual incompatibility has prevented hybridization of these species with cantaloupe or cucumber.

The nature of the incompatibility between *C. metuliferus* and *C. melo* has been studied and the results indicate that the two species were compatible at a very low frequency, but no interspecific hybrid plants were grown (Fassuliotis, 1977a; b). The hybridization of *C. metuliferus* and *C. melo* has been reported, but to this date no hybrid melons have been released to plant breeders (Norton, 1969).

There were no differences in penetration by *M. incognita* juveniles of the resistant *C. metuliferus* and *C. ficifolius* and susceptible *C. melo* roots (Fassuliotis, 1970). Resistance in these two species was associated with arrested development of juveniles to adults, and increased stimulation toward maleness. Fewer juveniles developed into mature females in the resistant species because the small giant cells apparently did not supply sufficient nourishment to the developing nematode.

M. incognita juveniles entered bitter roots of cucumber cultivars and produced fewer juveniles than nematodes that entered nonbitter roots (Haynes and Jones, 1976). They attribute the reduction to an effect by cucurbitacins present in the bitter roots.

B. Reniform Nematode (*Rotylenchulus reniformis*)

The reniform nematode causes quality and yield loss in cantaloupe (Heald, 1980). *R. reniformis* penetrated the cortex of cantaloupe roots perpendicular to the stele and fed on the endodermis in young roots (Heald, 1975). During feeding, the pericycle on either side of the endodermal feeding cells became enlarged and granular in appearance. In older roots, the nematode fed directly on the pericycle. Pathogenicity experiments showed that the length of the life cycle was dependent on the soil temperature. Nematodes had penetrated roots within 24 hours equally at 21 and 27°C, but developed faster at 27°C. Eggs were deposited in a gelatinous matrix within 10 days and hatched three days later.

C. *Pratylenchus thornei* Sher and Allen

Watermelon was found to be a moderately good host of *Pratylenchus thornei* in the Northern Negev (Orion et al., 1979). Females, males, and juveniles invaded the cortical root tissue, leading to root collapse and reduction in shoot weight.

V. NEMATODE PARASITES OF CRUCIFEROUS VEGETABLES

Most crucifers grow best in a cool environment. In the southern United
States, cabbage, brussels sprouts, cauliflower, broccoli, turnip, kale, and
others, are grown as winter crops since they can withstand light frosts
(Walker, 1952). Fifteen nematode species of 11 genera have been associated
with crucifer cultivation. Two cyst nematodes, *Heterodera schachtii* and *H.
cruciferae*, are important parasites of crucifers.

A. Root-Knot Nematodes (*Meloidogyne* spp.)

M. incognita is apparently of little importance in spring cabbage in coastal
South Carolina and North Carolina. However, in late summer plantings, when
soil temperatures are favorable for nematode activity, a direct-seeded fall
crop can be severely damaged (Fassuliotis and Rau, 1969). *M. incognita* may
be a severe problem in cabbage transplant production in Georgia (Johnson et
al., 1979). *M. hapla* is present throughout muck and light mineral soils in
the commercial vegetable growing areas of Ontario and New York. Cabbage
and cauliflower yields were suppressed by 9% and 24%, respectively, and cauliflower curd maturity was delayed by three days at a population density of
18,000 per kilogram of soil (Olthof and Potter, 1972). No resistance to root-
knot nematodes was found in 11 cabbage cultivars (Winstead, 1959).

B. Sugar Beet Nematode (*Heterodera schachtii* Schmidt)

This nematode was first observed in 1859 on the roots of sugar beets in Germany and is now known to occur in most of the regions of the world where
sugar beets are grown. In the United States, *H. schachtii* occurs in most of
the sugar beet-growing areas from Michigan to California (Thorne, 1952).
Isolated infestations have been found in New York State, Canada, and Florida,
attacking crucifers (Fig. 12).

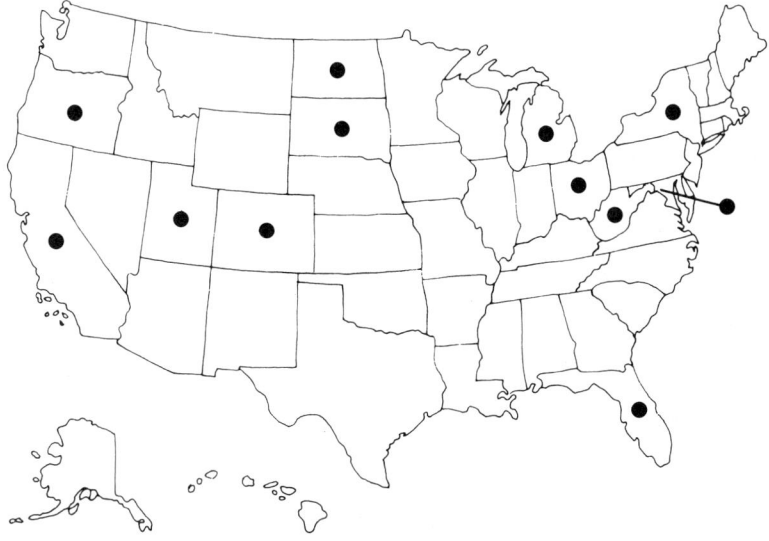

Fig. 12 Distribution of *H. schachtii* in the United States (Courtesy of Committee on Nematode Distribution, Society of Nematologists.)

H. schachtii has a very wide host range involving 218 plant species in 95 genera. Over 80% of the cruciferae are hosts to *H. schachtii*.

In New York State, the nematode causes significant yield losses to cabbage. Infected plants are severely stunted and chlorotic and most older leaves die. Marketable yields of cabbage were reduced by 21, 28, 46, and 54% by an initial population density of 9, 18, 34, and 64 eggs and juveniles per gram of soil (Abawi and Mai, 1980). Similar losses were reported by Olthof et al. (1974).

C. Cabbage Cyst Nematode (*Heterodera cruciferae* Franklin)

1. History

Heterodera cruciferae was first reported on brussels sprouts in California occurring either as a pure population in infested fields or as a mixed population with *H. schachtii* (Raski, 1952). Both nematodes alone or together cause significant yield losses. The known distribution of this nematode is confined to California.

H. cruciferae has many hosts in common with *H. schachtii* but its host range is more limited. *Brassica* seedlings infested with *H. cruciferae* produce extra lateral roots and stunted shoots with either reddened leaves or interveinal chlorosis (McCann, 1981). A marked reduction in curd quality of cauliflowers resulted when nematode densities exceeded 75 eggs per gram soil (Sykes and Winfield, 1966), and cabbage seedlings were severely stunted at a nematode density of 20 cysts per 100 g soil (McCann, 1981). No tolerance was found to *H. cruciferae* in six cultivars of brussels sprouts tested (Lear, 1971).

2. Life Cycles of H. schachtii *and* H. cruciferae

The life cycle of the sugar beet nematode and cabbage cyst nematode are similar. *H. schachtii* eggs within mature females and newly formed cysts hatch in the presence of growing host plants (Raski, 1950). Upon emerging from the cyst, the juveniles invade host plant roots, develop to maturity, and complete the life cycle in about 30 days.

The nematodes are amphimictic. Soon after fertilization, the adult females appear as white lemon-shaped bodies attached to the host roots. A gelatinous matrix surrounds the posterior end of the adult females in which are deposited a few to 130 eggs. After a short period of growth, the female dies and its cuticular body forms a tough, saclike protective cyst containing about 500 eggs.

Studies on the development of *H. cruciferae* on cabbage cv. Winningstadt in South Wales indicated that second-stage juveniles are stimulated to hatch by root diffusates from *Brassica* spp., but not from diffusates of other cruciferous hosts (Lewis, 1971). Two complete generations occurred between April and late September. Juveniles invading *Brassica* roots in September did not develop beyond the immature female stage by December. The gelatinous matrix in which eggs are deposited is produced by the uterine wall (Mackintosh, 1960), unlike root-knot nematodes in which it is produced by rectal glands.

D. Root Lesion Nematode (*Pratylenchus penetrans*)

Cabbage plants in Massachusetts have been reported to harbor high populations of *Pratylenchus penetrans*. Roots with dark brown lesions contained eggs, juveniles, and adults of this species (Acedo and Rohde, 1971). Injury is caused by mechanical and enzymatic destruction of cells, causing discoloration of nearby intact cells and subsequent accumulation of phenolic compounds.

E. Stunt Nematode (*Tylenchorhynchus brassicae* Siddiqi)

The stunt nematodes are root parasites that may feed ecto- or endoparasiti-cally. In India, poor areas in cabbage and cauliflower fields are always asso-ciated with high populations of *Tylenchorhynchus brassicae* (Khan, 1969).

1. Life History and Biology

T. brassicae penetrates throughout the cortical region. The nematodes are confined to the outer cortical layers with their bodies lying parallel to the longitudinal axis of the roots and the anterior portion of the body curved toward the conducting tissue.

Of 22 vegetables inoculated with 1000 *T. brassicae*, cabbage Golden Acre and cauliflower Snowball were the most suitable hosts and significant damage was incurred when the reproductive rate was five times above the initial popu-lation. The most favorable temperature for reproduction is 30°C and the most favorable soil moisture is 25-30% (Khan, 1969).

Several other nematodes are associated with cabbage cultivation. Patho-genicity studies showed that *Tylenchorhynchus dubius* and *Paratylenchus pro-jectus* reproduced on cabbage but caused no permanent damage (Brzeski, 1971a, b). *Belonolaimus longicaudatus* is extremely virulent to direct-seeded cabbage (Fig. 13) in Florida (Rhoades, 1971).

2. Interactions with Other Organisms

M. incognita did not reduce the resistance of cabbage cultivars to *Fusarium* wilt, but only increased the susceptibility of cabbage cultivars to low levels of the fungus (Fassuliotis and Rau, 1969).

Rhizoctonia solani and *T. brassicae* are often associated with roots of cabbage and cauliflower. *R. solani* alone suppressed the emergence of cauli-flower seedlings by 81%. When the two organisms occurred together, 97% of the seed failed to germinate (Khan and Saxena, 1969).

Fig. 13 Response of cabbage to chemical control of *B. longicaudatus*. (Cour-tesy of H. L. Rhoades.)

VI. NEMATODE PARASITES OF ROOT AND BULB CROPS

A. Onion and Garlic

1. *Historical*

Ditylenchus dipsaci and *Pratylenchus penetrans* are important nematode parasites of onion and garlic. *D. dipsaci* is also an important parasite of broad
bean, peas, carrots, parsnip, and potato. The nematode also attacks many
field crops and ornamental bulbs. It has been known in Europe since 1877.
In Germany it was referred to as Kroefziekte and in France as "*maladie vermiculaire de l'oignon*" (Thorne, 1961). In the United States, the disease was
first recognized in onions from Canastota, New York (Steiner, 1931), and
appears to be limited to that state. The term *onion bloat* was coined from the
symptoms exhibited by infested onions raised from sets (Chitwood et al.,
1940). *D. dipsaci* is widespread from New Hampshire to California and is a
potential pathogen in onion- and garlic-growing areas (Fig. 14).

Much of the garlic grown in California is infested with *D. dipsaci*, which
is spread by planting infected sections of the bulb called cloves.

In certain regions of Russia as much as 60% of the onion or garlic crop
was lost due to *D. dipsaci* (Kirjanova and Krall, 1980).

2. *Life Cycle and Biology*

The life cycle of *D. dipsaci* on onion seedlings is completed in 19-23 days
with four molts and four juvenile stages (Yuksel, 1960). The first molt takes
place within the egg, resulting in a second-stage juvenile. Upon hatching
from the egg, about two days after oviposition, development to the adult stage
is rapid, taking 4-5 days. Mating is necessary for reproduction and the fe-

Fig. 14 Distribution of *D. dipsaci* in the United States (Courtesy of Committee
on Nematode Distribution, Society of Nematologists.)

male begins to oviposit four days after the final molt. Females lay 8-10 eggs each day for 25-50 days for a total of 200-500 eggs. Males and females live for 45-73 days.

D. dipsaci feeds on stems, leaves, and bulbs. It is rarely found in the soil, but can be recovered from tissue remnants of bulb scales, leaves, and shoots. The nematode is very tolerant to desiccation, surviving anabiotically in the dried state for several years. During heavy rains, the nematodes become active and climb upward along the plant and penetrate the shoot stomates. In dry weather, the nematodes penetrate directly into the bulb (Kirjanova and Krall, 1980).

D. dipsaci penetrates the germinating seed and lives in the parenchyma tissue beneath the hilum or in the cotyledon (Newhall, 1943). The nematode caused much destruction of cortical parenchyma as it migrated through the tissue. Damage to onion seedlings was attributed to direct feeding activity. Cells were punctured and the contents were sucked out until the cells were so weakened that they collapsed under the pressure of the nematode's body. There is little evidence that the nematode is capable of inflicting much mechanical injury. Most of the damage results from the effects of its salivary secretion on the surrounding tissues. One effect of this secretion is to dissolve the middle lamella, causing the cells near the region of invasion to separate. In storage tissue these separated cells become rounded, which usually gives the structure a whitish appearance and a mealy texture. This separation of the cells is caused by a pectinase found in the salivary secretion produced by *D. dipsaci*. The dissolution of the middle lamella is necessary for the parasite's survival. In plants where this does not occur *D. dipsaci* will not reproduce and fails to survive.

D. dipsaci is composed of a complex of races that are differentiated from each other by their respective host ranges. Many races are host specific, whereas others have a wide host range. More than 20 biological races have been identified and named after their preferred hosts or those plants on which they were first detected (Sturhan, 1971). Morphological variations among the races are extremely minor, being based mostly on body length, body width, and position of the vulva. Races appear to be able to interbreed freely. Hybrids of onion race (female) × white clover race (male) have been produced (Eriksson, 1974). A giant triploid race that occurs on broad bean exists that cannot be crossed with the other races.

Symptoms of injury differ for different plants and different parts of the same plant. Infested seedlings become twisted, enlarged, and deformed and frequently die in severely infested areas of the field. Infested plants become stunted and irregular in form, and their leaves are short and thickened and frequently contain brown or yellowish spots (Fig. 15). As the season progresses, the foliage collapses and a softening of the bulb begins at the neck and gradually proceeds downward; the scales become soft and pale gray. In dry seasons, the bulbs become desiccated and very light in weight. There is an increase in number of split and double onions in infested fields. If adequate moisture is present, a soft rot completes the process of destruction and is accompanied by an offensive odor. Many of the diseased onions decay at the base and contain varied types of secondary invaders. The decay is usually caused by bacteria, fungi, and onion maggots in *D. dipsaci*-infested fields. The nematode has been eradicated from loose garlic cloves with a hot water, formalin, and detergent treatment (Lear and Johnson, 1962).

Fig. 15 Onions infected with *D. dipsaci* (Courtesy of Dr. W. F. Mai, Cornell University, New York.)

3. *Root Lesion Nematode* (Pratylenchus penetrans)

Onion yields in northern Indiana were severely reduced by *P. penetrans* (Bergeson, 1962). The nematode enters the delicate root system and begins to reproduce soon after the rootlets are formed. Eggs were recovered from roots of seedlings seven days after the seed germinated. The nematode caused long, narrow, greyish, and opaque root lesions initially. A single nematode caused a lesion 1-3 mm long and one to two cells wide. Commercial cultivars with suitable resistance to be used as an alternative to chemical control have not been found (Bergeson, 1962). The wild onion, *Allium canadense*, was resistant. Onion seedlings showed characteristic symptoms of injury where more than 100 *P. penetrans* were present per gram of root (Ferris, 1962).

B. Sweet Potato

Nematodes found associated with sweet potato (*Ipomoea batata*) cultivation are *Meloidogyne* spp., *R. reniformis*, *B. longicaudatus*, *Tylenchus* spp., *Helicotylenchus* spp., *Pratylenchus* spp., *Criconemella* spp., *Longidorus* spp., *Tylenchorhynchus* spp., *Aphelenchoides* spp., *Ditylenchus destructor*, *Pratylenchus coffeae*, *P. zeae*, *Radopholus similis*, *Tylenchorhynchus martini*, *Hemicriconemoides cocophilus*, *Xiphinema americanum*, *Paraphelenchus* spp., and *Aphelenchus* spp. (Román, 1978). Of these nematodes, however, the pathogenicity of only *Meloidogyne* spp. and *R. reniformis* has been established on sweet potato.

1. *Root-Knot Nematodes* (Meloidogyne spp.)

Sweet potatoes are subject to heavy damage by *M. incognita* (Nielsen and Phillips, 1973). Injury caused by root-knot nematodes results in malformed roots, stunted plants, scabbing, and cracking.

Sweet potatoes are planted in the field as sprouts cut from bedding roots. The growing season of the sweet potato is four to six months from the transplanting of the sprouts to digging. Infection by root-knot juveniles can occur throughout the growing season. There are three main sites of entry for the juveniles. Many juveniles enter through the root tips, resulting in a cessation of root elongation and development of small inconspicuous knots. This infection limits the enlargement of the root system and suppresses plant growth. The second infection site is through the rupture caused by lateral rootlets growing from the enlarging root. The juveniles migrate to the vascular cylinder, where the nematodes feed and develop to maturity. This invasion of lateral roots is probably responsible for root cracking, common to root-knot nematode-infected sweet potatoes. The third infection court is through cracked surfaces. When the roots crack, the exposed tender tissue is readily penetrated by the juveniles. This invasion occurs late in the growing season (Nielsen et al., 1957).

Feeding juveniles stimulate the production of giant cells, abnormal xylem, hyperplastic parenchyma, and cork (Fig. 16). Swollen roots, harvested late in the season, often contain females and egg masses surrounded by cork (Krusberg and Nielsen, 1958).

M. hapla, M. javanica, and *M. arenaria* also attack the sweet potato but usually cause only a light galling response, and juveniles do not usually develop to maturity. Sweet potato appears to be a nonhost of *M. arenaria* (Giamalva et al., 1963).

M. incognita was transmitted from infected bedding roots to the field as sprouts; however, the nematodes did not multiply sufficiently to impair plant productivity or to infect storage roots in large numbers (Nielsen and Phillips, 1973).

Interaction with Other Organisms Root-knot nematodes had no significant adverse effects on *Fusarium* wilt development in either wilt-resistant or wilt-susceptible cultivars (Giamalva et al., 1962).

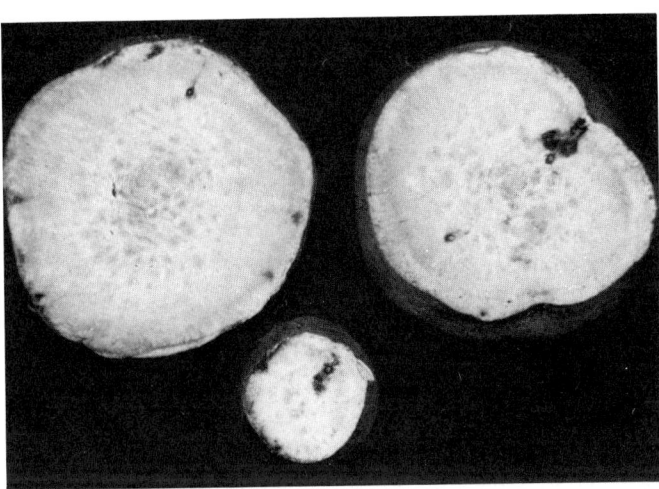

Fig. 16 Slices of sweet potato, *Ipomoea batatas*, rhizomes showing internal damage from infection by *M. incognita* (Courtesy of P. D. Dukes.)

Resistance Resistance to root-knot nematodes has been incorporated into several sweet potato cultivars. The major sources of resistance originally came from the Jersey types and Tinian, a wild type, which was not resistant to all races of *M. incognita*. The genetic base for root-knot nematode resistance was broadened from open pollinated seedlings from polycross nurseries of parental types developed through mass selection. Recent releases of sweet potato germ plasm from the U. S. Vegetable Laboratory are being used by plant breeders throughout the world for improving local cultivars. Many of these lines are highly resistant to *M. javanica*, *M. hapla*, and several races of *M. incognita* (Jones et al., 1980).

2. *Reniform Nematode* (Rotylenchulus reniformis)

Sweet potato is highly susceptible to reniform nematode in Puerto Rico (Román, 1978), and this nematode species is a potential problem in Louisiana (Martin, 1960). Plants inoculated with nematodes at various dosage levels had a sparse root system that was necrotic and devoid of feeder roots. The medium weight of roots inoculated with 2000 nematodes was 10.1 g, with 8000 nematodes it was 3.8 g, and without nematodes, 22.5 g. A possible source of resistance was found in the cultivar Goldrush, which did not support good reproduction of the reniform nematode (Martin et al., 1966).

The reniform nematode completed its life cycle in the sweet potato feeder roots in 17-21 days after planting. Infected roots had thick-walled cells in the cortex and endodermis, pitted tracheids, and giant cells (Brathwaite and Duncan, 1974).

C. Table Beets

In New York State, *Heterodera schachtii* causes significant yield losses to table beets. Marketable yields were suppressed by 23, 25, 42, and 54% by an initial population density of 9, 18, 34, and 64 eggs and juveniles per gram of soil (Abawi and Mai, 1980).

D. Carrots (*Daucus carotae*)

Almost 90 species of nematodes have been associated with carrot cultivation (Brzeski, 1970; Philis, 1976).

1. *Root-Knot Nematode* (Meloidogyne hapla)

M. hapla is a serious pathogen of carrots in Michigan. Carrots infected with *M. hapla* often exhibit galling, forking, stubbing and fasciculation of the roots (Fig. 17) and yield reduction. Only 55% of infected carrots were suitable for fresh market, compared with 97% of those grown in nematode-free soil (Slinger and Bird, 1978). Three breeding lines, MSU 872, 5988, and 1475, are tolerant to *M. hapla* (Yarger and Baker, 1981).

2. *Spiral Nematodes* (Rotylenchus spp.)

Rotylenchus uniformis Thorne is a parasite of carrots in The Netherlands (Seinhorst and Kuniyasu, 1969). The tolerance of carrots to the nematode is much lower at 8°C than at 18°C. Tolerance to the nematode was considered to be due to increased root density during development of the plants (Seinhorst and Kozlowska, 1977).

Rotylenchus laurentinus was found in carrot roots in sandy soils of the South Adriatic coast of Italy. The nematode is semiendoparasitic on the

Fig. 17 Carrot infected by *M. hapla*. (Courtesy of H. J. Jensen.)

Fig. 18 Distribution of *B. longicaudatus* in the United States. (Courtesy of Committee on Nematode Distribution, Society of Nematologists.)

Fig. 19 Corn field severely infested with sting nematode, *B. longicaudatus.*

feeder roots and causes lesions and cavities in epidermal and cortical tissues
(Vovlas et al., 1980).

Rotylenchus robustus (de Man) Filipjev is a destructive disease of carrots
characterized by a severe reduction in roots and stunting of top growth (Kuiper and Drijfhout, 1957).

E. Celery

The pathogenic nematodes of celery are *B. longicaudatus, D. heterocephalus,*
M. hapla, Paratylenchus hamatus, and *Longidorus vineacola.*

1. Historical

The genus *Belonolaimus* was established with *B. gracilis* Steiner as the type
species (Steiner, 1949). Since that time, *B. longicaudatus, B. euthychilus,*
B. maritimus, and *B. nortoni* have been described (Rau, 1958; 1963).

The sting nematodes appear to be indigenous to the light sandy soils of
the southeastern Coastal Plain of the United States (Fig. 18). These nematodes are highly pathogenic, and if only one specimen is found in a soil sample it indicates that crop damage can be expected.

B. longicaudatus is the most widespread species commonly found on vegetable crops. The sting nematode has a wide host range, and is highly pathogenic to sweet corn (Fig. 19), lima bean, southern pea, cabbage, onion, and
carrot. In the Sanford area of Florida it is a major pest of celery.

Symptoms of injury caused by sting nematodes include poor stands, erratic uneven plant growth, and severe stunting.

2. Life History and Biology

Little is known about the life history of the sting nematode. They are ectoparasites, feeding mostly on the surface of the roots at the root tips and along
the sides of succulent roots and other subterranean parts of a plant. Root
tips are stunted and stubby, and taproots are often malformed. Three populations of *B. longicaudatus* from North Carolina and one from Georgia have

been studied (Robbins and Barker, 1973). The North Carolina sting nematode populations were very similar in host suitability comparisons. When the Tifton, Georgia, population was compared with the North Carolina populations, significant differences were detected. A major difference was the ability of the Georgia nematodes to reproduce on cucumber compared with the inability of the North Carolina nematodes to do so. Also, the reproductive rate of the Georgia nematodes was greater than that of the North Carolina nematodes. Cucumber, okra, and watermelon are nonhosts for the North Carolina sting nematode populations. Carrot is a good host of populations of *B. longicaudatus* from North Carolina, Georgia, and Florida. Most of the carrots produced in these states are grown in fine sandy soils.

In the Sanford area of Florida, celery can be severely damaged by the awl nematode, *D. heterocephalus* Cobb (Steiner, 1949). This nematode also causes considerable damage to beans, sweet corn, and other vegetable crops in wet, sandy soils in Florida.

Pathogenicity of *D. heterocephalus* was first demonstrated on celery plants in greenhouse experiments (Tarjan et al., 1952). Plants growing in pots infested with 1000 nematodes were visibly stunted three weeks after planting. At the end of the experiment, roots from these plants weighed only about one-fourth as much as those from controls. Experiments indicated that *D. heterocephalus* is a factor in "red root" disease of celery.

Symptoms of injury include either stubby root or coarse root, or both, often with considerable discoloration and tissue destruction. *D. heterocephalus* is an ectoparasite that feeds on the root tips, along the sides of succulent roots, and at the base of the hypocotyl causing discoloration and necrosis. Yield of celery from infested fields may be suppressed 50% or more by *D. heterocephalus*.

Fig. 20 Parsley, *Petroselinum crispum*, infested with pin nematode, *Paratylenchus* spp. on a truck farm in Staten Island, New York. The heavy infestation was located in the two central rows causing dwarfed chlorotic plant growth.

M. hapla is a severe pest of celery in New York State (Starr and Mai, 1976). Another nematode, *Paratylenchus hamatus* Thorne and Allen, was found in the Northeast associated with severe stunting and chlorosis of celery and parsley (Fig. 20) when field populations of the nematode were as high as 10,000 per pound of soil (Lownsbery et al., 1952).

Damage to celery by *Longidorus vineacola* in Israel has been reported (Cohn and Auscher, 1971).

3. Interaction with Other Organisms

A common soil-borne fungus, *Trichoderma viride* Pers. ex Fr., normally a weak parasite, when present with *P. penetrans*, caused a greater reduction in plant growth than either organism alone. More *P. penetrans* were recovered from plants grown in soil infested with *T. viride* than in soil not infested with the fungus. However, populations of *P. penetrans* were reduced when an undescribed species of *Trichoderma* was mixed with soil containing the nematode (Mai et al., 1977).

Celery exhibited severe necrosis when *M. hapla* and *Pythium polymorphon* coinhabited the roots than when either pathogen occurred alone (Starr and Mai, 1976).

F. Lettuce

1. Root-Knot Nematode (Meloidogyne hapla)

M. hapla is an important pathogen of head lettuce in organic soils in New York. Infected lettuce plants are frequently left unharvested in the field because they fail to produce heads of marketable size.

The pathogenicity of *M. hapla* to lettuce was influenced by inoculum level, age of plant at inoculation, and temperature (Wong and Mai, 1973a, b). Top weight of Minetto lettuce was suppressed 32% when two-week-old lettuce plants each were inoculated with five egg masses. Higher inoculum levels did not further suppress top weight significantly. Inoculation at seeding suppressed top growth more than inoculation of one, two, or three-week-old seedlings. *M. hapla* suppressed growth more at the intermediate (21.1°C night and 26.7°C day) than at the low (15.5°C night and 21.1°C day) or high (26.7°C night and 32.2°C day) temperature regimes. Growth, development, and reproduction of *M. hapla* were significantly reduced as temperature decreased.

Low oxygen limits movement, invasion, and hatch of *M. hapla*. These activities can be manipulated by the level of CO_2 present (Wong and Mai, 1973c).

M. hapla juveniles need 627 degree days to develop to egg-laying females (Starr and Mai, 1976).

2. Spiral Nematode (Rotylenchus robustus)

Rotylenchus robustus has been found in Rhode Island, New Jersey, New York, and California. Stunting, yellowing, and loss in yield of lettuce have been associated with large populations of *R. robustus* (Lear et al., 1969). In field studies, yields of marketable heads of lettuce were doubled when *R. robustus* was controlled with nematicides. These nematodes also reproduce on cabbage, red beet, spinach, broccoli, celery, brussels sprouts, and cauliflower.

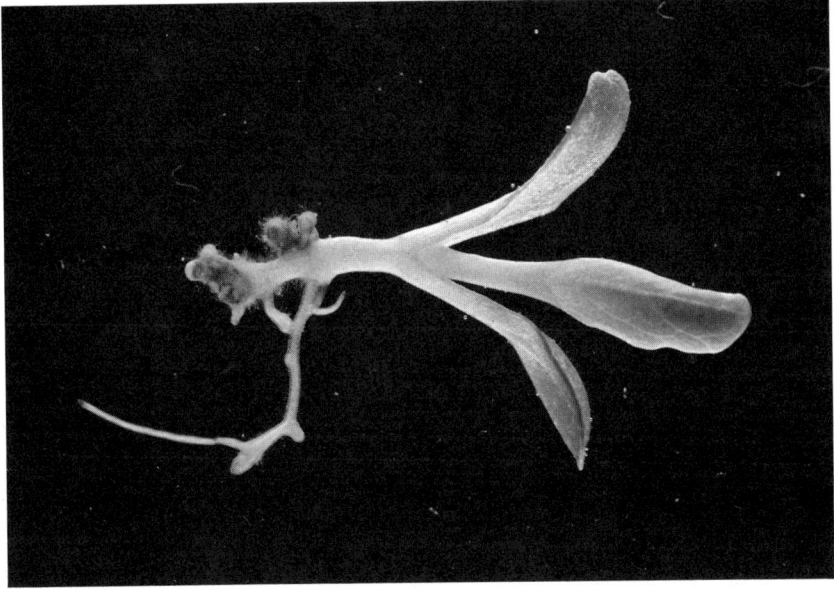

Fig. 21 Roots of lettuce damaged by *Longidorus africanus* (Courtesy of Dr. J. D. Radewald, University of California, Riverside.)

3. Needle Nematode (Longidorus africanus Merny)

Longidorus africanus was first reported to be a pathogen on head lettuce in the Imperial Valley of Southern California (Radewald et al., 1969). The disease was evident at the time of plant emergence, with distribution of diseased seedlings spotty in some fields and uniform in others. The cotyledon leaves of the stunted seedlings were cupped downward as if they were wilting, though there was ample moisture for normal growth. Leaves of infected plants were a grayish green color and the margins were often chlorotic.

Taproots of affected seedlings failed to elongate normally and were short with terminal swellings (Fig. 21). Necrotic areas on the swollen root tips were sometimes evident, and it appeared that the meristematic tissue had been destroyed. Diseased plants frequently compensated for the loss of the main taproot by lateral root proliferation; often these newly formed roots were similarly affected.

VII. CONTROL

The major objective in control of plant-parasitic nematodes is economic; to increase quantity and quality of vegetable crops. The procedures always involve reduction of nematode populations, or making them less infective than they would be otherwise.

Because the objective is economic, care and calculation are needed to be sure that expenses, including money, labor, and loss of income due to changes in customary farm practices, do not exceed the value of probable benefits. Because of farming risks, such as unfavorable weather, diseases, pests, or a poor market, the expected benefits should exceed the expenses by a ratio of at least 3:1, and preferably more. Current philosophy regarding control measures is centering around integrated pest management systems involving the minimal use of chemicals and maximum use of cultural methods.

A. Crop Rotation

In crop rotation for control of *Meloidogyne* species, susceptible crops are rotated with immune or resistant crops. Usually, the susceptible crop is the most profitable and the rotation crops less profitable.

For example, tomato is a profitable crop, but susceptible to all common species of *Meloidogyne*. After a tomato crop is harvested, the root-knot nematode population in the soil is usually high. A second crop of tomatoes or other susceptible vegetables would be severely damaged.

Studies are being conducted at the Coastal Plain Experiment Station, Tifton, Georgia, to develop intensive cropping systems utilizing vegetable crops and agronomic crops in rotations to manage populations of plant-parasitic nematodes. Two-year intensive cropping systems including sweet corn-soybean-wheat-soybean-spinach and turnip-peanut-cucumber-turnip-cucumber-soybean were more effective than an annual intensive cropping system including turnip-field corn-southern peas in suppressing populations of *M. incognita* race 1 (Johnson et al., 1976). Many other combinations of crops may also be used (Johnson et al., 1981). The first and most important requirement for effective nematode control is that the rotation crops be immune or highly resistant to the species and race of *Meloidogyne* present. As more nematode-resistant vegetables are developed, they will be integrated into cropping systems to further suppress nematode populations usually provided by the rotation crop.

The second requirement for effective management of nematode populations is weed control. Many nematode species reproduce on weeds, and the presence of weeds in a field can prevent success of a rotation.

With good selection of resistant rotation crops and good weed control, the reduction is enough so that the profitable susceptible crop can be grown next in succession or the next season.

Compared with the root-knot nematodes, cyst nematodes are difficult to control because the eggs in the body of the dead females are protected from desiccation and other adverse conditions.

Crop rotation is the most practical method of controlling *H. schachtii*. Cereals, corn, alfalfa, clover, potatoes, tomatoes, and other important crops are poor to nonhosts of *H. schachtii*. Susceptible vegetable crops, e.g., table beets, cabbage, cauliflower, broccoli, brussels sprouts, turnips, rutabagas, and radishes can be grown once every three years if the infestation is confined to a few areas in a field, but only once every four or five years if the infestation is extensive.

To control *H. cruciferae*, rotations with noncruciferous plants for three or more years should be sufficient to produce one good *Brassica* crop.

Rotation with small grains, *Brassica*, or other nonhost crops for three or four years gives control of *D. dipsaci* if clean fallow is practiced and weeds are eliminated, but much depends on the host preference of the particular nematode race present.

The increase of *P. penetrans* on different cultivars of such important crops in the Northeast as beans, tomatoes, and potatoes varied considerably. Because resistance is an economical and desirable control measure, there is a need for a greatly expanded research program in the Northeast to find or develop cultivars with resistance and tolerance to *P. penetrans*. Resistant cultivars are needed not only for cash crops, but for cover crops and plants used in gardens by homeowners. Crop rotation cannot be used effectively to control *P. penetrans* unless suitable resistant and tolerant crops are available. Present information indicates that crop rotations will be of no value because of the wide host range of *P. penetrans*.

The most extensive studies of crop rotations on population dynamics of ectoparasitic nematodes on vegetable crops have been conducted in Georgia (Johnson et al., 1981). Data from a four-year study on six intensive cropping sequences [turnip (T)-corn (C)-snapbean (SNB), T-peanut (P)-SNB, T-C-T, T-P-T, SNB-soybean (SB)-cabbage, and T-cucumber (CU)-southern pea (SP)-T] indicated that numbers of *Paratrichodorus minor* were < 20/150 cm^3 soil on all crops except corn. When corn appeared in the cropping sequences numbers of *P. minor* increased to 80-150 cm^3 soil. Numbers of *Criconemella ornata* remained low (< 60/150 cm^3 soil) on vegetable crops, but increased on corn and peanuts. Numbers of *C. ornata* were consistently suppressed to lowest levels (< 15/150 cm^3 soil) in the T-CU-SP-T cropping sequence.

Although crop rotation is probably the most effective single control measure for most ectoparasitic nematode problems at present, it is by no means completely satisfactory. Crop rotation will not control all ectoparasitic nematodes that injure vegetable crops because of the overlapping host suitability. Rotations that include poor host or nonhost crops can effectively reduce the damage caused by ectoparasitic nematodes and, where all crops have a low per-acre value, this may be the only economical method of ectoparasitic nematode control.

B. Cultural Methods

1. Land Preparation

The value and subtle effects of land preparation methods on nematode populations have never been fully investigated. Tilling land in preparation for planting reduces nematode populations, but provides little long-term crop protection. The destruction of nematode host plants after harvest reduces nematode populations and has been recommended for nematode control for many years. Tillage practices (disk harrow, subsoil-bed, subsoil-plant, and moldboard) did not signficantly affect nematode populations in intensive cropping systems including several vegetable crops (Johnson et al., 1981). Bed construction and planting, allowing maximum utilization of dried surface soil, may minimize nematode damage.

2. Flooding

Where water is abundant and fields are level, it is sometimes possible to control nematodes by flooding the land to a depth of 10 cm or more for several months. Flooding does not necessarily kill the eggs and juveniles of root-knot nematodes. Flooding does not inhibit "infection" and reproduction of nematodes on any plants that grow while the field is flooded. Flooding experiments are best evaluated by measuring yields of a subsequent crop, not by survival of juveniles. Juveniles may survive flooding but not be infective (Taylor and Sasser, 1978).

3. Fallow

Summer and winter fallow are effective in controlling nematodes. The best nematode control is obtained when fallow is maintained during hot, dry weather—either by withholding irrigation (which is easily accomplished in arid areas) or by preventing plant growth by repeated plowing, harrowing, or use of nonselective herbicides. In some areas of the United States, the *Meloidogyne* population of fields can be reduced by plowing at intervals of two to four weeks during the dry season. Plowing exposes eggs and juveniles to desiccation, and many in the upper layers of soil are killed. This practice may be sufficient to increase yield of a subsequent susceptible vegetable crop.

4. Time of Planting

In long-season, warm climate regions, certain vegetable crops may be planted during the winter months and harvested before injury occurs in the spring. For example, potato production in the San Joaquin Valley of California is based on this method of farming. Plantings are made early in the year and harvested before June without visible infestation. If allowed to remain in the soil a month or two longer, the entire crop would be unsalable. Also, in Georgia, turnips for greens are planted in February and harvested in April. During this time root-knot nematodes are less active and cause less damage than in warmer months.

5. Trap Crops

The use of trap crops has been evaluated as a method for controlling endoparasitic nematodes. A highly susceptible, quick growing crop is planted on the land, allowed to grow a short time, then plowed under or otherwise destroyed. Control is based on the fact that the nematodes, after entering roots, become sedentary parasites incapable of locomotion and are destroyed with the plants before reproducing. The method looks good in theory, but has not been very effective in practice. It involves careful timing, for if the crop grows too long the infestation may be increased instead of decreased. A new idea demonstrated that root-knot nematode juveniles freely entered the roots of some resistant plants like *Crotalaria spectabilis* but failed to survive (Barrons, 1940). The effect of *Tagetes* (marigold) species on populations of *Meloidogyne* species is highly variable (Belcher and Hussey, 1977). The effectiveness depends on the combination of species of *Tagetes* and *Meloidogyne*. Reduction of populations of *M. incognita* by *Tagetes patula* was primarily due to an antagonistic or trap crop effect. Second-stage juveniles entered roots and died because of a hypersensitive necrotic root reaction. In 12 weeks, populations in flats in a greenhouse were reduced 97% by marigold, but only 70% by peanut (*Arachis hypogaea*), which is also a nonhost of *M. incognita*.

C. Resistant Cultivars

The least expensive and often the only practical means of controlling root-knot nematodes is by the use of resistant cultivars. Resistant cultivars not only prevent nematode damage to the resistant crop, but lower population densities and extend protection to the following susceptible crop.

Resistant cultivars usually provide nematode control and increase crop yields equal to that obtained from use of nematicides provided resistance is adequate for the types of nematodes inhabiting the soil (Wyatt, et al., 1980). For low-value crops, use of a resistant cultivar often is the only practical control method because of the current high cost of nematicides.

"Resistance-breaking" races or pathotypes of root-knot nematodes frequently limit the continued usefulness of a resistant cultivar. Rotations must be designed to maximize the use of resistant cultivars in order to limit the development of resistance-breaking races. Resistance-breaking races of several nematode species have evolved on a number of resistant cultivars when monocropping was practiced (Sauer and Giles, 1959). The selection of resistance-breaking types has been minimized by planting resistant tomatoes with susceptible tomatoes and after five years resistance was still retained (Giles and Hutton, 1958).

Several lines of cowpea, sweet potato, and tomato have multiple root-knot nematode resistance and are more valuable for planting over a larger geographical area than crops having resistance to only one species, e.g., beans.

D. Chemical

Nematicides are one of the most important and reliable means of controlling a wide variety of nematodes. Commercial use of nematicides began about 1945 and has increased rapidly during the last 15 years. The principal use of nematicides is for control of nematode populations in soil before planting annual crops. Suitable application methods have been devised for applying about 25 highly effective nematicides. These chemicals are used on approximately 1.7 million acres, which is less than 1% of the cultivated agricultural land in the United States. Nematicides are used to protect a number of highly susceptible vegetables from nematode attack. The recent action by EPA to suspend the use of DBCP (1,2-dibromo-3-chloropropane) on all crops except pineapple in Hawaii (1977) and EDB (ethylene dibromide) (1983) reduces the number of effective and economical soil fumigants that can be used for control of nematodes on vegetable crops. Nematicides available on world markets are listed in Table 2, but many of these are not registered for use on vegetable crops.

There are two broad types of nematicides for control of nematodes on vegetable crops.

Soil fumigants: The older nematicides are liquids, which are injected beneath the soil surface. These nematicides volatilize to produce vapors that kill nematodes. Vapors from soil fumigants diffuse through the soil, are dissolved in soil water, and enter nematode bodies through the cuticle.

Nonfumigants and systemics: Newer nematicides are water soluble and are called "nonfumigants." Nonfumigant nematicides are distributed through the soil by percolation of water and also enter nematode bodies through the cuticle. The newest types of nematicides are systemic. They can be taken up by plants through the roots after application to the soil; or through foliage after spray application, then translocated to roots and kill nematodes feeding on the plants. Presumably they are taken up by nematodes in feeding, but might also enter nematode bodies through the cuticle in contact with soil and plant tissue.

E. Biological

The literature on biological control of plant-parasitic nematodes has been reviewed (Sayre, 1971; Webster, 1972). Both authors discuss various soil organisms antagonistic to nematodes.

Predators include fungi, nematodes, turbellarians, enchytraeids, insects, and mites. Parasites include viruses, protozoa, bacteria, and fungi. A great diversity of microorganisms attack nematodes under natural conditions in agricultural soils. However, biological control of nematodes by natural enemies per se is not yet practical under field conditions.

Table 2 Nematicides Available on World Markets*

Common name	Registered trade name and manufacturer	Chemical name	Formulation and classification
Aldicarb	(TEMIK) Union Carbide Corp.	2-Methyl-2-(methylthio) propionaldehyde O-(methylcarbamoyl) oxime	Granular nematicide-insecticide
Carbofuran	(FURADAN) Niagara Chem. Div., FMC Corp.	2,3-Dihydro-2,2-dimethyl-7-benzofuranyl methylcarbamate	Granular and flowable nematicide-insecticide
Chloropicrin	Great Lakes Chem. Corp.	Trichloronitromethane	Liquid fumigant nematicide-insecticide
DBCP[a]	(FUMAZONE) Dow Chem. Co. (NEMAGON) Shell Dev. Co.	1,2-Dibromo-3-chloropropane	Emulsifiable and nonemulsifiable liquid nematicide
1,3-D	(TELONE) Dow Chem. Co.	1,3-Dichloropropene and related chlorinated C_3 hydrocarbons	Liquid fumigant nematicide
DD Mixture	(DD) Shell Dev. Co. (VIDDEN-D) Dow Chem. Co.	1,3-Dichloropropene and 1,2-dichloropropane and related chlorinated hydrocarbons	Liquid fumigant nematicide

EDB[b]	(DOWFUME W-85) Dow Co.	Ethylene dibromide	Liquid fumigant nematicide
Ethoprop	(MOCAP) Mobil Chemical Co.	O-Ethyl S,S-dipropyl phosphorodithioate	Granular or emulsifiable liquid nematicide-insecticide
Fenamiphos	(NEMACUR) Chemagro Agri. Div., MOBAY Chem. Co.	Ethyl 4-(methylthio)-m-tolyl isopropylphosphoramidate	Granular or emulsifiable liquid nematicide
Fensulfothion	(DASANIT) Chemagro Agri. Div., MOBAY Chem. Co.	O,O-diethyl O-[p-methylsulfinyl) phenyl] phosphorothioate	Granular nematicide
MBR	(DOWFUME MC-2) Dow Chem. Co.	Methyl bromide	Gas fumigant nematicide
Oxamyl	(VYDATE) E. I. duPont de Nemours and Co.	Methyl-N',N'-dimethyl-N-[(methylcarbamoyl)oxy]-1-thioxamidate	Granular or emulsifiable liquid nematicide-insecticide

[a]DBCP has recently (1977) been removed from the market and manufacture discontinued.
[b]EDB was suspended by the Environmental Protection Agency – 1983. Stocks in dealers' hands will be recalled. Any EDB in growers' hands may be used until September 1984.

ACKNOWLEDGMENTS

Thanks are due to Dr. K. R. Barker, Dr. J. D. Gay, Professor Amegda Overman, Dr. J. N. Sasser, Dr. D. R. Sumner, and Mr. A. L. Taylor for reviewing the chapter and making many helpful suggestions with reference to its organization and content, and to nematologists and other scientists who have contributed directly and indirectly to the information presented.

Special thanks are due to Mrs. Carol Ward, who did the preliminary and final typing of the manuscript, and to Mr. Tom Girardeau, for preparing the figures on nematode distribution.

REFERENCES

Abawi, G. S., and Mai, W. F. (1980). Effects of initial population densities of *Heterodera schachtii* on yield of cabbage and table beets in New York State. *Phytopathology 70*: 481-485.

Acedo, J. R., and Rohde, R. A. (1971). Histochemical root pathology of *Brassica oleracea capitata* L. infected by *Pratylenchus penetrans* (Cobb) Filipjev and Schuurmans-Stekhoven (Nematoda: Tylenchidae). *J. Nematol. 3*: 62-68.

Barrons, K. C. (1940). Root-knot resistance in beans. *J. Hered. 31*: 35-38.

Belcher, J. V., and Hussey, R. S. (1977). Influence of *Tagetes patula* and *Arachis hypogaea* on *Meloidogyne incognita*. *Plant Dis. Rep. 61*: 525-528.

Bergeson, G. B. (1962). A report on the testing of onion varieties for resistance to lesion nematodes (*Pratylenchus penetrans*). *Plant Dis. Rep. 46*: 535-536.

Bergeson, G. B. (1971). Response of muskmelon to fumigation for control of *Meloidogyne incognita* following one year of a nonhost crop. *Plant Dis. Rep. 55*: 55-56.

Bergeson, G. B., Van Gundy, S. D., and Thomason, I. J. (1970). Effect of *Meloidogyne javanica* on rhizosphere microflora and Fusarium wilt of tomato. *Phytopathology 69*: 1245-1249.

Berkeley, M. J. (1855). [Vibrio forming excrescences on the roots of cucumber plants]. *Gardeners Chron.* Apr., 1855, p. 220.

Bird, A. F. (1959). Development of the root-knot nematodes *Meloidogyne javanica* (Treub) and *Meloidogyne hapla* Chitwood in the tomato. *Nematologica 4*: 31-42.

Bird, A. F. (1972). Quantitative studies on the growth of syncytia induced in plants by root-knot nematodes. *Int. J. Parasitol. 2*: 157-170.

Bird, A. F. (1974). Plant response to root-knot nematode. *Ann. Rev. Phytopathol. 12*: 69-85.

Bird, A. F. (1978). Root-knot nematodes in Australia. *CSIRO Aust. Div. Hort. Res. Tech. Pap. No. 2*, pp 1-26.

Blazey, D. A., Smith, P. G., Gentile, A. G., and Miyagawa, S. T. (1964). Nematode resistance in the common bean. *J. Hered. 55*: 20-22.

Bowman, P., and Bloom, J. R. (1966). Breaking the resistance of tomato varieties to Fusarium wilt by *Meloidogyne incognita* (abstract). *Phytopathology 56*: 871.

Brathwaite, C. W. D., and Duncan, E. J. (1974). Development and histopathology of *Rotylenchulus reniformis* in sweet potato roots. *Trop. Agric. (Trinidad) 51*: 437-441.

Brzeski, M. W. (1970). Plant parasitic nematodes associated with carrot in Poland. *Roczniki Nauk Rolniczych Ser.* E TOM 1 Z. *1*: 93-102.

Brzeski, M. W. (1971a). Nematodes associated with cabbage in Poland. IV. Experiment with *Tylenchorhynchus dubius* Bütschli. *Zeszyty Prob. Post. Nauk Rolniczych 121*: 107-111.

Brzeski, M. W. (1971b). Nematodes associated with cabbage in Poland. V. Experiments with *Paratylenchus projectus* s.1. *Zeszyty Prob. Post. Nauk Rolniczych 121*: 113-119.

Chitwood, B. G. (1949). Root-knot nematodes. I. A revision of the genus *Meloidogyne* Goeldi 1887. *Proc. Helm. Soc. Wash. 16*: 90-104.

Chitwood, B. G., Newhall, A. G., and Clement, R. L. (1940). Onion bloat or eelworm rot, a disease caused by the bulb or stem nematode, *Ditylenchus dipsaci* (Kühn) Filipjev. *Proc. Helm. Soc. Wash.* 7: 44-51.

Christie, J. R. (1936). Development of root-knot nematode galls. *Phytopathology 26*: 1-22.

Christie, J. R. (1946). Host-parasite relationships of the root-knot nematode, *Heterodera marioni*. II. Some effects of the host. *Phytopathology 36*: 340-352.

Christie, J. R., and Albin, F. E. (1944). Host-parasite relationships of the root-knot nematode, *Heterodera marioni*. I. The question of race. *Proc. Helm. Soc. Wash. 11*: 31-37.

Christie, J. R., and Perry, V. G (1951). A root disease of plants caused by a nematode of the genus *Trichodorus*. *Science 113*: 491-493.

Clark, S. A. (1967). The development and life history of the false root-knot nematode, *Nacobbus serendipiticus*. *Nematologica 13*: 91-101.

Cohn, E., and Auscher, R. (1971). Seasonal occurrence of *Longidorus vineacola* on celery in Israel and its control. *Israel J. Agric. Res. 21*: 23-25.

Cohn, Eli, and Minz, G. (1960). Nematodes and resistance to Fusarium wilt in tomatoes. *Hassadeh 40*: 1347-1349.

Conroy, J. H., Green Jr., R. J., and Ferris, J. M. (1972). Interaction of *Verticillium albo-atrum* and the root lesion nematode, *Pratylenchus penetrans*, in tomato roots at controlled inoculum densities. *Phytopathology 62*: 362-366.

Crop Losses Committee. (1971). Estimated crop losses due to plant-parasitic nematodes in the United States. Society of Nematologists. *Special Publication No. 1. Supplement to J. Nematol. 4.*

Dhawan, S. C., and Sethi, C. L. (1978). Interrelationship between root-knot nematode *Meloidogyne incognita* and little leaf of brinjal. *Indian Phytopathol. 39*: 55-63.

Dropkin, V. H. (1969). Cellular responses of plants to nematode infections. *Ann. Rev. Phytopathol.* 7: 101-122.

Dunn, V. H. (1972). Importance of depth in soil, presence of host roots, and role of eggs as compared to vermiform stages in overwintering of *Pratylenchus penetrans* at Ithaca, New York (abstract). *J. Nematol.* 4: 221-222.

Eriksson, K. B. (1974). Intraspecific variation in *Ditylenchus dipsaci* I. Compatibility tests with races. *Nematologica 20*: 147-162.

Estores, R. A., and Chen, T. A. (1972). Interactions of *Pratylenchus penetrans* and *Meloidogyne incognita* as coinhabitants in tomato. *J. Nematol.* 4: 170-174.

Fassuliotis, G. (1967). Species of *Cucumis* resistant to the root-knot nematode, *Meloidogyne incognita acrita*. *Plant Dis. Rep. 51*: 720-723.

Fassuliotis, G. (1970). Resistance of *Cucumis* spp. to the root-knot nematode, *Meloidogyne incognita acrita*. *J. Nematol. 2*: 174-178.

Fassuliotis, G. (1971). Susceptibility of *Cucurbita* spp. to the root-knot ne-
    matode, *Meloidogyne incognita*. *Plant Dis. Rep. 55*: 666.
Fassuliotis, G. (1975). Regeneration of whole plants from isolated stem paren-
    chyma cells of *Solanum sesymbriofolium*. *J. Am. Soc. Hort. Sci. 100*:
    636-638.
Fassuliotis, G. (1977a). Self-fertilization of *Cucumis metuliferus* Naud. and
    its cross-compatibility with *C. melo* L. *J. Am. Soc. Hort. Sci. 102*:
    336-339.
Fassuliotis, G. (1977b). Embryo culture of *Cucumis metuliferus* and the in-
    terspecific hybrid with *C. melo* (abstract). In *4th Ann. Colloquium
    Plant Cell and Tissue Culture*. Sept. 6-9, 1977. Ohio State Univ. Co-
    lumbus, Ohio.
Fassuliotis, G. and Deakin, J. R. (1973). Stem galls on root-knot nematode
    resistant snapbeans. *J. Am. Soc. Hort. Sci. 98*: 425-427.
Fassuliotis, G., Deakin, J. R., and Hoffman, J. C. (1970). Root-knot ne-
    matode resistance in snapbeans: breeding and nature of resistance.
    J. Am. Soc. Hort. Sci. 95: 640-645.
Fassuliotis, G., Nelson, B. V., and Bhatt, D. P. (1981). Organogenesis in
    tissue culture of *Solanum melongena* cv. Florida Market. *Plant Sc. Ltr.
    22*: 119-125.
Fassuliotis, G. and Rau, G. J. (1969). The relationship of *Meloidogyne in-
    cognita acrita* to the incidence of cabbage yellows. *J. Nematol. 1*: 219-
    222.
Ferris, J. M. (1962). Some observations on the number of root lesion nema-
    todes necessary to cause injury to seedling onions. *Plant Dis. Rep. 46*:
    484-485.
Fery, R. L. and Dukes, P. D. (1980). Genetics of root-knot resistance in
    the southern pea (*Vigna unguiculata* (L.) Walp). *J. Am. Soc. Hort.
    Sci. 105*: 671-674.
Gay, J. D. (1976). 'Worthmore' Southernpea. *HortScience 11*: 621-622.
Giamalva, M. J., Martin, W. J., and Hernandez, T. P. (1962). Relationship
    of root-knot nematodes to the development of Fusarium wilt in the sweet
    potato (abstract). *Phytopathology 52*: 733.
Giamalva, M. J., Martin, W. J., and Hernandez, T. P. (1963). Sweet potato
    varietal reaction to species and races of root-knot nematodes (*Meloi-
    dogyne Phytopathology 53*: 1187-1189.
    Sacc. in brinjal (*Solanum melongena* L.). *Indian Phytopathol. 23*: 587-
    589.
Giles, J. E., and Hutton, E. M. (1958). Combining resistance to the root-
    knot nematode, *Meloidogyne javanica* (Treub) Chitwood, and Fusarium
    wilt in hybrid tomatoes. *Aust. J. Agric. Res. 9*: 182-192.
Goswami, B. K., and Chenulu, V. V. (1974). Interaction of root-knot nema-
    tode, *Meloidogyne incognita*, and tobacco mosaic virus in tomato. *Indian
    J. Nematol. 4*: 68-80.
Goswami, B. K., Singh, D. V., Sethi, and Gupta, J. N. (1970). Studies on
    asssociation of root-knot nematodes, *Meloidogyne incognita* (Kofoid and
    White) Chitwood and *Sclerotium rolfsii* Sacc. in brinjal (*Solanum melongena*
    L.). *Indian Phytopathol. 23*: 587-589.
Hare, W. W. (1967). A combination of disease resistance in a new cowpea,
    Mississippi Silver. *Phytopathology 57*: 460.
Harrison, A. L. (1960). Breeding of disease resistant tomatoes with special
    emphasis on resistance to nematodes. In *Proc. Plant Science Seminar*,
    Campbell Soup Company, Camden, New Jersey, pp. 57-78.
Hastings, R. J., and Bosher, J. E. (1938). A study of the pathogenicity of

the meadow nematode and associated fungus *Cylindrocarpon radicicola* Wr. *Can. J. Res. 16:* 225-229.

Haynes, R. L., and Jones, C. M. (1976). Effects of the Bi Locus in cucumber on reproduction, attraction and response of the plant to infection by the southern root-knot nematode. *J. Am. Hort. Sci. 101:* 422-424.

Heald, C. M. (1975). Pathogenicity and histopathology of *Rotylenchulus reniformis* infecting cantaloup. *J. Nematol. 7:* 149-152.

Heald, C. M. (1980). Effect of reniform nematode on cantaloup yields. *Plant Dis. Rep. 64:* 282-283.

Högger, C. H. (1973). Preferred feeding site of *Trichodorus christiei* on tomato roots. *J. Nematol. 5:* 228-229.

Jenkins, W. R., and Coursen, B. W. (1957). The effect of root-knot nematodes, *Meloidogyne incognita acrita* and *M. hapla*, on Fusarium wilt of tomatoes. *Plant Dis. Rep. 41:* 182-186.

Jensen, H. J. (1972). Nematode pests of vegetable and related crops. In *Economic Nematology* (J. M. Webster, ed.). Academic, New York, pp. 377-408.

Johnson, A. W. (1967). Ecological associations between certain species of plant parasitic nematodes in mixed populations. Ph.D. Thesis, North Carolina State University, Raleigh.

Johnson, A. W., McCarter, S. M., Jaworski, C. A., and Williamson, R. E. (1979). Chemical control of nematodes and soil-borne plant pathogenic fungi on cabbage transplants. *J. Nematol. 11:* 138-144.

Johnson, A. W., Dowler, C. C., Glaze, N. C., and Sumner, D. R. (1983). Effects of intensive cropping systems and pesticides on nematodes and crop yields. U.S. Dept. Agr., ARS, ARR-S-14 pp 36.

Jones, A., Dukes, P. D., Schalk, J. M., Mullen, M. A., Hamilton, M. G., Paterson, D. R., and Boswell, T. E. (1980). W-71, W-115, W-119, W-149, and W-154 sweet potato germplasm with multiple insect and disease resistances. *HortScience 15:* 835-836.

Johnson, A. W., Sumner, D. R., Dowler, C. C., and Glaze, N. C. (1976). Influence of three cropping systems and four levels of pest management on populations of root-knot and lesion nematodes (abstract). *J. Nematol. 8:* 290-291.

Khan, A. M. (1969). Studies on plant parasitic nematodes associated with vegetable crops in Uttar Pradesh. *Final Tech. Report. Grant No. FG-In-225, Proj. No. A7-CR-65.* Aligarh Muslim Univ. Aligarh, India.

Khan, M. W, and Saxena, S. K. (1969). Effect of *Rhizoctonia solani* and *Tylenchorhynchus brassicae* on the emergence of cauliflower seedlings. *All Indian Nematology Symposium, 1967,* The Indian Agricultural Research Institute, New Delhi - 12.

Kirjanova, E. S., and Krall, E. L. (1980). Plant-Parasitic Nematodes and their Control. Nauka Publishers, Leningrad Section, Leningrad. Transl. Amerind Publishing Co., PVT. LTD., New Delhi.

Krusberg, L. R.. and Nielsen, L. W.. (1958). Pathogenesis of root-knot nematodes to the Puerto Rico variety of sweetpotato. *Phytopathology 48:* 30-39.

Kuiper, K., and Drijfhout, E. (1957). Bestrijding van het worterlaaltje *Hoplolaimus uniformis* Thorne, 1949, bij de Teelt van peen. *Overdruck uit Meded. Landbouwhogesch. Opzoekingsst. Staat Gent 22:* 419-426.

Lear, B. (1971). Reproduction of the sugarbeet nematode and the cabbage root nematode on several cultivars of Brussels sprouts. *Plant Dis. Rep. 55:* 1005-1006.

Lear, B., and Johnson, D. E. (1962). Treatments for eradication of *Ditylenchus dipsaci* in cloves of garlic. *Plant Dis. Rep. 46*: 635-639.

Lear, B., Johnson, D. E., and Miyagawa, S. T. (1969). A disease of lettuce associated with an ectoparasitic nematode, *Rotylenchus robustus. Plant Dis. Rep. 53*: 952-954.

Lewis, S. (1971). Observations on the development of the Brassica cyst eelworm. *Plant Pathol. Lond. 20*: 144-148.

Linford, M. B., and Oliveira., J. M. (1940). *Rotylenchulus reniformis* nov. gen., n. sp., a nematode parasite of roots. *Proc. Helm. Soc. Wash. 7*: 35-42.

Linford, M. B., and Yap, F. (1940). Some host plants of the reniform nematode in Hawaii. *Proc. Helm. Soc. Wash. 7*: 42-44.

Loof, P. A. A. (1960). Taxonomic studies on the genus *Pratylenchus* (Nematoda). *T. Pliekt. 66*: 29-90.

Loof, P. A. A. (1978). The genus *Pratylenchus* Filipjev, 1936. (Nematoda: Pratylenchidae): A review of its anatomy, morphology, distribution, systematics and identification. *Swedish Univ. of Agric. Sci. Upsala.*

Lownsbery, B. F., Stoddard, E. M., and Lownsbery, J. W. (1952). *Paratylenchus hamatus* pathogenic to celery. *Phytopathology 42*: 651-653.

Mackintosh, G. M. (1960). The morphology of the Brassica root eelworm *Heterodera cruciferae* Franklin, 1945. *Nematologica 5*: 158-165.

Mai, W. F., Bloom, J. R., and Chen, T. A. (1977). Biology and ecology of the plant-parasitic nematode *Pratylenchus penetrans. The Pennsylvania State University College of Agriculture, Agricultural Experiment Station Bulletin 815.* 64 pp.

Martin, W. J. (1960). The reniform nematode may be a serious pest of the sweetpotato. *Plant Dis. Rep. 44*: 216.

Martin, W. J., Birchfield, Wray, and Hernandez, T. P. (1966). Sweetpotato varietal reaction to the reniform nematode. *Plant Dis. Rep. 50*: 500-502.

Maung, O. and Jenkins, W. R. (1959). Effects of root-knot nematode, *Meloidogyne incognita acrita* Chitwood, 1949, and a stubby-root nematode *Trichodorus christiei* Allen, 1957 on the nutrient status of tomato, *Lycopersicon esculentum* hort. var. Chesapeake. *Plant Dis. Rep. 43*: 791-796.

McCann, J. (1981). Threshold populations of *Heterodera cruciferae* and *H. schachtii* causing damage to cabbage seedlings. *Plant Dis. Rep. 65*: 264-266.

McGuire, D. C., Allard, R. W., and Harding, J. A. (1961). Inheritance of root-knot nematode resistance in lima beans. *Proc. Soc. Hort. Sci. 78*: 302-307.

McKeen, C. D., and Mountain, W. B. (1960). Synergism between *Pratylenchus penetrans* (Cobb) Filipjev and Stekhoven and *Verticillium alboatrum* R & B in eggplant wilt. *Can. J. Bot. 38*: 789-794.

Mountain, W. B., and McKeen, C. D. (1962). Effect of *Verticillium dahliae* on the population of *Pratylenchus penetrans. Nematologica 7*: 261-266.

Mountain, W. B., and McKeen, C. D. (1965). Effects of transplant injury and nematodes on incidence of Verticillium wilt of eggplant. *Can. J. Bot. 43*: 619-624.

Nath, R. P., Sinha, B. K., Haider, M. G. (1979). Nematodes of vegetables in Bihar. II. Combined effect of *Meloidogyne incognita* and *Ozonium texanum* var. *parasiticum* on germination of eggplant. *Indian J. Nematol. 6*: 177-179.

Nath, R. P., Swarup, G., and Rama Rao, G. V. S. V. (1969). Studies on the reniform nematode, *Rotylenchulus reniformis* Linford and Oliveira, 1940. *Indian Phytopathol. 22*: 99-104.

Newhall, A. G. (1943). Pathogenesis of *Ditylenchus dipsaci* in seedlings of *Allium cepa*. *Phytopathology 33*: 61-69.

Ngundo, B. W., and Taylor, D. P. (1974). Effects of *Meloidogyne* spp. on bean yields in Kenya. *Plant Dis. Rep. 58*: 1020-1023.

Nielsen, L. W., and Phillips, D. V. (1973). Relevance of *Meloidogyne incognita*-infected sweetpotato bedding roots on sprout transmission of the nematode to the succeeding crop. *Plant Dis. Rep. 57*: 291-294.

Nielsen, L. W., Sasser, J. N., and Krusberg, L. R. (1957). Root-knot nematode problems on sweet potatoes. In *Proc. Shell Nematology Workshop*, Columbia, SC, pp. 44-52.

Norton, D. C. (1969). Incorporation of resistance to *Meloidogyne incognita acrita* into *Cucumis melo* (abstract). *Proc. Assoc. Agr. Workers 66*: 212.

Olthof, T. H. A. (1979). The use of beans and Kentucky blue grass for rearing *Pratylenchus neglectus, P. projectus* and *Helicotylenchus digonicus. Can. J. Plant Sci. 59*: 897-898.

Olthof, T. H. A., and Potter, J. W. (1972). Relationship between population densities of *Meloidogyne hapla* and crop losses in summer maturing vegetables in Ontario. *Phytopathology 62*: 981-986.

Olthof, T. H. A. and Potter, J. W. (1977). Effects of population densities of *Meloidogyne hapla* on growth and yield of tomato. *J. Nematol. 9*: 296-300.

Olthof, T. H. A., Potter, J. W., and Peterson, E. A. (1974). Relationship between population densities of *Heterodera schachtii* and losses in vegetable crops in Ontario. *Phytopathology 64*: 549-554.

Olthof, T. H. A., and Reyes, A. A. (1969). Effect of *Pratylenchus penetrans* on *Verticillium* wilt of pepper (abstract). *J. Nematol. 1*: 21-22.

Orion, D., Krikun, J., and Sullami, M. (1979). The distribution, pathogenicity, and ecology of *Pratylenchus thornei* in the Northern Negev. *Phytoparasitica 7*: 3-9.

Orton, W. A. (1913). The development of disease resistant varieties of plants. *Comp. Rend. Rapp. IV Conf. Int. Genetique. Paris 1911*: 247-265.

Oyekan, P. O., Blake, C. D., and Mitchell, J. E. (1972). Histopathology of pea roots axenically infected by *Pratylenchus penetrans. J. Nematol. 4*: 32-35.

Perry, V. G. (1953). The awl nematode, *Dolichodorus heterocephalus*, a devastating plant parasite. *Proc. Helm. Soc. Wash. 20*: 21-27.

Philis, J. (1976). Occurrence and control of nematodes affecting carrot crops in Cyprus. *Nematol. Medit. 4*: 7-12.

Potter, J. W., and Olthof, T. H. A. (1974). Yield losses in fall-maturing vegetables relative to population densities of *Pratylenchus penetrans* and *Meloidogyne hapla. Phytopathology 64*: 1072-1081.

Potter, J. W., and Olthof, T. H. A. (1977). Analysis of crop losses in tomato due to *Pratylenchus penetrans. J. Nematol 9*: 290-295.

Prasad, S. K., and Webster, J. M. (1967). Effect of temperature on the rate of development of *Nacobbus serendipiticus* in excised tomato roots. *Nematologica 13*: 85-90.

Prot, J. C., and Netscher, C. (1979). Influence of movement of juveniles on detection of fields infested with *Meloidogyne*. In *Root-knot Nematodes (Meloidogyne species) Systematics, Biology and Control* (F. Lamberti and C. E. Taylor, eds.). Academic, New York, pp. 193-203.

Radewald, J. D., Osgood, J. W., Mayberry, K. S., Paulus, A. O., and
    Shibuya, F. (1969). *Longidorus africanus* a pathogen of head lettuce in
    the Imperial Valley of southern California. *Plant Dis. Rep. 53:* 381-384.
Raski, D. J. (1950). The life history and morphology of the sugar-beet ne-
    matode, *Heterodera schachtii* Schmidt. *Phytopathology 40:* 135-152.
Raski, D. J. (1952). The first record of the Brassica-root nematode in the
    United States. *Plant Dis. Rep. 36:* 438-439.
Rau, G. J. (1958). A new species of sting nematode. *Proc. Helm. Soc. Wash.*
    25: 95-98.
Rau, G. J. (1963). Three species of *Belonolaimus* (Nematoda: Tylenchida)
    with additional data on *B. longicaudatus* and *B. gracilis*. *Proc. Helm.*
    Soc. Wash. 30: 119-128.
Razak, A. R., and Evans, A. A. F. (1976). An intracellular tube associated
    with feeding by *Rotylenchulus reniformis* on cowpea root. *Nematologica*
    22: 182-189.
Rebois, R. V., Eldridge, B. J., Good, J. M., and Stoner, H. J. (1973).
    Tomato resistance and susceptibility to the reniform nematode. *Plant*
    Dis. Rep. 57: 169-172.
Rhoades, H. L. (1971). Chemical control of the sting nematode, *Belonolaimus*
    longicaudatus, on direct-seeded cabbage. *Plant Dis. Rep. 55:* 412-414.
Robbins, R. T., and Barker, K. R. (1973). Comparisons of host range and
    reproduction among populations of *Belonolaimus longicaudatus* from
    North Carolina and Georgia. *Plant Dis. Rep. 57:* 750-754.
Rhode, R. A., and Jenkins, W. R. (1957). Host range of a species of *Tri-*
    chodorus and its host-parasite relationship on tomato. *Phytopathology*
    47: 295-298.
Román, J. (1978). *Fitonematologia Tropical*. Estac. Exp. Agric., Univ. de
    P. R.
Sasser, J. N. (1977). Worldwide dissemination and importance of the root-
    knot nematodes (*Meloidogyne* spp.). *J. Nematol. 9:* 26-29.
Sasser, J. N., and Kirby, M. F. (1979). Crop cultivars resistant to root-
    knot nematodes, *Meloidogyne* species with information on seed sources.
    Int. Meloidogyne Proj. Contract No. AID/ta-c-1234.
Sasser, J. N., Taylor, A. L., and Nelson, L. A. (1980). Ecological factors
    influencing survival and pathogenicity of *Meloidogyne* species (abstract).
    J. Nematol. 12: 237.
Sauer, M. R., and Giles, J. E. (1959). A field trial with a root-knot resis-
    tant tomato variety. *Irrigation Research Stations Technical Paper No. 3.*
Sayre, R. M. (1971). Biotic influences in soil environment. In: *Plant Para-*
    sitic Nematodes, vol. I, (B. M. Zuckerman, W. F. Mai, and R. A.
    Rhode, eds.). Academic, New York, pp. 235-256.
Schilt, H. G., and Cohn, E. (1975). Pathogenicity and population increase
    of *Paratrichodorus minor* as influenced by some environmental factors.
    Nematologica 21: 71-80.
Schwartz, H. F., and Galvez, G. E. (1980). Bean production problems.
    CIAT Series Number 09EB-1.
Seinhorst, J. W. (1954). Een ziefte in erwten, veroorzaakt door het saltje
    Hoplolaimus uniformis Thorne. *Tijdschr. Plantenziekten 60:* 262-264.
Seinhorst, J. W., and Kozlowska, J. (1977). Damage to carrots by *Rotylen-*
    chus uniformis with a discussion on the cause of increase of tolerance
    during the development of the plant. *Nematologica 23:* 1-23.
Seinhorst, J. W., and Kuniyasu, K. (1969). *Rotylenchus uniformis* (Thorne)
    on carrots. *Neth. J. Plant Pathol. 75:* 205-223.
Shafiee, M. F., and Jenkins, W. R. (1962). Effect of some single element

deficiencies on pathogenicity of *Pratylenchus penetrans* to pepper plants. *Plant Dis. Rep. 46*: 472-475.

Sher, S. A., and Allen, M. W. (1953). Revision of the genus *Pratylenchus* (Nematoda: Tylenchidae). *Univ. Cal. Pub. Zool. 57*: 441-470.

Siddiqi, M. R. (1973). Systematics of the genus *Trichodorus* Cobb 1913 (Nematoda: Dorylaimida), with descriptions of three new species. *Nematologica 19*: 259-278.

Singh, R. V., and Khera, S. (1978). Culturing and life history studies of *Rotylenchulus reniformis* Linford and Oliveira on brinjal (Nematoda). *Bull. Zool. Surv. India 1*: 115-128.

Sivakumar, C. V., and Seshadri, A. R. (1972). Histopathology of infection by the reniform nematode, *Rotylenchulus reniformis* Linford and Oliveira, 1940 on castro, papaya, and tomato. *Indian J. Nematol. 2*: 173-181.

Slinger, L. A., and Bird, G. W. (1978). Ontogeny of *Daucus carotae* infected with *Meloidogyne hapla*. *J. Nematol. 10*: 188-194.

Starr, J. L., and Mai, W. F. (1976). Effect of soil microflora on the interaction of three plant-parasitic nematodes with celery. *Phytopathology 66*: 1224-1228.

Steiner, G. (1931). Two interesting findings of *Tylenchus dipsaci* the bulb or stem nema (mimeographed). *U.S. Bur. Plant. Indus. Plant Dis. Rep. 15*: 92-93.

Steiner, G. (1949). Plant nematodes the grower should know. *Proc. Soil Sci. Soc. Fla.* (1942) *4-B*: 72-117.

Sturhan, D. (1971). Comparative investigations on the host plants of stem eelworms (*Ditylenchus dipsaci*) from beets of different origin. *Meded. Land. 30*: 1468-1474.

Sykes, G. B., and Winfield, A. L. (1966). Studies on Brassica cyst nematode *Heterodera cruciferae*. *Nematologica 12*: 530-538.

Tarjan, A. C., and Frederick, J. J. (1978). Intraspecific morphological variation among populations of *Pratylenchus brachyurus* and *P. coffeae*. *J. Nematol. 10*: 152-160.

Tarjan, A. C., Lownsbery, B. F. and Hawley, W. O. (1952). Pathogenicity of some plant-parasitic nematodes from Florida soils. I. The effect of *Dolichodorus heterocephalus* Cobb on celery. *Phytopathology 42*: 131-132.

Taylor, A. L., and Sasser, J. N. (1978). Biology, identification and control of root-knot nematodes (*Meloidogyne* species). A Cooperative Publication of the Department of Plant Pathology, North Carolina State University and the United States Agency For International Development.

Thomason, I. J., and McKinney, H. E. (1959). Reaction of some Cucurbitaceae to root-knot nematodes (*Meloidogyne* spp.) *Plant Dis. Rep. 43*: 448-450.

Thomason, I. J., Rich, J. R., and O'Melia, F. C. (1976). Pathology and histopathology of *Pratylenchus scribneri* infecting snapbean and lima bean. *J. Nematol. 8*: 347-352.

Thorne, G. (1952). Control of the sugar beet nematode. *U.S. Dept. Agr. Farmers Bull. 2054*.

Thorne, G. (1961). *Principles of Nematology*. McGraw-Hill, New York.

Toler, R. W., and Wester, R. E. (1966). A survey of lima bean diseases in the south. 1965. *Plant Dis. Rep. 50*: 316-317.

Townsend, C. R. (1937). Development of the root-knot nematode on beans as affected by soil temperature. *Fla. Agr. Exp. Bull. 309*.

Townshend, J. L., Tarté, R. and Mai, W. F. (1978). Growth response of three vegetables to smooth- and crenate-tailed females of three species of *Pratylenchus*. *J. Nematol. 10*: 259-263.

Triantaphyllou, A. C., and Hirschmann, H. (1960). Post infection development of *Meloidogyne incognita* Chitwood (1949) Nematoda: Heteroderidae. *Ann. Inst. Phytopathol. Benaki 3*: 3-11.

Triantaphyllou, A. C., and Sasser, J. N.(1960). Variation in perineal patterns and host specificity of *Meloidogyne incognita*. *Phytopathology 50*: 724-735.

Valdez, R. B. (1978). Nematodes attacking tomato and their control. In *1st International Symposium on Tropical Tomato*, AVRDC Publ. 78-59, pp. 136-152.

Van Gundy, S. D., Kirkpatrick, J. D., and Golden, J. (1977). The nature and role of metabolic leakage from root-knot nematode galls and infection by *Rhizoctonia solani*. *J. Nematol. 9*: 113-121.

Verma, T. S., Choudhury, B., and Swarup, G. (1979). Root-knot nematodes—nature of damage and relationship of initial larval penetration to resistance in brinjal. *Indian J. Hortic. 36*: 105-109.

Vovlas, N., Cham, S., and Hooper, D. J. (1980). Observations on the morphology and histopathology of *Rotylenchus laurentinus* attacking carrots in Italy. *Nematologica 26*: 302-307.

Walker, J. C. (1952). *Diseases of Vegetable Crops*. McGraw-Hill, New York.

Webber, H. J., and Orton, W. A. (1902). Some diseases of cowpea. II. A cowpea resistant to root-knot (*Heterodera radicicola*). *U.S. Dept. Agric. Bur. Plant Industry Bull.*, No. 17: 23-28.

Webster, J. M. (1972). Nematodes and biological control. In *Economic Nematology* (J. M. Webster, ed.). Academic, London, pp. 469-496.

Winstead, N. N. (1959). Reaction of cabbage varieties and clubroot-resistant lines to root-knot nematodes. *Plant Dis. Rep. 43*: 1280-1287.

Winstead, N. N., and Riggs, R. D. (1963). Reaction of watermelon varieties to root-knot nematodes. *Plant Dis. Rep. 43*: 909-912.

Wong, T. K., and Mai, W. F. (1973a). Pathogenicity of *Meloidogyne hapla* to lettuce as affected by inoculum level, plant age at inoculation, and temperature. *J. Nematol. 5*: 126-129.

Wong, T. K., and Mai, W. F. (1973b). Effect of temperature on growth, development and reproduction of *Meloidogyne hapla* in lettuce. *J. Nematol. 5*: 139-142.

Wong, T. K., and Mai, W. F. (1973c). *Meloidogyne hapla* in organic soil: Effects of environment on hatch, movement and root invasion. *J. Nematol. 5*: 130-138.

Wyatt, J. E., Fassuliotis, G., and Johnson, A. W. (1980). Efficacy of resistance to root-knot nematode in snap beans. *J. Amer. Soc. Hort. Sci. 105*: 923-926.

Wyatt, J. E., Fassuliotis, G., Hoffman, J. C., and Deakin, J. R. (1983). 'NemaSnap' snap bean. *HortScience 18*: 776.

Yarger, L. W., and Baker, L. R. (1981). Tolerance of carrots to *Meloidogyne hapla*. *Plant Dis. 65*: 337-339.

Yu, P. K., and Viglierchio, D. R. (1964). Plant growth substances and parasitic nematodes. I. Root-knot nematodes and tomato. *Exp. Parasitol. 15*: 242-248.

Yuksel, H. S. (1960). Observations on the life cycle of *Ditylenchus dipsaci* on onion seedlings. *Nematologica 5*: 289-296.

Zimmer, R. C., and Walkof, C. (1968). Occurrence of the northern root-knot nematode *Meloidogyne hapla* on field-grown cucumber in Manitoba. *Can. Plant Dis. Surv. 48*: 154.

Chapter 10
Nematode Parasites of Peanuts

Norman A. Minton *USDA, Agricultural Research Service, Coastal Plain Experiment Station, Tifton, Georgia*

I. INTRODUCTION

The peanut, *Arachis hypogaea* L., a legume native of South America, is cultivated in all six continents encompassing about 80 countries. The total world
production in 1979-1980 was estimated at 17.4 million metric tons on 18.2 million
hectares (Anon., 1981). India, the People's Republic of China, the United
States, Senegal, and Sudan produce approximately 65% of the total world production. In world agriculture, peanuts rank 13 among edible crop plants.
Edible food is the chief use of peanuts in the United States, compared with its
use as edible oil elsewhere in the world. Since peanuts average 25% highly
digestible protein and contain 5.6 cal/g, their use as an important high-protein food crop is rapidly gaining acceptance in a world deficient of protein
(Mottern, 1973). It is the number 2 edible legume worldwide. Peanuts rank
ninth in acreage among the major row crops in the United States and second
in dollar value per acre.

Most of the available information on damaging effects of nematodes on peanuts is limited to research done in the United States. Loss of yields due to
nematodes may range from a negligible level to that approaching 100%. Field
populations may be monospecific or, as is the usual case, several species may
be present (Minton and Morgan, 1974; Sasser et al., 1975a, b; Ingram and
Rodriguez-Kabana, 1980). Therefore, loss estimates are usually based on the
average loss in a given region for all species. Estimated losses in the United
States were set at 10% for the 1962-1968 period (Anon, 1971a). Based on a loss
value of 10%, total losses due to nematodes in the United States in 1979 would
have been estimated at 200,791,000 kg, or $91,550,120. However, the percent-
age loss in 1979 may not have been as great as it was during the 1962-1968
period because of improved control measures used in 1979.

The major genera damaging peanuts are *Meloidogyne*, *Pratylenchus*,
Belonolaimus, and *Criconemella*. Several additional genera have been reported
to be associated with peanuts, but their relationship may be of negligible
economic importance.

II. ROOT-KNOT NEMATODES

A. Nematode Species

Root-knot nematodes known to damage peanuts are *Meloidogyne arenaria* (Neal
1889) Chitwood 1949, *M. hapla* Chitwood 1949, and *M. javanica* (Treub 1885)
Chitwood 1949. Neal (1889) reported a root-knot nematode that produced severe galling on peanuts in Florida which he described as *Anguillula arenaria*,
and which Chitwood (1949) renamed *Meloidogyne arenaria*. Wilson (1948) found
root-knot nematodes on peanuts in Alabama. Sasser (1954) reported that peanuts were susceptible to *M. arenaria* and *M. hapla* and resistant to *M. incognita*,

M. incognita acrita, and *M. javanica*. However, Martin (1958) indicated that peanuts in South Rhodesia were infected by *M. javanica* and Minton et al. (1969) found a population of *M. javanica* damaging peanuts in Georgia.

B. Associated Symptoms

Root-knot nematodes enter and damage peanut roots, gynophores (pegs), and pods (Schenck, 1961). Larvae, which enter roots, cause slight mechanical injury, except when large numbers enter in a limited area. Most of the effects on the surrounding plant tissues are caused by the secretion ejected through the stylet while the juveniles are feeding. Sometimes root tips are devitalized and their growth stopped.

Galls formed on peanut roots are of two general types depending on the nematode species. Sasser (1954) found that peanut roots infected with *M. hapla* developed small galls and extensive root proliferation, and the roots often formed a dense mat when infection was severe (Fig. 1). Production of lateral roots just above the gall was found to be typical of plants infected with *M. hapla*. Galls caused by *M. arenaria* (Fig. 2) and *M. javanica* on peanut roots are larger than those caused by *M. hapla*. Galls caused by *M. javanica* are very similar in size and form to those caused by *M. arenaria*. Infected plants tend to have fewer small rootlets than normal, and symptoms might be characterized as a combination of galls and coarse root. The galls tend to be large and involve the main roots. Nodules on peanut roots may be diagnosed by an inexperienced person as root-knot galls (Fig. 3). Root-knot galls are

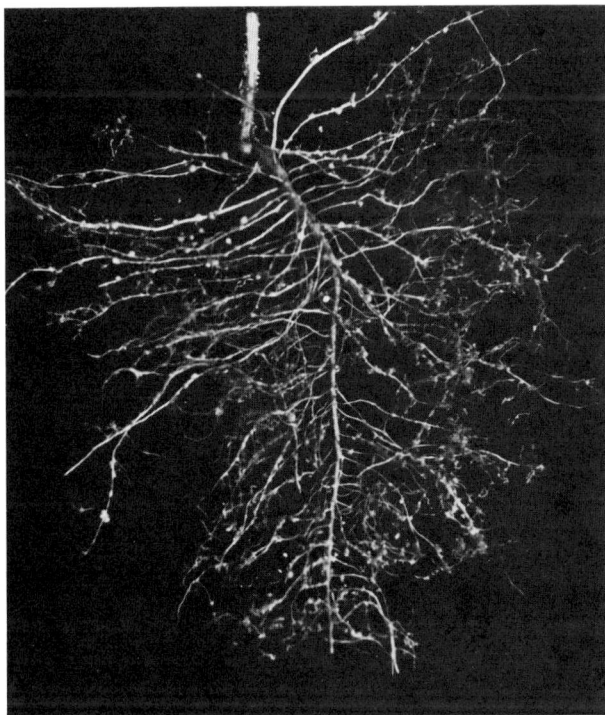

Fig. 1 *M. hapla* galls on peanut roots. Root proliferation in the vicinity of galls results in a matted root system.

Fig. 2 *M. arenaria* galls on peanut roots, pods, and pegs.

of a woody consistency and are swellings of the roots themselves, whereas nodules are of a spongy consistency and occur on the surface of the roots and can be easily rubbed off.

M. *arenaria* and M. *javanica* may also cause extensive damage to peanut fruiting structures (Fig. 4). They enter and become established in the devel-

Fig. 3 Peanut plant with no nematode damage but with numerous nitrogen-fixing nodules attached to roots.

Fig. 4 *M. arenaria* galls on peanut pods and pegs. In some instances, the seed burst through the shell.

oping pegs and pods, often preventing seed development. Pegs may become weak, causing the pod to be detached from the vine before or during digging.

Above-ground symptoms for all root-knot nematode species include stunting and, in extreme cases, death of the plant (Fig. 5). The foliage may appear light green to yellow, indicative of nutrient deficiencies. Infected plants are more susceptible to drought damage than noninfected plants.

Root-knot nematodes have been reported to interact with fungi on peanuts. Garcia and Mitchell (1975) observed a synergistic interaction in damping off in peanut that had been inoculated with *M. arenaria* and *Pythium myriotylum*. *M. hapla* enhanced the development of *Cylindrocladium* black rot of peanut caused by *Cylindrocladium crotalariae* (Diomande and Beute, 1981a, b). Also, *M. arenaria* enhanced the development of *C. crotalaria* on a peanut cultivar resistant to both pathogens (Diomande et al., 1981).

Fig. 5 A peanut field in Georgia with severe *M. arenaria* damage in the foreground.

C. Life Cycle

All root-knot nematodes have a similar life cycle. Usually, a one-celled egg
is deposited into a gelatinous matrix. Embryogenesis begins within a few
hours after deposition, resulting in two cells, four, eight, and so on, until a
fully formed juvenile with a visible stylet lies coiled in the egg membrane (Taylor and Sasser, 1978). This is the first juvenile stage. The first molt takes
place in the egg and the second-stage juvenile, which is eelworm shaped,
emerges from the egg membrane. The second-stage juveniles are infective and
move through the soil in search of roots on which to feed. Penetration usually
occurs just above the root cap. The juveniles position their heads in the developing stele near the region of cell elongation and with their bodies in the
cortex become sedentary. Secretions from the esophageal glands cause enlargement of cells in the vascular cylinder and increase the rate of division in
the pericycle. This leads to formation of giant cells (syncytia) formed by enlargement of cells (hypertrophy), possible dissolution of cell walls, enlargement of nuclei, and changes in composition of the cell content (Minton, 1963).
At the same time there is active cell multiplication (hyperplasia) around the
juvenile head. These changes usually result in enlargement of the root to form
galls.
     While the giant cells and galls are developing, the juveniles undergo three
additional molts. They become oval in shape with a width about half their
length. The male is a sedentary parasite only during its juvenile development
and emerges after two to three weeks as a slender worm with a typical nema-
toid shape. The female remains sedentary throughout her entire adult life,
and continues to grow, increasing in girth, and may become almost spherical.
If a plant is a suitable host and the temperature is warm, females begin to lay
eggs 20-30 days after penetrating roots as juveniles.

D. Distribution

One or more of the three *Meloidogyne* species that infect peanuts occur in the
major peanut-producing areas of the world. Sasser (1977) indicated that *M.
arenaria, M. hapla*, and *M. javanica* occur in the following continents or regions: North America, Central and South America, Africa, India, and Europe
and the Mediterranean region. Only *M. arenaria* and *M. javanica* were listed
as occurring in Southeast Asia, Australia, and Fiji Islands. Both *M. hapla*
and *M. arenaria* were reported to occur in Queensland, Australia, but only *M.
hapla* reproduced on peanuts in host studies (Colbran, 1958). Taylor and
Sasser (1978) stated that the region of the world between 35° S and 35° N latitudes is widely infested by *M. arenaria* and *M. javanica* and that north of 35°
latitude in the northern hemisphere the most common *Meloidogyne* species is
M. hapla.

E. Losses

M. arenaria and *M. hapla* are the most important nematode species affecting
peanuts. In addition to their inflicting heavy damage (Minton and Morgan,
1974; Rodriguez-Kabana et al., 1979) they are widely distributed (Taylor and
Sasser, 1978). Losses in infested fields may exceed 50%. However, infestations in most fields are unevenly distributed and average yield losses may be
less than 50%.

F. Control

The ideal method of controlling nematodes on peanuts would be the use of resistant cultivars. However, there is currently no cultivar available with a
high level of resistance to *M. arenaria* or *M. hapla*. Extensive efforts (Miller,
1972a, Minton and Hammons, 1975) to locate germ plasm with resistance to the
most prevalent race (race 1) of *M. arenaria* (Taylor and Sasser, 1978) in the
United States have been unsuccessful. However, Minton (1963) reported significant differences in infectivity and pathogenicity to peanuts between morphologically identical populations of *M. arenaria*. Populations of *M. arenaria*
that do not damage peanuts were designated as race 2 by Taylor and Sasser
(1978). Sasser (1966) also found that some populations of *M. arenaria* from
various countries did not reproduce on peanuts. Kirby et al. (1975) found
that only three of six Florida populations of *M. arenaria* severely galled Florunner peanuts. Although *M. arenaria* occurs throughout West Africa, Netscher (1975) reported that there has never been a root-knot nematode problem
on peanuts grown in Senegal. Seven Institute Research Agronomique Tropi-
cale peanut cultivars inoculated with isolates of *M. arenaria* from Senegal exhibited a high degree of resistance to the nematode. In a second experiment,
roots of peanut plants inoculated with *M. arenaria* were necrotic, but serious
damage was not observed. Castillo et al. (1973) reported that eight peanut
lines were only moderately susceptible to *M. hapla* and four wild peanuts exhibited resistance. Even though certain populations of root-knot nematodes
do not damage peanuts, a large percentage of peanuts are grown in areas infested with damaging populations. The probability of developing commercial
peanut cultivars resistant to the damaging populations in the near future seems
remote.

Fig. 6 A peanut field in Alabama infested with *M. arenaria*. Left, untreated;
right, treated with ethylene dibromide at the rate of 35.8 kg a.i./ha. (Courtesy of R. Rodriguez-Kabana.)

Rotation of peanut with nonhost or poor host crops can effectively reduce root-knot nematode damage (Cooper, 1950; Thames and Langley, 1967). Corn, wheat, barley, and watermelon are nonhosts for *M. hapla*, and cotton is a nonhost for both *M. hapla* and *M. arenaria* (Sasser, 1954).

In many instances chemicals are the only reliable means to control root-knot nematodes of peanuts. Two types of nematicides are widely used, fumigants and nonfumigants with contact or systemic properties. Effective fumigants that have been used contained 1,3-dichloropropenes (DD, 1,3-D), ethylene dibromide (EDB), or 1,2-dibromo-3-chloropropane (DBCP) (Fig. 6). However, DBCP is no longer available for use on peanuts because of governmental regulations. Some of the nonfumigant nematicides available, such as aldicarb, carbofuran, and phenamiphos, function as both contact poisons in the soil and systemic poisons after entering the plants. Others, such as fensulfothion and ethoprop, function primarily as contact poisons. These fumigant and nonfumigant materials have been shown to be nematicidal against root-knot, lesion, sting, ring, and other nematodes (Sturgeon and Russell, 1971; Minton and Morgan, 1974; Dickson and Mitchell, 1974; Sasser et al., 1975b; Chhabra and Mahajan, 1976; Dickson and Waites, 1978; Rodriguez-Kabana, et al., 1979, 1980). However, their effectiveness varies with the material and the target nematode.

III. LESION NEMATODES

A. Nematode Species

Pratylenchus brachyurus (Godfrey 1929) Goodey 1951 is the major lesion nematode that damages peanuts in the United States. Steiner (1949) first reported this nematode on peanuts from Alabama in 1942 at the Annual Meeting of the Soil Science Society of Florida; however, the proceedings of this meeting were not published until 1949. He described this nematode as a new species and named it *P. leiocephalus*. Sher and Allen (1953) synonymized *P. leiocephalus* with *P. brachyurus*. Steiner (1945) illustrated conspicuous shell lesions on the outer shell of the pods from both Spanish and runner peanuts from Virginia which he attributed to this nematode. Boyle (1950) later found large numbers of this nematode in dark-colored lesions on pods of Spanish peanuts in Georgia.

B. Associated Symptoms

Steiner (1945) and Boyle (1950) described conspicuous lesions on the pods of peanuts infected with *P. brachyurus*. Later Good et al. (1958) indicated that *P. brachyurus* were found in the roots, pegs, and shells of mature pods of peanuts, but were most numerous in the shells, where they colonized in dark-colored necrotic lesions (Fig. 7). Several hundred nematodes may occur in a single lesion. Infection of the pegs was correlated with a peg rot resembling that caused by *Sclerotium rolfsii*. The weakening and rotting of the mature pegs was thought to be responsible for a loss of pods in the soil at harvest. *S. rolfsii* and *P. brachyurus* are frequently found occurring together as pathogens. In addition to producing lesions on the roots, pods, and pegs, this nematode may also reduce the size of the root system. The nematode may be well established in the roots of the plant without visual above-ground symptoms to indicate disease. However, yield reductions may occur in this case.

C. Life Cycle

P. brachyurus are migratory parasites with all developmental stages occurring in plant tissue. Adults and juveniles may migrate into and out of tissues.

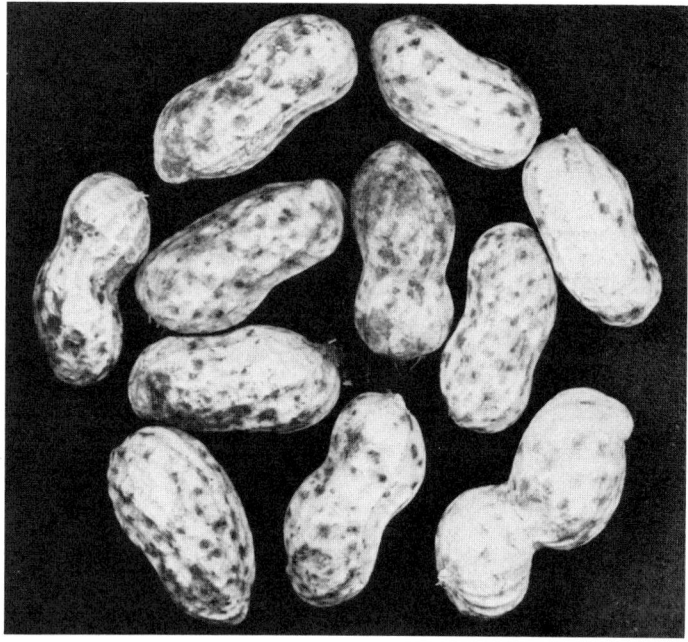

Fig. 7 Lesions on peanut pods caused by *Pratylenchus brachyurus*. (Courtesy of L. I. Miller.)

Boswell (1968), being unable to culture peanut pegs and pods on an artificial medium, made a thorough study of *P. brachyurus* on corn root tissue cultures. He observed external feeding upon root hairs, root cap cells, and epidermal cells, but juveniles did not develop further unless they had successfully penetrated the root tissue. The nematodes migrated at random through the root tissue, feeding on cells. Most of the cytoplasm was removed from the cell, leaving the nucleus and the remaining cytoplasm adhering to the periphery of the cell. Roots that have been fed on extensively become attenuated as if much of the cell contents had been removed, giving a clearing effect. Microscopic examination of this tissue revealed that the cortical parenchyma was extensively damaged. Godfrey (1929) noted that the most common point of entry into pineapple roots was basipetal to the zone of elongation in the region of root hairs; however, root tips also were invaded and destroyed.

Good et al. (1958) found *P. brachyurus* in roots, elongating pegs, mature pegs, and pods. In shells of pods, nematodes were located in the tissue between the vascular network of the pericarp. Orientation of the nematodes within the tissue suggested that they were feeding on cells that lay along the vascular tissue. The nematode has never been reported to enter the peanut seed.

Graham (1955) reported that *P. brachyurus* reproduced more rapidly at soil temperatures of 26.7-32.2°C than at lower temperatures and optimum conditions for development occurred in middle or late summer in South Carolina. Boswell (1968) reported that reproduction in root and shell tissue of peanut was greatest at 26°C.

Boswell (1968) observed hatching of an egg in culture 41 days after it was deposited. Olowe and Corbett (1976) found that the generation time for

P. brachyurus at 15 and 30°C was 14 weeks and 4 weeks, respectively. The optimum temperature for egg deposition was 35°C. Development requires four molts, the first occurring in the egg. Second-stage juveniles hatch from the egg and molt three times between feeding intervals to become adults. Eggs are laid in soil as well as in the plant tissue. In the southeastern United States, these nematodes overwinter in the soil, dead peanut roots, weeds, and grasses, as well as in the peanut shell. In South Africa, Koen (1967), working with potato and maize, found that at the end of winter 66.1% of *P. brachyurus* were found in the organic matter in the soil although the organic matter constituted only 0.29% of the soil. Since this is a polyphagous parasite, it may also overwinter on live roots of winter crops and weeds. In order to properly assess population levels of this nematode, assays of soil as well as subterranean plant parts should be made.

D. Distribution

P. brachyurus is distributed chiefly in the warmer zones of the world (Loof, 1964). In addition to its widespread occurrence in peanut production areas of the southeastern United States (Graham, 1951; Good et al., 1954; Norton, 1959; Boswell, 1968; Sturgeon and Russell, 1971; Fox and Phillips, 1980), it has been found in 81.2% of the peanut fields sampled in Egypt (Oteifa, 1962) and it is a problem on peanuts in Africa (Anon., 1971b) and Australia (Saint-Smith et al., 1972). However, its host range is not limited to peanuts, as it also occurs on other crops in many countries with warm climates.

E. Losses

Control of lesion nematodes with nematicides has resulted in peanut yield increases ranging up to 20-26% (Good and Steele, 1959; Boswell, 1968; Jackson and Sturgeon, 1973). Although most infestations usually do not inflict losses of great magnitude, this nematode must be considered an important pest of peanuts because of its widespread distribution, the high percentage of infested fields, and wide host range (Alexander, 1963; Boswell, 1968; Motsinger et al., 1976; Ingram and Rodriguez-Kabana, 1980).

F. Control

At present, there are no commercial cultivars resistant to *P. brachyurus*, although there is some evidence to suggest the existence of sources of resistance (Boyle, 1950; Minton et al., 1970; Smith et al., 1978). Because of the wide host range of *P. brachyurus*, crop rotations may be of little value for its control. However, this nematode is readily controlled by nematicides (Good and Steele, 1959; Boswell, 1968; Jackson and Sturgeon, 1973; Minton and Morgan, 1974), which can be used where the economic benefits justify the cost.

IV. STING NEMATODES

A. Nematode Species

Owens (1951) reported that the sting nematode, *Belonolaimus gracilis* Steiner 1949, injured peanuts in Virginia and later Graham and Holdeman (1953) and Holdeman (1955) indicated that it had also been found parasitizing peanuts in North Carolina, South Carolina, and Georgia. Rau (1958) described *B. longicaudatus* from Florida and suggested that *B. longicaudatus* was probably the common sting nematode species of the southeastern United States. Since Rau's

Fig. 8 Peanut plant with root system greatly reduced by *Belonolaimus longi-caudatus*. (Courtesy of L. I. Miller.)

publication, *B. longicaudatus* has been the species referred to most often on peanuts

B. Associated Symptoms

Peanut plants infected by sting nematodes are severely stunted and chlorotic and have stubby, sparse roots (Fig. 8). Small dark necrotic spots, which are caused by feeding, may occur on roots and pods (Owens, 1951). The sting nematode is an ectoparasite and is rarely found inside the roots or pods.

C. Life Cycle

A detailed life cycle of *B. longicaudatus* on peanuts has not been reported. They are migratory nematodes and feed mostly at root tips, along young roots, and on other subterranean parts of plants. Reproduction is rapid and may reach high levels. In pots planted to Florunner peanuts and inoculated with 600 nematodes, populations as high as 2100 nematodes per pot were attained in nine weeks (Robbins, 1972).

Fig. 9 A peanut field in North Carolina showing damaging effect of *Belonolai-mus longicaudatus*. Center row was untreated; rows to right and left of center were treated with different nematicides. (Courtesy of A. W. Johnson.)

D. Distribution

B. longicaudatus occurs in sandy soils along the Atlantic Coastal Plain from New Jersey to Florida and westward to Texas and Arkansas. Although *B. longicaudatus* has been associated with peanuts in most of the peanut-producing states (Brooks and Christie, 1950; Owens, 1951; Holdeman, 1955; Rau, 1958), it is most severe on peanuts in Virginia and North Carolina. Populations in Georgia and Florida do not damage peanuts appreciably in the field. Coarse, sandy soils are the optimum habitat of this nematode; consequently, soil type is a major limiting factor in its distribution (Miller, 1972a; Robbins and Barker, 1974). This nematode has not been reported on peanuts outside the United States.

E. Losses

B. longicaudatus can seriously damage peanuts (Fig. 9). Miller (1952) reported that control of sting nematodes on peanuts in Virginia was highly profitable. Cooper et al. (1959) found that control of this nematode in North Carolina increased yields as much as 400% or approximately 3300 kg/ha. In subsequent experiments in North Carolina (Sasser et al., 1960), yields were also increased approximately 3300 kg/ha. Based on 1980 peanut prices, the value of these yield increases could be as much as $1600/ha. Fortunately the distribution of this nematode is limited, so that losses of this magnitude are not widespread.

F. Control

Miller (1972b) indicated peanuts resistant to *B. longicaudatus* are not available. Crop rotations for managing this nematode are usually of little value because the crops (corn, cotton, and soybean) commonly grown in rotation are also susceptible (Holdeman and Graham, 1953; Robbins and Barker, 1973). Nematicides are effective, since populations of sting nematodes are easily reduced by fumigants and many of the nonfumigant nematicides (Cooper 1959; Sasser et al., 1960, 1967; Johnson and Chalfant, 1972).

V. RING NEMATODES

A. Nematode Species

Criconemella ornata (Raski, 1958) (formerly *Criconemoides ornatus*) are present in many peanut fields in the southeastern United States (Machmer, 1953; Graham, 1955; Minton et al., 1963; Motsinger et al., 1976; Ingram and Rodriguez-Kabana, 1980). Relatively large numbers of nematodes may be associated with peanut roots, yet with little discernible loss in yield or quality. However, occasionally peanuts appear to be affected by the nematode. Results of the limited research that has been done to date suggest that *C. ornata* is not a severe pathogen on peanuts but may inflict damage where large populations occur (Machmer, 1953; Graham, 1955; Sasser et al., 1968; Minton and Bell, 1969).

B. Associated Symptoms

Obvious damage to peanuts are seldom caused by *C. ornata*, and large populations are necessary to produce symptoms. Machmer (1953) described a chlorotic condition of peanuts growing in Georgia in soil heavily infested with a species of *Criconemella*, and Graham (1955) reported that peanuts in South Carolina when inoculated with a species of *Criconemella*, were stunted and roots were decayed. Both were probably working with the ectoparasitic nematode, *C. ornata*, that feeds on the roots, pods, and pegs of peanuts. In microplots (Minton and Bell, 1969) heavily infested with this nematode, roots, pods, and pegs of Argentine and Starr peanut cultivars were severely dis-

Fig. 10 *Criconemella ornata* lesions on peanut pods (A) compared with undamaged pods (B). (From Minton and Bell, 1969.)

colored with brown necrotic lesions (Fig. 10). Small necrotic lesions were
often superficial, but necrosis in large lesions usually extended deep into the
tissues. Many lateral root primordia and young roots were killed, resulting
in reduced numbers of lateral roots. Pod yields from nematode-infected plants
were reduced by about one-half.

C. Life Cycle

The life cycle of *C. ornata* has not been studied. However, the life cycle of
C. xenoplax has been observed in the laboratory (Seshadri, 1965) and the
life cycle of *C. ornata* is probably similar. The life cycle of *C. xenoplax* from
egg to egg was found to be 25-34 days. There were four molts, including one
inside the egg prior to hatching. Adults began to lay eggs two to three days
after their final molt and deposited 8-15 eggs during a two to three day period,
at the rate of one to eight eggs per day.

D. Distribution

C. ornata is widely distributed in the peanut-producing regions of the United
States (Boyle, 1950; Machmer, 1953; Graham, 1955; Anon., 1960; Minton et al.,
1963; Motsinger et al., 1976; Ingram and Rodriguez-Kabana, 1980; Fox and
Phillips, 1980).

E. Losses

Losses due to *C. ornata* have not been well defined. Damage in the field is
subtle, and low levels of damage may often go undetected. Sasser et al.
(1975a) obtained significant negative correlations between *C. ornata* population levels and peanut growth index, and also between population levels and
peanut yields. Diomande and Beute (1981b) obtained correlations between *C.
ornata* and *Cylindrocladium* black rot symptoms of peanuts in field experiments.
In greenhouse tests, *C. ornata* increased the severity of the disease on the
Florigiant cultivar, but failed to affect the syndrome on NC 3033 (Diomande
and Beute, 1981a).

F. Control

There are no known commercial peanut cultivars with resistance to *C. ornata*.
Some of the crops grown in rotation with peanuts, such as cotton, soybeans,
corn, and sorghum may reduce population levels of this nematode (Good, 1968;
Johnson et al., 1974; Kinloch and Lutrick, 1975). Most of the nematicides in
use are effective against this nematode.

VI. TESTA NEMATODE

A. Nematode Species

Aphelenchoides arachidis Bos 1977, an endoparasite of peanuts, was described
from northern Nigeria (Bos, 1977a). This nematode causes discoloration of the
peanut seed tissues, reduces seed size, and causes seeds to shrivel (Bridge
et al., 1977). No yield reductions have been reported.

B. Associated Symptoms

A. arachidis is a facultative endoparasite of peanuts (Bridge et al., 1977). It
occurs within the tissues of the pods, testas, roots, and hypocotyls, but not

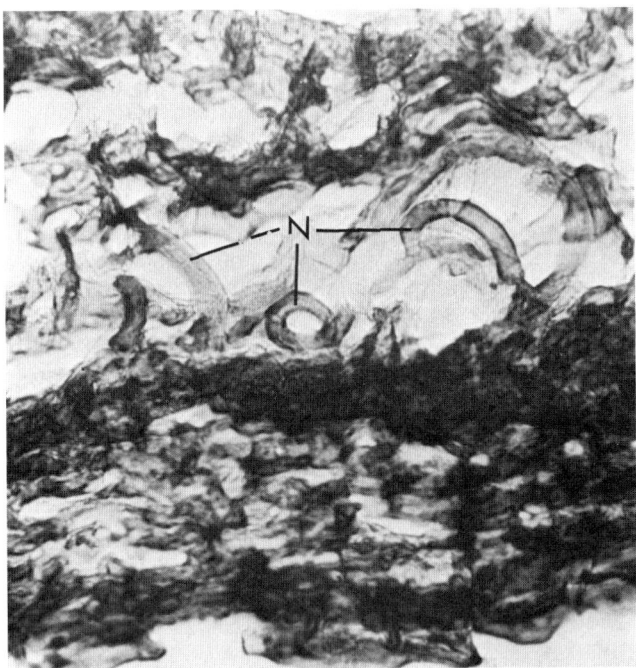

Fig. 11 Transverse section of peanut testa infected with *Aphelenchoides arachidis* (N = nematodes). (From J. Bridge, W. S. Bos, L. J. Page, and D. McDonald, 1977.)

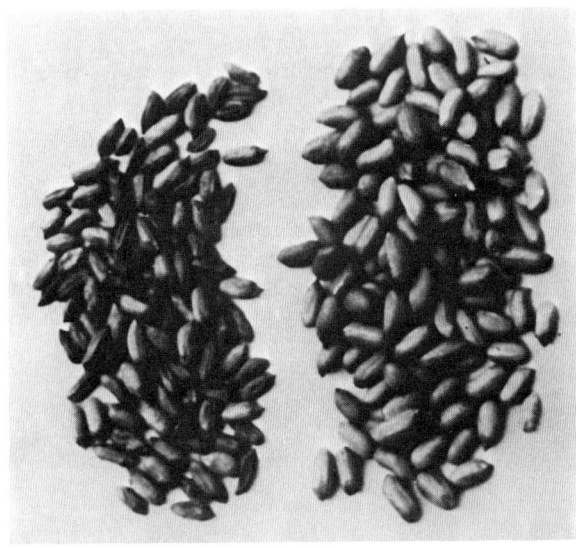

Fig. 12 Seeds of peanut infested with *Aphelenchoides arachidis* (left) and healthy (right). (From Bridge et al., 1977.)

within the cotyledons, embryos, or other parts of the plant (Fig. 11). Seed coats were discolored when more than 2000 nematodes per testa were present and more than 25,000 nematodes are often found. Heavily infected seeds, immediately after removal from fresh, mature pods, had translucent testas, were light brown in color, and had dark vascular strands within the testas. Infected testas of dry seeds were often wrinkled and dark brown (Fig. 12). Testas infected with *A. arachidis* were thicker and more uneven than normal testas. The epidermal layer of the testa was reduced in infected testas and the basal tissues, including the aleurone layer, were disorganized. No necrosis or other symptoms could be attributed to the presence of *A. arachidis* in the roots or hypocotyl. Infected seeds weighed less than healthy seeds, and seedling emergence was reduced (McDonald et al., 1979).

C. Life Cycle

Little has been written about the life cycle of *A. arachidis*. However, observations have been made relative to the biology of this nematode (Bridge et al., 1977). It is a facultative parasite of the seed testa, pod shell, roots, and hypocotyl of peanuts. It has also been observed feeding ectoparasitically on roots and on two fungi, *Macrophomina phaseoli* and *Botrytis cinerea*. It appeared that the nematodes may move from the soil or infested seeds to infect plants. In mature infested pods, many nematodes migrate into the spaces between shell, testa, and cotyledons and are released when moisture is absorbed before germination and the pods and testas burst. At this stage they can penetrate the young roots and hypocotyls and reproduce in very young plants. Low numbers of the nematode can survive dessication in stored pods of peanuts under certain circumstances. All juvenile stages and some adults can be revived from stored seeds.

D. Distribution

A. arachidis was found at a significant level of infestation in a limited area around Samaru, North Nigeria (Bos, 1977b). It was also found at a low level of infestation in peanut fields at Kadawa and Gwoza. Results of a survey of Nigeria indicated that the nematode is ubiquitous over a large area, and was found in large numbers in roots of maize, millet, sorghum, sugar cane, rice, and some wild grasses.

E. Losses

A. arachidis, a facultative endoparasite of the seed testa, pod shell, root, and hypocotyl, devalues the confectionary peanuts because it causes shriveled and discolored seeds. However, it has not been shown to decrease yields (Bridge et al., 1977). Severe infestation of peanuts with *A. arachidis* not only has an adverse effect on the appearance and size of seed but it also predisposes seeds to invasion by fungi (*Rhizoctonia solani*, *Sclerotium rolfsii*, *Macrophomina phaseoli*, and *Fusarium* spp.), which can lead to reduced seed emergence (McDonald et al., 1979). Because of its very limited distribution (Nigeria) this nematode is probably not inflicting appreciable loss. However, should it become established in other peanut-producing regions of the world, it could become a major economic pest.

F. Control

There are no reports concerning the control of *A. arachidis* in the field. Since this nematode appears to have a very limited distribution, restricting its spread is of paramount importance. Infested seeds should not be planted in noninfested areas. However, if it should become necessary to plant infested seeds they should be hot water treated in order to eradicate the nematodes. All stages of the nematode were killed when infested peanut seeds were emersed in 60°C water for five minutes after the seeds had been soaked first in cold water for 15 minutes (Bridge et al., 1977). Germination or growth of seedlings was not affected. Sun drying pods after harvesting in very dry conditions, as occurs in northern Nigeria, usually controls the nematodes. In southern regions of Nigeria and other countries, where sun-drying of pods is less efficient and conditions are more humid, the nematodes can survive and be disseminated in the seeds. Also, shelling before planting eliminates the tissues in which most nematodes occur and where they survive best.

VII. OTHER NEMATODES

Several other species of nematodes have been reported in association with peanuts, but their economic importance is not clear. Perry and Nordon (1963) and Norden et al. (1977) reported *Hoplolaimus galeatus* (Cobb 1913) parasitizing and reducing yields of peanuts in Florida. In addition, they also found relatively high populations of *Criconemella curvata* (Raski 1952) on peanuts. Chhabra and Mahajan in India (1976) reported *Pratylenchus coffeae* (Zimmermann 1898) Goodey 1951 on peanuts and indicated that fensulfothion, aldicarb, and dazomet controlled them effectively. The banana race of *Radopholus similis* (Cobb 1893) Thorne 1949 was found by O'Bannon et al. (1971) to be pathogenic on peanuts but it has not been recognized as a problem. The stubby-root nematode, *Paratrichodorus minor* (Colbran 1956) was reported as a parasite on peanuts by Coursin et al. (1958), but other studies have indicated that the nematode declines or fails to increase on peanuts (Johnson et al., 1974; Kinloch and Lutrick, 1975). Schindler (1954) reported that galls and curly tips developed on roots of peanuts growing in soil infested with *Xiphinema diversicaudatum* (Micoletzky 1927) Thorne 1939; however, this is the only report of this nematode damaging peanuts.

REFERENCES

Alexander, P. M. (1963). Stylet-bearing nematodes associated with various crop plants in South Carolina, 1962-1963. *Plant Dis. Rep. 47*: 978-982.

Anon. (1960). Distribution of plant-parasitic nematodes in the south. *South Coop. Ser. Bull. 74.*

Anon. (1971a). Estimated crop losses due to plant-parasitic nematodes in the United States. *Special Pub. No. 1, Soc. Nematol. (USA).*

Anon. (1971b). Plant Nematology. In *Report of the Secretary for Agriculture for the period 1 October 1970 to 30 September 1971., Salisbury, Rhodesia.*

Anon. (1981). World crop production. *For. Agric. Cir. WCP-1-80. U.S. Dept. Agric. For. Agric. Serv. Econ., Stat. And Coop. Serv.*

Bos, W. S. (1977a). *Aphelenchoides arachidis* n. sp. (Nematoda: Aphelenchoidea), an endoparasite of the testa of groundnuts in Nigeria. *Z. Pfl. Kranksh. Pfl. Schutz. 84*: 95-99.

Bos, W. S. (1977b). A preliminary report on the distribution and host range of the nematode *Aphelenchoides arachidis*, in the north of Nigeria. *Samaru Newslett. 19*: 21-23.

Boswell, T. (1968). Pathogenicity of *Pratylenchus brachyurus* to Spanish peanut. Ph.D. Dissertation, Texas A.&M. University, College Station.

Boyle, L. W. (1950). Several species of parasitic nematodes on peanuts in Georgia. *Plant Dis. Rep. 34:* 61-62.

Bridge, J., Bos, W. S., Page, L. J., and McDonald, D. (1977). The biology and possible importance of *Aphelenchoides arachidis*, a seed-borne endoparasitic nematode of groundnuts from Northern Nigeria. *Nematologica 23:* 253-259.

Brooks, A. N. and Christie, J. R. (1950). A nematode attacking strawberry roots. *Proc. Fla. State Hort. Soc. 63:* 123-125.

Castillo, M. B., Morrison, L. S., Russell, C. C., and Banks, D. J. (1973). Resistance to *Meloidogyne hapla* in peanut. *J. Nematol. 5:* 281-285.

Chhabra, H. K. and Mahajan, R. (1976). *Pratylenchus coffeae*, the root-lesion nematode in groundnut and its control by granular nematicides. *Nematol. Medit. 4:* 241-242.

Chitwood, B. G. (1949). Root-knot nematodes – Part I. A revision of the genus *Meloidogyne* Goeldi, 1887. *Proc. Helm. Soc. Wash. 16:* 90-104.

Colbran, R. C. (1958). Studies of plant and soil nematodes. 2. Queensland host records of root-knot nematodes (*Meloidogyne* spp.). *Queensland J. Agric. Sci. 15:* 101-136.

Cooper, W. E. (1950). Root-knot of peanuts (abstract). *Phytopathology 40:* 786.

Cooper, W. E., Wells, J. C., Sasser, J. N., and Bowery, T. G. (1959). The efficacy of preplant and postplant applications of 1,2-dibromo-3-chloropropane for control of the sting nematode, *Belonolaimus longicaudatus*. *Plant Dis. Rep. 43:* 903-908.

Coursin, B. W., Rohde, R. A., and Jenkins, W. R. (1958). Additions to the host lists of the nematode *Pratylenchus projectus* and *Trichodorus christiei*. *Plant Dis. Rep. 42:* 456-460.

Dickson, D. W. and Mitchell, D. J. (1974). Nematode and soil-borne disease control on peanut (abstract). *J. Nematol. 6:* 138-139.

Dickson, D. W. and Waites, R. E. (1978). Efficacy of at-plant and additional at-pegging applications of nematicides for control of *Meloidogyne arenaria* on peanut (abstract). *Am. Peanut Res. Educ. Assn. Proc. 10:* 51.

Diomande, M. and Beute, M. K. (1981a). Effects of *Meloidogyne hapla* and *Macroposthonia ornata* on Cylindrocladium black rot on peanut. *Phytopathology 71:* 491-496.

Diomande, M. and Beute, K. M. (1981b). Relations of *Meloidogyne hapla* and *Macroposthonia ornata* populations to Cylindrocladium black rot in peanuts. *Plant Dis. Rep. 65:* 339-342.

Diomande, M., Black, M. C., Beute, M. K., and Barker, K. R. (1981). Enhancement of *Cylindrocladium crotalaria* root rot by *Meloidogyne arenaria* (race 2) on a peanut cultivar resistant to both pathogens. *J. Nematol. 13:* 321-327.

Fox, J. A. and Phillips, P. M. (1980). The role of a predictive nematode assay program in defining nematode problems on peanut in Virginia (abstract). *Am. Peanut Res. Educ. Assoc. Proc. 12:* 35.

Garcia, R., and Mitchell, D. J. (1975). Interactions of *Pythium myriotylum* with *Fusarium solani*, *Rhizoctonia solani*, and *Meloidogyne arenaria* in a pre-emergence damping-off of peanut. *Plant Dis. Rep. 59:* 665-669.

Godfrey, G. H. (1929). A destructive root disease of pineapple and other plants due to *Tylenchus brachyurus*, n. sp. *Phytopathology 19:* 611-629.

Good, J. M. (1968). Relation of plant parasitic nematodes to soil management practices. In *Tropical Nematology* (G. C. Smart and V. G. Perry, eds.).

University of Florida Press, Gainesville.

Good, J. M., Boyle, L. W., and Hammons, R. O. (1958). Studies of *Pratylenchus brachyurus* on peanuts. *Phytopathology 48:* 530-535.

Good, J. M., Robertson, W. K., and Thomason, L. G. Jr. (1954). Effects of crop rotation on the population of meadow nematodes, *Pratylenchus leiocephalus* in Norfolk loamy fine sand. *Plant Dis. Rep. 38:* 178-180.

Good, J. M. and Steele, A. E. (1959). Evaluation of methods of applying 1.2-dibromo-3-chloropropane for controlling root-lesion nematodes on Spanish peanuts (abstract). *Phytopathology 49:* 317.

Graham, T. W. (1951). Nematode root rot of tobacco and other plants. *S. C. Agric. Exp. Sta. Bull. 390.*

Graham, T. W. (1955). Pathogenicity of *Rotylenchus brachyurus* on tobacco and of *Criconemoides* sp. on tobacco and peanuts (abstract). *Phytopathology 45:* 347.

Graham, T. W. and Holdeman, Q. L. (1953). The sting nematode *Belonolaimus gracilis* Steiner: A parasite on cotton and other crops in South Carolina. *Phytopathology 43:* 434-439.

Holdeman, Q. L. (1955). The present known distribution of the sting nematode, *Belonolaimus gracilis*, in the Coastal Plain of the Southeastern United States. *Plant Dis. Rep. 39:* 5-8.

Holdeman, Q. L. and Graham, T. W. (1953). The effect of different plant species on the population trends of the sting nematode. *Plant Dis. Rep. 37:* 497-500.

Ingram, E. G. and Rodriguez-Kabana, R. (1980). Nematodes parasitic on peanuts in Alabama and evaluation of methods for detection and study of population dynamics. *Nematropica 10:* 21-30.

Jackson, E. E., and Sturgeon, R. V., Jr. (1973). Effect of nematicides upon root lesion nematode populations. *J. Am. Peanut Res. Educ. Assoc. 5:* 178-181.

Johnson, A. W. and Chalfant, R. B. (1972). Control of nematodes and corn earworm on sweet corn. *J. Nematol. 4:* 227-228.

Johnson, A. W., Dowler, C. C., and Hauser, E. W. (1974). Seasonal population dynamics of selected plant-parasitic nematodes on four monocultured crops. *J. Nematol. 6:* 187-190.

Kinloch, R. A. and Lutrick, M. C. (1975). The relative abundance of nematodes in an established field crop rotation. *Soil Crop Sci. Soc. Fla. Proc. 34:* 192-194.

Kirby, M. F., Dickson, D. W., and Smart, G. C. (1975). Physiological variation within species of *Meloidogyne* occurring in Florida. *Plant Dis. Rep. 59:* 353-356.

Koen, H. (1967). Notes on the host range, ecology, and population dynamics of *Pratylenchus brachyurus*. *Nematologica 13:* 118-124.

Loof, P. A. A. (1964). Free-living and plant-parasitic nematodes from Venezuela. *Nematologica 10:* 201-300.

Machmer, J. H. (1953). *Criconemoides* sp., a ring nematode associated with peanut "yellows". *Plant Dis. Rep. 37:* 156.

Martin, G. C. (1958). Root-knot nematodes (*Meloidogyne* spp.) in the Federation of Rhodesia and Nyasaland. *Nematologica 3:* 332-349.

McDonald, D., Bos, W. S., and Gumel, M. H. (1979). Effects of infestations of peanut (groundnut) seed by the testa nematode, *Aphelenchoides arachidis*, on seed infection by fungi and on seedling emergence. *Plant. Dis. Rep. 63:* 464-467.

Miller, L. I. (1952). Control of the sting nematode on peanuts in Virginia (abstract). *Phytopathology 42:* 470.

Miller, L. I. (1972a). The influence of soil texture on the survival of *Belono-laimus longicaudatus* (abstract). *Phytopathology 62*: 670-671.

Miller, L. I. (1972b). Resistance of plant introductions of *Arachis hypogaea* to *Meloidogyne hapla, Meloidogyne arenaria*, and *Belonolaimus longicaudatus* (abstract). *Virg. J. Sci. 23*: 101.

Minton, N. A. (1963). Effects of two populations of *Meloidogyne arenaria* on peanut roots. *Phytopathology 53*: 79-81.

Minton, N. A. and Bell, D. K. (1969). *Criconemoides ornatus* parasitic on peanuts. *J. Nematol. 1*: 349-351.

Minton, N. A., Cairns, E. J., and Hopper, B. E. (1963). Occurrence of plant-parasitic nematodes in Alabama. *Plant Dis. Rep. 47*: 743-745.

Minton, N. A., McGill, J. F., and Golden, A. M. (1969). *Meloidogyne javanica* attacks peanuts in Georgia. *Plant Dis. Rep. 53*: 688.

Minton, N. A. and Hammons, R. O. (1975). Evaluation of peanut for resistance to the peanut root-knot nematode, *Meloidogyne arenaria*. *Plant Dis. Rep. 59*: 944-945.

Minton, N. A., Hammons, R. O., and Parham, S. A. (1970). Infection of shell and peg tissues of six peanut cultivars by *Pratylenchus brachyurus*. *Phytopathology 60*: 472-474.

Minton, N. A. and Morgan, L. W. (1974). Evaluation of systemic and nonsystemic pesticides for insect and nematode control on peanuts. *Peanut Sci. 1*: 91-98.

Motsinger, R. E., Crawford, J. L., and Thompson, S. S. (1976). Nematode survey of peanuts and cotton in Southwest Georgia. *Peanut Sci. 3*: 72-74.

Mottern, H. H. (1973). Peanuts and human nutrition. In *Peanuts, Culture and Uses, a Symposium*. American Peanut Research and Education Association, Inc., Stone Printing, Roanoke, Virginia.

Neal, J. C. (1889). The root-knot disease of the peach, orange, and other plants in Florida, due to the work of *Anguillula*. *U.S. Dept. Agric. Div. Ent. Bull. 20*.

Netscher, C. (1975). Studies on the resistance of groundnut to *Meloidogyne* spp. in Senegal. *Cah. ORSTOM Ser. Biol. 10*: 227-232.

Norden, A. J., Perry, V. G., Martin, F. G., and Nesmith, J. (1977). Effect of age of bahiagrass sod on succeeding peanut crops. *Peanut Sci. 4*: 71-74.

Norton, D. C. (1959). Plant parasitic nematodes in Texas. *Texas Agric. Exp. Sta. Misc. Pub. 321*.

O'Bannon, J. H., Yuhl, W. A., and Tomerlin, A. T. (1971). Pathogenicity of two races of *Radopholus similis* to six peanut cultivars. *Soil Sci. Soc. Fla. Proc. 31*: 264-265.

Olowe, T. and Corbett, D. C. M. (1976). Aspects of the biology of *Pratylenchus brachyurus* and *P. zeae*. *Nematologica 22*: 202-211.

Oteifa, B. A. (1962). Species of root-lesion nematodes commonly associated with economic crops in the delta of the U.A.R. *Plant Dis. Rep. 46*: 572-575.

Owens, J. V. (1951). The pathological effects of *Belonolaimus gracilis* on peanuts in Virginia (abstract). *Phytopathology 41*: 29.

Perry, V. G., and Norden, A. J. (1963). Some effects of a cropping sequence on populations of certain plant nematodes. *Soil Crop Sci. Soc. Fla. Proc. 23*: 116-121.

Rau, G. J. (1958). A new species of sting nematode. *Proc. Helm. Soc. Wash. 25*: 95-98.

Robbins, R. T. (1972). Morphology and ecology of the sting nematode, *Belonolaimus longicaudatus*. Ph.D. Dissertation, North Carolina State University, Raleigh, North Carolina.

Robbins, R. T. and Barker, K. R. (1973). Comparison of host range and reproduction of *Belonolaimus longicaudatus* from North Carolina and Georgia. *Plant Dis. Rep.* 57: 750-754.

Robbins, R. T. and Barker, K. R. (1974). The effect of soil type, particle size, temperature, and moisture on reproduction of *Belonolaimus longicaudatus*. *J. Nematol.* 6: 1-6.

Rodriguez-Kabana, R., King, P. S., Penick, H. W., and Ivey, H. (1979). Control of root-knot nematodes on peanuts with planting time and post-emergence applications of ethylene dibromide and an ethylene dibromide-chloropicrin mixture. *Nematropica* 9: 54-61.

Rodriguez-Kabana, R., Mawhinney, P. G., and King, P. S. (1980). Efficacy of planting time injections to soil of liquid formulations of three systemic nematicides against root-knot nematodes in peanuts. *Nematropica* 10: 45-49.

Saint-Smith, J. H., McCarthy, G. J. P., Rawson, J. E., Langford, S., and Colbran, R. C. (1972). Peanut Growing. *Queensland Agric. J.* 98: 639-644.

Sasser. J. N. (1954). Identification and host-parasite relationships of certain root-knot nematodes (*Meloidogyne* spp.) *Univ. Maryland Agric. Exp. Sta. Tech. Bull.* A-77.

Sasser, J. N. (1966). Behavior of *Meloidogyne* spp. from various geographical locations on ten host differentials (abstract). *Nematologica* 12: 97-98.

Sasser, J. N. (1977). Worldwide dissemination and importance of the root-knot nematodes, *Meloidogyne* spp. *J. Nematol.* 9: 26-29.

Sasser, J. N., Barker, K. R., and Nelson, L. A. (1975a). Correlations of field populations of nematodes with crop growth responses for determining relative involvement of species. *J. Nematol.* 7: 193-198.

Sasser, J. N., Barker, K. R., and Nelson, L. A. (1975b). Chemical soil treatment for nematode control of peanut and soybean. *Plant Dis. Rep.* 59: 154-155.

Sasser, J. N., Cooper, W. E., and Bowery, T. G. (1960). Recent developments in the control of sting nematode, *Belonolaimus longicaudatus* on peanuts with 1,2-dibromo-3-chloropropane and EN 18133. *Plant Dis. Rep.* 44: 733-737.

Sasser, J. N., Wells, J. C., and Nelson, L. A. (1967). Correlations between sting nematode populations at three sampling dates following nematicide treatments and the growth and yield of peanuts (abstract). *Nematologica* 13: 152.

Sasser, J. N., Wells, J. C., and Nelson, L. A. (1968). The effect of nine parasitic nematode species on growth, yield and quality of peanuts as determined by soil fumigation and correlation of nematode populations with host response (abstract). *Nematologica* 14: 15.

Schenk, R. U. (1961). Development of the Peanut Fruit. *Ga. Agric. Exp. Sta. Tech. Bull. N.S.* 22.

Schindler, A. F. (1954). Root galling associated with dagger nematode, *Xiphinema diversicaudatum* (Micoletzky, 1927) Thorne, 1939 (abstract). *Pytopathology* 44: 389.

Seshadri, A. R. (1965). Investigations on the biology and life cycle of *Criconemoides xenoplax* Raski, 1952 (Nematoda: Criconematidae). *Nematologica* 10: 540-562.

Sher, S. A., and Allen, M. W. (1953). Revision of the genus *Pratylenchus* (Nematoda: Tylenchidae). *Univ. Ca. Pub. Zool. 57*: 441-470.

Smith, O. C., Boswell, T. E., and Thames, W. H. (1978). Lesion nematode resistance in peanuts. *Crop. Sci. 18*: 1008-1011.

Steiner, G. (1945). Meadow nematodes as the cause of root destruction. *Phytopathology 35*: 935-937.

Steiner, G. (1949). Plant nematodes the grower should know. *Proc. Soil Sci. Soc. Fla. 1942 4-B*: 72-117.

Sturgeon, R. V., Jr., and Russell, C. C. (1971). Spanish peanut yield response to nematicide-soil fungicide combinations. *Am. Peanut Res. Educ. Assoc. Proc. 3*: 29-30.

Taylor, A. L., and Sasser, J. N. (1978). Biology, identification and control of root-knot nematodes (*Meloidogyne* species). Department of Plant Pathology, North Carolina State University, and U.S. Agency for International Development, Printed by North Carolina State University Graphics, Raleigh, North Carolina.

Thames, W. H. and Langley, B. C. (1967). Effects of sorghum rotations on yield of Spanish peanuts from plots infested with *Meloidogyne arenaria* (abstract). *Phytopathology 57*: 464.

Wilson, C. (1948). Root-knot nematodes on peanuts in Alabama. *Plant Dis. Rep. 52*: 443.

Chapter 11

Nematode Parasites of Citrus

Armen C. Tarjan *University of Florida, Gainesville, Florida*

John H. O'Bannon *USDA, Agricultural Research Service, Irrigated Agriculture Research and Extension Center, Prosser, Washington*

I. INTRODUCTION

Citrus is a crop appreciated by human and nematode alike, as judged by the astronomical numbers of both fed by citrus trees the world over. From the tree's point of view both human and animal are tolerable parasites since they do not kill the tree but, in the case of the latter parasite, only debilitate it, albeit often severely.

The various species of citrus originated in the Orient and India and were carried westward by humans. Columbus is credited with having first planted citrus seeds on the Caribbean Island of Haiti.

Commercially important species are sweet orange (*Citrus sinensis* Osbeck), sour orange (*C. aurantium* L.), lemon (*C. limon* Burm.), lime (*C. aurantifolia* Swing.), grapefruit (*C. paradisi* Macf.), pummelo (*C. grandis* Osbeck), and citron (*C. medica* L.). These are grown in the tropics and subtropics, between 40° N and S latitudes, where they thrive in a frost- and freeze-free climate along with at least 1200 mm of annual natural and/or applied irrigation (Chapot, 1975).

It had been estimated that production of oranges, grapefruit, tangerines, and lemons in 1978-1979 was 35,953 million metric tons among the major citrus-producing countries of the world. The United States was the largest producer and grew 28% of the orange crop, 75% of the grapefruit crop, and 29% of the lemon crop. The U.S. production of citrus fruits in 1978-1979 was 13.3 million tons, of which Florida produced 76% of the crop (Anon., 1979). In all, there are about 1.3 million ha of citrus grown commercially in the world; about 480,000 ha are in the United States.

Citrus is enjoyed primarily as either fresh fruit or in its juice form. Frozen concentrated orange juice, a blend of different varieties, is shipped the world over in 200 liter steel drums. Canned single-strength juice and chilled pasteurized juice are also saleable products depending on consumer preference. Among some of the remaining citrus products, perhaps not so well known to the average consumer, are essential oils, dried citrus pulp cattle feed, citrus molasses, bioflavonoids and pectins, and citrus peel products, such as candied peels and marmalades (Kesterson and Braddock, 1975).

II. SLOW DECLINE DISEASE

The first report of an association between a nematode and citrus appeared in 1889 (Neal, 1889). However, it was not until 1912, when *Tylenchulus semipenetrans* Cobb was discovered by J. R. Hodges on roots of citrus trees in California, that a nematode was found to cause a diseased condition of citrus, appropriately called "slow decline." The causal organism (Fig. 1), the citrus root nematode (later shortened to the citrus nematode) was first reported by Thomas (1913). About the same time, according to Cohn (1969), it was independently discovered in Israel. Cobb (1913) named and described the nematode in an extensive account of the pest (Cobb, 1914). At that time is was reported in Malta, Spain, Israel, Australia, and South America. In the United States, it was reported in Florida in 1913, Alabama in 1914, Arizona in 1926, Texas in 1950, and Louisiana and Hawaii in 1954. It has now been found in all citrus-growing areas worldwide. Until recently most citrus rootstocks used commercially were attacked by the citrus nematode (O'Bannon and Ford, 1977). The pest is often unnoticed because it causes no obvious root symptoms on nursery stock to indicate its presence, which probably accounts for its worldwide distribution.

A. Economic Importance

Surveys of major citrus-growing areas of the world report that from 50 to 100% of the citrus examined was infested with the citrus nematode (Van Gundy and Meagher, 1977). Since citrus nematode-debilitated trees are not killed, but slowly decrease in productivity, they eventually become marginally nonproductive. The economic importance of the nematode as a pest of citrus was proved

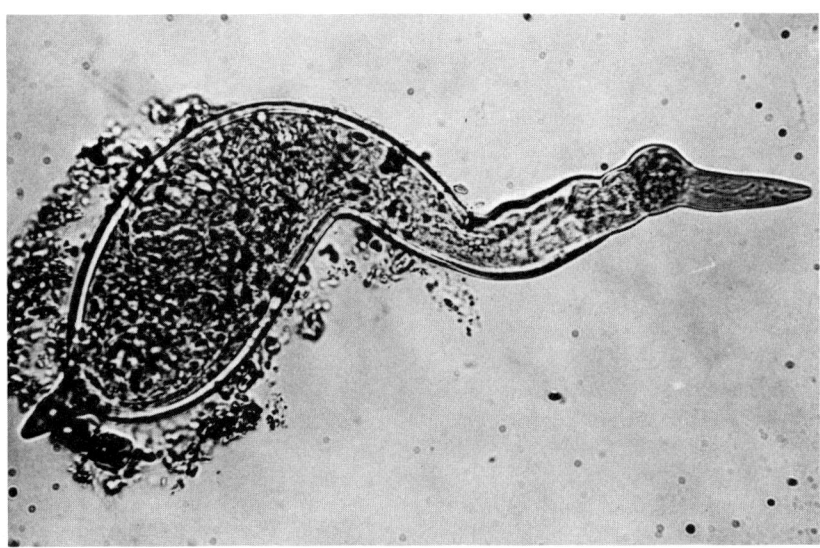

Fig. 1 Adult female citrus nematode, *Tylenchulus semipenetrans*.

when it was demonstrated that growth of young lemon and orange trees was reduced from 10 to 60%, and that yields of mature trees were decreased by 30-50% due to the nematode (Baines, 1950; Baines and Clarke, 1952; Baines and Martin, 1953; Baines et al., 1956, 1962). Control of the nematode with nematicides in the United States has resulted in yield increases from 10 to 300% (Baines, 1964; Heald, 1972; Reynolds and O'Bannon, 1963a; Tarjan and O'Bannon, 1974). Research in other citrus areas of the world have also demonstrated the importance of the citrus nematode as an economic pest (Cohn and Minz, 1965; Davide and dela Rosa, 1976; Laborda and Bello, 1975; Meagher, 1969; Milne, 1977; Mukhopadhyaya and Dalal, 1971; Oteifa et al., 1965; Philis, 1969; Scaramuzzi and Perrotta, 1962; Scotto La Massese, 1965a; Vilardebo, 1963a; Yokoo, 1964).

According to Cohn (1972), estimates of loss can be attained by measuring yield increases as a result of successful nematode control, and by comparing the performance of infected trees with that of uninfected trees. Data available on yield increases from citrus nematode control in various citrus-growing countries suggest a world average increase of 20-30% in citrus yield. Cohn (1972) suggests that actual reduction in world citrus yields due to citrus nematode could be estimated at 8.7-12.2% because all infected citrus trees are not economically damaged by this nematode.

B. Life Cycle

The life cycle consists of several stages, with the complete cycle from egg to egg requiring four to eight weeks, depending upon temperature (Cohn, 1965b; Macaron, 1972; O'Bannon et al., 1966; Van Gundy, 1958). Nematodes emerge from eggs after having been second-stage juveniles for 12-14 days. The males undergo four molts, changing in length and width within seven to ten days and becoming slightly smaller than second-stage juveniles. The male stylet becomes indistinct and the esophagus degenerates; males can neither feed on nor infect roots. Juveniles destined to become females also undergo four molts

Fig. 2 Colony of citrus nematode females on a citrus root.

and become embedded in the root, where they feed on cortical cells and develop to maturity. About one-fourth of the anterior portion of the female body is within the root, usually four to five cells deep, but it never penetrates beyond the cortex. The head of the nematode is located in a cell that is void of content and from which the nematode feeds on surrounding cells, called "nurse" cells (Van Gundy and Kirkpatrick, 1964). After a feeding site is established the body of the nematode becomes immobile, with its posterior portion exterior to the root. This portion, outside the root, becomes greatly enlarged at maturity (Fig. 2). The adult female excretes a gelatinous matrix from the excretory pore (Maggenti, 1962) in which the eggs are deposited, forming an egg mass about the female body.

Van Gundy (1958) observed that hand-picked second-stage juvenile females developed and reproduced in the absence of males. Both male and female juveniles were produced by these unfertilized individuals.

C. Associated Symptoms

1. Roots

The citrus nematode does not cause galling or knots on roots; however, the gelatinous matrix exuded by the female in which nematode eggs are embedded allows soil to adhere to roots at the point of infection. Roots thus look larger in diameter than healthy roots (Fig. 3). Actually, infected roots are only slightly enlarged at these places, owing to the surface being rough instead of smooth. This condition can be detected readily with a microscope. The feeding of nematodes in the cortex results in cell breakdown at feeding sites. Secondary microorganisms invade infected tissue, causing dark necrotic lesions within the cortex. Thus, the cortex of heavily infected feeder roots decays and readily separates from the axial portion of the infected area. When this happens, the cortical tissues slough off, exposing the central cylinder, and resulting in eventual death of that portion of the root (Cohn, 1965a; Schneider and Baines, 1964; Van Gundy and Kirkpatrick, 1964). Such feeding kills many roots over a period of years causing a gradual "dieback" of the tree canopy which is known as slow decline.

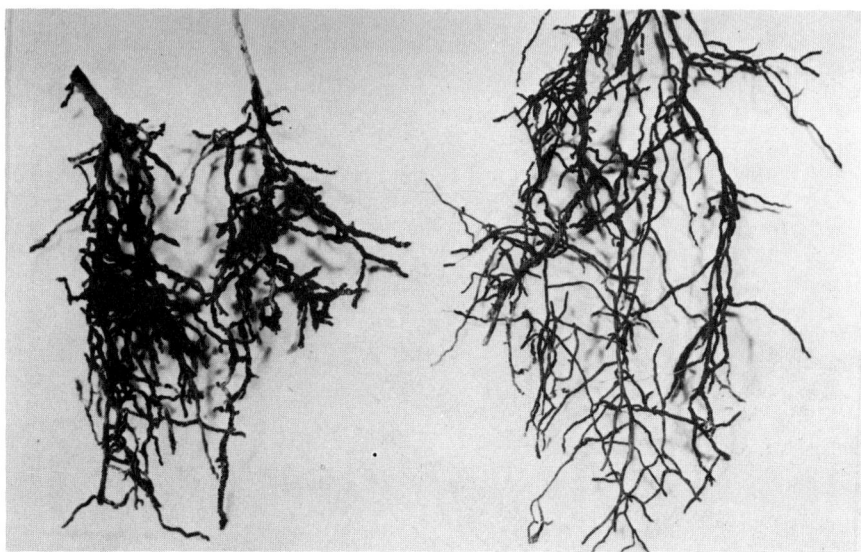

Fig. 3 Left, citrus roots heavily infected with citrus nematodes. Soil particles and root debris adhere to gelatinous matrix deposited on the root surface
by egg-laying female citrus nematodes. Right, appearance of roots from a
tree to which nematode control measures were applied.

2. Above Ground

Not all citrus trees having a large citrus nematode population parasitizing roots
show above-ground symptoms. Symptom expression may not be noticeable
until several years after the peak of the nematode population is reached.
Severity of symptoms is determined by the care the grove received and overall
vigor of trees. When above-ground symptoms finally occur they consist of
yellowing of leaves and general symptoms of malnutrition (sparse foliage, small,
nonuniform fruit, and defoliated branch ends). The symptoms are particularly
noticeable in the uppermost portion of a tree (Fig. 4A). Infected trees shed
more leaves than uninfected trees, or trees that have been treated with nematicides (Fig. 4B). This is particularly evident during late winter or early
spring, and during periods of environmental stress. Degree of decline and
symptom expression may vary considerably from tree to tree within a grove
and from one grove to another.

D. Progress of Disease

Influence of the nematode on its host develops in two ways. If citrus nematode-
infected nursery trees are planted in grove soil free of citrus nematodes, trees
generally grow and will produce fruit satisfactorily for several years while
nematode populations are increasing on the roots. Eventually, heavily infected
trees will decline, which results in a reduction of fruit yield and quality. Cohn
et al. (1965) reported that in Israel this level is approximately 40,000 larvae
per 10 g of feeder roots and is usually attained in orchards in 12-17 years
after planting infested seedlings. Conversely, noninfected nursery trees
planted in well-infested grove soil may show symptoms (Fig. 5) as early as five
years after planting (O'Bannon and Tarjan, 1973).

Fig. 4 (A) Citrus nematode-infected grapefruit tree on sour orange rootstock showing typical symptoms of defoliated branch ends, small leaves, and leaf yellowing mainly in upper portions of tree. (B) Same tree two years after treatment for control of the citrus nematode.

Fig. 5 Five-year-old citrus tree infected with citrus nematodes growing in nonfumigated soil. (Note typical wilt symptoms during drought.)

The condition of infected trees depends on severity of nematode attack and can vary among trees from year to year. Presently there is little evidence to suggest that toxic substances or other metabolic products that enhance tree decline are introduced into trees by the nematode as it feeds, but the possibility exists that other than physical factors may alter host response. We do know that decline results from the debilitative effect of nematode numbers on feeder roots. Reynolds and O'Bannon (1963a), Cohn et al. (1965), and Bindra et al. (1967) correlated nematode infestation rates with decline symptoms by studying the relationship between tree performance and nematode populations. General tree vigor was correlated with the condition of the root system. Trees in early stages of decline still have a rather vigorous root system that enables the roots to support a large nematode population. Trees in advanced stages of decline have deteriorated root systems that generally support fewer nematodes. Reynolds and O'Bannon (1963a) found that if nematode numbers are low some tree recovery occurs as new roots develop; an increase in nematodes to damaging numbers is followed by tree decline. Thus, nematode population and root system reach an unsteady equilibrium. With general health of the tree depending on the condition of the root system, the tree can recover only to a limited extent because of the increase in nematode population. On the other hand, the nematode population can increase only to a limited extent because of root damage. Since the citrus nematode alone will not kill trees, this cyclic condition continues indefinitely.

The role of secondary organisms in influencing disease progress is an important part of the disease complex. Cobb (1914) was the first to observe that other organisms may be associated with the citrus nematode disease syndrome; Thomas (1923) showed an association with *Fusarium solani*. Later, Van Gundy and Tsao (1963) demonstrated that reduction in growth of citrus seedlings due to citrus nematode and *F. solani* combined was greater than either

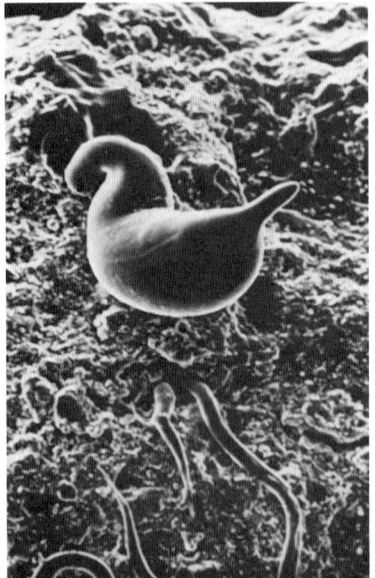

A

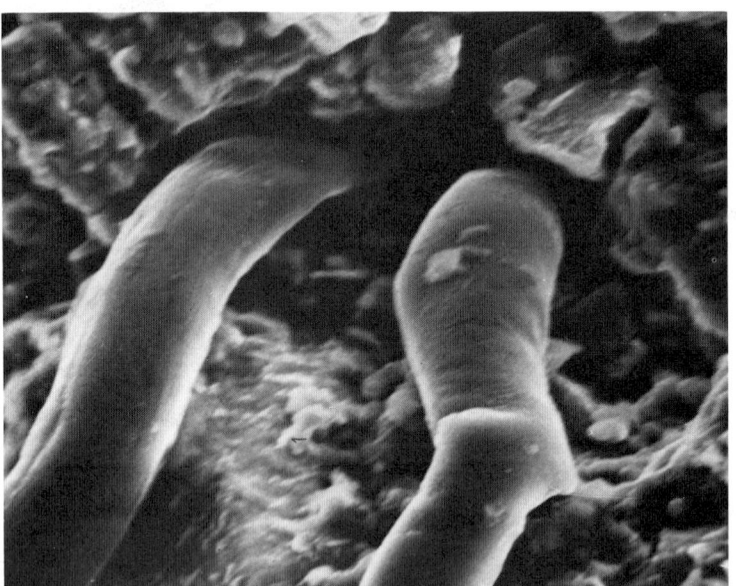

B

Fig. 6 (A) Scanning electron photomicrograph of citrus nematode adult female and juveniles invading citrus root. (B) Close-up showing cavity created by nematode entry. (Courtesy of R. Inserra and N. Vovlas.)

alone. The deleterious effect of other microorganisms invading nematode-infected roots (Fig. 6) resulting in tissue necrosis was reported by Cohn (1965a).

E. Environmental Effects on Nematode Infection, Development, and Survival

1. *Effect of Moisture and Temperature*

Soil moisture and temperature favorable for growth of citrus are generally favorable for nematode development and reproduction. However, certain environmental conditions may directly or indirectly influence development. Gutierrez (1947) demonstrated the extreme sensitivity of the nematode to lack of moisture; when he exposed infected roots to air and sunlight for 24 hours, juveniles and most of the eggs were killed. Feldmesser and Rebois (1963) found that eggs and second-stage juveniles in root debris can withstand temperatures as high as 45°C for several hours. Van Gundy et al. (1964) found that nematode population levels varied with soil moisture. In fine-textured soils populations were favored by water suctions between 0 and 60 centibars, whereas in coarse-textured soils they developed best between 0 and 10 centibars. Labanauskas et al. (1965) reported that more nematodes were found around roots from dry soils (60 centibars) than in comparable samples from wet soils (9 centibars). According to Stolzy et al. (1963), nematode reproductive processes in "wet" soils were slowed down because of lower oxygen diffusion rates.

Soil temperatures most favorable for infectivity and development of the citrus nematode on citrus are between 25°C and 30°C, whereas at 20°C and 35°C only slight infection occurs (Baines, 1950; O'Bannon et al., 1966). Active larvae were recovered by Baines (1950) from moist soil held at 15°C for 2.5 years, but at 33°C they survived no longer than 2.5 months. Van Gundy et al. (1967) found that a decrease in infectivity was associated with a corresponding decrease in motility and in body contents of second-stage juveniles held at 27°C in soil and in water. Survival of infective juveniles was more pronounced in soil than in water and was associated with retention of body contents. Juveniles stored in vitro were motile and infective for 128 days. Cohn (1966) reported that approximately 70% of citrus nematode juveniles stored in vitro at 10°C were infective after 24 months.

Reynolds et al. (1970) studied the effect of moisture and temperature on citrus nematode survival under field conditions for 18 months and reported a 95% reduction in recoverable nematode populations during the first two months. Nematode longevity was enhanced when soil moisture was near optimum for citrus growth and when average daily soil temperatures did not exceed 30°C. High summer soil temperatures around young trees aided in reducing citrus nematode populations, according to Reynolds and O'Bannon (1963b). They reported that temperatures in unshaded moist soil around young trees at depths of 2.5-50 cm were greater than 35°C and that nematode populations remained low. Populations increased when conditions became more favorable for the nematodes, as the trees became larger, providing more shade and thus changing the soil environment. Even though nematode-infected trees are removed and a grove is not replanted, roots may survive deep in the soil for many years. Baines et al. (1962) found that citrus nematodes survived up to nine years after infected trees were removed.

2. Influence of Soils on Infection and Reproduction

The citrus nematode occurs over a wide range of soil conditions throughout the world. Certain soil factors readily influence infection and reproduction. Van Gundy et al. (1964) found that growth and reproduction of citrus nematodes occurred on citrus seedlings in soils containing 5-50% clay. The rate of reproduction, however, was significantly lower in soils of 50% clay than in soils containing 5, 15, or 30% clay. In general, citrus nematode invasion and reproduction is slower in very sandy, coarse-textured soils than in other soil types (Baines, 1974b; O'Bannon, 1968; Van Gundy et al., 1964; Vilardebo, 1963b).

Since nematode infection of roots occurs more slowly in coarse sand (O'-Bannon, 1968), it takes longer for a population to increase; thus, trees are

able to establish a more vigorous root system before severe attack by the nematode. In time, nematode numbers will increase to damaging proportions on
the extensive root system. Conversely, soils containing organic matter (up
to 9%) favored infection and rapid increase in nematodes that resulted in early
damage. Van Gundy (1958) found that organic debris created a thin protective cover over citrus roots that enhanced nematode infectivity.

Soil type apparently has little influence on citrus nematode migration.
Baines (1974b) and Tarjan (1971) found that nematode mobility in various
soils was limited. Tarjan (1971) reported that maximum citrus nematode
movement in two soil types was 53.3 cm in two years. Many years of field observations by the authors indicate that nematode migration from infected to
noninfected trees rarely occurs except where roots overlap or the nematodes
are moved by drainage or irrigation water (Meagher, 1967).

Citrus nematodes survive and reproduce best at pH 6 to 7.5 (Martin and
Van Gundy, 1963; Van Gundy et al., 1964). Although soil pH does affect
population levels, populations will exist at pH extremes and will influence tree
growth (Van Gundy and Martin, 1961; Reynolds et al., 1974).

An important aspect of the influence that citrus nematodes have on tree
debilitation is the parasite's effect on root cells. It seems to alter the semi-
permeable nature of cells, allowing trees to imbibe greater concentrations of
some elements and less of others. When high concentrations of salts occur in
the soil, particularly when trees are irrigated with water of high salt content,
leaf concentrations detrimental to the tree often result (Baines et al., 1962;
Van Gundy and Martin, 1961). With nematode control a healthier root system
develops, resulting in an improvement in the aerial portions of the tree (Reynolds and O'Bannon, 1963a).

3. Seasonal Variation

Environmental influences of climate, soil, food supply, and competition are
principal factors regulating nematode activity. Superimposed on these is population variation due to seasonal growth habits of the host. In Florida, O'Bannon et al. (1972) found that peak populations developed during corresponding
periods of increased root growth that occurred in April-May and November-
December. Infection and subsequent population cycles are restricted to primary roots because citrus nematodes feed only in the cortex of primary roots
(Van Gundy and Kirkpatrick, 1964). Macaron (1972), Toung (1963), and
Vilardebo (1964) also recognized two high and low population periods. Several
other investigators observed some seasonal variation (Prasad and Chawla, 1965;
Yokoo, 1964); however, Cohn (1966) in Israel found no correlation between
population fluctuations.

F. Host Differentiation and Biotypes

Cobb (1914) first reported that trifoliate orange (*Poncirus trifoliata*), grapefruit (*Citrus paradisi*), sweet orange (*C. sinensis*), and sour orange (*C. aurantium* were susceptible to citrus nematodes. Since then, 75 rutaceous
species, 50 of which are *Citrus* or citrus hybrids, and nine nonrutaceous
hosts have been reported (Vilardebo and Luc, 1961; Cohn and Milne, 1977;
Cohn and Milne, 1977).

DuCharme (1948) first found that a selection of trifoliate orange was resistant to the citrus nematode. Others (Baines et al., 1948, 1958) also re-

ported certain selections of trifoliate orange, and some other plants botanically close to citrus, were highly resistant or immune to the citrus nematode. Later, Cameron et al. (1954) found the F_1 hybrid seedlings from crosses with susceptible citrus species and resistant strains of *Poncirus trifoliata* were resistant to the citrus nematode. Cohn (1965b) and Feder (1968) found variation in degree of resistance among selections of *P. trifoliata*. Swingle (1946) developed two commercially acceptable hybrids (Troyer and Carrizo citranges) considered resistant to the citrus nematode. Troyer and Carrizo have been found to be susceptible by Hutchison et al. (1972), but Troyer is considered only moderately resistant according to Baines et al. (1969), Meagher (1969), Muthukrishnan et al. (1975), and Scotto La Massese (1965b).

Van Gundy and Kirkpatrick (1964) studied host resistance and identified three reactions in cultivars resistant to the citrus nematode. These are: (1) a hypersensitive cell reaction to feeding of the nematode, (2) a formation of wound periderm in the root cortex, and (3) a toxic factor in the root juice. Baines et al. (1969) reported host preference of field populations of the citrus nematode and demonstrated the existence of biotypes on citrus. Studies based on host reaction by others (Lamberti et al., 1976; Stokes, 1969) confirm the existence of other *T. semipenetrans* biotypes. In Japan, a citrus nematode biotype readily attacks *P. trifoliata* (O'Bannon and Ford, 1977). A similar biotype was found in California (Baines et al., 1974) and was named the *Poncirus* biotype (O'Bannon et al., 1977). Due to host differentiation, four biotypes, Poncirus, Citrus, Mediterranean, and Grass, were recognized by Inserra et al. (1980).

Trifoliate orange and their hybrids continue to be the major source of resistance to citrus nematodes (O'Bannon and Ford, 1977; McCarty et al., 1979); however, Baines (1974a) reported that a selection of sour orange also was highly tolerant.

G. Nonchemical Control

1. *Cultural*

Management of nematodes without pesticides has resulted in investigation of several methods to reduce detrimental effects of the pest, thus enabling infected trees to successfully cope with the root parasites. Methods that have been investigated are application of inorganic and organic amendments to soils, growing antagonistic plants, use of electricity, gamma rays, or sound, flooding, fallowing, burning, soil stirring, and rotation (Tarjan and O'Bannon, 1977).

2. *Biological*

Soil organisms known to be parasites and predators of nematodes include a number of bacteria, viruses, fungi, protozoa, turbellarians, tardigrades, collembola, mites, and predacious nematodes. A recent comprehensive review (Stirling and Mankau, 1977) has indicated the potential value of biological organisms in regulating nematode populations. At present, however, there are no such organisms the use of which offer reliable economic control. Characteristics required by a successful natural enemy are mobility and the ability to seek prey, adaptability to the environment, host specificity, synchronization with the host, and ability to survive host-free periods (Huffaker et al., 1971).

H. Chemical Control

1. Preplant Treatment

Trees planted in infested soil from an old grove are readily infected with the surviving soil nematode population. To avoid this, preplant fumigation is important for the establishment of the young resets. This is also true for nematode-infested citrus nurseries, or for individual tree sites within a grove. In California and Florida, preplanting applications of nematicides have been found effective for citrus nematode control (Baines et al., 1956; O'Bannon and Tarjan, 1973). Applications by chisel of such chemicals as 1,3-dichloropropene nematicides (D-D and Telone) give effective nematode control in most soils. Other compounds, such as methyl bromide, chloropicrin, ethylene dibromide, Vapam, and Vorlex, when properly applied, have provided effective nematode control, particularly in nurseries or as tree site treatments (Baines et al., 1957; Baines et al., 1966; Cohn et al., 1968; Hannon, 1964; O'Bannon and Bistline, 1969).

It is now known that certain fumigants have a deleterious effect on vesicular-arbuscular mycorrhizal fungi, which are beneficial to citrus (Kleinschmidt and Gerdemann, 1972). Citrus mycorrhizae are widely distributed (Menge et al., 1975), and citrus plants grow poorly, or not at all, when mycorrhizae are not present (Kleinschmidt and Gerdemann, 1972; Marx et al., 1971).

Recently a number of workers have reported effects of certain fumigants on citrus mycorrhizae (Menge et al., 1977; Milne, 1974; Nemec and O'Bannon, 1979; O'Bannon and Nemec, 1978; Timmer and Leyden, 1978), which has resulted in the suppression of symbiotic endomycorrhizae. Precautions must be observed when certain of these fumigants are used in replant or in nursery fumigation to prevent seedling growth suppression.

2. Postplant Treatment

Treating living trees in place with chemical compounds that are not toxic to the tree but that provide effective nematode control has spurred the testing of many compounds. DBCP (1,2-dibromo-3-chloropropane) has received widespread use by the citrus industry for citrus nematode control. The chemical has been injected into the soil and applied as an aqueous emulsion, either with perforated irrigation pipe (Fig. 7) or sprayed on the soil (Fig. 8). It also has been pressure injected into the soil (Fig. 9) and used as a transplant drench (Fig. 10). Since the chemical has been found detrimental to humans, it no longer is approved for use. A search for alternate compounds has shown that several nonvolatile nematicidal compounds provided nematode control, were not phytotoxic, and/or resulted in yield increase (Baines and Small, 1969; Baines et al., 1977; Elgindi et al., 1976; Heald, 1970, 1972; Milne, 1977; Milne and DeVilliers, 1977; Radewald et al., 1973; Tarjan, 1976; Timmer, 1977; Vilardebo et al., 1975; Vovlas et al., 1977).

Materials most actively investigated include aldicarb, carbofuran, diazinon, fensulfothion, oxamyl, and phenamiphos. Only aldicarb is currently registered for use on orange trees.

Nonfumigants can be incorporated with the soil, applied in irrigation water, or sprayed on foliage. Some of these compounds, such as aldicarb, oxamyl, and phenamiphos, have systemic properties. Most of these compounds are highly toxic to warm-blooded animals and must be handled with extreme caution. Many questions still arise when using the newer compounds on established citrus or as preplant treatments. With the loss of DBCP these problems will have to be addressed.

Fig. 7 Application of DBCP to citrus nematode-infected grapefruit trees by using perforated irrigation pipe.

Fig. 8 DBCP sprayed on soil in which nematode-infected citrus trees are growing.

Fig. 9 Pressure injection of DBCP into a citrus nematode-infested grove.

Fig. 10 Basin treatment of a citrus replant with DBCP emulsion.

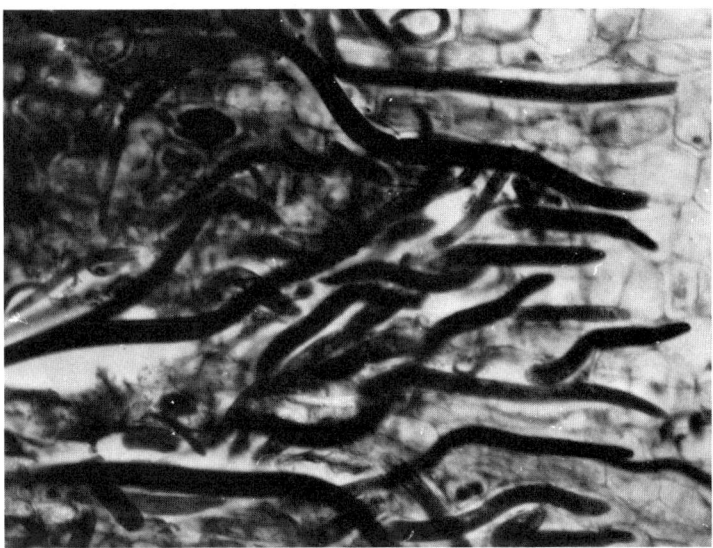

Fig. 11 Burrowing nematodes (*R. similis*) colonizing a citrus rootlet.

III. SPREADING DECLINE DISEASE

This disease is aptly named because it is capable of spreading at an average rate of 15.2 m/year. It is, for all practical purposes, a disease limited mainly to the deep, sandy soils of the central "ridge" area of Florida. Although the causal organism, *Radopholus similis* (Cobb), can occur in heavier citrus soils in other areas of Florida, it is not as destructive to citrus trees as in the lighter, well-drained sandy soils of central Florida.

The malady was first noted in 1928, but it was not until two decades later that concerted investigations were begun to find the cause (Suit, 1947). In 1953, *R. similis* was shown to be the pathogen (Suit and DuCharme, 1953). It has been shown that spreading decline causes a reduction in yields of grapefruits and oranges of 40-80% (Suit and DuCharme, 1967).

A. Associated Symptoms

1. Below Ground

The nematode is migratory and can penetrate citrus root tips from the region of the root cap back to where suberization of epidermal cells occurs (Fig. 11). Destruction of the apical meristem will prevent terminal growth, resulting in stubby, somewhat swollen root tips. DuCharme (1968) reported that all types of root parenchymatous tissue could be attacked. He noted that the phloem-cambium ring was a preferred feeding site and could be destroyed accordingly. At the sites of nematode infection, brown-black lesions can be formed which may coalesce to form a canker. Whereas only about 30% of feeder roots are killed at depths between 25 and 75 cm, 90% of the roots are destroyed at depths below 75 cm (Ford, 1953). An infected tree can have a reduction of functional feeder roots of 50% and more.

Fig. 12 Citrus tree infected with burrowing nematodes and showing typical spreading decline symptoms.

Fig. 13 Sparse foliage in canopy of tree affected by spreading decline.

Fig. 14 Healthy citrus tree showing full foliar canopy.

2. Above Ground

In general, infected trees show symptoms of debilitation (Fig. 12) best illustrated by the sparse foliage in the tree canopy (Fig. 13). One can stand by
the trunk of a declined tree and look up through the thin foliage at patches of
sky; this is not possible with normal healthy trees (Fig. 14). Though these
symptoms can easily be applied to most trees sustaining debilitating nemic infections, spreading decline is unique because infected trees show about the
same degree of decline, and the margin of declining trees spreads each year.
Infected trees wilt more readily than healthy trees during droughts and will
react adversely to environmental stress conditions much more than normal
trees. Leaves are small and sparse, branches show twig dieback, and fruit
is small sized and of lowered yield. Feldman and Hanks (1964; 1965) reported
that nutrient levels in leaves of infected trees are affected. They found that
more protein amino acids occurred in roots of decline plants than in healthy,
and that arginine accumulated in decline roots.

   Infected trees do not always show symptoms; on several occasions the
authors have observed trees with root infections of *R. similis* that do not have
a decline appearance. It can be speculated that this may be due to several
factors of which grove care, soil moisture, and tree vitality are some.

B. Nematode Species

R. similis is a migratory, endoparasitic nematode closely related to the root
lesion nematodes, *Pratylenchus* spp. Adults of *Radopholus* show sexual dimorphism and females have a vulva medially situated (Fig. 15), whereas *Pratylenchus* adults are not sexually dimorphic and the vulva is posterior on the body
of the female. The life cycle of *Radopholus* in citrus roots requires 18 to 20
days at a soil temperature of 24-27°C. It has been reported that first- and
second-generation females can produce ova parthenogenetically (DuCharme
and Price, 1966). The nematode feeds on parenchyma cells of the cortex and
stele, where eggs are also deposited. Eggs hatch in three to seven days, and

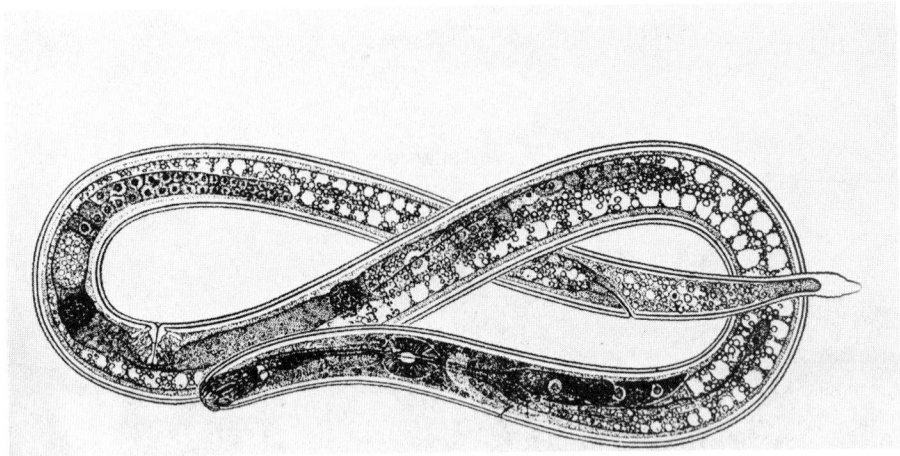

Fig. 15 A female of *R. similis*, the burrowing nematode. (From N. A. Cobb, 1915.)

the juveniles can feed on the same sites of infestation if the tissues remain free of decay caused by other invading microorganisms. It has been determined that as many as 739 nematodes can be dissected from root lesions on mature trees, but as many as 30,000 individuals were estimated in large lesions on roots grown in monoexenic culture (DuCharme and Price, 1966).

C. Nematode Behavior

Longevity of the nematode in sandy soil free of citrus roots has been determined to be six months at most (Tarjan, 1961b). *R. similis* was found experimentally to be more motile, particularly in light-textured soil, than the two other important citrus nematodes *Tylenchulus semipenetrans* (Cobb) and *Pratylenchus coffeae* (Zimmermann) (Tarjan, 1971). The nematode can move in all directions, including in an upward direction through the soil. Dispersal of the nematode was found to be aided by the flow of soil water (DuCharme, 1955), by the intergrove movement of mechanical equipment (Tarjan, 1956), and by the movement of infected nursery stock and ornamental plants (Poucher, 1967b).

D. Host Plants

This polyphagous nematode can attack 1275 species and relatives of citrus, and 250 tropical and subtropical plants as well (Ford et al., 1960). However, not all populations found may be pathogenic to citrus since biotypes (e.g., physiologic races) are known that attack only bananas, only citrus, or both bananas and citrus (DuCharme and Birchfield, 1956). Whereas such biotypes had been morphologically indistinguishable from one another, recent studies with the scanning electron microscope on two biotypes of burrowing nematodes have revealed minute differences in head annulations and vulvar patterns (Baldwin et al., 1978). Recent cytogenetic studies have shown that the chromosome number of the banana biotype is n = 4; that of the citrus biotype is n = 5

(Huettel and Dickson, 1981). According to Huettel (1982), there are sufficient biological differences between the banana race and citrus race to consider them genetically separate. She elevated the citrus race to species rank and renamed it *Radopholus citrophilus* for its well established association with *Citrus* spp. in Florida. Since it is indistinguished morphologically from *R. similis*, it satisfies the definition of a sibling species. It is the first true sibling species to be described among the plant-parasitic nematodes (Huettel, 1982).

E. Environmental Factors

Burrowing nematodes attacking citrus are mainly found at soil depths of 60-150 cm (DuCharme, 1967) and can invade roots and reproduce at temperatures of 12- 30°C. Their most rapid population increases occur at 24°C (DuCharme, 1969). Spreading decline is found only in the sandy ridge zone of Central Florida, where soils are well drained and have a relatively deeply situated clay layer. The type of soil in which infection occurs can influence symptomatology. O'Bannon and Tomerlin (1971) found that *R. similis* was less pathogenic to greenhouse-grown citrus seedlings in a Leon sandy loam than in Lakeland fine sand. Hughes (1955) observed that citrus growing in soil with more than 3% clay would not be affected by the nematode. The nematode has been found in soils having 0.5% moisture (Feldmesser and Feder, 1957) or, under greenhouse conditions, when soil moisture was kept between 75 and 100% of field capacity (O'Bannon and Tomerlin, 1971). Yet, the nematode frequently cannot be found in known infested groves during droughts occurring from January to May when its population can drop to imperceptible levels.

F. Decline Complexes

The malady known as spreading decline is caused primarily by the burrowing nematode, but not wholly by it. Various fungi, oligochetes, and mites invariably are associated with the root lesions (DuCharme, 1968). These organisms must be considered as important secondary root invaders that contribute to spread of the disease by rendering unsuitable the nematode feeding sites, thus forcing the migratory nematodes to invade healthy root tissues. An interesting, yet controversial, test was reported by Feldmesser et al. (1959) who applied captan (trichloromethyl-mercapto-cyclohexene-dicarboximide) at 50 lb/acre (56 kg/ha) to sour orange seedlings infected primarily with burrowing nematodes. After six months they recovered almost twice as many burrowing nematodes from the roots as from the controls. The captan-treated plants were of a darker green coloration and had root systems twice as heavy and aerial parts more than three times as heavy as the controls. Their modest interpretation of the results stated: "Apparently application of captan controlled some factor other than nematodes affecting growth of the plants."

G. Control

Spreading decline has been unique by the amount of concentrated research effort that has been directed toward its control. During the period from the mid-1950s through the 1960s, teams of Florida nematologists and plant pathologists at the University of Florida Citrus Experimental Station at Lake Alfred, U.S. Department of Agriculture Horticultural Research Laboratory at Orlando, and Florida Department of Agriculture, Division of Plant Industry at Winter Haven and Gainesville, worked almost continuously on the problem.

Fig. 16 Bulldozer pushing out trees affected by spreading decline. (Courtesy of Division of Plant Industry and Consumer Services, Florida Department of Agriculture – "Florida DPI", Gainesville, Florida.)

Fig. 17 Fumigating a nematode-infested citrus grove site. (Courtesy of Florida DPI.)

Fig. 18 A replanted section of citrus grove amid older trees. (Courtesy of Florida DPI.)

Fig. 19 A barrier protecting a healthy citrus grove from trees infected with burrowing nematodes which surround it. (Courtesy of Florida DPI, Gainesville, Florida.)

The most recommended means of control was the "push and treat method," which involved removing all infected trees (Fig. 16), plus a two tree margin around the infected area, burning the trees, deep plowing the remaining roots, treating the soil with dichloropropene-dichloropropane fumigant (D-D) (Fig. 17), maintaining the treated area free of plants for six months, and waiting for two years before replanting with citrus (Fig. 18) (Poucher, 1967a). This method is not always effective because some roots remaining in the soil are not always killed by the fumigant and these serve as a breeding site for surviving nematodes. The barrier or buffer zone method (Suit and Feldman, 1965) is also employed for protecting a healthy grove from adjacent infested groves. In this method a barrier zone measuring 16 feet (4.9 m) to 100 feet (30.5 m) wide is fumigated with ethylene dibromide (EDB) every six months while the surface of the barrier is maintained free of vegetation (Fig. 19). Such methods may be effective, but they are not infallible, since a small percentage of "escapes" do occur.

The best possible control for any disease is its prevention, which can only be accomplished by excluding the pathogen from its host. The regulatory program of the Florida Division of Plant Industry that has been in effect since the control program was started has been instrumental in preventing excessive and uncontrolled spread of the disease.

A number of citrus, citrus relatives, and hybrid rootstocks have been tested for resistance, but only a limited few have been worthy of release to growers. Milam, a lemon, and Ridge Pineapple, a sweet orange, were two of three rootstocks released in 1964 (Ford and Feder, 1964); the former has been rather extensively used. Carrizo citrange and Algerian navel orange were also recommended (Ford and Feder, 1969).

Chemotherapy of infected citrus trees has received considerable attention. Suit (1969) described a number of trials with various chemicals dating back to 1954 but suggested ineffectiveness of control since his goal was *eradication* of the pest. Although DBCP (dibromochloropropane) has been used with favorable results for the control of the citrus nematode, it was not as effective against the burrowing nematode (Feldmesser and Feder, 1956). Under experimental field conditions, oxamyl and phenamiphos provided nematode control and improved growth of trees, which resulted in increased yield (O"Bannon and Tomerlin, 1977; O'Bannon and Tarjan, 1979).

Attempts at biological control of burrowing nematodes using nematophagous fungi (Tarjan, 1961a) and African marigolds (Tarjan, 1960) (Fig. 20) were unsuccessful. Applications of kelp derivatives (Tarjan, 1977b) and municipal refuse compost (Tarjan, 1977a) (Fig. 21) have been found to be mildly beneficial. A large experiment was conducted from 1964 to 1967 testing the effect of compost, nutritive foliar spray, root pruning, lime, and organic fertilizer on burrowing nematode-infected citrus (Tarjan and Simmons, 1967). Both favorable and unfavorable responses resulted from each combination of treatments. Still another, performed by the authors, investigated the effect of ammonium nitrate fertilizer and calcium nitrate fertilizer, but was terminated with inconclusive results (Tarjan and O'Bannon, 1980). The prospects of biological control of nematodes attacking citrus are discussed by Tarjan and O'Bannon (1977) and O'Bannon and Tarjan (1978).

At present, no chemicals are approved for use on burrowing nematode-infected citrus trees. Although phenamiphos and oxamyl have been found effective against the burrowing nematode, neither of these is approved for use. Aldicarb is used in Florida on *Tylenchulus semipenetrans*-infected orange trees (Fig. 22). It also has been reported as effective on Meyer lemons infected primarily with *Belonolaimus longicaudatus* Rau (Tarjan, 1980). It

Fig. 20 African marigolds (*Tagetes erecta*) growing in a grapefruit grove affected by spreading decline.

Fig. 21 Application of municipal refuse compost to a grapefruit grove suffering from spreading decline.

Fig. 22 Application of aldicarb to infested grove soil using a Gandy granule applicator.

would seem unreasonable to assume that aldicarb would be without effect against the burrowing nematode.

IV. CITRUS SLUMP DISEASE

The first published reports on association of lesion nematodes, *Pratylenchus* spp., with citrus were by Suit and DuCharme (1953) in Florida and Jensen (1953) in California. Of the several species of lesion nematodes associated with citrus, only *Pratylenchus brachyurus* (Godfrey), *P. coffeae* (Zimmermann), and *P. vulnus* Allen and Jensen are known to adversely affect citrus. The pathogenicity of *P. brachyurus* on a citrus rootstock was first reported by Brooks and Perry (1967) in Florida, where it was found in 90% of the groves surveyed (Tarjan and O'Bannon, 1969). The pathogenicity of *P. coffeae* was demonstrated in greenhouse tests by Feldmesser and Hannon (1969) and in the field (Fig. 23) by O'Bannon and Tomerlin (1973). A severe stunting of *Citrus aurantium* seedlings infected with *P. vulnus* was detected in citrus nurseries in Italy (Inserra and Vovlas, 1974) but not in established plantings.

   Although *P. brachyurus* was reported to cause growth reduction to citrus seedlings in greenhouse tests (Brooks and Perry, 1967), field studies have shown that young trees were more adversely affected than older ones, and that damage usually decreased with plant age (O'Bannon et al., 1974). *P. coffeae*, however, was found to be much more pathogenic to citrus (Radewald et al., 1971) and was also reported to cause citrus tree decline in other parts of the world (Huang and Chiang, 1976; Siddiqi, 1964; Yokoo and Ikegemi, 1966).

   P. coffeae invades extensively the cortex of feeder roots (Fig. 24) in much the same manner as *R. similis*, where it creates cavities and cell necrosis during feeding and migration. The vascular region is not usually invaded

Fig. 23 Influence of *P. coffeae* on tree growth. (A) Five-year-old infected tangelo tree showing decline symptoms. (B) Five-year-old uninfected tree. Both trees on sour orange rootstock.

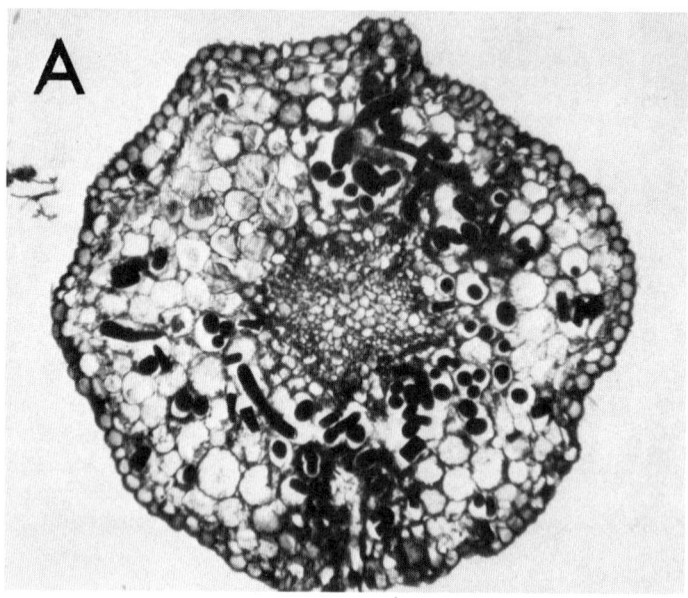

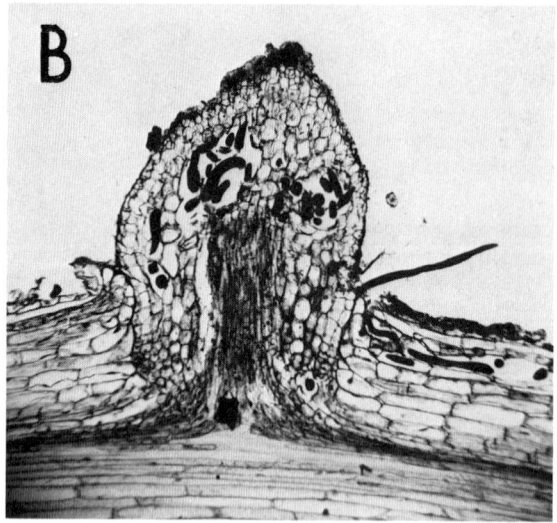

Fig. 24 (A) Cross section of citrus root heavily infested by *P. coffeae*. (B) Longitudinal section of citrus root showing invasion and infection by *P. coffeae*.

unless infection is massive in that area. Population densities can increase rapidly, reaching peak populations greater than 10,000 nematodes per gram of root in just a few months (O'Bannon et al., 1976). It can survive in, and cause growth retardation to, citrus plants growing in several soil types and can survive without host roots more than four months (Radewald et al., 1971). Host testing in the greenhouse showed that 125 citrus species, hybrids, and relatives are known hosts of *P. coffeae*. Four *Microcitrus australis* × *M. australasica* hybrids and one *Poncirus trifoliata* selection appeared to have some resistance (O'Bannon and Esser, 1975).

In Italy, *P. vulnus* has been detected only in citrus nurseries, where it was the cause of poor seedling growth. Invasion and subsequent root damage are similar to that reported for *P. coffeae*. Initial population densities of 100 *P. vulnus* per gram of root resulted in growth retardation of seedlings in two soil types. Only *C. aurantium* has been shown to support high populations of *P. vulnus* (Inserra and Vovlas, 1977a). Baines et al. (1959) reported that *P. vulnus* disappeared from soils two years after planting rough lemon and sweet orange trees. A later report (Baines et al., 1978) suggested *P. vulnus* reproduced on, and decreased the growth rate of, orange trees with *P. trifoliata* or Troyer citrange rootstock.

V. OTHER NEMATODES PATHOGENIC TO CITRUS

Numerous surveys for nematodes associated with citrus roots and soil have reported a wide variety of known and suspected genera (O'Bannon et al., 1975). Yet, it would be fallacious to accept the presence around roots of any known or suspected pathogenic nematode as indicative of pathogenicity, or even parasitism, without adequate proof. Accordingly, many reported associations of nematodes with citrus will not be reviewed in this chapter.

There exist two excellent accounts of nematodes pathogenic to citrus besides those previously discussed. These are by Baines et al. (1978) and Inserra and Vovlas (1977b). The reader should refer to these publications for amplification of some of the material hereafter presented. Cohn (1972) presents a table listing nematodes known to attack citrus. Among these he lists sting nematodes, *B. longicaudatus* Rau 1958; sheath nematodes, *Hemicycliophora arenaria* Raski 1958; root-knot nematodes (*Meloidogyne* spp.), stubby-root *Paratrichodorus minor* (Colbran 1956); and two species of dagger nema-

Fig. 25 Sting nematode (*B. longicaudatus*) injury on grapefruit seedlings. Left, healthy plant. Center and right, parasitized plants. (Courtesy of F. W. Bistline and V. G. Perry.)

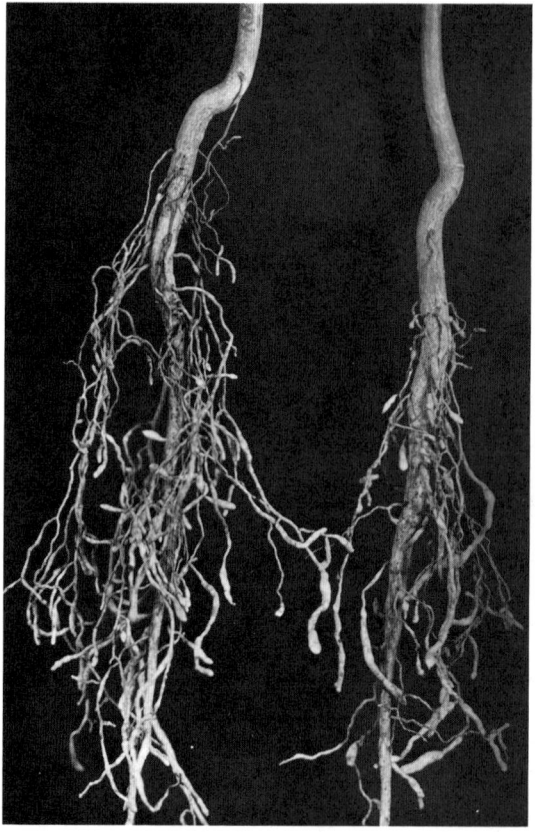

Fig. 26 Troyer citrange roots parasitized by root-knot nematodes, *M. javanica*. (Courtesy of R. Inserra.)

todes (*Xiphinema*) as proven pathogens. Of the above-named nematodes, only *Meloidogyne* spp. are endoparasites; the others have an ectoparasitic feeding habit.

Sting nematodes (*Belonolaimus longicaudatus*) are elongate, polyphagous nematodes that affect citrus only in Florida. The effects of parasitism on citrus were reported by Standifer and Perry (1960), who observed lesions and swellings on roots of grapefruit seedlings (Fig. 25). The senior author has seen similar root symptoms on mature Meyer lemon trees on the West Coast of Florida, where sting nematodes were the predominant soil nematodes.

Sheath nematodes (*Hemicycliophora* spp.) are often found associated with citrus. Only two species are proven parasites that produce plant symptoms. Van Gundy and Rackham (1961) found that *H. arenaria* causes root tip gall formation and can affect growth of rough lemon seedlings. Similar galls on citrus in Australia were reported as being caused by *H. nudata* Colbran 1963 by that author. This genus of nematodes also parasitizes grass roots, which, often being present in citrus plantings, may account for several of the reported associations of sheath nematodes with citrus.

Root-knot nematodes (*Meloidogyne* spp.) are rarely found parasitizing citrus in the United States, but have been reported from other parts of the

world. Chitwood and Toung (1960) reported infections of citrus in Taiwan
and India by a *Meloidogyne* species referred to as the "Asiatic pyroid citrus
nema," and Colbran (1958) reported the occurrence of *M. incognita* Kofoid and
White 1919) on citrus in Queensland, Australia. These and other accounts
report the formation of knots on roots following penetration of tissues by the
second-stage infective juveniles (Fig. 26), but further development into reproductive females does not always occur (Inserra et al., 1978). The authors
on numerous occasions have isolated infective juveniles from citrus root and
soil samples, the roots from which showed no evidence of galling. The presence of these juveniles has been regarded as indicative of the presence nearby of a noncitrus host.

Stubby-root nematodes (*Trichodorus* and *Paratrichodorus* spp.) are
often found in citrus soil, but only a report by Standifer and Perry (1960)
describes the effect of parasitism on roots by *P. minor* to be a reduction of
root elongation. A personal communication (McLeod, in Meagher, 1969) states
that a *Trichodorus* species was associated with stubby-root symptoms and poor
growth of citrus seedlings and trees in Australia. Response to preplant fumigation of the soil by citrus seedlings subsequently planted suggested the
probable pathogenicity of the species to citrus. High populations of *Paratrichodorus porosus* (Allen, 1957) have caused a devitalization of root tips,
cessation of growth, and a stubby-root symptom in greenhouse experiments
(Baines et al., 1978).

Dagger nematodes (*Xiphinema* spp.) are truly cosmopolitan animals with
several species that can impair growth of citrus. High populations of *X. americanum* Cobb 1913 were reported as causing small lesions and swellings close to
the root tip and necrotic shriveling on the older root tissues by Baines et al.
(1978) in California. Greenhouse tests proved that *X. brevicolle* Lordello and
daCosta 1961 and *X. index* Thorne and Allen 1950 could reduce growth of sour
orange seedlings by about 45% (Cohn and Orion, 1970). Species of *Xiphinema*
(and also *Longidorus* and *Trichodorus*) are notorious in their ability to transmit viruses. The senior author spent several years in the early 1960s unsuc-
cessfully attempting transfer of tristeza, exocortis, and xyloporosis viruses
by using nematodes from the roots of the virus-infected trees (Tarjan and
Hannon, 1963; 1964). Transmission of exocortis virus was attempted, and
similar failure was experienced by Nigh and Allen (1967) using *X. americanum*
and *L. elongatus* (deMan 1876); by Madaluni et al. (1972) using *Criconema*,
Criconemoides, *Tylenchulus*, and *Xiphinema*; and by workers in California
using *X. americanum* (Baines et al., 1978). Stokes (1977) gives an excellent
review of this subject.

Some relatively minor occurrences of plant nematodes on citrus involve
Paratylenchus spp. that have been found in California infesting orange and
lemon trees. Baines noted a 20-29% increase in growth when lemon trees were
planted in fumigated soil as compared with those planted in the *Paratylenchus*-
infested soil (Baines et al., 1978). *Rotylenchulus* is a genus of great importance in tropical and subtropical areas. Talamé et al. (1970) found *R. macrodoratus* Dasgupta, Raski, and Sher 1968 parasitizing roots of sweet orange in
Italy. The nematode also has been found in close association with citrus roots
in the southern part of Florida; however, documented accounts of its pathogenic behavior are still lacking.

Two additional genera of nematodes are prevalent in Florida where citrus
is grown. These are the ring nematodes, *Criconemella* and *Hemicriconemoides*.
The former genus is almost invariably isolated from citrus soil and root samples when sugar elutriation techniques are employed. Although no pathogeni-

city tests have been conducted, it is a foregone conclusion that the ring nematodes are citrus root parasites.

REFERENCES

Anon. (1979). *Citrus Summary, 1979.* Florida Crop and Livestock Reporting
    Service.
Baines, R. C. (1950). Nematodes on citrus. *Calif. Agric.* 4: 7.
Baines, R. C. (1964). Controlling citrus nematode with DBCP increases
    yields. *Calif. Citrogr.* 49: 222, 233.
Baines, R. C. (1974a). Susceptibility and tolerance of eight citrus rootstocks
    to the citrus nematode, *Tylenchulus semipenetrans* (abstract). *J. Nema-
    tol.* 6: 135.
Baines, R. C. (1974b). The effect of soil type on movement and infection rate
    of larvae of *Tylenchulus semipenetrans.* *J. Nematol.* 6: 60-62.
Baines, R. C., Cameron, J. W., and Soost, R. K. (1974). Four biotypes of
    Tylenchulus semipenetrans in California identified, and their importance
    in the development of resistant citrus rootstocks. *J. Nematol.* 6: 63-66.
Baines, R. C., and Clarke, O. F. (1952). Some effects of the citrus root ne-
    matode on the growth of orange and lemon trees (abstract). *Phytopath-
    ology 42*: 1.
Baines, R. C., Clarke, O. F., and Bitters, W. P. (1948). Susceptibility of
    some citrus species and other plants to the citrus-root nematode, *Tylen-
    chulus semipenetrans* (abstract). *Phytopathology 38*: 912.
Baines, R. C., Clarke, O. F., and Cameron, J. W. (1958). A difference in
    the pathogenicity of the citrus nematode from trifoliate and sweet orange
    roots. *Phytopathology 48*: 391.
Baines, R. C., Foote, F. J., and Martin, J. P. (1956). Fumigate soil before
    replanting citrus for control of the citrus nematode. *Citrus Leaves 36*:
    6-8; 25; 27.
Baines, R. C., Klotz, L. J., DeWolfe, T. A., Small, R. H., and Turner, G.
    O. (1966). Nematocidal and fungicidal properties of some soil fumigants.
    Phytopathology 56: 691-698.
Baines, R. C., and Martin, J. P. (1953). Effect of soil fumigation on growth
    and yield of Valencia orange trees (abstract). *Phytopathology 43*: 465-
    466.
Baines, R. C., Martin, J. P., DeWolfe, T. A., Boswell, S. B., and Garber,
    M. J. (1962). Effect of high doses of D-D on soil organisms and the
    growth and yield of lemon trees (abstract). *Phytopathology 52*: 723.
Baines, R. C., Miyakawa, T., Cameron, J. W., and Small, R. H. (1969). In-
    fectivity of two biotypes of the citrus nematode on citrus and some other
    hosts. *J. Nematol. 1*: 150-159.
Baines, R. C., and Small, R. H. (1969). Efficacy of some new compounds for
    control of the citrus nematode, *Tylenchulus semipenetrans* on established
    trees. *Proc. 1st Int. Citrus Symp., Riverside, Calif. 2*: 973-977.
Baines, R. C., Small, R. H., DeWolfe, T. A., Martin, J. P., and Stolzy, L.
    H. (1957). Control of the citrus nematode and *Phytophthora* sp. by
    Vapam. *Plant Dis. Rep. 41*: 405-414.
Baines, R. C., Van Gundy, S. D., and DuCharme, E. P. (1978). Nematodes
    attacking citrus. In *The Citrus Industry, vol. IV, Crop Protection* (W.
    Reuther, E. C. Calavan, and G. E. Carman, eds.). University of Cali-
    fornia Division of Agricultural Science, pp. 321-345.

Baines, R. C., Van Gundy, S. D., and Sher, S. A. (1959). Citrus and avocado nematodes. *Calif. Agric. 13*: 16-18.

Baines, R. C., Van Gundy, S. D., and Small, R. H. (1977). Efficacy of nonfumigant and low volatile nematicides for control of *Tylenchulus semipenetrans* on navel and Valencia oranges (abstract). *J. Nematol. 9*: 262.

Baldwin, J. G., O'Bannon, J. H., and Huettel, R. N., (1978). Scanning electron microscopy of *Radopholus similis* (abstract). *Nematropica 8*: 2-3.

Bindra, O. S., Chhabra, H. K., Chadha, K. L., and Mehrotra, N. K. (1967). A study on the correlation of citrus nematode population with decline of citrus. *J. Res. Punjab Agric. Univ. 4*: 543-546.

Brooks, T. L., and Perry, V. G. (1967). Pathogenicity of *Pratylenchus brachyurus* to citrus. *Plant Dis. Rep. 51*: 569-573.

Cameron, J. W., Baines, R. C., and Clarke, O. F. (1954). Resistance of hybrid seedings of the trifoliate orange to infestation by the citrus nematode. *Phytopathology 44*: 456-458.

Chapot, H. (1975). The citrus plant, In *Citrus*, Ciba-Geigy Chemicals, Tech. Monog. No. 4., pp. 6-13.

Chitwood, B. G., and Toung, M. C. (1960). Host-parasite interaction of the asiatic pyroid citrus nema. *Plant Dis. Rep. 44*: 848-854.

Cobb, N. A. (1913). Notes on *Mononchus* and *Tylenchulus*. *J. Wash. Acad. Sci. 3*: 287-288.

Cobb, N. A. (1914). Citrus-root nematode. *J. Agric. Res. 2*: 217-230.

Cohn, E. (1965a). On the feeding and histopathology of the citrus nematode. *Nematologica 11*: 47-54.

Cohn, E. (1965b). The development of the citrus nematode on some of its hosts. *Nematologica 11*: 593-600.

Cohn, E. (1966). Observations on the survival of free-living stages of the citrus nematode. *Nematologica 12*: 321-327.

Cohn, E. (1969). The citrus nematode, *Tylenchulus semipenetrans* Cobb, as a pest of citrus in Israel. *Proc. 1st Int. Citrus Symp., Riverside, Calif. 2*: 1013-1017.

Cohn, E. (1972). Nematode diseases of citrus. pp. 215-244. In *Economic Nematology* (J. M. Webster, ed.). Academic, New York, pp. 215-244.

Cohn, E., Feder, W. A., and Mordechai, M. (1968). The growth response of citrus to nematocide treatments. *Israel J. Agric. Res. 18*: 19-24.

Cohn, E., and Milne, D. L. (1977). New host plants of the citrus nematode from South Africa. *Plant Dis. Rep. 61*: 466-467.

Cohn, E., and Minz, G. (1965). Application of nematicides in established orchards for controlling the citrus nematode, *Tylenchulus semipenetrans* Cobb. *Phytopathol. Medit. 4*: 17-20.

Cohn, E., Minz, G., and Monselise, S. P. (1965). The distribution, ecology and pathogenicity of the citrus nematode in Israel. *Israel J. Agric. Res. 15*: 187-200.

Cohn, E., and Orion, D. (1970). The pathological effects of representative *Xiphinema* and *Longidorus* species on selected host plants. *Nematologica 16*: 423-428.

Colbran, R. C. (1958). Studies of plant and soil nematodes. 2. Queensland host records of root-knot nematodes (*Meloidogyne* spp.). *Queensland J. Agr. Sci. 15*: 101-135.

Colbran, R. C. (1963). Studies of plant and soil nematodes. 6. Two new species from citrus orchards. *Queensland J. Agr. Sci. 20*: 469-474.

Davide, R. G., and dela Rosa, A. G. (1976). Survey, host-parasite relationships and control of the citrus nematodes in the Philippines. I. The association of plant parasitic nematodes with the citrus decline in the Philippines. *Univ. Philipp. Coll. Agric. Laguna, NSDB Tech. J.* 1: 21-31.

DuCharme, E. P. (1948). Resistance of *Poncirus trifoliata* rootstock to nematode infestation in Argentina. *Citrus Ind.* 29: 9; 15.

DuCharme, E. P. (1955). Sub-soil drainage as a factor in the spread of the burrowing nematode. *Proc. Fla. State Hort. Soc.* 68: 29-31.

DuCharme, E. P. (1967). Annual population periodicity of *Radopholus similis* in Florida citrus groves. *Plant Dis. Rep.* 51: 1013-1034.

DuCharme, E. P. (1968). Burrowing nematode decline of citrus. A review. In *Tropical Nematology* (G. C. Smart and V. G. Perry, eds.). Univ. Florida Press, pp. 20-37.

DuCharme, E. P. (1969). Temperature in relation to *Radopholus similis* (Nematoda) spreading decline of citrus. *Proc. 1st Int. Citrus Symp., Riverside, Calif.* 2: 979-983.

DuCharme, E. P., and Birchfield, W. (1956). Physiologic races of the burrowing nematode. *Phytopathology* 46: 615-616.

DuCharme, E. P., and Price, W. C. (1966). Dynamics of multiplication of *Radopholus similis*. *Nematologica* 12: 113-121.

Elgindi, A. Y., Ahmed, S. S., and Oteifa, B. A. (1976). Effects of nonfumigant nematocides on root populations and manganese and zinc levels in rough lemon seedlings infected with the citrus nematode, *Tylenchulus semipenetrans*. *Plant Dis. Rep.* 60: 682-683.

Feder, W. A. (1968). Differential susceptibility of selections of *Poncirus trifoliata* to attack by the citrus nematode, *Tylenchulus semipenetrans*. *Israel J. Agric. Res.* 18: 175-179.

Feldman, A. W., and Hanks, R. W. (1964). Quantitative changes in the free and protein amino acids in roots of healthy *Radopholus similis*-infected, and recovered grapefruit seedlings. *Phytopathology* 54: 1210-1215.

Feldman, A. W., and Hanks, R. W. (1965). Quantitative determination of the free amino acids and amides in roots and leaves of healthy and exocortis-infected *Citrus sinensis* Osbeck on *Poncirus trifoliata*, Raf. In *Third Conf. Intern. Org. Citrus Virologists Proc.* (W. C. Price, ed.). University of Florida Press, Gainesville, pp. 285-290.

Feldmesser, J., and Feder, W. A. (1956). Use of 1,2-dibromo-3-chloropropane in living citrus trees infected with the burrowing nematode. *Proc. Fla. State Hort. Soc.* 69: 105-106.

Feldmesser, J., and Feder, W. A. (1957). Survival of *Radopholus similis* in field soil subjected to drying and to elevated temperature (abstract). *Phytopathology* 47: 1.

Feldmesser, J., and Hannon, C. I. (1969). Susceptibility of two citrus rootstocks to *Pratylenchus* spp. *Plant Dis. Rep.* 53: 603-607.

Feldmesser, J., and Rebois, R. V. (1963). Effects of soil temperature and moisture on *Tylenchulus semipenetrans* (abstract). *Phytopathology* 53: 875.

Feldmesser, J., Rebois, R. V., and Taylor, A. L. (1959). Progress report on growth response of burrowing nematode infected citrus following chemical treatments under greenhouse conditions. *Plant Dis. Rep.* 43: 261-263.

Ford, H. W. (1953). Effect of spreading decline disease on the distribution of feeder roots of orange and grapefruit trees on rough lemon rootstock. *Proc. Am. Soc. Hort. Sci.* 61: 68-72.

Ford, H. W., and Feder, W. A. (1964). Three citrus rootstocks recommended for trial in spreading decline areas. *Univ. Fla. Agric. Exp. Sta. Circ. S-151.*

Ford, H. W., and Feder, W. A. (1969). Development and use of citrus rootstocks resistant to the burrowing nematode, *Radopholus similis. Proc. 1st Int. Citrus Symp., Riverside, Calif. 2:* 941-948.

Ford, H. W., Feder, W. A., and Hutchins, P. C. (1960). Citrus varieties, hybrids, species and relatives evaluated for resistance to the burrowing nematode, *Radopholus similis. Plant Dis. Rep. 44:* 405.

Gutierrez, R. O. (1947). The nematode of citrus roots, *Tylenchulus semipenetrans* in Argentina. *Rev. Invest. Agric. 1:* 119-146.

Hannon, C. I. (1964). Control of the citrus nematode, *Tylenchulus semipenetrans*, in microplot experiments. *Plant Dis. Rep. 48:* 471-475.

Heald, C. M. (1970). Distribution and control of the citrus nematode in the Lower Rio Grande Valley of Texas. *J. Rio Grande Val. Hortic. Soc. 24:* 32-35.

Heald, C. M. (1972). Control of the citrus nematode on Valencia orange in Texas. *J. Rio Grande Val. Hortic. Soc. 26:* 38-42.

Huang, C. S., and Chiang, Y. C. (1976). Pathogenicity of *Pratylenchus coffeae* on sunki orange. *Plant Dis. Rep. 60:* 957-960.

Huettel, R. N. (1982). Genetic bases for identification and separation of the two Florida races of *Radopholus similis* (Cobb) Thorne. Ph.D. Dissertation, University of Florida, Gainesville.

Huettel, R. N., and Dickson, D. W. (1981). Karyology and oogenesis of *Radopholus similis* (Cobb) Thorne. *J. Nematol. 13:* 16-20.

Huffaker, C. B., Messenger, P. S., and DeBach, P. (1971). The natural enemy component in natural control and the theory of biological control. In *Biological Control*, Proceedings of AAAS Symposium on Biological Control, Boston, Mass., December, 1969, C. B. Huffaker, (ed.), pp. 16-67.

Hughes, T. J. (1955). Citrus burrowing nematode dislikes clay soils. *Fla. Grow. Rancher 63:* 11; 49.

Hutchison, D. J., O'Bannon, J. H., Grimm, G. R., and Bridges, G. D. (1972). Reaction of selected citrus rootstocks to foot rot, burrowing and citrus nematodes. *Proc. Fla. State Hort. Soc. 85:* 39-43.

Inserra, R. N., Perotta, G., Vovlas, N., and Catara, A. (1978). Reaction of citrus rootstocks to *Meloidogyne javanica. J. Nematol. 10:* 181-184.

Inserra, R. N., and Vovlas, N. (1974). Damage by *Pratylenchus vulnus* to sour orange in Apulia. *Nematol. Medit. 2:* 183-185.

Inserra, R. N., and Vovlas, N. (1977a). Effects of *Pratylenchus vulnus* on the growth of sour orange. *J. Nematol. 9:* 154-157.

Inserra, R. N., and Vovlas, N. (1977b). Nematodes other than *Tylenchulus semipenetrans* Pathogenic to citrus. *Proc. Int. Soc. Citricult. 3:* 826-831.

Inserra, R. N., Vovlas, N., and O'Bannon, J. H. (1980). A classification of *Tylenchulus semipenetrans* biotypes. *J. Nematol. 12:* 283-287.

Jensen, H. J. (1953). Experimental greenhouse host range studies of two root-lesion nematodes, *Pratylenchus vulnus* and *Pratylenchus penetrans. Plant Dis. Rep. 37:* 384-387.

Kesterson, J. W., and Braddock, R. J. (1975). Citrus fruit processing. In *Citrus*, Ciba-Geigy Chemicals, Tech. Monog. No. 4, pp. 75-80.

Kleinschmidt, G. D., and Gerdemann, J. W. (1972). Stunting of citrus seedlings in fumigated nursery soils related to the absence of endomycorrhizae. *Phytopathology 62*: 1447-1453.

Labanauskas, C. B., Baines, R. C., and Stolzy, L. H. (1965). Effects of citrus nematode and irrigation on growth and nutrient concentrations of leaves and roots of navel orange trees. *Calif. Citrogr. 50*: 432-437.

Laborda, E., and Bello, A. (1975). On the problem of phytoparasitic nematodes on citrus crops of the Mediterranean region. *IV Cong. Mediterr. Phytopathol. Union, Zadar, Yugosl.* p. 181.

Lamberti, F., Vovlas, N., and Tirro, A. (1976). An Italian biotype of the citrus nematode *Tylenchulus semipenetrans*. *Nematol. Medit. 4*: 117-120.

Macaron, J. (1972). Contribution to the study of the plant-parasitic nematode *Tylenchulus semipenetrans* Cobb 1913 (Nematoda-Tylenchida). Ph.D. Thesis, University of Science and Technology, Languedoc, Montpellier.

Madaluni, A. L., Scognamiglio, A., and Talamé, M. (1972). Failure to transmit citrus exocortis and crinkly-leaf virusus by nematodes. In *Proc. 5th Conf. Intern. Organ. Citrus Virol.* (W. C. Price, ed.), University of Florida Press, Gainesville, pp. 285-286.

Maggenti, A. R. (1962). The production of the gelatinous matrix and its taxonomic significance in *Tylenchulus*. *Proc. Helm. Soc. Wash. 29*: 139-144.

Martin, J. P., and Van Gundy, S. D. (1963). Influence of soil phosphorus level on the growth of sweet orange seedlings and the activity of the citrus nematode (*Tylenchulus semipenetrans*). *Soil Sci. 96*: 128-135.

Marx, D. H., Bryan, W. C., and Campbell, W. A. (1971). Effect of endomycorrhizae formed by Endogone mosseae on growth of citrus. *Mycologia 63*: 1222-1226.

McCarty, D. D., Bitters, W. P., and Van Gundy, S. D. (1979). Susceptibility of rootstocks to the citrus nematode. *Calif. Citrogr. 64*: 129, 144.

Meagher, J. W. (1967). Observations on the transport of nematodes in subsoil drainage and irrigation water. *Aust. J. Exp. Agric. Anim. Husb. 7*: 577-579.

Meagher, J. W. (1969). Nematodes as a factor in citrus production in Australia. *Proc. 1st Int. Citrus Symp., Riverside, Calif. 2*: 999-1006.

Menge, J. A., Gerdemann, J. W., and Lembright, H. W. (1975). Mycorrhizal fungi and citrus. *Citrus Ind. 61*: 16-18.

Menge, J. A., Munnicke, D. E., Johnson, E. L. V., and Carnes, D. W. (1977). Dosage response of the vesicular-arbuscular mycorrhizal fungus *Glomus fasciculatus* to methyl bromide. *Proc. Am. Phytopathol. Soc. 4*: 155-156.

Milne, D. L. (1974). Citrus seedbeds: Methyl bromide and mycorrhizae. *Citrus Subtrop. Fruit J. 488*: 9-11.

Milne, D. L. (1977). Impact of new nematicides and irrigation practices on methods of citrus nematode control. *Proc. Int. Soc. Citricult. 3*: 835-838.

Milne, D. L., and DeVilliers, E. A. (1977). Soil application of systemic pesticides for control of thrips and nematodes on citrus. *Citrus Subtrop. Fruit J. 518*: 9, 18.

Mukhopadhyaya, M. C., and Dalal, M. R. (1971). Effects of two nematicides on *Tylenchulus semipenetrans* and on sweet lime yield. *Indian J. Nematol. 1*: 95-97.

Muthukrishnan, T. S., Santhakumar, T., Rajendran, G., Chandrasekaran, J., Sankaranarayanan, R., and Rajagopalan, P. (1975). Studies on the varietal susceptibility of certain rootstocks to the citrus nematode, *Tylenchulus semipenetrans* Cobb, 1913. *S. Indian Hortic. 23*: 91-93.

Neal, J. C. (1889). Root-knot disease of peach, orange, and other plants in Florida, due to the work of *Anguillula*. *USDA Bur. Entomol. Bull. No. 20*: pp. 1-31.

Nemec, S., and O'Bannon, J. H. (1979). Response of *Citrus aurantium* to *Glomus etunicatus* and *G. mosseae* after soil treatment with selected fumigants. *Plant Soil 53*: 351-359.

Nigh, E. L., Jr., and Allen, R. M. (1967). Failure of nematodes to transmit citrus exocortis virus (abstract). *Phytopathology 57*: 100.

O'Bannon, J. H. (1968). The influence of an organic soil amendment on infectivity and reproduction of *Tylenchulus semipenetrans* on two citrus rootstocks. *Phytopathology 58*: 597-601.

O'Bannon, J. H., and Bistline, F. W. (1969). A simple device for injecting methyl bromide into a replant site. *Plant Dis. Rep. 53*: 799-802.

O'Bannon, J. H., Chew, V., and Tomerlin, A. T. (1977). Comparison of five populations of *Tylenchulus semipenetrans* to Citrus, Poncirus, and their hybrids. *J. Nematol. 9*: 162-165.

O'Bannon, J. H., and Esser, R. P. (1975). Evaluation of citrus, hybrids, and relatives as hosts of the nematode *Pratylenchus coffeae*, with comments on other hosts. *Nematol. Medit. 3*: 113-122.

O'Bannon, J. H., Esser, R. P., and Inserra, R. N. (1975). Bibliography of nematodes of citrus. *U.S. Dept. Agr. ARS-S-68.*

O'Bannon, J. H., and Ford, H. W. (1977). Resistance in citrus rootstocks to *Radopholus similis* and *Tylenchulus semipenetrans*. *Proc. Int. Soc. Citricult. 2*: 544-549.

O'Bannon, J. H., and Nemec, S. (1978). Influence of soil pesticides on vesicular-arbuscular mycorrhizae in a citrus soil. *Nematropica 8*: 56-61.

O'Bannon, J. H., Radewald, J. D., and Tomerlin, A. T. (1972). Population fluctuation of three parasitic nematodes in Florida citrus. *J. Nematol. 4*: 194-199.

O'Bannon, J. H., Radewald, J. D., Tomerlin, A. T., and Inserra, R. N. (1976). Comparative influence of *Radopholus similis* and *Pratylenchus coffeae* on citrus. *J. Nematol. 8*: 58-63.

O'Bannon, J. H., Reynolds, H. W, and Leathers, C. R. (1966). Effects of temperature on penetration, development and reproduction of *Tylenchulus semipenetrans*. *Nematologica 12*: 483-487.

O'Bannon, J. H., and Tarjan, A. C. (1973). Preplant fumigation for citrus nematode control in Florida. *J. Nematol. 5*: 88-95.

O'Bannon, J. H., and Tarjan, A. C. (1978). Effective pest management of the burrowing nematode on citrus. *Citrus Ind. 59*: 18-20, 23.

O'Bannon, J. H., and Tarjan, A. C. (1979). Management of *Radopholus similis* infecting citrus with DBCP or phenamiphos. *Plant Dis. Rep. 63*: 456-460.

O'Bannon, J. H., Tarjan, A. C., and Bistline, F. W. (1974). Control of *Pratylenchus brachyurus* on citrus and tree response to chemical treatment. *Soil Crop Sci. Soc. Fla. Proc. 33*: 65-67.

O'Bannon, J. H., and Tomerlin, A. T. (1971). Response of citrus seedlings to *Radopholus similis* in two soils. *J. Nematol. 3*: 255-260.

O'Bannon, J. H., and Tomerlin, A. T. (1973). Citrus tree decline caused by *Pratylenchus coffeae*. *J. Nematol. 5*: 311-316.

O'Bannon, J. H., and Tomerlin, A. T. (1977). Control of the burrowing nematode, *Radopholus similis*, with DBCP and oxamyl. *Plant Dis. Rep. 61*: 450-454.

Oteifa, B. A., Shafiee, V. A., and Eissa, F. M. (1965). Efficacy of DBCP flood irrigation in established citrus. *Plant Dis. Rep. 49*: 598-599.

Philis, J. (1969). Control of citrus nematode, *Tylenchulus semipenetrans*, with DBCP in established Cyprus citrus groves. *Plant Dis. Rep. 53*: 804-806.

Poucher, C. (1967a). Control. In *Burrowing Nematode in Citrus*. (C. Poucher, H. W. Ford, R. F. Suit, and E. P. DuCharme, eds.) Florida Department of Agriculture Bulletin 7, pp. 30-40.

Poucher, C. (1967b). Regulatory measures. In *Burrowing Nematode in Citrus*. (C. Poucher, H. W. Ford, R. F. Suit, and E. P. DuCharme, eds.). Florida Department of Agriculture Bulletin 7, pp. 41-48.

Prasad, S. K., and Chawla, M. L. (1965). Observations on the population fluctuations of citrus nematode, *Tylenchulus semipenetrans* Cobb, 1913. *Indian J. Entomol. 27*: 450-454.

Radewald, J. D., O'Bannon, J. H., and Tomerlin, A. T. (1971). Temperature effects on reproduction and pathogenicity of *Pratylenchus coffeae* and *P. brachyurus* and survival of *P. coffeae* in roots of *Citrus jambhiri*. *J. Nematol. 3*: 390-394.

Radewald, J. D., Rosedale, D., Shibuya, F., and Nelson, J. (1973). Control of the citrus nematode, *Tylenchulus semipenetrans* with foliar Vydate sprays on Valencia oranges in southern California (abstract). *Phytopathology 63*: 1217.

Reynolds, H. W., and O'Bannon, J. H. (1963a). Decline of grapefruit trees in relation to citrus nematode populations and tree recovery after chemical treatment. *Phytopathology 53*: 1011-1015.

Reynolds, H. W., and O'Bannon, J. H. (1963b). Factors influencing the citrus replants in Arizona. *Nematologica 9*: 337-340.

Reynolds, H. W., O'Bannon, J. H., and Nigh, E. L. (1974). The citrus nematode and its control in the Southwest. *USDA Tech. Bull. No. 1478*.

Reynolds, H. W., O'Bannon, J. H., Tomerlin, A. T., Nigh, E. L., Jr., and Rodney, D. R. (1970). The influence of various ecological factors on survival of *Tylenchulus semipenetrans*. *Soil Crop. Sci. Soc. Fla. Proc. 30*: 366-370.

Scaramuzzi, G., and Perrotta, G. (1962). Research on the control of citrus nematode in the Mediterranean area. *Proc. 1st Int. Citrus Symp., Riverside, Calif. 2*: 957-960.

Schneider, H., and Baines, R. C. (1964). *Tylenchulus semipenetrans*: Parasitism and injury to orange tree roots. *Phytopathology 54*: 1202-1206.

Scotto La Massese, C. (1965a). Studies on citrus nematodes in Algeria. *C.R. 1 J. Phytiatr. Phytopharm. Circum. Mediterr., Marseille, France*, pp. 48-51.

Scotto La Massese, C. (1965b). Susceptibility of some citrus rootstocks to *Tylenchulus semipenetrans* Cobb, 1913. *C.R. 1 J. Phytiatr. Phytopharm. Circum. Mediterr., Marseille, France*, pp. 59-68.

Siddiqi, M. R. (1964). Studies on nematode root-rot of citrus in Uttar Pradesh, India. *Proc. Zool. Soc. (Calcutta) 17*: 67-75.

Standifer, M. S., and Perry, V. G. (1960). Some effects of sting and stubby root nematodes on grapefruit roots. *Phytopathology 50*: 152-156.

Stirling, G. R., and Mankau, R. (1977). Biological control of nematode parasites of citrus by natural enemies. *Proc. Int. Soc. Citricult. 3*: 843-847.

Stokes, D. E. (1969). *Andropogon rhizomatus* parasitized by a strain of *Tylenchulus semipenetrans* not parasitic to four citrus rootstocks. *Plant Dis. Rep. 53*: 882-885.

Stokes, D. E. (1977). The role of nematodes as vectors of citrus viruses. *Proc. Int. Soc. Citricult. 3*: 848-853.

Stolzy, L. H., Van Gundy, S. D., Labanauskas, C. K., and Szuszkiewicz, T. E. (1963). Response of *Tylenchulus semipenetrans* infected citrus seedlings to soil aeration and temperatures. *Soil Sci. 96*: 292-298.

Suit, R. F. (1947). Spreading decline of citrus in Florida. *Proc. Fla. State Hort. Soc. 60*: 17-23.

Suit, R. F. (1969). Treatment of citrus trees for burrowing nematode control. *Proc. 1st Int. Citrus Symp., Riverside, Calif. 2*: 961-968.

Suit, R. F., and DuCharme, E. P. (1953). The burrowing nematode and other parasitic nematodes in relation to spreading decline of citrus. *Plant Dis. Rep. 37*: 379-383.

Suit, R. F., and DuCharme, E. P. (1967). Spreading decline of citrus. In *Burrowing Nematode in Citrus*. (C. Poucher, H. W. Ford, R. F. Suit, and E. P. DuCharme, eds.). Florida Department of Agriculture Bulletin, pp. 1-20.

Suit, R. F., and Feldman, A. W. (1965). Barriers for spreading decline control. *Proc. Fla. State Hort. Soc. 77*: 52-56.

Swingle, W. T. (1946). The botany of citrus and its wild relatives of the orange subfamily. In *The Citrus Industry*, vol. 1 (H. J. Webber and L. D. Batchelor, eds.). University of California Press, Berkeley, pp. 129-474.

Talamè, M., Brzeski, M. W., Scognamiglio, A., and D'Errico, F. P. (1970). The nematode *Rotylenchulus macrodoratus* Dasgupta, Raski and Sher, 1968 (Tylenchida) on orange trees in Abruzzo. *Boll. Lab. Entomol. Agrar. Portici 28*: 229-235.

Tarjan, A. C. (1956). The possibility of mechanical transmission of nematodes in citrus groves. *Proc. Fla. State Hort. Soc. 69*: 34-37.

Tarjan, A. C. (1960). Some effect of African marigold on the citrus burrowing nematode, *Radopholus similis* (abstract). *Phytopathology 50*: 577.

Tarjan, A. C. (1961a). Attempts at controlling citrus burrowing nematodes using nematode-trapping fungi. *Soil Crop Sci. Soc. Fla. Proc. 21*: 17-36.

Tarjan, A. C. (1961b). Longevity of *Radopholus similis* (Cobb) in host free soil. *Nematologica 6*: 170-175.

Tarjan, A. C. (1971). Migration of three pathogenic citrus nematodes through two Florida soils. *Soil Crop Sci. Soc. Fla. Proc. 31*: 253-255.

Tarjan, A. C. (1976). Application of systemic nematicides to trunks of trees (abstract). *J. Nematol. 8*: 803.

Tarjan, A. C. (1977a). Application of municipal solid waste compost to nematode-infected citrus. *Nematropica 7*: 53-56.

Tarjan, A. C. (1977b). Kelp derivatives for nematode-infected citrus trees (abstract). *J. Nematol. 9*: 287.

Tarjan, A. C. (1980). Increased yield from Meyer lemons following nematicide applications (abstract). *Nematropica 10*: 73.

Tarjan, A. C., and Hannon, C. I. (1963, 1964). The biology of nematodes associated with citrus. *Univ. Florida Agr. Exp. Sta. Ann. Rep.*, pp. 232,237.

Tarjan, A. C., and O'Bannon, J. H. (1969). Observations on meadow nematodes (*Pratylenchus* spp.) and their relation to decline of citrus in Florida. *Plant Dis. Rep. 53*: 683-686.

Tarjan, A. C., and O'Bannon, J. H. (1974). Postplant fumigation with DBCP for citrus nematode control in Florida. *J. Nematol. 6*: 41-48.

Tarjan, A. C., and O'Bannon, J. H. (1977). Nonpesticidal approaches to nematode control. *Proc. Int. Soc. Citricult. 3*: 848-853.

Tarjan, A. C., and O'Bannon, J. H. (1980). Comparison of calcium and ammonium nitrate fertilizers for treating citrus with spreading decline caused by *Radopholus similis*. *Absts. XV Intern. Nematol. Symp.*, European Soc. Nematol., Bari, Italy, pp. 49-50.

Tarjan, A. C., and Simmons, P. N. (1967). The effect of interacting cultural practices on citrus trees with spreading decline. *Soil Crop Sci. Soc. Fla. Proc. (1966) 26*: 22-31.

Thomas, E. E. (1913). A preliminary report of a nematode observed on citrus roots and its possible relation with the mottled appearance of citrus trees. *Calif. Agric. Exp. Sta. Circ. 85*.

Thomas, E. E. (1923). The citrus nematode, *Tylenchulus semipenetrans*. *Univ. Calif. Agric. Exp. Sta., Coll. Agric. Tech. Paper 2*.

Timmer, L. W. (1977). Control of citrus nematode *Tylenchulus semipenetrans* on fine-textured soil with DBCP and oxamyl. *J. Nematol. 9*: 45-50.

Timmer, L. W., and Leyden, R. F. (1978). Relationship of seedbed fertilization and fumigation to infection of sour orange seedlings by mycorrhizal fungi and *Phytophthora parasitica*. *J. Am. Hort. Soc. 103*: 537-541.

Toung, M. C. (1963). A study of seasonal influence in quantitative variation of the citrus nema, *Tylenchulus semipenetrans* Cobb. *J. Soc. Plant. Prot. (Taiwan) 5*: 323-327.

Van Gundy, S. D. (1958). The life history of the citrus nematode, *Tylenchulus semipenetrans* Cobb. *Nematologica 3*: 283-294.

Van Gundy, S. D., Bird, A. F., and Wallace, H. R. (1967). Aging and starvation in larvae of *Meloidogyne javanica* and *Tylenchulus semipenetrans*. *Phytopathology 57*: 559-571.

Van Gundy, S. D., and Kirkpatrick, J. D. (1964). Nature of resistance in certain citrus rootstocks to citrus nematode. *Phytopathology 54*: 419-427.

Van Gundy, S. D., and Martin, J. P. (1961). Influence of *Tylenchulus semipenetrans* on the growth and chemical composition of sweet orange seedlings in soils of various exchangeable cation ratios. *Phytopathology 51*: 146-151.

Van Gundy, S. D., Martin, J. P., and Tsao, P. H. (1964). Some soil factors influencing reproduction of the citrus nematode and growth reduction of sweet orange seedlings. *Phytopathology 54*: 294-299.

Van Gundy, S. D., and Meagher, J. W. (1977). Citrus nematode (*Tylenchulus semipenetrans*) problems worldwide. *Proc. Int. Soc. Citricult. 3*: 823-826.

Van Gundy, S. D., and Rackham, R. L. (1961). Studies on the biology and pathogenicity of *Hemicycliophora arenaria*. *Phytopathology 51*: 393-397.

Van Gundy, S. D., and Tsao, P. H. (1963). Growth reduction of citrus seedlings by *Fusarium solani* as influenced by the citrus nematode and other soil factors. *Phytopathology 53*: 488-489.

Vilardebo, A. (1963a). Nematodes parasitic on citrus in Morocco. *A. Awamia (Rabat) 8*: 57-59.

Vilardebo, A. (1963b). Studies on *Tylenchulus semipenetrans* Cobb in Morocco. I. *Al Awamia (Rabat) 8*: 1-23.

Vilardebo, A. (1964). Study on *Tylenchulus semipenetrans* Cobb in Morocco. II. *Al Awamia (Rabat) 11*: 31-49.

Vilardebo, A., and Luc, M. (1961). Slow decline of citrus caused by the nematode *Tylenchulus semipenetrans*. *Fruits (Paris) 16*: 445-465.

Vilardebo, A., Squalli, A., and Devaux, R. (1975). Possible use of DBCP, phenamiphos and prophos against *Tylenchulus semipenetrans* in orchards in Morocco. *Fruits (Paris) 30*: 313-327.

Vovlas, N., Lamberti, F., and Inserra, R. (1977). Results of glasshouse experiments with new nematicides against the citrus nematode, *Tylenchulus semipenetrans* Cobb. *Proc. 1973 Int. Citrus Congr., Murcia, Valencia, Spain 2*: 687-691.

Yokoo, T. (1964). Studies on the citrus nematode (*Tylenchulus semipenetrans* Cobb, 1913) in Japan. *Agric. Bull. Saga Univ. 20*: 71-109.

Yokoo, T., and Ikegemi, Y. (1966). Some observations on growth of the new host plant, snapdragon (*Antirrhinum majus* L.) attacked by root lesion nematode, *Pratylenchus coffeae* and control effect of some nematicides. *Agric. Bull. Saga Univ. 22*: 83-92.

Chapter 12
Nematode Parasites of Peach and Other Tree Crops

E. J. Wehunt *USDA, Agricultural Research Service, Booneville, Arkansas*

I. INTRODUCTION

The people of the United States consumed more than 24 billion pounds of
fruits and 424 million pounds of nuts in 1981. The 1981 farm value of deciduous tree fruit and nut crops in the United States was more than 2.4 billion
dollars (U.S. Dept. Agriculture, 1982). In addition, large quantities of these
crops utilized locally are not reported in the crop reporting service figures.
Thousands of people are employed in the growing, harvesting, processing,
and marketing of these crops. Fruit and nut crops are important items of
world commerce, enhancing the United States balance of payments position.

Fruit and nut crops as trees are permanent and their proper culture
discourages soil erosion, which is a major contribution to loss of productive
land. Costs of fruit and nut production are increasing because costs of
energy, machinery, fertilizers, pesticides, and labor are rising. The increased production cost magnifies the losses due to weeds, pests, diseases,
and nematodes.

Nematode parasitism of fruit and nut trees reduces production by 5-15%
(Anon., 1971), which converted to 1981 values is more than 256 million dollars.
Further losses are sustained because nematodes interact in disease complexes.

II. NEMATODES OF PEACH AND OTHER STONE FRUIT

A. History

In the United States the value of stone fruits — almonds, apricots, cherries,
nectarines, peaches, and plums — was more than $880 million in 1981 (U. S.
Dept. Agriculture, 1982). Peaches, and probably other stone fruits, were
cultured in China for thousands of years, but when found in Persia shortly
after world trade began they were mistakenly thought to be native and hence
were named *Prunus persica*. The plants were transported, probably as pits,
throughout Europe. They were brought to the Western Hemisphere, probably
by Spanish priests, more than 400 years ago.

The modern peach industry in the United States is a highly developed,
complex system. Several technological advances were necessary before today's
commercial peach industry was possible. The Elberta cultivar, discovered
near Marshallville, Georgia, in 1885, was of high quality and could be shipped
for long distances. Development of a suitable shipping container — the familiar
peach basket — rapid rail transportation, iced rail cars, and later mechanically
refrigerated cars all contributed to promote the peach industry. Similar technological developments enhanced the growth of other stone fruit enterprises.

B. Culture

In the United States stone fruits are usually propagated by budding the desired scion variety onto a seedling rootstock in a nursery. The budded trees
are removed from the nursery soil, transported to the orchard site where they
are planted directly or "heeled in" until they are planted in the orchard. The
heeling-in area is usually a plowed furrow where the trees are placed in the
furrow and the soil firmed around the roots to prevent them from drying or freezing. Growers in the Southeast prefer trees budded in June, which are 18-36
inches high by the following winter when they are removed from the nursery.
During the first year after the planted trees begin growth, they are headed
back to about 18 inches and are shaped with three to five future scaffold limbs.

Cultural practices for peach in the Southeast vary from place to place,
but the 10 point program outlined by the peach tree short-life work group of

the southeastern peach workers organization is generally recommended
(Ritchie and Clayton, 1981):

1. Apply lime before planting to adjust pH to at least 6.5.
2. Subsoil before planting to break up hardpans and promote improved root
   development.
3. In sandy soils where peach trees have been planted previously and in
   other soils where nematodes are a problem, fumigate the soil before planting.
4. Plant trees that have been grown in fumigated soil or are certified to be
   free of nematodes.
5. Plant trees propagated on Lovell rootstock. (Halford rootstock has also
   performed well in southeastern research plots.)
6. Apply nutrients and lime as needed based on soil tests, foliar analyses,
   and local recommendations.
7. Prune as late as possible, never before January 1 and preferably after
   February 1. If earlier pruning is unavoidable, prune older trees first.
   Early pruning is especially hazardous on old peach tree land. Discontinue
   summer pruning (including topping and hedging) by September 15.
8. Use recommended herbicides for weed control. Keep cultivation (if needed)
   shallow to avoid root injury.
9. In sites where preplant fumigation is necessary, postplant fumigate at
   approximately two-year intervals or as indicated by nematode populations.
10. *Promptly remove and burn all dead or dying trees.*

   With the exception of point 9 this program is currently recommended not
only to control PTSL (Peach Tree Short Life) but also to "establish and maintain a long-lived, productive peach orchard." PTSL is a "disease syndrome
following freeze injury and/or bacterial canker (caused by *Pseudomonas
syringae* Van Hall)".

C. Root-Knot Nematodes (*Meloidogyne* spp.)

Root-knot nematodes were first reported on peaches in Germany in 1885 and
in the United States in 1889 (Neal, 1889). The early recommendation for control was the use of clean plants planted in clean soil, although the means to
eliminate nematodes from plants and soil were not available. In the early 1920s,
workers noted differences in susceptibility among rootstocks (Tufts and Day,
1934). Variation in reaction of peach roots to root-knot nematodes was common,
and when Chitwood revised the classification of root-knot nematodes in 1949,
the variation in reaction was found to be due to different species of *Meloidogyne*.
   Peach roots are susceptible to four species of *Meloidogyne* — *M. incognita*,
(Kofoid and White 1919) Chitwood 1949, *M. javanica* (Treub 1885) Chitwood
1949, *M. hapla* Chitwood 1949, and *M. arenaria* (Neal 1889) Chitwood 1949.
Root-knot nematodes are found wherever peaches are grown.
   Root-knot nematodes have been implicated in peach disease complexes with
fungi and bacteria in which the nematode creates an avenue of entry or breaks
resistance to another organism. Esser et al. (1968) reported simultaneous
occurrence of root-knot nematodes and crown gall bacteria in peach and other
plants. The Javanese root-knot nematode (*M. javanica*) increased the incidence of crown gall of peach roots caused by *Agrobacterium tumefaciens* (E.
F. S. M. and Towns.) Conn. (Nigh, 1966). In Canada, injury by *M. hapla*
increased crown gall of peach in a test lasting 62 days, but in a later test

Fig. 1 (A) Peach roots infected with the root-knot nematode, *M. incognita* (Kofoid and White 1919) Chitwood 1949. (B) Peach orchard infested with the root-knot nematode, *M. incognita* (Kofoid and White 1919) Chitwood (note uneven size of the trees). (Courtesy of USDA.)

lasting 157 days, incidence of crown gall was not affected by the nematodes *M. incognita, M. hapla*, or *Pratylenchus penetrans* (Cobb 1917) Filip. and Sch. Stek. 1941 (Dhanvantari et al., 1975).

1. Symptoms

Invasion of susceptible peach roots by second-stage juveniles of *Meloidogyne* spp., and their subsequent development, elicit the galls or knots that are distinctive symptoms of root-knot nematode (Fig. 1A). When many juveniles invade a root the root tip is devitalized and elongation of the root ceases. Sometimes the root branches in response to parasitism by the nematodes. The development of the vascular system is altered, and short irregular xylem elements develop. Peach trees infected with root-knot nematodes are usually stunted. In periods of drought infected trees wilt more readily than do non-infected trees. Because root-knot nematodes are seldom uniformly distributed in the field, infested orchards are usually composed of unevenly sized trees (Fig. 1B). The effects of root-knot nematodes on peach trees are reduced growth, vigor, and yield, and the ability to withstand environmental stress is impaired.

2. Economic Loss

Reduced yield is an important economic factor in production. The effect of individual nematodes on peaches is difficult to assess because the monospecific populations are almost impossible to maintain in the field long enough for the trees to grow and produce fruit. Foster et al. (1972) assessed the economic loss to peaches. Interpolation of their data shows use of resistant rootstock (S-37) increased yield about 403.2 kg/ha, preplanting applications of 202 kg of ethylene dibromide increased yield about 420 kg, and preplanting applications of 134 kg of DBCP increased yield by about 306 kg/ha. Profits of $723, $729, and $510 per acre accrued in three years of harvesting.

3. Distribution

M. incognita, *M. incognita acrita*, and *M. javanica* occur throughout the United States where the temperature permits completion of the life cycle. Although *M. hapla* occurs on peanut and strawberry in the South, reports of occurrence on peaches are limited to the northern states and Canada. *M. arenaria* sometimes occurs on peaches in the southern states.

4. Life Cycle

The juvenile molts four times, once in the egg and three times after hatching. Second-stage juveniles enter the root near the root tip and move through the root until they find suitable tissue. When a site is reached the eel-shaped juvenile begins to swell. Feeding is accomplished by inserting the stylet into the cell, and digestive fluids that dissolve the cell contents are injected. Cell contents are then sucked out by the nematode. Giant cells, called syncytia, which serve as a food source, are formed in response to the nematode invasion. It has been generally believed that the digestive fluids dissolve the cell walls, allowing the protoplasm and nuclei to coalesce to form a multinucleate giant cell or syncytium. Recent studies, however, indicate that the cell wall does not dissolve but the syncytium is formed by enlargement of a single cell. The multinucleate condition arises from repeated mitosis without cytokinesis (Jones and Payne, 1978).

The swollen juvenile undergoes the fourth and final molt and becomes an adult. The male is eel shaped and motile and has a strong stylet and a large median esophageal bulb. Most species are parthenogenetic although males sometimes are abundant. The females continue in place, swell further, and, except for the anterior end, are immobile.

The female matures and deposits several hundred eggs in gelatinous material. The gelatinous material resists drying and serves as a survival mechanism for the eggs. Eggs hatch without delay when the moisture and temperature conditions are favorable. When the soil solution is concentrated by drought, eggs do not hatch.

5. Control

a. Genetic Variation in reaction to root-knot nematode among peach seedlings in the early part of the twentieth century led to a search for cultivars that were resistant to the nematode (Tufts and Day, 1934). By the 1950s, several cultivars were available for use in orchards infested with root-knot nematodes. The most popular, S-37, was sometimes unsuitable, probably because of different *Meloidogyne* species. Breeders sought a cultivar that was resistant to several nematodes. After 1949, when Chitwood revised the genus, breeding programs for individual species began. Nemaguard was released from the USDA breeding program at Fort Valley, Georgia, in 1961. Okinawa, another cultivar resistant to *M. incognita* and *M. javanica*, is used in Florida. Yunnan, no longer popular, is resistant to *M. incognita* but susceptible to *M. javanica*. Resistance to *M. incognita* is controlled by one gene with resistance dominant, and resistance to *M. javanica* is controlled by two or more genes (Sharpe et al., 1969).

Nemaguard is a popular rootstock in California, but in the Southeast it is used sparingly because trees on Nemaguard rootstock are often susceptible to cold injury. The close scrutiny of pesticide use patterns by the Environmental Protection Agency will undoubtedly increase interest in developing resistant varieties that are more acceptable to growers.

b. Cultural practices where applicable are effective controls for root-knot nematodes. Host-free fallow, long enough for the nematodes to starve, is effective. This practice is not suitable for perennial crops, however. Usually a nonhost crop is grown during the fallow period to lessen the economic impact and to reduce soil erosion. Varying the planting time and trap cropping are not always effective because timing must be exact and need for expert control precludes general use.

c. Chemical control of root-knot nematodes is most commonly accomplished by use of nematicidal chemicals. Because the nematodes are protected by root tissue, control measures must be applied before planting. For peaches and other perennials, nematicidal protection must be furnished throughout the life of the plant.

Nematicidal chemicals are either fumigants, which are halogenated hydrocarbons, or nonfumigants, which are organic phosphates or carbamates. Some of the nonfumigant nematicides are systemic and some are also insecticidal.

The fumigants are applied by injection beneath the soil surface with special equipment or in irrigation water. The chemicals volatilize and move through the soil pore spaces, where they dissolve in the water film around the soil particles, and kill the nematodes.

The nonfumigants, or contact nematicides, must be throughly mixed with the soil or transported to the nematode in water. Except for the systemics, the contact nematicide will not move independently.

D. Lesion Nematodes (*Pratylenchus* spp.)

1. History

Lesion nematodes were first observed in 1865, but it was not until 1898 that damage to plants was documented. Lesion nematodes were reported from potatoes, cotton, camphor, and violets in 1917, cereals in 1924, pineapple in 1929, and apples in 1936.

Lesion nematode damage to peach roots was reported in Connecticut in 1949, New York in 1950, California in 1953, and Georgia in 1966.

2. Causal Organisms

In Canada and in the northeastern United States, the lesion nematode *Pratylenchus penetrans* may be involved in a specific replant problem of peach (Mountain and Boyce, 1958). The syndrome occurs when peaches follow peaches on the same land. The nematode parasitizes apples and pears, but the syndrome does not develop when peaches follow these crops.

In Georgia, high populations of *P. vulnus* Allen and Jensen 1953 were associated with root destruction and reduced growth (Fliegel, 1969).

When Barker and Clayton (1969) tested the reaction of six peach cultivars to six species of lesion nematodes there was no significant difference in reaction among cultivars. *P. vulnus* and *P. penetrans* injured peach seedlings, but *P. brachyurus* (Godfrey 1929) Filip. and Sch. Stek. 1941, *P. coffeae* (Zimmerman 1898) Filip. and Sch. Stek. 1941, *P. scribneri* Steiner in Sherbakoff and Stanley 1943, and *P. zeae* Graham 1951 were of little importance. In Florida *P. brachyurus* parasitized peaches (Stokes, 1966).

3. Symptoms

The activities of *Pratylenchus* spp. in roots cause lesions and discolorations of roots that are usually reddish brown at first but turn dark and ultimately are black. The lesion grows as the nematodes and their progeny feed and eventually the root is girdled. The lesion could furnish an avenue of entry for other organisms, but such interactions in peach have not been reported.

4. Life Cycle

P. penetrans and *P. vulnus* have similar life histories and habits. The nematodes enter the roots wherever the tissue is immature and the nematodes can penetrate. The nematodes can enter or leave the root as juveniles or as adults. They move inter- and intracellularly, feeding primarily on cortical cells. Eggs are laid singly as the nematode migrates through the root and soil. There is no delay in hatching of the eggs.

5. Economic Losses

The economic loss in stone fruit production caused by lesion nematodes has not been determined. Lesion nematodes parasitize stone fruits in many areas including California and the Southeast but they have not yet been implicated in the PTSL problem. In Canada and in the northeastern United States, *P. penetrans* is the cause of a specific replant problem (Arneson and Mai, 1976; Mai et al., 1970; Mountain and Boyce, 1958).

6. Distribution

The lesion nematodes *P. penetrans* and *P. vulnus* occur in stone fruit orchards throughout the United States, Canada, and Europe.

7. Control

a. *Genetic* Although there have been tests for resistance to *P. vulnus*, none has been found in stone fruits. In Canada (Layne, 1974), evaluation of peach breeding stocks indicates that a source of resistance to *P. penetrans* is available in some Russian clones and that a U.S. clone (Nemaguard x Okinawa) is tolerant to *P. penetrans* and retains resistance to *M. javanica* and *M. incognita*.

b. *Cultural* Cultivation of nonhost crops reduces nematode numbers; however, when culture of susceptible perennial crops, such as stone fruits, is resumed, populations soon return to damaging numbers. The control of lesion nematodes has not been obtained with cultural practices.

   c. *Chemical* At present, lesion nematodes in stone fruits can be controlled only by nematicidal chemicals. Preplanting nematicides are effective
in reducing populations, but postplanting applications are necessary to maintain control in perennial crops.

E. Ring Nematodes (*Criconemella* spp.)

The relation of ring nematode and peach trees has been intensively studied
because of their possible role in a complex syndrome known as peach decline
or peach tree short life.

1. *History*

Chitwood (1949) reported large numbers of ring nematodes from peach orchards
in North Carolina and Maryland in 1949 and suggested that they might be involved in the peach decline problem. The ring nematodes are widespread and
are associated with many plants. Very high populations are often found in
peach orchards.

2. *Symptoms*

Parasitism of *Criconemella curvata* (Raski 1952) Luc and Raski 1981 (*Criconemoides curvatum* Raski 1952) on peach roots caused pits and lesions under
sterile conditions in New Jersey (Hung and Jenkins, 1969). Under nonsterile
conditions the pits and lesions were invaded by other microorganisms that
caused discoloration and low vigor of the root system. The control of ring
nematodes by nematicides in replant sites enhanced growth and tree survival.
Based on their research they proposed that nematodes and other cultural or
disease factors interact in the peach decline complex.
   Criconemella xenoplax (Raski 1952) Luc and Raski 1981 (*Criconemoides
xenoplax* Raski 1952) caused chlorosis and leaf drop under greenhouse conditions, and soil in pots infested with *C. xenoplax* tended to be waterlogged
(Lownsbery et al., 1973). In South Carolina fewer feeder roots were present
on peach trees growing in infested soil than in nematicide-treated soil (Fig. 2).
Trees treated with nematicide annually in Georgia were more vigorous (Fig. 3)
than trees in nontreated soil. In an experiment at Byron, Georgia, involving
irrigation, soil manipulation, and fumigation, plots receiving DBCP required
more irrigation water to maintain tensiometer readings below 25 bars than did
nonfumigated plots (Horton et al., 1981). In a greenhouse pot test at Byron,
Georgia, peach seedlings were grown in *C. xenoplax* infested with noninfested
soil to study the phytohormonal response of the seedlings to the nematode.
All pots were fertilized and watered alike. After six months the seedlings
growing in the infested soil were chlorotic and stunted compared to seedlings
growing in noninfested soil. Analysis of the soil showed higher content of
phosphorous, potassium, and magnesium in the infested soil than in the noninfested soil (Wehunt and Edwards, unpublished). Abscisic acid (ABA) content
was higher in stem tissue of seedlings growing in infested soil than in noninfested soil (Wehunt and Yadava, unpublished). In South Carolina parasitism
by *C. xenoplax* increased indole acetic acid (IAA) content of peach stem bark
(Nyczepir and Lewis, 1980). The phytohormonal evidence is somewhat conflicting: higher concentrations of IAA indicates growth, whereas higher concentrations of ABA indicates dormancy or stress. *C. xenoplax* feeding on *Prunus
cerasifera* produced an IAA-inactivating agent (Viglierchio and Mjuge, 1975).
   Collectively the evidence indicates that parasitism of peach roots by *C.
xenoplax* impairs root development or root function, which imposes a stress
condition on the plants.

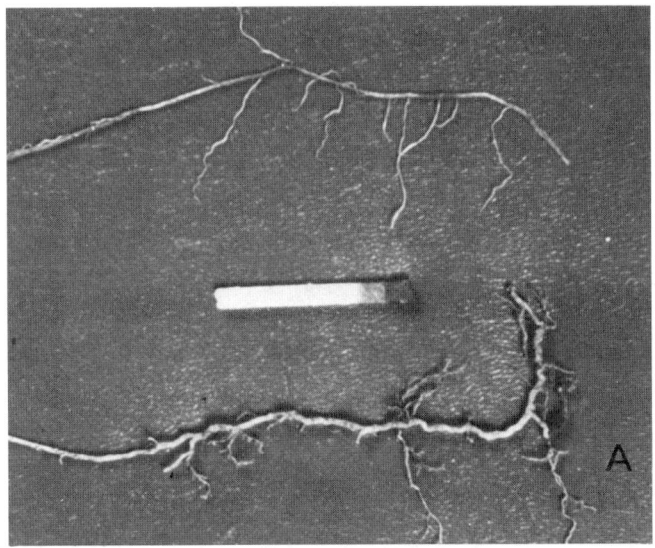

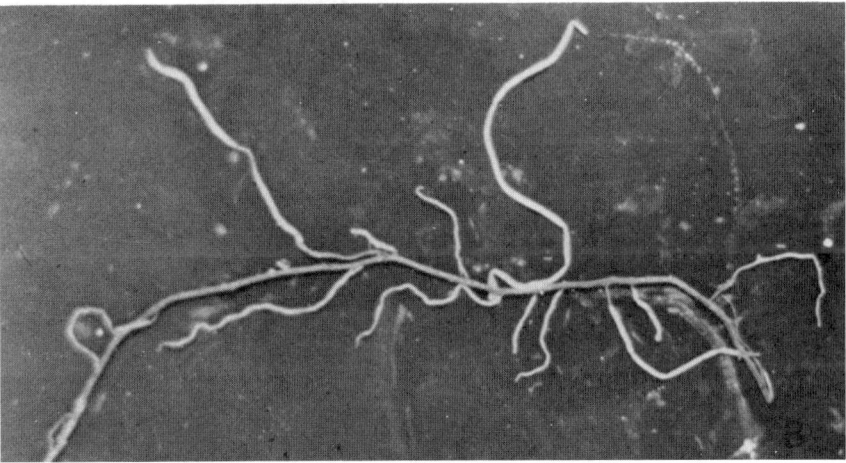

Fig. 2 Peach roots from nonfumigated soil infested with *C. xenoplax* (A), and from fumigated soil (B) in South Carolina. (Courtesy Dr. E. I. Zehr, Clemson University, Clemson, South Carolina.)

    C. xenoplax has been strongly implicated in the peach tree short-life (PTSL) condition in California and the Southeast. The implication, however, is circumstantial, from results of field trials, and does not preclude the influence of other factors. The complete PTSL syndrome has not been produced under controlled conditions. Peach tree short life is characterized by sudden death of peach trees in the spring shortly after foliation (Fig. 4). Control by fumigation, in particular postplanting fumigation, reduced tree death in California and the Southeast (Lownsbery et al., 1973; Nesmith et al., 1981; Taylor et al., 1970; Wehunt et al., 1980; Wehunt and Weaver, 1982; Zehr et al., 1976).
    Peach tree short life is a "disease syndrome characterized by collapse and death of trees above the soil line in late winter and spring following freeze

Fig. 3 Peach trees of equal age in a peach tree short-life site in Georgia: (A) nontreated; (B) fumigated annually with DBCP. (Courtesy of USDA.)

injury and/or bacterial canker (caused by *Pseudomonas syringae* Van Hall). Bacterial canker-damaged or freeze-injured bark is invariably invaded and colonized by Cytospora canker fungi (*Leucostoma persoonii* [Nits] Hohmel and *L. cincta* [Pers. ex Fr.] Hohmel)." (Ritchie and Clayton, 1981).

In North Carolina, studies were made on the influence of the fungi *Cytospora* and *Clitocybe* (or *Armillaria*), rootstock, and nematodes on PTSL. Nema-

Fig. 4 Peach orchard in Georgia with peach tree short life (A); compared with adjacent healthy trees (B). (Courtesy of USDA.)

todes, rootstock, *Cytospora*, and winter temperature regimes (cold injury) influenced tree survival and were therefore involved in the complex (Ritchie and Clayton, 1981). Soil fumigation prevented tree death, and trees on Lovell rootstock lived longer than did trees on Nemaguard. In South Carolina (Dowler and Peterson, 1966) and in Georgia (Prince and Horton, 1972) trees pruned in the fall were more susceptible to bacterial canker than were trees pruned in late winter. More carbohydrates were converted to sugars in the fall-pruned trees than in winter-pruned trees (Dowler and Peterson, 1966). Carter (1976) found that fall-pruned trees were higher in IAA than were winter-pruned trees. In Georgia the influence of nematodes and soil organisms on PTSL were studied. The PTSL syndrome could be controlled by soil fumigation. Furthermore, the tests indicated that the root injury caused by disking increased tree death. The research also indicated that fungi, in particular *Pythium* spp., reduced tree survival (Hendrix et al., 1965; Hendrix and Powell, 1968; Taylor et al., 1970). The research in California was concentrated on bacterial canker and the influence of nematodes on the incidence of the disease. Soil fumigation reduced susceptibility of trees to bacterial canker. Parasitism by the ring nematode, *C. xenoplax*, predisposed trees to bacterial canker (Lownsbery et al., 1968; 1973).

This complex syndrome is responsible for death of several thousand peach trees each year in the Southeast and in California. The role the interacting factors play in the syndrome is not clear.

3. Life Cycle

The life cycle of *C. xenoplax* on peach has been intensively studied (Lownsbery, 1959; Seshadri, 1964a). The nematode lives in the soil, feeding from the outer few layers of cells of the root. Eggs are deposited singly in the soil and hatch without delay. There can be several generations in a year.

4. Economic Loss

The economic loss due to parasitism of ring nematodes has not been assessed; however, the loss due to the involvement of ring nematodes in the peach tree short-life problem is estimated to be several million dollars per year. In a replicated experiment in Georgia, nematicides applied to a short-life site decreased incidence of peach tree short life from 70 to 30% after five years (Wehunt et al., 1980). In another experiment, annual postplanting application of nematicide increased average tree life on a severe site from 2.8 to 6.8 years during a seven year test period. Incidence of bacterial canker was reduced from 74 to 6% in the latter test (Wehunt and Weaver, 1982). Similar results have been reported from South Carolina (Nesmith et al., 1981; Zehr et al., 1976), North Carolina (Ritchie and Clayton, 1981), and California (Lownsbery et al., 1968). Based on income potential, the value of a healthy eight-year-old peach tree was estimated at $74 in 1978 (Bauer, 1978). Based on 109 trees per acre, loss from short life would be 76%, or 80 trees, which would be approximately $5600-5900 per acre loss in income potential.

5. Distribution

C. curvata is involved in "decline" of peach orchards in New Jersey (Hung and Jenkins, 1969). In the Southeast and in California, *C. xenoplax* is widely distributed. In Georgia (Hendrix et al., 1965) the nematode was found in 86% of the orchards surveyed. I found the nematodes in 85% of the orchards surveyed in Georgia, Alabama, South Carolina, North Carolina, and Florida. The nematode was found in all orchards surveyed in which peach tree short life occurred.

6. Control

   a. *Genetic* In Georgia pot tests, Halford seedlings supported fewer *C. xenoplax* than did HW-208 (Wehunt et al., 1976). In other tests Lovell supported fewer than did Nemaguard or Elberta but no resistance was found in more than 160 seedlings tested. In South Carolina, although not significant, field populations increased less on Lovell than on Elberta or Nemaguard (Zehr et al., 1976).

   b. *Cultural* The population of *C. xenoplax* is reduced by cultural practices, such as application of hydrated lime, and soil manipulation (Wehunt et al., 1980; Wehunt and Weaver, 1982). Soil moisture, temperature, and pH affect the population of *C. xenoplax*. The effect of soil factors, such as chemical composition, texture, structure, and other soil organisms, on ring nematodes is unknown. These ectoparasitic nematodes have not been studied sufficiently to provide this information.

c. Chemical *C. xenoplax* can be controlled by preplanting application of fumigant nematicides, such as 1,3-D, D-D Mencs, or ethylene dibromide. Preplanting and postplanting applications of DBCP (1,2-dibromo-3-chloropropane) increased growth of peach trees in soil infested with *C. xenoplax* (Fig. 4). Postplanting applications at reduced rates of several fumigants normally used as preplanting nematicides have been tried with little success in South Carolina (Zehr et al., 1982).

F. Pin Nematodes (*Paratylenchus* spp.)

Pin nematodes (*Paratylenchus* spp.) have been reported from stone fruit plantings in California and Connecticut (Wehunt and Good, 1975).

In California, *P. hamatus* Thorne and Allen 1950 parasitizes peach, plum, and prune trees, reducing vigor and yield (Lownsbery et al., 1974). In Connecticut, *P. projectus* Jenkins 1956 has been reported from peach orchards, but damage has not been demonstrated.

Population levels of pin nematodes can be altered by cover crop culture in peach orchards; however, control of pin nematodes in peach, plum, and prune orchards can best be accomplished by the use of nematicides.

G. Dagger Nematodes (*Xiphinema* spp.)

Dagger nematodes (*Xiphinema* spp.) were reported from stone fruit orchards in Connecticut, West Virginia, California, and Georgia.

The only dagger nematode shown to be a pathogen of stone fruits is *X. americanum* Cobb 1913, which reduces vigor and yield of peach trees. An important effect of *X. americanum* in stone fruits is the transmission of viruses causing *Prunus* stem pitting and peach yellow bud mosaic, which are debilitating and sometimes fatal diseases of stone fruits (Wehunt and Good, 1975).

Fig. 5 Effect of *Hoplolaimus galeatus* and *Fusarium oxysporum* on the growth of Elberta peach seedlings. From left to right, *H. galeatus* alone, *H. galeatus* plus *F. oxysporum*, *F. oxysporum* alone, and noninfested control. (Courtesy of USDA.)

H. Miscellaneous Nematodes of Stone Fruits

Scutellonema brachyurum (Steiner 1938) Andrassy 1958 attacks peach in South
Carolina (Nesmith et al., 1981). *Hoplolaimus galeatus* (Cobb 1913) Thorne
1935 increased incidence of *Fusarium* root rot of peach seedlings in a small
greenhouse test at Byron, Georgia (Fig. 5). Nematodes that are parasitic on
stone fruits but for which pathological studies have not been made are: *Helicotylenchus dihystera* (Cobb 1893) Sher 1961; *Tylenchorhynchus claytoni*
Steiner 1937; *Belonolaimus longicaudatus* Rau 1958; and an unidentified *Gracilacus* species (Wehunt and Good, 1975).

Chemical control measures used against other nematodes will also control
the miscellaneous nematodes listed above.

III. NEMATODES OF APPLE

Based on production, apple, *Malus domestica* Bork, is the most important deciduous tree fruit in the world, with about 14 million metric tons produced.
The production in the USA is about half of the production of Europe. Apples
are an important crop in world commerce. In 1981, U.S. apple production
was almost 8 billion pounds, valued at $864 million (U.S. Dept. Agriculture,
1982).

A. History

The early immigrants brought apple trees and seed to the new world from
Europe, where the crop has been grown for more than 2000 years. Settlers
aided by missionaries and the legendary Johnny Appleseed (John Chapman)
spread apples westward. The states of Washington, New York, Michigan,
Pennsylvania, California, Virginia, and North Carolina account for more than
75% U.S. production.

B. Culture

Apples are propagated by budding or grafting the desired scion onto the
seedling rootstock in the nursery. This practice, with apples as with most
other fruit trees, is hazardous, since the soil-borne pests, diseases, and
nematodes acquired in the nursery may be introduced into the orchard.

C. Lesion Nematodes (*Pratylenchus* spp.)

The most important nematode on apples is the lesion nematode, *Pratylenchus
penetrans* (Cobb 1917) Filip. and Sch. Stek. 1941. This nematode is the cause
of "soil sickness" of nurseries and orchards in Europe. In the United States,
the nematode causes decline in apple and other tree fruit orchards (Arneson
and Mai, 1976; Mai et al., 1970).

The migratory endoparasitic nematodes enter the root in the meristematic
region as larvae or adults. The life cycle is completed inside or outside the
root.

Another lesion nematode, *P. vulnus*, causes damage to apple trees in
California (Siddiqui et al., 1973). The habits and biology of *P. vulnus* and
P. penetrans are similar.

1. Distribution

P. penetrans has been reported from apple orchards throughout the United
States and Canada.

2. Control

There are no reports of resistance to *P. penetrans* in apples. Performance of apple trees is enhanced by soil fumigation. In New York (Arneson and Mai. 1976) net returns from apple trees planted in fumigated soil were more than twice the net returns from nonfumigated soil.

IV. NEMATODES OF PEAR

A. History

The pear, *Pyrus communis* L., is apparently native to the Northern Hemisphere and has been cultivated for thousands of years (Hedrick, 1921). Pear is the second most important deciduous tree fruit, based on production. Pears are grown commercially in almost every temperate country on earth. Pears, like other tree fruits, are propagated by budding the desired scion onto a rootstock in the nursery. Both seedling pear and quince (*Cydonia elongata* Mill.) are used as rootstocks. In 1981 more than 1.7 billion pounds were produced in the United States and the crop was valued at more than $174 million (U.S. Dept. Agriculture, 1982). Among all fruit crops produced in the United States, pears are exceeded in importance only by oranges, apples, grapefruit, grapes, peaches, and lemons. The major producing states are California, Oregon, and Washington, which account for more than 90% of production.

B. Nematodes

More than 100 nematode species have been reported from pear orchards throughout the world, but for only four has evidence of parasitism been presented. In Europe, *Longidorus elongatus* (DeMan 1876) Thorne and Swanger 1936 and *P. penetrans* (Cobb 1917) Filip. and Sch. Stek. 1941 are considered of major economic importance, and in Japan the root-knot nematodes *M. hapla* Chitwood 1949 and *M. incognita* (Kofoid and White 1919) Chitwood 1949, are parasitic on pears (Wehunt and Golden, 1982).

Root-knot nematodes are not considered to be serious pathogens of pears in the United States. Pears were reported resistant to root knot (Tufts and Day, 1934).

The only nematodes that are known to damage pears in the United States are the lesion nematodes, *Pratylenchus* spp.

P. penetrans is implicated in nursery soil sickness and in a replant problem in Europe (Hoestra and Oostenbrink, 1962). In the United States and Canada the nematode is a part of a pear replant problem (Wehunt and Golden, 1982). *P. vulnus* attacks roots of pear in the western United States (Siddiqui et al., 1973). The importance of this nematode in pear production is unknown.

1. Symptoms

Symptoms of lesion nematodes on pears are dark lesions on roots that increase in size as reproduction, feeding, and migration of the nematode continues.

2. Life Cycle

The life cycle and life history of lesion nematodes on pears are similar to those on other crops.

3. Economic Loss

The estimate of the loss of pear production due to nematodes is 5% (Anon., 1971), which would be almost $6 million.

4. Distribution

P. penetrans is found throughout the world where fruit trees are grown. P. vulnus is less widespread, being found in California, Oregon, Washington, Georgia, and North Carolina.

5. Control

a. *Genetic* In California results of trials indicate that a source of resistance to root-knot nematodes in pear is available (Tufts and Day, 1934). No source of resistance has been reported for P. penetrans.

b. *Cultural* There are no reports of attempts to control nematodes of pears by cultural means.

c. *Chemical* Soil fumigation with ethylene dibromide, 1,3-D, and DBCP has resulted in increases in growth, vigor, and yield of pears. Symptoms of "pear decline," however, were not alleviated by nematicide treatment (Wehunt and Golden, 1982).

In New York application of oxamyl increased growth and yield of pears infested with P. penetrans (Arneson and Mai, 1976).

V. NEMATODES OF PECAN

A. History

The pecan [*Carya illinoensis* (Wangenh.) Koch.] is a large tree nut native to the Mississippi Valley (Woodruff, 1976). Improved varieties produce higher yields, but seedlings are also cultivated and harvested. The pecan trees, both seedling and older improved varieties, require about 10 years for full production. The U.S. Department of Agriculture pecan-breeding program at Brownwood, Texas, has produced several varieties that produce nuts at a much younger age.

The U.S, production of pecans was 339 million pounds in 1981, valued at $117 million. Georgia, New Mexico, and Texas are the leading states in pecan production (U.S. Dept. Agriculture, 1982).

B. Culture

Pecans are propagated by budding or grafting the desired scion variety onto a rootstock in a nursery. The taproot is cut in the digging operation and may affect anchorage of the tree.

In the United States pecans are grown primarily in the Southeast and Southwest. In the Southwest, irrigation must be provided for growing pecans. In recent years irrigation has been profitable in the Southeast.

Pecans exhibit irregular or alternate bearing with a high production year followed by a light production year. Control of insects and diseases and proper fertilization tend to moderate the irregular bearing. Root pathogens and nematodes have not been implicated in the irregular bearing.

C. Nematodes

Little research has been done on the nematodes of pecans. Several nematodes were found in Georgia and Texas pecan orchards (Hendrix and Powell, 1968; Johnson et al., 1975; Orr, 1976). The nematodes most likely to be pathogenic were root-knot nematodes (*Meloidogyne* spp.). The symptoms of attack are blinded and branched root tips. Trees in a heavily infested orchard in Georgia were sparsely foliated and the foliage was chlorotic. Trees were stunted with some branches barren.

In Texas (Johnson et al., 1975) distorted, yellow, chlorotic foliage with zinc deficiency symptoms accompanied root-knot infection. Galls on roots were from 1/16 to 1/4 in. in size. The reduction in growth and yield of pecans due to root-knot nematodes is undetermined. From limited observations in Texas, Choctaw, and Burkett seedlings appear to have some resistance, whereas Riverside, Mahan, Sioux, Hollis, and Shoshoni were heavily infested. Control of nematodes and other organisms by methyl bromide treatment of nursery soil increases seedling growth. The factors responsible for the reduction of growth have not been investigated.

In Georgia the root-knot nematode, *M. incognita* (Kofoid and White 1919) Chitwood 1949, was present and reproducing on pecan roots of three unthrifty groves near Waycross, Georgia. In a survey of 60 other unthrifty groves throughout the state, *Criconemella* was most widespread, followed in order of occurrence by *Trichodorus* spp. and *Paratrichodorus* spp., *Helicotylenchus* spp., *Pratylenchus* spp., *Tylenchorhynchus* spp., and *Meloidogyne* spp. Except for *Meloidogyne* spp. the nematodes were not positively associated with pecan roots, since weeds and grass were present. Several fungi were isolated from roots and soil of the same groves. *Pythium* spp. were isolated from 55 of the 60 groves, with *P. irregulare* (Buism.) most often found. In greenhouse experiments several *Pythium* spp. tested reduced root weights of pecan seedlings (Hendrix and Powell, 1968).

Criconemella rusium (Khan, Ghawla, and Saha 1976) Luc and Raski 1981, the most frequently found nematode in Georgia pecan groves, was tested for pathogenicity and for its influence on feeder root necrosis caused by *P. irregular* and *Fusarium solani* (Mart.) App. and Wr. emend Snyd. and Hans. The nematode alone did not affect root weight, whereas both fungi reduced root weight. However, there was a synergistic effect of the nematode in combination with either and both of the fungi (Hsu and Hendrix, 1973).

A lance nematode, *Hoplolaimus stephanus* Sher 1963, is found in pecan orchards of South Georgia. The nematode survived on pecans in greenhouse pots, indicating parasitism of the nematode, but pathogenicity studies have not been made.

A pin nematode, *Gracilacus*, found in pecan orchards throughout Georgia, survives on pecans in the greenhouse. There is no information on pathogenicity of this nematode to pecans.

Occasionally species of *Longidorus, Xiphinema, Pratylenchus, Helicotylenchus*, and *Tylenchorhynchus* are found in Georgia pecan orchards, but there is no information on parasitism or pathogenicity to pecans.

VI. NEMATODES OF WALNUT

A. History

Three species of nut trees are commonly known as walnuts. Black walnut, *Juglans nigra* L., is a forest tree of temperate zones (Woodruff, 1976). Although the nuts are harvested, black walnut trees are more important for lumber. The butternut, *J. cinerea* L., also known as white walnut, is not an important crop. The principal walnut grown for nuts is the English or Persian walnut, *J. regia* L. Black walnut and butternut are native to the eastern United States. English walnut has been cultivated in Europe for many years. In the United States, the major areas of the production of English walnut are the western states of California and Oregon (U.S. Dept. Agriculture, 1982). Two types of English walnut are grown. The Santa Barbara group of cultivars are grown in southern California where temperatures are warm. The French group of cultivars are more hardy and are grown from central California to Oregon. Deep, fertile, well-drained soils are best for walnut production, and pure water is required for irrigation. Traditionally English walnuts do not do well in the southern or eastern states (Woodruff, 1976).

The United States is the leading producer of English walnut. The farm value of walnuts in the United States was $224 million in 1981 (U.S. Dept. Agriculture, 1982), with an estimated 15% loss due to nematodes, which equals $34 million (Anon., 1971).

B. Nematodes

English walnuts are parasitized by lesion nematodes, *P. vulnus* (Lownsbery, 1956; Siddiqui et al., 1973). Symptoms are black lesions on the epidermis and cortex of roots, which grow as the nematodes feed and multiply. Feeder roots are decreased by the nematode. The nematode freely enters roots both as juvenile and adults. The life cycle is completed inside the root and in the soil. Reduction in seedling growth (Lownsbery, 1956) and yield (Lownsbery et al., 1968) due to lesion nematodes has been demonstrated in California.

In California seedlings of *J. hindsii* (Jeps.) Rehder, *J. major* (Torr.) ex (Stisgr.) Hellar, *J. nigra*, *J. regia*, and *J. microcarpa* Berl. were susceptible to *P. vulnus* ; Paradox hybrid (*J. hindsii* x *J. regia*) was intermediate in reaction, and the Chinese wingnut, *Pteracarya stenoptera* C. DC, which is often used as a rootstock, was tolerant to *P. vulnus* (Lownsbery et al., 1974).

Cacopaurus pestis Thorne 1943 parasitizes roots of walnut in California. The nematode is a sedentary ectoparasite feeding from epidermal cells, and the body of the female stays on the outside of the root (Thorne, 1943). Eggs are deposited in a gelatinous matrix exuded from the posterior end of the female. No delay in hatching or requirement for hatching stimulants has been recorded.

Thorne (1943) described *C. pestis* from material from an orchard of Mayette English walnut on seedling rootstock near Santa Clara, California. The original trees, planted in 1902-1903, made normal growth until 1936 when a few trees that had died from cold injury were replaced with Mayette on Hinds black walnut (*J. hindsii*) rootstock. Symptoms described by Thorne (1943) were "a dieback developed in 1937 which progressed rapidly and by 1940 several trees were in various stages of decline. The injury was first manifest by a reduction in the size and numbers of leaves, twigs, and nuts and, in a few years, culminated in complete defoliation and death of the trees." Roots of the affected trees were dead or had extensive lesions. High populations of the nematode, *C. pestis* was found associated with the damaged roots.

The trees propagated on *J. hindsii* rootstock were not affected, although they were growing in what should have been infested soil.

In Tulare county, California, *Gracilacus epacris* (Allen and Jensen 1950) Raski 1962 (*G. epacris* Allen and Jensen 1950) was found in a declining orchard of black walnut. The contribution of *G. epacris* could not be determined because the roots were also infested with root-lesion nematodes (Allen and Jensen, 1950).

REFERENCES

Allen, M. W., and Jensen, H. J. (1950). *Cacopaurus epacris*, new species (Nematoda: Criconematidae), a nematode parasite of California Black Walnut roots. *Proc. Helm. Soc. Wash. 17*: 10-14.

Anon. (1971). Estimated crop losses due to plant-parasitic nematodes in the United States. *Soc. Nematol. Spec. Publ. No. 1.*

Arneson, P. A., and Mai, W. F. (1976). Root diseases of fruit trees in New York state. VII. Costs and returns of preplant soil fumigation in a replanted apple orchard. *Plant Dis. Rep. 60*: 1054-1057.

Barker, K. R., and Clayton, C. N. (1969). Relative host suitability of peach cultivars to six species of lesion nematodes. *Phytopathology 59*: 1017.

Bauer, L. L. (1978). Costs and returns of producing peaches in South Carolina. *S. C. Agric. Exp. Sta. Bull. SB 617.*

Carter, G. E. Jr. (1976). Effect of soil fumigation and pruning date on the indoleacetic acid content of peach trees in a short life site. *Hort. Sci. 11*: 594-595.

Chitwood, B. G. (1949). Ring nematodes (Criconematinae) a possible factor in decline and replanting problems of peach orchards. *Proc. Helm. Soc. Wash. 16*: 6-7.

Dhanvantari, B. N., Johnson, P. W., and Dirks, V. A. (1975). The role of nematodes in crown gall infection of peach in southwestern Ontario. *Plant Dis. Rep. 59*: 109-112.

Dowler, W. M., and Peterson, D. H. (1966). Infection of bacterial canker in the field. *Phytopathology 56*: 989-990.

Esser, R. P., Martine, A. P., and Longdon, K. R. (1968). Simultaneous occurrence of root-knot nematode and crown gall bacteria. *Plant Dis. Rep. 52*: 550-553.

Fliegel, P. (1969). Population dynamics and pathogenicity of three species of *Pratylenchus* on peach. *Phytopathology 59*: 120-124.

Foster, H. H., Gambrell, C. E., Jr., Rhodes, W. H., and Byrd, W. P. (1972). Effects of preplant nematicides and resistant rootstocks on growth and fruit production of peach trees in *Meloidogyne* spp. infested soil of South Carolina. *Plant Dis. Rep. 56*: 169-173.

Hedrick, U. P. (1921). *The Pears of New York*. Report of the New York Agricultural Experiment Station for the year 1921, vol. II. J. B. Lyon Company, Albany, New York.

Hendrix, F. F., Jr., and Powell, W. M. (1968). Nematode and *Pythium* species associated with feeder root necrosis of pecan trees in Georgia. *Plant Dis. Rep. 52*: 334-335.

Hendrix, F. F., Jr., Powell, W. M., Owen, J. H., and Campbell, W. A. (1965). Pathogens associated with diseased peach roots. *Phytopathology 55*: 1061.

Hoestra, H., and Oostenbrink, M. (1962). Nematodes in relation to plant growth. IV. *Pratylenchus penetrans* (Cobb) on orchard trees. *Neth. J. Agric. Sci. 10*: 286-296.

Horton, B. D., Wehunt, E. J., Edwards, J. H., Bruce, R. R., and Chesness, J. L. (1981). The effect of drip irrigation and soil fumigation on "Redglobe" peach yields and growth. *J. Am. Soc. Hort. Sci. 106*: 438-443.

Hsu, Dienshe, and Hendrix, F. F. (1973). Influence of *Criconemoides quadri-*

cornis on pecan feeder root necrosis caused by *Pythium irregulare* and *Fusarium solani* at different temperatures. *Can. J. Bot.* *51*: 1421-1424.

Hung, Cia-Ling Pi, and Jenkins, W. R. (1969). *Criconemoides curvatum* and the peach tree decline problem. *J. Nematol.* *1*: 12.

Johnson, J. D., Smith, H. P., Thames, W. H., Jr., Smith, L. R., Brown, M. H., and Henderson, W. C. (1975). Nematodes, *Meloidogyne incognita*, can be a problem in pecans. *Pecan Q.* *9*: 6.

Jones, M. G. K., and Payne, H. L. (1978). Early stages of nematode-induced giant-cell formation in roots of *Impatiens halsomina*. *J. Nematol.* *10*: 70-84.

Layne, R. E. C. (1974). Breeding peach rootstocks for Canada and the northern United States. *Hort. Sci.* *9*: 364-366.

Lownsbery, B. F. (1956). *Pratylenchus vulnus*, primary cause of the root-lesion disease of walnut. *Phytopathology* *46*: 376-379.

Lownsbery, B. F. (1959). Studies of the nematode *Criconemoides xenoplax* on peach. *Plant Dis. Rep.* *43*: 913-917.

Lownsbery, B. F., English, H., Moody, E. H., and Schick, F. J. (1973). *Criconemoides xenoplax* experimentally associated with a disease of peach trees. *Phytopathology* *63*: 994-997.

Lownsbery, B. F., Martin, B. C., Forde, H. I., and Moody, E. H. (1974). Comparative tolerance of Walnut species, Walnut hybrids and Wingnut to the root-lesion nematode, *Pratylenchus vulnus*. *Plant Dis. Rep.* *58*: 630-633.

Lownsbery, B. F., Mitchell, J. R., Hart, W. H., Charles, F. M., Gertz, M. H., and Greathead, A. H. (1968). Responses of postplanting and pre-planting soil fumigation in California peach, walnut and prune orchards. *Plant Dis. Rep.* *52*: 890-894.

Lownsbery, B. F., Moody, E. H., and Braun, A. J. (1974). Plant-parasitic nematodes in California prune orchards. *Plant Dis. Rep.* *58*: 633-636.

Mai, W. F., Parker, K. G., and Hickey, K. D. (1970). Root diseases of fruit trees in New York State. II. Populations of *Pratylenchus penetrans* and growth of apple in response to soil treatment with nematicides. *Plant Dis. Rep.* *54*: 792-795.

Mountain, W. G., and Boyce, H. R. (1958). The peach replant problem in Ontario. IV. The relation of *Pratylenchus penetrans* to the growth of young peach trees. *Can. J. Bot.* *36*: 135-151.

Neal, D. C. (1889). The root-knot disease of peach, orange and other plants in Florida due to the work of *Anguillula*. *USDA Div. Ent. Bull. No. 20*.

Nesmith, W. C., Zehr, E. I., and Dowler, W. M. (1981). Association of *Macroposthonia xenoplax* and *Scutellonema brachyurum* with the peach tree short life syndrome. *J. Nematol.* *13*: 220-225.

Nigh, E. L., Jr. (1966). Incidence of crown gall infection in peach as affected by the Javanese root-knot nematode. *Phytopathology* *56*: 150.

Nyczepir, A. P., and Lewis, S. A. (1980). The influence of *Macropostonia xenoplax* Raski on indole-3-acetic acid (IAA) and absisic acid (ABA) in peach. *J. Nematol.* *12*: 234.

Orr, C. C. (1976). Nematode diseases in pecan culture. *Pecan South 3*: 428-429.

Prince, V. E., and Horton, B. D. (1972). Influence of pruning at various dates on peach tree mortality. *J. Am. Soc. Hort. Sci.* *97*: 303-305.

Ritchie, D. F., and Clayton, C. N. (1981). Peach tree short life: A complex of interacting factors. *Plant Dis.* *65*: 462-469.

Seshadri, A. R. (1964a). Histological investigations on the ring nematode

Criconemoides xenoplax Raski, 1952 (Nematoda: Criconematidae). *Nematologica 10*: 519-539.

Seshadri, A. R. (1964b). Investigations on the biology and the life cycle of *Criconemoides xenoplax* Raski, 1952. (Nematoda: Criconematidae). *Nematologica 10*: 540-562.

Sharpe, R. H., Hess, C. O., Lownsbery, B. F., Perry, V. G., and Hansen, C. J. (1969). Breeding peaches for root-knot resistance. *J. Am. Soc. Hort. Sci. 94*: 209-212.

Siddiqui, I. A., Sher, S. A., and French, A. M. (1973). Distribution of plant parasitic nematodes in California. *State Calif. Dept. Food Agric. Div. Plant Ind. Bull.*

Stokes, D. E. (1966). Parasitism by *Pratylenchus brachyurus* on three peach rootstocks. *Nematologica 13*: 153.

Taylor, J., Biesbrock, J. A., Hendrix, F. F., Jr., Powell, W. M., Danniell, J. W., and Crosby, F. L. (1970). Peach tree decline in Georgia. *Ga. Agric. Exp. Sta. Bull. 77*.

Thorne, G. (1943). *Cacopaurus pestis*, nov. gen., nov. spec. (Nematoda: Criconematinae) a destructive parasite of the walnut, *Juglans regia* Linn. *Proc. Helm. Soc. Wash. 10*: 78-83.

Tufts, W. P. and Day, L. H. (1934). Nematode resistance of certain deciduous fruit tree seedlings. *Proc. Am. Soc. Hort. Sci. (Suppl.) 31*: 75-82.

U.S. Dept. Agriculture. (1982). *Agricultural Statistics – 1982*. U.S. Govt. Printing Office, Washington, D.C.

Viglierchio, D. R., and Mjuge, S. G. (1975). Auxin inactivation systems of nemic origin. *Nematologica 21*: 471-475.

Wehunt, E. J. and Edwards, J. H., unpublished.

Wehunt, E. J. and Golden, A. M. (1982). Nematodes of Pears. In *The Pear* (N. F. Childers, ed.). Communications Dept., Cook College, Rutgers University, New Jersey, pp. 377-387.

Wehunt, E. J. and Good, J. M. (1975). Nematodes on peaches. In *The Peach: varieties, culture, marketing, and pest control*, 3rd ed. (N. F. Childers, ed.). Communications Dept., Cook College, Rutgers University, New Jersey, pp. 377-387.

Wehunt, E. J., Horton, B. D., and Prince, V. E. (1980). Effects of nematicides, lime and herbicide on a peach tree short life site in Georgia. *J. Nematol. 12*: 183-189.

Wehunt, E. J., and Weaver, D. J. (1982). Effect of planting site preparation, hydrated lime and DBCP (1,2-dibromo-3-chloropropane) on populations of *Macroposthonia xenoplax* and peach tree short life in Georgia. *J. Nematol. 14*: 567-571.

Wehunt, E. J., Weaver, D. J., and Doud, S. L. (1976). Effect of peach rootstock and lime on *Criconemoides xenoplax*. *J. Nematol. 8*: 304.

Wehunt, E. J. and Yadava, U. L., unpublished.

Woodruff, J. G. (1976). *Tree Nuts: Production, Processing, Products*, 2nd ed. AVI Publishing Co., Westport, Connecticut.

Zehr, E. I., Lewis, S. A., and Gambrell, C. E., Jr. (1982). Effectiveness of certain nematicides for control of *Macropostonia xenoplax* and short life of peach trees. *Plant Dis. 66*: 225-228.

Zehr, E. I., Miller, R. W., and Smith, F. H. (1976). Soil fumigation and peach rootstock for protection against peach tree short life. *Phytopathology 66*: 689-694.

Chapter 13

Nematode Parasites of Grapes and Other Small Fruits

Dewey J. Raski *University of California, Davis, California*

Lorin R. Krusberg *University of Maryland, College Park, Maryland*

I. GRAPES

The culture of grapes goes back to our earliest records. According to Winkler et al. (1974), Egyptian mosaics of 2440 B.C. give details of grape growing. Also, many Roman Empire accounts describe varieties, types of wine, and cultural practices, as well as wine production.

World production of grapes totals about 22 million acres, most of which are varieties of *Vitis vinifera* L. The origin of vinifera grapes is in areas south of the Caucasus Mountains and Caspian Sea and of Asia Minor, where it is believed grape culture began (Hewitt and Raski, 1967).

Viticulture was introduced to western United States by the Spanish missionaries, according to Winkler et al. (1974), the first vines being planted about 1697 by Father Juan Ugarte at Mission San Francisco Xavier in Baja, California. The culture of vines spread northward to other missions but was largely abandoned by 1834 after secularization of the church. Commerical plantings began in the Los Angeles area in 1824 primarily for the fresh market and based largely on the Mission variety. From 1850 onward many new and different varieties were introduced for wine and raisin production as well as fresh fruit.

By 1969-1971, the bearing acreage of grapes in California totaled 449,682 acres, with an average annual value of $259,508,000. The current plantings total 596,354 bearing acres plus 56,173 nonbearing acres and an estimated crop value of $1,016,261,000 in 1979.

In the eastern United States, grape growing started much earlier, but cultivation of *V. vinifera* varieties failed due largely to the excessively cold winters that occur occasionally in the North and to the infections of Pierce's disease in southern areas. The successful vineyards in the eastern United States are mostly varieties of *Vitis labrusca* L., native to New England.

The first report of nematodes attacking grapevines in the United States was by Neal (1889). He described root-knot nematode (*Anguillula*) affecting many cultivated and uncultivated plants in Florida, including various species of *Vitis*. Bessey (1911) gave a more extensive report on root-knot but added little new information concerning grapes. Other workers (Milbrath, 1923; Nougaret, 1923; Brown, 1931; Tyler, 1933a, b, c; Whittle and Drain, 1935; Snyder, 1936; Tyler, 1941) contributed much to our knowledge of root-knot and grapevines, but it was not until 1954 (Raski, 1954) that many other plant-parasitic nematode species were found associated with declining vineyards, suggesting a much wider array of pathogens threatened that industry. *Xiphinema* spp. (*index*, *americanum*, and others); *Pratylenchus vulnus* and others; *Criconemella xenoplax* and others; *Paratylenchus hamatus* and others; and *Trichodorus* spp. were implicated at that time; some of them have since proved to be major pests of grape. Two years later the citrus nematode, *Tylenchulus semipenetrans*, was found attacking grape roots (Raski et al., 1956).

A. Root-Knot Nematodes (*Meloidogyne* spp.)

1. *History*

Much of the research reported before 1949 has limited value because it is doubtful identification to species of the *Meloidogyne* populations under study are dependable. Previous work was reported as *Heterodera radicicola* or more often as *Heterodera marioni*. Chitwood's revision of the genus (1949) distinguished five species and presented morphological differences that permit relatively easy and accurate identification.

Since then three species of root-knot nematode have been recognized as major pests of grapevine: *Meloidogyne incognita* (Kofoid and White 1919), *M. javanica* (Treub 1885), and *M. arenaria* (Neal 1889).

2. Symptoms

The above-ground parts of grapevines do not show symptoms diagnostic for damage by any of the nematode species attacking the root system. Often, weak areas of unthrifty growth are mistakenly attributed to water stress, low fertility, salt excess, or other pathogens. Nematode damage may appear in localized spots or areas (Fig. 1) but has been known to affect entire plantings. This is especially true of cuttings or rootings planted in heavily infested soils. The plants present a generally unthrifty appearance and poor color and most importantly have lowered fruit production. Evidence of nematode damage to roots by some nematode species, on the other hand, can be specifically and reliably identified.

In most cases of root-knot nematode there are swellings or galls on the feeder rootlets and young secondary roots (Fig. 1A). Individual nematodes produce small galls, but multiple infections of four to six or more females cause much larger, elongated swellings. It is also characteristic of heavy infections that ultimately most feeder rootlets are killed and their paucity is quite striking and noticeable. It has also been reported (Nougaret, 1923) that numerous females were found attacking the below-ground internodal trunk tissues.

3. Causal Organisms

The genus *Meloidogyne* Goeldi 1887 comprises the group of plant pests commonly referred to as root-knot nematodes. The three species that parasitize grapevines in California, *M. incognita, M. javanica,* and *M. arenaria,* are widespread, and are also well-known throughout the world. Interestingly, the only species so far identified from grape in Washington is *M. hapla* (G. S. Santo, personal communication) for which there are few records on grape in California.

The life habits and general morphology of root knot are quite similar. Specific identifications are made essentially from comparisons of specialized preparations of perineal sections from adult females, but differences in host ranges are also helpful (Sasser, 1954). Roots of grape are invaded only by second-stage larvae, usually near the tip of new rootlets. The larvae penetrate to the cortex and orient their head end in contact with the vascular system. As a result of nematode feeding some of the host plant cells are induced to form "giant cells," which provide a permanent feeding site for the developing larvae. Once the nematode is located in the feeding site, it is sedentary and completes its cycle there. The nematode undergoes four molts; the first occurs inside the egg and the larva, which hatches is the second-stage, mobile, infective stage. During development inside the root the second-stage female larva quickly molts three times and grows remarkably in girth and

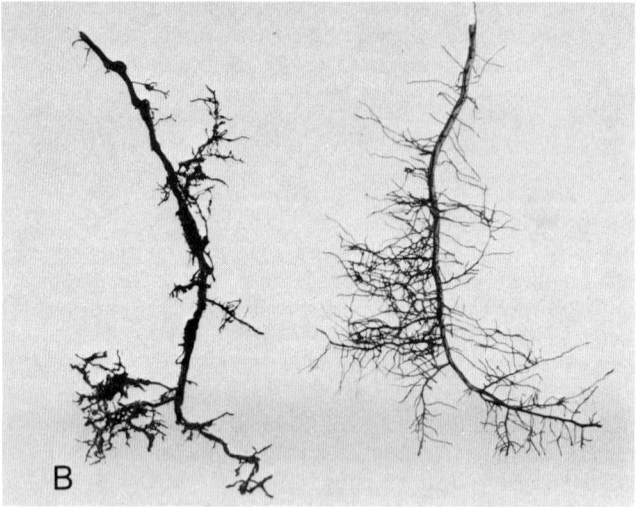

Fig. 1 (A) Typical area of weak vine growth in a vineyard infested with nematodes. (B) Root-knot nematode damage on grape roots. Right, healthy
roots. Left, infected roots.

slightly in length (Fig. 2). Eggs occur free in the soil but usually are deposited in a gelatinous matrix (Fig. 3) produced by rectal glands in the posterior end of the females. The life cycle from egg to egg at 27°C is reported
to be 25 days (Tyler, 1933a). The males develop by swelling only slightly,
then begin to elongate and complete their molting and development inside the
second-stage larval cuticle. These three principal species found on grape
are parthenogenetic, and males are not required for reproduction.

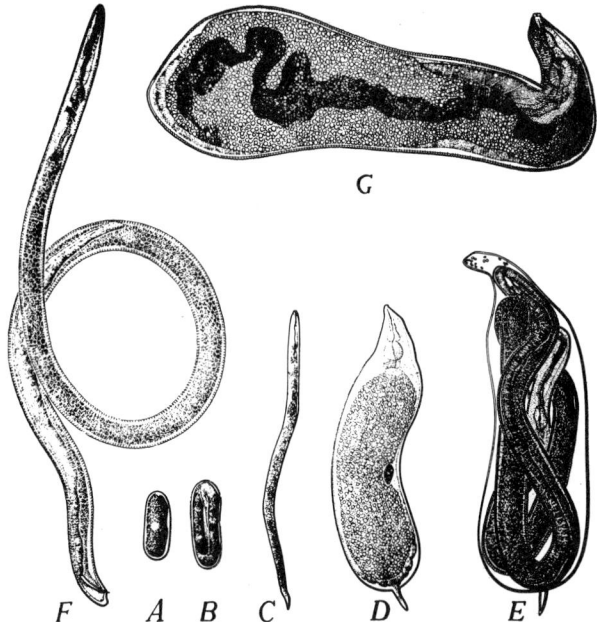

Fig. 2 Life stages of a root-knot nematode. (A) Unsegmented egg; (B) larva before hatching; (C) second-stage, migratory larva; (D) spike-tail larva in root; (E) fully developed male in larval cutical; (F) adult male; (G) adult but young female.

4. *Percentage and Dollar Loss Estimates*

Nematode problems on grapevines share the same status as many other pests of crops. Initially information on the problem is generated from survey data, collection, and identification, mostly from soil samples. The next step is to establish parasitism, which is relatively easy by collection of roots and incubation or dissection of the roots to recover nematodes from host tissues.

Ultimate determination of pathogenicity is much more difficult. Axenic cultures maintained under sterile conditions provide fundamental facts of histological as well as gross pathology. Few such tests have been accomplished on nematodes of grape. Even then the criticism remains that such information reflects laboratory conditions only and is not a true picture of expectations under field conditions.

Plant responses to side-dressing chemical treatments or preplant soil fumigation have been valuable as indicators of nematode effects when compared with untreated controls. However, even here some doubt is held as to effects of the chemicals on other organisms that may be important as contributors to a disease complex.

Percentage and dollar loss estimates thus are largely subjective. There have been no attempts in recent years to estimate total losses due to nematodes, but national studies on the impact of suspending DBCP have produced one measure of estimated loss. It is judged from treatment data that 165,200 acres of grapes are treated on a two year cycle using 3,200,000 lb DBCP annually in California. Due to suspension of use of DBCP it is expected plant vigor and production will decline and the first cumulative three year loss will exceed $65,000,000.

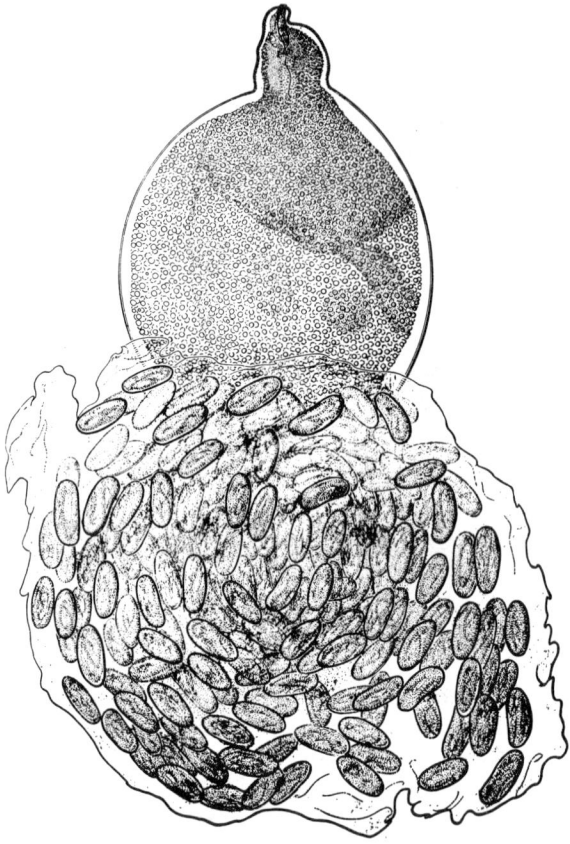

Fig. 3 Adult female of root-knot nematode with attached egg mass in gelatinous matrix.

In 1958, the Department of Nematology, University of California, prepared estimates of losses due to nematodes on all affected crops in California. At that time the total value of production was reported to be $168,292,000 and losses due to nematodes as 20% for $33,658,000. Today the production is valued at $1,016,261,000 and the same percentage loss would total $203,252,000.

Dr. G. S. Santo estimates losses in Washington at 10% of a $39,000,000 crop for a total loss of approximately $3,900,000.

5. Distribution

The four species of *Meloidogyne* mentioned above are worldwide in distribution (Taylor and Sasser, 1978). *M. incognita* is the most common species in California and on grapes is particularly prevalent throughout the interior valleys (Siddiqui et al., 1973). In southern counties it is also known in some coastal areas. *M. javanica* is almost coincident with *M. incognita* in the interior valleys but is commonly found on grapes only in the southern areas of the state. It is also reported from Australia (Seinhorst and Sauer, 1956; Sauer, 1962). *M. arenaria* is reported from San Joaquin Valley plantings (Ferris and McKenry, 1974), but its prevalence in vineyards is not yet known. *M. incognita*, *M. javanica*, and *M. arenaria* are also reported from Chile (Allen et al., 1971). *M. hapla* is rare on vines in California but is more frequently found in vineyards of Washington state (G. S. Santo, personal communication).

6. Control

The first line of defense against nematodes in vineyards is to establish healthy, vigorous, productive plants. To achieve this it is imperative the problem be identified to know with precision the kinds of nematodes present in the soils, their density and distribution. If there is a history of root-knot nematode in a given field, there is no alternative except to plant resistant rootstocks on which to graft scion *V. vinifera* varieties or to follow preplant soil fumigation after which own-rooted *V. vinifera* can be grown successfully.

a. Resistant Varieties. Interest in plant genetic sources of resistance or immunity to root-knot was reported as early as 1889 (Neal, 1889). He described *V. vinifera* grafted on *V. cordifolia* or *V. vulpina* made "superb growth free of the disease." Considerable time elapsed before Bessey made a more detailed report on root knot (Bessey, 1911). In the 1930s and early 1940s the works of Tyler and Snyder were published (Tyler, 1933a-c; 1941; Snyder, 1936), the last being a detailed study of susceptibility of grape rootstocks to root-knot nematode.

Unfortunately, most of that information has limited value because all the reports were made under the name *Heterodera marioni* (Cornu). Only after 1949 was it possible to distinguish the differences in species of root knot based on the classic revision of that group of nematodes by Chitwood (1949).

Soon after, Lider (1954) reported his studies on resistance to *M. incognita* var. *acrita* in *Vitis* spp. Since then four rootstocks have been the major stocks recommended in California (Lider, 1959; Raski, Hart, and Kasimatis, 1973). Dogridge (*V. champini*) and Salt Creek "Ramsey" (*V. champini*) are very vigorous and used only in very sandy soils of low fertility where nematode damage is usually severe. Both tend to have excessive growth but a high degree of resistance to root knot. The rootstock 1613 (*V. solonis* × Othello) and Harmony (1613 seedling × Dogridge seedling) are better adapted to the fertile sandy and loam soils of the San Joaquin Valley, but have the drawback of damage by root-knot populations in some local areas. Harmony also has good resistance to phylloxera as well.

b. Preplant Soil Fumigation. Early attempts at soil fumigation for replanting grapes in nematode-infested land (Raski, 1954; 1955) were largely unsuccessful because nematode control was inadequate. The materials available at that time were 1,3-D (1,3-dichloropropene), EDB (ethylene dibromide), and DBCP (1,2 dibromo-3-chloropropane) and were tested at rates up to 67 kg/ha, at conventional depths of 24-30 cm and in most cases were applied the same year the old infested vines were removed. The vines made excellent growth the first year but excessive nematode buildup recurred during the first year after planting and the vines failed to continue vigorous growth or to achieve satisfactory productivity.

Continued tests with 1,3-D (Raski et al., 1971) established effective control by deep placement (60-100 cm) on wide spacing (Fig. 4) (60-100 cm) of high dosages (up to 325 kg/ha). At the same time methyl bromide (CH_3Br) was developed (Raski and Schmitt, 1972; Raski, Hart, and Kasimatis, 1973) for field applications using 1 mil polyethylene covers (Fig. 5) to contain this highly volatile gas in the soil. The material was applied 60-100 cm deep, spaced 3.3 m apart with dosages of approximately 450 kg/ha. It has been found by using dosages of 560 kg/ha that the plastic cover can be omitted, but the soil must be roller packed as soon as possible after treatment (Raski et al., 1975; 1976). Eradication with these chemicals is not possible, but nema-

Fig. 4 Deep-placement soil fumigation equipment for preplant nematode control with 1,3-dichloropropene. (A) Deep chisels set 1.0 m apart, 0.75-1.0 m deep. (B) Detail of land wheel for metering fumigant.

nematode buildup is deterred for four to five years or more and in the nematode-free soils, the vines can establish as strong, healthy plants. Ultimate longevity after nematodes again attack the vines is yet to be determined.

    c. Treatment of Established Vines. Once vines are established and nematodes attack the roots, little can be done to offset the damage. For more than 20 years DBCP was successfully used (Fig. 6) at 18-36 liter ai/ha. It was low in phytotoxicity to grapevines, highly nematicidal, and dispersed effectively through the soil as a fumigant. However, its use has been suspended since 1977 and there are no alternative materials available. Several

Fig. 5 Methyl bromide application under polyethylene cover.

nonfumigant compounds, some of them systemic in activity, applied as granular or water-soluble formulations, are being tested, but to date none has achieved registration for use.

   d. Cultural Practices. With special care vineyards can remain productive in many areas despite nematode attacks on the roots, especially if the vines are given a healthy start. The single most important factor is stress:

Fig. 6 Thompson Seedless var. grape response to nematicidal treatment. Center row and partial row to right untreated; rows to left treated. Nematode infestation predominantly root-knot and dagger nematodes.

lack of timely irrigations, insect and mite damage, insufficient fertilizers and overcropping, are all critical practices that stress vines. Pruning to produce crops within the capability of vines is especially difficult with temptations to leave excess spurs, buds, and clusters.

3. Hot Water Treatments. The safest way to produce clean rootings is to plant cuttings in fumigated nursery soil. At times it becomes important to save nematode-infected rootings. Chemical dips have never succeeded without excessive phytotoxicity. Hot water treatments have been reported effective for eradication of *M. incognita* and *M. javanica* on 16 commercial varieties and 4 experimental selections of grape rootings (Lear and Lider, 1959).

B. Dagger Nematodes (*Xiphinema* spp.)

1. History

Dagger nematodes belong to the genus *Xiphinema*, and two species are the most important to grape culture. By far the most widespread is *X. americanum* Cobb 1913, which is present in all the grape growing areas of California and reported generally spread throughout the United States. Lamberti and Bleve-Zacheo recently (1979) revised a group of species related to *X. americanum*, which leaves doubt as to exactly what species are in California. Until this is clarified, use of *X. americanum* will be continued.

The second species, *X. index* (Fig. 7), was described by Thorne and Allen (1950) from fig, *Ficus carica* L., suffering "leaf drop" near Planada, Merced County, California. Since then it has been reported widely distributed in the central valley south of Sacramento and throughout the northern coastal grape-growing regions. It is one of the species reportedly worldwide in distribution (Martelli, 1978).

2. Symptoms

Similar to root knot, there are no symptoms on the foliage or fruit of grape caused by dagger nematode damage to the root system. *X. index* reproduces well under greenhouse conditions and has been studied in considerable detail. Gross symptoms (Raski and Radewald, 1958; Radewald, 1962) on roots (Fig. 8) show (1) terminal swellings with necrosis, (2) cessation of root elongation and extensive necrosis of main roots resulting in a witches' broom effect from lateral proliferation, (3) unequal swelling on one side of rootlets, which produces a curvature or bending of 45-90° or more. According to Fisher and Raski (1967), feeding by larvae is in the piliferous region as well as at root tips, but galls or swellings are produced only at the root tip. Females feed only at the root tip and galls are evident within 24 hours. This is at variance with Cohn (1970), who found *X. index* fed at several sites along the root but only occasionally near the tip.

Histologically, cells near feeding sites undergo hypertrophy, which accounts for the swelling or gall formation. Multinucleate giant cells and mononucleate hypertrophied cells develop at feeding sites (Rumpenhorst and Weischer, 1978). Incomplete cell separation and irregular wall formation also occur in giant cells. Sometimes the cell walls are ruptured and void of cytoplasm. Necrosis was found at varying depths in the cortex with underlying phellogen tissue. In general, plants attacked by *X. index* produce few feeder roots. Rootings inoculated with 500 *X. index* were greatly reduced in size after one year (Pinochet, Raski, and Goheen, 1976) and in addition appear to have auxin relationships altered in the plant (Van Gundy et al., 1968).

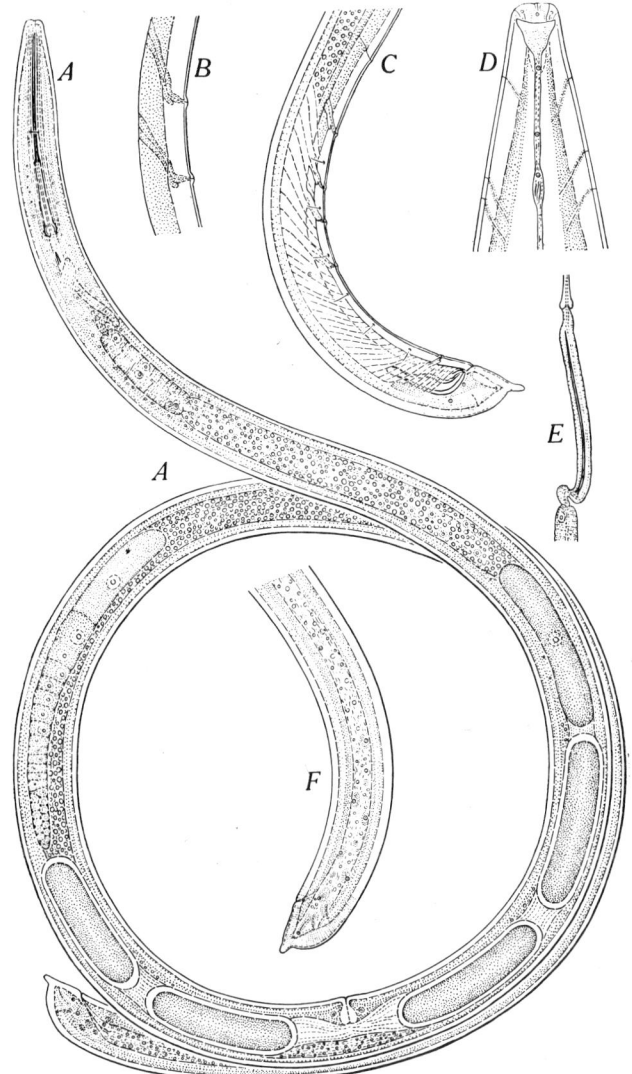

Fig. 7 The dagger nematode, *X. index*, pathogen of grapevines and vector of fanleaf virus. (A) Female, full length; (B) male supplements; (C) male tail; (D) female cephalic region; (E) developing odontostyle in larval esophagus; (F) female tail.

X. americanum has been studied in much less detail because it is difficult to increase or even maintain populations under controlled conditions of the greenhouse. Root damage is principally cessation of elongation and absence of feeder roots and occasionally may show distortion, but necrosis and gross malformations such as caused by *X. index* are lacking in vines attacked by *X. americanum*. Generally, poor root system with few feeder roots is associated with *X. americanum* presence.

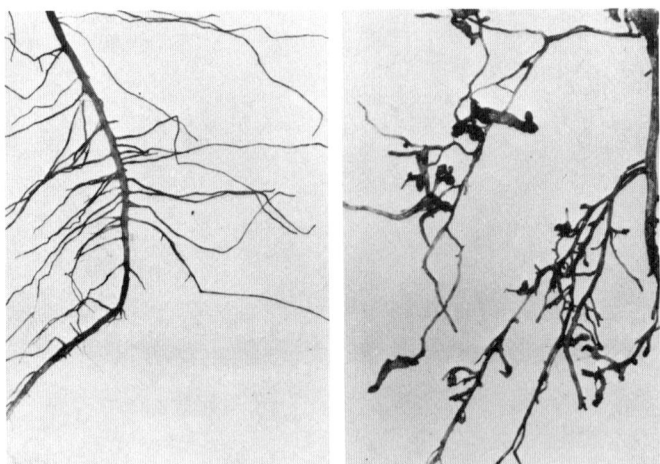

Fig. 8 Healthy grape roots, left. Grape roots damaged by *X. index* right.

3. *Causal Organism*

X. index is a typical member of the genus *Xiphinema*. The adults are long
(2.6-3.6 mm), slender, almost cylindrical in shape, tapering only slightly
anteriorly to a rounded head region. The tail of both sexes narrows or tapers
very little, is short, bluntly rounded with prominent concavity on the dorsal
side, and with a distinct digitate extension ventrally. The adults bear an
odontostyle 119-133 μm long and basal extensions 64-80 μm long (range of
combined odontostyle plus extension is 183-213 μm). The vulva is located at
37-43% and both ovaries are well developed. The life cycle as reported by
Radewald (1962) begins with eggs laid singly in the soil. Development to
ecdysis takes place in six to seven days and the first molt takes place soon
after. Three more molts follow in the next 15-18 days, with successive increase in size averaging 0.72 mm for the first-stage larvae and 1.25 mm, 1.62
mm, and 2.06 mm for second, third, and fourth-stage larvae, respectively.
The average spear length likewise increases from 48 μm, to 62 μm, 88 μm,
and 114 μm for the four stages. Adult males are similar to females except for
sexual morphology, the males bearing paired spicules, a pair of adanal supplements, and four or more ventromedian supplements.

   Reproduction is parthenogenetic in most cases. Males are rare but can
be found usually single or in low numbers. More rarely still, the males may
be more numerous perhaps due to prolonged unfavorable conditions.

   Host range records of *X. index* have been extended considerably in recent years, but compared with many plant-parasitic species it has a relatively restricted range. Besides grape (*V. vinifera*) and fig (*F. carica* L.) it is
known to attack *Pistacia vera*, *P. mutica*, *Ampelopsis aconitifolia* (Weiner and
Raski, 1966); rose (*Rosa* sp.), Boston ivy (*Parthenocissus tricuspidata*),
Virginia creeper (*Parthenocissus quinquefolis*) (Radewald, 1962); sour orange
(*Citrus aurantium*), bur marigold (*Bidens tripartita*), dwarf nettle (*Urtica
urens*), strawberry (*Fragaria* × *ananassa*) (Cohn and Mordechai, 1969); and
tomato (research of author unpublished).

   X. americanum measuring 1.4-2.0 mm in length is slightly smaller and
more slender than *X. index* with similar proportions and aspect. The odontostyle measures 68 (63-73) μm, and the odontophore 42 (38-50) μm. The tail

of both sexes is short and conoid with a rounded terminus. Vulva is located at 46-55%, and both ovaries are well developed. The life cycle and juvenile stages have not been described in detail. As with *X. index*, reproduction seems to be parthenogenetic and males are seldom found.

This species has been reported from a wide range of plants, both herbaceous and perennial, but experimental data proving host-parasite relationships are lacking in most records.

a. Dagger Nematodes as Vectors of Grapevine Viruses. Some species of dagger nematodes have a unique and important role as pathogens of plants because they also transmit certain viruses to their host plants. In fact, in 1958, *X. index* was the first nematode proved to be the vector of a plant virus, in that case grapevine fanleaf (Hewitt et al., 1958).

Prior to 1958 the group of plant viruses classified as "soil-borne" were known to have two basic characteristics: (1) either infectivity was retained after air drying the soil or (2) infectivity was lost upon air drying (Cadman, 1963). The former group later was proved to be transmitted by fungi (Teakle, 1967; 1969). The latter group have nematode vectors, all the species of which belong to the Adenophorea, either to the family Longidoridae or Diphtherophoridae.

The species of Diphtherophoridae that transmit viruses belong to two genera, *Trichodorus* and *Paratrichodorus*. These transmit rod-shaped, rigid particles classified as Tobra- (or previously Netu-) viruses. In the Longidoridae two genera are represented, *Xiphinema* and *Longidorus*. These species transmit Nepoviruses, which are polyhedral, isodiametric particles about 30 nm in diameter.

The most recent review of this subject (Martelli, 1978) lists 12 viruses of grapevine known or suspected to be nematode transmitted. All vectors of grapevine viruses are longidorids, four species of *Xiphinema* (*X. index*, *X. italiae*, *X. diversicaudatum*, *X. americanum*) in addition to *Longidorus attenuatus*.

In the United States, *X. index* and *X. americanum* are the principal species found commonly in vineyards. *X. americanum* is by far the more widespread throughout the United States and Canada in all types of crops and soils, including vineyard soils. It is a proved vector of tomato ringspot (grape yellow vein virus strain) to herbaceous plants (Teliz et al., 1966) and to peach, apricot, and plum roots (Teliz et al., 1967). So far, however, there is evidence only of a high degree of association of *X. americanum* with tomato ringspot virus in grapes in the field. Definite proof of transmission directly to grapevines is still lacking.

In the eastern United States, the French hybrids with tomato ringspot have foliar symptoms somewhat different from the *V. vinifera* grapes in California infected with yellow vein. Also, the French hybrids could not be infected by bud grafts of yellow vein infected Carignane var. (A. C. Goheen, personal communication). These differences may be due to cultivar reactions. However, both areas have in common the poor fruit set caused by the virus (Hewitt et al., 1962; Uyemoto and Gilmer, 1972; Dias, 1977). From the evidence, Dias (1977) believes tomato ringspot virus and its vector are native to the Niagara peninsula of Canada and is a potentially serious threat to the grape industry. Over 20% of the vineyards there are infected, causing losses in production up to 95% (Dias, 1977), and it spreads actively in the field (Uyemoto and Gilmer, 1972). Most California infections probably came from contaminated planting stocks because spread in the field is very slow (Teliz et al., 1966). Yellow vein of grape has never been commonly encountered in

California, and the clean planting stock program there has essentially eliminated the disease.

X. *index* has been studied in much more detail and is a known pathogen of major importance to the grape industry of California. Grapevine fanleaf virus complex is a devastating disease with at least three distinct strains distinguished by foliar symptoms: (1) in early stages of *fanleaf* the leaves are malformed with chlorosis in shapes of rings scattered about the leaf or arcs on the margins; later the leaves have an open sinus, sharply serrated edges and veins radiating like the ribs of a fan; (2) *yellow mosaic virus* is characterized by extensive deep-yellow chlorosis; (3) *veinbanding* has chlorosis closely following the veins (Fig. 9). As the disease progresses internodes shorten on new growth, double nodes appear, growth is stunted, fruit set is decreased, "shot-berries" form instead of normal fruit (Fig. 10), and productivity drops.

Presence of the vector X. *index* adds nematode damage. The combined effects of nematode and virus ultimately can kill the vine; more commonly the plants survive but are economic failures. The source of virus in a given plant may be via (1) rooting of cuttings taken from an infected plant; (2) grafting infected scions to healthy rootstocks or, conversely, grafting healthy scions to infected rootstocks; or (3) by infective nematodes feeding on roots of healthy transplants. The virus can also be transmitted mechanically to healthy host or indicator plants.

b. *Virus-Vector relationships.* It is now firmly established that single nematode vectors can acquire and transmit virus. Acquisition can occur in five minutes or less (Alfaro and Goheen, 1974) and inoculation within 24 hours (Raski and Hewitt, 1960). This was reduced to one hour for inoculation of tomato ringspot virus (yellow vein of grape) to cucumber by X. *americanum* (Teliz et al., 1966). Retention of infectivity by nematodes is reported to be three months when feeding on virus nonhost fig (Das and Raski, 1968) or up

Fig. 9 Veinbanding symptoms of fanleaf virus on grapevine, Cabernet sauvignon var.

Fig. 10 Fanleaf virus damage to Cabernet sauvignon var. grape. Left, cluster from healthy vine; right, cluster from diseased vine showing shot berry and poor fruit set.

to eight months in absence of host plants (Taylor and Raski, 1964). There is no evidence of multiplication of virus in the vector host, and infectivity is lost on molting.

Location of fanleaf virus has been determined as arrangement in monolayers in the lumen of the esophagus (Fig. 11) (Taylor and Robertson, 1970; Raski, Maggenti, and Jones, 1973). The entire cuticular lining of the esophagus is cast off on molting, which explains failure of transstadial passage. No evidence of transovarial passage has been found.

Peach rosette mosaic virus (PRMV) was first observed on peach in Michigan in 1917 (Dias, 1975) and in New York in 1941 (Hildebrand, 1941). More recently (Dias and Cation, 1976) it has been reported in Concord grape plantings in Michigan, which followed removal of peach trees infected with that virus. The vector of that virus has been reported to be *X. americanum* (Klos et al., 1967), based on the close association of nematode/virus in the field. Experimental transmission from *Chenopodium quinoa* to *C. quinoa* has been accomplished, but transmission directly to grapevine has not been achieved. PRMV symptoms on concord include delayed bud break, asymmetric and mottled leaves, shelling of berries, and stunting of vines, which may die.

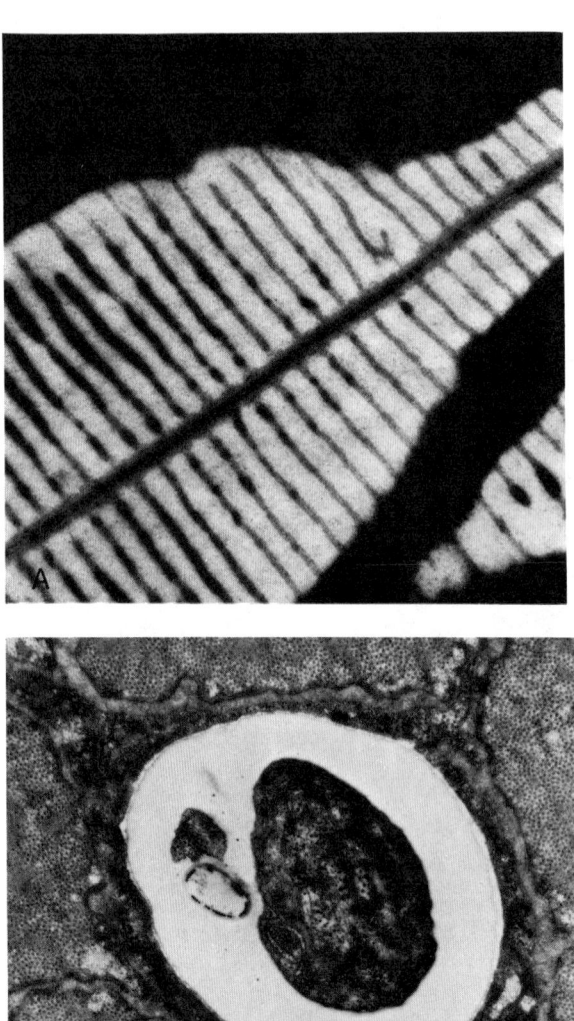

Fig. 11 Location of fanleaf virus in nematode vector. (A) Virus particles in lumen of bulbar region of esophagus. (B) Polyhedral virus particles on wall of lumen of odontophore (stylet extension).

The other species of nematodes established experimentally as vectors of grapevine viruses are: *X. italiae* (fanleaf, Israel), *X. diversicaudatum* (arabis mosaic, Europe), and *L. attenuatus* (tomato blackring, English strain, Europe).

4. Percentage and Dollar Loss Estimates

No specific estimates of losses caused by *Xiphinema* spp. are available but are included in the estimates of total losses referred to earlier. One complication in judging losses exists when *X. index* infestations are coincident with fanleaf virus. Both are serious pathogens causing major damage to vine growth and productivity.

5. Distribution

Typically the most prevalent species of *Xiphinema* in vineyards is *X. americanum*. This is true throughout the United States, particularly in California, but also in Australia and Chile. *X. index* is especially widespread in the north coastal vineyards of California but is also found in scattered plantings in the San Joaquin Valley. It is also reported from Chile.

6. Control

Chemical control measures are much the same for *Xiphinema* spp. as for *Meloidogyne* reported earlier. However, infections of both *X. index* and fanleaf virus have never responded to side-dressing treatments of established vineyards because such chemicals have no effect on the systemic virus disease. Also, some of the resistant rootstocks currently available (such as St. George, A × R#1, Dogridge) are attacked by *X. index* and are not suitable for use where this nematode is present.

C. Lesion Nematodes (*Pratylenchus* spp.)

1. History

Nematodes of this group belong to the genus *Pratylenchus* Filipjev 1936 and earlier were referred to as the root-lesion or meadow nematodes. *Pratylenchus pratensis* was the first species to be described in 1880 by de Man and by 1930 nine more had been added. At present more than 60 species have been described, but taxonomically this is a very difficult group. About half the species are judged by Loof (1978) to be synonyms of other named species or have been inadequately diagnosed and assigned *species inquirendae*. The taxonomy of this genus was not established on a sound and orderly basis until the classic revision by Sher and Allen (1953).

Allen (1949) also was the first to report *Pratylenchus* from vineyard soils but at that time could not identify them to species. *P. vulnus* was described in 1951 by Allen and Jensen (1951) (Fig. 12) and grapes were listed as a host. That species is the most important *Pratylenchus* damaging grapes, is widespread in California and also reported from Australia (Seinhorst and Sauer, 1956; Sauer, 1962). Other species found in vineyard soil include *P. scribneri*, *P. neglectus*, *P. brachyurus*, and *P. thornei*.

2. Symptoms

Except for paucity of rootlets there are no clearly defined symptoms on vine roots growing under field conditions. Pinochet et al. (Pinochet, Raski, and Goheen, 1976; Pinochet and Raski, 1977) studied the effects of *P. vulnus* on grape roots under greenhouse conditions. Necrotic lesions of varying size and color were found throughout the root system 362 days after inoculation with 1000 nematodes. Secondary and adventitious roots were extensively colonized by *P. vulnus* and associated with cavities in the cortical parenchyma

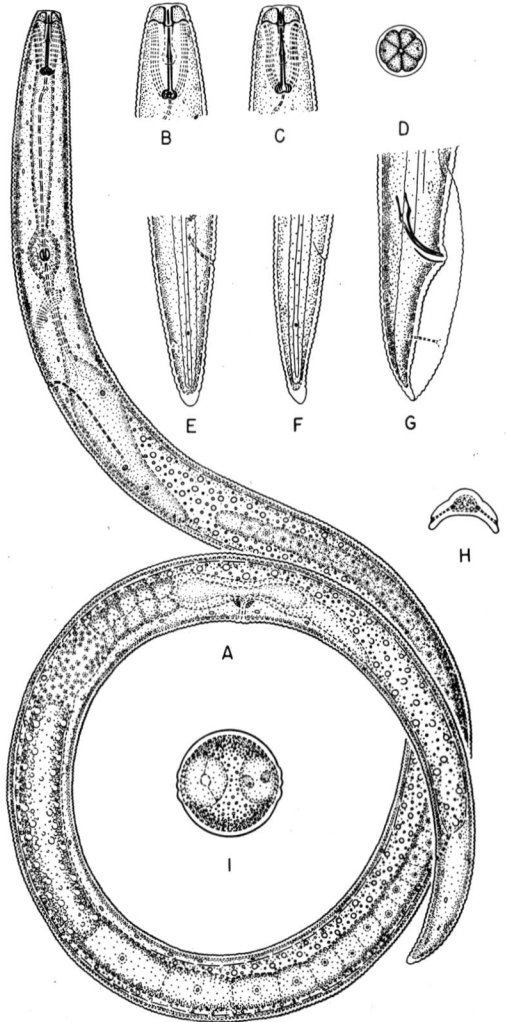

Fig. 12 Lesion nematode, *P. vulnus*. (A) Adult female; (B, C) cephalic regions of female; (D) face view of female; (E, F) female tails; (G) male tail; (H) transverse section through male tail; (I) transverse section of female at midbody.

but not in the meristematic or vascular tissues. Lesions initially appear as light brown spots, then turn dark, enlarge, and ultimately girdle the roots. Total root growth was reduced almost 40% by *P. vulnus* compared to uninoculated check plants. Marked stunting of the tops was another result of *P. vulnus* infection reducing top fresh weights by more than 50% (Fig. 13).

Seinhorst and Sauer (1956) reported similar symptoms caused by *P. vulnus* on roots in the field and also described damage by *P. scribneri*. In contrast the roots showed extensive proliferation, forming tangled masses of strongly branched roots with many very short lateral rootlets covered with small lesions. Severe damage to root tips was also present.

P. minyus (*P. neglectus*) is often found in vineyard soils in small numbers and has been incubated from the roots of grape, occasionally in large

Fig. 13 Damage to Thompson Seedless var. grape rootings grown in lesion nematode infested soil. Four plants on left grown in soil treated with a nematicide.

numbers (Raski, Hart, and Kasimatis, 1973), but no studies on host-parasite relationships have been reported (Sauer, 1962).

    P. brachyurus has been noted once in California in roots of grape being propagated in a nursery but no further information is known.

    P. thornei is frequently found in vineyard soils but has never been proved a parasite of grape. Most likely it feeds on herbaceous plants and is not a factor in vineyards.

    Control studies on *P. vulnus* and *P. minyus* were reported (Raski, 1954) with significant improvements of growth after preplant soil fumigation. The role of the lesion nematodes in that control is not established because root-knot nematode was a complicating infestation in each test.

3. Causal Organisms

P. vulnus is typical of this genus. The adults are of moderate size, females measuring 0.46-0.91 mm in length. Males are slightly smaller and more slender but similar in aspect except for the sexual apparatus. The nematodes are slender, except fully mature females become somewhat more broad and sluggish in movement. The head is almost continuous with the body with heavily sclerotized framework and three to four annules on the surface. The stylet is short (15-18 μm), but stout with rounded knobs. The esophagus has a distinct median bulb with valve and ends by overlapping the intestine as a long ventral lobe. The vulva of the female is located posteriorly at 78-84%, the gonad is monoprodelphic, with postvulval sac about two anal body widths long. The lateral field has four equidistant, longitudinal incisures. The tail tapers regularly to a rounded terminus without incisures around the tip. Slightly curved spicules, small, simple gubernaculum, and leptoderan caudal alae are similar on almost all males of *Pratylenchus*.

4. Life Cycle

The life cycle of lesion nematodes is fairly simple and consistently similar for all the species studied so far. They are obligate, migratory endoparasites of plants. Reproduction is sexual where males are common (*P. vulnus*), but for many species males are rare and reproduction is parthenogenetic (*P. neglectus*, *P. brachyurus*, *P. thornei*). Eggs are deposited singly in roots or soil (Fig. 14). There are four larval stages, which appear similar to the adults only smaller. The first stage molts before ecdysis and emerges as second-stage larva. After three more molts, the adults appear. Penetration of host roots is by all stages but principally by fourth-stage larvae and adults. Overwintering is by fourth-stage larvae and adults, but eggs can survive in roots (Dunn, 1972). The life cycle is completed in 30-86 days; the shortest time is at temperatures of 30-38°C, depending on host. Most species seem to prefer sandy loam soils.

5. Percentage and Dollar Loss Estimates

No estimates are available for losses caused by *Pratylenchus* spp. but are included in the estimates of total losses caused by all nematodes referred to earlier.

6. Distribution

P. vulnus is widely spread throughout the San Joaquin Valley in California and occasional records are known from coastal plantings. It is also known in Chile (Allen et al., 1971) and from many vineyards in Australia (Sauer, 1962).

7. Control

Similar to *Meloidogyne* spp., but *P. vulnus* is known to attack some resistant rootstocks (Raski, Hart, and Kasimatis, 1973). This needs further study and clarification to fully understand the role of *P. vulnus* in the performance of those resistant rootstocks.

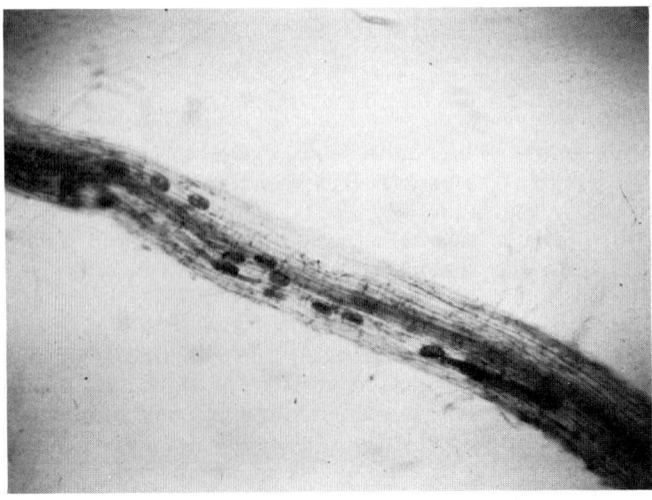

Fig. 14 Lesion nematodes and eggs in cortex of host root.

D. Citrus Nematode (*Tylenchus semipenetrans*)

1. History

Mr. J. R. Hodges, a Horticultural Inspector at Covina, California, was the first to observe this nematode on citrus roots. This observation was reported by Thomas (1913), and the animal was described the same year by Cobb (1913) with more complete illustrations and observations the following year (Cobb, 1914). Since then it has been found in citrus plantings throughout the world.

It was first recorded on grape roots in 1956 (Raski et al., 1956; Seinhorst and Sauer, 1956) both in California and in Australia.

2. Symptoms

There are no distinct symptoms associated with citrus nematode attack on roots of grape. Some extra branching of roots seemed noticeable, and soil particles cling persistently to the affected roots as it does with citrus roots (Seinhorst and Sauer, 1956). Irregular bends and some extensive necrosis have also been noted.

3. Causal organism

Tylenchulus semipenetrans is a relatively small nematode. Immature, infective adult females measure 0.25-0.36 mm in length and are slender (a = 15-20) vermiform in aspect. It has a rounded head region only lightly sclerotized and bears a short (12-15 μm) but robust stylet with rounded knobs. The medium bulb is well set off and is provided with a prominent valve. The posterior glandular region is enclosed and distinctly set off from the intestine. The excretory pore is unusual in its posterior location (77-84% of body length) just anterior to the vulva (V = about 90%). Gonad is single, prodelphic. Rectum and anus are obscure, nonfunctional. Tail posterior to vulva is short, tapering slightly to blunt rounded terminus. Mature females are sedentary, partly embedded in root tissue, and that part of the esophageal region is highly distorted in conformity with host cell outlines. Outside the root the body swells markedly (a = 3.8-5.1), except for blunt tail.

Males are slightly longer than immature females (L = 0.33-0.40 mm) and remain slender. Males do not feed and have a reduced stylet (10-12 μm) with slight swellings for basal knobs. The esophagus also is degenerate. Excretory pore is posterior to middle of body. Bursa is absent; spicules are 14-18 μm long. Tail is slender with a rounded terminus.

Second-stage larvae are slender (female L = 0.28-0.36 mm; slightly less in male) with rounded head similar to immature female. Stylet is robust, 12-14 μm long in female, slightly less in male. Excretory pore is about midway on body (52-56% of body length). Tail narrows to slender, elongate, conoid shape with finely rounded terminus (Van Gundy, 1958; Siddiqi, 1974).

4. Life Cycle of Nematode

The life cycle begins when eggs are deposited in a gelatinous matrix outside the host root (Fig. 15) and hatch in 12-14 days at 24°C. First-stage larvae molt to second stage inside eggs before ecdysis. Male larvae usually molt to third stage before leaving the matrix and molt twice more to become adults. Males and male larvae do not feed. Although this species is bisexual, reproduction may be amphimictic or parthenogenetic (Siddiqi, 1974). Hand-picked second-stage female larvae reproduced on citrus roots in the absence of males.

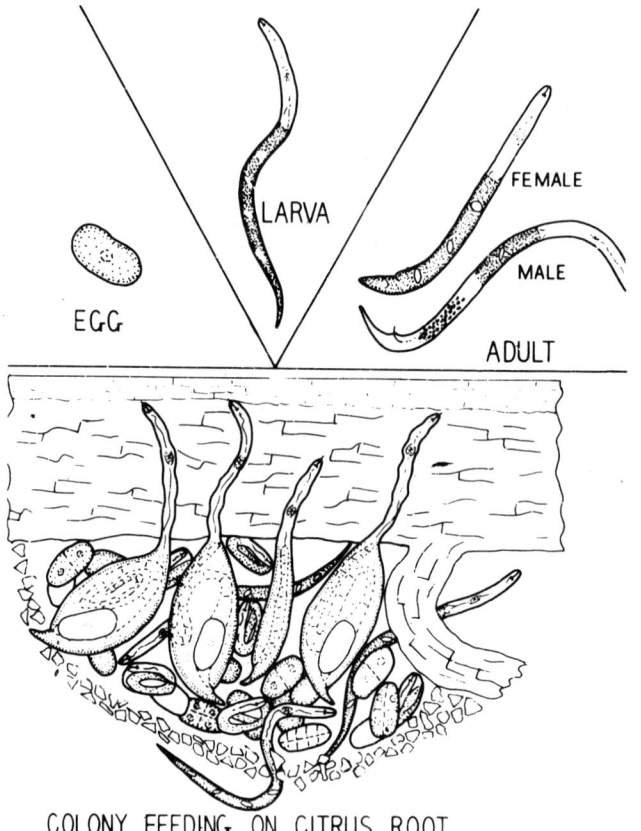

Fig. 15 Life stages of citrus nematode, *Tylenchulus semipenetrans*.

Female larvae feed on surface cells, and immature females then penetrate to the cortex, and within a week may reach the pericycle and develop 3/4 to fully swollen size (Fig. 16) (Van Gundy, 1958). The complete life cycle from egg is 6-8 weeks.

E. Miscellaneous Ecto- and Semiendoparasitic Species

Reported from Australia (Sauer, 1962) are many species, including: *Tylenchorhynchus (brevidens ?)*, *Paratylenchus hamatus*, *Criconemella xenoplax*, *Scutellonema brachyurum*, *Paratrichodorus minor*, *Helicotylenchus dihystera*, *Hemicycliophora conida*, and *Rotylenchus gracilidens*.

Other species found in California are *Nothocriconema mutabilis*, *Paratylenchus neoamblycephalus*, *Helicotylenchus erythrinae*, and *H. dihystera*.

Results of an inoculation trial suggest *Helicotylenchus pseudorobustus* is a weak pathogen of Thompson Seedless grape. It was observed feeding ecto- and endoparasitically in the cortical parenchyma of the roots (Pinochet, Raski, and Jones, 1976).

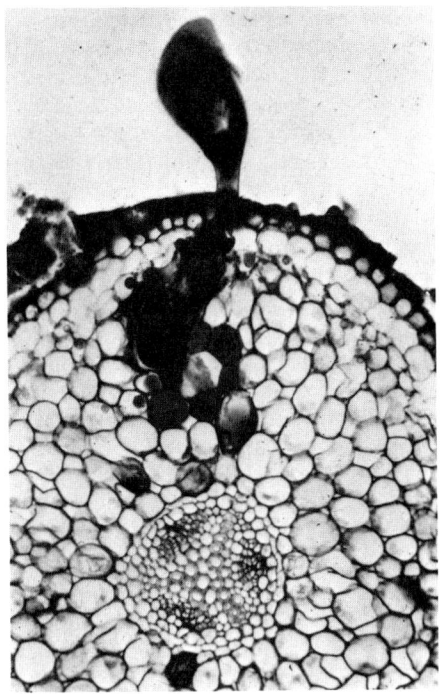

Fig. 16 Female of citrus nematode partially embedded in root of sweet orange.

II. STRAWBERRY

According to Wilhelm and Sagan (1974) humans of all ages enjoyed strawberries of the wild type found in forests. Transplanted to cultivated plots they became known as garden strawberries. Some of the earliest varieties were first developed in France in the sixteenth century. The modern strawberries as we know them today came from explorations of North and South America. It is from crosses of *Fragaria virginiana* from Virginia and *Fragaria chiloensis* of Chile that the large-fruited variety, first known as the pineapple strawberry, has become extensively cultivated throughout the world. The commercial strawberry often goes under the scientific name *Fragaria* × *ananassa* Duch. and is a hybrid of two variable octaploid species. The United States leads the world in strawberry production, followed by Japan, Mexico, and Poland (Childers, 1978). In the United States, California produces about 72% of the total crop, and around 42% of the U.S. crop is processed.

A. Bud and Leaf Nematodes (*Aphelenchoides* spp.)

1. History and Distribution

In California the principal nematode problem on strawberries is the spring dwarf nematode, *Aphelenchoides fragariae* (Ritzema Bos). One of its earliest records was made by Raski and Allen (1948). It is widely distributed along the coastal regions of California wherever strawberries are grown (Siddiqui et

al., 1973) and is also recorded from Oregon and Washington. On the East
Coast it is widely distributed along the Atlantic Seaboard from New England to
Georgia (Christie, 1959; USDA, 1978c). *A. fragariae* is a common and impor-
tant pest of strawberry in the eastern USSR (Szczygiel, 1977); in southern
Poland it infested 34% of 726 fields surveyed (Szczygiel, 1970). It has been
reported from Holland (Klinkenberg, 1955), and recently for the first time in
India (Sharma, 1977). *A. ritzemabosi* (Schwartz) Steiner and Buhrer has
also been reported in strawberry nurseries as well as production fields
(Siddiqui et al., 1973), but only rarely, and is not considered a problem on
strawberry in California. This nematode was also reported on strawberry in
Poland (Szczygiel, 1970) and Holland (Klinkenberg, 1955). Perhaps *A. besseyi* Christie is second in importance to *A. fragariae* as a parasite of strawberry. *A. besseyi* has been found in the United States on strawberry south
of Virginia and Arkansas, but can survive overwinter as far north as Delaware
and Illinois (USDA, 1978c). In Australia *A. besseyi* has caused severe yield
losses in some strawberry plantings (McCulloch, 1974; 1978).

2. Symptoms

These nematodes do not invade strawberry plant tissues, but feed on external
leaf surfaces in developing buds (Raski and Allen, 1948) so that when leaves
expand they are crinkled, distorted, and often reduced in size (Fig. 17).
Nematode feeding activity may also blind fruit buds causing reductions in fruit
yields. Plants may be killed by nematode feeding.

Nematodes of *A. fragariae* and/or *A. ritzemabosi* may also interact with
the plant pathogenic bacterium *Corynebacterium fascians* on strawberry plants
to produce unique "cauliflower" disease symptoms. These symptoms are unlike those caused by either nematodes or bacteria alone and are characterized
by the plant crowns being reduced to stunted fleshy rosettes about 1 cm high
(Crosse and Pitcher, 1952; Pitcher and Crosse, 1958).

Fig. 17 Spring crimp caused by *Aphelenchoides fragariae* in strawberry, cv.
Premier.

3. *Causal Organism*

These nematodes are bisexual species with females measuring 0.45-0.80 mm and males 0.48-0.65 mm in length (Allen, 1952). The sexes are very similar in appearance except for sexual organs. The adults are very slender (a = 45-63) with a short delicate stylet about 10 µm long. The head has very lightly sclerotized framework and is rounded and set off with a slight constriction. The metacorpus is strongly developed with prominent sclerotized valve. The posterior glandular region is long and slender, overlapping the intestine which joins the esophagus immediately behind the median bulb. The dorsal gland is connected with the lumen of the esophagus immediately anterior to the valve of the metacropus. The female gonad is monoprodelphic with the vulva located posteriorly (64-70%). The postuterine branch is long and slender and contains spermatozoa. The tail is slender and conoid. Thorne (1961) provides a key including these species of *Aphelenchoides*. The male tail curves 45-90% ventrad when relaxed with gentle heat. Spicules are paired, typically thornlike in shape; gubernaculum and caudal alae are lacking. Larval forms resemble adults, but are smaller.

4. *Life History*

These nematodes develop within buds of strawberry plants and have the typical four larval stages separated by molts. The life cycle of *A. besseyi* is completed in about two weeks (Christie, 1959), and it survives winter and adverse conditions in the soil and in plant debris in California, although in Poland *A. fragariae* persisted in plant residues under optimum conditions for no more than three months (Szczygiel and Hasior, 1971).

5. *Percentage and Dollar Loss Estimates*

Few reliable estimates have been made. In southern Poland *A. fragariae* and *A. ritzemabosi* during two years of tests decreased yields by 32-61% in the seven varieties of strawberries tested (Szczygiel, 1967). *A. besseyi* decreased yields of infested plants in Australia by up to 50% (McCulloch, 1978).

6. *Control*

In California there is no satisfactory control method for *A. fragariae* in the field. Parathion did not penetrate tightly folded buds and reach nematodes, and was phytotoxic to the strawberry plants (Raski and Allen, 1948). Propagation of nematode-free stock for fruit producers has been intensively developed under state supervision and is largely successful in California. However, outbreaks occur periodically and *A. fragariae* continues to be an important pathogen on strawberry in the state.

In Maryland, *A. fragariae* caused damage to strawberries in the 1940s and 1950s, but has been virtually unseen either in strawberry nurseries or fruiting fields in the past 20 years. Nurserymen have been using parathion in their insect control programs for many years, and the near disappearance of bud and leaf nematodes has been coincidental. Strawberry nurserymen are currently using aldicarb in their plant production schemes, making it unlikely that these nematodes will make a resurgence as important pests in the near future.

In Russia several methods of control are used singly or in various combinations, including sanitation, hot water treatment of dormant stock plants, chemical dips, and chemical sprays of established plants with mixtures containing parathion (Kirjanova and Krall, 1971).

A recent review of hot water treatments states that immersion of dormant plants in water at 46.1°C for 10 minutes will eliminate bud and leaf nematodes (Bridge, 1975).

In Australia several grasses common around strawberry fields were found to harbor *A. besseyi* and to serve as a source of nematodes when the grasses were plowed under and strawberries planted (McCulloch, 1978). The suggested control method was to apply 1 of phenamiphos 10G granules into the crown of each plant showing symptoms at one month after planting.

B. Root-Knot Nematodes (*Meloidogyne* spp.)

1. History and Distribution

Meloidogyne hapla Chit. is the only root-knot nematode species known to attack strawberries in California, and it is only found occasionally. This nematode is fairly widely distributed on strawberry, especially in the northern two-thirds of the United States (USDA, 1978c), and although it may not be a serious problem in some areas (Bridge, 1975), it severely reduces production of plants in strawberry plant nurseries on the Delmarva Peninsula when not controlled. It is the most serious nematode pest of strawberry in the Middle Atlantic States, and was also reported to cause damage in Louisiana (Horn et al., 1956). *M. hapla* is thought to be indigenous to areas of Minnesota as it was found in uncultivated land adjacent to strawberry plantings, as well as in strawberry beds causing damage (Crow and MacDonald, 1978).

A strain of *M. javanica* (Treub) Chit., presumably originating in Japan, has been reported causing damage to strawberry plantings in Israel (Minz, 1958; Strich-Harari and Minz, 1961) and Zambia, and more recently was found causing severe galling of plants in Senegal (Taylor and Netscher, 1975). On differential hosts this latter isolate of *M. incognita*, presumably identified by perineal patterns, was discovered to be a severe pathogen of strawberry in Florida, and when inoculated, reproduced on the roots of Florida 90 strawberry plants (Perry and Zeikus, 1972).

2. Symptoms

M. hapla causes mostly small inconspicuous galls on the roots of strawberry, with proliferation of lateral feeder rootlets off the galls (Fig. 18). Under heavy attack in sandy soils, root systems may be virtually lacking with severe loss of plants.

3. Causal Organism

The causal organism is *Meloidogyne hapla*.

4. Life History

The life history is typical for root-knot nematodes (see Section I).

5. Percentage and Dollar Loss Estimates

No reliable estimates exist for losses in berry production fields, but root-knot nematode infestations in strawberry plant nurseries make the infected plants unsalable.

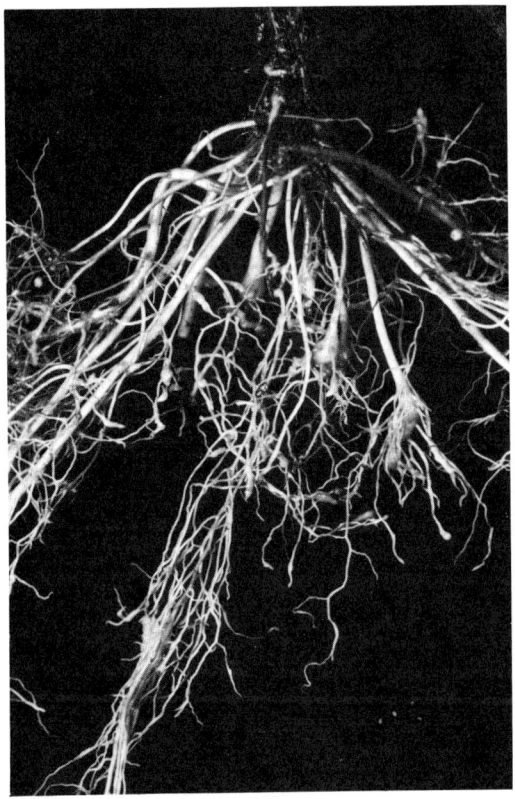

Fig. 18 Strawberry root system showing galling caused by *M. hapla.*

6. Control

Planting nematode-free stock plants in nematode-free soil is the ideal situation for controlling root-knot nematodes on strawberry. In California, apparently, according to some workers, root-knot nematodes, or other nematodes, are no longer a problem in strawberry nursery or fruit production fields because of nearly 100% fumigation with mixtures of methyl bromide and chloropicrin that have been used since about 1960 (Wilhelm and Paulus, 1980). According to the authors all other diseases of strawberry have also been virtually eliminated by this practice. Before a new field is planted to strawberry for the first time it is broadcast treated with methyl bromide-chloropicrin, 1:1, at 420-471 kg/ha, and subsequently before each successive strawberry crop the soil is fumigated with methyl bromide-chloropicrin, 2:1, at somewhat lower rates. Some fields have been repeatedly fumigated up to 15 times with no reported buildup of toxic residues or destruction of favorable soil microflora. Apparently nematodes have never been a problem in California strawberry production as these authors mention the importance of control of soil-borne pathogenic fungi by the soil fumigation, but never mention nematodes.

M. hapla is a serious problem in both strawberry fruit and nursery plant production in certain other areas of the United States, especially on the East Coast (Fig. 19). Control must start in the plant nursery, because the fruit producer must avoid introducing nematodes to his fields in the planting stock, especially if he treats his fields with nematicides. A few nurseries on the

Fig. 19 Growth of strawberry, cv. Guardian, planted in soil heavily infested with *M. hapla*: left row, mother plants heavily galled when planted five months earlier; right row, mother plants free of galls when planted. (Courtesy of D. E. Babineau.)

Delmarva Peninsula produce some 25% of the strawberry plants sold in the United States. *M. hapla* has been a recurring problem in these nurseries at times. Early efforts using the fumigants DD, EDB, and DBCP in various pre-plant and postplant combinations were quite successful, and DBCP became a mainstay in the production of essentially nematode-free plants (Kantzes and Morgan, 1962; Potter and Morgan, 1956). Then, in the 1970s, growers began experiencing occasional outbreaks of root knot in some of their fields, and also DBCP came under fire for environmental and health reasons. Five years of experiments were conducted at the University of Maryland Vegetable Research Farm at Salisbury, Maryland, to evaluate nonfumigant nematicides for control of nematodes in strawberry plant nurseries (Babineau, 1978; 1981). The nematicide sequence that gave effective and consistent nematode control was fall broadcast application of a fumigant nematicide followed by four monthly alternating sidedress applications of granular formulations of phenamiphos, then aldicarb, starting at the time of planting (Fig. 20).

In Florida, thionazin provided good control of root-knot nematodes, whereas DBCP failed to give adequate control (Smart et al., 1967). In other experiments these workers found that DD plus methyl bromide under clear or black plastic increased berry yields by about 38% over DD alone under clear plastic (Smart et al., 1968); plastic mulch is routinely used for strawberry production in Florida. No mention was made of the economics of these control methods. Tests in Arkansas demonstrated that soil fumigation with methyl bromide-chloropicrin and methyl bromide-chloropicrin-propargyl bromide mixtures gave good weed control the first year, increased plant production by 200%, and increased fruit yields by 50% (Riggs and Hamblen, 1962). When plants of 33 cultivars of strawberries were exposed to nematodes of one isolate of *M. hapla* the cultivars varied significantly in resistance although none were immune (Dickstein and Krusberg, 1978). On a root-knot indexing system of 1 (no galls) to 6 (75-100% galled) the indices for the cultivars ranged from

Fig. 20 Growth of strawberry, cv. Guardian, in soil treated with aldicarb at 3.36 kg active per hectare in furrow 8 cm deep immediately preplant followed by a postplant application at 3.36 kg/ha in a 40 cm band over the row with light soil incorporation seven weeks later. (Courtesy of D. E. Babineau.)

2.4 to 4.8. In another study only plants of *Fragaria vesca* of 11 species or subspecies of *Fragaria* challenged with *M. hapla* remained free of egg masses eight weeks after inoculation, but *F. vesca* is not a commercial strawberry (Orchard and Andrichem, 1961).

The USDA *Farmers' Bulletin* recommendations for control of *M. hapla* by fruit producers start by recommending the planting of nematode-free plants (USDA, 1978c). If this nematode is already present in a field to be planted with strawberries it is suggested that preplant soil fumigation may be profitable. If root-knot nematodes are present in a planting, the following is recommended: (1) shallow cultivation, (2) control weeds, (3) maintain high soil fertility, and (4) irrigate during dry periods.

Various methods have been tested to disinfect strawberry plants of nematodes. Different investigators have experimented with hot water for disinfection of nematode-infected plants, and some have experienced that treatments that killed the nematodes often injured the plants severely (Colbran, 1957; 1959). However, one report claimed that completely dormant plants exposed to water at 52°C for 120 seconds, then placed in cold storage, resulted in killing of both *M. hapla* and *Pratylenchus penetrans* in the roots, excellent plant survival, and good subsequent growth of plants and runner production in the field (Goheen and McGrew, 1954). A recent review indicates that dipping dormant plants in water at 52.8°C for 15 minutes will eliminate both root-knot and lesion nematodes (Bridge, 1975). In another report root-knot nematodes infecting plants were greatly reduced in numbers when the plants were dipped in solutions of thionazin or fensulfothion at 300 PPM (Smart and Locascio, 1968).

C. Lesion Nematodes (*Pratylenchus* spp.)

1. History and Distribution

Several species of *Pratylenchus* have been reported as attacking strawberry. *P. penetrans* seems to be the most ubiquitous species on strawberry worldwide, having been reported from both the East and West Coasts of the United States, Canada, and several European countries. Other species reported include *P. vulnus* in California and France (Scotto la Massese, 1973), *P. crenatus* in Poland (Szczygiel and Hasior, 1972) and Canada, *P. tenuis* in Minnesota (Crow and MacDonald, 1978), *P. scribneri* in Kentucky (Chapman, 1956), *P. brachyurus* in Kentucky (Chapman, 1956), *P. zeae* in Kentucky (Chapman, 1956), *P. minyus* in France (Scotto la Massese, 1973) and Holland (Schindler and Braun, 1957), *P. pratensis* in Florida (Steiner, 1931), Canada (Bosher, 1954), and Holland (Klinkenberg, 1955), and *P. coffeae* in Arkansas (Riggs et al., 1956).

2. Symptoms

Lesion nematode infection causes young strawberry roots to become wiry and brown (USDA, 1978c). *P. penetrans* caused extensive blackening of roots of sterile strawberry seedlings growing on agar medium by 60 to 75 days after surface-sterilized nematodes were introduced, illustrating that the nematodes alone are fully capable of causing symptoms of black-root rot (Chen and Rich, 1962; Townshend, 1963). In the field this nematode causes elliptical amber to dark brown lesions on strawberry roots (Townshend, 1962). It was suggested that in the field black-root rot symptoms are probably caused by a complex of lesion nematodes and soil fungi.

Above-ground symptoms of lesion nematode feeding on roots are similar to those caused by other root-feeding nematodes, namely, stunting of plants, lack of vigor, low productivity, and increased proneness to drought (Goheen and Smith, 1956; USDA, 1978c) (Fig. 21). In Ontario, Canada, 70% of the strawberry plantings examined were infested with *P. penetrans* (Townshend, 1962). Areas of weakly growing plants were correlated with the highest soil and root populations of lesion nematodes. Plants in these areas also produced few runner plants and suffered from root rot. All 15 strawberry cultivars tested harbored *P. penetrans* in their roots.

3. Causal Organism

This is *P. penetrans* (Cobb) Filip. and Schur. Stek. most often. See Section I for general description of species in this genus.

4. Life Cycle

See Section I for general life cycle.

5. Percentage and Dollar Loss Estimates

No estimates exist.

Fig. 21 Field damage caused by the lesion nematode, *P. penetrans*, to strawberry, cv. Sparkle.

6. Control

Lesion nematodes in general are more easily controlled on strawberry than are root-knot nematodes. In the recent studies in Maryland on chemical control both *M. hapla* and *P. penetrans* were present in these fields. Many nematicide applications that failed to adequately control *M. hapla* gave excellent control of *P. penetrans*, and every application that controlled *M. hapla* also controlled *P. penetrans*. In Canada, where a fine sandy loam soil containing *P. penetrans* was treated with chloropicrin (472 liters/ha), DD (326 liters/ha), or DD-Mencs (135 liters/ha), plants of Earlidawn and Sparkle cultivars subsequently grown produced higher yields of fruit and more runner plants than plants grown in untreated soil; DBCP (37 liters/ha) was not effective (Townshend et al., 1966). In another study preplant applications in the plow furrow of DD gave the best reduction of root rot and the greatest increase in strawberry yields as compared to EDB, DBCP, or Metam-sodium (Morgan, 1964). All the materials except DBCP were applied at the rate of 169 liters/ha, except that DBCP was at 11 liters/ha.

Soil incorporation of composted timothy hay gave the largest decrease in soil populations of *P. penetrans* in studies with several soil amendments in Canada (Morgan and Collins, 1964). However, timothy sod resulted in the greatest increase in soil *P. penetrans*, whereas growing beans resulted in low nematode populations. Soil applications of DD, phenamiphos, or dazomet provided good control of *P. penetrans* and two other species of phytoparasitic nematodes on strawberry at two locations in Switzerland (Klingler and Kunz, 1978).

One of the problems complicating control of lesion nematodes on strawberries is that they have wide host ranges. In Canada, roots of 55 species of weeds collected from areas of strawberry plantations where root rot was a problem were infected with *P. penetrans* (Townshend and Davidson, 1960). Some of the best weed hosts were common mouse-eared chickweed (*Cerastu vulgatum* L.), creeping yellow cress [*Rorippa sylvestris* (L.) Bess.], annual daisy fleabane [*Erigeron annuus* (L.) Pers.], and yellow hawkweed (*Hieracium pratense* Tansch).

A unique method of producing nematode-free strawberry plants was developed in Canada (Townshend, 1965). Runner plants were trained to root and grow in protected pots of sterile soil medium placed in the field and then were cut loose and nurtured under mist until established.

D. Stem Nematode (*Ditylenchus dipsaci*)

1. History and Distribution

The stem nematode, *Ditylenchus dipsaci* (Kühn) Filip., has been reported from a number of countries as attacking and causing damage to strawberries. This nematode is a serious pest in parts of Russia (Kirjanova and Krall, 1971), Holland (Klinkenberg, 1955), Italy (Tacconi, 1976), and in the Pacific Northwest of the United States (USDA, 1978c). There are no records of it on strawberry in other parts of the United States. It occurs sometimes on strawberries in Poland but is not considered a serious pest (Szczygiel, 1970, 1977).

2. Symptoms

D. dipsaci causes swelling of leaf petioles, twisting and distortion of leaves, galling of flowers and fruit, and stunting of plants (Filipjev and Schuurmans-Stekhoven, 1941; Goodey, 1951; Szczygiel, 1977). These symptoms are very similar to those caused by foliar nematodes (*Aphelenchoides* spp.) on strawberry; therefore, nematodes must be identified from plants showing such symptoms (Klinkenberg, 1955; Slack and Pitcher, 1970). *D. dipsaci* is an endoparasite and will be found in largest numbers inside the distorted plant tissues, living in the swollen cortical or parenchyma tissues (Filipjev and Schuurmans-Stekhoven, 1941). This nematode is most damaging in cool wet weather. In the U.S. Pacific Northwest it is only damaging early in the spring (USDA, 1978c).

3. Causal Organism

Strawberry is a host plant for perhaps six physiological races of *D. dipsaci* (Hesling, 1966).

4. Life Cycle

The life cycle of all races of the stem nematode is similar (Jenkins and Taylor, 1967). Eggs hatch about seven days after oviposition. Molts occur at intervals of two to seven days, the entire life cycle requiring 19 to 23 days to complete on a suitable host plant at 15°C. Stem nematodes are bisexual and, after mating, females lay eight to ten eggs per day, laying a total of between 200 and 500 eggs. Individuals may live 45 to 70 days. *D. dipsaci* is an endoparasite of plant tissues, and nematodes in all developmental stages may be found inside swollen, distorted host tissues. Fourth-stage larvae are quite resistant to desiccation, especially when they congregate in masses as "wool," and can survive for years in dried infected plants.

5. Percentage and Dollar Loss Estimates

No monetary estimates of losses caused by stem nematode to strawberry exist. In eastern Russia it has been reported to decrease strawberry yields by up to 85% (Szczygiel, 1977).

6. Control

In Russia methods used to control stem nematode on strawberry are the same as those used to control bud and leaf nematodes (Kirjanova and Krall, 1971). These include sanitation, hot water and chemical dips of dormant plants, and chemical sprays of established plants with mixes containing parathion. Dipping dormant strawberry plants in water at 46.1°C for 10 minutes was reported in a recent review to eliminate stem nematodes (Bridge, 1975). Another recent review indicates that this nematode has not been effectively controlled in Russia using chemicals (Szczygiel, 1977). It is suggested that propagation of nematode-free plants in the nursery, a difficult task, is the best approach.

Three years of studies involving eight nematicides were conducted in Italy (Tacconi, 1976). The most effective nematicides were miral, oxamyl, and thionazin when applied as foliar sprays in the fall. Foliar sprays with oxamyl in the spring were nearly as effective as fall sprays. Phenamiphos, when applied around plants 19 days after planting, also effectively controlled *D. dipsaci* and led to increased fruit production.

Apparently little resistance to *D. dipsaci* exists among strawberry cultivars. In Russia, 27 of 29 cultivars tested were susceptible to the stem nematode (Szczygiel, 1977).

E. Dagger and Needle Nematodes (*Xiphinema* **spp.**, *Longidorus elongatus*)

1. History and Distribution

At least four species of dagger nematodes have been reported from strawberry. *Xiphinema diversicaudatum* is common in various parts of Europe, England (Cotten, 1975; Pitcher and Jha, 1961; Thomas, 1969), Belgium (d'Herde and van den Brande, 1964), and Switzerland (Klinger and Kunz, 1978), to name a few, and is occasional in the United States (Schindler, 1954; Schindler and Braun, 1957). *X. americanum* is the most common species associated with strawberry in North America (Bosher, 1954; Chapman, 1956; Braun, 1958; Perry, 1958; White, 1960; Nemec and Malek, 1975; Crow and MacDonald, 1978). *X. bakeri* has been reported from Canada (Williams, 1961) and *X. chambersi* from the United States (Perry, 1958) on strawberry.

The needle nematode, *Longidorus elongatus*, has been reported from strawberry in several European countries, at least England (Cotten, 1975), Germany (Wyss, 1970b), Holland (Seinhorst, 1966), Sweden (Andersson, 1974), and Belgium (d'Herde and van den Brande, 1964), and from Canada (McElroy, 1971). Other species of needle nematodes may also parasitize strawberries.

2. Symptoms

X. diversicaudatum and *L. elongatus* cause sparse root systems to form and induce blackening of the roots and distortion and galling of the root tips through ectoparasitic feeding (Andersson, 1974; McElroy, 1971; Thorne, 1961). *X. americanum* and *X. chambersi* induce similar symptoms except for the root tip galls (Perry, 1958). All these nematodes cause a decline of plantings through reduction of plant growth and severly limited runner production (Perry, 1958; McElroy, 1971; Thorne, 1961). With their long stylets these nematodes can penetrate to the vascular tissue in the tips of the fine fibrous feeder roots (Trudgill, 1976). They also vector certain plant viruses, which will be dealt with in a succeeding section.

3. Causal Organisms

These are primarily *X. diversicaudatum* (Micol.) Thorne, *X. americanum* Cobb, and *L. elongatus* (de Man) Thorne and Swanger.

4. Life Cycle

The life cycles of these nematodes in general are quite long and the time required for development from egg to adult may require from 22 days to more than one year to complete (Cotten, 1976; Jenkins and Taylor, 1967; Wyss, 1970a). Eggs and all developmental stages of these nematodes occur in soil as they are ectoparasites of plants.

5. Percentage and Dollar Loss Estimates

None are available.

6. Control

Many chemical control studies have been performed towards dagger and needle nematodes in connection with their acknowledged role as vectors of a number of viruses that cause disease in plants. In Scotland, application to soil of 67 kg/ha of quintozene or 448 kg/ha of DD killed 95% of the *L. elongatus* infesting the soil and largely prevented transmission of raspberry ringspot and tomato black ring viruses to strawberry (Taylor and Murant, 1968). Fruit yields increased by four- to sixfold the first year after treatment and by 9- to 12-fold the second year. In England, application to soil of 1.0 kg per 10 m^2 of DD or methyl bromide eliminated 99% of the *X. diversicaudatum* to a depth of 70 cm and nearly stopped transmission of arabis mosaic virus to strawberry plants (Harrison et al., 1963). Dazomet, Mencs, metam-sodium, DBCP, or tetramethyl thiuram disulfide killed too few nematodes to prevent virus transmission. This elegant series of experiments showed that soil treated with DD or methyl bromide would probably not require treatment again for several years. Recent field studies at two locations in Switzerland demonstraed that DD, dazomet, or phenamiphos gave good control of *X. diversicaudatum* on strawberry (Klingler and Kunz, 1978). In the United States, preplant fall soil treatments with methyl bromide or DD-Mencs gave the best control of *X. americanum* (Nemec and Malek, 1975); DD or EDB were a little less effective.

Studies in England (Cotten, 1975) and Scotland (Taylor and Murant, 1968) have shown that cultural methods, such as crop rotations or fallow, are inadequate for controlling dagger and needle nematodes. Their wide host ranges and long survival in soil make the use of such methods impractical.

F. Other Nematodes

Many other genera and species of nematodes have been reported as associated with and sometimes pathogenic to strawberry around the world (Chapman, 1956; Horn et al., 1956; Braun, 1958; Szczygiel, 1966; Wyss, 1970; Scotto la Massese, 1973; Szczygiel, 1977; USDA, 1978c; Szczygiel, 1980). These include various species within the genera *Tylenchorhynchus*, *Helicotylenchus*, *Pratylenchus*, *Criconemella*, *Belonolaimus*, *Trichodorus*, and *Rotylenchus*. Little work has been done for the most part to determine how important these nematodes are to strawberry production.

G. Interactions of Nematodes with Other Plant Disease-Causing Agents

Nematodes are known to interact with several other agents that cause diseases in strawberry. The nematodes may act as vectors, as in the case of several viruses that affect strawberries, they may act to lessen resistance and/or to provide infection courts for other pathogens, or they may become involved as a component of a complex disease. Sometimes these categories are difficult to differentiate in a given disease situation.

The interaction of *Aphelenchoides* sp. with the bacterium, *Corynebacterium fascians*, to cause the unique cauliflower disease complex in strawberry is a classic complex disease (Pitcher and Crosse, 1958). Cauliflower symptoms are produced only in the presence of both the nematode and bacterium.

Nematodes are known to interact with several pathogenic fungi in strawberry. Lesion, root-knot, and needle nematodes have been implicated as interacting with the fungus *Verticillium dahliae* to cause wilt in strawberry plants (Abu-Gharbieh et al., 1962; Meagher and Jenkins, 1970; Szczygiel, 1980; Townshend et al., 1966). The incidence of verticillium wilt can be controlled by controlling the nematodes (Meagher and Jenkins, 1970; Townshend et al., 1966). Lesion nematodes have frequently been associated with a condition called black-root rot of strawberry (Riggs et al., 1956; Townshend, 1962, 1963). Probably fungi, but perhaps other organisms too, are involved in this disease. However, in one study lesion nematodes could not be implicated as a factor in causing black-root rot of strawberry (Raski, 1956).

Several viruses causing diseases in strawberry are known to be transmitted by a number of species of nematodes feeding on strawberry. This topic has been reviewed thoroughly recently and will not be covered here, so the reader is directed to these reviews (Milne, 1974; McElroy, 1977).

III. RASPBERRY

A. History

Raspberries are perennial plants with a biennial growth and fruiting pattern (Childers, 1978). The red and black raspberries are the major types planted. The United States is the world's leading producer, followed by Germany, Britain, Yugoslavia, Hungary, Poland, and Canada. Brambles are sensitive to extremes of heat, cold, and moisture, hence in the United States the major producing states are Michigan, Oregon, Washington, and New York.

B. Nematodes

The various nematodes found associated with or causing damage to raspberry will be covered collectively since not a great deal of information exists on this topic. Probably the greatest amount of interest and research has centered on the dorylaimoid nematodes because of their vectoring of a number of viruses causing diseases in raspberry (Forer et al., 1975; McElroy, 1975b; McElroy, 1977a; Taylor and Alphey, 1973; Taylor and Thomas, 1968). These nematodes are included in the genera *Longidorus, Paralongidorus, Paratrichodorus, Trichodorus*, and *Xiphinema*, and since this subject has been recently reviewed (McElroy, 1977a) it will not be covered here. Nematodes in the genera *Xiphinema* and *Longidorus* are commonly associated with raspberry in various parts of Europe (Cotten, 1975; Lamberti, 1981; McElroy, 1977a).

Lesion nematodes, *Pratylenchus* spp., have been widely reported as associated with and causing damage to raspberry (USDA, 1978b; Golden and Converse, 1965; McElroy, 1977a). Although several species have been detected, the most common species found has been *P. penetrans*. In British Columbia, Canada, *P. penetrans* was found in 90% of the raspberry acreage checked, caused a 24% mortality rate in new plantings, inhibited first year plant growth by 66%, caused the number of second year canes to be reduced by 23-30%, and reduced yields by 47% (McElroy, 1977b). In the United States, lesion and dagger nematodes are the most widespread and are considered the most damaging nematodes on raspberry (USDA, 1978b). Nematodes in several other genera have also been reported as associated with or causing damage to raspberry, including *Helicotylenchus* (Golden and Converse, 1965; USDA, 1978b), *Tylenchorhynchus* (Golden and Converse, 1965), *Meloidogyne* (USDA, 1978b), and *Criconemella* (USDA, 1978b). When nematodes cause plant damage directly, symptoms of this damage are spindly, thin stands of canes, small cane size, small fruit, and poor yields (USDA, 1978b). *Agrobacterium tumefaciens* caused crown galls on the roots of two cultivars of raspberry only when *M. hapla* was also present (Griffin et al., 1968). Nematodes did not occupy the crown galls, but were in nematode-galled root tissues adjacent to the crown galls. Plants of a third cultivar were not galled by either pathogen whether alone or in combination.

C. Control

The use of nematicides is the primary method used to control nematodes on raspberry because most of these nematodes have wide host ranges and some can survive for long periods of time in the absence of host plants (Cotten, 1975). Nematode control must start in the nursery. If analysis of nursery soil demonstrates the presence of plant-parasitic nematodes, this soil should be treated with a nematicide before planting (Lamberti, 1981; USDA, 1978b). In the United States some suggested nematicides for nursery application include methyl bromide, DD, chloropicrin, and methyl isothiocyanate, or certain mixtures of these fumigants. Disease-free plantings can last 10-15 years, producing 9000-11,000 kg/ha annually, whereas with diseases a planting may last only seven to eight years, yielding only 45000-6700 kg/ha each year (McElroy, 1975). Replanting of raspberry fields where tomato ringspot virus was present in plants and *X. americanum* occurred in the soil without applying nematode control measures necessitated replanting of fields every three to four years (McElroy, 1975). DD applied at 386 kg active per hectare to 15 cm deep failed to give adequate control of *Longidorus macrosoma* in a fine sandy loam soil in England (Cotten, 1975). In Canada, four successive biweekly foliar applications of oxamyl at 0.45 or 0.90 kg per 378 liters starting at full leaf in May held *P. penetrans* and *X. bakeri* populations to levels below the controls for at least two years and increased yields by over 4500 kg/ha for the two year test period (McElroy, 1975). In Scotland, application of oxamyl to soil at sublethal levels prevented *L. elongatus* from acquiring and transmitting tomato black ring virus for at least six weeks and *X. diversicaudatum* from transmitting arabis mosaic virus (Forer et al., 1975). Nematicides are considered to be the best tool for controlling virus vectoring by nematodes, and thus far the fumigants have outperformed the nonfumigants (Lamberti, (1981). Exposure of roots of dormant raspberry plants to hot water at 46.7-48.3°C for 15 minutes controlled infecting *P. penetrans* (Bridge, 1975).

IV. PINEAPPLE

A. History

Pineapple [*Ananas comosus* (L.) Merr] production is important to the economies of many tropical countries (Guerout, 1975). It is a tropical bromeliad that grows best where the monthly rainfall is at least 60 mm and the average temperature is 20-30°C. Pineapple grows for 16-40 months once planted, producing two and sometimes three crops from the mother plant without replanting. It is officially estimated that 4×10^6 tons of pineapples are produced annually, with over half being processed. Pineapples are grown in more than 20 countries. Hawaii is the major pineapple producer, followed by Brazil, Malaysia, Formosa, and Mexico.

B. Nematodes

The major nematode pests of pineapple in the 18 primary producing countries are *Meloidogyne* spp., *Pratylenchus* spp., and *Rotylenchulus reniformis* (Guerout, 1975). Nematodes can be so serious that the first crop may be reduced by up to 42%, and there is no second crop. On a worldwide basis it has been estimated that nematodes cause yield losses of 13.7% to pineapple. The particular nematodes that are most important vary from country to country. In Brazil and the Ivory Coast *P. brachyurus* is the most frequent nematode found attacking pineapple (Guerout, 1975; Rocha Monteiro and Lordello, 1972), in South Africa it is *P. brachyurus* and *M. javanica* (Keetch, 1976; Milne, 1976), in Australia *M. javanica* (Colbran, 1969), and in Jamaica *R. reniformis* (Hutton, 1975). A large number of other nematodes have been found associated with pineapples, some of them causing damage, but not widespread damage (Guerout, 1971, 1975; Rocha Monteiro and Lordello, 1972; Hutton, 1975; Milne et al., 1977).

C. Control

The most widely used method for controlling nematodes on pineapple is chemical soil treatment (Guerout, 1975). Nematicide use varies from nearly all the pineapple soil in Hawaii being treated to none of the soil in most of the smaller producing countries, due to economics. On a worldwide basis less than 25% of the land used to raise pineapples is treated. In South Africa several studies have involved a combination of soil treatment with nematicides and chemical dips to control nematodes (Horn et al., 1956; Milne and de Villiers, 1975a, b; Milne et al., 1977). Several different treatment regimes and chemicals seemed to provide adequate nematode control and increased yields. Preplant fumigation with EDB combined with seed piece dips using oxamyl or fenamiphos gave good control of *M. javanica* and *Helicotylenchus dihystera* and increased yields, although it was suggested that sidedress application of a systemic nematicide three to six months after planting might have been additionally beneficial (Horn et al., 1956). Oxamyl dips followed by monthly foliar applications of oxamyl without preplant fumigation gave good nematode control and high yields in another study (Milne and de Villiers, 1975a). Further studies demonstrated that preplant oxamyl or phenamiphos dips followed by postplant treatments with these nematicides gave extended nematode control and highly significant yield increases (Milne and de Villiers, 1975a; Milne et al., 1977). However, conventional preplant fumigation with a mixture of DD and DBCP still gave the best profit margin (Milne et al., 1977). Various pre- and postplant applications of DBCP, and preplant applications of DBCP + DD,

Fig. 22 Pineapple plants grown in soil fumigated with D-D (right) and untreated soil (left). (Courtesy of E. Pattimore.)

DBCP + EDB, or DD, gave yield increases of 19-50%, averaging 34% (Keetch,
1976). Similar studies in Jamaica confirmed that preplant soil fumigation with
DD-DBCP mixture followed by postplant applications of phenamiphos (20 kg
active per hectare) at 4, 8, and 12 months controlled *Pratylenchus* sp. and
Helicotylenchus multicinctus and gave excellent yield increases (Hutton, 1978).
In the Ivory Coast the response of pineapples grown in soil treated preplant
with 22.5 liters active DBCP per hectare followed by 11.2 liters in four months
controlled *P. brachyurus* and *Helicotylenchus* spp. (Guerout, 1976); supple-
mentary applications of phenamiphos improved nematode control, but not plant
growth. The recommendation of the Pineapple Research Institute for nematode control in pineapple is preplant application of 375 liters active DD + 28.1
liters active DBCP per hectare (Guerout, 1975); although nonfumigants gave
good nematode control they were too expensive to use commercially (Fig. 22).
      Several methods of nematode control on pineapple not involving nemati-
cides have been and are being investigated (Guerout, 1975). In the Ivory
Coast, 23 consecutive monthly plowings failed to adequately reduce *P. brachyurus* populations in the soil. Trap crops have experimentally controlled root-
knot nematodes in soil but are not very practical and therefore are not used
commercially. There is some resistance in certain pineapple cultivars to *M.
incognita* and *R. reniformis*, but this resistance has not been utilized in the
field. Rotation of crops using plants that are nonhosts of nematodes attacking
pineapple can sometimes be utilized, but if mixtures of nematodes are present
in the soil sometimes one or more of those nematodes will multiply on the rotation plant. In Australia, allowing a mixture of siratro (*Phaseolus atropur-
pureus* DC.) and green panic to grow for at least 12 months before planting
pineapples gave excellent control of *M. javanica* (Colbran, 1969).

V. CRANBERRY

A. History

The cranberry (*Vaccinium macrocarpon*) is native to North America, and the crop is mostly consumed in the United States and Canada, with only a small percentage being exported (Childers, 1978). Climate and suitable bog sites largely dictate where cranberries can be grown successfully. The major producing states in the United States are Massachusetts, Wisconsin, and New Jersey.

B. Nematodes

Little is known concerning nematodes parasitizing cranberry, and all the literature that could be found on this topic was published before 1970. In Wisconsin, 45 bogs were surveyed and *Hemicycliophora typica* and *Helicotylenchus pseudorobustus* were detected in nine of the bogs (Barker and Boone, 1966); populations of *H. typica* as high as 530 per liter of soil were found. In New Jersey a survey was made of 49 bogs (Bird and Jenkins, 1964). Nineteen species of stylet-bearing nematodes were detected in which the most common genera were *Helicotylenchus* (82%), *Hemicycliophora* (74%), *Criconemella* (57%), and *Trichodorus* (24%). Four species of nematodes were pathogenic of which *Paratrichodorus minor* caused the most severe cessation of root growth and degree of root discoloration. Of the 16 species of nematodes tested, 14 increased on cranberry plants. The most extensive survey of nematodes associated with cranberry was conducted in Massachusetts, where 355 soil and root samples were examined from 160 bogs (Zuckerman and Coughlin, 1960). *Hemicycliophora* spp. were found in 73% of the bogs and *Trichodorus* spp. in 42%. At least seven other genera of plant-parasitic nematodes were detected in lesser frequency. *Paratrichodorus minor* was again the most pathogenic species detected (Zuckerman, 1961).

C. Control

Although nematodes can be controlled by application of nematicides to cranberry bogs, there are no data showing that this practice is economically feasible. Application of thionazin at 9 or 18 kg active per hectare gave yield increases in one study (Zuckerman, 1964b); however, at 36 kg/ha yields were decreased due to phytotoxicity, although nematode control was best. Other workers found that thionazin at 18 and 32 kg active per hectare provided 89 and 93% reduction in soil nematode populations, respectively, and a concomitant increase in plant growth with no evidence of phytotoxicity (Bird and Jenkins, 1963). Constant flooding of bogs for two years failed to kill the plant-parasitic nematodes present (Bird and Jenkins, 1965).

VI. BLUEBERRY

A. History

Blueberry (*Vaccinium* spp.) is native to North America and because of its abundance, efforts to improve varieties were not started until 1906 (Childers, 1978). In the United States most blueberries are produced in the eastern half of the country, with the major cultivated acreages in Michigan, New Jersey, and North Carolina. A blueberry industry is now developing in Western Europe.

B. Nematodes

All the literature that was found concerning nematodes associated with blueberries was published before 1970, and all is from North America. Many nematodes have been found in association with blueberries (Goheen and Braun,
1955; Hutchinson et al., 1960; Morgan and Wood, 1962; Griffin et al., 1963;
Zuckerman, 1964a; Fox, 1967; USDA, 1978a). The most pathogenic of these
seems to be *Paratrichodorus minor* (Zuckerman, 1962). A root-knot nematode,
Meloidogyne carolinensis Eisenback 1982, seems to be widespread on blueberries in North Carolina, but its pathogenicity has not been evaluated (Fox,
1967; Eisenback, 1982). *P. penetrans* appears not to be a pathogen of blueberry (McCrum and Hilborn, 1962). *X. americanum* seems not in itself to be
pathogenic but transmits tobacco ringspot virus, which causes a necrotic
ringspot disease in blueberry (Griffin et al., 1963). Nematodes have been
reported to stunt and slow the growth of blueberry plants, cause poorly developed root systems and poor stands of plants, and reduce fruit size and
yields (Hutchinson et al., 1960; USDA, 1978a). Nematodes are reported to
cause damage both in nurseries and producing fields.

C. Control

The application of nematicides is the best means for controlling nematodes on
blueberry (USDA, 1978a). Several fumigants are recommended for this purpose including DD, EDB, DD-Mencs, methyl bromide, and metam-sodium
(USDA, 1978a). Dichlofenthion at rates of 113-225 liters/ha was reported to
provide excellent nematode control in both rooting beds and blueberry nursery
fields (Hutchinson et al., 1960). If they can be obtained, blueberry growers
should plant only nematode-free plants (USDA, 1978a). There are no nematode-resistant cultivars of blueberries at present (USDA, 1978a).

REFERENCES

Abu-Gharbieh, W., Verney, E. H., and Jenkins, W. R. (1962). Relationship
of meadow nematode to Verticillium wilt of strawberries (abstract). *Phytopathology 52*: 921.

Alfaro, A., and Goheen, A. C. (1974). Transmissions of strains of grapevine
fanleaf virus by *Xiphinema index*. *Plant Dis. Rep. 58*: 549-552.

Allen, M. W. (1949). Root-lesion nematodes. *Calif. Agric. 3*: 8, 14.

Allen, M. W. (1952). Taxonomic status of the bud and leaf nematodes related
to *Aphelenchoides fragariae* (Ritzema Bos, 1891). *Proc. Helm. Soc. Wash.
19*: 108-120.

Allen, M. W., and Jensen, H. J. (1951). *Pratylenchus vulnus*, new species
(Nematoda: Pratylenchinae), a parasite of trees and vines in California.
Proc. Helm. Soc. Wash. 18: 47-50.

Allen, M. W., Noffsinger, E. M., and Valenzuela, A. (1971). Nematodos en
huertos y vinedos de Chile. *Agric. Tech. 31*: 115-119.

Allen, M. W., and Raski, D. J. (1952). Nematodes on strawberries. *Calif.
Agric. 6*: 3, 14.

Andersson, S. (1974). Skador av *Longidorus elongatus* in jordgubbar. *Vaxt-
skyddsnotiser 38*: 14-18.

Babineau, D. E. (1978). Efficacy of nematicides for control of *Meloidogyne
hapla* and *Pratylenchus penetrans* on strawberry. Master's Thesis, University of Maryland, College Park.

Babineau, D. E. (1981). Efficacy of nonfumigant nematicides for control of *Meloidogyne hapla* and *Pratylenchus penetrans* in strawberry plant nurseries, Ph.D. Dissertation, University of Maryland, College Park.

Barker, K. D., and Boone, D. M. (1966). Plant-parasitic nematodes on cranberries in Wisconsin. *Plant Dis. Rep. 50*: 957-959.

Bessey, E. (1911). Root-knot and its control. *USDA Bur. Plant Ind. Bull. 217*: 1-89.

Bird, G. W., and Jenkins, W. R. (1963). Nematode control in cranberry (abstract). *Phytopathology 53*: 347.

Bird, G. W., and Jenkins, W. R. (1964). Occurrence, parasitism, and pathogenicity of nematodes associated with cranberry. *Phytopathology 54*: 677-680.

Bird, G. W., and Jenkins, W. R. (1965). Effect of cranberry bog flooding and low dissolved oxygen concentrations on nematode populations. *Plant Dis. Rep. 49*: 517-518.

Bosher, J. E. (1954). Root-lesion nematodes associated with root decline of small fruits and other crops in British Columbia. *Can. J. Agric. Sci. 34*: 429-431.

Braun, A. J. (1958). Plant parasitic nematodes found in association with strawberry roots in the United States. *Plant Dis. Rep. 42*: 76-83.

Bridge, J. (1975). Hot water treatment to control plant parasitic nematodes of tropical crops. *Meded. Fac. Landbouwiw. Univ. Gent. 40*: 249-259.

Brown, J. G. (1931). Root-knot in Arizona. *Plant Dis. Rep. 15*: 148.

Cadman, C. H. (1963). Biology of soil-borne viruses. *Ann. Rev. Phytopathol. 1*: 143-172.

Chapman, R. A. (1956). Plant parasitic nematodes associated with strawberries in Kentucky. *Plant Dis. Rep. 40*: 179-181.

Chen, T. A., and Rich, A. E. (1962). The role of *Pratylenchus penetrans* in the development of strawberry black root-rot. *Plant Dis. Rep. 46*: 839-843.

Childers, N. F. (1978). *Modern Fruit Science*, Hort. Publ., Rutgers Univ., New Brunswick, N. J.

Chitwood, B. G. (1949). Root-knot nematodes — Part 1, Revision of the genus *Meloidogyne* Goeldi, 1887. *Proc. Helm. Soc. Wash. 16*: 90-104.

Christie, J. R. (1959). Plant nematodes, their bionomics and control. *Univ. Fla. Agric. Exp. Sta.*

Cobb, N. A. (1913). Notes on *Mononchus* and *Tylenchulus*. *J. Wash. Acad. Sci. 3*: 287-288.

Cobb, N. A. (1914). Citrus-root nematode. *J. Agric. Res. 2*: 217-230.

Cohn, E. (1970). Observations on the feeding and symptomatology of *Xiphinema* and *Longidorus* on selected host roots. *J. Nematol. 2*: 167-173.

Cohn, E., and Mordechai, M. (1969). Investigations on the life cycles and host preference of some species of *Xiphinema* and *Longidorus* under controlled conditions. *Nematologica 15*: 295-302.

Colbran, R. C. (1957). The strawberry root-knot nematode. *Queensland Dept. Agric. Stock Div. Plant Ind. Advisory Leaflet 462*.

Colbran, R. C. (1959). Strawberry root-knot nematode investigations in Queensland. *Queensland J. Agr. Sci. 16*: 265-370.

Colbran, R. C. (1969). Cover crops and nematode control in pineapple. *Queensland Agric. J. 95*: 658-661.

Cotten, J. (1975). Virus vector species of *Xiphinema* and *Longidorus* in relation to certification schemes for fruit and hops in England. In *Nematode Vectors of Plant Viruses* (F. Lamberti, C. E. Taylor and J. W. Seinhorst, eds.). Plenum, New York, pp. 283-285.

Cotten, J. (1976). Observations of life-cycle, population development and vertical distribution of *Longidorus macrosoma* on raspberry and other crops. *Ann. Appl. Biol. 83*: 407-412.

Crosse, J. E., and Pitcher, R. S. (1952). Studies in the relationship of eelworms and bacteria to certain plant diseases. I. The etiology of strawberry cauliflower disease. *Ann. Appl. Biol. 39*: 475-484.

Crow, R. V., and MacDonald, D. H. (1978). Phytoparasitic nematodes adjacent to established strawberry plantations. *J. Nematol. 10*: 204-207.

Das, S., and Raski, D. J. (1968). Vector-efficiency in *Xiphinema index* in the transmission of grapevine fanleaf virus. *Nematologica 14*: 55-62.

De Man, J. G. (1880). Die einheimischen, frei in der reinen Erde und im süssen Wasser lebenden Nematoden. *Tijdschr. Ned. Dierk. Ver. 5*: 1-104.

D'Herde, J., and van den Brande, J. (1964). Distribution of *Xiphinema* and *Longidorus* sp. in strawberry fields in Belgium and a method for their quantitative extraction. *Nematologica 10*: 454-458.

Dias, H. F. (1975). Peach rosette mosaic. *C.M.I./A.A.B. Descriptions of Plant Viruses*, vol. 150.

Dias, H. F. (1977). Incidence and geographic distribution of tomato ringspot virus in De Chaunac vineyards in the Niagara peninsula. *Plant Dis. Rep. 61*: 24-28.

Dias, H. F., and Cation, D. (1976). The characterization of a virus responsible for peach rosette mosaic and grape decline in Michigan. *Can. J. Bot. 54*: 1228-1239.

Dickstein, E. R., and Krusberg, L. R. (1978). Reaction of strawberry cultivars to the northern root-knot nematode, *Meloidogyne hapla*. *Plant Dis. Rep. 62*: 60-61.

Dunn, R. A. (1972). Importance of depth in soil, presence of host roots, and role of eggs as compared to vermiform stages in overwintering of *Pratylenchus penetrans* at Ithaca, New York. *J. Nematol. 4*: 221-222.

Eisenback, J. D. (1982). Description of the blueberry root-knot nematode, *Meloidogyne carolinensis*, n. sp. *J. Nematol. 14*: 303-317.

Ferris, H., and McKenry, M. V. (1974). Seasonal fluctuations in the spatial distribution of nematode populations in a California vineyard. *J. Nematol. 6*: 203-210.

Filipjev, I. N., and Schuurmans-Stekhoven, J. H. (1941). *A. Manual of Agricultural Helminthology*. E. J. Brill, Leiden.

Fisher, J. M., and Raski, D. J. (1967). Feeding of *Xiphinema index* and *X. diversicaudatum*. *Proc. Helm. Soc. Wash. 34*: 68-72.

Forer, L. B., Trudgill, D. L., and Alphey, J. W. (1975). Some effects of oxamyl on the virus-vector nematodes *Longidorus elongatus* and *Xiphinema diversicaudatum*. *Ann. Appl. Biol. 81*: 207-214.

Fox, J. A. (1967). *Diss. Abstr. 28*: 1311-1312.

Goheen, A. C., and Braun, A. J. (1955). Some parasitic nematodes associated with blueberry roots. *Plant Dis. Rep. 39*: 908.

Goheen, A. C., and Braun, A. J. (1956). Some parasitic nematodes associated with wild strawberry plants in woodlands in Maryland. *Plant Dis. Rep. 40*: 43.

Goheen, A. C., and McGrew, J. R. (1954). Control of endoparasitic root nematodes in strawberry propagation stock by hot-water treatments. *Plant Dis. Rep. 38*: 818-826.

Goheen, A. C., and Smith, J. B. (1956). Effects of inoculation of strawberry roots with meadow nematodes, *Pratylenchus penetrans*. *Plant Dis. Rep. 40*: 146-149.

Golden, A. M., and Converse, R. (1965). Nematodes on raspberry in the Eastern United States. *Plant Dis. Rep. 49*: 987-991.

Goodey, J. B. (1951). Observations on the attack by the stem eelworm, *Ditylenchus dipsaci*, on strawberry. *Ann. Appl. Biol. 38*: 618-623.

Griffin, G. D., Anderson, J. L., and Jorgenson, E. C. (1968). Interaction of *Meloidogyne hapla* and *Agrobacterium tumefaciens* in relation to raspberry cultivars. *Plant Dis. Rep. 52*: 492-493.

Griffin, G. D., Huguelet, J. E., and Nelson, J. W. (1963). *Xiphinema americanum* as a vector of necrotic ringspot virus of blueberry. *Plant Dis. Rep. 47*: 703-704.

Guerout, R. (1971). Importance relative des champignons du sol et des nematodes sur la croissance des ananas. *Fruits 26*: 287-293.

Guerout, R. (1975). Nematodes of pineapple: a review. *Pest Articles and News Summaries (PANS) 21*: 123-140.

Guerout, R. (1976). Usefulness of plots without nematodes in pineapple experiments (abstract). *Nematropica 6*: 1.

Harrison, B. D., Peachey, J. E., and Winslow, R. D. (1963). The use of nematicides to control the spread of arabis mosaic virus by *Xiphinema diversicaudatum* (Micol.). *Ann. Appl. Biol. 52*: 243-255.

Hesling, J. J. (1966). 3. Biological races of stem eelworm. *Rep. Glasshouse Crops Res. Inst. 1965*, pp. 132-141.

Hewitt, W. B., Goheen, A. C., Raski, D. J., and Gooding, Jr., G. V. (1962). Studies on virus diseases of the grapevine in California. *Vitis 3*: 57-83.

Hewitt, W. B., and Raski, D. J. (1967). Factors limiting crop production: 6 grapes. *Span. 10*: 56-59.

Hewitt, W. B., Raski, D. J., and Goheen, A. C. (1958). Nematode vector of soil-borne fanleaf virus of grapevine. *Phytopathology 48*: 586-595.

Hildebrand, E. M. (1941). A new case of rosette mosaic on peach. *Phytopathology 31*: 353-355.

Horn, N. L., Martin, W. J., Wilson, W. F., Jr., and Giamalva, M. J. (1956). The relation of nematodes in strawberry culture in Louisiana. *Plant Dis. Rep. 40*: 790-797.

Hutchinson, M. T., Reed, J. P., and Race, S. R. (1960). Nematodes stunt blueberry plants. *New Jersey Agric. 42*: 12-13.

Hutton, D. G. (1975). Pineapple nematodes in Jamaica and relationship between their populations and rainfall in two areas. *Nematropica 5*: 23-24.

Hutton, D. G. (1978). Response of pineapple plants growing in nematode-infested soil to after-planting nematicidal treatments. *Nematropica 8*: 29-49.

Jenkins, W. R., and Taylor, D. P. (1967). *Plant Nematology*. Reinhold, New York.

Kantzes, J. G., and Morgan, O. D. (1962). Comparison of nematicides for control of root-knot nematodes on strawberries (abstract). *Phytopathology 52*: 164.

Keetch, D. P. (1976). Nematodes of pineapples in South Africa. *Gewasprod. Crop Prod. 5*: 51-54.

Kirjanova, E. S., and Krall, E. L. (1971). *Plant-Parasitic Nematodes and Their Control*, volume II (Translated from Russian, 1980). Amerind Publ. Co. Pat. Ltd., New Delhi.

Klinger, J., and Kunz, P. (1978). Beobachtungen und Versuche mit wurzelparasitischen und virusubertragenden Nematoden an Erdbeeren. *Schweiz. Z. Obst-u. Weinbau. 114*: 342-350.

Klinkenberg, C. H. (1955). Nematode diseases of strawberry in the Netherlands. *Plant Dis. Rep. 39*: 603-606.

Klos, E. J., Fronek, F., Knierim, J. A., and Cation, D. (1967). Peach rosette mosaic transmission and control studies. *Q. Bull. Mich. State Univ. Agr. Exp. Sta. 49*: 287-293.

Lamberti, F. (1981). Combating nematode vectors of plant viruses. *Plant Dis. 65*: 113-117.

Lamberti, F., and Bleve-Zacheo, T. (1979). Studies on *Xiphinema americanum sensu lato* with descriptions of fifteen new species (Nematoda, Longidoridae). *Nematol. Medit. 7*: 51-106.

Lear, B., and Lider, L. A. (1959). Eradication of root-knot nematodes from grapevine rootings by hot water. *Plant Dis. Rep. 43*: 314-317.

Lider, L. A. (1954). Inheritance of resistance to a root-knot nematode (*Meloidogyne incognita* var. *acrita* Chitwood) in *Vitis* spp. *Proc. Helm. Soc. Wash. 21*: 53-60.

Lider, L. A. (1959). Nematode resistant rootstocks for California vineyards. *Calif. Agric. Exp. Sta. Leaflet 114*.

Loof, P. A. A. (1978). The Genus *Pratylenchus* Filipjev, 1936 (Nematoda: Pratylenchidae): A Review of its anatomy, morphology, distribution, systematics and identification. *Vaxtskyddsrapporter Jordbruk 5*: 1-50.

Martelli, G. P. (1978). Nematode-borne viruses of grapevine, their epidemiology and control. *Nematol. Medit. 6*: 1-27.

McCrum, R. C., and Hilborn, M. T. (1962). Nonpathogenicity of *Pratylenchus penetrans* to sterile low-bush blueberry seedlings. *Plant Dis. Rep. 46*: 84-85.

McCulloch, J. S. (1974). Grass hosts of the strawberry "crimp" nematode. *Aust. P. Path. Soc. Newsletter 3*: 69.

McCulloch, J. (1978). Strawberry crimp. *Queensland Agric. J. 104*: 345-347.

McElroy, F. D. (1971). *Longidorus elongatus* damaging strawberry in British Columbia. *Plant Dis. Rep. 55*: 266-267.

McElroy, F. D. (1975a). Nematode control in established red raspberry plantings. In *Nematode Vectors of Plant Viruses* (F. Lamberti, C. E. Taylor, and J. W. Seinhorst, eds.). Plenum, New York, pp. 445-446.

McElroy, F. D. (1975b). Nematode transmitted viruses in British Columbia, Canada. In *Nematode Vectors of Plant Viruses* (F. Lamberti, C. E. Taylor, and J. W. Seinhorst, eds.). Plenum, New York, pp. 287-288.

McElroy, F. D. (1977a). Nematodes as vectors of plant viruses – a current review. Symposium on Nematode Transmission of Viruses. Amer. Phytopath. Soc. 69th Ann. Meeting, East Lansing, Mich., 8/16/77.

McElroy, F. D. (1977b). Effect of two nematode species on establishment, growth, and yield of raspberry. *Plant Dis. Rep. 61*: 277-279.

Meagher, I. W., and Jenkins, P. T. (1970). Interaction of *Meloidogyne hapla* and *Verticillium dahliae*, and the chemical control of strawberry wilt. *Aust. J. Exp. Agric. Anim. Husb. 10*: 493-496.

Milbrath, D. G. (1923). The root-knot nematode in relation to deciduous fruit trees and grapevines. *Calif. Dept. Agric. Monthly Bull. 12*: 127-135.

Milne, D. L. (1974). Dips for the control of pineapple eelworms. *Citrus and Sub-Trop. Fruit J. 484*: 11-13.

Milne, D. L. (1976). News from South Africa. *OTAN Newsl. 8*: 4-5.

Milne, D. L., and deVilliers, E. A. (1975a). Foliar systemic nematicides for pineapple nematode control (abstract). *J. Nematol. 7*: 327.

Milne, D. L., and deVilliers, E. A. (1975b). Nematodes of citrus, banana, mango, papaya, litchi, granadilla and pineapple in South Africa and their control (abstract). *Nematropica 5*: 25-26.

Milne, D. L., deVilliers, E. A., and Smith, B. L. (1977). Response of nematode populations and pineapple yields to foliar applications of systemic nematicides. *Citrus Sub-Trop. Fruit J. 525*: 8-14.

Minz, G. (1958). *Meloidogyne javanica* in strawberry roots. *F.A.O. Plant Prot. Bull. 6.*

Morgan, G. T. (1964). Effects of spring preplanting fumigation in strawberry soil in New Brunswick. *Can. J. Plant Sci. 44*: 170-174.

Morgan, G. T., and Collins, W. B. (1964). The effects of organic treatments and crop rotation on soil populations of *Pratylenchus penetrans* in strawberry culture. *Can. J. Plant Sci. 44*: 272-275.

Morgan, G. T., and Wood, G. W. (1962). An ectoparasitic root nematode on lowbush blueberry (*Vaccinium* spp.). *Plant Dis. Rep. 46*: 800.

Neal, J. C. (1889). The root-knot disease of the peach, orange, and other plants in Florida, due to the work of *Anguillula*. *USDA Div. Ent. Bull. 20*: 1-31.

Nemec, S., and Malek, R. B. (1975). Effects of nematicides and strawberry growth on nematodes, especially *Xiphinema americanum*, in root-rot sites in Illinois. *J. Nematol. 7*: 328.

Nougaret, R. L. (1923). Rootknot on grape. *Calif. Dept. Agric. Monthly Bull. 12*: 139-150.

Orchard, W. R., and Andrichem, M. C. J. (1961). Relative susceptibility of *Fragaria* spp. to the root-knot nematode, *Meloidogyne hapla* Chitw. *Plant Dis. Rep. 45*: 308.

Perry, V. G. (1958). Parasitism of two species of dagger nematodes (*Xiphinema americanum* and *X. chambersi*) to strawberry. *Phytopathology 48*: 420-423.

Perry, V. G., and Zeikus, J. A. (1972). Host variations among populations of the *Meloidogyne incognita* group (abstract). *J. Nematol. 4*: 231-232.

Pinochet, J., and Raski, D. J. (1977). Observations on the host-parasite relationship of *Pratylenchus vulnus* on grapevine, *Vitis vinifera*. *J. Nematol. 9*: 87-88.

Pinochet, J., Raski, D. J., and Goheen, A. C. (1976). Effects of *Pratylenchus vulnus* and *Xiphinema index* singly and combined in vine growth of *Vitis vinifera*. *J. Nematol. 8*: 330-335.

Pinochet, J., Raski, D. J., and Jones, N. O. (1976). Effect of *Helicotylenchus pseudorobustus* on Thompson Seedless grape. *Plant Dis. Rep. 60*: 528-529.

Pitcher, R. S., and Crosse, J. E. (1958). Studies on the relationship of eelworms and bacteria to certain plant diseases. II. Further analysis of the strawberry cauliflower disease complex. *Nematologica 3*: 244-256.

Pitcher, R. S., and Jha, A. (1961). On the distribution and infectivity with Arabis mosaic virus of a dagger nematode. *Plant Pathol. 10*: 67-71.

Potter, H. S., and Morgan, O. D. (1956). Nemagon control of root-knot nematode on strawberries. *Plant Dis. Rep. 40*: 187-189.

Radewald, J. D. (1962). The biology of *Xiphinema index* and the pathological effect of the species on grape. Ph.D., Dissertation, University of California, Davis.

Raski, D. J. (1954). Soil fumigation for the control of nematodes on grape replants. *Plant Dis. Rep. 38*: 811-817.

Raski, D. J. (1955). Additional observations on the nematodes attacking grapevines and their control. *Am. J. Enol. Vitic. 6*: 29-31.

Raski, D. J. (1956). *Pratylenchus penetrans* tested on strawberries grown in black-root-rot soil. *Plant Dis. Rep. 40*: 690-693.

Raski, D. J., and Allen, M. W. (1948). Spring dwarf nematode. *Calif. Agric.* 2: 23-24.

Raski, D. J., Hart, W. H., and Kasimatis, A. N. (1973). Nematodes and their control in vineyards (revised). *Calif. Agr. Exp. Sta. Circ. 533:* 1-20.

Raski, D. J., and Hewitt, W. B. (1960). Experiments with *Xiphinema index* as a vector of fanleaf of grapevines. *Nematologica 5:* 166-170.

Raski, D. J., Hewitt, W. B., and Schmitt, R. V. (1971). Controlling fanleaf virus – dagger nematode disease complex in vineyards in soil fumigation. *Calif. Agric. 25:* 11-14.

Raski, D. J., Jones, N. O., Hafez, S. L., Kissler, J. J., and Luvisi, D. A. (1981). Systemic nematicides tested as alternatives to DBCP. *Calif. Agric. 35:* 11-12.

Raski, D. J., Jones, N. O., Kissler, J. J., and Luvisi, D. A. (1975). Futher results from deep-placement fumigation for control of nematodes in vineyards. *Plant Dis. Rep. 59:* 345-349.

Raski, D. J., Jones, N. O., Kissler, J. J., and Luvisi, D. A. (1976). Soil fumigation: One way to cleanse nematode-infested vineyard lands. *Calif. Agric. 30:* 4-6.

Raski, D. J., Maggenti, A. R., and Jones, N. O. (1973). Location of grapevine fanleaf and yellow mosaic virus particles in *Xiphinema index*. *J. Nematol. 5:* 208-211.

Raski, D. J., and Schmitt, R. V. (1972). Progress in control of nematodes by soil fumigation in nematode-fanleaf infected vineyards. *Plant Dis. Rep. 56:* 1031-1035.

Raski, D. J., and Radewald, J. D. (1958). Reproduction and symptomatology of certain ectoparasitic nematodes on roots of Thompson Seedless grape. *Plant Dis. Rep. 42:* 941-943.

Raski, D. J., Sher, S. A., and Jensen, F. N. (1956). New host records of the citrus nematode in California. *Plant Dis. Rep. 40:* 1047-1048.

Riggs, R. D., and Hamblen, M. L. (1962). Soil fumigation increases strawberry plant and berry production in Arkansas. *Down to Earth 18:* 5-7.

Riggs, R. D., Slack, D. A., and Fulton, J. P. (1956). Meadow nematode and its relation to decline of strawberry plants in Arkansas (abstract). *Phytopathology 46:* 24.

Rocha Monteiro, A., and Lordello, L. G. E. (1972). Nematoides parasitos do abacaxizeiro. *Rev. Agr. Piracicaba 47:* 163.

Rumpenhorst, H. J., and Weischer, B. (1978). Histopathological and histochemical studies on grapevine roots damaged by *Xiphinema index*. *Rev. Nematol. 1:* 217-225.

Sasser, J. N. (1954). Identification and host parasite relationships of certain root-knot nematodes (*Meloidogyne* spp.). *Univ. Md. Agric. Exp. Sta. Tech. Bull. No. A-77.*

Sauer, M. R. (1962). Distribution of plant parasitic nematodes in irrigated vineyards at Merbein and Robinvale. *Aust. J. Exp. Agric. Anim. Husb.* 2: 8-11.

Schindler, A. F. (1954). Root galling associated with dagger nematode, *Xiphinema diversicaudatum*. *Phytopathology 44:* 389.

Schindler, A. F., and Braun, A. J. (1957). Pathogenicity of an ectoparasitic nematode, *Xiphinema diversicaudatum* on strawberries. *Nematologica 2:* 91-93.

Scotto la Massese, C. (1973). Les nematodes phytophages associes a la culture des petits fruits en France. *Bull. Tech. SICANAPF 5:* 6-14.

Seinhorst, J. W. (1966). *Longidorus elongatus* on *Fragaria vesca*. *Nematologica 12*: 275-279.

Seinhorst, J. W., and Sauer, M. R. (1956). Eelworm attacks on vines in the Murray Valley irrigation area. *J. Aust. Inst. Agric. Sci. 22*: 296-299.

Sharma, N. K. (1977). Incidence of 'spring dwarf' disease caused by the nematode *Aphelenchoides fragariae* of strawberry from India. *Curr. Sci. 46*: 566.

Sher, S. A., and Allen, M. W. (1953). Revision of the genus *Pratylenchus* (Nematoda: Tylenchidae). *Univ. Calif. Publ. Zool. 57*: 441-470.

Siddiqui, M. R. (1974). *Tylenchulus semipenetrans*. *C.I.H. Descriptions of Plant-parasitic nematodes* Set 3. No. *34*: 1-4.

Siddiqui, I. A., Sher, S. A., and French, A. M. (1973). Distribution of plant parasitic nematodes in California. *Calif. Dept. Food Agric. Div. Plant Ind.*

Slack, D. A., and Pitcher, R. S. (1970). Bud and leaf nematodes of strawberry. In *Vine Diseases of Small Fruits and Grapevines* (N. W. Frazier, J. P. Fulton, J. M. Thresh, R. H. Converse, E. H. Varney, and W. B. Hewitt, eds.). University of California Division of Agric. Sci., Berkeley, pp. 56-59.

Smart, G. C., Jr. and Locascio, S. J. (1968). Influence of nematicides and polyethylene mulch color on the control of nematodes on strawberries. *Proc. Soil Crop Sci. Soc. Fla. 28*: 292-299.

Smart, G. C., Jr., Locascio, S. J., and Rhoades, H. L. (1967). Root-knot nematode control on strawberry (abstract). *Nematologica 13*: 152-153.

Snyder, E. (1936). Susceptibility of grape rootstocks to root-knot nematode. *USDA Bur. Plant Ind. Circ. 405*: 1-15.

Steiner, G. (1931). *Tylenchus pratensis* deMan, on tobacco, tomato, and strawberry. *Plant Dis. Rep. 15*: 106-107.

Strich-Harari, D., and Minz, G. (1961). A strain of *Meloidogyne javanica* attacking strawberry. *Israel J. Agric. Res. 11*: 75-77.

Szczygiel, A. (1966). Studies on the present population dynamics of nematodes occuring in strawberry plantations. *Ekol. Pol. 14*: 651-709.

Szczygiel, A. (1967). Preliminary estimation of the harmfulness of nematodes of the genus *Aphelenchoides* to strawberries in South Poland (Polish). *Prace Inst. Sadownictwa 11*: 211-224.

Szczygiel, A. (1970). Distribution of leaf and bud nematodes (*Aphelenchoides* spp.) and stem nematode (*Ditylenchus dipsaci*) in strawberry fields in Poland. *Proc. IX Int. Nema. Symp.* (Warsaw, 1967), pp. 321-329.

Szczygiel, A. (1977). Review of Soviet literature on plant parasitic nematodes associated with strawberries. *Res. Inst. Pomol.* Brzezna, Poland.

Szczygiel, A. (1980). Pathogenicity of plant parasitic nematodes to strawberry plant as affected by host plant conditions and environmental factors. Final report, *Res. Inst. Pomol.*, Skierniewice, Poland, 6/1/73-12/3/79.

Szczygiel, A., and Hasior, H. (1971). Possibility of persistence of leaf and bud nematodes (*Aphelenchoides fragariae*) on strawberry plants and in the soil. *Zesz. Probl. Post. Nauk Roln. 121*: 101-106.

Szczygiel, A., and Hasior, H. (1972). Vertical distribution of plant parasitic nematodes in the soil of strawberry plantations. *Ekologia Polska 20*: 493-506.

Tacconi, R. (1976). Prove di lotta control il *Ditylenchus dipsaci* su fragola. II. *Ins. Fitopatol. 26*: 29-36.

Taylor, A. L., and Sasser, J. N. (1978). Biology, identification and control of root-knot nematodes (*Meloidogyne* species). North Carolina St. Univ. Graphics, Raleigh.

Taylor, C. E., and Alphey, T. J. W. (1973). Aspects of the systemic nematicidal potential of Du Pont 1410 in the control of *Longidorus* and *Xiphinema* virus vector nematodes. *Ann. Appl. Biol.* 75: 464-467.

Taylor, C. E., and Murant, A. F. (1968). Chemical control of raspberry ringspot and tomato black ring viruses in strawberry. *Plant Pathol.* 17: 171-178.

Taylor, C. E., and Raski, D. J. (1964). On the transmission of grape fanleaf by *Xiphinema index*. *Nematologica* 10: 489-495.

Taylor, C. E., and Robertson, W. M. (1970). Sites of virus retention in the alimentary tract of the nematode vectors *Xiphinema diversicaudatum* (Micol.) and *X. index* (Thorne et Allen). *Ann. Appl. Biol.* 66: 375-380.

Taylor, C. E., and Thomas, P. R. (1968). The association of *Xiphinema diversicaudatum* (Micoletzky) with strawberry latent ringspot and arabis mosaic viruses in a raspberry plantation. *Ann. Appl. Biol.* 62: 147-157.

Taylor, D. P. and Netscher, C. (1975). Occurrence in Senegal of a biotype of *Meloidogyne javanica* parasitic on strawberry. *ORSTOM, ser Biol.* 10: 247-250.

Teakle, D. S. (1967). Fungus transmission of plant viruses. In *Methods in Virology* (K. Maramorosch and H. Koprowski, eds.). Academic, New York, pp. 369-391.

Teakle, D. S. (1969). Fungi as vectors and hosts of viruses. In *Viruses, Vectors and Vegetation* (K. Maramorosch, ed.). John Wiley and Sons, New York, p. 666.

Teliz, D., Grogan, R. G., and Lownsbery, B. F. (1966). Transmission of tomato ringspot, peach yellow bud mosaic, and grape yellow vein viruses by *Xiphinema americanum*. *Phytopathology* 56: 658-663.

Teliz, D., Lownsbery, B. F., Grogan, R. G., and Kimble, K. A. (1967). Transmission of peach yellow bud mosaic virus to peach, apricot, and plum by *Xiphinema americanum*. *Plant Dis. Rep.* 51: 841-843.

Thomas, E. E. (1913). A preliminary report of a nematode observed on citrus roots and its possible relation with the mottled appearance of citrus trees. *Calif. Agr. Exp. Sta. Circ.* 85: 1-14.

Thomas, P. R. (1969). Population development of *Longidorus elongatus* on strawberry in Scotland with observation on *Xiphinema diversicaudatum* on raspberry. *Nematologica* 15: 582-590.

Thorne, G. (1961). *Principles of Nematology*. McGraw-Hill, New York.

Thorne, G., and Allen, M. W. (1950). *Paratylenchus hamatus* n. sp. and *Xiphinema index* n. sp., two nematodes associated with fig roots, with a note on *Paratylenchus anceps* Cobb. *Proc. Helm. Soc. Wash.* 17: 27-35.

Townshend, J. L. (1962). The root-lesion nematode, *Pratylenchus penetrans* (Cobb, 1917) Filip. and Stek., 1941, in strawberry in the Niagara Peninsula and Norfolk County in Ontario. *Can. J. Plant Sci.* 42: 728-736.

Townshend, J. L. (1963). The pathogenicity of *Pratylenchus penetrans* to strawberry. *Can. J. Plant Sci.* 43: 75-78.

Townshend, J. L. (1965). Production of nematode-free strawberry plants. *Can. J. Plant Sci.* 45: 201-203.

Townshend, J. L., and Davidson, T. R. (1960). Some weed hosts of *Pratylenchus penetrans* in Premier strawberry plantations. *Can. J. Bot.* 38: 267-273.

Townshend, J. L., Ricketson, C. L, and Wiebe, J. (1966). The effect of Spring application of nematicides on strawberry in the Niagara Peninsula. *Can. J. Plant Sci. 46:* 111-114.

Trudgill, D. L. (1976). Observations of the feeding of *Xiphinema diversicaudatum*. *Nematologica 22:* 417-423.

Tyler, J. (1933a). The root-knot nematode (revised 1944). *Calif. Agric. Exp. Sta. Circ. 330:* 1-30.

Tyler, J. (1933b). The root-knot nematode. *Calif. Agric. Exp. Sta. Circ. 330:* 1-34.

Tyler, J. (1933c). Development of the root-knot nematode as affected by temperature. *Hilgardia 7:* 389-415.

Tyler, J. (1941). Plants reported resistant or tolerant to root-knot nematode infestation. *USDA Misc. Public. 406:* 1-91.

USDA (1978a). Commercial blueberry growing (revised). *SEA Farmers' Bull. No. 2254.*

USDA (1978b). Controlling diseases of raspberries and blackberries (revised). *Farmers' Bull. No. 2208.*

USDA (1978c). Strawberry diseases (revised). *Farmers' Bull. No. 2140.*

Uyemoto, J. K., and Gilmer, R. M. (1972). Spread of tomato ringspot virus in "Baco Noir" grapevines in New York. *Plant Dis. Rep. 56:* 1062-1064.

Van Gundy, S. D. (1958). The life history of the citrus nematode *Tylenchulus semipenetrans* Cobb. *Nematologica 3:* 283-294.

Van Gundy, S. D., Kirkpatrick, J. D., and Martin, J. P. (1968). Nutritional response of *Vitis vinifera* "Carignane" infested with *Xiphinema index*. Rep. 8th International Symp. Nematol., Antibes, 8-14 Sept. 1965, E. J. Brill, Leiden.

Weiner, A., and Raski, D. J. (1966). New host records for *Xiphinema index* Thorne and Allen, 1950. *Plant Dis. Rep. 50:* 27-28.

White, V. (1960). Host-parasite relationship of *Xiphinema americanum* Cobb, 1913, on apple, corn and strawberry. Ph.D. Thesis, University of Wisconsin, Madison.

Whittle, W. O., and Drain, B. D. (1935). The root-knot nematode in Tennessee. Its prevalence and suggestions for control. *Tenn. Agric. Exp. Sta. Circ. 54:* 1-8.

Wilhelm, S., and Paulus, A. O. (1980). How soil fumigation benefits the California strawberry industry. *Plant Dis. 64:* 264-270.

Wilhelm, S., and Sagen, J. E. (1974). A history of the strawberry. *Univ. Calif. Div. Agr. Sci.*

Williams, T. D. (1961). *Xiphinema bakeri* n. sp. (Nematoda: Longidorinae) from the Fraser River Valley, British Columbia, Canada. *Can. J. Zool. 39:* 407-412.

Winkler, A. J., Cook, J. A., Kliewer, M. W., and Lider, L. A. (1974). *General Viticulture*. Univ. Calif. Press, Berkeley.

Wyss, U. (1970a). Parasitierungsvorgang und Pathogenität Vorndernder Wurzelnematoden an *Fragaria vesca* var. *semperflorens*. *Nematologica 16:* 55-62.

Wyss, U. (1970b). Untersuchungen zur populationsdynamik von *Longidorus elongatus*. *Nematologica 16:* 74-84.

Zuckerman, B. M. (1961). Parasitism and pathogenesis of the cultivated cranberry by some nematodes. *Nematologica 6:* 135-143.

Zuckerman, B. M. (1962). Parasitism and pathogenesis of the cultivated highbush blueberry by the stubby root nematode. *Phytopathology 52:* 1017-1019.

Zuckerman, B. M. (1964a). Studies of two nematode species associated with roots of the cultivated highbush blueberry. *Plant Dis. Rep.* 48: 170-172.

Zuckerman, B. M. (1964b). The effects of zinophos on nematode populations and cranberry yields. *Plant Dis. Rep.* 48: 172-175.

Zuckerman, B. M., and Coughlin, J. W. (1960). Nematodes associated with some crop plants in Massachusetts. *Agric. Exp. Sta. Univ. Mass. Bull.* 521.

Chapter 14

Nematode Parasites of Sugar Beet

Arnold E. Steele *USDA, Agricultural Research Service, Salinas, California*

I. HISTORICAL BACKGROUND

In 1747 the chemist Andreas Siegmund Marggraf, director of the Mathematical Physical Class of the Royal Prussian Academy of Science and protégé of Frederick the Great, found sucrose in the plants *Beta alba*, *B. rubra*, and *Sium sisaram*, which are endemic to Europe (Coons, 1949). Franz-Carl Achard, pupil of and successor to Marggraf in the Academy, in 1784 initiated experiments with beets as a source of sugar. He selected a variety that was closely

related to the mangel-wurzel (*Beta vulgaris* L.). Through his program of
plant breeding and selection, Achard developed the White Silesian beet, considered by some to be "ancestress of all the sugarbeets of the world." Achard
learned to grow the crop, developed the processes for sugar extraction, puri-
fication, and crystallization, and established seed and production programs.

Disruption of the supply of cane sugar from the West Indies to France
by the British naval blockade of continental ports, prompted Napoleon in 1811
to issue a series of decrees requiring beets to be grown in France, and en-
couraged the establishment of beet sugar factories through government sub-
sidies. The sugar beet industry, which eventually failed in France and disappeared in Germany, was revived about 1840 with the introduction of the
Imperial beet, a more vigorous variety with higher sugar content than the
White Silesian from which it was derived. Soon after, beet sugar factories
were established throughout Central Europe. The first sugar factory in the
United States was started in Northampton, Massachusetts, in 1837, but after
three years it was abandoned. In 1870 the first successful factory was established at Alvarado, California.

The world production of raw sugar from beets and cane is estimated at
87.1 million metric tons for the 1980 sugar crop year. Raw sugar produced
from 475,071 ha of sugar beet in 18 states of the United States is estimated to
be 2.7 million metric tons, or about 3.1% of the worldwide production (Anon,
1980).

Although sugar beet is a susceptible host for a number of nematode species, the sugar beet nematode, *Heterodera schachtii* Schmidt, probably accounts
for more than 90% of the damage to sugar beet caused by nematodes. The
average annual loss in yield of sugar beet attributed to nematodes in the
United States is estimated to be as high as 10% (Anon., 1970; Good, 1968).
Total crop failures of beets grown in heavily infested soil are not uncommon
in the United States, and in Central Europe yield losses exceeding 25% occur
when sugar beets are cropped too frequently (Weischer and Steudel, 1972).

II. SUGAR BEET NEMATODE

The sugar beet nematode, *Heterodera schachtii* A. Schmidt 1871, variously
called the beet eelworm or the beet cyst nematode, was discovered by Schacht
(1859a,b) in 1859 near Halle, Germany. In 1871 Schmidt named the nematode
after its discoverer and gave a comprehensive description of its morphology
(1871, 1872). The intensive cultivation of sugar beet was accompanied by the
rapid spread of the sugar beet nematode, which soon produced severe yield
losses. Research by Kühn (1881, 1882) and Liebscher (1878) established the
nematode as the true cause of the alleged exhaustion of the soil termed "Ruben-
mudgkeit" or "beet weariness" that in 1876 forced the closure of at least 24
beet sugar factories in Germany. The devastating impact of this pest to the
sugar beet industry prompted other European researchers to seek additional
information on the nematode and its hosts. Shortly thereafter, thorough studies of the anatomy and morphology were accomplished by Strubell (1888) and
Chatin (1888; 1891), and later by Raski (1949). Nemec (1910, 1911, 1933) investigated the histopathology of nematode-infected sugarbeet, and Bergman
(1958) and Hijner (1952) evaluated the host reactions and nematode development in susceptible and resistant wild *Beta* spp.

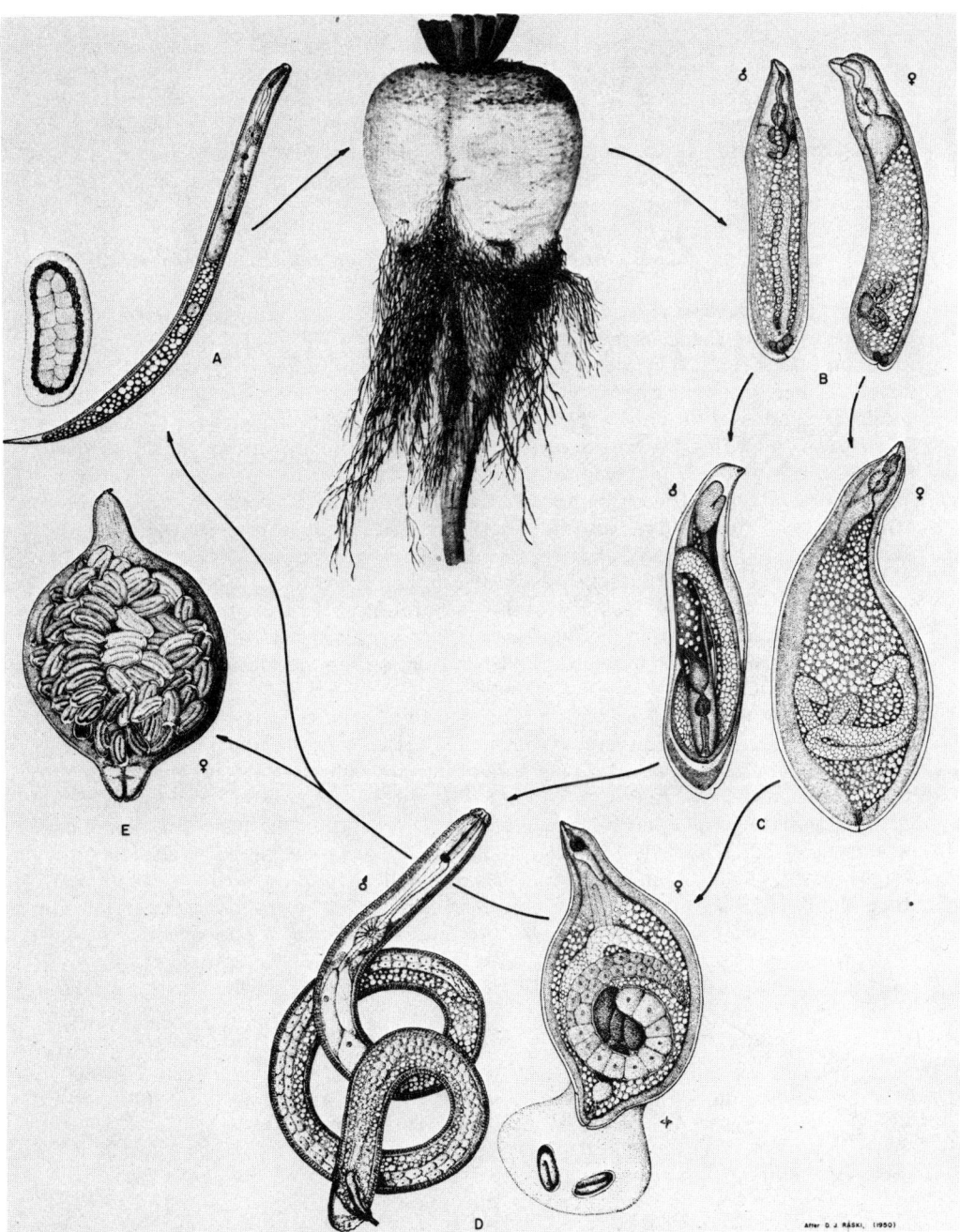

Fig. 1 Life cycle of the sugar beet nematode, *H. schachtii*: (A) Second-stage infective larvae hatch from eggs and invade roots of sugar beet (0.435-0.492 mm). (B) Third-stage larvae develop within roots (0.324-0.377 mm). (C) Fourth-stage larvae within roots. Female breaks through root surface, molts, and becomes fifth-stage adult. Adult male develops in fourth-stage cuticle and emerges from root (1.119-1.438 mm). (D) Fertilized female begins egg production (0.626-0.890 mm). (E) Cyst. (Courtesy of C. S. Papp.)

A. Life Cycle

The life cycle of *H. schachtii* (Fig. 1) is essentially typical of all known species of the genus. In the presence of growing host plants, eggs hatch and larvae escape from protective cysts and invade host plant roots where they develop to maturity. Soon after fertilization the adult females can be seen as white lemon-shaped bodies attached to external surfaces of host roots. After a short period of growth the female dies and its cuticular remains form a tough saclike reddish brown cyst that protects the enclosed eggs from desiccation and, to a certain extent, from predators.

The following account of the life cycle of *H. schachtii* is summarized from Raski's (1949) extensive account of the life history and morphology of this nematode. The periods for development given are those that would be expected at or near the optimum temperature of 24°C.

The embryonic development of *H. schachtii* is typical of that of other nematode species. Soon after fertilization, the ovum undergoes cell division, which continues at a rapid pace. Cellular differentiation leads to development of an embryo, which increases in length, becomes more slender, and begins to move. Progressive development and elongation of the embryo results in a larva having two or three flexures. The first molt occurs within the egg soon after the embryo is fully grown and the cast-skin separates from the second-stage larva (L2). The fully developed L2 (Polychronopoulos 1969) may hatch in the presence of adequate soil moisture or become quiescent until hatching is initiated by root exudates containing a hatching stimulus. A comprehensive review of hatching of cyst nematodes is given by Shepherd (1962), and Clarke and Perry (1977) have reviewed the literature on the hatching phenomenon published since 1971.

After hatching and emerging from the cyst, the infective L2 (Fig. 1A) migrate to and penetrate nearby roots of a suitable host plant, move a short distance in the cortex and become sedentary parasites. Size ranges for larval stages and adults are listed in Table 1. At six to seven days after penetration the larva undergoes a second molt, and by the end of this molt, the sexes can be distinguished. Female L3 (Fig. 1B) develop two ovaries and males have a single testis. Male L3 enlarge and become flask shaped, whereas females are more slender. The third molt of male L3 begins 10-11 days after penetration, and the L4 (Fig. 1C) are fully developed and vermiform in appearance at 14 days when the fourth molt occurs. Adult males (Fig. 1D) begin to emerge from roots about 16 days after penetration. With low inoculum levels, the rate of emergence rapidly increases and is constant from about 21-43 days (Steele, 1973). L3 females (Fig. 1B) are fully grown 11-12 days after penetration, and after the third molt L4 females develop the typical flask shape. Females (Fig. 1C) undergo the fourth molt at about day 15 to 17 of development. After day 18 adult females (Fig. 1D) increase in size, grow to the typical lemon shape, and ova develop to fully formed eggs, which by day 30 completely fill the body cavity. Variable numbers of eggs are deposited in a gelatinous matrix or egg sac that is secreted from the posterior of the female. The sac remains attached to the female. The body wall of the female hardens, and is transformed into a reddish-brown cyst that may contain from a few to more than 600 eggs (Raski, 1949; Steele, 1972a). Soon after formation of the cyst the internal organ systems of the female collapse and disintegrate. The time required for development from penetration to cyst formation is highly variable and is dependent upon temperature and other environmental factors.

Table 1 Size Ranges for Larvae and Adults of *Heterodera schachtii*

Stage of development		Body length (mm)	Body width (mm)	Stylet length (μm)
L2[a]		0.435-0.492	0.021-0.022	25
L3	Males	0.324-0.377	0.057-0.085	16-19
L3	Females	0.320-0.366	0.044-0.070	22
L4	Males[b]	—	—	—
L4	Females	0.396-0.419	0.115-0.076	
Adult males		1.119-1.438	0.028-0.042	29
Adult females		0.626-0.890	0.361-0.494	27

[a]Sex of L2 is not distinguishable.
[b]Measurements of L4 males are not given.
Source: Data extracted from Raski (1949).

B. Reproduction and Sex Ratios

Males of *H. schachtii* are attracted to their females by a chemical(s) secreted by the female, which elicits klinokinetic aggregation behavior and klinotaxic orientation (Green, 1966). The attractant acts in solution and not as airborne gases (Green, 1967). Green concludes that in soil males move along gradients of attractants produced in the water film surrounding soil particles in close proximity to the female. A study of the male attractants emitted by 10 *Heterodera* species led researchers (Green and Plumb, 1970) to place *H. schachtii*, *H. glycines*, and *H. trifolii* within a subgeneric group. Females within this group are attractive to males of their own or other species within the group. Additional studies revealed the females secrete male attractants all over the body (Green and Greet, 1972) and the attractant includes volatile components that are able to diffuse through the air (Greet et al., 1968). *H. schachtii* males can inseminate up to 10 females. Green et al. (1970) suggested that multiple mating is advantageous in species in which the sex ratio ranges greatly and females often greatly outnumber males.

Impregnation of female *H. trifolii*, a triploid parthenogenic species, by *H. schachtii* males has been reported (Mulvey, 1958), but no males were observed in the F_1 progeny. Mulvey (1960) suggested that giant larvae observed in the progeny of mixed matings may have been polyploid progeny having 36 chromosomes, produced by *H. schachtii* sperm entering *H. trifolii* oocytes. Giant larvae of *H. schachtii*, however, were observed under conditions that precluded interspecific hybridization (Johnson and Viglierchio, 1970). The authors suggested "giantism" may be produced by environmental stress or by chemical additives.

H. schachtii and *H. glycines* are morphologically similar diploid (n = 9), amphimictic species (Triantaphyllou, 1973, Triantaphyllou and Hirschmann, 1962). Viable male and female F_1 progeny, which readily developed on table beet (*B. vulgaris* L.), were obtained by hybridizing *H. glycines* females and *H. schachtii* males (Potter and Fox, 1965). Reciprocal crosses were not attempted in these studies.

The number of males versus females developing on a host plant is influenced by the physiological condition of the host and the density of invading larvae. Molz (1920) suggested the sex of *H. schachtii* may be determined by the environment within host plant roots. Shifts in sex ratio were attributed to differential death rates (Senbusch, 1927), and numbers of maturing females correlated with increases in dead larvae in all stages of development. This suggests that the main factors influencing sex ratio are density of infections and thickness of roots (Kerstan, 1969). Another explanation offered for unbalanced sex ratios in the genus *Heterodera* is failure of female larvae to reach maturity under adverse conditions (Kerstan, 1969). Experiments showed development of male and female *H. schachtii* are disproportionately influenced by the nematode inoculum level and root size, which together determine the density of invading larvae, and differential population changes of host-selected *H. schachtii* races were attributed to selective development of males and females (Steele, 1975).

As early as 1888, Strubell (1888) and later Berliner and Busch (1914) reported that it was not unusual to find larvae in various stages of development external to the fibrous root surfaces of sugar beet and attached by their heads only (Fig. 2). Steele (1971c) found there was a wide variation in degree of larval penetration, ranging from completely endoparasitic to nearly completely ectoparasitic with only the cephalic region of the nematode buried in the root (Figs. 3 and 4). With the exception of a single L3 female, all of the sexually differentiated larvae external to the roots were males, but the number of ectoparasitic males never exceeded 10% of the total population parasitizing a given plant.

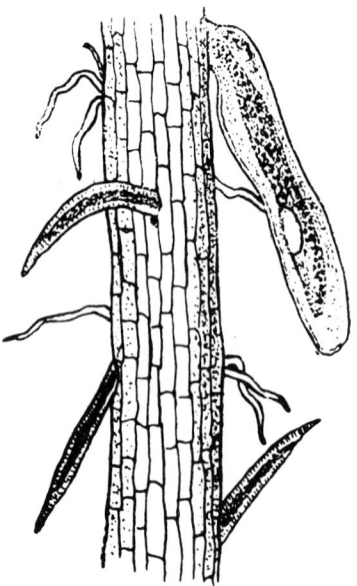

Fig. 2 Developing nematodes in various positions on sugar beet root (After Strubell, 1888.)

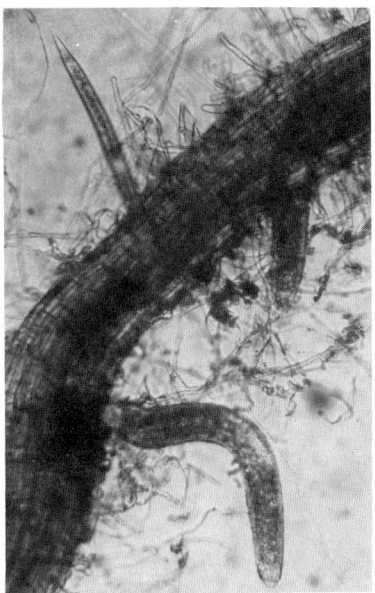

Fig. 3 Second-stage larvae and third-stage male of *H. schachtii* attached to sugar beet root.

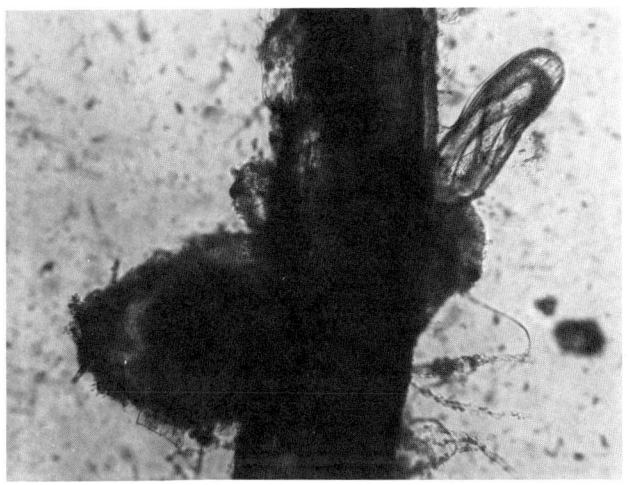

Fig. 4 Adult male within larval integuments (left) and adult female *H. schachtii* (right) on fibrous root of sugar beet.

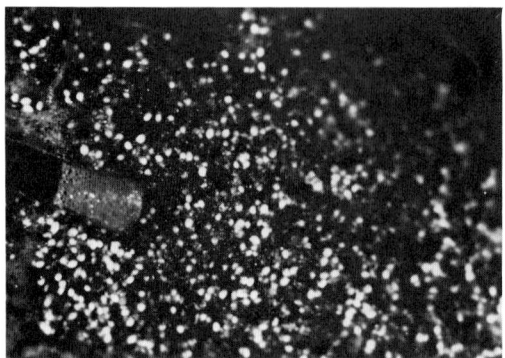

Fig. 5 *H. schachtii* parasitizing storage root of sugar beet. (Courtesy of L. R. Faulkner.)

Although *H. schachtii* normally parasitizes only the fibrous roots of sugar beet, development of a few to several hundred nematodes per square inch on the fleshy surfaces of storage roots of field-grown sugar beet has been reported (Fig. 5). Bergman (1958) reported formation of large multi-nucleate cells (syncytia) in the tissue of the secondary cortex under the cell layers of the storage root periderm. Although Steele (1972b) found that *H. schachtii* could be easily cultured on storage root slices of sugar beet, table beet (Fig. 6), radish, turnips, and rutabaga, he was unable to infect unin-jured storage root surfaces of these plants. He suggested the periderm might act as a barrier to nematode penetration and that destruction of peridermal tissue by fungi or bacteria, or both, may facilitate invasion of the deeper cortical tissues. Greco and Vovlas (1979) maintained that the nematode's feed-ing activity is confined to the periderm and the outer two layers of secondary xylem. They found the nematode invaded both storage roots and lateral root-lets equally and concluded the nematode did not have a preferred feeding

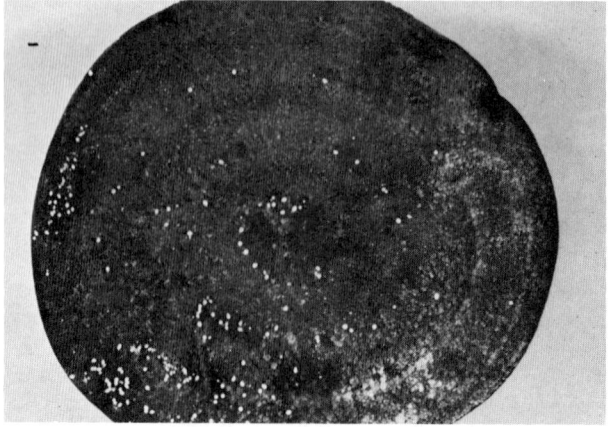

Fig. 6 Root slice of red table beet (*B. vulgaris* L.) 21 days after inoculation of larvae. White objects on slice are developing nematodes.

site. The populations used in these studies were investigated at Salinas, California, by this author. Although the nematode readily parasitized fibrous roots of greenhouse-grown sugar beet, no development occurred on storage roots.

C. Distribution

The sugar beet nematode has been found in at least 40 countries within North and South America, Europe, Africa, and the Middle East, and has been reported in the subcontinent of Australia, in New Zealand, and in Hawaii. Of these countries, 64% lie between 30° and 60° N latitude, 23% between 30° and 60° S latitude, and 13% in the equatorial region between 30° N latitude and 30° S latitude.

The means by which *H. schachtii* gained entry to the United States is not known; however, the nematode may have been introduced with imported sugar beet seed contaminated with infested soil (Shaw, 1915; Triffit, 1935).

Although damage to sugar beet by *H. schachtii* was first observed in the United States as early as 1895, the causal agent was not known until 1905, when E. G. Titus of the U.S. Bureau of Entomology discovered white females on roots of diseased plants (Thorne, 1949). According to Shaw (1915), Dr. Albert Schneider observed the nematode in a field of the Martin's Island ranch near Spreckels, California, in 1906, and his report probably constitutes the first record of *H. schachtii* in the United States. By 1907 the nematode was well established and caused severe losses in several beet-growing areas in California and Utah (Bessey, 1911), and in 1920, a total 31,238 ha in California, Utah, Colorado, and Idaho were infested (Thorne and Giddings, 1922). A report of a survey in 1958 (Caveness, 1958) listed 96 counties in beet-growing areas within 15 states infested. Although by 1925 crop rotations for control of *H. schachtii* were in widespread use (Thorne, 1949), analysis of available crop sequence data showed that sugar beet growers were following unsatisfactory rotations, and Caveness (1958) recommended four years of nonhost crops between crops of sugar beet. Although *H. schachtii* also occurs in the states of New York (Mai, 1961), Ohio (Partyka, 1969), and Florida (Rhoades, 1970), these states are not presently within the beet-growing areas. Today, in California alone, the total area of infested land in sugar beet growing areas totals 148,347 ha (Cooke and Thomason, 1978).

The vertical distribution of *H. schachtii* adult females and cysts was studied in three fields of sandy loam soil in which beets were growing (Thorne and Giddings, 1922). Nematodes were recovered from all samples taken in 5 cm increments to the maximum depth sampled of 76 cm. Vertical distribution of *H. schachtii* expressed as a percentage of the total population from 0-35 cm was as follows: 0-5 cm, 6.8; 6-10 cm, 30.6; 11-15 cm, 38.1; 16-20 cm, 15.4; 21-25 cm, 6.2; 26-30 cm, 2.6; and 31-35 cm, 0.2. Of all nematodes, 98% occurred in the soil profile from 0-35 cm depth and only 2% from 35-76 cm depth. Whitehead (1977), however, reported that in two fields of peaty loam soils following crops of sugar beet, *H. schachtii* populations were similar between the 20-40 cm and 0-20 cm soil depths, and in a third field highest populations were recovered at the 40-60 cm depth. The low populations in the upper soil levels were attributed to a drought that influenced nematode invasion and reproduction. A similar study established that vertical distribution of *H. schachtii* was influenced by the age of the infestation (Goffart, 1954). Nearly 83% of the cysts were found in the upper 10 cm of soil in a young infestation, whereas cysts were uniformly distributed throughout the vertical soil profile to a depth of 50 cm in an older infestation. In Germany, the maximum number of cysts are found in the upper 10 cm of soil when *Brassica rapa* and *Raphanus oleiferous* are sown at the end of July, and nearly the same numbers are reached in the 10-20 cm zone when plants are sown at the end of August (Thielemann, 1978). The greatest numbers of viable cysts were found at 30-40 cm depth, and other cysts were recovered from soil taken at a depth of 130-140 cm (Korab, 1929). It is apparent that soil type, moisture, temperature, age of the infestation, depth of cultivation, and perhaps the growth characteristics of various host-plant root systems are all factors that determine the vertical distribution of the sugar beet nematode.

D. Dissemination

In the absence of factors that produce directed movements of nematodes in soil the rate of spread by active migration probably varies from 1 to 2 mm (Wallace, 1963) to several centimeters per year. Larvae of *H. rostochiensis* can migrate up to 45 cm to reach host-plant roots and 40 cm in a temperature gradient (Rode, 1962). Dispersal of *Heterodera* within a field, or from field to field, is primarily by passive transportation. Nematode dispersal can occur when infested soil is moved or transferred in routine farming practices and by natural dissemination by birds, livestock, and flooding. Eight genera of plant-parasitic nematodes, including *Heterodera*, were recovered from irrigation water (Faulkner and Bolander, 1966) and from seed contaminated with infested soil (Epps, 1968; 1969). *Heterodera glycines* can survive in the digestive tracts of blackbirds (Epps, 1971) and swine (Smart, 1963), but similar experiments have not been attempted for *H. schachtii*. Manure of four of five steers fed on beet roots infected with *H. schachtii* contained viable larvae, which subsequently parasitized sugar beet seedlings (Kontaxis et al., 1971). Cysts of various *Heterodera* spp. are carried by wind (Chitwood, 1951; Kumaraswami, 1966; Meagher, 1977; Whitehead, 1977), and this may well be a primary means of rapid dissemination of the sugar beet nematode.

E. Persistence and Survival

The length of time *H. schachtii* can survive in field soil in the absence of host plants is not known. A small percentage of eggs within protective cysts can remain viable for more than six years under Utah conditions (Thorne, 1923, 1926). Microplot studies by the author (unpublished) demonstrated low populations of *H. schachtii* survived in fallowed soil 12 years after removal of sugar beet.

The annual rate of decline of *H. schachtii* under fallow or nonhost crops varies from 40 to 60% and depends upon climatic and edaphic factors influencing hatching and survival. Decay of populations in soil fallowed or cropped to nonhosts in England was about 20, 40, and 50% per annum for cysts, cysts with contents, and eggs, respectively (Jones, 1956), and decline was independent of population level (Jones, 1956; Moriarity, 1963). In England, eggs declined at the rate of 48% per annum in fallowed peaty clay soil (Moriarity, 1961), in Holland a decline rate of 38% for eggs under nonhost crops was observed. Hijner (1952) reported total cyst numbers declined 13% per annum during an eight year period. Under nonhost plants in Germany, populations declined by 60% the first year after sugar beet and 45% the following years (Kühn, 1877). With new cysts decline was negligible the first year, then about 30% thereafter. Egg numbers decreased 36% the second year and 60% per annum for subsequent years (Olthof et al., 1974).

Winslow (1956) was unable to observe seasonal decline in the hatchability of *H. schachtii*, whereas Oostenbrink (1967) concluded from his studies that there were clear and distinct influences of diapause, although diapausal suppression of hatching is not complete, and is broken by warmth in September to November. Spontaneous hatching of *H. schachtii* in the absence of host plants (Wallace, 1956c, 1958), however, results in a seasonal decline in populations in soil. The stimulus for cessation of hatching may derive from metabolites of soil biota rather than from physiological factors of the environment, but the causes of diapause have not been clearly established and may result from a number of interacting conditions that include physical, chemical, and biological factors.

Important factors affecting persistence and survival of *H. schachtii* are soil temperature and moisture, susceptibility and availability of host plants, soil type, and predators and parasites of the nematode.

Shepherd (1962) reported that air drying *H. schachtii* cysts greatly lowered hatching and emergence of larvae. Lowering the humidity from saturation to about 98% decreased hatching (Wallace, 1955). Viglierchio (1961) obtained good hatches from cysts stored one month at relative humidity of 92, 92.5, and 98%, but cysts stored at 84% RH showed much lower hatches. He also found that rapid drying or storing of cysts one month at 18°C reduced hatches to low numbers. Ellenby (1968) found that only 50% of second-stage larvae survived 30 minutes at 86% RH or 20 minutes at about 82% RH. Moriarity (1962) demonstrated that multiplication of *H. schachtii* was inhibited by dry conditions in sandy soil and suggested that irrigation of such soils increased nematode development.

Golden and Shafer (1960) found about half of the L2 larvae surviving after three months' storage in tap water at room temperatures (about 24°C) and none surviving after seven months of such storage. The L2 survived up to 10 months in soil.

Fuchs (1911) [cited by Triffit and Hurst (1935)] reported that prolonged heat retarded development of *H. schachtii*, whereas cyst contents were destroyed by exposures of 48 hours at 50°C, 24 hours at 55°C, or 6 hours at 60°C. In dry soil encysted L2 did not survive 10 minutes at 60°C, five minutes at 62°C, or one minute at 63°C. Steele (1973) exposed cysts of *H. schachtii* in a water bath at constant temperatures ranging from 45 to 62.5°C for one second to 28 hours. Within the temperature range of 49-54°C the minimum lethal temperature was proportional to the log time of treatment. No larvae survived when they were exposed 10 minutes to 60°C. Exposure of cysts for 8 hours at 45°C significantly reduced emergence, but increasing the treatment period to 28 hours did not completely suppress emergence of larvae.

Although soil moisture and temperature directly affect hatching, development, and survival of the nematode, these factors also affect nematode populations through their influence on plant growth and food availability (Santo and Bolander, 1979; Whitney and Doney, 1973b). Steele (1975) found that delaying inoculation of two cultivars of sugar beet by 30 days resulted in a 10-fold increase in the number of *H. schachtii* parasitizing the plants. Inoculation of 10-, 20-, and 30-day-old tomato plants with *H. schachtii* resulted in development of 13, 275, and 537 females per plant, respectively, and the number of females on tomato roots was proportional to plant weights 30 days after inoculation. Griffin (1981) observed root weight depression of sugar beet grown in soil infested with *H. schachtii* was inversely correlated with seedling planting age in weeks.

Nebel (1926) found that infection of *B. rapa* by *H. schachtii* larvae did not occur at 6-9.5°C, and only a few larvae invaded at 10-14°C. Maximum

invasion occurred at 18-20°C, the highest range tested. Jones (1975) reported
that plotting accumulated temperature above an assumed basal development
temperature of 4.4°C gave curves that closely resembled curves for development of *H. schachtii* on sugar beet. From studies carried out in controlled
temperature tanks, Raski and Johnson (1959) concluded that although *H.
schachtii* could develop over a wide temperature range (18-29.5°C) the optimum
range for development was relatively narrow (21-26.5°C). Similar studies by
Santo and Bolander (1979) showed the nematodes could develop at temperatures
as low as 16°C, and development at 24°C was significantly greater than at 21,
18, or 16°C. Johnson and Viglierchio (1969) observed the optimum temperature
for development of *H. schachtii* on axenic *B. vulgaris* root explants was 25°C,
and only males were observed in cultures maintained at 30°C. Thomason and
Fife (1962) found the maximum reproduction of *H. schachtii* at 27.5°C was
significantly greater than at 25°C. Griffin (1981) reported a correlation between sugar beet root weight loss in nematode-infested fields and soil temperature at planting within the range of 6-24°C.

Inoculation with 1000 larvae per plant delayed emergence of sugar beet
seedlings grown under monoxenic conditions in an environmental chamber for
one to two days. Ten thousand larvae caused a three to four day delay. During a day length of 12 hours, the chamber was maintained at 26°C and plants
received a light intensity of 2000 ft C, whereas at night temperatures were
held at 20°C (Polychronopoulos and Lownsbery, 1968).

The longevity of eggs and larvae within *Heterodera* cysts is also affected
by predators and parasites. In Canada, amoeboid organisms have been observed attacking larvae of *H. schachtii* (Winslow and Williams, 1957), and in
Utah, the predacious nematodes, *Mononchus papillatus* and *M. sigmaturus*, fed
on larvae and males (Thorne, 1927). Larvae of *H. schachtii* are destroyed by
enchytraeid larvae, which enter host plant roots (Schaerffenberg 1951;
Schaerffenberg and Tendl, 1951). The enchytraeid *Fridericia* spp. (Murphy
and Doncaster, 1957) and the collembolans *Orychiurus armatus, Isotoma viridis, Orchesella villosa*, and an *Achorutes* species feed on *H. cruciferae*. The
collembolan *Folsemia*, symphylids, the predacious mite *Pergamasus crossipes*,
larvae of several staphylinid beetles (*Philonthus, Omalium*, and *Trechus*),
and a centipede *Lithobius dubosqui* also feed on the contents of *Heterodera*
cysts (Doncaster, 1962).

Parasitic fungi probably cause the greatest natural decline in populations
of *Heterodera*. As early as 1877, Kühn (1877) reported that *Tarichium auxi-
liare* [redescribed by Tribe (1977a,b) as *Catenaria auxilieris*] parasitized females
of *H. schachtii*, and Korab (1927) later reported that eggs and larvae of this
nematode were destroyed by *Torula heteroderae*. The fungi *Cylindrocarpon
radicicola* (Goffart, 1932) and *Isaria destructor* (Rademacher and Schmidt,
1933) parasitize cysts of *H. schachtii*. *Trichosporium populenum* destroys
eggs and larvae and *Oepidium nematodae* and *Protomycopsis* spp. inhabit
H. schachtii cysts (Rozsypal, 1934). *Verticillium chlamydosporium* and
a nonsporulating "contortion fungus" are the principal egg pathogens of *H.
schachtii* (Bursnall and Tribe, 1974), and *H. schachtii* was one of six species
within the genus *Heterodera* experimentally infected with *Nematophthora gynophila* (Kerry and Crump, 1980). Tribe (1979) reported that some 14% of 112
populations of *H. schachtii* were diseased, but less than 50% of the diseased
cysts were parasitized by recognizable fungal pathogens. Where beet monoculture was practiced, overall disease was approximately doubled. The "black
yeast," the "crystal-forming fungus," and *Cylindrocarpon destructans* (Zinss-
meister) Scholten are minor egg pathogens of cyst nematodes, and *Glomus* sp.,

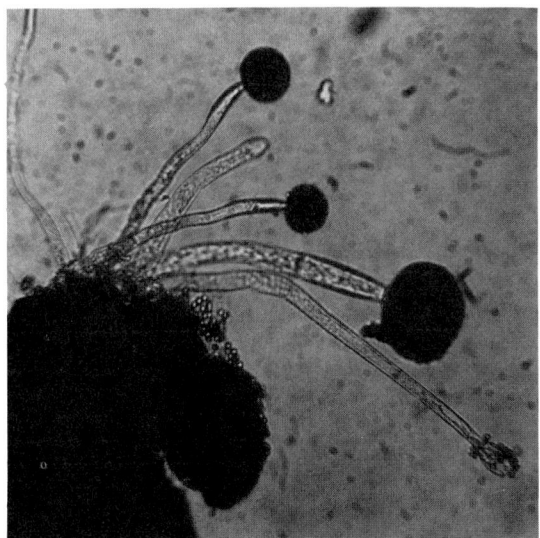

Fig. 7 Fungal sporangiophores with sporangia containing sporangiospores (probably *Rhizopus* sp.) protruding from cyst of *H. schachtii*.

Pythium spp., *Stachybotrys chararum*, *Fusarium*, *Phoma*, and *Penicillium* are found associated with cysts (Kerry and Crump, 1977). Various saprophytic fungi frequently inhabit cysts (Fig. 7).

The author has frequently observed large numbers of eggs and larvae in cysts from widely separated locations in the Salinas Valley parasitized by *F. oxysporum*, and on one occasion nematode-trapping fungi were observed feeding in vitro on larvae of this nematode. A culture of *H. schachtii* infested with a mycorrhizal fungus, *Glomus* sp. (Figs. 8 and 9), has been maintained several years on tomato. Although the fungus may be feeding on dead cysts rather than live larvae, hatching of larvae is greatly depressed from cysts heavily infested with the fungus. The cyst nematode, *Globodera solanacearum*, and the fungus, *Endogone gigantea* Nicol. and Gerd. [(*Gigaspora gigantea* Gerdemann and Trappe (1974)], mutually suppress reproduction of the other on tobacco (Fox and Spasoff, 1972).

Damage by *H. schachtii* to sugar beet is less in the presence of *F. oxysporum*, and this fungus decreases the numbers of nematodes developing to maturity to one-third (Jorgenson, 1970) of the original population. Eggs of *H. schachtii* on agar or in soil were parasitized by *F. oxysporum* and/or *Acremonium strictium*, or both, and both fungi grew saprophytically in eggs killed by heat (Nigh et al., 1980). In studies of natural infestations in separate fields in California, researchers found an apparent correlation between decline in *H. schachtii* egg viability and the number of eggs parasitized by *A. strictum* and *F. oxysporum* (Roberts et al., 1981). They suggested that these and other fungi may be a major cause of death of *H. schachtii* eggs.

The current interest in integrated approaches to pest management coupled with the recent loss of one of the most effective nematicides has directed attention to the employment of fungi as agents for biocontrol of nematodes. According to Mankau (1980), no fungal species has yet controlled nematodes to the degree achieved with nematicides. Kerry (1980), however, asserts that once fungal parasites are established in soil they are as effective as resistant cultivars or efficient nematicides in limiting nematode numbers.

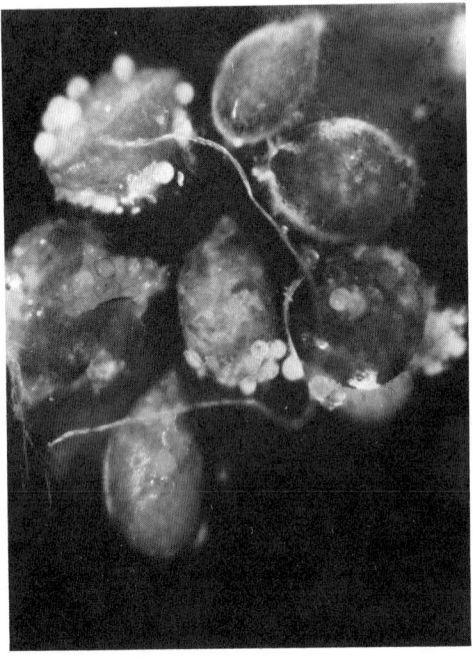

Fig. 8 *Glomus* sp. attached to external surface of cyst wall and within cysts of *H. schachtii*.

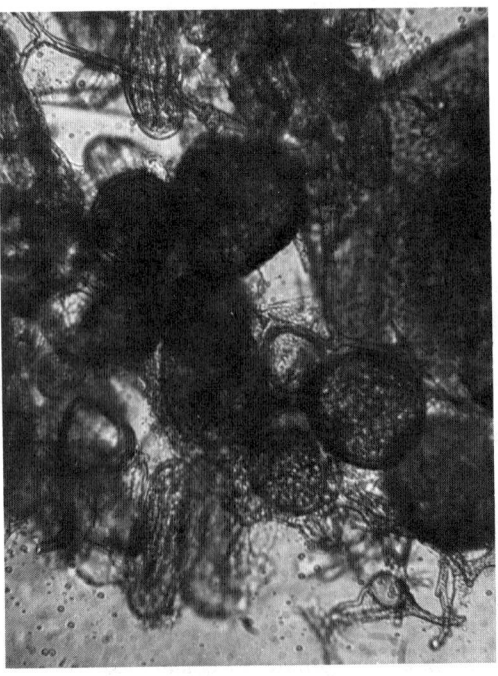

Fig. 9 Crushed cyst of *Heterodera schachtii* showing eggs and chlamydospores of *Glomus*.

Fig. 10 View of sugar beet field heavily infested with *Heterodera schachtii*, Salinas Valley, California.

Fig. 11 View of sugar beet field heavily infested with *H. schachtii*, Salinas Valley, California.

F. Field Symptoms and Yield Losses

The first indication of an infestation of *H. schachtii* in fields of sugar beet is usually the appearance of one or more well-defined circular to oval areas of reduced growth or poor stands. In irrigated soils poor growth may extend down and be restricted to one or more rows. Frequently, a single row of two row beds may show symptoms of nematode injury. Nematode damage is most severe in the seedling stage. In heavily infested soil seedlings may not emerge, emergence may be delayed (Polychronopoulos and Lownsbery, 1968), or plants may die after emergence (Figs. 10 and 11). In midseason the foliage of lightly infected plants may remain green or may show a lighter green color than is normal. Young sugar beet plants infected with *H. schachtii* may show a significant increase in petiole length that is correlated with increased size of spongy parenchymal cells of the petiole (Doney, Whitney, and Steele, 1970). Within heavily infested areas, beets often show irregular growth, with larger plants dark green in color and smaller plants varying from lighter shades of green to yellow. Plants that survive may remain stunted throughout the season. The inner leaves of older plants may remain green, but are undersized. The contrast between the dark inner leaves and the yellow-to-brown outer leaves is indicative of sugar beet nematode infestations. The outer leaves of infected plants typically wilt or droop during the hot period of the day in spite of adequate soil moisture (Fig. 12). In the evening, leaves regain their turgidity and assume an upright posture. The wilting phenomena can be attributed, at least in part, to pathological changes in the water-conducting elements within infected roots; however, extracts of tissues of infected sugarbeet and rape cause curling, rolling, discoloration, and folding of leaves of otherwise healthy sugar beets (Nolte and Kohler, 1952). Plants showing wilting symptoms may not survive severe moisture stress. Affected beets show excessive development of white, light brown, or dark brown fibrous roots (Fig. 13). These fibrous roots may or may not show slight swellings,

Fig. 12 View of severely wilted sugar beet plant growing in field heavily infested with *H. schachtii*.

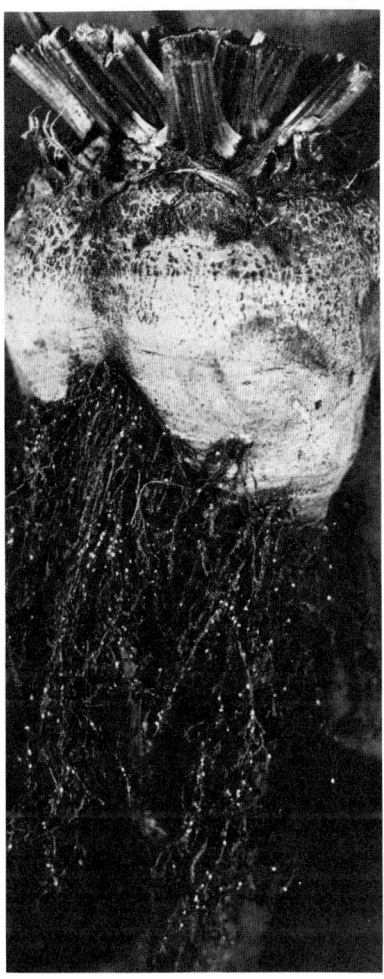

Fig. 13 Nematode-infected sugar beet showing typical proliferation of fibrous roots and malformed storage root. Small white bodies on the fibrous roots are female *H. schachtii*. (Courtesy of the U.S. Department of Agriculture.)

with localized lesions at sites of penetration of nematodes. Early attack by nematodes often causes severe branching or "sprangling" of the storage root.

Secondary invasions by pathogenic soil fungi may induce light to severe rotting of the storage root. Beets within an infested area of the field may be so severely stunted or damaged that machine harvesting is not possible.

Sugar beets grown in lightly infested soils may not show symptoms of nematode damage, and yields may closely approximate average yield for the area. At harvest, however, fibrous roots of these beets may be lined with adult females and cysts of *H. schachtii* and planting beets the following year could result in total loss of the crop.

The severity of damage to sugar beet by *H. schachtii* is dependent upon the nematode population density at time of seeding, and upon the interacting effects of soil moisture and temperature on both nematode and plant during the early weeks of growth. According to Seinhorst (1967b), Hellinga (1942) was the first to show a proportional relationship between cyst density at planting and yield loss of sugar beet. Preplant density in larvae per gram of soil on a logarithmic scale is related to sugar beet yields (Heijbroek, 1973; Johnson and Viglierchio, 1970). Such results were presented as straight regression lines by Jones (1956), but were transformed to partial sinusoids by Seinhorst (1965). Griffin (1981) found that sugar beet yields were inversely correlated with population densities at planting as measured by either larvae per gram soil (p = 0.01) or viable cysts per gram soil (p = 0.05). Experiments in the United States (Radewald et al., 1971) and in Germany (Steudel and Thielemann, 1970, 1979) demonstrated that early-planted beets were less severely damaged by *H. schachtii* than were late-planted beets. The tolerance limit (the level of the initial population below which damage is not measurable) when planting at the end of March or at the beginning of April was about 2000 eggs and larvae per 100 cm^3, and in May the tolerance limit was 250 per 100 cm^3.

The relation between sugar beet root dry weight, preplanting populations of *H. schachtii*, the sugarbeet tolerance limit, and soil temperature can be described in an equation (Cooke and Thomason, 1979). In greenhouse studies the tolerance limits in eggs per 100 g soil for various temperatures is 65 at 23 and 27°C and 430 at 19°C. In the Imperial Valley the tolerance limit is 100 eggs per 100 g soil.

When sugar beet is grown at varying initial population densities, the final population tends to rise to a "ceiling," which varies with soil and season (Jones, 1956). The equilibrium density of *H. schachtii* is reported as 200 eggs and larvae per gram of soil (Seinhorst, 1967b). More than 15,000 larvae per 100 cm^3 of soil is rarely found when sugar beet is grown in alternate years, and populations vary between 5000 and 10,000 larvae per 100 cm^3 under continuous cropping of sugar beet (Stelter, 1976).

In Germany (Thielemann and Steudel, 1973), the total numbers of cysts increase continuously during 10 years of monoculture of sugar beet, whereas the numbers of eggs and larvae increase during the first year only, and stabilize at 3000 larvae per milliliter soil. Continuous cropping of sugar beet does not result in further decline in yields of beets or sugar.

Marketable yields of rutabagas (Swedes), cabbage, table beet, and spinach, but not cauliflower, are inversely correlated with preplant population densities of *H. schachtii* (Lear et al., 1966). At a density of 18 larvae per gram soil, losses in marketable weight yields are: rutabagas, 35%; cabbage, 24%; table beet, 30% and spinach, about 40%. In contrast to these findings, other research has shown that row treatment of 30 gal/A of 1,3-dichloropropenes and related C_3 hydrocarbons (1,3-D) (Telone) increases quality and yields of cauliflower by nearly 86% of that of untreated controls (Olthof et al., 1974). Red table beet appears to be only slightly affected by *H. schachtii*. Preplant population densities in this experiment were not given; however, in untreated plots, damage to cabbage, mustard, and Swiss chard was so severe that there were not marketable yields, whereas broccoli yields in the nematicide-treated plots increased 186%. Total marketable yields of table beet and cabbage (direct-seeded and transplanted), but not sweet corn (a nonhost), were inversely correlated with population density of *H. schachtii* at planting (Abawi and Mai, 1980). Initial densities as low as six to nine viable eggs and larvae per gram soil decreased marketable yields of table beets and cabbage, whereas the highest density of 68 per gram decreased marketable yields of direct seeded cabbage and table beet by 51.5 and 53.3%, respectively.

The number of generations of *H. schachtii* that complete development in a season depends upon the growing period of the host crop and soil moisture and temperature. In England the number of generations may vary from two to three (Whitehead, 1977) and in The Netherlands, Germany (Nejad and Dern, 1979), and the coastal valleys of California from one to two. The higher soil temperatures of the Imperial Valley of California permit completion of five generations during the growing season (Thomason and Fife, 1962); however, hatching, penetration, and development are greatly influenced by conditions such as fertilizer and irrigation, which affect production of new plant rootlets. Thus, the rate of cyst production during the growing season under field conditions can vary from year to year. During some years adult precystic females are present on host plant roots throughout the growing season. In other years this stage may not be observed in a given field at the time selected for examination of plant roots.

G. Interactions

Nematodes act in concert with pathogenic viruses, bacteria, and fungi in disease complexes producing greater damage to plants than the single action of either component. The effects produced by these associations may be either additive or multiplicative. From results of investigations of the effects of fungicides, S-ethyl [3-(dimethylamino) propyl] carbamothioate monohydrochloride (prothiocarb) (Previcur) + the insecticide dimethyl (3-[(dimethoxyphosphinyl)oxy]-2-2-pentendioate (CG 3707) (Bomyl) on control of unspecified soil fungi and the insecticide-nematicide 2-methyl-2-(methylthio) propionaldehyde O-(methyl carbamoyl) oxime aldicarb on *H. schachtii* it was concluded that these pathogens in combination produced a negative synergistic effect on growth of sugar beet (Muller, 1980).

Infection with *H. schachtii* can aggravate yield losses of sugar beet due to leaf spot (caused by *Cercospora beticola*) or virus yellows (*Beta* virus 4) (Weischer and Steudel, 1972). Sugar beet yields were reduced by the interacting effects of *H. schachtii* and *Rhizoctonia solani* (Price and Schneider, 1965). The separate and combined effects of the pathogens on percentage loss were: fungus alone, 12.4; nematode, 26.9; and fungus and nematode, 45.8. Polychronopoulos et al. (1969) investigated the penetration and development of *R. solani* in sugar beet seedlings infected with *H. schachtii*. They found that the fungus enters wounds made by the nematode and by lateral roots. There are more lateral roots on nematode-infected plants. Syncytia were more favorable substrates for the fungus than normal cells. Necrosis was synergistically increased by the presence of both pathogens.

Death of sugar beet seedlings results from high population levels of *H. schachtii* and high levels of *Aphanomyces cochlioides* (Whitney and Doney, 1973a). Although *Pythium ultimum* and *H. schachtii* readily produce a synergistic effect in pre- and postemergence damping-off of sugar beet, the complex only rarely produces this effect on later root rot of sugar beet. The effect of *P. aphanidermatum* and the nematode is additive for both damping-off and root rot (Whitney, 1974).

Although mixed populations of *H. schachtii* and other plant-parasitic nematodes sometimes occur (Caveness, 1959), the only interrelationship that has been studied extensively is that between *H. schachtii* and *Meloidogyne hapla*. Simultaneous inoculation of these nematodes does not alter population dynamics of either parasite (Jatala and Jensen, 1972; 1976a). When inoculation of sugar beets with *H. schachtii* precedes inoculation with *M. hapla*, the production of root-knot galls is reduced. This reduction was interpreted as the manifestation of an amensal relationship, with *M. hapla* the amensal and *H. schachtii* the inhibitor. Conversely, when inoculation with *M. hapla* precedes *H. schachtii* there is a significant increase (three to five fold) in numbers of cysts formed. The presence of both parasites on a single feeding site resulted in formation of two distinctive pathological tissues typical of both nematodes (Jatala and Jensen, 1976b). In most infections xylem elements separate the two pathological tissues, but occasionally they are separated by a single cell wall. Because discrete infection foci were maintained, *M. hapla* and *H. schachtii* must be considered as competitors for available feeding sites, with *M. hapla* being the weaker antagonist.

The combination of *M. hapla* and *H. schachtii* significantly reduced tomato root weights and top growth below that of single inoculations of either pathogen, but *H. schachtii* did not increase galls produced by *M. hapla* (Griffin, 1980a).

H. The Hatching Phenomenon

Baunacke (1922) was the first to discover that eggs of *H. schachtii* were stimulated to hatch by sugar beet root exudates. Germinating sugar beet seed (Golden and Shafer, 1959b) and leaves of vigorously growing sugar beet (Golden, 1958b) also give off a substance or substances [hatch factor(s)] that stimulate hatching and emergence of larvae from cysts of *H. schachtii*. Triffit (1930) demonstrated that low levels of hatch occurred in water. Later it was shown that hatches of *H. schachtii* in water vary from about 10 to 40% (Shepherd, 1962).

Hatching is readily stimulated by most host plants, and a number of non-host plants, such as *Hesperis matronalis* (dames violet), *Matthiola incuna* (stock), and *Coronopus squamatus* (swine cress) have active diffusates (Ouden, 1956a). Certain hosts, however, i.e., *Lycopersicon esculentum* (tomato) and *Linaria vulgaris* (butter and eggs), do not stimulate hatching (Golden and Shafer, 1959a).

The concentration of hatch factors and activity of root exudates vary within wide limits. Maximum production of the hatch factor occurs during periods of optimum root growth (about three to six weeks for sugar beet). Short days (eight to ten hours) favor production of more active sugar beet diffusate than do long days (14-16 hours), and temperatures of 15.6-21.1°C are superior to 9.4-26.7°C, suggesting activity is related to root growth (Winslow and Ludwig, 1957). A 10-fold dilution of sugar beet root diffusate does not decrease activity (Steele and Fife, 1964, 1969). Cumulative hatch of *H. schachtii* is proportional to log concentration of diffusate until a maximum hatch is reached. With concentrations above optimum, hatch declines (Steele, 1971a). Activity of sugar beet root diffusate is not affected by freezing or drying, is lost only slowly by boiling, and more rapidly by heating at 15 lb pressure. The loss in activity is proportional to log time of treatment (Steele, 1973). Concentration of diffusate 50-fold by vacuum distillation does not result in loss of activity, and the active principle is dialyzable with a diffusion rate between those of inorganic salts and compounds with molecular weights greater than 15,000 (Steele and Fife, 1969).

When *H. schachtii* cysts were placed in sand held at a constant suction of 15 cm at various distances from growing cress plants, the rate of larval emergence decreased proportionately with distance from the plants, until at a distance of 4.5 cm the rate of emergence was similar to emergence in the absence of plants (Shepherd and Wallace, 1959). The rate of larval emergence in coarse sand decreases with depth, but the relationship is not linear and may be related to O_2 diffusion through sand (Wallace, 1956a). Soil moisture relations probably exert their greatest influence through their effect on aeration, for little hatching occurs at suctions greater or less than 15 cm. Hatching in vitro increases with increasing oxygen concentration (Wallace, 1955). Hatching of *H. schachtii* occurs at field capacity when soil pores are empty of water except where particles are in contact (Wallace, 1954).

A number of inorganic and organic chemicals inhibit hatching of *H. schachtii*. Other compounds in concentrations of 1-10 mM/liter stimulate hatching, and a few materials (anhydrotetronic acid, picrolonic acid, nicotinic acid, and aminobenzoic acid) are more active than sugar beet root diffusate (Clarke and Shepherd, 1964). Clarke and Shepherd (1965, 1966) tested 26 inorganic salts for hatching activity and found that 21 stimulated and 5 inhibited hatching. Salts showing the best activity in decreasing order are: zinc chloride, cadmium chloride, zinc sulfate, and zinc nitrate.

Extracts of certain fungal fruiting bodies stimulate hatching, whereas others inhibit hatching of *H. schachtii* (Vinduška, 1973). Bacteria isolated from the rhizospheres of host plants produce metabolites attractive to the sugar beet nematode (Bergman and Van Duuren, 1959). The fungicide disodium ethylenebis[dithiocarbamate]nabam (Steele, 1961) and the herbicide (S-(2,3-Dichlorallyl-diisopropylthiocarbamate) (diallate) (Kraus and Sikora, 1980) increase hatching, whereas hatching is inhibited by the herbicide 5-amino-4-chloro-2-phenyl-3-(2H)-pyridazinone (pyridazon) (chloridazon).

The optimum concentration of most organic and inorganic chemicals for inducing hatching of *H. schachtii* is equivalent to 10^{-2} M sodium chloride (Wallace, 1956b). The optimum osmotic pressure of sugar beet root diffusate obtained by diluting diffusate 1/16 was 0.48 atm. At 0.6 M concentrations, no hatching occurs, whereas cysts exposed for five weeks to molar solutions of urea or sodium chloride recover completely. Immersion of cysts in 3-4 M solutions, however, is lethal to *H. schachtii* (Wallace, 1956b). Thirty percent sucrose solutions completely suppress hatching of *H. schachtii*, but hatching resumes when cysts are later transferred to diffusate, and pretreatment of cysts for 96 hours with 60% sucrose is not lethal (Steele, 1962). Wallace (1963) suggests that plant nematodes can tolerate osmotic pressures up to 10 atm. Beyond this level activity may be inhibited although the nematodes may not be killed. He notes that the osmotic pressure of soil solutions in most agricultural soils rarely exceeds 2 atm, even at the wilting point of plants.

Steiner (1952) suggests that pH may be unimportant to nematode ecology, and Wallace (1963) observes that evidence for the influence of pH on plant nematodes is contradictory. In vitro hatch of *H. schachtii* occurs between pH 5 and 8 (author's unpublished results). Although pH 3.0 sometimes stimulates hatch, repeated experiments gave radically different results and proportional relationships between pH and hatching numbers could not be established.

I. Host Range

Unlike most of the *Heterodera* species, *H. schachtii* has a rather wide host range that includes weeds, cultivated vegetable, and field crops, and ornamentals occurring in 23 plant families (Steele, 1965). Although at least 218 plant species within 95 genera are hosts, experimental infections vary from a few to many nematodes per plant. Agriculturally important host species are listed in Table 2.

Many nonhost plants are readily invaded by *H. schachtii* larvae (Golden, 1958a; Golden and Schafer, 1959a; Steele, 1971b), and occasional development to maturity of *H. schachtii* on plants that are not ordinarily hosts has been

Table 2 Partial List of Cultivated and Weed Plant Hosts of *Heterodera schachtii*

Latin name	Common name or cultivar
Amaranthaceae	
Amaranthus blitoides S. Wats.	Tumbling pigweed
A. graecizans L.	Tumbleweed
A. retroflexus L.	Redroot pigweed
A. tricolor L	Josephs-coat
Capparidaceae	
Cleome spinosa Jacq.	
var. *rosea*	Giant pink queen
Caryophyllaceae	
Dianthus barbatus L.	Large flowered sweet william
D. deltoides L.	Maiden pink
Dianthus sp.	Carnation
Dianthus spp.	Pinks
Stellaria media L. Vill.	Common chickweed
Chenopodiaceae	
Atriplex confertifolia	Sheep-fat spiny saltbush
A. confertifolia (Torr. & Fum.) S. Wats.	Shadscale
A. hortensis L. var. *rubra* Moq.	Garden orache
A. lentiformis	Quailbush
A. polycarpa	Cattle spinach
Beta atriplicifolia	
B. atriplicifolia Rouy × *Beta vulgaris* L. (Fa)	
B. corolliflora Zoss.	Wild beet
B. intermedia Bunge	Wild beet
B. lomatogona Fisch. and Meyers	Wild beet
B. macrocarpa Guss.	Wild beet
B. macrorrhiza	Wild beet
B. maritima L.	Wild beet
B. patula Ait.	Wild beet
B. patula Soland	Wild beet
B. trigyna Wald. et Kitt.	Wild beet
B. vulgaris L.	Mangel
B. vulgaris L.	Red table beet
B. vulgaris L.	Spinach beet
B. vulgaris L.	Sugar beet
B. vulgaris L.	Swiss chard
B. vulgaris L. var. *cicla* L.	Leaf beet
Chenopodium album L.	Fat hen
C. album L.	Lambsquarters
C. album L.	White pigweed or goosefoot

Table 2 (Continued)

Latin name	Common name or cultivar
C. amaranticolor	
C. ambrosioides var. chilensis (Schrad.) Spegaz.	
C. bonus henricus L.	Good-King-Henry
C. capitatum (L.) Aschers	Strawberry-blite
C. ficifolium Sm.	Fig-leaved goosefoot
C. glaucum L.	Oak-leaved goosefoot
C. hybridum L.	Sowbane
C. murale L.	Nettle-leaved goosefoot
C. murale L.	Sowbane
C. polyspermum L.	Many-seeded goosefoot
C. rubrum L.	Red goosefoot
C. schraderianum Roem and Schult.	
Chenopodium sp.	Pigweed
C. urbicum L.	
C. vulvaria L.	Stinking goosefoot
Obione portulacoides (L.) Moq.	Sea purslane
Spinacia glabra Mill.	Summer spinach
S. oleracea L.	Spinach
Spinacia oleracea L.	Winter spinach
Cruciferae	
Alyssum alpestre	
A. maritimum (L.) Lam.	Sweet alyssum var. carpet of snow
A. montanum	
A. saxatile L.	Yellow saxatile compactum alyssum
Alyssum sp.	Sweet alyssum
A. spinosum	
Arabis arenosa Scop.	
A. caucasia Willd.	Garden arabis
A. muralis Bertol	
A. turrita L.	
A. verna R. Br.	
Armoracia lapathifolia Gilib.	Horseradish
Barbarea vulgaris R. Br.	Yellow-rocket
Brassica campestris L.	Common mustard
B. campestris L.	Wild turnip
B. caulorapa Pasq.	Kohlrabi
B. juncea	So. gt. curled mustard
B. napobrassica Mill.	Rutabaga
B. napus L.	Coleseed or rape
B. napus L.	Turnip
B. nigra (L.) Koch.	Black mustard
B. oleracea L.	Broccoli
B. oleracea L.	Brussels sprouts
B. oleracea L.	Cabbage

Table 2   (Continued)

Latin name	Common name or cultivar
B. oleracea L.	Red cabbage
B. oleracea L.	Cauliflower
B. oleracea L.	Collards
B. oleracea L.	Kale
B. oleracea L.	Kohlrabi
B. pekinensis (Lour.)Rupr.	Chinese cabbage
B. rapa L.	Turnip
Capsella bursa-pastoris L. Medic.	Shepherds-purse
Cardaria pubescens (C.S. Mey) Roll.	Hoarycress
Cheiranthus allionii Hort.	Siberian wallflower
Cochlearia amoracia L.	Horseradish
C. officinalis L.	Scurvy grass
Erysimum capitatum	Wallflower
E. insulare var. grandifolum	Wallflower
Iberis umbellata L.	Globe candytuft
Lepidium sativum L.	Garden cress
Lunaria annua L.	Honesty
L. biennis	Money plant
Malcolmia maritima (L.) R. Br.	Virginia stock
Nasturtium microphyllum (Boenn.) Reichenb.	One-rowed watercress
N. officinale R. Br.	Watercress
Raphanus maritimus Sm.	Sea radish
R. raphanistrum L.	Wild radish
R. sativus L.	Radish
R. sativus L.	Wild radish
Rorippa amphibia (L.) Besser	Great yellowcress
R. islandica (Oeder) Borkas	Marsh yellowcress
R. nasturtium	True watercress
Sinapis alba L.	White mustard
Sisymbrium irio L.	London rocket
Thlaspi arvense L.	Field pennycress
Labiatae	
Galeopsis speciosa Mill.	Large flowered hemp-nettle
G. tetrahit agg.	Common hemp-nettle
Leguminosae	
Glycine max (L.) Merr.	Soybean cv. Earlyana
Lespedeza stipulacea Maxim.	Korean lespedeza cv. Auburn
Medicago hispida Gaertn.	Burclover
Sesbania exaltata (Raf.) Rydb.	
S. macrocarpa	Sesbania
Phytolaccaceae	
Phytolacca americana L.	Pokeweed

Table 2 (Continued)

Latin name	Common name or cultivar
Polygonaceae	
Polygonum convolvulus L.	
P. pensylvanicum L.	
Polygonum sp.	Smartweed
Rheum rhaponticum L.	Rhubarb
Rumex acetosella L.	
R. alpinus L.	Monks-rhubarb
R. confertus	
R. crispus L.	Curled dock
R. hydrolapathum Huds.	Great water dock
R. maritimus	
R. obtusifolius L.	Broad-leaved dock
R. palustris Sm.	Marsh dock
R. patientia L.	
R. pulcher L.	Fiddle dock
R. sanguineus L.	Red-veined dock
Portulacaceae	
Portulaca oleracea L.	Purslane
Primulaceae	
Anagallis arvensis L.	Scarlet pimpernel
Resedaceae	
Reseda odorata	Red goliath mignonette
Scrophulariaceae	
Chaenorrhinum minus (L.) Lange	Small toadflax
Collinsia heterophylla	
Linaria vulgaris Hill	Butter-&-eggs
Solanaceae	
Lycopersicon esculentum Mill.	Pearson A-1 tomato
L. esculentum Mill.	Pearson XL tomato
Physalis sp.	Yellow pear tomato
Salpiglossis sinuata	Painted tongue
Solanum douglasii	Nightshade
S. nigrum L.	Black nightshade
Salpiglossis sinuata	Painted tongue
Solanum douglasii	Nightshade
S. nigrum L.	Black nightshade
Tropaeolaceae	
Tropaeolum pereginum	Canary nasturtium
Tropaeolum spp.	Calif. giants mixed nasturtium

Table 2   (Continued)

Latin name	Common name or cultivar
Umbelliferae	
Anethum graveolens L.	Dill
Conium maculatum L.	Poison hemlock
Pastinaca sativa L.	Holly crown parsnip
Urticaceae	
Urtica gracilis Ait.	Stinging nettle

Source: Information extracted from Steele (1965).

reported (Steele, 1971b; Golden and Shafer, 1959a). Infrequent development of the sugar beet nematode on highly resistant crop and weed species could maintain localized infestations at low levels that become detectable only after continuous cropping of susceptible host plants.

There is increasing evidence of the existence of biotypes (some authors prefer the terms *race* or *pathotype*) of *H. schachtii* that differ either qualitatively in host specificity or quantitatively in population dynamics on susceptible host plants. Steele (1964) isolated a population that increased 194.6-fold on tomato over that of a population transferred from sugar beet to tomato. Similar results with isolates of *H. schachtii* and tomato have been reported by others (Graney and Miller, 1980; Griffin, 1980b; Lear and Miyagawa, 1972). These isolates of *H. schachtii* are pathogens of tomato (Griffin, 1980; Lear and Miyagawa, 1972). Resistance of tomato (*Lycopersicon esculentum*) and wild tomato (*L. peruvianum*) varies with nematode and plant biotypes (Steele, 1977a) and is correlated with resistance of tomato to *Rotylenchulus reniformis* (Rebois et al., 1977). Selections of pigweed (*Chenopodium album*) differ in susceptibility to *H. schachtii* (Griffin, 1980b; Golden and Shafer, 1958). Isolates of *H. schachtii* differ in ability to develop egg-bearing females on certain cultivated and weed plants (Griffin, 1980b; Graney and Miller, 1980).

J. Histopathology of Susceptible Sugar Beet

Nemec (1911; 1933) was the first to describe the anatomical changes in fibrous roots of *H. schachtii*-infected sugar beet. Reports of his thorough investigations have been repeatedly consulted by those concerned with the pathology of diseases caused by *Heterodera*. The discussions that follow are based in part upon Nemec's contributions.

Although the sugar beet nematode may invade at any point along the length of fibrous roots, preferred sites of entry are the zone of elongation behind growing root tips and at wounds or ruptures produced by emerging lateral roots. After a short intercellular migration, the L2 larvae take up a position within the cortex, usually parallel to the root axis, with their anterior ends directed toward the stele and lip regions pressed upon the endodermis (Fig. 14). The larvae orient themselves with anterior ends toward the hypocotyl or the root tip (Steele, 1971c). Larvae may remain semiendoparasitic and are often wrapped around the root. L2 and L4 males are frequently observed external to the root attached by their head only.

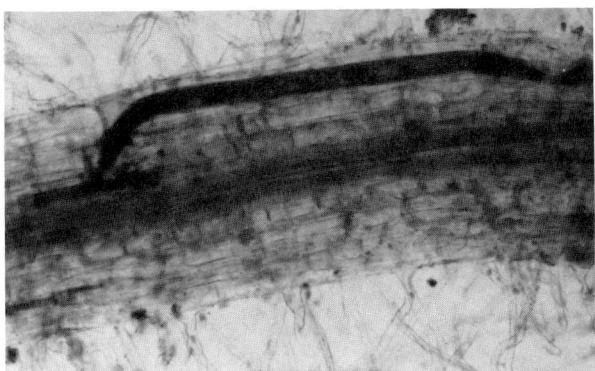

Fig. 14 Unstained whole root of sugar beet with second stage larvae of *H. schachtii*. Note small syncytia at anterior end of nematode.

Soon after the larvae establish a feeding site, usually adjacent to the protoxylem poles, host cells increase in size and contain finely granulated, densely strained cytoplasm and enlarged nuclei and nucleoli. Dissolution of cell walls is evident within a few days after penetration. Transverse walls disappear between cells to a greater extent than do longitudinal walls. The cytoplasm, initially interspersed with vacuoles, becomes densely granular and increases in volume, eventually joining cytoplasm of other cells to form a single mass that fills the entire space of the syncytial unit (Fig. 15). In young tap roots expansion of the syncytium results in the arrested formation of metaxylem and metaphloem. Wall boundaries of syncytia become thickened, especially in areas proximal to the nematode and in regions bordering xylem elements, and typically stain densely with fast green (Fig. 16). The multi-

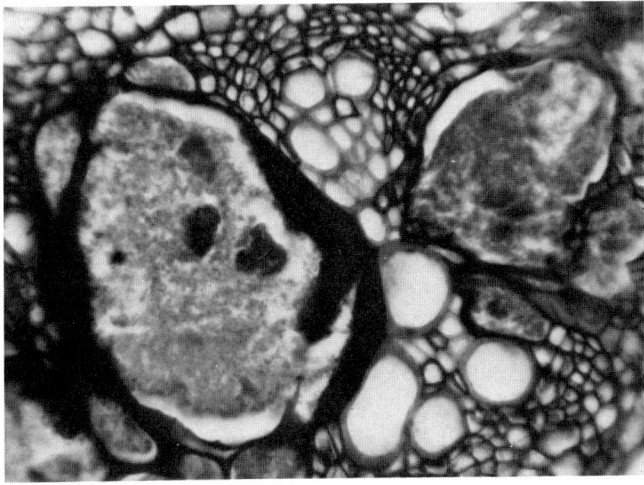

Fig. 15 Transverse section through sugar beet root infected with *H. schachtii*, 20 days after inoculation. Note multinucleate cell, thickening of cell walls, and displacement of xylem.

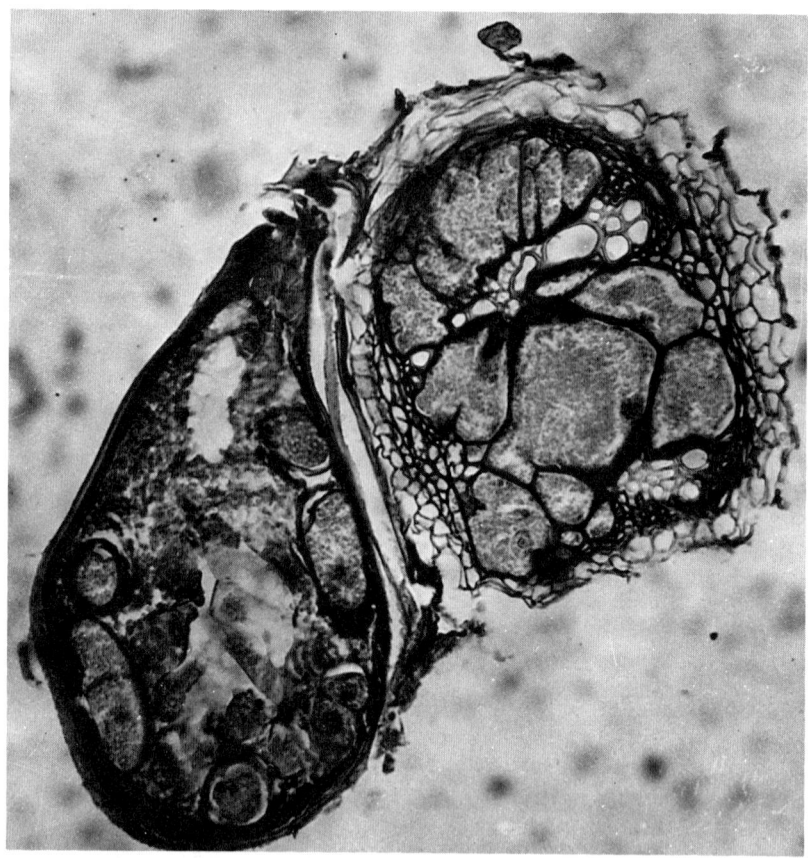

Fig. 16 Transverse section through sugar beet 25 days after inoculation with *H. schachtii*. Female nematode attached to left side of root. Note dense granular cytoplasm of syncytia and thickening of cell walls adjacent to xylem.

nucleate condition of syncytia results from loss of cellular integrity through breakdown of cell walls rather than from nuclear division as occurs in giant cells associated with *Meloidogyne*. As the syncytium expands, it progressively incorporates parenchyma cells distal to the feeding site. In young tap roots the syncytium interrupts the cambial ring for some distance and secondary xylem and phloem may fail to develop in the affected areas. The protoxylem is compressed between giant cells in heavy infections. The entire cross-sectional root area may be occupied by syncytia, which results in missing, crushed, or displaced xylem vessels. Although the phloem does not undergo radical changes, the number of phloem elements may be reduced or phloem may be displaced from usual positions. Syncytia may extend for a distance of 1-2 mm along the root axis. As the syncytium ages, the cytoplasm becomes turbid and heavily stained. Cessation of feeding by the adult nematode is concomitant with deterioration and collapse of the syncytial unit, and areas vacated by the receding syncytia are invaded by rejuvenative parenchymatous tissue. Syncytia sustaining male larvae are comparatively small in size and are frequently limited to tissues of the cortex or pericycle.

Syncytia are considered by some researchers to be examples of transfer cells induced in the presence of a metabolic sink (Jones and Northcoat, 1972). According to this view, the parasite continuously removes and utilizes metabolites for growth and development and induces modified cells (the syncytia) that have a specialized function of massive short-distance transport of solutes. These transfer cells are thought to be transitory and nonpathologic and are sustained only as long as there is a continuous drain of metabolites. Indeed, it has been shown that fibrous roots parasitized by *H. schachtii* contain significantly increased amounts of total amino acids, aspartic acid, glutamic acid, and glutamine (Doney, Fife, and Whitney, 1970). However, there is considerable evidence that nematode-induced transfer cells exhibit both morphologic and physiologic differences from naturally occurring transfer cells frequently found in plants, and plants apparently have several biochemical mechanisms operating separately and concurrently to provide resistance to nematode parasitism (Gommers, 1981). Leaf secretions of nematode-infected plants, however, induce wilting symptoms in otherwise healthy sugar beet (Ouden, 1956b). Although it is not known whether the wilting factor is of plant or nematode origin, tolerance to wilting in sugar beet appears to be genetically controlled (Heijbroek et al., 1977).

K. Control

Perhaps the first method of control of *H. schachtii* was devised by Kühn in 1881 (Kühn, 1881). His method consisted of repeated planting of rape between rows of beet (three to four times in a season). The rape was then destroyed after nematodes had hatched and invaded the roots and before there was sufficient time for larvae to complete development. Instead of rape, Fuchs (1911) used white mustard as a trap crop and Baunacke (1922) recommended planting beets, turnip, or *Barbarea*, which were plowed under eight days later. With the latter method nematode hatching was stimulated, but the crop was destroyed before nematodes could invade host-plant roots. Hijner (1952) found that growing the resistant wild beet, *Beta patellaris*, for two months reduced nematode populations by 50%, whereas a growing period of five months resulted in a decline of 90%. Nebel (1926) reported that sugar beets were more heavily invaded after use of trap crops. Under Polish conditions sowing of early cabbage and radish resulted in mean declines of 51 and 48%, respectively, in populations of *H. schachtii* (M. W. Brzeski, personal communication). The use of trap crops is not recommended today because its effectiveness is too variable and unpredictable. Unforeseen problems could delay destruction of the trap crop and result in increased rather than decreased populations. In addition, productivity of land is lost during implementation of trap cropping.

In the early years sugar beets were cropped continuously in Europe, often 8 to 10 years. In Germany, in 1901, Spiegler devised and employed a method where sugar beets were rotated with nonhost crops for different lengths of time depending on the degree of nematode infestation. Shaw (1915) adapted Spiegler's rotations for use in the widely scattered beet-growing areas of the United States. Today, crop rotation has worldwide acceptance as the most practical, economical means of obtaining profitable yields on nematode-infested land.

In nematode-infested fields, recommended rotational schemes may permit growing sugar beet once in three to seven years, depending on severity of infestation and local conditions that influence population dynamics. For the Imperial Valley of California, three years out of beets is considered a minimum rotation in an infested field, and in noninfested fields, beets can be grown

Fig. 17 View of experimental test plots infested with *H. schachtii* and treated with chlorinated hydrocarbon nematicides, then planted to sugar beet. Note two nontreated plots showing nearly complete loss of stand six weeks after planting.

consecutively two years in five. In Germany, the average yield of sugar beet in a rotation with beets every third year was 15% less than the yield obtained with rotations where beets were planted every eighth year (White, 1953).

The effects of rotations are enhanced by planting sugar beet as early as possible when low soil temperatures greatly reduce rates of nematode hatching, migration, and invasion. Under these conditions young beets can acquire a well-established root system that can withstand later attack by nematodes.

The halogenated aliphatic hydrocarbons, particularly those with a high content of 1,3-dichloropropene, will effectively control the sugar beet nematode (Jorgenson, 1969a; Kontaxis and Thomason, 1978; Lear et al., 1966, 1961; Radewald et al., 1971; Steele et al., 1971) (Figs. 17, 18). Consistent performance of the fumigant nematicides is difficult to achieve (Whitehead, 1973), however, because of the influences of factors such as depth of application,

Fig. 18 View of experimental plots shown in Fig. 17. Nontreated plot is to the left of the treated plot.

soil temperature and moisture, soil type, compaction, and organic matter content of soil at time of application. Although fumigants are usually applied as preplanting treatments in the spring, fall applications of 1,2-dichloropropane-1,3 dichloropropene mixtures have given good control of *H. schachtii* in Colorado (Altman and Fitzgerald, 1960; Altman and Thomason, 1971). The recommended minimum interval between treating with fumigant nematicides and planting is one week for each 15.3 liters/ha of nematicide. In light soils of low soil moisture and high soil temperatures, however, the interval between treatment and planting can be shortened to achieve the added advantage of early planting (Jorgenson, 1969b).

In recent years oximecarbamates and organophosphates have been tested extensively under laboratory, greenhouse, and field conditions for control of *Heterodera*. Research has shown that these nonfumigant materials act in several ways to control nematodes. Oximecarbamates and carbofuran (2,3-dihydro-2,2-dimethyl-7-benzofuranyl methycarbamate) inhibit hatching of *H. schachtii*, but this action is reversed by removing the chemicals; suppression of hatching by organophosphates, however, is irreversible (Steele, 1977b). Aqueous solutions of 1 g/ml aldicarb [2-methyl-2-(methylthio) propionaldehyde O-(methylcarbamoyl)oxime] partially reduce hatching of *H. schachtii* and 5 μg/ml completely suppress hatching; while 10 μg/ml was found to adversely affect movement of larvae (Steudel, 1972). Concentrations of 5 μg/ml or greater of aldicarb or aldicarb sulfoxide [aldicarb + 1,2-(methylenedioxy)-4-[2-(octylsulfinyl)propyl]benzene] inhibit hatching, but hatching resumes after removal of the inhibitor and exposure to a hatching agent (Steele and Hodges, 1975). Aldicarb sulfone (aldoxycarb) [2-methyl-2-methylsulfonyl) propanal O-[(methylamine)carbonyl]oxime] does not affect hatching. Control of L2 is proportional to the concentration and duration of treatment with aldicarb, and the material acts systemically in sugar beet to suppress development of parasitic larvae (Steele and Hodges, 1975).

Systemic levels of toxic carbamates (aldicarb and aldicarb sulfoxide) that suppress development of *H. schachtii* females and males are 0.35 and 0.8 g/μg of root tissue, respectively (Steele, 1979). Aldoxycarb and oxamyl [methyl N^1,N^1-dimethyl-N-[methylcarbamoyl)oxy]-1-thioamimidate] do not have systemic activity (Steele, 1976). Aldicarb sulfoxide and aldoxycarb affect coordinated movement of *H. schachtii* larvae, but aldicarb oxime sulfoxide does not affect movement (Batterby, 1979). Male *H. schachtii* exposed to 0.01 μg/ml aldicarb fail to migrate toward nubile females (Hough and Thomason, 1975).

Steudel et al. (1978) and Thieleman and Steudel (1973) investigated the effects of aldicarb on population dynamics of *H. schachtii* on sugar beet during a nine year period of monoculture. They found that aldicarb retarded development of the nematode during the first months after application, but that the treatment had no effect on final population density, which stabilized at 3000 eggs and larvae per 100 cm^3 of soil. After plowing, nematode population levels were similar for treated and untreated plots. Aldicarb increased sugar beet yields by 7%, foliage by 5%, and sugar content by 0.11%. More importantly, yields of roots and sugar showed no tendency to decline over the nine year period.

Hatching of *H. schachtii* is depressed 88% by 4-8 μg/ml phenamiphos [ethyl (methylthio)-*m*-tolyl isopropylphosphoramidate] (Greco and Thomason, 1980), and foliar applications of this material or oxamyl retard development of the nematode (Griffin, 1975). In a field experiment at least 5 kg/ha of phenamiphos was required to reduce infection by *H. schachtii*, and higher rates did not increase control (Greco and Thomason, 1980). Pretransplant drench and foliar sprays of oxamyl (DPX 1410) reduce numbers of *H. schachtii* invading or developing in roots of cabbage (Potter and Marks, 1971).

Treatments combining soil fumigants (1,3-D) with systemic nematicides (aldicarb or carbofuran) have increased yields of sugar beet grown in heavily infested soil by as much as 266% (Kontaxis and Thomason, 1978).

Resistance to plant pests and disease has long been considered the preferred method of crop protection. Early attempts to detect resistance or tolerance to *H. schachtii* in *B. vulgaris* were unsuccessful (Hülsenberg, 1935; Hulsfeld, 1926; Molz, 1917). More recent attempts to find resistance (Curtis, 1970; Doney and Whitney, 1970; Finkner and Swink, 1956) or tolerance (Rietberg, 1954; Swink, 1954) have not yielded genetically transferable genes. Tolerance to wilt caused by *H. schachtii* appears to be controlled by complete and incomplete dominant genes (Heijbroek et al., 1977). The apparent absence of genes for resistance to *H. schachtii* in *B. vulgaris* has directed the attention of researchers to wild *Beta* spp. Extensive research has demonstrated that of 13 wild *Beta* spp. investigated, only *B. procumbens*, *B. patellaris*, and *B. webbiana* are resistant (Golden, 1958a; Golden, 1959; Hijner, 1952; Shepherd, 1957; Steele and Savitsky, 1962, 1974). All resistant species have similar morphologic characters and are members of the taxonomic section *Patellares* (Transhel, 1927).

Bergman (1951, 1958) investigated the penetration and development of *H. schachtii* in roots of resistant *B. patellaris* and *Hesperis matronalis* and subsequent histopathology. He found that although L2 sometimes penetrated roots of these plants in large numbers, they did not develop beyond the L3 stage. Necrosis of *B. patellaris* roots occurred soon after penetration, and larvae often induced a restricted formation of giant cells, whereas no giant cells were initiated in *H. matronalis*. Bergman concluded that failure of *H. schachtii* to develop in *B. patellaris* was due to an inability of the nematode to initiate syncytia formation or to maintain them once they are formed.

Crosses between *B. vulgaris* and three resistant *Beta* spp. were accomplished by Savitsky (1975, 1978; Savitsky and Price, 1965), and she continued her investigations of the cytogenetics and resistance to *H. schachtii* in one of these, *B. vulgaris* × *B. procumbens*.

Tetraploid sugar beet (*B. vulgaris*) was first crossed with *B. procumbens*, which served as the pollinator. The F_1 progeny of this cross were triploids with 27 chromosomes (9 from *B. procumbens*). Pollination of these triploid hybrids with sugar beet results in viable trisomic progeny having 19 chromosomes (18 from *B. vulgaris* and 1 bearing the gene for resistance from *B. procumbens*). A gene for resistance was incorporated into a chromosome of sugar beet by crossover transfer of a segment of a chromosome from *B. procumbens*. In the transfer, dominant genes linked to the gene for resistance and controlling undesired features such as bolting, narrow leaves (Figs. 19 and 20), and tumor formation were deleted. Recently, Savitsky (1980) reported that she obtained two F_3 self-sterile hybrids that transmitted genes for resistance to all of their progeny.

Steele and Savitsky (1974, 1981) investigated the resistance of trisomic and diploid hybrids of *B. vulgaris* and *B. procumbens* to *H. schachtii*. Resistance in these hybrids is not due to failure of larvae to enter roots, but is due to failure of larvae to reach maturity. Although the hybrids exhibited different degrees of resistance to populations of the nematode, the differences were extremely small and not due to selections of "resistance-breaking" biotypes. Greater numbers of males than females matured on root slice cultures of *H. schachtii* on resistant hybrids. Examination of the histopathology of

Fig. 19 Diploid (left) and trisomic (right) hybrids of *B. vulgaris* L. and *B. procumbens* Chrys. Sm. resistant to *H. schachtii*.

Fig. 20 Leaf blades and petioles of diploid (A) and trisomic (B) hybrids of *B. vulgaris* and *B. procumbens*, *B. procumbens* (C) is shown for comparison.

roots of resistant hybrids (Yu and Steele, 1981) revealed the presence of necrotic lesions at sites of larval penetration as early as four days after inoculation. Often lesions were accompanied by slight swelling of the roots. Phase illumination of plant sections showed crystalline cytoplasmic granules in forming syncytia within the stele. Dead L2 (Fig. 21), living L4 males (Fig. 22), and dead female larvae (Figs. 23 and 24) were frequently observed within roots 10-15 days after inoculation, which showed extensive necrosis and disorganization of stelar tissues.

After 10 days, necrosis of infected root tissues became more severe and resulted in the total collapse of syncytia. Syncytial cytoplasm eventually became turbid and heavily stained. Although rejuvenated parenchymatous tissues often invaded cavities formed by receding boundaries of syncytia, frequently the entire stele was obscured by heavily stained necrotic tissue (Fig. 25). The severity of the hypersensitive reaction of the hybrids appeared to depend upon the number of larvae invading a given root locus,

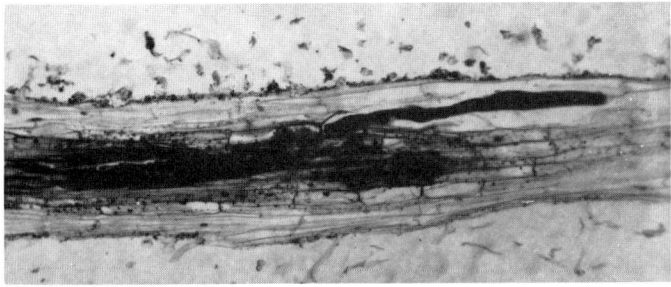

Fig. 21 Longitudinal section through root of interspecific hybrid of *B. vulgaris* and *B. procumbens* 10 days after inoculation showing dead second-stage larvae and necrotic area.

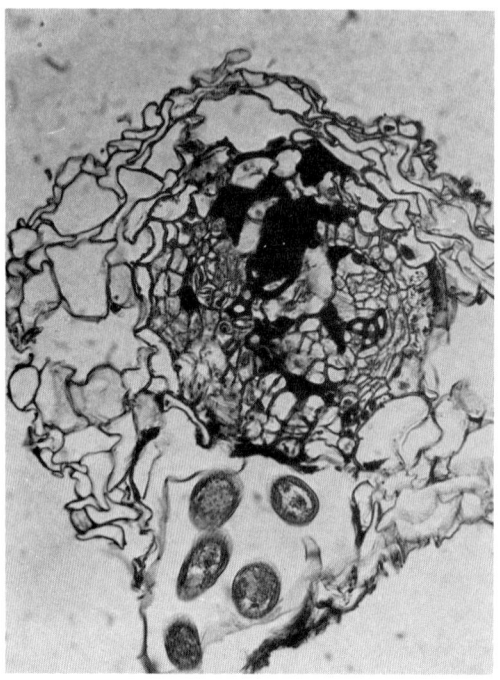

Fig. 22 Transverse section through root of resistant hybrids of *B. vulgaris* and *B. procumbens* 15 days after inoculation with *H. schachtii*. Note necrotic (dark stained) areas, disorganization of root tissues, and fifth-stage male nematode to the left of the stele.

the sex of developing larvae, and the length of time between initiation of syncytial formation and death of larvae leading to disintegration of giant cells. The occurrence of severe hypersensitivity of resistant hybrids to nematode invasion suggests that the growth, and hence, the yields of the hybrids, may be drastically reduced when the hybrids are grown in fields heavily infested with *H. schachtii*. Consequently, it is likely that growing resistant hybrids in nematode-infested areas will at least initially require the integration of rotational and chemical methods of control. It should be noted that neither *B. procumbens* nor hybrids of *B. vulgaris* and *B. procumbens* are resistant to species of *Meloidogyne* that attack sugar beet.

Historically, the development of resistance to species within the genus *Heterodera* has been followed soon by the appearance of resistance-breaking biotypes. As yet there is no evidence that biotypes of *H. schachtii* may develop to reduce resistance transferred from *B. procumbens* to sugar beet; however, resistant beets have not been planted in widely scattered beet-growing areas and natural populations have not been subjected to selection pressures imposed by intensive cropping of the resistant hybrids. Whether populations of *H. schachtii* contain sufficient genetic diversity to enable development of biotypes has yet to be determined.

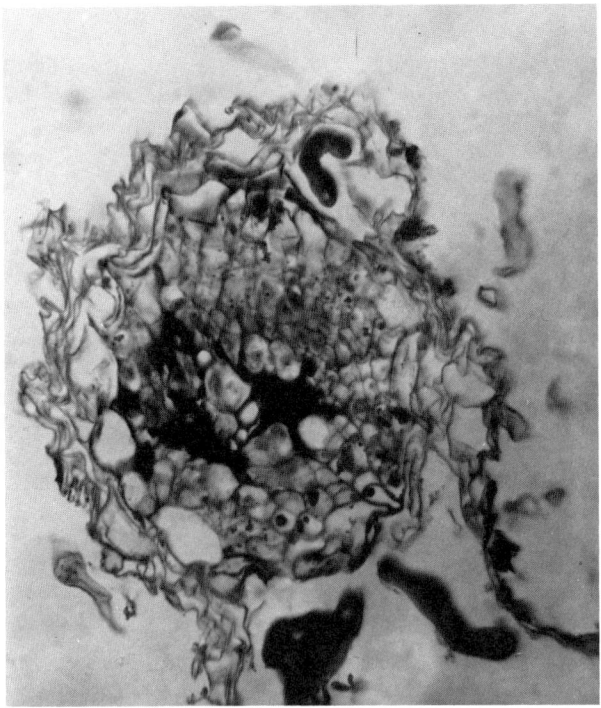

Fig. 23 Transverse section through root of resistant hybrid 10 days after inoculation with *H. schachtii*. Note the presence of three shrunken, irregularly shaped female larvae in the periphery of the root.

Fig. 24 Whole root of resistant interspecific hybrid showing dead female, necrosis of syncytium, and displaced xylem.

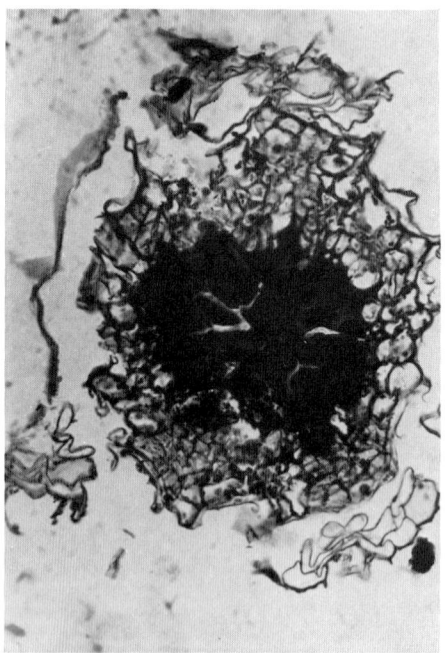

Fig. 25 Transverse section through root of resistant hybrid of *B. vulgaris* and *B. procumbens* 25 days after inoculation with *H. schachtii*. Note almost total necrosis of stelar area and absence of xylem vessels.

III. CLOVER CYST NEMATODE

After 1970 the cultivation of sugar beet in southeastern Holland was increased, which, except for rather small areas of reclaimed peat subsoils and alluvial loams, is almost entirely alluvial sandy soil. Before 1970 the clover cyst nematode, *Heterodera trifolii* Goffart 1937, was known to be widespread in this area.

Within a few years after the increase of the area cropped to sugar beet, yield losses became apparent in fields in which the beets appeared to be severely damaged by nematodes. Routine sampling revealed high populations of *H. trifolii* in many of the fields, and in 1975 it was discovered that beets were infected by a cyst nematode which had an intermediate yellow stage. The nematode was subsequently identified as a pathotype of *H. trifolii*. Soon after, Willem Heijbroek of the Institute for Sugarbeet Research at Bergen op Zoom and Paul Maas of the Plant Protection Service in Wageningen initiated studies of the distribution, biology, and host range of the "yellow beet cyst nematode (YBCN)" pathotype. The following is a summary of their findings.

The pathotype is widespread throughout the area, with 238 fields infested. Of these fields, 24 were also infested with *H. schachtii*. Twenty-one fields were infested with *H. schachtii* only. Although the infested area is of alluvial sandy soil, greenhouse studies have shown the nematode will multiply on beets grown in heavy clay soils. As with other species of the *H. trifolii* complex (*H. galeopsidis* and *H. lespedezae*), no males have been observed, and it is presumed the YBCN reproduces parthenogenetically. If all invading larvae become egg-producing females, populations would be expected to increase and spread more rapidly than the amphimictic species *H. schachtii*. In tests where larvae of *H. schachtii* or the YBCN were inoculated on sugar beet seedlings, newly formed cysts contained 238 and 292 eggs and larvae, respectively, and the multiplication rate for the YBCN was 92×, nearly three times that of the *H. schachtii*. In a field trial, end-of-season counts of 98 eggs and larvae per milliliter of soil were accompanied by reduction in stand count of 30%, increased numbers of branched roots, and at-harvest root weights were decreased by 35%. In another experiment, inoculation of 18 larvae per milliliter of soil decreased sugar beet root yields by 35%, and populations of the YBCN increased by 4.6 times during the seven month test period.

Heijbroek and Maas also investigated the penetration and development of the YBCN and the sugar beet nematode on selected legume crops. Inoculation of 800 larvae per plant gave the following mean numbers of YBCN per gram of root: Pea, 91; bean, 21; white clover, 120; red clover, 48; oil radish, 20; and red beet, 165. A mean count of 61 YBCN cysts per plant was found on Perfection, the only pea cultivar on which cysts developed, whereas a mean of 23 cysts per plant were recovered for 13 bean cultivars inoculated with the

Fig. 26 Views of a field of sugar beet heavily parasitized by a biotype of *H. trifolii* Goffart.

Fig. 27 View of a field of sugar beet infected with *H. trifolii* and showing severe wilting.

YBCN. White clover averaged seven cysts per plant and oil radish and red clover one cyst per plant, whereas red beet averaged 54 cysts per plant.

Although eight pea cultivars were invaded by a mean of three *H. schachtii* larvae per plant, no cysts developed. Cysts developed on six of ten bean cultivars inoculated with the sugar beet nematode, whereas no cysts developed on white or red clover. Ten cysts per plant developed on oil radish, and 100 cysts per plant developed on red beet.

Of interest is the finding that in Hawaii, *H. trifolii* heavily parasitized white clover (*Trifolium repens*), Swiss chard (*Beta vulgaris* var. *cicla*), and spinach (*Spinacea oleracea*), but only one female was recovered from a single plant of Early Wonder table beet (*B. vulgaris*) (Holtzmann and Aragaki, 1963). Swiss chard, beet, and spinach are efficient hosts of *H. schachtii*.

Five populations of the *H. trifolii* complex were evaluated for morphological differences and host preferences. The host plant appeared to influence larval length, position of the excretory pore, median valve, and genital primordium. No reliable differences between the populations were observed for body width, lip region, stylet knobs, lateral fields, shape, underbridge, bullae, and shape of fenestra of cysts.

Although the optimum temperature of hatching was 25°C for both YBCN and *H. schachtii*, only the latter hatched at temperatures as low as 10°C. At 25°C, hatching of both nematodes was about 80% in cole-seed leachate or picric acid. An additional test revealed that about 90% of the larvae hatched in picric acid with no seasonal effects evident. Increased hatching in tap water occurred from YBCN cysts collected in May and December, however, compared with hatches from cysts collected in August and March.

High populations of the YBCN biotype produce severe damage, including stand loss, uneven growth, yellowing of outer leaves, and severe wilting (Fig. 26), even when there is adequate soil moisture (Figs. 27 and 28). Infected beets lack well-formed storage roots, and root systems are branched with excessive growth of fibrous roots. High populations throughout the field can result in a total loss of the sugar beet crop.

Fig. 28 Sugar beets infected with *H. trifolii* showing discoloration and severe wilting.

Current recommendations in The Netherlands on planting sugar beet in infested fields, based upon population levels of viable eggs and/or larvae per 100 cm^3 of soil, are as follows: less than 100, sugar beets can be grown with no waiting period; 105-400, beets may not be planted for one year; with an infestation of 405-1000, beets may not be grown for one to two years without application of a chemical nematicide. Very heavy infestations of greater than 1000 preclude planting beets for two to four years if nematicides are applied.

IV. ROOT-KNOT NEMATODES

In 1885, Professor A. B. Frank reported that a number of cultivated plants in Germany, including *B. vulgaris*, were attacked by root-knot nematodes (*Meloidogyne* spp.). In 1911 Bessey included beets in a list of hosts of the root-knot nematode in the United States (Bessey, 1911). Of some 50 described species of *Meloidogyne*, only a few parasitize sugar beet, and fewer still are of economic importance to sugar beet production. Important species include *M. arenaria, M. incognita, M. javanica,* and *M. hapla*. The wild beet species, *B. patellaris, B. procumbens,* and *B. webbiana,* are highly resistant to *H. schachtii,* but are susceptible to all of the aforementioned *Meloidogyne* species (Golden, 1959). *M. hapla* is widely distributed in beet-producing areas of the United States (Caveness, 1959), but this species does not appear to be a major factor in sugar beet production. A recently described species, *M. chitwoodi,* shows a greater affinity for sugar beet than *M. hapla,* but this root-knot nematode is limited to the vicinity of the Columbia River in Washington, Oregon, and Idaho (Santo et al., 1980). *Meloidogyne naasi* can severely damage sugar beet in the seedling stage, and in Belgium, heavy infestations can reduce sugar beet yield by nearly 60% and sugar content by 13% (Weischer and Steudel, 1972). In England, infection of sugar beet is rare (Franklin, 1965), and in The Netherlands sugar beet is a poor host for *M. naasi* (Maas and Maenhout, 1978). Although *M. naasi* has been reported from Illinois and Kansas, and occurs in 13 widely scattered counties in California, apparently this nematode does not damage sugar beet in the United States. Species

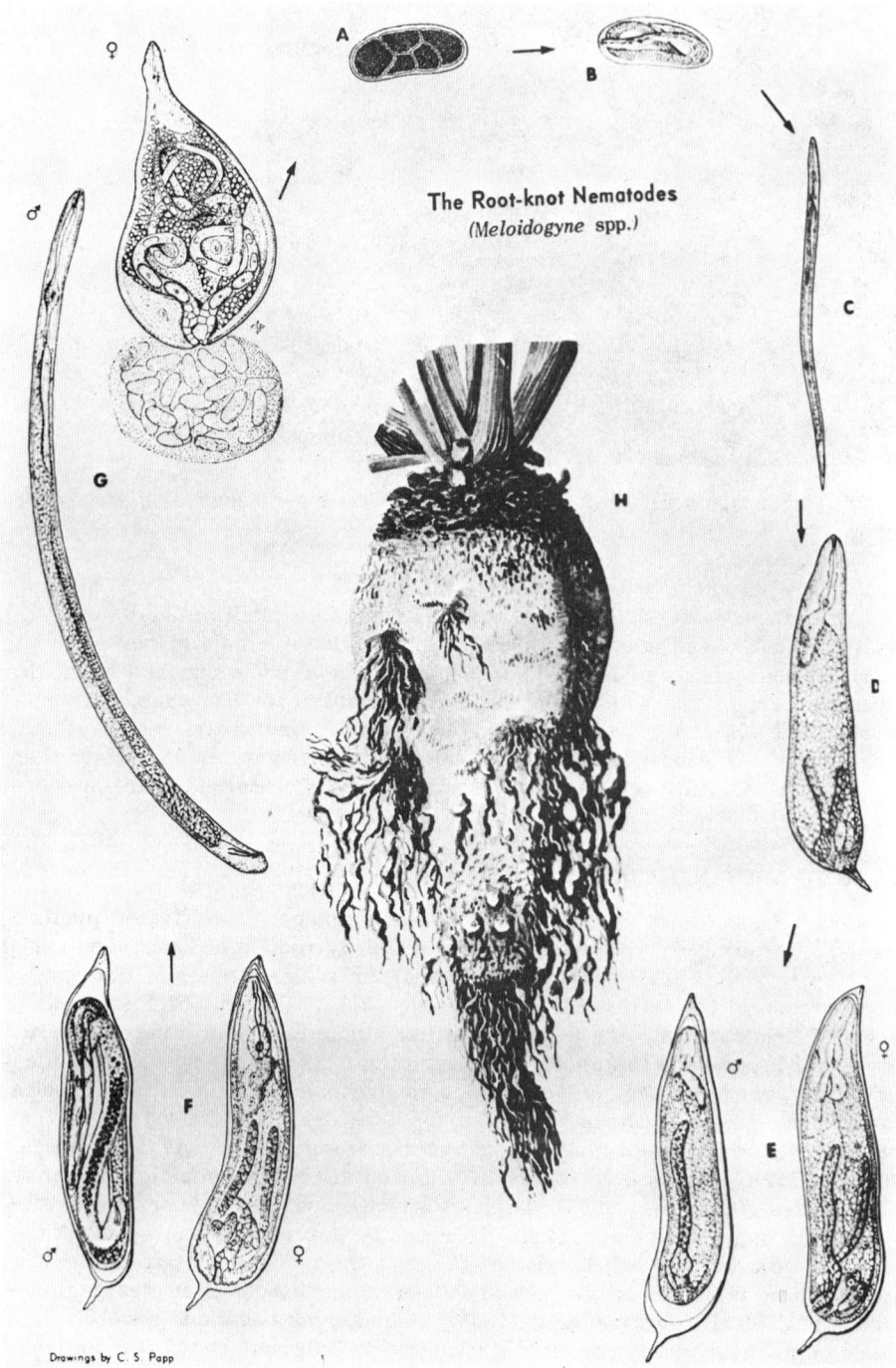

Fig. 29 The life cycle of the root-knot nematode, *Meloidogyne* spp.: (A, B) egg; (C) second-stage larvae; (D) spike-tail stage; (E) fourth-stage male and female; (F) early fifth-stage male and female; (G) mature adult male and female; (H) sugar beet with root galls caused by *Meloidogyne*. (Courtesy of C. S. Papp.)

causing the greatest losses of sugar beet in the United States are *M. incog-nita* and *M. javanica*.

The life cycle of the root-knot nematode is illustrated in Fig. 29. As with other nematodes, development is marked by the occurrence of four molts and five stages. After molting within the egg, L2 hatch and invade host-plant roots, where at optimum temperatures development is completed in 20-25 days, each female producing as many as 1000 eggs. In warmer climates four or five generations may develop in a single growing season.

Field symptoms produced by root-knot nematodes are similar to those of *H. schachtii*. As with *H. schachtii*, damage by *Meloidogyne* spp. may appear as localized areas of plants with chlorotic leaves, or plants may be stunted or missing. Alternatively, damage may extend throughout the entire field and severely affected plants may wilt and collapse in warm dry climates. Although *Meloidogyne* spp. can attack beets grown in any type of soil, highest popula-tions of the nematode and most severe damage occur in well-drained coarse-textured soils.

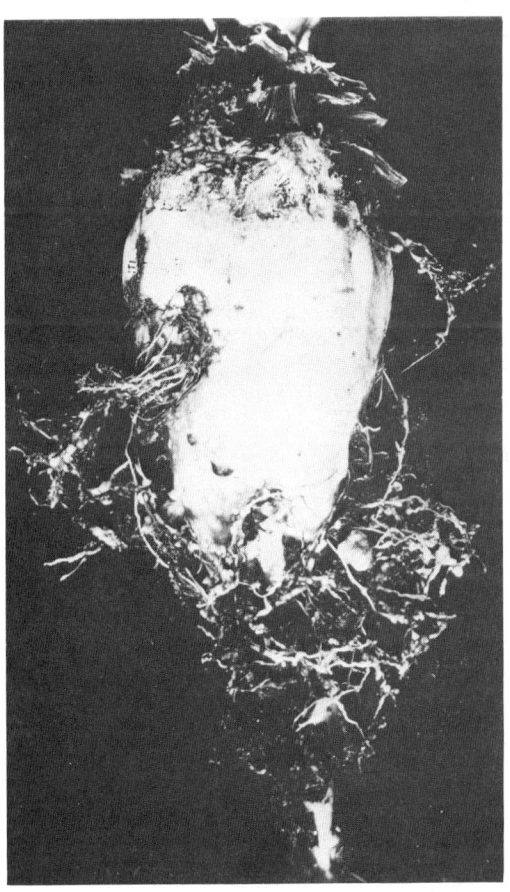

Fig. 30 Galls on roots of sugar beet infected with *Meloidogyne javanica*. (Courtesy of I. J. Thomason).

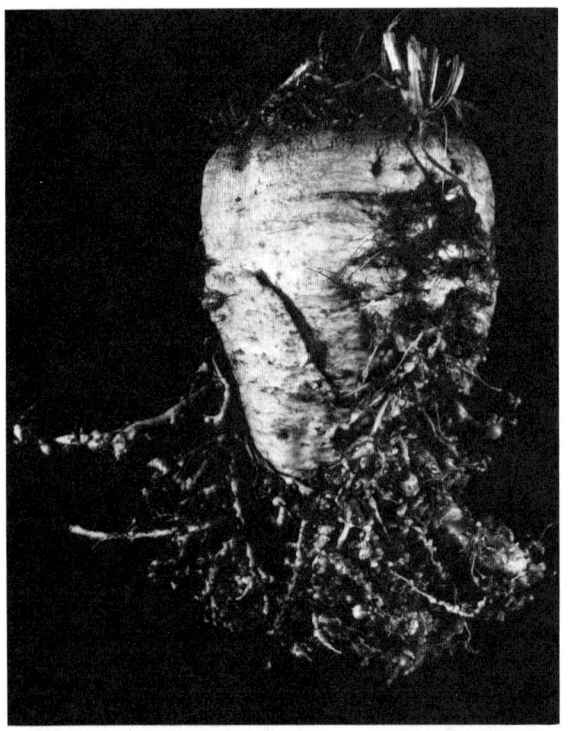

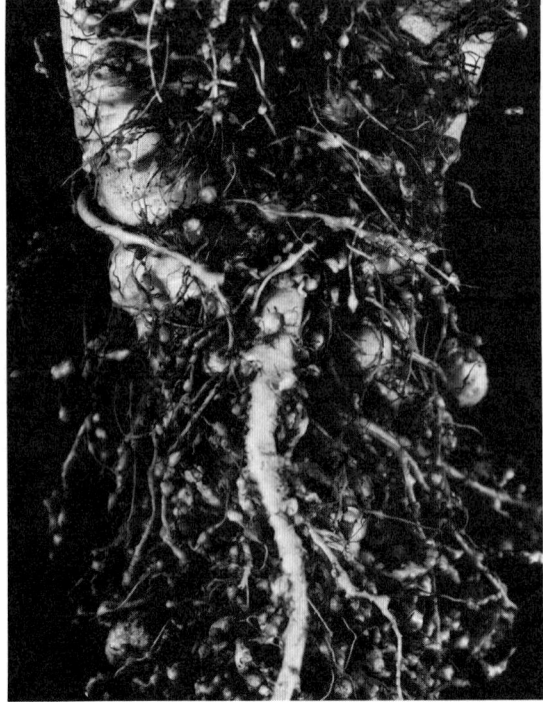

Fig. 31 Galls on roots of sugar beet caused by *Meloidogyne* sp. (probably *M. hapla*.) (Courtesy of I. J. Thomason.)

The root-knot nematodes attack the smaller roots and the tap root, inducing formation of galls (Figs. 30 and 31) that may contain one to several nematodes. Galls formed by hypertrophic and hyperplastic changes in host root tissues may be typically rounded, as in *M. hapla*, may be elongate or spiral shaped, as in *M. naasi*, or galls may fuse to form irregular clublike swellings, as in *M. incognita* (Fig. 32). The extent of lateral root growth, branching of roots, and degree of galling is dependent upon root size and population levels of invading larvae. Seedlings are most vulnerable to attack, and early attacks leading to death or severe stunting of seedlings are favored by high soil temperatures. Severe root rot of sugarbeet, which can result in complete collapse of roots and major problems in processing because of increased levels of nitrogenous compounds, is often associated with *Meloidogyne* infections.

The root-knot nematodes may overwinter as eggs or as second-stage larvae in the soil or in galls and root tissues of the previous crop, sometimes encased in clods of soil (Steele and Hodges, 1974).

Because of the wide host ranges of *Meloidogyne* spp., control by rotations that exclude host species is extremely difficult and requires identification of the root-knot nematode species for the employment of economic and practical rotation schemes. The organophosphate and carbamate nematicides appear to be less effective for control of *Meloidogyne* spp. on sugar beet than fumigants containing 1,3-dichloropropenes (Lear and Raski, 1958). Although applications of 5 kg/ha ai of granular formulation of phenamiphos are more effective against *M. javanica* than *H. schachtii* (Greco and Thomason, 1980), the nematicide is not registered for control on sugar beet in the United States.

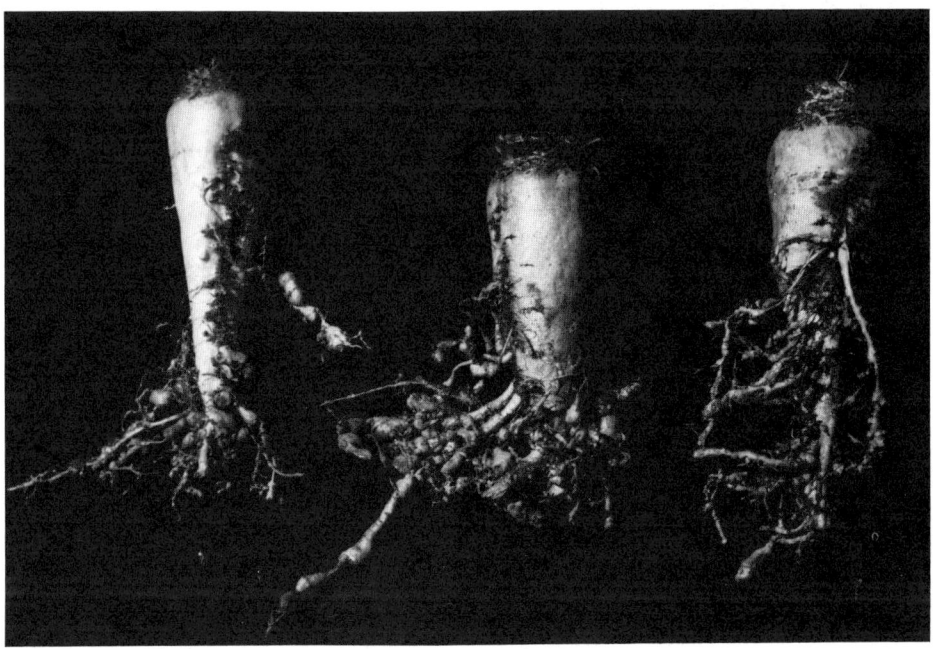

Fig. 32 Galls on roots of sugar beet infected with *M. incognita*.

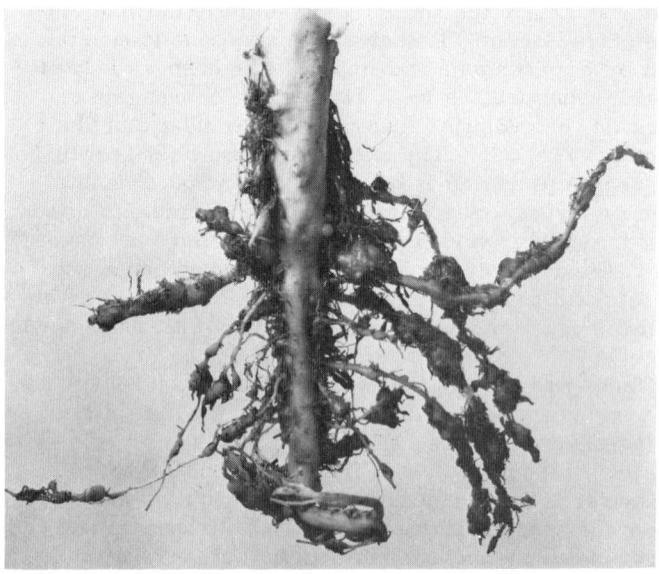

Fig. 33 Sugar beet roots infected by *Nacobbus aberrans*. (Courtesy of M. W. Schuster.

V. FALSE ROOT-KNOT NEMATODES

Thorne and Schuster (1956) described *Nacobbus batatiformis*, a nematode parasite of sugar beet that produced distinctive gall-like swellings (Fig. 33) on infected roots. In a revision of the genus *Nacobbus*, Sher (1970) synonymized *N. batatiformis* with *N. aberrans*. Surveys of 125 sugar beet fields in western Nebraska during 1953 and 1954 revealed 32% were infested with this nematode. *N. aberrans* has also been found in limited areas in Montana, Wyoming, South Dakota, Colorado, and Kansas (Caveness, 1959). Larvae frequently feed on root hairs as well as fibrous and storage roots. In contrast to *Meloidogyne* spp. and *H. schachtii*, *N. aberrans* typically is found in cortical tissues of sugar beet (Schuster et al., 1965). Necrosis and hypertrophy of the epidermal and cortical cells occur within a few days after surface feeding or penetration. Giant cells induced by the feeding of *N. aberrans* are a highly granular, deeply staining multinucleate mass formed by merging of protoplasts after gradual dissolution of cell walls and are usually located entirely within the cortex. The increase in the number of nuclei in syncytia appears to be due to amitotic division. *N. aberrans* alters the metabolism of the syncytial unit to induce formation of starch, which first appears as grains in the periphery of nuclei, and later, near the feeding area of the nematode.

In addition to sugar beet, important economic hosts of *N. aberrans* include broccoli, cabbage, carrot, cucumber, lettuce, pumpkin, pea, radish, rutabaga, tomato, and turnip. Three species of cacti native to Nebraska, *Corypantha vivipara* Nutt., *Opuntia fragilis* Nutt., and *O. tortispina* Nutt, and the weeds *Kochia scoparia* (L.) Schrad. and *Chenopodium album* L., are also hosts for *N. aberrans* (Schuster and Thorne, 1956).

In 1978, sugar beets in a 40.5 ha field near King City, California, were so severely damaged by *Nacobbus dorsalis* Thorne and Allen 1944 that the beets were not harvested. The nematode was identified by nematologists of the California Department of Food and Agriculture and University of California at Riverside. Although *N. dorsalis* occurs in widely scattered locations in California (Sher, 1970), the aforementioned finding is the only report of this nematode on sugarbeet.

VI. STUBBY-ROOT AND NEEDLE NEMATODES

The stubby-root nematodes, *Trichodorus* spp. and *Paratrichodorus* spp., and the needle nematode, *Longidorus* spp., are economically important pests of sugar beet in England and The Netherlands but are rarely a factor in production of sugar beet in the United States. In England the disease of sugar beet caused by either or both of these nematodes is termed "Docking disorder" (Fig. 34) after the parish in northwestern Norfolk where it was first reported (Whitehead, Dunning, and Cooke, 1970). Seven species of stubby-root nematodes (*Paratrichodorus anemones* Loof, *T. cylindricus* Hooper, *P. pachydermus* Seinhorst, *T. primitivus* de Man, *P. teres* Hooper, and *T. viruliferus* Hooper) and four species of needle nematodes (*Longidorus attenuatus* Hooper, *L. elongatus* de Man, *L. caespiticola* Hooper, and *L. leptocephalus* Hooper) have been associated singly or in combination with Docking disorder. These nematodes feed as migratory external parasites on the fibrous roots of sugar beet. *Trichodorus* and *Paratrichodorus* spp. cause stubby-ended lateral roots that may eventually turn dark brown and die (Fig. 35). High nematode populations may kill the tip of the growing tap root early in the seedling stage.

Fig. 34 Field with docking disorder caused by *Paratrichodorus* spp. and/or *Longidorus* spp. (Courtesy of D. A. Cooke.)

Other roots thicken to produce a "fangy" storage root (Fig. 36). *Longidorus* spp. severely stunt seedlings, which have many short lateral roots (Fig. 37) and swollen tips, but tap roots of parasitized beets appear to be relatively un- affected. In fields affected with Docking disorder, affected areas appear as diffuse patches of poorly growing plants with plants often lacking uniform size and shape and showing symptoms of nitrogen or manganese deficiencies. Although the nematodes alone rarely kill the seedlings, as many as 9000 stub- by-root nematodes or 2000 needle nematodes per liter of soil may be found in the root zones of stunted plants (Dunning and Cooke, 1972), and, as a rule, population levels and yield losses show no close relationship. Docking disorder is most severe after wet springs or when high numbers of the nematodes are active at the time of germination of beet seed. The disease is confined to alkaline sandy soils of low organic content.

Two soilborne viruses, tobacco rattle virus (TRV) transmitted by tricho- dorid nematodes and tomato black ring virus transmitted by *Longidorus* spp.,

Fig. 35 Sugar beet infected with *Paratrichodorus* spp. (Courtesy D. A. Cooke.)

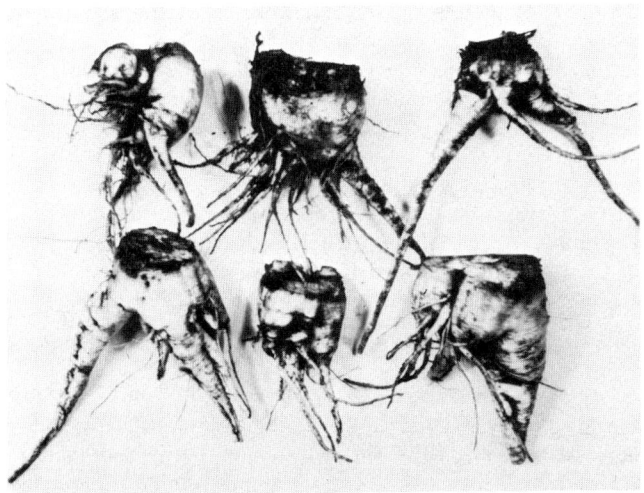

Fig. 36 Roots of sugar beet showing severe branching and stunted growth of root tips caused by *Paratrichodorus* spp. (Courtesy D. A. Cooke.)

Fig. 37 Sugar beet seedling showing root damage caused by *Longidorus* spp.
(Courtesy of D. A. Cooke.)

are associated with Docking disorder, but neither virus causes serious disease
of sugar beet in England (Heathcote, 1973).

Addition of granular fertilizers that increase nitrogen content of soil and
organic matter, such as farmyard manure, decrease losses due to Docking disorder. Whitehead, Dunning, and Cooke (1970) recommend avoiding practices
that weaken plant growth, such as early or deep sowing, applying too much
herbicide, and damaging the soil structure. They maintain that the amount of
nematode feeding can be minimized by interrow cropping, close sowing of
sugar beet seed, addition of marl, establishing a firm seedbed, and killing or
repelling nematodes from beet roots. Measures should include control of weed
hosts of stubby-root and needle nematodes. Fumigation with 1,3-dichloropro-
pene nematicides will effectively control the nematodes, and treatment with
carbamates and organophosphates are promising alternatives. In England,
early fall fumigation with 68-135 liters/ha of dichloropropane-dichloropropene
mixture or 100 liters/ha ethylene dibromide was effective in controlling *Trichodorus*, *Paratrichodorus*, and *Longidorus* and increased sugar beet yield
without affecting sugar percentage or juice purity of the roots (Whitehead,
Tite, and Fraser, 1970).

VII. *DITYLENCHUS* SPP.

The stem and bulb nematode, *Ditylenchus dipsaci* (Kühn 1857) Filipjev 1936
(T.) is a serious pest of sugar beet, fodder beet, and mangels in a number of
European countries, but disease of sugar beet caused by this nematode is un-
known in the United States. Damage to *B. vulgaris* has been reported from
Belgium, the CSSR, England, Germany, Ireland, Italy, Poland, Switzerland,

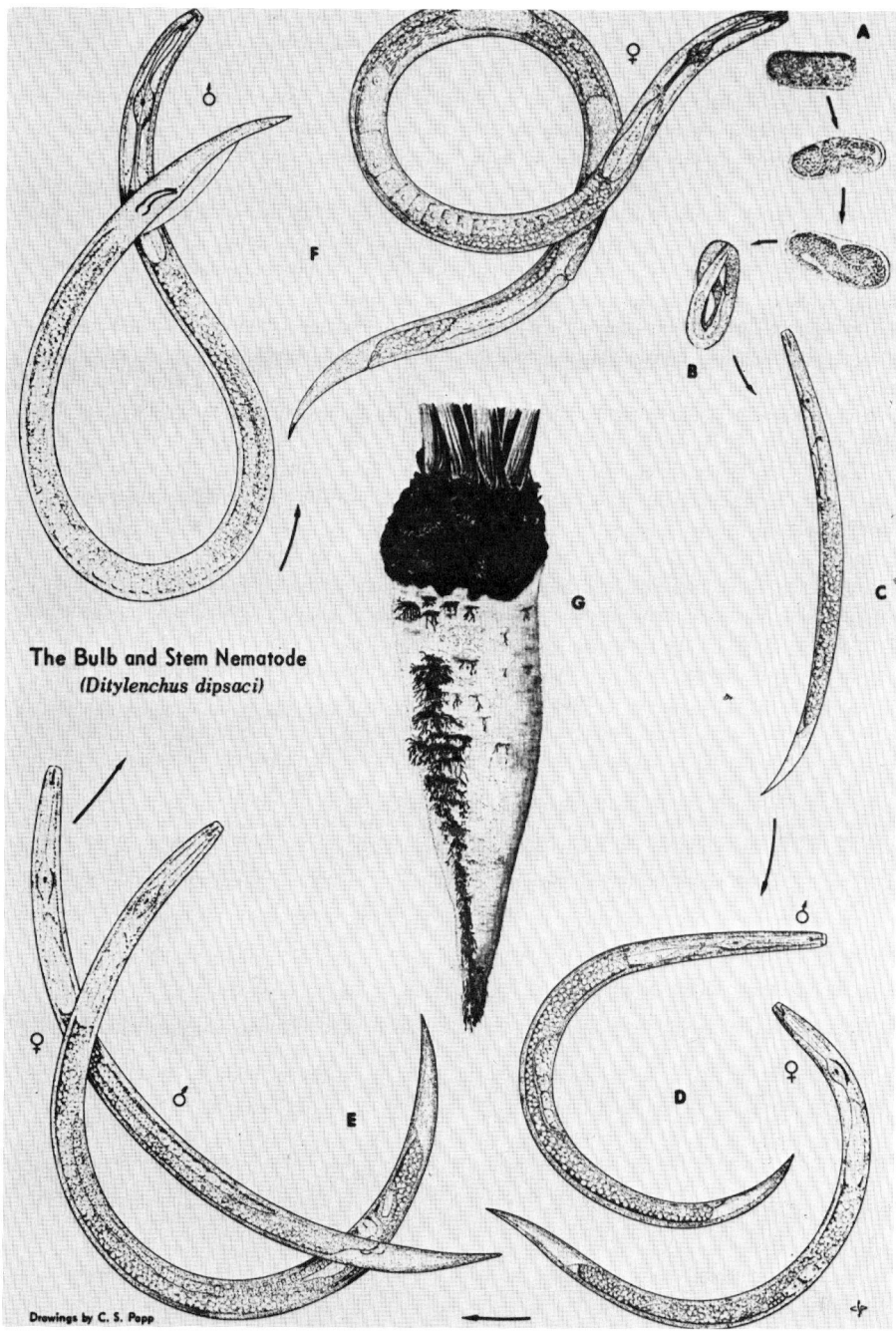

The Bulb and Stem Nematode
(Ditylenchus dipsaci)

Drawings by C. S. Papp

Fig. 38 Life cycle of *Ditylenchus dipsaci*: (A) egg; (B) larvae within egg membranes; (C) second-stage larva; (D) third-stage larva; (E) fourth-stage male and female; (F) mature adult male and female; (G) crown canker of sugar beet caused by *D. dipsaci*. (Courtesy of C. S. Papp.)

The Netherlands, and the USSR, but damage seldom occurs in Sweden and Denmark. The life cycle of this nematode is illustrated in Fig. 38.

Investigations by Dunning (1957) established two types of injury to sugar beet produced by *D. dipsaci*. Early attack of seedlings in the cotyledon stage causes swelling of stalks, midribs, and main veins of the leaves, and sometimes formation of galls. The growing point may be injured or killed, and the plant may be severely stunted or may die. Plants that survive may develop a typical multiple crowned appearance. Invasion of young seedlings is favored by cool and rainy weather in the spring. During the summer damage is less apparent and plants appear to show some recovery from the spring attack. In late summer or fall, crown canker may develop at scars of infected petioles and healthy plants that have escaped the spring attack may be invaded. The canker usually occurs above soil level, and gradually spreads to form a continuous girdle around the crown. Secondary pathogens enter the damaged tissue and cause rotting that may eventually destroy the entire storage root.

The primary sources of infection appear to be infested plant material of preceding susceptible crops, or weed hosts, or both. Although seedborne infection seems possible, survival and infection by this means have not been proven (Dunning, 1957). Experimental infection of sugar beet by the potato rot nematode *D. destructor* Thorne 1945 was accomplished by Dallimore and Thorne (1951), who described symptoms similar to crown canker produced by *D. dipsaci*. Field infection of sugar beet has been reported from the USSR (Kirjanova and Krall, 1971), but positive information on occurrence of the disease elsewhere has not been published.

Table 3 List of Other Nematodes Parasitizing *B. vulgaris* L.[a]

Species	Reference
Belonolaimus gracilis Steiner 1949	Christie et al. (1952)[b]
Helicotylenchus microlobus Perry 1959	Taylor (1960)
Hemicycliophora similis Thorne 1955	Khera and Zuckerman (1963)
Neotylenchus abulbosus Steiner 1931	Thorne and Price (1935)
Paratylenchus projectus Jenkins 1956	Coursen et al. (1958)
Pratylenchus scribneri Steiner 1943	Thomason and O'Melia (1962)
Radopholus similis (Cobb 1893) Thorne 1949	Feder and Feldmesser (1957)
Rotylenchulus reniformis Linford and Oliveria 1940	Linford and Yap (1940)
Paratrichodorus minor (Colbran 1956)	Coursen et al. (1958)
Tylenchorhynchus dubius (Bütschli 1873) Filipjev 1936	Kirjanova and Krall (1971)

[a]Species not discussed in text.
[b]See literature cited.

Rotation, weed control, and sanitation will effectively control *Ditylenchus* spp. Susceptible crops should not be included in rotations for at least two years. Sugar beet should not be grown after rye, oats, maize, onion, carrot, bean, cucumber, sunflower crops (Weischer and Steudel, 1972), or potato when *D. destructor* is present in the area. Because the nematode may persist on common weeds and debris of host crops, weed control and sanitation to prevent the spread of nematode-infested plant materials are required. Control of *Ditylenchus* spp. with organophosphate nematicides appears to be effective when economics justify their use.

VIII. NEMATODES OF LESSER IMPORTANCE

Table 3 lists 10 additional nematode species that attack *B. vulgaris* but are not known to be important to sugar beet production. Goodey, Franklin, and Hooper (1966), Caveness (1958), and Mathur and Varaprasad (1980) list other species found in association with sugar beet.

REFERENCES

Abawi, G. S., and Mai, W. F. (1980). Effects of initial population densities of *Heterodera schachtii* on yield of cabbage and table beets in New York State. *Phytopathology 70*: 481-485.

Altman, J., and Fitzgerald, B. T. (1960). Late fall application of fumigants for the control of sugar beet nematodes, certain fungi and weeds. *Plant Dis. Rep. 44*: 868-971.

Altman, J., and Thomason, I. J. (1971). Nematodes and their control. In *Advances in Sugarbeet Production: Principles and Practices* (R. T. Johnson, J. T. Alexander, G. E. Rush, and G. R. Hawkes, eds.). Iowa State University Press, Ames, Iowa, pp. 335-370.

Anon. (1970). Estimated crop losses from plant-parasitic nematodes in the United States. *Report of the Soc. Nem. Comm. on Crop Losses - 1970.*

Anon. (1980). U.S.D.A. Sugar and Sweetner Report, December 1980.

Batterby, A. (1979). Toxic effects of aldicarb and its metabolites on second stage larvae of *Heterodera schachtii. Nematologica 25*: 377-384.

Baunacke, W. (1922). Untersuchungen zur Biologie und Bekämpfung des Rübennematoden, *Heterodera schachtii* Schmidt. *Arb. Biol. Abt. Reich-Anst. Berlin 11*: 185-288.

Bergman, B. H. H. (1951). Degevoeligheid van wildebieten voor het bieten-cystenaaltje (*Heterodera schachtii*). *Med. Inst. Rationale Suikerprod. 21*: 1-13.

Bergman, B. H. H. (1958). Het bietencystenaltje en zijn bestrijding. V. Enige microscopische waarnemingen betreffende de ontwikkeling van larven van *Heterodera schachtii* in de wortels van vatbare en resistente planten. *Med. Inst. Suikerbiet. Bergen-O.-Z. 28*: 151-168.

Bergman, B. H. H., and Van Duuren, A. J. (1959). Het bietencystenaaltje en zijn bestrijding. VII. De werking van stofwisselingsproducten van sommige micro-organismen op de larven van *H. schachtii. Med. Inst. Suikerbiet. Bergen-O.-Z. 29*: 27-52.

Berliner, E., and Busch, K. (1914). Uber die Züchtung des Rübennematoden (*Heterodera schachtii* Schmidt) auf *Agar. Biol. Zbl. 34*: 349-356.

Bessey, E. A. (1911). Root-knot and its control. *U.S. Dept. Agric. Bur. Plant Ind. Bull. 217.*

Bursnall, L. A., and Tribe, H. T. (1974). Fungal parasitism in cysts of *Heterodera*. II Egg parasites of *Heterodera schachtii*. *Trans. Br. Mycol. Soc. 62*: 595-601.

Caveness, F. E. (1958). A study of nematodes associated with sugarbeet production in selected northwest and north central states. *Beet Sugar Devel. Found.*, Fort Collins, Colorado.

Caveness, F. E. (1959). Distribution of cyst and gall-forming nematodes of sugarbeets in the United States. *J. Am. Soc. Sugar Beet Technol. 10*: 544-552.

Chatin, J. (1888). Sur la structure des téguments de l'*Heterodera schachtii* et sur les modifications qu'ils présentent chez les femelles fécondées. *C.R. Hebd. Seanc. Acad. Sci. Paris 107*: 139-141.

Chatin, J. (1891). L'angiullule de la betterave (*Heterodera schachtii*). *Bull. Minist. Agric. Paris 10*: 457-506.

Chitwood, B. G. (1951). The golden nematode of potatoes. *U. S. Dept. Agric. Circ. No. 875*.

Christie, J. R., Brooks, A. N., and Perry, V. G. (1952). The sting nematode, *Belonolaimus gracilis*, a parasite of major importance on strawberries, celery, and sweet corn in Florida. *Phytopathology 42*: 173-176.

Clarke, A. J., and Perry, R. N. (1977). Hatching of cyst nematodes. *Nematologica 231*: 350-368.

Clarke, A. J., and Shepherd, A. M. (1964). Synthetic hatching agents for *Heterodera schachtii* Schm. and their mode of action. *Nematologica 10*: 431-453.

Clarke, A. J., and Shepherd, A. M. (1965). Zinc and other metallic ions as hatching agents for the beet cyst nematode, *Heterodera schachtii* Schm. *Nature (Lond.) 208*: 502.

Clarke, A. J., and Shepherd, A. M. (1966). Inorganic ions and the hatching of *Heterodera* spp. *Ann. Appl. Biol. 58*: 497-508.

Cooke, D. A., and Thomason, I. J. (1978). The distribution of *Heterodera schachtii* in California. *Plant Dis. Rep. 62*: 989-993.

Cooke, D. A., and Thomason, I. J. (1979). The relationship between population density of *Heterdera schachtii*, soil temperature and sugar beet yields. *J. Nematol. 11*: 124-128.

Coons, G. H. (1949). The sugar beet: Product of science. *The Scientific Monthly 68*: 149-164.

Coursen, B. W., Rhode, R. A., and Jenkins, W. R. (1958). Additions to the host lists of the nematodes *Pratylenchus projectus* and *Trichodorus christiei*. *Plant Dis. Rep. 42*: 456-460.

Curtis, G. J. (1970). Resistance of sugar beet to the cyst-nematode *Heterodera schachtii* Schm. *Ann. Appl. Biol. 66*: 169-177.

Dallimore, C. E., and Thorne, G. (1951). Infection of sugar beets by *Ditylenchus destructor* Thorne, the potato rot nematode. *Phytopathology 41*: 872-874.

Doncaster, C. C. (1962). Natúrliche Feinde wirtschaftlich bedeutsamer Schadnematoden. Ein Einblick in das Räuber-Beute-Verhältnis im Erdboden. *Umschau 62*: 443-446.

Doney, D. L., Fife, J. M., and Whitney, E. D. (1970). Effect of the sugarbeet nematode *Heterodera schachtii* on the free amino acids in resistant and susceptible *Beta* species. *Phytopathology 60*: 1727-1729.

Doney, D. L., and Whitney, E. D. (1970). Genetic diversity in sugar beet lines selected for nematode resistance. *J. Am. Soc. Sugar Beet Technol. 16*: 219-224.

Doney, D. L., Whitney, E. D., and Steele, A. E. (1970). Effect of *Hetero-dera schachtii* infection on sugarbeet leaf growth. *Phytopathology 61:* 40-41.

Dunning, R. A. (1957). Stem eelworm invasion of seedling sugar beet and development of crown canker. *Nematologica II (Suppl.),* pp. 362-368.

Dunning, R. A., and Cooke, D. A. (1972). Docking disorder of sugar beet Min. Agric. Fish. and Foods, Middlesex, England.

Ellenby, C. (1968). The survival of desiccated larvae of *Heterodera rosto-chiensis* and *H. schachtii. Nematologica 14:* 544-548.

Epps, J. M. (1968). Survival of soybean cyst nematode in seed bags. *Plant Dis. Rep. 52:* 45.

Epps, J. (1969). Survival of the soybean cyst nematode in seed stocks. *Plant Dis. Rep. 53:* 403-405.

Epps, J. M. (1971). Recovery of soybean cyst nematodes (*Heterodera gly-cines*) from the digestive tracts of blackbirds. *J. Nematol. 3:* 417-419.

Faulkner, L. R., and Bolander, W. J. (1966). Occurrence of large nematode populations in irrigation canals of South Central Washington. *Nemato-logica 12:* 591.

Feder, W. A., and Feldmesser, J. (1957). Additions to the host list of *Rado-pholus similis,* the burrowing nematode. *Plant Dis. Rep. 41:* 33.

Finkner, R. E., and Swink, J. F. (1956). Breeding sugar beets for resis-tance to nematodes. *J. Agron. 48:* 389-392.

Fox, J. A., and Spasoff, L. (1972). Interaction of *Heterodera solanacearum* and *Endogone gigantea* on tobacco (abstract). *J. Nematol. 4:* 224-225.

Franklin, M. T. (1965). A root-knot nematode, *Meloidogyne naasi* n. sp., on field crops in England and Wales. *Nematologica. 11:* 79-86.

Fuchs, O. (1911). Beiträge zur Biologie des Rübennematoden *Heterodera schachtii, Z. Landw. Ver Wes. Ost. 14:* 923-952.

Gerdemann, J. W., and Trappe, J. M. (1974). The *Endogonaceae* in the Pa-cific Northwest. *Mycol. Mem. 5:* 1-76.

Goffart, H. (1932). Untersuchungen am Hafernematoden *Heterodera schachtii* Schm. unter besonderer Berücksichtigung der schleswigholsteinischen Verhältnisse I. III. Beitrag zu: Rassenstudien an *Heterodera schachtii* Schm. *Arb. Biol. Bund Anst. Land-u. Forstw. 20:* 1-26.

Goffart, H. (1954). Gegenwartsfragen zum Rübennematodenproblem. *Zucker* 7: 130-137.

Golden, A. M. (1958a). Interrelationships of certain *Beta* species and *Hete-rodera schachtii,* the sugar-beet nematode. *Plant Dis. Rep. 42:* 1157-1162.

Golden, A. M. (1958b). Influence of leaf diffusate of sugar beet on emergence of larvae from cysts of the sugar beet nematode (*Heterodera schachtii*). *Plant Dis. Rep. 42:* 188-193.

Golden, A. M. (1959). Susceptibility of several *Beta* species to the sugar-beet nematode (*Heterodera schachtii*) and root-knot nematodes (*Meloi-dogyne* spp.). *J. Am. Soc. Sugar Beet Technol. 10:* 444-447.

Golden, A. M., and Shafer, T. (1958). Differential response of *Heterodera schachtii,* the sugar-beet nematode, to selections of *Chenopodium album. Plant Dis. Rep. 42:* 184-187.

Golden, A. M., and Shafer, T. (1959a). Host-parasite relationships of various plants and the sugar-beet nematode (*Heterodera schachtii*). *Plant Dis. Rep. 43:* 1258-1262.

Golden, A. M., and Shafer, T. (1959b). Influence of germinating seeds of sugar beet (*Beta vulgaris*) on emergence of larvae from cysts of the sugar-beet nematode (*Heterodera schachtii*). *Plant Dis. Rep. 43:* 1103-1104.

Golden, A. M., and Shafer, T. (1960). Survival of the sugar-beet nematode (*Heterodera schachtii*) in water and in soil. *Nematologica 5*: 32-36.

Gomers, F. J. (1981). Biochemical interactions between nematodes and plants and their relevance to control. *Helm. Abstr. Ser. B. Plant Nematol. 50*: 9-24.

Good, J. M. (1968). Assessment of crop losses caused by nematodes in the United States. *FAO Plant Prot. Bull. 16*: 37-40.

Goodey, J. B., Franklin, M. T., and Hooper, D. J. (1966). The nematode parasites catalogued under their hosts. *Commonw. Agric. Bur. Farnham Royal*, Bucks, England.

Graney, L. S., and Miller, L. I. (1980). Differentiation of five isolates of *Heterodera schachtii* as races of the sugar beet cyst nematode (abstract). *J. Nematol. 12*: 223.

Greco, N., and Thomason, I. J. (1980). Effect of phenamiphos on *Heterodera schachtii* and *Meloidogyne javanica*. *J. Nematol. 12*: 91-96.

Greco, N., and Vovlas, N. (1979). Infestation of *Heterodera schachtii* on sugar beet storage roots. *Nematol. Medit. 7*: 1-5.

Green, C. D. (1966). Orientation of male *Heterodera rostochiensis* Woll. and *H. schachtii* Schm. to their females. *Ann. Appl. Biol. 58*: 327-339.

Green, C. D. (1967). The attraction of male cyst-nematodes by their females. *Nematologica 13*: 172-173.

Green, C. D., and Greet, D. N. (1972). The location of the secretions that attract *Heterodera schachtii* and *H. rostochiensis* to their females. *Nematologica 18*: 347-352.

Green, C. D., Greet, D. N., and Jones, F. G. W. (1970). The influence of multiple mating on the reproduction and genetics of *Heterodera rostochiensis* and *H. schachtii*. *Nematologica 16*: 309-326.

Green, C. D., and Plumb, S. C. (1970). The interrelationships of some *Heterodera* spp. indicated by the specificity of the male attractants emitted by their females. *Nematologica 16*: 39-46.

Greet, D. N., Green, C. D., and Poulton, M. E. (1968). Extraction, standardization and assessment of the volatility of the sex attractants of *Heterodera rostochiensis* Woll. and *H. schachtii* Schm. *Ann. Appl. Biol. 61*: 511-519.

Griffin, G. D. (1975). Control of *Heterodera schachtii* with foliar application of nematicides. *J. Nematol. 7*: 347-351.

Griffin, G. D. (1980a). The pathogenicity of a combination of *Meloidogyne hapla* and *Heterodera schachtii* on tomato (abstract). *J. Nematol. 12*: 223-224.

Griffin, G. D. (1980b). The effect of genetic variability on the susceptibility of certain weed hosts to *Heterodera schachtii* (abstract). *J. Nematol. 12*: 224.

Griffin, G. D. (1981). The relationship of plant age, soil temperature and population density of *Heterodera schachtii* on growth of sugar beet. *J. Nematol. 13*: 184-190.

Healthcote, G. D. (1973). Nematode-transmitted viruses of sugar beet in England, 1965-1972. *Plant Pathol. (Lond.) 22*: 156-160.

Heijbroek, W. (1973). Forecasting incidence of and issuing warnings about nematodes, especially *Heterodera schachtii* and *Ditylenchus dipsaci*. *Meded. Inst. Rat. Suikerprod. 6*: 76-86.

Heijbroek, W. J., McFarlane, J. S., and Doney, D. L. (1977). Breeding for tolerance to beet-cyst eelworm *Heterodera schachtii* A. Schm. in sugar beet. *Euphytica 26*: 557-564.

Hellinga, J. J. A. (1942). De invloed van het bietenaltje op de opbrengst en de samenstelling van suikerbieten. *Meded. Inst. SuikBietTeelt. Bergen-O.-Z. 12:* 163-182.

Hijner, J. A. (1952). Devgevoeligheid van wildevbieten voor het bietencystenaaltje (*Heterodera schachtii*). *Meded. Inst. Rat. Suik. Prod. Bergen-O.-Z. 21:* 1-13.

Holtzmann, O. V., and Aragaki, M. (1963). Clover cyst nematode in Hawaii. *Plant Dis. Rep. 47:* 886-889.

Hough, A., and Thomason, I. J. (1975). Effects of aldicarb on the behavior of *Heterodera schachtii* and *Meloidogyne javanica*. *J. Nematol. 7:* 221-229.

Hülsenberg, H. (1935). Beitrag zur Züchtung einer nematodenfesten Zuckerrübe. *Landw. Jb. 81:* 505-523.

Hulsfeld, B. (1926). Beitrag zur Züchtung von nematoden immunen Zuckerube. *Ill. Landro. Ztg.*, p. 18.

Jatala, P., and Jensen, H. J. (1972). Interrelationships of *Meloidogyne hapla* and *Heterodera schachtii* populations on *Beta vulgaris* (abstract). *J. Nematol. 4:* 226.

Jatala, P., and Jensen, H. J. (1976a). Self-interactions of *Meloidogyne hapla* and *Heterodera schachtii* on *Beta vulgaris*. *J. Nematol. 8:* 43-48.

Jatala, P., and Jensen, H. J. (1976b). Histopathology of *Beta vulgaris* to individual and concomitant infections by *Meloidogyne hapla* and *Heterodera schachtii*. *J. Nematol. 4:* 336-341.

Johnson, R. N., and Viglierchio, D. R. (1969). Sugar beet nematode (*Heterodera schachtii*) reared on axenic *Beta vulgaris* root explants II. Selected environmental and nutritional factors affecting development and sex-ratio. *Nematologica 15:* 144-152.

Johnson, R. N., and Viglierchio, D. R. (1970). Incidence of aberrancy in *Heterodera schachtii*. *Nematologica 16:* 33-38.

Jones, F. G. W. (1956). Soil populations of the beet eelworm (*Heterodera schachtii* Schm.) in relation to cropping. 2. Microplot and field plot results. *Ann. Appl. Biol. 44:* 25-56.

Jones, F. G. W. (1975). Accumulated temperature and rainfall as measures of nematode development and activity. *Nematologica 21:* 62-70.

Jones, M. G. K., and Northcote, D. H. (1972). Multinucleate transfer cells induced in *Coleus* roots by the root-knot nematode, *Meloidogyne arenaria*. *Protoplasma 75:* 381-395.

Jorgenson, E. C. (1969a). Control of the sugar beet nematode, *Heterodera schachtii* on sugar beets with organophosphate and carbamate nematicides. *Plant Dis. Rep. 53:* 625-628.

Jorgenson, E. C. (1969b). Influence of planting time on phytotoxicity of dichloropropenes to sugar beets. *Plant Dis. Rep. 53:* 629-630.

Jorgenson, E. C. (1970). Antagonistic interaction of *Heterodera schachtii* Schmidt and *Fusarium oxysporum* (Woll.) on sugar beets. *J. Nematol. 2:* 393-398.

Kerry, B. (1980). Biocontrol-fungal parasites of female cyst nematodes. *J. Nematol. 12:* 253-259.

Kerry, B. R., and Crump, D. H. (1977). Observations of fungal parasites of females and eggs of the cereal cyst-nematode *Heterodera avenae* and other cyst nematodes. *Nematologica 23:* 193-201.

Kerry, B. R., and Crump, D. H. (1980). Two fungi parasitic on female cyst nematodes (*Heterodera* spp.) *Trans. Br. Mycol. Soc. 74:* 119-125.

Kerstan, U. (1969). Die Beeinflussung des Geschlechterverhältnisses in der Gattung *Heterodera*. II. Minimallebensraum-selektive Absterberate der Geschlechterverhaltnis (*Heterodera schachtii*). *Nematologica 15:* 210-228.

Khera, S., and Zuckerman, B. M. (1963). In vitro studies of host-parasite relationships of some plant-parasitic nematodes. *Nematologica 9*: 1-6.

Kirjanova, E. S., and Krall, E. L. (1971). *Plant-Parasitic Nematodes and Their Control*. Part 2 Izd. "Nauka," Leningrad. English translation published by Amerind Publ. Co. Pvt. Ltd., New Delhi, 1980.

Kontaxis, D. G., Lofgreen, G. P., Thomason, I. J., and McKinney, H. E. (1971). Survival of the sugar beet cyst nematode in the alimentary canal of cattle. *Calif. Agric. 30*: 15.

Kontaxis, D. G., and Thomason, I. J. (1978). Chemical control of *Heterodera schachtii* and sugar beet production in Imperial Valley, California. *Plant Dis. Rep. 62*: 79-82.

Korab, I. I. (1927). Some data on the problem of control of the sugar beet nematode. *Zakhyst Rosl. Ruspub. Mizhvid Lemat. Nauk. Zb. 2*: 17-38. Cited by I. N. Filipjev and J. H. S. Stekhoven, Jr., in *A Manual of Agricultural Helminthology*. E. J. Brill, Leiden, 1941.

Korab, I. I. (1929). Materials for the study of the beet nematode *Heterodera schachtii* Schm. from data of investigations carried out at the nematode laboratory of the Velotserkov Selection Station S.S.U. Preliminary report. *Ukrainian Res. Inst. Sugar Indus. Kiev 8*: 20-67.

Kraus, R., and Sikora, R. A. (1980). Herbicides as cyst nematode hatching factors and their subsequent effect on *Heterodera schachtii* population levels and sugar beet growth (abstract). *Institute fur Pflanzenkrankheiten, Ninth Int. Congr. Plant Protection and 71st Ann. Meeting Am. Phytopathol. Soc.*, Washington, D.C., Aug. 5-11, 1979.

Kühn, J. (1877). Vorläufiger Bericht des Herrn Prof. Dr. Kuhn, Directors des Landwirthschaftlichen Institutes in Halle, über die bisherigen Ergebnisse der seit dem Jahre 1875 im Auftrage des Vereins für Rübenzucker-Industrie ausgeführten Versuche zur Ermitelung der Ursache der Rübenmüdigkeit des Bodens und zur Erforschung der Natur der Nematoden. *Z. Ver. RubenZuckInd. Zollverein 27*: 452-457.

Kühn, J. (1881). Bericht über die Ergebnisse der im Auftrage des Vereins für Rübenzucker-Industrie des Deutschen Reiches ausgefuhrten Versuche zur Ermitelung der Ursache der Rübenmüdigkeit des Bodens und zur Erforschung der Natur der Nematoden. *Ber. Physiol. Lab. Inst. Univ. Hallevd. 3*.

Kühn, J. (1882). Die Wirksamkeit der Nematoden-Fangpflanzen nach den Versuchsergebnissen des Jahres 1881. *Ber. Physiol. Lab. Landw. Inst. Univ. Halle 4*: 1-14.

Kumaraswami, T. (1966). Preliminary studies on the wind-borne dispersal of cysts of the golden nematode of potatoes *Heterodera rostochiensis* Woll., 1923 in the Nilgiris. *Madras Agric. J. 53*: 41.

Lear, B., Miyagawa, S. T., Johnson, D. E., and Atlee, C. B. (1966). The sugar beet nematode associated with reduced yields of cauliflower and other vegetable crops. *Plant Dis. Rep. 50*: 611-612.

Lear, B., and Miyagawa, S. T. (1972). Development of a strain of the sugar beet nematode as a potential pest of tomato. *J. Nematol. 4*: 296-297.

Lear, B., and Raski, D. J. (1958). Control by soil fumigation of root-knot nematodes affecting sugar beet production. *Plant Dis. Rep. 42*: 861-864.

Lear, B., Sciaroni, R. H., Atlee, C. B., and Hart, W. H. (1961). Yield response of Brussels sprouts to soil fumigation for control of sugar beet nematode, *Heterodera schachtii*. *Plant Dis. Rep. 45*: 739-741.

Liebscher, G. (1878). Die Rübenmüdigkeit des Ackers hervorgerufen durch *Heterodera schachtii*. *Z. Ver. Ruben Zuc Ind. Zoll Ver. 28*: 893-895.

Linford, M. B., and Yap, F. (1940). Some host plants of the reniform nematode in Hawaii. *Proc. Helm. Soc. Wash.* 7: 42-44.

Maas, P. W. T., and Maenhout, C. A. A. A. (1978). Hetgraswortel knobbe-laaleje (*Meloidogyne naasi*) bij suikerbieten. *Gewasbescherming* 9: 159-166.

Mai, W. F. (1961). Sugar beet nematode found in New York State. *Plant Dis. Rep. 45*: 151.

Mankau, R. (1980). Biocontrol: Fungi as nematode control agents. *J. Nematol. 12*: 244-252.

Mathur, V. K., and Varaprasad, K. G. (1980). Nematodes associated with sugar beet in India. *Indian J. Nematol. 8*: 75-77.

Meagher, J. W. (1977). World distribution of the cereal-cyst nematode (*Heterodera avenae*) and its potential as a pathogen of wheat. *J. Nematol. 9*: 9-15.

Molz, E. (1917). Über die Züchtung widerstandsfähiger Sorten unserer Kulturpflanzen. *Z. PflZücht. 5*: 121-124.

Molz, E. (1920). Versuche zur Ermittlung des Einflusses änsserer Faktoren auf das Geschlechtsverhältnis des Rübennematoden *Heterodera schachtii*. *A. Schmidt. Landw. Jrb. Schweiz. 54*: 769-791.

Moriarty, F. (1961). The effects of red table beet and of *Hesperis matronalis* L. on a population of *Heterodera schachtii*. *Nematologica 6*: 214-221.

Moriarty, F. (1962). The effect of water-table on beet eelworm, *Heterodera schachtii* Schm., in a sandy loam. *Ann. Appl. Biol. 50*: 693-701.

Moriarty, F. (1963). The decline of a beet eelworm (*Heterodera schachtii* Schm.) population in microplots in the absence of host plants. *Nematologica 9*: 24-30.

Muller, J. (1980). Wechselwirkungen zwischen *Heterodera schachtii* und Bodenpilzen an Zuckerrüben. *Phytopathol. Z. Berlin 97*: 357-363.

Mulvey, R. H. (1957). Chromosome number in the sugar beet nematode *Heterodera schachtii* Schmidt. *Nature (Lond.) 180*: 1212-1213.

Mulvey, R. H. (1958). Impregnation of *Heterodera trifolii* by males of *H. schachtii* (Nematoda:Heteroderidae). *Can. J. Zool. 36*: 839-841.

Mulvey, R. H. (1960). Giant larvae of the clover cyst-nematode *Heterodera trifolii* (Nematoda:Heteroderidae) *Nematologica 5*: 53-55.

Murphy, P. W., and Doncaster, C. C. (1957). A culture method for soil meiofauna and its application to the study of nematode predators. *Nematologica 2*: 202-214.

Nebel, B. (1926). Ein Beitrag zur Physiologie des Rübennematoden *Heterodera schachtii* vom Standpunkt der Bekämpfung. *Küehn-Atch. 12*: 38-103.

Nejad, S., and Dern, R. (1979). Über die populationsentwicklung von Rübennematoden (*Heterodera schachtii*) nach anbau von Zuckerruben in Hessen-Nassau. *Gesunde Pflanzen 31*: 73-75.

Nemec, B. (1910). Das problem der Befruchtingsvorgange under anders Zytologishe fragen. VI. Vielkernige Riesenzellen in *Heterodera* Gallen. *Gebrüder Borntrager Berlin*, pp. 151-173.

Nemec, B. (1911). Über die nematodenkrankheit der Zukerrübe. *Z. Pfl Krankr. 21*: 1-10.

Nemec, B. (1933). Über die Gallen von *Heterodera schachtii* auf der Zuckerrübe. *Mem. Soc. R. Sci. Bohéme 6*: 1 14.

Nigh, E. A., Thomason, I. J., and Van Gundy, S. D. (1980). Identification and distribution of fungal parasites of *Heterodera schachtii* eggs in California. *Phytopathology 70*: 884-889.

Nolte, H. W., and Kohler, H. (1952). Pflanzenschädigungen bie Nematoden-
befall und ihre kausalen Ursachen. Nachr. Bl. dt. PflSchutzdienst Berlin 6: 24-28.

Olthof, T. H. A., Potter, J. W. and Peterson, E. A. (1974). Relationship
between population densities of Heterodera schachtii and losses in vegetable crops in Ontario. Phytopathology 64: 549-554.

Oostenbrink, M. (1967). Studies on the emergence of encysted Heterodera
larvae. Meded. Fac. Landbouwwet Rijksfac. Gent 32: 503-539.

Ouden, H. Den. (1956a). The influence of hosts and non-susceptible hatching
plants on populations of Heterodera schachtii. Nematologica 1: 138-144.

Ouden, H. Den. (1956b). Het bietencystinaaltje en Zijn bestrijding. IV.
Enige proeven over de invioed van door bietencystenaaltjes-larven afge-
scheiden stoften. Med. Inst. Rationele Suikerprod. 26: 119-125.

Partyka, R. E. (1969). Sugar beet nematode found on garden red beet in
Ohio. Plant Dis. Rep. 53: 118.

Polychronopoulos, A. G., Houston, B. R., and Lownsbery, B. F. (1969).
Penetration and development of Rhizoctonia solani in sugar beet seedlings
infected with Heterodera schachtii. Phytopathology 59: 482-485.

Polychronopoulos, A. G., and Lownsbery, B. F. (1968). Effect of Heterodera
schachtii on sugar beet seedlings under monoxenic conditions. Nematologica 14: 526-534.

Potter, J. W., and Fox, J. A. (1965). Hybridization of Heterodera schachtii
and H. glycines. Phytopathology 55: 800-801.

Potter, J. W., and Marks, C. F. (1971). Effect of duPont 1410-X on rate of
development of Heterodera schachtii on cabbage (abstract). J. Nematol.
3: 325.

Price, C., and Schneider, C. L. (1965). Heterodera schachtii in relation to
damage from root rot of sugar beets. J. Am. Soc. Sugar Beet Technol.
13: 604-606.

Rademacher, B., and Schmidt, O. (1933). Die bisherigen Erfahrungen in der
Bekaumpfung des Rubcnnematoden (Heterodera schachtii Schm.) auf dem
Wege der Reizbeeinflussung. Arch. PflBau, 10: 237-296. (Helm. Abstr.,
2, No. 145a).

Radewald, J. D., Hall, B. J., and Shibuya, F., and Nelson, J. (1971). Results of preplant fumigation trial for the control of sugar beet nematode
on cabbage. Plant. Dis. Rep. 55: 841-845.

Raski, D. J. (1949). The life history and morphology of the sugar beet nematode, Heterodera schachtii. Phytopathology 40: 135-152.

Raski, D. J., and Johnson, R. T. (1959). Temperature and activity of the
sugar-beet nematode as related to sugar-beet production. Nematologica
4: 136-141.

Rebois, R. V., Steele, A. E., Stoner, A. K., and Eldridge, B. J. (1977).
Rotylenchulus reniformis resistance and a possible correlation with Heterodera schachtii resistance in tomatoes (abstract). J. Nematol. 9: 280-
281.

Rhoades, H. L. (1970). Occurrence of the sugar beet nematode, Heterodera
schachtii in Florida. Plant Dis. Rep. 54: 635.

Rietberg, H. (1954). Possibilities of breeding for tolerance against virus
yellows and beet eelworm. Proc. Am. Soc. Sugar Beet Technol. 8: 104-
108.

Roberts, P. A., Thomason, I. J., and McKinney, H. E. (1981). Influence of
nonhosts, crucifers and fungal parasites on field populations of Heterodera schachtii. J. Nematol. 13: 164-171.

Rode, H. (1962). Untersuchungen über das Wandervermögen von Larven des Kartoffelnematoden (*Heterodera rostochiensis* Woll.) in Modellversuchen mit verschiedenen Bodenarten. *Nematologica* 7: 74-82.

Rozsypal, J. (1934). Houby na hád'átku repném *Heterodera schachtii* Schmidt v moravských pudách. (Pilze in Cysten von *Heterodera schachtii* Schmidt usa mährischen Rübenboden.) (In Czech, with German summary.) *Bull. Acad. Tchecoslovaque Agric.* 10: 413-422.

Santo, G. S., and Bolander, W. J. (1979). Interacting effects of soil temperature and type on reproduction and pathogenicity of *Heterodera schachtii* and *Meloidogyne hapla* on sugar beet. *J. Nematol.* 11: 289-291.

Santo, G. S., O'Bannon, J. H., Finley, A. M., and Golden, A. M. (1980). Occurrence and host range of a new root-knot nematode (*Meloidogyne chitwoodi*) in the Pacific Northwest. *Plant Dis. Rep.* 64: 951-952.

Savitsky, H. (1975). Hybridization between *Beta vulgaris* and *B. procumbens* and transmission of nematode (*Heterodera schachtii*) resistance to sugar beet. *Can. J. Genet. Cytol.* 17: 197-209.

Savitsky, H. (1978). Nematode (*Heterodera schachtii*) resistance and meiosis in diploid plants from interspecific *Beta vulgaris* × *B. procumbens* hybrids. *Can. J. Genet. Cytol.* 20: 177-186.

Savitsky, H. (1980). Nematode resistance transmission of diploid *Beta vulgaris-procumbens* hybrids and the production of homozygous nematode-resistant plants (abstract). *Genetics* 94: s93.

Savitsky, H., and Price, C. (1965). Resistance to the sugar beet nematode (*Heterodera schachtii*) in f_1 tetraploid hybrids between *Beta vulgaris* and *Beta patellaris*. *J. Am. Soc. Sugar Beet Technol.* 13: 370-373.

Schacht, H. (1859a). Ueber einige Feinde der Rübenfelder. *Ztschr. Ver. Rübenzucher-Ind. Zollver.* 9: 175-179.

Schacht, H. (1859b). Ueber einige feinde and Krankheiten der Zucherrube. *Ztschr. Ver. Rubenzucher-Ind. Zollver.* 9: 239-250.

Schaerffenberg, B. (1951). Untersuchungen über die bedeutung der enchytraiden als nematoden feinde. *Mitt. Biol. Zen. Anst. Berl.* 70: 55-58.

Schaerffenberg, B., and Tendl, H. (1951). Untersuchungen über das *Heterodera schachtii* (Schm.). *Z. Angew. Ent.* 32: 476-488.

Schmidt, A. (1871). Über den Rubennematoden. *Z. Ver. Rubenzukerindustr. Zollver* 21: 1-19.

Schmidt, A. (1872). Zweiter Bericht über den Rübennematoden. *Z. Ver. Rubenzukerindustr. Zollver* 22: 67-75.

Schuster, M. L., Sandstedt, R., and Estes, L. W. (1965). Host-parasite relations of *Nacobbus batatiformis* and the sugar beet and other hosts. *J. Am. Soc. Sugar Beet Technol.* 13: 523-537.

Schuster, M. L., and Thorne, C. (1956). Distribution, relation to weeds, and histology of sugar beet root galls caused by *Nacobbus batatiformis*. Thorne and Schuster. *J. Am. Soc. Sugar Beet Technol.* 9: 193-197.

Seinhorst, J. W. (1965). The relation between nematode density and damage to plants. *Nematologica* 11: 137-154.

Seinhorst, J. W. (1967a). Review of methods for measuring damage caused by nematodes. *FAO Symp. Crop Losses, Rome*, pp. 311-312.

Seinhorst, J. W. (1967b). The relationships between population increase and population density in plant parasitic nematodes. III. Definitions of the terms host, host status and resistance. IV. The influence of external conditions on the regulation of population density. *Nematologica* 13: 429-442.

Senbusch, R. V. (1927). Beitrag zur biologie des Rübennematoden *Heterodera schachtii: Z. Pflanzenkr. Pflanzenpathol. Pflanzenschutz* 37: 86-102.

Shaw, H. B. (1915). The sugar-beet nematode and its control. *Sugar-Chicago*, vol. 17.

Shepherd, A. M. (1957). Development of the beet eelworm *Heterodera schachtii* Schmidt, in the wild beet *Beta patellaris*. *Nature (Lond.) 180*: 341.

Shepherd, A. M. (1962). Emergence of larvae from cysts in the genus *Heterodera*. *Tech. Comm. No. 32. Commonw. Bur. of Helm.*, England.

Shepherd, A. M. and Wallace, H. R. (1959). A comparison of the rate of emergence and invasion of beet eelworm, *Heterodera schachtii* Schmidt and pea root eelworm *Heterodera göttingiana* Liebscher. *Nematologica 4*: 227-235.

Sher, S. A. (1970). Revision of the genus *Nacobbus* Thorne and Allen 1944 (Nematoda: Tylenchoidea). *J. Nematol. 2*: 228-235.

Smart, G. C. (1963). Survival of encysted eggs and larvae of the soybean cyst nematode, *Heterodera glycines* ingested by swine. *Phytopathology 53*: 889-890.

Steele, A. E. (1961). Effect of nabam solutions on the emergence of larvae from cysts of *Heterodera schachtii* Schmidt. *J. Am. Soc. Sugar Beet Technol. 11*: 528-532.

Steele, A. E. (1962). Effects of pretreatment of *Heterodera schachtii* cysts with sugar solutions on emergence of larvae in sugar beet root diffusate. *Plant Dis. Rep. 46*: 43-44.

Steele, A. E. (1964). Influence of prolonged association of sugar beet nematode and tomato on intensity of parasitism. *J. Am. Soc. Sugar Beet Technol. 13*: 170-176.

Steele, A. E. (1965). The host range of the sugar beet nematode *Heterodera schachtii* Schmidt. *J. Am. Soc. Sugar Beet Technol. 13*: 573-603.

Steele, A. E. (1971a). Influence of dilution on the hatching activity of sugar-beet-root diffusate. *J. Am. Soc. Sugar Beet Technol. 16*: 575-576.

Steele, A. E. (1971b). Invasion of non-host plants by larvae of the sugar beet nematode, *Heterodera schachtii*. *J. Am. Soc. Sugar Beet. Technol. 16*: 457-460.

Steele, A. E. (1971c). Orientation and development of *Heterodera schachtii* larvae on tomato and sugar beet roots. *J. Nematol. 3*: 424-426.

Steele, A. E. (1972a). Evaluation of cyst selection as a means of reducing variation in sugar beet nematode inocula. *J. Am. Soc. Sugar Beet. Technol. 17*: 22-29.

Steele, A. E. (1972b). Development of *Heterodera schachtii* on large rooted crop plants and the significance of root debris as substratum for increasing field infestation. *J. Nematol. 4*:: 250-256.

Steele, A. E. (1973). The effects of hot water treatments on survival of *Heterodera schachtii*. *J. Nematol. 5*: 81-84.

Steele, A. E. (1975). Population dynamics of *Heterodera schachtii* on tomato and sugar beet. *J. Nematol. 7*: 105-111.

Steele, A. E. (1976). Effects of oxime carbamate nematicides on development of *Heterodera schachtii* on sugar beet. *J. Nematol. 8*: 137-141.

Steele, A. E. (1977a). Inheritance of resistance to *Heterodera schachtii* in *Lycopersicon* spp. (abstract). *J. Nematol. 9*: 285.

Steele, A. E. (1977b). Effects of selected carbamate and organophosphate nematicides on hatching and emergence of *Heterodera schachtii*. *J. Nematol. 9*: 149-154.

Steele, A. E. (1979). Residues of aldicarb and its oxides in *Beta vulgaris* L. and systemic control of *Heterodera schachtii*. *J. Nematol. 11*: 42-46.

Steele, A. E., and Fife, J. M. (1964). Factors affecting the hatching activity of sugarbeet-root diffusate. *Plant Dis. Rep. 48*: 229-233.

Steele, A. E., and Fife, J. M. (1969). Effect of liquid nutrient culture, vacuum distillation and dialysis on hatching activity of sugar beet root diffusate for *Heterodera schachtii*. *J. Nematol. 1*: 223-226.

Steele, A. E., and Hodges, L. R. (1974). Overwintering of *Meloidogyne incognita* in root galls of sugar beet in the Salinas Valley of California. *Plant Dis. Rep. 58*: 88-90.

Steele, A. E., and Hodges, L. R. (1975). *In vitro* and *in vivo* effects of aldicarb on survival and development of *Heterodera schachtii*. *J. Nematol. 7*: 305-312.

Steele, A. E., and Savitsky, H. (1962). Susceptibility of several *Beta* species to the sugar beet nematode (*Heterodera schachtii* Schmidt). *Nematologica 8*: 242-243.

Steele, A. E., and Savitsky, H. (1974). Quantitative and qualitative evaluation of resistance of interspecific hybrids *Beta vulgaris* × *B. procumbens* to *Heterodera schachtii* (abstract). *J. Nematol. 6*: 153.

Steele, A. E., and Savitsky, H. (1981). Resistance of trisomic and diploid hybrids of *Beta vulgaris* and *B. procumbens* to the sugar beet nematode, *Heterodera schachtii*. *J. Nematol. 13*: 352-357.

Steele, A. E., Thompson, J., and Wheatley, G. (1971). Depth of application and efficacy of soil nematicides for controlling *Heterodera schachtii*. *Plant Dis. Rep. 55*: 1101-1105.

Steiner, G. (1952). Soil in its relationship to plant nematodes. *Proc. Soil Sci. Soc. Fla. 12*: 24-29.

Stelter, H. (1976). Zur Populationsdynamik von *Heterodera schachtii* Schm. *Arch. Phytopath. Pflshutz. 12*: 393-400.

Steudel, W. (1972). Versuche zum Einfluss von Aldicarb auf den Schüpfvorgang bei Zysten von *Heterodera schachtii* nach längerer Einwirkung. *Nematologica 18*: 270-274.

Steudel, W., and Thielemann, R. (1970). Weitere Untersuchungen zur Frage der Empfindlichkeit von Zuckerruben gegen den Rübennematoden (*Heterodera schachtii* Schmidt). *Zucker 4*: 106-109.

Steudel, W., and Thielemann, R. (1979). Uber die Prognose von Schäden durch den Rübennematoden (*Heterodera schachtii* Schmidt) bei Zuckerrüben mittels Untersuchungen des Vorbefalls. *Nachrichtenbl. Dtsch. Pflanzenschutzdienstes (Braunschw.) 31*: 179-181.

Steudel, W., Thielemann, R., and Haufe, W. (1978). Der einfluss von Aldicarb auf die vermehrung des Rübenzystenälchens (*Heterodera schachtii* Schmidt) und den ertrag von Zukerrüben in der Köln-Aachener Bucht. *Nematologica 24*: 361-375.

Strubell, A. (1888). Untersuchungen über den Bau und die Entwicklung des Rubennematoden *Heterodera schachtii* Schmidt. *Cassel Bibliotheca Zoologica*, Heft 2.

Swink, J. F. (1954). Breeding for resistance to the sugar beet nematode. *Proc. Am. Soc. Sugar Beet Technol. 8*: 109-111.

Taylor, D. P. (1960). Host range study of the spiral nematode, *Helicotylenchus microlobus*. *Plant Dis. Rep. 44*: 747-750.

Thielemann, R. (1978). Zystenentwicklung des Rübennematoden *Heterodera schachtii* Schmidt an Cruciferen-Stoppelfruchten. *Z. Pflanzenkrankr Pflanzenschutz 85*: 657-665.

Thielemann, R., and Steudel, W. (1973). Neunjährige Erfahrugen mit Monokultur von Zukerrüben auf mit *Heterodera schachtii* (Schmidt) verseuchtem Boden. *Nachrichtenbl. Dtsch. Pflanzenschutzdienstes. Braunschw. 25*: 145-149.

Thomason, I. J., and Fife, D. (1962). The effect of temperature on develop-
    ment and survival of *Heterodera schachtii* Schm. *Nematologica* 7: 139-
    145.

Thomason, I. J., and O'Melia, F. C. (1962). Pathogenicity of *Pratylenchus
    scribneri* to crop plants. *Phytopathology* 52: 755.

Thorne, G. (1923). Length of the dormancy period of the sugar beet nema-
    tode in Utah. *U.S. Dept. Agric. Circ. 262.*

Thorne, G. (1926). Control of the sugar beet nematode by crop rotation.
    U.S. Dept. Agric. Farmers' Bull. No. 1514.

Thorne, G. (1927). The life history, habits and economic importance of some
    Mononchs. *J. Agric. Res. 34:* 265-283.

Thorne, G. (1949). Thirty years of nematological research in Utah. *Utah
    Acad. Sci. Arts Letters 17:* 26-31.

Thorne, G., and Giddings, L. A. (1922). The sugar beet nematode in the
    western states. *U.S. Dept. Agric. Farmers' Bull. No. 1248.*

Thorne, G., and Price, C. (1935). The nematode *Neotylenchus abulbosus*
    Steiner (Anguillulinidae) as a parasite of sugar beets. *Proc. Helm. Soc.
    Wash. 2:* 46.

Thorne, G., and Schuster, M. L. (1956). *Nacobbus batatiformis* n. sp. (Ne-
    matoda:Tylenchidae), producing galls on the roots of sugar beets and
    other plants. *Proc. Helm. Soc. Wash. 23:* 128-134.

Transhel, B. A. (1927). Review of *Beta* species L. *Bull. Appl. Bot. Genet.
    Breeding 18:* 203-223.

Triantaphyllou, A. C. (1973). Environmental sex differentiation of nema-
    todes in relation to pest management. *Annu. Rev. Phytopathol. 11:*
    441-462.

Triantaphyllou, A. C., and Hirschmann, H. (1962). Oogenesis and mode of
    reproduction in the soybean cyst nematode, *Heterodera glycines. Ne-
    matologica 7:* 235-241.

Tribe, H. T. (1977a). Pathology of cyst nematodes. *Biol. Rev. 52:* 477-507.

Tribe, H. T. (1977b). A parasite of white cysts of *Heterodera: Catenaria
    auxilieris. Trans. Br. Mycol. Soc. 69:* 367-376.

Tribe, H. T. (1979). Extent of disease in populations of *Heterodera* with
    special reference to *H. schachtii. Ann. Appl. Biol. 92:* 61-72.

Triffit, M. J. (1930). On the bionomics of *Heterodera schachtii* on potatoes
    with special reference to the influence of mustard on the escape of larvae
    from the cysts. *J. Helm. 8:* 19-48.

Triffit, M. J. (1935). The origin of strains of *Heterodera schachtii* occurring
    in Britain, with special reference to the beet-strain. *J. Helm. 13:* 149-
    158.

Triffit, M. J., and Hurst, R. H. (1935). On the thermal death-point of *Hete-
    rodera schachtii. J. Helm. 4:* 219-222.

Viglierchio, D. R. (1961). Effects of storage environment on "in vitro" hatch-
    ing of larvae from cysts of *Heterodera schachtii* Schmidt. 1871. *Phyto-
    pathology 51:* 623-625.

Vinduska, L. (1973). The effect of some substances on the hatching activity
    of the larvae of the beet nematode *Heterodera schachtii. S. B. Uvti
    (Ustav Vedeckotech Inf.) Ochr. Rostl. 9:* 247-252.

Wallace, H. R. (1954). Hydrostatic pressure-deficiency and the emergence
    of larvae from cysts of the beet-eelworm. *Nature (Lond.) 173:* 502-503.

Wallace, H. R. (1955). Factors influencing the emergence of larvae from
    cysts of the beet eelworm, *Heterodera schachtii* Schm. *J. Helm. 29:* 3-
    16.

Wallace, H. R. (1956a). Soil aeration and the emergence of larvae from cysts of the beet eelworm, *Heterodera schachtii* Schm. *Ann. Appl. Biol. 44*: 57-64.

Wallace, H. R. (1956b). The emergence of larvae from cysts of the beet eelworm, *Heterodera schachtii* Schmidt in aqueous solutions of organic and inorganic substances. *Ann. Appl. Biol. 44*: 274-282.

Wallace, H. R. (1956c). The seasonal emergence of larvae from cysts of the beet eelworm, *Heterodera schachtii* Schmidt. *Nematologica 1*: 227-238.

Wallace, H. R. (1958). Observations on the emergence from cysts and the orientation of larvae of three species of the genus *Heterodera* in the presence of host plant roots. *Nematologica 3*: 236-243.

Wallace, H. R. (1963). *The Biology of Plant Parasitic Nematodes*. Edward Arnold Ltd., London.

Weischer, B., and Steudel, W. (1972). Nematode diseases of sugar beet. In *Economic Nematology* (J. M. Webster, ed.). Academic, London, pp. 49-65.

White, J. H. (1953). Wind-borne dispersal of potato root eelworm. *Nature (Lond.) 172*: 686.

Whitehead, A. G. (1973). Control of cyst-nematodes (*Heterodera* spp.) by organophosphates, oximecarbamates and oil fumigants. *Ann. Appl. Biol. 75*: 439-453.

Whitehead, A. G. (1977). Vertical distribution of potato, beet and pea cyst nematodes in some heavily infested soils. *Plant Pathol. (Lond.) 26*: 85-90.

Whitehead, A. G., Dunning, R. A., and Cooke, D. A. (1970). Docking disorder and root ectoparasitic nematodes of sugar beet. *Rep. Rothamsted Exp. Sta. for 1970*, pp. 219-236.

Whitehead, A. G., Tite, D. J., and Fraser, J. E. (1970). The effect of small doses of nematicides on migratory root-parasitic nematodes and on the growth of sugar beet and barley in sandy soils. *Ann. Appl. Biol. 65*: 361-375.

Whitney, E. D. (1974). Synergistic effect of *Pythium ultimum* and the additive effect of *P. aphanidermatum* with *Heterodera schachtii* on sugar beet. *Phytopathology 64*: 380-383.

Whitney, E. D., and Doney, D. L. (1973a). The effects of *Heterodera schachtii* and *Aphanomyces cochlioides* on root rot of sugar beet. *J. Am. Soc. Sugar Beet Technol. 17*: 240-245.

Whitney, E. D., and Doney, D. L. (1973b). The effects of soil types, inoculum levels, fertilizers and water regimes on the development of *Heterodera schachtii* on lines of sugar beet. *J. Am. Soc. Sugar Beet Technol. 17*: 309-314.

Winslow, R. D. (1956). Seasonal variations in the hatching responses of the potato-root eelworm, *Heterodera rostochiensis* Wollenweber and related species. *J. Helm. 30*: 157-164.

Winslow, R. D., and Ludwig, R. A. (1957). Studies on hatch stimulation in the beet nematode *Heterodera schachtii* Schmidt. *Can. J. Bot. 35*: 619-634.

Winslow, R. D., and Williams, T. D. (1957). Amoeboid organisms attacking larvae of potato root eelworm (*Heterodera rostochiensis* Woll.) in England and the beet eelworm (*H. schachtii* Schm.) in Canada. *T. Pl. Ziekten 63*: 242-243.

Yu, M. H., and Steele, A. E. (1981). Host-parasite interaction of resistant sugar beet and *Heterodera schachtii*. *J. Nematol. 13*: 206-212.

Chapter 15

Nematode Parasites of Sugarcane

Wray Birchfield *USDA, Agricultural Research Service, Louisiana State University, Baton Rouge, Louisiana*

I. INTRODUCTION

Sugarcane was among the first host plants of plant-parasitic nematodes. According to Bell (1929), Soltwedel described sugarcane diseases caused by nematodes in Java in 1887. He described a new nematode species from sugarcane, and named it *Tylenchus sacchari*. Cobb (1893, 1915, 1931, 1935) published classic papers on nematodes parasitizing sugarcane in Hawaii. He described a nematode on sugarcane, which he named *Tylenchus similis*. This nematode is now known as *Radopholus similis*, a pathogen of many crop plants, including sugarcane. Cobb believed nematodes caused sugarcane diseases alone, and indirectly, in complex with fungi and bacteria.

Root-knot nematode is recorded in the early literature as the most common nematode on sugarcane. Cobb reported the first root-knot disease of sugarcane in 1887. Other early records of root-knot associated with sugarcane were those of Matz (1925), Cassidy (1930), and Muir (1926).

Van Zwaluwenburg (1926, 1930) summarized the observations and research on sugarcane nematode problems published between 1905 and 1932. He considered the nematodes *Tylenchus similis* (*Radopholus similis* and *Heterodera* (probably a root-knot nematode) of greatest importance to sugarcane at that time.

Muir, with Henderson (1926) and Van Zwaluwenburg, (1927) reported *Heterodera schachtii* Schmidt 1871; the stem and bulb nematode, *Tylenchus dipsaci* Kuehn 1858 (*Ditylenchus dipsaci*); and the meadow nematode, *Pratylenchus pratensis* de Man 1881, associated with sugarcane root rot in Hawaii. Other genera associated with sugarcane, according to Cobb (1893), are *Aphelenchus*, *Criconema*, *Anguina*, *Actinolaimus*, *Dorylaimus*, *Discolaimus*, and *Xiphinema*.

Rands (1929) reported *Heterodera* spp., *Radopholus similis*, and *Hoplolaimus* spp. associated with fungi causing sugarcane root rots in Louisiana. His was the first report of the burrowing nematode in the United States. Flor (1930) reported on the meadow nematode, *Pratylenchus pratensis*, associated with sugarcane root disease in Louisiana. Neither of these nematodes is now known in Louisiana sugarcane fields.

Evidence that nematodes cause sugarcane diseases was based on critical observations of early scientists. Stewart (1926, 1927) and Stewart and Hanson (1928) inoculated sugarcane with *T. similis* in Hawaii, but bacteria and fungi were associated with the nematode inoculum. The inoculations resulted in poor sugarcane growth attributed to nematodes, compared to good growth in nontreated check plants in his experiments.

Flor (1930) published results of controlled inoculation experiments with root-knot nematodes on sugarcane in Louisiana. He concluded that galling and sugarcane stunting, caused by root-knot nematode, were not correlated with moisture and temperature within limits favorable to the host plant.

The history and distribution, disease symptoms, pathogenicity, and host-parasite relations, economic importance, and control are reviewed for those genera on sugarcane for which we have information. Additional genera of nematodes found associated with sugarcane are mentioned.

II. SUGARCANE ROOT-KNOT NEMATODE

A. History and Distribution

Root-knot disease of sugarcane, described from Hawaii and Louisiana, prior to Chitwood's classic revision of the genus *Meloidogyne* (1949), was attributed to nematodes in the genus *Heterodera*. They most probably belonged in the *Meloidogyne incognita* group. New root-knot species, described by Golden and Birchfield (1965), from grass hosts, were likely involved with sugarcane diseases at that time. Muir and Henderson (1926) recorded a root-knot disease of sugarcane and referred to the causal nematodes as *Heterodera schachtii* Schmidt 1871, but their description was probably that of a root-knot nematode now placed in the genus *Meloidogyne*. Root-knot disease of sugarcane was reported by Cobb (1893), followed by Matz (1925), Cassidy (1930), Muir (1926), Rands (1929), Rands and Abbot (1938), Flor (1930), Martin and Field-ing (1956), David (1959), and Winchester (1968, 1969). We now know that root-knot disease of sugarcane caused by *Meloidogyne* spp. is common, and a serious sugarcane disease in the sandy and muck soils worldwide wherever sugarcane is grown. *M. incognita* and *M. graminicola* are probably the most important species on sugarcane.

B. Associated Symptoms

Sugarcane disease symptoms caused by *Meloidogyne* spp. are similar to disease symptoms caused by root-knot nematode on other hosts. Infected roots show small galls in a beadlike arrangement (Fig. 1). Brown, jellylike egg masses appear on the root surfaces. The egg masses are hard and are easily pulled away from the roots. The round, posterior parts of the females under the egg masses glisten like water bubbles. The infected roots decay, due to the combined action of nematodes and root-rotting fungi. The sugarcane becomes stunted due to the depleted root system. Diseased sugarcane stalks are thin and short; the foliage has a yellowish color that resembles that of nitrogen-deficient plants.

C. Pathogenicity and Host-Parasite Relations

Second-stage larvae, after molting and leaving the egg, infect sugarcane roots. They penetrate the roots and completely embed in the soft cortical parenchyma. The larval entry path into the root is marked by a red color reaction of the host plant. The larvae orient with the head turned toward the main root axis and feed in or near vascular tissues. After feeding, they begin to enlarge and molt a second, third, and fourth time inside the roots. After the fourth molt, young females become round to pear shaped.

The males become sausage shaped in the second and third stages, but after the fourth and final molt, inside the root, they retain the eel-shaped form and leave the roots. The function of males in fertilization is not well understood; few males are observed. Females are parthenogenetic in some

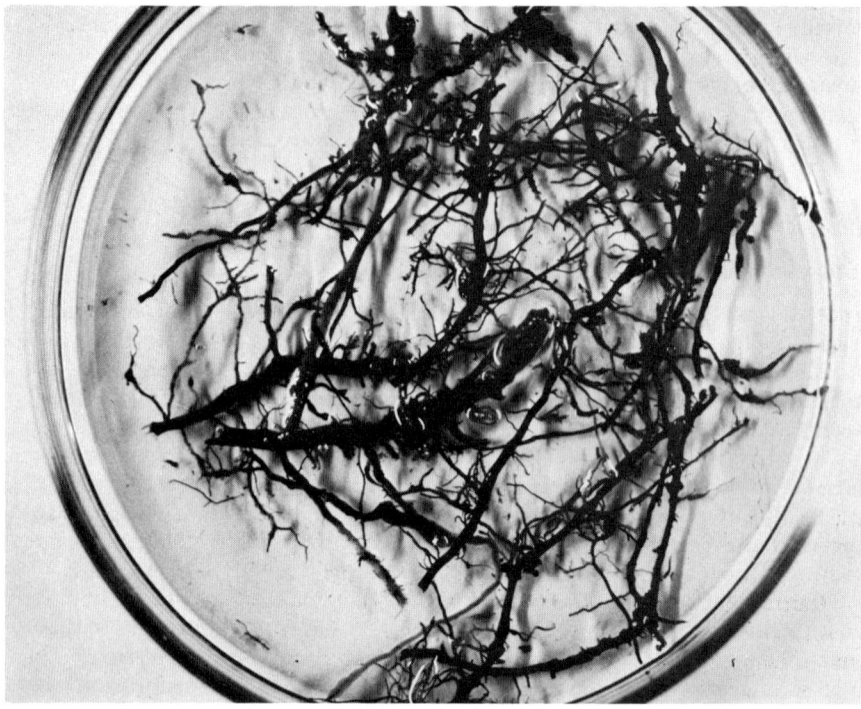

Fig. 1 Root-knot galls on sugarcane roots.

species, but may alternate with fertilization by the male in other cases. Male sperm, in some cases, are stored in the ovaries of the female. Fertilization, when it occurs, probably takes place on the epidermal roots surface where the female vulva may be exposed due to lysis of the epidermis by the vulva secretions. Females deposit 300-500 eggs per egg mass. These hatch to cause secondary infection. The life cycle is 21-35 days, depending on temperature and growing conditions of the sugarcane.

D. Economic Importance

Root-knot nematode sugarcane disease worldwide is the most economically important nematode disease of this crop. Estimated crop losses due to this nematode, based on controlled experiments in replicated plots, average about 5 tons/acre, or about 20% of the crop in infested fields according to Birchfield (1969) and the Society of Nematologists Committee on Crop Losses (1971). About 30% of the sugarcane-growing areas of the world are infested with root-knot nematodes.

E. Control

Few commercial sugarcane varieties are known to have root-knot nematode resistance. Marcano (1971) discovered root-knot resistance in Canal Point Variety 48-103 and moderate resistance in varieties L-62-98, L-62-88, and C.P.-36-105. He listed 15 additional varieties susceptible to *M. incognita*. Some varieties selected for vigor probably have moderate resistance. Crop rotation and fallow plowing, normally practiced in sugarcane culture for weed control, controls root knot, if a nonhost crop is used in the rotation. Carbamate nematicides, such as Furadan and Temik, are highly effective in root-knot control, and other plant-parasitic sugarcane nematodes. Also, the organophosphates, such as Mocap, Dasanit, and Nemacur, are equally effective in low, safe dosages applied in-furrow treatment on sugarcane at planting. Nematicide control is not cost effective in many parts of the world, and critical judgments must be made on the economic advantages of their usage.

Biological control of root-knot nematodes of sugarcane may eventually be realized by use of a protozoan that has an affinity for root-knot larvae. Williams (1960) described the life cycle of *Duboscqia penetrans* found in sugarcane fields of Mauritius. Birchfield and Antonopoulos (1975) found it common in Louisiana sugarcane fields.

III. LESION OR MEADOW NEMATODE

A. History and Distribution

Several lesion or meadow nematode species, *Pratylenchus* spp., are reported parasitic on sugarcane. The most common and destructive are *P. brachyurus* and *P. zeae*.

Soltwedel, according to Filipjev and Stekhoven (1941), described *P. sacchari* in sugarcane roots from Hawaii in 1888. This species is now *species inquirenda*, having not since been reported. Filipjev and Stekhoven listed *P. pratensis* de Man as a parasite of sugarcane and other plants. Jensen et al. (1959) first reported *P. brachyurus* (Godfrey 1929) Goodey 1951 (Syn. *P. leiocephalus* Steiner 1949), commonly occurring on sugarcane in Hawaii. Birchfield and Martin (1956) and Fielding and Hollis (1956) found *P. brachyurus*, the most common lesion or meadow nematode, in Louisiana sugarcane fields. According to Khan (1959), 76% of soil samples collected from Louisiana sugarcane fields were infested with *P. brachyurus* and/or *P. zeae*.

B. Pathogenicity and Host-Parasite Relations

Lesion-meadow nematodes occur as sugarcane root endoparasites. All stages are infective on sugarcane: males, females, second-, third-, and fourth-stage larvae. All stages are found inside sugarcane roots, including the eggs. Eggs hatch, and larvae feed and mature in lesions made in the cortical parenchyma. Copulation and fertilization take place in root lesions where eggs are layed. The life cycle is completed in 35-65 days on sugarcane, depending on temperature and moisture. When sugarcane roots begin to decay, the nematodes leave the roots, become vagrant, and find new roots for food. Specimens are generally present in the soil near sugarcane roots. They are highly resistant to desiccation and adverse growing conditions, and persist several months in fallow soil without food. Khan (1959) observed the pathogenicity of *P. zeae* on sugarcane in greenhouse cultures. He reported severe weight losses to tops and roots caused by this nematode. He also studied the effects of *P. zeae* on sugarcane with the fungus *Phytophthora megasperma*. They caused sugarcane root rot separately, but no synergistic effects were observed when placed in combination.

C. Associated Symptoms

Pratylenchus species cause lesions or blotches on sugarcane roots. These lesions are red in color on young roots, but later turn to a dark brown and then to black. The lesions are elliptically shaped spots 5-10 mm in diameter,

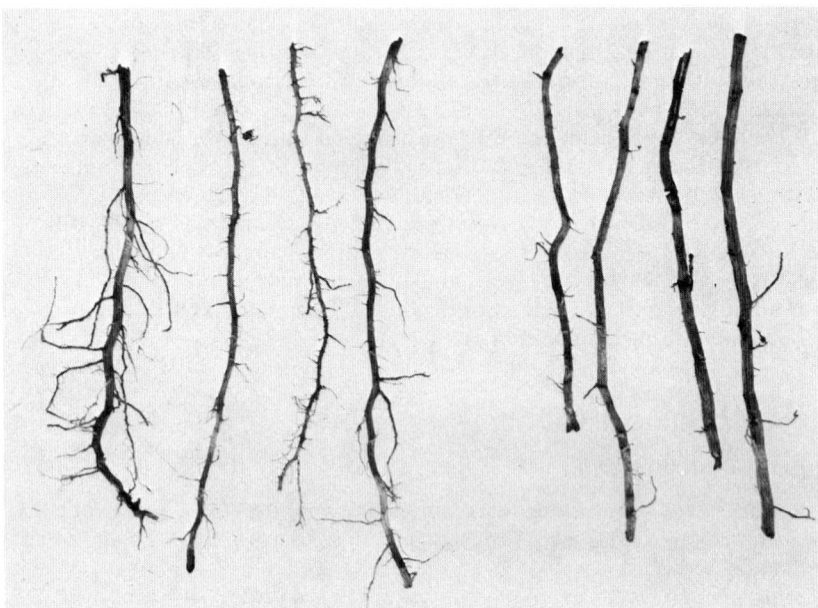

Fig. 2 Left, sugarcane roots grown in steam-sterilized soil. Right, sugarcane roots exposed to the lesion nematode, *P. zeae*.

located in the young sugarcane root cortex. Eggs, larvae, and adults are found inside these lesions. Infection and lesion formation on sugarcane roots are not confined to root tips or any particular root area, but the parasite infects and burrows only in young, white adventitious roots. Infected roots respond to the parasite by forming additional roots; these added roots are stubby in appearance, a condition caused by secondary infection. Normal sugarcane roots are white and succulent, but naturally become brown to black and corky with age. The dark color hides lesion-nematode symptoms on older roots. Khan described yellowing of sugarcane leaves, suppressed growth, reduced internode numbers, and reduced green and dry top and root weights associated with *P. zeae*. Sugarcane roots inoculated with *P. zeae* become sparse and blunt compared to normal roots that were parasite free (Fig. 2).

D. Economic Importance

The wide distribution of *P. brachyurus* and *P. zeae*, the root rot they cause, and presence of other lesion nematode species make this group one of the most damaging economically important to sugarcane. Exact sugarcane tonnage and sugar losses caused by these nematodes have not been estimated. However, an 11-18% increase in sugarcane tonnage was recorded in Louisiana with the use of Temik, Mocap, and Dasanit where *P. brachyurus* was present in combination with several other plant-parasitic genera, according to Birchfield (1969). This increase amounted to 4-5 tons/acre. *Pratylenchus* spp. damage to sugarcane is believed to be second only in economic importance to root-knot nematodes.

E. Control

No sugarcane varieties are known with resistance to *Pratylenchus* spp. Sugarcane varietal selection has traditionally been based on higher sucrose yields. Varieties decline rapidly due to disease susceptibility, and new varieties rapidly replace them. This unstable condition in the industry has not contributed greatly to nematode and other disease control with resistant varieties.

Pratylenchus spp. are easily controlled with nematicides. Their slow rate of reproduction and low population levels in sugarcane render them susceptible to chemical control. About a 5 ton increase can be realized with the organophosphate and carbamate nematicides placed in granular form in the furrow at planting time.

Crop rotation is ineffective in controlling *Pratylenchus* spp. because of its wide host range and holdover in weed hosts in the field. Biological control is not a reality for controlling this nematode at present.

IV. LANCE NEMATODES

A. History and Distribution

Several lance nematode species, *Hoplolaimus* spp., have been reported from sugarcane, including *H. columbus*, *H. tylenchiformis*, *H. galeatus*, and *H. seinhorsti*. Cobb (1923) reported *H. coronatus (H. galeatus)* on sugarcane roots. Flor (1930) listed a *Hoplolaimus* species among factors influencing root-rot disease of sugarcane in Louisiana. Birchfield et al. (1978) found *H. tylenchiformis* common on sugarcane in Louisiana. Astudillo (1979) was the first to find and report *H. columbus* causing damage to sugarcane. This species is known to cause extensive damage to soybeans, according to Fassuliotis (1975). Soybeans are often grown in rotation with sugarcane.

B. Pathogenicity and Host-Parasite Relations

Astudillo (1979) made critical inoculation experiments with *H. columbus*, proved its pathogenicity to sugarcane, and recorded the host-parasite relations. He discovered that females completely enter and feed in the cortical parenchyma of sugarcane. They deposit eggs inside the roots, which hatch and produce larvae that continue to feed and complete the life cycle. All stages were found inside the roots. The endoparasitic habit of this nematode on soybeans was first reported by Fassuliotis (1975), who assigned the common name, the Columbia nematode. According to Astudillo, the Columbia nematode population increased on sugarcane and significantly reduced fresh and dry top and root weights. Stained sugarcane roots showed all nematode stages inside the roots. Many nematodes were observed in 2 mm of sugarcane tissue near the vascular vessels. They caused parenchyma cell necrosis at the feeding site and mechanical damage to the cortex resulting from entrance wounds. *H. columbus* fed on the primary, secondary, and tertiary sugarcane roots, but not on root tips. According to Astudillo, first-stage larvae developed 11 days after eggs of the Columbia nematode were layed on sugarcane at 27°C. Astudillo's observations on the embryology of *H. columbus* agreed with studies reported by Fassuliotis on *H. columbus* on soybeans and observations of Dasgupta et al. (1970) with *H. indicus*.

Complete life cycles of other *Hoplolaimus* species on sugarcane are unknown.

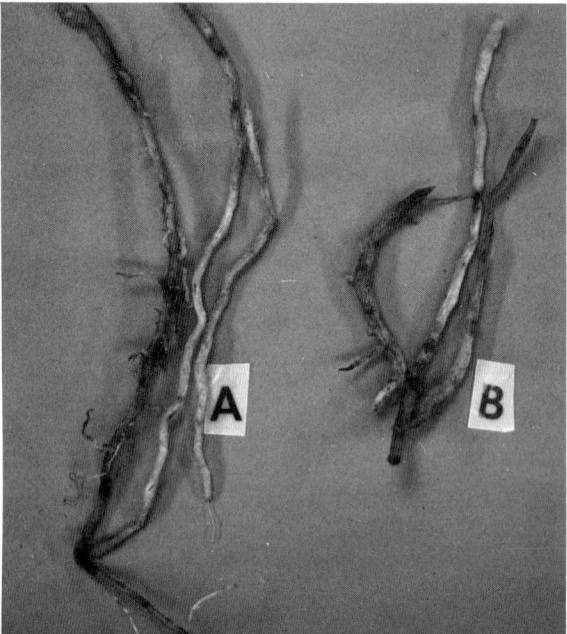

Fig. 3 (A) Healthy sugarcane roots. (B) Sugarcane roots grown in the presence of the Columbia nematode, *H. columbus.*

C. Associated Symptoms

Astudillo (1979) observed stubby, sparse, and decayed sugarcane roots associated with *H. columbus* in the field (Fig. 3). It caused cell necrosis and damage to root parenchyma where the nematode fed in sugarcane roots. Fresh and dry root and shoot weights were reduced by this nematode, but visible top symptoms were not observed on sugarcane.

D. Economic Importance

The economic importance of lance nematodes on sugarcane under field conditions is not known. Most nematicide tests have been done in sugarcane fields where *Hoplolaimus* occurs in combination with *Pratylenchus* spp., *Helicotylenchus* spp., *Criconemella* spp., and *Meloidogyne* spp. Lance nematodes are believed to cause economic losses to sugarcane production based on greenhouse tests that show weight reduction of tops and roots. With this indirect evidence, plus the common occurrence of several *Hoplolaimus* spp. in many areas where sugarcane is grown, it is believed to cause extensive production losses in sugarcane tonnage and sucrose.

E. Control

Resistant sugarcane varieties to lance nematodes are unknown. Most crops grown in rotation with sugarcane are lance nematode hosts. Fallow plowing is an effective control measure, but impractical, because of increasing land and crop values. Soybeans and corn should be avoided in rotation with sugarcane where lance nematodes are present because these crops are good hosts and keep the population high.

According to Birchfield (1965), nematicides control lance nematodes and increase sugarcane production, according to experimental tests where lance nematodes occur with other nematode genera. When resistant sugarcane varieties and crops that can be rotated profitably with sugarcane are developed, these should be used, in combination with nematicides, for control. Fields should be kept free of Johnson grass, a weed that is a good host for lance nematodes.

V. STUNT-STYLET NEMATODES

A. History and Distribution

The first report of a stunt-stylet nematode on sugarcane was by Birchfield (1953). He showed the life cycle and pathogenicity of a new *Tylenchorhynchus* species, later named *T. martini* in honor of W. J. Martin (Fielding, 1956). Host-plant studies showed all Louisiana sugarcane varieties are susceptible. A large host list was recorded for this nematode, including several noxious weeds found in sugarcane fields. Since then, *T. martini* has been reported as a sugarcane pathogen in all sugarcane-growing areas worldwide and on many other crop plants. Other species that feed on sugarcane are: *T. claytoni, T. acutus, T. nudus, T. dactylurus, T. curvus, T. crassicaudatus, T. brevilineatus,* and *T. elegans,* according to Prasad (1972).

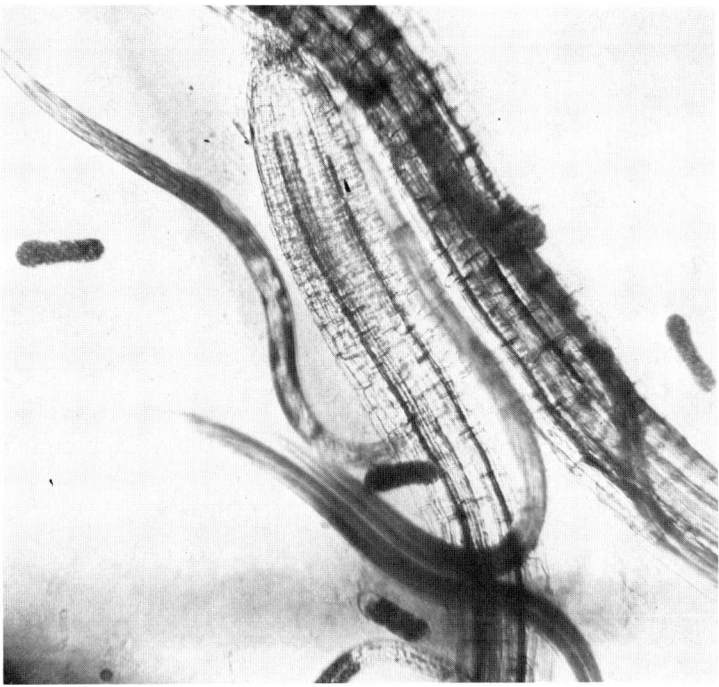

Fig. 4 Stunt or stylet nematode feeding on sugarcane roots.

B. Pathogenicity and Host-Parasite Relations

Only the pathogenicity and host-parasite relations of *T. martini* on sugarcane are known. Birchfield and Martin (1956) showed this nematode feeds on the root surfaces of sugarcane roots, root hairs, and epidermal cells. It occurs in large numbers in the soil around sugarcane roots. *T. martini* is a mild sugarcane parasite. It requires 500 or more per pint of soil to cause measurable damage. Males are unknown for this species. The parthenogenetic female lays eggs near feeder roots in the soil; these hatch in about 8-10 days and feed on the young sugarcane roots (Figure 4). All larval stages and the adult female feed.

The life cycle of *T. martini* on sugarcane under greenhouse conditions took 14-21 days from egg to egg. Eggs hatch in a few hours. The second-, third-, and fourth-stage larvae, and females feed on sugarcane roots as ectoparasites. They survive long periods without food in fallow sugarcane fields.

C. Associated Symptoms

Disease symptoms on sugarcane tops are not evident, even where large numbers of *T. martini* feed on the roots. However, roots exposed to the nematode show poor development. The roots are stubby and sparse and have a coarse appearance caused by this nematode (Fig. 5). No internal root symptoms are caused by *T. martini*, which feeds on root hairs and epidermal tissue from outside the roots. Single sugarcane nodes planted and inoculated in replicated tests in the greenhouse showed a high reproduction of *T. martini* on

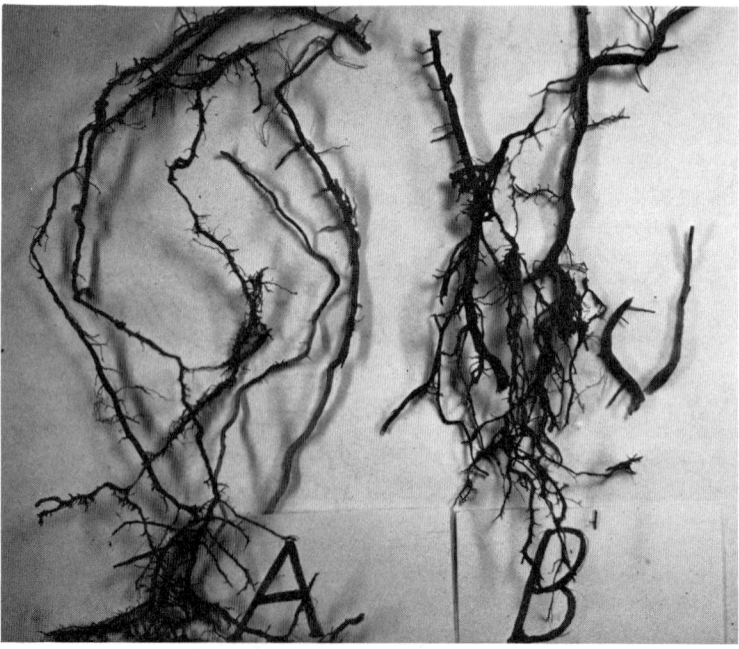

Fig. 5 (A) Healthy sugarcane roots. (B) Sugarcane roots grown in presence of the stunt-stylet nematode, *T. martini*.

sugarcane. Heights and green weights of plants grown in infested soil were lower than normal plants grown in noninfested soil. However, stunt-stylet nematodes are generally considered mild pathogens of most host plants, and sugarcane is no exception with *T. martini*.

D. Economic Importance

The economic importance of stunt-stylet nematodes on sugarcane is not great, although it is generally considered that they do minor damage to this crop. Sugarcane responds to soil fumigation and granular applications of nematicides where this nematode is present in large numbers. Thus, ethylene dibromide at 2 gal/acre increased sugarcane production 3.6 tons/acre or 15% more than the nontreated where this nematode was present, according to Birchfield (1965). Recent field experiments with newer organophosphate and carbamate nematicide granules applied on plant cane to control this nematode confirmed these results.

E. Control

Resistant sugarcane varieties are unknown for control of *Tylenchorhynchus* spp. This nematode group has a wide host range, and immune or resistant crops of economic importance are unknown for rotation with sugarcane. Nematicides are available that will control this nematode group effectively. However, the margin between use of nematicides and making a profit is narrow with this nematode, and therefore use of chemicals for control is questionable. Crop rotation is not effective because these nematodes can live a long time without food, and have a wide host range. Better methods for economic control of this group of parasites must await further research.

VI. SPIRAL NEMATODES

A. History and Distribution

Cobb (1893) reported *Helicotylenchus dihystera* from sugarcane in Hawaii. Jensen et al. (1959) reported *H. dihystera* again on sugarcane in Hawaii. This is among the most common nematodes associated with sugarcane roots in Louisiana. It is generally associated with sugarcane wherever this crop is grown. Other genera, referred to as spiral nematodes, *Rotylenchus* spp., and *Scutellonema* spp., have been described from sugarcane roots in southern Rhodesia and Kenya according to Prasad (1972).

B. Pathogenicity and Host-Parasite Relations

Spiral nematodes are ectoparasitic feeders, and although obligate parasites, are mild host pathogens. They feed from outside the roots by repeatedly jabbing the stylet into the epidermis of young succulent roots. Rarely do these nematodes completely embed in sugarcane roots. Eggs are layed in the soil close to young roots. These hatch in two to three days, molt in the soil, and feed on the roots. Mature males and females also feed on sugarcane and other host plants.

C. Associated Symptoms

Spiral nematodes do not ordinarily cause pronounced disease symptoms on sugarcane roots. Occasionally, *H. dihystera* buries the anterior in young adventitious roots, and leaves brownish red lesions where it feeds. No top symptoms on sugarcane are visible when spiral nematodes feed on the roots. Prasad (1972) reported *H. dihystera* completely embed in the cortex and feed near the stele. He reported disorganization and collapse of cortical tissues infected with *H. dihystera*. Other symptoms were sloughing of epidermal cells and necrosis (death) of cells in the feeding area, extending far from the feeding site of the parasite. Jensen et al. (1959) observed *H. dihystera* with the head embedded in sugarcane roots. Secondary infection occurred at the feeding site. Apt and Koike (1962a) reported blunt, malformed roots, and small branch root reduction caused by *H. dihystera*.

D. Economic Importance

The damage that spiral nematodes do to sugarcane and other host plants is not considered extensive. On sugarcane, where these may occur in high numbers, the damage cannot be measured with nematicide treatment. The damage is considered below the level of detection.

The mildness of their parasitic habit has not prompted researchers to identify resistant nematode germplasm, and consequently no resistant sugarcane varieties to spiral nematodes are known.

E. Control

It is generally not considered economically feasible to control spiral nematodes on sugarcane, since the damage they do is too small to measure. Their impact on the industry is not great.

VII. STUBBY-ROOT NEMATODE

A. History and Distribution

Stubby-root nematodes (*Trichodorus* and *Paratrichodorus* spp.) were first recognized as plant parasites by Christie (1953). Christie (1959) reported injury to sweet corn and vegetable crops in Florida caused by this nematode. Several *Trichodorus* and *Paratrichodorus* species, which cause stubby-root disease of corn and vegetable crops, are found abundantly associated with sugarcane wherever this crop is grown. Hawaii, South Africa, Australia, the Dominican Republic, Puerto Rico, Cuba, Florida, and Louisiana are only a few places that stubby-root nematodes occur. It is believed to cause much damage to sugarcane, but few experiments have been made to prove the pathogenicity of this genus.

Jensen et al. (1959) noticed *P. minor*, and *P. porosus* among other nematodes associated with sugarcane decline in Hawaii. Martin and Birchfield (1955) reported the association of this nematode with sugarcane in Louisiana. Prior to this, no record of sugarcane as a host of stubby-root nematodes was known.

B. Associated Symptoms

Critical inoculation tests by Apt and Koike (1962b) with *Paratrichodorus minor* on sugarcane seedlings showed severe stunting and lack of feeder roots on seedlings grown in nematode cultures. It preferred to feed on root tips. Top growth was suppressed, and leaves were reduced with high nematode populations. Dry weights of roots in stubby-root nematode-infested soil was less than dry weights of roots grown in noninfested soil. Also, dry weights

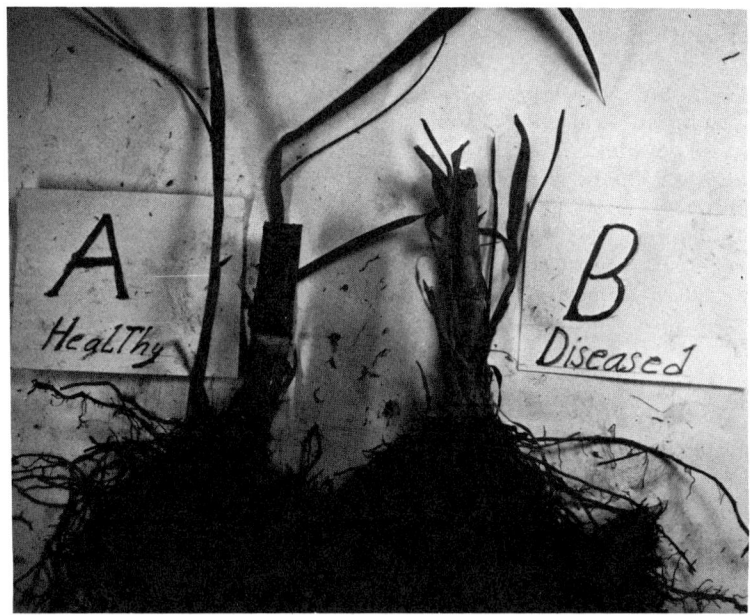

Fig. 6 (A) Healthy sugarcane root systems. (B) Diseased sugarcane caused by stubby root nematode, *P. minor*.

of tops were reduced by large stubby-root nematode populations. In Louisiana, large stubby-root nematode populations are believed responsible for tonnage reduction and sugarcane foliage discoloration (yellowing).

C. Pathogenicity and Host-Parasite Relations

Stubby-root nematodes cause a stubby appearance of sugarcane roots. They prefer to feed on young root tips. When they feed they secrete digestive enzymes into the root tissues that inhibit or stop root elongation. This results in stubby, blunt, coarse roots resembling herbicide or chemical damage to the roots (Fig. 6). *Paratrichodorus* spp. feed ectoparasitically on sugarcane and other hosts. All stages feed vagrantly. Eggs are layed near the roots; these hatch in a few days during the growing season. The life cycle from egg to egg on sugarcane is believed to be short, since large populations build up quickly during the growing season. Rhode and Jenkins (1958) found that *P. minor* completed its life cycle in 16-17 days at 30°C and 21-22 days at 22°C on tomatoes. Most likely the life cycle on sugarcane from egg to egg is similar, but detailed life cycle studies have not been reported on this host.

D. Economic Importance

The amount of damage to sugarcane caused by *Paratrichodorus* spp. is unknown. Root symptoms caused by these nematodes are not obvious; no root lesions are present, but leaf discolorations are seen on sugarcane where the parasite feeds. Critical basic research is needed to determine accurately the amount of tonnage and sucrose losses to sugarcane caused by *Paratrichodorus* spp.

E. Control

No experiments to control stubby-root nematodes on sugarcane have been reported. Christie (1959) experienced difficulty in controlling species of stubby-root nematodes on corn and vegetables in Florida. These pests were easy to kill with soil fumigation, but reestablished rapidly after about six weeks and built up to large populations on suitable hosts. Limited field observations suggested that fallow or fallow and tillage may be effective against stubby-root nematodes, but that flooding is not very effective.

P. minor soil counts after nematicide treatments with organophosphates and carbamates were reduced effectively in Louisiana sugarcane fields. Yields were increased with nematicides where *P. minor* occurred with other spear-bearing types.

VIII. OTHER NEMATODES REPORTED FROM SUGARCANE

Many other nematode genera are commonly associated with sugarcane, but we know little about the biology and damage they do.

Rotylenchulus spp. have been reported on sugarcane roots from several parts of the world, but in general fail to reproduce on plants in the family Graminae. *Rotylenchulus reniformis* does not parasitize sugarcane in Louisiana, although according to Roman (1961) it is a serious parasite of sugarcane in Puerto Rico. There may be races of this nematode involved.

Sting nematode, *Belonolaimus gracilis*, was reported from sugarcane in Louisiana. It produced disease symptoms on sugarcane in greenhouse tests. However, is not widely distributed in Louisiana sugarcane fields or other sugarcane areas of the world.

Cyst nematode, *Heterodera* spp., are associated with sugarcane roots in India and Africa, but remain of academic interest in regard to the damage caused on sugarcane.

The burrowing nematode, *Radopholus similis*, is known to parasitize sugarcane as reported by Van Zwaluwenburg (1932) and Williams (1959). It causes diseases of citrus, black pepper, and several tropical crop plants. Experimental evidence supporting it as a serious pathogen to sugarcane is lacking.

Other nematode genera from various sugarcane areas associated with sugarcane according to Prasad (1972) are: *Rotylenchus, Longidorus, Criconema, Hemicycliophora, Dolichodorus, Criconemella, Hemicriconemoides, Xiphinema*, and various members of the Dorylaimidae. The economic importance of these on sugarcane has not been determined.

IX. NEMATODE-FUNGAL AND BACTERIAL ASSOCIATES

The hypothesis that a single pathogen causes a single disease is seldom tenable in discussions of soil-borne plant diseases; so it is with nematodes. Seldom, if ever, in the plant root zone do nematodes occur alone. Nematodes may cause plant diseases by several methods. Often they provide avenues of entrance for bacteria that are not capable of causing plant disease unless they have a natural or artificial opening to incite the disease. Some fungi are capable of penetrating root tissue directly, but others need a natural opening or avenue of entrance. Nematodes may provide avenues of entrance for weak fungi that otherwise would not infect the host. Some nematode genera are known to vector plant viruses.

The classic example of a nematode-fungal complex involves *Fusarium* wilt of cotton and root-knot nematodes, *M. incognita*. Cotton *Fusarium* wilt seldom occurs unless root knot is present in the soil. Without root knot, it is a mild disease, if it occurs at all. Root knot will cause the most *Fusarium* wilt-resistant cotton varieties to be susceptible. The exact relationship involved is not definitely known. Such relationships involving bacteria, fungi, nematodes, and sugarcane viruses have not been proven. However, there is strong evidence that nematodes are involved in a disease of sugarcane called Rattoon stunting disease (RSD), thought by some to be caused by a virus and others a bacterium. When soil is methyl bromide fumigated or when other nematicides are used, symptoms of RSD are suppressed until the nematode population builds up again. It sometimes takes six months or a year for sugarcane, known to have RSD, to express the symptoms after nematicide treatment. Work with combinations of *Tylenchorhynchus martini* and *Pythium arrhenomanes* showed that no synergistic effects were involved. Both organisms produced diseased sugarcane roots separately, according to Birchfield and Martin (1956). Khan (1959) was unable to show any relationships between *Pratylenchus zeae* and *Phytophthora megasperma*. Apt and Koike (1962a) investigated the pathogenicity of *Helicotylenchus nannus* and its relation with *Pythium graminicola* on sugarcane in Hawaii. When plants were grown in the presence of *H. nannus* the fungus reduction in dry weight was greater than that caused by either agent alone.

Many possible combinations of different nematodes, root-rotting fungi, bacteria, and insects may occur in the same field. In Louisiana seldom is a sugarcane stubble examined that does not have six to eight genera of plant-parasitic nematodes, plus the root-rotting organisms—*Pythium* spp., *Phytophthora* spp., and the red rot fungus *Colletotrichum falcatum*. Consequently, stubble deterioration caused by diseases is one of the main problems in sugarcane production. Louisiana farmers are likely to raise only two to three crops of sugarcane per planting, one seed crop, and one to two stubbles. If nematodes and root diseases are controlled, a single planted crop will yield profitably. Soil types have a great deal to do with the type of nematode-fungal complexes encountered. The sandier soils have mostly nematode problems, but the heavier soils are most apt to have *Pythium*, *Phytophthora*, and *Colletotrichum* involved in nematode complexes.

REFERENCES

Apt, W. J., and Koike, H. (1962a). Pathogenicity of *Heliocotylenchus nannus* and its relation with *Pythium graminicola* on sugarcane in Hawaii. *Phytopathology 52*: 798-802.

Apt, W. J., and Koike, H. (1962b). Influence of stubby-root nematode and growth of sugarcane in Hawaii. *Phytopathology 52*: 963-964.

Astudillo, E. E. (1979). Pathology of *Hoplolaimus columbus* on sugarcane. M.S. thesis, Louisiana State University, Baton Rouge.

Bell, A. F. (1929). A key for the identification of sugarcane diseases. *Bur. Sugar Exp. Sta. Queensland Bull. 2*: 55-61.

Birchfield, W. (1953). Parasitic nematodes associated with diseased roots of sugarcane. *Plant Dis. Rep. 37*: 38.

Birchfield, W. (1965). Effects of soil fumigation and organic amendments on plant-parasitic nematodes and sugarcane yields (abstract). *Phytopathology 55*: 1051-1052.

Birchfield, W. (1969). Nematicides for control of plant-parasitic nematodes of sugarcane in Louisiana. *Plant Dis. Rep. 53:* 530-533.

Birchfield, W., and Antonopoulos, A. A. (1975). Scanning electron microscopic observations of *Duboscqia penetrans* parasitizing root-knot larvae. *J. Nematol. 8:* 272-273.

Birchfield, W., and Martin, W. J. (1956). Pathogenicity and host plant
studies of a species of *Tylenchorhynchus. Phytopathology 46:* 277-280.

Birchfield, W., Martin, W. J., and Hollis, J. P. (1978). A list of nematodes
associated with some Louisiana plants. *LA State Univ. Tech. Bull. 101.*

Cassidy, G. (1930). Nematodes associated with sugarcane in Hawaii. *Hawaiian Planters Record 34:* 374-381.

Chitwood, B. G. (1949). Root-knot nematodes. Part I-A revision of the
genus *Meloidogyne* Goeldi 1887. *Proc. Helm. Soc. Wash. 16:* 90-104.

Christie, J. R. (1953). Ectoparasitic nematodes of plants. *Phytopathology
43:* 295-297.

Christie, J. R. (1959). *Plant Nematodes, Their Bionomics and Control.* W.
B. Drew Co., Jacksonville, Florida.

Cobb, N. A. (1893). Nematodes attacking sugarcane. *Agr. Gazette N. S.
Wales,* pp. 808-833.

Cobb, N. A. (1915). *Tylenchus similis,* the cause of a root disease of sugarcane and banana. *J. Agric. Res. 4:* 561-588.

Cobb, N. A. (1923). An amendation of *Hoplolaimus* Daday, 1905 nec ductores.
Contr. Sci. Nematol. 13: 363-370.

Cobb, N. A. (1931). *Tylenchus similis. J. Parasitol. 21:* 315-322.

Cobb, N. A. (1935). A key to the genera of free-living nematodes. *Proc.
Helm. Soc. Wash. 2:* 1-40.

Dasgupta, D. R., Naud, S., and Seshadri, A. R. (1970). Culturing, embryology, and life history studies on the lance nematode, *Hoplolaimus indicus.
Nematologica 16:* 235-248.

David H. (1959). The root-knot nematode in sugarcane. *Indian J. Sugar
Res. Dev. 3:* 234-236.

Fassuliotis, G. (1975). Feeding, egg laying, and embryology of the Columbia
lance nematode, *Hoplolaimus columbus. J. Nematol. 7:* 152-158.

Fielding, M. J. (1956). *Tylenchorhynchus martini,* a new nematode found in
sugarcane and rice fields of Louisiana and Texas. *Proc. Helm. Soc.
Wash. 23:* 47-48.

Fielding, M. J., and Hollis, J. P. (1956). Occurrence of plant-parasitic nematodes in Louisiana soils. *Plant Dis. Rep. 40:* 402-405.

Filipjev, I. M., and Stekhoven, J. H. S. (1941). *A Manual of Agricultural
Helminthology.* E. J. Brill Co., Leiden, The Netherlands.

Flor, H. H. (1930). Factors influencing the severity of the root rot troubles
of sugarcane. *LA Bull.,* p. 212.

Golden, A. M., and Birchfield, W. (1965). *Meloidogyne graminicola* (Heteroderidae), a new species of root-knot nematode from grass. *Proc. Helm.
Soc. Wash. 32:* 228-231.

Jensen, H. J., Koike, H., Martin, J. P., and Wismer, C. A. (1959). Nematodes associated with varietal decline of sugarcane in Hawaii. *Plant Dis.
Rep. 43:* 253-260.

Khan, S. A. (1959). Studies of *Pratylenchus zeae* (Nematoda: Tylenchida)
on sugarcane in Louisiana. Ph.D. dissertation, Louisiana State University, Baton Rouge.

Marcano, J. C. (1971). Resistance in sugarcane varieties to root-knot nematode (*Meloidogyne incognita* group). M.A. thesis, Louisiana State University, Baton Rouge.

Martin, W. J., and Birchfield, W. (1955). Notes on plant-parasitic nematodes in Louisiana. *Plant Dis. Rep. 39*: 3-4.

Martin, W. J., and Fielding, M. J. (1956). The root-knot nematode, *Meloidogyne incognita acrita* on sugarcane in Louisiana. *Plant Dis. Rep. 40*: 406.

Matz, J. (1925). Root-knot on sugarcane in Puerto Rico. *Phytopathology 15*: 539-563.

Muir, F. (1926). Nematodes considered in relation to root-rot of sugarcane in Hawaii. *Rep. Hawaiian Sugarcane Technologists*, pp. 14-18.

Muir, F., and Henderson, G. (1926). Nematodes in connection with sugarcane root rot in the Hawaiian islands. *Hawaiian Planters' Record 30*: 233-250.

Muir, F., and van Zwaluwenburg, R. H. (1927). A generic list of spear-bearing nematodes with a revised dichotomous table. *Hawaiian Planters' Record 31*: 354-361.

Prasad, S. K. (1972). Nematode diseases of sugarcane. In *Economic Nematology* (J. M. Webster, ed.). Academic, New York, pp. 144-158.

Rands, R. D. (1929). Fungi associated with root rots of sugarcane in the Southern United States. *Proc. 3rd Cong. Int. Soc. Sugarcane Technologists*, pp. 119-131.

Rands, R. D., and Abbott, E. V. (1938). Sugarcane diseases in the United States. *Proc. Int. Sugarcane Technologists*: 202-212.

Rhode, R. A., and Jenkins, W. R. (1958). The chemical basis of resistance of asparagus to the nematode *Trichodorus christiei* (abstract). *Phytopathology 48*: 463.

Roman, J. (1961). Pathogenicity of five isolates of root-knot nematodes (*Meloidogyne* spp.) to sugarcane roots. *J. Agr. Univ. Puerto Rico 45*: 55-84.

Society of Nematologists Committee on Crop Losses. (1971). Estimated crop losses due to plant-parasitic nematodes in the United States. Special publication No. 1.

Soltwedel, F. (1887). *Tylenchus sacchari* n. sp. *Bylage Archief voor de Java Suberindustrie*, pp. 7-12.

Stewart, G. R. (1926). The possible relation between nematode injury to plant roots and soil conditions. *Rep. Hawaiian Sugarcane Technologists*, pp. 10-15.

Stewart, G. R. (1927). Further studies of the relation between soil conditions and nematodes in cane roots. *Proc. 2nd Cong. Sugarcane Technologists*, pp. 16-21.

Stewart, G. R., and Hanson, F. (1928). A study of the effect of nematodes upon cane roots in sterilized soils. *Hawaiian Planters' Record 32*:217-223.

Van Zwaluwenburg, R. H. (1926). The soil fauna of Hawaiian sugarcane fields. *Hawaiian Planters' Record 30*: 250-255.

Van Zwaluwenburg, R. H. (1930). Summary of investigations of the soil fauna of sugarcane fields in Hawaii. *Int. Soc. Sugarcane Tech. Proc. 3rd Cong.*, pp. 216-225.

Van Zwaluwenburg, R. H. (1932). The nematodes attacking sugarcane roots in Hawaii. *Proc. Int. Sugarcane Technologists. 4th Cong. Bull*: 5.

Williams, J. R. (1959). Nematode investigations. *Mauritius Sugarcane Ann. Rep.*, p. 60.

Williams, J. R. (1960). Nematode investigations. *Mauritius Sugarcane Ann. Rep.*, p. 69.

Winchester, J. A. (1968). Some effects of root-knot nematodes on sugarcane. *Nematologica 14*: 18-19.
Winchester, J. A. (1969). Nematicide trials in Florida sugarcane. *Pathologist Newsletter 2*: 41.

Chapter 16
The Pine Wood Nematode

Yasuharu Mamiya *Forestry and Forest Products Research Institute, Ibaraki, Japan*

I. HISTORY

A. Pine Wilt in Japan

The wilting disease of pine trees caused by the pine wood nematode, *Bursaphelenchus xylophilus* (Steiner and Buhrer 1934) Nickle 1970, is widespread in stands of Japanese red pine (*Pinus densiflora* Sieb. and Zucc.) and Japanese black pine (*P. thunbergii* Parl.) throughout Japan.

The first occurrence of pine wilt in Japan goes back to the early part of this century. Yano (1913), a forest entomologist, reported the death of pine trees in Nagasaki of Kyushu in epidemic proportions. The damage had spread since 1905, and the causal agent of the disease had not been determined at that time. According to his report, however, disease symptoms of pine trees correspond well to those symptoms induced currently by the pine wood nematode. Since then this dreadful disease had spread throughout the coastal areas in Kyushu. In 1914, pine wilt was recorded for the first time in Honshu at a locality in Hyogo Prefecture. Since the early 1930s, infested areas had progressively spread throughout Kyushu and along the coast of the Seto Inland Sea in Honshu, which is a warm temperature zone. In most cases, pulp factories and shipyards in which large numbers of pine logs were accumulated played an important role as starting points of disease spread in those areas. The recent occurrences of the disease in scattered points in the northern region of Honshu are also attributed to bringing infected pine logs into those noninfested areas.

In the 1930s, the annual loss of pines increased from 30,000 m^3 to 200,000 m^3. In the 1940s, during World War II, timber losses were estimated at 400,000 m^3 annually and the nematode caused a huge loss of 1,230,000 m^3 in 1948 (Fig. 1). Such a rapid increase of timber losses was mainly attributed to the abandoned attempt of control and to the social disorder resulting from the war. Extensive efforts to fight against such devastating damage resulted in reducing the annual loss to 400,000 m^3. Eradication of dead trees by means of felling and burning was basically adopted as the control method at that time. In the 1970s, the total volume of annual loss increased again and has exceeded 1,000,000 m^3 annually since 1973. Extensive spread of the infested area was pointed out as one of the most characteristic features of the epidemic during that period. The heaviest loss of timber, 2,400,000 m^3, was recorded in 1979. This amount of timber loss represented almost 1% of the growing stock of pine in Japan. At the present time about 500,000 ha of pine forest are thought to be infested with pine wilt disease out of a total of 2.6 million ha of pine forest in Japan. This increase in damage, despite efforts to control the disease, is due to several reasons. One of the reasons is unfavorable weather conditions, such as drought and unusually high temperatures. Also, in Japan, many more dead trees are left untreated in the forests than before due to labor shortages in forestry and because of the drastic changes in demand for pine for firewood because of the use of oil for cooking and home heating.

This is the most serious pest of forest trees in Japan (Fig. 2). The Japanese Government appropriated $35 million in the 1980 fiscal year to fight against pine wilt. In addition to this figure, additional amounts of money are spent by local governments and private owners of forests to control this disease.

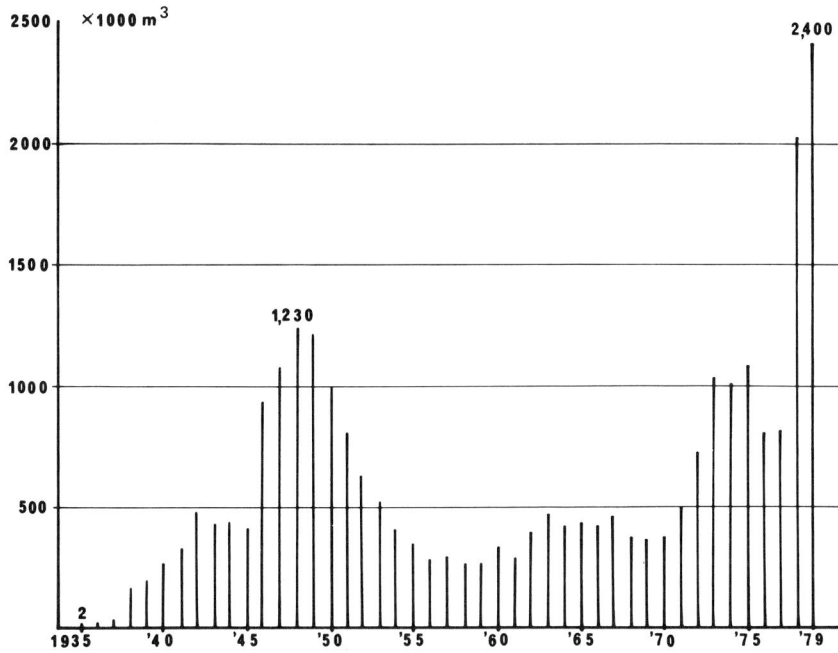

Fig. 1 Annual loss of pine trees in Japan. (Ministry of Agriculture, Forestry and Fisheries, Forest Agency, 1980.)

Fig. 2 Pine trees (*P. densiflora*) killed by the pine wood nematode in Japan.

B. Discovery of the Pine Wood Nematode in Association with Pine Wilt

A number of research and control programs were established. Numerous investigations had been conducted mainly on insects associated with deteriorated
pine trees as suspected causal agents of tree mortality up to the early 1960s.
Many species of bark and wood borers, which belonged mostly to the Cerambycidae, Curculionidae, and Scolytidae, were found to be associated with dead
pine trees. Among them about 10 species received the attention of entomologists and were extensively studied as to their roles in tree mortality. Entomologists had not yet been successful in demonstrating a primary role of insects in the deterioration of pine trees. They came to a conclusion that, at
the time of insect attack by oviposition, the tree had already become diseased
even though it showed a healthy appearance. To detect such diseased trees,
a technique measuring the amount of oleoresin exudation from the sapwood
was developed by entomologists. A high degree of association between the
reduction of oleoresin exudation and the incidence of a lethal effect on pine
trees was demonstrated.

A research project aimed at investigating the cause of disease occurrence
prior to insect attacks was initiated in 1968 at the Government Forest Experi-

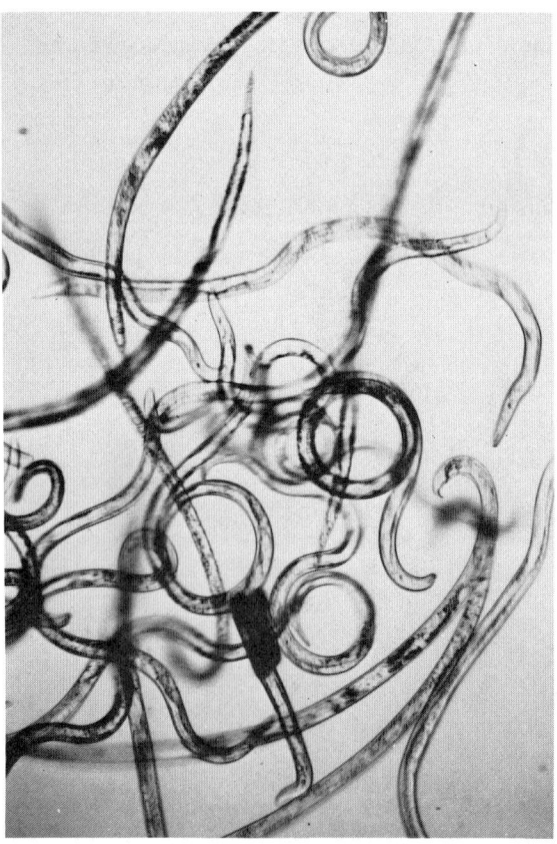

Fig. 3 *B. xylophilus* extracted from the wood of a dead pine tree.

Table 1 Inoculation of Pine Trees with *B. xylophilus*

| | 16-24-year-old *P. densiflora*[a] | | 14-year-old *P. thunbergii*[b] | |
	No. trees tested	No. trees dead	No. trees tested	No. trees dead
Inoculated[c]	20	20	40	34
Control[d]	20	0	15	0

[a]Inoculation on June 27, 1970.
[b]Inoculation on July 9, 1970.
[c]Injection of a water suspension of 30,000 nematodes per tree.
[d]Injection of sterilized water.
Source: From Kiyohara and Tokushige, 1971.

ment Station (now Forestry and Forest Products Research Institute). Scientists from various disciplines, such as entomology, pathology, soil science, silviculture, physiology, meteorology, wood chemistry, joined in this project.

In 1969, Tokushige and Kiyohara found a large number of nematodes (Fig. 3) living in wood of dead pine trees while conducting research on fungal association with deterioration of pine trees. They demonstrated a widespread distribution of this nematode throughout Kyushu in close association with the

Fig. 4 Inoculation of a pine tree with the pine wood nematode. (A) Making a hole on the trunk. (B) Injection of a water suspension of nematodes.

Table 2   Inoculation of 25-Year-Old *P. densiflora* and *B. xylophilus*

	Inoculation on May 18		Inoculation on July 12	
	No. trees tested	No. trees dead	No. trees tested	No. trees dead
Inoculated[a]	40	36	66	59
Control[b]	20	1[c]	30	0

[a]Injection of a water suspension of 30,000 nematodes per tree.
[b]Injection of tap water.
[c]*B. xylophilus* was not found in wood.
Source: From Mamiya, 1972a.

pine epidemic. This nematode, the pine wood nematode, was described in 1972 as *Bursaphelenchus lignicolus* (Mamiya and Kiyohara, 1972). Later, however, based upon typical morphological characters of original specimens of *B. xylophilus* that were rediscovered in the USDA Nematode Collection and genetic crosses among the Japanese and American nematode populations, *B. lignicolus* was placed as a synonym of *B. xylophilus* (Nickle et al., 1981). Since its discovery in Kyushu, *B. xylophilus* was found in many other areas from southwestern to central Japan, indicating a clear coincidence with the disease distribution. Kiyohara and Tokushige (1971) first suggested that *B. xylophilus* was the causal organism of the pine wilt disease. They demonstrated the drastic effect of this nematode on pine trees as the result of their inoculation tests (Table 1). Since then, many investigators have presented experimental evidence that showed a strong correlation between nematode inoculation and the death of pine trees (Mamiya, 1972a, b) (Table 2). Symptoms similar to those occurring in naturally infected trees were produced by introducing nematode suspensions (Fig. 4) to healthy trees. Death of trees followed in 40-60 days after inoculation when experiments were conducted in summer.

II. CAUSAL ORGANISM

B. xylophilus was first described as *Aphelenchoides xylophilus* in 1934 from long leaf pine, *Pinus palustris* Mill., at a sawmill in Bogalusa, Louisiana, by Steiner and Buhrer (1934). Nickle (1970) transferred *A. xylophilus* to the genus *Bursaphelenchus*.

B. xylophilus can be distinguished from other species in the genus by a distinct vulval flap and the shape of the spicules.

All species of *Bursaphelenchus* described so far have a phoretic relationship with insects, especially bark beetles and wood borers, and are mycophagous (Rühm, 1956; Nickle, 1970). Although many species of Aphelenchoididae are also associated with insects, none of these have been noted as having any pathological effect on plants. The one exception is *Rhadinaphelenchus cocophilus*, which causes red ring disease of coconut palms. Most nematodes of Aphelenchoididae, having a phoretic association with bark and wood borers, were observed on the body of their vectors and in a wood habitat limited to insect niches. When Steiner and Buhrer described *A. xylophilus*, they emphasized the apparent specialization of that nematode to a life in wood, and named

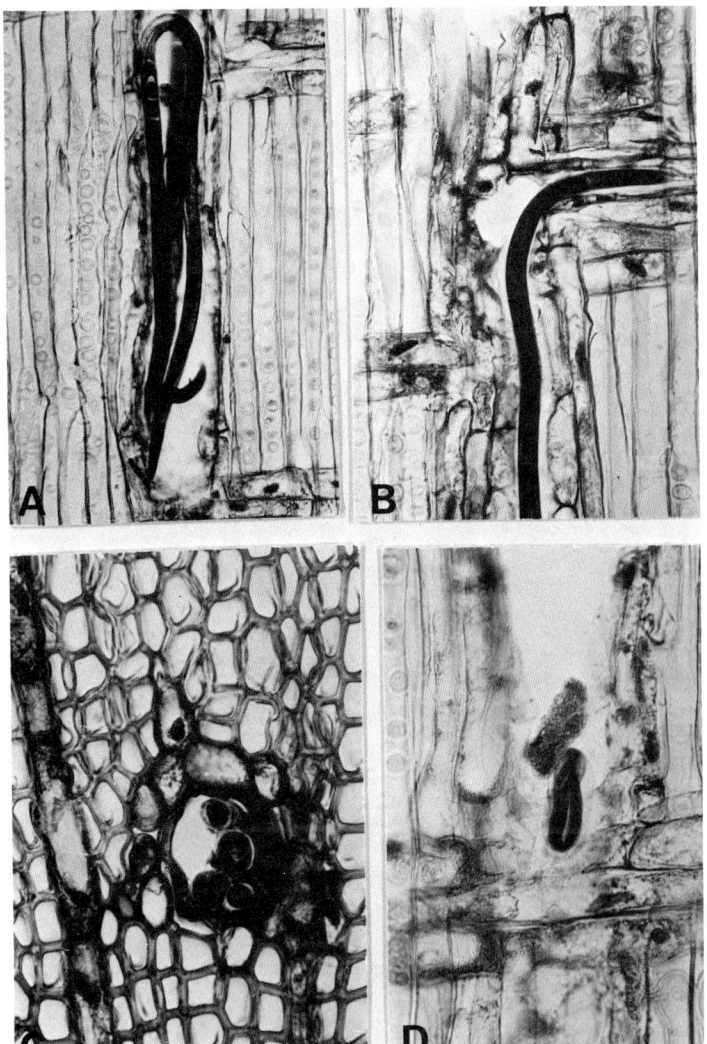

Fig. 5 *B. xylophilus* in resin canals of pine tree. Epithelial cells are destroyed. (A) Radial section; nematodes in an axial resin canal. (B) Radial section; a nematode passing from an axial resin canal to a radial resin canal. (C) Cross section; nematodes in an axial resin canal. (D) Radial section; eggs in an axial resin canal.

the species the *timber nema*. Since Steiner and Buhrer's observation, there has been little information on recovering nematodes from wood. It would appear that finding *B. xylophilus* located in resin canals (Fig. 5) was the first case in regard to habitat among nematodes of this group (Mamiya and Kiyohara, 1972).

   B. xylophilus easily reproduces on cultures of *Botrytis cinerea*, *Pestalotia* spp., *Ceratocystis* spp., and many other fungi (Dozono and Yoshida, 1974; Kobayashi et al., 1974, 1975; Mamiya and Kiyohara, 1972). The nematode also

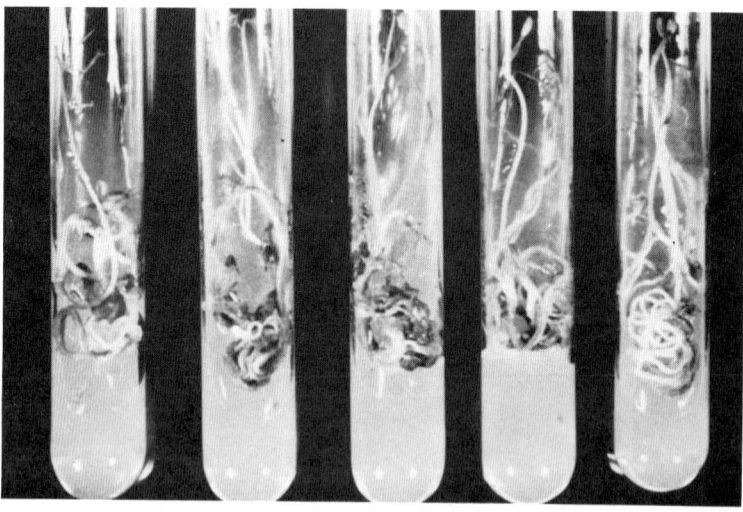

Fig. 6 Reproduction of the pine wood nematode on pine callus tissues. (Courtesy of H. Tamura.)

is successfully cultured on callus tissues (Fig. 6) of pines, *P. densiflora* and
P. thunbergii, and alfalfa, *Medicago sativa*, grown on artificial medium (Tamura and Mamiya, 1976, 1979). Successful reproduction of *B. xylophilus* cn
callused pine tissues indicates the possibility of nematode feeding on parenchyma cells of pine wood. Epithelial cells of resin canals are most likely the
main source of food for nematodes, especially in the early stage of disease
development. Anatomical observation demonstrated that nematodes used the
axial and radial resin canals, which formed a network in the wood as passages
for spreading throughout the wood (Mamiya and Kiyohara, 1972).

Marked reduction in virulence to pine trees was demonstrated on *B. xylophilus* populations that were subcultured successively on *Botrytis cinerea* during more than five years (Kiyohara, 1976). Variations of virulence were also
revealed among cultures of *B. xylophilus* started from different populations
(Kiyohara, 1977).

While conducting a survey of pine wilt, a species very similar to *B. xylophilus* was found living in the wood of dead pine trees (Mamiya and Enda,
1973). This species differs in having a distinct mucro at the tail terminus of
females and larvae from *B. xylophilus*, and was described as a new species,
B. mucronatus (Mamiya and Enda, 1979). It was also found not to mate with
B. xylophilus. Since the beginning of the survey, it has come to our attention that this species has a wider distribution than *B. xylophilus*. It was
usually recovered from a few dead trees scattered over pine forests located
in areas beyond the distribution boundary of *B. xylophilus*. Inoculation experiments showed that *B. mucronatus* was not very pathogenic to pine trees.
The geographical distribution of *B. mucronatus* and absence of the epidemic
spread of the disease is also highly correlated with nonpathogenicity of this
nematode to pine trees. The principal insect vector of this nematode is also
Monochamus alternatus. *B. xylophilus* is successfully transmitted by *M. alternatus* to healthy pine trees, as indicated by its ability to kill trees after
invading wood through wounds produced by maturation feeding of beetles.
Further studies concerning transmission and infestation of *B. mucronatus* to

pine trees are necessary to determine whether successful transmission of the nematode results from maturation feeding of *M. alternatus*, or whether oviposition of the beetle on pine trees weakened by other causes predisposes the trees to nematode infection. The fact that *B. mucronatus* is distributed over a larger area than *B. xylophilus* but is not associated with the epidemic disease suggests that *B. mucronatus* is an aboriginal inhabitant whereas *B. xylophilus* has successfully extended its distribution area by being very pathogenic to pine trees.

III. VECTORS

A. Vectors of the Pine Wood Nematode

Most species of *Bursaphelenchus* are associated with beetles and apparently are disseminated by them. Beetles of various genera in the Cerambycidae, Curculionidae, and Scolytidae, all of which were associated with deteriorated pine trees, were examined for the presence of *B. xylophilus* (Mamiya and Enda, 1972; Morimoto and Iwasaki, 1972). Dauerlarvae of *B. xylophilus* have been recovered from adults of eight species of Cerambycidae so far (Table 3). They occurred most frequently and in greatest numbers in *Monochamus alternatus*, the pine sawyer (Fig. 7); other cerambycids carried dauerlarvae less commonly.

In a severely damaged pine forest, more than 75% of the examined adults *M. alternatus* were contaminated with *B. xylophilus*. An average of 15,000 dauerlarvae (Fig. 8) per insect were recovered. A maximum was 230,000 dauerlarvae per insect. Dauerlarvae are held in the tracheae throughout the body of the *M. alternatus* adult. None of the *M. alternatus* adults collected from pine forests in which there were only a few deteriorated trees, and in which *B. xylophilus* had not been detected from dead trees, carried dauerlarvae of *B. xylophilus*.

Although no dauerlarvae were detected on pupae and larvae, they were first observed on the body surface and around spiracles of the *M. alternatus* adult just after emergence. Dauerlarvae enter the body of the *M. alternatus* adult through the abdominal spiracles and are held only in the tracheae (Fig. 9) throughout the body, even in the tracheae of the antenna and legs. The largest numbers are usually found in the metathorax, followed by those in the

Table 3 Cerambycid Beetle Species Transmitting Dauerlarvae of *B. xylophilus*

Acalolepta fraudatrix Bates

Acanthocinus griseus Fabricius

Arhopalus rusticus Linne

Corymbia succedanea Lewis

Monochamus alternatus Hope.

Monchamus nitens Bates

Spondylis buprestoides Linne

Uraecha bimaculata Thomson

Fig. 7 Maturation feeding of *M. alternatus* on a twig of healthy pine tree. (Courtesy of N. Enda.)

Fig. 8 Dauerlarvae of *B. xylophilus* extracted from the body of *M. alternatus*.

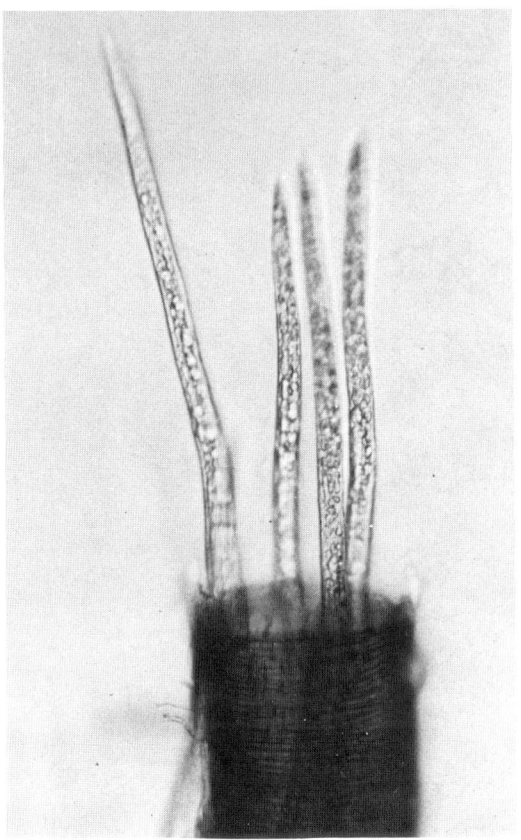

Fig. 9 Dauerlarvae in tracheae dissected from antenna of *M. alternatus*. (Courtesy of N. Enda.)

abdomen. The number of dauerlarvae in the beetle's body decreases gradually in relation to time after emergence of the *M. alternatus* adult from dead pine trees. Under experimental conditions, it was observed that dauerlarvae began to leave the beetle's body three to six days after the emergence of *M. alternatus* adults from dead trees. Enda (1972) showed that more than 90% of dauerlarvae held by the beetle left the beetle's body within 30 days after the beetle's emergence, and of these 50% invaded the twig tissues. Kishi (1978) also provided data that 40% of the dauerlarvae held by the beetle still remained in the body at the time of the beetle's death. Only 10-20% of the dauerlarvae leaving the beetle's body were successful in entering twig tissues of a pine tree. When the *M. alternatus* adults, just after emerging from dead trees, were caged on healthy pine seedlings, maturation feeding of beetles during the first week did not cause the death of seedlings (Mineo and Kontani, 1975). Seedlings caged with the beetles that had fed on other seedlings more than one week after beetle emergence became diseased and died. This result indicates that the transfer of dauerlarvae from the beetle's body to pine seedlings occurred at least one week after the beetle emerged from the wood of dead trees.

According to the result of direct observation on the bodies of *M. alternatus*, dauerlarvae were found on the body surface, mostly on the abdomen,

Fig. 10 Dauerlarvae gathering at the tail tip of *M. alternatus*. (Courtesy of N. Enda.)

with the largest numbers at the tail tip (Fig. 10). In regard to the pathway of nematode transfer from the beetle's body to the pine twig, it can be supposed that dauerlarvae leaving the tracheae through spiracles crawl on the body surface toward the tail tip and then leave the beetle to wounds on twigs during maturation feeding by the adult *M. alternatus*.

Under natural conditions the number of dauerlarvae recovered from matured *M. alternatus* decrease markedly when the beetles had been collected at the time of egg laying (Mamiya and Enda, 1972).

When nematode-contaminated adults of *M. alternatus* were caged in a bag net covering a branch of healthy pine tree, disease transmission occurred in 100% of tested trees. No pine trees on which nematode-free adult beetles were caged were affected (Enda and Mamiya, 1972; Enda, 1973).

B. Biology of *Monochamus alternatus*

The life cycle of *M. alternatus* is annual. The adult of *M. alternatus* emerges during May to July, though the beginning of adult emergence is earlier in the southern region than in the northern region. It is from early to late May in

Kyushu, and during early June in Kanto. Half of the adult beetles have emerged from mid-June to early July, and the period of adult emergence is completed by late July. The threshold temperature for development from the overwintering larvae to the adult is about 11°C. After the adult beetle emerges in the pupal chamber, it remains quiescent for five to seven days and then leaves the pupal chamber by making a hole of 6-10 mm diameter through the wood and bark. The newly emerged adults attain maturity by feeding on living twigs of healthy pine trees. After maturation feeding for a period of three to four weeks, the adult beetle is fully developed and is ready for egg laying. A female beetle lays eggs successfully only on weakened, suppressed, or dead trees. The oviposition period is from June until August, and most eggs are laid during late July to late August. A female lays one to three eggs per day, with a maximum of 188 eggs laid by a female during a three month period. Eggs are laid one by one through each slitlike scar made on the bark surface to the inner bark and hatch after one week's incubation period. Hatched larvae bore into the inner bark. Third- and fourth-instar larvae (Fig. 11) bore tunnels through the sapwood. They then

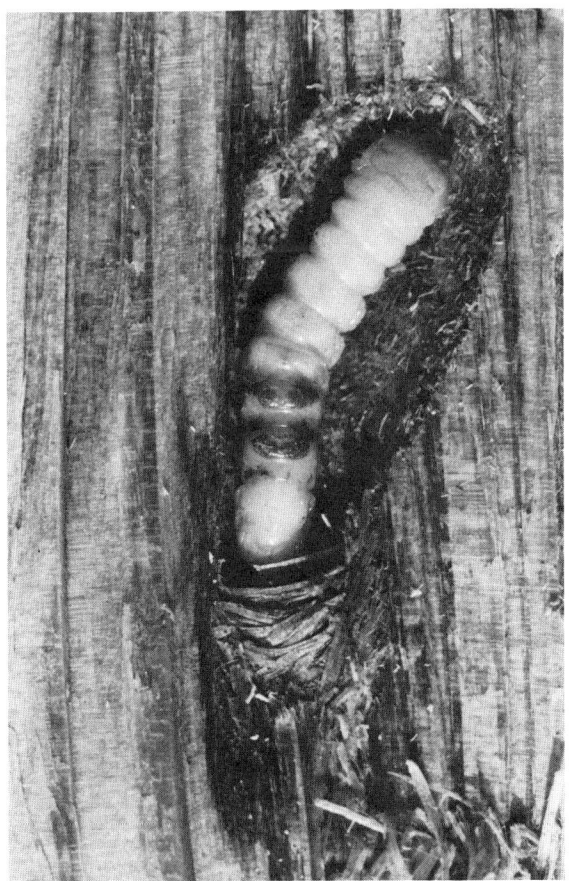

Fig. 11 A larva of *M. alternatus* in the pupal chamber. (Courtesy of N. Enda.)

Fig. 12 A pupa of *M. alternatus* in the pupal chamber. (Courtesy of N. Enda.)

turn around and bore back toward the surface, thereby forming a characteristic U-shaped tunnel. A pupal chamber is formed at the outer end of the tunnel, where larvae enter the period of diapause around October. After overwintering, the larvae pupate (Fig. 12) between April and July. The length of the pupal period depends on the temperature and requires 46 days at 15°C, 20 days at 20°C, and 12 days at 25°C.

The life cycle of *M. alternatus* corresponds well with the chronological development of pine wilt under natural conditions.

M. alternatus is distributed over a larger area than *B. xylophilus* throughout Honshu, Shikoku, and Kyushu. Before the epidemic spread of pine wilt, this beetle was regarded as a rare species and secured the survival of its offspring on pine trees weakened by some other causes, such as overmaturing, injuries caused by wind and lightning, and suppression. *M. alternatus* is also known to occur in China, Taiwan, Laos, and Vietnam.

Based on the results of feeding tests with *M. alternatus* on 40 pine species, Furuno and Uenaka (1979) concluded that all pine species might be infested with *M. alternatus* when the occasion occurred, because no feeding preference of *M. alternatus* was observed among the tested pine species.

IV. SYMPTOMS

A. Disease Development

Quick death of affected trees is characteristic of the pine wilt disease. Affected pine trees, which appear healthy in early summer, die in late summer showing reddish brown foliage. The only outstanding external symptom before the sudden death of the pine tree is the yellowing and subsequent death of the needles, as if they had been subjected to rapid desiccation. Disease development of the nematode-infected pine tree can be summarized as follows:

1. Reduction and cessation of oleoresin exudation. No oleoresin exudation flow from a wound artificially made on trunk is observed on diseased trees at an early stage of disease development. This is characteristic of the disease as the first detectable internal symptom, and occurs as early as mid-July. Oleoresin exudation in inoculated pine trees comes to a complete stop within two weeks after inoculation.
2. Transpiration reduction. Transpiration from foliage decreases and then stops. In the case of the inoculated trees, transpiration reduction occurs 20-30 days after inoculation. During these periods no external symptoms appear.
3. Wilting and yellowing of foliage. Foliage wilting develops on the diseased trees subsequent to transpiration reduction. Sapwood moisture decreases rapidly, and desiccation of the wood is characteristic of this stage.
4. The diseased tree eventually dies from late August to October, just 30-40 days after the first symptoms appear. The brown foliage is retained on the dead tree until early the next summer.

Huge populations of pine wood nematode build up in a dead or advanced diseased tree and are found throughout the trunk, branches, and roots. Nematodes can be easily recovered from wood pieces by using a Baermann funnel (Fig. 13) or by just placing the wood in a beaker of water over night.

The chronological development of the disease under natural conditions coincides with the life cycle of the vector, *M. alternatus*. During maturation

Fig. 13 Baermann funnels for nematode extraction from pine wood pieces.

feeding of the adult beetles on fresh twigs of healthy pine trees, nematodes leave the beetle and enter pine tissues through these feeding injuries. Therefore, nematode infection occurs from early June to late July, which is the maturation feeding period of *M. alternatus*. Adult beetles oviposit successfully in diseased trees that have not exuded oleoresin from wounds, but still show a healthy appearance. This occurs from July to late August. In other words, the pine trees that are attacked or oviposited on by *M. alternatus* have already become diseased even though they show a healthy appearance. When the external symptoms become apparent on the trees, damage is also caused by the feeding of hatched beetle larvae developing on the inner bark. Nematodes multiply and are spread throughout the dead tree. They are successfully transmitted to a new healthy tree in the next early summer by the beetle.

Just before finding the pine wood nematode, forest entomologists had investigated the conditions under which these beetles attack pine trees. They demonstrated, based on experimental results on felling living trees in early summer in disease-infested forests, that beetle oviposition was found on only the trees that showed no oleoresin exudation from the cross section of the cut trunk. A healthy pine tree would produce much oleoresin, which would engulf the larval insects and kill them. They found also that all dead trees within the season were developed from those on which beetles oviposited. A very close relationship between reduced and ceased oleoresin exudation from a pine tree in early summer, and the death of such a tree in late summer was suspected by entomologists. Oda (1967) developed a simple technique to measure the oleoresin exudation flow, aiming to differentiate diseased trees at the time that no external symptoms were observed. A hole of 10-15 mm in diameter (Fig. 14) is made in such a manner as to remove both the outer and inner bark at the basal part of the trunk by using a punching tool and hammer. Oleore-

Fig. 14 A punched hole made on the trunk of pine tree for observation of oleoresin exudation.

sin exudation through the sapwood in the hole is observed and classified as follows:

1. Abundant and overflowing from the hole
2. Abundant but not overflowing
3. Only appreciable amounts as tiny drops
4. No exudation
5. Desiccation of sapwood

Trees at grades 3, 4, and 5 are diagnosed as being diseased. Extensive studies carried out by using this technique revealed that a very high rate of mortality was proven among such diseased trees.

An investigation was conducted to follow the chronological development of the disease under natural conditions in a regrowth stand of 25-year-old *Pinus densiflora* located in Chiba Prefecture in Honshu (Fig. 15) (Mamiya et al., 1973). Observations on oleoresin exudation from a punched hole were made on each pine tree at monthly intervals, and disease development was diagnosed. In late June, the first occurrence of diseased trees was observed. From mid-July to mid-August, diseased trees appeared progressively. Of total dead trees throughout this season, 83% were diagnosed as diseased trees during this period. Most of them, more than 90%, died from late August to mid-October. In late July to mid-August, oviposition by *M. alternatus* was usually observed on diseased trees that had not yet shown external symptoms. A few trees became diseased only after September. No oviposition of *M. alternatus*

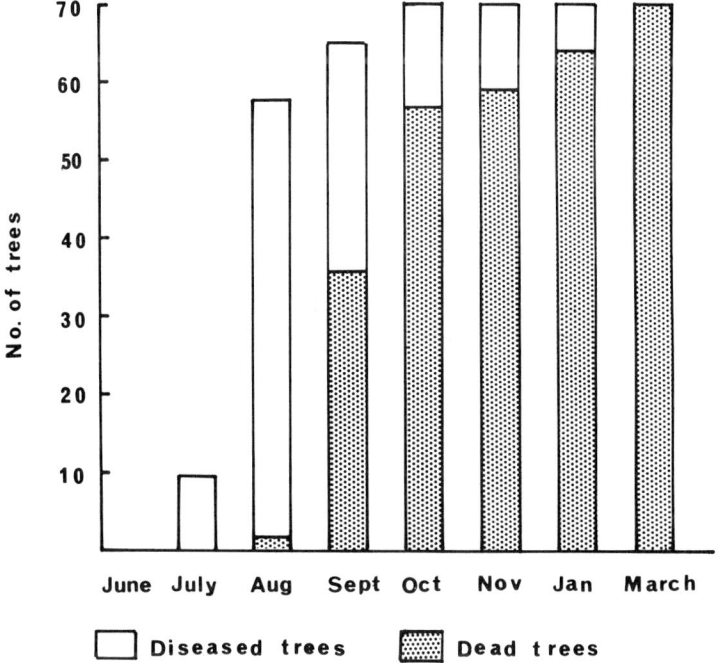

Fig. 15 Natural infection of pine trees with *B. xylophilus* and disease development. A total of 322 of 25-year-old *P. densiflora* were examined. (After Mamiya et al., 1973).

was observed on such trees, because it was beyond the oviposition period of the beetle. The mortality rate of a total of 322 trees examined was 21% in the experimental stand throughout that year.

Similar investigations carried out at several different sites and in different years provided the same results on chronological disease development, although the mortality rate of trees at each site and year was different. Kishi (1980) reported that more than 30% of total dead trees throughout the season appeared during January to May of the following year in Ibaraki Prefecture. The similar trends of disease development was shown in newly infested areas, mostly located in the northern regions, which is a low temperature zone. According to observations by Kishi, most trees that displayed browning of the foliage and died in the next spring, had been diagnosed as diseased in September. This indicates the delayed development of disease in those areas after infection occurred.

B. Pathological Responses of the Pine Tree to Nematode Infection

Under natural conditions nematode infection occurs via wounds made by *M. alternatus* on living twigs through maturation feeding. Dauerlarvae enter the wood tissues immediately after transmission to pine trees. Without too much delay they become adults after molting, and reproduction takes place in wood. The results of inoculation experiments showed that the rapid movement of nematodes throughout the trunk, branches, and roots was observed during the first 24 hours after infection.

Studies have been conducted to demonstrate pathological responses of pine trees to nematode infection, especially during the early stages of pathogenesis (Mamiya, 1975a, 1980a, b). According to the results of histological observations on the nematode-inoculated pine seedlings, cell death, signaled by a granulation of cytoplasm, yellowed and browned cell contents, and deformed or missing nuclei of ray and axial parenchyma cells, was detected as the first visible pathological responses of pine tissues to nematode infection at an initial stage of pathogenesis. Pine seedlings, three-year-old *P. densiflora* and *P. thunbergii*, were inoculated with *B. xylophilus* at the cross section of a cut branch. At the inoculation site, nematodes directly entered most resin canals immediately after inoculation and then invaded areas, gradually extending deeper inside the branch. In the inoculated branch, severity of parenchyma cell death and extent of destruction of epithelial cells directly caused by nematode attack preceded those of any other parts throughout the seedling. Almost all nematodes extracted from a seedling were from the inoculated branch until the ninth day after inoculation, although the rapid spread of nematodes from the inoculation site was shown as early as 24 hours after inoculation. Cell death (Fig. 16) was first observed on axial parenchyma cells surrounding epithelial cells of an axial resin canal and ray parenchyma cells of stem, even at the part far from inoculation site, as early as 24 hours after inoculation. This cellular response developed in relation to the time after infection occurred. It might be concluded that a steady increase in numbers and extent of dead parenchyma cells reflected the disease development of pine seedlings. When inoculated seedlings ceased to exude oleoresin six to nine days after inoculation, death of the ray cells and axial parenchyma cells had prevailed markedly in wood tissues. It was demonstrated that cell death occurred and became advanced prior to the nematode population increase and distribution in the wood of inoculated seedlings. This indicates that pathological reactions of pine tissues to some chemicals might be involved in the mechanism of cell death. Oku and his coworkers (1979, 1980) reported that some

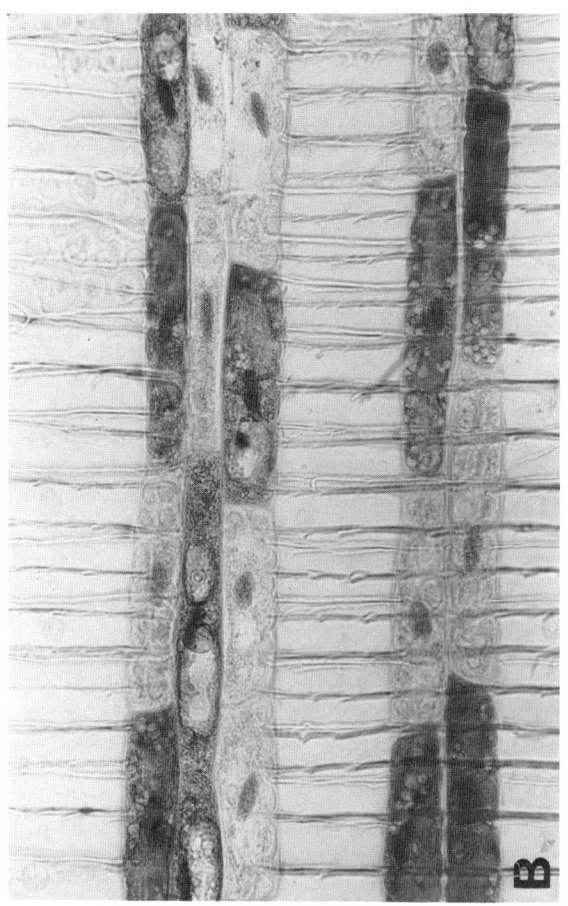

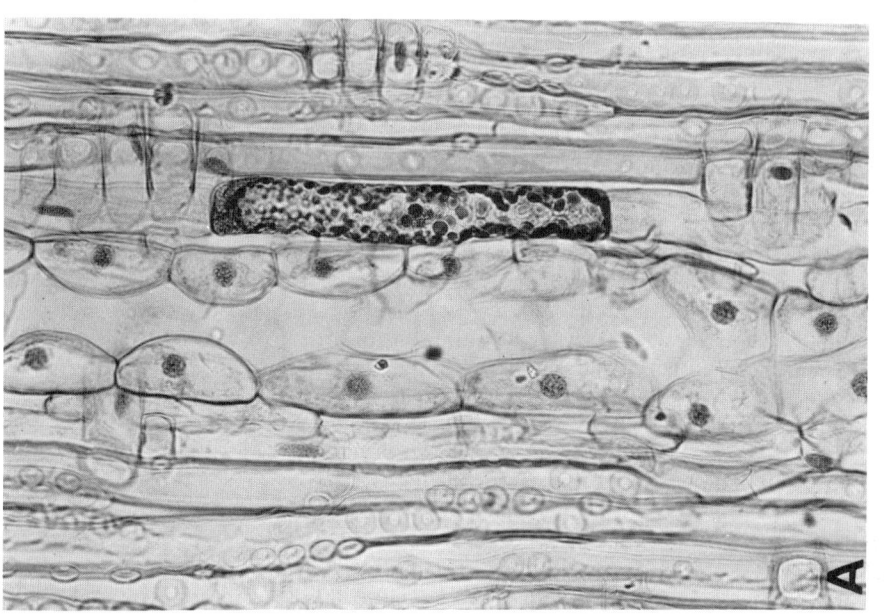

Fig. 16 Pathological response of pine wood tissues to nematode infection. (A) 24 hours after inoculation: Cell death of axial parenchyma cells. (B) Nine days after inoculation: Cell death of ray parenchyma cells.

metabolites of the pine wood nematode or bacterium associated with the nematode had toxic effects on a pine tree.

As mentioned above, nematode feeding and reproduction in resin canals starts at the inoculation site as early as infection occurs. This settlement of invading nematodes at the inoculation site at an early stage of infection has an effect on disease development of a tree thereafter. When an inoculated branch was cut off 24 hours after inoculation, it took longer for these seedlings to show disease symptoms and to come to an eventual death than those on which the inoculated branch remained. Seedlings on which the inoculated branch had been cut off three days after inoculation did not show any difference in disease development from the untreated control seedlings. The number of nematodes in the parts of the seedling other than the inoculated branch were nearly the same in each treatment in which the inoculated branch was cut after 24 hours and in treatments in which the inoculated branch was cut after three days. The results indicate that the pathological responses of living cells appeared at an early stage of nematode infection and showed enough symptoms of disease occurrence to develop disease thereafter without the inoculated branch. At the same time, the extent of the pathological responses depended on the nematode activity at the inoculation site (Mamiya, 1976).

C. Environmental Influences on the Disease

Environmental factors modify susceptibility to disease in pine. High temperature and dry growing seasons greatly favor the intensification and spread of pine wilt. This explains the fact that the rapidly increasing annual loss of pine trees (Fig. 1) has been closely correlated with such weather conditions as higher temperatures and less precipitation during the growing seasons examined so far. For instance, the heaviest annual loss in volume and number of pine trees reported in the history of the disease was recorded in 1978. This was attributed to unusually hot and dry conditions during the summer. In Ibaraki Prefecture, in the central part of Honshu, the annual loss of pine trees in 1978 reached the record loss of 742,000 m^3, which was nearly 30 times that of the previous year's loss. In 1978, the mean temperature of June, July, and August was 2.3°C, 2.7°C, and 1.8°C higher than those of the normal year, respectively, and total precipitation of these three months was only 35% of a normal year (Kishi, 1980). The pine wilt in Ibaraki was particularly destructive in pine stands located in areas with a dry type of soil. Pine forests showed nearly a 90% mortality rate during only that season, and pine wilt was widely distributed throughout the region (Kishi, 1980).

As a result of extensive studies on the relationship between the disease incidence and climatic factors, Takeshita et al. (1975) demonstrated that poor precipitation in the summer (less than 30 mm during 40 days of the summer season) and higher temperature (more than 55 days whose mean daily temperature is above 25°C throughout season) were highly responsible for the incidence of severe damage. The mean annual temperature is one of the most useful factors related to the extent of the incidence and distribution of pine wilt in an area. In southwestern Japan, the areas most severely infested since the early stages of disease spread are located within zones of the mean annual temperature of 15-16°C. At the present time the disease occurs commonly throughout areas where the mean annual temperature is higher than 14°C. Since 1975, the pine wilt disease has spread in scattered localities to the northern region. In most cases, introduction of the disease to those localities resulted from transfer of nematode-infected pine logs from diseased areas. The mean annual temperature, in the areas newly invaded by pine wood nematode, is 10-12°C.

Actual damage to pine trees in these areas is not so severe, and usually it can be regarded as the introductory stage. During this stage of spread, pine infection centers are established. According to the results of surveys on the disease occurrence in these areas, apparent differences in regard to disease development of each infected tree and disease spread patterns from those of the southern region can be indicated. Since this may be attributed to lower temperature, it is doubtful that rapid increase in damage of pine trees and progressive spread of the disease will occur in the future throughout the areas.

No damage to pine is observed in forests located at the higher elevation of mountains, even in the midst of disease-spread areas of southwestern Japan. In Kyushu, incidence of the disease decreases gradually as the elevation becomes higher, and no actual damage is apparent at higher elevation than 700 m (Hashimoto et al., 1974).

Experimental results also indicated a close relationship between disease incidence and environmental factors, such as temperature and humidity. Pine seedlings inoculated with *B. xylophilus* and kept at 25°C and 30°C became diseased and died 30-45 days after inoculation, whereas those kept at 20°C and 15°C did not show any disease symptoms (Kiyohara, 1973). Suzuki and Kiyohara (1978) demonstrated that conditions of water stress had a favorable influence on inducing disease in pine trees. Nematode inoculation caused death in 1 seedling of 35 in the group that received daily watering, and 100% mortality in the group that received a one day watering and a three day drying cycle.

V. BIOLOGY AND LIFE CYCLE

A. Life Cycle

B. xylophilus completes its life cycle in four to five days at 25°C on *Botrytis cinerea* (Mamiya, 1975b). The threshold temperature for development is 9.5°C, and at a temperature higher than 33°C no reproduction occurs. A female produces an average of 79 eggs during a 28 day oviposition period. Females need to be inseminated repeatedly to continue egg laying. It is theoretically demonstrated that the population of *B. xylophilus* in an unlimited environment increases from a female adult to 263,000 in 15 days (Mamiya and Furukawa, 1977).

B. xylophilus has a resting stage in its life cycle. This stage, which appears in wood after the nematode population reaches the highest level, is regarded as the third stage but different in its morphological and biological features from the usual third-stage larva. It is adapted to surviving unfavorable conditions, such as dry conditions, low temperatures, and lack of food (Mamiya, 1972a, 1975b; Ishibashi and Kondo, 1977; Shoji, 1979). In fact, larvae of this stage can survive for a longer period than the other propagative-stage larvae and adults. Larvae of this stage are tentatively called the dispersal third-stage larvae, to differentiate them from the usual third-stage larvae. The dispersal third-stage larvae (Fig. 17) have densely packed materials in their bodies, which are lipid droplets deposited deeply in the intestine. The dispersal third-stage larvae also have the thickest cuticle of all stages in both propagative and dispersal forms (Kondo and Ishibashi, 1978). They molt to dauerlarvae in wood. When they are placed on fungal culture, such as *B. cinerea*, they molt to propagative fourth-stage larvae and begin to increase their population. Dauerlarvae are adapted to being carried by the insect vector to a new habitat. The dauerlarvae begin to molt immediately after craw-

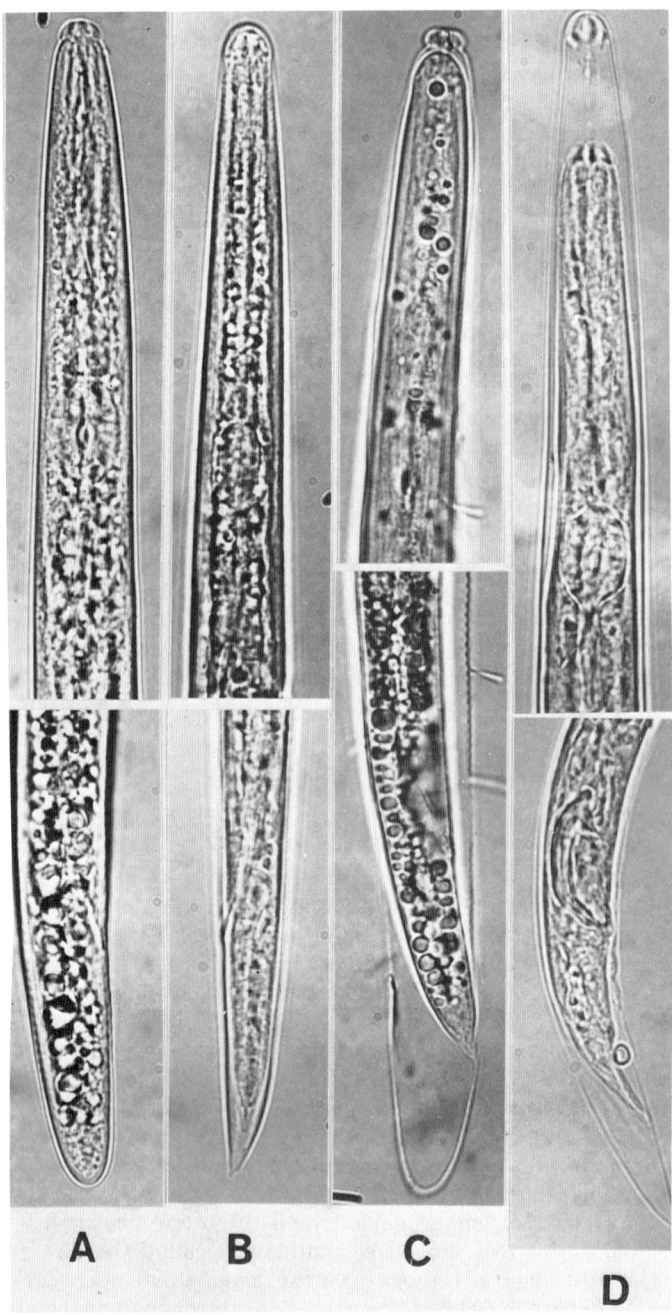

Fig. 17 Dispersal stages of *B. xylophilus*. (A) Dispersal third-stage larvae
(L3). (B) Dauerlarva (DL). (C) Molting; L3-DL. (D) Molting; DL-adult.

ling off the insect's body onto pine twigs, which are the feeding site of the vector. The dauerlarvae can be easily differentiated from other stage larvae by their morphological features, such as a dome-shaped head, lack of stylet, degenerate esophagus and esophageal glands, and subcylindrical tail with a digitate terminus. Their bodies are covered with a protective sticky substance that seems to play a role in attaching these nematodes to the insect's body and to help them move on the insect's body, and to leave the insect. Dauerlarvae that have adapted to dissemination are designated as the third stage in most other species of phoretic nematodes. In the case of *B. xylophilus*, however, dauerlarvae are the fourth stage, and the dispersal third-stage larvae have a biological significance as the stage preceding the dauerlarvae (Mamiya, 1975b).

B. Population Increase of Nematodes in the Host

At the infection site, which are wounds on twigs of a healthy pine tree made by *M. alternatus*, the transmitted dauerlarvae invade the wood tissues immediately. After molting to the adult stage, they begin to feed and reproduce in the resin canals. Death of the parenchyma cells is the first visible pathological response of pine to the nematode infection. This occurs and becomes advanced prior to nematode population increase throughout the pine tree. Re-

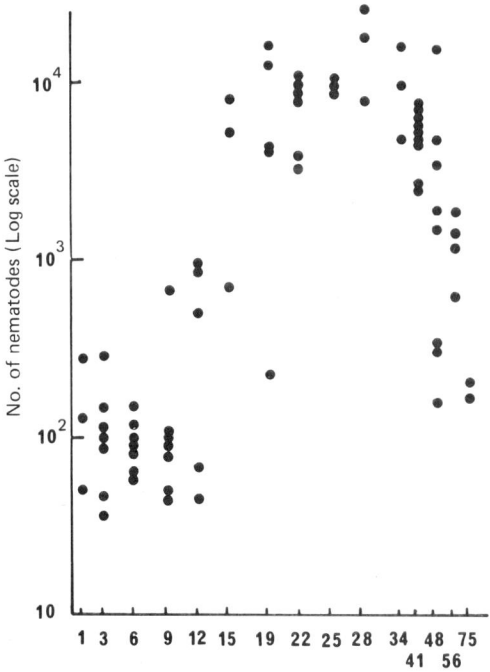

Days after Inoculation

Fig. 18 Population of *B. xylophilus* in wood of inoculated pine seedlings. Seedlings were inoculated with 5000 nematodes per seedling (average 240 nematodes per gram of dry wood). Nematodes were extracted from a whole body of each seedling and then total numbers were converted into numbers per gram of dry wood. (After Mamiya, 1974).

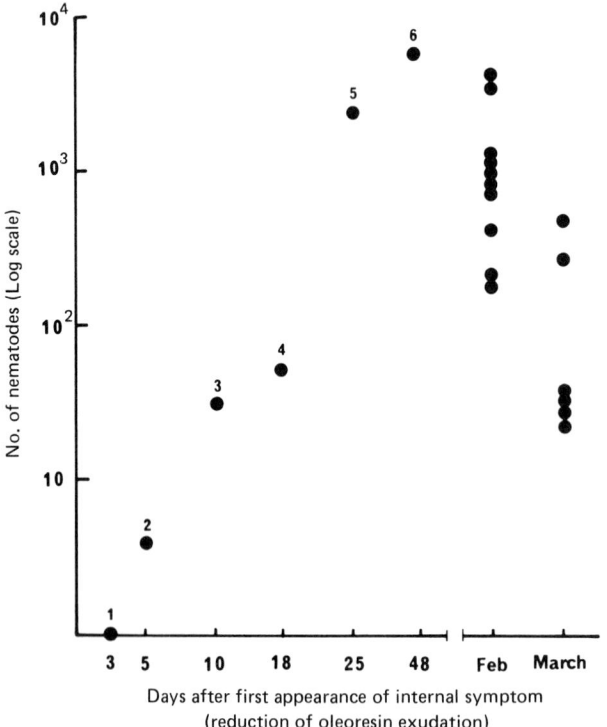

Fig. 19 Population of *B. xylophilus* in wood of inoculated pine trees (per gram of dry wood). Pine trees (7-year-old *P. thunbergii*) were inoculated on July 5, 1971. Nematodes were extracted from pieces of wood sampled at each internode of a tree. Mean numbers for each tree are shown. (1) Healthy appearance. (2) Healthy appearance. (3) Healthy appearance, transpiration reduction. (4) Healthy appearance, transpiration reduction. (5) Wilting of foliage. (6) Wilting and yellowing of foliage. (After Mamiya, 1972b.)

duction and cessation of oleoresin exudation as the characteristic symptom of the disease following cell death also precedes nematode population growth and distribution in wood. According to the results of the inoculation test to three-year-old pine seedlings, seedlings ceased to exude oleoresin six to nine days after inoculation. Almost all nematodes inoculated into the seedling were still in the branch used as the inoculation site. The anatomical observation also gave evidence that destruction of epithelial cells of the resin canals, resulting from the presence of nematodes, had not occurred in the seedling. After the cessation of the oleoresin exudation, the nematode population increases rapidly. From the ninth to the twelfth day after inoculation, the apparent increase of the nematode population is observed on the seedlings showing the symptoms (Fig. 18) (Mamiya, 1974). The same result was obtained from inoculation tests on the first-year seedlings (Mamiya, 1980a).

The population of nematodes in the wood increases as the disease becomes more advanced. At the time when the cessation of oleoresin exudation begins, however, *B. xylophilus* is rarely detected in wood of a diseased tree (Figs. 19 and 20). If nematode populations are compared on the basis of population density per unit of weight of wood between seedlings and young or matured

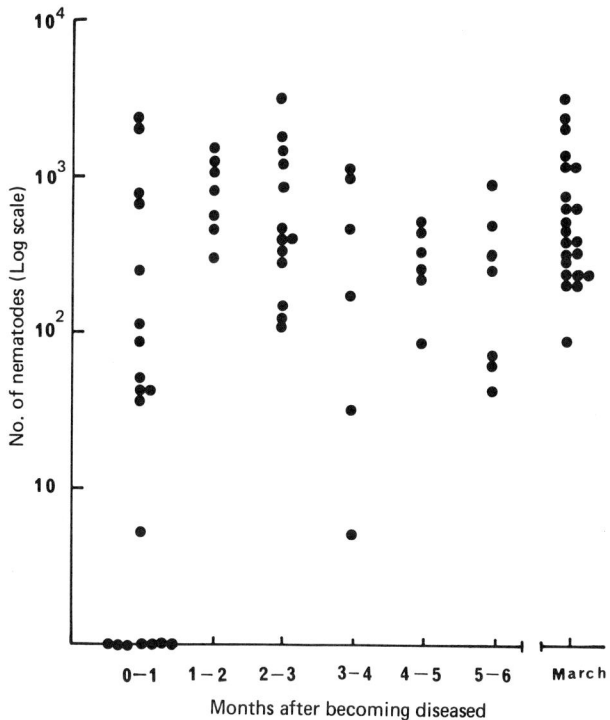

Fig. 20 Population of *B. xylophilus* in wood of diseased pine trees, 25-year-old *P. densiflora* (per gram of dry wood). Nematodes were extracted from pieces of wood sampled at 2 m intervals of the trunk. Mean numbers for each tree are shown. (After Mamiya et al., 1973.)

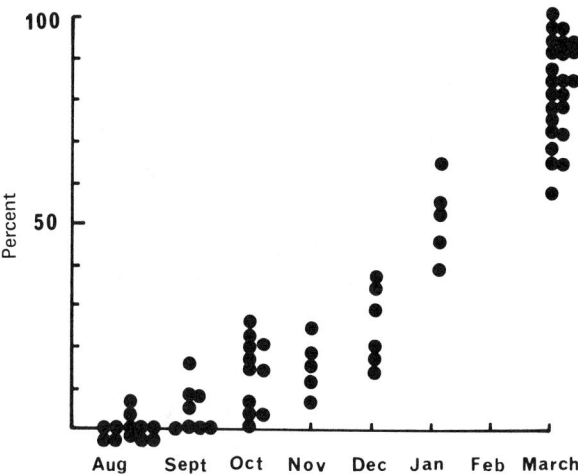

Fig. 21 Proportion of the dispersal third-stage larvae to total numbers of *B. xylophilus*. Mean proportional numbers were obtained from nematode numbers of each pine tree, 25-year-old *P. densiflora*, on which samples were taken at 2 m intervals of the trunk. (After Mamiya et al., 1973.)

trees, population density in the latter is much lower than in the former for a fairly long time after the symptom appears. Since it can be postulated that total numbers of the nematode population growing in wood may be the same at that stage of disease development, this result is mainly attributed to the difference of the volume of wood. Regarding disease development, it is noteworthy that there is not much difference between the seedling and matured tree at the time of the cessation of oleoresin exudation, after nematode infection occurs. Large numbers of nematodes are easily extracted from wood collected at any part of a diseased tree as external symptoms, wilting and yellowing of the foliage, became noticeable. Trees with advanced disease, which are dying or dead, contain tens of millions of living nematodes in the wood throughout the tree. After reaching a maximum level, the population decreases gradually as deterioration of the tree becomes advanced. In this stage of population dynamics, dispersal third-stage larvae appear, and the proportional number of this stage larvae to the whole population of *B. xylophilus* in wood increases gradually as time passes (Mamiya et al., 1973) (Fig. 21). Almost no larvae except dispersal third-stage larvae are observed in winter and the following spring.

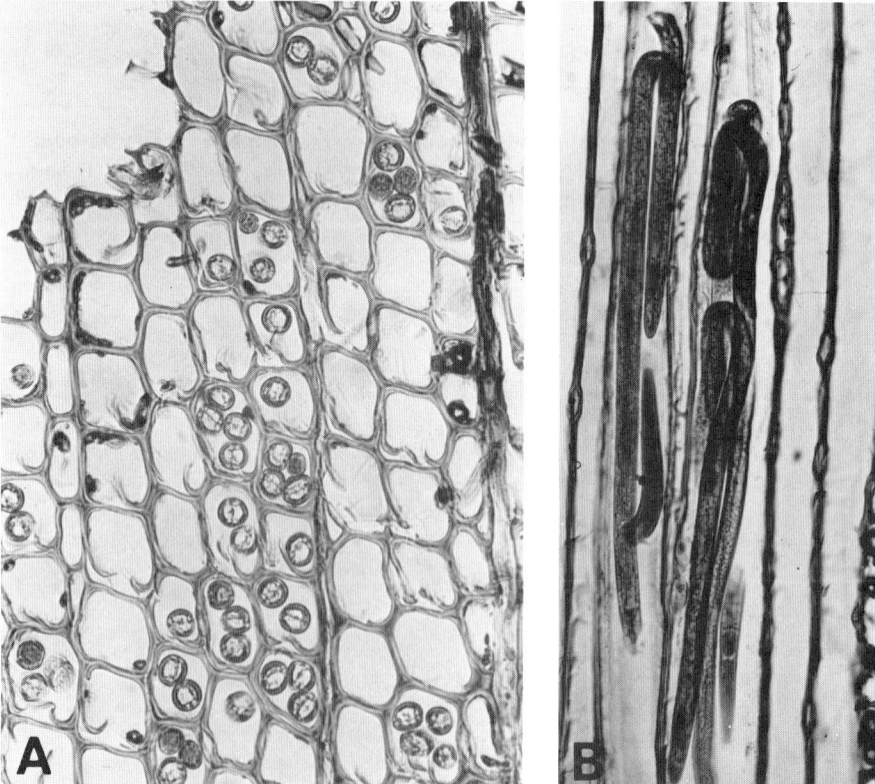

Fig. 22 Dispersal third-stage larvae of *B. xylophilus* located in tracheids around pupal chamber of *M. alternatus*. (A) Cross section. (B) Tangential section.

C. Behavioral Aspects of Nematode Transmission

Migration of *B. xylophilus* occurs in wood of the dead tree during winter to spring, and nematodes consequently accumulate around pupal chambers (Fig. 22) of *M. alternatus* in the early spring. All the aggregated nematodes are in the dispersal third stage, and located within 1-2 mm of the wood tissues surrounding the pupal chamber (Mamiya, 1972a). Most nematodes are found in tracheids and more are found in spring wood than in summer wood. In late spring, the dispersal third-stage larvae become dauerlarvae after molting there. The beginning of this molt coincides with the time of pupation of *M. alternatus*, early May, and the proportion of dauerlarvae in the population becomes larger as the time of emergence comes closer. Dauerlarvae appear on the wall surface of the pupal chamber, and climb up the long perithecial neck (Fig. 23) of blue stain fungi, *Ceratocystis* spp. At the tip of the neck they attach to spore masses of blue stain fungi with sticky substances. Contamination with dauerlarvae occurs only on adults just after emerging from pupae (Fig. 24). The number of dauerlarvae that move onto the body of *M. alternatus* varies with environmental conditions of the pupal chamber. Dry conditions in the pupal chamber markedly reduce the numbers of dauerlarvae held by the beetle coming out of such pupal chambers. Nematode-free adult beetles are easily obtained from pine logs that have been placed under extensively dry conditions.

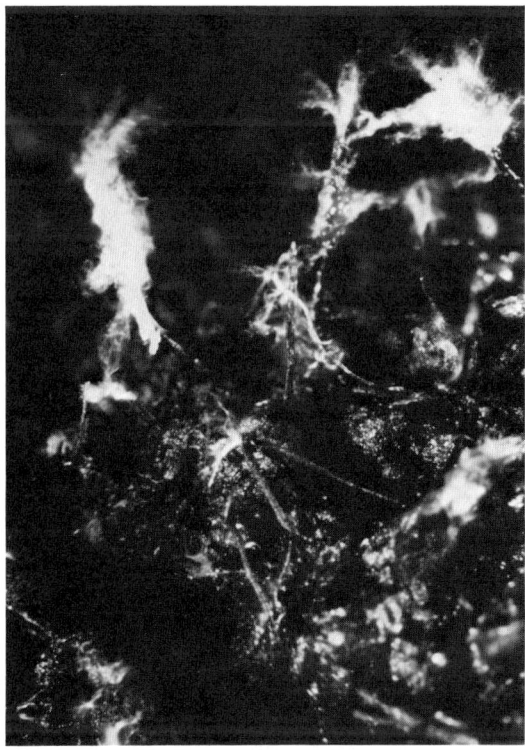

Fig. 23 Dauerlarvae of *B. xylophilus* at the tip of the perithecial necks of blue stain fungi on the wall surface of pupal chamber of *M. alternatus*.

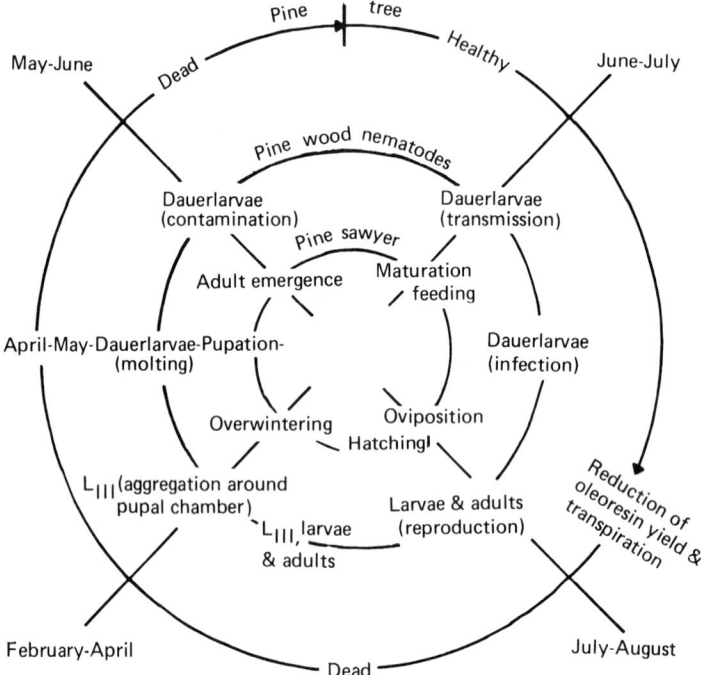

Fig. 24 Biological relationship of *B. xylophilus*, *M. alternatus*, and pine.

Unsaturated fatty acids, such as linoleic acids and oleic acids, originating from the body of larval *M. alternatus*, were suspected as one of the substances involved in the mechanism of nematode accumulation around the pupal chamber (Miyazaki et al., 1977). The attractive effect of carbon dioxide on the behavior of dauerlarvae also was observed positively (Miyazaki et al., 1978).

VI. DISTRIBUTION

A. Japan

The pine wilt was first noticed in Nagasaki of Kyushu early this century. Since then it has spread throughout Kyushu, Shikoku, and Honshu (Fig. 25). The disease spread along coastal areas and to inland areas of low elevation. In Honshu, infection now extends into the northern regions, which are cooler areas. Outlying infection, probably caused by the introduction of infected pine logs, was found at Mito of Ibaraki Prefecture in 1971. At that time the disease distribution showed its apparent boundary south of Ibaraki. The disease rapidly spread in all directions from Mito at the speed on an average of 8 km/year, and has covered almost all areas of Ibaraki. In 1974, the annual loss of pine trees in Ibaraki was estimated at 9000 m³, and enormously increased to 742,000 m³ in 1978 (Kishi, 1980). The pine wood nematode is now known to be causing much damage in Okinawa since the first occurrence in 1973.

Although the origin of the pine wood nematode is obscure, it seems significant that certain pine species distributed in parts of the world other than Japan are resistant to the nematode, and pines native to Japan are quite susceptible. The possible introduction of the nematode from outside Japan is suspected.

40°
HONSHU
SHIKOKU
KYUSHU
30°

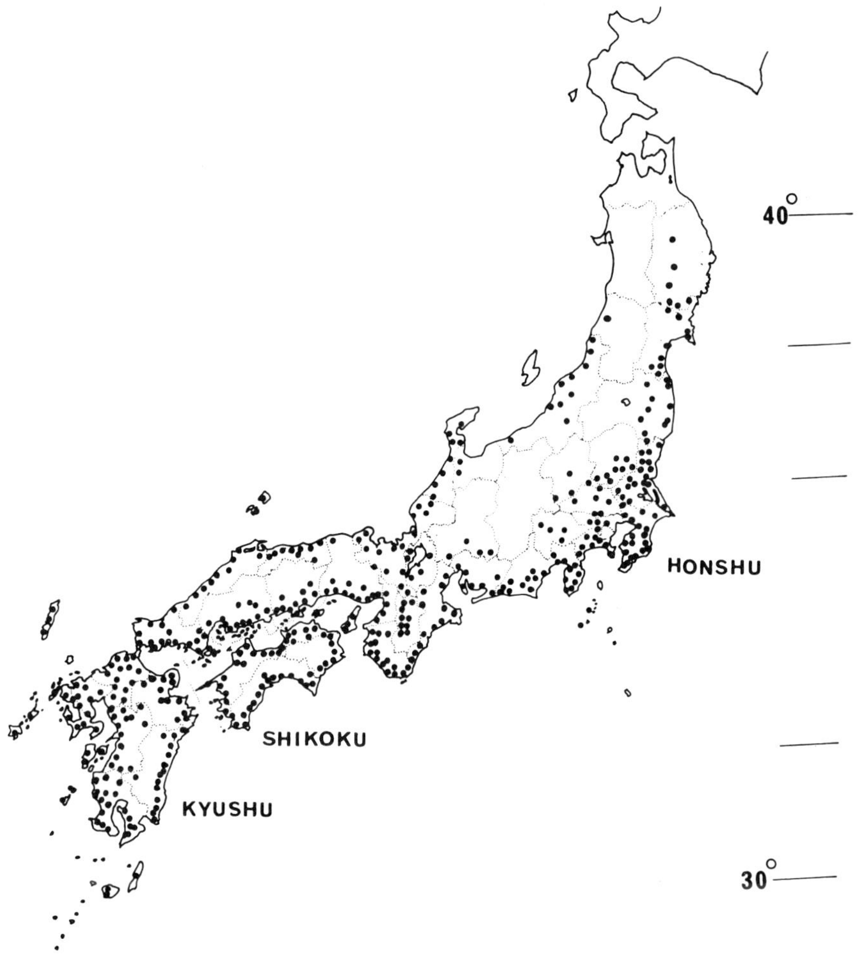

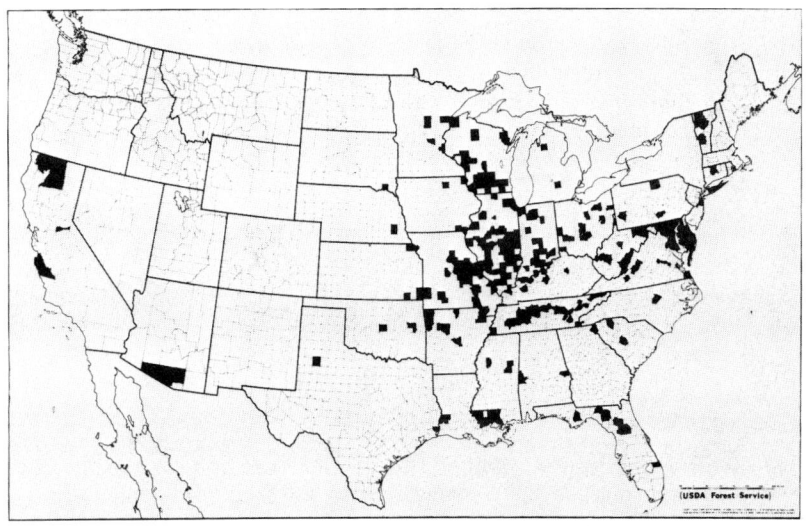

Fig. 25 Distribution of *B. xylophilus* in Japan, 1980.

(USDA Forest Service)

Fig. 26 Pine wood nematode distribution in the United States, March 1982. (Courtesy of Kathryn Robbins, U.S. Forest Service.)

B. United States

The first report of the pine wilt disease in the United States was made by Dropkin and Foudin (1979) based upon finding the pine wood nematode on an Austrian pine (*Pinus nigra* Arnold) in Columbia, Missouri. Dr. S. Ouchi, a visiting plant pathologist from Japan, saw the similarity in symptoms on the tree during the summer of 1978 to the Japanese situation and suggested that the tree be checked for nematodes in the wood. This tree proved positive for the nematode and the identification of Dropkin was confirmed by Dr. Y. Mamiya from Japan and W. Friedman from USDA, APHIS, Beltsville, Maryland. Since that time (Dropkin et al., 1981) many more pine trees, mainly ornamentals, have been found killed in Missouri and several other states, including Kansas, Louisiana, Florida, Maryland, Pennsylvania, Iowa, Illinois, California, and others (Fig. 26). Other tree species found to be killed by this nematode include Japanese black pine (*Pinus thunbergii*), scotch pine (*P. sylvestris* L.), loblolly pine (*P. taeda*), slash pine (*P. elliottii* Engelm.), and others.

Nickle (1981) reported losses of Japanese black pine along the Atlantic coast from Long Island to North Carolina. This tree is a favorite pine planted along the sandy beaches. This loss was especially heavy at Henlopen State Park in Delaware. He also reported that plantations of scotch, red, and sand pine are dying in Maryland and Florida. He has seen 7 acres of red pine and 5 acres of scotch pine dying at Loch Raven Reservoir just north of Baltimore, Maryland. Over half the trees were dead. Eight rows of white pine planted between the scotch and red pine were not affected. Dr. Tarjan found the pine wood nematodes killing sand pine in plantations in the Appalachicola and Ocala National Forests southwest of Gainesville, Florida.

One of the concerns of workers in the United States is the effect of this pest on Christmas tree plantations. Scotch pine is the preferred tree for Christmas tree production, and millions are grown for this purpose annually. Nickle (1981) reports that some of these trees have been killed by this nematode in Missouri and Maryland.

Though the main Japanese vector, *Monochamus alternatus*, is not present in the United States, three American vectors have been reported in the Cooperative Plant Pest Report. *M. carolinensis* has been reported from Florida by Wilkinson and Tarjan (1980) to be a vector of the pine wood nematode from sand pine (*P. clausa* Engelm.). Nickle et al. (1980), in the same issue, reported *M. scutellatus* collected from Virginia pine (*P. virginiana* Mill.) to be vectoring the pine wood nematode in Maryland. Later from Iowa, Williams (1980) reported *M. titillator* as a vector of the pine wood nematode collected from white pine (*P. strobus* L.).

C. France

The pine wood nematode has also been reported in maritime pine (*P. pinaster* Ait.) from France as reported by Scotto la Massese et al. (1978), and by Scotto la Massese (1979).

VII. CONTROL

A. Chemical Control of Vectors

To prevent pine disease, sanitation against the beetles that spread the nematode is necessary. The emergence of beetles from dead pine trees can be prevented by peeling or burning infested wood of felled trees by May. Infested wood with the bark on can be made safe by spraying with chemicals that will destroy the beetles in their larval stages. Chemicals used for this purpose are fenitrothion, fenthion, diazinon, and carbaryl in emulsion or dissolved in fuel oil. The spray should be applied so as to wet the bark surface completely, 400-600 ml of emulsion of fenitrothion will usually treat 1 m^2 of bark surface. Treatment in late summer and fall is effective. This is when the insect larvae are still under the bark. Later, larvae bore their tunnels deep into sapwood and form the pupal chamber to overwinter and to pupate and the treatment in winter or spring is not as effective as in fall. During the winter season, chemicals dissolved in fuel oil are usually used to enhance their effectiveness. Although felling-peeling, felling-burning, and felling-spraying of infested trees have been adopted to eradicate beetles since the time when the actual cause of the disease was not known, sanitation against the vectors is still the most effective and basic method to control the disease, especially in newly infested areas to eliminate the vectors.

Effective control of the disease can be achieved by preventing maturation feeding of *M. alternatus* on living twigs. Just prior to maturation feeding, the crowns of healthy pine trees are sprayed with insecticides, such as fenitrothion, fenthion, and carbaryl. In spraying tree crowns from the ground, a 0.5% emulsion of insecticide is usually sprayed at the rate of 2-3 liters per tree once during late May to June. To spray pine trees on a large scale, aerial applications of chemicals are now practical. An emulsion of fenitrothion at 3% concentrate is commonly used for aerial control, and this chemical is applied at the rate of 60 liters/ha twice during late May to mid-June. In 1979, 121,000 ha were treated by aerial spray in Japan. The total area of pine forests in Japan is roughly estimated at 2,660,000 ha, and those of 450,000 ha have been infested with pine wilt.

B. Chemical Control of Nematodes

Trunk infection (Fig. 27) and soil application of chemicals are being investigated to obtain practical methods for direct control of the nematodes. Systemic chemicals, such as fensulfothion, disulfton, and thionazin, are effective when applied on pine trees two weeks to three months before nematode infection occurs.

Trunk infection of 50% fensulfothion in methyl isobutyl ketone was successful in controlling the disease when applied two weeks before nematode inoculation. Injection of 100 ml of fensulfothion, 50 g of active ingredient, per tree of 25 cm diameter at chest height resulted in 100% control of nematode inoculated trees. The trees treated with fensulfothion survived a series of inoculations in each of three successive years without additional injection of the chemical (Matsuura et al., 1975, 1976). Diffusion of fensulfothion throughout the wood was shown by its preventive effect on population growth of the nematode in the wood of a chemically injected pine tree that was inoculated with nematodes after felling (Mamiya and Tamura, 1976b). This diffusion was also shown by chemical analysis of fensulfothion in wood (Matsuura, 1976).

This nematicide application to individual trees is being developed as a useful technique for responding to the increasing demands for the protection of valuable trees in parks and gardens from the disease. Morantel-tartrate which is a vermicide widely used in veterinary medicine and mesulfenphos have been developed (Matsuura, 1981), and are put to practical use in Japan after being officially registered by the government.

Fig. 27 Trunk injection of nematicide.

There is little value in trying to cure diseased trees, because irreparable damage has already been done to the tissues by the time external symptoms become visible.

C. Other Control Methods

Researches on natural enemies of the pine wood nematode and the vector are being conducted for biological control against the disease. Several species of bacteria and fungi, such as *Serratia* sp. and *Beauveria bassiana* isolated from bodies of dead *M. alternatus*, have been tested for their pathogenicity to *M. alternatus* (Katagiri and Shimazu, 1980).

Three species of mesostigmatid mites have been found in the pupal chamber of *M. alternatus* with high frequency. They have a phoretic association with *M. alternatus*, and two species among them feed on nematodes. According to the results of our observations, however, there was no high correlation between numbers of phoretic mites and nematodes on the body of *M. alternatus* (Tamura and Enda, 1980a, b).

Nematode-trapping fungi (Fig. 28) also have been detected around the pupal chamber of *M. alternatus* (Mamiya and Tamura, 1976a) and in the wood (Yoneda et al., 1980).

Behavior-regulating chemicals, such as an oviposition attractant for *M. alternatus*, are being extensively studied. Yamazaki et al. (1980) demonstrated an interesting effect of paraquat injection to healthy pine trees on the attraction of *M. alternatus*. They suggested that the part of the trunk secreting large quantities of oleoresin resulting from paraquat injection probably released

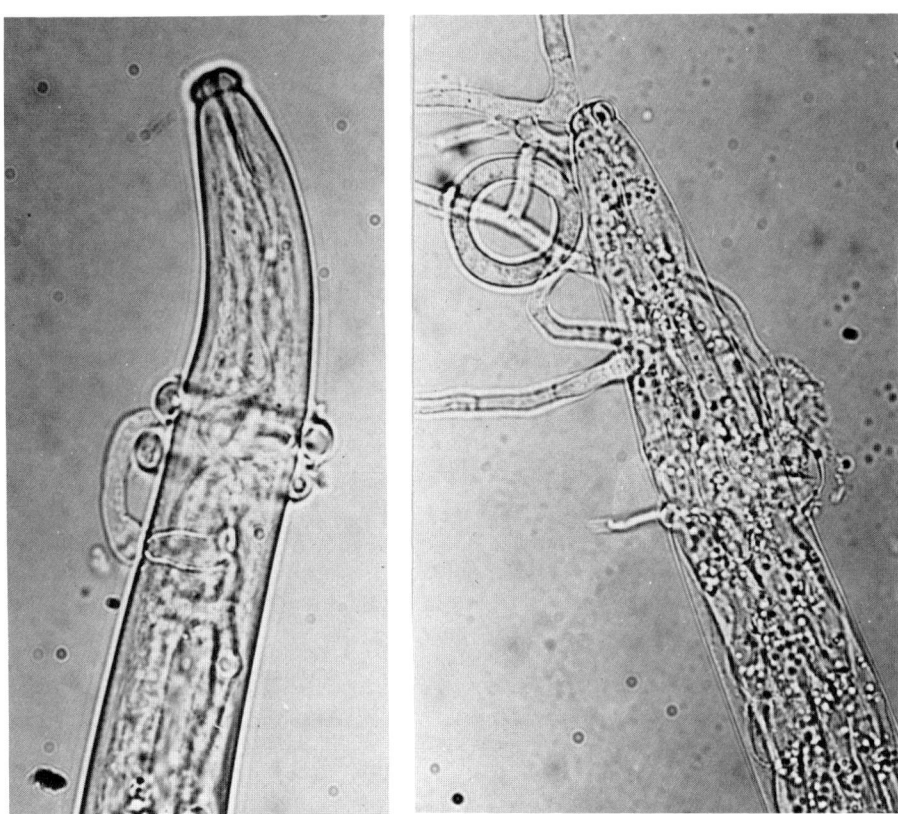

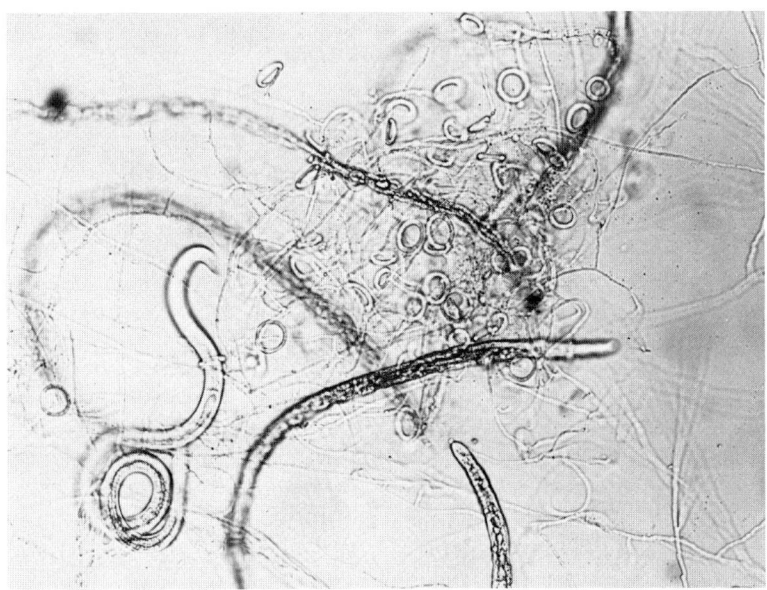

Fig. 28 Nematode-trapping fungus, *Dactylella leptospora*, isolated from pupal chambers. Trapped dispersal third-stage larvae of *B. xylophilus*.

powerful oviposition attractants. Ikeda et al. (1980) found that the combined bait of ethanol and monoterpene hydrocarbons was highly attractive to the beetle.

D. Breeding of Resistant Trees

Since within the species of pines susceptible to the nematode there is a variation in the reaction of individual trees to the disease, it may eventually be controlled by developing a resistant tree. Extensive studies for breeding resistant trees are in progress as the most promising method for controlling the disease (Ohba, 1980).

According to the results of inoculation tests, pine species vary in their resistance to nematode infection. Native pines in Japan, *P. densiflora, P. thunbergii*, and *P. luchuensis*, of which the first two species are very important conifers, are highly susceptible to the disease. Results on other pine species, which are mostly exotic in Japan, are summarized as follows:

Resistant species are:
P. taeda*,†,‡, *P. rigida**, *P. elliottii**‡, *P. taiwanensis**, *P. excelsa**, *P. strobus*,*†, *P. massioniana**, *P. thunbergii* × *P. massoniana**, *P. resinosa**, *P. tabulaeformis**, *P. banksiana**,†, *P. contorta**, *P. echinata*,‡ *P. palustris*‡, *P. pungens*†, *P. bungeana*‡, *P. caribaea*†.

Susceptible species (Futai and Furano, 1979) are: *P. monticola, P. pentaphylla, P. strobiformis, P. nigra, P. pinaster, P. sylvestris, P. ponderosa, P. rudis, P. engelmanni, P. oocarpa, P. radiata, P. koraiensis, P. leiophylla, P. mugo, P. muricata.*

REFERENCES

Dozono, Y., and Yoshida, N. (1974). Application of the logistic curve for the population growth of pinewood nematode, *Bursaphelenchus lignicolus*, on the culture of *Botrytis cinerea* (Japanese). *Jpn. J. For. Soc. 56*: 146-148.

Dropkin, V. H., and Foudin, A. S. (1979). Report of the occurrence of *Bursaphelenchus lignicolus*-induced pine wilt disease in Missouri. *Plant. Dis. Rep. 63*: 904-905.

Dropkin, V. H., Foudin, A., Kondo, E., Linit, M., Smith, M., and Robbins, K. (1981). Pinewood Nematode: A Threat to U.S. Forests? *Plant Dis. 65*: 1022-1027.

Enda, N. (1972). Removing dauerlarvae of *Bursaphelenchus lignicolus* from body of *Monochamus alternatus* (Japanese). *Trans. 24th Mtg. Kanto Branch Jpn. For. Soc.*, p. 32.

Enda, N. (1973). The effect of maturation feeding of *Monochamus alternatus* contaminated with *Bursaphelenchus lignicolus* on pine trees. II. (Japanese. *Trans. 84th Mtg. Jpn. For. Soc.*, pp. 319-321.

Enda, N., and Mamiya, Y. (1972). The effect of maturation feeding of *Monochamus alternatus* contaminated with *Bursaphelenchus lignicolus* on pine tree. I. (Japanese). *Trans. 83rd Jpn. For. Soc.*, pp. 320-322.

---

*Futai and Furuno (1979).
†Kiyohara and Tokushige (1971).
‡Ogawa et al. (1973).

Furuno, T., and Uenaka, K. (1979). Studies on the insect damage upon pine species imported in Japan (No. 6). On the feeding of Japanese pine sawyer adult, *Monochamus alternatus* Hope (Japanese with English summary). *Bull. Kyoto Univ. Forests 51:* 12-22.

Futai, K., and Furuno, T. (1979). The variety of resistances among pine-species to pinewood nematode, *Bursaphelenchus lignicolus*. *Bull. Kyoto Univ. Forests 51:* 23-36.

Hashimoto, H., Kiyohara, T., Dozono, Y., Takizawa, Y., Miyazaki, T., Kawabata, K., Katsu, Z., and Taniguchi, A. (1974). Relationship between the distribution of *Bursaphelenchus lignicolus* and *Monochamus alternatus* in relation to elevation and the incidence of pine wilt (Japanese). *Trans. 85th Mtg. Jpn. For. Soc.* pp. 253-256.

Ikeda, T., Enda, N., Yamane, A., Oda, K., and Toyoda, T. (1980). Attractants for the Japanese pine sawyer, *Monochamus alternatus* Hope (Coleoptera: Cerambycidae). *Appl. Ent. Zool. 15:* 358-361.

Ishibashi, N., and Kondo, E. (1977). Occurrence and survival of the dispersal forms of pinewood nematode, *Bursaphelenchus lignicolus* Mamiya and Kiyohara. *Appl. Ent. Zool. 12:* 293-302.

Katagiri, K., and Shimazu, M. (1980). Search for pathogenic bacteria and fungi to *Monochamus alternatus* (Japanese). *Forest Pests 29:* 28-33.

Kishi, Y. (1978). Invasion of pine trees by *Bursaphelenchus lignicolus* M. & K. from *Monochamus alternatus* (Japanese). *J. Jpn. For. Soc. 60:* 179-182.

Kishi, Y. (1980). Mortality of pine trees by *Bursaphelenchus lignicolus* M. & K. (Nematoda: Aphelenchoididae) in Ibaraki Prefecture and its control (Japanese with English summary). *Bull. Ibaraki Pref. For. Exp. Sta. 11:* 1-83.

Kiyohara, T. (1973). Effect of temperature on the disease incidence of seedlings inoculated with *Bursaphelenchus lignicolus* (Japanese). *Trans. 84th Mtg. Jpn. For. Soc.,* pp. 334-335.

Kiyohara, T. (1976). The decrease of pathogenicity of pinewood nematode, *Bursaphelenchus lignicolus*, induced by the extended subculturing on the fungal mat of *Botrytis cinerea* (Japanese with English summary). *Jpn. J. Nematol. 6:* 56-59.

Kiyohara, T. (1977). Differences of pathogenicity and reproduction rate among cultures of *Bursaphelenchus lignicolus* (Japanese). *Trans. 30th Mtg. Kyushu Branch Jpn. For. Soc.,* pp. 241-242.

Kiyohara, T., and Tokushige, Y. (1971). Inoculation experiments of a nematode, *Bursaphelenchus* sp., onto pine trees (Japanese with English summary). *J. Jpn. For. Soc. 53:* 210-218.

Kobayashi, T., Sasaki, K., and Mamiya, Y. (1974). Fungi associated with *Bursaphelenchus lignicolus*, the pinewood nematode. I. (Japanese with English summary). *J. Jpn. For. Soc. 56:* 136-145.

Kobayashi, T., Sasaki, K., and Mamiya, Y. (1975). Fungi associated with *Bursaphelenchus lignicolus*, the pinewood nematode. II. (Japanese with English summary). *J. Jpn. For. Soc. 57:* 184-193.

Kondo, E., and Ishibashi, N. (1978). Ultrastructural differences between the propagative and dispersal forms in pinewood nematode, *Bursaphelenchus lignicolus*, with reference to the survival. *Appl. Ent. Zool. 13:* 1-11.

Mamiya, Y. (1972a). Pinewood nematode, *Bursaphelenchus lignicolus* Mamiya and Kiyohara, as a causal agent of pine wilting disease. *Rev. Plant Protec. Res. 5:* 46-60.

Mamiya, Y. (1972b). Reproduction of pine lethal wilting disease by the inoculation of young trees with *Bursaphelenchus lignicolus* (Japanese with

English summary). *Jpn. J. Nematol.* 2: 40-44.

Mamiya, Y. (1974). Population increase of *Bursaphelenchus lignicolus* in wood of pine seedlings in relation with the time after inoculation (Japanese). *Trans. 85th Mtg. Jpn. For. Soc.*, pp. 249-251.

Mamiya, Y. (1975a). Behaviour of *Bursaphelenchus lignicolus* in the wood of pine seedlings and pathological responses of pine to nematode infection (Japanese). *Trans. 86th Mtg. Jpn. For. Soc.*, pp. 285-286.

Mamiya, Y. (1975b). The life history of the pinewood nematode, *Bursaphelenchus lignicolus* (Japanese with English summary). *Jpn. J. Nematol.* 5: 16-25.

Mamiya, Y. (1976). The numbers of *Bursaphelenchus lignicolus* associated with disease development of pine seedlings (Japanese). *Trans. 87th Mtg. Jpn. For. Soc.*, 225-226.

Mamiya, Y. (1980a). Inoculation of the first year pine (*Pinus densiflora*) seedlings with *Bursaphelenchus lignicolus* and the histopathology of diseased seedlings (Japanese with English summary). *Jpn. J. For. Soc.* 62: 176-183.

Mamiya, Y. (1980b). Pine wilt and pinewood nematode; histopathological aspects of disease development. *Proceedings of Workshop on Genetics of Host-Parasite Interactions in Forestry*, Wageningen, The Netherlands, September 11-21, 1980.

Mamiya, Y., and Enda, N. (1972). Transmission of *Bursaphelenchus lignicolus* (Nematoda: Aphelenchoididae) by *Monochamus alternatus* (Coleoptera: Cerambycidae). *Nematologica* 18: 159-162.

Mamiya, Y., and Enda, N. (1973). Occurrence of a species of *Bursaphelenchus* closely related to *Bursaphelenchus lignicolus* in wood of dead pine trees (Japanese). *Trans. 84th Mtg. Jpn. For. Soc.*, pp. 328-330.

Mamiya, Y., and Enda, N. (1979). *Bursaphelenchus mucronatus* n. sp. (Nematoda: Aphelenchoididae) from pine wood and its biology and pathogenicity to pine tree. *Nematologica* 25: 353-361.

Mamiya, Y., and Furukawa, M. (1977). Fecundity and reproductive rate of *Bursaphelenchus lignicolus*. *Jpn. J. Nematol.* 7: 6-9.

Mamiya, Y., and Kiyohara, T. (1972). Description of *Bursaphelenchus lignicolus* n. sp. (Nematoda: Aphelenchoididae) from pine wood and histopathology of nematode-infested trees. *Nematologica* 18: 120-124.

Mamiya, Y., Kobayashi, T., Zinno, Y., Enda, N., and Sasaki, K. (1973). Disease development of pine trees naturally infected with *Bursaphelenchus lignicolus* (Japanese). *Trans. 84th Mtg. Jpn. For. Soc.*, pp. 332-334.

Mamiya, Y., and Tamura, H. (1976a). A kind of nematode-trapping fungi, *Dactylella leptospora*, found in wood around pupal chambers of *Monochamus alternatus* (Japanese). *Forest Pests* 25: 147-149.

Mamiya, Y., and Tamura, H. (1976b). Trunk injection of chemicals for control of pine wilting disease caused by *Bursaphelenchus lignicolus*. III. Biological test of the effect of injected chemicals on nematodes (Japanese). *Trans. 87th Mtg. Jpn. For. Soc.*, pp. 271-272.

Matsuura, K. (1976). Trunk injection of chemicals for the control of pine wilting disease caused by *Bursaphelenchus lignicolus*. II. Fate of fensulfothion and thiophanatemethyl injected into the wood of pine trunk (Japanese). *Trans. 87th Mtg. Jpn. For. Soc.*, pp. 267-269.

Matsuura, K. (1981). Detectable amount of trunk infected nematicides and their protective effects against pine wilt disease. *Proc. 17th IUFRO World Congr., Div. 2.* pp. 593-596, Japanese IURFO Congress Committee. 636 pp.

Matsuura, K., Fujishita, A., and Kishi, Y. (1975). The effect of trunk injection of chemicals on the pine wilting disease caused by *Bursaphelenchus lignicolus* (Japanese). *Trans. 86th Mtg. Jpn. For. Soc.*, pp. 309-310.

Matsuura, K., Kawasaki, T., Kobayashi, T., Zinno, Y., Mamiya, Y., Tamura, H., and Sasaki, K. (1976). Trunk injection of chemicals for the control of pine wilting disease caused by *Bursaphelenchus lignicolus*. I. Effect of injected chemicals on the disease development of nematode inoculated pine tree (Japanese). *Trans. 87th Mtg. Jpn. For. Soc.*, pp. 265-266.

Mineo, K., and Kontani, S. (1975). Removing process of *Bursaphelenchus lignicolus* from *Monochamus alternatus* (Japanese). *Trans. 86th Mtg. Jpn. For. Soc.*, pp. 307-308.

Miyazaki, M., Oda, K., and Yamaguchi, A. (1977). Behaviour of *Bursaphelenchus lignicolus* to unsaturated fatty acids (Japanese with English summary). *Mokuzai Gakkaishi* 23: 255-261.

Miyazaki, M., Yamaguchi, A., and Oda, K. (1978). Behaviour of *Bursaphelenchus lignicolus* in response to carbon dioxide released by respiration of *Monochamus alternatus* pupa (Japanese with English summary). *J. Jpn. For. Soc.* 60: 249-254.

Morimoto, K., and Iwasaki, A. (1972). Role of *Monochamus alternatus* (Coleoptera: Cerambycidae) as a vector of *Bursaphelenchus lignicolus* (Nematoda: Aphelenchoididae) (Japanese with English summary). *J. Jpn. For. Soc.* 54: 177-183.

Nickle, W. R. (1970). A taxonomic review of the genera of the Aphelenchoidea (Fuchs, 1937) Thorne, 1949 (Nematoda: Tylenchida). *J. Nematol.* 2: 375-392.

Nickle, W. R. (1981). Research on the pine wood nematode in the United States. XVII IUFRO World Congress—Japan, Division 2: 269-271.

Nickle, W. R., Friedman, W., and Spilman, T. J. (1980). *Monochamus scutellatus*: Vector of pinewood nematode on Virginia pine in Maryland. *Cooperative Plant Pest Rep.* 5: 383.

Nickle, W. R., Golden, A. M., Mamiya, Y., and Wergin, W. P. (1981). On the taxonomy and morphology of the pinewood nematode, *Bursaphelenchus xylophilus* (Steiner & Buhrer, 1934) Nickle, 1970. *J. Nematol.* 13: 385-392.

Oda, K. (1967). The applicability of measurement of the oleoresin yield in determining the susceptibility of pine trees to beetle infestation (Japanese). *Forest Pest News* (Tokyo) 16: 263-266.

Ogawa, S., Nakajima, Y., and Hagihara, Y. (1973). A preliminary report on the distribution and biology of *Bursaphelenchus lignicolus* in Fukuoka Prefecture (Japanese). *Research Note No. 2. Fukuoka Forest Exp. Sta.*, pp. 3-27.

Ohba, K. (1980). Breeding of pines for resistance to the wood nematode, *Bursaphelenchus lignicolus* Mamiya et Kiyohara. *Proceedings of Workshop on Genetics of Host-Parasite Interactions in Forestry*, Wageningen, The Netherlands. September 11-12, 1980.

Oku, H., Shiraishi, T., and Kurozumi, S. (1979). Participation of toxin in wilting of Japanese pines caused by a nematode. *Naturwissenschaften* 66: 210.

Oku, H., Shiraishi, T., Ouchi, S., Kurozumi, S., and Ohta, H. (1980). Pine wilt toxin, the metabolite of a bacterium associated with a nematode. *Naturwissenschaften* 67: 198-199.

Rühm, W. (1956). Die Nematoden der Ipiden. *Parasit. Schriftenreihe 6*: 1-487.

Scotto la Massese, C. (1979). Report of European and Mediterranean Plant Protection Organization 3632, Item 3.3, Paris.

Scotto la Massese, C., Boulbria, A., and Baujard, P. (1978). Epigeal nema-
tofauna associated with *Pinus pinaster* decay. *Abstr. Papers presented
at 3rd Intn. Congr. Plant. Pathol.*, Munchen, August 16-23, 1978.

Shoji, T. (1979). Resistance of pinewood nematode, *Bursaphelenchus lignicolus*, to low temperatures (Japanese with English summary). *Jpn. J.
Nematol. 9:* 5-8.

Steiner, G., and Buhrer, E. M. (1934). *Aphelenchoides xylophilus* n. sp. a
nematode associated with blue-stain and other fungi in timber. *J. Agric.
Res. 48:* 949-951.

Suzuki, K., and Kiyohara, T. (1978). Influence of water stress on development of pine wilting disease caused by *Bursaphelenchus lignicolus*. *Eur.
J. For. Path. 8:* 97-107.

Takeshita, K., Hagihara, Y., and Ogawa, S. (1975). Environmental analysis
to pine damage in Western Japan (Japanese with English summary). *Bull.
Fukuoka For. Exp. Sta. 24:* 1-45.

Tamura, H., and Enda, N. (1980a). Mesostigmatid mites associated with
Japanese pine sawyer beetle (Japanese with English summary). *Jpn. J.
Appl. Ent. Zool. 24:* 54-61.

Tamura, H., and Enda, N. (1980b). Life histories of three species of nema-
tode-feeding mesostigmatid mites associated with the pine sawyer beetle,
Monochamus alternatus (Japanese with English summary). *J. Jpn. For.
Soc. 62:* 301-307.

Tamura, H., and Mamiya, Y. (1976). Reproduction of *Bursaphelenchus lignicolus* on alfalfa callus tissues. *Nematologica 21:* 449-454.

Tamura, H., and Mamiya, Y. (1979). Reproduction of *Bursaphelenchus lignicolus* on pine callus tissues. *Nematologica 25:* 149-151.

Tokushige, Y., and Kiyohara, T. (1969). *Bursaphelenchus* sp. in the wood
of dead pine trees (Japanese). *J. Jpn. For. Soc. 51:* 193-195.

Wilkinson, R. C., and Tarjan, A. C. (1980). *Monochamus carolinensis:* Vector
of pinewood nematode on sand pine in Florida. *Cooperative Plant Pest
Rep. 5:* 383.

Williams, D. J. (1980). *Monochamus titillator:* Vector of pinewood nematode on
red pine in Iowa. *Cooperative Plant Pest Rep. 5:* 627.

Yamazaki, T., Hata, K., and Okamoto, H. (1980). Luring of Japanese pine
sawyer *Monochamus alternatus* Hope by paraquat-treated pine trees. I.
The effect of paraquat and ethephon on pine trees. *J. Jpn. For. Soc.
62:* 99-102.

Yano, M. (1913). Reports of investigation on the death of pine trees. (Japanese). *Sanrin Koho 4:* 1-14.

Yoneda, K., Katsumata, J., Saiki, H., Tsubaki, K., and Tatsumi, S. (1980).
A nematode-trapping fungus detected in pine sap. *J. Jpn. For. Soc.
62:* 227-229.

Chapter 17

History, Development, and Importance
of Insect Nematology

William R. Nickle *USDA, Agricultural Research Service, Beltsville, Maryland*

Harold E. Welch *University of Manitoba, Winnipeg, Manitoba, Canada*

I. HISTORY

Entomophilic (*entos*, insect; *philos*, to like) nematodes occur in some 27 families in nine major groups of nematodes; namely, rhabditoid, tylenchoid, aphelenchoid, strongyloid, oxyuroid, ascaridoid, spiruroid, filarioid, and mermithoid. The first eight belong to the Secernentea, and the ninth to the Adenophorea, the two primary taxonomic divisions of the Nematoda. Entomophilic nematodes vary greatly in size and shape, have insects as intermediate or as definitive hosts, may be facultative or obligatory in their host relations, and often involve other microorganisms in their relationships with their host.

In a historical review we need to search for the first recognition of these worms, the establishment of their life cycle, and their association with other microorganisms.

A. Pre-Rudolphi Period

Rudolphi is often named the "Father of Helminthology." Studies before his time are often anecdotal, but nevertheless of interest. Aldrovandus (1527-1605) found dead grasshoppers with worms emerging from their bodies (1623). The account is recorded in his *De Animalibus Insectis*. In an earlier study (1602), he coined the word *vermes*. Lister (1672) described similar worms from a plant in his garden and compared his findings to that of Aldrovandus. In 1742, Reaumur, an insect physiologist, described a worm that was undoubtedly *Sphaerularia bombi* Dufour. Gould (1747) graphically described the emergence of worms, probably mermithids, from ants. The great Carl von Linnaeus (1707-1778), in his *Systema Naturae*, listed eight genera in the Vermes Intestini (1758). Two of these were truly parasitic worms, and the name of the third, *Gordius*, is associated with Linnaeus. A Lutheran pastor, J. A. E. Goeze (1731-1793), was the first to study nematodes seriously under the microscope and described the vinegar eelworm (1782). He began to distinguish between the various kinds of worms. He also described the emergence of mermithids from soil following a heavy rain (1782). Zeder (1800) used Goeze's notes and distinguished several types of worms and gave to them the common name, roundworms.

B. Rudolphi and the Nineteenth Century Researchers

In the first half of the eighteenth century, the great improvement in microscopes led to a new interest in microscopic worms. Rudolphi (1771-1832) was born in Sweden of German parents (Fig. 1). He was really an anatomist and adopted a philosophy that the phenomena of life must be based on a real knowledge of the organs in different animals. His philosophy was to reject the mysticism of natural history, and this gave his work a sound basis akin to modern research. His search for knowledge was humane, and he would not perform vivisections. He was Professor of Anatomy at Berlin and founded the Berlin Zoological Museum. Despite his abandonment of the concept of spontaneous generation he believed that intestinal worms were spontaneously produced from fluid in an animal. Pallas (1741-1811) opposed Rudolphi's view on the origin of worms, and believed that parasites or their eggs entered the host from outside. Rudolphi (1819), in *Entozoorum Synopsis*, widened knowledge to include 350 species belonging to 11 genera. He also gave us the scientific name, Nematoidea.

Ernst von Siebold (1804-1885) studied under Rudolphi in Berlin, and like Pallas, could not accept Rudolphi's view of the spontaneous generation of intestinal worms or endoparasites. He argued that the presence of large quanti-

ties of eggs indicated that the parasites reproduced in the same way as other
animals. But the question remained: How did the worms gain access to the
tissues of the animal? Neither the experimental design nor the required knowledge was available to perform the decisive experiment. An idea of a Swedish
worker, Steenstrup, concerning the alternation of generations of marine animals, provided both the knowledge and the stimulus for von Siebold (Fig. 2)
to carry out the critical test. He showed how a parasite, now called *Coenurus
cerebralis*, insinuated in sheep brain to cause "gid." He further showed that
when the parasitized tissue was fed to a dog, the adult stages of the tapeworm
developed in the intestine of the dog. This led to the concept of a life cycle
involving different kinds of hosts, giving helminthology and parasitic nematology a more rational basis (1842, 1843, 1848, 1850).

We are also indebted to von Siebold for the recognition of the Gordiacea.
His own studies, plus those of Charvet (1834), Berthold (1843), and Dujardin
(1842), led him in 1843 to establish the Gordiacea, though he still included
mermithids in the group. Under the title "Ueber die Fadenwürmer der Insekten" in six works (1842-1858), von Siebold described and noted 233 nematode
species from insects. Von Siebold held his first appointment as Professor
of Anatomy and Physiology at Erlangen, a university to which we will refer
again.

Leuckart (1822-1898) clarified Linnaeus' groups and established Vermes
on a firm basis akin to that which we use today (Fig. 3). Rudolphi (1809),
with true perspicacity, recognized Nematoidea as separate from Acanthocepha-
lea, Trematodea, and Cestoidea, but Leuckart (1887) established them as
separate groups. He, in competition with von Siebold, worked out the life
cycles of *Taenia solium* and *T. saginata*. The concept of life cycles and the
role of different hosts was now well established.

In 1839, Hope wrote "The genera and species of insects infected by filariae" and complicated an already complex puzzle concerning the identity of
filariids, mermithids, and gordiids. Bremser (1824), in his "New Atlas of
Intestinal Worms," recorded Leblond's discussion of the finding of Audouin of
mermithid worms in cockchafers in France. Another French worker, F. Dujardin (1801-1860), was also a pioneer in the study of nematodes in insects in
France. He described *Mermis nigrescens* in 1842 and *Mermis aquatilis* in 1845.
In Scotland in 1861, Sir John Bulloch described a worm that was obviously
Sphaerularia bombi. In 1853, Meissner (1829-1905) described *Mermis albicans*
in detail.

In 1851, Karl Diesing (1800-1867) published *Systema Helminthum*, with
175 insect nematode records and involving five entomophilic genera. He listed
118 species of *Gordius*, 17 of *Mermis*, including *M. nigrescens* and *M. albicans*
and *Sphaerularia bombi* as members of the same suborder and tribe. Not surprisingly, he considered *Sphaerularia* as a *genus inquirendum*. He assigned
twelve species to *Anguillula*, nine of which had been assigned to *Oxyuris*
spp., and found in intestines of insects. Diesing recognized two genera,
Gordius and *Mermis*, but kept them in the same suborder and tribe. It
remained for Vejdovsky (1886) to separate Nematoda and Nematomorpha.
Remarkably, Max Braun in 1883 defined the Mermithidae, a definition that
still stands. Linstow (1842-1916), in Berlin, initiated a series of papers
that extended from 1860 to 1914. The admirable practice of naming new
genera by the use of a new prefix with *-mermis* originated with von Linstow and was first used in establishing the name *Paramermis* in 1898. In 1878,
he published a *Compendium der Helminthologie*, which lists several entomophilic nematodes.

Fig. 1 K. A. Rudolphi (1771-1832).

Fig. 2 C. von Siebold (1804-1885).

Fig. 3 R. Leuckart (1822-1898).

Fig. 4 I. N. Filipjev (1889-1940).

Schneider (1831-1890) provided one of the first classifications of nematodes (1866). He accepted only *M. nigrescens* and named a new species *M. lacinulata*. He separated *Mermis* from *Gordius*, but left *Sphaerularia bombi* with *Gordius* in the same group. In the same monograph, he reveals his curiosity about *S. bombi* by including a chapter on its development.

Late in the nineteenth century, three Russian researchers, Fedchenko (1874, 1886), Keppen (1870, 1881, 1882), and Radkevitch (1869), gave us information on the entomophilic nematode fauna of that vast country. They created a tradition that was carried into the next century by several researchers. Dr. E. S. Kirjanova (1900-1976) was one of these. She worked on plant, animal, and insect nematodes, but had a special interest in the Nematomorpha. She supervised 25 dissertations, and described 100 new species of nematodes and Nematomorpha. I. N. Filipjev will be discussed later.

C. Cobb to Present (1900-1982)

During the first 30 years of this century, many faunal lists were prepared. In 1900, Schultz published "Filarein in Palaarktischen Lepidopteren." Other lists included van Zwaluwenburg's compendium (1928), exclusive of medical and veterinary titles, up to 1926, and LaRivers (1949) updated van Zwaluwenberg's compendium until 1946. Hall (1929) presented "Arthropods as an intermediate host for helminths," but regrettably had no references. Stiles' and Hassall's 1920 compendium of generic and specific names was praised when published as much as it is esteemed today. It is a valuable volume filled with names and dates, and remarkably free of errors. It is known as the *Index Catalogue of Medical and Veterinary Zoology, Roundworms*. Checklists of more recent dates include Polozhentsev and Negrobov (1967), who listed 400 insect species that are intermediate and definitive hosts for trematodes, cestodes, acanthocephalans, and nematodes. Another Russian worker, Ipatyeva (1970), provided a list of nematodes associated with the Scarabaeoidea. Finally, Gilbert Fuchs published a series of papers on bark beetle nematodes. The series extended from 1914 to 1938. Yatsenkowsky (1924) in the USSR provided evidence that a small number of nematodes could cause castration of the bark beetle hosts, and heavy infections killed the beetles.

In the United States, Cobb (1927), from his experience with three entomophilic nematodes, stressed the potential of nematodes in controlling insects. In Britain, Oldham (1933), on the basis of his own experience and that of T. Goodey (1930) on *Tylenchinema oscinellae*, made the same recommendation. Both T. Goodey (1951) and J. B. Goodey (1963) produced systematic books containing nematodes associated with insects.

If Rudolphi is recognized as the Father of Helminthology, then Ivan Nikolaevich Filipjev (1889-1940) may certainly be named the father of Insect Nematology (Fig. 4). Not only did he contribute indirectly through his work on nematode classification, but directly by bringing together the scattered information on insect nematodes and incorporated his own findings into the synthesis. This formed the second section (about 80 pages) of a monograph published in Russian in 1934 entitled "Nematodes that are harmful and useful in Agriculture" (Filipjev, 1934a). Sections of this book were sent to Dr. Prof. J. H. Schuurmans-Stekhoven in Belgium who, as junior author, had the original Russian translated into French and then into the "American Language" (see foreword)! This was published in Leiden under the title *A Manual of Agriculture Helminthology* in 1941.

Also in 1934, Filipjev's "Classification of the free-living nematodes and their relation to parasitic ones" was published in English by the Smithsonian

Miscellaneous Collections (1934b). Micoletzky (1925) and Cobb (1919) dropped their classification in favor of Filipjev's. Western workers lost contact with Filipjev, and Kirjanova (1959) reported his death on October 22, 1940.

Filipjev's book on harmful and helpful helminths in agriculture combines both plant and insect nematodes and provides a pattern for this book.

Filipjev (1934b) included those nematodes that have insects as "intermediate" hosts, the Spirurida and Filariida as well as Nematomorpha. The limits of a scientific field have no sharp boundaries. Poinar (1975), for example, includes the Spirurids, Filarids, and Nematomorphs, whereas this book combines the insect-parasitic nematodes and the plant-parasitic nematodes.

Detailed morphological and taxonomic studies were strongly pursued in the early decade of this century. Dadai (1911), Daday* (1913), and Hagmeier (1912) established many of the type species for future genera. Cobb (1859-1932) (Fig. 5), Steiner (1886-1961) (Fig. 6), and Christie (1889-1978) (Fig. 7), three Americans, established insect nematology in North America in single, dual, and triple authorships in a long series of papers. Their "*Agamermis decaudata* Cobb, Steiner, and Christie 1923; a nema parasite in grasshoppers and other insects" is recognized as a classic, as pointed out by Gerald Thorne (1961). Two other Americans, Rudolf William Glaser and Norman R. Stoll, richly deserve our gratitude. Glaser (1888-1947) was the first to glimpse the potential of *Neoaplectana* nematodes for biological control, and gave much effort to mass-culture techniques and was the first researcher to culture *N. glaseri* on an artificial culture in vitro. Norman Stoll (1892-1976) continued Glaser's work on vermiculture (Stoll, 1959). His breadth of scientific knowledge was great and, perhaps unknown to many, he was a member of the International Commission of Zoological Nomenclature and Chairman of the Editorial Committee. A reading of his "Introduction to the Code" is time well spent.

The Dane, P. Bovien (1933, 1937, 1944), made three excellent contributions and suggested the interrelationship among bacteria, nematodes, and insects (1937). The Americans, S. R. Dutky and W. S. Hough (1955), and the Czechoslovakian, J. Weiser (1955), simultaneously found a nematode in the Codling moth that the former designated by its accession number, DD-136, and the latter named the worm *Neoaplectana carpocapsae*. Dutky, and later Weiser, confirmed the presence of an associated bacterium.

J. R. Christie (1974) reviewed the parasites of invertebrates, noting many newly discovered species, chiefly in insects.

In the United States, insect pathology was beginning to develop, and Steinhaus (1949) devoted a chapter to nematodes. Welch (1956, 1963, 1965) revived interest and reviewed progress in the field, which he named entomophilic nematology.

Reference was made to von Siebold's sojourn at the University of Erlangen, but the major contribution of this institution was under Professor H. J. Stammer, who established a school of insect nematology and graduated such notables as Drs. H. Korner, E. Liebersperger, G. Osche, W. Rühm, and F. Wachek.

In Brazil, I. Travassos has a long series of publications (1925-1965) contributing to our knowledge of the oxyurids and thelastomatids in Diplopoda, Chilopoda, and Insecta. His 1953 paper is a typical example of his work. Both Nickle and Poinar have contributed to the identification of entomophilic

---

*German spelling of the Hungarian "Dadai."

Fig. 5 N. A. Cobb (1859-1932).

Fig. 6 G. Steiner (1886-1961).

Fig. 7 J. R. Christie (1889-1978).

nematodes. The former provided lengthy papers (1967a, 1972, 1973, 1974) showing worms in situ and reviews on taxonomy and biology of insect parasitic nematodes, and the latter published books in 1975 and 1979, and a key in the Commonwealth Agricultural Bureaux Series (1977). Another CIH contribution of considerable value is the compendium of abstracts produced by Shephard (1974). Massey (1974) produced a monograph on the taxonomy of nematodes of bark beetles in the United States.

We are fortunate in having a strong group of contemporary workers in the field of entomophilic nematodes in addition to those mentioned above. Perhaps this represents the largest number that we have ever had at any given period. They include, from the United States, J. J. Petersen, G. O. Poinar, A. A. Johnson, H. K. Kaya, J. G. Stoffolano, C. J. Geden, E. G. Platzer, D. P. Molloy, and R. Gaugler; from Canada, J. M. Webster, C. H. S. Thong, J. R. Finney, and R. Gordon; from England, W. M. Hominick and P. N. Richardson; from Austria, H. Kaiser; from France, C. Laumond; from the Soviet Union, P. A. Polozhentsev, A. K. Artyukovsky, I. A. Rubtsov, S. L. Lazarevskaya, and G. V. Veremchuk; from New Zealand, W. Wouts; R. Bedding and A. Akhurst from Australia; and from Brazil, G. R. Kloss.

II. HIGHLIGHTS IN THE DEVELOPMENT OF INSECT NEMATOLOGY

Our first records of mermithids from insects can be found in the fossil record, where *Heydonius antiquus* and *H. matutinus* were collected from Rhine lignite (Eocene) and Baltic amber (Taylor, 1935). There are a few scattered references to entomophilic nematodes in the seventeenth and eighteenth centuries. The field developed more dramatically during the nineteenth century, but the last 30 years have certainly been the most productive period for the development of insect nematology as determined by the number of workers and papers in the literature.

Many different kinds of insects are parasitized by nematodes. Some of the insects parasitized by mermithids can be found in Fig. 8, and some insects known to be parasitized by tylenchid-sphaerulariid nematodes can be found in Fig. 9.

A. Grasshopper Mermithids

Probably the earliest and best known nematode parasite of insects is *Mermis nigrescens* from grasshoppers. It was found over 350 years ago and was described in 1842. The grasshopper, being an important agricultural pest, was the subject of early studies, and dissections revealed these long parasites, which were noted in the early European literature. *M. nigrescens* and *Agamermis decaudata* are two mermithids that have always been of interest to entomologists, and many surveys and life history studies have been made on these nematodes. This and other information will be taken up later in a whole chapter on this subject. Noteworthy at this time is the fact that the adult nematodes spend some time in the soil and that they or their young infective stages are capable of climbing grasshopper host plants to lay eggs on the foliage or to infect grasshoppers.

B. Chironomid Mermithids

Though they are not important agronomic or medical insects, chironomids perhaps have the largest variety and number of mermithid nematode parasites of any insect group. The aquatic habitat and concentrations of large numbers of

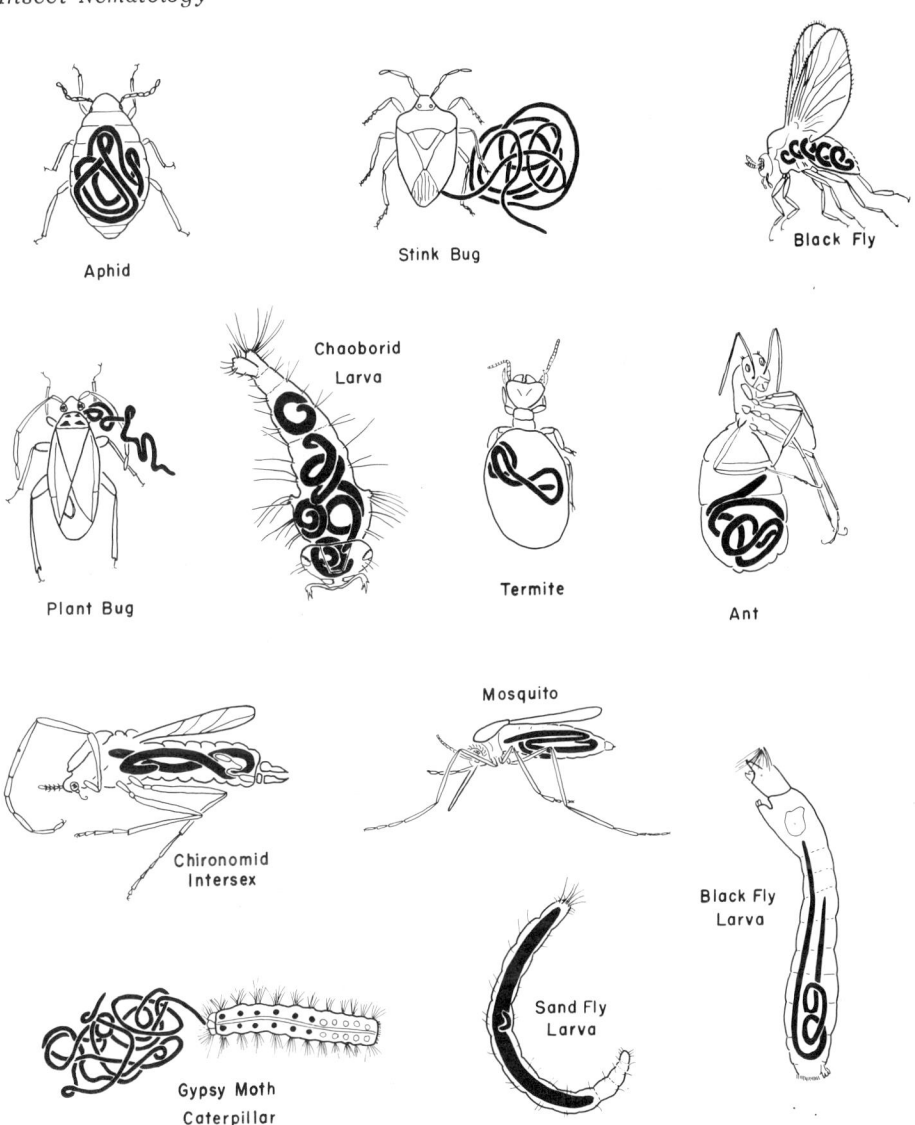

Fig. 8 Some insects known to be parasitized by mermithids, showing location and relative size of the nematode parasites.

larval insects make them very susceptible to mermithid parasitism. In addition to the ultimate death of either the parasitized larval or adult chironomid, we find here parasite-induced intersexes. Wülker (1961, 1963, 1964) has contributed much to our knowledge of these interesting chironomid parasites. At certain times of the year he found 100% parasitism of *Tanytarsus* larvae in Germany. As to variety of mermithids on chironomids, we can find in the literature at least seven different genera of mermithids from *Chironomus plumosus* (L.).

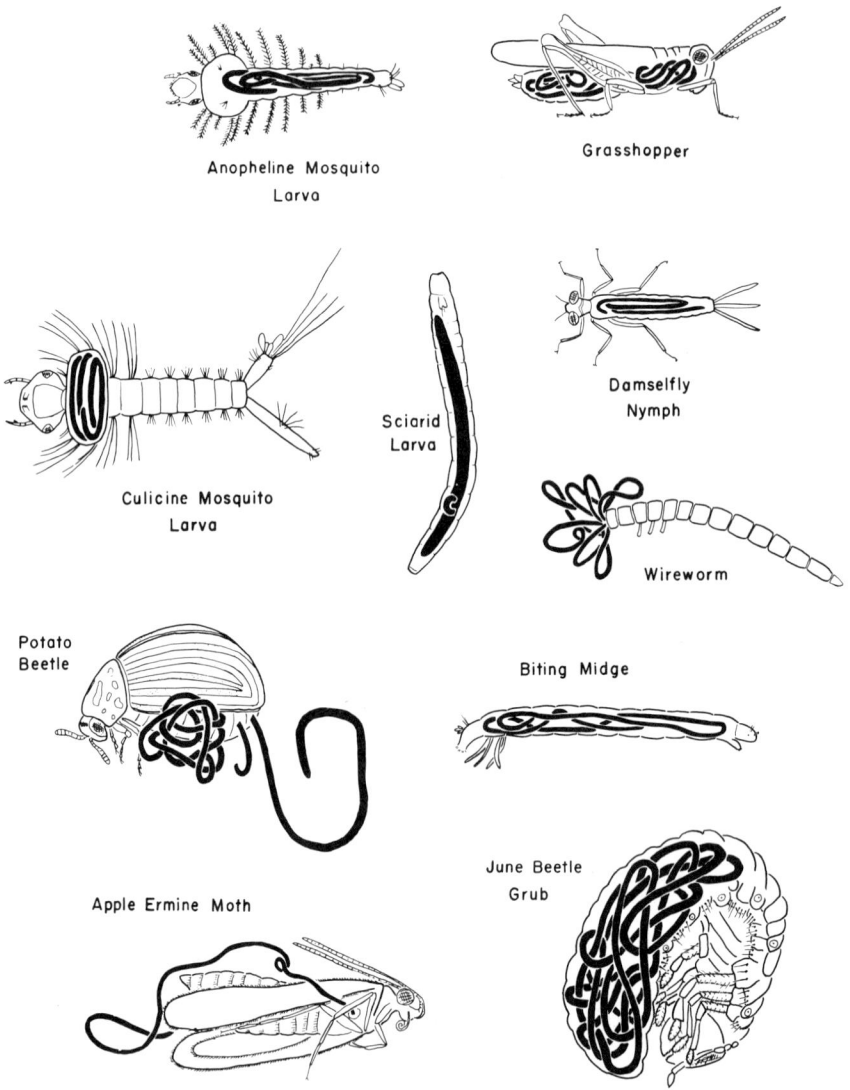

Fig. 8 Continued.

C. Scarabid Beetle Mermithids

It is not too surprising to find that some soil-dwelling mermithids have successfully parasitized the large soil cutworms. These May or June beetle grubs,
and also the Japanese beetle grub, make fine hosts for a number of mermithid
species on a worldwide basis. Polozhentsev (1952) and his student, Artyuk-
hovsky, have studied mermithids from the June beetle, *Melolontha hippocastani*
(F.), from the Soviet Union and have recorded up to 60% mortality by two mermithids in sandy soil of a pine forest. Recently, Klein et al. (1976) found a
similar type of parasitism in which up to 60% of the Japanese beetle grubs
taken from the public green in Brattleboro, Vermont. and other Northeastern
states were infected.

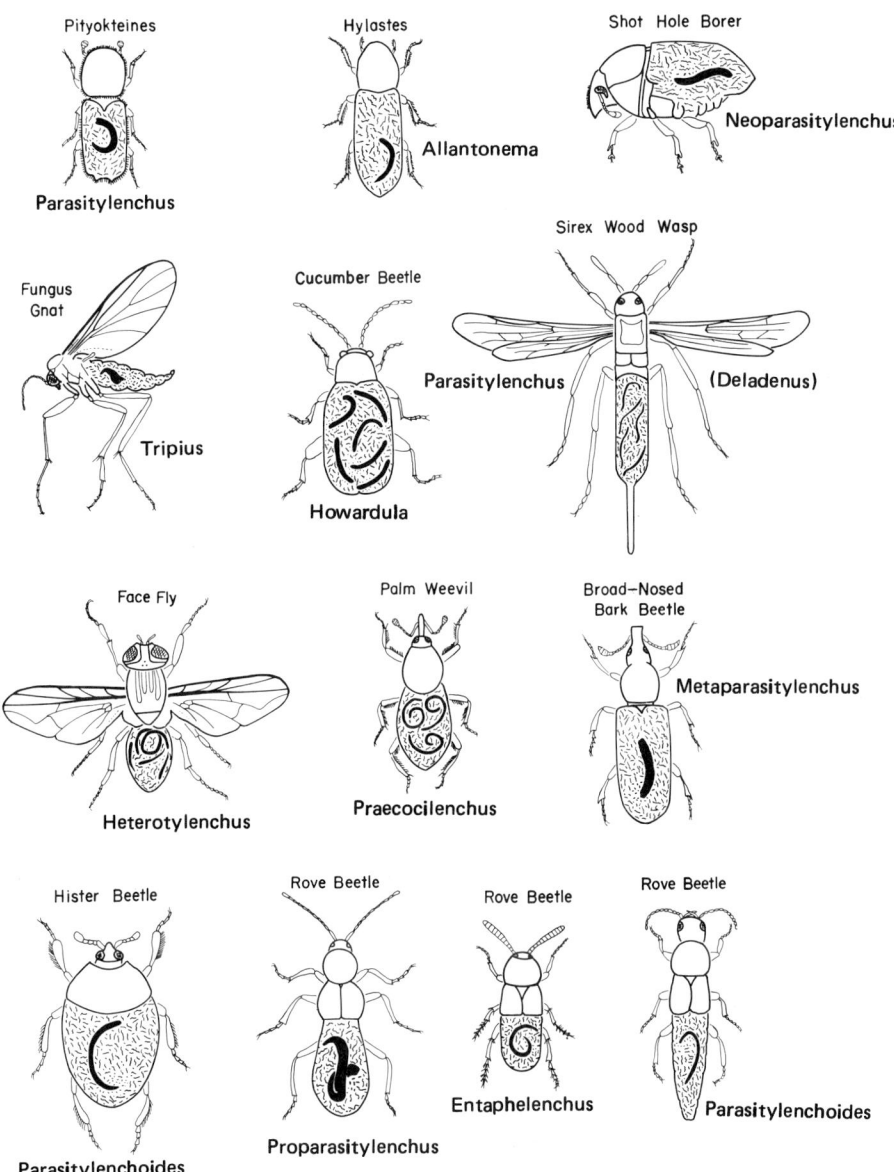

Fig. 9 Some insects known to be parasitized by sphaerulariids and entaphelen-chids, showing location and relative size of the nematode parasites.

D. Ant Mermithids

Classic studies on ant mermithids can be found in the works of Wheeler (1928), Gosswald (1930), and Vandel (1930a, b, 1934). They found distinct morpho-logical changes on the ants induced by mermithid parasitism. They classified these forms as mermithogynes, mermithergates, and mermithostratiotes. They described mermithogynes as forms that are intermediary between workers and fecundated females. Mermithostratiotes are female forms that have the shape

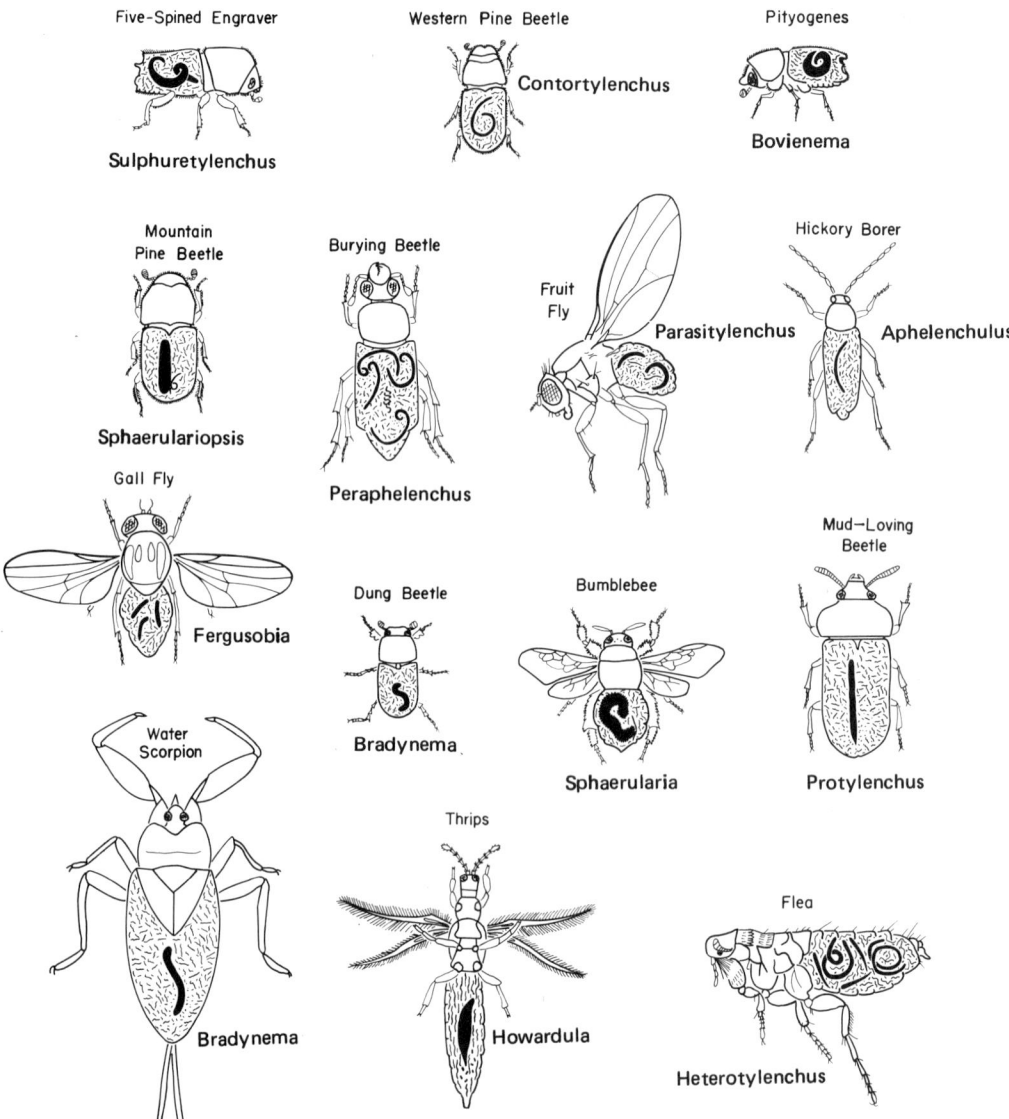

Fig. 9 Continued.

of soldiers. Both mermithogynes and mermithergates have swollen abdomens, though their heads and thoraces are smaller than normal. It was observed by Wheeler (1928) that ants parasitized by mermithids appeared to be in a constant state of hunger. The soil habitat of the ants would again be favorable for high levels of parasitism. Also, ants are concentrated in their colonies.

E. Black Fly Mermithids

Though black fly larvae live in one of the harshest habitats, their numbers are often ravaged by mermithid nematode parasites. One would not think that infective stages of mermithid nematodes would be capable of infecting black fly larvae in rapidly flowing cold water (2°-13°C). However, they have adapted to this situation quite well and often parasitize 80-95% of the black fly larvae that are attached to rocks and debris on the bottoms of streams. Classic studies on black fly mermithids can be found by Professor DeFoliart and his students Anderson and Phelps at the University of Wisconsin (Anderson and Dicke, 1960; Anderson and DeFoliart, 1962; Phelps, 1962; Phelps and DeFoliart, 1964).

Black flies are important medical insects that make many thousands of acres of rich farm land unusable due to onchocerciasis debilitation of residents of West Africa.

F. Chrysomelid Beetle Mermithids

Some important agricultural pest beetles spend a part of their lives in the soil to feed or pupate. Here they become susceptible to infection by soil-dwelling mermithids that have adapted to these insects. Cucumber beetles, corn root worms, and Colorado potato beetles are the best known of this group. Cuthbert (1968) found a mermithid nematode parasite of the banded cucumber beetle, *Diabrotica balteata* Lec, in a sweet potato field in Charleston, South Carolina. He reported that 50-100% of the grubs in the soil were parasitized by this mermithid and indicated that few larval *Diabrotica* would be able to develop to maturity during late August and September. It has become apparent that after the Colorado potato beetle invaded Europe, it became parasitized by a mermithid nematode of the genus *Hexamermis* (Kaiser, 1972).

G. *Pheromermis* Wasp-Ant Mermithids

Though there are six to eight known types of life cycles in mermithid nematode parasites of insects, none is more intricate than the one we now know for members of the genus *Pheromermis*. Poinar and Thomas (1976) and Kaiser (1978) have studied the life cycles of two members of this genus. The former authors found that the yellow jacket (*Vespula pensylvanica* Saussuri) probably became infected by the *Pheromermis* by ingesting other insects that contained infective-stage nematodes in their intestinal cells and muscles. Some of these insects were: caddisflies, craneflies, beetle larvae, and ephemerid nymphs. Kaiser (1978) worked with *Pheromermis myrmecophila* from the ants *Lasius niger* and *L. flavus* in Austria. He found that infective-stage mermithid larvae occurred most frequently in turbellarians (*Polycelis niger*) and Tricoptera, but less frequently in Plecoptera and Tubificida. This type of life cycle allows for higher rates of parasitism in the definitive host yellow jacket or ant.

H. Mosquito Mermithids

Probably the best known and most successful insect mermithid association has been that of *Romanomermis culicivorax* (*Reesimermis nielseni*) on various mosquito species. It was first found at the USDA Lake Charles Mosquito Control Laboratory by Drs. H. C. Chapman and J. J. Petersen. Nickle (1972) separated out the various mermithids found at Lake Charles on mosquitoes and described the infective stage. Fortunately, the workers at Lake Charles added these infective stages to the mosquito mass-rearing pans and they were able to mass rear the parasitic mermithid. Nickle and Chittick of the Fairfax Biological Laboratory approached the Environmental Protection Agency for a ruling on the commercialization of this nematode for the production of a product called "Skeeter Doom." EPA responded in 1976 stating that "the mermithid

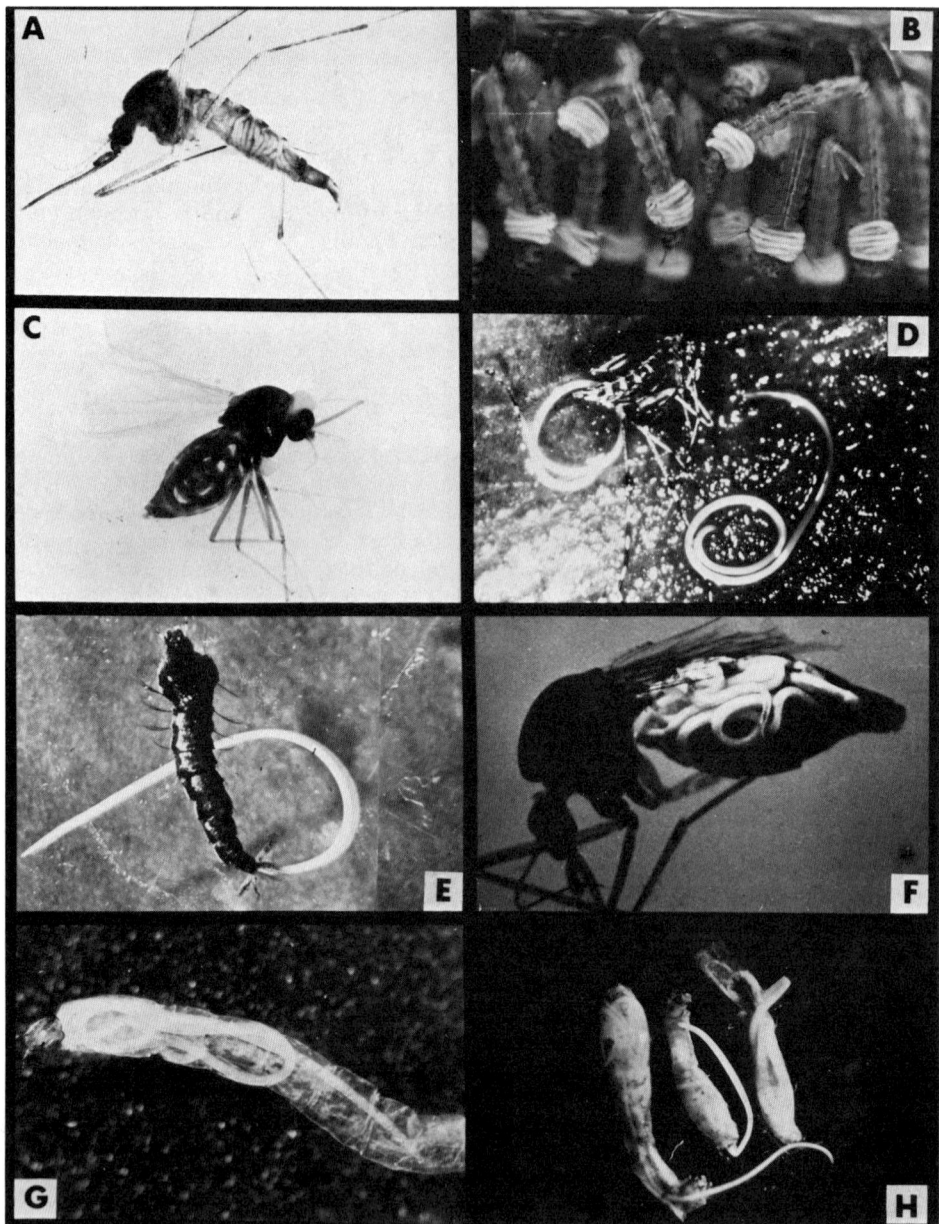

Fig. 10 Mosquitoes and other aquatic insects parasitized by mermithids. (A) *Aedes stimulans* parasitized by a mermithid. (Courtesy of D. L. Haynes.) (B) *Culex pipiens quinquefasciatus* larvae parasitized by *Romanomermis culicivorax*. (Courtesy of J. J. Petersen.) (C) Adult chironomid parasitized by *Hydromermis contorta*. (Courtesy of R. P. Esser and J. B. Mac Gowan.) (D) Adult *Aedes sollicitans* parasitized by *Perutilimermis culicis*. (Courtesy of J. J. Petersen.) (E) *Anopheles crucians* larvae with *Diximermis peterseni* emerging. (Courtesy of J. J. Petersen). (F) *Aedes vexans* with mermithid parasite. (Courtesy of Trpis and Shemanchuk.) (G) Chironomid larvae (*Psectrocladius* sp.) parasitized by *Limnomermis bathybia*. (Courtesy of V. J. E. McCauley.) (H) Blackfly larvae (*Simulium vittatum*) with mermithid parasite. (Courtesy of G. R. De Foliart.)

640

nematode, *Reesimermis nielseni* is not a pesticide as defined in the Federal Insecticide, Fungicide, and Rodenticide Act (FIFRA). Therefore, it is not in our jurisdiction. . . ." This allowed the sale of "Skeeter Doom," which sold for three years. Later, a similar product, Q-Licide, was promoted by Petersen, who will discuss these subjects at length in another chapter. Unfortunately, no Q-Licide was ever sold. The aquatic habitat allows this mermithid to reach high levels of parasitism. Also, it kills the mosquito before it pupates, which eliminates the pest adult stage as the larval mosquito does no economic damage. Mermithid parasites of mosquitoes and other aquatic insects can be seen in Fig. 10.

I. Bark Beetle Sphaerulariids

Early in this century, workers in Germany and other countries began to notice that bark beetles were being parasitized by nematodes. Dissections of parasitized beetles are rather spectacular, exposing thousands of larval nematodes and one or more females free in the hemolymph of the body cavity of each adult bark beetle. The moist habitat in the phloem area under the bark is suitable for infections by nematodes. Also, high concentrations of bark beetle grubs are present here. Rühm (1956) did the classic work on bark beetle nematodes. In 1963, Nickle did his Ph.D. thesis on bark beetle nematodes and worked on the infective-stage female of *Contortylenchus*, which he favored over the genus *Aphelenchulus* in use at the time. Kaya at the University of California and others are currently working in this area.

J. *Sphaerularia* Nematodes

The everted uterus found in insect-parasitic nematodes, such as *Sphaerularia*, *Sphaerulariopsis*, and *Tripius*, are interesting because of their physiological adaptation to life bathed in the hemolymph of the body cavity of a host insect. *S. bombi* was first described by Dufour (1837) from bumblebees. After penetration of the bumblebee by the infective-stage female nematode into the body cavity, the uterus everts out of the body of the female nematode. The uterus, containing the ovaries and sperm, then obtains nourishment from the insect hemolymph and grows to 15,000 times the volume of the original female body, which remains as a minute appendage to the uterus. The genus *Sphaerulariopsis* also has a similar adaptation to the habitat of being free in the hemolymph of the hemocoel of host insects in this case being the coleopteran genera, *Dendroctonus*, *Pissodes*, and *Ernobius*, and the hymenopteran genus, *Coeloides*.

K. Sirex Woodwasps

Deladenus siricidicola was described by Bedding (1968) and has two life cycles, each of which can continue indefinitely without the intervention of the other. One life cycle is free-living, involving a neotylenchid form of female that reproduces oviparously while feeding on the symbiotic fungus of *Sirex* woodwasps. The other cycle is parasitic, with a sphaerulariid form of female reproducing ovoviviparously within the hemocoel. This wood borer habitat is moist and quite suitable for high populations of nematodes. The mycetophagous cycle has been utilized for mass rearing this nematode, and already hundreds of millions of nematodes have been reared and distributed throughout many of the *Sirex*-infested forests of Australia with encouraging early results (Bedding, 1981). If we assume that sphaerulariid nematodes evolved from fungus-feeding nematodes, then this form has not gone all the way to obligate parasitism and has the best of two worlds.

L. *Fergusobia*-Eucalyptus Gallfly

Fergusobia curriei was described by Currie (1937) and came from inside Eucalyptus gallflies in Australia. Later, Fisher and Nickle (1968) found the infective-stage female in the gall, which showed the typical sphaerulariid life cycle for this nematode. Though *Fergusobia* is similar to *Heterotylenchus* in having an alternation of gametogenetic and parthenogenetic generations, it differs in that the parthenogenetic generation is outside the insect, where it may feed on plant or fungal tissue similar to the *Sirex-Deladenus* nematode. Two female stages occur outside the insect, inside Eucalyptus galls. More work is needed on this parasite to determine whether certain stages feed on fungi within the galls, which would place it close to the *Sirex* nematode, both being from Australia.

M. *Heterotylenchus*-Flies

This group of nematodes was first found by Bovien (1937) when he described the intricate life cycle of *Heterotylenchus aberrans* from the onion fly, *Hylemyia antiqua* Meig. He concluded that this nematode sterilized the female flies because no eggs were found in infected females. The ovaries and ovarial ducts were replete with nematodes. More recently, Stoffolano and Nickle (1966) and Nickle (1967b) found a similar nematode *H. autumnalis* from the face fly, *Musca autumnalis* De Geer. Parasitism averaged about 25%. Female insects were found to be sterile when parasitized by this nematode, similar to the one Bovien described. Dissection of a parasitized adult face fly reveals thousands of nematodes of three sizes in the body cavity, ovaries, and thorax of the insect, where one or more adult-parasitic females, 12-24 parthenogenetic females, and thousands of nematode eggs and larvae can be found. Closer examination of the insect's ovaries reveals packets of males and unmated females that have developed at sites normally occupied by eggs. During mock oviposition, the nematodes are deposited in manure, then the nematodes mate; the male dies; and the impregnated young female enters the body cavity of the fly maggot, apparently through the body wall. Once inside the maggot, the small female nematode develops into the adult parasitic stage, which lays eggs in the hemolymph. These eggs develop into parthenogenetic females, which lay larger numbers of eggs, which, in turn, develop into the small males and females. These males and females penetrate the insect ovaries, completing the life cycle.

N. *Neoaplectana carpocapsae (DD-136)*

This nematode vectors a bacterium that builds up in the body cavity of the insect, causing a septicemia and death of over 1000 species of insects on a worldwide basis. The eggs of the nematode are laid in the hemocoel of host insects. They hatch and produce larvae, which mature, mate, and lay eggs in the hemocoel of the same insect. Larvae by the thousands are produced within one insect cadaver (Fig. 11). What makes this life cycle so efficient is that the ensheathed nematode larva, after being ingested by the insect grub, penetrates the gut and releases a mutualistic bacterium from its intestine into the insect's hemolymph. This bacterium then multiplies rapidly, producing an abundance of food for the nematode and its progeny. Therefore, this life cycle has three different forms, the male, the female, and many larvae. The adult nematodes, which are produced later, are usually smaller than the early brood. Dutky and Weiser, as mentioned earlier, found this nematode parasite and its associated bacterium. One of us (WRN) sprayed leaves of potato and bean with ensheathed larvae of this nematode for Colorado potato beetle and

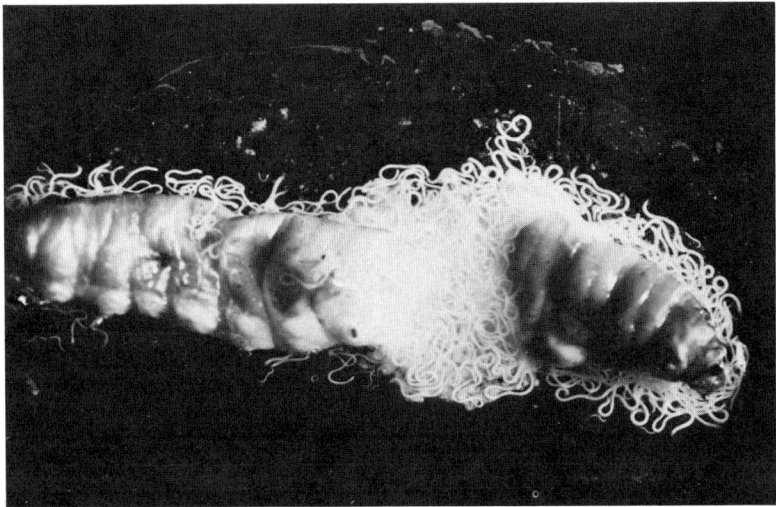

Fig. 11 Wax moth larva replete with *N. carpocapsae* (DD-136). (Courtesy of S. R. Dutky.)

Mexican bean beetle control. These larval insects stopped feeding in 36 hours and died.

In December 1981, the Environmental Protection Agency had exempted the macrobiological agent *Neoaplectana carpocapsae* from the requirements of registration under FIFRA. They felt that the bacterium, *Xenorhabdus nematophilus*, could not survive and replicate independently of the nematode, *N. carpocapsae*, which made this agent unique and helped to influence the EPA decision to exempt this macrobial pesticide. The current regulations provide that certain macrobiologicals are exempt from FIFRA because they are adequately controlled by other Federal Agencies.

Another important development in the practical use of these neoaplectanid and heterorhabditid nematodes was reported by Bedding (1981). It concerned the monoxenical mass rearing of these nematodes economically at a cost of less than two cents per million for biological control purposes. The medium consisted of pork kidney-fat homogenate on polyether polyurethane crumbed sponge preinoculated with the primary form of the bacterial symbiont.

Recently there has been considerable taxonomic activity concerning the name of *Neoaplectana carpocapsae* (DD-136). It has always been somewhat controversial. Stanuszek (1972, 1974) synonymized it under *N. feltiae* Filipjev 1934 after finding some new specimens from the type host in Eastern Europe. Because this nematode has over 1000 known hosts and is worldwide in distribution, there is a good possibility that his action is the correct one. Two contributors to this book (Wouts and Bedding) have accepted Stanuszek's work.

There is also strong evidence that even the genus *Neoaplectana* is a synonym of *Steinernema* (see Chap. 21), and soon this change will be presented in a separate concurrent work.

O. Pine Wood Nematode

Though this nematode, *Bursaphelenchus xylophilus*, is vectored by long horned beetles of the genus *Monochamus* to its pine tree hosts, it had escaped our attention so long that we wonder how many other such associations exist today. This devastating disease is discussed thoroughly by Mamiya in Chap. 16. Dauerlarvae of *B. xylophilus* were found in the tracheae throughout the body of the adult *M. alternatus* in a severely damaged forest. An average of 15,000 nematode dauerlarvae per insect were recovered, with a maximum of 230,000 found on one insect. These larvae leave the tracheae through the spiracles, crawl on the body surface toward the tail tip, and then leave the beetle to wounds on twigs during the maturation feeding of the adult beetle. Up to 75% of the beetles are vectors in a severe outbreak. Though the nematode apparently causes no harm to the insect, the association is devastating to pine trees both in Japan and in the United States on certain ornamental pines.

P. Taxonomic Highlights

1. *Hagmeir*

Hagmeir (1912) made the first extensive revision of the family Mermithidae. His article is a recognized cornerstone for workers dealing with mermithid taxonomy. His spicule drawings are especially well done and are holding up well even under current-day scrutiny. The aquatic mermithid groups and terrestrial mermithid groups, though artificial, are still used and help in separating some genera.

2. *Filipjev*

Filipjev (1934), in his *Manual of Agricultural Helminthology*, dealt extensively with the taxonomy of insect-parasitic nematodes. He states that the classification of the Mermithidae has not yet been established on a sound base. He did not like many of the older papers written by entomologists because, though they used the name of a nematode, they had no description or figure of the parasite. Therefore, he concluded that the practical thing to do is to reject all unrecognizable old species and to describe as new with new names all species that remain open to doubt even if they might have been discovered earlier.

3. *Wachek and Welch*

The non-bark beetle tylenchid parasites were reviewed taxonomically by Wachek (1955). His goal was to set up a natural classification based on descriptions of 33 new endoparasitic tylenchid species and seven new endoparasitic aphelenchoid species from the vicinity of Erlangen, West Germany. Welch (1959) studied the complicated taxonomy and life cycle of *Parasitylenchus diplogenus* on fruit flies of the genus *Drosophila*. He found five different adult forms, i.e., the two sexual generations and one larger swollen female.

4. *Rühm*

Rühm (1956) has done the largest and best work on taxonomy of nematodes of bark beetles. He described one new family, two new subfamilies, five new genera, seven new subgenera, seventy-eight new species, and four new subspecies. He well understood the life cycles of the bark beetle parasites and other associated nematodes and presented the information in an easily understood manner.

5. *Nickle*

Nickle (1967a) studied the taxonomy of specimens of representatives of 15 of the 22 genera of the Sphaerulariidae and presented a classification system of this family. He found the dorsal gland orifice of *S. bombi* to be in the tylenchoid position and made the necessary shifts of taxa. Later, Nickle (1972) reviewed 16 genera of the Mermithidae that were well documented and illustrated pertinent stages of these parasites. Some new mermithids were described, including three new genera. The mosquito mermithids from Louisiana were separated into five different genera.

6. *Rubtsov*

Rubtsov (1971, 1974) continues to do the bulk of the taxonomic work produced on the Mermithidae. He is located in Leningrad, USSR, and has described over 20 new genera and dozens of new species. His work points out the great variety of mermithid fauna that can be found by dedication to this area of research. Many new host insects and much interesting biology can be found here. Three of his larger books have been translated into English, which makes Rubtsov's work more available to Western scientists.

III. IMPORTANCE OF INSECT NEMATOLOGY

A. General Considerations

The importance of insect nematology will be presented in the chapters that follow, and we only attempt here to introduce the subject to our readers. Entomophilic nematodes are a group of parasites that cause debilitation, sterility (partial or complete), or death of a large number of insects in several insect orders and families. They seem to do best in the soil and water environments, though they are not limited to these habitats. Interestingly enough, these environments are two in which it is not easy to utilize parasitic Hymenoptera, viruses, or even chemical pesticides, and therefore, they serve a useful role in suppressing pest insects, often in a self-perpetuating manner. The soil and water insects that are killed by nematodes disintegrate quickly, making it difficult to survey for nematode parasitism. Often the nematodes move out and away from the dead insect to molt and mate deeper in the soil, as is the case with mermithids. So, timing of the sampling or exploration for entomophilic nematodes is important and often this timing is not possible. If the survey is taken too early, the infective stage nematode may not have entered the larval insect yet or may be too small to see. Sometimes, if the survey is taken too late, one might obtain too high a percentage of parasitism because the healthy members of the population have already emerged and left the larval insect environment. This situation occurs because insects with nematode parasites are often behind in their development and, in the case of many mermithid parasites, the insects remain in the larval stage until death occurs, when the large worm emerges from its body.

B. Pathological and Biological Importance of Nematode Parasitism of Insects

1. *Mermithoidea*

a. Tetradonematidae The pathology of *Tetradonema plicans* on the fly, *Sciara coprophila* Lint., in greenhouse cultures was studied by Hungerford (1919). He found that all the insects were parasitized in flower pots that had been in the greenhouse for a long time. The normal fat bodies were lacking in the larval insect, leaving the body very clear so that the parasitic worm could easily be seen. The head capsule was often found to be small, indicating that the maggot failed to make its normal molts. Heavily parasitized maggots often died and disintegrated, leaving only a mass of several thousand nematode eggs. Parasitized adult insects lacked reproductive organs and had swollen abdomens. The importance of this type of parasitism is that maggots that had fed upon plant roots were destroyed or rendered less active, and the insect larvae that transformed to the adult fly stage were rendered incapable of reproduction.

Another tetradonematid nematode parasite, *Corethrellonema grandispiculosum* Nickle, has been found by Chapman et al. (1967) in Louisiana on *Corethrella brakeleyi* (Coquillett). Infected late-instar larvae (about 2.9 mm) contain at least one male and one female nematode and may possess a total of six or more nematode parasites. The nematodes evidently mate shortly before the chaoborid larva is ready to pupate, since the emerging nematode female is usually replete with eggs. Escape of the adult female nematode is always fatal to the larval insect; male nematodes almost always remain in the cadaver.

b. *Mermithidae* The emergence of a mermithid from the body cavity of an insect host produces a large hole in the chitinous exoskeleton of the insect, and death usually follows quickly due to loss of essential body fluids. The fourth-stage mermithid larva uses its lancelike tooth to perforate the insect exoskeleton from the inside. Therefore, it can be said that parasitism by mermithids usually results in death of the host insect. Sometimes, interesting morphological and biological changes occur in insects parasitized by mermithids. These changes appear as swollen abdomens, brown spots on the integument, insects with intersexual characters, such as females with male genitalia, intercastes in social insects, inability to pupate, and other similar deleterious changes. Mermithids parasitize many kinds of invertebrates, including leeches, snails, crustaceans, spiders, and insects, in just about all taxonomic groups. Entomologists often find larval mermithids in insects, place them in 95% alcohol, and send them in for identification. Proper identification can only be done from adult mermithids. Therefore, it is best to allow the mermithid to exit naturally from the insect into a suitable medium, such as soil, and allow the nematode to molt to adult stage, usually requiring four weeks. These specimens can then be fixed in alcohol or 3% formalin, 2% glycerin, and sent to an expert for identification.

2. *Sphaerulariids*

Insects parasitized by sphaerulariid nematodes exhibit pathological symptoms, such as reduced egg production, total sterility, and even death. These pathological effects occur because of the presence of one or more large parasitic female nematodes found free in the hemolymph of a parasitized insect along with her progeny of 2000-7000 smaller nematodes. All these nematodes derive their complete nourishment from the hemolymph of the insect host. Nickle (1971) confirmed the observations of Schvester (1957) concerning a nematode-induced change in oviposition behavior. In this case, the parasitized female bark beetle made a gallery pattern horizontally rather than vertically and did not lay any eggs.

The face fly nematode, *Heterotylenchus autumnalis*, was found by Stoffolano and Nickle (1966) to sterilize from 25-50% of these flies in the United States. This fly was introduced into the East Coast of the United States from Europe, and because the first wave of flies moving West were healthy, they caused considerable difficulties to cattle and horses. Later, about two to three seasons, this nematode and other parasites and predators reduced the population levels of the face fly, and we are now more able to cope with the problem. We in the USDA had leap-frogged some nematodes to Montana and California to help in disseminating this parasite. The nematodes invade the gonads in large numbers, and the infested female fly mock oviposits a ball of infective-stage nematodes in the cow patty. These nematodes then are able to search out and sterilize healthy fly maggots, thus continuing the self-perpetuating biological control process. These and other similar nematodes will be dealt with in other chapters of this book.

3. Neoaplectana *and* Heterorhabditis

Dutky (1959) described a complex involving a species of the nematode genus, *Neoaplectana*, vectoring a bacterium. It is now known as *Neoaplectana carpocapsae*, though it was also known as DD-136 and *N. dutkyi*. The pathology is caused after the nematode larva reaches the insect body cavity. Here it releases the bacterium, *Xenorhabdus nematophilus* into the hemolymph, causing a septicemia and the death of the host insect. The nematodes then feed on bacteria in the insects' cadaver and build up population levels of several thousand within the dead insect. This nematode has thousands of host insects, and there are few insects that are not killed by this nematode, which can be used as a biological insecticide. Members of the nematode genus *Heterorhabditis* cause a similar pathology to insects. It also vectors a bacterium that is actually the cause of the insects' death. These nematodes will be dealt with at length in later chapters.

IV. CONCLUSIONS

Our current status with regard to entomophilic nematodes shows that we are making some progress. We now are able to identify the main groups of insect-parasitic nematodes, though one important genus was described less than five years ago. We now have a voluminous and impressive accumulation of data and field evidence illustrating that nematodes play an important role in natural control of many pest insect populations. One can only speculate that if we transported these nematode parasites across oceans or mountain ranges, thereby introducing these parasites into otherwise nematode-free populations, or if we did something to enhance the existing nematode parasites by inundation or manipulation, the benefit to humanity would be significant. The writers feel that the potential for use of entomophilic nematodes as self-perpetuating biological control agents lies in areas where chemical pesticides are too expensive, not practical, or most noxious to humans and the environment. These situations, such as near water, streams, ponds; near livestock and other animals; near our dwellings; and in soil situations like concentrated irrigated or river bottoms where agricultural production occurs, are also fortunately ideal habitats for high nematode parasitism of insects.

REFERENCES

Aldrovandus, U. (1602). *De animalibus insectis libri septum, cum singulorum iconibus ad vivum expressis.*

Aldrovandus, U. (1623). *De animalibus insectis libri septum.* fol. Francofurti.

Anderson, J. R., and DeFoliart, G. R. (1962). Nematode parasitism of black fly (Diptera: Simuliidae) larvae in Wisconsin. *Ann. Ent. Soc. Am. 55:* 542-546.

Anderson, J. R., and Dicke, R. J. (1960). Ecology of the immature stages of some Wisconsin Black flies (Simuliidae: Diptera). *Ann. Ent. Soc. Am.* 53: 386-404.

Bedding, R. A. (1968). *Deladenus wilsoni* n. sp. and *D. siricidicola* n. sp. (Neotylenchidae), entomophagous-mycetophagous nematodes parasitic in Siricid woodwasps. *Nematologica 14*: 515-525.

Bedding, R. A. (1981). Low cost *in vitro* mass production of *Neoaplectana* and *Heterorhabditis* species (Nematoda) for field control of insect pests. *Nematologica 27*: 109-114.

Berthold, A. A. (1843). Uber den Bau des Wasserkalbes (*Gordius aquaticus*). *Abhandl. K. Gesellsch Wissensch. Gottingen Math-Phys. Cl. 1*: 1-18.

Bovien, P. (1933). On a new nematode, *Scatonema wülkeri* gen. et. sp. n. parasitic in the body cavity of *Scatopse fuscipes* Meig. (Diptera, Nematocera). *Vidensk Medd. Dansk Naturh. Foren. Köben'havn. 94*: 13-32.

Bovien, P. (1937). Some Types of Association between Nematodes and Insects. *Vidensk. Medd. fra Dansk Naturh. Foren. 101.*

Bovien, P. (1944). *Proatractonema sciarae* n. g., n.sp., a parasitic nematode from the body cavity of a dipterous larva. *Vidensk Medd. Dansk Naturh. Foren. Köben'haven 108*: 1-14.

Braun, M. (1883). Die thierishchen Parasiten des Menschen nebst einer Anleitung zur praktischen Beschaftigung mit der Helminthologie fur Studierende und Aerzte.

Bremser, J. G. (1824). Nouvel Atlas, compose de 15 planches in -4° dissinees par A. Foucaud, avec un texte explicatif renfermant des observations inedites par Charles Leblond. Paris.

Bulloch, J. (1861). On *Sphaerularia bombi*. *Natural. Hist. Review 1*: 44-57.

Chapman, H. C., Woodard, D. B., and Petersen, J. J. (1967). Nematode parasites of Culicidae and Chaoboridae in Louisiana. *Mosquito News 27*: 490-492.

Charvet, P. (1834). Observations sur deux especes du genre dragonneau qui habitent dans quelques eaux courantes aux environs de Grenoble. *Nouvelles Ann. Mus. Hist. Nat. Paris 3*: 37-46.

Christie, J. R. (1941). Chapter XVIII, Life History (Zooparasitica). Parasites of Invertebrates. In *Introduction to Nematology* (B. G. Chitwood and M. B. Chitwood, eds.). Reprinted, University Park Press, 1974.

Cobb, N. A. (1919). The orders and classes of nemas. *Contrib. Sci. Nematol 8*: 213-216.

Cobb, N. A. (1927). Nemas and recent progress in nematology research. *Agric. Yearbook USDA* (1926), pp. 540-543.

Cobb, N. A., Steiner, G., and Christie, J. R. (1923). *Agamermis decaudata* Cobb, Steiner, and Christie, a nema parasite of grasshoppers and other insects. *J. Agric. Res. 28*: 921-926.

Currie, G. A. (1937). Galls on Eucalyptus trees. A new type of association between flies and nematodes. *Proc. Linn. Soc. N.S. Wales 62*: 147-174.

Cuthbert, F. P., Jr. (1968). Bionomics of a mermithid (Nematode) parasite of soil-inhabiting larvae of certain Chrysomelids (Coleoptera). *J. Invert. Path. 12*: 283-287.

Dadai, J. (1911). Adatok a mermithidae – csalad edes vizben elo fajainak ismerethehez. Math. es Termeszettud. Ertesito Magyar Tudoman Akad., Budapest *29*: 450-514.

Daday, J. (1913). Beitrage zur Kenntnis der in Susswassern lebenden Mermithiden. *Math. Naturw. Berg. Ungarn* (1909) *27*: 214-272; 4: 273-281.

Diesing, K. M. (1851). *Systema helminthum. 2.*

Dufour, L. (1837). Recherches sur quelques entozoaires et larves parasites des insectes orthopteres et hymenopteres. *Ann. Sci. Natur. Zool.* 7: 5-20.

Dujardin, F. (1842). Memoire sur les *Gordius* et les *Mermis*. *Compt. Rend. Acad. Sci. Paris 15*: 117-119.

Dujardin, F. (1845). Histoire naturelle des helminthes ou vers intestinaux. Paris.

Dutky, S. R. (1959). Advances in applied microbiology. *Insect Microbiology 1*: 175-200.

Dutky, S. R., and Hough, W. S. (1955). Note on a parasitic nematode from Codling moth larvae (*Carpocapsa pomonella*) (Lepidoptera, Olethreutidae). *Proc. Entomol. Soc. Wash.* 57: 24.

Fedchenko, A. P. (1874). Zoological observations III. The anatomy of round worms (Russian). *Izvest. Imp. Obsh. Liub Estestvozn.* Moskva. (1872-1873) *10*: 51-68.

Fedchenko, A. P. (1886). Travels in Turkestan II. *Zoogeographical Investigations, Roundworms and Trematodes.* (Russian). Moscow.

Filipjev, I. N. (1934a). *Harmful and Useful Nematodes in Rural Economy* (Russian). Figs. 1-333. Moskva, Leningrad.

Filipjev, I. N. (1934b). The classification of the freeliving nematodes and their relation to the parasitic nematodes. *Smithsonian Misc. Coll.* (Publ. 3216) *89*: 1-63.

Filipjev, I. N., and Schuurmans-Stekhoven, J. H. Jrs., (1941). *A Manual of Agricultural Helminthology.* E. J. Brill, Leiden.

Fisher, J. M., and Nickle, W. R. (1968). On the classification and life history of *Fergusobia curriei* (Sphaerulariidae: Nematoda). *Proc. Helm. Soc. Wash.* 35: 40-46.

Fuchs, A. G. (1915). Die Naturgeschichte der Nematoden und einiger anderer Parasiten. 1. Des *Ips typographus* L., 2. Des *Hylobius abietis* L. *Zool. Jahrb. Jena Abt. Syst. 38*: 109-222.

Goeze, J. A. E. (1782). Versuch einer Naturgeschichte der Eingeweidewurmer thierischer Korper.

Goodey, J. B. (1963). *Soil and Freshwater Nematodes.* Methuen, London.

Goodey, T. (1930). On a remarkable new nematode *Tylenchinema oscinellae* gen. et sp. nov., parasitic in the fritfly, *Oscinella frit* L., attacking oats. *Phil. Trans. R. Soc. London 218*: 315-343.

Goodey, T. (1951). *Soil and Freshwater Nematodes.* Methuen, London.

Gosswald, K. (1930). Weitere Beitrage zur Verbreitung der Mermithiden bei Ameisen. *Zool. Anz. Bd. 90*: 13-27.

Gould, W. (1747). *An Account of English Ants.* London.

Hagmeier, A. (1912). Beitrage zur Kenntnis der Mermithiden. 1. Biologische Notizen un systematische Beschreibung einiger alter und neuer Arten. *Zool. Jahrb. Abt. Syst. 32*: 521-612.

Hall, M. C. (1929). Arthropods as intermediate hosts of helminths. *Smithsonian Misc. Coll. (Publ. 3024). 81*: 77 pp.

Hope, F. W. (1839). Lists of the genera and species of insects infested by Filariae. *Tr. Entom. Soc. London* (1837-40) 2: 256-271.

Hungerford, H. B. (1919). Biological notes on *Tetradonema plicans* Cobb, a nematode parasite of *Sciara coprophila* Lintner. *J. Parasitol.* 5: 186-192.

Ipatyeva, G. V. (1970). On the knowledge of geohelminths (Nematoda) of Lamellicorn beetles (Scarabaeidae, Coleoptera) (Russian). *Tr. Saratov Zootekh-Veterin. Inst. 19*: 158-177.

Kaiser, H. (1972). Mermithidae (Nematoda) als Parasiten des Kartoffelkafers (*Leptinotarsa decemlineata* Say.) in der Steiermark. Ph.D. Thesis, University of Graz, Austria.

Kaiser, H. (1978). The ecological position of 'parasitoid' nematodes with a change of hosts (Nematoda: Mermithidae). *Zentralbl. Bakteriol. Parasitenkunde Infektionskrankh. Hygiene 257*: 29-30.

Keppen, P. (1870). On locusts and other pest orthopterans in the family *Acridioidea* mainly in relation to Russia (Russian). *Russian Entomolog. Soc.* SPb. *5*: 1-345.

Keppen, P. (1881). *Pest Insects* (Russian). SPb. *I*: 1-374.

Keppen, P. (1882). *Pest Insects* (Russian). SPb. *II*: 1-585.

Kirjanova, E. S. (1959). The 70th Anniversary of the birth of the outstanding Soviet scientist, Professor Ivan Nikolaevich Filipjev. *Izv. Akad. Nauk. Tadzhik. SSSR 2*: 51-55.

Klein, M. G., Nickle, W. R., Benedict, P. R., and Dunbar, D. M. (1976). *Psammomermis* sp. (Nematoda: Mermithidae): A new nematode parasite of the Japanese beetle, *Popillia japonica* (Coleoptera: Scarabaeidae). *Proc. Helm. Soc. Wash. 42*: 235-236.

LaRivers, I. (1949). Entomic nematode literature from 1926 to 1946, exclusive of medical and veterinary titles. *Wassman Collect. 7*: 177-206.

Leuckart, K. G. F. R. (1887). Neue Beitrage zur Kenntniss des Baues und der Lebensgesichte der Nematoden. *Abhand Math-Phys. Cl. K. Sachs Gesellsch. 13*: 565-704.

Linnaeus, Carl von. (1758). Systema naturae regna tria naturae, secundum classes, ordines, genera, species, cum characteribus differentiis, synonymis, locis. *Editio decima reformata. 1.*

Linstow von, O. F. B. (1878). *Compendium of Helminthologie.* Ein Verzeichniss der bekannt Helminthen, die frei oder in thierischen Korpern leven, geornet nach ihren Wohnthieren, unter der Organe, in denen sie gefunden sind, und mit Beifugung der Litteraturquellen. Hannover.

Linstow von, O. F. B. (1898). Das Genus *Mermis. Arch. Mikrosk. Anat. 53*: 149-168.

Lister, M. (1672). An extract of a letter written from York, April 12,1672, concerning animated horse hairs, rectifying a vulgar error. *Phil. Trans. R. Soc. London 83 (7)*: 4064-4065.

Massey, C. L. (1974). Biology and Taxonomy of Nematode Parasites and Associates of Bark Beetles in the United States. *Agriculture Handbook No. 446.* Forest Service. USDA, Washington, D.C.

Meissner, G. (1853). Beitrage zur Anatomie und Physiologie von *Mermis albicans. Ztschr. Wissensch. Zool. 5*: 207-284.

Micoletzky, H. (1925). Die freilebenden Susswasser – und Moornematoden Denemarks. *K. Danske Vidensk. Selsk. Skr. Natury.* 8th ser. *10*: 57-310.

Nickle, W. R. (1963). The endoparasitic nematodes of California bark beetles with descriptions of *Bovienema* n.g. and *Neoparasitylenchus* n. subg. and with the presentation of new information on the life history of *Contortylenchus elongatus* n. comb. Ph.D. thesis, University of California.

Nickle, W. R. (1967a). On the classification of the insect parasitic nematodes of the Sphaerulariidae Lubbock, 1861 (Tylenchoidea: Nematoda). *Proc. Helm. Soc. Wash. 34*: 72-94.

Nickle, W. R. (1967b). *Heterotylenchus autumnalis* sp. n. (Nematoda: Sphaerulariidae), a parasite of the face fly, *Musca autumnalis* de Geer. *J. Parasitol. 53*: 398-401.

Nickle, W. R. (1971). Behavior of the shothole borer, *Scolytus rugulosus*, altered by the nematode parasite *Neoparasitylenchus rugulosi*. *Ann. Entomol. Soc. 64*: 751.

Nickle, W. R. (1972). A contribution to our knowledge of the Mermithidae (Nematoda). *J. Nematol. 4*: 113-146.

Nickle, W. R. (1973). Identification of insect parasitic nematodes – a review. *Exp. Parasitol. 33*: 303-317.

Nickle, W. R. (1974). Nematode infections. In *Insect Diseases*, volume 2 (G. E. Cantwell, ed.). Dekker, New York, pp. 327-376.

Oldham, J. N. (1933). Helminths in the biological control of insect pests. *Imp. Bur. Agric. Parasitol. Notes Mem. 9*: 6 pp.

Phelps, R. J. (1962). Nematode parasitism of larval Simuliidae. *Diss. Abstr. 23*: 1128-1129.

Phelps, R. J., and DeFoliart, G. R. (1964). Nematode parasitism of Simuliidae. University of Wisconsin Experimental Station, *Res. Bul. 245*.

Poinar, G. O., Jr. (1967). Description and taxonomic position of the DD-136 nematode (Steinernematidae: Rhabditoidea) and its relationship to *Neoaplectana carpocapsae* Weiser. *Proc. Helm. Soc. Wash. 34*: 199-209.

Poinar, G. O., Jr. (1975). *Entomogenous nematodes. A Manual and Host List of Insect-Nematode Association*.

Poinar, G. O., Jr. (1977). CIH Key to the groups and genera of nematode parasites of invertebrates. *Commonwealth Agricultural Bureaux*.

Poinar, G. O., Jr. (1979). *Nematodes for Biological Control of Insects*. CRC Press Inc., Boca Raton, Florida.

Poinar, G. O., Jr., and Thomas, G. M. (1976). Biology and redescription of *Pheromermis pachysoma* (V. Linstow) n. gen., n. comb (Nematoda: Mermithidae), a parasite of yellowjackets (Hymenoptera: Vespidae). *Nematologica 22*: 360-370.

Polozhentsev, P. A. (1952). New mermithidae of sandy soil of pine forests. *Trudy Helm. Lab. 6*: 376-382.

Polozhentsev, P. A., and Negrobov, V. P. (1967). On insects; hosts for parasitic worms of man and animals (Russian). In *Harmful and Beneficial Insects*. Voronezh.

Radkevitch, G. (1869). On roundworms parasitizing in the oesophageal canal of *Blatta* (Russian). *Proc. Verb. of the University of Khar'kov (3)* annere 1-18 and 1-111.

Reaumur de, R. A. F. (1742). *Memoires pour servir a l'histoire des insectes. 6*.

Rubtsov, I. A. (1971). New species of mermithids. *Parasitologia 5*: 458-461.

Rubtsov, I. A. (1974). Anatomy and diagnosis of mermithid larvae. *Zoologicheskii Zhurnal 44*: 660-675.

Rudolphi, C. A. (1809). *Entozoorum sive vermium intestinalium histoiria naturalis*, Amsterdam. *2*.

Rudolphi, C. A. (1819). *Entozoorum synopsis cui accedunt mantissa duplex et indices locu pletissiimi*. Berlin.

Ruhm, W. (1956). Die Nematoden der Ipiden. *Parasitologische Schriftenreihe 6*: 1-437.

Schneider, A. (1866). *Monographie der Nematoden*. Reimer, Berlin.

Schultz, O. (1900). Filarien in Palaarktischen Lepidopteren. Illust. *Ztschr. Entom. 5*: 148-152; 164-168; 183-185; 199-201; 264-265; 279-280; 292-297.

Schvester, D. (1957). Contribution a l'etude de coleopteres scolytides. *Serie C. Ann. Epiphyties*: 1-162.

Shephard, M. R. M. (1974). *Arthropods as final hosts of nematodes and nematomorphs*. An annotated Bibliography. 1900-1972. Tech. Comm. 45. Commonwealth Agricultural Bureaux. 248 pp.

Siebold von, C. T. E. (1842; 1843; 1848; 1850). Ueber die Fadenwurmer der Insekten. *Entomol. Z. 3*: 146-161; *4*: 78-84; *9*: 290-300; *11*: 329-336.

Stanuszek, S. (1972). Revision of the genus *Neoaplectana* Steiner, 1929. (Rhabditoidea: Steinernematidae). *Abst. 10th Int. Symp. Nemat. Readings*: 69-70.

Stanuszek, S. (1974). *Neoaplectana feltiae* complex (Nematoda: Rhabditoidea. Steinernematidae) its taxonomic position within the genus *Neoaplectana* and intraspecific structure. *Zesz. Probl. Postep. Nauk soln. 154*: 331-360.

Steinhaus, E. A. (1949). *Principles of Insect Pathology*. McGraw-Hill, New York.

Stiles, C. H. W., and Hassall, A. (1920). *Index-catalogue of Medical and Veterinary Zoology*. Subjects: Roundworms (Nematoda, Gordiacea and Acanthocephali) and the Diseases That They Cause. *USPHS Hygenic Laboratory-Bulletin No. 114*.

Stoffolano, J. G., and Nickle, W. R. (1966). Nematode parasite (*Heterotylenchus* sp.) of face fly in New York State. *J. Econ. Entomol. 59*: 221-222.

Stoll, N. R. (1959). Conditions favoring the axenic culture of *Neoaplectana glaseri*, a nematode parasite of certain insect grubs. *In Axenic Cultures of Invertebrate Metazoa: A Goal*. Ed. E. C. Dougherty. *Ann. N.Y. Acad Sci. 77*: 126-136.

Taylor, A. L. (1935). A review of fossil nematodes. *Proc. Helm. Soc. Wash. 2*: 47-49.

Thorne, G. (1961). *Principles of Nematology*. McGraw-Hill, New York.

Travassos, L. (1953). Nematodeos parasitos de *Gryllotalpa*. G. S. Thapar, *Commemoration Volume Lucknow*, 277-288.

Vandel, A. (1930a). La production d'intercastes, chez la Fourmi, *Pheidole pallidula*, sous l'action de parasites du genre *Mermis*. *Compt. Rend. Sceances, Acad. Sci. 190*: 770-772.

Vandel, A. (1930b). La production d'intercastes chez la fourmi *Pheidole pallidula* suns l'action de parasites du genre *Mermis*. I. Etude morphologique des individus parasites. *Bull. Biol. France Belgique 64*: 457-494.

Vandel, A. (1934). Le cycle evolutif d-*Hexamermis* sp., parasite de la fourmi (*Pheidole pallidula*). *Ann. Sci. Nat. Zool. 17*: 47-58.

Vejdovsky, F. (1886). Zur Morphologie der Gordiiden. *Z. Wissensch. Zool. 43*: 369-433.

Wachek, F. (1955). Die entoparasitischen Tylenchiden. *Parasitol. Schriftenreihe 3*: 1-119.

Weiser, J. (1955). *Neoaplectana carpocapsae* n. sp. (Anguillulata, Steinernematidae) novy Cizopasnik housenek obatece jablecneho *Carpocapsa pomonella* L. *Vestnik Cesk. Zool. Spolecnosti 19*: 44-52.

Welch, H. E. (1956). Review of recent work on nematodes associated with insects with regard to their utilization as biological control agents. *Proc. 10th International Congress of Entomology 4*: 863-868.

Welch, H. E. (1959). Taxonomy, life cycle, development, and habits of two new species of Allantonematidae (Nematoda) parasitic in drosophilid flies. *Parasitology 49*: 83-103.

Welch, H. E. (1963). Nematode infections. In *Insect Pathology* (E. A. Steinhaus ed.). Academic, New York, pp. 363-392.

Welch, H. E. (1965). Entomophilic nematodes. *Ann. Rev. Entomol. 10*: 275-302.

Wheeler, W. M. (1928). *Mermis* parasitism and intercastes among ants. *J. Exp. Zool. 50*: 165-237.

Wouts, W. M. (1981). *Hexamermis truncata* (Rudolphi, 1809) new combination, the valid name for *Hexamermis albicans* (Von Siebold, 1848). *Systematic Parasitol. 3*: 127-128.

Wülker, W. (1961). Untersuchungen uber die Intersexualitaet der Chironomiden (Dipt.) nach *Paramermis* Infektion. *Archiv fur Hydrobiologie. Suppl. 25, 4*: 127-181.

Wülker, W. (1963). Prospects for biological control of pest Chironomidae in the Sudan. *World Health Organ. 11*: 23 pp.

Wülker, W. (1964). Parasite-induced changes of internal and external sex characters in insects. *Exp. Parasitol. 15*: 561-597.

Yatsenkowsky, A. W. (1924). The castration of *Blastophagus* of pines by roundworms and their effect on life phenomena of the Ipidae (Russian). *Publ. Agric. Inst. Western White Russian 3*: 1-19.

Zeder, J. G. H. (1800). *Erster Nachtrag zur Naturgeschichte der Eingeweiderwurmer, mit Zufassen un Anmerkungen herausgeben.* Leipzig.

Zwaluwenburg van, R. H. (1928). The interrelationships of insects and roundworms. *Bull. Exp. Sta. Hawaiian Sugar Planters Assoc. Entomol. Series 20.*

Chapter 18
Nematode Parasites of Lepidopterans

Wilhelmus M. Wouts *Mt. Albert Research Center, Auckland, New Zealand*

I. INTRODUCTION

Many species of nematodes parasitize insects, but only mermithids and bacteria-transmitting rhabditids parasitize lepidopterans. The unidentified rhabditid and two tylenchids, including *Mikoletzkya aerivora* (Cobb 1916), reported in the literature (Poinar, 1975) were found in dead hosts and are not proven parasites of lepidopterans.

Mermithids are associated with their host for up to a month and occur in a wide range of insect hosts and habitats. They are large animals and can be easily recognized as a group. The diagnosis of the individual species is difficult because detailed comparative studies and comprehensive keys are not available. The stage most commonly encountered in nature is the parasitic stage, and diagnostic characters are confined to the free-living adults. That these large animals require laborious sectioning for identification further complicates the situation and has caused major misidentifications and the perpetuation of obvious errors. In this chapter only those records that include an accurate diagnosis are discussed.

Bacteria-transmitting nematodes belong to the families Steinernematidae Filipjev 1934 and Heterorhabditidae Poinar 1976. They resemble microscopic, free-living species. They are occasionally encountered in field collections of insects and can be obtained from infected host cadavers or from material extracted from infested soil. They infect a wide range of lepidopteran hosts. The bacteria they transmit are symbiotically associated with the nematodes and are lethal to the host. Transmitted to a host, the bacteria rapidly multiply in the body cavity, and together with the decomposing tissues of the host serve as food for the nematode.

Parasitism by mermithids and bacteria-transmitting rhabditids is not restricted to lepidopterans, and for a full account of each individual species the sections on the same species in other chapters in this book should also be considered.

II. AUTHENTIC RECORDS OF LEPIDOPTERAN PARASITES

The nematode species reported as lepidopteran parasites in the literature include: *Mermis nigrescens* Dujardin 1842; *M. indica* Schultz 1899; *Hexamermis truncata* (Rudolphi 1809) Wouts 1981; *H. microamphidis* Jaynes 1933; *H. ferghanensis* Kirjanova, Karavaeva, and Romanenko 1959; *H. cavicola* Welch 1963; *H. arvalis* Poinar and Gyrisco 1962; *Agamermis decaudata* Cobb, Steiner, and Christie 1923; *Amphimermis elegans* (Hagmeier 1912) Welch 1963; *A. bogongae* Welch 1963; *A. zuimushi* Kaburaki and Imamura 1923, and species of the genera *Neoaplectana* Steiner 1929 and *Heterorhabditis* Poinar 1976.

In a critical search to sift out the properly documented lepidopteran parasites, only a small number of these records proved valid. *Mermis nigrescens* larvae have been reported from lepidopterans repeatedly, but it is unlikely that they are correctly identified for they have never been maintained to maturity to check their identity; experiments by Hagmeier (1912), Baylis (1944), and myself (unpublished data), in which *M. nigrescens* eggs were fed to lepidopterans, failed to result in infections. *M. nigrescens* eggs have to be ingested for the species to become parasitic. In this respect the species resembles oxyurid nematodes. Oxyurids commonly parasitize insects whose digestive processes involve a more or less prolonged period of bacterial fermentation; they do not parasitize lepidopterans. As is the case for oxyurids (Dale, 1970), the passage of food through the simple digestive system of caterpillars is too fast for *M. nigrescens* eggs to hatch and for the nematodes to become established. *Mermis indica* and *Hexamermis arvalis* are both misidentifications. In the original description of *M. indica* the postparasitic larva is illustrated with the distinctly attenuated tail tip typical of *Hexamermis* rather than *Mermis* larvae, and the S-shaped vagina, which was claimed to separate *H. arvalis* from *H. truncata*, is diagnostic for the latter (Nickle, 1972). Like *Hexamermis microamphidis* (Poinar and Gyrisco, 1962), *H. ferghanensis* is a *nomen nudum* because the detailed description mentioned by the original authors was never published. Identifications of *Agamermis decaudata* and *Amphimermis elegans* were based on postparasitic larvae that were not reared to maturity. As specific characters are located in the adult females, the true identity of these larvae is uncertain. Thus, only four mermithid species definitely parasitize lepidopterans: *H. truncata, H. cavicola, Amphimermis bogongae*, and *A. zuimushi*.

The genus *Neoaplectana* contains about 20 nominal species, but few are considered to be valid. On the basis of the body length of the infective larvae, Stanuszek (1972, 1974) distinguished three species, *N. glaseri* Steiner 1929, *N. bibionis* Bovien 1937, and *N. feltiae* Filipjev 1934, and suggested that no further species should be recognized until it has been established that they do not interbreed with any of these three species. [Since the manuscript of this chapter was submitted, Wouts et al. (1982) have officially established that only these three species are valid.]

Two species of the genus *Heterorhabditis*, *H. bacteriophora* Poinar 1976 and *H. heliothidis* (Khan, Brooks, and Hirschmann 1976) Poinar, Thomas, and Hess 1977, have been found associated with lepidopterans. *H. hambletoni* (Pereira 1934) is the only other nominal species in the genus.

III. HISTORY OF LEPIDOPTERAN PARASITES

A. Mermithidae

Of the four mermithid species that parasitize lepidopterans, *Hexamermis truncata* has been known the longest (about 170 years) and has been studied in greatest detail. *Amphimermis zuimushi*, *A. bogongae*, and *Hexamermis cavicola* have been known for less than 50 years, and the only information on them is provided in their original descriptions. *H. truncata* occurs in the literature under various names. Rudolphi (1809) described it as *Filaria truncata* from material isolated from caterpillars of the moth *Yponomeuta padella* (L.). von Siebold (1848) changed the name to *Mermis albicans*, and Steiner (1924) transferred the species to the genus *Hexamermis*. The name *H. albicans* remained valid until recently, when the specific name *truncata* was reestablished (Wouts, 1981a).

Rudolphi (1809) characterized *H. truncata* by its elongated filiform body, truncated head, and wide tail with short, more or less blunt tip. von Siebold and Rosenhauer, in the 1840s, established that the name was based on and had always been applied to immature forms, placed the species in the genus *Mermis* Dujardin 1842, and worked out the life cycle. Meissner (1854) used von Siebold's accumulated material for detailed studies of the morphology of this nematode. He distinguished most of the morphological features now used to identify the species. Hagmeier (1912) revealed that the esophagus does not function as a digestive system but that food was absorbed through the cuticle by endosmosis.

Since the beginning of the twentieth century, several more mermithids have been named. Their different characteristics provided a better understanding of the morphology of the group in general, and this in turn stimulated the study of the functions of the organs recognized. This general activity turned attention away from *H. truncata*. Hagmeier's (1912) comparative study and the more recent life cycle studies of Couturier (1950) and Rathke (1953) have been the only major contributions in the last 80 years.

A. zuimushi parasitizes the rice borers *Chilo simplex* Butler (Kaburaki and Imamura, 1932) and *Chilo suppressalis* (Wlk) (Tateishi et al., 1955). The life cycle worked out by Kaburaki and Imamura (1932) revealed that the infective larvae actively find the host by entering the rice plant through the holes left by the young caterpillar when it bored into the plant. *A. bogongae* and *H. cavicola* parasitize the bogong moth, *Agrotis infusa* (Bois.), estivating in large congregations on the walls of mountain caves during the summer. Their life cycles were partially worked out by Welch (1963). The two species are unusual in that they infect an adult lepidopteran; usually mermithids attack immature stages.

Hosts lists of *H. truncata* presented by earlier workers were compiled by Schultz (1900) and Poinar (1975). An updated list is presented in Appendix I.

In the literature a large number of lepidopterans have been observed as hosts of *Hexamermis* species and unidentified mermithids. They are presented in Appendixes II and III. In the majority of these cases the parasite involved was probably *H. truncata*.

B. Steinernematidae

Studies of *Neoaplectana* species began 50 years ago with the description of the type species *N. glaseri* Steiner 1929, a parasite of the Japanese beetle (*Popillia japonica*) in the United States. This species killed its host in the laboratory very effectively, and rearing techniques were quickly developed to test its potential as a biological control agent (Glaser, 1931, 1940). The results of field tests were promising, but continued research was not justified, as better results were being obtained with the milky disease organism *Bacillus popilliae* (Welch, 1962). Consequently, the discovery of *N. feltiae* Filipjev 1934 in the USSR and *N. bibionis* Bovien 1937 in Denmark made little impact.

Interest in biological control revived after the discovery by Dutky and Hough (1955) and Dutky et al. (1962) of a *N. feltiae* population in codling moth (*Cydia pomonella*) (L.) in the United States. In the initial experiments with this population simple spray applications gave 60-70% control of codling moth, and this was soon repeated by different workers all over the world. These results, reviewed by Dutky (1968), Poinar (1971), and Benham and Poinar (1973), generated an interest in this nematode that continues today.

With the emphasis on field evaluation the nematode was simply referred to as strain DD-136. In 1967, Poinar stated that DD-136 was identical to *N. carpocapsae* Weiser 1955, a species originally obtained from codling moth in Czechoslovakia. *N. carpocapsae* has since been synonymized with *N. feltiae* by Stanuszek (1972) and transferred to Steinernema by Wouts et al. (1982).

The life cycle and gross morphology of *N. glaseri* and *N. bibionis* were studied by Glaser (1932) and Bovien (1937), respectively, and have since been found to be very similar to that of *N. feltiae*. Detailed descriptions of the individual stages of the life cycle were presented recently (Wouts, 1980). Up-to-date host lists were presented by Poinar (1979) and Laumond et al. (1979).

The significance of a symbiotic bacterium in the life cycle of the nematode was first suggested by Bovien (1937) and later confirmed by Dutky and Hough (1955). The symbiont was described by Poinar and Thomas (1965) and redescribed as *Xenorhabdus nematophilus* (Thomas and Poinar, 1979). The knowledge that the nematodes feed on the bacterium enabled improvements in culture media for the mass production of *Neoaplectana*. Initially nematodes were produced in axenic cultures with the help of expensive components, such as liver extract of pregnant rabbits (Glaser et al., 1942; Stoll, 1954a, b). Now large numbers are produced on cheap homogenized animal tissues inoculated with the symbiont (House et al., 1965). Increases in surface area by coating the medium onto inert carriers further improved yields (Bedding, 1976).

C. Heterorhabditidae

The genus *Heterorhabditis* is of very recent date. *H. hambletoni* is the oldest species, but its characteristics were not understood until *H. bacteriophora* and *H. heliothidis* were described. Detailed descriptions of the developmental stages and the life cycle of *H. heliothidis* were given by Wouts (1979), and a host range was published by Milstead and Poinar (1978). The associated symbiont was recently described as *Xenorhabdus luminescens* (Thomas and Poinar, 1979). Though detailed studies of the pathology and biological control potential of this group of nematodes have not yet been made, preliminary tests showed that in these respects they are similar to *Neoaplectana*.

IV. CLASSIFICATION AND DIAGNOSIS OF LEPIDOPTERAN PARASITES

A. Mermithidae

Although mermithids have been recognized as an independent genus for more than a century, and have been studied by many workers, their classification remains unsatisfactory. The only comprehensive, comparative study of their morphology and taxonomy is that of Hagmeier (1912). A useful summary of the genera with a key was provided by Filipjev and Schuurmans Stekhoven (1941) and extended by Rubtzov (1972, 1978) and Nickle (1972). A general classification was presented by Andrassy (1976). The following definitions and identifications were extracted from these works.

1. Classification

Phylum: Nematoda Chitwood 1950
Class: Adenophorea von Linstow 1905
Order: Dorylaimida (de Man 1876) Pearce 1942
Suborder: Mermithina Andrassy 1976
Superfamily: Mermithoidea (Braun 1883) Wülker 1924
Family: Mermithidae Braun 1883
Genus: *Hexamermis* Steiner 1924
  Species: *H. truncata* (Rudolphi 1809) Wouts 1981
          H. cavicola Welch 1963
Genus: *Amphimermis* Kaburaki and Imamura 1932
  Species: *A. zuimushi* Kaburaki and Imamura 1932
          A. bogongae Welch 1963

2. Diagnosis

    a. The Family Mermithidae Mermithids are parasites of invertebrates. The larvae gain access to the body cavity of the host by means of a piercing tooth. In the body cavity the mermithid absorbs food through its smooth skin and stores it as fat globules in the pseudointestine (trophosome). Inside the host mermithids can grow up to 50 cm in length. The preadult emerges from the host to mature in the soil to a free-living nonfeeding adult. The adults have six cephalic papillae and two amphids. The reproductive tract is paired

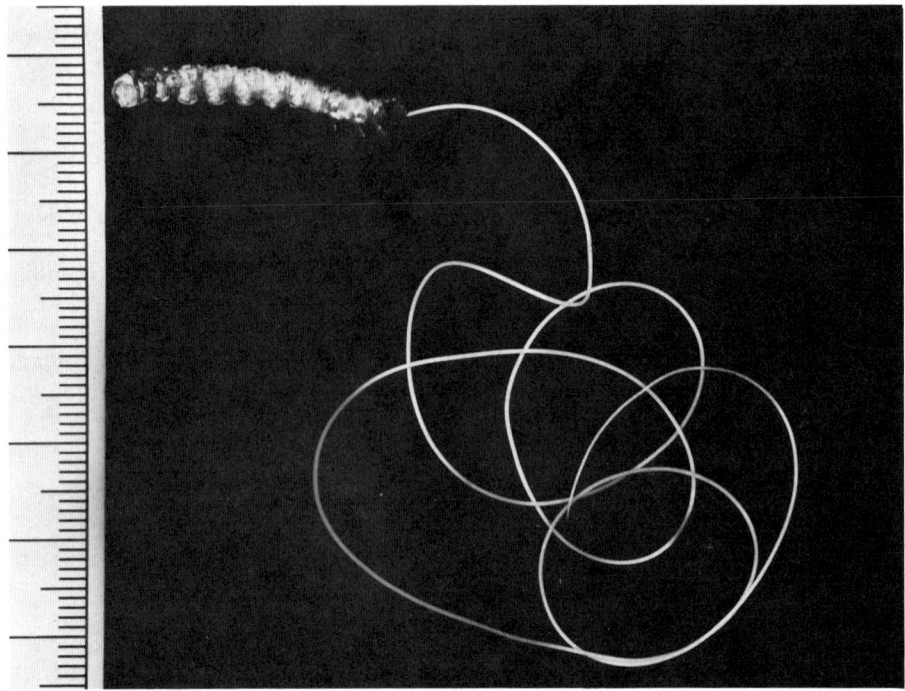

Fig. 1 Postparasitic *H. truncata* juvenile and *Wiseana cervinata* (Walker) caterpillar, a serious pest of pastures in New Zealand (scale of centimeters and millimeters).

in both sexes. The vulva is equatorial in position. The muscular vagina is S-shaped or barrel shaped. In mature females the uteri are filled with eggs. The males are shorter than the females. The cephalic papillae and the constriction of the esophageal tissue immediately below the cephalic papillae are more distinct in the male than in the female. The spicules are paired or single, and the tail carries numerous papillae arranged in three or more rows.

b. *The Genus Hexamermis* Species of the genus *Hexamermis* are characterized by their length (30-200 mm) (Fig. 1), the terminal mouth opening (Fig. 2A, D), small amphids, blunt tail in both sexes (Fig. 2B, E), and their parasitic development in terrestrial insects. The males have spicules that are less than 1.0 mm long and tail papillae in four to six rows (Fig. 2G) in a characteristic pattern. In the female the vagina is S-shaped and vulval lips are present (Fig. 2C). Postparasitic larvae have a terminal mouth opening (Fig. 2H) and a digitate appendage on the tip of the tail (Fig. 2I).

H. *truncata*, the type species, is characterized by spicules that are less than 0.3 mm long (Fig. 2F). *H. cavicola* differs from the type species in that the spicules are more than 0.4 mm long (Welch, 1963).

c. *The Genus Amphimermis* Species of the genus *Amphimermis* differ from those of *Hexamermis* by the large amphids and the slightly ventral mouth opening in both sexes. The males have twisted spicules that are more than 1.0 mm long, and three single rows of tail papillae of which the central row bifurcates around the spicule opening.

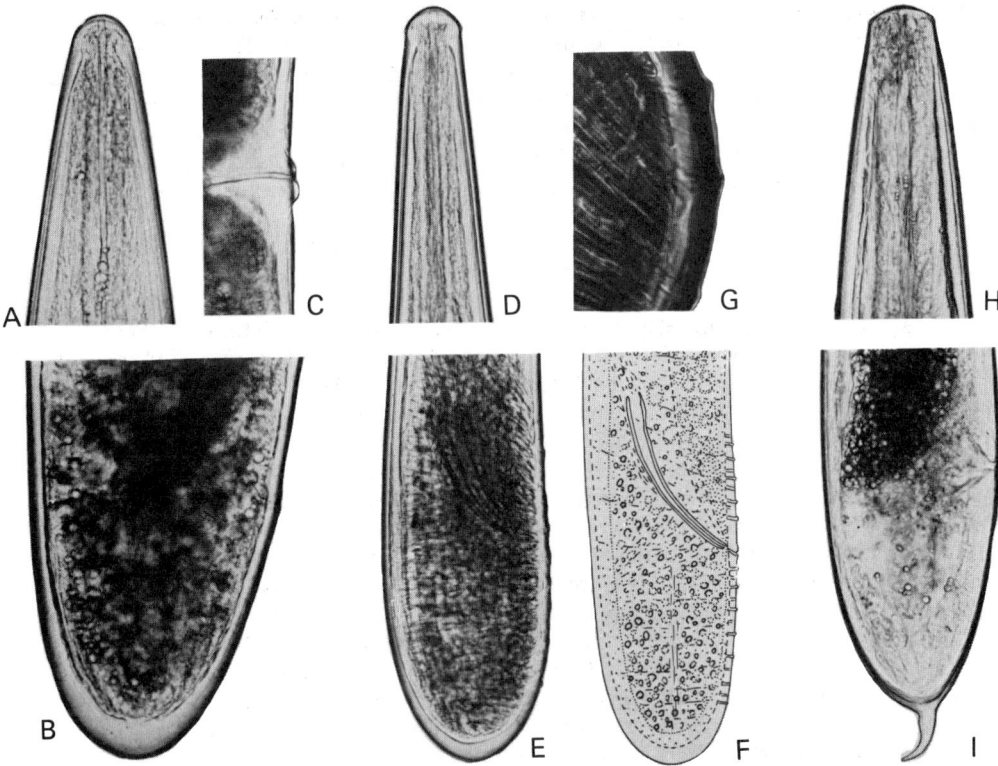

Fig. 2 *H. truncata.* (A-C) Female. (A) Anterior region. (B) Tail, lateral aspect. (C) Vulval region. (D-G) Male. (D) Cephalic region. (E) Tail showing one row of genital papillae. (F) Tail, lateral aspect, to show spicule. (G) Cross section through the three longitudinal rows of genital papillae located on one side of the ventral line of the nematode. (H-I) Postparasitic juvenile. (H) Anterior region. (I) Tail, lateral aspect. (Except for G all same magnification; drawn spicules 0.13 mm long.)

A. zuimushi, the type species, is characterized by kidney-shaped amphids and by spicules that are less than 2.0 mm long. In *A. bogongae* the amphids are cup shaped and the spicules more than 2.0 mm long (Welch, 1963).

B. Steinernematidae and Heterorhabditidae

Travassos (1927) placed *Aplectana kraussei* Steiner 1923 in a new genus for which he introduced the name *Steinernema.* Filipjev (1934) raised *Steinernema* to subfamily level, and Chitwood and Chitwood (1937) gave it family status. Currently the family Steinernematidae contains the type genus *Steinernema* and the genus *Neoaplectana.*

Steinernema kraussei is the only species in the genus *Steinernema.* According to the original description of this species (Steiner, 1923), no valve is present in the basal bulb of the esophagus. This characteristic would clearly separate *S. kraussei* from species of the genus *Neoaplectana,* but apparently Steiner's specimens in general did not show this feature well, for he also reported this valve as indistinct in the original description of the type material of *Neoaplectana* and separated *Neoaplectana* from *Steinernema* by the number of cephalic and genital papillae (Steiner, 1929). Mráček (1977). in a redescription of *S. kraussei*, illustrated a distribution of genital papillae similar to that in *Neoaplectana* so that the lower number of cephalic papillae in *Steinernema* is the only remaining characteristic distinguishing the genera. As these papillae are small and easily overlooked, both genera may be the same. [Since submission of this paper, Wouts et al. (1982) proved that they are synonyms.]

The family Heterorhabditidae was introduced by Poinar in 1976 for nematodes with a bacterial association similar to that of *Neoaplectana*, but with an heterogonic life cycle and a different morphology.

Bacteria associated with *Neoaplectana* and *Heterorhabditis* were recently described and placed in the new genus *Xenorhabdus* Thomas and Poinar 1979 (Enterobacteriaceae). They are defined as large (0.8-2.0 by 4.0-10.0 µm) gram-negative, rod-shaped, facultative anaerobic, entomopathogenic bacteria. Each species occurs in two forms (Akhurst, 1980): a "healthy" or primary form and a "weakened" or secondary form. Under normal conditions only the primary form exists. In the laboratory, subcultured at monthly intervals, this form is stable, but in old cultures the secondary form regularly develops. The secondary form does not seem to revert back to the primary form, is not suitable for nematode mass production, does not kill the host effectively, and is not suitable for biological control purposes. In nature it probably does not survive. It can be prevented from developing in vitro by subculturing the bacteria at least once a month. In the mass-rearing methods discussed later, the occurrence of the weak form is reduced by inoculating the mass-rearing medium with the symbiont two to three days before the nematodes are introduced, in combination with monthly subculturing.

Xenorhabdus nematophilus (Thomas and Poinar 1965), the bacterium associated with *Neoaplectana* species, is the type species of the genus *Xenorhabdus* and is characterized by colonies on nutrient agar that are smooth, moist, and somewhat granular in appearance. On Tergitol media the colonies become blue in color with maroon centers, or blue green with dark centers and clear zones around them (Thomas and Poinar, 1976).

Xenorhabdus luminescens Thomas and Poinar 1979, the bacterium associated with *Heterorhabditis* species, is the only other species in the genus, and is characterized by colonies on nutrient agar that are smooth and mucoid in appearance and with irregular margins. Initially they are pale yellow but change to deep yellow and usually red with age. They are bioluminescent, and the light they produce can be detected by the human eye after 10 minutes of adjusting in the dark (Fig. 3). It is sufficiently strong to show through the cuticle of an infected host and to expose photographic film (Fig. 4).

Descriptions of the Steinernematidae and Heterorhabditidae have been presented by Poinar (1975, 1976), Khan et al. (1976), and Wouts (1979, 1980). The following definitions are extracted from these works.

1. Classification

Phylum: Nematoda Chitwood 1950
Class: Secernentea von Linstow 1905
Order: Rhabditida (Oerly 1880) Chitwood 1933
Suborder: Rhabditina (Oerly 1880) Chitwood 1933
Superfamily: Rhabditoidea (Oerly 1880) Travassos 1920
Family: Steinernematidae (Filipjev 1934) Chitwood and Chitwood 1937

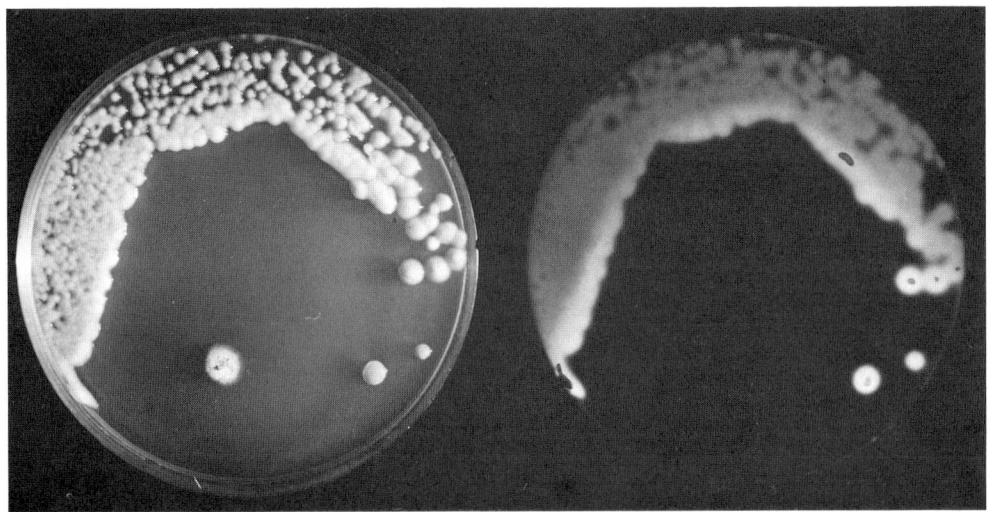

Fig. 3 *Xenorhabdus luminescens* culture photographed in daylight (left) and in the dark (right). A nonluminescent colony contaminates the lower center of the plate.

Fig. 4 *Galleria mellonella* larvae infected with *X. luminescens* photographed in daylight (left) and in the dark (right).

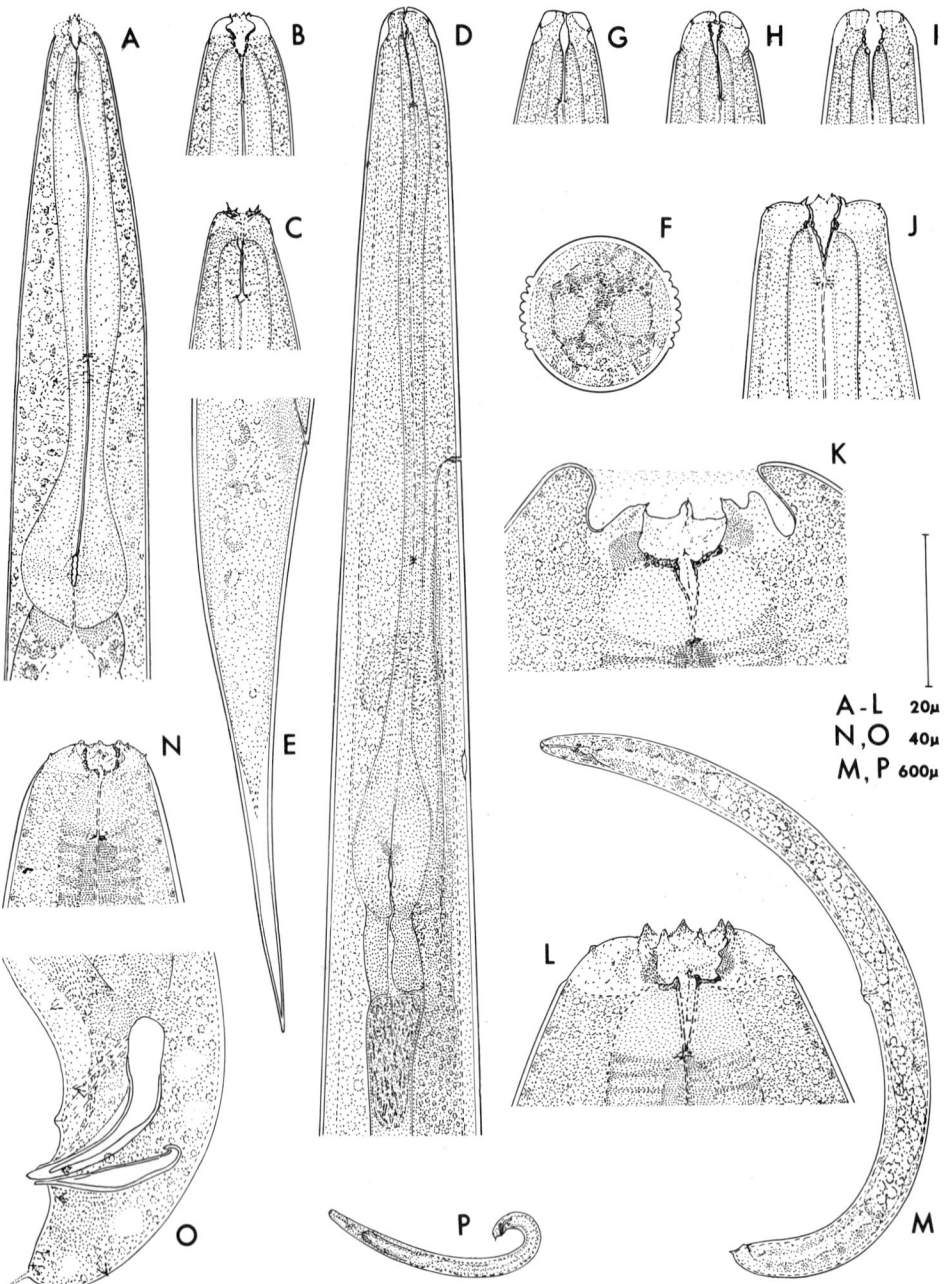

Fig. 5 *N. bibionis*. (A-B) First larval stage. (A) Esophageal region of a young specimen. (B) Cephalic region of an old specimen. (C) Cephalic region of a second-stage larvae. (D-F) Infective larva. (D) Esophageal region. (E) Tail, lateral aspect. (F) Cross section at midbody. (G-I) Cephalic region of the parasitic third-stage larva showing progressive development after feeding has started. (J) Cephalic region fourth-stage larva. (K-L) Female. (K) Cephalic region retracted. (L) Cephalic region not retracted. (M) Female whole. (N-P) Male. (N) Cephalic region. (O) Tail, lateral aspect. (P) Male whole.

Genus: *Neoaplectana* Steiner 1929 [for latest status see Wouts et al (1982)]
   Species: *N. glaseri* Steiner 1929
          N. feltiae Filipjev 1934
          N. bibionis Bovien 1937
Family: Heterorhabditidae Poinar 1976
  Genus: *Heterorhabditis* Poinar 1976
    Species: *H. bacteriophora* Poinar 1976
          H. heliothidis (Khan, Brooks, and Hirschmann 1976)
          Poinar, Thomas, and Hess 1977

2. Diagnosis

a. The Family Steinernematidae Steinernematids are parasites of invertebrates. The first-stage larvae are parasitic. Initially their lip region is very delicate (Fig. 5A), but it soon develops its distinct outline with six labial papillae (Fig. 5B). The second larval stage has a closed stoma (Fig. 5C); it develops into the free-living infective third larval stage. Infective larvae are characterized by a closed stoma (Fig. 5D), a sharp tail (Fig. 5E), and nine lines in the lateral field (Fig. 5F); in the parasitic larvae the stoma gradually opens (Fig. 5G-I). The fourth larval stage (Fig. 5J) resembles young females and males. The cuticle of the female is smooth, without lateral lines. The labial region has six lips, each with a papilla. In large females the labial region may be retracted (Fig. 5K). Four cephalic papilla are located near the base of the lips (Fig. 5L) (Mrácek and Weiser, 1979). The amphids are obscure. The stoma is as wide as it is long and tapers into the narrow esophagus. Relative to the length of the body the esophagus is short, ending in a valved basal bulb. A distinct esophageal-intestinal valve is present. Large glands that open out into the excretory duct displace the anterior part of the intestine dorsally. The ovaries are paired, opposite, and reflexed. The rectum and the anus are distinct (Fig. 5M). Except for size (Fig. 5L, and N) and sexual characteristics the males are identical to the females (Fig. 5M, P). The testis is single and reflexed, the spicules are paired, a gubernaculum is present, and several pairs of genital papillae (including a pronounced preanal ventral papilla) are located in the caudal region. There is no bursa (Fig. 5O).

Neoaplectana is the only genus in this family reported from lepidopterans. The type species *N. glaseri* is characterized by infective larvae that are longer than 1 mm. *N. bibionis* and *N. feltiae* can be distinguished from the type species by their shorter infective larvae (0.7-1.0 mm for *N. bibionis* and 0.4-0.7 mm for *N. feltiae*).

b. The Family Heterorhabditidae Heterorhabditids are similar to neoaplectanids in general life cycle and gross morphology, and their first larval stages are almost indistinguishable (Fig. 6A). The second larval stage is the infective stage. The infective larvae are characterized by a closed stoma (Fig. 6B), a sharply pointed tail (Fig. 6C), and longitudinal lines on the cuticle (Fig. 6D). The third-stage larvae possess a characteristic, sclerotized labial tooth above the stoma (Fig. 6E, F). The tooth is supported by rays extending between the lips. An independent ventral sclerotized labial plate is present (Fig. 6G). The lip region of fourth-stage larvae resembles that of young females (Fig. 6I) and young males (Fig. 6L). The lip region of the female has six papillae, which in large specimens are formed into flaps (Fig. 6H) that may have a function in food gathering. Males are produced in the second generation only. They do not seem to feed. Fully grown males are

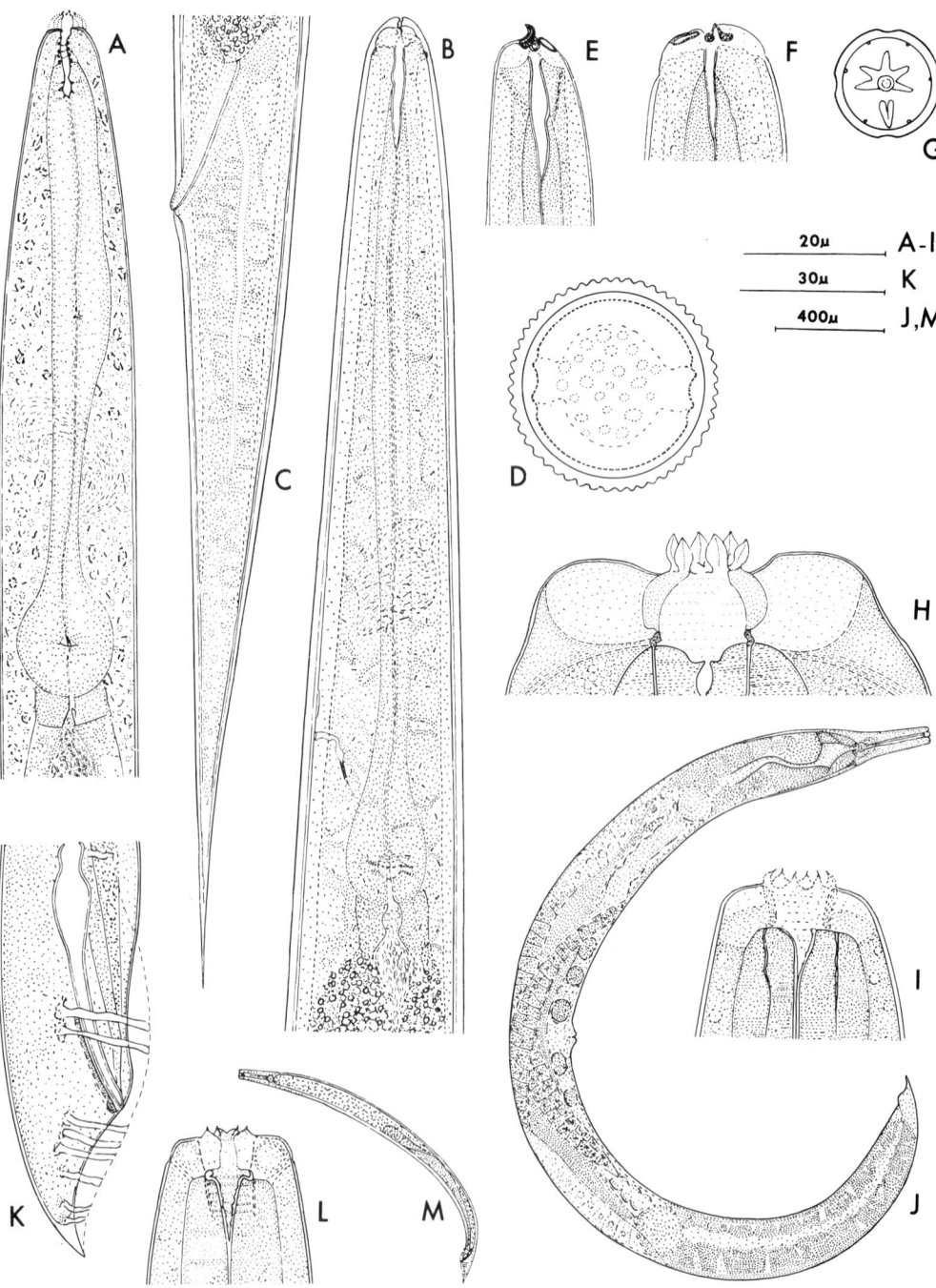

Fig. 6 *Heterorhabditis heliothidis*. (A) Esophageal region of a first-stage larva. (B-D) Infective larvae. (B) Esophageal region. (C) Tail, lateral aspect. (D) Cross section at midbody. (E-G) Third-stage larvae. (E) Cephalic region of a young specimen. (F) Cephalic region of an old specimen. (G) Face view. (H-I) Anterior region female. (H) Large specimen. (I) Small specimen. (J) Female whole. (K-M) Male. (K) Tail, lateral aspect. (L) Cephalic region. (M) Male whole.

considerably smaller than fully grown females (Fig. 6M, J). The spicules
are slender and genital papillae are absent but a bursa supported by nine
ribs is present (Fig. 6K).

   Heterorhabditis bacteriophora, the type species, is characterized by
infective larvae that are less than 0.6 mm long and have a tail length of about
0.09 mm. *H. heliothidis* can be distinguished from the type species by infective larvae that are more than 0.6 mm long and have a tail length of about
0.1 mm.

V. BIOLOGY AND LIFE CYCLES AND MODE OF INFECTION

A. Mermithidae

The life cycle of *H. truncata* as described by Rosenhauer (1847) and von
Siebold (1843, 1848, 1850, 1855) and extended by Couturier (1950) is schematically presented in Fig. 7. The infective larvae develop in the soil and can
survive there for several months. They move actively in search of a suitable,
soil-dwelling host. During rainy periods or humid early hours in the spring
when plants are covered with a layer of moisture they also move up plants to
infect hosts feeding on the foliage. The infective larvae penetrate the host
with the help of a mouth stylet and settle in the body cavity; the initial
growth rate is fast. After four days the larval length has doubled, and
after 11 days it has increased 12-fold. The full length of up to 30 cm is
reached after about one month. If several specimens infect the same host
(Fig. 8), they do not grow as long and more males are produced. During
development of the parasitic stage the characteristic digitate appendage of
the postparasitic larva is formed. It is generally assumed that this appendage
is the result of extensive enlargement of the body in which the tail tip does
not take part (Hagmeier, 1912).

   The morphology and function of the esophagus of mermithids differ
from that of all other nematodes and reflect the unusual habitat of this
group of nematodes. Muscle attachments for active food intake are lacking,
and there is no open connection with the intestine. A layer of small cells,
the episome, surrounds the length of the esophagus. They probably have
a function in the formation of a new cuticular esophageal canal during molting. Some cells of the episome, mainly located in the anterior portion, are of
glandular character. Other cells increase considerably in size and form the
stichocytes that are homologous to the large glandular cells along the esophagus in free-living nematodes. The stichocytes surround the esophagus in
one or two rows and regulate osmosis and nutrition. The most anterior pair
of stichocytes, the homorocytes, are large in size and regulate osmosis and
excretion (Rubtzov, 1972). Where and how food is digested has not been
established with certainty. According to Rathke (1953), it takes place in the
esophagus after the nutrients have entered by capillary action. Digested
food is absorbed through the esophageal wall and transported through the
body cavity of the parasite to the intestine. The intestine acts as a storage
body, called a *trophosome*. According to Müller (1931) and Rubtzov (1972),
no food enters the esophagus and digestion takes place in the body cavity of the
host. Müller (1931) suggests that digestive enzymes excreted into the esophagus are released in the body cavity of the host, where digestion then
takes place. Rubtzov (1972) agrees that the glandular cells of the episome
excrete matter into the esophageal canal and that this matter is discharged
through the oral aperture into the body cavity of the host but does not
assign any enzymatic significance to it. According to him digestion in the

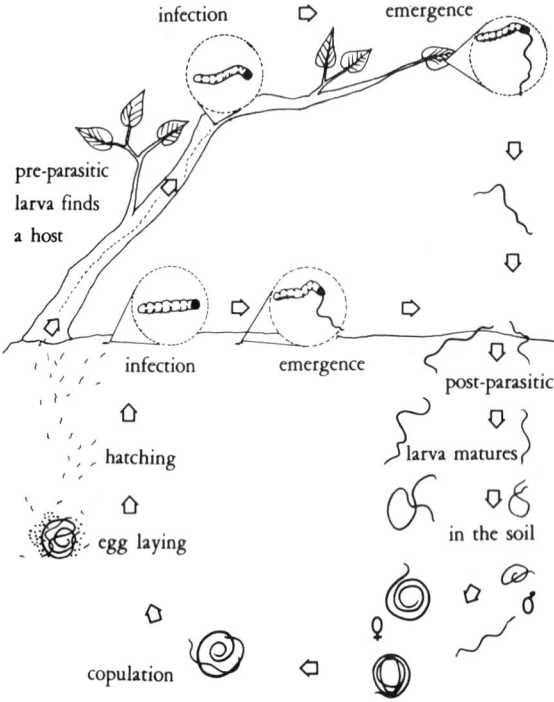

Fig. 7 Schematic presentation of the life cycle of *H. truncata*.

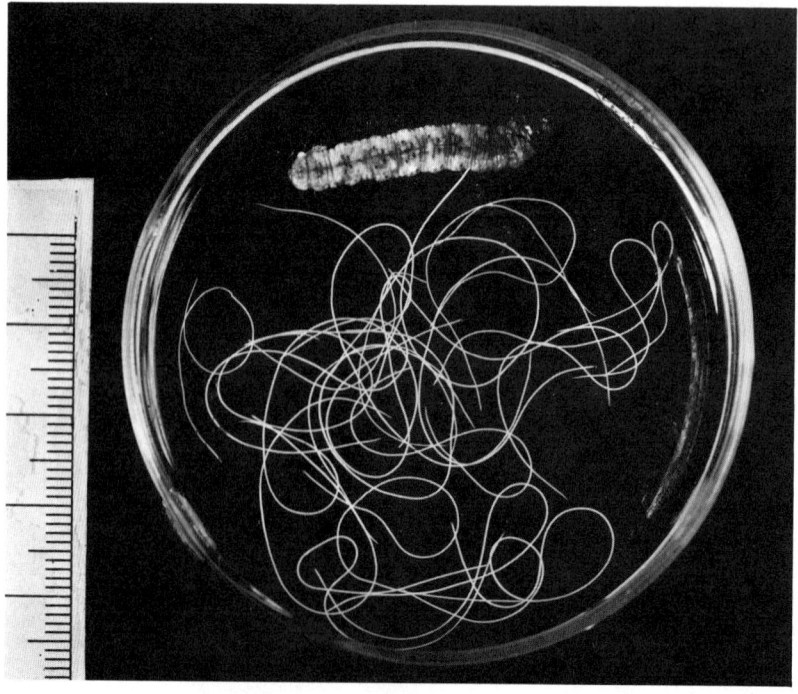

Fig. 8 *H. truncata* juveniles extracted from two *Wiseana cervinata* caterpillars.

body cavity of the host takes place by enzymes from the glandular cells in the longitudinal hypodermal cords. Digested food is absorbed through pores in the mermithid's cuticle (Poinar and Hess, 1977), a process regulated by the trichocytes, and is transported through the body cavity of the nematode to the trophosome, where it is stored until it is needed for development of the free-living phase and for reproduction (Rubtzov, 1972). Excretion of waste products takes place through the excretory pore and apparently is under direct control of the homorocytes. Some waste products are not ex-creted but are stored in crystal form, either in the trophosome or in the hypodermal cords (Rathke, 1953).

It has been suggested that *H. truncata* leaves the host as the result of the depletion of essential substances and the accumulation of metabolic wastes in the host (Couturier, 1950). It is more likely, however, that it is determined by the availability of food in general, because field-collected infected hosts without food often release the nematode within 48 hours. Emergence from the host is usually completed within a few minutes. The host dies within hours from loss of body contents and invading microorgan-isms. The rapid deterioration of some dead hosts may prevent the remaining mermithids from emerging from the host, though if freed artificially these individuals are capable of completing development. Six specimens reared in one *Galleria* larva all emerged (unpublished data). Parasitic larvae that emerge from a host in a moist environment move into the soil to a depth of 10-30 cm, coil up (Fig. 9), and mature. Maturation may take from three weeks to two months and may involve two molts (Rathke, 1953). Mature males move about in search of the females. The copulating partners form a tight cluster, which is maintained for the rest of their lives.

Egg-laying starts 20-40 days after copulation at a rate of 10-20 eggs per day, accumulating around the cluster to 1000-2000 eggs in eight months (Fig. 10). For continuous egg-laying, multiple copulation is required. The males die in the spring; the majority of females die in late summer. Unfer-tilized females lay no eggs and can live up to four years (Couturier, 1950). The eggs measure up to 180 × 200 μm and at 16-18°C they hatch after about

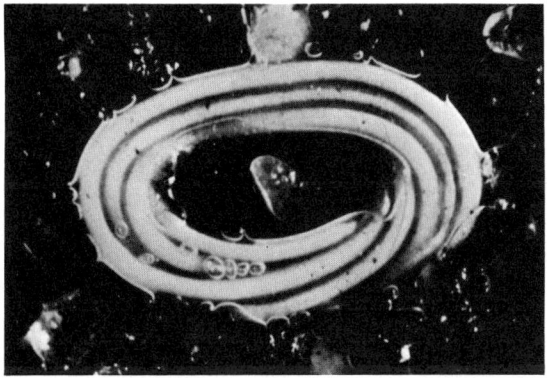

Fig. 9 *H. truncata*. The postparasitic juvenile maturing in the soil. (Repro-duced from Couturier, 1950, by courtesy of Director Institute Nationale de la Recherche Agronomique.)

Fig. 10 *H. truncata* adults surrounded by eggs. (Reproduced from Couturier, 1950, by courtesy of Director Institute Nationale de la Recherche Agronomique.)

three months. The newly hatched larvae are 2.8-3.5 mm long and 18-23 μm wide. They have completed one molt in the egg and are infective.

The life cycles of *A. bogongae* and *H. cavicola* are similar to that of *H. truncata*, but they differ in that they take place in an unusual host and habitat. The host is the adult bogong moth, estivating in large congregations on the walls of mountain caves in Australia. The infective nematode larvae develop in pools on the floor of the cave and reach the host by moving up through condensation on the cave wall, a phenomenon first reported by von Siebold (1854).

The life cycle of *A. zuimushi* is also similar to that of *H. truncata* (Kaburaki and Imamura, 1932). The free-living larva is positively phototropic, apogeotropic, rheotropic, and xenotropic and thermotropic to an adequate warm temperature. It appears on the field surface, swims through the irrigation water, and follows the host through its entry hole in the rice plant. Penetration of the host and the subsequent development of the nematode is as for *H. truncata*.

B. Steinernematidae

The biology and life cycle of *Neoaplectana* was studied by Bovien (1937) and reexamined in detail recently (Wouts, 1979). There are six distinct stages in the life cycle: egg, four morphologically distinct larval stages, and adults.

The life cycle of *N. bibionis* in caterpillars of the greater waxmoth, *Galleria mellonella*, is schematically presented in Fig. 11. The free-living infective third-stage larvae are resistant to desiccation and can survive in damp soil for several months. Under artificial conditions, in 0.1% formalin, at 5°C they survive for several years. Attracted to chemostimulants on the exterior of the host (Schmidt and All, 1978), they enter the host through the mouth or anus. Penetration through the spiracles has not been observed.

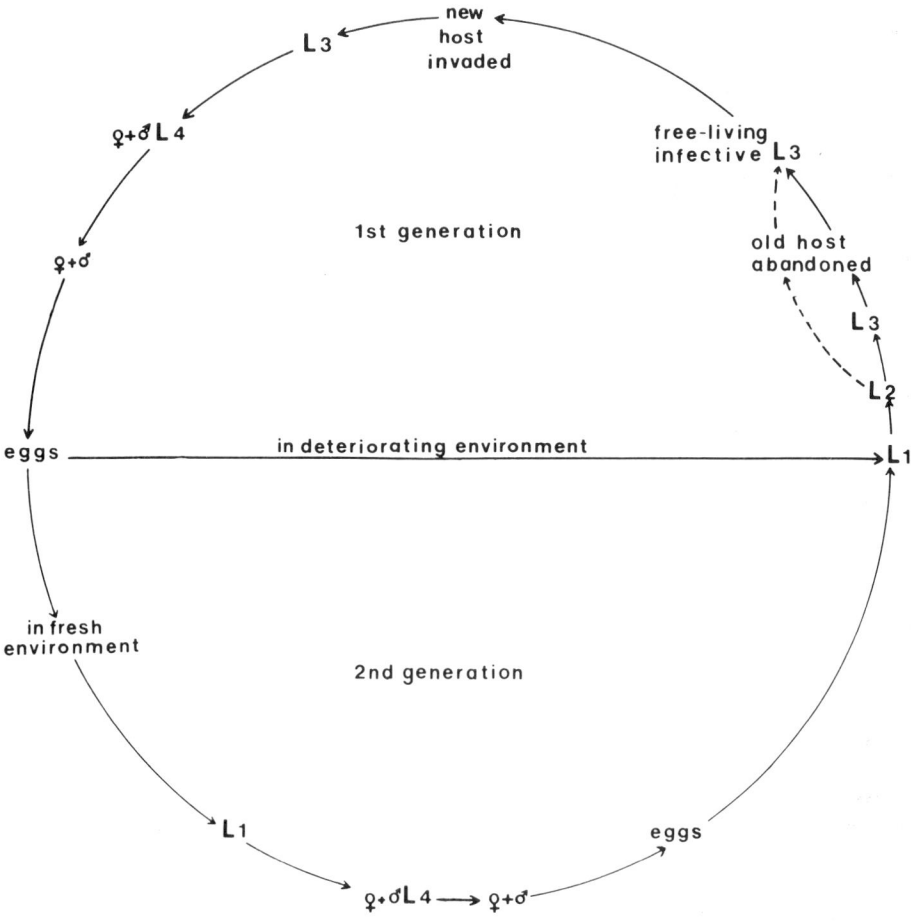

Fig. 11 Schematic presentation of the life cycle of *N. bibionis* in a natural host (solid lines, usual two cycles; broken lines, alternative pathway).

Once inside the host they penetrate the wall of the alimentary canal and enter the body cavity. Initially they seem to congregate in the head capsule. Feeding starts immediately. The host hemolymph taken up by the nematodes accumulates in the bacterial pouch in the anterior part of their intestine. The bacteria in the pouch multiply and are released, first into the lumen of the nematode intestine and then via the anus into the body cavity of the host. In the host the bacteria multiply and cause a lethal septicemia. The nematodes live and develop on the bacteria and the decomposed tissues of the host. To facilitate food intake the stoma gradually opens wide and the esophagus expands. At the same time the excretory glands increase in size and their three nuclei become clearly discernible. The third-stage larva grows very little in length but doubles in width and, about 48 hours after penetrating the host, develops into a fourth-stage larva, which again doubles in width and increases somewhat in length. The genital primordia develop substantially at this stage. Males and females are present 72 hours after penetration of the host by the infective larvae. Both sexes live for about five days. Males appear before females and are essential for reproduction. Copulation takes place soon after the females are formed. Initially both sexes are of equal size. Males change very little, but females in a freshly invaded host grow considerably, reaching up to five times the original length. The first females lay eggs. First-stage larvae hatching from these eggs develop rapidly, through one molt, into fourth-stage larvae and into adults that initially may lay eggs. As the condition of the host deteriorates, egg laying ceases and unlaid eggs develop and hatch inside the female (*endotokia matricida*). First-stage larvae now develop into second-stage larvae that do not feed. Their stoma closes and their body grows thinner. Concurrently the esophagus reduces in width and the lumen of the intestine closes starting at the anus, gradually moving forward until only a small pouch remains immediately below the basal bulb of the esophagus. This pouch retains bacteria to infect the next host. While the lumen of the intestine is closing, the cuticle of the third-stage larva is formed and separates from the cuticle of the second-stage larva. The nematode then increases in length, ruptures the old cuticle, and emerges as the slender, free-living infective stage. The late second-stage larva or the infective third-stage larva leaves the old host in search of a new one. One successful larva is sufficient to kill a host, but in the field a concentration of about 10 infective larvae per 100 g of soil is required for successful penetration to occur (Bedding and Akhurst, 1975). Infective larvae are partially resistant to gradual desiccation and are able to survive dry summer conditions. They have been observed to position themselves on dry soil particles and wave the anterior portion of their body through the air, reaching out for a new host.

The nematode life cycle is completed in about 10 days. Usually two generations are produced in one host, but this varies with the size and condition of the host as well as the number of invading infective larvae. The eventual size of the females is determined by the amount of food available. They attain their greatest length when they develop individually in a host in the absence of a male.

C. Heterorhabditidae

A brief account of the biology and life cycle of *Heterorhabditis* was first published by Pereira (1937). Further information was added by Poinar (1967), Khan et al. (1976), and Wouts (1979).

Figure 12 schematically presents the heterogonic life cycle of *H. heliothidis* as it takes place in a natural host. As in the genus *Neoaplectana* there are six distinct stages. The free-living, infective, second-stage larvae live in the soil where they can survive for several months. In the laboratory, stored in 2.5% Ringers solution at 7°C, they survive for more than a year (Milstead and Poinar, 1978). Until they locate a host the bacterial contents of the lumen of the intestine are gradually digested, but a small quantity of bacteria remains immediately below the basal bulb of the esophagus, sufficient to inoculate the next host. In storage in the laboratory the fat reserves in the wall of the intestine are gradually used up and development into third-stage larvae takes place. These third-stage larvae remain infective, but do not live long. In nature it is probably always the developing third-stage larva, still in the cuticle of the second stage, that infects the new host. Soon after having penetrated the host the infective larva develops into a short-lived third-stage parasitic larva. This stage is very vulnerable to changes in osmotic pressure. A possible function of the characteristic, sclerotized tooth above the stoma of these larvae may be to tear host tissue in the initial phase of infection. Third-stage larvae feed on host hemolymph, release their sym-

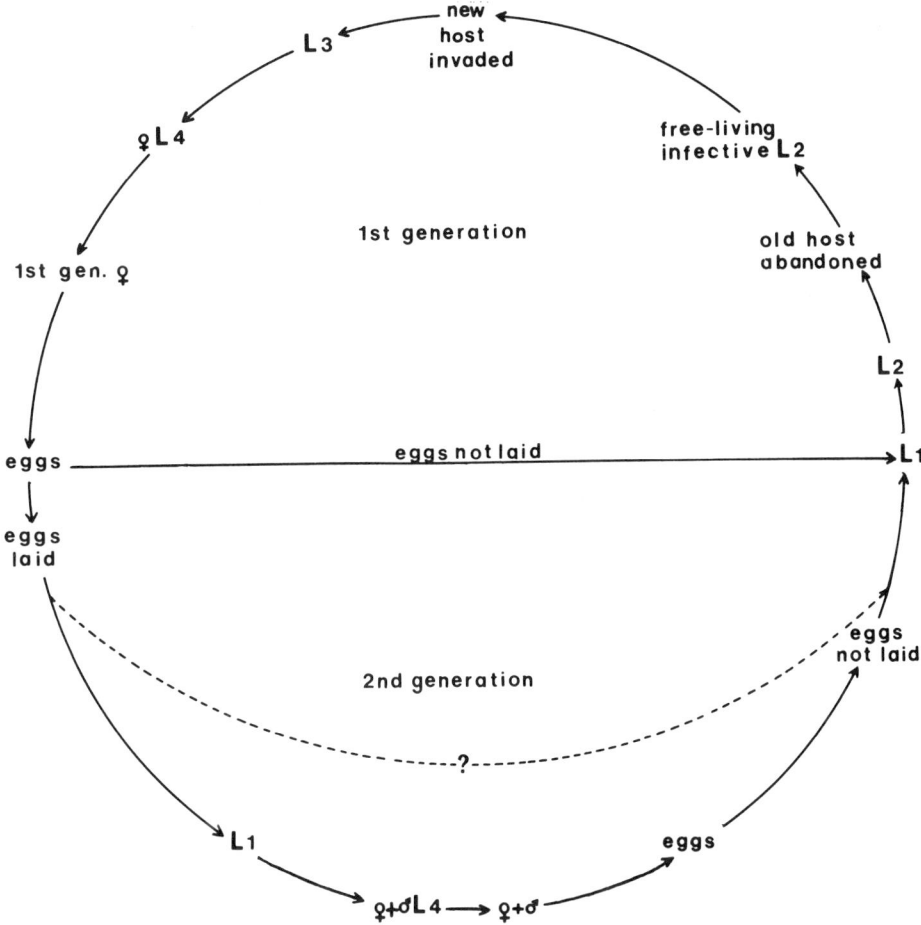

Fig. 12 Schematic presentation of the life cycle of *Heterorhabditis heliothidis* in a natural host (solid lines, usual two cycles; broken lines, alternative pathway).

biotic bacteria, double in width while changing little in length, and molt within two days. Fourth-stage larvae further increase in width and almost double their length. Their genital primordia develop substantially to become functional genital systems at the next molt. At 20°C first-generation, hermaphroditic females are present in the host about four days after penetration. The females may grow up to 6 mm in length and produce more than 1000 eggs. As in *Neoaplectana* they lay eggs initially, but larvae develop by *endotokia matricida* as the population increases. First-stage larvae hatch from the eggs. In a relatively fresh environment they grow rapidly through one molt into fourth-stage larvae and then to males and females of the second generation. By then competition for food has intensified and conditions in the host have deteriorated to the extent that newly developing females remain small and lay no eggs. Unlaid eggs hatch inside the females where, because of the restricted food supply, the majority develop into second-stage infective larvae. These larvae usually remain inside the female for two or three days and give the

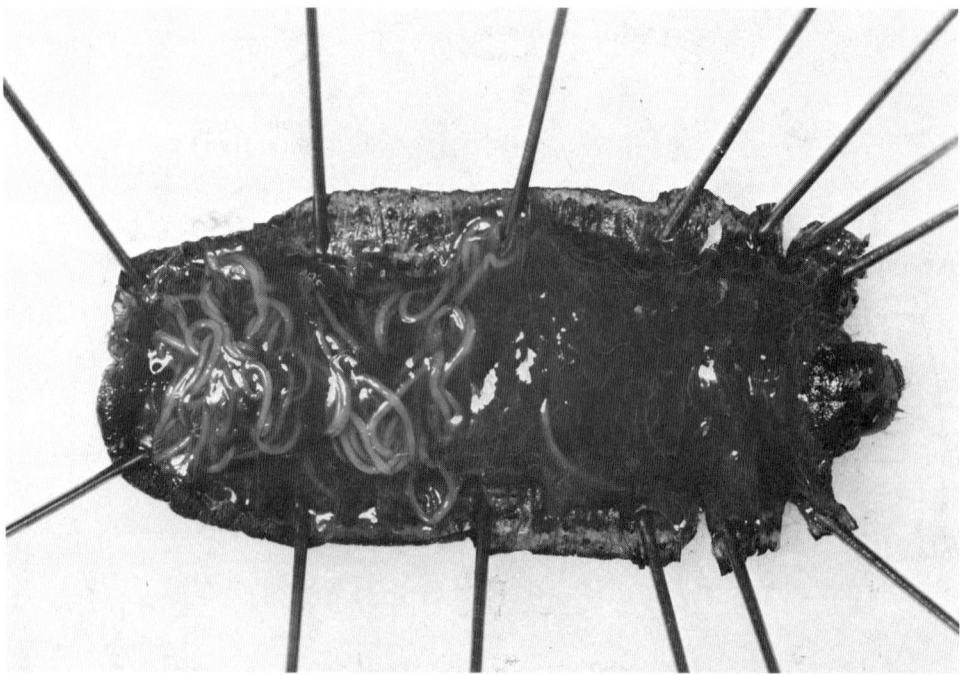

Fig. 13 Large, white *Heterorhabditis* females in a dissected host.

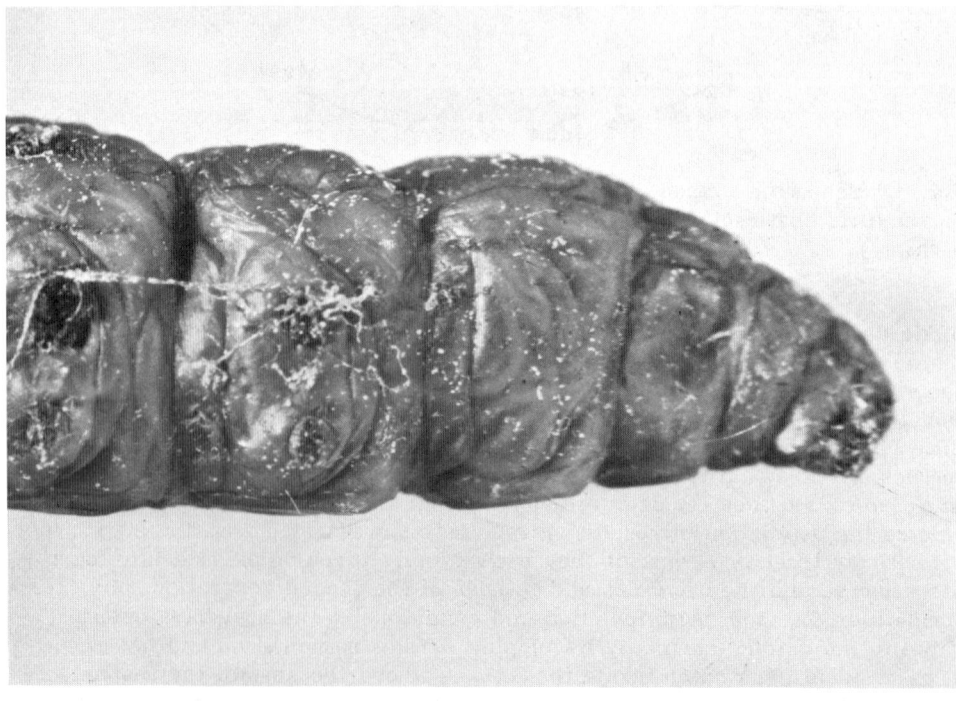

Fig. 14 Large, white *Heterorhabditis* females showing through the cuticle of an infected host.

parent a white appearance (Fig. 13) by which they can be recognized through the transparent cuticle of the host (Fig. 14). Males are produced only in the second generation. They are present for only a short period and can easily be overlooked. They do not feed and are not essential for reproduction. Copulation has not been observed.

VI. PATHOLOGY AND ECONOMIC SIGNIFICANCE

A. Mermithidae

The parasitic larvae of mermithids develop on reserves that normally provide the fat body of the host. Retarded development and delayed pupation of the host are the only external symptoms of infection, although each host is eventually killed when the fully grown mermithid larva departs. In field populations incidence of mermithid infection can be very high. Chatterjee and Singh (1965) reported a 73% infection of *Antigastra catalaunalis* Dup. population on *Sesamum indicum* and a 92.5% infection of *Achara janata* L. on cotton. Kaburaki and Imamura (1932) found 76% of a population of the rice borer *Chilo simplex* Butler affected. Although the percentage of parasitism reduces with the distance of the host from the ground and is nil in insects feeding above 5 m (Chatterjee and Singh, 1965), the high incidence of parasitism in some lepidopteran species clearly demonstrates the importance of mermithids as regulators of pest populations. Detailed studies on how this is reflected in reduced damage have not yet been made, and there are no examples of lepidopteran pests either naturally or artificially controlled by mermithids.

B. Steinernematidae and Heterorhabditidae

The parasitic larvae of *Neoaplectana* and *Heterorhabditis* species are extremely damaging to their host because of the lethal bacteria they transmit. Although not essential to the nematode in laboratory cultures, these bacteria provide advantages without which the nematode could not survive in nature. Despite the defense reactions of the host manifested by rapid proliferation of phagocytes and free hemocytes, reaching a maximum 8-12 hours after infection (Seryczynska et al., 1974), the bacteria have sufficiently multiplied within 24 hours to damage all major internal organs of the host. The host phagocytes are overwhelmed, and movement and feeding are considerably impeded. Normally the host dies about 48 hours after infection. The bacteria developing in the host produce antibiotics that prevent invasion of the host by other microorganisms and delay breakdown of the host cuticle.

The eventual breakdown of the host cuticle proceeds differently for *Neoaplectana* and *Heterorhabditis*. *Neoaplectana* and its associated bacteria gradually break down the cuticle of the host until it ruptures, about two days before the development of the infective larvae, and the contents flow out as a liquid mass of bacteria and nematodes (Fig. 15). *Heterorhabditis* and its symbiotic bacteria preserve the cuticle of the host until after the infective larvae have emerged. Only then do the antibiotics dissipate and other organisms invade and decompose the host remains.

Of the *Neoaplectana* species studied, *N. feltiae* has the widest host range, but because of the lethal effect on their host all species are suitable for consideration as a biological insecticide against lepidopterans. Effective control programs are being developed (Bedding and Miller, 1981), and results should be promising where physical conditions suitable for nematode infection occur.

The temperature range for *Neoaplectana* activity (10-32°C) is similar to that of lepidopterans (Dutky, 1959; Danilov, 1976) and is usually not a

Fig. 15 *N. bibionis* emerging through the cuticle of an infected *Galleria* larva.

limiting factor for nematode infection. Humidity and radiation (Gaugler and Bousch, 1978) are both limiting factors, but the effect of radiation generally diminishes when high humidity is maintained. Humidity can be manipulated by applying the nematodes in large quantities of water and by adding gels (Nash and Fox, 1969) or by using oils (Bedding, 1976), but no successful field results with these methods have yet been reported.

Promising field results were obtained with aqueous suspensions of infective nematode larvae applied against lepidopterans that live in environments that naturally experience periods of high relative humidity. Damp soil and tunnels of wood-boring insects provide such environments. For insects feeding on foliage, extended periods of light rain and heavy dew, especially in combination with dense foliage such as the canopy of trees and closely planted agricultural crops, have been most inducive. Applications of *Neoaplectana* infective larvae under these conditions led to reduced damage of the fall army worm *Spodoptera frugiperda* (J. E. Smith) on maize (Landazabal et al., 1973), the tobacco hornworm *Protoparce sexta* Johan. and the tobacco budworm *Heliothis virescens* (Fab.) on tobacco (Chamberlin and Dutky, 1958), and caterpillars of the swallowtail *Papilio demodocus* Esper on trees (Srivastava, 1978). Applications on humid soil allowed some control of soil-dwelling caterpillars of the turnip moth *Agrotis segetum* (Schiff) (Simons, 1978), the leaf roller *Pseudexentera mali* (Freeman), and the winter moth *Operophtera brumata* (L.) (Jaques et al., 1968). Complete control has been obtained with applications against the wood-boring caterpillars of the sesiids *Sciapteron tabaniformis* (Rottemburg 1775) and *Synanthedon tupiliformis* (Clerck), the former by injection of a nematode suspension into their tunnels in poplars (Simons, 1978), the latter by surface spraying of tightly stacked infected black currant cuttings (Bedding and Miller, 1981).

All wood-boring caterpillars can probably be controlled by injecting infective nematode larvae into their tunnels. In practice this method of application will generally be too time consuming to be practical. Therefore, Bedding and Miller's (1981) finding that infective *N. bibionis* larvae are actively attracted to the tunnels of *S. tipuliformis* larvae opens promising perspectives.

VII. MASS REARING OF *NEOAPLECTANA* AND *HETERORHABDITIS* SPECIES

A. Rearing on a Natural Host

Mass production on a suitable host, such as larvae of the greater wax moth, *Galleria mellonella*, or any other large caterpillar, yields sufficient numbers of infective larvae to permit small-scale field trials. A culture can be started with a population obtained from established cultures at other laboratories, or with field-collected infective larvae. In general the low natural incidence of parasitized insects makes collecting of hosts in the field an inefficient method for obtaining *Neoaplectana* or *Heterorhabditis* species. More efficient is extracting infective larvae from soil, either directly by traditional nematode extraction methods or by using a trap host, but requires a thorough knowledge of the morphology of the nematodes as large numbers of different species are obtained.

The number of *Neoaplectana* and *Heterorhabditis* specimens obtained by direct extraction of soil is usually small. To ensure their survival they should be injected into caterpillars at a rate of about 10 per caterpillar. The caterpillars are maintained on damp filter paper in a Petri dish at about 20°C for 10 days and then transferred to a nematode extracting device, such as the White trap described below.

Once large numbers of infective larvae are available, mass production becomes easier as natural infection of caterpillars can take place. Twenty caterpillars of a noncannibalistic species are placed on a filter paper disk in a Petri dish. About 20,000 infective nematode larvae, surface sterilized in 0.4% Hyamine 10X (Poinar, 1975) or 0.1% Merthiolate (Bedding, 1981) for half an hour and thoroughly washed, are pipetted onto the filter paper. The quantity of water used should be sufficient to make the paper evenly wet without leaving free water. The dish is closed without being air tight and kept in the dark at room temperature. The infective larvae penetrate the caterpillars and kill them within 48 hours. Ten days after the first exposure to nematodes the caterpillars are transferred to an extraction device in which the emerging infective larvae are trapped. Suitable devices include: the modified Baerman funnel of Carne and Reed (1964) (Fig. 16) and White's (1927) trap (Fig. 17). The White trap consists of a small inverted Petri dish in a large dish partly filled with water. The small dish is covered with a filter paper disk onto which the infected caterpillars are placed. When large numbers of infected caterpillars have to be processed, photographic developing trays can be used successfully. In these trays the caterpillars are placed on strips of filter paper between the water-filled grooves in the bottom of the tray (Fig. 18). In each of these systems the caterpillars are either directly or indirectly, through the filter paper that supports them, in contact with water. This wet environment stimulates the movement of the nematode larvae, which emerge and are sooner or later trapped in the surrounding water. The nematode larvae are collected daily and after being washed are placed in flasks of aerated water. To enable all nematodes extracted to develop into infective larvae the nematodes should remain in aerated water for at least two weeks before being used for experiments or for further rearing. Infective larvae can be stored for extended periods in 0.1% formalin at 5-10°C.

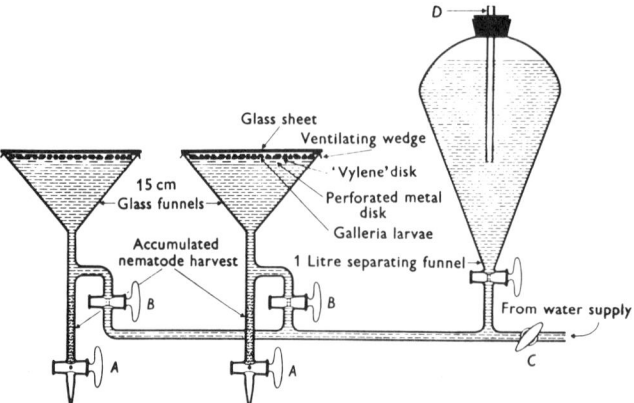

Fig. 16 Nematode extraction apparatus as developed by Carne and Reed (1964). (Courtesy of Cambridge University Press, New York.)

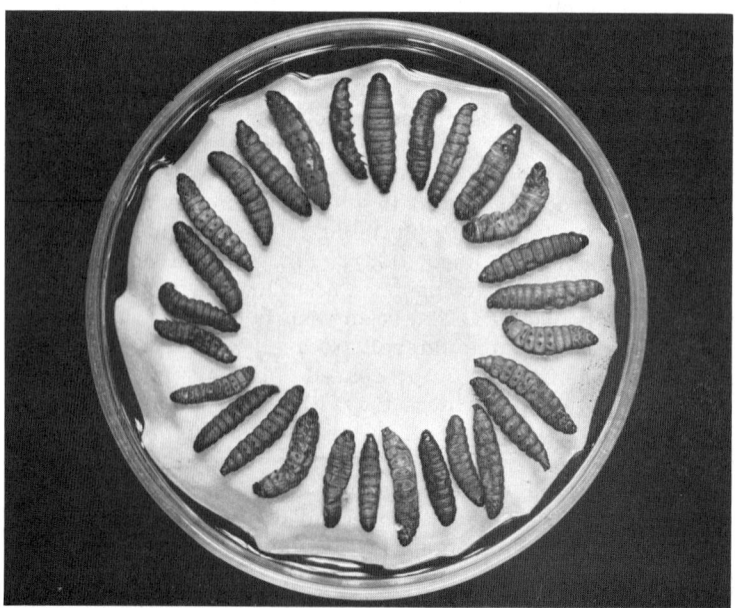

Fig. 17 White's water trap with *Heterorhabditis*-infected *Galleria* larvae.

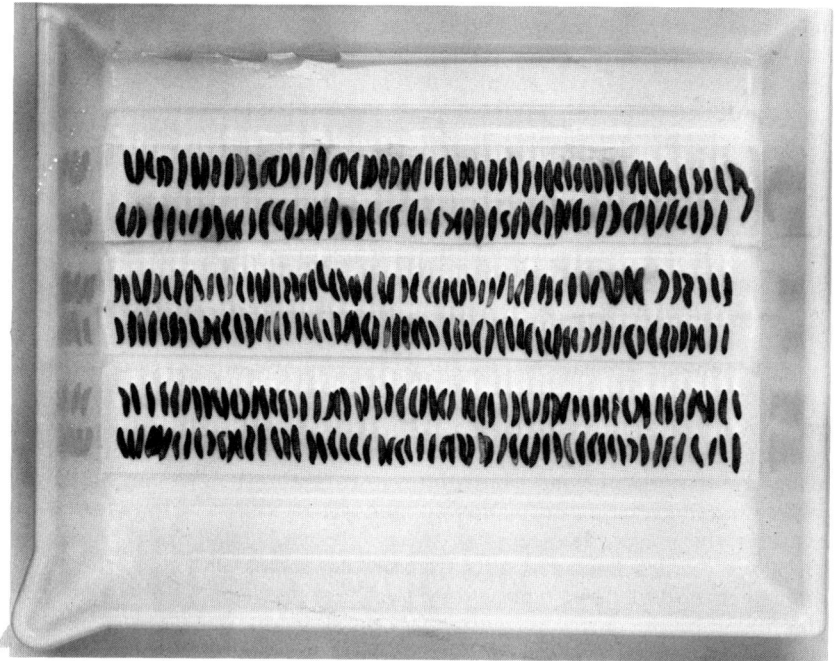

Fig. 18 Photographic developing tray with *Neoaplectana*-infected *Galleria* larvae on filter paper strips.

B. Rearing on Artificial Media

Infective larvae reared on a natural host may be used to start mass production on artificial media. First the symbiotic bacterium is isolated and then a monoxenic culture is developed.

The bacterium is located in the anterior part of the gut of the infective nematode larva. Infective nematode larvae injected into the body cavity of a *Galleria* larva start feeding, and the bacteria they release can be isolated from the host's blood 24 hours later. The original blood smear usually requires further subculturing to eliminate contaminants. Pure cultures can be stored for up to two months at 6°C. For long-term storage they may be freeze dried or deep frozen (Akhurst, 1980).

The monoxenic nematode inoculum for mass production on artificial media can be developed from one of the feeding stages of the nematode, first-generation females being most suitable. Nonfeeding stages, such as infective larvae, should not be used as they contain contaminating bacteria and viruses (Kaya, 1980) that cannot easily be removed. First-generation females are usually free of contaminants as they develop in the host when contaminating bacteria are effectively suppressed by the antibiotics produced by the active symbiont. First-generation females are obtained by exposing *Galleria* larvae to surface-sterilized infective nematode larvae. After three days the *Galleria* larvae are surface sterilized, placed in a small dish of sterile 50% Ringers solution, and carefully opened up. The emerging nematode females are washed in 50% Ringers solution and transferred to a small quantity of medium on an agar plate and incubated at 25°C. Suitable media include autoclaved homogenized dog-food biscuits (House et al., 1965), homogenized animal tissues (Bedding, 1976),

and double-strength Difco nutrient-broth-agar enriched with 0.4% yeast extract and 20-30% vegetable oil (Wouts, 1981b). The symbiont transferred with
the nematode will develop on the medium and provide a substrate suitable for
nematode reproduction. When the developing nematodes become visible to the
naked eye the purity of the cultures is checked and monoxenic cultures are
transferred to autoclaved mass-rearing vessels.

Rearing vessels contain a medium that for maximum surface area has been
coated onto an inert carrier. Polyether polyurethane sponge at a rate of 0.25
g/g of medium, with water added to facilitate impregnation of the sponge has
been found an excellent carrier (Bedding, 1981). Nutrient broth is a liquid
but can be coated onto sponge when cooked with flour into a smooth roux,
soybean flour being most suitable. It may yield slightly less than that obtained with homogenized animal tissues, but is a realistic alternative as it is
more pleasant and easier to work with. Yields are improved by inoculating
the culture flasks with the bacterial symbiont three days before introduction
of the monoxenic nematode culture. Rearing vessels are incubated at 20-25°C
and are ready for harvesting one month later. Infective larvae can be extracted by placing the sponge on sieves suspended in dishes with water. In
this environment the nematodes increase their activity, move out of the sponge
through the sieve, and collect at the bottom of the dish from where they
should be removed and cleaned once every two hours.

Besides providing a suitable medium for mass production, broth-yeast-
oil agars can conveniently be used to develop monoxenic females from contaminated infective larvae (Wouts, 1981b). For this purpose culture plates
containing double the normal strength of Difco nutrient broth, 0.4% yeast extract, 1% vegetable oil, and 1.5% agar are inoculated with a suspension of
symbiotic bacteria and incubated at 25°C for three days. In the center of
the plate, which is now covered with an even layer of bacterial growth, a
small square of agar is removed and replaced by a concentrated suspension
of surface-sterilized infective nematode larvae. At 20-25°C the infective larvae will spread out over the plate and develop into females in three days'
time. Produced in an environment in which only the symbiont was present
they are generally free of contaminating bacteria.

VIII. CONCLUSION

Nine nematode species definitely parasitize lepidopterans: *Hexamermis truncata*, *H. cavicola*, *Amphimermis zuimushi*, *A. bogongae*, *Neoaplectana glaseri*,
N. feltiae, *N. bibionis*, *Heterorhabditis bacteriophora*, and *H. heliothidis*.
The shared characteristics that enable these species to parasitize and kill
lepidopterans are: (1) a resistant infective larva that in an atmosphere of
high humidity can find and penetrate a host; (2) an efficient conversion of
host energy reserves and tissues to parasite; in the rhabditids symbiotic
bacteria aid in the process.

H. truncata and *A. zuimushi* are frequent parasites of lepidopterans, at
times killing more than 90% of the population. This reduces the insect populations and undoubtedly also the damage the insect causes in subsequent
crops. The infected caterpillar's immediate capacity for damaging host plants,
however, seems undiminished, discouraging attempts to employ these mermithids in biocontrol programs.

Symbiotic bacteria make neoaplectanids and heterorhabditids lethal to
their insect host. Mammals are not affected (Gaugler and Bousch, 1979) but
many lepidopterans, especially the larger species, can be parasitized by these
nematodes. As death of the host usually occurs within two days of infection,

the nematodes can be used as biological insecticides. *N. feltiae* seems to have the widest host range and greatest biological control potential, but the activity of the other species should not be ignored. They may each have their individual preference for specific hosts and climatic conditions. The recent development of cheap mass-rearing methods, made possible by the finding that the bacterial symbiont serves as food for the nematode, and the discovery of polyurethane sponge as inert carrier to increase the surface area of the medium, now makes field applications, even at a very high dose, economical. The success of these nematodes in biological control programs against susceptible hosts is almost solely determined by environmental conditions, humidity being the critical factor. Best results in field trials were obtained against hosts in damp environments, such as soil, orchards during periods of rain or heavy dew, and tunnels of wood-boring species. In general, excellent results may be expected in any humid environment where a spatial and temporal coincidence of parasite and host can be achieved.

APPENDIX I Lepidopteran Hosts of *Hexamermis truncata*[a]

Arctiidae
  Euprepiae cajae (L.)
  pr.n. *Arctia caja* (L.)
  Setina aurita Esp.
  pr.n. *Endrosia aurita* Esp.
  Spilosoma lubricipeda (L.)
  pr.n. *Diacrisia lubricipeda* (L.)

Cossidae
  Cossus ligniperda Fabr.
  pr.n. *Cossus cossus* (L.)

Geometridae
  Abraxas marginata (L.)
  pr.n. *Lomaspilis marginata* (L.)
  Acidalia impluviaria Hb (von Siebold, 1858)
  pr.n. *Hydriomena impluviata* (Hb.)
  Amphidasis betularia (L.)
  pr.n. *Biston betularius* (L.)
  Cabera exanthemaria Esp.
  pr.n. *Cabera exanthemata* (Scop.)
  Cheimatobia brumata (L.) (von Linstow, 1898)
  pr.n. *Operophtera brumata* (L.)
  Cidaria berberata Hb
  pr.n. *Pareulype berberata* (Denis + Schiff.)
  C. sordidata Fabr.
  pr.n. *Hydriomena furcata* (Thun.)
  C. trifasciata (Bkh.) (Schultz, 1900)
  pr.n. *Hydriomena coerulata* Fab.
  Corythea juniperata (L.)
  pr.n. *Thera variata* Denis + Schiff.
  Ennomos alniaria (Denis + Schiff.)
  pr.n. *E. autumnaria* (Wern.)
  E. illunaria Hb.
  pr.n. *Selenia bilunaria* Esper

Hepialidae
  Hepialus humuli (L.)
  pr.n. *H. humuli* (L.)

Lasiocampidae
  Gastropacha neustria (L.) (von Siebold, 1853)
  pr.n. *Malacosoma neustria* (L.)
  G. pruni (L.)
  pr.n. *Odonestis pruni* (L.) (von Siebold, 1855)
  G. rubi (L.) (von Linstow, 1898)
  pr.n. *Macrothylacia rubi* (L.)
  Ondonestis potatoria (L.)
  pr.n. *Philudoria potatoria* (L.)

Lymantriidae
  Dasychira salicis (L.)
  pr.n. *Leucoma salicis* (L.)
  Liparis chrysorrhoeae L.
  pr.n. *Euproctis chrysorrhoea* (L.) (von Siebold, 1853)
  Ocneria dispar (L.) (von Linstow, 1898)
  pr.n. *Lymantria dispar* (L.)

Noctuidae
  Agrotis ipsilon (Hufn.)
  pr.n. *A. ipsilon* (Hufn.)
  A. linogrisea (Schiff.)
  pr.n. *Epilecta linogrisea* (Schiff.)
  Calpe libatrix S.V.
  pr.n. *Scoliopterix libatrix* (L.)
  Catocala nupta (L.) (von Siebold, 1842)
  pr.n. *C. nupta* (L.)
  C. paranympha S.V.
  pr.n. *Mormonia antinympha* (Hb.)
  C. sponsa (L.)
  pr.n. *C. sponsa* (L.)
  Cucullia scrophulariae (Denis + Schiff.)
  pr.n. *C. scrophulariae* (Denis + Schiff.)
  C. tanaceti S.V.
  pr.n. *C. tanaceti* Denis + Schiff.
  C. verbasci S.V. (von Siebold, 1858)
  pr.n. *Cucullia verbasci* (L.)
  Diloba caerulocephala (L.)
  pr.n. *D. caerulocephala* (L.)
  Episema graminis L.
  pr.n. *Cerapterix graminis* (L.)
  Feltia subgothica (Haworth)
  pr.n. *F. subgothica* (Haworth)
  Hadena polyodon (L.) (von Siebold, 1858)
  pr.n. *Polia serratilinea* Ochsenh.
  Lacinipolia renigera (Stephens)
  pr.n. *Standfussiana lucernea* (L.)
  Leucania l. album L.
  pr.n. *Mythimna (Aletia) 1-album* (L.)
  L. pallens (L.)

Noctuidae (con't.)
  pr.n. *Mythimna (Aletia) pallens* (L.)
  Lithocampa ramosa Esp.
  pr.n. *Calliergis ramosa* (Esp)
  Mamestra persicariae L.
  pr.n. *M. persicariae* L.
  M. pisi L.
  pr.n. *M. pisi* L.
  Naenia typica (L.) (von Linstow, 1892)
  pr.n. *N. typica* (L.)
  Noctua brunnea S.V.
  pr.n. *Diarsa brunnea* (Denis + Schiff.)
  Plusia gamma L.
  pr.n. *Autographa gamma* (L.)

Notodontidae
  Notodonta dromedarius (L.)
  pr.n. *N. dromedarius* (L.)
  N. ziczac (L.)
  pr.n. *Eligmodonta ziczac* (L.)
  Ptilophora plumigera (Esp.)
  pr.n. *P. plumigera* (Esp.)
  Pygaera bucephala (L.)
  pr.n. *Phalera bucephala* (L.)

Nymphalidae
  Melitaea athalia L.
  pr.n. *Mellicta athalia* (Rott.)
  Vanessa antiopae (L.) (von Siebold, 1850)
  pr.n. *Nymphalis antiopia* (L.)
  V. io L. (von Siebold, 1842)
  pr.n. *Inachis io* (L.)
  Vanessa album (L.)
  pr.n. *Polygonia c. album* (L.)

Pieridae
  Pontia crataegi (L.) (von Siebold, 1853)
  pr.n. *Aporia crataegi* (L.)

Pyralidae
  Diatraea saccharalis (Fabr.) (Jaynes, 1933)
  pr.n. *D. saccharalis* (Fabr.)
  Hypsipyla grandella Zell. (Nickle and Grijpma, 1974)
  pr.n. *H. grandella* Zell.

Satyridae
  Epinephele tithonus (L.) (von Linstow, 1898)
  pr.n. *Pyrania tithonus* (L.)

Sphingidae
  Smerinthus tiliae (L.)
  pr.n. *Mimas tiliae* (L.)

Tortricidae
  Carpocapsa pomonata (L.) (von Siebold, 1850)
  pr.n. *Cydia pomonella* (L.)
  Penthina salicana (Schiff.)
  pr.n. *Argyroploce salicana* (Schiff.)
  Tortrix heparana W.V.
  pr.n. *Pandemis heparana* (Denis + Schiff.)
  T. textana Hubs. (von Siebold, 1848)
  pr.n. *Pandemis corylana* (Fab.)
  T. viridana L.
  pr.n. *T. viridana* L.

Yponomeutidae
  Hyponomeuta evonymi Zeller (Rosenhauer, 1847)
  pr.n. *Yponomeuta evonymella* (L.)
  H. malinellus Zeller
  pr.n. *Yponomeuta malinella* Zeller
  Yponomeuta cognatella Treitschke (von Siebold, 1848)
  pr.n. *Y. cagnagalla* (Hb.)
  Y. padella (L.) (Rudolphi, 1809)
  pr.n. *Y. padella* (L.)

Zygaenidae
  Zygaena minos (Denis + Schiff.) (von Siebold, 1848)
  pr.n. *Z. purpuralis minos* (Denis + Schiff.)

---

[a]Hosts are presented in alphabetical order of the names used in the original publications. The reference to each of the host records can be found in Poinar (1975), unless otherwise stated. pr.n. = present name.

APPENDIX II Lepidopteran Hosts of Unidentified *Hexamermis* Species[a]

Agonoxenidae
  Agonoxena pyrogramma Meyr.
  pr.n. *A. pyrogramma* Meyr.

Arctiidae
  Isia isabella (Abbot + Smith)
  pr.n. *Pyrrharctrea isabella* (Abbot + Smith)

Ctenuchidae
  Amsacta moorei Butler
  pr.n. *Eressa aperiens* Wlk.

Gelechiidae
  Gnormoschema operculella (Zeller) (Usman, 1956)
  pr.n. *Phthorimaea operculella* (Zeller)

Geometridae
  Dilinia medardaria Schaeffer
  pr.n. *Petelia medardaria* (Schaeffer)

Hesperiidae
  Cephrenes angiades Selb.
  pr.n. *C. angiades* Selb.

Hyblaeidae
  Hyblaea puera Cram. (Mathur, 1959)
  pr.n. *H. puera* Cram.

Noctuidae
  Achaea janata (Fabr.)
  pr.n. *A. janata* (Fabr.)
  Brithys crini Fabr.
  pr.n. *B. crini* Fabr.
  Busseola fusca (Fuller) (Mahyuddin and Greathead, 1970)
  pr.n. *B. fusca* (Fuller)
  Cirphis sp.
  pr.n. *Mythimna (Pseudaletia)* sp.
  Episparis sp.
  pr.n. *Episparis* sp.
  Fodina stola Guen. (Gokulpure, 1970)
  pr.n. *F. stola* Guen.
  Helicoverpa armigera (Hb.) (Nickle and Grijpma, 1974)
  pr.n. *Heliothis armiger* (Hb.)
  Hypena iconicalis Wlk.
  pr.n. *H. iconicalis* Wlk.
  Hypocala rostrata Fabr.
  pr.n. *H. rostrata* Fabr.
  Laphygma frugiperda (J. E. Smith)
  pr.n. *Spodoptera frugiperda* (Smith + Abbott)
  Plathypena scabra (Fabr.)
  pr.n. *P. scabra* (Fabr.)
  Sesamia calamistis Hamps.
  pr.n. *S. calamistis* Hamps.
  Spodoptera exigua (Hb.)
  pr.n. *S. exigua* (Hb.)
  Spodoptera mauritia (Boisd.)
  pr.n. *S. mauritia* (Boisd.)
  Thiacidas postica Wlk. (Mehra et al., 1968)
  pr.n. *T. postica* Wlk.
  Trichoplusia ni (Hb.)
  pr.n. *T. ni* (Hb.)

Notodontidae
  Neopheosia excurvata Hamps. (Mathur, 1959)
  pr.n. *Poecilopheosia excurvata* (Hamps.)

Nymphalidae
  Polygonia comma Harris
  pr.n. *P. comma* Harris
  P. interrogationis F.
  pr.n. *P. interrogationis* F.
  Proclassiana eunomia Esp.
  pr.n. *P. eunomia* Esp.
  Vanessa atalanta (L.)
  pr.n. *V. atalanta* (L.)

Papilionidae
  Papilio helenus L.
  pr.n. *P. helenus* L.

Pieridae
  Pieris rapae (L.)
  pr.n. *P. rapae* (L.)
  Terias blanda silhetana Wall.
  pr.n. *T. blanda silhetana* Wall.

Pyralidae
  Chilo partellus (Swinhoe)
  pr.n. *C. partellus* (Swinhoe)
  Chilotroea auricilia Ddgn.
  pr.n. *Chilo auricilia* Ddgn.
  Diaphania pyloalis Wlk.
  pr.n. *Margaronia pyloalis* Wlk.
  Eldana saccharina Wlk.
  pr.n. *E. saccharina* Wlk.
  Hapalia machaeralis Wlk.
  pr.n. *Pyrausta machaeralis* Wlk.
  Hypsipila robusta Moore
  pr.n. *H. robusta* Moore
  Proceras indicus Kapur
  pr.n. *Chilo indicus* Kapur
  Scirpophaga nivella (Fab.)
  pr.n. *S. nivella* (Fab.)
  Tryporyza incertulas (Wlk.)
  pr.n. *Scirpophaga incertulas* (Wlk.)

Sphingidae
  Cephonodes sp.
  pr.n. *Cephonodes* sp.

Thyrididae
  Rhodoneura myrtaea Drury
  pr.n. *R. myrtaea* Drury

---

[a]Hosts are presented in alphabetical order of the names used in the original
publications. The reference to each of the host records can be found in
Poinar (1975), unless otherwise stated. pr.n. = present name.

APPENDIX III Lepidopteran Hosts of Unidentified Mermithids[a]

Arctiidae
  Halisidota tesselaris Smith in Abbott + Smith
  pr.n. *H. tesselaris* Smith in Abbott + Smith
  Hyphantria cunea (Drury)
  pr.n. *H. cunea* (Drury)

Attacidae
  Attacides sp.
  pr.n. *Attacides* sp.

Brassolidae
Caligo sp.
pr.n. *Caligo* sp.

Drepanidae
Drepana harpagula (Esp.)
pr.n. *Palaeodrepana harpagula* (Esp.)
Platypteryx falculae (Schiff.)
pr.n. *Drepana falcataria* Denis + Schiff.

Epiplemidae
Epiplema sp. (Mathur, 1959)
pr.n *Epiplema*

Geometridae
Acidalia dilutata (Hb.)
pr.n. *Oporinia dilutata* (Denis + Schiff.)
Ascotis infixaria Wlk.
pr.n. *Glena cognataria* Hb.
Cusialia raptaria Wlk.
pr.n. *C. raptaria* Wlk.
Hybernia defoliaria L.
pr.n. *Erannis defoliaria* (Clerck)
Hyposidra successaria Wlk.
pr.n. *H. talaca* Wlk.
Eupithecia insigniata (Hb.)
pr.n. *E. venosata* (Fabr.)

Hesperiidae
Hidara irava Moore
pr.n. *H. irava* Moore

Hypsidae
Euprepia jacobae (L.)
pr.n. *Tyria jacobaeae* (L.)

Lasiocampidae
Bombyx catax L.
pr.n. *Eriogaster catax* (L.)
Gastropacha quercus (L.) (Schultz, 1900)
pr.n. *Lasiocampa quercus* (L.)
Gastropachae trifolii Denis + Schiff. (von Siebold, 1842)
pr.n. *Lasiocampa trifolii* (Denis + Schiff.)
Malacosoma americanum (Fabr.)
pr.n. *M. americana* (Fabr.)

Limacodidae
Thosea cana Wlk.
pr.n. *T. cana* Wlk.

Lycaenidae
Lycaena betulae (L.)
pr.n. *Thecla betulae* (L.)

Lymantriidae
  Orgya mixta Snellen
  pr.n. *O. mixta* Snellen

Noctuidae
  Alabama argillacea Hb.
  pr.n. *Anomis argillacea* Hb.
  Anomis flava Fab. (Mundiwale et al., 1968)
  pr.n. *A. flava* Fab.
  Anomis fulvida Guen. (Mathur, 1959)
  pr.n. *Rusicada fulvida* Guen.
  Arcyophora patricula Hamp. (Büttiker and Nicolet, 1975)
  pr.n. *Arcyophora patricula* Hamp.
  Busseola sacchariphaga Fletch.
  pr.n. *B. sacchariphaga* Fletch.
  Catocala flebilis Grt. (Schaffner and Griswold, 1934)
  pr.n. *Mormonie flebilis* Grt.
  C. neogama Smith + Abbott
  pr.n. *Mormonia piatrix* Grt.
  Diparopsis watersi Roths. (Galichet, 1961)
  pr.n. *D. watersi* Roths.
  Earias insulana Boisd. (Galichet, 1961)
  pr.n. *E. insulana* Boisd.
  Eutelia favillatrix Wlk. (Mathur, 1959)
  pr.n. *E. favillatrix* Wlk.
  Lobocraspis griseifusa Hamps.
  pr.n. *Plecoptera griseifusa* Hamps.
  Mamestra oleracea (L.) (Negrobov, 1962)
  pr.n. *Lacanobia oleracea* (L.)
  M. suasa (Denis + Schiff.) (Negrobov, 1962)
  pr.n. *Lacanobia suasa* (L.)
  Parallelia algira L.
  pr.n. *Dysgonia algira* L.
  Platypteryx sicula S.V. (von Siebold, 1858)
  pr.n. *Mythimna (Mythimna) sicula* (Treitschke)
  Plusia chrysitis L.
  pr.n. *P. chrysitis* (L.)
  Rhesala imparata Wlk.
  pr.n. *R. imparata* Wlk.
  Risoba obstructa L.
  pr.n. *Illatia octo* (Guen.)
  Sesamia inferens (Wlk.)
  pr.n. *S. inferens* (Wlk.)
  Trachea piniperda Panzer (von Siebold, 1858)
  pr.n. *Panolis flammea* (Denis + Schiff.)

Notodontidae
  Gargetta curvaria Hamps.
  pr.n. *Porsica curvaria* Hamps.
  Harpyia furcula (Clerck.) (von Siebold, 1858)
  pr.n. *H. furcula* (L.)
  Pygaera fulgurita Wlk.
  pr.n. *Clostera fulgurita* Wlk.
  Schizura concinna Smith + Abbott
  pr.n. *S. concinna* Smith + Abbott

Nymphalidae
  Atella phalanta Drury
  pr.n. *A. phalanta* Drury
  Vanessa polychlori (L.)
  pr.n. *Nymphalis polychloros* (L.)
  V. urticae (L.)
  pr.n. *Aglais urticae* (L.)

Oecophoridae
  Tonica niviferana (Wlk.)
  pr.n. *T. niviferana* (Wlk.)

Papilionidae
  Papilio helenus L.
  pr.n. *P. helenus* L.

Pieridae
  Pieris brassicae (L.) (Mathur, 1959)
  pr.n. *P. brassicae* (L.)

Pyralidae
  Agrotera sp.
  pr.n. *Agrotera* sp.
  Antigastra catalaunalis Dup.
  pr.n. *A. catalaunalis* Dup.
  Bapara sp.
  pr.n. *Bapara* sp.
  Chilo aleniella (Strand)
  pr.n. *C. aleniella* (Strand)
  C. partellus (Swinhoe)
  pr.n. *C. partellus* (Swinhoe)
  Chilotraea infuscatellus Snell.
  pr.n. *C. infuscatellus* Snell.
  Diaphania laticostalis Guen.
  pr.n. *Margaronia laticostalis* Guen.
  Dichocorcis evaxalis Guen.
  pr.n. *D. evaxalis* Guen.
  D. leptalis Hamps.
  pr.n. *D. leptalis* Hamps.
  Etiella venustella Hamps.
  pr.n. *Phyala venustella* Hamps.
  Hypsipila robusta Moore
  pr.n. *H. robusta* Moore
  Lamida carbonifera Meyr.
  pr.n. *Macalla carbonifera* Meyr.
  L. nubilalis Hamps.
  pr.n. *Macalla? nubilalis* Hamps.
  Margaronia hilaralis Wlk.
  pr.n. *Arthroschista hilaralis* Wlk.
  M. pyloalis Wlk.
  pr.n. *M. pyloalis* Wlk.
  Nacoleia octasema (Meyr.)
  pr.n. *Lamprosema octasema* (Meyr.)
  Nephopteryx rhodobasalis Hamps.

Pyralidae (con't.)
  pr.n. *N. rhodobasalis* Hamps.
  Psara stultalis Wlk.
  pr.n. *P. stultalis* Wlk.
  Sylepta balteata Fab.
  pr.n. *S. balteata* Fab.
  S. derogata Fab.
  pr.n. *S. derogata* Fab.
  S. lunalis Guen.
  pr.n. *S. lunalis* Guen.
  Tyspanodes linealis Moore
  pr.n. *T. linealis* Moore

Saturniidae
  Anisota rubicunda Fab.
  pr.n. *Dryocampa rubicunda* Fab.
  Antheraea mylitta Drury
  pr.n. *A. mylitta* Drury
  Antheraea pernyi (Guen.)
  pr.n. *A. pernyi* (Guen.)
  Hemileuca lucina H. Edw.
  pr.n. *H. lucina* H. Edw.
  H. maia (Drury)
  pr.n. *H. maia* (Drury)

Sphingidae
  Deilephila nerii L.
  pr.n. *Daphnis nerii* (L.)
  Smerinthus ocellatus (L.)
  pr.n. *S. ocellatus* (L.)
  Sphingis euphorbiae (L.)
  pr.n. *Hyles euphorbiae* (L.)

Thyatiridae
  Thyatira derasa L.
  pr.n. *Habrosyne pyritoides* (Hufn.)

Thyrididae
  Striglina scitaria Wlk.
  pr.n. *S. scitaria* Wlk.

Tineidae
  Opogona (Paine, 1964)
  pr.n. *Opogona* sp.

Tortricidae
  Argyroploce cellifera Meyr. (Mathur, 1959)
  pr.n. *Hedya cellifera* (Meyr.)
  A. leucacaspis Meyr.
  pr.n. *Olethreutes leucacaspis* (Meyr.)

Yponomeutidae
  Atteva sp.
  pr.n. *Atteva* sp.

---

[a]Hosts are presented in alphabetical order of the names used in the original publications. The reference to each of the host records can be found in Poinar (1975), unless otherwise stated. pr.n. = present name.

ACKNOWLEDGMENT

I thank the staff members of Entomology Division, DSIR, Auckland, for their suggestions and comments on the manuscript.

REFERENCES

Akhurst, R. J. (1980). Morphological and functional dimorphism in *Xenorhabdus* spp., bacteria symbiotically associated with the insect pathogenic nematodes *Neoaplectana* and *Heterorhabditis*. *J. Gen. Microbial. 121*: 303-309.

Andrassy, I. (1976). *Evolution as a Basis for the Systematisation of Nematodes*. Pitman Publishing, London.

Baylis, H. A. (1944). Observation on the nematode *Mermis nigrescens* and related species. *Parasitology 36*: 122-132.

Bedding, R. A. (1976). New methods increase the feasibility of using *Neoaplectana* spp. (Nematoda) for the control of insect pests. *Proc. 1st Intern. Colloquium Invert. Path. Kingston.* Canada: 250-254.

Bedding, R. A. (1981). Low cost, in vitro mass production of *Neoaplectana* and *Heterorhabditis* species (Nematoda) for field control of insect pests. *Nematologica 27*: 109-114.

Bedding, R. A., and Akhurst, R. J. (1975). A simple technique for the detection of insect parasitic rhabditid nematodes in soil. *Nematologica 21*: 109-110.

Bedding, R. A., and Miller, L. A. (1981). Disinfesting blackcurrant cuttings of *Synanthedon tipuliformis* using the insect parasitic nematode *Neoaplectana bibionis*. *Envir. Ent. 10*: 449-453.

Benham, G. S., and Poinar, G. O. (1973). Tabulation and evaluation of recent field experiments using the DD-136 strain of *Neoaplectana carpocapsae* Weiser: a review. *Exp. Parasitol. 33*: 248-252.

Bovien, P. (1937). Some types of association between nematodes and insects. *Vidensk. Meddr. Dansk Naturh. Foren. 101*: 1-114.

Büttiker, W., and Nicolet, J. (1975). Observations complementaires sur les lépidoptères ophtalmotrapes en Afrique Occidentale Revue d'Elevage Méd. *Vet. Pays Trop. 28*: 319-329.

Carne, P. B., and Reed, E. M. (1964). A simple apparatus for harvesting infective stage nematodes emerging from their insect host. *Parasitology 54*: 551-553.

Chamberlin, F. S., and Dutky, S. R. (1958). Tests of pathogens for the control of tobacco insects. *J. Econ. Ent. 51*: 560.

Chatterjee, P. N., and Singh, P. (1965). Mermithid parasites and their role in natural control of insect pests. *Indian Forester 91*: 714-721.

Chitwood, B. G., and Chitwood, M. B. (1937): *An Introduction to Nematology*. Monumental Printing Co., Baltimore, Maryland.

Couturier, A. (1950). Biologie d'un *Hexamermis* parasite des insectes defolia-
teurs de l'osier. *Ann. Epiphyt. 1:* 13-37.

Dale, P. S. (1970). Dispersal and phylogeny of some oxyuroid nematodes.
Proc. Int. Nematol. Symp. Warsaw 1967, pp. 333-338.

Danilov, L. G. (1976). The effect of temperature on the susceptibility of
Galleria mellonella larvae to infection by *Neoaplectana carpocapsae*
Weiser, 1955 "agriotis" strain, and the development of the nematode in
the insect host. *Byull, Vses. Nauchno-issled. Inst. Zash. Rast. 37:*
17-21.

Dutky, S. R. (1959). Insect microbiology. In *Advances in Applied Microbiology*, vol. I (W. W. Umbriet, ed.). Academic, New York, pp. 175-
200.

Dutky, S. R. (1968). An appraisal of the DD-136 nematode for the control
of insect populations and some biochemical aspects of its host-parasite
relationships. *Proc. Joint U.S.-Japan Seminar on Microbial Control of
Insects*, Tokyo, pp. 139-140.

Dutky, S. R., and Hough, W. S. (1955). Note on a parasitic nematode from
codling moth larvae *Carpocapsa pomonella* (Lepidoptera, Olethreutidae).
Proc. Ent. Soc. Wash. 57: 244.

Dutky, S. R., Thompson, J. V., and Hough, W. S. (1962). A new nematode
parasite of codling moth showing promise in insect control. *8th Int.
Congr. Microbial. Montreal.*

Filipjev, I. N. (1934). Miscellanea Nematologica 1. Eine neue Art der Gattung
Neoaplectana Steiner nebst Bemerkungen über die systematische Stellung
der letzteren. *Tr. Parazit. lab. Zool. Inst. Akad. Nauk SSSR. 4:* 229-
240.

Filipjev, I. N., and Schuurmans Stekhoven, J. H. (1941). *Agricultural
Helminthology*. Brill, Leiden.

Galichet, P. F. (1961). Parasitisme multiple chez *Diparopsis watersi* Roths.
(Lep. Agrotidae). *Entomophaga 6:* 203-205.

Gaugler, R., and Bousch, G. M. (1978). Effects of ultraviolet radiation and
sunlight on the entomogenous nematode *Neoaplectana carpocapsae*. *J.
Invert. Pathol. 32:* 291-296.

Gaugler, R., and Bousch, G. M. (1979). Non susceptibility of rats to the
entomogenous nematode *Neoaplectana carpocapsae*. *Envir. Entom. 8:*
658-660.

Glaser, R. W. (1931). The cultivation of a nematode parasite of an insect.
Science 73: 614-615.

Glaser, R. W. (1932). Studies on *Neoaplectana glaseri*, a nematode parasite
of the Japanese beetle (*Popillia japonica*). *New Jersey Dept. Agric. Circ.
211:* 3-34.

Glaser, R. W. (1940). The bacteria-free culture of a nematode parasite.
Proc. Soc. Exp. Biol. Med. 43: 512-514.

Glaser, R. W., McCoy, E. E., and Girth, H. B. (1942). The biology and culture of *Neoaplectana chresima*, a new nematode parasite in insects. *J.
Parasitol. 28:* 123-126.

Gokulpure, R. S. (1970). Some new hosts of a juvenile mermithid of the
genus *Hexamermis. Indian J. Entomol. 32:* 387-389.

Hagmeier, A. (1912). Beiträge zur Kenntnis der Mermithiden. 1. Biologische Notizen und systematische Beschreibung einiger alter und neuer
Arten. *Zool. Jb. Abt. Syst. 32:* 521-612.

House, H. L., Welch, H. E., and Cleugh, T. R. (1965). A food medium
of prepared dog biscuit for the mass-production of the nematode DD-136
(Nematoda-Steinernematidae) (correspondence). *Nature (Lond.) 206:* 847.

Jaques, R. P., Stultz, H. T., and Huston, F. (1968). The mortality of the pale apple leafroller and winter moth by fungi and nematodes applied to soil. *Can. Ent. 100*: 813-818.

Jaynes, H. A. (1933). The parasites of the sugar cane borer in Argentina and Peru and their introduction into the USA. *USDA Tech. Bull. 363*.

Kaburaki, T., and Imamura, S. (1932). A new mermithid-worm parasitic in the rice borer, with notes on its life history and habits. *Proc. Imp. Acad. Tokyo 8*: 109-112.

Kaya, H. K. (1980). Granulosis virus in the intestinal lumen of *Neoaplectana carpocapsae*: Retention of infectivity after treatment with formaldehyde or high pH. *J. Invert. Pathol. 35*: 20-25.

Khan, A., Brooks, W. M., and Hirschmann, H. (1976). *Chromonema heliothidis* n.gen., n.sp. (Steinernematidae: Nematoda), a parasite of *Heliothis zea* (Noctuidae, Lepidoptera), and other insects. *J. Nematol. 8*: 159-168.

Landazabal, A. J., Fernandez, A. F., and Figueroa, P. A. (1973). Control biologico de *Spodoptera frugiperda* (J. E. Smith), con el nematodo *Neoaplectana carpocapsae* en maize (*Zea mays*). *Acta Agron. Columbia 23*: 41-70.

von Linstow, O. F. B. (1892). Ueber *Mermis nigrescens* Duj. *Arch. Mikrosk. Anat. 40*: 498-512.

von Linstow, O. F. B. (1898). Das Genus *Mermis*. *Arch. Mikrosk. Anat. 53*: 149-168.

Mahyuddin, A. I., and Greathead, D. J. (1970). An annotated list of the parasites of graminaceous stem borers in East Africa with a discussion of their potential in biological control. *Entomophaga 15*: 241-274.

Mathur, R. N. (1959). *Mermis* sp. (Mermithidae, Ascaroidea, Nematoda) and its insect host. *Curr. Sci. (India) 28*: 255-256.

Mehra, B. P., Gokulpure, R. S., and Sah, B. N. (1968). New records of *Hexamermis* spp. (Nematoda: Mermithidae) from Neuroptera and Lepidoptera in India. *Bull. Ent. Loyola Coll. 9*: 83-84.

Meissner, G. (1854). Beiträge zur Anatomie u. Physiologie von *Mermis albicans*. *Z. Wiss. Zool. 5*: 207-284.

Milstead, J. E., and Poinar, G. O. (1978). A new entomogeneous nematode for pest management systems. *Calif. Agric.*: 12.

Mráček, Z. (1977). *Steinernema kraussei*, a parasite of the body cavity of the sawfly, *Cephalea abietis*, in Czechoslovakia. *J. Invert. Pathol. 30*: 87-94.

Mráček, Z., and Weiser, J. (1979). The head papillae of the Steinernematidae. *J. Invert. Pathol. 34*: 310-311.

Müller, G. W. (1931). Über Mermithiden. *Z. Wiss. Biol. A* (superseded by *Z. Morph. Okol. Tiere*) *24*: 82-147.

Mundiwale, S. K., Radke, S. G., and Borle, M. N. (1978). Natural control of *Heliothis armigera* Hub. and *Anomis flava* Fab. on cotton by nematodes (Mermithids). *Indian J. Entomol. 4*: 101.

Nash, P. F., and Fox, R. C. (1969). Field control of the Nantucket pine tip moth by the nematode DD-136. *J. Econ. Ent. 62*: 660-663.

Negrobov, V. P. (1962). Observations on the effect of variations in humidity in the environment on *Complexomermis ghilarovi* Polozhentsev & Artyukovski 1958 (Nematoda: Mermithidae). *Tr. vses Soveshoh. Fitogelmint* (5th) *Samarkand*, pp. 173-179.

Nickle, W. R. (1972). A contribution to our knowledge of the Mermithidae (Nematoda). *J. Nematol. 4*: 113-146.

Nickle, W. R, and Grijpma, P. (1974). Studies on the shootborer *Hypsipyla grandella* (Zeller) Lep. Pyralidae) XXV. *Hexamermis albicans* (Siebold) (Nematoda: Mermithidae) a parasite of the larva. *Turrialba* 24: 222-226.

Paine, R. W. (1964). The banana scab moth *Nacoleia octasema* (Meyrick) its distribution, ecology and control. *South Pacific Commission Technical Paper 145*.

Pereira, C. (1937). *Rhabditis hambletoni* n.sp. nema apparentemente semi-parasito da "Broca do algodoeiro" (*Gasterocercodes brasiliensis*). *Argos. Inst. Biol. Sao Paulo* 8: 215-230.

Poinar, G. O. (1967). Description and taxonomic position of the DD-136 nematode (Steinernematidae, Rhabditoidea) and its relationship to *Neoaplectana carpocapsae* Weiser. *Proc. Helm. Soc. Wash.* 34: 199-209.

Poinar, G. O. (1971). Use of nematodes for microbial control of insects. In *Microbial Control of Insects and Mites* (H. D. Burges and N. W. Hussey, eds.). Academic, New York, pp. 181-203.

Poinar, G. O. (1975). *Entomogenous Nematodes*. E. J. Brill, Leiden.

Poinar, G. O. (1976). Description and biology of a new insect parasitic rhabditoid *Heterorhabditis bacteriophora* n.gen. n.sp. (Rhabditida; Heterorhabditidae n. fam.). *Nematologica* 21:463-470.

Poinar, G. O. (1979). *Nematodes for Biological Control of Insects*. CRC Press, Boca Raton, Florida.

Poinar, G. O., and Gyrisco, G. G. (1962). A new mermithid parasite of the alfalfa weevil *Hypera postica* (Gyllenhal). *J. Insect Pathol.* 4: 201-206.

Poinar, G. O., and Hess, R. (1977). *Romanomermis culicivorax*: Morphological evidence of transcuticular uptake. *Exp. Parasitol.* 42: 27-33.

Poinar, G. O., and Thomas, G. M. (1965). A new bacterium, *Achromobacter nematophilus* sp. nov. (Achromobacteriaceae; Eubacteriales) associated with a nematode. *Int. Bull. Bact. Nomencl. Taxon.* 15: 249-252.

Rathke, B. (1953). Zur Biologie von *Hexamermis albicans* (Siebold). *Veröff. Überseemus. Bremen* 2: 161-210.

Rosenhauer (1847). Mitteilungen 1. Ueber die in den Raupen der *Hyponomeuta evonymi* Zell. (*Tinea cognatella* Tr) lebende *Filaria truncata* Rud. *Ent. Z. Stettin* 8: 318-320.

Rubtzov, I. A. (1972). *Aquatic Mermithidae of the Fauna of the USSR, vol. I*. Nauka Publishers Leningrad. Translated from Russian, published for ARS, USDA, and NSF, Washington D.C., by Amerind Publishing Co. New Delhi, 1977.

Rubtzov, I. A. (1978). *Mermithids: Classification, Importance, and Utilization*. Publishing House Nauka, Leningrad.

Rudolphi, C. A. (1809). *Entozoorum sive Vermium Intestinalium Historia Naturalis* Vol. II. Amstelaedami.

Schaffner, J. V., and Griswold, C. L. (1934). Microlepidoptera and their parasites reared from field collections in the northeastern part of the United States. USDA Miscellaneous Publications 188.

Schmidt, J., and All, J. N. (1978). Chemical attraction of *Neoaplectana carpocapsae* (Nematoda: Steinernematidae) to insect larvae. *Envir. Ent.* 7: 605-607.

Schultz, O. (1900). Filarien in paläarktischen Lepidopteren. *Ill. Z. Entomologie* (superceded by *Z. wiss Insect Biol.*) 5: 148-152, 164-166; 183-185; 199-201; 264-265; 279-280; 292-297.

Seryczynska, H., Kamionek, M., and Sandner, H. (1974). Defense reactions of caterpillars of *Galleria mellonella* L. in relation to bacteria *Achromobacter nematophilus* Poinar and Thomas (Eubacteriales: Achromobacteriaceae) and bacteria-free nematodes *Neoaplectana carpocapsae* Weiser (Nematoda: Steinernematidae). *Bull. Acad. Pol. Sci.* 22: 193-196.

von Siebold, C. T. E. (1842). Ueber die Fadenwürmer der Insekten. *Ent. Z. Stettin 3*: 146-161.

von Siebold, C. T. E. (1843). Ueber die Fadenwürmer der Insekten. Nachtrag 1. *Ent. Z. Stettin 4*: 78-84.

von Siebold, C. T. E. (1848): Ueber die Fadenwürmer der Insekten. Nachtrag II. *Ent. Z. Stettin 9*: 290-301.

von Siebold, C. T. E. (1850). Ueber die Fadenwürmer der Insekten. Nachtrag III. *Ent. Z. Stettin 11*: 329-336.

von Siebold, C. T. E. (1853). Beiträge zur Naturgeschichte der Mermithen. *Z. Wissensch. Zool. 5*: 201-206.

von Siebold, C. T. E. (1854). Ueber die Fadenwürmer der Insekten. Nachtrag IV. *Ent. Z. Stettin 15*: 103-121.

von Siebold, C. T. E. (1855). Zusatz. *Z. Wissensch. Zool. 7*: 141-144.

von Siebold, C. T. E. (1858). Ueber die Fadenwürmer der Insekten. Nachtrag V. *Ent. Z. Stettin 19*: 325-344.

Simons, W. R. (1978). Preliminary research on entomophagous nematodes in particular on *Neoaplectana* species in the Netherlands. *Meded. Faculteit Landbouwwetenschappen Gent 43*: 765-768.

Srivastava, R. P. (1978). Studies on the feasibility of the use of microbial pesticides for the integrated control of *Papilio demodocus* Esper. *Abstracts XXth International Horticultural Congress, Sydney, Austr.* International Society for Horticultural Sciences, Abs. No. 1624.

Stanuszek, S. (1972). Revision of the genus *Neoaplectana* Steiner, 1929. (Rhabditoidea: Steinernematidae). *Abst. 10th Int. Symp. Nemat. Reading*: 69-70.

Stanuszek, S. (1974). *Neoaplectana feltiae* complex (Nematoda: Rhabditoidea. Steinernematidae) its taxonomic position within the genus *Neoaplectana* and intraspecific structure. *Zesz. Probl. Postep. Nauk soln. 154*: 331-360.

Steiner, G. (1923). *Aplectana kraussei* n.sp. eine in der Blattwespe *Lyda* sp. parasitierende Nematodenform, nebst Bemerkungen über das Seitenorgan der parasitischen Nematoden. *Zentralbl. Bakteriol. Parasitenkd. Infektionskr. Hyg. Abt 1. 59*: 14-18.

Steiner, G. (1924). Beitrage zur Kenntnis der Mermithiden. 2. Teil. Mermithiden aus Paraguay in der Sammlung des Zoologischen Museums zu Berlin. *Zentralbl. Bakteriol. Parasitenkd. Infektionskr. Hyg. 62*: 90-110.

Steiner, G. (1929). *Neoaplectana glaseri*. n.g. n.sp. (Oxyuridae), a new nemic parasite of the Japanese beetle (*Popillia japonica* Newm.). *J. Wash. Acad. Sci. 19*: 436-440.

Stoll, N. R. (1954a). Further progress in the axenic cultivation in fluid media of *Neoaplectana glaseri*, a parasitic nematode. *Anat. Rec. 120*: 745-746.

Stoll, N. R. (1954b). Improved yields in axenic fluid cultures of *Neoaplectana glaseri* (Nematoda). *J. Parasitol. 40*: 14.

Tateishi, E., Murata, T., and Tyotoku, N. (1955). On the parasites of the rice stem borer (*Chilo suppressalis*) (Japanese). *Kyushu Agric. Res. 16*: 105.

Thomas, G. M., and Poinar, G. O. (1979). *Xenorhabdus* gen. nov., a genus of entomopathogenic nematophilic bacteria of the family Enterobacteriaceae. *Int. J. System. Bact. 29*: 352-360.

Travassos, L. (1927). Sobre o genera *Oxysomatium*. *Bot. Biol. Sao Paulo 5*: 20-21.

Usman, S. (1956). Some field parasites of the potato tuber moth *Gnorimoschema opercullella* Zell in Mysore. *Indian J. Ent. 18*: 463-468.

Welch, H. E. (1962). Nematodes as agents for insect control. *Proc. Ent. Soc. Ont. 92:* 11-19.

Welch, H. E. (1963). *Amphimermis bogongae* sp. nov. and *Hexamermis cavicola* sp. nov. from the Australian bogong moth, *Agrotis infusa* (Boisd.) with a review of the genus *Amphimermis* Kaburaki and Imamura, 1932 (Nematoda, Mermithidae). *Parasitology 53:* 55-62.

White, G. F. (1927). A method for obtaining infective nematode larvae from cultures. *Science 66:* 302-303.

Wouts, W. M. (1979). The biology and life cycle of a New Zealand population of *Heterorhabditis heliothidis* (Heterorhabditidae). *Nematologica 25:* 191-202.

Wouts, W. M. (1980). Biology, life cycle and redescription of *Neoaplectana bibionis* Bovien, 1937 (Nematoda: Steinernematidae). *J. Nematology 12:* 62-72.

Wouts, W. M. (1981a). *Hexamermis truncata* (Rudolphi, 1809) new combination, the valid name for *Hexamermis albicans* (von Siebold, 1848). *Systematic Parasitol. 3:* 127-128.

Wouts, W. M. (1981b). Mass production of the entomogenous nematode *Heterorhabditis heliothidis* (Nematoda: Heterorhabditidae) on artificial media. *J. Nematol. 13:* 467-469.

Wouts, W. M., Mracek, Z., Gerdin, S., and Bedding, R. A. (1982). *Neoaplectana* Steiner, 1929, a junior synonym of *Steinernema* Travassos, 1927 (Nematoda; Rhabditida). *Systematic Parasitol. 4:* 147-154. (Taxonomic changes proposed in this paper could not be included here as the paper was published after the final draft of this chapter was submitted.)

Chapter 19
Nematode Parasites of Orthopterans

John M. Webster and Cyril H. S. Thong *Simon Fraser University, Burnaby, Vancouver, British Columbia, Canada*

I. INTRODUCTION

The economic importance of the Orthoptera, especially locusts, grasshoppers, and cockroaches, is large, indisputable, persistent, and indeed legendary. There is no group of insects about which the public, growers, or governments have known so much and yet tried for so long to exterminate. These same insect pests are infected by nematode parasites that are frequently lethal to their host and are often debilitating to the insect population. It behooves us, therefore, to understand fully the nature of these nematodes and their relationship with the insect host in view of the possibility of utilizing them as biological control agents of these insect pests.

The relative economic and social importance of locusts, grasshoppers, and cockroaches in the Orthoptera has resulted in a greater research effort and greater fund of knowledge than for other members of this insect order. As a corollary to this, the Mermithoidea and the Oxyuroidea are the most extensively studied nematode parasites of the Orthoptera. Hence, this chapter gives these insect and nematode groups special attention, and emphasizes the biology and host-parasite relationships of the nematode parasites. The systematics and taxonomy of these nematodes is not considered in detail, and readers are advised to consult the reviews by Basir (1956), Corbel (1967), Kloss (1959), Leibersperger (1960), Nickle (1972), and Poinar (1975) for such information.

II. BIOLOGY AND ECOLOGY OF PARASITES

A. Mermithids in Locusts and Grasshoppers

> Early in September last I visited my brother entomologist who resides
> in Montreal, and . . . on a few occasions in the evenings, I accompanied
> [a] friend in his boat to the Rapids opposite the city, where we fished.
> The bait generally used for still-fishing are grasshoppers, freshly
> collected and kept in a bottle. On one occasion I selected a specimen
> measuring about 14 lines (probably an *Oedipoda*) [Dr. Kevan states that
> this is probably *Encoptolophus sordidus sordidus*] found commonly on
> the island of St. Hellens. It had been a short time in the water and I
> had indication of a "nibble"; shortly afterwards, on examining the bait,
> the posterior part of its body had been bitten off, and something pro-
> truded, having a resemblance to white thread, and which, at first sight,
> I took to be its intestines. I disengaged the thread-like substance, and
> discovered it to be an *intestina* measuring at least *nineteen* inches in
> length. (Couper, 1855).

This report by Couper (1855) in Toronto and that of Leidy (1851) in Philadel-
phia are probably the first records of mermithids in North America and were
made only a few years after Dujardin (1842) described *Mermis nigrescens* in
France. *Mermis subnigrescens* was described subsequently by Cobb (1926)
in the United States and since that time there has been a continual debate as
to its synonymy with *M. nigrescens*. There is no convincing evidence for defining two species, and certainly specimens found in western Canada closely
resemble the description of *M. nigrescens* from Europe. The pros and cons
of the synonymy debate are discussed in some detail by Baylis (1944) and
clarified by Nickle (1972). There are many less well-known species of *Mermis*
recorded from Orthoptera, several of them from Africa, which possess obvious
species-specific characters. Unfortunately, some of these species descriptions
are based on single specimens.

 M. nigrescens is a common parasite of locusts and grasshoppers (Acrididae and Tettigoniidae) and may also infect other insect species. Poinar (1979)
listed 48 species of grasshoppers and locusts as hosts of *M. nigrescens*.
These insects become infected with *M. nigrescens* by feeding on vegetation to
which the dark brown, embryonated nematode eggs (40 μm diameter) are
attached. These hosts are vulnerable to infection by this nematode species
throughout their life from the first nymphal to the adult stage.

 The eggs containing the L_2 larvae pass unchanged to the proventriculus
region of the gut (Craig and Webster, 1978). Here, and in the midgut, the
L_2 larvae hatch within an hour of egg ingestion (Fig. 1). These larvae use
their stylets (odontostyle) to probe and penetrate the egg shell; at this stage
of development the nematode possesses an anus (Poinar and Hess, 1974b).
The L_2 larvae immediately penetrate the host gut wall in the same general location where hatching occurs by using the stylet and secretions from the two
penetration glands. Although several factors, such as temperature, size, and
species of hosts, influence the duration of the parasitic phase in the host
hemocoel, the larvae commonly take three to six weeks to develop from the ingested egg to the preadult emerging larval stage. Developing female nematodes remain longer in the host and reach a greater body length than do
developing males. The duration of parasitism is inversely proportional to the
number of parasites in the host and directly proportional to the size of the
host (Christie, 1937), but this may be modified by the ambient temperature.
In locusts, *Schistocerca gregaria*, reared at 35°C, *M. nigrescens* develops
faster than those parasitizing locusts reared at 22°C. Further, nematode

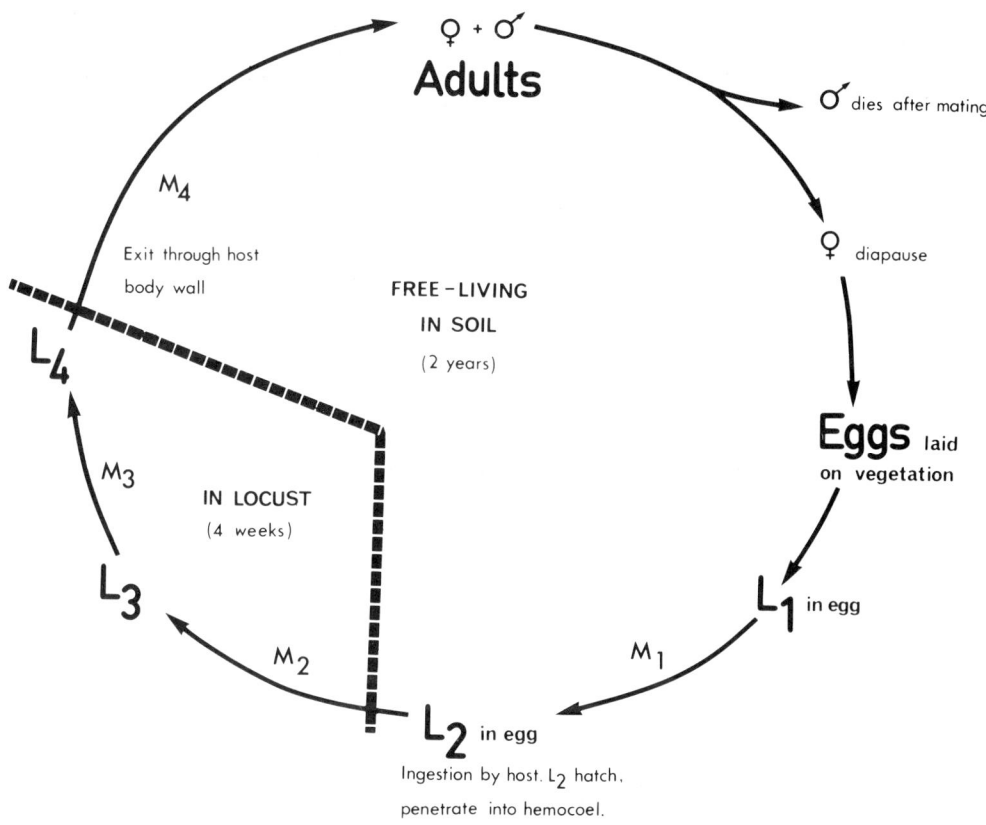

Fig. 1 Life cycle of *M. nigrescens* in locusts (L = larval stage; M = molt).

larvae grow larger and remain for a longer period in mature hosts than in nymphal or young adult hosts (Craig, 1973).

In Europe and North America, *M. nigrescens* L_4 larvae emerge from the host's hemocoel in the summer, usually by penetrating the cuticle and occasionally through the natural orifices (Fig. 2). Larval emergence kills the host and also any larvae remaining in the host. The L_4 larvae burrow 6-8 in. down into the soil where they molt into adults and may mate. Adult females remain loosely coiled in the soil until the following June, when they crawl to the surface (Christie, 1937) and, under suitably wet conditions, such as after a thunderstorm or in the early morning dew, they climb the vegetation to lay their eggs (Fig. 3). Under such moist conditions, the mermithid is able to resist desiccation by virtue of its thick multilayered cuticle. In adult *M. nigrescens* the cortical layer of the cuticle is penetrated by regular rows of canals, which Lee (1970) suggested may secrete a mucoprotein that assists in retarding water loss. Female *M. nigrescens* are very active during egg laying, and there is evidence to suggest that migration up the vegetation and the rate of uterine contractions are influenced by light. Only mature females possess a reddish pigment, which is concentrated in the chromatrope, a region of the hypodermal chords anterior to the cerebral ganglion (Cobb, 1929; Croll, 1966). In view of the apparent positive phototaxis exhibited by

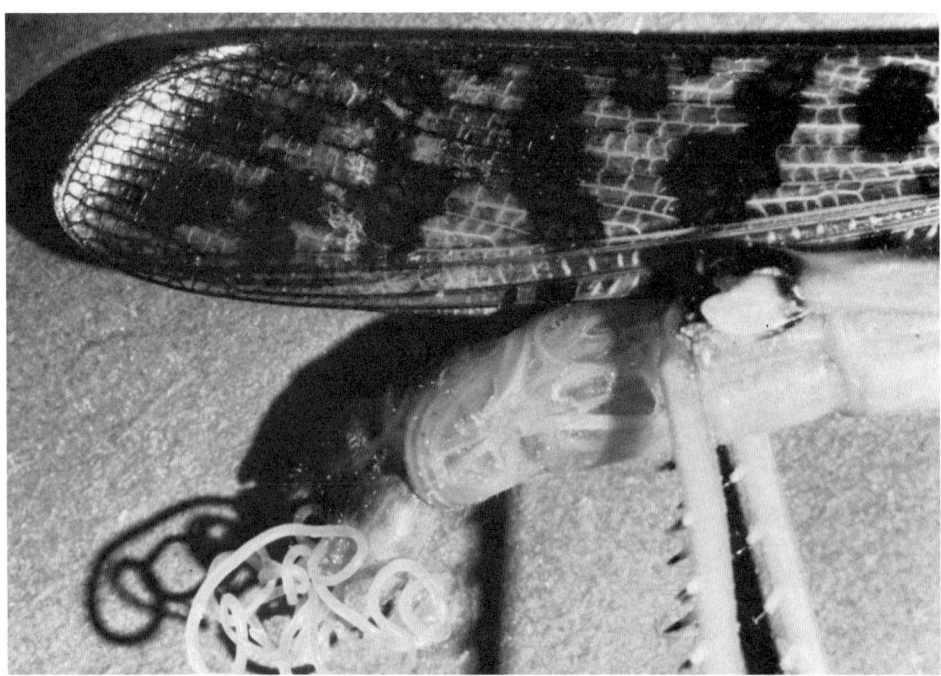

Fig. 2 Postparasitic larva of *M. nigrescens* emerging from the desert locust, *Schistocerca gregaria*. (Courtesy of R. Gordon.)

Fig. 3 Female *M. nigrescens* in the process of egg laying on alfalfa foliage. (Courtesy of T. A. Rutherford.)

females, this chromatrope may serve as a shading pigment for light-sensory organs similar to the role shown by Burr (1979) for the melanin pigments of *Oncholaimus vesicarius*. Ellenby and Smith (1966) considered the pigment more likely to assist in supplying oxygen to the tissue. The hemoglobin composition of the pigment in *M. nigrescens* was confirmed by Burr et al. (1975), who believe that it is still an open question whether the chromatrope pigment serves for oxygen storage (presumably for use during periods of low oxygen concentration in the soil) or for shading light-sensory organs during the female's migration.

The eggs may develop parthenogenetically or as a result of fertilization. Many thousands of shiny, dark brown eggs are laid on the higher vegetation (Fig. 3) in the ground cover, to which they adhere by byssi until swallowed by a foraging insect. Their viability is undiminished by 15 months of storage in the laboratory at 5°C (Craig and Webster, 1978).

North American grasshoppers and other insects are frequently parasitized by another mermithid, *Agamermis decaudata*. The free-living, whitish adults of this parasite are often found where the eggs (166 μm in diameter) are laid, coiled 5-15 cm below the soil surface. They differ markedly, therefore, from *M. nigrescens* in laying their plain, colorless eggs in these subterranean chambers. Within 30 days of laying, the eggs develop into coiled juveniles, each with an odontostyle. The first molt occurs within the egg, and the L_2 larva emerges in the following May to July and migrates to the soil surface. Infection of newly hatched nymphal grasshoppers can occur below the soil surface, but most grasshopper nymphs come in contact with the larval nematodes as they climb the vegetation (Christie, 1936). The nematode larvae penetrate the host integument, and frequently the posterior portion of the nematode, behind the node, breaks off due to the host's vigorous activity. Entry is completed within 10 minutes. The larvae develop, slowly at first, and eventually occupy a large volume within the hemocoel. The larvae emerge from the host after one to three months of parasitic development. There is no evidence of a molt during the parasitic phase of development, and, unlike *M. nigrescens*, there is no evidence of parthenogenesis (Christie, 1936). Females are receptive to fertilization for at least 14 months after emergence, and each male is capable of fertilizing several females. There is considerable morphological variability among adults, especially in the structure and arrangement of the cephalic sense organs. However, these appear to be intraspecific variations rather than species differences. Intersexes of *A. decaudata* are very common (Steiner, 1923).

Two other genera of mermithids, namely *Hexamermis* and *Amphimermis*, parasitize the Orthoptera on a worldwide scale but they are not reported as frequently as *Mermis* or *Agamermis* (Nickle, 1972). *Hexamermis albicans* adults are recorded up to 20 cm long and when found on food crops, have caused considerable public anguish and concern. The outcry and exaggerated claims were so great in 1908 that the then Secretary of Agriculture in the United States approved a public clarification of the problem in a USDA circular by Chittenden (1908) that concluded with the statement "the cabbage snake [cited as *Mermis albicans*] is entirely harmless, and . . . public rumors and superstitions are fallacies without semblance of foundation."

Mermithids are characterized by a trophosome (modified intestine) and by a group of large cells closely associated with the esophagus, which is a nonmuscular, thin cuticular tube. Steiner (1933) described three types of these cells, namely, the stichocytes (often grouped together into a stichosome), homorocytes (large, oval cells anterior to the stichocytes), and episomes (small cells closely applied to the cuticular tube). The function of these three types of cells is unclear, but they are probably important for the nematode's parasitic development since they are smaller in the free-living adults. They might have an excretory or osmoregulatory function, or may be used for protein digestion and tissue penetration (Baylis, 1947). *M. nigrescens* is unusual in having eight rather than sixteen stichocytes, and because of the large amount of endoplasmic reticulum within them, Poinar and Hess (1974b) suggest that they are a center of protein synthesis. The trophosome is the major food-storage organ that enables the nematode to survive the nonfeeding, postparasitic phase of its life.

B. Other Nematode Parasites of Locusts and Grasshoppers

Grasshoppers and locusts are the intermediate hosts for several nematode parasites of birds and mammals. These nematodes are confined almost entirely to families of the Ascaridida and Spirurida (Table 1).

A detailed study of *Diplotriaena tricuspis*, a parasite of the air sacs of American crows, showed that the development of this species in an intermediate host exemplifies the general pattern of development of many heteroxenous nematode species (Cawthorn and Anderson, 1980). When the grasshopper, *Melanoplus sanguinipes*, and the locust, *S. gregaria*, were infected at 30°C with the thick-walled, oval embryonated eggs, the eggs hatched in the midgut and the first nematode molt occurred four days and the second molt eight days after ingestion by the intermediate host. Between 12 and 21 hours after infection the hatched larvae had penetrated the midgut wall, traversed the

Table 1 Some Heteroxenous Nematodes That Use Orthopteran Intermediate Hosts, Listed to Show Their Possible Phylogenetic Relationships

Order	Family	Species	Definitive Host
Ascaridida	Subuluridae	*Maupasina weissi*	Rodents
		Allodapa suctoria	Gallinaceous birds
		Subulura elongata	Rodents
	Seuratidae	*Seuratum cadarachense*	Rodents
Spirurida	Rictulariidae	*Pterygodermatites hispanica*	Rodents
		P. affinis	Carnivores
	Physalopteridae	*Abbreviata caucasica*	Mammals
	Spiruridae	*Spirura guianensis*	Mammals
	Diplotriaenidae	*Diplotriaena tridens*	Birds
		D. tricuspis	Birds
	Tetrameridae	*Tetrameres cardinalis*	Birds
		T. americana	Gallinaceous birds
	Acuariidae	*Acuaria anthuris*	Birds
	Gongylonematidae	*Gongylonema brevispiculum*	Rodents

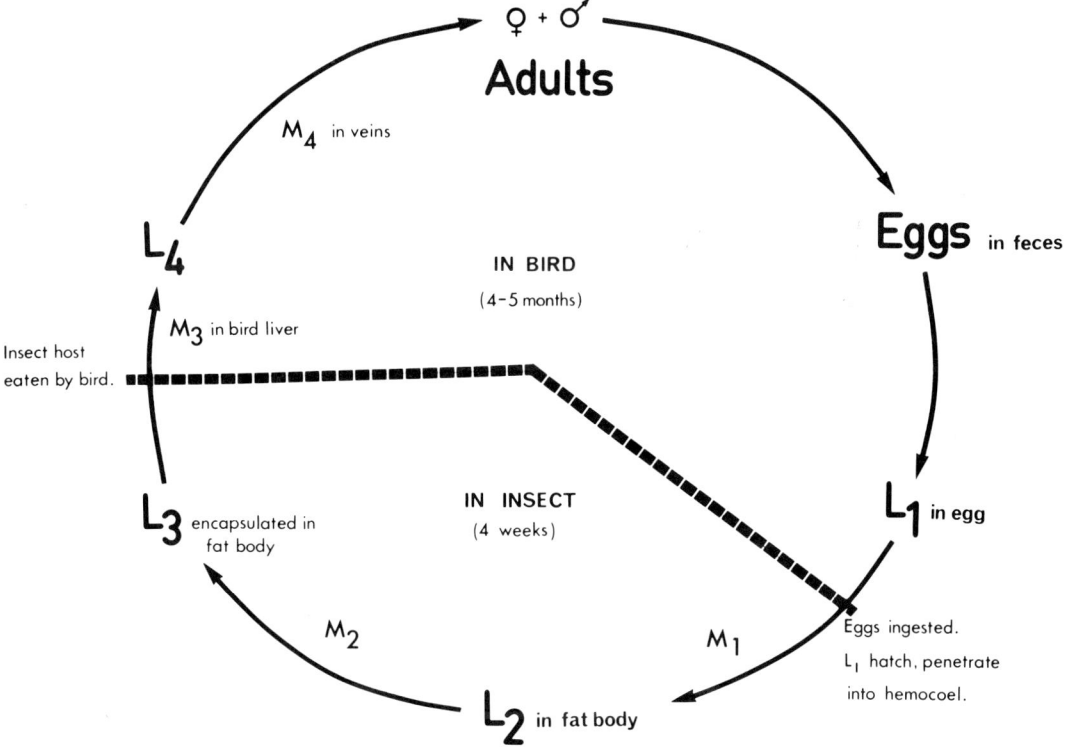

Fig. 4 Life cycle of *Diplotriaena tricuspis* in grasshopper and crow (L = larval stage; M = molt).

hemocoel, and entered the fat body (Fig. 4). The disproportionately long and bimorphic (muscular and glandular) esophagus, which is typical of the genus, becomes apparent in the L_2 stage and has compressed the small intestine to a small fraction of the body length by the L_3. Larval migration from the gut toward the hemocoel and subsequent encystment on or in one of the body tissues, or encapsulation in the hemocoel, appears to be common to all ascaridids and spirurids that use insects as intermediate hosts (Figs. 4 and 5). However, in a series of papers, Seureau, Quentin, and colleagues showed that the precise migratory path, the speed of migration, and the site and type of encystment of the larva vary with the species of nematode in the experimental intermediate host, *Locusta migratoria* (Fig. 5). Furthermore, this larval behavior is different in different species of intermediate host.

Seureau and Quentin (1977) reviewed the larval migrations of 17 species of subulurids, spirurids, and associated groups in the experimental intermediate host, *L. migratoria*. All larvae hatch in the intenstine, traverse the gut wall in the posterior midgut, and follow one of three patterns of migration (see also Figs. 4 and 5), namely,

1. A relatively long period of migration that is confined to traversing the midgut wall and entering the hemocoel where they are encapsulated (Subuluridae and Seuratidae)
2. A relatively short period of intraepithelial migration along the wall of the midgut, hindgut and sometimes the rectum prior to encystment in the gut wall (Rictulariidae and Physalopteridae)

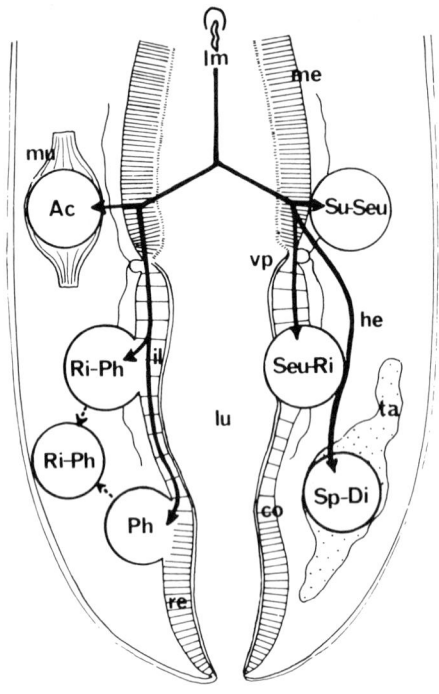

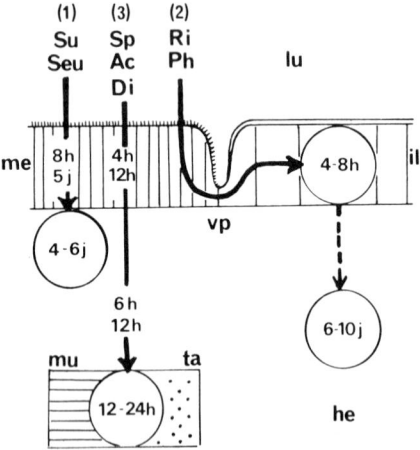

Fig. 5 Diagrammatic representation of (above) the migratory routes and sites of encystment and (below) the length of time of migration of the larvae of seven families of heteroxenous nematodes in an experimental intermediate host, *Locusta migratoria*. Key: Seu = Seuratidae; Ph = Physalopteridae; Di = Diplotriaenidae; Su = Subuluridae; Ri = Rictulariidae; Sp = Spiruridae; Ac = Acuariidae; me - midgut; il = hindgut; re = rectum; mu = muscle; lu = gut lumen; lm = hatched, migrating larva; vp = proctodeal valve; co = colon; ta = fat body; he = hemocoel. (Courtesy of Seureau and Quentin, 1977).

3. A short period of migration through the midgut wall, across the hemocoel and penetration and encystment in the fat body and muscle fibers (Spiruridae, Acuariidae, and Diplotriaenidae)

Larval nematodes belonging to the Subuluridae and Seuratidae migrate the shortest distance but take a relatively long time, e.g., *Subulura elongata*, a parasite of gerbils, takes three days to migrate across the gut wall of *L. migratoria* (Seureau and Quentin, 1977). The subulurid *Maupasina weissi*, a parasite of the elephant shrew, shows a more primitive phylogenetic relationship in requiring egg maturation in the external environment prior to ingestion by an intermediate host (Quentin and Verdier, 1979).

The development in the laboratory of *Pterygodermatites hispanica*, a parasite of the field mouse, in the experimental intermediate host *L. migratoria*, shows the larval behavior characteristic of the Rictulariidae (Quentin and Seureau, 1974). The larvae hatch in the midgut of the locust, penetrate the midgut wall, and migrate rapidly intraepithelially in the gut wall to the outer wall of the hindgut, where they are encysted within six hours of egg ingestion. The cysts and contained larvae are often subsequently released into the hemocoel. This migration within the gut tissue (see Fig. 5) is seen by Seureau and Quentin (1981) to be a way of delaying or avoiding hemocytic encapsulation and also suggests a more advanced phylogenetic standing of the Rictulariidae as compared with the Subuluridae, the larvae of which are in the hemocoel for a longer period and become encapsulated. The larval migration and development of *Seuratum cadarachense* (Seuratidae) support the hypothesis implied by adult morphological characteristics that within the Ascaridida (Table 1), the Seuratidae is a phylogenetic group between the Subuluridae and members of the Spirurida (Quentin and Seureau, 1975). In this instance the L_1 of *S. cadarachense* occur in the posterior midgut of *L. migratoria* 24 hours and in the epithelial lining of the midgut wall 48 hours after ingestion of the eggs; encapsulation in the hemocoel on the surface of muscle fibers occurs six days after experimental intake of the eggs. The migration of the subulurids and seuratids in Orthoptera is somewhat similar to that of larval cestodes in Coleoptera, larval trematodes in the Coleoptera and Hymenoptera, and larval Acanthocephala in Dictyoptera.

The fat body is a common site for encysted spirurids. The larvae of *Diplotriaena tridens* may be found encysted in the fat body as early as 24 hours after ingestion of the embryonated eggs by *L. migratoria* (Bain and Vaucher, 1973; Seureau and Quentin, 1977). In the same species of experimental host, larval *Spirura guianensis*, a parasite in neotropical primates and marsupials, took 23 days to become encysted in the fat body (Quentin, 1973) but the speed of migration was influenced by temperature. The larvae of *Tetrameres americana*, a parasite of the stomach glands of gallinaceous birds, migrate rapidly through the gut wall, but remain active in the hemocoel for about 10 days before encysting in the muscle tissue. Encysted larvae may occur throughout the hemocoelic musculature of the intermediate host and cause the insect to become less active and, consequently, easier prey for the definitive host (Cram, 1931). *Tetrameres cardinalis*, a parasite of cardinals, is regarded by Quentin and Barre (1976) to be a more primitive form of the genus because the larvae encyst in the fat body of *L. migratoria* rather than in the musculature as does *T. americana*.

The influence of temperature on parasitic development is demonstrated also by *Acuaria anthuris*, a parasite in the gut of magpies (Quentin et al., 1972). At 22-24°C the hatched larvae occur in the lumen of the midgut three hours after ingestion of the eggs by *L. migratoria*. After 48 hours they occur in the hemocoel, and two weeks later these larvae are found encysted in the muscle fibers of the abdominal tergites (Fig. 5). However, at higher temperatures (e.g., 28°C) the development and migration occur considerably faster (Seureau and Quentin, 1977) and the encysted larvae occur in the muscles within hours of host ingestion of the embryonated eggs. Undoubtedly, both the ambient temperature and the species of intermediate host have a significant effect on the larval development time of these nematodes, and the effects of both these factors are shown in the development of *Allodapa suctoria*, a common parasite of gallinaceous birds, that experimentally infects the intermediate host *Conocephalus brevipennis* (Baruš, 1970 a, b).

An additional perspective on the phylogenetic relationship of some of the heteroxenous subulurid and spirurid nematodes is given by Quentin and Poinar (1973) in their detailed comparison of the larval development of several of these nematodes. The development of each species is unique and is characterized by the number and growth of the intestinal cell initials, the number and rate of multiplication of the mesenchymal cells, and the position and nature of the rectal cells in relation to the R_1 mesenchymal initial. In the more primitive forms, e.g., *A. suctoria* (Subuluridae) and *S. cadarachense* (Seuratidae), the number of tissue initials is relatively high but the development is much slower than in the more evolved species, e.g., *A. anthuris* (Acuariidae).

C. Nematode Parasites of Cockroaches

The nematode parasites that use cockroaches as their definitive host belong mainly to the family Thelastomatidae. These nematodes are intestinal "parasites" although "commensalism" would more appropriately describe their association with cockroaches, since they feed mostly on the gut contents and there is little evidence that they are detrimental to the health of their host (Roth and Willis, 1960; Jarry, 1964).

Adult and nymphal stages of the host may be infected with thelastomatids. Bhatnagar and Edwards (1970), working on *Gromphadorhina portentosa* infected with *Leidynema appendiculata*, and Tsai and Cahill (1970), working on *Blatella germanica* infected with *Hammerschmidtiella diesingi*, reported no difference in the incidence of infection between host nymphs and adults and between males and females. However, Leong and Paran (1966) reported that adult *Blatta orientalis* and *Periplaneta americana* were more heavily infected with five thelastomatid species than were their nymphal stages and Hammerschmidt (1847) found adults more often than nymphs of *P. americana* infected with *H. diesingi*. Hominick and Davey (1972a) found adults but not nymphs of *P. americana* infected with *L. appendiculata*. The different findings may be the result of host differences and/or different sampling methods.

Thelastomatids usually occupy the proximal region of the hindgut of cockroaches, although *L. appendiculata* has been reported to infect the rectum of the Madagascar cockroach, *G. portentosa* (Bhatnagar and Edwards, 1970). They feed almost entirely on digested and partially digested food materials and on the microorganisms in the host hindgut (Jarry, 1964). It has been reported, for example, that starved *P. americana* tend to have a reduced parasite burden of *Thelastoma attenuatum* (Peregrine, 1974a), which suggests that the nematode depends on the host's food materials for its nutrition.

H. diesingi and *L. appendiculata*, the two most common thelastomatids in cockroaches, have similar life cycles (Fig. 6). Eggs are laid in the hindgut of the host, they pass out with the host feces, and usually the first larval molt occurs outside the host. Eggs containing L_2 larvae are infective to new hosts that ingest them. They pass unchanged through the host gut until they reach

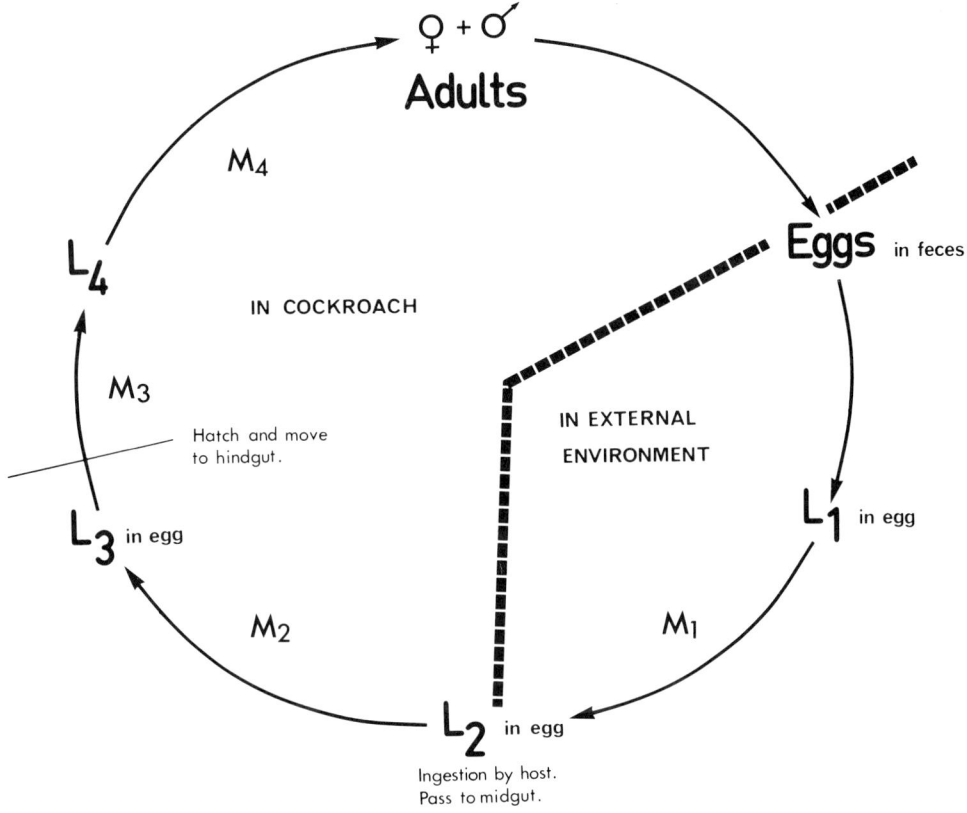

Fig. 6 Life cycle of *Hammerschmidtiella diesingi* and/or *Leidynema appendiculata* in cockroaches (L = larval stage; M = molt).

the posterior midgut, where a second molt occurs within the egg and the L_3 larvae then hatch. Two more molts occur within the host hindgut giving rise to females and males (Dobrovolny and Ackert, 1934; Todd, 1941, 1944).

The nematode *Blatticola blattae*, which commonly occurs in the German cockroach, *B. germanica*, has a slightly different life cycle from that described above in that only one molt occurs in the egg and it is the L_2 rather than the L_3 larva that hatches in the host gut. The remainder of the development to the adult occurs in the anterior hindgut (Bozeman, 1942; Cali, 1964; Cali and Mai, 1965).

Mating rarely has been observed in the thelastomatids. Jarry (1964) reported a single observed mating of *H. diesingi*. He attributes the rareness of such observations to the fact that in the related *L. appendiculata*, the males mature 15 days after infection whereas the females mature 20-30 days after infection. The males have a short life span not exceeding 20-30 days. Therefore, mating must occur within 10-15 days of female maturation. He concluded that although egg production occurs throughout the life of the female, she probably mates only once.

Thelastoma bulhoesi is typical of the thelastomatids in having eggs with a sticky coat, the active constituent of which is an acid mucopolysaccharide (Lee, 1961). This sticky coat causes the eggs to stick together and to other

objects, such as the feet of cockroaches, and consequently the eggs are ingested when the cockroaches clean themselves. This helps to ensure the infection of new hosts with high infective doses of the nematode. The hygro-
scopic nature of the acid mucopolysaccharide coat increases the eggs' resistance to desiccation. These sticky eggs readily adhere to the oothecae of
their hosts and to the exoskeleton of nymphs (Peregrine, 1974b) and the
only way to obtain nematode-free B. orientalis is to thoroughly clean the
oothecae in detergent (Fay, 1961). Fay claimed that thelastomatid eggs may
be carried within the oothecae, but this seems unlikely, since this would
necessitate the nematode eggs' gaining entry into the genital tract. This
process has not been observed and it is not characteristic of this taxonomic
group.

Cockroaches also act as intermediate hosts for several species of nematodes parasitic in vertebrates. Young and Babero (1975) experimentally
tested four species of cockroach as potential transmitting agents for 21 species of helminths including 11 nematodes. Most of the nematode eggs were
voided by the cockroaches unharmed, but embryonated eggs of Ascaridia
galli and Toxocara canis, when subsequently fed to chicks and rats, respectively, established viable infections in these vertebrate hosts. Abbreviata
caucasica (Physalopteridae) hatches in the midgut of B. germanica and the
L_1 larva enters the colon wall where it develops to an L_3 (Poinar and Quentin, 1972). Further larval development is delayed until after the cockroach
is ingested by the appropriate vertebrate host. B. germanica serves as an
intermediate host for several species of Physaloptera (Alicata, 1937; Hobmaier,
1941; Petri, 1950; Petri and Ameel, 1950; Schell, 1952). The ingested eggs
of this genus hatch in the gut and then develop to L_3 larvae within cysts located in the wall of the cockroach hindgut. These L_3 larvae are infective to
several species of carnivorous mammals, which are the definitive hosts.

Cockroaches are also intermediate hosts to other spirurids (Alicata,
1935; Brumpt, 1931; Seurat, 1911), some rictulariids (Oswald, 1958), and
a few thelaziids (Schwabe, 1951). The reaction of cockroaches to these nematodes has been reviewed by Poinar (1969).

D. Nematode Parasites of Crickets and Mantids

Crickets serve as intermediate and definitive hosts for a few species of nematodes. Ackert and Wadley (1921) described Cephalobium microbivorum (Diplogasteridae) from the hindgut of the cricket, Gryllus assimilis. Diplogasterids are widespread in the soil, and many species have phoretic relations
with insects. They are typically bacterial feeders, and C. microbivorum
feeds on the bacterial flora of the hindgut. Eggs are laid and pass out with
the host feces and, after a brief period of larval development in the outside
environment, the larvae are ingested by new cricket hosts. The infective
larvae, of unknown developmental stage, mature in the host hindgut. The
average parasite load is higher for female than for male hosts of C. microbivorum, and adult female nematodes appear to outnumber the males within the
insect host. These nematodes are not known to cause deleterious effects to
crickets.

Another common intestinal parasite of crickets is the thelastomatid,
Protrellatus alii, which has been reported from Gryllus domesticus (Farooqui,
1970). Not more than one female was found in the parasite population of each
infected host, and male nematodes were rare.

Chitwoodiella ovofilamenta (Oxyuridae), a parasite of the mole cricket,
Gryllotalpa africana, has a novel way of ensuring multiple infections in new

hosts. Eggs are laid in strings, with each egg connected to the next by filaments (Basir, 1948). It seems that once egg laying commences, each uterus
must be emptied completely in turn to ensure the stringlike arrangement of
the eggs. This nematode is unusual also in having an annulated stoma. A
similar egg arrangement occurs in the nematode *Cameronia multiovata* from
the gut of *Gryllotalpa gryllotalpa*. Apparently, this egg arrangement together with the ability of such egg strings to stick to the exoskeleton of the
host facilitates autoreinfection through the cleaning habit of crickets, and
cross infection through the ingestion of molted exoskeletons by other mole
crickets (Jarry, 1964).

Crickets act as intermediate hosts for a number of vertebrate parasites.
Lincoln and Anderson (1975), in studying the development of the spirurid
nematode *Physaloptera maxillaris* in the common field cricket *Acheta pennsylvanicus*, found that nematode larval development occurred in capsules in the
wall of the host hindgut. After being ingested by the cricket, the eggs were
encapsulated by host tissues in the wall of the hindgut. The eggs hatch as
L_1 larvae within the capsules, and the first molt occurs four days and the
second molt eight days after infection. L_3 larvae were recovered up to 25
days after infection. The L_3 larva does not develop further until the cricket
is ingested by its definitive host, the striped skunk, wherein the larva is
released from the insect tissue and attains maturity in the host gut. The
ambient temperature affects the growth rate of the nematode within the cricket (Cawthorn and Anderson, 1976), with the optimum being at 30°C.
Furthermore, older nymphs harbored, on average, shorter nematode larvae
than did younger nymphs independent of parasite load. When *P. maxillaris*
uses the German cockroach, *B. germanica*, as an intermediate host, the larvae develop in capsules in the host colon rather than in the hindgut (Cawthorn and Anderson, 1977).

Apart from a few reports of mermithid infections, nematodes are rarely
reported from mantids. However, the diplogasterid, *Gynopoecilia pseudovipara*, occurs in the esophagus and intestines of the mantids *Polyspilota aeru-
ginosa* and *Mantis viridis*. Chabaud et al. (1965) described three types of
females in this nematode species. These authors suggested that the three
types represented different ages of the females. Hence, the predominance
of a female type at any one time is probably dependent on the age of the infection in the host.

III. HOST-PARASITE RELATIONSHIPS

A. Mermithoidea

Gut penetration by infective *M. nigrescens* larvae does not cause any significant behavioral or tissue response by the grasshopper even when up to 70
eggs are fed experimentally to a single host nymph (Denner, 1976).

The developing *M. nigrescens* may be located throughout the host hemocoel, from the head capsule to the genital sclerites, and during its development, the host's morphology and physiology can be greatly modified. Larvae
that occur in the mandibular adductor muscle may cause the muscle bundles
to become atrophied and highly compressed. Individual host muscle fibers
are sometimes destroyed. Larvae in the brain of grasshoppers may cause the
protocerebrum and/or deutocerebrum to become atrophied or misshapened,
which may, in turn, affect host mobility. Penetration of the salivary glands
by the larvae may reduce the size of these structures. The fat body may
also decrease in size in heavy infections. The gut and lateral oviducts some-

times become compressed and distorted due to pressure from the growing mermithid larvae in the hemocoel (Denner, 1976). The ovaries are diminished in size in mermithid-infected hosts, and Kevan et al. (1962) found that the number of eggs in the ovaries of naturally infected *Metrioptera roeseli* decreased with increasing parasite load. Male grasshoppers infected with either *A. decaudata* or *M. nigrescens* retain the capability of producing sperm, and the testes are "not materially diminished" in size (Christie, 1936).

Mermithid parasitism sometimes discolors and softens the cuticle and may deform the wings of the host. All these phenomena are also symptomatic of the absence of juvenile hormone, though there is no evidence that mermithids cause such a change in the Orthoptera. The deformations in grasshoppers and locusts, such as crumpled and shortened wings, are not correlated with parasite load but with the timing of infection (Sugiyama, 1956; Webster, 1972). Adult grasshoppers that became infected as nymphs by mermithids are likely to develop deformed wings. The softened cuticle may facilitate egress of the larva from its host.

Developing mermithids in their insect hosts grow very rapidly in a relatively short time; e.g., preparasitic larvae of *M. nigrescens* measuring 370 µm develop into postparasitic larvae measuring 10 cm within three weeks in *S. gregaria*. The developing female mermithid takes up sufficient food to enable it to survive without feeding for one to three years and for the production and development of many thousands of eggs. It is reasonable to suppose, therefore, that large quantities of amino acids, lipids, and carbohydrates are taken up by the parasitic larvae.

Amino acids are taken up in large quantities by developing *M. nigrescens* larvae and incorporated into protein at varying rates during mermithid development. Dipeptides and polypeptides are not taken up by the developing larvae. Under experimental conditions protein was synthesized most rapidly from [^{14}C] leucine by 17-day-old larvae. This timing of maximum synthesis coincides with the rapid increase in total dry weight and in protein level that occurs mostly between 17 and 21 days after infection (Gordon and Webster, 1972) as the developing nematode accumulates stored proteins and lipids in the trophosome (Chitwood and Jacobs, 1938; Gordon and Webster, 1972) prior to emergence from the host. Parasitic *M. nigrescens* larvae preferentially take up glucose from the hemolymph of *S. gregaria* rather than trehalose, the host's main blood sugar. They do this faster in the midphase (i.e., about 14 days after egg ingestion) than in the closing phase (i.e., after 21 days) of their parasitic development. Glucose absorption through the cuticle is probably mediated by an active transport system in the trophosome, which sets up a diffusion gradient across the pseudocoel, hypodermis, and cuticle (Rutherford and Webster, 1974). It was shown experimentally (Rutherford et al., 1977) using this host-parasite model that glucose transport is saturable in 14-day-old larvae, and not in 21-day-old larvae. It may be significant, therefore, that the cuticle is about 11 times thicker in 21-day-old larvae than in 14-day-old larvae, that glucose uptake is about 10 times faster at about day 14 of parasitism than at day 21. Amino acid transport from the *S. gregaria* host to the *M. nigrescens* parasite is by a stereospecific transport system probably located in the nematode cuticle (Rutherford et al., 1977).

M. nigrescens significantly depletes a *S. gregaria* host of carbohydrates, proteins, and amino acids after one week of parasitism, and this depletion is not compensated for by increased food consumption by the host (Gordon and Webster, 1971). Despite the large glucose and amino acid requirements of the developing *M. nigrescens*, the glucose levels in the hemolymph of parasitized locusts is not significantly decreased (Rutherford and Webster, 1978), and the levels of various amino acids in parasitized hosts vary with the stage of parasitism. The fatty acid composition of the host hemolymph is not significantly changed by mermithid parasitism, but the levels of cholesterol and cholestanol appear to be increased (Rutherford and Webster, 1976). However, the lipid content of fat body and flight muscle of *L. migratoria* is diminished by *M. nigrescens* infections (Jutsum and Goldsworthy, 1974).

Trehalose in the host hemolymph is maintained at a constant level for the first seven days of parasitism then drops to about a third of its initial level, but rises again during the closing phase of parasite development. As mermithids utilize glucose rather than trehalose the host's metabolism is modified to produce more glucose. The decrease in host hemolymph trehalose reflects indirectly the host's inability, during the parasite's most rapid growth phase, to compensate for the large uptake of glucose by the parasite without affecting the trehalose level. At the same time (two weeks after infection with *M. nigrescens*), there is a significant reduction in the level of glycogen and non-glycogen carbohydrate in the fat body of *S. gregaria* together with a progressive depletion of active and inactive glycogen phosphorylases (Gordon et al., 1971). The rapid and persistent uptake of glucose modifies the fat body-hemolymph balance of carbohydrates. This causes a reduction in glycogenesis in the host fat body that thus ensures the availability of more glucose in the hemolymph for the parasite (Rutherford and Webster, 1978). The depletion of the glycogen phosphorylases in the fat body prevents further glycogenolysis there and helps maintain a constant low level of glycogen in the host fat body.

Although Rubtsov (1967) and others have suggested that parasitic mermithids may digest the protein reserves of the host fat body by secreting hydrolytic enzymes, it is more likely that mermithids indirectly utilize host fat body proteins by inducing changes in the host's metabolism. The level of hemolymph proteins and amino acids remains relatively constant in adult *S. gregaria* parasitized by *M. nigrescens*, but a significant decrease in fat body proteins and amino acids occurs (Gordon and Webster, 1971). Vitellogenic and nonvitellogenic proteins in the fat bodies of parasitized locust are depleted within two weeks of mermithid parasitism (Gordon et al., 1973). The breakdown of fat body proteins and the subsequent release of amino acids into the hemolymph provides the rapidly developing mermithid larvae with a good nitrogen source, while, at the same time, maintaining a relatively constant level of amino acids in the blood. This level of hemolymph amino acids is diminished only during periods of very rapid mermithid growth in adult hosts, and in nymphal locust hosts where the mermithid's protein requirements cannot be satisfied by resorption of vitellogenic proteins from the ovaries.

It appears, therefore, that the mermithids stimulate catabolism and/or inhibit anabolism of host fat body proteins, which thus provides the necessary dietary amino nitrogen via the host hemolymph. Cessation of vitellogenesis in the host can be attributed to an inability of the oocytes to sequester available proteins from the hemolymph. Subsequent oocyte resorption probably results from a depletion of vitellogenic proteins in the hemolymph due to the nutrient requirements of the developing mermithids. This results in sterilization of the female host.

The older the instar of *S. gregaria* at infection, the greater the parasitic burden of *M. nigrescens* required to inhibit the next-but-one molt. However, this inhibition of molting is due not to a change in the ecdysone level but to the decreased availability of fat body proteins (Craig and Webster, 1974), an abundant supply of which in the host's epidermal cells is an essential prerequisite to molting (Wigglesworth, 1970). Since *M. nigrescens* decreases the catabolism or increases the anabolism of fat body proteins, there are sufficient proteins in the hemolymph for only one molt, and further molts are precluded by the developing *M. nigrescens* taking up large quantities of amino acids (Craig and Webster, 1974).

Mermithid infections diminish the efficiency of the excretory system (Gordon and Webster, 1971). Condon and Gordon (1977) showed that a heavy (80 ova per host) *M. nigrescens* infection of *L. migratoria* second- and third-instar nymphs increased the uric acid level of the hemolymph five times over that of the controls. Furthermore, the concentration of fecal uric acid was reduced to one-quarter that of uninfected controls. The results of similar experiments by Jutsum and Goldsworthy (1974) do not show this effect of mermithids on the locust host, probably because they used a parasite burden of only three eggs per host. The accumulation of toxic waste products in the host's hemocoel or the lack of some nutrients may be the trigger that induces the larval mermithid to leave the host.

S. gregaria (Weis-Fogh, 1956) and *L. migratoria* (Jutsum and Goldsworthy, 1974) parasitized by *M. nigrescens* fly more slowly than nonparasitized locusts. This may be due to a diminished ability to mobilize lipid in the shrunken fat body for the flight muscles (Jutsum and Goldsworthy, 1974) or to the diminished ability of the flight muscles to metabolize trehalose (Rutherford and Webster, 1978).

Many of the locust's physiological processes are influenced by mermithid parasitism, and most of these processes are normally under control of the host's endocrine system. However, there is, as yet, no convincing evidence that mermithid parasites of Orthoptera modify the host's physiological processes by directly altering the endocrine regime.

Glaser and Wilcox (1918) showed a difference in the percentage of mermithid parasitism between male and female grasshoppers, and Denner (1968) implied that this was due to differential ingestion of mermithid eggs by male and female grasshoppers. It is more probable, however, that males and females are equally susceptible to mermithid infection and that the higher percentage of female grasshoppers Glaser and Wilcox (1918) found to be mermithid infected in a field population was due to the longer duration of mermithid parasitism in females, and to the corollary that more male hosts had been removed from the field population by the relatively more rapid mermithid development and subsequent emergence.

The period of time that a host is capable of supporting a mermithid infection, and the size and sex ratio of the mermithids, probably depends on the total available nutrients in the host. *M. nigrescens* larvae grow fastest in adult *S. gregaria* between days 11 and 21 after egg ingestion (Craig, 1973), which is when they take up most nutrients. Adult female locusts have higher hemolymph protein and carbohydrate concentrations than do males, and females are about 50% heavier than males. Under such conditions of "abundant food supply" in female hosts, *M. nigrescens* larvae have a longer period of parasitic development and attain a greater body length than in male or nymphal hosts (Fig. 7).

Species of grasshoppers differ in weight and, therefore, in total available nutrients. This, in turn, affects the sex ratio of a mermithid parasite. From field and laboratory observations (Christie, 1929) it is apparent that a balanced sex ratio in *M. nigrescens* and *A. decaudata* is readily modified by such factors as host size, sex, and age, and especially by the number of parasites per host. All these factors also affect quantitatively and qualitatively the nutrients available in a host for a developing mermithid. Christie (1929) showed experimentally that heavy infections of *M. nigrescens* or of *A. decaudata*

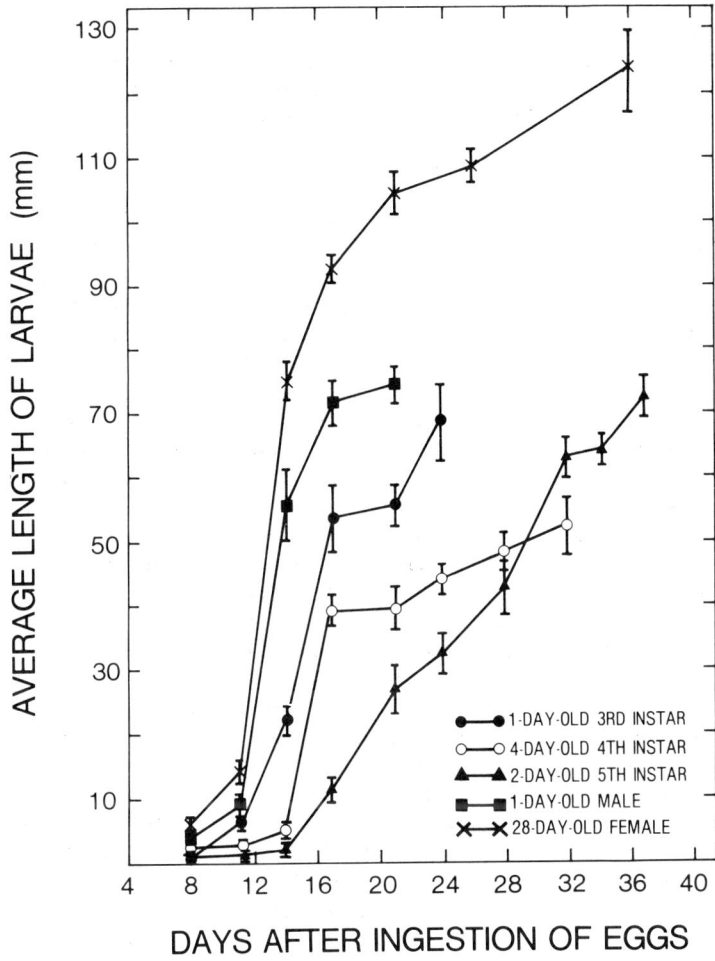

Fig. 7 Graph showing the rate of growth of *M. nigrescens* larvae parasitic in *S. gregaria* of different age and sex. (Courtesy of S. M. Craig.)

caused a marked swing in the sex ratio to maleness in heavily infected *Chortophaga viridifasciata* and *Melanoplus femurrubrum*, with *A. decaudata* doing so more readily than *M. nigrescens*.

Where there is an abundance of food, the parasitic development of *M. nigrescens* is longer, and the resulting parasites, which are nearly all females, are larger and presumably lay more eggs. With a progressive increase in food stress mermithid larvae could be expected to develop into smaller females, to have a shorter duration of parasitism, or to develop into males. Under laboratory conditions large numbers (i.e., more than 50) of mermithid larvae are required per host in order to significantly shift the sex ratio (Craig, 1973). It is probably only rarely that an adult locust would pick up doses of over 50 eggs in the field. However, a shift in the sex ratio of *M. nigrescens* in a field population could occur if, instead of locust adults, young nymphs became heavily infected or if the food available to the host was severely limited due to environmental factors.

The implications of nutrient stress affecting sex ratio, rate of development, adult size, and fecundity of mermithids are considerable in view of the possible in vitro culture of these organisms for subsequent application as biological control agents. A detailed knowledge of the affect of diet on the culturing of mermithids is a prerequisite to successful, economic in vitro culture of each species.

B. Oxyuroidea

Todd (1944) maintains that the second molt of *H. diesingi* and *L. appendiculata* occurs under conditions of partial anaerobiosis after the egg has entered the host midgut. Furthermore, this molt can be stimulated in vitro by solutions containing peptone and tryptone. Eggs of *H. diesingi* differ from those of *L. appendiculata* with respect to the kinds of amino acids that stimulate the second molt in the egg. *H. diesingi* eggs are stimulated by dl-methionine, l-histidine HCl, l-proline, and dl-valine, whereas those of *L. appendiculata* respond to dl-valine, l-proline, and l-arginine HCl. Other amino acids only stimulated molting of larval *L. appendiculata* in the eggs in binary combinations with each other. Although Todd's experiments were conducted in vitro, he found that bacteria from the cockroach hindgut had to be added to the in vitro cultures in order for any of the amino acids to induce the second molt. No molting occurred in the absence of the bacteria. He postulated that the ammonia produced by the bacteria in the host gut determines, at least in part, the completion of the second molt and the hatching of the eggs of these two nematodes.

Host diet appears to significantly affect the development of these nematode commensals despite conflicting reports. Starved cockroaches characteristically have a lower nematode burden (Peregrine, 1974a), and egg production by the nematodes is reported to be diminished (Nadakal and Nayar, 1968) or to increase (Peregrine, 1974a). Hominick and Davey (1972b) found that a reduction in nutrition of *P. americana* did not affect the total parasite number of adult *H. diesingi* and *L. appendiculata*. However, Peregrine (1974a, b) found that significantly greater numbers of *T. attenuatum* and *H. diesingi* occurred in hosts fed normal diets in comparison to those fed artificial diets. The artificial diets, although apparently adequate for normal growth of the host, *P. americana*, were disadvantageous to the parasites. The greatest hatch of infective eggs of *H. diesingi* and *T. attenuatum* was achieved in high-carbohydrate host diets and least in high-protein, vitamin-deficient, high-roughage, and starvation diets. The establishment of larvae of these two nematode species in the gut of *P. americana* is decreased by the same diets that diminish egg hatch, which suggests that the two processes may be related.

The position of *H. diesingi* and *T. attenuatum* in the hindgut of *P. americana* changes slightly with host diet (Peregrine, 1974a, b). Nematode larvae were found more anterior than normal in the hindgut of cockroaches fed a high-protein diet and more posterior than normal in hosts fed a high-roughage diet. There were also differences in the optimum nutritional conditions for each of the developmental stages of *H. diesingi*. There were greater numbers of females than males in hosts fed balanced or carbohydrate diets and greater numbers of males than females in hosts fed high-protein diets. L_4 larvae were more abundant in hosts fed balanced or high-fat diets. Peregrine suggests that these differences reduce competition for food in the hindgut. This conclusion is questionable because it would necessitate each developmental stage selectively ingesting food molecules, which is highly unlikely.

Host diet also affects the relative proportions of *H. diesingi* and *L. appendiculata* in those cockroach hindguts containing mixed populations of these nematode species. Hominick and Davey (1972a, 1975) observed that with normal dietary intake by the host, *P. americana*, the population of *H. diesingi* was lower than that of *L. appendiculata*, whereas at low levels of food intake by the host the reverse was true. Since *H. diesingi* feeds on fine food particles and bacteria, and *L. appendiculata* feeds on larger food particles, the observed population changes with host diet were explained by the fact that cockroaches fed normal diets had much larger particles of food material in their hindguts than those of low food intake. However, it does not explain why *L. appendiculata* does not increase its intake of smaller food particles during the period of low food intake by the host. These observations on *L. appendiculata* are contrary to those of Fay (1961), who observed that this species did not ingest particulate materials and postulated that nutrients were absorbed through the cuticle. This hypothesis has not been substantiated for *L. appendiculata*. A probable explanation for the effects of host diet on nematode numbers is that it alters the composition of the bacterial population on which these nematodes feed in the host hindgut (Peregrine, 1974a).

Differences also exist between *H. diesingi* and *L. appendiculata* in the relative population size of these nematodes in nymphal and in adult cockroaches. The number of *H. diesingi* increased progressively from the early seventh-instar to a maximum in the ninth-instar host and then decreased in adults of both sexes (Hominick and Davey, 1972a). *L. appendiculata* does not occur in nymphs but occurs in large numbers in adults of both sexes. The absence of a nymphal infection by *L. appendiculata* is apparently due to the inability of the nematode to withstand the molting processes of the host. *H. diesingi* adults and larvae apparently are able to withstand the osmotic changes that accompany the molting process (Lee, 1960). Differences in the relative population sizes of these species in nymphal and in adult cockroaches may reflect differences in host hormonal regimes.

Several studies have attempted to elucidate the relationship between the host cockroach hormonal system and the nematode parasite population. Removal of the retrocerebral complex or the median neurosecretory cells of the host does not affect the total food intake of *B. orientalis* but significantly decreases the number of adult *H. diesingi* in the gut. This suggests that the host median neurosecretory cells-corpora cardiaca complex is in some way essential for successful nematode population development (Gordon, 1968, 1970). In contrast, Hominick and Davey (1972b) found that adult female *P. americana*, with their corpora allata and corpora cardiaca removed, decreased their food intake, but the population size of adult *H. diesingi* was unchanged. They attributed the differences between the two studies to differences in hosts and to the differential effects of allatectomy and cardiacectomy on food intake between the hosts.

There have been some reports of detrimental effects of oxyuroid infections on orthopterans. Taylor (1968) reported that *L. appendiculata* caused lesions in the hindgut of *Leucophaea maderae*, which apparently were initiated as tissue damage followed by a hemocytic response and melanization on the part of the host. However, the nematodes were not encapsulated and the hosts were otherwise quite healthy. Total lipids are reduced in nymphal and in adult *P. americana* by *L. appendiculata* infections, and male insects are affected more than the females (Majumdar, 1970). There is some evidence that *L. appendiculata* uses lipids as an endogenous source of energy and as a storage product in the female reproductive organs.

C. Other Nematodes

Certain pathological conditions arise when orthopterans act as intermediate hosts of nematodes, but histopathological damage is rarely significant except during heavy infestations when the epithelium of the midgut and proctodeum (hindgut and/or rectum) may be damaged by larval penetration and migration. Reactions of cockroaches to some nematodes that use them as intermediate hosts has been reviewed by Poinar (1969).

Larvae of *P. maxillaris* develop within host-produced capsules in the insect hindgut and have been observed to ingest the capsule contents. These capsules are thin in cockroaches, and frequently hemocytes invade the capsules and melanization destroys the nematode larvae. In contrast, such capsules in crickets are thick walled and so there is little invasion by hemocytes or melanization, and less destruction of the nematode larvae. In cockroaches and crickets the larvae usually develop in the hindgut, but occasionally they penetrate into the hemocoel, which results in a hemocytic encapsulation and melanization response and the consequent death of the larvae (Cawthorn and Anderson, 1977). Cawthorn and Anderson (1976) found that crickets previously infected with *P. maxillaris* exhibited no immunity to further infections, and accumulations of large numbers of larvae by continual reinfection was common.

Larvae of the spirurid nematode *A. caucasica* enter the colon wall of *B. germanica* and develop into L_3 larvae before being infective to skunk, their definitive host (Poinar and Quentin, 1972). The larval invasion of the colon wall causes cells of the colon to break down and coalesce into giant syncytial cells (Poinar and Hess, 1974a). The nematodes move freely within these giant cells, which in turn become surrounded by hemocytes as the cockroach responds to the tissue disruption. Developing nematodes have been observed ingesting cytoplasmic material from the syncytial giant cells, and there is evidence that successful development of the nematode is dependent on the disrupted epithelial cells forming a giant syncytial cell that protects and nourishes the parasite.

Penetration of the locust midgut wall by larval *D. tricuspis* causes little damage other than hypertrophy of the regenerative crypts. Once within the fat body, a fibrous wall develops around the larva, probably by conversion of adipocytes to fibrous tissue. The connective tissue membrane of the lobule of the fat body remains attached to the capsule, and helps to protect the larva from the hemocytes. The developing larva derives nutrients from the adipocytes and is protected from melanization by the capsule (Cawthorn, 1980). Although the intensity of infection of *M. sanguinipes* was greater in experimental infections than that of *S. gregaria*, the average body length of the parasitic larvae from the two intermediate hosts was not significantly different (Cawthorn and Anderson, 1980).

The ascaridid and spirurid nematodes have sites of encystment in the intermediate host, and a cyst structure that is characteristic of particular nematode families (Fig. 5) (Seureau, 1973). Thus, larvae of the Subuluridae induce a "granuloma" type of hemocytic reaction, larvae of the Rictulariidae, Spiruridae, and Diplotriaenidae induce cellular hypertrophy of the fat body or hindgut tissue with proliferation of fibrillar material, and larvae of the Acuariidae induce a localized lysis of the muscle fibers. The phylogenetically more primitive forms (Subuluridae and Seuratidae) induce a host response in the hemocoel that is hemocytic and extracellular, which is similar to the normal defense response of the host insects. Larvae of the more advanced forms (Rictulariidae, Physalopteridae, Spiruridae, Diplotriaenidae, Acuariidae,

Gongylonematidae, and Tetrameridae) enter a host cell in the gut wall, fat body, or muscle tissue, where they generally induce the formation of a syncytium without the participation of hemocytes (Seureau and Quentin, 1981). This intracellular encystment has been progressively acquired, so as to isolate the larva from the host's defense mechanism. A summary table of possible phylogenetic relationships of heteroxenous nematodes is given in Table 1.

Cellular hypertrophy in the epithelium of the hindgut is particularly extensive in the experimental intermediate host *L. migratoria* in response to the spirurid *A. caucasica*. The resulting thick-walled capsule surrounding the larva consists entirely of epithelial cells of the gut, and they are modified to form a syncytium containing a large number of microtubules (Seureau, 1977). There are fewer microtubules found in the syncytia of the natural intermediate host of this parasite, *B. germanica*, than in those of the experimental host. *Pterygodermatites affinis* also induces a weaker tissue response in its natural host, the beetle *Tachyderma hispida*, than in the locust experimental intermediate host (Quentin et al., 1976).

Gongylonema brevispiculum parasitizes the gut of the elephant shrew and utilizes several insect species as intermediate hosts. After hatching in the gut of the experimental host, *L. migratoria*, the larvae migrate through the midgut wall to the hemocoel and are encysted in the muscle tissue in a similar fashion to the Acuariidae. There is no accumulation of melanin or hemocytes associated with the intramuscular cyst, and the larva is in direct contact with the sarcoplasm (Quentin and Seureau, 1978).

Few records exist of nematode parasites of Orthoptera, other than those of ascaridids, spirurids, oxyurids, and mermithids, though van Waerebeke (1969) records a *Parasitylenchus* species in the hemocoel of longhorn grasshoppers, *Mastododera nodicollis*, in Madagascar. These nematodes caused atrophy of the gonads and fat body of the host resulting in sterility and diminished survival. A recent report from Pakistan (Khan, 1979) suggests that an increased mortality of *S. gregaria* eggs was probably caused by the saprophytic nematode, *Paroigolaimella coprophages*, associated with the egg pods.

IV. BIOLOGICAL CONTROL POTENTIAL OF MERMITHIDS

Mermithids are common parasites of grasshoppers, and consequently it has been claimed that they are significant factors in controlling wild grasshopper populations. However, the incidence of mermithid parsitism is unpredictable and varies from place to place and from year to year. Glaser and Wilcox (1918) recorded high mortality among populations of two species, *Melanoplus atlanis* and *M. bivittatus*, in Vermont, where mermithid infection ranged from 2 to 75% in samples of up to 100 insects. Similar variations in the incidence of parasitism have been recorded from grasshopper populations in eastern Canada, where the incidence of parasitism by *M. nigrescens* was greater than that by all other parasites (Briand and Rivard, 1964). *M. nigrescens* and *A. decaudata* are widely distributed in southern and western Canada as parasites of grasshoppers (Smith, 1958). In a survey of western Canada, the incidence of mermithid parasitism in the samples ranged widely to a maximum of 36% in *Melanoplus mexicanus*. Although *M. nigrescens* infections of grasshoppers are very common in the United States, infections by *A. decaudata* and several instances of multiple infections by both *A. decaudata* and *M. nigrescens* have been recorded. In a survey done by Hayes and DeCoursey (1938) in the summer of 1937 in Illinois and Indiana, mermithid (*M. nigrescens* and *A. de-*

caudata) infections occurred in 13.9% of the adults and 22.6% of the nymphs, mostly of *Melanoplus differentialis* and of *M. femurrubrum*. Mermithids have been found in the overwintering nymphs of *C. viridifasciata* in late fall and early spring in Missouri (Blickenstaff and Sharifullah, 1962). *M. nigrescens* is probably the more important of the two mermithid parasites in the United States in that it is able to withstand a wider range of soil and climatic conditions and is able to maintain itself even when grasshopper populations are consistently low (Christie, 1937). The infective mobile larval phase of *A. decaudata* is less able to withstand dry environmental conditions than is the infective embryonated egg of *M. nigrescens*.

There is substantial evidence for mermithids causing mortality among field populations of grasshoppers and locusts, but there are very few data on the effect of mermithids on overall population size and development. The most comprehensive study of a long-term relationship of a grasshopper population and its parasites is that of Smith (1965). He provides data on a complex of about 10 parasites on a field population of *M. sanguinipes* in Ontario over 10 years. The average percentage parasitism of grasshoppers by mermithids was greater than for any other parasite. However, he found no evidence of a spatial relationship between the abundance of grasshoppers and numbers of mermithids (probably *M. nigrescens*), but there was a relationship from year to year between grasshopper abundance and numbers parasitized. The peak of the host population coincided with the maximum incidence of mermithid parasitism. This may be expected, as the chances of mermithid infection are greater when the grasshoppers are more abundant. In only 2 of the 10 years of the survey was the population of mermithids significantly higher than the 10 year average, and they coincided with the population peaks of the host. After determining the seasonal index of parasitism and the aggregate percentage mortality, Smith concluded that there was no evidence that any of the parasites, including mermithids, affected the trend in grasshopper abundance. Nevertheless, he recognized weather as a factor affecting the development and survival of grasshoppers, especially the seasonal distribution of rain. There was a positive correlation between the number of grasshoppers parasitized with mermithids and the precipitation in July (the period of egg laying by *M. nigrescens*) and a negative correlation between the hours of sunshine in July and the number of grasshoppers containing mermithids. However, despite this detailed survey, the extent to which mermithid parasites diminish grasshopper populations remains unknown.

Mongkolkiti and Hosford (1971) reported the decline and disappearance of a natural population of *Hesperotettix viridis pratensis* in North Dakota and attributed it to a severe infestation by *M. nigrescens*. This claim was not supported by as complete a set of data as that of Smith (1965) and, hence, is not as convincing. Nevertheless, the data show that the highest incidence of mermithid parasitism in *H. viridis pratensis* occurred during the early spring rains, and there is strong support for the theory that moist habitats increase the number of infected insects.

A perceptive report by Baker (1979) gives a preliminary analysis of the joint influence of weather and mermithid parasitism on populations of the wingless grasshopper, *Phaulacridium vittatum*. Populations of this grasshopper decline with average or above-average rainfall because of the proliferation of unfavorable food plants. Populations increase greatly in spring periods of below-average rainfall. He maintains that the gravid mermithids remain quiescent in the soil during dry conditions, and this probably explains the low mortality of the grasshopper population during dry springs. The drought-breaking rains initiate nymphal grasshopper emergence and mermithid oviposition on relatively sparse foliage. The consequent high concentration of mermithid eggs per unit area of vegetation results in a high incidence of parasitism (more than 80%) and major host population crashes (more than 70% over two weeks). He suggests that the high mermithid parasitism late in the season may not significantly diminish the grasshopper population, as most breeding has occurred prior to parasitism. Average rainfall facilitates a protracted mermithid oviposition on lush vegetation and, hence, a low concentration of eggs per unit area of vegetation. This results in a modest level of mermithid parasitism. Several successive seasons of average rainfall result in a low number of hosts and parasites. Few hosts and poor pasture militate against the ingestion of mermithid eggs by the small acridids with a low capacity for food intake. This results in changes in the incidence and level of multiple parasitism within different host species, associated with a different vegetation pattern following prolonged periods of average rainfall. Baker states that local variation in the density of *P. vittatum* reflects differences in the efficiency of mermithid parasitism. Hence, high stocking rates denude pastures and result in an effect analogous to drought in the population dynamics of the wingless grasshopper and of the mermithids.

As mermithid parasitism is always fatal to the host, the incidence of parasitism among field populations of insects provides an assessment of the induced mortality, and some estimate of the contribution of such parasites to the natural regulation of insect populations. Despite our limited knowledge of certain aspects of the biology of mermithids their use as a biological control agent of some insect pests has considerable potential except in areas of persistent low moisture (Webster, 1972).

The monsoon areas of Southwest Asia appear to be appropriate locations for mermithid control of grasshoppers, as are also the irrigated areas of Europe, North America, and Australia. Irshad et al. (1977) in Pakistan showed that the incidence of mermithid parasitism in grasshoppers (*Oxya multidentata, Shirakiacris shirakii*, and *Hieroglyphus banian*) was low in paddy fields but, nevertheless, was at its highest during or soon after the monsoon season in the cooler regions. Nickle (1977) maintains that a wide range of grasshopper species throughout the world are parasitized by *M. nigrescens*, especially on agricultural land with over 100 cm per annum of rainfall. He also cites reports from the Soviet Union, where eggs milked from adult *M. nigrescens* were sprayed on cabbage plants as a grasshopper control measure.

The influence of climatic factors on the distribution of *M. nigrescens* can be explained by the fact that the females only come to the surface to lay eggs during or after substantial rainfall or in the presence of significant moisture. It is likely that females prevented from laying their eggs in one year by the absence of rain may be successful in the subsequent year, thereby helping to ensure the survival of the species. If mermithids are to be used successfully to control grasshoppers and locusts, special consideration must be given to environmental tolerance of the mermithid species and to the biological competitiveness and mass culture of mermithids (Webster, 1980).

ACKNOWLEDGMENTS

We much appreciate the help of several colleagues and especially of Dr. G. Baker of the New South Wales Department of Agriculture and Dr. D. K. McE. Kevan of the Lyman Entomological Museum, McGill University. An operating grant from the Natural Sciences and Engineering Research Council of Canada was held by J.M.W.

REFERENCES

Ackert, J. E., and Wadley, F. M. (1921). Observations on the distribution and life history of *Cephalobium microbivorum* Cobb and of its host *Gryllus assimilis* Fabricius. *Trans. Am. Microsc. Soc. 40*: 97-115.

Alicata, J. E. (1935). Early developmental stages of nematodes occurring in swine. *Tech. Bull. No. 489. USDA*: 1-96.

Alicata, J. E. (1937). Larval development of the spirurid nematode *Physaloptera turgida* in the cockroach *Blatella germanica* L. *Papers on Helminths, 30 year Jubileum*, K. I. Skrjabin, pp. 11-14.

Bain, O., and Vaucher, C. (1973). Développement larvaire de *Diplotriaena tridens* (Nematoda: Filarioidea) chez *Locusta migratoria*. *Ann. Parasitol. Hum. Comp. 48*: 81-89

Baker, G. L. (1979). The adverse effects of below-average rainfall on parasites of the post-embryonic stages of the wingless grasshopper. *Australian Applied Entomological Research Conference (Working Paper), Queensland*: No. 6, 3 pp.

Baruš, V. (1970a). Studies of the nematode *Subulura suctoria*. II. Development in the intermediate host. *Folia Parasitol. Praha 17*: 49-59.

Baruš, V. (1970b). Studies on the nematode *Subulura suctoria*. IV. Intermediate hosts. *Folia Parasitol. Praha. 17*: 191-199.

Basir, M. A. (1948). *Chitwoodiella ovofilamenta* gen. et sp. nov., a nematode parasite of *Gryllotalpa*. *Can. J. Res. 26D*: 4-7.

Basir, M. A.(1956). Oxyuroid parasites of Arthropoda. A monographic study. 1. Thelastomatidae. 2. Oxyuridae. *Zoologica, Stuttgart 38*: No. 106, 79 pp.

Baylis, H. A. (1944). Observations on the nematode *Mermis nigrescens* and related species. *Parasitology 36*: 122-132.

Baylis, H. A. (1947). The larval stages of the nematode *Mermis nigrescens*. *Parasitology 38*: 10-16.

Bhatnagar, K. N., and Edwards, L. J. (1970). Parasites of the Madagascar cockroach *Gromphadorhina portentosa*. *Ann. Ent. Soc. Am. 63*: 620-621.

Blickenstaff, C. C., and Sharifullah, M. (1962). Infestation of overwintering nymphs of *Chortophaga viridifasciata* by mermithids. *J. Econ. Ent. 55*: 268.

Bozeman, Jr., W. B. (1942). An experimental investigation into the life history of *Blatticola blattae*, a nematode found in *Blatella germanica Trans. Kans. Acad. Sci. 45*: 304-310.

Briand, L. J., and Rivard, I. (1964). Observations sur *Mermis subnigrescens* Cobb (*Mermithidae*), nématode parasite des criquets au Québec. *Phytoprotection 45*: 73-76.

Brumpt, E. (1931). Nemathelminthes parasites des rats sauvages (*Epimys norvegicus*) de Caracas. I. *Protospirura bonnei*. Infections expériméntales et spontanees. Formes adultes et larvaires. *Ann. Parasitol. Hum. Comp. 9*: 344-358.

Burr, A. H. (1979). Analysis of phototaxis in nematodes using directional statistics. *J. Comp. Physiol. 134*: 85-93.

Burr, A. H., Schiefke, R., and Bollerup, G. (1975). Properties of a hemoglobin from the chromatrope of the nematode *Mermis nigrescens*. *Biochim. Biophys. Acta 405*: 404-411.

Cali, C. T. (1964). Studies on the life cycle of the nematode *Blatticola blattae* (Graeffe, 1860) Schwenk, 1926 with suggested laboratory exercises. *Diss. Abstr. 25*: 3755.

Cali, C. T., and Mai, W. F. (1965). Studies on the development of *Blatticola blattae* (Graeffe, 1860) Chitwood, 1932 within its host, *Blatella germanica* L. *Proc. Helm. Soc. Wash.* 32: 164-169.

Cawthorn, R. J. (1980). The cellular responses of migratory grasshoppers (*Melanoplus sanguinipes* F.) and African desert locusts (*Schistocerca gregaria* L.) to *Diplotriaena tricuspis* (Nematoda: Diplotriaenoidea). *Can. J. Zool.* 58: 109-113.

Cawthorn, R. J., and Anderson, R. C. (1976). Effects of age, temperature and previous infection on the development of *Physaloptera maxillaris* (Nematoda: Physalopteroidea) in field crickets (*Acheta pennsylvanicus*). *Can. J. Zool.* 54: 442-448.

Cawthorn, R. J., and Anderson, R. C. (1977). Cellular reactions of field crickets (*Acheta pennsylvanicus* Burmeister) and German cockroaches (*Blatella germanica* L.) to *Physaloptera maxillaris* Molin (Nematoda: Physalopteroidea). *Can. J. Zool.* 55: 368-375.

Cawthorn, R. J., and Anderson, R. C. (1980). Development of *Diplotriaena tricuspis* (Nematoda: Diplotriaenoidea), a parasite of Corvidae, in intermediate and definitive hosts. *Can. J. Zool.* 58: 94-108.

Chabaud, A. G., Golvan, Y., Bain, O., and Brygoo, E. R. (1965). *Gynopoecilia pseudovipara* n. gen., et cycles endoxènes chez les nématodes zooparasistes. *C. R. Hebd. Seanc. Acad. Sci.* 260: 4602-4604.

Chittenden, F. H. (1908). The cabbage hair-worm. *Circular No. 62. USDA Bureau of Entomology*: 1-6.

Chitwood, B. G., and Jacobs, L. (1938). Stored nutritive materials in the trophosome of the nematode, *Agamermis decaudata* (Mermithidae). *J. Wash. Acad. Sci.* 28: 12-13.

Christie, J. R. (1929). Some observations on sex in the Mermithidae. *J. Exp. Zool.* 53: 59-76.

Christie, J. R. (1936). Life history of *Agamermis decaudata*, a nematode parasite of grasshoppers and other insects. *J. Agric. Res.* 52: 161-198.

Christie, J. R. (1937). *Mermis subnigrescens*, a nematode parasite of grasshoppers. *J. Agric. Res.* 55: 353-364.

Cobb, N. A. (1926). The species of *Mermis*. *J. Parasitol.* 13: 66-72.

Cobb, N. A. (1929). The chromatropism of *Mermis subnigrescens*, a nemic parasite of grasshoppers. *J. Wash. Acad. Sci.* 19: 159-166.

Condon, W. J., and Gordon, R. (1977). Effects of the mermithid nematode *Mermis nigrescens* on the levels of hemolymph and fecal uric acid in its host, the migratory locust *Locusta migratoria*. *Can. J. Zool.* 55: 690-692.

Corbel, J.-C. (1967). Les parasites des Orthoptères. *Ann. Biol.* 6: 391-426.

Couper, W. (1855). Vermes in grasshopppers. *Can. J.* 3: 355.

Craig, S. M. (1973). The host-parasite relationship between *Mermis nigrescens* Dujardin and the desert locust, *Schistocerca gregaria* Forskål. M.Sc. Thesis, Simon Fraser University, Burnaby, British Columbia.

Craig, S. M., and Webster, J. M. (1974). Inhibition of molting of the desert locust, *Schistocerca gregaria*, by the nematode parasite *Mermis nigrescens*. *Can. J. Zool.* 52: 1535-1539.

Craig, S. M., and Webster, J. M. (1978). Viability and hatching of *Mermis nigrescens* eggs and subsequent larval penetration of the desert locust *Schistocerca gregaria*. *Nematologica* 24: 472-474.

Cram, E. B. (1931). Developmental stages of some nematodes of the Spiruroidea parasitic in poultry and game birds. *Tech. Bull. No. 227. USDA*: 1-27.

Croll, N. A. (1966). A contribution to the light sensitivity of the "chroma-trope" of *Mermis subnigrescens*. *J. Helm. 40*: 33-38.

Denner, M. W. (1968). Biology of the nematode *Mermis subnigrescens* Cobb [sic]. Ph.D. Thesis, Iowa State University, Ames, Iowa.

Denner, M. W. (1976). Preliminary studies on the pathology caused by *Mermis nigrescens* Duj. in Orthoptera. *Proc. Indiana Acad. Sci. 85*: 258-261.

Dobrovolny, C. G., and Ackert, J. E. (1934). The life history of *Leidynema appendiculata* (Leidy), a nematode of cockroaches. *Parasitology 26*: 468-480.

Dujardin, F. (1842). Mémoire sur la structure anatomique des *Gordius* et d'un autre helminthe, le *Mermis*, qu'on a confundu avec eux. *Ann. Sci. Nat. 18*: 129-151.

Ellenby, C., and Smith, L. (1966). Haemoglobin in *Mermis subnigrescens* (Cobb), *Enoplus brevis* (Bastian) and *E. communis* (Bastian). *Comp. Biochem. Physiol. 19*: 871-877.

Farooqui, M. N. (1970). Some known and new genera and species of the family Thelastomatidae Travassos, 1929. *Riv. Parassit. 31*: 195-214.

Fay, C. M. (1961). On the ecology of *Leidynema appendiculata* (Leidy, 1850) (Nematoda: Oxyuroidea). *J. Parasitol. 47*: 518.

Glaser, R. W., and Wilcox, A. M. (1918). On the occurrence of a *Mermis* epidemic amongst grasshoppers. *Psyche 25*: 12-15.

Gordon, R. (1968). Observations on the effect of the neuro-endocrine system of *Blatta orientalis* L. on the midgut protease activity of the adult female and the level of infestation with the nematode *Hammerschmidtiella diesingi* (Hammerschmidt, 1838). *Gen. Comp. Endocr. 11*: 284-291.

Gordon, R. (1970). A neuroendocrine relationship between the nematode *Hammerschmidtiella diesingi* and its insect host, *Blatta orientalis*. *Parassitology 61*: 101-110.

Gordon, R., and Webster, J. M. (1971). *Mermis nigrescens*: Physiological relationship with its host, the adult desert locust *Schistocerca gregaria*. *Exp. Parasitol. 29*: 66-79.

Gordon, R., and Webster, J. M. (1972). Nutritional requirements for protein synthesis during parasitic development of the entomophilic nematode *Mermis nigrescens*. *Parasitology 64*: 161-172.

Gordon, R., Webster, J. M., and Hislop, T. G. (1973). Mermithid parasitism, protein turnover and vitellogenesis in the desert locust, *Schistocerca gregaria* Forskål. *Comp. Biochem. Physiol. 46B*: 575-593.

Gordon, R., Webster, J. M., and Mead, D. E. (1971). Some effects of the nematode *Mermis nigrescens* upon carbohydrate metabolism in the fat body of its host, the desert locust *Schistocerca gregaria*. *Can. J. Zool. 49*: 431-434.

Hammerschmidt, K. E. (1847). Beschreibung einiger Oxyurisarten. *Naturw. Abh. Wien 1*: 379-388.

Hayes, W. P., and DeCoursey, J. D. (1938). Observations of grasshopper parasitism in 1937. *J. Econ. Entomol. 31*: 519-522.

Hobmaier, M. (1941). Extramammalian phase of *Physaloptera maxillaris* Molin 1860 (Nematoda). *J. Parasitol. 27*: 233-235.

Hominick, W. M., and Davey, K. (1972a). The influence of host stage and sex upon the size and composition of the population of two species of thelastomatids parasitic in the hindgut of *Periplaneta americana*. *Can. J. Zool. 50*: 947-954.

Hominick, W. M., and Davey, K. (1972b). Reduced nutrition as the factor controlling the population of pinworms following endocrine gland removal in *Periplaneta americana* L. *Can. J. Zool.* *50*: 1421-1432.

Hominick, W. M., and Davey, K. (1975). The effect of nutritional level of the host on space and food available to pinworms in the colon of *Periplaneta americana* L. *Comp. Biochem. Physiol.* *51A*: 83-88.

Irshad, M., Mazhar, R. A., and Ghani, M. A. (1977). Grasshoppers associated with paddy and their natural enemies in Pakistan. *Agriculture Pakist.* *28*: 55-64.

Jarry, D. T. (1964). Les oxyuroides de quelques arthropodes dans le Midi de la France. *Ann. Parasitol. Hum. Comp.* *39*: 381-508.

Jutsum, A. R., and Goldsworthy, G. J. (1974). Some effects of mermithid infection on metabolic reserves and flight in *Locusta*. *Int. J. Parasitol.* *4*: 625-630.

Kevan, D. K., LeRoux, E. J., and d'Ornellas, C. (1962). Further observations on *Metrioptera (Roeseliana) roeseli* (Hagenbach, 1822) in Quebec, with notes on the genus *Metrioptera* Wesmael, 1938 (Orthoptera: Tettigoniidae: Decticinae). *Ann. Ent. Soc. Queb.* *7*: 70-86.

Khan, A. A. (1979). Mortality of desert locust eggs in the laboratory by a nematode. *Pakis. J. Zool.* *11*: 51-55.

Kloss, G. R. (1959). Nematoides parasitas de Gryllotalpidae (Orthoptera) do Brasil. *Mems Inst. Oswaldo Cruz 57*: 137-170.

Lee, D. L. (1960). The effect of changes in the osmotic pressure upon *Hammerschmidtiella diesingi* (Hammerschmidt, 1838) with reference to the survival of the nematode during moulting of the cockroach. *Parasitology 50*: 241-246.

Lee, D. L. (1961). Studies on the origin of the sticky coat on the eggs of the nematode *Thelastoma bulhoesi* (Magalhaes, 1900). *Parasitology 51*: 379-384.

Lee, D. L. (1970). The ultrastructure of the culticle of adult female *Mermis nigrescens* (Nematoda). *J. Zool. 161*: 513-518.

Leibersperger, E. (1960). Die Oxyuroidea der europäischen Arthropoden. *Parasitol. Schriftenr., Jena, No. 11*.

Leidy, J. (1851). Contribution to helminthology. *Proc. Acad. Natl. Sci. Phila.* *5*: 262-263.

Leong, L., and Paran, T. P. (1966). A study of the nematode parasites of cockroaches in Singapore. *Med. J. Malaya 20*: 349.

Lincoln, R. C., and Anderson, R. C. (1975). Development of *Physaloptera maxillaris* (Nematoda) in the common field cricket (*Gryllus pennsylvanicus*). *Can. J. Zool. 53*: 385-390.

Majumdar, G. (1970). Host-parasite relationships. I. Interference in the lipid metabolism of the cockroach, *Periplaneta americana*, following infection of a thelastomatid nematode, *Leidynema appendiculata*. *Proc. Zool. Soc. Calcutta 23*: 161-168.

Mongkolkiti, S., and Hosford, Jr., R. M. (1971). Biological control of the grasshopper *Hesperotettix viridis pratensis* by the nematode *Mermis nigrescens*. *J. Nematol. 3*: 356-363.

Nadakal, A. M., and Nayar, K. K. (1968). Neural and hormonal influence on the fecundity and egg laying of certain oxyurid nematodes inhabiting the hind gut of the cockroach, *Periplaneta americana* L. *Indian J. Exp. Biol. 6*: 29-32.

Nickle, W. R. (1972). A contribution to our knowledge of the Mermithidae (Nematoda). *J. Nematol. 4*: 113-146.

Nickle, W. R. (1977). Taxonomy of nematodes that parasitize insects, and their use as biological control agents. *Biosystematics in Agriculture, Beltsville Symposium 2*: 37-51.

Oswald, V. H. (1958). Studies on *Rictularia coloradensis* Hall, 1916 (Nematoda: Thelaziidae). I. Larval development in the intermediate host. *Trans. Am. Microsc. Soc. 77*: 229-240.

Peregrine, P. C. (1974a). The effects of host diet on *Thelastoma attenuatum* (Nematoda: Thelastomatidae) populations in cockroaches. *J. Helm. 48*: 47-57.

Peregrine, P. C. (1974b). Host dietary changes and the hindgut fauna of cockroaches. *Int. J. Parasitol. 4*: 645-656.

Petri, L. H. (1950). Life cycle of *Physaloptera rara* Hall and Wigdor, 1918 (Nematoda: Spiruroidea) with the cockroach, *Blatella germanica*, serving as the intermediate host. *Trans. Kans. Acad. Sci. 53*: 331-337.

Petri, L. H., and Ameel, D. J. (1950). Studies on the life cycle of *Physaloptera rara* Hall and Wigdor, 1918, and *Physaloptera praeputialis* Linstow, 1889. *J. Parasitol. 36*: 40.

Poinar, Jr., G. O. (1969). Arthropod immunity to worms. In *Immunity to Parasitic Animals*, vol. 1 (G. Jackson, R. Herman, and I. Singer, eds.). Appleton-Century-Crofts, New York, pp. 173-210.

Poinar, Jr., G. O. (1975). *Entomogenous Nematodes*. E. J. Brill, Leiden.

Poinar, Jr., G. O. (1979). *Nematodes for Biological Control of Insects*. CRC Press, Boca Raton, Florida.

Poinar, Jr., G. O., and Hess, R. (1974a). An ultrastructural study of the response of *Blatella germanica* (Orthoptera: Blattidae) to the nematode *Abbreviata caucasica* (Spirurida: Physalopteridae). *Int. J. Parasitol. 4*: 133-138.

Poinar, Jr., G. O., and Hess, R. (1974b). Structure of the pre-parasitic juveniles of *Filipjevimermis leipsandra* and some other Mermithidae (Nematodea). *Nematologica 20*: 163-173.

Poinar, Jr., G. O., and Quentin, J.-C. (1972). The development of *Abbreviata caucasica* (von Linstow) (Spirurida: Physalopteridae) in an intermediate host. *J. Parasitol. 58*: 23-28.

Quentin, J.-C. (1973). Présence de *Spirura guianensis* (Ortlepp, 1924) chez des Marsupiaux néotropicaux. Cycle évolutif. *Ann. Parasitol. Hum. Comp. 48*: 117-133.

Quentin, J.-C., and Barre, N. (1976). Description et cycle biologique de *Tetrameres (Tetrameres) cardinalis* n. sp. *Ann. Parasitol. Hum. Comp. 51*: 65-81.

Quentin, J.-C., and Poinar, Jr., G. O. (1973). Comparative study of the larval development of some heteroxenous subulurid and spirurid nematodes. *Int. J. Parasitol. 3*: 809-827.

Quentin, J.-C., and Seureau, C. (1974). Cycle biologique de *Pterygodermatites hispanica* Quentin, 1973 (Nematoda: Rictulariidae). *Ann. Parasitol. Hum. Comp. 49*: 701-719.

Quentin, J.-C., and Seureau, C. (1975). Sur l'organogenèse de *Seuratum cadarachense* Desportes, 1947 (Nematoda: Seuratoidea) et les réactions cellulaires de l'Insecte *Locusta migratoria*, hôte intermédiaire. Z. Parasitkde. 47: 55-68.

Quentin, J.-C., and Seureau, C. (1978). Identification et biologie du Gongylonéme parasite du Macroscélide en Tunisie. *Ann. Parasitol. Hum. Comp. 53*: 631-640.

Quentin, J.-C., Seureau, C., and Gabrion, C. (1972). Cycle biologique d'*Acuaria anthuris* (Rudolphi, 1819), Nématode parasite de la Pie. Z. Parasitkde. 39: 103-126.

Quentin, J.-C., Seureau, C., and Vernet, R. (1976). Cycle biologique du Nématode Rictulaire *Pterygodermatites (Multipectines) affinis* (Jagerskiold, 1904). *Ann. Parasitol. Hum. Comp. 51*: 51-64.

Quentin, J.-C., and Verdier, J.-M. (1979). Cycle biologique de *Maupasina weissi* Seurat, 1913 (Nématoda: Subuluroidea), parasite du Macroscélide. Ontogenèse des structures céphaliques. *Ann. Parasitol. Hum. Comp. 54*: 621-635.

Roth, L., M., and Willis, E. R. (1960). The biotic associations of cockroaches. *Smithson. Misc. Collns. 141*: 190-211.

Rubtsov, I. A. (1967). Scheme and organs of the extraintestinal digestion of mermithides. *Izv. AN SSSR, Ser. biol. 6*: 883-891.

Rutherford, T. A., and Webster, J. M. (1974). Transcuticular uptake of glucose by the entomophilic nematode, *Mermis nigrescens*. *J. Parasitol. 60*: 804-808.

Rutherford, T. A., and Webster, J. M. (1976). Effects of the nematode *Mermis nigrescens* on some chemical components of the insect host's hemolymph. *Proc. Int. Colloquium Invert. Pathology, Kingston, Ontario, 1976*, pp. 272-275.

Rutherford, T. A., and Webster, J. M. (1978). Some effects of *Mermis nigrescens* on the hemolymph of *Schistocerca gregaria*. *Can. J. Zool. 56*: 339-347.

Rutherford, T. A., Webster, J. M., and Barlow, J. S. (1977). Physiology of nutrient uptake by the entomophilic nematode *Mermis nigrescens* (Mermithidae). *Can. J. Zool. 55*: 1773-1781.

Schell, S. C. (1952). Studies on the life cycle of *Physaloptera hispida* Schell (Nematoda: Spiruroidea), a parasite of the cotton rat (*Sigmodon hispidus littoralis* Chapman). *J. Parasitol. 38*: 462-472.

Schwabe, C. W. (1951). Studies on *Oxyspirura mansoni*, the tropical eyeworm of poultry. II. Life history. *Pac. Sci. 5*: 18-35.

Seurat, L. G. (1911). Sur l'habitat et les migrations du *Spirura talpae* Gmel. (= *Spiroptera strumosa* Rud.). *C.R. Soc. Biol. Paris 71*: 606-608.

Seureau, C. (1973). Réactions cellulaires provoquées par les Nématodes Subulures et Spirurides chez *Locusta migratoria* (Orthoptère): localisation et structure des capsules. *Z. Parasitkde. 41*: 119-138.

Seureau, C. (1977). A cytopathological accumulation of microtubules in the epithelial cells of the gut of an insect parasitized by a heteroxenic nematode. *J. Invert. Pathol. 29*: 240-241.

Seureau, C., and Quentin, J.-C. (1977). Migrations larvaires des Nématodes Subulures et Spirurides chez *Locusta migratoria* (Insecte, Orthoptere), hôte intermédiaire expérimental. *Ann. Parasitol. Hum. Comp. 52*: 457-470.

Seureau, C., and Quentin, J.-C. (1981). Évolution de l'adaptation des Nématodes hétéroxènes à leur hôte intermédiaire: passage progressif d'un parasitisme extracellulaire à un parasitisme intracellulaire. *C.R. Acad. Sc. Paris 292*: 421-425.

Smith, R. W. (1958). Parasites of nymphal and adult grasshoppers (Orthoptera: Acrididae) in Western Canada. *Can. J. Zool. 36*: 217-262.

Smith, R. W. (1965). A field population of *Melanoplus sanguinipes* (FAB.) (Orthoptera: Acrididae) and its parasites. *Can. J. Zool. 43*: 179-201.

Steiner, G. (1923). Intersexes in nematodes. *J. Hered. 14*: 147-158.

Steiner, G. (1933). Some morphological and physiological characters of the mermithids in their relationship to parasitism (abstract). *J. Parasitol. 19*: 249-250.

Sugiyama, K. (1956). Effects of the parasitism by a nematode on a grass-
    hopper, *Oxya japonica*. I. Effects of the parasitism on wing length,
    pronotal length and genitalia. *Zool. Mag. 65*: 382-385.

Taylor, R. L. (1968). Tissue damage induced by an oxyuroid nematode,
    Leidynema sp., in the hindgut of the Madeira cockroach, *Leucophaea
    maderae*. *J. Invert. Pathol. 11*: 214-218.

Todd, A. C. (1941). An addition to the life history of *Leidynema appendicu-
    latum* (Leidy, 1850) Chitwood, 1932, a nematode parasitic in cockroaches.
    J. Parasitol. 27: 34-35.

Todd, A. C. (1944). On the development and hatching of the eggs of *Ham-
    merschmidtiella diesingi* and *Leidynema appendiculata*, nematodes of
    roaches. *Trans. Am. Microsc. Soc. 63*: 54-67.

Tsai, Y. H., and Cahill, K. M. (1970). Parasites of the German cockroach
    (*Blatella germanica* L.) in New York City. *J. Parasitol. 56*: 375-377.

van Waerebeke, D. (1969). Quelques cas d'association entre nématodes et
    insectes à Madagascar. *Revue Agric. Sucr. Ile Maurice 48*: 274-276.

Webster, J. M. (1972). Nematodes and biological control. In *Economic Nema-
    tology* (J. M. Webster, ed.). Academic, London, pp. 469-496.

Webster, J. M. (1980). Biocontrol: The potential of entomophilic nematodes
    in insect management. *J. Nematol. 12*: 270-278.

Weis-Fogh, T. (1956). Biology and physics of locust flight. II. Flight
    performance of the desert locust (*Schistocerca gregaria*). *Phil. Trans.
    R. Soc. Series B239*: 459-510.

Wigglesworth, V. B. (1970). *Insect Hormones*. Oliver and Boyd, Edinburgh.

Young, P. L., and Babero, B. B. (1975). Studies on the transmission of
    helminth ova by cockroaches. *Proc. Okla. Acad. Sci. 55*: 169-174.

Chapter 20

Nematode Parasites of Bark Beetles

Harry K. Kaya *University of California, Davis, California*

I. INTRODUCTION

Bark beetles are among the most destructive insects in the forest, not only in the United States (Baker, 1972), but also throughout the world (Berryman, 1974). In the western United States, the genera *Dendroctonus* and *Ips* are the most destructive scolytids of coniferous forests (Furniss and Carolin, 1977). These bark beetles can destroy millions of board feet of valuable timber in a relatively short period of time. Trees attacked by bark beetles sustain structural damage by gallery construction and from introduced microorganisms, primarily fungi, which hasten the death of trees and devalue the wood. Blue-stain fungi in the genus *Ceratocystis* are the primary pathogens associated with bark beetles in pines and cause devaluation of wood (Carter, 1973). In addition, bark beetles in the genera *Scolytus* and *Hylurgopinus* are vectors of the Dutch elm disease fungus, *Ceratocystis ulmi*, which is the most destructive pathogen of elms in the world (Carter, 1973).

Factors responsible for the rise and decline of bark beetle populations are not completely understood, but parasitic nematodes are believed to be major biotic factors in population reduction (Massey, 1974). The majority of parasitic nematodes in bark beetles are sphaerulariids, although a few aphelenchs and rhabditids are parasitic. These nematodes, generally, do not kill their insect hosts, but are capable of altering host behavior, reducing fecundity, reducing longevity, reducing flight ability, or delaying emergence.

Besides parasitic nematodes, many species of nematodes are phoretically associated with bark beetles. These phoretic nematodes are usually not harmful to the beetles. This chapter will review the bark beetle parasites and some of their associates.

II. HISTORY OF BARK BEETLE NEMATODES

The early work on bark beetle nematodes was conducted primarily by European scientists. von Linstow (1890) was the first to record and describe the parasitic nematode, *Contortylenchus (Allantonema) diplogaster*, from the bark beetle, *Ips typographus*, in Germany. Although he described the parasitic female from a beetle host, he mistook the free-living female of a diplogasterid in the beetle's gallery as belonging to the parasitic group. Subsequently, Fuchs (1914, 1915) initiated studies on bark beetle nematodes, which included parasitic nematodes and nematode associates of several European bark beetle species. He published further works on bark beetle parasites and associates, including taxonomic, biological, and ecological studies that have been important contributions to this field (Fuchs, 1929, 1930, 1937, 1938). His studies showed that parasitic nematodes killed their hosts, or reduced longevity and fecundity, and that microclimate was important to the nematodes.

Other scientists during Fuch's time demonstrated that parasitic nematodes could be detrimental to bark beetles. Yatsenkovskii (1924), in Russia, reported that nematode-infected bark beetles were sterilized or killed if large numbers of nematodes occurred in the hemocoel of their hosts. Oldham (1930), an English scientist, observed that a tylenchid could sterilize the elm bark beetles, *Scolytus multistriatus* and *S. scolytus* (*S. destructor*). (Later, this observation would be questioned by a number of scientists.) In the United States, Steiner (1932) and Thorne (1935) published on the nematode parasites and associates of the mountain pine beetle, *Dendroctonus ponderosae* (*D. monticolae*). In 1937, Bovien, a Danish scientist, recovered *Bovienema tomici* from several species of *Pityogenes* bark beetles. Furthermore, he speculated that nematodes in the gut of wood-feeding insects were evolving toward parasitism.

Only a few studies on parasitic nematodes of bark beetles were reported between 1940 and 1955. Some of these studies included works by Hetrick (1940), Schvester (1950), Théodoridés (1950), and Hirschmann and Rühm (1953). In 1956, Rühm published a monumental monograph "Die Nematoden der Ipiden," which summarized much of the older literature and added new taxonomic, biological, and ecological information on nematode parasites and associates of bark beetles. Since 1956, a number of workers have published significant taxonomic and biological studies on nematode parasites and associates of bark beetles. Moreover, Massey (1974) summarized the biology and taxonomy of nematode parasites and associates of bark beetles in the United States, and Laumond and Ritter (1971) summarized the phoretic and parasitic relationships existing between nematodes and bark beetles. Host lists of parasitic nematodes and associates of insects including scolytids have been published by Poinar (1975). However, as Massey (1974) stated, "Much remains to be learned on the biology, ecology and life histories of nematode parasites; only the surface has been scratched."

III. CLASSIFICATION OF NEMATODE PARASITES OF BARK BEETLES

The reader should be aware that the classification of parasitic nematodes of bark beetles is confused. Confusion stems from the lack of taxonomic and biological studies of this group and the difficulty in finding good morphological characters for separation into genus and species. Biological studies are difficult because of the obligate nature of the nematodes and problems of rearing some of the bark beetles in the laboratory. Therefore, when parasitic nematodes are found, infectivity tests are difficult to conduct. In recent years, certain techniques have beeen developed to rear bark beetles on artifical diets (Bridges, 1979) and on tissue culture of pine callus (Mott et al., 1978), which may facilitate research with parasitic nematodes.

Nematode parasites of bark beetles belong to the families Sphaerulariidae, Aphelenchoididae, and Rhabditidae (Table 1). The greatest number of species occurs in the family Sphaerulariidae. Nickle (1967) presented a taxonomic revision of the sphaerulariids that includes four subfamilies. The family diagnosis was emended, and generic diagnoses were given along with a listing of species and synonymies. Parasitic bark beetle nematodes are found in the subfamilies Allantonematinae and Sphaerulariinae. The reader is referred to this paper by Nickle (1967), which provides basic information to do further taxonomic studies with this group.

Researchers differ in the classification they use. Rühm (1956) and Massey (1974) recognize Contortylenchidae as a family of parasitic bark beetle nematodes, whereas Nickle (1967) and Poinar (1975) do not give this genus family status and place it in the subfamily Allantonematinae. Poinar (1975), on the other hand, gives family status to the allantonematids, but Nickle (1967) considers the allantonematids as a subfamily within the Sphaerulariidae. Furthermore, Nickle (1967) and Massey (1974) place the genus *Sphaerulariopsis* in the family Sphaerulariidae. However, Massey (1974) retains the genus *Sphaerularia* as a valid generic name, and Poinar (1975) uses the genus *Sphaerulariopsis* as a valid generic name, but places this genus into the allantonematids. In this chapter, *Contortylenchus* is in the subfamily Allantonematinae, and *Sphaerulariopsis* is in the subfamily Sphaerularinae. Both subfamilies are in the family Sphaerulariidae.

At the generic level, Massey (1974) synonymized *Sulphuretylenchus* and *Neoparasitylenchus* into the genus *Parasitylenchus*. However, Nickle (1967),

Table 1 Classification of Insect-Parasitic
Nematodes of Bark Beetles

Family: Sphaerulariidae
Subfamily: Sphaerulariinae
Genus: *Sphaerulariopsis*
Subfamily: Allantonematinae
Genus: *Allantonema*
    Bovienema
    Contortylenchus
    Neoparasitylenchus
    Parasitylenchus
    Sulphuretylenchus

Family: Aphelenchoididae
Subfamily: Aphelenchoidinae
Genus: *Parasitaphelenchus*

Family: Rhabditidae
Subfamily: Protorhabditinae
Genus: *Parasitorhabditis*

Poinar and Caylor (1974), and Poinar (1975) recognize *Sulphuretylenchus*, *Neoparasitylenchus*, and *Parasitylenchus* as valid genera. In this chapter, these three genera are considered to be valid and are listed in Table 1.

The family Aphelenchoididae contains the genus *Parasitaphelenchus* which is parasitic in bark beetles. Other genera in this group may be associated phoretically with bark beetles. Nickle (1970) reviews this group of nematodes, and the reader is referred to this paper for a description of the family and genera.

The family Rhabditidae contains the genus *Parasitorhabditis*, which sometimes can cause pathological effects in its bark beetle host and sometimes can be found in the hemocoel of its host. Most parasitorhabditids appear to be phoretics of bark beetles.

In addition to the families listed in Table 1, nematodes in the families Steinernematidae and Heterorhabditidae have been used as biological control agents against bark beetles on a limited scale. Brief consideration will be given to this group and only to their pathological effects on bark beetles.

IV. BIOLOGY, LIFE CYCLE, MODE OF INFECTION

The life cycles of parasitic nematodes of bark beetles are closely synchronized with those of their hosts (Ashraf and Berryman, 1970a; MacGuidwin, Smart, and Allen, 1980; Massey, 1956, 1962; Nickle, 1963a; Rühm, 1956; Saunders and Norris, 1961; Thong and Webster, 1973). Thus, *Contortylenchus elongatus* develops in one to eight months, depending upon the generation of beetles infected (Massey, 1962); *C. reversus* takes two years to complete its life cycle because its host, *D. rufipennis* (*engelmanni*), has a two year life cycle (Massey, 1956); and *Parasitaphelenchus oldhami* becomes quiescent during the winter when its host, *S. multistriatus*, is inactive (Saunders and Norris, 1961). The following gives a brief description of the biology, life cycle, and mode of infection of the parasitic nematode genera infecting bark beetles.

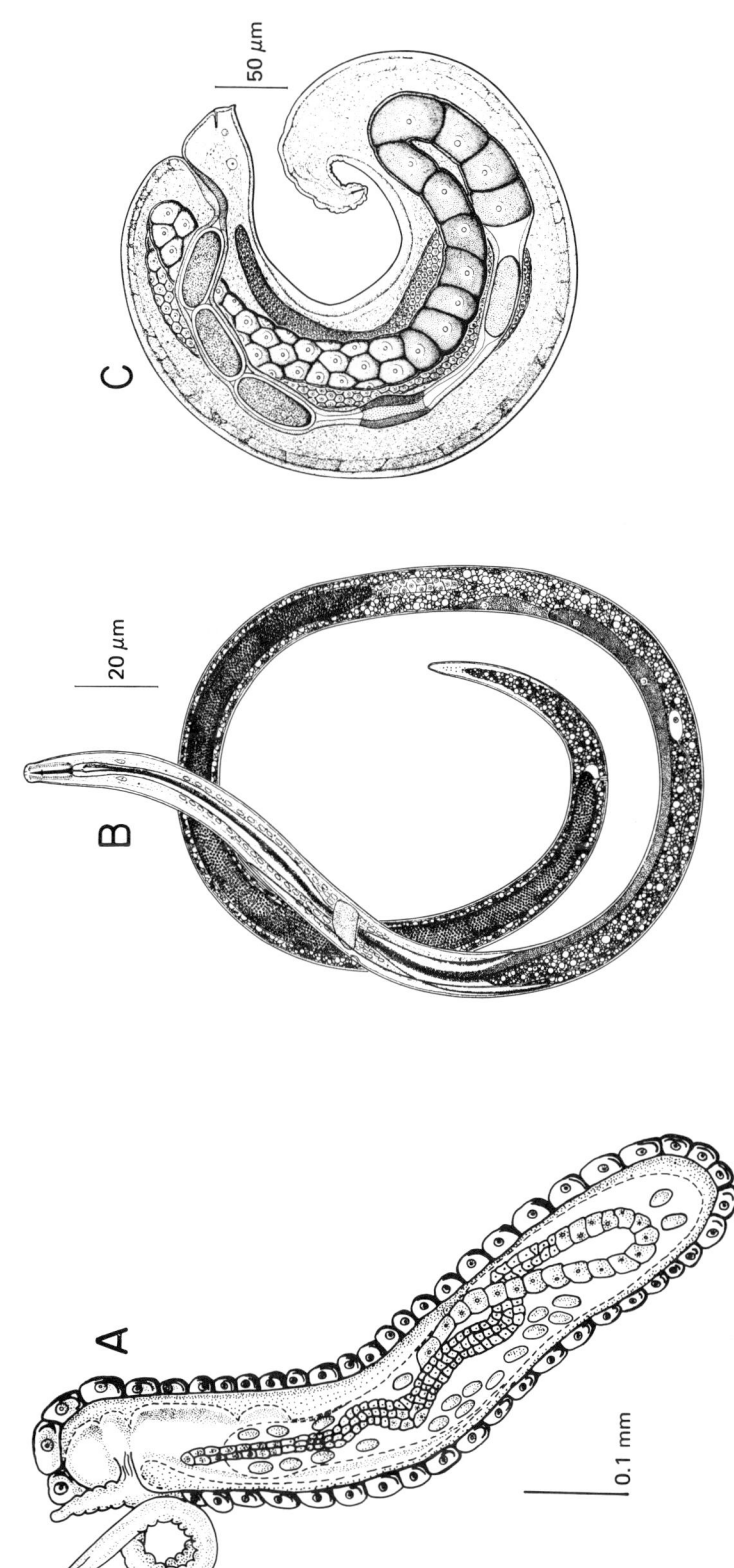

Fig. 1 (A) Adult female of *Sphaerulariopsis* with everted uterus. (Redrawn from Massey, 1974, by R. M. Giblin.) (B) Mated infective female of *Contortylenchus*. (Courtesy of W. R. Nickle.) (C) Parasitic *Bovienema* female adult from hemocoel of bark beetle. (Courtesy of W. R. Nickle.)

A. Sphaerulariopsis

The life cycle of *Sphaerulariopsis* in bark beetles is not completely known. Massey (1956) found *S. dendroctoni* only in the adults of the spruce beetle, *D. rufipennis*. The mode of infection, the stage of beetle infected, and the method of leaving the host have not been observed in bark beetles. Khan (1957a, b) and Massey (1956) observed adult females with everted uteri inside the hemocoel of adult beetles (Fig. 1A). According to Massey (1956), as many as 50 juvenile females and many hundreds of juveniles and eggs are present in the hemocoel. In the case of *S. ungulacauda*, adult females and preadult juvenile males are observed in the beetle, but adult males and females can be found outside the host (Khan, 1957b). Mated females are observed outside the host. Thus, Khan (1957b) suggested that mating takes place outside the host's body and that the mated female infects a new host. The mode of infection is either through the cuticle or by ingestion (Massey, 1956). So far, *Sphaerulariopsis* has been isolated only from *Dendroctonus* bark beetles in North America.

B. Allantonema

The life cycle of *Allantonema* is not completely known. This genus is rare in bark beetles in the United States (Massey, 1974) and, even in Europe, has been isolated only from a few species of bark beetles (Poinar, 1975). Rühm (1956) found *A. morosa* in the bark beetle, *Hylastes ater*, and unlike other *Allantonema* species that mature in adult insects, *A. morosa* matures in the beetle larvae. This nematode is ovoviviparous, and eggs and juveniles can be found in the hemocoel of beetle larvae. It is not known whether the beetle larvae are killed by the nematode or whether infected beetle larvae develop into adults. One of the reasons for the lack of information is the low incidence of infection in *H. ater* populations. On the average, only 3% of the beetle population are infected, and in many cases most beetle populations are free of infection (Rühm, 1956).

C. Contortylenchus and Bovienema

The life cycle of *Contortylenchus diplogaster* in *Ips typographus* has been studied by Rühm (1956) and of *C. elongatus* in *I. confusus* by Massey (1960, 1962) and Nickle (1963a, b), respectively. The contortylenchs are oviparous and have four juvenile stages and an adult stage (Nickle, 1963b; Thong and Webster, 1973; MacGuidwin, Smart, and Allen, 1980). Rühm (1956) reported only three juvenile stages in *C. diplogaster* because he did not observe the molt in the egg stage (Nickle, 1963a).

The life cycles of contortylenchs and bovienemas are similar. The one described for *C. elongatus* by Nickle (1963a, b) is given in detail (Fig. 2). Mated infective female adults (Fig. 1B) occur in the galleries of its host and seek beetle larvae. Usually second- and third-instar larvae are infected, but all larval and pupal stages are probably susceptible to infection. The usual mode of infection is believed to occur by direct penetration through the insect's integument. The possibility of infection via the oral or anal opening and subsequent penetration through the gut into the hemocoel cannot be ruled out. However, the infective females that are ingested are injured by the mouthparts of beetle larvae. Once inside the hemocoel, the female nematodes are usually found in the abdomen or thorax. Infected beetle larvae continue their development, pupate, and become adults. Parasitic female nematodes

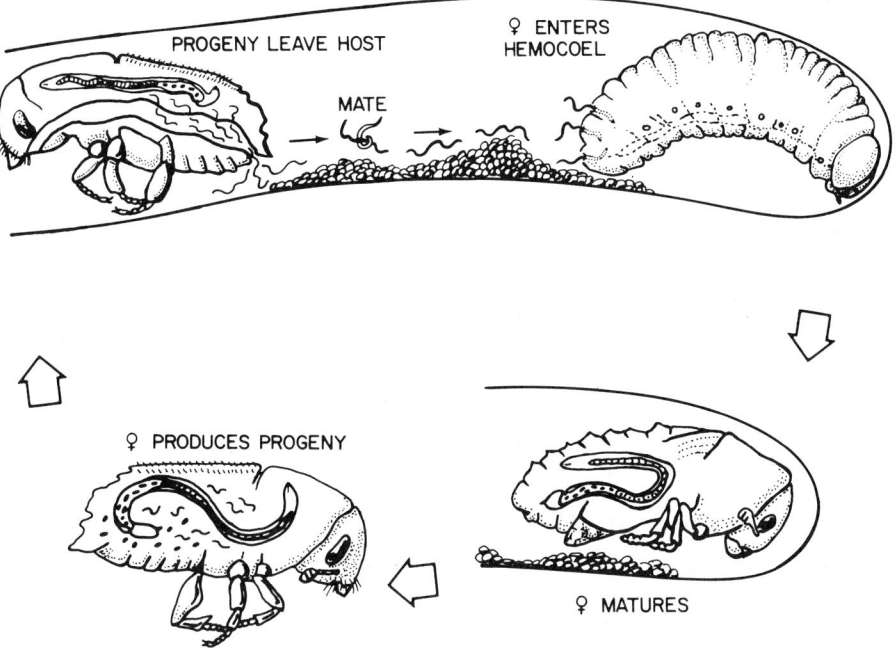

PROGENY LEAVE HOST   ♀ ENTERS HEMOCOEL

MATE

♀ PRODUCES PROGENY

♀ MATURES

Fig. 2 Life cycle of *Contortylenchus*.

also continue to develop. When infected bark beetle pupae are dissected, female nematodes with fully developed eggs in the uterus are found, but the
nematodes do not deposit eggs until the beetle emerges as an adult. Dissections of infected beetles may reveal the presence of 7500 larvae and eggs of
C. elongatus in the hemocoel. Adult female parasites are about 4 mm long
and deposit single-celled eggs inside the hemocoel of the beetle. One molt
occurs within the egg, and the nematodes emerge as second-stage juveniles.
These undergo two more molts to the fourth stage. These fourth-stage nematodes attain a size of about 0.55 mm, migrate to the gut region, and penetrate
into the hindgut and are defecated into the gallery (Fig. 3). Rühm (1956)
states that the penetration of *C. diplogaster* into the hindgut occurs in waves.
The fourth-stage juveniles molt into adult males and females in the galleries.
After mating, males die and mated infective females seek larval and pupal bark
beetle hosts.

The life cycle of *C. brevicomi* in the southern pine beetle, *Dendroctonus
frontalis*, is very similar to *C. elongatus*. MacGuidwin, Smart, and Allen,
(1980) reported that second-stage juveniles molt to third-stage juveniles in
the adult bark beetles, but the nematodes do not molt to the fourth stage until the beetles attack a host tree. Fourth-stage *C. brevicomi* apparently exit
from their hosts through the digestive or reproductive tracts. The life cycle
of *C. reversus* in the Douglas fir beetle, *D. pseudotsugae*, is similar to the
basic life cycle outlined for *C. elongatus* (Thong and Webster, 1973). However, the route of leaving the hosts by the fourth-stage juvenile has not been
resolved. In some nematode-infected adult beetles, the hemocoels contain
many nematodes in the fourth stage, but no juveniles or adults of *C. reversus*
are found in the gallery until the death of the beetle.

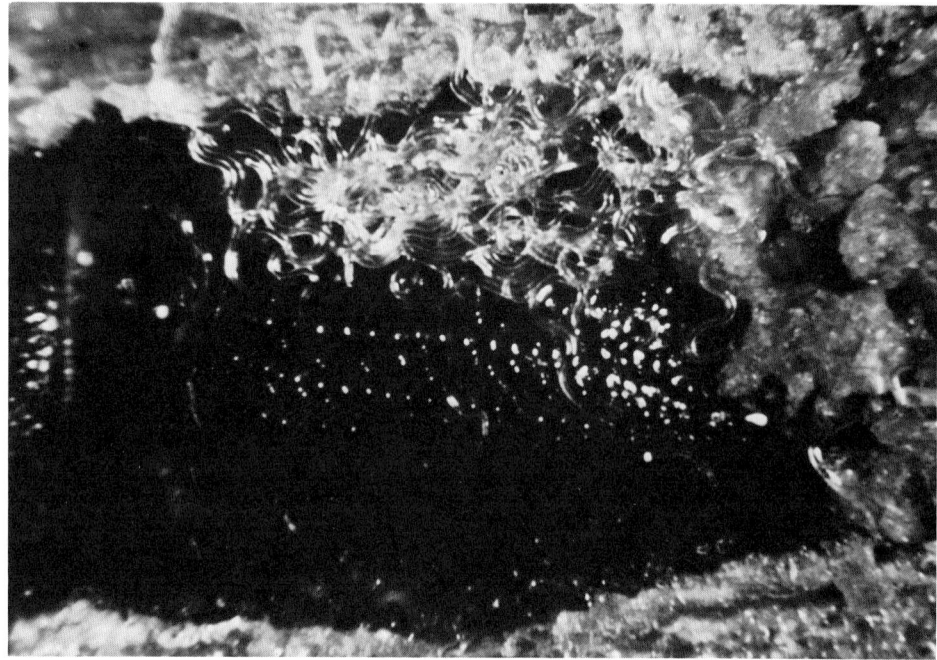

Fig. 3 Emergence of *Contortylenchus* juveniles from *Ips* adult. (Courtesy of D. N. Kinn.)

The biology of *Bovienema*, a nematode parasite of *Pityogenes* beetles, is similar to that of *Contortylenchus* (Nickle, 1963b). The mated female, which measures from 0.30 to 0.32 mm, is small in comparison to the infective female of *C. elongatus*, which measures 0.52-0.65 mm. The parasitic *Bovienema* female adult from the hemocoel of the bark beetle host is also much smaller (0.48-0.73 mm) (Fig. 1C) than that of *C. elongatus* from *Ips*, which measures 1.33-6.7 mm. This difference in size may be associated with the size of the beetle host because *Pityogenes* is much smaller than *Ips*.

D. Parasitylenchus

The biology of *Parasitylenchus* (*Polymorphotylenchus*) involves two sexual generations within the same beetle host (Rühm, 1956). The mated infective female in the frass seeks and infects a beetle larva (Fig. 4). The mode of infection into the hemocoel is not known, but presumably infection occurs by direct penetration through the cuticle or through the digestive tract into the hemocoel. The infected beetle continues its development to adulthood. In the adult beetle, the female nematode produces progeny ovoviviparously. These progeny mature in the hemocoel of the bark beetle. Mating takes place in the hemocoel of the bark beetle, and this mated female produces progeny. The progeny from the second-generation female adult molt twice in the hemocoel before they bore through the intestinal wall of the beetle and are defecated into the frass. Nematodes become sexually mature after molting in the frass.

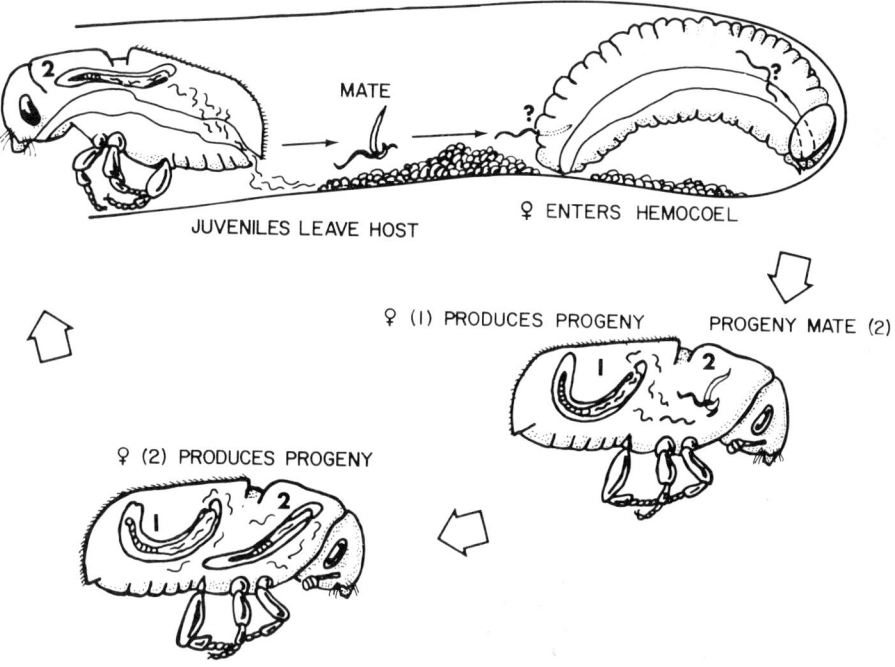

Fig. 4 Life cycle of *Parasitylenchus*.

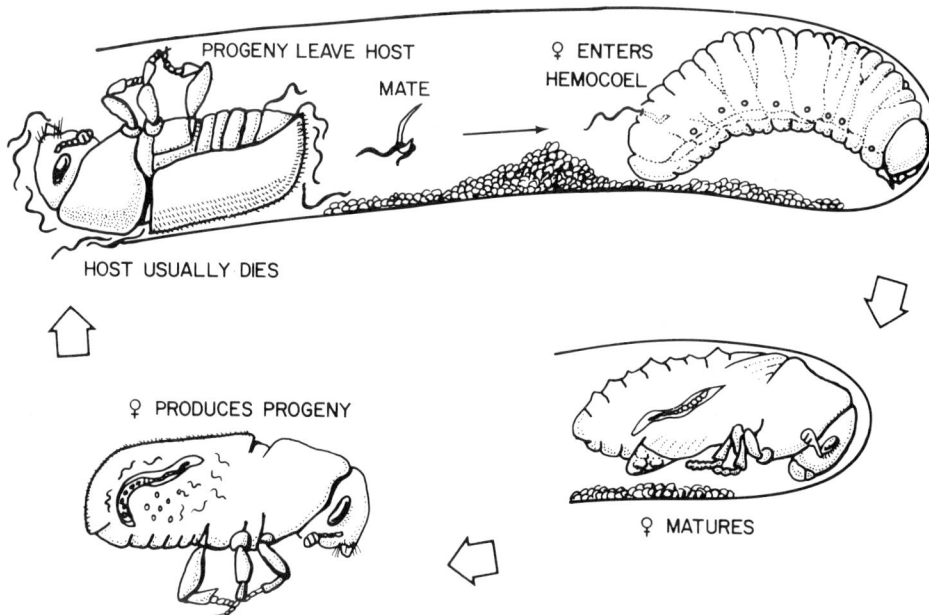

Fig. 5 Life cycle of *Sulphuretylenchus*.

E. Neoparasitylenchus and Sulphuretylenchus

The life cycles of neoparasitylenchs and sulphuretylenchs are similar and will
be considered together. The biology of *Sulphuretylenchus elongatus* has been
studied in its host, *Scolytus ventralis*, by Ashraf and Berryman (1970a) and
Massey (1974) (Fig. 5). Mated infective females seek beetle larvae and penetrate directly through the cuticle into the hemocoel. Although the mechanism
for invasion is not known, the stylet, and possibly enzymes, are important
in the penetration of the cuticle, which takes about two hours to complete.
There is no evidence for penetration through the midgut. Males are not infective and die in the host gallery after mating with the females. After an
infective female enters its larval host, the lip region degenerates and the stylet becomes nonfunctional and displaced. The nematode continues its development, but usually no progeny are produced until the beetle emerges as an
adult. *S. elongatus* is ovoviviparous, and a mature female nematode produces
an average of 446 ± 196 eggs. Emergence of sulphuretylenchs and neoparasitylenchs from their hosts occurs by penetrating the gut of the host and defe-
cation into the gallery (Rühm, 1956; Massey, 1974). Ashraf and Berryman
(1970a) stated that *S. elongatus* usually emerges from dead adult beetles.
Emergence occurs through intersegmental membranes and natural body openings of the host. According to Ashraf and Berryman (1970a), *S. elongatus*
may also emerge from living beetles, but the emergence site was not specified
though it is presumed to be through the anal opening.

Ashraf and Berryman (1970a) identified only two developmental stages
of *S. elongatus*. Rühm (1956) observed three juvenile stages of parasitylenchs
in adult beetles and stated that the third-stage juveniles bore through the
intestinal wall into the gut and may molt in the rectum of the beetle into fourth-
stage juveniles. The old cuticle remains until the nematodes are defecated into
the gallery. Rühm (1956) observed only three juvenile stages with contortylenchs, but Nickle (1963a) demonstrated that there was a molt in the egg and
thus accounted for four juvenile stages. Accordingly, the development of
these tylenchs seems to be similar to contortylenchs and four juveniles stages
probably occur in the insect host.

The life cycle of *Neoparasitylenchus rugulosi* in the shothole borer,
Scolytus rugulosus, is similar to that of *S. elongatus* (Nickle, 1971). Fourth-
stage male and female juveniles molt in the gallery of the beetle. After mating,
the infective female seeks a beetle larva. Once inside the hemocoel of the
beetle, the nematode takes up nourishment from the hemolymph and transforms
into a large, swollen, egg-producing form (Fig. 6). The infected beetle continues its development and becomes an adult. Shortly after adult emergence,
the female nematode deposits hundreds of eggs or larvae into the hemocoel.
The female beetle dies after constructing a short horizontal gallery (Fig. 7).
The nematodes exit from the head end of the dead beetle into the gallery as
fourth-stage juveniles (Nickle, 1971; Schvester, 1957). The infected beetles
are sterile, and therefore, the infective female nematodes must migrate to
adjacent galleries containing beetle larvae (Fig. 7).

F. Parasitaphelenchus

The members of the genus *Parasitaphelenchus* seem to be host specific to various bark beetles and have been recovered from scolytids, which attack conifers and deciduous trees (Hunt and Hague, 1974a). Members of this genus
appear to be distributed worldwide.

The life cycle of parasitaphelenchs as it occurs in the bark beetle, *Scolytus*, is shown in Fig. 8. The third-stage juvenile is the infective stage and

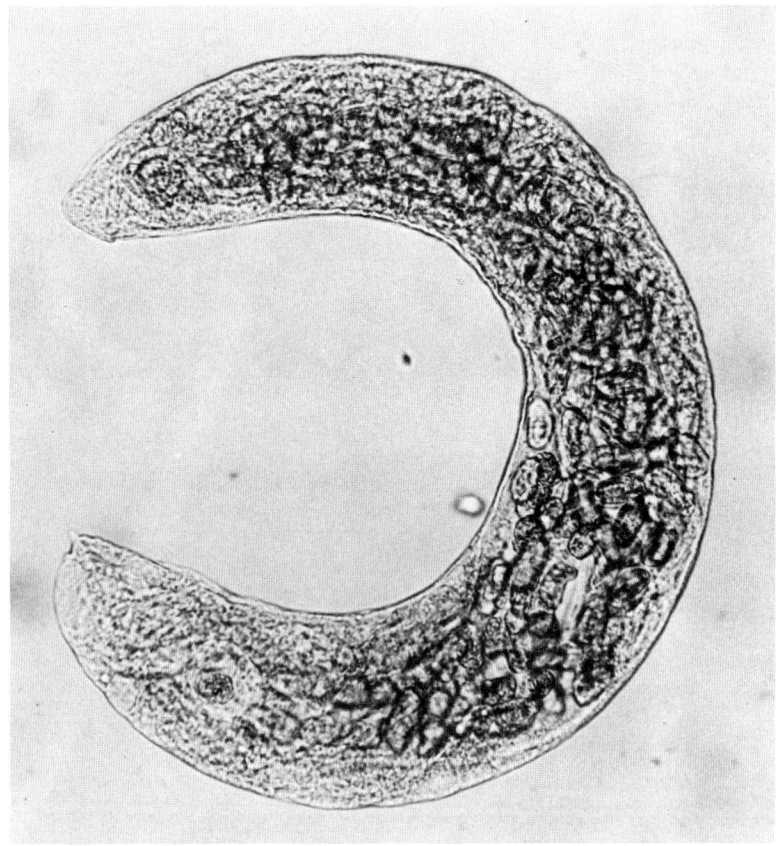

Fig. 6 Parasitic *Neoparasitylenchus* female from hemocoel of bark beetle.
(Courtesy of D. N. Kinn.)

only the fourth stage is parasitic. Remaining stages are freeliving. This
fourth-stage juvenile occurs in the hemocoel of bark beetles and can be found
in larvae, pupae, and adults. It is also found in the gut and frass. The
other stages are found in the frass in the beetle's gallery. The mode of infection of the third-stage juvenile is not known. Fuchs (1938) and Rühm
(1956) suggested that juveniles are ingested by beetle larvae and then penetrate through the gut into the hemocoel. Hunt and Hague (1974a) did not
observe gut penetration, but suggested that the nematodes may penetrate
very rapidly through the gut into the hemocoel or the nematodes may invade
through the cuticle. Saunders and Norris (1961) dissected many beetles but
were unable to detect gut penetration and suggested that gut penetration may
not be the mode of invasion.

Once inside the hemocoel, the third-stage juveniles molt to the fourth
stage and grow rapidly (Hunt and Hague, 1974a). Nematodes are found in
the abdomen, around the flight muscle in the thorax, and in the head of adult
Scolytus beetles; in beetle larvae and pupae, the juveniles are scattered
among the fat bodies throughout the hemocoel. The fourth-stage juveniles
emerge from the beetles from anal openings. If the beetles die, the juveniles
emerge from the mouth or anus. After three to four days, these nematodes

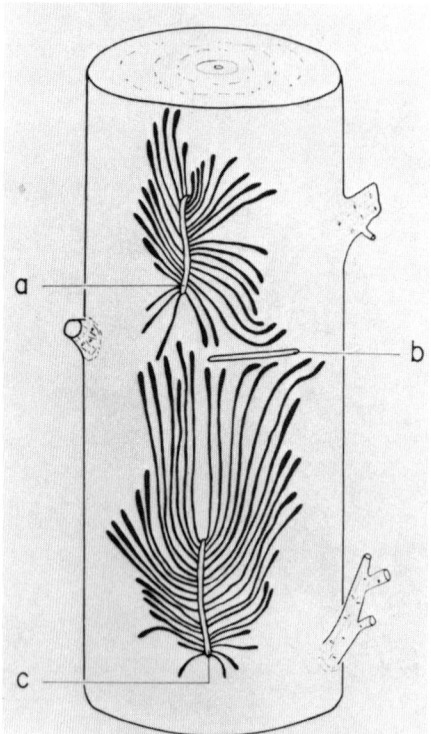

Fig. 7 Horizontal gallery system (a) of female *Scolytus rugulosus* infected
with *Neoparasitylenchus* contrasted with gallery systems (a and c) of normal
females. (Courtesy of W. R. Nickle.)

molt into adults. These adult males and females feed on the mycelia of several
fungal species, including *Ceratocystis ulmi*, the causative agent of Dutch elm
disease. After mating, the females produce eggs that hatch three to four days
later. The first-stage juveniles molt rapidly to the second- and then to the
third-stage juveniles without feeding. The infective third-stage juveniles do
not develop further until infection of a beetle host occurs.

It is not unusual to find more than 100 juveniles of *Parasitaphelenchus*
in *S. multistriatus* in the United States (Saunders and Norris, 1961). An
average of 41 parasitaphelenchs in *S. scolytus* and an average of 10 in *S.
multistriatus* have been found in England (Hunt and Hague, 1974a). In heavily parasitized beetles, nematodes are easily found, but if only a few occur in
the hemocoel, they can be overlooked. Both male and female beetles are infected, but in male *Scolytus*, the nematodes are in a dead-end host, because
only female beetles construct new gallery systems (Hunt and Hague, 1974a).

Parasitaphelenchus is not restricted to the genus *Scolytus*, and has been
isolated from a number of bark beetle genera (Poinar, 1975). The life cycle
of parasitaphelenchs may differ in other hosts. In *Blastophagus destruens*,
the third-stage juvenile of *P. papillatus* occurs in the hemocoel and no molt to
the fourth stage has been observed (Laumond and Carle, 1971). Laumond
and Carle (1971) counted juveniles remaining in the hemocoel of *B. destruens*
as a function of gallery length and found that juvenile nematodes appear to

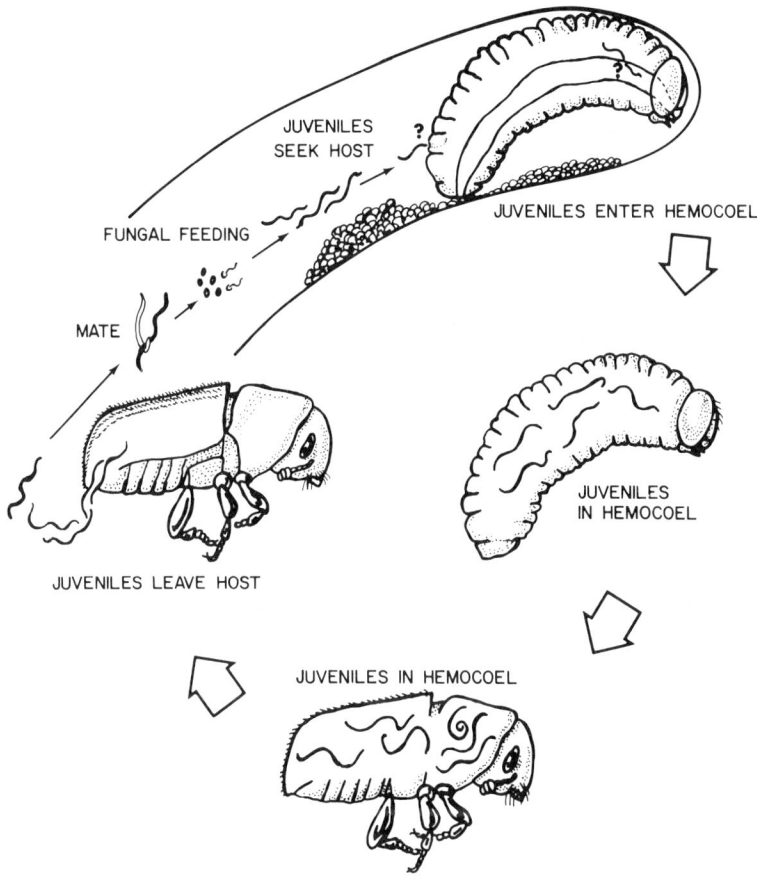

Fig. 8 Life cycle of *Parasitaphelenchus* in *Scolytus* bark beetle.

leave the host shortly after gallery construction. At the end of the beetle's oviposition, the beetles contained a reduced number of nematodes in the hemocoel.

G. Parasitorhabditis

Parasitorhabditis can be considered as being phoretic or parasitic depending upon the effect it has upon its host. Some detrimental effects by certain parasitorhabditid species in certain bark beetles have been reported (see Section VII). Juveniles of *Parasitorhabditis* are commonly found in the intestinal lumen and sometimes in the Malpighian tubules of wood-boring insects, especially among bark beetles. In some instances, the juveniles of certain species are found in the hemocoel of the bark beetle hosts (Fuchs, 1937; Rühm, 1956; Laumond and Carle, 1971).

The life cycle of this nematode genus is not complex, but may have several alternatives (Fig. 9). It may occur in the galleries as a free-living organism, being transported externally on bark beetles. In another situation, third-stage juveniles of *P. ipsophila* are found in the intestinal lumen of *Ips sexdentatus* (Lieutier and Laumond, 1978). In an infested adult beetle, 1-92

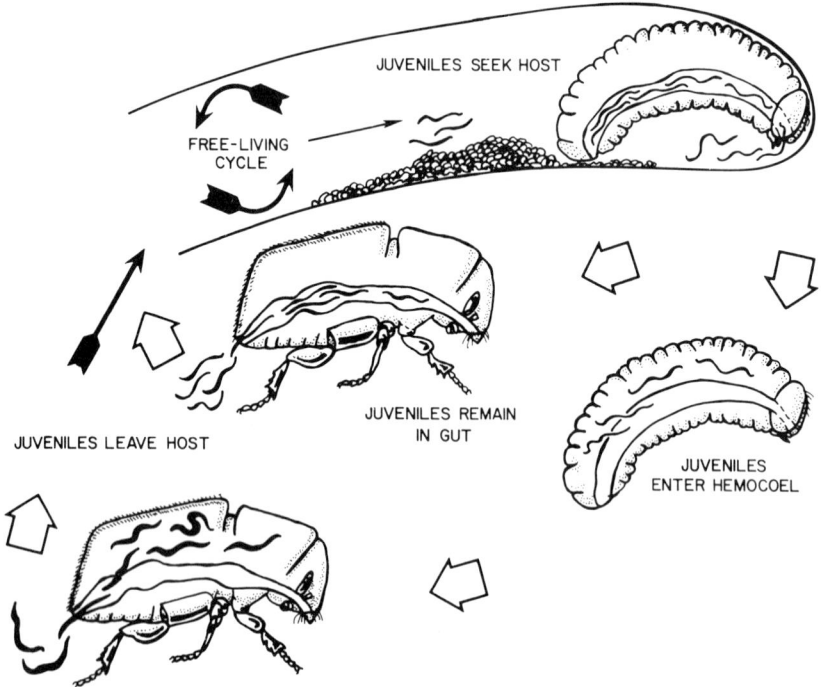

Fig. 9 Life cycle of *Parasitorhabditis* showing several alternatives.

juveniles can be found in the intestine. Juveniles leave via the anus and feed
on fungi in the beetle's gallery and develop to adulthood. After mating, the
female oviposits and other juvenile stages are found in the gallery. The third-
stage juveniles seek new hosts (probably larval beetles) and apparently enter
the host via the oral opening.

In some cases, the juveniles of *P. piniperdae* are found in the larval
hemocoel of the bark beetle, *Myleophilus piniperda* (Rühm, 1956). The third-
stage juveniles apparently penetrate through the intestinal wall and enter the
hemocoel. They can be found in the fat body tissues, molt to the fourth stage,
reenter the intestine, and leave via the anus. Laumond and Carle (1971)
stated that third-stage juveniles of *P. piniperdae* undergo considerable development in the hemocoel of *Blastophagus destruens*, leave their hosts, and then
molt to the fourth-stage juvenile. The length of time the nematodes stay in
the hemocoel is varied. Once nematodes leave the beetle, adults can be found
in the frass in the gallery and, after mating, females deposit eggs. Juveniles
are also found in the frass. Rühm (1956) states that it is the third- and
fourth-stage juveniles that seem to seek new hosts. It is not clear whether
fourth-stage juveniles are also capable of penetrating into the hemocoel.

In other cases, the insect host does not seem to be obligatory for the
development of parasitorhabditids. Hunt and Poinar (1971) successfully cultured *Parasitorhabditis* spp. from the bark beetle, *Dendroctonus valens*, on
the blue-stain fungus, *Ceratocystis minor*, for several generations. Massey
(1956) found *P. obtusa* in the intestinal lumen of *D. rufipennis* and believed
that infestation occurred through ingestion. Eggs, juveniles, and adults of
this nematode were found in the frass, and juveniles isolated from the beetles'
intestines can be reared to adults on malt agar.

V. HOST SPECIFICITY AND RESISTANCE

The host range of most entomophilic nematodes, including those infecting bark beetles, is usually narrow. This specificity may be related to the effects of host habitat selection or host limitation in the form of resistance (Stoffolano, 1973). An examination of insect associates of bark beetles that occur in the galleries may reveal that bark beetle nematodes may have a wider host range. In one instance, Khan (1957a) found that *Sphaerulariopsis hastata* not only infects *Dendroctonus ponderosae* and *D. pseudotsugae*, but also their hymenopterous parasites, *Coeloides dendroctoni*. Occasionally, a species of nematode parasite of one bark beetle species has been recovered from other bark beetle species. Thus, *Contortylenchus reversus* infects *D. rufipennis, D. ponderosae, D. pseudotsugae, Ips borealis*, and *I. pilifrons*.

Host resistance or immunity to nematode infection has not been adequately studied in bark beetles. Host escape and encapsulation are two mechanisms of arthropod immunity to nematodes. Host escape, which consists of physical barriers of the host, such as thickness of cuticle or behavioral responses, probably occurs with bark beetles but has not been examined critically. Encapsulation, which is the accumulation of hemocytes around the nematode to form a capsule, has been observed in certain bark beetles. Eggs and juveniles of *Sulphuretylenchus elongatus* were frequently encapsulated in first to third instar *Scolytus ventralis* (Ashraf and Berryman, 1970a). In *Ips sexdentatus* infected with *Contortylenchus*, 31% of the adult beetles contained encapsulated nematodes (Nelmes and Hussain, 1972). The capsules were attached to tracheoles, body wall and muscles, and Malpighian tubules of the host. Furthermore, tracheoles of the insect infiltrated the capsules. The nematodes were not killed by encapsulation and produced viable progeny. Rühm (1956) and Fuchs (1915) made similar observations with other bark beetle nematodes. Generally, encapsulation is more apt to occur when nematodes infect an insect other than its normal host (Salt, 1963).

VI. SUPERPARASITISM AND MULTIPLE PARASITISM

Superparasitism, defined here as more than one infective nematode of the same species occurring within the hemocoel of an insect, is of common occurrence in bark beetles as well as in other insects. Many researchers noted upon dissection that more than one infective female or infective juvenile nematode can be found in the hosts. A few examples are given to illustrate this point. In *Ips confusus*, 83 *Contortylenchus elongatus* adults have been recovered from an adult beetle (Nickle, 1963b). With *C. reversus*, the average number of nematode adults in *D. rufipennis* was 5.5 with a range from 1 to 20 (Massey, 1974), while with *Bovienema*, the average number per beetle was two with a maximum of five (Nickle, 1963c).

Lieutier (1979) found an average of three *C. diplogaster* in *Ips sexdentatus* with a range of 1-26. The distribution of infective *C. diplogaster* females suggests that infections may be made at random with nematodes collected in the galleries. However, the widespread superparasitism of *I. sexdendatus* by *C. diplogaster* may be explained by an "attractiveness" of parasitized hosts by the nematodes. It is not unusual to find six or seven adult nematodes of sulphuretylenchs and neoparasitylenchs in a single host (Ashraf and Berryman, 1970a; Poinar and Caylor, 1974). With parasitorhabditids and parasitaphelenchs, the number of juveniles found in the bark beetle hosts is usually greater than one and often more than 50 occur in a given host (Rühm, 1956; Hunt and Hague, 1974a; Saunders and Norris, 1961; Lieutier and Laumond, 1978).

Multiple parasitism, which is defined as two or more infective nematodes of different species occurring within the hemocoel of an insect host, has also been observed in bark beetles. In *D. rufipennis*, 2% of the adults were infected with *Contortylenchus reversus* and *Sphaerulariopsis dendroctoni* (Massey, 1956). Similarly, these two nematodes have been recovered from the same individual of *D. pseudotsugae* (Furniss, 1967). Moreover, 14% of *Conophthorus monophyllae* were found infected with a parasitaphelench and a neoparasitylench (Poinar and Caylor, 1974).

VII. PATHOLOGICAL EFFECTS OF NEMATODE INFECTIONS IN BARK BEETLES

A. Sulphuretylenchus, Neoparasitylenchus, and Parasitylenchus

True nematode parasites of bark beetles generally do not kill their host immediately even when more than one infective nematode invades the hemocoel. Sulphuretylenchs, neoparasitylenchs, and parasitylenchs are the most virulent of the true bark beetle nematodes. Many of the pathological effects have been reported for this particular group.

The nematode *S. elongatus* can directly penetrate and kill about 8% of the eggs and 2-4% of the larval stages of the fir engraver, *S. ventralis* (Ashraf and Berryman, 1970a). Many larvae survive nematode infection, but nematode-infected fir engraver beetles show a delayed emergence of adults compared to uninfected ones. *S. elongatus* significantly reduces the longevity of male and female bark beetle adults (Ashraf and Berryman, 1970a; Massey, 1964). In most cases, death occurs after the female beetle constructs a short gallery. The cause of death of larval and adult beetles is probably related to the mass production of nematode progeny.

Reduced flight by beetles infected with sulphuretylenchs has been reported (Ashraf and Berryman, 1970a). Ashraf and Berryman (1970a) categorized *S. ventralis* infected with sulphuretylenchs into lightly infected (1-500 juveniles or one adult nematode), moderate (500-1500 juveniles or two to three adults), heavy (1500-3000 juveniles or four to five adults), and very heavy (3000+ juveniles or 6+ adults). Using this classification, they found that heavily infected beetles were completely absent from flight traps and that male beetles were more affected than the females. Ultrastructural pathology of the flight muscles of nematode-infected beetles showed that the mitochondria were disoriented and reduced in numbers, and, in some cases, had ruptured membranes (Ashraf et al., 1971).

The most pronounced effect of these nematodes is on the reproductive organs and fat body. Sulphuretylenchs and neoparasitylenchs may sterilize their hosts (Ashraf and Berryman, 1970a; Nickle, 1971; Schvester, 1957). Reduced fecundity may occur and is associated with the intensity of nematode infection. Thus, uninfected *S. ventralis* beetles average 51 eggs, lightly infected females average 37 eggs, and heavily infected females produce no eggs (Ashraf and Berryman, 1970a). Histologically, the developing oocytes in infected females are less numerous, show signs of cellular disintegration, and are separated from the follicular epithelium compared to uninfected beetles (Ashraf and Berryman, 1970b). In heavily infected males, the epithelial sheaths around the testes are disintegrated and the testes are reduced in size. (Fat bodies of infected beetles lack vacuoles and are reduced in size.) In other tissues, the fore- and hindgut epithelial cells and muscle layers of infected *S. ventralis* are greatly reduced and midgut muscle fibers are separated and disintegrated.

Bark beetles infected with parasitylenchs and neoparasitylenchs are also known to show aberrant gallery construction, reduced longevity, reduced flight activity, reduced fat body, and sterilization (Fuchs, 1915; Schvester, 1950, 1957; Rühm, 1956; Nickle, 1971; Poinar and Caylor, 1974).

The bark beetle, *S. rugulosus*, infected with neoparasitylenchs dies soon after making a short horizontal gallery (Nickle, 1971; Schvester, 1957) (Fig. 7), and *C. monophyllae* infected with neoparasitylenchs shows poor flight (Poinar and Caylor, 1974). Rühm (1956) noted that 6% of the population of *Ips (Pityokteines) curvidens* infected with parasitylenchs are sterile or have reduced gonads. Oldham (1930) reported that *S. multistriatus* and *S. scolytus* infected with *Neoparasitylenchus scolyti* were sterilized. However, he found only one adult neoparasitylench in the beetles and remaining beetles in his samples apparently were infected with *Parasitaphelenchus oldhami* (Rühm, 1956). Oldham's report is considered in greater detail in the *Parasitaphelenchus* section.

There are some reports that infection by this group of nematodes has no adverse effects on bark beetles. Hoffard and Coster (1976) reported that *Ips avulsus* adults showed no difference in size, shape, color, or texture of ovaries and testes between beetles infected with neoparasitylench and uninfected beetles. Dale (1967) concluded that *Hylastes ater* was not adversely affected by a neoparasitylench infection.

B. Contortylenchus

Contortylenchs in general do not seem to kill their hosts, and the effects are less dramatic than reported for other parasitic tylenchs of bark beetles. The most pronounced effects of this nematode group are on reproduction, gallery construction, and fat body reduction. *I. confusus* infected with *C. elongatus* has reduced brood to about one-third the normal number of uninfected beetles (Nickle, 1963a, d). Similarly, Massey (1962) reported a 70% reduction in brood produced by *C. elongatus*-infected *I. confusus*. The average brood per infected female was 14, whereas uninfected beetles averaged 48. However, the developmental stage of the beetle at the time of nematode infection and the number of mature adult nematodes in the beetle affected the number of beetle progeny produced. The earlier the infection of the beetle larva or the greater the number of nematodes per beetle, the less progeny is produced. In concurrence with these observations, Nickle (1963d) suggested that the presence of several adult *C. elongatus* females and up to 7500 of their eggs and juveniles in the hemocoel (Figs. 10 and 11) of *I. confusus* will probably cause a drain of the reserve food supply of the host, and may cause further damage because of catabolic toxins. Fewer fat cells are found in infected beetles compared to uninfected beetles. In addition, delayed development and adult emergence of nematode-infected bark beetles have been hypothesized because beetles that emerged earlier had a lower percentage of infection than those that emerged later.

With other bark beetles, MacGuidwin, Smart, Wilkinson, and Allen (1980) reported that *D. frontalis* females infected with *C. brevicomi* produced fewer eggs than uninfected ones, but the data were not significantly different. These observations were made for a one week period, and perhaps longer observations may show significant differences in fecundity. However, fertility of uninfected *D. frontalis* mated with *C. brevicomi*-infected males is reduced, suggesting that sperm viability or production may be adversely affected. In the Douglas fir beetle, *D. pseudotsugae*, infected with *C. reversus*, females laid 33-50% fewer eggs than uninfected beetles (Thong and Webster, 1975a).

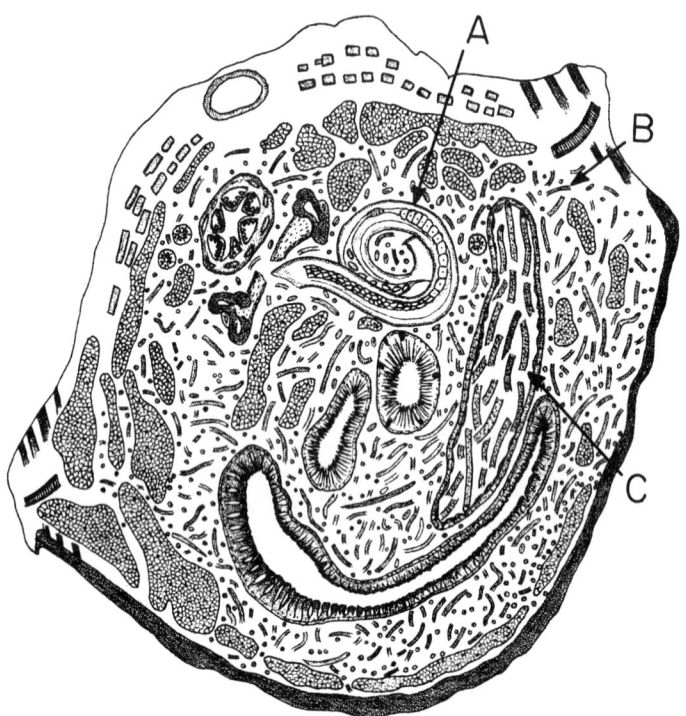

Fig. 10 Cross section through abdomen of *Ips* adult parasitized by *Contorty-lenchus* and *Parasitorhabditis*. (Courtesy of W. R. Nickle.) (A) Adult *Contortylenchus*. (B) Juvenile *Contortylenchus*. (C) Juveniles of *Parasitorhab-ditis* in midgut.

Parasitism by *C. reversus* does not affect fertility of male *D. pseudotsugae* (Thong and Webster, 1975a), and sperm production of *I. confusus* is not reduced in males parasitized by *C. elongatus* (Massey, 1962).

The effect on flight of bark beetles infected with contortylenchs is not drastic. The duration of initial flight of *D. pseudotsugae* is reduced by 40-50% by nematode infection (probably contortylenchs), but the total flight duration or the number of rests is not affected (Atkins, 1960; 1961). In the case of *D. frontalis* infected with *C. brevicomi*, the incidence of infection of flying beetles trapped at different heights is similar (Kinn and Stephens, 1981). However, *D. frontalis* males infected with *C. brevicomi* seem to respond less to pheromone traps baited with Frontalure-33 than males not infected with the nematode (Atkinson and Wilkinson, 1979). Reasons for the lower incidence of nematode infection may be due to the adverse effect of *C. brevicomi* on the flight ability of males, alteration of response to the phero-mones, or increased susceptibility to other mortality factors. Another possi-bility is that this particular population of *D. frontalis* was infected with a pro-tozoan parasite (microsporidian), and the interactions between the nematode and the microsporidian may have affected the results.

Beetles infected with contortylenchs also construct aberrant galleries. *D. pseudotsugae* infected with *C. reversus* has about 25% reduction in the length of the primary egg gallery built by the infected female (Thong and Webster, 1975a). Nematode parasitism did not affect gallery shape or egg

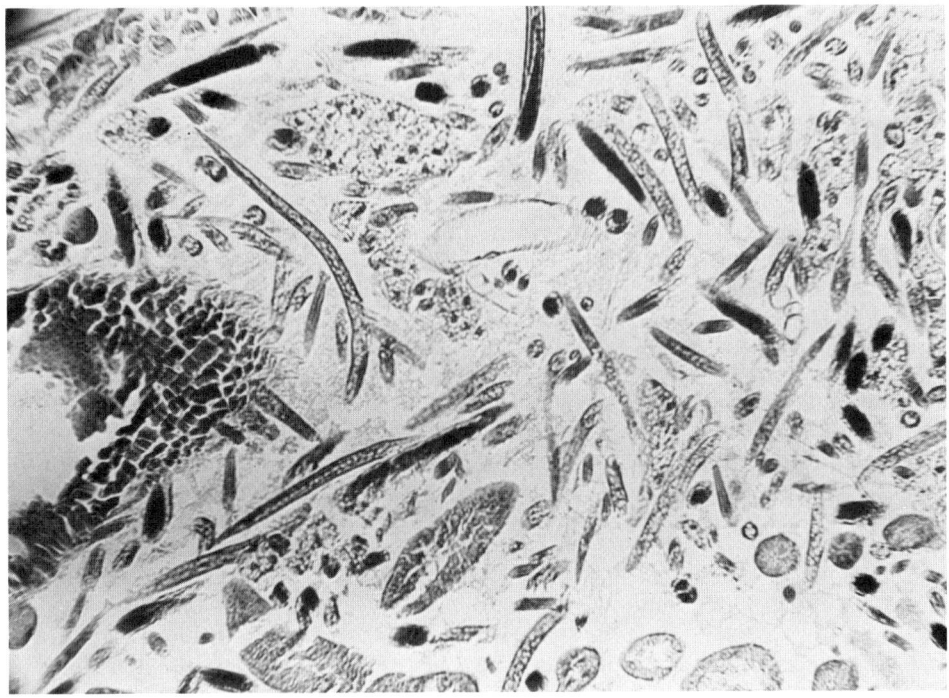

Fig. 11 Juveniles of *Contortylenchus* in hemocoel of *Ips*. (Courtesy of S. Barras and T. Perry.)

viability. Comparison of gallery length made by *D. frontalis* infected with *C. brevicomi* and uninfected beetles showed significant differences in gallery length. In *D. frontalis*, both male and female beetles are involved in gallery construction. Therefore, when both male and female beetles were infected, gallery lengths were consistently shorter and fewer eggs were laid than observed for uninfected pairs (MacGuidwin, Smart, Wilkinson, and Allen, 1980). Similarly, gallery lengths of *I. confusus* infected with *C. elongatus* average 11.3 cm, whereas those of uninfected beetles average 17.8 cm (Massey, 1962).

Comparisons of carbohydrate, protein, and amino acids between *D. pseudotsugae* adults infected with *C. reversus* and uninfected adults indicate that nematode infection does not affect trehalose or amino acid levels, but protein levels are less in infected beetles (Thong and Webster, 1975b). A consequence of this protein reduction is 20% smaller oocytes in infected beetles.

C. Sphaerulariopsis

Sphaerulariopsis dendroctoni is an obligate parasite and does not kill its bark beetle host, *D. rufipennis*. As with other bark beetle nematodes, egg production is affected by *Sphaerulariopsis* infection (Massey, 1956). The average number of eggs produced by infected beetles is 29, whereas the average for uninfected beetles is 78 eggs. In some cases, infected beetles are completely sterilized. Reid (1958) reported that infected adults of mountain pine beetle, *D. ponderosae*, show lethargic movements and tremor in the antennae and legs. Effects of this nematode on bark beetles have not been studied in greater detail, primarily because of its limited occurrence to a few beetle species.

D. Parasitaphelenchus and Parasitorhabditis

Although the parasitaphelenchs occur in the hemocoel of many bark beetle species and one to several hundred juveniles can be found in a bark beetle host, the effect on the host appears to be minimal (Hunt and Hague, 1974a; Saunders and Norris, 1961). Oldham (1930), who mistook the juveniles of parasitaphelenchs as being a tylenchid parasite (*Neoparasitylenchus*) reported that 60% of *S. multistriatus* and *S. scolytus* were infected and of these about 40% were sterile. Rühm (1956) does not agree with these figures and suggested that Oldham may have dissected beetles with underdeveloped gonads and classified them as sterile. In Rühm's study, a low percentage (less than 8%) appeared to be sterile or had reduced gonads as a result of parasitaphelench infection in *Myelophilus piniperda*, *S. scolytus*, *I. acuminatus*, and *H. ater*. On the other hand, Hunt and Hague (1974a) found no noticeable adverse effect on size of adult beetles, gonads, fat body, flight muscles, or sex ratio in *S. scolytus* and *S. multistriatus* infected with *Parasitaphelenchus oldhami*.

Many species of the parasitorhabditids are phoretically associated with bark beetles, and only a few species may cause damage to their hosts. Juveniles of *P. hectographi*, which are found in the mid- and hindgut of *Dryocoetes hectographus*, cause the cytoplasm of the intestinal cells to become reduced and vacuolated and the microvilli of these cells to disappear (Rühm and Chararas, 1957). Similarly, juvenile parasitorhabditids reduce the intestinal cells of *Ips confusus* (Nickle, 1963d) (Fig. 10). *Parasitorhabditis piniperdae* can invade the hemocoel of *Myelophilus piniperda* and *Blastophagus destruens* (Fuchs, 1937; Rühm, 1956; Laumond and Carle, 1971). Interestingly, Laumond and Carle (1971) indicated that when *B. destruens* is heavily infected with *P. piniperdae* and *Parasitaphelenchus papillatus* the adult beetles have less fat body tissue and underdeveloped gonads in comparison to uninfected ones. In addition, heavily infected beetles seem to require a longer developmental period, which translates into later emergence and longer beetle galleries.

E. Neoaplectana and Heterorhabditis

Neoaplectanids and heterorhabditids have a wide host range and infect a number of insect species including bark beetles. *Neoaplectana* has been used against *S. scolytus* and *D. frontalis* (Finney and Mordue, 1976; Finney and Walker, 1977, 1979; Moore, 1970). *Neoaplectana* infects mature beetle larvae of *D. frontalis* in their galleries, but few young larvae or adults and no pupae are infected (Moore, 1970). Larvae and adults of *S. scolytus* are susceptible to *Neoaplectana* (Finney and Mordue, 1976; Finney and Walker, 1977, 1979). Infected beetle larvae become flaccid and turn brown, but dead adult beetles show no outward signs of neoaplectanid infection.

VIII. PARASITIC NEMATODES IN BARK BEETLE POPULATIONS

The impact of parasitic nematodes in bark beetle populations has not been adequately assessed. Numerous reports are available on the percentage of bark beetles infected with nematodes in a given population, but long-term studies are generally lacking. The percentage of infected beetles in a given species in a given population may vary from 0 to 90% and fluctuate drastically from year to year (Massey, 1974). Rühm (1956) also observed differences in nematode infection in bark beetle populations from generation to generation. It appears that dense beetle populations have higher numbers of infected individuals than smaller populations, but the percentage of infection may be greater in a smaller population.

During a five year (1958-1962) study on the Douglas fir beetle, *D. pseudotsugae*, collected from Idaho and Utah, the incidence of nematode infection by *Contortylenchus reversus* fluctuated from 10 to 31% and that of *Parasitaphelenchus* spp. varied from 49 to 60% (Furniss, 1967). The highest infection in field populations seems to occur most commonly with *Parasitaphelenchus* spp. (Massey, 1974; Oldham, 1930; Hunt and Hague, 1974a, b; Saunders and Norris, 1961). However, as previously mentioned, the impact of parasitaphelenchs on individual beetles is negligible.

Kinn and Stephens (1981) found that the incidence of infection by *Contortylenchus brevicomi* in *D. frontalis* populations declined during the summers of 1976 and 1977. The infection rate of emerging beetles in May and June was 20-25% whereas during July to September, the infection rate fluctuated between 0 and 14%. The reason for this summer decline is not known, but may be related to predation of the free-living juveniles and adults of the contortylenchs in the frass from mites and/or higher summer temperatures that reduced survival of stressed beetles. Hetrick (1940) also reported a higher incidence of nematode infection in the overwintering brood of *D. frontalis* than in the first spring brood. Conversely, Rühm (1956) found that nematode infection of *Ips typographus* is higher in the second generation than in the first. Hoffard and Coster (1976) reported that nematode infection showed a consistent trend toward higher infection levels during the summer and fall months, with *Ips avulsus* infected with *Neoparasitylenchus avulsi*, *I. grandicollis* infected with *Contortylenchus grandicollis*, and *I. calligraphus* infected with *C. elongatus* and *Parasitaphelenchus* sp. in eastern Texas. In laboratory studies, the incidence of infection by *Bovienema tomici* on *Pityogenes bidentatus* increased from 30 to 60% in a population living in logs of *Pinus contorta* (Nickle, 1963a). Similar studies with *I. confusus* and *C. elongatus* under controlled conditions indicate that incidence of nematode infection can be increased over several generations. Initial nematode infection by *C. elongatus* in *Ips* populations was 13%, after six months was 34%, and by 21 months had risen to 83%. The infection rate by *Sulphuretylenchus californicus* in *I. confusus* was 5% initially and still 5% at the end of six months, but at 21 months this nematode was not present in the population. The more virulent sulphuretylenchs apparently cannot compete with the less virulent contortylenchs under controlled conditions in *Ips* populations.

IX. NEMATODE ASSOCIATES OF BARK BEETLES

The number of nematode species associated with bark beetles is greater than the number of parasitic species. Nematode associates are, for the most part, phoretics. They are carried from gallery to gallery and from tree to tree externally on various parts of the beetle's body, such as beneath the elytra, between intersegmental folds, and on tarsal and tibial joints of the legs and internally in the gut, Malpighian tubules, reproductive tract, and trachea. After being transported to a new environment, the nematodes feed on microorganisms in the beetles' galleries. Nematode associates are found among many different nematode families (Table 2). Massey (1974) and Lieutier and Laumond (1978) describe many species of nematode associates and parasites of bark beetles of the world.

The occurrence of nematode associates on bark beetles is an important consideration because in some cases these nematodes may be mistaken for parasitic forms and may result in erroneous conclusions on the impact of nematodes on bark beetles. Consequently, before any dissections of bark beetles are

Table 2 Families of Nematodes Associated with Bark Beetles

Aphelenchidae	Mononchidae
Aphelenchoididae	Neotylenchidae
Cephalobidae	Panagrolaimidae
Chambersiellidae	Plectidae
Cylindrocorporidae	Rhabditidae
Diplogasteridae	Tylenchidae
Dorylaimidae	

made for parasitic nematodes, the external surface of the beetle should be carefully examined for phoretic ones. Phoretic nematodes that occur internally are more difficult to find and may be easily mistaken as parasites. Care should be taken to prevent rupturing the midgut and Malpighian tubules. The location of nematodes should be noted and specimens preserved for future identification. The majority of phoretic nematodes occurs as juveniles on or in bark beetles. Identification in such cases can only be made to family and, in some cases, to generic level. Usually, phoretic nematodes can be reared on fungal or bacterial isolates from the beetles' galleries. If definitive identification is needed, the adult nematodes must be reared or isolated from the gallery. Generally, tylenchs and aphelenchoidids can be reared on a variety of fungal species; diplogasterids and rhabditids can be reared on bacterial species. However, a variety of microorganisms should be tried because some rhabditids can be reared on fungi (Hunt and Poinar, 1971).

Nematode associates of bark beetles (and of insects in general) have adapted to their host so that only a certain stage is transported to a new environment. This stage is often a resistant form known as dauer juvenile. Sometimes, dauer juveniles occur anhydrobiotically on the beetles. A few examples are given to illustrate the phoretic relationships between nematodes and bark beetles.

Diplogasterids are usually transported as dauer juveniles, which are enclosed by an oily cuticle. Dauer juveniles can be found under the elytra and between thoracic folds (Rühm, 1956). Often, the juveniles are arranged in series with one another or occur in clumps or bundles. The oily cuticle provides a means of adhesion and protects the nematode from desiccation. In some cases dauer juveniles, such as *Goodeyus* spp., are found under the elytra as well as in the genital area (bursa copulatrix) of female *Scolytus scolytus*. Another diplogasterid, *Mikoletzkya pinicola*, is reported to be predacious on eggs of the mountain pine beetle, *Dendroctonus ponderosae* (Massey, 1974). In galleries with high populations of *Mikoletzkya*, eggs of the bark beetle failed to hatch, but eggs in a light population of this nematode hatched normally.

Although some parasitorhabditids are parasitic (see Section VII), others occur in the midgut and Malpighian tubules and cause no damage and are considered to be phoretics (Lieutier and Laumond, 1978). Other rhabditids, such as *Cephaloboides* and *Mesorhabditis*, are phoretics of bark beetles (Massey, 1974).

Aphelenchoidids contain a large group of phoretic nematodes that are associated with bark beetles. Juveniles are found on the external body regions as described for the diplogasterid. Dauer juveniles of *Bursaphelenchus* can occur anhydrobiotically on various parts of the beetle's body (Rühm, 1956). Some species of aphelenchoidid juveniles are found in the reproductive tract (colleterial glands) of female beetles. Juveniles of at least three species of *Cryptaphelenchus* occur in the Malpighian tubules, and juveniles of a single species, *Cryptaphelenchus viktoris*, occur in the hemocoel of *Myelophilus piniperda* (Rühm, 1956).

In *Cryptaphelenchoides (Ektaphelenchus) scolyti*, mated females occur on the surface of the fourth- and fifth-instar larvae and beneath the elytra of pupae and adult *Scolytus scolytus* (Hunt and Hague, 1976). Female nematodes are attached beneath the inner anterior angle of the elytra by means of a sheathlike invagination of the insect cuticle. This cuticular sheath encloses the anterior end of the nematode to about the position of the median bulb. The nematodes occur anhydrobiotically until the beetles die. Water droplets condense beneath the elytra, and female nematodes become active and move into the frass. Presumably, this nematode feeds upon fungi in the galleries. Rühm (1956) suggested that *C. scolyti* is an ectoparasite of bark beetles, apparently because of the formation of the cuticular sheath. It is questionable that this nematode is parasitic because the nematode occurs anhydrobiotically on the beetle. However, Kurashvili et al. (cited in Poinar, 1972) provided evidence to indicate that *C. (Ektaphelenchus) dendroctoni*, which occurs under the elytra of *D. micans*, obtained hemolymph through the insect cuticle. Nematodes under the elytra of beetles exposed to radioactive phosphorus were also found to be radioactive, and the authors concluded that the nematodes obtained the radioactivity by ingesting radioactive hemolymph.

The tylenchs also contain a number of nematodes phoretically associated with bark beetles. For example, juveniles of *Neoditylenchus* have been observed in the trachea of *Dendroctonus* adults in California (unpublished data). A number of genera and species of nematode associates of bark beetle have not been covered. Bionomics of these nematode associates, in many cases, remain unknown and they offer a fertile field of study.

X. ASSESSMENT

Many factors affect the level of nematode infection in bark beetle populations. Factors such as host populations, predation of infective-stage nematodes, competition with other nematodes and microorganisms, interaction with other biological control agents, quality of the host trees for beetles, and nematode survival and sampling procedures for nematode infection have not been adequately examined. Moreover, the effects of abiotic factors, such as humidity and temperature, on the nematode have not been critically examined. A better understanding of the biology of parasitic nematodes is also needed to develop artificial means of in vivo and in vitro production of nematodes for biological control programs. It has been shown that, under controlled conditions, nematode parasites can have a great impact upon the bark beetle population and thus show promise as biological control agents.

Massey (1974) is optimistic that nematode parasites can be effective biological control agents and stresses the need for more research on these nematodes. He states that populations of beetles can be eradicated by breeding nematode-infected males and females or infected females to noninfected males. Although eradication of bark beetles under field conditions does not appear to be feasible, the use of parasitic nematodes may reduce bark beetle populations to levels below the economic threshold.

Rühm (1956) does not feel that parasitic nematodes can be manipulated for biological control of bark beetles because (1) the development of culture media for the parasitic nematodes is difficult, (2) artificial infection of the bark beetle has not been achieved because the conditions required for infection cannot be produced, and (3) the economics of introducing infective nematodes into the breeding galleries are too great because many nematodes will not find hosts and die.

The utilization of bark beetle nematodes as biological control agents has yet to be realized under field conditions. However, artificial infection of bark beetles has been made in the laboratory (Nickle, 1963a). Furthermore bark beetles have been reared on artificial diets, which may someday be used to infect them with nematodes. These nematode-infected beetles can be introduced into natural populations. Although we are far from the practical aspect of manipulating nematodes in beetle populations, the importance of certain parasitic forms on individual beetles has shown that they can sterilize or reduce fecundity, reduce longevity, and cause aberrant behavior. These nematodes may play an important role in the natural control of bark beetle populations and should be assessed as factors in population reduction of bark beetles.

ACKNOWLEDGMENTS

I thank D. N. Kinn, R. M. Giblin, and H. E. Welch for their suggestions and comments on the manuscript. Special thanks are extended to D. N. Kinn for making unpublished data available for this manuscript, to R. M. Giblin for doing much of the line drawings for the life cycles, and to Jennifer Swan for translating some of the articles from French to English.

REFERENCES

Ashraf, M., and Berryman, A. A. (1970a). Biology of *Sulphuretylenchus elongatus* (Nematoda: Sphaerulariidae) and its effect on its host, *Scolytus ventralis* (Coleoptera: Scolytidae). *Can. Entomol. 102*: 197-213.

Ashraf, M., and Berryman, A. A. (1970b). Histopathology of *Scolytus ventralis* (Coleoptera: Scolytidae) infected by *Sulphuretylenchus elongatus* (Nematoda: Sphaerulariidae). *Ann. Entomol. Soc. Am. 63*: 924-930.

Ashraf, M., Mayr, W., and Sybers, H. D. (1971). Ultrastructural pathology of the flight muscles of *Scolytus ventralis* (Coleoptera: Scolytidae) infected by a nematode parasite. *J. Invert. Pathol. 18*: 363-372.

Atkins, M. D. (1960). A study of the flight of the Douglas-fir beetle *Dendroctonus pseudotsugae* Hopk. (Coleoptera: Scolytidae). II. Flight movements. *Can. Entomol. 92*: 941-954.

Atkins, M. D. (1961). A study of the flight of the Douglas-fir beetle *Dendroctonus pseudotsugae* Hopk. (Coleoptera: Scolytidae). III. Flight capacity. *Can. Entomol. 93*: 467-474.

Atkinson, T. H., and Wilkinson, R. C. (1979). Microsporidan and nematode incidence in live-trapped and reared southern pine beetle adults. *Fla. Entomol. 62*: 169-175.

Baker, W. L. (1972). *Eastern Forest Insects*. USDA Miscell. Publ. No. 1175.

Berryman, A. A. (1974). Dynamics of bark beetle populations: Towards a general productivity model. *Environ. Entomol. 3*: 579-585.

Bovien, P. (1937). Some types of association between nematodes and insects. *Vid. Meddel. Dansk Naturh. Forening. Kobenhavn 101*: 1-114.

Bridges, J. R. (1979). An artificial diet for rearing the southern pine beetle, *Dendroctonus frontalis* Zimm. (Coleoptera: Scolytidae). *J. Ga. Entomol. Soc. 14:* 278-279.

Carter, W. (1973). *Insects in Relation to Plant Disease.* John Wiley and Sons, New York.

Dale, P. S. (1967). Nematodes associated with the pine-bark beetle, *Hylastes ater*, in New Zealand. *N.Z. J. Sci. 10:* 222-234.

Finney, J. R., and Mordue, W. (1976). The susceptibility of the elm bark beetle *Scolytus scolytus* to the DD-136 strain of *Neoaplectana* sp. *Ann. Appl. Biol. 83:* 311-312.

Finney, J. R., and Walker, C. (1977). The DD-136 strain of *Neoaplectana* sp. as a potential biological control agent for the European elm bark beetle, *Scolytus scolytus. J. Invert. Pathol. 29:* 7-9.

Finney, J. R., and Walker, C. (1979). Assessment of a field trial using the DD-136 strain of *Neoaplectana* sp. for the control of *Scolytus scolytus. J. Invert. Pathol. 33:* 239-241.

Fuchs, G. (1914). *Tylenchus dispar curvidentis* m. und *Tylenchus dispar cryphali* m. *Zool. Anz. 45:* 195-207.

Fuchs, G. (1915). Die Naturgeschichte der Nematoden und einiger anderer Parasiten des *Ips typographus* L. 2. des *Hylobius abietis* L. *Zool. Jahrb. (Syst.) 38:* 109-222.

Fuchs, G. (1929). Die Parasiten einiger Rüssell-und Borkenkäfer. *Z. Parasitenkd. 2:* 248-285.

Fuchs, G. (1930). Neue an Borken-und Rüsselkäfer gebundene Nematoden, halbparasitische und Wohnungseinmieter. *Zool. Jahrb. Abt. Syst. 59:* 505-646.

Fuchs, G. (1937). Neue parasitische und halbparasitische Nematoden bei Borkenkäfern und einige andere Nematoden. I. Teil *Zool. Jahrb. Abt. Syst. 70:* 291-330.

Fuchs, G. (1938). Neue Parasiten und halbparasiten bei Borkenkäfern und einige andere Nematoden. II., III., u. IV. Teil *Zool. Jahrb. Abt. Syst. 71:* 123-190.

Furniss, M. M. (1967). Nematode parasites of the Douglas-fir beetle in Idaho and Utah. *J. Econ. Entomol. 60:* 1323-1326.

Furniss, R. L., and Carolin, V. M. (1977). *Western Forest Insects.* USDA Miscell. Publ. No. 1339.

Hetrick, L. A. (1940). Some factors in natural control of the southern pine beetle, *Dendroctonus frontalis* Zimm. *J. Econ. Entomol. 33:* 554-556.

Hirschmann, W., and Rühm, W. (1953). Milben und Fadenwürmer als Symphoristen und Parasiten des Buchdruckers. *Mikrokosmos 43:* 7-10.

Hoffard, W. H., and Coster, J. E. (1976). Endoparasitic nematodes of *Ips* bark beetles in eastern Texas. *Environ. Entomol. 5:* 128-132.

Hunt, D. J., and Hague, N. G. M., (1974a). A redescription of *Parasitaphelenchus oldhami* Rühm, 1956 (Nematoda: Aphelenchoididae) a parasite of two elm bark beetles: *Scolytus scolytus* and *S. multistriatus*, together with some notes on its biology. *Nematologica 20:* 174-180.

Hunt, D. J., and Hague, N. G. M. (1974b). The distribution and abundance of *Parasitaphelenchus oldhami*, a nematode parasite of *Scolytus scolytus* and *S. multistriatus*, the bark beetle vectors of Dutch elm disease. *Plant Pathol. 23:* 133-135.

Hunt, D. J., and Hague, N. G. M. (1976). The bionomics of *Crytaphelenchoides scolyti* n. comb., syn. *Ektaphelenchus scolyti* Rühm, 1956 (Nematoda: Aphelenchoididae) a nematode associate of *Scolytus scolytus* (Coleoptera: Scolytidae). *Nematologica 22:* 212-216.

Hunt, R. S., and Poinar, G. O. (1971). Culture of a *Parasitorhabditis* sp. (Rhabditida: Protorhabditinae) on a fungus. *Nematologica 17*: 321-322.

Khan, M. A. (1957a). *Sphaerularia bombi* Duf. (Nematoda: Allantonematidae) infesting bumblebees and *Sphaerularia hastata* sp. nov. infesting bark beetles in Canada. *Can. J. Zool. 35*: 519-523.

Khan, M. A. (1957b). *Sphaerularia ungulacauda* sp. nov. (Nematoda: Allantonematidae) from the Douglas fir beetle, *Dendroctonus pseudotsugae* Hopk., with key to *Sphaerularia* species (Emended). *Can. J. Zool. 35*: 635-639.

Kinn, D. N. and Stephen, F. M. (1981). The incidence of endoparasitism of *Dendroctonus frontalis* Zimm. (Coleoptera: Scolytidae) by *Contortylenchus brevicomi* (Massey) Rühm (Nematoda: Sphaerulariidae). *Z. Ang. Entomol. 9*: 452-458.

Laumond, C., and Carle, P. (1971). Némalodes associés et parasites de *Blastophagus destruens* Woll. (Col. Scolytidae). *Entomophage 16*: 51-66.

Laumond, C., and Ritter, M. (1971). Les némalodes parasites des insectes Xylophages. *Ann. Zool. Ecol. Anim.* (no hors-série) 195-205.

Lieutier, F. (1979). Le parasitisme d' *Ips sexdentatus* (Coleoptera: Scolytidae) par *Contortylenchus diplogaster* (Nematoda: Allantonematidae). *Rev. Nematol. 2*: 143-151.

Lieutier, F., and Laumond, C. (1978). Némalodes parasites et associés á *Ips sexdentatus* et *Ips typographus* (Coleoptera, Scolytidae) en région Parisienne. *Nematologica 24*: 184-200.

Linstow, D. von (1890). Über *Allantonema* und *Diplogaster*. *Zentralbl. Bakt. Paraskde. 8*: 489-493.

MacGuidwin, A. E., Smart, Jr., G. C., and Allen, G. E. (1980). Redescription and life history of *Contortylenchus brevicomi*, a parasite of the southern pine beetle *Dendroctonus frontalis*. *J. Nematol. 12*: 207-212.

MacGuidwin, A. E., Smart, Jr., G. C., Wilkinson, R. C., and Allen, G. E. (1980). Effect of the nematode *Contortylenchus brevicomi* on gallery construction and fecundity of the southern pine beetle. *J. Nematol. 12*: 278-282.

Massey, C. L. (1956). Nematode parasites and associates of the Engelmann spruce beetle (*Dendroctonus engelmanni* Hopk.). *Proc. Helm. Soc. Wash. 23*: 14-24.

Massey, C. L. (1960). Nematode parasites and associates of the California five-spined engraver, *Ips confusus* (Lec.). *Proc. Helm. Soc. Wash. 27*: 14-22.

Massey, C. L. (1962). Life history of *Aphelenchulus elongatus* Massey (Nematoda), an endoparasite of *Ips confusus* LeConte, with a description of the male. *J. Insect Pathol. 4*: 95-103.

Massey, C. L. (1964). The nematode parasites and associates of the fir engraver beetle, *Scolytus ventralis* LeConte, in New Mexico. *J. Insect Pathol. 6*: 133-155.

Massey, C. L. (1974). *Biology and Taxonomy of Nematode Parasites and Associates of Bark Beetles in the United States*. USDA Agric. Handbook No. 446.

Moore, G. E. (1970). *Dendroctonus frontalis* infection by the DD-136 strain of *Neoaplectana carpocapsae* and its bacterium complex. *J. Nematol. 2*: 341-344.

Mott, R. L., Thomas, H. A., and Namkoong, G. (1978). In vitro rearing of southern pine beetle larvae on tissue-cultured loblolly pine callus. *Ann. Entomol. Soc. Am. 71*: 564-566.

Nelmes, A. J., and Hussain, W. I. (1972). The response of the bark boring beetle *Ips sexdentatus* infected by a nematode parasite *Contortylenchus* sp. *Intern. Symp. Nematol., Reading, U.K. 3-8 Sept. 1972*, p. 48.

Nickle, W. R. (1963a). The endoparasitic nematodes of California bark beetles with descriptions of *Bovienema* n. g. and *Neoparasitylenchus* n. subg. and with the presentation of new information on the life history of *Contortylenchus elongatus* n. comb. Ph.D. thesis, University of California, Davis.

Nickle, W. R. (1963b). Notes on the genus *Contortylenchus* Rühm, 1956, with observations on the biology and life history of *C. elongatus* (Massey, 1960) n. comb., a parasite of a bark beetle. *Proc. Helm. Soc. Wash. 30*: 218-223.

Nickle, W. R. (1963c). *Bovienema* (Nematoda: Allantonematidae), a new genus parasitizing bark beetles of the genus *Pityogenes* Bedel, with notes on other endoparasitic nematodes of scolytids. *Proc. Helm. Soc. Wash. 30*: 256-262.

Nickle, W. R. (1963d). Observations on the effect of nematodes on *Ips confusus* (LeConte) and other bark beetles. *J. Insect Pathol. 5*: 386-389.

Nickle, W. R. (1967). On the classification of the insect parasitic nematodes of the Sphaerulariidae Lubbock, 1861 (Tylenchoidea: Nematoda). *Proc. Helm. Soc. Wash. 34*: 72-94.

Nickle, W. R. (1970). A taxonomic review of the genera of the Aphelenchoidea (Fuchs, 1937) Thorne, 1949 (Nematoda: Tylenchida). *J. Nematol. 2*: 375-392.

Nickle, W. R. (1971). Behavior of the shothole borer, *Scolytus rugulosus*, altered by the nematode parasite *Neoparasitylenchus rugulosi*. *Ann. Entomol. Soc. Am. 64*: 751.

Oldham, J. N. (1930). On the infestation of elm bark beetles (Scolytidae) by a nematode, *Parasitylenchus scolyti* n. sp. *J. Helm. 8*: 239-248.

Poinar, G. O., Jr. (1972). Nematodes as facultative parasites of insects. *Annu. Rev. Entomol. 17*: 103-122.

Poinar, G. O., Jr. (1975). *Entomogenous Nematodes*. E. J. Brill, Netherlands.

Poinar, G. O., Jr., and Caylor, J. N. (1974). *Neoparasitylenchus amvlocercus* sp. n. (Tylenchida: Nematodea) from *Conophthorus monophyllae* (Scolytidae: Coleoptera) in California with a synopsis of the nematode genera found in bark beetles. *J. Invert. Pathol. 24*: 112-119.

Reid, R. W. (1958). Nematodes associated with the mountain pine beetle. *Can. Dep. Agr. For. Biol. Div. Bi.-Mon. Prog. Rep. 14*: 3.

Rühm, W. (1956). Die Nematoden der Ipiden. *Parasitol. Schrift. 6*: 1-437.

Rühm, W., and Chararas, C. (1957). Description, biologie et histologie de quatre espéces nouvelles de nématodes parasites de *Dryocoetes hectographus* Reit. (Col. Scolytidae). *Entomophaga 2*: 253-269.

Salt, G. (1963). The defense reactions of insects to metazoan parasites. *Parasitology 53*: 527-642.

Saunders, J. L., and Norris, Jr., D. M. (1961). Nematode parasites and associates of the smaller European elm bark beetle, *Scolytus multistriatus* (Marsham). *Ann. Entomol. Soc. Am. 54*: 792-798.

Schvester, D. (1950). Sur un nématode du groupe des *Parasitylenchus dispar* Fuchs, parasite nouveau du Xylebore disparate (*Xyleborus dispar* F.). *Ann. Epiphyt. 1*: 1-6.

Schvester, D. (1957). Contribution à l'étude de coléopteres scolytides. *Ann. Epiphyt. 8*: 1-62.

Steiner, G. (1932). Some nemic parasites and associates of the mountain pine beetle (*Dendroctonus monticolae*). *J. Agric. Res. 45*: 437-444.

Stoffolano, J. G., Jr. (1973). Host specificity of entomophilic nematodes—A review. *Exp. Parasitol. 33*: 263-284.

Théodoridés, J. (1950). Les nématodes des Coléoptéres Scolytides de France. *Vie Milieu: 1*: 53-68.

Thong, C. H. S., and Webster, J. M. (1973). Morphology and post-embryonic development of the bark beetle nematode *Contortylenchus reversus* (Sphaerulariidae). *Nematologica 19*: 159-168.

Thong, C. H. S., and Webster, J. M. (1975a). Effects of the bark beetle nematode, *Contortylenchus reversus*, on gallery construction, fecundity, and egg viability of the Douglas fir beetle, *Dendroctonus pseudotsugae* (Coleoptera: Scolytidae). *J. Invert. Pathol. 26*: 235-238.

Thong, C. H. S., and Webster, J. M. (1975b). Effects of *Contortylenchus reversus* (Nematoda: Sphaerulariidae) on hemolymph composition and oocyte development in the beetle *Dendroctonus pseudotsugae* (Coleoptera: Scolytidae). *J. Invert. Pathol. 26*: 91-98.

Thorne, G. (1935). Nemic parasites and associates of the mountain pine beetle (*Dendroctonus monticolae*) in Utah. *J. Agric. Res. 51*: 131-144.

Yatsentkovskii, A. V. (1924). The castration of pine borers by roundworms (Nematodes) and their influence on the life of bark beetles (Ipidae) (in Russian). *Rep. Belorussian Agr. Inst. 3*: 278-296.

Chapter 21
Nematode Parasites of Hymenoptera

Robin A. Bedding   *Commonwealth Scientific Industrial Organisation Tasmanian Regional Laboratory, Hobart, Tasmania, Australia*

I. INTRODUCTION

There are about 100,000 described species in the order Hymenoptera, which include some of the most highly specialized and successful insects. These range from the primitive phytophagous Symphyta, to the ants, bees, and wasps, many of which exhibit an extraordinary degree of social organization and caste differentiation. They include many of the most important insect parasitoids. Although some Hymenoptera are serious pests to humans, no other order contains as many species that are of direct benefit.

What was possibly the earliest irrefutable record of an entomophagous nematode (Reamur, 1742) concerns the most unusual *Sphaerularia bombi*, which commonly parasitizes queen bumblebees. This was followed a short while later by the observation of Gould (1747) of mermithid parasitism of ants. These two topics, both equally fascinating in their own, but different ways, have since been the subject of many dozens of publications.

Of more economic importance and of no less biological interest, are the nematode parasites of woodwasps, which were not discovered until 1962. These nematodes were the first to be successfully used for large-scale biological control of an insect pest.

Taken as a whole, the nematode parasites of Hymenoptera, though not necessarily the most economically important, must certainly rate as among the most varied and interesting.

II. ANTS

A. Rhabditids

Many free-living rhabditoid nematodes, which feed saprozoically, have resistant, long-lived third-stage juveniles that are produced when the food supply is depleted or exhausted. These survivor-larvae or "dauer larvae," using surface tension forces, stand on their posterior ends and weave and wave the rest of their bodies free of the substrate, particularly when mechanically disturbed. They can then rapidly attach to passing insects, make their way to regions most protected from desiccation and mechanical damage, and be transported (phoresy) to a fresh environment and new food supply. Also, since every insect must of course eventually die, those nematodes already present have an advantage over other saprozotes in colonizing the cadaver.

An interesting, but slight transition from this, toward a parasitic mode of life, is illustrated by a few species of such nematodes associated with ants. Janet (1893, 1894, 1909) was first to observe, from *Formica rufa*, that frequently one of the four pairs of ant head glands, the pharyngeal glands, were inhabited by one or more tiny nematodes (Fig. 1). Janet was able to show that the juvenile nematodes (originally described as *Pelodera janeti*, but now known to be *Caenorhabditis dolichura*) roughly doubled their size after entry into the gland, but would not grow in water, indicating that the nematodes obtained nourishment from their host. However, the worms never reached sexual maturity in the live ant; they only completed development and reproduced after feeding saprozoically in refuse heaps within the ants' nest. The nematodes could also complete many reproductive cycles without invading an ant.

Wahab (1962) made a comprehensive study of *C. dolichura* and three other species of nematodes he found while examining the pharyngeal glands of 25 species from six genera of ants, mainly from *Frankonia* in Germany. After dissecting some 14,000 ants from 550 nests and making 1752 hanging-drop

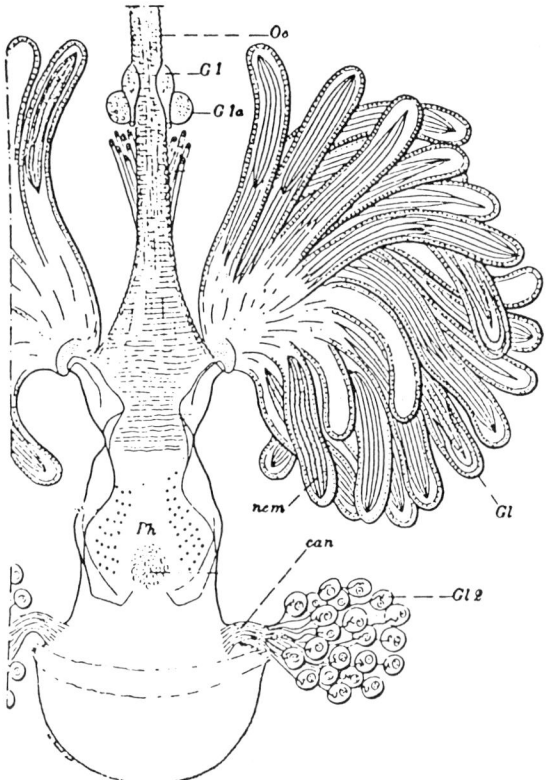

Fig. 1 Dissection of the head of the ant, *Formica rufa*, showing pharyngeal glands invaded by *Caenorhabditis dolichura*; (Gl) shows tubes of pharyngeal glands containing nematodes (nem). (After C. Janet, 1894).

cultures, he found that whereas most species of the ant subfamily Camponotinae were parasitized, with over 40% of nests and 20% of individual ants infected in nature, parasitism of various species of Myrmicinae was very rare. Since Wahab was readily able to infect myrmicine ants with nematodes in the laboratory, he concluded that occurrence in the field was dependent on ant feeding behavior and type of habitat. Thus Camponotinae are much more readily infected because, besides being predatory, they feed on dead and decaying substances on which nematodes reproduce, whereas myrmicine ants are rarely exposed to infection because they are strictly predatory or herbivorous. Wahab also found that nests in moist situations were more susceptible to infection, doubtless because free-living stages of nematodes require high humidity for infection.

Wahab found two nematode species to be common in ant nests. *Diploscapter lycostoma* (in 127 nests) preferred cultivated soils, and *Caenorhabditis dolichura* (in 58 nests) preferred forest soils. *Pristionchus lheritieri* was found in only five nests and *Eudiplogaster histophorus* in three. Two nematode species were often found in the same nest and occasionally even in the same ant.

In the laboratory, artificially reared nematodes mixed with ant food readily infected all ant species tested. Although those nematodes entering the intestine soon died, most entered the pharyngeal glands, so that these

glands often contained more than a hundred nematodes (although infected ants
from field populations usually contain only from 1 to 10).

Even heavily infested ants appeared to suffer no other harm than expansion of the gland tubules and thinning of the epithelium lining these tubules.
Wahab concluded that natural infection was *per os*, occurring when the ant
fed saprozoically, during the cleaning process, and possibly during trophyl-
laxis. Wahab found that although infective nematodes would not develop in
water prior to infecting the ant, they would, after spending some time within
the ants' glands, then mature to adults when placed in water. This observation confirmed the parasitic nature of the relationship.

Wahab tried without success to infect ants with other rhabditid nematodes
and Bedding (unpublished) was unable to infect ants either with rhabditids
associated with scarabeid larvae or with a *Pseudodiplogasteroides* species obtained from the head of the termite *Mastotermes darwiniensis*.

The first record of nematodes from the pharyngeal glands of ants in
North America was provided by Nickle and Ayre (1966). *C. dolichura* was
found in *Camponotus herculeanus* and *Acanthomyops claviger*, but not in
Formica fusca or *F. integra*; the authors were not able to artificially infect
Formica species. Nickle and Ayre noted that attachment of the waving nematode juveniles to ants was aided by an oily secretion covering the nematodes
and that it was during the ants' grooming activities that nematode juveniles
became transferred to the ants' buccal cavity. After this, nematodes were
presumed to make their way actively to the pharyngeal glands.

The Argentine ant *Iridomyrmex humilis*, well known as a pest species,
was found to be parasitized by *C. dolichura* (Markin and McCoy, 1968).

B. Mermithids

Commencing with the observation by Gould (1747) of a mermithid parasite of
the ant *Lasius flavus* E., there has possibly been more written about mermithid parasitism of the Formicidae than nematode parasitism of any other family
of insects. That ants, which have long fascinated entomologists and are
readily encountered in large numbers, can sometimes become conspicuously
changed in their morphology, caste differentiation, and behavior by mermithid
parasitism, has intrigued many researchers. The description by Wasmann
(1909) of *Pheidole symbiotica* "a remarkable, new parasitic ant" from a nest of
P. pallidula illustrates the degree to which ants can be changed by mermithid
parasitism [parasitized *P. pallidula* were later found by Wheeler (1928) to be
identical with Wasmann's description of "*Ph. symbiotica*"]. Likewise, Emery
(1890) described specimens of *Odontomachus haematoda* as a new variety
"microcephalus" and of *O. chelifer*, as a new variety "leptocephalus" before
later (Emergy, 1904) finding these to be parasitized by mermithids.

Wheeler (1901) was first to describe morphological and behavioral modifications brought about by mermithid parasitism of an ant (*P. commutata*)
and followed this up with a series of papers on the subject culminating in an
excellent review (Wheeler, 1928). During the same period, many other workers described various aspects of the parasitism of a wide variety of ant species from many parts of the world. Unfortunately the mermithid parasites involved were often not accurately identified. A recent work (Passera, 1976)
has done much to clarify the mechanisms producing the spectrum of modifications caused in ants by mermithid parasites.

1. Life Cycle

Three genera and two species have been identified: *Hexamermis* sp., *Allomermis myrmecophila*, *Mermis racovitzai*, and another *Mermis* species. The life cycle of each parasitic species doubtless differs from the others to a greater or lesser degree, but this cannot be discerned from the literature.

In general, parasitic mermithids are found within adult ants from early summer to late autumn (Wheeler, 1928; Gosswald, 1930; Taylor, 1933; Vandel, 1934; Passera, 1974, 1976), although Gosswald (1930) reports finding infected *Solenopsis fugax* in early spring. Fully fed, postparasitic juvenile mermithids leave their hosts, usually via the anus, between late autumn and midwinter (Vandel, 1934; Gosswald, 1930) and burrow into the soil of the ant nest. Vandel (1934), in a study of *Hexamermis* sp. from *P. pallidula*, found that postparasitic juveniles, which are strongly thigmotactic, never occurred within the galleries or chambers but at a depth of from 10 to 25 cm within soil, usually lying coiled at soil/stone interfaces. Gosswald (1930) and Vandel (1934) observed that postparasitic juveniles molted twice over a period of two to three months before becoming adults, and Vandel (1934) observed mating at this stage, which was followed shortly afterward (during the winter months) by oviposition. Despite the contention of Gosswald (1930) that he had obtained parthenogenetic oviposition (based on very limited evidence), Vandel (1934) showed quite clearly that in *Hexamermis* from *P. pallidula* only mated females would oviposit. Oviposition occurred over a period of several months while the worms changed from a pure white color to become quite clear just before death. Interestingly, virgin female *Hexamermis* survived two to three years, compared with the survival time of fertile females of six months at the most, and so could presumably mate with males produced during the next year. Vandel (1934) observed the ovoid eggs (30-60 µm in length) at various stages of development and stated that juveniles (probably after one month in the egg) hatch from the eggs and burrow into the soil where they may survive for several months.

Entry of infective juvenile mermithids into ants has never been observed, although Vandel (1934) achieved this experimentally by adding soil, in which mermithid oviposition had previously occurred, to an unparasitized ant colony. Gosswald (1930) conducted a similar experiment, but added honey to the soil and when ants became infected, considered that he had confirmed that infection of ants was *per os*. Poinar (1975) states that "In all detailed studies of mermithids that deposit eggs in the environment, the infective juvenile enters the host by direct penetration of the insects' cuticle," and it seems more likely that such is the case in infection of ants.

Considerable controversy existed between Vandel and Wheeler as to whether ant larvae could become infected. Vandel was never able to find an infected ant larva, but found a small infective juvenile within a pronymph and concluded this was the earliest stage at which ants were infected. Although Wheeler found no ant larvae infected, he was of the opinion that because of the considerable modifications induced in some ant specimens, well-developed mermithid juveniles must have been present by the time differentiation within the ant pupa was occurring. Wheeler (1933) therefore correctly surmised that ant larvae could become infected. Hagmeier (1912) actually found young mermithids in *Myrmica rubra* larvae in early spring, and Gosswald (1929, 1930) found well-grown (0.5 cm long) mermithids in larvae of both *S. fugax* and *Leptothorax tuberum*. However, Gosswald notes that these ant larvae were unable to complete their development. Passera (1976) determined that in *P. pallidula*, first- and second-stage and early third-stage larvae were never infected (or

at least never reached an adult stage if they were), but that differentiating
third-stage larvae could be. If, as seems likely, infection is by direct penetration rather than *per os*, it is doubtful whether adult ants could commonly
become infected because of their sclerotization and lack of any intimate contact with the soil. Further, since most mermithized ants show some structural
modifications (if only in some cases an increase in their length), mermithid
penetration must have occurred at least before completion of sclerotization.

2. Effect on the Host

It is this aspect of mermithid parasitism of ants that has received most attention, particularly in the early literature, although it was not until the work of
Passera (1976) that certain aspects were clarified.

In common with other Hymenoptera, haploid (unfertilized) ant eggs produce males; male ants are usually winged and exhibit no polymorphism. Fertilized eggs produce females, but depending upon how the resulting larvae
are fed and possibly upon exposure to pheromones, either queens or workers
result. Virgin queens exhibit no polymorphism, are winged, have a well-developed pterothorax and flight muscles, large compound eyes and three ocelli,
and are usually larger than workers of the same species. Workers are wingless and often monomorphic, but in many species there is a soldier caste which
is larger than the worker and has a disproportionately large head; in some
species, there is a series of intermediate forms between workers and soldiers.

In order to understand the array of differing modifications produced by
mermithids on ants, several variables should be considered. First, many
different subfamilies, genera, and species of ants and an indeterminate number
of mermithid genera and species are involved; second, mermithized juveniles
can enter ant larvae or pupae before, during, or after the process of differentiation of larvae or pupae already destined or partly destined to become
males, queens, workers, or soldiers; third, a variable number of mermithids
enter each ant.

In the simplest situation, if a mermithid juvenile enters an ant after its
process of differentiation has been completed, obviously, little morphological
modification apart from abdominal and gaster distension can occur and this
may not be readily apparent until the mermithid parasite is nearly full grown
and often several times the length of its host. Thus, Gosswald (1930) found
large numbers of infected workers and males of *Lasius alienus*, a male *L. niger*,
females of *L. flavus*, workers of *Formica fusca*, *F. rufibarbis*, and *F. rufopro-
tensis*, and many males and neuters of *Myrmica scabriosis* to be relatively unchanged by parasitism except for matt-grey coloration. At the other extreme
are the mermithergate, a term coined by Wheeler (1907) for parasitized workers, intercastes produced by mermithid infection occurring early in the differentiation of third-stage worker larvae. These are essentially small-headed
soldiers with certain worker and queen characteristics; the phenomenon is
well illustrated by Emery (1904), whose figures of *P. absurda* are reproduced
here (Fig. 2). The presence of ocelli in these intercastes is a queen characteristic that may be absent or poorly or well developed. Passera (1976) has
shown that in *P. pallidula*, the more worms there are and the earlier the infection, the more likely it is for the mermithergates to have well-developed ocelli.

Apart from external morphological changes, mermithid parasitization
affects behavior. Wheeler (1907) graphically describes a study of mermithergates:

> On exposure to the sunlight they hurried, like the females, to the dark
> chamber, thereby evincing a much higher degree of negative phototro-

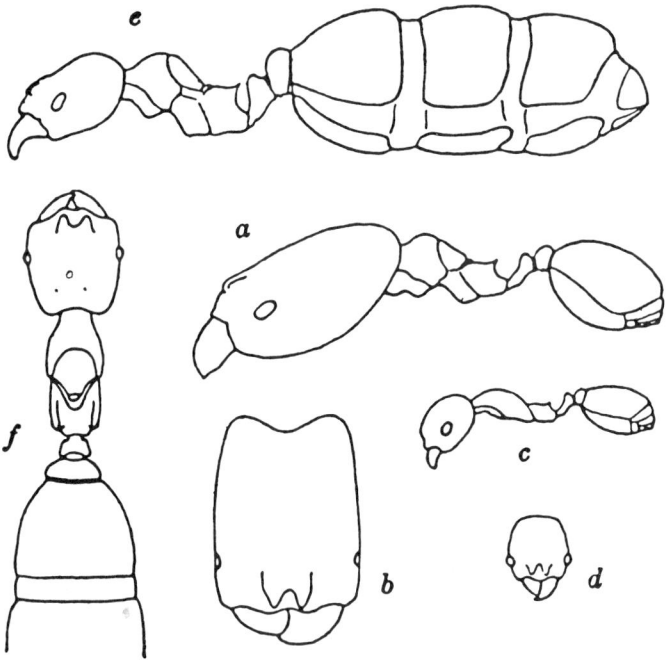

Fig. 2 The ant, *Pheidole absurda*, showing (a) soldier, in profile; (b) head of soldier; (c) worker; (d) head of worker; (e) mermithergate, lateral; (f) mermithergate, dorsal. (After C. Emery, 1904.)

pism than the workers and soldiers. They never carried the brood, although even the soldiers were seen to do this occasionally. They never fed the larvae, workers or females, and, had earth been present in the nest, it is very probable that they would never have shown any inclination to excavate. They never visited the manger of the nest but were fed exclusively by regurgitation. As befitted animals containing such enormous parasites, they were in a chronic state of hunger. It was impossible at any time to uncover the nest without finding one or more of them either being fed by a worker or eagerly begging for food. And as soon as one was offered food, three or four of them would rush up and put out their tongues for a share of it. Once I saw a single worker trying to feed five of her huge parasitized sisters simultaneously. While imbibing their food the mermithergates stridulated, either continuously or at regular intervals. Sometimes they were so impatient to be fed that they would hold down a worker's head with their large fore feet and compel the little creature to regurgitate.

Gosswald (1930) made the interesting observation that mermithergates of *L. niger*, even though difficult to distinguish morphologically from healthy workers, were readily identified because, unlike normal workers, they were unable to secrete poison when handled.

The origin of profoundly modified intercastes is a matter of some dispute; Vandel (1927, 1930, 1934) considered them to have arisen from immatures already destined to become soldiers, but depleted of nutrition by parasitism and accordingly termed them "mermithostratiotes," whereas Wheeler (1928) thought of them as modified workers or mermithergates. Passera (1976), working with *Ph. pallidula*, produced circumstantial, but significant evidence to show that the true mermithostratiotes were very rare and that intercastes were in fact mermithergates. Passera examined 18,695 nymphs from 28 infected nests (another 21 were uninfected), found an average parasitism of 2.1%, and that the percentage of healthy soldiers in both uninfected and infected nests was almost identical at 1.94 and 1.96% of the total population. Based on the assumption that all larvae should be similarly parasitized, he could have expected about eight (although Passera calculated 3.66) and in fact found three parasitized, but relatively unmodified soldiers. However, 369 intercastes were found and a similar number (367) of soldiers. Had intercastes derived from ant larvae already predestined to be soldiers, it would be necessary to explain, first, why there were twice as many soldier nymphs in infected nests as in uninfected, and, second, why when only 0.78% of the worker nymphs were parasitized, there should be a 50% parasitism of soldier nymphs.

It is of interest to note that in some other species of *Pheidole*, very similar intercastes occur without the intervention of mermithid parasitism. That mermithid parasitism can produce far greater modifications in ants than in any other group of insects is doubtless related to the plasticity of female larvae, which are naturally liable to produce profoundly differing images depending on cultural conditions. Parasitization may well interfere with the hormonal balance leading to production of intercastes, but since intercastes are larger than workers they must have received more food. It may be that part of the explanation is that parasitization of worker larvae causes these to be treated by the workers as though they were to become soldiers, but that depletion of nutrition by the parasite holds back their full development.

Other castes are less dramatically altered by parasitism. Passera (1976) seems to be alone in observing true mermithostratiotes and found little modification induced. One of the more obvious features of mermithized queens, termed *mermithogynes* by Mrázek (1908), is the tendency to shortening of the wings (see Fig. 3), but there is also often a distinct narrowing of the head and thorax. Crawley and Baylis (1921) sectioned mermithogynes of *Lasius* spp. and found atrophied ovaries and nurse cells, an absence of fat body, but hypertrophy of the tracheae. Taylor (1933) observed both mermithogynes and mermithanders (parasitized males) of *Acanthomyops flavus* and recorded that whereas the wings of the former were only about 0.6 the length of unparasitized females (average 11 as compared to 19 mm), the wings of mermithanders were only about 1 mm shorter than normal males. Taylor also noted that in mermithanders, distension of the abdomen resulted in protrusion of their genitalia. Wheeler (1928) records, as compared with normal queens, that the mermithogynes of *L. americanus* Emery and *L. brevicornis* Emery were thicker, more opaque, and more densely pilose and, although the venation was normal, the veins were darker and stouter. Wheeler (1933) describes a *Myrmica rubra* mermithogyne as being almost completely apterous although little modified in other features.

3. *Occurrence and Incidence*

The association between ants and mermithids may well have been a long one. Parasitized ants have been discovered widely in Europe from Russia to Portugal, in North, Central, and South America, the Antilles, Jamaica, Taiwan, and Australia. Species of 19 genera of ants are known as hosts.

Several authors record that mermithized ant nests are much more common in some areas than others; Gosswald (1930) noted different frequencies of in-

Fig. 3 Mermithized queen ant, left; normal queen, right. (Courtesy of H. Kaiser.)

festation of nests for different ant species and found *Lasius* spp. and *Myrmica* spp. more often affected than *Formica* spp. However, Kloft (1950) has provided the most comprehensive data on mermithid incidence in ants from his studies in a 21 km^2 area northwest of Wurtzburgmer in Germany. He found 118 infested nests out of a total of 1050 nests examined (11.2%). Of particular interest was the significant difference in the percentage of nests infected for different species of ants even of the same genus. Thus, over 200 nests of each of *L. flavus*, *L. niger*, and *L. alienus* were examined, and the percentages infested were 2, 8.5, and 31.6, respectively. Percentages of nests of other species infested varied from 3.15 for *M. rubra scabrinoidis rugulosa* to 37.9 for *M. rubra laevinoidis*, and the mean percentage parasitism in infected *Lasius* nests ranged from 5.2 in *L. flavus* to 14.3 in *L. niger*. As a result of this study, Kloft also established, as opposed to Vandel (1930), that ant nests on limestone were no more susceptible to infestation than in other areas.

The work of Passera (1976) in the Quercy district of France is notable because although he found 57% of nests of *P. pallidula* infected, average parasitism of nymphs was only 2.7% with a maximum of 7.5% in one nest (compared with a maximum of 83% of adult ants in a nest of *M. rubra laevinoidis* found by Kloft).

It seems doubtful whether mermithid parasitism has any major effect in controlling ant populations, which is perhaps fortunate, since ants are considered rather more beneficial than detrimental.

C. Steinernematids

Although it is relatively easy to infect ants in the laboratory with *Neoaplectana* spp., it is surprising that there are no reports of natural parasitism, particu-

larly from species nesting in soil. This may be because infected ants are de-
voured by workers as soon as they become moribund or die, as was observed
in laboratory tests with *Myrmica* spp. and *Camponotus* spp. (Bedding, unpublished).

III. BEES

A. Mermithids

Records of mermithids from honeybees, which commenced with that of Assmuss
(1858), are rare, and when honeybees have been found parasitized, infection
has usually been at a very low level. Milum (1938) found a larval *Mermis subnigrescens* in a single *Apis mellifera* worker from Illinois and suggested that
infection may have occurred during water uptake from foliage (on which this
nematode oviposits); Morse (1955) obtained two larvae of an *Agamomermis*
species from over 2000 honeybees in the New York area. However, honeybees
can sometimes be heavily infected. Vasiliadi (1970) reports that, in 1966, T.
A. Atakishiev found 2017 worker bees out of 3254 (61% parasitized by mermithids in bee colonies on flood plains and river banks whereas parasitization
was totally absent in arid areas. In 1969, Vasiliadi found 15% of worker bees
alighting at the entrance of one hive at the Moscow Agricultural Academy to
be mermithized. Such levels of parasitism may have an adverse affect on
honey production.

B. Sphaerularia

With the exception of a single record of an unidentified mermithid from *Bombus*
(Maclean, 1966) there has been only one species of nematode found in bumblebees. However, this nematode, *Sphaerularia bombi* (Tylenchida: Sphaerulariidae), is relatively common and is in many respects one of the most remarkable
of the nematode parasites of insects. First described and illustrated by Reaumur (1742), it provides one of the earliest authenticated records of an entomogenous nematode. The parasitic stage is so unusual that Dufour, in 1836,

> . . . at first supposed that it was a dipterous larva, but soon saw that
> it belonged to the Entozoa; and as it certainly could not be referred to
> any other genus, he gave it the appropriate name of *Sphaerularia*.

(Lubbock, 1861). von Siebold (1838) classified *S. bombi* as a nematode after
finding juveniles in the hemocoels of parasitized bees but was still unable to
explain the structure of the "parent" parasite. Of this Lubbock (1861) re-
marked that:

> In turning to the internal anatomy, one can, to some highly important
> organs, and systems of organs, only parody Van Troil's celebrated
> chapter on the snakes in Iceland, and say simply that there are in
> *Sphaerularia*, no muscles, no nervous or circulatory systems and no
> intestinal canal.

The confusion of these early investigators was hardly surprising for they
were observing a unique and hitherto unknown phenomonon. What they mistook for the parent parasite was in fact the everted tubular and highly hypertrophied female reproductive system (Schneider, 1885) sometimes over
20 mm long and nearly 2 mm in diameter (Lubbock, 1861). Lubbock (1861)
actually saw and illustrated the tiny (1-2 mm long) parent female nematode
attached to the everted reproductive system, but considered this to be a male
in permanent copulation with the giant female.

Accounts of various aspects of parasitism by *S. bombi* have been particularly numerous since Leuckart (1887), although this author produced a comprehensive and accurate treatise.

1. Life History

The Bombini (bumble or humble bees) include some of the most familiar insects in temperate regions of the Northern Hemisphere. Fertilized queen *Bombus* spp. hibernate usually in the ground, from six to nine months, depending on species. Some species begin hibernation in midsummer, others in autumn. Emergence from hibernation may occur from early to late spring or even early summer. Each unparasitized queen builds a nest soon after the end of hibernation, collects a mass of pollen, and lays a batch of eggs in a cell over this. The first eggs produce workers, which then take over pollen collected while the queen remains in the nest. After the queen has produced from 200 to 400 worker eggs she lays other eggs that give rise to males and queens at the end of the season. The new queens hibernate after copulation; males, workers, and the old queen die. Bumblebees of the genus *Psythirus* are parasitic in the nests of *Bombus* spp. and hibernate in similar situations.

S. bombi has only ever been found in queens. When parasitized queens emerge from hibernation, the huge uterine sacs of *S. bombi* are readily apparent on dissection, as can be seen from Figs. 4 and 5 from Poinar and van der Laan (1972). These authors found that most queens collected at various

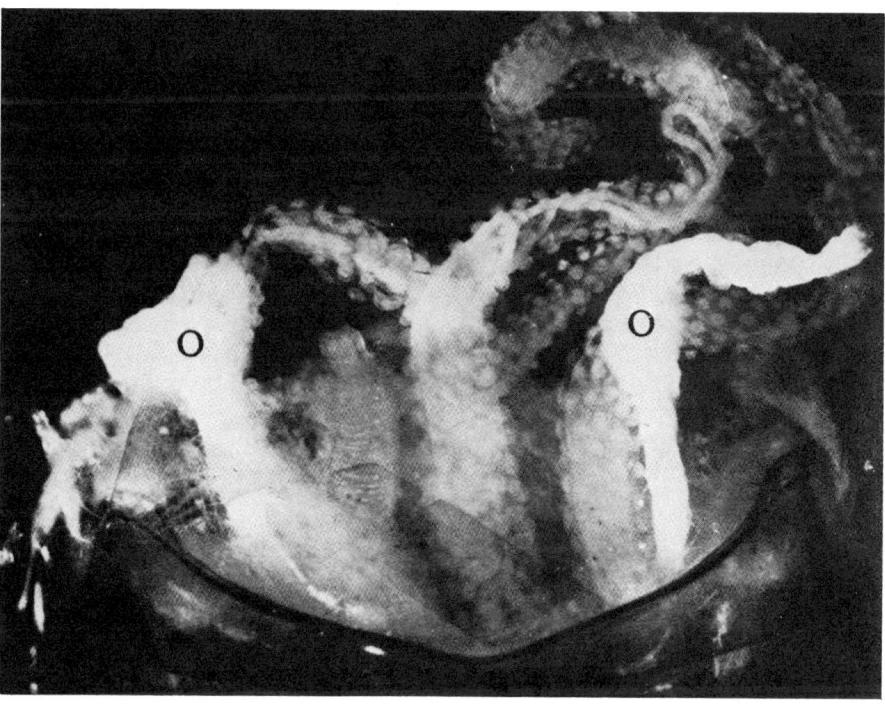

Fig. 4   Abdomen of a parasitized *Bombus* species opened to show the uterine sacs of *S. bombi*; O shows atrophied ovaries. (After G. O. Poinar and P. A. Van der Laan, 1972.)

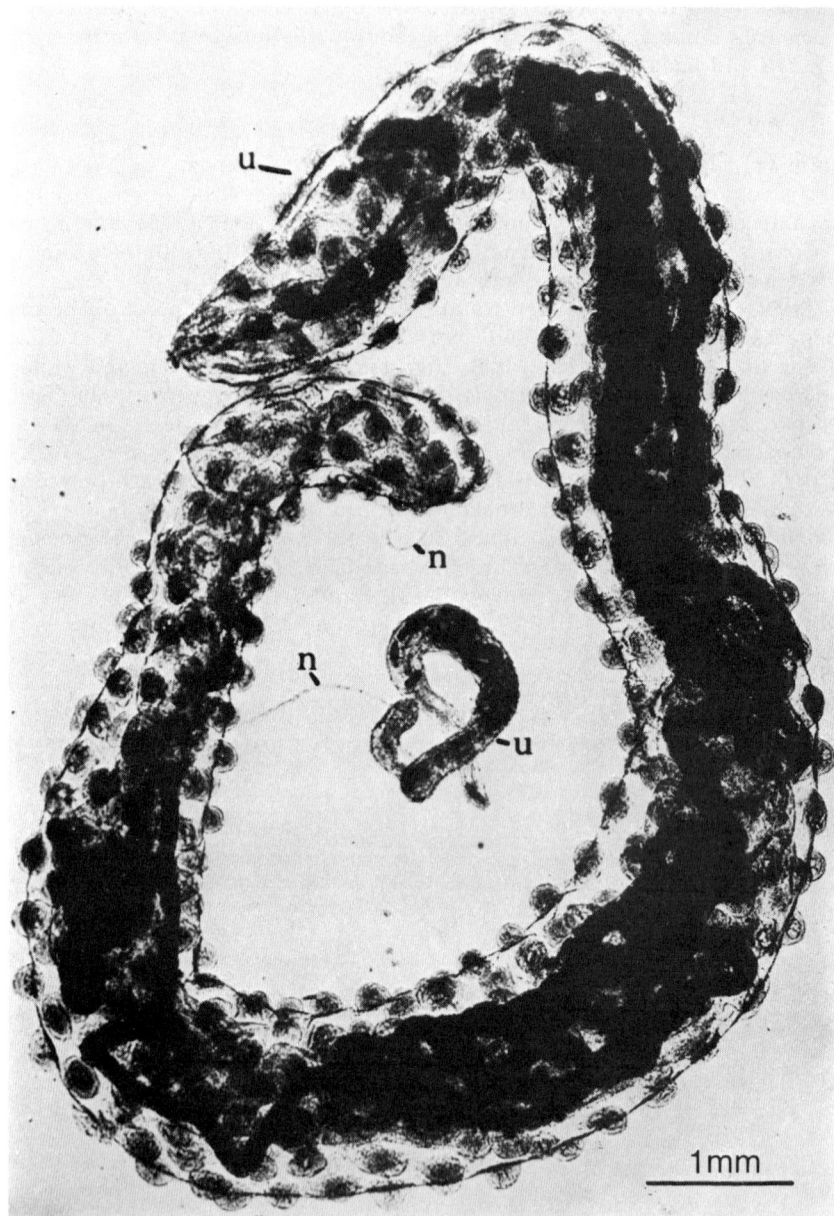

Fig. 5 Size variation in egg-producing uterine sacs of two individuals of *S. bombi*; (n) shows body of female nematode without reproductive organs; (u) shows uterine sacs. (After G. O. Poinar and P. A. van der Laan, 1972.)

localities in The Netherlands contained only a single parasite, but recovered 72 parasites from one *B. terrestris* queen. Lubbock (1861) found most parasitized bees contained from five to eight parasites, and Alford (1969) notes that most hosts have fewer than six parasites. From two locations in England the latter author found means of 5.8 and 3.5 parasites per host from 81 and 26 parasitized queens, respectively. Only one to four parasites were found in *Bombus* queens from California (Poinar, 1974). As might be expected, there appears to be a correlation between the percentage parasitism of a population and the degree of superparasitism. Generally, heavily superparasitized hosts produce smaller parasites although there is often considerable variation in the size of parasites from a single host (Cumber, 1949). Poinar and van der Laan found a variation of from 2 to 20 mm in the length of mature egg-producing uterine sacs depending on the number of parasites present, and Lundberg and Svensson (1975) obtained uterine sacs as long as 30 mm from queens with few parasites.

When parasitized bumblebees first emerge from hibernation, their behavior is little modified and like unparasitized bees they collect nectar and pollen. Production of eggs within *S. bombi* uteri usually occurs within a few weeks of parasitized queens emerging from hibernation. Cumber (1949) recorded that the first release of eggs into the hemocoel of a *B. lucorum* collected near London occurred at the end of April and found newly hatched juveniles in another of the same species five days later. Stein (1956) records that brood reached full development during April and May.

The parasite eggs are usually passed into the host's hemocoel, but they may hatch within the uterine tube, resulting in this becoming packed with juveniles. Poinar and van der Laan (1972) found that two molts occur within the egg of *S. bombi* so that all juveniles within living hosts are of the third stage. However, Khan (1957) notes a significant increase in length (with no reduction of breadth) of juveniles in the hemocoel, which indicates that feeding probably occurs at this stage.

Although few authors record the number of juvenile nematodes produced within a single host, Lubbock (1861) estimated that one queen examined had 50,000 juveniles, three had 60,000 each, and one over 100,000 juveniles swarming within the hemocoel of their host.

Juvenile *S. bombi* penetrate from the hemocoel into the rectum of their hosts and to a lesser extent into the oviducts (Pouvreau, 1964). Pouvreau observed expulsion, through the bee's anus, of whitish balls consisting of several thousand juveniles and also found many juveniles in the feces. Cumber (1949), Stein (1956), and Poinar and van der Laan (1972) have noted that large numbers of juveniles can be observed swarming up the sides of small glass containers soon after parasitized bees have been confined. It is likely, therefore, that wherever bees in the later stages of parasitism alight, numbers of third-stage juveniles are deposited. Poinar and van der Laan (1972) reported that almost every queen of *B. terrestris, B. lucorum, B. lapidarius*, and *B. hypnorum* caught at one site after the beginning of summer contained third-stage juveniles ready to leave their hosts. At an even earlier stage of parasitism the behavior of *Bombus* spp. queens becomes profoundly modified in a manner that later leads to the wide dispersal of third-stage juvenile *S. bombi*.

Unlike unparasitized queens, which establish nests soon after the end of hibernation, parasitized queens never do this, although Alford (1969) found two foundress queens containing dead *S. bombi*. Parasitized queens exhibit a behavior pattern somewhere between the nest seeking and later hibernacula seeking of unparasitized queens, but with the addition of considerable disorientation.

Poinar and van der Laan (1972) describe infected queens in woodlands in The Netherlands flying close to the ground, often alighting and crawling under fallen leaves and digging small holes up to 2 cm deep. After remaining in the soil for only a few minutes, the bees flew a few meters and repeated the process, gradually working their way to the top of a hill before flying down and starting again. These workers were able to find third-stage *S. bombi* larvae in the soil where bees had been digging. Similarly, Lundberg and Svensson (1975) observed *Bombus* spp. in a sub- to midalpine region in northern Sweden. Parasitized queens were particularly attracted to small wind-exposed hillocks and approached these in a fast, high, and direct flight before descending to the base of the hillock and working their way to the top in a manner already described. The flight of these queens was unsteady, and the wing beat produced a tone easily distinguished from that of healthy queens. During landing, parasitized bees often hit grass stems and fell to the ground. Although unparasitized queens were readily able to select hibernation sites in sheltered places on the wind-protected lower slopes of these hillocks, parasitized individuals were apparently undiscriminating and dug both in these places and in the most exposed areas at the top of hillocks. There was nevertheless an overlap of areas that would later be visited by unparasitized queens seeking hibernation sites. Bols (1939) found holes originally dug by infected queens were sometimes used later by hibernating unparasitized *Bombus* queens, and Lundberg and Svensson (1975) suggest that unparasitized queens are likely to be attracted visually to the preliminary diggings of parasitized queens.

After juvenile nematodes have been deposited by parasitized bees they take about seven weeks to molt into the fourth stage and another two weeks to become adult (Poinar and van der Laan, 1972). There is no feeding by the nematodes after leaving their host. Males and females mate and the males die, but female nematodes may survive for many months before entering a host. Stein (1956) was able to keep free-living *S. bombi* alive on agar plates for nine months, revived desiccated free-living *S. bombi* two years after keeping them on agar plates together with activated charcoal, and found that they would survive in a refrigerator for several years.

Leuckart (1887) originally suggested that infection of each new generation of bumblebee queens occurred after they entered soil previously contaminated with *S. bombi* from parasitized queens of the previous generation. This contention has since been supported by many workers (Cumber, 1949; Alford, 1969; Poinar and van der Laan, 1972; Lundberg and Svensson, 1975), although it has also been suggested that infective-stage nematodes may enter bee larvae or adults in the nest (Schneider, 1885; Stein, 1956; Nickle, 1967). Leuckart (1887) and Madel (1966) showed that when queen bees were kept in soil containing free-living *S. bombi*, they became infected. There is very strong, though admittedly only circumstantial, evidence that most if not all parasitism observed in field-collected bumblebees is a result of infection during hibernation. However, it must be remembered that parasitized queens, aptly called *de eeuwige zoekers* (the eternal seekers) by Minderhoud (1951), fly from dawn to dusk for several months of the year depositing juvenile nematodes wherever they alight. Lundberg and Svensson (1975) even recorded parasitized queens exploring vole holes, which may be used for nest building by unparasitized queens. Juvenile nematodes are also likely to be deposited on flowers visited by parasitized bees, and these may be picked up when unparasitized bees visit the same flower shortly afterward. Both these possibilities could occasionally lead to the presence in nests of juvenile *S. bombi*, but some 10 weeks would need to elapse before infective female *S. bombi* resulted, and a relative humidity of near 100% would be required for infection.

Penetration by *S. bombi* has never been observed, but Madel (1973) provided the first evidence that infection is, at least in part, *per os*, when he found in sections of queens, accumulations of *S. bombi* infectives in the musculature of the anterior region of the midintestine.

This gives some significance to the observation by Lubbock (1864) on young parasites that:

> I have found that they lie at the anterior part of the stomach; the female (everted uterus) indeed being free — but the male (female nematode) being attached by its cephalic extremity.

It seems that infection may occur at any time during the long period (six to nine months) of hibernation of queen bees. If young queens become infected when temperatures are still warm, full development of parasitism may take only four to five weeks (Stein, 1956). Occasionally queens enter hibernation early but may leave it again if conditions become warmer. Poinar and van der Laan found a hibernating *B. hypnorum* already infected in mid-August. The possibility exists that there may occasionally be a second generation of parasitism during a single year.

2. Effect on Host

Apart from the behavioral effects already noted in connection with the life cycle of *S. bombi*, parasitism also sterilizes the queen. Cumber (1949) noted that the presence of even a single parasite inhibited ovarial development so that no eggs developed although ovaries appeared similar to those of a normal queen freshly emerged from hibernation. When several parasites were present, however, the ovaries were reduced to translucent filaments, and the fat bodies, which were little affected when parasitism was light, were greatly reduced, translucent, and watery. Palm (1948) reported the significant finding that the corpora allata of parasitized bees were inactive, a feature not found in unparasitized bees until the onset of hibernation. The continual searching and digging activities of parasitized bees may possibly be connected with this.

3. Physiology of Parasitic S. bombi

Fertilized female *S. bombi* average 1.32 mm in length and 0.025 mm in greatest diameter (Poinar and van der Laan, 1972). The everted uterus may reach a length of 30 mm and a diameter of 2 mm. Considering each as of basically cylindrical shape, there may therefore be up to a 36,000-fold increase in volume. However, the original female after eversion of the reproductive organs is nonfunctional, motionless, and does not enlarge after entering the hemocoel (Poinar and van der Laan, 1972). From an early stage, the uterus has had to take over the functions of nutrition, excretion, and gaseous exchange in order to accommodate this huge size increase and fulfill its main function of producing many thousands of offspring. It has no internal digestive apparatus, and its physiology is unusual.

Poinar and Hess (1972), who examined the ultrastructure of the everted uterus, found that the outer surface area of the uterus is vastly increased by being modified into a network of saccular indentations separating off fine cytoplasmic extensions (see Fig. 6). Also, each nucleus and its associated

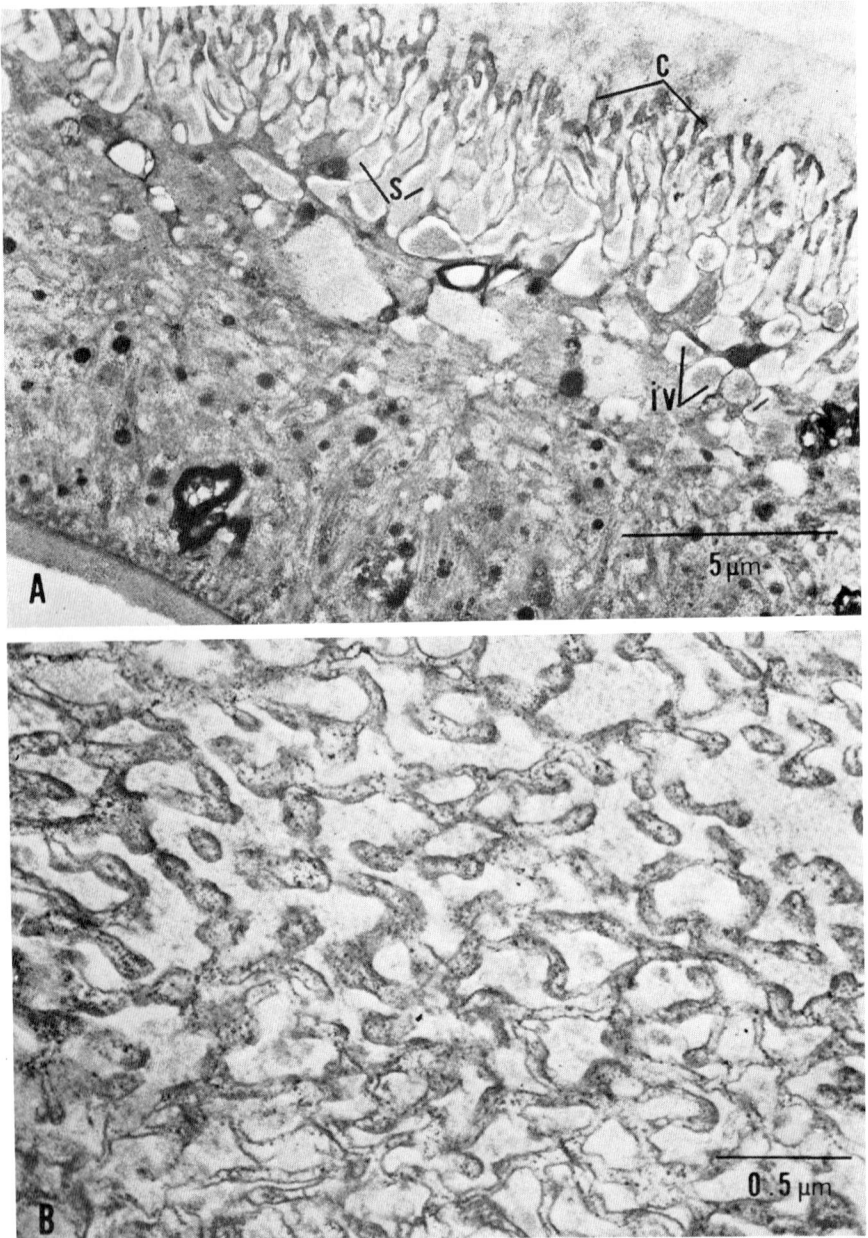

Fig. 6 (A) Outer surface of the uterus of *S. bombi* showing saccular indentations (S) surrounded by cytoplasmic extensions (C) and the formation of intracellular vacuoles (IV). (B) Tangential section through cytoplasmic extensions. (After G. O. Poinar and P. A. van der Laan, 1972.)

cytoplasm is greatly enlarged to produce the pronounced knobs that give *Sphaerularia* its name; this further increases the absorptive area. Poinar and Hess (1972) suggest that the presence of pinocytoticlike vacuoles in the cytoplasmic extensions and intracellular vacuoles at the base of the saccular indentations indicate that nutrition is brought about by pinocytosis. However, Madel and Scholtyseck (1976), who also studied the ultrastructure of *Sphaerularia*, do not agree that this is necessarily so and report a very rapid direct absorption of neutral red dye by the knob cells. Both groups identify an outer layer of mucopolysaccharide and believe that this may act as a protective layer preventing the host's blood cells attaching to the uterus.

4. Occurrence and Incidence

S. bombi has been recovered from some 37 species of *Bombus* and five species of the parasitic bumblebee *Psythirus*. It has been found in Europe as far north as northern Sweden, in Canada and in the United States and has been accidentally introduced into New Zealand, where it infects *B. terrestris* (Macfarlane, 1975).

Several workers have noted the difficulties of ascertaining accurately the levels of parasitization, particularly later in the season when unparasitized queens have commenced nest building but parasitized queens remain flying. Thus, Cumber (1949) found 25% of queen *B. agrorum* from Putney Heath, London, to be parasitized in early spring, whereas over 80% of this species were parasitized in early summer, and Stein and Lohmar (1972) found 26% parasitization early in the season with 57% later.

Cumber (1949) recorded that levels of parasitism in spring queens varied considerably with the *Bombus* species concerned; of 146 *B. lucorum* queens, 100 were parasitized (69%) and in another four species (with fewer numbers dissected) parasitization ranged from 11 to 28%. Alford (1969) examined at least 20 queens each of six species of *Bombus* some 50 km away from Cumber's site and also found *B. lucorum* most heavily parasitized (37%), with the five other species ranging from 2.4 to 31% parasitization. However, whereas Cumber (1949) found *B. pratorum* next most heavily parasitized (28%), Alford found this species least so (2.4%). Stein and Lohmar (1972) found considerable variation in the percentage of parasitization within different species and between each of six localities near Bonn. The most common species, *B. terrestris*, was most heavily parasitized (average, 30%) and this was attributed to a hibernation environment with high soil moisture content sheltered by trees and high bushes, whereas *B. agrorum* (average 12% parasitism) preferred arable land where temperature variations were greater and soil moisture content lower. Medler (1962), examining 13 species of *Bombus* over six years in Wisconsin, found an overall average of 12% parasitization, with parasitization ranging from 1.2 to 28.2% depending upon the species, and Fye (1966) examined four species in Ontario, Canada, over four years and obtained an average parasitization of 15%.

In general, it is likely that the localized hibernation environment will influence degree of parasitization. However, the suitability for parasitization of any one site will vary from year to year according to fluctuations in climate. It may also be significant that the most common species in each of the three areas studied was the most highly parasitized. Presumably, parasitized individuals of each species may tend to overlap more with the hibernation sites of unparasitized members of their own species and density-dependent factors may be of some importance.

It is quite probable that *S. bombi* exerts some influence on the abundance of its hosts, and since bumblebees must be considered as beneficial insects because of their pollinating activities, *S. bombi* may be one of the few harmful insect-parasitic nematodes. Although probably little can be accomplished in the way of reducing parasitism where bumblebees are already found, countries such as Australia which might profitably introduce these insects will of course have to exercise strict quarantine of any imported insects, and in New Zealand, where *S. bombi* may still be localized, steps should be taken to reduce to a minimum the spread of this nematode.

IV. WASPS

A. Mermithids

Only a single species of mermithid has been identified from the Vespidae. This was first discovered by Baird (1853) from an adult *Vespula vulgaris* and named *Gordius vespae vulgaris*, renamed *Mermis pachysoma* by von Linstow (1905), and placed in the genus *Agamomermis* by Welch (1958). Welch examined the type material of the previous authors and material from Blackith and Stevenson (1958), and found all material identical. Only well-developed parasitic juveniles of this species have been observed. In Welch's material these ranged in length from 22 to 77 mm, but Beck (1937) found one 90 mm long.

How, and at what stage, *A. pachysoma* enters its host remain unknown. The observation by Blackith and Stevenson (1958) of an infected *Vespula sylvestris* nest within an old beehive, which had been kept entirely out of contact with the ground, suggests that mermithid eggs or larvae must have been brought into the nest by foraging workers. No larval or pupal wasps have been found infected, and since no external modifications (apart from increase in size) have been reported in parasitized adults such as occur in ants, it is likely that *A. pachysoma* infects only adult wasps. Blackith and Stevenson (1958) found 65 parasitized queens out of a total of 710 (9% parasitization) in a nest of *V. vulgaris*, but none of a sample of 135 larval and pupal queens from the same nest were parasitized; none of the 226 adult males or 218 adult workers sampled from this nest were parasitized. In a nest of *V. germanica* these authors found 35% of a sample of nearly 200 queens to be parasitized, but only 0.25% of nearly 400 males and 4.5% of 67 workers parasitized from the same nest. However, Fox-Wilson (1946) quotes H. Latter finding 100% parasitization in several hundred male wasps and that parasitization caused castration. Kloft (1951) reported atrophy of wing musculature, reduction of fat body, and ovarial degeneration in parasitized *V. germanica* workers.

It seems unlikely that mermithid parasitism would have any significant effect on wasp populations.

B. Sphaerularia

The occurrence of *S. bombi* in wasps was recorded by Cobbold (1883), who reports that queen *V. rufa* and *V. vulgaris* sent to him by a Dr. Omerod contained this parasite. It seems likely that these queen wasps had hibernated in soil previously visited by parasitized *Bombus* queens (see Section III).

C. Steinernematids

Although the author has found the cadaver of a single queen *V. germanica* containing many thousands of infective juvenile *Neoaplectana feltiae* (*carpocapsae*), which had been hibernating beneath the loose back of a *Eucalyptus* tree in Tasmania, there have been no other records of wasps being naturally infected with steinernematids. However, Poinar and Ennick (1972) were readily able to infect *Vespula* workers from a bait of fruit extract, containing juvenile *N. feltiae*. Whether these nematodes could be used for the biological control of *Vespula* is doubtful, because although individual workers can be readily killed by the nematodes, spread of infection within the colony could only occur at relative humidities near 100%, which rarely occur in wasps nests.

V. FIG WASPS

All species of the Agaonidae live in figs, where in some cases (e.g., Capri figs) they are essential for pollination and production of mature fruit.

In 1864, Gasparrini first described an aphelenchid nematode, *Schistonchus caprifici*, as an associate of *Blastophaga psenes*, and Poinar (1975) reported that many fig wasps carry juvenile aphelenchid nematodes in abdominal pouches or abdominal folds and in some cases even internally.

The first evidence of an at least partially parasitic nematode in fig wasps was provided by Poinar (1979). Although there has been no previous record of adult diplogasterids within insects, Poinar found adults and larvae of a new genus, *Parasitodiplogaster sycophilon*, within the hemocoel of *Elisabethiella stuckenbergi*. It seems likely that juvenile nematodes enter the larva or pupa of a host and by the time adult female wasps leave one fig to enter another, the nematodes have matured and are ready to leave their host.

VI. SAWFLIES

A. Mermithids

von Siebold (1854) found and described the mermithid *Hexamermis albicans* in five species of tenthredinid sawflies. Since then there have been isolated records of another four species of this family of sawflies parasitized by other mermithids.

B. Steinernematids

The first steinernematid nematode to be described was found in the pamphilid sawfly *Cephaleia abietis*. It was named *Aplectana draussei* by Steiner (1923), but in 1927, Travassos erected a new genus *Steinernema* for this species. Subsequently, Chitwood and Chitwood (1937) established the family Steinernematidae, which is composed of the genera *Steinernema* and *Neoaplectana*. Mráček (1977) reported the rediscovery of *S. kraussei* from a dense infestation (800-1000/3 m^2) of *C. abietis* in forest soils in Czechoslovakia where 10% of the larvae were infected. Although the biology of this species was found to be like that of *Neoaplectana*, Mráček reported that a *Flavobacterium* species that he isolated from the *C. abietis* cadavers was the symbiotic bacterium, whereas the bacterial symbiont of any *Neoaplectana* species is a *Xenorhabdus* species. However, repeated examination by Mr. R. J. Akhurst (personal communication) of the infective juveniles of *S. kraussei* from cultures given the author by Mráček indicated that these are unable to carry *Flavobacterium* internally: recently reisolated *S. kraussei* kindly sent by Mráček have been found by Akhurst (unpublished data) to be symbiotically associated with a *Xenorhabdus* species.

Poinar (1979) points out that the only characters separating *Steinernema* from *Neoaplectana* are the number and arrangement of head papillae. Scanning

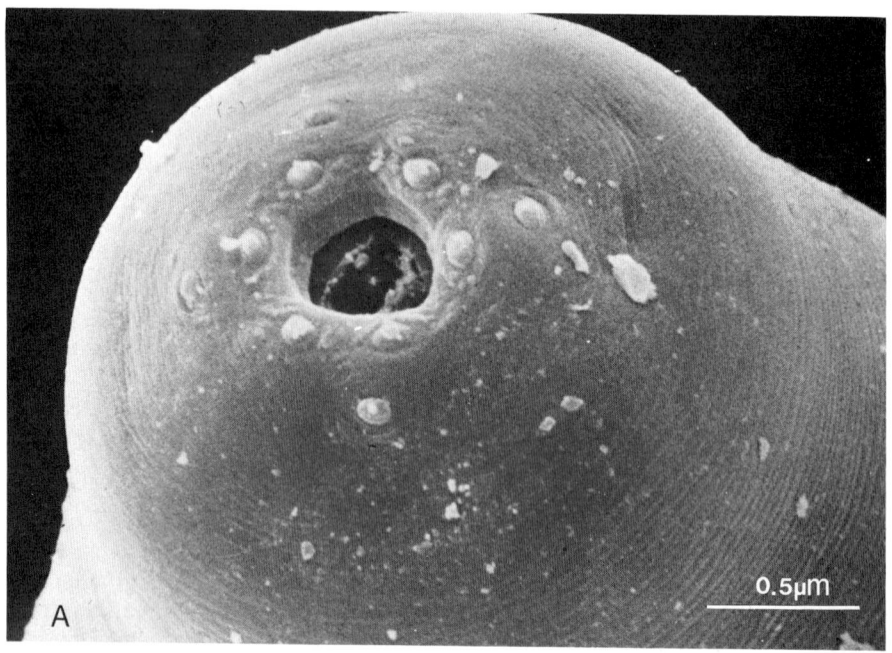

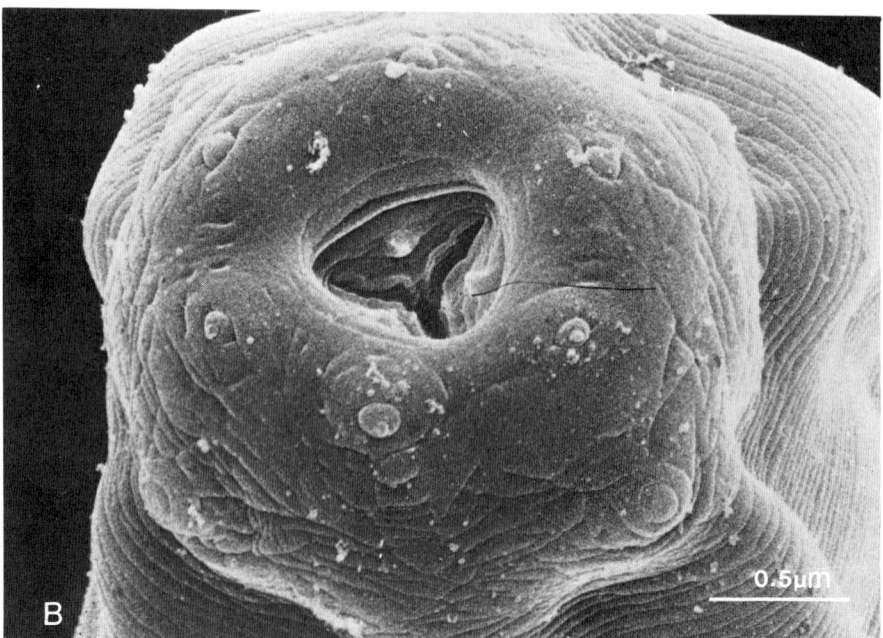

Fig. 7 Scanning electron micrographs of the head end of (A) *Neoaplectana bibionis*, (B) *N. glaseri*, (C) *Steinernema kraussei*.

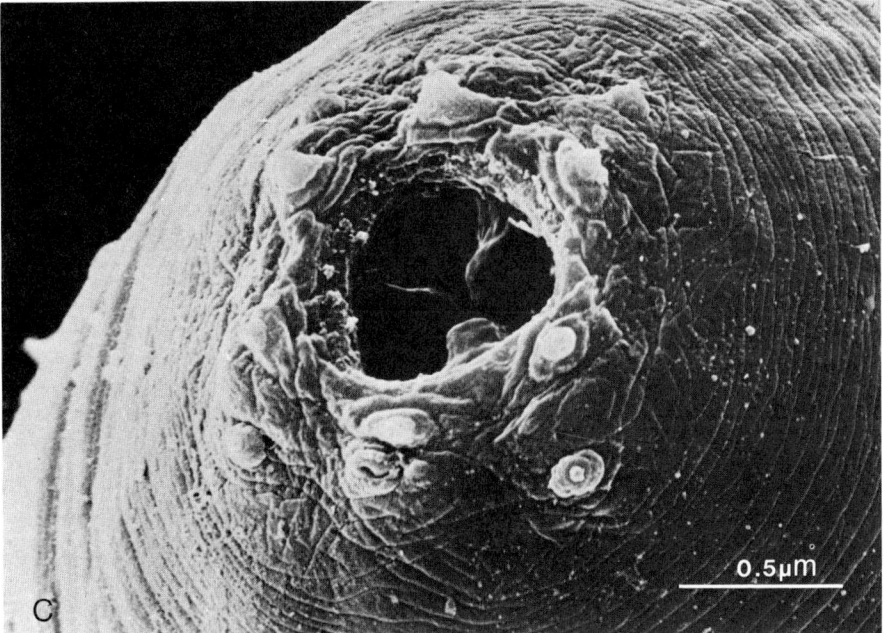

Fig. 7 (Con't.)

electron micrographs of the heads of adult female specimens of *S. kraussei* and some *Neoaplectana* species show in all species an inner circle of six papillae and an outer four papillae (Fig. 7). Previous descriptions of *Neoaplectana* spp. report an outer circle of six papillae whereas they have in fact an outer circle of only four as originally described for *S. kraussei*. The present author therefore considers that *Steinernema* and *Neoaplectana* spp. belong to the same genus, with *Steinernema* Travassos (1927) having precedence over *Neoaplectana* Steiner (1929).

N. *janickii* was first discovered in another pamphilid, *Acantholyda nemoralis* (Weiser and Kohler, 1955) from fir forests in Silesia (Poland). Parasitization was usually less than 1% but reached 28% in one forest (67% of the 42% dead larvae were parasitized) and the insect cadavers showed characteristic signs of infection with symbiotic bacteria. This nematode has never been recovered again, although Weiser (1976) originally considered the *S. kraussei* he found in *C. abietis* to be *N. janickii*.

The only other record of steinernematids naturally occurring in sawflies is the recent report by Georgis and Hague (1979) of a *Neoaplectana* species similar to *N. feltiae* (*carpocapsae*) in the larch sawfly *Cephalcia lariciphila* in Britain. Parasitism of the prepupae of this insect ranged from 8 to 15% in one forest. Since infestations by *C. lariciphila* have increased during 1972-1978 with some 2500 ha being affected in 1977, Georgis and Hague (1979) have considered the possibility of using the nematode for biological control, particularly since the sawfly larvae can be found at high densities, but localized within a forest.

Although no steinernematids have been found naturally infecting tenthredinid sawflies, a number of economically important species of this family have proved susceptible to *N. feltiae* (*carpocapsae*) in laboratory experiments (Schmiege, 1963; Laumond et al., 1979). The author has found that *N. bibio-*

nis, *N. glaseri*, and several species of *Heterorhabditis* are far more effective against insects in the soil than is *N. feltiae*, and it may well be possible to use these and *S. kraussei* in the biological control of sawflies as they enter the soil to pupate. The use of these nematodes for large-scale control has now been facilitated by Bedding's (1981) method of mass rearing; by this method *Neoaplectana* species and *Heterorhabditis* species can be produced at a cost of little more than 2¢ (Australian 1981 values) per million. A more recent development of this technique (Bedding, unpublished data) has permitted the production of 2000 million nematodes per container at a cost of 0.5¢ per million, and several billion nematodes have already been produced for commercial use and field trials. Methods of cheaply storing and transporting thousands of millions of infective nematodes have also been developed (Bedding, unpublished data). These techniques are applicable to *S. kraussei*.

VII. WOODWASPS

Woodwasps of the family Siricidae are widely distributed throughout the world. These large and often spectacular insects drill 1 cm or more into coniferous trees to oviposit and at the same time deposit mucus and spores of a symbiotic fungus; the fungus (*Amylostereum areolatum* or *A. chailletii*, depending on the siricid species) grows, drying out the timber, which stimulates hatching of woodwasp eggs; the larvae tunnel through the wood feeding on the fungus. Woodwasp larvae, just prior to and just after hatching, are subject to attacks from the internal parasitoids, *Ibalia* species (Hymenoptera: Cynipidae) and well-developed siricid larvae are attacked by some of the largest ichneumonids known (*Megarhyssa*, *Rhyssa*, and *Schlettererius* spp.).

Although most siricids attack only dead, dying, or damaged trees, one species, *Sirex noctilio*, aided by a phytotoxic mucous and the symbiotic fungus *A. aerolatum*, is able to kill relatively healthy pine trees. *Pinus radiata*, one of the exotic pines introduced into Australia during the nineteenth century, flourished under Australian conditions and now forms the bulk of commercial softwood plantations with an annual return exceeding $250 million Australian. However, in 1952, *S. noctilio* was discovered in Tasmania, having been introduced probably via a shipment of timber from New Zealand. Thus, what is probably the most susceptible tree species (from California) was exposed to the most virulent siricid (originally from Europe) in the absence of any of its natural biological control agents, and in a country of prevalent droughts. (Droughts are known to predispose trees to *Sirex* attack.) Within a decade *S. noctilio* had killed some 40% of radiata pine trees in the first forest it infested.

Fortunately, when *S. noctilio* arrived in New Zealand, it must have brought with it a nematode parasite. Zondag (1962) discovered this nematode, and later attributed to it the collapse of *Sirex* populations in New Zealand. This tylenchid nematode, later described as *Deladenus siricidicola* (Bedding, 1968), was the first to be found in any siricid, although it and related nematode species occur abundantly in many siricids throughout the world (Bedding and Akhurst, 1978).

When it was found in 1961 that *S. noctilio* had reached the mainland of Australia, the National Sirex Fund was established, and one of the most extensive worldwide searches for natural enemies was instituted. This provided an ideal opportunity for a thorough study of the nematode parasites of all the siricids and of the insect parasitoids. Such a study was commenced by CSIRO in 1965.

It was soon found (Bedding, 1967) that these nematodes not only had great potential for the biological control of *Sirex* since they effectively steri- lized the females of many species, but that the nematodes' life history was quite extraordinary. Bedding (1967, 1972) described the biology of *D. siri- cidicola*, which is basically the same for all *Deladenus* species associated with siricids and their hymenopterous parasitoids.

1. *Biology of* D. siricidicola

Parasitized adult *Sirex* may contain up to 100 or more inactive, cylindrical adult female nematodes that are 325 mm long, 0.1-0.5 mm wide, and often bright green. The adult nematodes occur in the hemocoel, mainly in the ab- domen, but also in the head, thorax, or even within the legs of their host. At this stage, the adult nematodes are usually depleted of juveniles, the lat- ter being found almost entirely within the hosts' reproductive organs. In parasitized female *S. juvencus, S. cyaneus*, and *S. noctilio* (most strains) all the woodwasp's eggs are nonviable and almost invariably contain 50-200 juvenile nematodes. Nevertheless, the female *Sirex* oviposit these eggs as though they were normal, thus transferring the nematodes to new trees. Al- though the testes of parasitized *Sirex* males are packed with juvenile nema- todes, these cannot enter the vesiculae seminales and are not transferred during copulation; the male *Sirex* is thus a "dead end" for the nematodes.

An extraordinary aspect of the life cycle of *D. siricidicola* was demon- strated experimentally by Bedding (1967); juvenile nematodes removed asep- tically from their insect hosts to cultures of *Sirex-symbiotic* fungus (*A. aero- latum*) fed on the fungus and grew into oviparous adults that also fed on the fungus and were remarkably different from the parasitic parent nematodes.

Fig. 8 Section of pine wood showing tracheids containing *D. siricidicola* eggs. (After R. A. Bedding, 1972.)

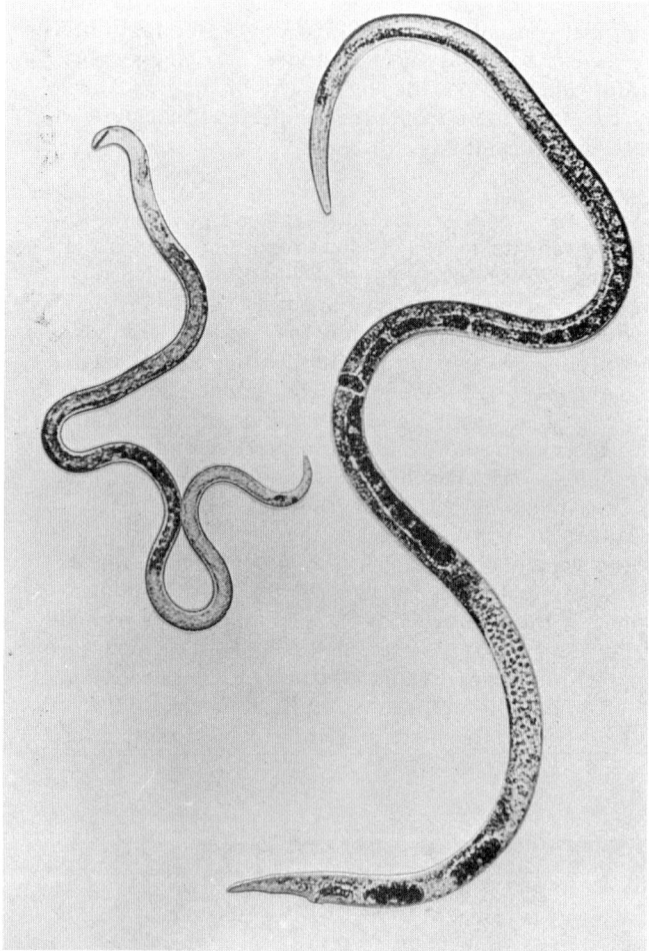

Fig. 9 Infective female (left) and free-living mycetophagous female (right) of *D. siricidicola.* (After R. A. Bedding, 1972.)

It was found that mycetophagous life cycles could continue indefinitely. Indeed, the author has cultures of *D. siricidicola* on *A. areolatum* that have been serially subcultured for more than 12 years and nearly 200 generations without the intervention of an insect-parasitic cycle.

In the natural situation, a *Sirex* female oviposits nematode-filled eggs and inserts spores of the symbiotic fungus and mucus into the tree. As the fungus grows in the wood, nematode juveniles emerge from the *Sirex* eggs, feed on the fungal hyphae, grow, and reproduce within tracheids (see Fig. 8), resin canals, and beneath the bark of the tree. Eventually, fungus and an ever-increasing nematode population spread throughout the now-dead tree. Since as a rule many *Sirex* females attack the same tree, the tree will often contain healthy *Sirex* eggs, and subsequently larvae, as well as nematodes. Prior to infection of healthy *Sirex* larvae, dramatic changes take place in the nematode populations; juvenile nematodes developing in the microenvironment around *Sirex* larvae, instead of becoming mycetophagous, egg-laying adults, become the strikingly different adult female infectives (Fig. 9).

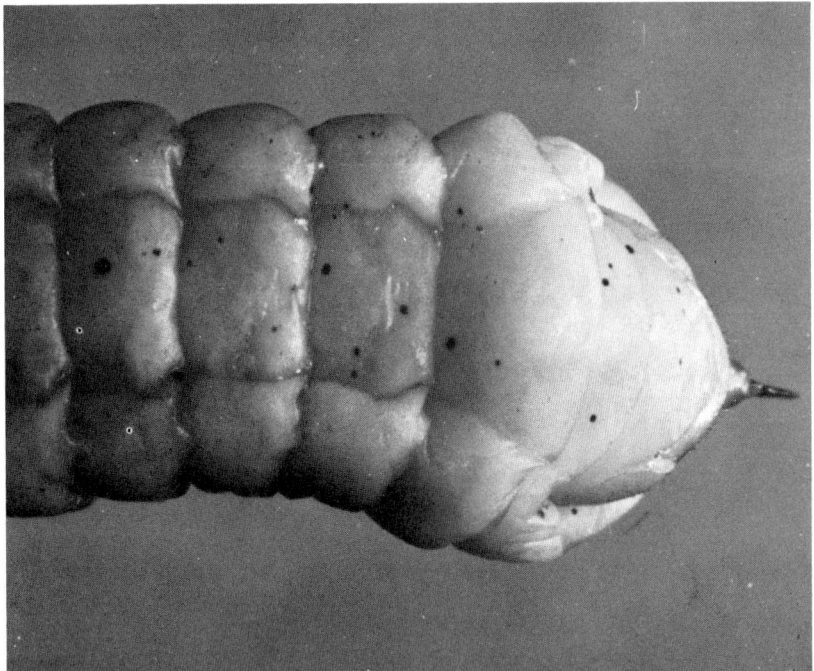

Fig. 10 Posterior end of a parasitized *S. noctilio* larva showing scarring after natural attack. (After R. A. Bedding, 1972.)

After mating, infectives actively bore through the cuticle of *Sirex* larvae (Fig. 10) and enter the larval hemocoel. After two or three days, they lose their adult cuticle and within a few weeks increase up to 1000-fold in volume (depending on the size of their host and number of infectives entering). However, the tiny reproductive system of the infective remains unchanged until the onset of host pupation (up to three years). At this time the nematode's reproductive system rapidly proliferates from 0.25 mm long to fill the adult parasite. Toward the end of host pupation, a single parasitic nematode may produce as many as 10,000 eggs, which soon hatch within the parent nematode. Initially, juveniles emerge from the parental vulva, but soon burst out from all over the nematode's body surface and into the blood of their host. Juveniles then migrate to the reproductive organs of their host.

D. siricidicola can be considered as a parasite able to breed in large numbers in the environment of the host, thus facilitating infection of further hosts, or it can be regarded as primarily a fungal-feeding nematode transported by an insect to a fresh environment with the added advantage that it also multiplies within the insect. The two cycles (Fig. 11) are bound together by the complete specificity of the mycetophagous nematode to the symbiotic fungus (*A. areolatum*) of its host.

Other species of *Deladenus* that parasitize siricids differ from *D. siricidicola* only slightly in their biology (Bedding, 1974); all of these, with the exception of *D. wilsoni*, will complete mycetophagous life cycles only in the presence of the fungus *A. chailettii* (the symbiont of most siricids). Only *D. wilsoni* is able to utilize either *A. areolatum* or *A. chailettii* to complete its life cycle.

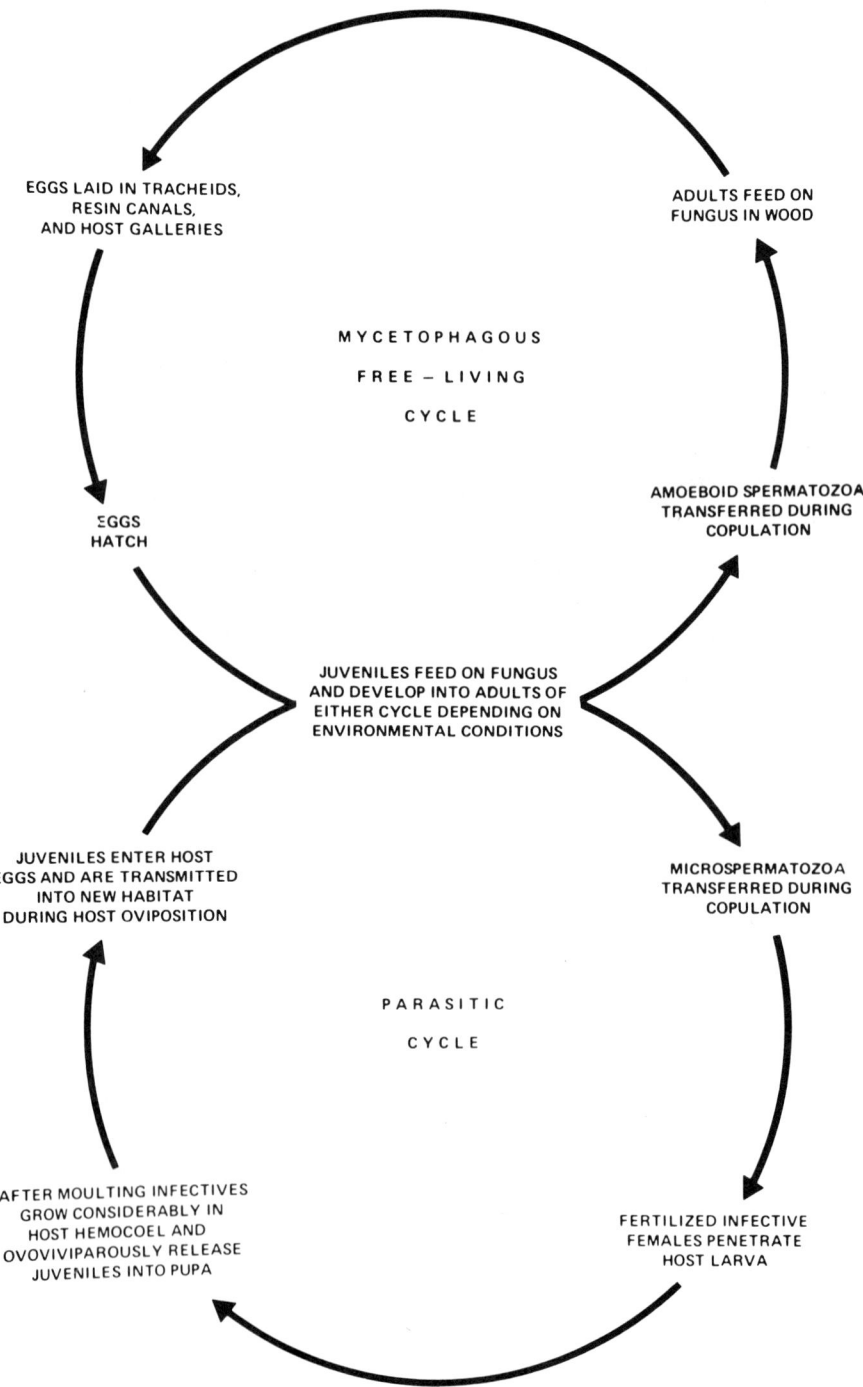

MYCETOPHAGOUS

FREE – LIVING

CYCLE

EGGS LAID IN TRACHEIDS,
RESIN CANALS,
AND HOST GALLERIES

ADULTS FEED ON
FUNGUS IN WOOD

EGGS
HATCH

AMOEBOID SPERMATOZOA
TRANSFERRED DURING
COPULATION

JUVENILES FEED ON FUNGUS
AND DEVELOP INTO ADULTS OF
EITHER CYCLE DEPENDING ON
ENVIRONMENTAL CONDITIONS

JUVENILES ENTER HOST
EGGS AND ARE TRANSMITTED
INTO NEW HABITAT
DURING HOST OVIPOSITION

MICROSPERMATOZOA
TRANSFERRED DURING
COPULATION

PARASITIC

CYCLE

AFTER MOULTING INFECTIVES
GROW CONSIDERABLY IN
HOST HEMOCOEL AND
OVOVIVIPAROUSLY RELEASE
JUVENILES INTO PUPA

FERTILIZED INFECTIVE
FEMALES PENETRATE
HOST LARVA

Fig. 11 Diagram of the life history of *D. siricidicola*. (After R. A. Bedding, 1972.)

2. Effect on Host

Depending on the strain of nematode involved and density of nematode populations within the tree, the mycetophagous nematodes may compete significantly for fungal food with siricid larvae so that the latter do not grow as large as those from uninfected trees.

During the parasitic phase, nematodes appear to have little effect on the larval host, except in heavily parasitized larvae where fat body is reduced.

In adult female hosts, the time of release of juvenile nematodes in relation to the stage of development of the host has an important influence upon the effect of parasitization on the host. This is dependent on both strain of nematode and species or strain of host concerned. It varies from juvenile nematodes entering the host's ovaries well before the end of pupation, causing suppression of ovarian development and greatly reducing the number and size of eggs produced (Australian *S. noctilio*), to juvenile nematodes not being released into the host's hemocoel until well after the host emerges from pupation when ovaries and eggs are fully developed so that no eggs contain nematodes and all juvenile nematodes are located in the ovaries and oviducts and are transmitted with the eggs during oviposition (Japanese strain of *D. siricidicola* in *S. nitobei* and *S. noctilio*, Belgian strain of *S. noctilio* with normal *D. siricidicola*). In between these two extremes, in *S. juvencus* and *S. cyaneus*, most juvenile nematodes are not released into the host hemocoel until just before the host emerges from pupation, at which stage the host's ovaries are fully developed and the eggs full grown, but not fully hardened; juvenile nematodes enter host eggs (Fig. 12) before the insect emerges from the wood,

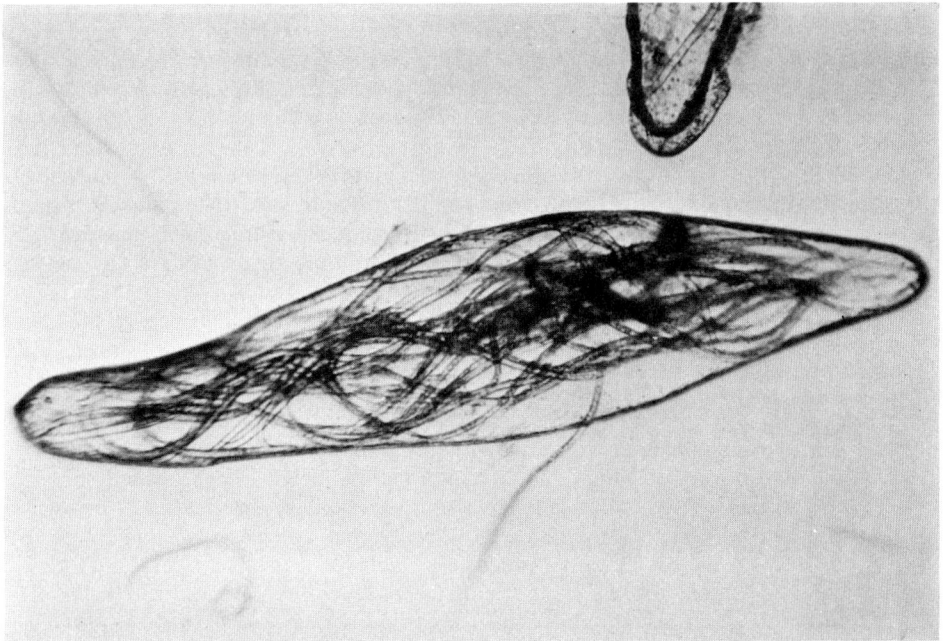

Fig. 12 *S. juvencus* egg containing juvenile nematodes. (After R. A. Bedding, 1972.)

Fig. 13 Testes from unparasitized (left) and parasitized (right) *S. noctilio*. (After R. A. Bedding, 1972.)

and all eggs are infected before oviposition. In the woodwasp *Xeris spectrum*, eggs are of normal size and only 10-30% of the eggs contain nematodes.

During pupation of male hosts, juvenile nematodes migrate from the hemocoel to the testes, which become filled with many thousands of juveniles and greatly hypertrophied (FIg. 13). However, the fertility of male *Sirex* is not impaired, since early in pupation and well before invasion of the testes by juveniles, spermatozoa are passed from the testes and into the vasiculae seminales. Parasitized male *Sirex* mate readily and transmit viable spermatozoa, but cannot transmit juvenile nematodes as these are unable to pass from the testes to the vesiculae seminales.

3. *Form Change and Functional Dimorphism of* Deladenus

Depending on environmental conditions, eggs laid by mycetophagous female *Deladenus* spp. may give rise to mycetophagous females and associated males or infective females and associated males. In monoxenic cultures on young fungus, all juveniles hatching from eggs invariably become mycetophagous egg-laying females or males containing large amoeboid spermatozoa. However, when monoxenic cultures of *D. siricidicola* on potato dextrose agar (PDA) become old and brown or when cultures are contaminated with bacteria, infective females and males with microspermatozoa are produced. Aseptic extracts of such cultures when incorporated into fresh PDA cultures also produce infective females and associated males from added nematode eggs, even with young fungus. Incorporation of 0.2% lactic or hydroxybutyric acid into media also results in production of infective females from eggs of some strains of

D. siricidicola, but these substances do not appear to be the primary causative agents.

In the natural situation it is believed that much of the wood of a *Sirex*-infested tree is an almost monoxenic culture of symbiotic fungus. In contrast, in the microenvironment around *Sirex* larvae, there are many other active microorganisms present (some producing lactic acid), and as on culture plates this leads to the production of infective females and associated males. In other words, infectives are produced mainly in situations where prospective hosts are available.

The morphological differences between the two forms of adult female are profound (see Figs. 9 and 14) and immediately obvious.

The mycetophagous females have characters that indubitably place them in an already established genus, *Deladenus*, within the Neotylenchidae although this family was previously thought to contain only free-living nematodes. However, the infective female has all the characters of another family of nematodes, the Sphaerulariidae (Allantonematidae), all of which are insect parasitic.

The striking differences between these two forms of female nematodes, which are both of the same species and yet exhibit such different characters as would place them in different families, is an example of profound functional dimorphism. Most immediately obvious is the difference in stylets: that of the mycetophagous female is a delicately tapered hypodermic syringelike apparatus adapted for insertion into fungal hyphae to withdraw fluids; the stylet of the infective form is twice as long, considerably more robust and spearlike, admirably adapted to puncturing the relatively thick cuticle of *Sirex* larvae. Associated with the greatly differing modes of feeding and penetrating to the food supply, the esophageal glands of each form are very differently developed. In the mycetophagous female the dorsal esophageal gland is expanded to lie in a long lobe beside the intestine and the subventral esophageal glands are almost vestigial. However, in infective females the dorsal gland is small and lies within the esophagus and the subventral glands are greatly expanded and may occupy half the body length.

As might be expected, the reproductive systems of the two forms also differ greatly, with that of the mycetophagous female (having no constraints on the timing of oviposition) being fully functional, occupying much of the adult nematode. In this form, oocytes pass through a thin-walled oviduct filled with large amoeboid spermatozoa and then through the thick-walled muscular uterus before being violently expelled from the protruding vagina. In infective females, the reproductive system is scarcely developed since several months or even one or more years elapse before reproduction occurs in synchrony with the host reaching maturity. In this form of female there is only a thin-walled oviduct, filled with microspermatozoa and with a few primordial cells at its distal end.

The dimorphism of spermatozoa may be unique in the animal kingdom. Presumably the large amoeboid spermatozoa are more primitive, but perfectly adequate for the free-living mycetophagous form which, although only able to retain 200-500, can have them repeatedly replenished. On the other hand, the infective female must contain, prior to entering a host, enough spermatozoa to fertilize the much larger number of eggs she will produce (up to 10,000), and this has been achieved by reducing the spermatozoon to little more than a nucleus.

Apart from other considerations, the example of *Deladenus* should serve as a warning to nematode taxonomists. The considerable differences between

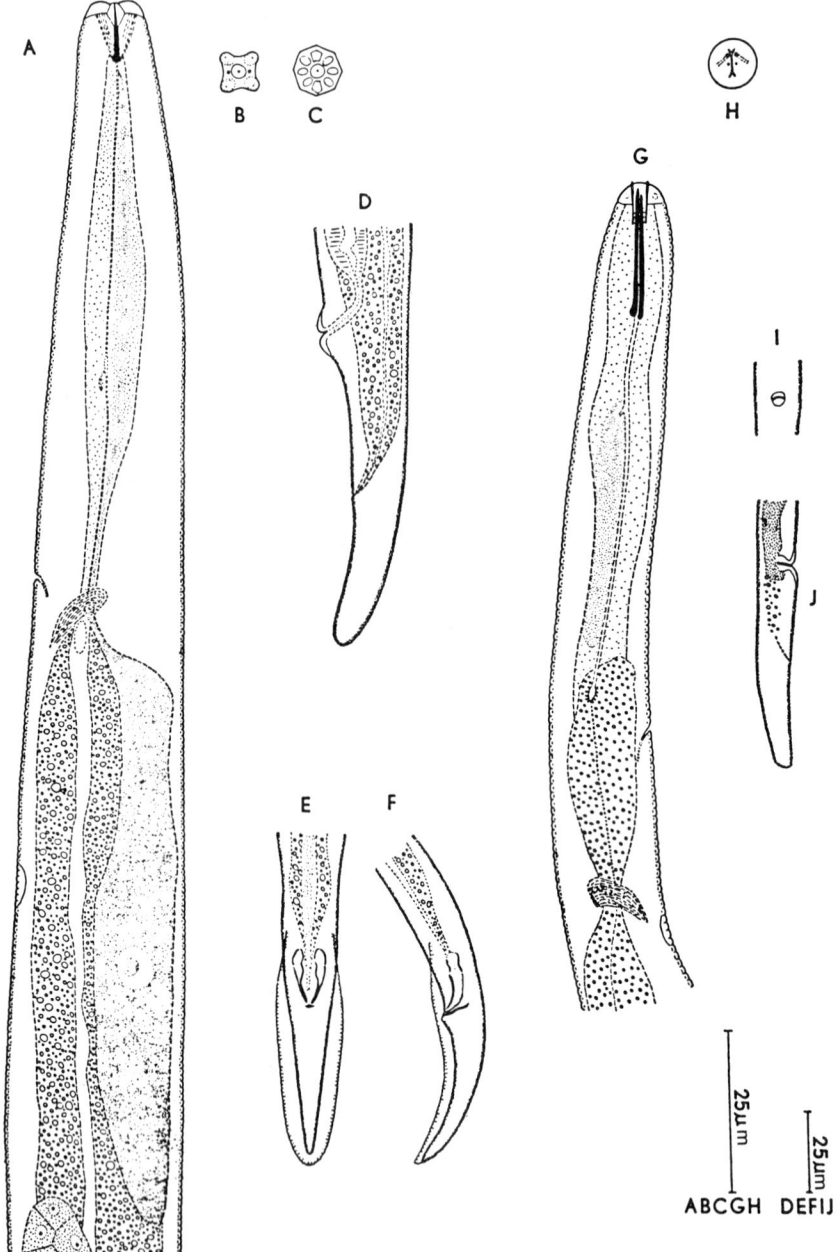

Fig. 14 *D. siricidicola*: (A) anterior end of mycetophagous female, lateral;
(B) *en face* view of lip region of mycetophagous female; (C) cephalic framework of mycetophagous female; (D) posterior end of mycetophagous female,
lateral; (E) posterior end of male, ventral; (F) posterior end of male, lateral;
(G) anterior end of infective female, lateral; (H) *en face* view of infective female; (I) vulva of infective female, ventral; (J) posterior end of infective female, lateral. (After R. A. Bedding, 1968).

the two forms constitute major adaptations of morphology to function and indicate that such characters may have little phylogenetic importance because they constrast so considerably within a single species.

4. *Distribution and Host Preferences of* Deladenus *spp.*

Bedding and Akhurst (1978) made a comprehensive study, involving over 22,000 dissections, on the distribution and host preferences of seven species of *Deladenus* parasitizing siricids. These nematodes were found parasitizing 11 species of *Sirex*, six species of *Urocerus*, and two species of *Xeris* from 27 countries and 30 tree species.

Although not particularly insect-host specific, *Deladenus* species are highly fungal specific; limited feeding may occur on several fungi, but nematode reproduction is restricted even within the genus *Amylostereum*. Only *D. wilsoni* will feed and reproduce significantly on either symbiotic fungus associated with siricids; *D. siricidicola* is specific to *A. areolatum*, and the other five species are specific to *A. chailettii*. It is considered that fungal specificity is important in determining which siricid species may be parasitized in the field, though there are obvious geographical considerations as well. This in turn is related to which tree species supports which nematode species.

The extent of geographical distribution ranges from that of *D. wilsoni*, which is found throughout the Holarctic and Asia wherever there are siricids and their parasitoids, to *D. proximus*, which was found only in the southeastern United States. Of the other five species, two are confined to the Neartic, two to the Paleartic and Asia, and a third was only found in Pakistan.

Incidence of parasitization ranged from a fraction of a percent in some localities to nearly 100% in others. Where large samples were available, a majority of localities were found to have nematode parasites in siricid populations.

5. *Utilizing* Deladenus *Species to Control* S. noctilio

Accounts of the use of *D. siricidicola* in the biological control of *S. noctilio* (Bedding and Akhurst, 1974; Bedding, 1979) demonstrate how essential a thorough knowledge of the biology of this nematode has been to allow it to have been successfully manipulated to control *S. noctilio* on a large scale.

The importance of being able to culture readily and therefore store nematodes after removal from their hosts cannot be overrated. During the search for parasites conducted by CSIRO, the author received nematode-parasitized siricids from hundreds of localities. These were dissected and the nematodes cultured (Bedding and Akhurst, 1974) so that an armory of many different species and strains of *Deladenus* could be stored for evaluation over several years. This made possible the rapid screening of hundreds of cultures for new species by crossbreeding experiments on fungal cultures (Akhurst, 1975). (It was not possible to separate species on the basis of insect-parasitic stages.) It also made feasible extensive testing to determine the effect of various nematode species and strains on important insect parasitoids, prolonged testing of otherwise suitable strains of nematodes for greatest efficacy against *Sirex*, and mass production of nematodes for field releases.

6. *Evaluation of Species and Strains of* Deladenus

Only two of the seven species of *Deladenus* could be considered for the control of *S. noctilio*, as only these two were able to utilize the fungal symbiont of this insect. Of these, only *D. siricidicola* gave good control of *S. noctilio*

without harming the parasitoids (see Section VIII. B). There was some variation among the many strains of *D. siricidicola* that had been collected from
many European countries, Japan, and New Zealand. After eliminating those
few that parasitized, but did not sterilize female *S. noctilio*, several strains
were tested in comprehensive experiments to determine the relative levels of
parasitization that could be reached. In these experiments, some strains were
obviously inferior to others and these were not considered for field liberation.

Since the nematode relies for its dispersal on *S. noctilio* females, the extent to which parasitization affects flying and oviposition capabilities is important. Comprehensive experiments using flight mills showed that although
flight capabilities were highly variable, there was no significant difference
between parasitized and unparasitized *S. noctilio* of the same size. However,
the largest female *S. noctilio* flew up to 50 times as far as the smallest, and
there is a direct relationship between decrease in size of *S. noctilio* and the
presence of nematodes. Various strains of *D. siricidicola* differed in this respect, and two strains that had little significant effect on the size of emerging *S. noctilio* as well as giving the highest levels of parasitization were therefore selected for liberation. Size, of course, is also important in relation to
the number of eggs laid and the number of juvenile nematodes produced within
the host.

7. Mass Rearing Nematodes

The use of host insects, which have a life cycle of from one to three years,
to rear large numbers of nematodes for release would have been both tedious
and expensive; taking advantage of the nematodes free-living cycle was far
more efficient. Although potato dextrose agar culture plates each produced
only a few thousand nematodes, when these were used to inoculate flasks containing autoclaved wheat (100 g wheat and 150 ml water per 500 ml flask),
yields of from 3 to 10 million juvenile nematodes were obtained within four to
six weeks after incubation at 24 C (Bedding and Akhurst, 1974). Since the
contents of a single flask were sufficient to inoculate about 100 m of *S. noctilio*-
infected tree trunk, no attempts were made to scale up the process.

8. Inoculation of Trees

In order to introduce *D. siricidicola* to *S. noctilio* infestations, a number of
easily accessible infested trees scattered throughout the area were usually
felled and inoculated where they lay. Most female *S. noctilio* emerging from
correctly inoculated trees (see below) are parasitized, and these oviposit on
trees, which usually are also attacked by wild unparasitized *S. noctilio*.

Early methods used for inoculation resulted in low, variable, or no parasitism, but later methods regularly produced over 99% parasitism (Bedding
and Akhurst, 1974). Both preparation of inoculation holes and the medium
used to inoculate were found to be critical: normal drilling of wood resulted
in twisted tracheids that impeded nematode entry; water suspensions were
rapidly absorbed, leaving the nematodes to desiccate. A wad punch mounted
to form a hammer proved the most efficient tool for making inoculation holes
(Fig. 15), aerated 12% gelatin with 4000 nematodes per milliliter the most
satisfactory medium, and a spacing of one inoculation per meter was found
adequate to produce almost 100% parasitization. Heavier inoculation resulted
in early competition between nematodes and *S. noctilio* larvae for fungal food
and thus smaller *S. noctilio* females in the next generation.

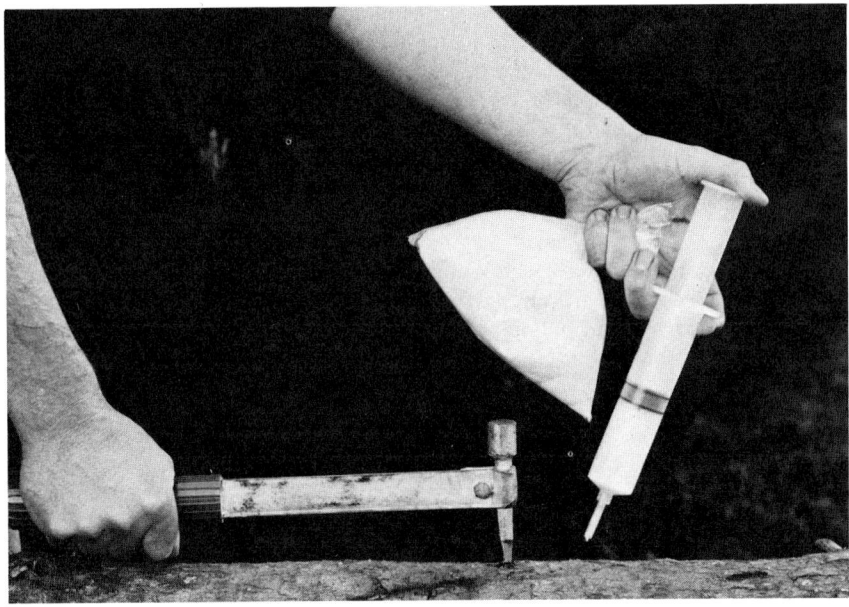

Fig. 15 Inoculation of *S. noctilio*-infested timber with *D. siricidicola* using punch and syringe. (After R. A. Bedding and R. J. Akhurst, 1974.)

9. Field Liberations

Experimental liberations in Tasmania (Bedding and Akhurst, unpublished data) have indicated that *D. siricidicola* establishes, spreads, and achieves high levels of parasitization rapidly. In a 400 ha forest in northern Tasmania with a low infestation of *S. noctilio*, some 50 parasitized females were allowed to emerge from a stack of logs in one corner of the forest during 1970. By 1972, nematodes had spread to 37% of siricid-infested trees in the whole forest and 92% in the compartment of liberation. (Over 70% of *S. noctilio* emerging from any nematode-infested tree were parasitized.) Two years later, over 70% of all trees contained nematodes (with over 90% of *S. noctilio* emerging from these parasitized), and the number of trees killed by *S. noctilio* in the next year dropped dramatically. No *Sirex*-killed trees have been found in the several years since then. In a nearby forest of similar size, several thousand trees were killed annually by *S. noctilio*, but after a heavy inoculation program designed to achieve 10% parasitization during the first year (1972), over 90% parasitization resulted in the following year. The year after this, only 200 trees were killed by *S. noctilio*; the next year only five, and no fresh *S. noctilio*-killed trees have been found either by ground or aerial survey in the several years since then. The nematode also spread naturally to other nearby forests 2, 7, 8, and 13 km away, and produced high levels of parasitization.

During 1970, some 1000 inoculated logs were sent from Tasmania to the Forest Commission, Victoria, for distribution in *S. noctilio*-infested areas. After this hundreds of millions of nematodes were sent to Victoria and, in a major program of liberation by the Forests Commission (which at one stage had 10 mobile crews searching for and inoculating trees in infested forests), nematodes were liberated throughout most of the *S. noctilio*-infested areas in

Victoria. *D. siricidicola* is now established over most of Victoria, and high levels of parasitization are in evidence. In Victoria, *D. siricidicola* is generally considered as the main natural control agent of *Sirex*, although cultural practice and in particular timing of thinning are very important.

Following the spread of *Sirex* infestations to South Australia and New South Wales during 1980, similar programs of nematode liberation have been undertaken.

The program of nematode control is complemented by and coordinated with a similar one for insect parasitoids (Taylor, 1976), and it is expected that the manipulation of both nematodes and parasitoids will become standard forestry practice. In this respect, nematodes are most easily and cheaply reared in the laboratory, but both kinds of agents can usually also be introduced to new *S. noctilio* infestations by transfer of infected logs from one forest to another.

VIII. PARASITOIDS

Although there are some 20,000 different species of hymenopterous parasitoids, records of nematode parasitism of these are rare.

A. Sphaerulariopsis

Khan (1957) described *Sphaerulariopsis hastata*, which he found in the pine bark beetle *Dendroctonus monticolae* and cocoons of its braconid parasite *Coeloides dendroctoni*. The effect, if any, of this nematode on its hymenopterous host is not known, although a related species, *S. dendroctoni*, reduces egg production in the bark beetle *Dendroctonus rufipennis*.

B. Deladenus

Several species of *Deladenus* have been discussed earlier as parasites of siricid woodwasps, but one species, *D. wilsoni*, is largely restricted to parasitization of the ichneumonid parasitoids of woodwasps. Bedding (1967) described the basic biology of this species and its effect on *Rhyssa persuasoria*, and Hocking (1967) described its effects on *R. amoena* and *R. himalayensis*. Bedding and Akhurst (1978) added nine more rhyssine host records.

D. wilsoni is found throughout the holarctic and part of Asia wherever rhyssine parasitoids of siricids occur. It has a similar free-living cycle to that of *D. siricidicola* (see earlier), but it alone is able to utilize as food either of the symbiotic fungi associated with siricids.

Unlike other species of *Deladenus*, the large adult female parasitic *D. wilsoni* continue to liberate juveniles into their host long after this becomes adult. Although parasitization has little effect on the reproductive system of the male host, it almost invariably results in the suppression of egg development in the female host; the juvenile nematodes accumulate in the oviduct and become densely packed within the accessory glands. Juvenile nematodes emerge into the ovipositor, and when parasitized female rhyssines drill into wood, the nematodes migrate into the tracheids and commence free-living mycetophagous cycles.

Parasitization of *Rhyssa* species is frequently about 30%. During importation of rhyssines into Australia for the biological control of *Sirex noctilio*, great care was taken to quarantine insects once it was known that these could be parasitized; parasitoids were individually caged on *Sirex*-infested logs, and where the parasitoid was found on death to be parasitized by nematodes, the logs were destroyed (Taylor, 1967).

The cynipid parasitoids of siricids, *Ibalia* species, are occasionally found to contain infective stages of *Deladenus* species, but these invariably fail to grow although they often remain alive within the adult insect for several weeks.

Each of the six species of *Deladenus* found only in siricids was tested in the laboratory against various insect parasitoids of siricids. In all cases where parasitoid larvae were exposed to dense infestations of infective nematodes, it was found (Bedding and Akhurst, 1978) that penetration could occur, but that infectives, although often surviving for considerable periods, were invariably unable to grow. However, when infectives were placed aseptically in hanging drops of rhyssine blood, development proceeded readily (Bedding and Akhurst, unpublished data).

C. *Neoaplectana* and *Heterorhabditis*

No natural infestation of parasitic Hymenoptera by *Neoaplectana* spp. or *Heterorhabditis* spp. has been recorded. Kaya (1978a) reported that artificially applied *N. feltiae* (*carpocapsae*) adversely affected the development of *Apanteles militaris* when the nematode killed its host, the army worm *Pseudaletia unipuncta*, before this parasitoid could complete its development. However, *A. militaris* was able to successfully complete its development and pupated if its host was infected at a late stage of parasitization, and once the parasitoid had spun a cocoon infection by *N. feltiae* was not possible.

Kaya (1978b) exposed emerging *A. militaris* larvae to dosages of 100-10,000 infectives of *N. feltiae* and *Heterorhabditis heliothidis* in Petri dishes, but even at dosages of 10,000 nematodes, mortality in this artificially favorable environment was only 32 and 22%, respectively. As pointed out by Kaya, the likelihood is remote that emerging parasitoids will encounter such numbers in the field.

The impact on hymenopterous parasitoids of mass releases of *Neoaplectana* and *Heterorhabditis* species in the field is likely to be minimal, apart from the effect of any reduction achieved in host population size. Nematode infection will only occur in situations where films of moisture are present, and this will frequently be within the soil; most insects usually living in the soil have few hymenopterous parasitoids, and those parasitoids of the many insects that enter the soil to pupate normally kill their hosts before this occurs. Nematode sprays, such as the oil wax encapsulation method suggested by Bedding (1976) used against leaf-eating insects, should have no direct impact on parasitoids because of the necessity for an insect to first eat the leaf together with nematodes in order to become infected.

IX. HOST LIST – ADDITIONS TO POINAR (1975)

Poinar (1975) provided a comprehensive host list of insects infected by nematodes. The following are the few additions to the list of Hymenoptera, adopting the same tabulation procedures.*

Agaonidae

Elisabethiella stuckenbergi Grandi	D - *Parasitodiplogaster sycophilon*	Poinar (1979)

Apidae

Bombus alpinus L.	Sp- *Sphaerularia bombi* Dug.	Lundberg and Svensson (1975)

B. balteatus Dahlb.	Sp- *S. bombi* Duf.	Lundberg and Svensson (1975)
B. jonellus K.	Sp- *S. bombi* Duf.	Hasselroth (1960)
B. lapponicus Fabr.	Sp- *S. bombi* Duf.	Lundberg and Svensson (1975)
B. pascuorum Scop.	Sp- *S. bombi* Duf.	Hasselroth (1960)
B. pratorum L.	Sp- *S. bombi* Duf.	Hasselroth (1960)
Formicidae		
Tapinoma erraticum	M - Unknown	Gosswald (1930)
Ichneumonidae		
Rhyssa alaskensis Ashm.	N - *Deladenus wilsoni* Bed.	Bedding and Akhurst (1978)
R. crevieri (Prov.)	N - *D. wilsoni* Bed.	Bedding and Akhurst (1978)
R. hoferi Rohwer	N - *D. wilsoni* Bed.	Bedding and Akhurst (1978)
R. howdenorum Townes	N - *D. wilsoni* Bed.	Bedding and Akhurst (1978)
R. jozana Mats.	N - *D. wilsoni* Bed.	Bedding and Akhurst (1978)
R. lineolata (Kirby)	N - *D. wilsoni* Bed.	Bedding and Akhurst (1978)
Megarhyssa nortoni (Cress).	N - *D. wilsoni* Bed.	Bedding and Akhurst (1978)
Psuedorhyssa sternata Merr.	N - *D. wilsoni* Bed.	Bedding and Akhurst (1978)
P. approximator (F.)	N - *D. wilsoni*	Bedding and Akhurst (1978)
Pamphiliidae		
Cephalcia lariciphila (Wachtl.)	St- Unknown	Georgis and Hague (1979)

*D = Diplogasteridae, M = Mermithidae, N = Neotylenchidae, St = Steinernematidae, Sp = Sphaerulariidae.

ACKNOWLEDGMENTS

I thank Raymond J. Akhurst for his encouragement and suggestions and comments on the manuscript, Vinu Patel and Ted Woolcock for photographic help, Dr. George Poinar for permission to use several of his photographs, Dr. H. Kaiser for permission to use his photograph of a mermithogyne, and the editor of *Nematologica* for permission to use material published in that journal. For help with scanning electron micrographs I am grateful to Colin Beaton, Tony Molyneux, Ms. Elizabeth Lockie, and Dr. Barry Filshie.

REFERENCES

Akhurst, R. J. (1975). Cross-breeding to facilitate the identification of *Deladenus* spp., nematode parasites of woodwasps. *Nematologica 21*: 267-272.

Alford, D. V. (1969). *Sphaerularia bombi* as a parasite of bumble bees in England. *J. Apicult. Res. 8*: 49-54.

Assmuss, E. P. (1858). Verzeichniss einiger Insecten in denen ich Gordiaceen antraf. Wien. *Entomol. Monatschur. 2*: 171-181.

Baird, W. (1853). *Catalogue of the Species of Entozoa, or Intestinal Worms, Contained in the Collection of the British Museum London.* British Museum.

Beck, R. (1937). *Mermis* thread worm (Nematode) in wasp (*Vespa vulgaris*). *Entomol. Rec. J. War. 49*: 65.

Bedding, R. A. (1967). Parasitic and free-living cycles in entomogenous nematodes of the genus *Deladenus*. *Nature (Lond.) 214*: 174-175.

Bedding, R. A. (1968). *Deladenus wilsoni* n. sp. and *Deladenus siricidicola* n. sp. (Neotylenchidae) entomophagous-mycetophagous nematodes parasitic in siricid woodwasps. *Nematologica 14*: 515-525.

Bedding, R. A. (1972). Biology of *Deladenus siricidicola* (Neotylenchidae) an entomophagous-mycetophagous nematode parasitic in siricid woodwasps. *Nematologica 18*: 482-493.

Bedding, R. A. (1974). Five new species of *Deladenus* (Neotylenchidae), entomophagous-mycetophagous nematodes parasitic in siricid woodwasps. *Nematologica 20*: 204-225.

Bedding, R. A. (1976). New methods increase the feasibility of using *Neoaplectana* spp. (Nematoda) for the control of insect pests. *Proc. Int. Colloq. Invert. Pathol.* 250-254.

Bedding, R. A. (1979). Manipulating the entomophagous-mycetophagous nematode, *Deladenus siricidicola* for biological control of the woodwasp *Sirex noctilio* in Australia. *Current topics in forest entomology, U.S. For. Serv. Gen. tech. rept. WO-8*, pp. 144-147.

Bedding, R. A. (1981). Low cost, in vitro mass production of *Neoaplectana* and *Heterorhabditis* species, (Nematoda) for field control of insect pests. *Nematologica 27*: 109-114.

Bedding, R. A., and Akhurst, R. J. (1974). Use of the nematode *Deladenus siricidicola* in the biological control of *Sirex noctilio* in Australia. *J. Aust. Ent. Soc. 13*: 129-137.

Bedding, R. A., and Akhurst, R. J. (1978). Geographical distribution and host preferences of *Deladenus* species (Nematoda: Neotylenchidae) parasitic in siricid woodwasps and associated hymenopterous parasitoids. *Nematologica 24*: 243-251.

Blackith, R. E., and Stevenson, J. H. (1958) Autumnal populations of wasp nests. *Insectes Soc. 5*, 347-352.

Bols, J. H. (1939). Un remarquable terrain d'hivernation de *Bombus* et de *Psithyrus* près Louvain, a Lubbeek, en Belgique. *Verh. VII Int. Kongr. Ent. 2*: 1048-1060.

Chitwood, B. G., and Chitwood, M. B. (1937). *An Introduction to Nematology*. Monumental Printing Co., Baltimore.

Cobbold, T. S. (1883). On *Simondsia paradoxa* and on its probable affinity with *Sphaerularia bombi. Trans. Linn. Soc. London 2*: 357-361.

Crawley, W. C., and Baylis, H. A. (1921). *Mermis* parasitic on ants of the genus *Lasius*. *J. R. Microsc. Soc. 257*: 353-372.

Cumber, R. A. (1949). Humble-bee parasites and commensals found within a thirty-mile radius of London. *Proc. Roy. Ent. Soc. London 24*: 119-121.

Emery, C. (1890). Studii sulle Formiche della Fauna Neotropica. *Bull. Soc. Ent. Ital. 22*: 38-40.

Emery, C. (1904). Zur Kenntnis des Polymorphismus der Ameisen. *Zool. Jahrb., Suppl. VII, S.*: 587-610.

Fox-Wilson, G. (1946). Factors affecting populations of social wasps, *Vespula* species, in England. *Proc. R. Ent. Soc. Lond. 21*: 17-27.

Fye, R. E. (1966). *Sphaerularia bombi* Dufour parasitizing *Bombus* queens in Northwestern Ontario. *Can. Entomol. 98*: 88-89.

Gasparrini, G. (1864). Sulla maturazione e la qualita dei fichi dei contorni di Napoli. *Atti. Accad. Pontaniana 9*: 99-118.

Georgis, R., and Hague, N. G. M. (1979). A steinernematid nematode in the web-spinning larch sawfly, *Cephalcia lariciphila* (Wachtl.). *Plant. Pathol. 28*: 98-99.

Gosswald, K. (1929). Mermithogynen von *Lasius alienus* gefunden in der Umgebungen von Würzburg. *Zool. Anz. 84*: 202-204.

Gosswald, K. (1930). Weitere Beiträge zur Verbreitung der Mermithiden bei Ameisen. *Zool. Anz. 90*: 13-27.

Gould, W. (1747). *An Account of English Ants*. A. Millar, London.

Hagmeier, A. (1912). Beiträge zur Kenntnis der Mermithiden. I. Biologische Notizen und systematische Beschreibung einiger alten und neuen Arten. *Zool. Jahrb., Abt. Syst. 32*: 521-612.

Hasselroth, T. B. (1960). Studies on Swedish bumble-bees (Genus *Bombus* Latr.), their domestication and biology. *Opusc. Ent. Suppl. 17*: 1-192.

Hocking, H. (1967). A nematode (*Deladenus* sp., Neotylenchidae) associated with *Rhyssa* spp. (Hymenoptera: Ichneumonidae), parasites of siricid woodwasps. *J. Aust. Ent. Soc. 6*: 52-56.

Janet, C. (1893). Sur les nématodes des glandes pharyngiennes des fourmis. (*Pelodera* sp.) *C.R. Acad. Sci. 117*: 700.

Janet, C. (1894). Etudes sur les fourmis. *Pelodera* des glandes pharyngiennes de *Formica rufa* L. *Mem. Soc. Zool. France 7*: 45-62.

Janet, C. (1909). Sur un nématode que se developpe dans la tête de la *Formica fusca*. *Mém. Soc. Acad. Archeol. Sci. and Arts, Dept. Oise 20*: 1072-1073.

Kaya, H. K. (1978a). Interaction between *Neoaplectana carpocapsae* (Nematoda: Steinernematidae) and *Apanteles militaris* (Hymenoptera: Braconidae), a parasitoid of the army worm, *Pseudaletra unipuncta*. *J. Insect Pathol. 31*: 358-364.

Kaya, H. K. (1978b). Infectivity of *Neoaplectana carpocapsae* and *Heterorhabditis heliothidis* to pupae of the parasite *Apanteles militaris*. *J. Nematol. 10*: 241-244.

Khan, M. A. (1957). *Sphaerularia bombi* Duf. (Nematoda: Allantonematidae) infesting bumble bees and *Sphaerularia hastata* sp. nov. infesting bark beetles in Canada. *Can. J. Zool. 35*: 519-523.

Kloft, W. (1950). Ökologische Untersuchungen zur Verbreitung der Mermithiden bei Ameisen. *Zool. Jahrb. Abt. System. Okol. Geogr. Tiere 78*: 526-30.

Kloft, W. (1951). Pathologische untersuchungen an einem Wespenweibschen, infiziert durch rinen Gordioiden (Nematomorpha). *Z. Parasitenk. 15*: 134-47.

Laumond, C., Mauleon, H., and Kermarrec, A. (1979). Données nouvelles sur le spectra d'hôtes et le parasitisme du nématode entomophage *Neoaplectana carpocapsae*. *Entomophaga 24*: 13-27.

Leuckart, R. (1887). Neue Beiträge zur Kenntnis des Baues und der Lebensgeschichte der Nematoden. *Abh. Kgl.-sächs. Ges. Wiss. 22*: 567-704.

Linstow, von, O. (1905). Helminthologische Beobachtungen. *Arch. Mikrosc. Anat. 66*: 355-366.

Lubbock, J. (1861). On *Sphaerularia bombi*. *Nat. Hist. Rev. 1*: 44-57.

Lubbock, J. (1864). Notes on *Sphaerularia bombi*. *Nat. Hist. Rev. 4*: 265-270.

Lundberg, H., and Svensson, B. G. (1975). Studies on the behaviour of *Bombus* Latr. species (Hym., Apidae) parasitized by *Sphaerularia bombi* Dufour (Nematoda) in an alpine area. *Norw. J. Ent. 22*: 129-134.

MacFarlane, R. P. (1975). The nematode *Sphaerularia bombi* (Sphaerulariidae) and the mite *Locustacurus buchneri* (Podapolipidae) in bumble bee queens *Bombus* spp. (Apidae) in New Zealand. *N.Z. Entomologist 6*: 79.

MacLean, B. K. (1966). Internal parasitism of *Bombus* spp. in Indiana. *Proc. North Central Br. ESA 20*: 94-95.

Madel, G. (1966). Beiträge zur Biologie von *Sphaerularia bombi* Leon Dufour 1837. I. *Mitteilung. Z. Parasitk. 28*: 99-107.

Madel, G. (1973). Zur Biologie des Hummelparasiten *Sphaerularia bombi* Dufour 1886 (Nematoda, Tylenchida). *Bonn. Zool. Beitr. 24*: 134-151.

Madel, G., and Scholtyseck, E. (1976). Licht und elektronenoptische Untersuchungen an den Schlauchzellen des Hummelparasiten *Sphaerularia bombi* (Tylenchida, Nematoda). *Z. Parasitenk. 49*: 81-92.

Markin, G. P., and McCoy, C. W. (1968). The occurrence of a nematode, *Diploscapter lycostoma*, in the pharyngeal glands of the Argentine ant, *Iridomyrmex humilis*. *Ann. Entomol. Soc. Am. 61*: 505-509.

Medler, J. T. (1962). Development and absorption of eggs in bumblebees (Hymenoptera: Apidae). *Can. Entomol. 94*: 825-833.

Milum, V. G. (1938). A larval mermithid, *Mermis subnigrescens* Cobb, as a parasite of the honey-bee. *J. Econ. Entomol. 31*: 460.

Minderhoud, A. (1951). Het telen van hommels in verband met *Sphaerularia bombi*. *Med. Directie Lanbouw 477-482*.

Morse, R. A. (1955). Larval nematode recorded from honeybee *Apis mellifera* L. *J. Parasitol. 41*: 553.

Mrácek, Z. (1977). *Steinernema kraussei*, a parasite of the body cavity of the sawfly, *Cephaleia abietis*, in Czechoslovakia. *J. Invert. Pathol. 30*: 87-94.

Mrázek, A. (1908). Myrmekologiske poxnamky. III. Brachypterni mermithogyny u *Lasius alienus*. *Act. Soc. Ent. Bohem. 5*: 1-8.

Nickle, W. R. (1967). On the classification of the insect parasitic nematodes of the Sphaerulariidae Lubbock, 1861 (Tylenchoidea: Nematoda). *Proc. Helm. Soc. Wash. 34*: 72-94.

Nickle, W. R., and Ayre, G. L. (1966). *Caenorhabditis dolichura* (A. Schneider, 1866) Dougherty (Rhabditidae, Nematoda) in the head glands of the ants, *Camponotus herculeanus* (L.) and *Acanthomyops claviger* (Roger) in Ontario. *Proc. Entomol. Soc. Ont. 96*: 96-98.

Palm, N. (1948). Normal and pathological histology of the ovaries in *Bombus* Latr. (Hymenoptera). *Opusc. Ent. Suppl. 7*: 1-101.

Passera, L. (1974). Présence d'*Hexamermis* sp. (Nematoda, Mermithidae) dans les reines vierges et les males de la fourmi *Pheidole pallidula* Nyl. (Formicidae, Myrmicinae). *Bull. Soc. Zool. Fr. 99*: 315-324.

Passera, L. (1976). Origine des intercastes dans les sociétés de *Pheidole pallidula* (Ny L.) (Hymenoptera, Formicidae) parasitées par *Mermis* sp. (Nematoda, Mermithidae). *Insectes Soc. 23*: 559-575.

Poinar, G. O. (1974). The presence of *Sphaerularia bombi* Dufour (Tylenchida: Nematoda), a nematode parasite of *Bombus* queens (Apidae: Hymenoptera), in California. *Pan-Pacif. Entomologist 50*: 304-305.

Poinar, G. O. (1975). *Entomogenous Nematodes*. E. J. Brill, Leiden.

Poinar, G. O. (1979). *Nematodes for Biological Control of Insects*. C.R.C. Press, Boca Raton, Florida.

Poinar, G. O., and Ennick, F. (1972). The use of *Neoaplectana carpocapsae* Weiser (Steinernematidae: Rhabditoidea) against yellow jackets (*Vespula* spp., Vespidae: Hymenoptera). *J. Invert. Pathol. 19*: 331-340.

Poinar, G. O., and Hess, R. (1972). Food uptake by the insect-parasitic nematode, *Sphaerularia bombi* (Tylenchida). *J. Nematol. 4*: 271-277.

Poinar, G., and van der Laan, P. (1972). Morphology and life history of *Sphaerularia bombi*. *Nematologica 18*: 239-252.

Pouvreau, A. (1964). Observations d'une infestation precoce des reines de bourdons (Hymenoptera, Apoidea, *Bombus*), par *Sphaerularia bombi* (Nematoda, Tylenchida, Allantonematidae). *Bull. Soc. Zool. France 89*: 717-719.

Réaumur, R. A. F. (1942). *Mémoires pour servir à l'histoire des insectes 6*: 22.

Schmiege, D. C. (1963). The feasibility of using a neoaplectanid nematode for control of some forest insect pests. *J. Econ. Entomol. 56*: 427-431.

Schneider, A. (1885). Ueber die Entwicklung der *Sphaerularia bombi*. *Zool. Beitrage 1*: 1-9.

Siebold, C. T., von (1838). Ueber geschlechtslose Nematoideen. Helminthologische Beiträge 4, *Arch. Naturgesch. 4*: 302-314.

Siebold, C. T., von (1854). Ueber die Fadenwurmer der Insekten. *Entomol. Z. 15*: 103-121.

Stein, G. (1956). Weitere Beiträge zur Biologie von *Sphaerularia bombi* Léon Dufour 1837. *Z. Parasitenk. 17*: 383-393.

Stein, G., and Lohmar, E. (1972). Uber de infektion verschiedener Hummelarten mit *Sphaerularia bombi* Léon Dufour 1837 in raum Bonn. Frühjahr 1970. *Decheniana 124*: 135-140.

Steiner, G. (1923). *Aplectana kraussei* n.sp., eine in der Blattwespe *Lyda* sp. parasitierende Nematodenform, nebst Bemerkungen über das Seitenorgan der parasitischen Nematoden, *Cent. Bacteriol. Parasitenk. 59*: 14-18.

Steiner, G. (1929). *Neoaplectana glaseri* n.g., n.sp. (Oxyuridae), a new nemic parasite of the Japanese beetle. *J. Wash. Acad. Sci. 19*: 436-440.

Taylor, J. N. (1933). Mermithogynes. *Entomol. Rec. J. Var. 45*: 162-164.

Taylor, K. L. (1967). The introduction, culture, liberation and recovery of parasites of *Sirex noctilio* in Tasmania 1962-67. *Tech. Paper Div. Entomol. CSIRO Aust.*, No. 8.

Taylor, K. L. (1976). The introduction and establishment of insect parasitoids to control *Sirex noctilio* in Australia. *Entomophaga 21*: 429-440.

Travassos, L. (1927). Sobre o genera *Oxysomatium*. *Bol. Biol. Sao Paolo 5*: 20-21.

Vandel, A. (1927). Modifications déterminées par un Nématode du genre *Mermis* chez les ouvrieres et les soldats de la fourmi *Pheidole pallidula* (Nyl.). *Bull. Biol. Fr. Belg. 61*: 38-47.

Vandel, A. (1930). La production d'intercastes chez la fourmi *Pheidole pallidula* sous l'action des parasite due genre *Mermis*. *Bull. Biol. Fr. Belg. 64*: 457-494.

Vandel, A. (1934). Le cycle évolutif d'*Hexamermis* sp. parasite de la fourmi *Pheidole pallidula*. *Ann. Sci. Nat. 17*: 47-58.

Vasiliade, G. (1970). A nematode of the honeybee. *Pchelovadstvo 11*: 16-17.

Wahab, A. (1962). Untersuchungen über Nematoden in den Drüsen des Kopfes der Ameisen (Formicidae). *Z. Morph. Okol. Tiere 52*: 33-92.

Wasmann, E. (1909). Ueber den Ursprung des sozialen Parasitismus, der Sklaverei und der Myrmekophilie bei den Ameisen. *Biol. Centralbl., Bd. 29*: 692.

Weiser, J. (1955). *Neoaplectana carpocapsae*, novy cizopasnik housenek obalece jableeneho, *Carpocapsa pomonella* L. *Cesk. Zool. Spol. 19*: 44-52.

Weiser, J. (1976). *Steinernema kraussei* St. as an insect pathogen. Proceedings of the First International Colloquium on Invertebrate Pathology, and IXth Annual Meeting Society for Invertebrate Pathology. Queen's University, Kingston, Canada, pp. 245-249.

Weiser, J., and Koehler, W. (1955). Hlistice jako cizopasnici larev ploskohrbetky *Acantholyda nemoralis* Thoms. v Polsku. *Cesk. Parasitol. 2*: 185-190.

Welch, H. E. (1958). *Agamomermis pachysoma* (Linstow, 1905). N. comb. (Mermithidae: Nematoda), a parasite of social wasps. *Insectes Soc. 5*: 353-5.

Wheeler, W. M. (1901). The parasitic origin of Macroërgates among ants. *Am. Natural. 35*: 877-886.

Wheeler, W. M. (1907). The polymorphism of ants, with an account of some singular abnormalities due to parasitism. *Bull. Am. Mus. Natl. Hist. 23*: 1-93.

Wheeler, W. M. (1928). *Mermis* parasitism and intercastes among ants. *J. Exp. Zool. 50*: 165-237.

Wheeler, W. M. (1933). *Mermis* parasitism in some Australian and Mexican ants. *Psyche 40*: 20-31.

Zondag, R. (1962). A nematode disease of *Sirex noctilio*. *N.Z. For. Serv. Interim Research Release*.

Chapter 22

Nematode Parasites of Mosquitoes

James J. Petersen* *USDA, Agricultural Research Service, Lake Charles, Louisiana*

Present affiliation: USDA, Agricultural Research Service, University of Nebraska, Lincoln, Nebraska

I. INTRODUCTION

With the exception of two or three reports, all entomogenous nematodes (growing in or on an insect) developing in mosquitoes are limited to the Mermithidae and Filariidae groups. Though the filarids (*Wuchereria*, *Dirofilaria*, and so on) are important as disease agents of humans and animals and are vectored by mosquitoes, mosquitoes serve only as intermediate hosts in the life cycles of these nematodes. Thus, they rarely kill their mosquito hosts and will not be considered here as potential biological control agents. Limited studies on the potential of the nonaquatic rhabditoid *Neoaplectana carpocapsae* showed that although death of the host sometimes occurred (due principally to a pathogenic bacteria carried by the nematode), the mosquito host usually encapsulated the nematode, which then was unable to complete its life cycle (Welch and Bronskill, 1962).

Mermithids occur in mosquitoes throughout much of the world and have been reported from at least 79 different host species (41 *Aedes*, 21 *Anopheles*, eight *Culex*, and nine from four additional genera). The actual number of species of mermithids responsible for these infections is not known. To date 15 species have been studied and described from adult forms. Of these, five have been described from the Old World, the remainder from the United States and Canada. Only one species, *Romanomermis culicivorax*, has been extensively studied. As a result, much of the information on the biological control potential of mermithids was derived from studies of this species.

Mermithids have drawn considerable attention from scientists, especially in the last 10 years, because they are generally (1) well adapted to the life cycle of their hosts, (2) are host specific, (3) kill their hosts, (4) provide high levels of infection, (5) at least one can be easily mass produced, (6) have a high reproductive potential, (7) can be easily disseminated using standard procedures and techniques, (8) have a high potential for establishment and recycling, and (9) offer no threat to nontarget organisms or the environment. This chapter will discuss further these characteristics as well as those that limit the usefulness of mermithids as control agents for mosquitoes.

II. HISTORY

Perhaps the earliest record of mermithids from mosquitoes was by Dujardin (1845), who reported a mermithid from *Anopheles* spp. mosquitoes in France. Few, if any, additional records are available for the next half century. From 1903-1909, mermithids were reported from mosquitoes in Germany (Stiles, 1903), New Jersey (Smith, 1903; Johnson, 1903), India (Ross, 1906), and French Guinea (Gendre, 1909). Between 1910 and 1950, 18 mosquito-mermithid associations were recorded, and from 1960-1969, 30 additional studies were reported covering various aspects of mosquito-mermithid research. This field of study reached its peak of activity in the past 10 years, when no less than 150 publications (omitting review articles) were published. Nearly 50% of these articles involved studies with *R. culicivorax*.

Prior to the late 1960s most of the references to mosquito-mermithid associations were limited to distribution records, incidence of parasitism, and host records. With few exceptions, taxonomic work was not attempted or improperly done using immature stages. Very few studies included data on life cycles, laboratory culture, or biocontrol potential. These omissions undoubtedly resulted because much of the early work was done by mosquito biologists with little or no nematological background. Noteworthy exceptions during the early years of this field of science were Iyengar (1927), who first worked out the life cycle of a mosquito mermithid; Muspratt (1947), who first reported a method for culturing of a mosquito mermithid in the laboratory; and Welch (1960), who worked out the life cycle and developed additional basic information on mosquito-mermithid associations.

Since 1968, our knowledge of mermithids parasitic on mosquitoes has expanded rapidly; taxonomic problems have been clarified (Nickle, 1972), and the life cycles of several species worked out. Data have been developed on host specificity, temperature, and the effects of chemical, physical, and other factors. Also, one species has been successfully mass produced, thus permitting extensive field testing of a mermithid for insect control for the first time.

III. TAXONOMY

Though a number of attempts have been made to describe mermithids from mosquitoes, until Welch (1960) described *Hydromermis churchillensis*, previous descriptions were incomplete or done using juvenile forms and thus considered to be *species inquirenda*. Only 15 described species are presently recognized from six genera. Of these, 11 have been described since 1970.

A. *Romanomermis* (*Reesimermis*, in Part)

The genus *Romanomermis* was first established by Coman (1961), when he reassigned *Pseudomermis cazanica* as the type species. The taxonomic status of the members of this genus was clarified by the work of Ross and Smith (1976) and Galloway and Brust (1979). All seven species assigned to this genus infect aquatic hosts, and six are parasites of larval mosquitoes.

Romanomermis iyengari (syn. *Reesimermis iyengari*, syn. *Eurymermis iyengari*), a parasite of *Anopheles* larvae, was first isolated by Iyengar (1927) from Bangalore, India. Recently this species was reisolated and is presently being extensively studied both in the field and laboratory (Chandrahas and Rajagopalan, 1979).

Romanomermis nielseni (syn. *Reesimermis nielseni*) is a parasite of spring *Aedes* mosquito larvae from Lone Tree, Wyoming. Several recent attempts have been made to mass rear this species, but efforts to date have been unsuccessful (Tsai and Grundmann, 1969).

Romanomermis communensis is a parasite of *Aedes communis* larvae collected from pools at Goose Creek, Manitoba. Parasitism was also observed in a sibling species, *Ae. churchillensis*, and a chaoborid, *Mochlonyx* sp. (Galloway and Brust, 1979).

Romanomermis culicivorax (syn. *Reesimermis nielseni*, in part) has been collected from mosquito larvae in the vicinity of Lake Charles, Louisiana, and later Gainesville, Florida. This parasite has a wide host range and is known to naturally infect at least 16 species of mosquitoes in six genera. *R. culicivorax* has been the most intensively studied of the mosquito mermithids, has been successfully mass produced, and has been seriously considered for commercial preparation as a biological control agent for mosquitoes. This species was referred to as *Reesimermis nielseni* during the early and mid-1970s. All references to *Reesimermis nielseni*, unless specifically referring to the Lone Tree, Wyoming, location, are *R. culicivorax*.

Romanomermis hermaphrodita is a parasite of larval *Ae. nigripes* collected from tundra pools on the coast of Hudson Bay. This nematode is unique among the known *Romanomermis* species in its high incidence of intersexuality; one-quarter to one-third of the worms from the original collections bore fully formed genitalia of both sexes (Ross and Smith, 1976).

Romanomermis kiktoreak is a parasite of *Ae. impiger* in tundra pools near Bear Lake, Northwest Territory, Canada. This species also develops in *Ae. communis* and *Ae. rempeli*, but parasitism in these hosts is less intense than in *Ae. impiger*. Juveniles are parasitic in the hemocoel of larval hosts and are occasionally parasitic in pupae and adults. Taxonomic keys and detailed descriptions of the species of *Romanomermis* are found in Ross and Smith (1976) and Galloway and Brust (1979).

B. *Octomyomermis*

The genus *Octomyomermis* was first established for a mermithid found in Chironomidae (Johnson, 1963). Later, two species of mosquito mermithids were assigned to this genus. Both are parasites of treehole breeding mosquitoes, one from Africa and the other from California.

    Octomyomermis troglodytis is a parasite of larvae, pupae, and adults of the tree-hole mosquito *Ae. sierrensis* in Marin County, California. Adult nematodes remain in the moist rotten organic matter in the bottom of the tree hole for most of the year. Infections occur in March, April, and May and may reach 38% (Poinar and Sanders, 1974).

    Octomyomermis muspratti (syn. *Romanomermis* sp., syn. *Reesimermis muspratti*) was originally collected from larvae of treehole breeding mosquitoes at Livingston, Zambia. Though this nematode is principally a larval parasite, it readily develops in pupal and adult hosts. *O. muspratti* has a wide host range and is presently being maintained in several laboratories around the world. Problems with asynchronous egg hatch have prevented the development of a large scale mass-rearing system for this promising parasite (Obiamiwe and MacDonald, 1973).

C. *Strelkovimermis peterseni* (syn. *Diximermis peterseni*)

This genus contains only one species parasitic of mosquitoes, *S. peterseni*, which was first isolated from larvae of *Anopheles* spp. in southwestern Louisiana and later from New York and southern Canada. *S. peterseni* is easily maintained in laboratory culture, but because of its specificity for *Anopheles* mosquitoes, high-yield mass-production systems have yet to be developed. Small-scale field releases have demonstrated its ability to establish and recycle (Petersen and Chapman, 1970).

D. *Perutilimermis*

The genus *Perutilimermis* contains only one species, *P. culicis* (syn. *Agamomermis culicis*), which parasitizes a single host species, *Ae. sollicitans*, and completes parasitic development only in adult hosts. This nematode apparently has a wide distribution, first being isolated in New Jersey in 1902 and later in Mississippi and Louisiana. *P. culicis* is probably found throughout its host's range and has been reported to parasitize as many as 91% of some *Ae. sollicitans* populations (Petersen et al., 1967).

E. *Culicimermis*

This genus contains only two species, which are parasites of univoltine (one brood per year) *Aedes* mosquitoes, complete their parasitic development in adult hosts, and have only been reported from Russia (Rubtsov and Isaeva, 1975).

    C. schakovii has been described from mountain *Aedes* from the Kiev region of Russia. Levels of parasitism have been reported as high as 63% in natural host populations. Unlike other mermithids that complete parasitic development in the adult host, this parasite appears capable of parasitizing a wide range of spring *Aedes* mosquitoes (Rubtsov and Isaeva, 1975).

C. culicivora was reported infecting females of *Ae. communis* in the vicinity of the mountain River Tyi, Russia (Yakubovich, 1978).

F. *Empidomermis*

Two species have recently been described and assigned to this genus. Both complete their development in adult hosts in a manner similar to members of the genera *Perutilimermis* and *Culicimermis*.

E. cozii has been recovered only from adult females of *Anopheles funestus* from Upper Volta, West Africa. Parasitism of host populations was not observed to exceed 17.6%. Limited studies of the life cycle of *E. cozii* report it to be similar to that of *P. culicis* (Poinar, 1977).

E. riouxi has been reported only from *Ae. detritus* in France. Like other mermithids emerging from adults, *E. riouxi* arrests ovarian development of the host female and inhibits male mating activity. Reported rates of parasitism were very high and seemed to fluctuate closely with the host's life cycle (Doucet et al., 1979).

G. *Hydromermis*

H. churchillensis is the only species of this genus described from mosquitoes and was the first mermithid parasite of mosquitoes to be described from adult forms (Welch, 1960). Recent attempts failed to more completely describe this species because of the poor condition of the type material and failure to recover the nematode from the type locality. Confusion with *R. communensis* collected from the same hosts and from the same type locality have led Galloway and Brust (1979) to consider *H. churchillensis* as a *species inquirenda*.

IV. LIFE CYCLE

Mermithid nematodes parasitic on mosquitoes have life cycles similar to those of other aquatic mermithids (Fig. 1). The eggs are nearly round, transparent, usually less than 100 μm (about 200 μm in *R. culicivorax*) in diameter, and are usually covered with a slightly adhesive material (Fig. 2A). Egg development usually commences soon after oviposition. Egg maturation is temperature dependent and requires as little as 7-10 days at 26°C, but may require several months at temperatures below 10°C. Mature eggs held in a moist environment devoid of free water do not hatch; eggs of *O. muspratti*, however, can remain viable for more than five years under these conditions (Petersen, 1981). When mature eggs are subjected to free water, they generally begin hatching immediately. However, in some species, even under optimum conditions, only a portion of the eggs will hatch, thus ensuring future survival. Eggs allowed to mature in a free-water environment at adequate temperatures will hatch as they mature.

A free-living, infective-stage nematode (preparasite) hatches from the egg and actively swims about seeking a suitable host. The life of the preparasite is short and temperature dependent; at 26°C the preparasite of *R. culicivorax* remains infective for about 36-48 hours (Petersen, 1975). The preparasitic stage measures 0.7-1 mm in length and appears to be nondirectional in its swimming behavior (Kurihara and Kazuhisa, 1979). Contact with the host appears to be by accidental collision. However, the nematodes are reported to be positively thigmotactic and negatively geotactic, which greatly increases

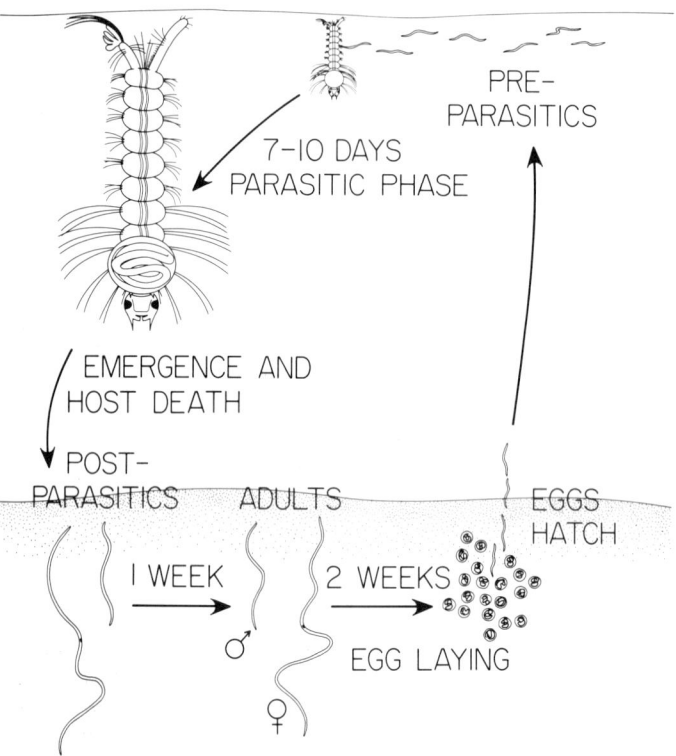

Fig. 1 Life cycle of *R. culicivorax*. (Courtesy of E. G. Platzer.)

their chance of contacting a suitable mosquito host (Petersen, 1973). Upon successful contact, the preparasite attaches to the host by means of a stylet (Fig. 2B) and enters the hemocoel through a hole made in the host's cuticle (Fig. 2C), an activity that requires about five minutes.

In mermithid species that characteristically develop in larval mosquitoes (*Romanomermis* spp., *S. pelerseni*, and so on), the parasitic stage begins development soon after the nematode enters the host. The parasitic stage grows slowly for the first three to four days and then rapidly increases in size, requiring a total of seven to eight days for completion at 26°C (Gordon et al., 1974). In the latter stages of parasitic development, the nematode can be readily seen inside the host, either wrapped transversely around the thorax (Fig. 2D) or folded longitudinally extending the entire length of the thorax and abdomen ventral to the alimentary canal (Fig. 3A). The nematode then molts to the postparasitic stage, ruptures the host's cuticle, and emerges (Figs. 2E and 3B). This emergence is always fatal to the larval host. In some mermithid species (*P. culicis, E. cozii, C. schakovii,* and others) the parasitic stage undergoes little development until the host becomes an adult; thereafter, the nematodes migrate to the host's abdomen, mature rapidly, cause sterility, and normally kill the adult mosquito at emergence (Petersen et al., 1967; Poinar, 1977; Isaeva, 1977). Still other species (i.e., *Octomyo-*

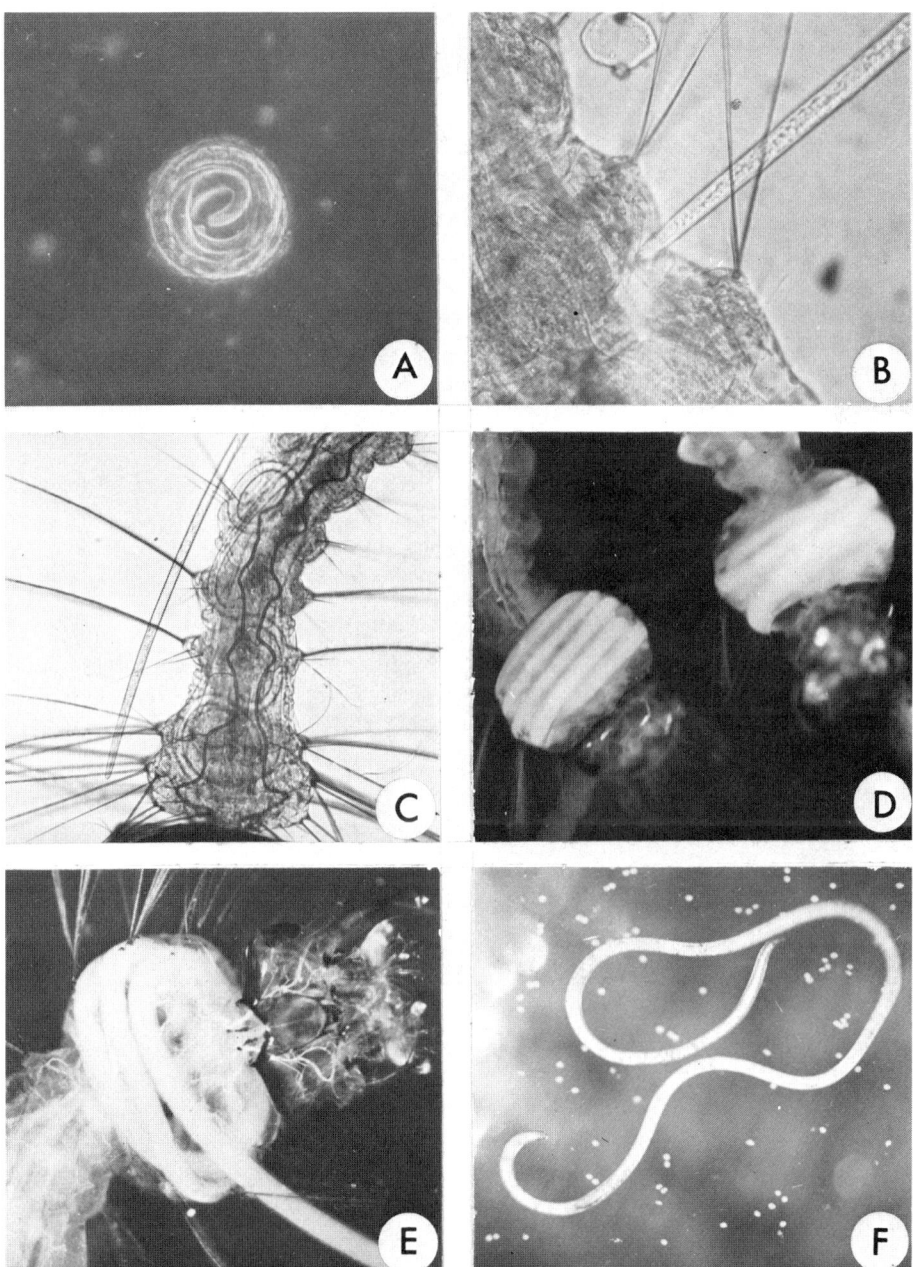

Fig. 2 Life cycle of *R. culicivorax*. (A) Mature egg. (B) Preparasite penetrating the cuticle of first instar mosquito larva. (C) Preparasite just after
entering host. (D) Nematodes in thorax of fourth-instar mosquito larvae.
(E) Postparasite emerging from thorax of host. (F) Adult female in the act
of oviposition. (Courtesy of USDA.)

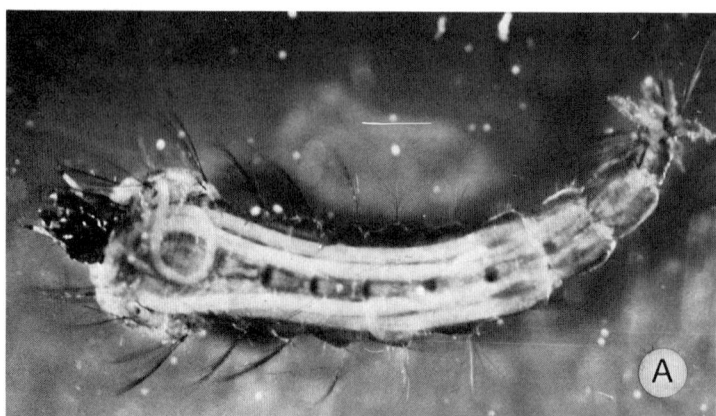

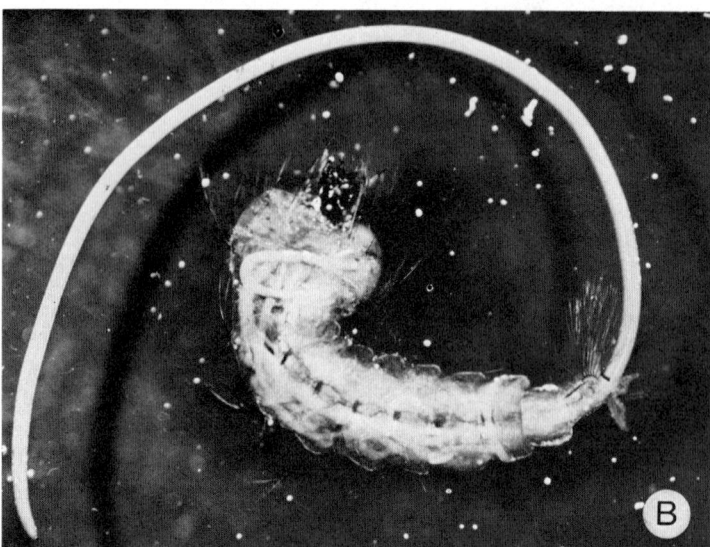

Fig. 3 Parasitism of *A. crucians* by *S. peterseni*. (A) Ventral view showing mature parasitic stage. (B) Postparasite emerging from host. (Courtesy of USDA.)

mermis spp.) that commonly develop in host larvae can be readily carried on into the adult if the preparasite invades the host in later larval instars (Petersen, 1977).

After emergence, the postparasitic nematodes burrow in the soil where they require seven or more days to mature to the adult stage. Soon after molting, adults mate and females begin to lay eggs. The number of eggs laid varies greatly with the species. *R. culicivorax* lays an average of about 2500 eggs over a 10-15 day period (Fig. 2F), but most other species lay considerably larger numbers. The life cycle of some species can be completed in as little as three to six weeks, but may require a year in species that are parasitic in univoltine mosquitoes.

V. FACTORS AFFECTING PARASITISM

A. Host Specificity

One of the most important requirements of an "ideal biological control agent" is its specificity for the target host. There are few biological agents that are as selective as aquatic mermithid nematodes. These mermithids apparently have been associated for a long time with the insects that serve as their normal hosts. Some mermithids are particularly well adapted to the temporary water conditions associated with the habitats of floodwater mosquitoes, and others are better adapted to the more permanent water habitats and associated mosquito species. Others, unique to tree-hole mosquitoes, have developed an egg stage that can persist in the environment for several years under unfavorable conditions. Still others have developed a life cycle to coincide with the univoltine mosquitoes that serve as their hosts.

With the exception of two species of Chaoboridae (closely related to mosquitoes) (Welch, 1960; Galloway and Brust, 1979), mermithids parasitic in mosquitoes have not been found parasitizing other organisms in nature. Coincidentally, both observations of Chaoboridae parasitism were from the same locality. Though two species of mermithids were reported to be involved, one might suspect only one species was responsible, especially since the original species as described (*H. churchillensis*) could not be recovered when the second species was studied (Galloway and Brust, 1979).

In the laboratory, extensive tests have been carried out with *R. culicivorax* against nontarget organisms and, with the exception of some early instar Chaoboridae, Chironomidae, and Simuliidae, no other invertebrates or vertebrates were found to be susceptible to attack, even under very heavy challenges (Ignoffo et al., 1973). Considerable research has been done with *R. culicivorax* as a possible biological control agent for blackflies (Finney, 1975; Hansen and Hansen, 1976; Poinar et al., 1979). Though early-instar simuliids could be infected under laboratory conditions, successful parasite development could not be demonstrated. As previously mentioned, even if successful development had taken place, it still remains that aquatic mermithids are adapted to the habitat of their natural hosts. Thus, it becomes impractical to attempt the control of an insect that inhabits a running water environment with a parasite adapted to the nonmoving water habitats of mosquitoes.

The past 10 years have provided considerable information about host specificity in mosquito mermithids. Generally, mermithids that complete their parasite development in the larval host exhibit a wide host range (i.e., *R. culicivorax*, *R. iyengari*, *O. muspratti*), but some are reported only from a single host (i.e., *O. troglodytis*, *R. kiktoreak*). The specificity of this latter group may not be as selective as studies indicate, since little is known except that observed in nature. One species, *S. peterseni*, was found to be generically specific for *Anopheles* mosquitoes and failed to develop in culicine mosquitoes under laboratory conditions (Petersen and Chapman, 1970). Mermithids that complete their parasitic development only in the adult host are usually specific for a single host (*P. culicis* for *Ae. sollicitans*, *C. culicivora* for *Ae. communis*, *E. cozii* for *An. funestus*, and *E. riouxi* for *Ae. detritus*). However, at least one species, *C. schakovii*, readily develops in a number of *Aedes* species.

Host susceptibility to a given mermithid species varies with each host species. *R. culicivorax* has been tested against at least 87 species of mosquitoes from 13 genera (Petersen and Chapman, 1979). Susceptibility to this nematode is generally highest in *Anopheles* mosquitoes, but at least one species is highly refractory and another moderately so in the form of encapsulation and melanization. Similar patterns are seen in the *Aedes, Culex,* and *Psoro-phora,* with at least one species in each genus exhibiting complete refractive-ness to *R. culicivorax.*

Though physiological defense mechanisms play a major role, they are by no means the only factor involved in host susceptibility. Many mosquito spe-cies exhibiting no physiological resistance to *R. culicivorax* vary in their sus-ceptibility to attack. This difference appears to be at least in part due to differences in behavior of the hosts. Further, age of the host has been shown to have a considerable effect on susceptibility (Petersen and Willis, 1970). Re-fractiveness because of age appears to be more physical than behavioral, prin-cipally because of the thicker cuticle in older larvae. The susceptibility of a given species can thus be difficult to determine because of the interaction of two or more of these forms of defense complicated by a given set of environ-mental factors that may or may not be favorable to the nematode.

In the laboratory, as a result of mass-rearing practices for *R. culicivorax,* where exposed but uninfected hosts were retained to supply the next genera-tion, the host mosquito species was found to have become measurably less sus-ceptible to the nematode after some 100 generations. No evidence of humoral defense mechanisms were found, and the mode of defense was not determined (Petersen, 1978a). Similar results were observed with the mass production of *S. peterseni* in *An. quadrimaculatus* (Woodard and Fukuda, 1977). Though the potential to develop a refractiveness for a mermithid nematode does exist, as demonstrated by these studies, it was slow to develop despite the pressure applied in the laboratory. Consequently, such changes in nature will probably require a very long period of time, if ever, to occur.

B. Effects of Parasites on Hosts

With the few exceptions where the host physiologically resists the nematode, mermithids that develop in larval hosts are always lethal to their host. As the nematodes develop, they place a heavy demand on the host's food reserves and thus prevent formation of pupal histoblasts (Gordon et al., 1974). The actual cause of death, however, is probably the result of excessive loss of body fluids at the time of nematode emergence. When the host is killed in the larval stage, the chances for the mermithids to become widely and uniformly distributed is greatly reduced. However, this apparent disadvantage can be considered an advantage as far as the biological control potential is concerned. Since these naturally occurring parasites do not effectively control large populations of mosquitoes, it becomes necessary to modify these agents in some way in order to make them more effective. With *R. culicivorax* we simply serve as the mecha-nism for distribution.

In some species of mermithids, the parasites are occasionally carried through the pupal stage into the adult host. These nematodes often create a lethal stress for the host if the parasite's development is advanced at the time of host pupation. The potential for dissemination of this latter group of mer-mithids, though better than that for nematodes that prevent host pupation, is still very limited because of the often weakened state of the parasitized adult host.

Mermithids that complete parasitic development only in adult hosts under-go little development while the host is in the larval stage and thus demand little of the host's food reserves. Because of the delayed parasite development, newly emerged adult hosts (unless heavily parasitized) can function normally for the first three to five days and readily take blood meal. In fact, it appears that a blood meal enhances the development of the parasite. Though the nematode does not prevent the blood-seeking activities of its host, it does prevent the host from reproducing, in most instances. When nematodes are ready to emerge, the host apparently seeks a natural oviposition site and attempts to oviposit. When she comes in contact with the moist substrate, the nematodes rupture the abdominal cuticle and emerge. This usually kills the host; however, a small percentage apparently do survive, but are capable of laying only a few eggs after subsequent blood meals. The desire of the host to return to an oviposition site places the nematodes in an ideal location for parasitism of future mosquito populations. As a result, these mermithids are usually widespread throughout their host's range. As an example, a survey in southwestern Louisiana showed infection levels of *P. culicis* in populations of *Ae. sollicitans* to range from 70 to 100% along a 100 mile stretch of the Gulf Coast (Petersen and Willis, 1969). A similar distribution of a *Culicimermis* sp. was reported from *Ae. vexans* in Manitoba and British Columbia (Harlos et al., 1980).

C. Physical Factors

Temperature is an important limiting factor, and its effect varies greatly with each mermithid species. Some species parasitize their hosts only once each year after a long severe cold period and are found as far North as the Arctic Circle; others are known only from the tropics of Africa. *S. peterseni* appears to have a uniquely broad tolerance for temperature. It was first discovered in southwestern Louisiana and was active the entire year (Petersen and Chapman, 1970). Later it was isolated in New York (Molloy and Wraight, 1980) and Canada (Ellis and Chapman, 1980). The importance of temperature on the potential use of mermithid nematodes as a biological control agent is exemplified by *R. culicivorax*. In its normal range, this species is active only during the summer when temperatures are above about 15°C (Petersen et al., 1968). Attempts to infect mosquitoes in the laboratory (Brown and Platzer, 1977) and the field (Galloway and Brust, 1977) at temperatures below 15°C were ineffective.

Desiccation does not appear to greatly affect mosquito mermithids in nature, even those that develop only in permanent water-breeding hosts. Habitats may remain "dry" for several months and continue to produce infected mosquitoes when conditions are right. Data indicate that the eggs of at least one species, *O. muspratti*, can remain viable for several years waiting for the proper conditions to stimulate hatch (Petersen, 1981). However, mosquito mermithids and their eggs cannot tolerate complete desiccation even for a few minutes. Thus, moisture becomes a critical factor in maintaining these parasites in the laboratory.

Since the infective stage (preparasite) is short lived, free swimming, and appears nondirectional in its attempt to locate a suitable host, dilution becomes an important factor in the use of these nematodes to control mosquitoes. In laboratory tests, increased volumes significantly reduced parasitism with *R. culicivorax* (Petersen and Willis, 1970). However, in large-volume outdoor experiments, *R. culicivorax* was found to be positively thigmotactic and negatively geotactic (Petersen, 1973). This behavior brought the nematodes to the surface of the water and along the edge of vegetation, which greatly increased their chance of contact with a suitable host. Thus, dilution under normal conditions was determined not to be an important factor in application of *R. culicivorax* for mosquito control.

Time of nematode applications is also very important when trying to control rapidly growing floodwater mosquitoes. As previously mentioned, hosts become less susceptible with age, and thus early nematode application is neces-

sary for effective control. With some species this timing is so critical that
effective use of R. culicivorax when applied for inundative control is impractical.

D. Chemical Factors

The chemical makeup of the aquatic environment can have a profound effect on
mosquito mermithids.

Tolerance to salinity varies greatly with each mermithid species. Petersen and Willis (1970) found that R. culicivorax was inhibited by mild salinity
(0.96%). This was confirmed by Brown and Platzer (1978a). Therefore, R.
culicivorax will be ineffective for mosquito control under conditions of increased
water salinity. In contrast, O. muspratti was shown to be tolerant of dilute
seawater containing 1.92% salt (Petersen, 1981). Also, the mermithid P. culicis
develops only in Ae. sollicitans, a saltmarsh-breeding mosquito. This mermithid, therefore, must be able to routinely tolerate very high salinity and has
adapted well to this type of environment.

Brown and Platzer (1978b) showed that transient exposure to low oxygen
tension increased the survival and thereby the infectivity of the preparasites
of R. culicivorax. More recently, Platzer (1981) reported that preparasites
of R. culicivorax stop moving within eight hours in water rich in organic content and low in oxygen content, and that lowered oxygen tensions in polluted
waters may be responsible for the inability of R. culicivorax to infect mosquitoes under such conditions.

Mermithid species also vary in their tolerance to pollution. Laboratory
and field studies have shown R. culicivorax to be ineffective in controlling
mosquitoes in heavily polluted environments. However, another species, O.
muspratti, has been shown to tolerate much higher levels of organically rich
tree-hole water than R. culcivorax. This again demonstrates the diversity in
tolerances of mermithid species that infect mosquitoes and suggests that there
may be mermithids in nature suited to many, if not most, mosquito habitats
(Petersen, 1981).

Mermithids apparently have a high tolerance for most pesticides. Mitchell
et al. (1974) in Taiwan found that levels of Abate, Dieldrin, and Gama-HCH,
normally used for mosquito control, did not adversely affect host parasitism
by R. culicivorax. Finney et al. (1977) reported that Altosid 5E, an insect
growth regulator, did not interfere with any phase of the parasite's development and host mortality was increased when a combination of Altosid and R.
culicivorax was used in laboratory experiments. Levy and Miller (1977a) reported that the infective stage of R. culicivorax could tolerate 0.1 ppm of
malathion, 0.003 ppm of Baytex, 0.001 ppm of Dursban, 0.001 ppm of Abate,
and 0.005 ppm of Dimilin with no apparent detrimental effects. Further,
Platzer and Brown (1976) reported that a variety of copper-based organic
algicides or copper sulfate failed to alter the infectivity of R. culicivorax at
concentrations used for algae and weed control.

Chen (1976) reported that optimum pH for infection by R. culicivorax
was 6.7-7.2; no infection was observed below pH 5.7 and above 8.7; and a
sharp drop in infection occurred below pH 6.2 and above 7.7. Chen's work
appears to be in error, since most aquatic habitats undergo a wide pH range
daily because of photosynthesis. Limited pH tolerances of this magnitude
would prevent the cycling of R. culicivorax in most aquatic environments.
Two recent studies actually substantiate that aquatic mermithids have a broad
tolerance for pH. Brown and Platzer (1978a) observed infections to occur
over a pH range of 3.6-8.6, and Petersen (1979) showed that host mosquitoes

were readily infected at all pH concentrations tested (5.4-7.9) and that infection increased at the lower pH ranges. Thus, under normal conditions, pH does not appear to be a limiting factor for *R. culicivorax* and probably most other aquatic mermithids.

VI. MASS PRODUCTION

A. In Vivo Rearing

To date, *R. culicivorax* is the only mermithid nematode that has been successfully mass produced. The primary techniques were developed during the early 1970s (Petersen and Willis, 1972a). At present *R. culicivorax* can only be produced in vivo (in an insect host). For mass production, 16,000-20,000 first-instar *Cx. pipiens* larvae are placed into rearing trays (136 × 52 × 5 cm) (Fig. 4A) and exposed to preparasites of *R. culicivorax* at a parasite-host ratio of 12:1. After seven days, the uninfected larvae begin to pupate and the nematodes become readily visible in the thoraxes of infected hosts (Fig. 4B); the mosquitoes are then removed from the rearing trays, washed, concentrated, and placed into nematode collection trays. The collection trays are constructed so that as the nematodes emerge from their hosts and drop to the bottom of the tray, they can pass through a screen divider (Fig. 4C). During this period, the pupae that results from uninfected hosts are separated from infected larvae by submerging the population in cold water (0-2°C). Chilled larvae sink to the bottom, are drawn off, and returned to the nematode collection trays (Fig. 4D). After emergence, the nematodes are concentrated, washed, and placed into culture pans (Fig. 4E). Each rearing tray produces about 10-15 g of postparasites.

Recent studies have shown that densities of the postparasites in culture pans had a significant effect on the yields of preparasites (Petersen, 1980). Cultures established in moist sand with a bottom surface area density of 15-25 postparasites per square centimeter (0.7-1.1 g nematodes per 100 cm^2) and a sand depth of 1-2 cm were most productive. Under these conditions, 5 g of postparasites per culture pan (22 x 32 x 5 cm) yielded 5-6 x 10^6 preparasites over three to four floodings.

Longevity of the nematode culture is an important limiting factor. Unlike chemicals, nematodes have a limited life span. Thus, there is an optimum period of preparasite hatch. At ambient temperatures (25-27°C) the maximum hatch will occur when the cultures are between 8 and 20 weeks old (Petersen, 1978b); this period can be extended about nine weeks by reducing the holding temperature to 15-20°C. Immediately upon hatching, the preparasites can then be applied directly to the mosquito habitat (Fig. 4F).

Sex in *R. culicivorax* is environmentally induced. When the host is either superparasitized (containing more than one *R. culicivorax*), inadequately fed, or overcrowded in the rearing tray, a preponderance of male nematodes will be produced (Petersen, 1972). Therefore, during the rearing procedure, it is important to control the level of infection (90-95%), provide adequate food, and limit the host populations to maintain maximum yields of female nematodes and, ultimately, preparasites.

B. In Vitro Rearing

Many inherent problems are associated with the mass production of parasites in vivo and, as a result, considerable attention has been directed toward the development of an in vitro rearing system for aquatic mermithids. Mermithids

Fig. 4 Mass rearing of *R. culicivorax*. (A) Rearing trays. (B) Mature parasitic nematodes in hosts. (C) Collection of emerging postparasitic nematodes. (D) Device for separation of host larvae and pupae. (E) Culture trays. (F) Application of nematodes. (Courtesy of USDA.)

are unique because they feed only during their relatively short parasitic phase and absorb nutrients directly (from the host's hemolymph) through their cuticle. Thus, they are very sensitive to mechanical damage, and hypo- and hypertonic changes in their feeding environment. Because of very selective needs in form and amounts of nutrients, in vitro rearing of mermithids has proven to be a difficult task. Again, *R. culicivorax* has been the most extensively studied species because of background information and availability as a test organism. Several researchers have succeeded in demonstrating limited growth by various means in *R. culicivorax*. To date the most successful attempt at in vitro rearing of a mermithid was reported by Finney (1977). She was able to get considerable growth and gonadal development in *R. culicivorax* over a six week period. However, these same nematodes lacked trophosome development (stored food reserves). Development of methodologies for in vitro rearing of nematodes is still at an early stage. It is unproven whether an in vitro rearing system can provide the ease of rearing, reduction of costs, and standardization currently found with the present in vivo system for *R. culicivorax*. However, even if a successful in vitro system proves less efficient than the present system for production of *R. culicivorax*, the methodology could be extremely valuable in providing a source for many potentially beneficial mermithids, which is not presently available.

VII. FIELD TRIALS

Though mermithids have been studied as potential control agents for insects for many years, few have progressed to the stage of field evaluation. With a few minor exceptions, only mosquito mermithids have been used for serious attempts to control host populations in nature.

The first recorded attempt to release mermithids for mosquito control was made on Nauru Island in the South Pacific in 1967 (Reynolds, 1972). A small number of eggs of *O. muspratti* (*Romanomermis* sp.) were introduced into several mosquito breeding sites; the results of this release, however, were never recorded. In 1970, hosts infected with *S. peterseni* were collected in the field and transferred to a new habitat. The nematodes became established and continued to produce high levels of parasitism in host populations (Petersen and Willis, 1974a). In 1971, the initial releases of *R. culicivorax* were made. Ten sites were treated with varied dosages of preparasites (at the time dosages were dependent on availability of nematodes). Mean levels of 65, 58, and 33% parasitism in second-, third-, and fourth-instar host larvae, respectively, were recorded (Petersen and Willis, 1972b). Though these initial treatments were applied on a small scale with many uncontrolled variables, the tests demonstrated that: (1) parasitism could be obtained in natural populations of mosquitoes, (2) high levels of parasitism could be obtained when preparasitic nematodes were applied at reasonable rates, (3) the infective-stage nematodes could be applied easily by many of the standard techniques used for application of insecticides, and (4) *R. culicivorax* could establish itself in many of the release sites for long-term control of many mosquito species.

In 1973, more extensive field trials were made with *R. culicivorax*. Fifteen sites treated with a dosage rate of 1200 preparasites per square meter of surface area produced a mean incidence of parasitism of 76%. Fifteen additional sites treated at 2400 preparasites per square meter of surface area averaged 85% parasitism in *Anopheles* mosquitoes (10 of the latter 15 treatments produced levels of infection in excess of 90%) (Petersen and Willis, 1974b). *R. culicivorax* became established and recycled, with high levels of parasitism in host populations in many of the sites treated in 1971 and 1973, and continued to produce

Fig. 5 Application of *R. culicivorax*. (A) Hand application using standard pesticide equipment. (Courtesy of USDA.) (B) Aerial application. (Courtesy of R. Levy.)

substantial levels of parasitism through 1975 (Petersen and Willis, 1975). Most of these habitats are now (1980) either inaccessible or no longer producing mosquitoes. However, many of those that are still producing mosquitoes have had continued *R. culicivorax* activity for seven to nine years. In these studies, preparasites of *R. culicivorax* were applied by standard compressed-air sprayers (Fig. 5a). Laboratory studies indicated that the nematodes were not significantly damaged when passed through these delivery systems. Studies by Levy et al. (1979) also indicated the *R. culicivorax* can be effective in controlling mosquitoes when applied by aerial spray from a helicopter equipped with a standard low-profile spray system (Fig. 5b).

Brown et al. (1977) reported that four species of mosquitoes were infected in three natural and four artificial sites in California when treated with preparasites of *R. culicivorax*. Infection levels were dependent on the mosquito subfamily, application rate, and test site. Anopheline mosquitoes were reported to be more susceptible than culicine mosquitoes, and sites with dense vegetation or algal mats reduced the effectiveness of the parasites. In Florida, Levy and Miller (1977b) reported that *R. culicivorax* effectively parasitized and developed in *Cx. quinquefasciatus* larvae which were in an abandoned sewage settling tank. Levy and Miller (1977c) also reported 96% parasitism of four floodwater mosquito species from eight potholes and ditches when these habitats were treated with *R. culicivorax* at a dosage rate of 3600 preparasites per square meter of surface area; when *R. culicivorax* was released into mosquito breeding sites in Maryland, it parasitized 50-100% of the mosquitoes, survived winter temperatures down to -19°C, and recycled the following year (Nickle, 1979).

The most extensive release of a pathogen or parasite for mosquito control to date was made on a lake in El Salvador. Nearly 110,000 m^2 of surface area were treated with *R. culicivorax* for the control of the malaria mosquito *An. albimanus*. Mean parasitism was only 39% for the first eight treatments. These treatments were applied in the morning at dosage rates that should have produced substantially higher levels of parasitism. Treatments were then changed to evening to avoid daily afternoon wave action in the lake. Parasitism averaged 92% in *Anopheles* spp. for five subsequent treatments. Even with the low infection levels during the early part of the study, a 94% reduction in host populations was achieved during the seven week study period (Petersen et al., 1978).

Though most of the field studies with *R. culicivorax* involved the release of the preparasitic stage in an inundative manner (immediate control), some work has been done using the inoculative approach (release of the postparasitic stage of the parasite). Limited studies in Louisiana showed that when postparasites of *R. culicivorax* were introduced into habitats of floodwater mosquitoes as a prehatch treatment, significant levels of parasitism were achieved (Petersen and Willis, 1976). Recent studies in California showed that early-season application of postparasites of *R. culicivorax* were effective in providing continuous partial control (weekly mean infection of 60%) of *An. freeborni* and *Cx. tarsalis*. Also, the nematodes successfully overwintered, survived rice culture practices, and continued to infect native host populations (Brown-Westerdahl et al., 1979).

Not all attempts to use *R. culicivorax* for mosquito control have been successful, especially those in environments known to be inhibitory to the nematode: attempts to infect *Cx. fatigans* (*quinquefasciatus*) in Taiwan were unsuccessful because of the polluted environment in which the tests took place (Mitchell et al., 1974); *R. culicivorax* was shown to be ineffective when tested against early spring *Aedes* mosquitoes in Canada because of low temperatures (Galloway and Brust, 1977); and tests against pasture *Aedes* mosquitoes in California were disappointing, presumably because of the running water conditions present in the test areas (Hoy and Petersen, 1973). Predation of the preparasitic stage of *R. culicivorax* by microcrustaceans (Platzer and MacKenzie-Graham, 1980) and other stages of the nematode by other invertebrate predators (Platzer and MacKenzie-Graham, 1978) may account for the lower than expected levels of parasitism observed following some releases of the mermithid in habitats that were otherwise assumed suitable.

These studies show that, if properly used, mermithids can be effective in controlling mosquito populations. However, as living organisms, they are subject to environmental limitations and thus cannot be used indiscriminately. A number of factors must be considered (water quality, temperature, movement, predation, host species, age and density, and makeup of the habitat) if these parasites are to be used effectively.

VIII. SAFETY

Certain representatives of the family Mermithidae have been reported as accidental human parasites. Poinar (1975) studied these reports and concluded the human infection by mermithid nematodes cannot be confirmed at this time and should be accepted as fact only when proven experimentally or when parasites are found developing in situ.

Few host specificity data are available for most mermithids. However, as previously mentioned, R. culicivorax has been reported only from larval mosquitoes and two species of Chaoboridae in nature. Laboratory studies showed limited penetration of early-instar larvae of several nematocerous Dipter under laboratory conditions (Ignoffo et al., 1973), but it is highly unlikely any of these dipterans would serve as suitable hosts in nature.

Ignoffo et al. (1974) reported that when suckling and adult mice and adult rats were subjected to either per os, intranasal, intraperitoneal, or dermal challenge of R. culicivorax, their body weight gain and histologies were identical to those of untreated animals. Immunodepressed rats also were not susceptible. Similar findings were reported for a Romanomermis species in India (Anon., 1978).

The selectivity shown by mosquito mermithids from the many field observations in recent years, the laboratory safety data available for R. culicivorax, and the questionable mermithid-human associations previously recorded, clearly demonstrate the safety of these organisms for nontarget organisms, the environment, and humans. In 1976, the Environmental Protection Agency determined that this nematode be classified as a parasite and not a pesticide, exempting it from pesticide regulations (Nickle, 1976).

IX. COMMERCIAL DEVELOPMENT

Although a number of nematode species show promise as biological control agents, only two species, Neoaplectana carpocapsae, a parasite of terrestrial insects, and R. culicivorax can be mass produced in sufficient quantities to be considered prospects for commercial production. Though R. culicivorax has been extensively studied and field tested, commercial development of this parasite has been slow to develop because of the environmental limitations, host specificity, and inherent problems associated with handling, storage, shipping, and application procedures. However, despite these problems, three companies have pursued commercial preparations of R. culicivorax (Fig. 6).

The first product "Skeeter Doom" was sold from 1975 to 1978 by the Fairfax Biological Laboratory, Clinton Corners, New York. It enjoyed early succes, but later it became apparent that a subsidy of some sort was needed to continue production until the product was sufficiently well-known to be profitable. This was not available, and the company ceased production. Shipping and packaging emerged as problems for this new type of product though some checks on material sold by the company showed that "Skeeter Doom" arrived in good condition and established in a subdivision in Maryland (Nickle, 1980).

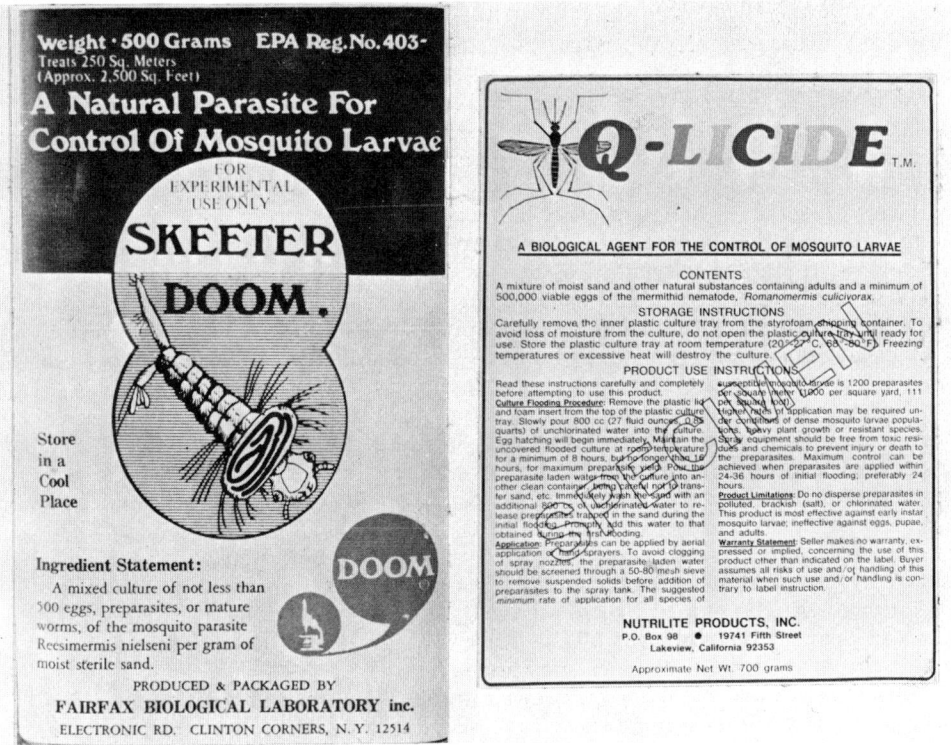

Fig. 6 Proposed labels for commercial preparations of *R. culicivorax*.

In 1977, Nutrilite Products, Inc., Buena Park, California, became interested in *R. culicivorax* because of its potential and because it was exempted from cost-prohibitive Environmental Protection Agency regulations. The company decided to allocate two years of effort (six person-years) to attempt the development of this parasite into a commercial product. After becoming familiar with *R. culicivorax* and its rearing procedures, a cost-effective mass-rearing facility was established. A major task was the development of shipping container that would keep the nematode eggs viable during shipment and storage prior to use. Several concepts and designs were studied. These were followed by extensive shipping studies before an effective container was developed that permitted quality-control guarantees for the new product (Cupello et al., 1982). Then, two months prior to test marketing of this product, the company changed its program and divested itself of all biological control research and development and, in so doing, terminated research on *R. culicivorax* (Petersen and Cupello, 1981). Though research at Nutrilite established that a marketable product could be developed and that a market potential was there for *R. culicivorax*, it became evident that the limited market volume puts production economics outside the interests of the larger manufacturer. Thus, the future for these types of biological control agents lies with so-called cottage industries (small corporations or partnerships) or with government-subsidized productions.

X. FUTURE PROSPECTS

Mermithid nematodes are the most promising of the true biological control agents for mosquitoes at the present time. *R. culicivorax* has drawn much of the attention because this species possesses the characteristics of an ideal biological control agent, as previously mentioned. However, because of these characteristics, this form of biological control finds itself in a "Catch-22" situation. Biological agents are desirable for many reasons, but perhaps the most significant is host specificity. This ability to selectively kill only target pests is their unique advantage over chemical pesticides. But, in being highly specific, their use is limited to, at most, only a few insect species. The result is that the total product volume sold per year will be considerably less than would be sold if the product were applicable to a larger range of pests. Couple host specificity with the potential for establishment and continued recycling and the potential market for these agents is further threatened.

Furthermore, biologicals, by definition, are living organisms (in this sense, sex pheromones, bacterial toxins, and insect growth regulator compounds are not true biologicals). Thus, shipping and storage become major problems for a producer. Also, the people who will use biologicals must be trained and develop some degree of skill in application.

We find ourselves in a serious dilemma. We are being constantly challenged by far-sighted academicians and environmentalists to embrace biological control. At the same time, however, laws and regulations are passed that delay or prevent development of the very biologicals that might provide alternatives to the use of chemical pesticides.

As things now stand, we still await the success of that first truly effective commercially available product to demonstrate the feasibility of nematodes as biocontrol agents for insect control. Although attempts to produce mermithid nematodes commercially are not currently underway, the methodologies for the commercial preparation of *R. culicivorax* have been developed. It now appears this nematode will remain "on the shelf" until increased pressures are applied for biological control, and cottage industries make it available to people involved in mosquito control.

REFERENCES

Anon. (1978). Pathogens and Parasites. Vector Control Research Center, Pondicherry. *Indian Council Med. Res. Ann. Rep.*, pp. 68-83.

Brown, B. J., and Platzer, E. G. (1977). The effects of temperature on the infectivity of *Romanomermis culicivorax*. *J. Nematol. 9*: 166-172.

Brown, B. J., and Platzer, E. G. (1978a). Salts and the infectivity of *Romanomermis culicivorax*. *J. Nematol. 10*: 53-64.

Brown, B. J., Platzer, E. G. (1978b). Oxygen and the infectivity of *Romanomermis culicivorax*. *J. Nematol. 10*: 110-113.

Brown, B. J., Platzer, E. G., and Hughes, D. S. (1977). Field trials with the mermithid nematode *Romanomermis culicivorax*, in California. *Mosq. News 37*: 603-608.

Brown-Westerdahl, B., Washino, R. K., and Platzer, E. G. (1979). Early season application of *Romanomermis culicivorax* provides continuous partial control of rice field mosquitoes. *Proc. Calif. Mosq. Vector Cont. Assoc. 47*: 55.

Chandrahas, R. K., and Rajagopalan, R. K. (1979). Mosquito breeding and the natural parasitism of larvae by a fungus, *Coelomomyces* and a mermithid nematode, *Romanomermis*, in paddy fields in Pondicherry. *Indian J. Med. Res. 69*: 63-70.

Chen, P. S. (1976). A study of *Reesimermis nielseni* for control of *Culex pipiens fatigans* in Taiwan. *Bull. Inst. Zool. Acad. Sin. 15*: 21-28.

Coman, D. (1961). Mermithidae. *Fauna Repub. Pop. Rom. 2*: 1-61.

Cupello, J. M., Petersen, J. J., and Chauthani, A. R. (1982). An improved method for long-distance shipping the mosquito parasite *Romanomermis culicivorax*. *J. Nematol. 14*: 121-125.

Doucet, M. M., Laumond, C., and Bain, O. (1979). *Empidomermis riouxi*, new species (Nematoda: Mermithidae) parasite from *Aedes detritus* (Diptera: Culicidae). *Ann. Parasit. Hum. Comp. 54*: 341-352.

Dujardin, F. (1845). Historic naturelle des helminthes ou vers intestinaux. *Paris Raret* 654 pp.

Ellis, R. A., and Chapman, H. C. (1980). Mermithid parasites of Canadian anophelines. *Mosq. News 40*: 115-116.

Finney, J. R. (1975). The penetration of three simuliid species by the nematode *Reesimermis nielseni*. *Bull. WHO 52*: 235.

Finney, J. R. (1977). The development of *Romanomermis culicivorax* in in vitro culture. *Nematologica 23*: 479-480.

Finney, J. R., Gordon, R., Condon, W. J., and Rusted, T. N. (1977). Laboratory studies on the feasibility of integrated mosquito control using an insect growth regulator and a mermithid nematode. *Mosq. News 37*: 6-11.

Galloway, T. D., and Brust, R. A. (1977). Effects of temperature and photoperiod on the infection of two mosquito species by the mermithid, *Romanomermis culicivorax*. *J. Nematol. 9*: 218-221.

Galloway, T. D. and Brust, R. A. (1979). Review of the genus *Romanomermis* (Nematoda: Mermithidae) with a description of *R. communensis* sp. n. from Canada. *Can. J. Zool. 57*: 281-289.

Gendre, E. (1909). Sur des larves de *Mermis* parasites des larves du *Stegomyia fasciata*. *Bull. Soc. Pathol. Exotique 2*: 106-108.

Gordon, R., Bailey, C. H., and Barber, J. M. (1974). Parasitic development of the mermithid nematode *Reesimermis nielseni* in the larval mosquito *Aedes aegypti*. *Can. J. Zool. 52*: 1293-1302.

Hansen, E. L., and Hansen, J. W. (1976). Parasitism of *Simulium damnosum* by *Romanomermis culicivorax*. *IRCS Med. Sci. 4*: 508.

Harlos, J., Brust, R. A., and Galloway, T. D. (1980). Observations on a nematode parasite of *Aedes vexans* (Diptera: Culicidae) in Manitoba. *Can. J. Zool. 58*: 215-220.

Hoy, J. B., and Petersen, J. J. (1973). Fish and nematodes – current status of mosquito control techniques. *Proc. Calif. Mosq. Control Assoc. 41*: 49-50.

Ignoffo, C. M., Biever, K. D., Johnson, W. W., Sanders, H. O., Chapman, H. C., Petersen, J. J., and Woodard, D. B. (1973). Susceptibility of aquatic vertebrates and invertebrates to the infective stage of the mosquito nematode *Reesimermis nielseni*. *Mosq. News 33*: 599-602.

Ignoffo, C. M., Petersen, J. J., Chapman, H. C., and Novotny, J. F. (1974). Lack of susceptibility of mice and rats to the mosquito nematode *Reesimermis nielseni* Tsai and Grundmann. *Mosq. News 34*: 425-428.

Isaeva, N. M. (1977). New data on the parasite-host relation of *Culicimermis schakhovii* Rubtsov et Isavea 1975 and *Aedes cantans* Mg (in Russian). *Vestnik Zool. 5*: 87-88.

Iyengar, M. O. T. (1927). Parasitic nematodes of *Anopheles* in Bengal. *Far Eastern Assoc. Trop. Med., 7th Congress British India 3*: 128-135.

Johnson, A. A. (1963). *Octomyomermis itascensis* gen. et sp. nov. (Nematoda: Mermithidae) a parasite of *Chironomus plumosus* (L). *Am. Microsc. Soc. 82*: 237-241.

Johnson, H. P. (1903). A study of certain mosquitoes in New Jersey, and a statement of the "mosquito-malaria theory". *N. J. Agric. Exp. Sta. Rep.* 23: 559-573.

Kurihara, T., and Kazuhisa, H. (1979). Two factors affecting parasitism by *Romanomermis culicivorax* in mosquito larvae. *Jpn. J. Sanit.* 30: 200-202.

Levy, R., Hertlein, B. C., Petersen, J. J., Doggett, D. A., and Miller, Jr., T. W. (1979). Aerial application of *Romanomermis culicivorax* (Mermithidae: Nematoda) to control *Anopheles* and *Culex* mosquitoes in southwest Florida. *Mosq. News* 39: 20-25.

Levy, R., and Miller, Jr., T. W. (1977a). Susceptibility of the mosquito nematode *Romanomermis culicivorax* (Mermithidae) to pesticides and growth regulators. *Environ. Entomol.* 6: 447-448.

Levy, R., and Miller, Jr., T. W. (1977b). Experimental release of a mermithid to control mosquitoes breeding in sewage settling tanks. *Mosq. News* 37: 410-414.

Levy, R., and Miller, Jr., T. W. (1977c). Experimental release of *Romanomermis culicivorax* (Mermithidae: Nematoda) to control mosquitoes breeding in Southwest Florida. *Mosq. News* 37: 483-486.

Mitchell, C. J., Chen, P., and Chapman, H. C. (1974). Exploratory trials utilizing a mermithid nematode as a control agent for *Culex* mosquitoes in Taiwan. *J. Formosan Med. Assoc.* 73: 241-254.

Molloy, D., and Wraight, S. P. (1980). Isolation of *Strelkovimermis peterseni*, a mermithid parasite of Anopheline mosquitoes in northeastern New York. *J. Nematol.* 12: 330-332.

Muspratt, J. (1947). The laboratory culture of a nematode parasite of mosquito larvae. *J. Entomol. Soc. South Africa.* 10: 131-132.

Nickle, W. R. (1972). A contribution to our knowledge of the Mermithidae (Nematoda). *J. Nematol.* 4: 113-146.

Nickle, W. R. (1976). Toward the commercialization of a mosquito mermithid *Proc. Int. Colloq. Invert. Pathol., Kingston, Canada*, pp. 241-244.

Nickle, W. R. (1979). Probable establishment and overwintering of a mermithid nematode parasite of mosquitoes in Maryland. *Proc. Helm. Soc. Wash.* 46: 21-27.

Nickle, W. R. (1980). Possible commercial formulations of insect-parasitic nematodes. *Biotech. Bioeng.* 22: 1407-1414.

Obiamiwe, B. A., and MacDonald, W. W. (1973). A new parasite of mosquitoes, *Reesimermis muspratti* sp. n. (Nematoda: Mermithidae), with notes on its life-cycle. *Am. Trop. Med. Parasitol.* 67: 439-444.

Petersen, J. J. (1972). Factors affecting sex determination in a mermithid parasite of mosquitoes. *Mosq. News* 32: 226-230.

Petersen, J. J. (1973). Relationship of density, location of hosts, and water volume to parasitism of larvae of the southern house mosquito by a mermithid nematode. *Mosq. News* 33: 516-420.

Petersen, J. J. (1975). Development and fecundity of *Reesimermis nielseni*, a nematode parasite of mosquitoes. *J. Nematol.* 7: 211-214.

Petersen, J. J. (1977). Biology of *Octomyomermis muspratti*, a parasite of mosquitoes as it relates to mass production. *J. Invert. Pathol.* 30: 155-159.

Petersen, J. J. (1978a). Development of resistance by the southern house mosquito to the parasitic nematode *Romanomermis culicivorax*. *Econ. Entomol.* 7: 518-520.

Petersen, J. J. (1978b). Observations on the mass production of *Romanomermis culicivorax*, a nematode parasite of mosquitoes. *Mosq. News* 38: 83-86.

Petersen, J. J. (1979). pH as a factor in parasitism of mosquito larvae by the mermithid *Romanomermis culicivorax*. *J. Nematol. 11*: 105-106.

Petersen, J. J. (1980). Mass production of the mosquito parasite *Romanomermis culicivorax*: Effect of density. *J. Nematol. 12*: 45-48.

Petersen, J. J. (1981). Observations on the biology of *Octomyomermis muspratti*, a nematode parasite of mosquitoes. *J. Invert. Pathol. 37*: 290-294.

Petersen, J. J., and Chapman, H. C. (1970). Parasitism of *Anopheles* mosquitoes by a *Gastromermis* sp. (Nematoda: Mermithidae) in southwestern Louisiana. *Mosq. News. 30*: 420-424.

Petersen, J. J., and Chapman, H. C. (1979). Checklist of mosquito species tested against the nematode parasite *Romanomermis culicivorax*. *J. Med. Entomol. 15*: 468-471.

Petersen, J. J., Chapman, H. C., Willis, O. R., and Fukuda, T. (1978). Release of *Romanomermis culicivorax* for the control of *Anopheles albimanus* in El Salvador. II. Application of the nematode. *Am. J. Trop. Med. Hyg. 27*: 1268-1273.

Petersen, J. J., Chapman, H. C., and Woodard, D. B. (1967). Preliminary observations on the incidence and biology of a mermithid nematode of *Aedes sollicitans* (Walker) in Louisiana. *Mosq. News 27*: 493-498.

Petersen, J. J., Chapman, H. C., and Woodard, D. B. (1968). Bionomics of a mermithid nematode of larval mosquitoes in southwestern Louisiana. *Mosq. News 28*: 346-352.

Petersen, J. J., and Cupello, J. M. (1981). Commercial development and future prospects for entomogenous nematodes. *J. Nematol. 13*: 280-284.

Petersen, J. J., and Willis, O. R. (1969). Incidence of *Agamomermis culicis* Stiles (Nematoda: Mermithidae) in *Aedes sollicitans* in Louisiana in 1967. *Mosq. News. 29*: 87-92.

Petersen, J. J., and Willis, O. R. (1970). Some factors affecting parasitism by mermithid nematodes in southern house mosquito larvae. *J. Econ. Entomol. 63*: 175-178.

Petersen, J. J., and Willis, O. R. (1972a). Procedures for the mass rearing of a mermithid parasite of mosquitoes. *Mosq. News 32*: 226-230.

Petersen, J. J., and Willis, O. R. (1972b). Results of preliminary field applications of *Reesimermis nielseni* (Mermithidae: Nematoda) to control mosquito larvae. *Mosq. News 32*: 312-316.

Petersen, J. J., and Willis, O. R. (1974a). *Diximermis peterseni* (Nematoda: Mermithidae): A potential biocontrol agent of *Anopheles* mosquito larvae. *J. Invert. Pathol. 24*: 20-23.

Petersen, J. J., and Willis, O. R. (1974b). Experimental release of a mermithid nematode to control *Anopheles* mosquitoes in Louisiana. *Mosq. News 34*: 316-319.

Petersen, J. J., and Willis, O. R. (1975). Establishment and recycling of a mermithid nematode for the control of mosquito larvae. *Mosq. News 35*: 526-532.

Petersen, J. J., and Willis, O. R. (1976). Experimental release of a mermithid nematode to control floodwater mosquitoes in Louisiana. *Mosq. News 36*: 339-342.

Platzer, E. G. (1981). Biological control of mosquitoes with mermithids. *J. Nematol. 13*: 257-262.

Platzer, E. G., and Brown, B. J. (1976). Physiological ecology of *Reesimermis nielseni*. *Proc. Int. Colloq. Invert. Pathol., Kingston, Canada*, pp. 263-267.

Platzer, E. G., and MacKenzie-Graham, L. L. (1978). Predators of *Romano-mermis culicivorax*. *Proc. Calif. Mosq. Vector Cont. Assoc. 46*: 93.

Platzer, E. G., and MacKenzie-Graham, L. L. (1980). *Cyclops vernalis* as a predator of the preparasitic stages of *Romanomermis culicivorax*. *Mosq. News 40*: 252-257.

Poinar, G. O., Jr. (1975). On the question of human infection by nematodes of the family Mermithidae (Dorylaimida: Adenophora). *WHO VBC* 564.

Poinar, G. O., Jr. (1977). *Empidomermis cozii* new genus, new species (Mermithidae: Nematoda), a parasite of *Anopheles funestus* (Culicidae: Diptera) in West Africa. *Can. J. Zool. 55*: 1475-1479.

Poinar, G. O., Jr., Hess, R., Hansen, E., and Hansen, J. (1979). Laboratory infection of blackflies (Simuliidae) and midge (Chironomidae) by the mosquito mermithid *Romanomermis culicivorax J. Parasitol. 65*: 613-615.

Poinar, G. O., Jr., and Sanders, R. D. (1974). Description and bionomics of *Octomyomermis troglodytis* sp. n. (Nematoda: Mermithidae) parasitizing the western treehole mosquito *Aedes sierrensis* (Ludlow) (Diptera: Culicidae). *Proc. Helm. Soc. Wash. 41*: 37-41.

Reynolds, D. G. (1972). Experimental introduction of a microsporidian into a wild population of *Culex pipiens fatigans* Wied. *Bull. WHO 46*: 807-812.

Ross, J. F., and Smith, S. M. (1976). A review of the mermithid parasites (Nematoda: Mermithidae) described from North American mosquitoes (Diptera: Culicidae) with descriptions of three new species. *Can. J. Zool. 54*: 1084-1102.

Ross, R. (1906). Notes on the parasites of mosquitoes found in India between 1895-1899. *Jour. Hygiene. 6*: 101-108.

Rubtsov, I. A., and Isaeva, N. M. (1975). *Culicimermis schakhovii* gen. et sp. n. (Mermithidae), a new parasite of mosquitoes (in Russian). *Vestn. Zool. 1*: 39-44.

Smith, J. B. (1903). Report on the mosquito investigations. *N. J. Agric. Exp. Sta. Rep. 23*: 509-512.

Stiles, C. W. (1903). A parasitic roundworm (*Agamomermis culicis* n.g., n. sp.) in American mosquitoes (*Culex sollicitans*). *Hygiene Lab. U.S. Public Health Marine Hosp. Serv. Bull. 13*: 15-17.

Tsai, Y., and Grundmann, A. W. (1969). *Reesimermis nielseni* gen. et. sp. n. (Nematoda: Mermithidae) parasitizing mosquitoes in Wyoming. *Proc. Helm. Soc. Wash. 36*: 61-67.

Welch, H. E. (1960). *Hydromermis churchillensis* n. sp. (Nematoda: Mermithidae) from Churchill, Manitoba with observations on its incidence and bionomics. *Can. J. Zool. 38*: 465-474.

Welch, H. E., and Bronskill, J. F. (1962). Parasitism of mosquito larvae by the nematoda, DD136 (Nematoda: Neoaplectanidae). *Can. J. Zool. 40*: 1263-1268.

Woodard, D. B., and Fukuda, T. (1977). Laboratory resistance of the mosquito *Anopheles quadrimaculatus* to the mermithid nematode *Diximermis peterseni*. *Mosq. News 37*: 192-195.

Yakubovich, V. Y. (1978). The finding of a new mermithid species *Culicimermis culicivora* n. sp. in *Aedes communis* De Geer in the Buryat USSR (in Russian). *Medits. Parazitol. 47*: 107-108.

Chapter 23

Nematode Parasites of Blackflies

Roger Gordon *Memorial University of Newfoundland, St. John's, Newfoundland, Canada*

I. INTRODUCTION

Blackflies (Diptera: Simuliidae) rank among the most serious pests of humans and domestic livestock. The major effects of simuliids in the Nearctic are socioeconomic, resulting from the voracity of the insect's feeding activity. In tropical Africa and Central America, blackflies are a serious human health problem, since they transmit to humans the filarioid nematode *Onchocerca volvulus*, causative agent of onchocerciasis, i.e., "river blindness."

The early Canadian missionaries and explorers provided colorful accounts of their encounters with blackflies, describing them as "little demons" that "stick to their prey like bulldogs" and "make one look like a leper, hideous to the sight" (Davies et al., 1962). Adult female blackflies are indeed persistent, attacking in swarms and crawling into the clothing, hair, ears, eyes, or nostrils of their victims. Their bites are extremely painful and itchy; vision may be obscured through swelling when the bites are located around the eyes. Secondary skin infections may result when the bites are scratched. Particularly sensitive individuals may incur "blackfly fever," an allergic reaction against the mass introduction of toxins into the circulatory system through large-scale blackfly attack. The disorder is characterized by nausea, headache, fever, and swollen neck glands; hospitalization may be required (Fredeen, 1969). Development of the Canadian north is hampered by dense indigenous populations of human-biting simuliids. In such numbers, blackflies cause a reduction in working efficiency, especially among those employed in the pulpwood industry, and pose a discouragement to tourists (Jamnback, 1973; Fredeen, 1977).

Cattle are especially vulnerable to simuliid attack, and this has been a long-standing problem for livestock producers in northern Alberta and north-central Saskatchewan. Outbreaks of blackflies disrupt grazing and breeding activities of cattle, reduce production of milk (by up to 50%) and beef, and cause fatalities among the herds (Rempel and Arnason, 1947; Fredeen, 1977). In a single year, monetary loss due to simuliid attack in Canada was estimated to be $105 million (Anon., 1982) and $600,000 for a survey area (53,245) ha) surrounding a single river system in northern Alberta (Charnetski and Haufe, 1981).

Blackflies act as vectors-intermediate hosts for a variety of filarioid nematodes and protozoans. Of major importance is the role that several species of blackflies play in transmitting *O. volvulus* to humans. Onchocerciasis is endemic throughout the greater part of tropical Africa, as well as in the Yemen and in part of the area of Latin America from northwestern Brazil to southern Mexico. The commonly cited figure of 20 million for the total number of people in the world infected with *O. volvulus* is now believed to grossly underestimate the severity of the problem.* The adult nematodes form nodules underneath the skin, causing disfigurement. The most serious clinical manifestations of the parasitemia, however, are inflicted by the microfilariae, first-stage juveniles that migrate into the peripheral blood circulation and into the eye tissues. Skin conditions include dermatitis, skin thickening, loss of skin elasticity, and resultant hernias (Nelson, 1970). The most debilitating aspect of the disease is blindness. Communities in the savannah regions of Africa often have blindness rates in excess of 10%, and 30% of the population may have impaired vision due to onchocerciasis (Nelson, 1970). Entire communities move away from the more fertile river valleys, where the simuliids breed, to the less arable upland areas, which become overcrowded, overcropped, and overgrazed. In 1974, the World Health Organization, in association with several other agencies, launched a long-term control program, centered around the Volta River Basin in West Africa. By 1978, continual spraying of the river systems with insecticide had reduced populations of the blackfly vector, *Simulium damnosum*, within the central part of the zone below a defined maximum permissible level for transmission of *O. volvulus* (Walsh et al., 1979). In addition to the well-documented role of simuliids as vectors of *O. volvulus*, recent evidence suggests that in Brazil, some species of blackflies may be involved in transmitting the nonpathogenic filarioid *Mansonella ozzardi* to humans (Shelley et al., 1980).

Blackflies transmit several filarioids to domestic livestock and wild game (Poinar, 1977; Omar et al., 1979; Addison, 1980), but there is no evidence that these nematodes cause ill health in their hosts. Of greater consequence is the role that ornithophilic blackflies play in transmitting species of the sporozoan *Leucocytozoon* to birds. Leucocytozoonosis is especially serious in turkeys; in South Carolina, it has resulted in substantial economic loss to commercial turkey producers (Barnett, 1977). The causative agent, *Leucocytozoon smithi*, is transmitted to turkeys by several ornithophilic simuliid spe-

---

*Epidemiology of onchocerciasis. Report of a WHO Expert Committee. WHO Technical Report Series No. 597, (1976).

cies and causes hemolytic anemia, gross hypertrophy of the liver and spleen, congested lungs, emaciation, and death (Fallis and Desser, 1977). In the Northern Hemisphere, simuliids transmit the highly pathogenic species *Leucocytozoon simondi* to young ducks and geese (Laird and Bennett, 1970; Fallis and Desser, 1977; Fredeen, 1977). Blackflies have not been directly implicated in the transmission of viral diseases, but such a role cannot be discounted. Eastern equine encephalitis virus has been isolated from blackflies in North America, and Venezuelan equine encephalitis, as well as a virus that causes vesicular stomatitis in livestock, from simuliids in Central America (Travis et al., 1974).

In addition to the filarioid nematodes, which utilize blackflies as intermediate hosts, other nematodes parasitize only blackflies and spend the remaining portion of their life cycles free-living in the environment. For the purposes of this discussion, the term *entomophilic* will apply to this latter category of nematode and the chapter will be devoted to them exclusively. Reviews of filarioid parasites of simuliids are available elsewhere (Poinar, 1975, 1977).

This chapter will review the major entomophilic nematode parasites of blackflies, their life cycles, host-parasite relations, and biocontrol potential. For other recent reviews, the reader is referred to Poinar (1979, 1981), Finney (1981a), Molloy (1981), and Rubtsov (1981b).

II. CLASSIFICATION OF NEMATODE PARASITES OF BLACKFLIES

In addition to the filarioids, several other nematode superfamilies may utilize blackflies as intermediate or paratenic hosts (Poinar, 1977).

All entomophilic nematodes thus far described from natural infections of simuliids belong to the family Mermithidae (Table 1). Rubtsov (1964, 1981a) recorded tetradonematid nematodes from simuliids in the USSR, but it is likely that the nematodes were mermithids, since taxonomic assignments were made on the basis of parasitic juveniles. It is possible, however, to experimentally infect larval blackflies with *Neoaplectana carpocapsae*, a steinernematid nematode with a host range that encompasses a variety of insect orders.

The morphology of mermithid nematodes is considerably modified from the usual nematode plan. Except for the infective preparasites, all stages

Table 1 Classification of entomophilic nematode parasites of blackflies

---

Superfamily: Mermithoidea
Family: Mermithidae
Genus: *Gastromermis*
    Hydromermis
    Isomermis
    Limnomermis
    Neomesomermis

Superfamily: Rhabditoidea
Family: Steinernematidae
Genus: *Neoaplectana*[a]

---

[a]These nematodes have not been recorded from natural infections of the host.

lack a digestive tube. Parasitic juveniles absorb nutrients from the host hemolymph across their outer cuticle and store them in a modified intestine, the trophosome, which occupies most of the pseudocoelom. Mermithids resemble trichurid and trichinellid nematodes in possessing a stichosome. In addition to possessing so atypical a structure, mermithids are difficult to speciate because they possess few readily distinguishable features. Compounding this difficulty, the literature abounds with inadequate descriptions of mermithid nematodes. Until relatively recently, the majority of recordings of mermithid parasitism in blackflies were provided by entomologists, who accidentally discovered juvenile mermithids during primary studies of the blackfly host. These nematodes were loosely referred to as *Mermis* (Strickland, 1911; Petersen, 1924; Wu, 1930; Twinn, 1936, 1939; Lewis, 1953; Carlsson, 1962), *Paramermis* (McComb and Bickley, 1959), *Limnomermis* (Sommerman et al., 1955), and *Hydromermis* (Anderson and Dicke, 1960; Peterson, 1960).

Blackfly-parasitic mermithids most commonly belong to the three genera: *Neomesomermis*, *Gastromermis*, and *Isomermis*. The genus *Neomesomermis* was erected by Nickle (1972) to replace *Mesomermis* on the grounds that the type species of the latter genus had been insufficiently described (Chitwood, 1935). Some authorities, believing such a substitution of generic name to be unnecessary, have retained the genus *Mesomermis* (Molloy, 1979; Poinar, 1979). Species of *Limnomermis* and *Hydromermis* have also been described from simuliids. There is disagreement over the legitimacy of the genus *Spiculimermis*. This genus was listed as one that includes species parasitic in blackflies (Rubtsov, 1977, 1981a; Molloy, 1981), but Poinar (1975) considered it to be one of several mermithid genera that are insufficiently described.

There is a division of opinion on the number of valid species of simuliid parasitic mermithids that have been thus far recorded. Most authorities recognize approximately 35-40 species worldwide. Rubtsov (1981b), however, considers there to be at least 80 species, plus many subspecies, in Eurasia alone. This difference results from a basic disagreement over what constitutes valid taxonomic criteria. Nickle (1972) and Poinar (1981) represent the mainstream of opinion that only the morphology of the adult and preparasitic stages should, at present, be considered for taxonomic purposes. By contrast, Rubtsov (1981b) contends that mermithids may be speciated on the basis of the structure of parasitic and postparasitic juveniles. A partial list of features that, in Rubtsov's view, are species characteristic include the structure of the trophosome, stichosome, longitudinal cords, and arrangement of anterior sense organs. He defends the practice of assigning specific taxa to juvenile mermithids (often single specimens collected from the host's fresh-water habitat or from the host itself) on the grounds that adult mermithids are rarely encountered in nature by nematologists and the likelihood that additional biological information of the nematodes may emanate from such descriptions. There is no doubt in this reviewer's mind that a wide range of studies should be done on juvenile mermithids as well as on the adult stages. The question still remains, however, as to whether taxonomic studies on juveniles are legitimate. More specifically, are the features listed by Rubtsov distinctive at the species level? Until this question has been satisfactorily answered, it is difficult to argue that judgment be withheld and juvenile mermithids either reared to adults in the laboratory for taxonomic assignment or assigned to the genus *Agamomermis*, created specifically for juvenile mermithids.

Unorthodox approaches toward the classification of blackfly-parasitic mermithids ought not to be discouraged, however, provided that their legitimacy has first been proven. Returning to the biological definition of what constitutes a species, experiments should be done to ascertain whether juveniles with, for example, differing trophosomal structure are, when reared to adults, capable of interbreeding. Provided that a sufficiently representative sampling of mermithid entities were examined in this way, the validity of the juvenile character could be established. Similarly, biochemical criteria may prove of assistance in classifying mermithid parasites of blackflies. Recent studies by Kaiser and Fachbach (1977) suggest that the protein composition of the trophosomes is of taxonomic value for mermithid parasites of chironomids.

III. LIFE HISTORY AND ECOLOGY OF PARASITES

A. Mermithoidea

Recently, Poinar (1981) listed 67 species of simuliids known to harbor mermithid nematodes and opined that "the actual number is probably several times greater than that." Mermithids are widely distributed throughout tropical and

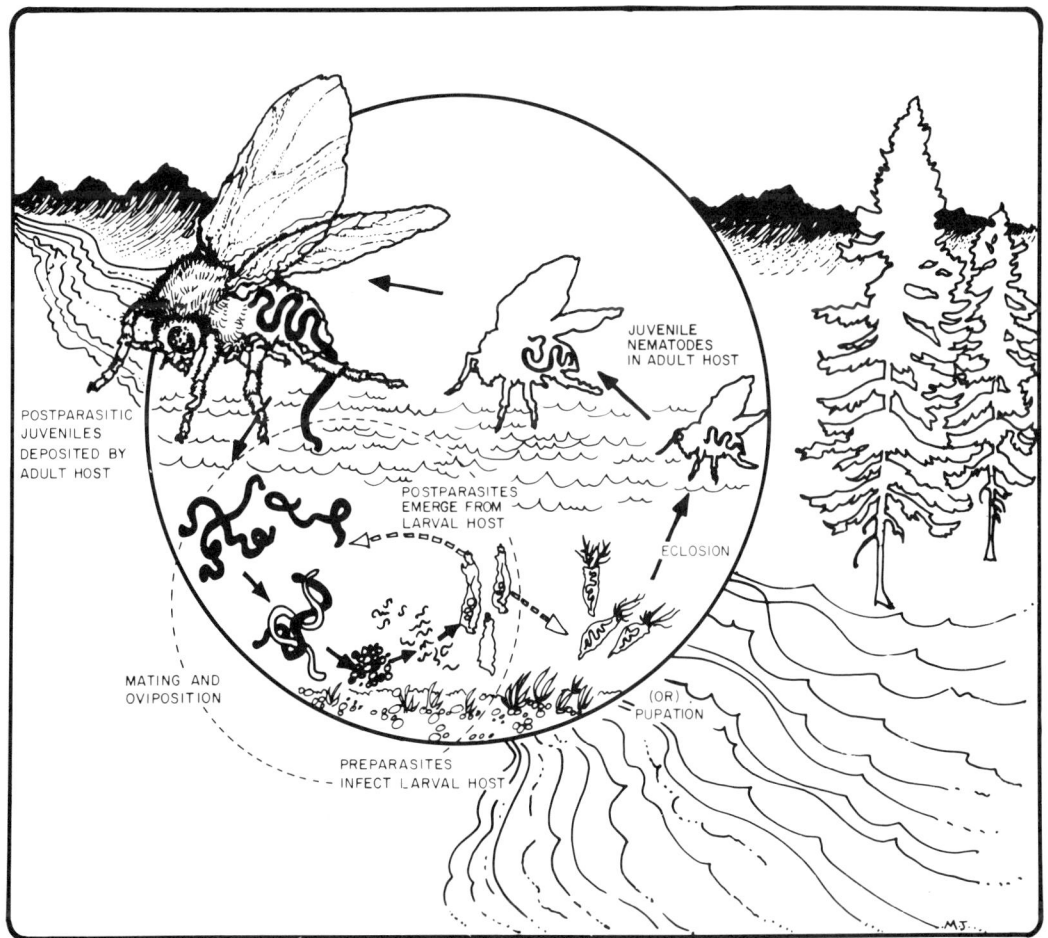

Fig. 1 Life cycle of mermithid parasites of blackflies, showing completion of parasitic development in the larval or adult host.

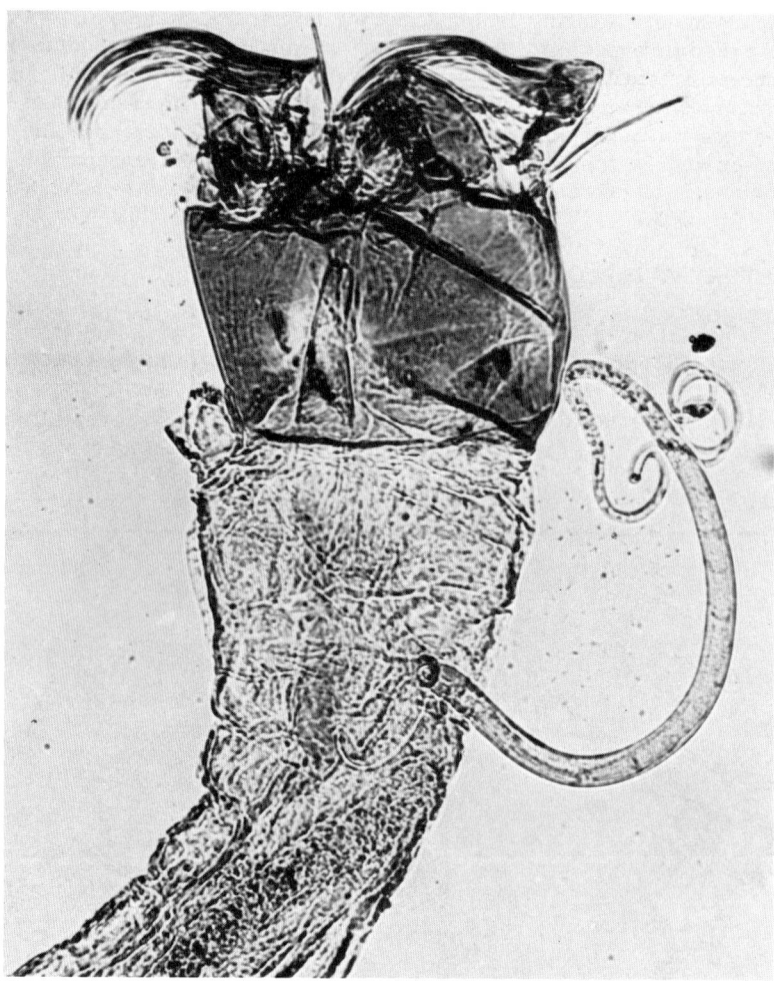

Fig. 2 Preparasite of *N. flumenalis* entering a larval blackfly, *Simulium venustum*. (Courtesy of J. E. Mokry.)

arctic regions of the New and Old Worlds, and their presence is coincident with that of the susceptible blackfly populations.

In outline, the life cycle of blackfly-parasitic mermithids follows the typical mermithid plan. The host becomes infected during the larval phase of its life history, spent attached to the substratum of a shallow body of fastflowing fresh water (Fig. 1). Information on the specific stage(s) at which the host becomes infected under natural conditions is sketchy, but available evidence suggests that this may vary among mermithids or even within the same host-parasite system. Thus, although early- or late-instar *S. damnosum* larvae may become infected by *Isomermis lairdi* (Mondet et al., 1976), this does not appear to be the case for simuliid parasitism by *Neomesomermis flumenalis*, since only first and second instars of *Simulium venustum* and *Prosimulium mixtum* were found to be susceptible to infection (Molloy and Jamnback, 1975; Bailey and Gordon, 1977). In both *N. flumenalis* (Anon., 1973; Molloy and Jamnback, 1975) and *I. lairdi* (Mondet, Berl, and Bernadou, 1977), the infective stage is a stylet-bearing preparasitic juvenile that penetrates the host's exoskeleton to gain access to the hemocoel (Fig. 2). Molloy (1981) observed that the host larvae became partially paralyzed during invasion by the preparasitic *N. flumenalis*, facilitating easier penetration by the nematode and rendering the host more vulnerable to infection by other preparasites. This author theorized that the preparasite's penetration glands could be the source of a muscle-paralyzing secretion. The theory that access to the host's hemolymph may be gained indirectly through accidental ingestion of the infective preparasites by the larval insect (Phelps and DeFoliart, 1964; Welch, 1964; Welch and Rubtsov, 1965; Rubtsov, 1971) has been shown to be untenable (Molloy and Jamnback, 1975; Nickle, 1979).

Parasitic development within the host's hemocoel involves considerable growth and organ development. Parasitic juveniles grow in length from a millimeter or so (size of the preparasitic stage) to 1-5cm. Once parasitic development has been completed, the nematodes resume free-living existence.

Among the relatively few species of cold temperature-adapted mermithids that have been studied, the postparasitic juveniles have been found to emerge most frequently from the late-instar larvae. The nematodes bore through the host's intersegmental membranes (Fig. 3) or through one of its natural openings and, in so doing, kill the host. Within the substratum of the stream, the postparasites (nonfeeding stages) molt to adults, mate, then oviposit. The life cycle is completed when the embryonated eggs hatch to release the infective preparasites. Sometimes, the parasitized host is permitted to pupate and undergo eclosion, so that a proportion of the parasites are harbored by the insect's adult stage. In certain mermithid-simuliid populations, the majority of the nematodes may be carried over to the host's adult stage (Colbo and Porter, 1980). Parasitized adult hosts fly upstream to oviposit (Fig. 1), but since they are sterilized through parasitism, postparasitic nematodes rather than host eggs are deposited at the head of the stream (Rubtsov, 1971; Wenk, 1976; Colbo and Porter, 1980). Parasitism of adult hosts would appear to provide a mechanism for ensuring that mermithid populations primarily emerging from larval hosts are not eventually swept progressively downstream and for disseminating the infection to hosts in nearby streams (Welch, 1964; Finney, 1981a). Poinar (1981), however, believes that recycling of nematodes within streams by parasitized adult hosts is minimal and that juvenile postparasites maintain their position by moving in protected regions within the substratum of the stream. Whatever the dispersal mechanisms of the parasite may consist of, they would appear to be of limited effectiveness. The mermithids characteristically display a discontinuous distribution among stream systems of a particular locality, with mermithid-infested streams frequently in close proximity to uninfested ones (Gordon et al., 1973; Finney, 1981a).

There are several gaps in our knowledge of the life cycles of blackfly-parasitic mermithids. Parasitic development under controlled laboratory conditions has been studied at the organ system level for only two species, *Gastromermis viridis* and *Isomermis wisconsinensis* (Phelps and DeFoliart, 1964). Based on studies done on *R. culicivorax* (Poinar and Otieno, 1974), it is likely that there are four molts in the life cycles of mermithids: the first in the egg, the second during parasitic development, and the final two molts in the postparasitic juvenile. Such a full sequence of molts has not been thus far demonstrated for any species of blackfly-parasitic mermithid, however. The entire sequence of events involved in postparasitic development has been gleaned from laboratory rearing attempts. Studies have not been undertaken to ascertain the pattern of development of the free-living stages under natural conditions.

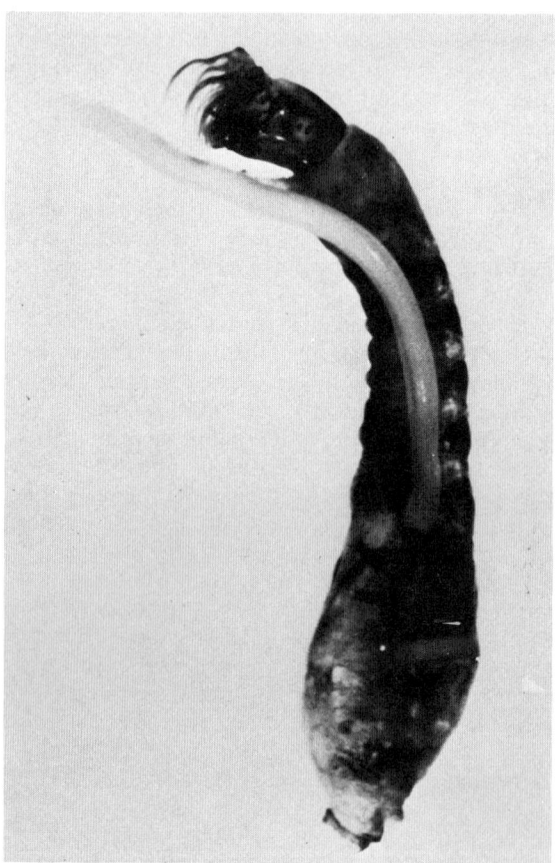

Fig. 3 Mermithid postparasitic larva *(N. flumenalis)* emerging from a larval simuliid to commence the free-living portion of its life cycle (Courtesy B. A. Ebsary. From Ebsary, 1973.)

Available evidence suggests that mermithid parasites of simuliids do not, under natural conditions, parasitize invertebrate fauna other than insects of the family Simuliidae. Curran (1981) recently found, however, that a species of *Gastromermis*, parasitic in British simuliids, was able under laboratory conditions to complete parasitic development in the tropical mosquito *Aedes aegypti*. From the host-parasite list provided by Poinar (1981), it would appear that blackfly-parasitic mermithids show variable host specificity, each nematode being capable of infecting 2-15 blackfly species. Finney (1981a), however, suggested that these nematodes may have a broader host range than is currently conceived and that under field conditions, the composition by species of the simuliid population largely determines which species become infected, because the mermithids exhibit preferences for certain hosts. Molloy (1981) pointed out that in view of present uncertainty surrounding the taxonomic status of many of the mermithids (and in some cases, their simuliid hosts), it is premature to speculate on their host range within the Simuliidae.

Such taxonomic problems notwithstanding, an attempt will be made below to summarize salient features of the life cycles and ecology of five mermithid species in which the postparasitic stages have been reared from newly emerged juveniles to infective preparasites.

1. Neomesomermis flumenalis

Neomesomermis is the focus of a considerable amount of taxonomic confusion. First described from simuliid larvae in Ontario, Canada (Welch, 1962), it has since been reported from blackfly populations in Wisconsin (Phelps and De-Foliart, 1964), New York State (Molloy and Jamnback, 1975), New Jersey (Bruder and Crans, 1979), the USSR (Rubtsov, 1963), Alberta (Abdelnur, 1968), the North West Territories (Mulvey and Nickle, 1978), Newfoundland (Ebsary and Bennett, 1974), and Costa Rica (Vargas and Fallas, 1976). There is a strong possibility, however, that at least some of the groups of ecologically and/or geographically distinct nematodes that have been thus far labeled N. flumenalis are actually other morphologically indistinguishable species or physiologically distinct strains of N. flumenalis. For example, nematodes classified as N. flumenalis were found in Wisconsin (Phelps and DeFoliart, 1964) and New Jersey (Bruder and Crans, 1979) to occasionally infect Stegopterna mutata, but such was found not to be the case for the Newfoundland N. flumenalis, even though sympatric simuliid species were infected (Ebsary and Bennett, 1975b; Bailey and Gordon, 1977; Colbo and Porter, 1980).

The nematode appears to be univoltine in Newfoundland (Ezenwa, 1974a; Ebsary and Bennett, 1975b), with two distinct cycles that are closely synchronized with those of their hosts, an overwintering P. mixtum generation and a spring generation of S. venustum. According to Ezenwa (1974a) and Ebsary and Bennett (1975b), postparasites emerge from P. mixtum in the spring, then molt, mate, and oviposit over the summer months, so that preparasites are released in the fall to infect the early instars of the overwintering host. Postparasites emerge from the first generation of the multivoltine S. venustum during the early part of the summer, but eggs that are eventually oviposited by these nematodes are unable to complete embryogenesis due to the arresting effect of the extremely low (close to 0°C) winter temperatures. Thus, overwintering eggs of N. flumenalis hatch in the spring, when stream temperatures approach those of the fall, releasing preparasites that infect the early instars of the first generation of S. venustum.

Bailey et al. (1977) felt that the nematode that infects P. mixtum is a physiological strain of N. flumenalis differing from the one that infects S. venustum, because laboratory studies showed that Prosimulium postparasites were found to be better adapted to lower temperatures (5°C), but less well adapted to higher temperatures (20°C) than Simulium postparasites; also, cross mating between the two groups of nematodes did not occur. There is a suggestion (Colbo and Porter, 1980) that the two entities may even be different species or constitute groupings of species.

The nematode designated as N. flumenalis in New Jersey infects an overwintering generation of Prosimulium magnum, Prosimulium fuscum, and P. mixtum and all three generations of Simulium tuberosum, which extend from late spring to early fall; it is present in simuliids throughout the year (Bruder and Crans, 1979). If all of the mermithid infections were due to a single strain of N. flumenalis, or even a winter and a summer-fall strain, the life cycle of the nematode must be either multivoltine or asynchronous with those of its hosts, to achieve such a year-round infection in four consecutive blackfly generations.

Whatever the taxonomic status of the various biological entities entitled N. flumenalis may be, its (their) geographical distribution indicates adapta-

tion to cold stream water temperatures. This capacity to tolerate cold temperatures enables the nematode to adapt to the extreme northern latitudes of
Churchill Falls, Labrador, where it was found to parasitize 11.2% of the larval
simuliid population (Ezenwa, 1973).

Laboratory rearing studies showed that postparasitic development was
optimal at 12°C, with least mortality occurring; 24°C was beyond the tolerance
range of the nematode (Ebsary and Bennett, 1973). At 12°C, postparasitic
juveniles molted (double molt) to adults 9-15 days after emergence from the
host, oviposition occurred 36-59 days after molting, and embryogenesis took
35-55 days for completion; the life span of the preparasites was found to be
only two to three days (Ebsary and Bennett, 1973).

Incidences of parasitism are extremely variable and are affected by such
factors as the location of the host-breeding sites, location within the host-
breeding sites, and the host-developmental stages. Incidences of infection
can be high (over 50%) when surveys are done at a time of year when most of
the uninfected larvae have pupated (Gordon; unpublished), because the larval
simuliid population will contain a disproportionately high number of infected
larvae that have been prevented from pupating by the parasitism. Earlier in
the season, host infection rates may be 5% or less (Bailey and Gordon, 1977).
Caution should be exercised in interpreting the numerous reports of incidences
of mermithid parasitism, many of which were based on individual collections
(Finney, 1981a).

Though *N. flumenalis* usually emerges from the larval host, conducting
its entire life cycle within the stream habitat, under as yet undetermined
circumstances, it can be carried over into the pupal and adult stages of the
simuliid. Mokry and Finney (1977) reported incidences of parasitism of adult
female *P. mixtum* and adult female *S. venustum* to be 11.7 and 22.2%, respectively. Though these authors found there to be no mermithid parasitism in
adult male hosts, a low proportion of the male pupae were infected. Several
mermithid species, in addition to *N. flumenalis*, were involved in these pupal-
adult infections.

In streams where the mermithids emerged mainly from larval hosts, ex-
tensity and intensity of infection increased progressively downstream. The
reverse proved true for streams in which the mermithids emerged from adult
hosts (Colbo and Porter, 1980). Presumably, the effect of upstream recoloni-
zation by adult blackflies would be minimal in the first instance, but in situations where large numbers of adult hosts are infected, it would compensate
for the downstream drift of preparasites and infected blackfly larvae known to
occur (Colbo, 1979; Colbo and Porter, 1980). The dynamics of the mermithid-
blackfly system(s) cannot be fully understood, however, on the basis of infection rates-levels alone. Changes in host population densities along the
stream will affect host-parasite ratios and, consequently, incidences and
levels of parasitism. Information on the distribution of the host population
and the postparasites along the stream is needed to complete the picture.

In view of the degree of taxonomic uncertainty surrounding the various
entities described as *N. flumenalis*, unequivocal comments cannot be made regarding its (their) host specificity, because there may be more than one species of nematode involved. To Poinar's (1981) list of hosts for the range of
nematode strains or species so designated, should be added *Prosimulium magnum, Simulium aureum*, and *St. mutata* (Bruder and Crans, 1979).

2. Mesomermis camdenensis*

This mermithid was recently described from *S. tuberosum* larvae in New York State (Molloy, 1979). It is morphologically very similar to *N. flumenalis* (Welch) and may be one of several species that have, in the past, been confused with it (Molloy, 1981). Host larvae, primarily *S. tuberosum*, were found to be infected from May through October. Hatching of the eggs appears to be staggered, occurring throughout most of this period. It is not known whether the life cycle of the nematode is univoltine or multivoltine. The minimum period of time between postparasite emergence and egg hatching was observed to be 12 weeks at 12°C. Theoretically, this would permit more than one generation to be accommodated per season, but since development will undoubtedly be temperature dependent and details of stream temperatures were not provided, the rate of postparasite development under natural conditions is not known. The nematode was found to (almost always) complete its development within the larval host, rarely being carried over to the pupal stage. Incidences of infection in *S. tuberosum* (10-25%) were found to be higher than in the sympatric *S. venustum* (<1%), indicating that *S. tuberosum* larvae are more susceptible to infection than *S. venustum* larvae.

3. Gastromermis viridis

This nematode was described from *Simulium vittatum* larvae in Wisconsin (Welch, 1962; Phelps and DeFoliart, 1964) where it is prevalent. It has since been reported from simuliids in northern Alberta, Canada (Abdelnur, 1968) and in Newfoundland, where it forms a minor component of the overall mermithid fauna (Ezenwa 1974a,b; Ebsary and Bennett, 1975a; Bailey and Gordon, 1977). The anatomy of the preparasitic juveniles has been described (Poinar and Hess, 1974). Postparasitic stages are easily recognized by their greenish color.

In Wisconsin, it was found to parasitize only the multivoltine *S. vittatum* and host infections were recorded in all months of the year except April (Phelps and DeFoliart, 1964). Incidences of parasitism are difficult to gauge from the data of Phelps and DeFoliart (1964), because cumulative rates resulting from *G. viridis* and *Isomermis wisconsinensis* infections were recorded. In a stream in which *G. viridis* was the dominant of the two mermithid species, infection rates of larval hosts varied seasonally from 14 to 91%. In northern Alberta, 7-47% of overwintering *S. vittatum* larvae were found to be infected with *G. viridis* (Abdelnur, 1968). Along with *I. wisconsinensis*, *G. viridis* can emerge from larval or adult hosts of either sex. High proportions (up to 100%) of adult *S. vittatum* were found to be infected with *G. viridis* and/or *I. wisconsinensis*. In Newfoundland, Mokry and Finney (1977) recovered *G. viridis* from mixed infections of adult female *P. mixtum* and *S. venustum*. Overall incidences of infection for *P. mixtum* and *S. venustum* were 11.7 and 22.2%, respectively, with *G. viridis* the least abundant of the four mermithid species identified.

Considering the almost year-round presence of this nematode in Wisconsin simuliids, it is probably multivoltine at this latitude. The overall time required from postparasite emergence to egg hatching was found to be 15-28 days at 23.9°C. At 14.4-15.6°C, oviposition occurred 25-31 days after nematode emer-

---

*Concerning the assignment of genus to this species, the reader is referred back to the previously mentioned difference of opinion among taxonomists concerning the validity of the genus *Neomesomermis*.

gence, but the eggs did not hatch at this temperature. Eggs were killed by storage in ice for three months, leading Phelps and DeFoliart (1964) to conclude that the nematode overwinters in the parasitic juvenile stage. In Newfoundland, the life cycle of *G. viridis* is incompletely known due to its sparse occurrence. However, it may be univoltine with two distinct cycles, one infecting the overwintering *Prosimulium* generation, the other infecting the first generation of *S. venustum* and *Simulium corbis* (Ezenwa, 1974a).

4. Isomermis wisconsinensis

This nematode was described from *S. vittatum* larvae in Wisconsin (Welch, 1962; Phelps and DeFoliart, 1964), where, along with *G. viridis*, it appears to be the codominant mermithid species parasitic in simuliids. It also infects the overwintering *P. mixtum-fuscum* and the first generation (Spring) of *S. venustum* and *S. vittatum* in Newfoundland, but it is not a common mermithid in that province (Ezenwa, 1974a, b; Ebsary and Bennett, 1975a; Bailey and Gordon, 1977). The postparasites are faintly pinkish brown in color.

In Wisconsin, *I. wisconsinensis* infections were recorded from *S. vittatum* larvae only between May and December. Thus, in contrast to *G. viridis*, this mermithid overwinters while undergoing free-living development. Poinar (1979) theorized that the nematode overwinters as a postparasitic juvenile, presumably because of the limited capability of the egg stage to withstand winter temperature conditions (Phelps and DeFoliart, 1964). In streams in which *I. wisconsinensis* was dominant over *G. viridis*, larval host infection rates varied from 0 to 95% according to season. The nematode may emerge from larval or adult hosts of either sex. In Wisconsin, 0-75% of adult simuliids, which emerged from streams in which *I. wisconsinensis* predominated, harbored mermithids (Phelps and DeFoliart, 1964).

From the limited data available, it is likely that *I. wisconsinensis* is multivoltine in Wisconsin. The period between emergence from the host and oviposition was found to be 8-17 days at 14.4-15.6°C and 4-10 days at 23.9°C. Egg hatching occurred 28-29 days (14.4-15.6°C) and 8-10 days (23.9°C) afterward (Phelps and DeFoliart, 1964). Thus, though no data on stream temperatures were provided by the authors, it would appear that the duration of the free-living portion of the life cycle is sufficiently short to permit more than one nematode cycle per year. At 23-27°C, development of the parasitic juveniles within the host was found to require 10-14 days (Phelps and DeFoliart, 1964).

Ezenwa (1974a) felt that this mermithid was univoltine in Newfoundland, but as with *G. viridis*, the nematode is of sparse occurrence in the province and the data obtained were insufficient to permit such an unequivocal assessment to be made.

Isomermis wisconsinensis appears to have a very narrow host range, since it is only known to infect the three abovementioned hosts.*

---

*Ezenwa (1974b), citing Anderson and DeFoliart (1962), stated that *I. wisconsinensis* infects *St. mutata*, *S. tuberosum*, *Simulium decorum*, and *Simulium latipes*. In fact, Anderson and DeFoliart only reared postparasites that emerged from *S. vittatum* to the adult stage. Identification of nematodes occurring in other hosts was based on examination of the parasitic juvenile stage(s) and is, therefore, of doubtful value.

5. Isomermis lairdi

This nematode is the most common parasite of *S. damnosum* in the Ivory Coast, West Africa; it has also been recovered from *S. damnosum* in northern Ghana (Poinar, 1979). Taxonomic description was provided by Mondet, Poinar, and Bernadou (1977).

The nematode is multivoltine with six to eight cycles per year. In the laboratory (23-28°C), the entire life cycle of *I. lairdi* was completed in one month, with a parasitic phase of 10-16 days and a free living phase of 15-23 days (Mondet, Berl, and Bernadou, 1977). Postparasitic juveniles and adults survive the West Africa dry season by burrowing into the moist sand in the river beds. Host larvae that become parasitized in their early instars are prevented from pupating, but those that become infected during later instars pupate and undergo eclosion. The rate of parasitism may reach 80% of the nulliparous females of *S. damnosum* and 35% of the entire population (Mondet et al., 1976). There is a suggestion that when the mermithid is carried over to the adult stage of the female host, a blood meal is required in order for parasitic development to be completed (Poinar, 1979).

Aside from *S. damnosum*, several other coexisting species of *Simulium* were found to be infected with *I. lairdi* (Mondet et al., 1976).

B. Rhabditoidea

Under laboratory conditions, the DD-136 strain of *N. carpocapsae* is highly lethal to simuliids. Infective third-stage juveniles are ingested by the host during filter feeding. In *S. vittatum*, engulfment of *N. carpocapsae* juveniles was found to be higher at 10°C than at a higher (20°C) temperature (Webster, 1973). After exsheathing their second-stage cuticles, the nematodes penetrate the gut wall to enter the insect's hemocoel; in *S. vittatum* (10°C; 20°C), nematodes were found to be present in the hemocoel 80 minutes after being ingested (Webster, 1973). A mutualistic bacterium, *Xenorhabdus nematophilus*, is then released from the nematode's anus into the hemocoel of the host (Thomas and Poinar, 1979). The bacterium kills the host, then multiplies within the hemolymph, to provide the nematode(s) with nutrition for development. In *S. verecundum* and *S. vittatum* (11°C), 54% of all blackfly mortality occurred within 24 hours after infection and all infected hosts died within six days posttreatment (Molloy et al., 1980). Though details are not known for simuliid infections, it can be surmised from other insect-*N. carpocapsae* relationships that the nematodes conduct one or several complete life cycles within the host cadaver. Ensheathed infective-stage juveniles are eventually produced, which either remain in the cadaver or migrate into the environment to reinitiate infection in another host.

Early simuliid instars are refractory to infection because their mouthparts are too small to capture and ingest the nematodes (Molloy et al., 1980; Gaugler and Molloy, 1981). Last (seventh) instar *S. vittatum* larvae were found to be more susceptible to infection than other stages, with greater mortality occurring in larger individuals than in smaller ones. The mouthparts of the smaller blackfly larvae inflicted greater damage to the nematodes than did those of larger hosts, so fewer viable nematodes were ingested (Gaugler and Molloy, 1981). The simuliid host range of this nematode is not known, but it is likely to be broad, considering that it is capable of infecting insect species pertaining to a wide variety of orders (Poinar, 1979).

IV. HOST-PARASITE RELATIONSHIPS–PHYSIOLOGICAL AND PATHOLOGICAL

Very little information is available on the physiology of mermithid parasites of simuliids. The mechanism of nutrition has not been studied for these nematodes, but it is probable that metabolites within the host's hemolymph are transported across the nematode's outer cuticle, as in mermithid parasites of grasshoppers (Rutherford and Webster, 1974; Rutherford et al., 1977) and mosquitoes (Poinar and Hess, 1977). The cuticle of simuliid-parasitic mermithids would appear on ultrastructural grounds to be well adapted for nutrient absorption. The cuticle of the parasitic stage(s) of *Gastromermis boophthorae* is very thin, contains fewer layers than subsequent free-living stages, and overlies a hypodermis replete with organelles indicative of high metabolic activity; the hypodermal surface is of a microvillous nature (Batson, 1979a). Rubtsov (1967, 1981a) theorized that mermithid parasites of aquatic Diptera predigest host nutrients before absorbing them. The cells of the longtitudinal cords are presumed by Rubtsov to secrete digestive enzymes to induce lysis of host fat body tissue; absorbed nutrients undergo metabolic transformations in the stichocytes before being further processed in the trophosome. This hypothesis remains to be proven. Predigestion of nutrients does not appear to occur in locusts parasitized by *Mermis nigrescens* (Gordon and Webster, 1972; Rutherford and Webster, 1974). However, digestive enzymes have been demonstrated in homogenates and exudates of mermithid parasites of chironomids (Spasskii et al., 1975).

Once absorbed, nutrients are transferred to the nematode's trophosome for metabolic conversion. Nutrients may enter the trophosome of *G. boophthorae* through cytoplasmic bridges that connect it to the hypodermal cords and/or through the pseudocoelom, since the outer surface of the trophosome is structurally well equipped for nutrient absorption (Batson, 1979b). Batson (1979b) showed that the trophosome of *G. boophthorae* is syncytial, without lumen, and contains globules of high and low electron density as well as reticulate granular inclusions. Gordon et al. (1979) showed that lipids are the predominant storage metabolite in the trophosomes of *N. flumenalis* and comprise, in order of prevalence, triacylglycerols, phospholipids, free sterols, and sterol esters. Almost all the fatty acid components of the trophosomal lipids were found to be available in the hemolymph of their simuliid hosts. By comparison with *Romanomermis culicivorax*, a mermithid parasite of warm water mosquitoes, the triacylglycerols and sterol esters in the trophosomes of *N. flumenalis* contain a higher proportion of unsaturated fatty acids (Gordon et al., 1979). It is possible that this high proportion of unsaturated fatty acids is of adaptive value in the boreally adapted *N. flumenalis*, permitting maintenance of physical state and consequent enzymatic activity at low temperatures (Fast, 1970; Alexandrov, 1977). A variety of sterols were tentatively identified from the free sterol and sterol ester fractions of the trophosomes of *N. flumenalis* (Gordon et al., 1980). These included a C26 sterol, which was not identified from the host's hemolymph and must, therefore, have been synthesized by the nematode from other sterols.

The metabolic fate of the trophosomal storage products is unknown. The large amounts of stored triacylglycerols suggest that the nematodes may utilize lipids for energy metabolism, at least during their free-living stages. In *R. culicivorax*, a functional β-oxidation pathway has been demonstrated (Gordon, Walsh, and Burford, 1981).

The first recordings of pathogenic effects of mermithids upon their simuliid hosts were made by Strickland (1911), who noted that mermithid parasites

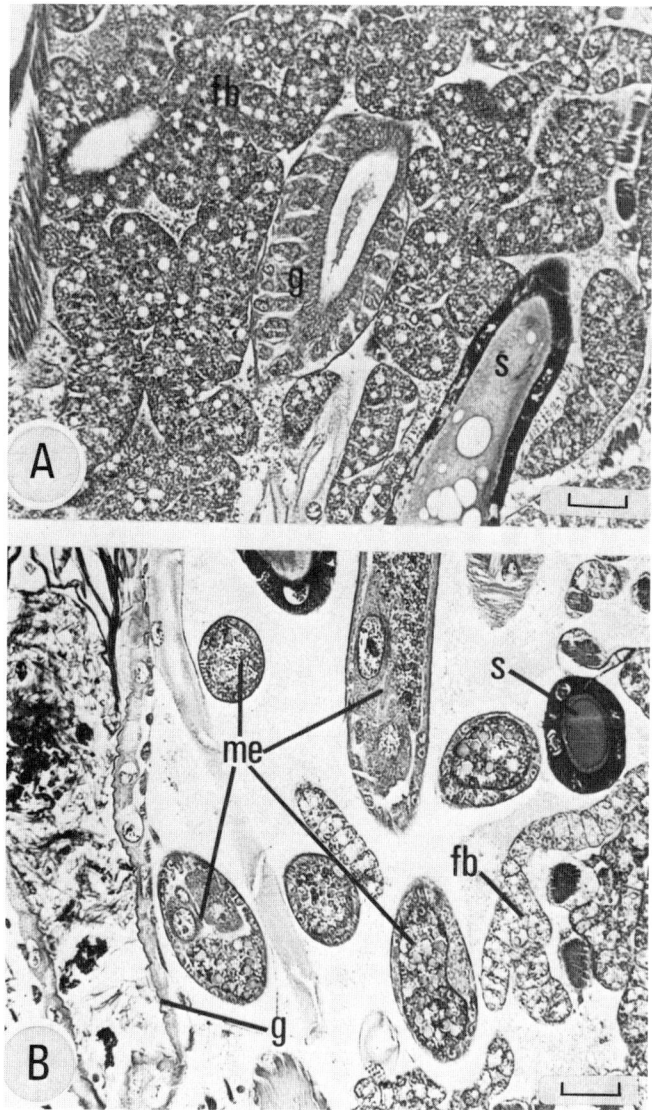

Fig. 4 Effect of mermithid parasitism on fat body of *P. mixtum*. (A) longitudinal section of an uninfected larva (abdominal region) showing densely packed fat body tissue throughout the hemocoel. (B) longitudinal section of an infected *P. mixtum* larva (abdominal region) showing the replacement of fat body tissue by a developing mermithid nematode. fb, fat body; g, gut; me, mermithid; s, salivary gland. The scale in (A) and (B) represents 30 μm. Stain: Periodic Acid-Schiff. (Courtesy W. J. Condon. From Condon 1975.)

prevented pupation, suppressed larval and pupal histoblast development, and caused degeneration of fat body tissue in *Prosimulium* sp. larvae. Numerous reports have since corroborated these findings and have shown that mermithids cause the degeneration of several tissues within the immature and adult simuliid host. The most dramatic of these recordings was by Hocking and

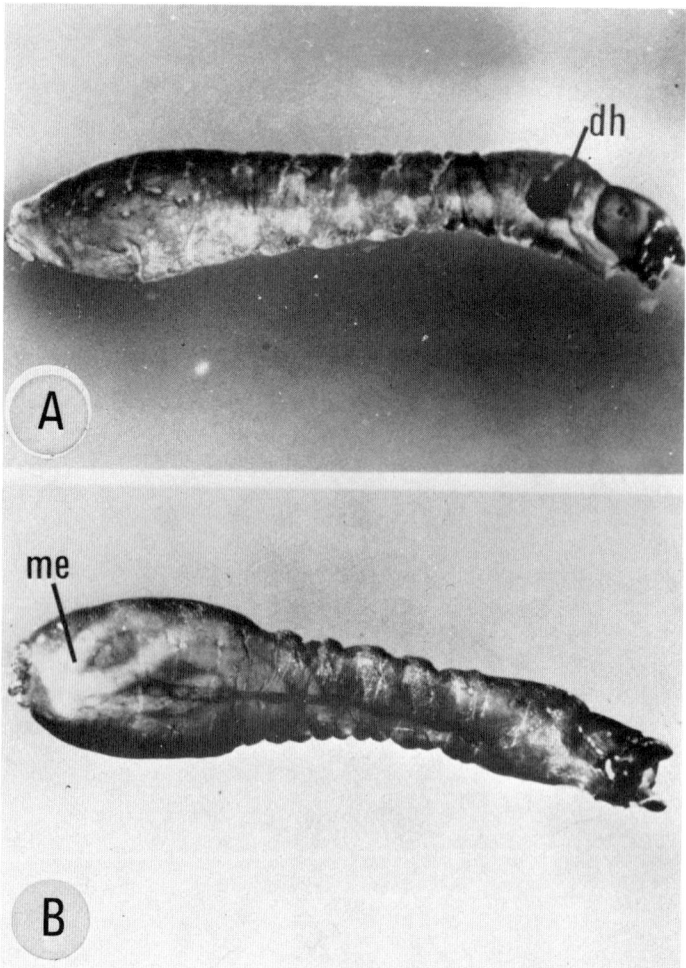

Fig. 5 Effect of mermithid parasitism on pupal histoblast development in *P. mixtum*. (A) A mature terminal-instar *P. mixtum* larva with black pupal histoblasts (dh) on the dorsolateral sides of its prothorax. (B) A mermithid-infected *P. mixtum* larva recognized by the prominent mermithid (me) coiled within the swollen abdomen of the blackfly larva. The infected larva lacks prominent histoblasts. (Courtesy W. J. Condon. From Condon 1975.)

Pickering (1954), who found that mermithid parasites caused the degeneration of the fat bodies as well as the autolysis of nervous, reproductive, and digestive systems of adult *S. venustum* in Manitoba, Canada.

Condon (1975) found that *N. flumenalis* caused almost total depletion of host fat body tissue (Fig. 4A, B). Moreover, glycogen concentrations within the fat body tissue that had not degenerated were found to be significantly lowered by the parasitism (Condon and Gordon, 1977).

Mermithids commonly prevent the simuliid host from pupating and suppress imaginal bud development (Fig. 5A, B). Since such processes are normally regulated by the insect's endocrine system it was thought that their disruption in parasitized insects resulted from an alteration in the host's hormonal status (Welch, 1965). Condon and Gordon (1977) found, by examination of naturally infected *S. venustum* larvae, that mermithid parasitism significantly increased the nuclear DNA-RNA activity of the host's corpus allatum gland as well as the volume and amount of stored neurosecretory material in the corpus cardiacum gland. In *P. mixtum*, however, mermithid parasitism did not induce any detectable changes in host endocrinology, yet this host was also prevented from pupating and developing normal pupal histoblasts. Condon and Gordon (1977) concluded, therefore, that disturbances in host development caused by mermithid parasites of simuliids are the result of depletion of host nutrient reserves by the developing nematode. The endocrinological disturbances recorded in *S. venustum* are best interpreted as stressful effects of the parasitemia induced by the nutritional disturbances caused by mermithid parasitism. The nematode develops more rapidly in the spring-developing *S. venustum* than in the overwintering *P. mixtum* larvae. As a consequence, the nematode exerts more intense nutritional demands upon *S. venustum* than upon *P. mixtum* and this is reflected in more pronounced changes in the host's endocrinology.

Gordon et al. (1978) found that mermithid parasitism resulted in a general depletion of hemolymph protein fractions in *P. mixtum* and *S. venustum*. These same authors found that mermithid parasitism in *P. mixtum* is characterized by a decrease in the majority of the hemolymph amino acids. In *S. venustum*, however, 14 of 32 ninhydrin-positive substances detected in the hemolymph were found to be increased by parasitism, 10 substances decreased in concentration, and eight substances were unaffected. The fact that the mermithid(s) do not appear to cause a general drain on the hemolymph amino acid pool of larval simuliids suggests that amino acids are recruited into the hemolymph by catabolism-autolysis of the host fat body, and possibly other tissues. The hemolymph of parasitized simuliids was found to be depleted in glucose, but not trehalose (Gordon et al., 1978). If the nutrition of black-fly parasitic mermithids resembles that of *M. nigrescens*, a disaccharide such as trehalose would be unable to traverse the nematode's cuticle, whereas glucose would (Rutherford and Webster, 1974).

Parasitized adult blackflies of both sexes are invariably rendered sterile. There are numerous reports in the literature of mermithid-parasitized adult female blackflies exhibiting partial or total suppression of ovarian development; these are reviewed by Finney (1981a). The cause of such parasitic castration is not known; it could be due to severe nutrient depletion and/or impairment of host endocrine activity. Occasionally, mermithids cause the production of intersexual adult blackflies (Welch, 1964; Fredeen, 1970; Hunter and Moorhouse, 1976). There is some evidence (Molloy, 1981) that the behavior of adult male simuliids may be modified by mermithid parasitism; such hosts may fly upstream to oviposition sites, simulating oviposition and depositing mermithid postparasites at the stream head.

It is worth pointing out that all the documented effects of mermithid parasitism on simuliids were observed on field-collected hosts. No studies have been undertaken to ascertain how the nematodes interfere with host tissue development and physiology under controlled laboratory conditions.

There is hardly any information at all on the reverse situation, i.e., the possibility that differences in physiology within or among hosts influence the parasitic development of the mermithids. Field-collected hosts have never been found to contain dead or dying mermithids, leading Molloy (1981) to conclude that physiological resistance involving host encapsulation and/or melanization reactions is of limited significance under natural conditions. Colbo and Porter (1980), however, reported the presence of a melanized mermithid in

S. vittatum larvae that were infected in the laboratory. More exhaustive laboratory and simulated stream infection studies are needed in order to fully evaluate the relative degrees to which host attractiveness, physiological resistance, and physical resistance (i.e., the inability of the preparasites to gain access to the hemocoel through structural barriers presented by the host, such as thickness of exoskeleton, mouthparts, and others) are involved in determining host susceptibility.

There is circumstantial evidence that, as with *R. culicivorax* (Petersen, 1972; Gordon, Squires, Babie, and Burford, 1981) sex is largely determined in mermithid parasites of simuliids by the availability of nutrients. As the degree of superparasitism (and presumably, competition for host nutrients) increases, the proportion of male to female nematodes increases (Phelps and DeFoliart, 1964; Rubtsov, 1971; Ebsary and Bennett, 1975b; Ezenwa and Carter, 1975). Male *I. lairdi* invariably developed in parasitized adult male *S. damnosum*, even when there was only one parasite per host. In parasitized female hosts, corresponding levels of infection resulted in the production of mostly (60%) female worms (Mondet et al., 1980). Such differences in nematode sex ratio may be a consequence of the higher level of nutrition afforded by the blood meal of the female host compared to the plant juice diet of the male blackfly.

No information is available on physiological interactions between *N. carpocapsae* and simuliids. Most of the research on neoaplectanid nematodes has focused on their biocontrol effectiveness, sensitivity of the infective stage to environmental stresses, and rearing procedures. In host-parasite relationships involving these nematodes, the host is alive for so short a period that physiological studies have not been considered worthwhile. Once the bacterium associated with the nematode kills the host, the nutrition of the nematode is essentially saprozoic.

V. CONTROL POTENTIALITIES

The recent discovery that one of the strains of *S. damnosum* in the WHO control zone in West Africa has, since the inception of the program, developed resistance to the insecticide Abate, underscores the need to develop alternate control methodologies (Guillet et al., 1980).

The potential of a neoaplectanid such as *N. carpocapsae* for simuliid biocontrol is difficult to assess, however. Since natural infections of simuliids by neoaplectanids are unknown, evaluation of their effectiveness can only be ascertained through extensive laboratory and field trials. In the only stream trial thus far conducted against blackfly larvae, Gaugler and Molloy (1981) obtained 50% mortality of *S. vittatum* larvae despite relatively low stream temperatures (9-12°C), but nematode establishment did not occur. *Neoaplectana carpocapsae* has the advantage of being readily mass cultivated by a variety of economically feasible procedures (Gaugler, 1981), but its host range is so broad that it could indiscriminately attack a wide range of stream insects, with undesirable consequences to the ecosystem as a whole.

Several reviewers have concluded that mermithid nematodes offer a promising alternative or adjunct to existing methods for controlling simuliids (Welch, 1964; Rubtsov, 1964, 1981b; Gordon et al., 1973; Poinar, 1979; Finney, 1981a). These nematodes kill and sterilize their hosts and, under natural conditions, appear only to infect simuliids. Moreover, the finding that parasitism of natural simuliid populations by mermithids was unimpaired by the larvacide Abate (Garris and Noblet, 1975) suggests promising integrated control potential. Extremely high incidences of mermithid parasitism, recorded in localized simuliid populations in the United States and the USSR, were found to lead to short-term (several months to two years) eradication of the blackflies involved (Phelps and DeFoliart, 1964; Rubtsov, 1964; Welch and Rubtsov, 1965). Aside from the numerous and widespread reports of mermithid parasitism in natural populations of Nearctic simuliids, two of the principal vectors of *O. volvulus* in tropical Africa and Central America, *S. damnosum* and *Simulium metallicum*, respectively, harbor mermithids (Dalmat, 1955; Mondet et al., 1976, 1980; Mondet, Berl, and Bernadou, 1977; Mondet, Prud'hom, Bellec, and Hebrard, 1980; Poinar and Takaoka, 1979; Takaoka, 1980; Vargas et al., 1980).

Only two trials have been carried out involving the application of preparasites of simuliid-parasitic mermithids to field populations of blackflies. In the USSR, Likhovoz (1978) introduced laboratory-reared postparasites, eggs, and preparasites into simuliid breeding places where mermithids had not been seen in five years of observation. Subsequent examination revealed the presence of larval simuliids infected with *Gastromermis boophthorae* and *Gastromermis likhovosi*. Molloy and Jamnback (1977) released preparasites, laboratory reared from field-collected postparasites, against blackfly larvae in a stream in Cambridge, New York. High infection rates (up to 71.4%) of *S. venustum* larvae were recorded near the point of introduction, but incidences of infection declined rapidly and progressively downstream. These authors calculated that it would be cost prohibitive to use mermithids for simuliid biocontrol by the method they had used for producing preparasites. To achieve a 90% rate of infection, 3.6×10^9 preparasites would have to be applied per mile, at a cost of $300 per million preparasites.

The eventual introduction of mermithid preparasites into field conditions can only be accomplished with probable success and on a commercially viable basis when the infective juveniles can be mass cultured by in vitro techniques. Avenues to be exploited for expediting the in vitro culture of mermithids in general have been reviewed (Finney, 1981b). Recently, Fassuliotis and Creighton (1982) succeeded in culturing *Filipjevimermis leipsandra*, a mermithid parasite of chrysomelid beetles through its entire life cycle, using a medium containing Schneider's *Drosophila* medium supplemented with fetal calf serum. The only recorded attempt at in vitro culture of blackfly-parasitic mermithids, however, is that of Finney (1976), who succeeded in culturing *N. flumenalis* preparasites to a length of 8-9 mm after 22 days in a medium containing Grace's tissue culture medium, Schneider's *Drosophila* medium, and fetal calf serum. Concerted efforts to culture these nematodes can only begin when in vivo laboratory colonization has been achieved and preparasites are thereby made available on a year-round basis for in vitro culture trials. Such an in vivo culture system would also enable studies to be done under controlled laboratory conditions on the host-parasite relationships, yielding information on the parasites' nutrition and growth requirements that could prove crucial in determining optimal conditions for in vitro culture. The lack of a procedure (in vivo or in vitro) for mass cultivating these nematodes has prevented exhaustive field and laboratory infection studies from being carried out. Such studies are vital in order to fully evaluate the biocontrol effectiveness of the mermithids and ascertain optimal application methodologies.

Some species of blackfly-parasitic mermithids have been cycled in vivo in the laboratory through a generation (Phelps and DeFoliart, 1964; Mondet, Berl, and Bernadou, 1977), but continuous cultivation has not been achieved. The major obstacle that has prevented the in vivo cultivation of mermithids has, until recently, been the inability to establish a permanent laboratory colony of blackflies. Simmons and Edman (1981), however, have succeeded in colonizing *Simulium decorum*, an autogenous simuliid with blood-feeding capabilities, through 16 generations, and this major breakthrough could pave the way for in vivo culture of blackfly-parasitic mermithids. It is highly unlikely that the mermithids could be experimentally maintained in a nonsimuliid host. The ability of *Gastromermis* spp. to complete parasitic development in *Ae. aegypti* (Curran, 1981) is an interesting finding, because it reinforces the concept that host specificity is largely determined by behavioral and/or physical phenomena. It does not appear to have special significance, however, as an in vivo culture device. Host mortality rates (prior to the time of post-parasite emergence) were high (85%) in parasitized *Ae. aegypti* and most (69%) of the nematodes were melanized. Nor would the converse approach, i.e., use of a mosquito-parasitic mermithid against blackflies, appear to hold very much promise. Nickle (1979) found, in laboratory tests, that *S. vittatum* larvae were refractory to infection by *R. culicivorax* preparasites. *R. culicivorax* preparasites were, however, found capable of penetrating several species of boreal blackfly larvae (Finney, 1975; Finney and Mokry, 1980; Poinar et al., 1979) as well as larvae of *S. damnosum* (Hansen and Hansen, 1976). Though parasitism was found always to be fatal to the hosts, the nematode was unable to complete parasitic development. The incidence of *S. damnosum* parasitism by *R. culicivorax* preparasites, introduced into blackfly-rearing systems, was found to be extremely low (Anon., 1978).

There are several problems to elucidate, in addition and consequent to in vitro culture, before mermithids could be effectively utilized in biocontrol programs. There appears to be general agreement that a multifaceted approach is necessary in order to implement integrated control methods that incorporate mermithids. The specific lines of inquiry that are needed have been detailed (Gordon et al., 1973; Finney, 1981a; Molloy, 1981); they include studies on host specificity, taxonomy (mermithids and simuliids), ecology of the candidate mermithid within the target biotope, its tolerance limits to environmental stresses, and optimal field and spraying conditions under which preparasite inocula should be applied. Colbo and Porter (1980) stressed the need to obtain information on the mermithids' free-living development under field conditions. These nematodes spend a greater period of time within their life cycles undergoing free-living development than they do as parasites. Thus, a univoltine mermithid may spend up to 11 months in the free-living phase of its cycle. Despite the undoubted importance of the free-living stages in controlling mermithid distribution among simuliids, virtually nothing is known about the location and nature of the microhabitat(s) in which they develop and the interrelationship between the nematode and its free-living environment. The time spent by the nematode(s) in each stage of development (postparasitic juvenile, adult, egg, preparasite) is not known, nor has the identity of the overwintering stages been established beyond doubt for any species. Conclusions on these points have been based on laboratory-rearing studies and, in some instances, on field infection rates of the simuliid host. There is a pressing need for direct studies to be done on the free-living stages in the field environment.

It is impossible to make an unequivocal statement of a general nature on whether it is desirable to control blackflies, since the potential consequences of achieving successful control must be separately considered for each biotype. Compelling arguments can be made for instituting control programs in situations where simuliids threaten the health of humans and livestock and where, through the density of their numbers, they cause severe socioeconomic problems. Where such reasons do not exist, however, the consequences of eliminating an important component of the food chain should be borne in mind. In eastern Canada, for example, larval simuliids are an important dietary component of salmon parr and speckled trout (Schiefer, 1969). Where blackfly control is considered to be a desirable objective, mermithid nematodes possess considerable potential for deployment in integrated control methodologies. Realization of this potential is to be considered a long-term goal, however, since a wide range of studies are necessary to establish successful mass-cultivation and field-application methodologies.

ACKNOWLEDGMENTS

I thank G. F. Bennett, M. H. Colbo, J. R. Finney, and M. Laird for their suggestions and comments on the manuscript. Thanks are also due to W. J. Condon and J. E. Mokry for permission to use their slides and prints, to R. Ficken for photographic services, to G. Campbell and M. Jones for preparing the line drawing of the life cycle, to A. V. Cowan and R. A. M. Cowan for translating one of the articles from Spanish to English, and to K. Wollenweber for translating relevant sections of one of the articles from Russian to English. I am grateful for continuing financial support from the Natural Sciences and Engineering Research Council of Canada (Grant A6679).

REFERENCES

Abdelnur, O. M. (1968). The biology of some black flies (Diptera: Simuliidae) of Alberta. *Quaest. Entomol.* 4: 113-174.

Addison, E. M. (1980). Transmission of *Dirofilaria ursi* Yamaguti, 1941 (Nematoda: Onchocercidae) of black bears (*Ursus americanus*) by blackflies (Simuliidae). *Can. J. Zool.* 58: 1913-1922.

Alexandrov, V. Ya. (1977). *Cells, Molecules and Temperature. Conformational Flexibility of Micromolecules and Ecological Adaptation* (translated from Russian by V. A. Bernstam). Springer, Berlin.

Anderson, J. R., and DeFoliart, G. R. (1962). Nematode parasitism of black fly (Diptera: Simuliidae) larvae in Wisconsin. *Ann. Entomol. Soc. Am.* 55: 542-546.

Anderson, J. R., and Dicke, J. R. (1960). Ecology of the immature stages of some Wisconsin black flies (Simuliidae: Diptera). *Ann. Entomol. Soc. Am.* 53: 386- 404.

Anon. (1973). *Third Biannual Report. Research Unit on Vector Pathology. Memorial University of Newfoundland.*

Anon. (1978). *Annual Report. Research Unit on Vector Pathology. Memorial University of Newfoundland.*

Anon. (1982). Health and economic consequences of biting flies in Canada. (M. Laird, Chairman), *Associate Committee on Scientific Criteria for Environmental Quality, National Research Council of Canada (mimeographed document)* NRC-19248, Ottawa.

Bailey, C. H., and Gordon, R. (1977). Observations on the occurrence and collection of mermithid nematodes from blackflies (Diptera: Simuliidae). *Can. J. Zool.* 55: 148-154.

Bailey, C. H., Gordon, R., and Mills, C. (1977). Laboratory culture of the free-living stages of *Neomesomermis flumenalis*, a mermithid nematode parasite of Newfoundland blackflies (Diptera: Simuliidae). *Can. J. Zool.* 55: 391-397.

Barnett, B. D. (1977). *Leucocytozoon* disease of turkeys. *Worlds Poult. Sci. J. 33*: 76-87.

Batson, B. S. (1979a). Body wall of juvenile and adult *Gastromermis boophthorae* (Nematoda: Mermithidae): Ultrastructure and nutritional role. *Int. J. Parasitol. 9*: 495-503.

Batson, B. S. (1979b). Ultrastructure of the trophosome, a food-storage organ in *Gastromermis boophthorae* (Nematoda: Mermithidae). *Int. J. Parasitol. 9*: 505-514.

Bruder, K. W., and Crans, W. J. (1979). The blackflies (Simuliidae: Diptera) of the Stony Brook watershed of New Jersey with emphasis on parasitism by mermithid nematodes (Mermithidae: Nematoda). *N. J. Agric. Exp. Sta. Bull. No. 851.*

Carlsson, G. (1962). Studies on Scandanavian black flies. *Opusc. Entomol. Suppl. 21*: 264 pp.

Charnetski, W. A., and Haufe, W. O. (1981). Control of *Simulium arcticum* Malloch in Northern Alberta, Canada. In *Blackflies: The Future for Biological Methods in Integrated Control* (M. Laird, ed.). Academic, London, pp. 117-132.

Chitwood, B. G. (1935). Nomenclatorial notes. *Proc. Helm. Soc. Wash. 2*: 51-54.

Colbo, M. H. (1979). Distribution of winter-developing Simuliidae (Diptera) in Eastern Newfoundland. *Can. J. Zool. 57*: 2143-2152.

Colbo, M. H., and Porter, G. N. (1980). Distribution and specificity of Mermithidae (Nematoda) infecting Simuliidae (Diptera) in Newfoundland. *Can. J. Zool. 58*: 1483-1490.

Condon, W. J. (1975). Some aspects of the host/parasite relations of Newfoundland blackflies and their mermithid parasites. M.Sc. thesis, Memorial University of Newfoundland, St. John's, Newfoundland, Canada.

Condon, W. J., and Gordon, R. (1977). Some effects of mermithid parasitism on the larval blackflies *Prosimulium mixtum/fuscum* and *Simulium venustum*. *J. Invert. Pathol. 29*: 56-62.

Curran, J. (1981). Infectivity and development of a blackfly (Simuliidae) mermithid in mosquitoes. *J. Invert. Pathol. 39*: 401-402.

Dalmat, H. T. (1955). The black flies (Diptera: Simuliidae) of Guatemala and their role as vectors of onchocerciasis. *Smithson. Misc. Collect. 125(1)*: 425 pp.

Davies, D. M., Peterson, B. V., and Wood, D. M. (1962). The blackflies (Diptera: Simuliidae) of Ontario. Part 1. Adult identification and distribution with descriptions of six new species. *Proc. Entomol. Soc. Ont. 92*: 71-154.

Ebsary, B. A. (1973). The mermithid (Nematoda) and other endoparasites of Simuliidae (Diptera) in insular Newfoundland. M.Sc. thesis, Memorial University of Newfoundland, St. John's, Newfoundland, Canada.

Ebsary, B. A., and Bennett, G. F. (1973). Molting and oviposition of *Neomesomermis flumenalis* (Welch, 1962) Nickle, 1972, a mermithid parasite of blackflies. *Can. J. Zool. 51*: 637-639.

Ebsary, B. A., and Bennett, G. F. (1974). Redescription of *Neomesomermis flumenalis* (Nematoda) from blackflies in Newfoundland. *Can. J. Zool. 52*: 65-68.

Ebsary, B. A., and Bennett, G. F. (1975a). The occurrence of some endoparasites of blackflies (Diptera: Simuliidae) in insular Newfoundland. *Can. J. Zool. 53*: 1058-1062.

Ebsary, B. A., and Bennett, G. F. (1975b). Studies on the bionomics of mermithid nematode parasites of blackflies in Newfoundland. *Can. J. Zool. 53*: 1324-1331.

Ezenwa, A. O. (1973). Mermithid and microsporidan parasitism of blackflies (Diptera: Simuliidae) in the vicinity of Churchill Falls, Labrador. *Can. J. Zool. 51*: 1109-1111.

Ezenwa, A. O. (1974a). Ecology of Simuliidae, Mermithidae, and Microsporida in Newfoundland freshwaters. *Can. J. Zool. 52*: 557-565.

Ezenwa, A. O. (1974b). Studies on host-parasite relationships of Simuliidae with mermithids and microsporidans. *J. Parasitol. 60*: 809-813.

Ezenwa, A. O. and Carter, N. E. (1975). Influence of multiple infections on sex ratios of mermithid parasites of blackflies. *Environ. Entomol. 4*: 142-144.

Fallis, A. M., and Desser, S. S. (1977). On species of *Leucocytozoon, Haemoproteus*, and *Hepatocystis*. In *Parasitic Protozoa*, vol. 3 (J. P. Kreier, ed.). Academic, New York, pp. 239-266.

Fassuliotis, G., and Creighton, C. S. (1982). In vitro cultivation of the entomogenous nematode *Filipjevimermis leipsandra*. *J. Nematol. 14*: 126-131.

Fast, P. G. (1970). Insect lipids. In *Progress in the Chemistry of Fats and Other Lipids*, vol. 11 (R. T. Holman, ed.). Pergamon, Oxford, pp. 179-242.

Finney, J. R. (1975). The penetration of three simuliid species by the nematode *Reesimermis nielseni*. *Bull. WHO 52*: 235.

Finney, J. R. (1976). The in vitro culture of mermithid parasites of blackflies and mosquitoes. *J. Nematol. 8*: 284.

Finney, J. R. (1981a). Potential of mermithids for control and in vitro culture. In *Blackflies: the Future for Biological Methods in Integrated Control* (M. Laird, ed.). Academic, London, pp. 325-333.

Finney, J. R. (1981b). Mermithid nematodes: In vitro culture attempts. *J. Nematol. 13*: 275-280.

Finney, J. R., and Mokry, J. E. (1980). *Romanomermis culicivorax* and simuliids. *J. Invert. Pathol. 35*: 211-213.

Fredeen, F. J. H. (1969). Outbreaks of the black fly *Simulium arcticum* Malloch in Alberta. *Quaest. Entomol. 5*: 341-372.

Fredeen, F. J. H. (1970). Sexual mosaics in the black fly *Simulium arcticum* (Diptera: Simuliidae). *Can. Entomol. 102*: 1585-1592.

Fredeen, F. J. H. (1977). A review of the economic importance of black flies (Simuliidae) in Canada. *Quaest. Entomol. 13*: 219-299.

Garris, G. I., and Noblet, R. (1975). Notes on parasitism of black flies (Diptera: Simuliidae) in streams treated with Abate. *J. Med. Entomol. 12*: 481-482.

Gaugler, R. (1981). The biological control potential of neoaplectanid nematodes. *J. Nematol. 13*: 241-249.

Gaugler, R., and Molloy, D. (1981). Instar susceptibility of *Simulium vittatum* (Diptera: Simuliidae) to the entomogenous nematode *Neoaplectana carpocapsae*. *J. Nematol. 13*: 1-5.

Gordon, R., Condon, W. J., Edgar, W. J., and Babie, S. J. (1978). Effects of mermithid parasitism on the haemolymph composition of the larval blackflies *Prosimulium mixtum/fuscum* and *Simulium venustum*. *Parasitology 77*: 367-374.

Gordon, R., Condon, W. J., and Squires, J. M. (1980). Sterols in the trophosomes of the mermithid nematodes *Neomesomermis flumenalis* and *Romanomermis culicivorax* relative to sterols in the host hemolymph. *J. Parasitol. 66*: 585-590.

Gordon, R., Ebsary, B. A., and Bennett, G. F. (1973). Potentialities of mermithid nematodes for the biocontrol of blackflies (Diptera: Simuliidae) – a review. *Exp. Parasitol. 33*: 226-238.

Gordon, R., Finney, J. R., Condon, W. J., and Rusted, T. N. (1979). Lipids in the storage organs of three mermithid nematodes and in the hemolymph of their hosts. *Comp. Biochem. Physiol. 64B*: 369-374.

Gordon, R., Squires, J. M., Babie, S. J., and Burford, I. R. (1981). Effects of host diet on *Romanomermis culicivorax*, a mermithid parasite of mosquitoes. *J. Nematol. 13*: 285-298.

Gordon, R., Walsh, D. J., and Burford, I. R. (1981). Beta-oxidation in the free-living stages of the entomophilic nematode *Romanomermis culicivorax*. *Parasitology 83*: 451-457.

Gordon, R., and Webster, J. M. (1972). Nutritional requirements for protein synthesis during parasitic development of the entomophilic nematode *Mermis nigrescens*. *Parasitology 64*: 161-172.

Guillet, P., Escaffre, M., Ouedraogo, M., and Quillévéré, D. (1980). Mise en évidence d'une résistance au téméphos dans le complexe *Simulium damnosum* (*S. sanctipauli et S. soubrense*) en Cote d'Ivoire (Zone du programme de lutte contre l'onchocercose dans la région du Bassin de la Volta). *Cah. ORSTOM Ser. Entomol. Med. Parasitol. 18*: 291-299.

Hansen, E. L., and Hansen, J. W. (1976). Parasitism of *Simulium damnosum* by *Romanomermis culicivorax*. *ICRS Med. Sci. 4*: 508.

Hocking, B., and Pickering, L. R. (1954). Observations on the bionomics of some northern species of Simuliidae (Diptera). *Can. J. Zool. 32*: 99-119.

Hunter, D. M., and Moorhouse, D. E. (1976). Sexual mosaics and mermithid parasitism in *Austrosimulium bancrofti* (Tayl.) (Diptera, Simuliidae). *Bull. Entomol. Res. 65*: 549-553.

Jamnback, H. (1973). Recent developments in control of blackflies. *Annu. Rev. Entomol. 18*: 281-304.

Kaiser, von H., and Fachbach, G. (1977). Polyacrylamiddiskelektrophoretische Untersuchungen an Organhomogenaten von Mermithiden (Nematoda). Ein Beitrag zum Problem der Trennung morphologisch schwer unterscheidbarer Arten. *Zool. Jahrb. Abt. Syst. Okol. Geogr. Tiere 104*: 72-79.

Laird, M., and Bennett, G. F. (1970). The subarctic epizootiology of *Leucocytozoon simondi*. *Proc. Second Int. Congr. Parasitol.*, Washington, D.C., p. 198.

Lewis, D. J. (1953). *Simulium damnosum* and its relation to onchocerciasis in the Anglo-Egyptian Sudan. *Bull. Entomol. Res. 43*: 579-644.

Likhovoz, L. K. (1978). Introduction of Mermithidae (Nematoda), parasites of larvae of blackflies (Simuliidae, Diptera) (in Russian with English summary). *Meditsinskaya Parazitologiya; Parazitarnye Balezni 47*: 90-94.

McComb, C. W., and Bickley, W. E. (1959). Observations on blackflies in two Maryland counties. *J. Econ. Entomol. 52*: 629-632.

Mokry, J. E., and Finney, J. R. (1977). Notes on mermithid parasitism of Newfoundland blackflies, with the first record of *Neomesomermis flumenalis* from adult hosts. *Can. J. Zool. 55*: 1370-1372.

Molloy, D. (1979). Description and bionomics of *Mesomermis camdenensis* n. sp. (Mermithidae), a parasite of black flies (Simuliidae). *J. Namtol. 11*: 321-328.

Molloy, D. (1981). Mermithid parasitism of blackflies (Diptera: Simuliidae). *J. Nematol. 13*: 250-256.

Molloy, D., Gaugler, R., and Jamnback, H. (1980). The pathogenicity of

Neoaplectana carpocapsae to blackfly larvae. *J. Invert. Pathol. 36*: 302-306.

Molloy, D., and Jamnback, H. (1975). Laboratory transmission of mermithids parasitic in blackflies. *Mosq. News 35*: 337-342.

Molloy, D., and Jamnback, H. (1977). A larval black fly control field trial using mermithid parasites and its cost implications. *Mosq. News 37*: 104-108.

Mondet, B., Berl, D., and Bernadou, J. (1977). Etude du parasitisme des Simulies (Diptera) par des Mermithidae (Nematoda) en Afrique de l'Ouest. III. Elevage de *Isomermis* sp. et infestation en laboratoire de *Simulium damnosum* s. 1. *Cah. ORSTOM Ser. Entomol. Med. Parasitol. 15*: 265-269.

Mondet, B., Pendriez, B., and Bernadou, J. (1976). Etude du parasitisme des Simulies (Diptera) par des Mermithidae (Nematoda) en Afrique de l'Ouest. I. Observations preliminaires sur un cours d'eau temporaire de savane. *Cah. ORSTOM Ser. Entomol. Med. Parasitol. 14*: 141-149.

Mondet, B., Poinar, G. O. Jr., and Bernadou, J. (1977). Etude du parasitisme des simulies (Diptera: Simuliidae) par des Mermithidae (Nematoda) en Afrique de l'ouest. IV. Description de *Isomermis lairdi*, n. sp. parasite de *Simulium damnosum*. *Can. J. Zool. 55*: 2011-2017.

Mondet, B., Prud'hom, J. M., Bellec, C., and Hebrard, G. (1980). Etude du parasitisme des simulies (Diptera: Simuliidae) par des Mermithidae (Nematoda) en Afrique de l'Ouest. V. Croissance et sex-ratio de deux especes parasites d'adultes de *Simulium damnosum* s. 1. *Cah. ORSTOM Ser. Entomol. Med. Parasitol. 18*: 49-57.

Mulvey, R. H., and Nickle, W. R. (1978). Taxonomy of mermithids (Nematoda: Mermithidae) of Canada and in particular of the Mackenzie and Porcupine river systems, and Somerset Island, N.W.T., with descriptions of eight new species and emphasis on the use of the male characters in identification. *Can. J. Zool. 56*: 1291-1329.

Nelson, G. S. (1970). Onchocerciasis. *Adv. Parasitol. 8*: 173-224.

Nickle, W. R. (1972). A contribution to our knowledge of the Mermithidae (Nematoda). *J. Nematol. 4*: 113-146.

Nickle, W. R. (1979). Probable establishment and overwintering of a mermithid nematode parasite of mosquitoes in Maryland. *Proc. Helm. Soc. Wash. 46*: 21-27.

Omar, M. S., Denke, A. M., and Raybould, J. N. (1979). The development of Onchocerca ochengi (Nematoda: Filarioidea) to the infective stage in *Simulium damnosum* s. 1. with a note on the histochemical staining of the parasite. *Tropenmed. Parasitol. 30*: 157-162.

Petersen, A. (1924). Contributions to the natural history of the Danish Simuliidae (in Danish with English summary). *Danske Videns Kabernes Selskab. Selskabs – Skrifter. Naturvidenskabelig og Matematisk Afdeling Series 8. 4*: 237-242.

Petersen, J. J. (1972). Factors affecting sex ratios of a mermithid parasite of mosquitoes. *J. Nematol. 4*: 82-87.

Peterson, B. V. (1960). Notes on some natural enemies of Utah blackflies (Diptera: Simuliidae). *Can. Entomol. 92*: 266-274.

Phelps, R. J., and DeFoliart, G. R. (1964). Nematode parasites of Simuliidae. *Univ. Wis. Madison Coll. Agric. Life Sci. Res. Div. Res. Bull. 245.*

Poinar, G. O., Jr. (1975). *Entomogenous Nematodes. A Manual and Host List of Insect-Nematode Associations.* E. J. Brill, Leiden.

Poinar, G. O., Jr. (1977). A synopsis of the nematodes occurring in blackflies (Diptera: Simuliidae). *Bull. WHO 55*: 509-515.

Poinar, G. O., Jr. (1979). *Nematodes for Biological Control of Insects*. CRC Press, Boca Raton, Florida.

Poinar, G. O., Jr. (1981). Mermithid nematodes of blackflies. In *Blackflies: the Future for Biological Methods in Integrated Control* (M. Laird, ed.). Academic, London, pp. 159-170.

Poinar, G. O., Jr., and Hess, R. (1974). Structure of the pre-parasitic juveniles of *Filipjevimermis leipsandra* and some other Mermithidae (Nematoda). *Nematologica 20*: 167-173.

Poinar, G. O., Jr., and Hess, R. (1977). *Romanomermis culicivorax*: Morphological evidence for transcuticular uptake. *Exp. Parasitol. 42*: 27-33.

Poinar, G. O., Jr., Hess, R., Hansen, E., and Hansen, J. W. (1979). Laboratory infection of blackflies (Simuliidae) and midges (Chironomidae) by the mosquito mermithid, *Romanomermis culicivorax*. *J. Parasitol. 65*: 613-615.

Poinar, G. O., Jr., and Otieno, W. A. (1974). Evidence of four molts in the Mermithidae. *Nematologica 20*: 370.

Poinar, G. O., Jr., and Takaoka, H. (1979). *Isomermis benevolus* sp. n. (Mermithidae, Nematoda), a parasite of *Simulium metallicum* (Diptera: Simuliidae) in Guatemala. *Jpn. J. Sanit. Zool. 30*: 305-307.

Rempel, J. G., and Arnason, A. P. (1947). An account of three successive outbreaks of the black fly *Simulium arcticum*, a serious livestock pest in Saskatchewan. *Sci. Agric. 27*: 428-445.

Rubtsov, I. A. (1963). On mermithids parasitizing simuliids (in Russian with English summary). *Zool. Zh. 42*: 1768-1784.

Rubtsov, I. A. (1964). Observations on mermithid nematodes parasitizing blackflies. *World Health Organization mimeographed document. WHO/ELB/19*.

Rubtsov, I. A. (1967). Scheme and organs of the extraintestinal digestion of mermithids (in Russian with English summary). *Izv. Akad. Nauk. SSSR Ser. Biol. 10*: 883-890.

Rubtsov, I. A. (1971). Protective and mutualistic relationships between the insect and its parasites (technical translation 1741, National Research Council of Canada). *Zh. Obshch. Biol. 32*: 193-201.

Rubtsov, I. A. (1977). *Aquatic Mermithidae of the Fauna of the USSR*, vol. 1. Nauka Publishers, Leningrad. Translated from Russian. Published for the USDA and the National Science Foundation, Washington, D.C., by Amerind Publishing Co., New Delhi.

Rubtsov, I. A. (1981a). *Aquatic Mermithidae of the Fauna of the USSR*, vol. 2. Nauka Publishers, Leningrad. Translated from Russian. Published for the USDA and the National Science Foundation, Washington, D.C., by Amerind Publishing Co., New Delhi.

Rubtsov, I. A. (1981b). Mermithidae: Taxonomic criteria for their juvenile stages and blackfly biocontrol prospects. In *Blackflies: the Future for Biological Methods in Integrated Control* (M. Laird, ed.). Academic, London, pp. 171-180.

Rutherford, T. A., and Webster, J. M. (1974). Transcuticular uptake of glucose by the entomophilic nematode, *Mermis nigrescens*. *J. Parasitol. 60*: 804-808.

Rutherford, T. A., Webster, J. M., and Barlow, J. S. (1977). Physiology of nutrient uptake by the entomophilic nematode *Mermis nigrescens* (Mermithidae). *Can. J. Zool. 55*: 1773-1781.

Schiefer, K. (1969). Ecology of Atlantic Salmon, *Salmo salar* L., in the Matamek River System. M.Sc. thesis, University of Waterloo, Ontario, Canada.

Shelley, A. J., Luna Dias, A. P. A., and Moraes, M. A. P. (1980). *Simulium* species of the amazonicum group as vectors of *Mansonella ozzardi* in the Brazilian Amazon, *Trans. R. Soc. Trop. Med. Hyg. 74*: 784-788.

Simmons, K. R., and Edman, J. D. (1981). Sustained colonization of the black fly *Simulium decorum* Walker (Diptera: Simuliidae). *Can. J. Zool. 59*: 1-7.

Sommerman, K. M., Sailer, R. I. and Esselbaugh, C. O. (1955). Biology of Alaskan blackflies (Simuliidae, Diptera) *Ecol. Monogr. 25*: 345-385.

Spasskii, A. A., Okopnyi, N. S., and Toderash, I. K. (1975). Character of the relationships between chironomids and mermithids parasitic on them. *Dokl. Akad. Nauk. SSSR Biol. Sci. Sect. (Engl. Transl.) 222*: 246-248.

Strickland, E. H. (1911). Some parasites of *Simulium* larvae and their effects on the development of the host. *Biol. Bull. 21*: 302-329.

Takaoka, H. (1980). Pathogens of blackfly larvae in Guatemala and their influence on natural populations of three species of Onchocerciasis vectors. *Am. J. Trop. Med. Hyg. 29*: 467-472.

Thomas, G. M., and Poinar, G. O., Jr. (1979). *Xenorhabdus* gen. nov., a genus of entomopathogenic, nematophilic bacteria of the family Enterobacteriaceae. *Int. J. Syst. Bacteriol. 29*: 352-360.

Travis, B. V., Vargas, M. V., and Swartzwelder, J. C. (1974). Bionomics of blackflies (Diptera: Simuliidae) in Costa Rica. I. Species biting man with an epidemiological summary for the Western Hemisphere. *Rev. Biol. Trop. 22*: 187-200.

Twinn, C. R. (1936). The blackflies of Eastern Canada (Simuliidae: Diptera). Part I. *Can. J. Res. Sect. D. Zool. Sci. 14*: 97-130.

Twinn, C. R. (1939). Notes on some parasites and predators of blackflies (Simuliidae, Diptera). *Can. Entomol. 71*: 101-105.

Vargas, M. V., and Fallas, F. B. (1976). Observaciones sobre *Neomesomermis flumenalis* (Welch, 1962) (Nematoda: Mermithidae) en una poblacion larval de simulidos en Costa Rica. *Resumenes de trabajos libres. Congreso (iv) Latinoamericano de Parasitologia, San Jose, Costa Rica, 7-11 Dec.*, p. 58.

Vargas, M. V., Rubtsov, I. A., and Fallas, F. B. (1980). Bionomics of black flies (Diptera: Simuliidae) in Costa Rica. V. Description of *Neomesomermis travisi* sp. n. (Nematoda: Mermithidae). *Rev. Biol. Trop. 28*: 73-89.

Walsh, J. F., Davies, J. B., and LeBerre, R. (1979). Entomological aspects of the first five years of the Onchocerciasis control programme in the Volta River basin. *Tropenmed. Parasitol. 30*: 328-344.

Webster, J. M. (1973). Manipulation of environment to facilitate use of nematodes in biocontrol of insects. *Exp. Parasitol. 33*: 197-206.

Welch, H. E. (1962). New species of *Gastromermis*, *Isomermis* and *Mesomermis* (Nematoda: Mermithidae) from black fly larvae. *Ann. Entomol. Soc. Am. 55*: 535-542.

Welch, H. E. (1964). Mermithid parasites of blackflies. *Bull. WHO 31*: 857-863.

Welch, H. E. (1965). Entomophilic nematodes. *Annu. Rev. Entomol. 10*: 275-302.

Welch, H. E., and Rubtsov, I. A. (1965). Mermithids (Nematoda: Mermithidae) parasitic in blackflies (Insecta: Simuliidae). I. Taxonomy and bionomics of *Gastromermis boophthorae* sp. n. *Can. Entomol. 97*: 581-596.

Wenk, P. (1976). Koevolution von Überträger und Parasit bei Simuliiden und Nematoden. *Z. Angew. Entomol. 82*: 38-44.

Wu, Y. F. (1930). A contribution to the biology of *Simulium* (Diptera). *Pap. Mich. Acad. Sci. Arts Lett. 13*: 543-599.

Chapter 24
Nematode Parasites of Other Dipterans

Christopher J. Geden and John G. Stoffolano, Jr. *University of Massachusetts, Amherst, Massachusetts*

I. INTRODUCTION

Bovien (1932), while describing a new nematode, *Scatonema wülkeri* Bovien, in the body cavity of the minute black scavenger fly, *Scatopse fuscipes* Meig., stated ". . . I found other specimens of undescribed nematodes in the body-cavity of dipterous insects and I feel sure that the nematode parasites of these insects offer a rich field for investigation." The Diptera, of all the insect orders, ranks third with respect to the size of the order in North America (16,130 species) (Stone et al. (1965) and comprises 17% of all described insect species from North America (93,728) (Borror et al., 1981). According to Poinar (1975), the Diptera comprised 22% of the 3142 citations he reported on insect-nematode related associations. The size of this order and its economic and medical-veterinary importance places this group in the top three with respect to significance of research priorities.

With the exceptions of the Culicidae and the Simuliidae (treated in other chapters), our coverage of the nematodes associated with flies was a difficult task. The large size of this order and the lack of information on many groups within the order complicated the literature search. We have attempted in Table 1 to give the reader a sense of the size of this group and, at the same

Table 1 Classification of the Diptera and the Associated Nematode Literature

Higher Classification of Diptera	Genera	Species
I. Suborder Nematocera	(417)	(5014)
A. Superfamily Tipuloidea	(62)	(1485)
1. Family Trichoceridae	3	27
2. Family Tipulidae	59	1458
B. Superfamily Psychodoidea	(17)	(107)
3. Family Tanyderidae	2	4
4. Family Psychodidae	12	87
5. Family Ptychopteridae	3	16
6. Family Nymphomyiidae	(1 unnamed)	(1 unnamed)
C. Superfamily Culicoidea	(121)	(1300)
7. Family Blephariceridae	5	22
8. Family Deuterophlebiidae	1	4
9. Family Dixidae	1	41
10. Family Chaoboridae	4	15
11. Family Culicidae	12	148
12. Family Thaumaleidae	2	5
13. Family Ceratopogonidae	27	348
14. Family Chironomidae	63	601
15. Family Simuliidae	6	116
D. Superfamily Anisopodoidea	(3)	(8)
16. Family Anisopodidae	3	8
E. Superfamily Bibionoidea	(8)	(80)
17. Family Bibionidae	6	78
18. Family Pachyneuridae	2	2
F. Superfamily Mycetophiloidea	(206)	(2034)
19. Family Mycetophilidae	53	612
20. Family Sciaridae	16	155
21. Family Hyperoscelididae	1	1
22. Family Scatopsidae	11	61
23. Family Cecidomyiidae	125	1205
II. Suborder Brachycera	(356)	(4408)
A. Superfamily Tabanoidea	(81)	(672)
24. Family Xylophagidae	7	26
25. Family Xylomyidae	2	9
26. Family Stratiomyidae	36	234
27. Family Pelecorhynchidae	1	1
28. Family Tabanidae	25	295
29. Family Rhagionidae	9	104
30. Family Hilarimorphidae	1	3

Common Names[a]	Category of Cited Literature[b]	Nematode Family[c]	References[d]
Long-horned flies			
Winter crane flies	N	—	—
Crane flies	BC	T,M,Syn	12,73,74,107
Primitive crane flies	N	—	—
Moth flies, sand flies, owl midges	BC,V	S,F,R,	9,60,70,158,159
Phantom crane flies	N	—	—
Nymphomyiid flies	N	—	—
Net-winged midges	BC	M	24
Mountain midges	N	—	—
Dixid Midges	N	—	—
Phantom midges	BC	M,Te	13,93,154
Mosquitoes			
Solitary midges	N	—	—
Biting midges, punkies, no-seeums	BC,V	S,M,F,Te,E	3,14,93,102,117, 119,131,156
Midges		M,S	15,54,106,107,157
Blackflies, buffalo gnats		—	
Wood gnats	N	—	
Marsh flies	BC	A,St,M	9,107,112,150
Pachyneurid gnats	N	—	
Fungus gnats	N	—	
Dark-winged fungus gnats, root gnats	BC	Te,Sp,St,M	10,22,55,56,68,103, 104,110
Hyperoscelid gnats	N	—	
Minute black scavenger flies	BC	Sp	8
Gall gnats or gall midges	BC	Sp	76
Straight-seemed flies			
Xylophagid flies	N	—	
Xylomyid flies	N	—	
Soldier flies	N	—	
Pelecorhyncid flies	N	—	
Horse flies, deer flies, greenheads	BC,V	F,M,D	1,21,48,61,81,98, 107,111,128,129
Snipe flies	N	—	
Hilarimorphid flies	N	—	

Table 1   (Con't.)

Higher Classification of Diptera	Genera	Species
B. Superfamily Asiloidea	(170)	(1885)
31. Family Therevidae	13	130
32. Family Scenopinidae	6	21
33. Family Apioceridae	2	29
34. Family Mydiadae	6	39
35. Family Asilidae	82	856
36. Family Nemestrinidae	3	6
37. Family Acroceridae	7	59
38. Family Bombyliidae	51	745
C. Superfamily Empidoidea	(105)	(1851)
39. Family Empididae	61	710
40. Family Dolichopodidae	44	1141
III. Suborder Cyclorrhapha	(1197)	(6702)
Division Aschiza	(153)	(1408)
A. Superfamily Lonchopteroidea	(1)	(4)
41. Family Lonchopteridae	1	4
B. Superfamily Phoroidea	(47)	(293)
42. Family Phoridae	38	226
43. Family Platypezidae	9	67
C. Superfamily Syrphoidea	(105)	(1111)
44. Family Pipunculidae	8	105
45. Family Syrphidae	88	939
46. Family Conopidae	9	67
Division Schizophora	(1044)	(5294)
Section Acalyptratae	(423)	(2351)
D. Superfamily Micropezoidea	(10)	(33)
47. Family Micropezidae	8	31
48. Family Neriidae	2	2
E. Superfamily Nothyboidea	(6)	(35)
49. Family Diopsidae	1	1
50. Family Psilidae	4	32
51. Family Tanypezidae	1	2
F. Superfamily Tephritoidea	(102)	(413)
52. Family Richardiidae	4	7
53. Family Otitidae	38	127
54. Family Platystomatidae	4	41
55. Family Pyrgotidae	3	5
56. Family Tephritidae	53	233
G. Superfamily Sciomyzoidea	(35)	(191)
57. Family Helcomyzidae	2	3
58. Family Ropalomeridae	1	1
59. Family Coelopidae	2	5
60. Family Dryomyzidae	2	8
61. Family Sepsidae	8	34
62. Family Sciomyzidae	20	140

Common Names*	Category of cited Literature	Nematode Family	References
	N	—	
Stiletto flies	N	—	
Window flies	N	—	
Flower-loving flies	N	—	
Mydas flies	N	—	
Robber flies	N	—	
Tangle-veined flies	N	—	
Small-headed flies	N	—	
Bee flies	N	—	
Dance flies	N	—	
Long-legged flies	N	—	
Circular-seamed flies	N	—	
	N		
Spear-winged flies	N	—	
Humpbacked flies	BC	A	57, 114, 115, 116
Flat-footed flies	N	—	
Big-headed flies	N	—	
Syrphid flies, flowerflies	N	Syr	75
Thick-headed flies	—	—	—
Muscoid flies			
Acalyptrate muscoid flies			
Stilt-legged flies	N	—	
Cactus flies	N	—	
	N	—	
Stalk-eyed flies	N	—	
Rust flies	N	—	
Tanypezid flies	N	—	
Richardiid flies	N	—	
Picture-winged flies	N	—	
Picture-winged flies	N	—	
Pyrgotid flies	N	—	
Fruit flies	BC	R	108, 113
Seabeach flies	N	—	
Ropalomerid flies	N	—	
Seaweed flies	N	—	
Dryomyzid flies	N	—	
Black scavenger flies	BC	Te	44
Marsh flies	BC	M	107
	N		

Table 1 (Con't.)

Higher Classification of Diptera	Genera	Species
H. Superfamily Lauxanioidea	(31)	(174)
63. Family Lauxaniidae	23	135
64. Family Chamaemyiidae	7	36
65. Family Periscelididae	1	3
I. Superfamily Pallopteroidea	(9)	(81)
66. Family Piophilidae	1	31
67. Family Thyreophoridae	1	2
68. Family Neottiophilidae	1	1
69. Family Pallopteridae	1	9
70. Family Lonchaeidae	5	38
J. Superfamily Milichioidea	(27)	(197)
71. Family Sphaeroceridae	3	117
72. Family Braulidae	1	1
73. Family Tethinidae	5	22
74. Family Milichiidae	15	52
75. Family Canaceidae	3	5
K. Superfamily Drosophiloidea	(86)	(534)
76. Family Ephydridae	65	347
77. Family Curtonotidae	1	1
78. Family Drosophilidae	17	179
79. Family Diastatidae	2	6
80. Family Camillidae	1	1
L. Superfamily Chloropoidea	(42)	(264)
81. Family Chloropidae	42	264
Unplaced Families of Acalyptratae	(73)	(426)
82. Family Odiniidae	3	11
83. Family Agromyzidae	16	189
84. Family Clusiidae	5	27
85. Family Acartophthalmidae	1	2
86. Family Heleomyzidae	23	113
87. Family Trixoscelididae	5	30
88. Family Rhinotoridae	1	1
89. Family Anthomyzidae	5	14
90. Family Opomyzidae	3	13
91. Family Chyromyidae	3	9
92. Family Aulacigastridae	1	1
93. Family Asteiidae	6	18
94. Family Cryptochetidae	1	1
Aclyptrate Genera of Uncertain Family Position	2	3
Section Calyptratae	(621)	(2943)
M. Superfamily Muscoidea	(131)	(1220)
95. Family Anthomyiidae	56	556

Common Names*	Category of cited Literature	Family	References
Lauxaniid flies	N		
Aphid flies	N	—	
Periscelidid flies	N	—	
Skipper flies	N	—	
Thyreophorid flies	N	—	
Neottiophilid flies	N	—	
Pallopterid flies	N	—	
Lonchaeid flies	N	—	
	N	—	
Small dung flies	N	—	
Bee lice	N	—	
Tethinid flies	N	—	
Milichiid flies	N	—	
Beach flies	N	—	
	—	—	
Shore flies	N	—	
Curtonotid flies	N	—	
Pomace flies	BC,V	A,S	25,123,153
Diastatid flies	N	—	
Camillid flies	N	—	
	N	—	
Chloropid flies	BC	A	43
Odiniid flies	N	—	
Leafminer flies	BC	N	26,35
Clusiid flies	N	—	
Acartophthalmid flies	N	—	
Heleomyzid flies	N	—	
Trixosclelidid flies	N	—	
Rhinotorid flies	N	—	
Anthomyzid flies	N	—	
Opomyzid flies	N	—	
Chyromyzid flies	N	—	
Aulacigastrid flies	N	—	
Asteiid flies	N	—	
Chryptochotid flies	N	—	

Calyprate muscoid flies

Anthomyiid flies	BC,V	A,M,St, P,S,Syn	5,7,9,108,120,143, 152

Table 1 (Con't.)

High Classification of Diptera	Genera	Species
96. Family Muscidae	58	622
97. Family Gasterophilidae	1	4
98. Family Hippoboscidae	12	28
99. Family Streblidae	3	5
100. Family Nycteribiidae	1	5
N. Superfamily Oestroidea	(490)	(1723)
101. Family Calliphoridae	23	78
102. Family Sarcophagidae	48	327
103. Family Tachinidae	414	1281
104. Family Cuterebridae	1	26
105. Family Oestridae	4	11
Unplaced Genus and Species of Diptera	1	6
Totals	1971	16130

[a]From Borror et al. (1981).
[b]From Poinar (1975).
[c]

A	Allantonematidae
Apl	Aphelenchidae
Ap	Aphelenchoididae
C	Carabonematidae
Ce	Cephalobidae
Ch	Chambersiellidae
Cy	Cylindrocorporidae
D	Diplogasteridae
Dr	Dorylaimidae
E	Entaphelenchidae
F	Filariidae
M	Mermithidae
Mc	Monochidae
N	Neotylenchidae
O	Oxyuridae
P	Panagrolaimidae
Pl	Plectidae
R	Rhabditidae
S	Spiruridae
Sp	Sphaerulariidae
St	Steinernematidae
Sb	Subuluridae
Syn	Syngamidae
Syr	Syrphonematidae
Te	Tetradonematidae
T	Thelastomatidae
Ty	Tylenchidae

Common Names*	Category of cited Literature	Family	References
Muscid flies	BC,V	A,M,F, S,Syn	20,23,32,36,37,51,86 90,91,92,108,120,122, 123,124,125,126,134, 136,146
Horse bot flies	N	−	
Louse flies	V	F	88
Bat flies	N	−	
Bat flies	N	−	
Blow flies	V	S,Syn	20,53
Flesh flies	V	S	63
Tachinid flies	N	−	
Robust bot flies	N	−	
Warble flies and bot flies	N	−	

time, to show the lack of information that exists for many families and, in some instances, superfamilies. Of the 105 dipteran families, 77 (73%) have not been reported to be associated with nematodes. Also, of the 23 superfamilies that comprise this order, nine have not been reported to be associated with nematodes. We have suggested that these groups have not been reported to be associated with nematodes either because they are not economically important, consequently little studied, or because the life cycle and ecological niche has placed them into environments unsuitable to nematodes or both. In Table 1, we have also attempted to give the reader some sense of the type of literature available for a given group. For each family of fly, we designated the literature associated with that group as either biocontrol studies, vector studies, or nothing reported. We have also reported which nematode groups have been reported with each dipteran group and have listed the associated references. Finally, we restricted our discussion to two major aspects of the dipteran-nematode relationship: flies as definitive hosts of nematodes and flies as vectors or intermediate hosts. In a few instances, however, we have discussed a relationship that does not fit into these categories. In these instances, the relationships were discussed either because of their historical significance or because of an interesting aspect of their biology.

II. FLY-NEMATODE ASSOCIATIONS

A. Suborder Nematocera (Crane Flies, Moth Flies, Sand Flies, Midges, Fungus Gnats)

1. *Superfamily Tipuloidea*

At present, no nematodes have been reported from the first family, the Trichoceridae (winter crane flies); however, the second family, the Tipulidae (crane flies) were shown to be associated with nematode parasites. The Tipulidae are easily recognized by their long legs and are not normally considered economically important. The larvae of some species feed on decaying vegetable matter in the soil and often attack various grasses, thus becoming a pest of lawns, turf, and potatoes.

In western Europe and British Columbia, one particular species, *Tipula paludosa* Meig., has become a serious pest. The larvae have a dark-colored, thickened cuticle, consequently the common name "leatherjackets." Larvae were reported to reduce dairy pasture yield up to 5022 kg/ha per year (Lam and Webster, 1971). Lam and Webster (1971) reported that of the tipulid species present in four study sites, *T. paludosa* was the most prevalent in three of the four and constituted 54-57% of the sample. Dissections revealed larvae parasitized by mermithids. The highest rate of infection was 12.2%, with death resulting as the nematodes exited from the third- or fourth- larval instar of the host. Since the damage has already been done to the crop by the third- or fourth- larval instar, the importance of these nematodes as effective biocontrol agents is questioned. Carter (1976) reported that a good biological control agent would destroy the host before damage levels were too severe; consequently, these mermithids appear to be ineffective against *T. paludosa*.

Lam and Webster (1971) evaluated the potential of two other nematodes as biocontrol agents of the leatherjacket. They discovered, however, that *Panagrolaimus tipulae* Lam and Webster (Panagrolaimidae) and *Rhabditis tipulae* Lam and Webster (Rhabditidae) survived only in insects that were debilitated or dead. Poinar (1975) provided a list of other nematodes associated with this family, but the majority of them belong to the Oxyuroidea and are not pathogenic to the insect host.

Since Dutky et al. (1962) reported that the DD-136 strain of *Neoaplectana carpocapsae* Weiser with its associated bacterium, *Achromobacter nematophilus* Poinar and Thomas, infected a *Tipula* species and Heimpel (1967) reported the β-exotoxin of *Bacillus thuringiensis* var. *thuringiensis* Berliner lethal to a wide variety of insects, Lam and Webster (1972) decided to evaluate the capability of these two agents in regulating the population of *T. paludosa*. Results suggested that the dosage of the nematode necessary to adequately reduce population levels would be too high and not practical or economical for large-scale control. Thus, one of the major problems in the use of nematodes as biocontrol agents was demonstrated: the practical and economic aspects of mass rearing.

2. Superfamily Psychodoidea

The only family, of the four present in this group (Table 1), reported to have nematodes associated with it is the Psychodidae; and, of the three subfamilies, only the Psychodinae (moth flies) and Phlebotominae (sand flies) were reported to be associated with nematodes.

Bovien (1937) reported a new rhabditid nematode, *Rhabditis dubia* Bovien, being carried from one dung pat to another by a *Psychoda* species. This phoretic association is necessary for the nematodes to move from one decaying and depleted resource to a new habitat. The "dauer" larval stage accomplish this by wrapping around the intersegmental membranes of the fly's abdomen. This casual relationship, however, could be confused as a parasitic one by investigators conducting dissections, since the nematodes will release when placed in saline and may appear to come from the body cavity of the fly.

The Phlebotominae (sand flies) are economically and medically important because of the nuisance created when the female seeks a blood meal and also because they are capable of vectoring human pathogens in the tropics and subtropics that cause visceral or dermal leishmaniasis (Harwood and James, 1979). Because of their worldwide importance, the World Health Organization sponsored two publications concerning the pathogens, parasites, and predators of medically important arthropod vectors (i.e., sand flies) (Jenkins, 1964a; Young and Lewis, 1977). These volumes reported several papers on nematodes associated with members of the genera *Phlebotomus*, *Lutzomyia*, and *Sergentomyia*, but all were casual reports and no identification of the nematodes were made.

In a laboratory study that probably has little relevance to the field, Yao et al. (1938) reported three species of *Phlebotomus* harboring microfilaria of *Wuchereria bancrofti* (Cobbold) after they were fed on microfilaremic patients. This nematode is normally vectored by mosquitoes. All nematodes exsheathed; however, they were unable to complete development because the sand flies died.

In France, Killick-Kendrick, et al. (1976) reported two spirurid nematodes from *Phlebotomus ariasi* Tonn. Both species, *Mastophorus muris* (Gmelin) and *Rictularia proni* Seurat, are normally parasites of small rodents. As a result, the authors concluded that the burrows of field mice may be the breeding sites of the sand fly larvae in southern France. To date, nematodes of only three groups have been found associated with this superfamily (Table 1). The importance of the phlebotominae and the type of habitat, moist decaying debris, of the larvae make them excellent candidates for nematode infections and future investigations.

3. Superfamily Culicoidea

The larvae and pupae of the nine families that comprise this superfamily are mainly aquatic or semiaquatic. Because they are not economically important, little or no information has been reported on nematode associations in the Deuterophlebiidae, Dixidae, and Thaumaleidae (Table 1). Craig (1963) reported on a mermithid (*Agamomermis* sp.) parasitizing the blepharicerid, *Peritheates turrifer* (Lumb). It is surprising that more nematodes are not reported in the above groups since these insects may often represent the dominant life form in many ecological studies.

The Chaoboridae or phantom midges resemble mosquitoes as adults, but do not bite. They obtained their common name because the larvae are almost transparent. Larvae live in pools and are predaceous on other aquatic insect larvae. It was this predaceous habit that caused Welch (1960) to hypothesize that the two *Chaborus* pupae and third-instar larvae of *Mochlonyx* he found parasitized by *Hydromermis churchillensis* Welch became infected by eating infected mosquito larvae of *Aedes communis* (DeG.). Chapman et al. (1967) reported 47% of the 260 larvae of *Corethrella brakeleyi* Coq. infected with a "mermithoid adult." They also noted that the last-instar larva of the fly was killed when the gravid female exited from the host. Later, Nickle (1969) identified this same nematode from *C. brakeleyi*, named it *Corethrellonema grandispiculosum* Nickle, placed it into the nematode family Tetradonematidae, and described the life cycle. Usually there are one to three male and one female adult nematodes per chaoborid larva. Adults mate in the host, and the host dies when the gravid female exits. The time from the onset of parasitism to exit from the host is about nine days; egg hatching is reported to take eight days following female exit from the host (Nickle, 1969). Infective-stage larvae are free-living forms prior to penetrating a host larva. The mode of entrance into the host was not reported. It is documented that parasitism of a biocontrol agent reduces the effectiveness of the agent; consequently, further information on this nematode may be important since its chaoborid hosts are predators of mosquito larvae.

The family Ceratopogonidae, commonly referred to as biting midges, punkies, or no-see-ums, includes four major genera, *Culicoides*, *Forcipomyia*, *Lasiohela*, and *Leptoconops*. The adult flies are small, the males do not bite, but the females are blood sucking and vicious biters. This feeding habit of the female makes them a serious nuisance to humans and other animals. Females also serve as vectors of various pathogenic agents of medical and veterinary importance (Kettle, 1965). Because of their importance as vectors, nematode biocontrol agents would be a viable alternative to insecticides since immature stages of many species occur in aquatic or semiaquatic areas considered extremely sensitive ecosystems (i.e., salt marshes); consequently, no insecticides are permitted.

The nematodes associated with this group of flies fall into two major categories: those serving as biocontrol agents (Tetradonematidae and Mermithidae) and those that are vectored by the adult fly and are parasites of vertebrates (Filariidae). Reviews concerning the use of various organisms, including nematodes, as biocontrol agents against this group of flies are provided by Bacon (1970) and Wirth (1977). Since the biting midge, *Culicoides variipennis* (Coquillett), has recently become an economically important fly in the southwestern United States, where it vectors a virus that is the causative agent of blue tongue disease of sheep, studies aimed at natural control of this fly may prove fruitful.

Nickle (1969) reported the only known nematode in the family Tetradonematidae parasitizing individuals in the Ceratopogonidae. *Aproctonema chap-*

mani Nickle was discovered in a field survey in southwestern Louisiana by Chapman et al. (1968), who were looking for nematode parasites of mosquitoes. They reported 1-10% of the tree-hole-inhabiting *Culicoides arboricola* Root and Hoffman parasitized by this nematode. Infected host larvae die just prior to pupation, and death may be attributed to the juvenile nematodes emerging from the host. When late-instar larvae of *C. arboricola* are dissected, one or more large females plus one or more male nematodes may be found filling the body cavity. Little more is known about this parasite and, as reported by Nickle (1969): "Little is known of the biology of this nematode. . .."

The major nematode parasites of the ceratopogonids are the mermithids. Phelps and Mokry (1976) listed five species of *Culicoides* infected with mermithids of unknown determination, while Wirth (1977) reported 33 references in which different biting midges served as hosts for undetermined mermithids. In general, the mermithids from *Culicoides* are undetermined. Rubtsov (1970), however, reported *Heleidomermis vivipara* Rub. from *C. nubeculosus* Mg. (a cattle feeder) and *C. stigma* Mg., and *Agamermis heleis* Rubt. from *C. pulicaris* L. (a human feeder). Recent research by Mullens (personal correspondence) has shown *Culicoides variipennis* larvae parasitized by *Heleidomermis* n. sp. (Fig. 1A, B). Sometimes, adult flies are parasitized (Fig. 1C). Without a doubt, there remains a considerable amount of research to be conducted on the mermithid parasites of this important group of flies.

To date, most research on nematodes of this group of flies has centered on those parasites that utilize them as intermediate, rather than definitive hosts.

Culicoides species have been shown to serve as intermediate hosts for three filarial nematodes of humans, *Dipetalonema perstans* (Manson), *D. streptocerca* M. and C., and *Mansonella ozzardi* (Mans.) and several filariae of animals, *Onchocerca cervicalis* R. and H. of horses, *O. gibsoni* (Clev. and John.) of cattle, *O. sweetae* Spratt and Moorhouse of buffalo, *Macacanema formosana* Sch. and And. from the Taiwan monkey, *Splendidofilaria quiscali* (*Chandlerella quiscali*) (v. Linst.) from the grackle, and *Icosiella neglecta* (Dies.) from the green frog.

A rather interesting relationship has been reported between the grackle, *Quiscalus quiscula versicolor* (Vieillot), and *Splendidofilaria quiscali*. Robinson (1971) fed *Culicoides crepuscularis* (Malloch) on an infected adult male bird and at given time intervals following feeding he dissected the flies to record nematode development. Exsheathing microfilariae were observed in the gnat 15 minutes following removal from traps. After one hour, some microfilariae were found in the abdomen but not the thorax. At 12 hours, both L_2 and L_3 nematodes were present, and between 120 and 132 hours, infective-stage larvae began migrating to the head and mouthparts of the host. At 13 days, 75% of the infective stages were reported in the head. No significant host response to the nematodes was reported.

The majority of filarial life cycles in ceratopogonids are very similar to the life cycle reported for the filarial nematode of the grackle. The exsheathing microfilariae are taken up by the arthropod host while blood feeding. These larvae then pass the gut barrier and enter the body cavity of the fly. They then make their way to the thoracic muscles where they molt and develop to L_3 larvae. These infective stages leave the muscles and enter the head of the host where they await the fly to feed. While feeding, the L_3 larvae exit from the mouth parts and enter the vertebrate host.

It was Steward (1935) who, after long searching by many previous workers for the intermediate host of *Onchocerca* in cattle, discovered that the genus *Culicoides* acted as the intermediate host for *O. cervicalis* in the horse. This

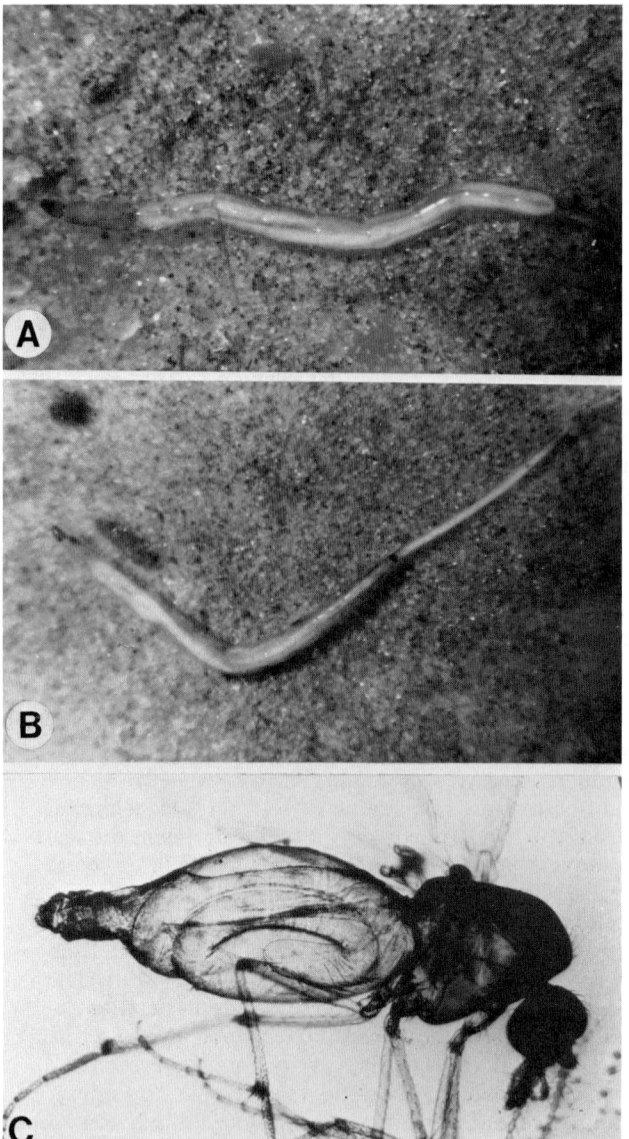

Fig. 1 Parasitism of *Culicoides variipennis* by *Heleidomermis* n.sp. Notice preemergence period (A) within insect hemocoel prior to emergence through the anus (B). (Courtesy of B. A. Mullens.) Adult *Culicoides* parasitized by a mermithid which fills the entire abdomen. (C). (Courtesy of W. R. Nickle.)

discovery was of paramount significance to the work of Buckley (1938), who had been unsuccessful in finding the insect host for *O. gibsoni*. The work of Buckley on *Culicoides* as vectors was considered by Nelson (1970) as ". . . one of the most outstanding publications in parasitology."

The microfilariae of various species of *Onchocerca* are intradermal (Fig. 2A) where they remain and are actively taken up by the adult female *Culicoides*

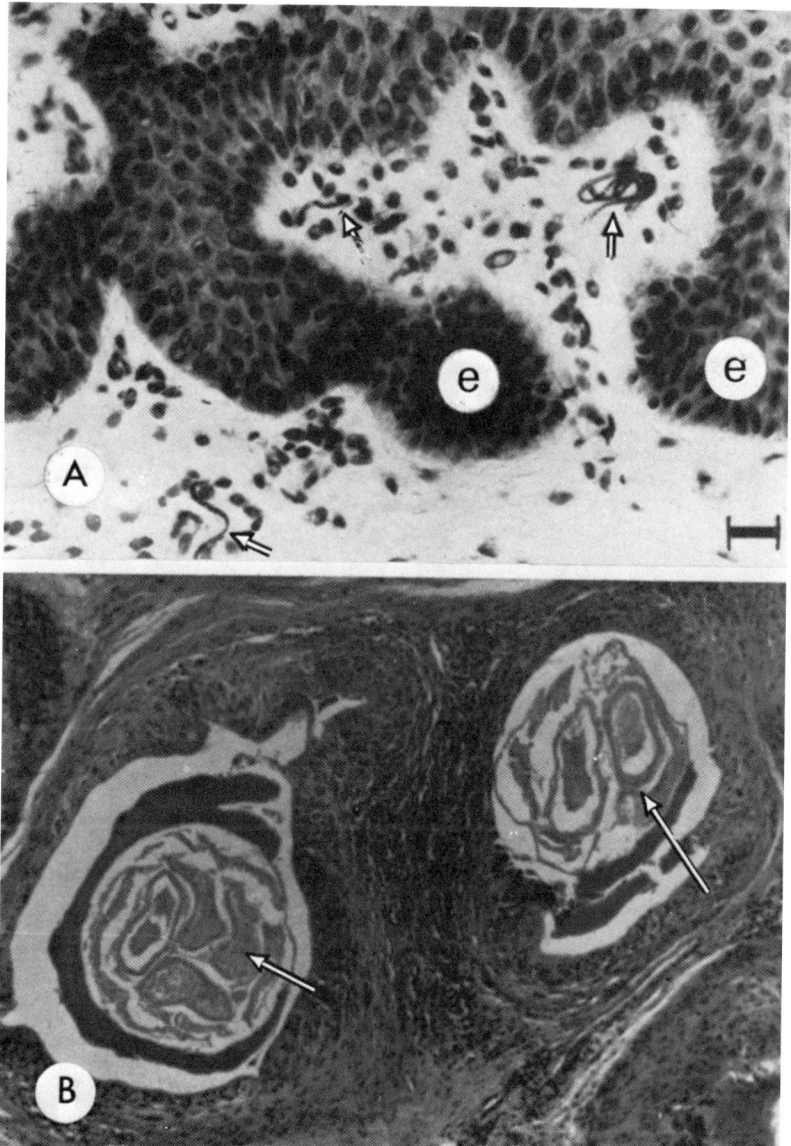

Fig. 2 (A) Microfilariae of *Onchocerca sweetae* (arrows) in the superficial layers of dermis of the water buffalo, *Bubalus bubalis* between and adjacent to epidermal papillae (e). (Courtesy of D. M. Spratt.) (B) Chronic inflammatory reaction around dead adult *Onchocerca cervicalis* (arrows) in the ligamentum nuchae of an infected horse. (Courtesy of P. S. Mellor.)

while feeding on these infected areas. Most studies to date show a close correlation between the site of the microfilariae in the host and the selected site of feeding preference by the intermediate host (Mellor, 1974, 1975). Diagnosis of vertebrate infection is performed by taking skin snips and observing them in saline placed under a microscope. If the vertebrate is infected, the microfilariae leave the tissue and are readily observed in the saline. Votava and Thompson (1978) showed that only 8 of 29 cattle known to be infected with *O. lienalis* Stiles showed microfilariae in skin biopsies. Unless skin biopsies were only taken in specific areas of the animal, their study raises some doubt as to the reliability of using just skin biopsies for diagnostic purposes in vertebrates.

Invertebrate vectors are identified by dissecting the fly in saline and locating nematodes in the thorax or head. The unsheathed microfilariae are taken up by the female fly during the act of feeding. The larvae enter the hemocoel where they make their way to the thoracic flight muscles. Development takes place there and the nematodes, now invasive L_3 forms, leave and enter the vertebrate host. Infected vertebrate hosts show varying signs of pathology. Varying degrees of economic loss have been attributed to bovine and equine onchocerciasis. Stannard and Cello (1975) reported *O. cervicalis* microfilarial invasion of the eyes in 60% of the infected horses; Mellor (1973) found only 2.4% of horses showing adult nematodes in the ligamentum nuchae with microfilariae in the eyes. Mellor (1973) concluded that no pathological conditions could be attributed to *Onchocerca* microfilariae. Figure 2B shows a chronic inflammatory reaction in horses around dead adults of *O. cervicalis* in the ligamentum nuchae (Mellor, personal correspondence).

Because of their increasing importance, mainly as nuisance pests and vectors of vertebrate pathogens, more emphasis should be given to finding nematode parasites of the aquatic stages of this group of flies. Since the use of many nematode biocontrol agents is often limited by the lack of moisture for survival, this factor certainly should not be the case with the aquatic stages of the ceratopogonids.

Two families of this superfamily, the Culicidae (mosquitoes) and Simuliidae (blackflies), are extremely important from a medical and veterinary standpoint. Both families include adults that are important vectors of parasites and pathogens of humans and animals; consequently, they have been subjects of intense investigation and are separately discussed in this book.

The Chironomidae (midges) are becoming more important as humans build houses closer to standing bodies of water and also as we attempt to use lagoons for water and/or waste holding. Both situations provide ideal breeding sites for the innocuous chironomid larvae. Upon emergence, however, the adults become serious problems as nuisance pests, even though they do not bite, because they are attracted to lights and will rest and get into dwellings by the thousands. In addition, larvae of these flies have been studied because they are food for fish and/or can be used as indicator species in environmental impact studies.

Poinar (1975) listed 62 chironomid hosts of nematodes, 58 of which were parasitized by mermithids while the remainder contained spirurids. The majority of research on nematode associations within the Chironomidae has been on the effects of the parasite on intersexual development of the adult host (Wülker, 1961). Hominick and Welch (1971), however, provided information on the topic of host-parasite synchronization, which is an extremely important characteristic of any biocontrol agent (Gordon and Webster, 1974).

Mermithid infection of chironomids usually takes place by infective-stage, parasitic juveniles penetrating the cuticle of early host larval instars (Poinar, 1968). Development occurs in the larva of the host (Fig. 3A), and mature parasitic juveniles exit through a hole made in the host's cuticle, thus killing the host. After a molt outside the host, the nematodes mate and the gravid female matures and releases eggs into the environment. The eggs hatch and the life cycle is completed. Sometimes, late-instar larvae are parasitized and this may result in the pupa or adult chironomid becoming infected (Fig. 3B).

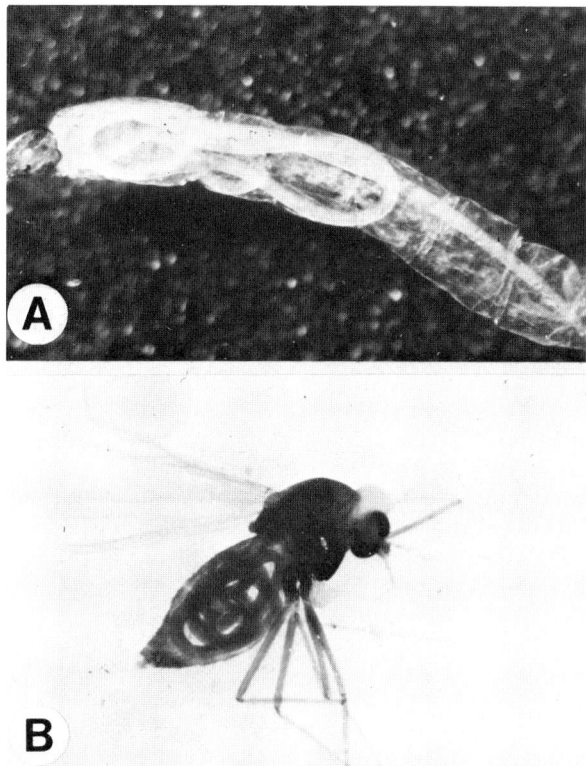

Fig. 3 (A) Chironomid larva (*Psectrocladius*) parasitized by *Limnomermis bathybia*. (Courtesy of V. J. McCauley.) (B) Adult chironomid parasitized by *Hydromermis contorta*. (Courtesy of R. P. Esser and J. B. MacGowan.)

During their investigation of percolation pond ecology in California, Chapman and Ecke (1969) found an inverse relationship between the percentage of parasitism and population density of the chironomid larvae. At the same time they reported that reduction in the host *Tanytarsus* population occurred and appeared to be directly associated with the high rate of parasitism by the mermithid, *Hydromermis conopophaga* Poinar. Further studies on this particular parasite-host association have not been reported; but this initial report suggested the possibility of using mermithids as biocontrol agents in ponds being used for supplemental drinking water where no pesticides could be applied.

4. Superfamily Anisopodoidea

Since the only family in this group is of insignificant economic importance, little research has been conducted on the associated parasites or pathogens. No nematodes have been reported associated with this group.

5. Superfamily Bibionoidea

Two families belong to this superfamily, Bibionidae and Pachyneuridae. No nematodes have been reported from the latter. The Bibionidae, or marsh flies, however, have been reported to host three families of parasitic nematodes: Mermithidae, Allantonematidae, and Steinernematidae. Poinar (1975) listed nine fly species associated with three species of nematodes: *Neoaplectana bibionis* Bov., *N. affinis* Bov., and *Bradynema bibionis* Wach., and only one mermithid of unknown determination.

The Pachyneuridae and Bibionidae are not very economically important, and this may account for lack of information on nematodes associated with these two families. Outbreaks of the marsh fly may occur and when they do they cause damage to grain and barley crops. Research on the nematodes of this group was initiated by Bovien (1937), who in 1931 received a sample of fly larvae from a neighborhood barley field. Of these *Bibio ferruginatus* L. and *Phlilia febrilis* (L.) larvae several died, and it was from these dead larvae that he described two new species of *Neoaplectana*.

Bovien (1937) was intrigued by the relationship of the *Neoaplectana* with larvae of *Bibio*. He stated, "The most important result of my investigations was the finding of a morphologically and biologically well-defined 'Dauerlarvae, a stage which is not obligatory, but only found under certain external conditions." Bovien was able to keep these dauer larvae under adverse conditions for up to 12 months and reported that this is how they survived adverse conditions. Transportation from one area to another could take place by the dauer larva clinging to the adult fly and was reported by Bovien as a means of spreading the infection. Forty years after Bovien's initial discovery, Poinar and Lindhardt (1971) reisolated *N. bibionis* from Danish bibionids and commented on their potential as biocontrol agents for this pest. Even though Bovien considered these nematodes generally saprozoic, latter work by Poinar and Lindhardt (1971) reported *N. bibionis* harboring a pathogenic bacteria, similar to *Achromobacter nematophilus* Poinar and Thomas in *N. carpocapsae* (Poinar, 1966), that causes lethal septicemia in the host and provides a bacterial population for the nematodes to feed upon. Because of their limited economic importance, however, the use of nematodes to control marsh flies has limited potential. In Florida, the bibionid called the "love bug" is a human nuisance pest and may be a potential candidate for a biocontrol attempt using one of these species of nematodes.

Our knowledge concerning the allantonematid, *Bradynema bibionis*, infecting various species of bibionids came from the research of Wachek (1955) and is scanty. If the bionomics of this species is similar to that of the *Bradynema* found by Hussey (1965), who reported up to 90% of a population of the mushroom phorid fly parasitized, it may have some potential as a biocontrol agent. Further information on the life cycle of *B. bibionis* is required prior to evaluating its potential as a biocontrol agent.

6. *Superfamily Mycetophiloidea*

Five families comprise this superfamily (Table 1). No reports have been made concerning nematodes associated with the Mycetophilidae or Hyperoscelididae. Nematodes have, however, been reported in the Sciaridae, Scatopsidae, and Cecidomyiidae.

The sciarids, or dark-winged fungus gnats, may become an economic pest, especially during the larval stage when they attack roots of plants, especially mushrooms in cultivation. Since these insects usually are associated with greenhouse crops, biocontrol agents may prove to be extremely effective because the ecosystem is fairly closed and should lend itself to greater human manipulation. Sciarid flies are parasitized by four groups of nematodes: Mermithidae, Tetradonematidae, Sphaerulariidae, and Steinernematidae (Poinar, 1975).

In his study on a new genus of mermithid parasitizing chironomid larvae, Poinar (1964) reported the same mermithid, *Orthomermis oedobranchus* Poinar, parasitizing larvae of the sciarid fly, *Lycoriella solani* (Winn.). When he placed infected host larvae in water, the mermithids exited through the neck or anus, thus killing the host. Since mermithids have already been successfully mass reared for mosquito larvae, the potential for manipulating a nematode parasite in a greenhouse, which is a rather well-defined ecosystem, should be ideal as an integral component of a pest management scheme for greenhouse agriculture.

The Tetradonematidae-sciarid fly relationship provides some very interesting life cycle adaptations between the host and parasite. Cobb (1919) described the nematode, *Tetradonema plicans* Cobb, infecting the fly, *Bradysia coprophila* (Lintner), and established the family whose name was based on the four large unicellular organs that occupy the greater part of the body. Hungerford (1919) worked out the bionomics of the host-parasite relationship. Death of an infected host results in the release of large numbers of eggs that hatch within about 24 hours when exposed to water. Infection probably takes place when eggs are eaten by the host larva. Presence in the host's gut probably initiates hatching. The larvae then make their way into the hemocoel, where both sexes become entangled in the Malpighian tubules and tracheal vessels of the host, thus making it difficult to extract them; consequently, the reason for their specific name, *plicans*. Unlike uninfected hosts that show a well-defined, shiny head capsule (Fig. 4A) and white fat body, infected larvae have a reduced head capsule and lack of fat body. Using transmitted light reveals nematodes in the hemocoel (Fig. 4B). The female nematode increases in size, and one to several males, considerably smaller in size, attach to her genital area (Fig. 4C). When inseminated, eggs mature within the female, which has become a large sac of eggs (Fig. 4D; 4F). Eggs (Fig. 4E) are released into the host hemocoel, and upon death and disintegration of the host become present in the soil. Sometimes adult insects become infected and may facilitate spreading the parasite. In these hosts the female fly is castrated; thus the nematode can have a detrimental effect on two life stages of the pest, the larva and adult.

In Britain the sciarid flies *Lycoriella solani* and *Lycoriella auripila* are pests in mushroom houses, and *Bradysia paupera* Tuom. is a pest in greenhouses. Hudson (1974) provided an excellent report of her work that was designed to manipulate the nematode *Tetradonema plicans* as a biological control agent of *B. paupera* and to evaluate its attributes as a biocontrol agent. Her study demonstrated that this nematode parasite satisfied many of the requirements of an efficient biocontrol agent. (1) It was highly pathogenic and specific to the target organism. (2) It could be mass-reared and the eggs stored up to one year at 10°C. (3) The eggs could be applied in water to the treated area. 4) The life cycles of both host and parasite were synchronized. (5) It had a high reproductive potential. In another study of the same pest, but with a different parasite, Poinar (1965) demonstrated that *Tripius sciarae* (Bovien) (Sphaerulariidae) was able to significantly decrease the population of *B. paupera* within four weeks of introduction. Failure of this nematode to have a resistant, free-living stage greatly reduces its effectiveness and potential as a biocontrol agent, since the free-living stage only remains viable for two weeks in the soil.

Another tetradonematid nematode reported to parasitize a sciarid fly is *Aproctema entomophagum* Keilin, a parasite of *Sciara pullala* (Winn.) that inhabits decaying wood. Even though the fly is not economically important, the nematode's life history is interesting enough to briefly mention. Adult

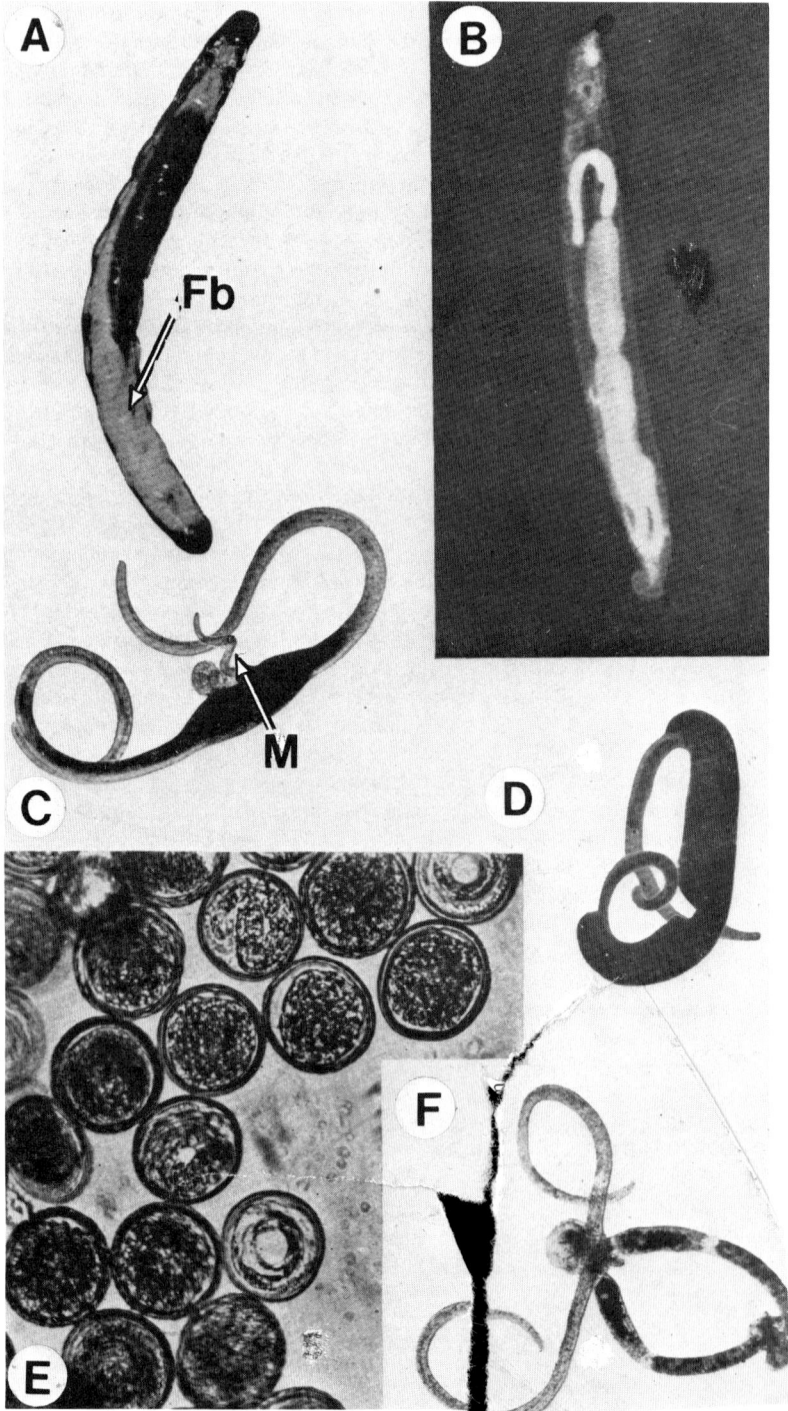

Fig. 4 (A) Uninfected, normal sciarid maggot showing the large white t
body (Fb). (Courtesy of the *J. Paras itol.*) (B) Parasitized sciarid ma ot
showing clearing of hemocoel and loss of fat body. (Courtesy of the *J. a-*

female nematodes usually mature in the host larva and kill it when they exit through the cuticle. In the environment of the dead host, the female nematode deposits about 216 eggs, which hatch after a few days into young larvae bearing a buccal stylet. These larvae have already molted once inside the egg shell and now penetrate the cuticle of the host. Presumably they use the buccal stylet to do this.

Keilin and Robinson (1933) reported "scars" on infected larvae very similar to those reported for face fly larvae infected with *Heterotylenchus autumnalis* Nickle (see Fig. 9). Stoffolano and Streams (1971) reported that these "scars" in the face fly larvae were produced by the host and were due to a host reponse and an attempt to repair the cuticle where the nematode entered the host. Sometimes the parasitic nematodes are carried over into the adult stage and, under these conditions, Keilin and Robinson (1933) reported the female nematodes exiting via the female fly's ovipositor at the time of a "mock-oviposition." This is very similar to that reported for *H. autumnalis*, whereby the normal exit route of male and female nematodes is via the ovipositor (Stoffolano and Nickle, 1966).

The only other report of a nematode parasitizing a sciarid fly was that of Bovien (1944), who originally described the parasite *Tripius sciarae*. Poinar (1965) later provided more information on the bionomics of the host-parasite relationship. The sciarid fly, *Bradysia paupera*, is a pest in the glasshouse industry in England and often the populations reach outbreak levels that have to be regulated by insecticidal treatment. Poinar (1965) conducted preliminary tests using this nematode as a biological control agent and was able to significantly reduce the pest populations to tolerable levels within four weeks following treatment. Little information, however, is available concerning the mode of parasite entrance into the host.

Poinar and Doncaster (1965) showed that the infective-stage, adult female, still ensheathed in the last juvenile cuticle, initially produces an adhesive mass about the head that aids in attachment to the host's cuticle. While penetrating the host's cuticle the nematode molts and sheds its cuticle, which acts as a plug to seal the penetration hole. Such a mechanism would not only facilitate the attachment to the host but would prevent the loss of the host's hemolymph and penetration of pathogenic organisms into the host. Often a late-stage larva or pupa is penetrated and the female nematode does not mature until the fly reaches the adult stage. Under these conditions, the female nematode completely everts the uterus which was never seen everted in parasitized fly larvae (Poinar, 1965). Poinar hypothesized that the uterine cells were absorptive in function and the ". . . dissolved nutrients may be less in the adult flies than in growing larvae. Thus, the remainder of the uterine cells may be extruded into the body cavity of the host as a reaction to diminishing nutrients." Being a sphaerulariid nematode, the infective stage is also the free-living stage.

---

sitol.) (C) Mating between two males (M) and one female *Tetradonema plicans*. (Courtesy of the *J. Parasitol.*) (D) Gravid female nematode parasite removed from host. (Courtesy of the *J. Parasitol.*). (E) Eggs of the nematode in various stages of development. (Courtesy of *J. Parasitol.*) (F) Female *Tetradonema plicans* with reproductive organs dissected out. (Courtesy of *J. Parasitol.*).

The use of nematodes that have a free-living stage for biocontrol agents may be hindered by the lack of resistance of this stage to fluctuating environmental conditions and may preclude disseminating them in the manner used for the neoaplectanid, DD-136. Hudson (1974) and Stoffolano (1973) both alluded to this problem. Stoffolano (1973) proposed that for certain nematodes, dispersal of the parasite in a biological control program may be best achieved by releasing infected adult hosts rather than spraying larval, free-living stages.

The minute black scavenger flies (family Scatopsidae) breed in decaying material and animal excrement. They are not considered economically important since they do not bother humans or animals. Bovien (1932) provided the only known study on nematodes parasitizing this group of flies. From specimens of *Scatopse fuscipes* Meig. sent to him he was able to elucidate the life cycle of *Scatonema wülkeri* (family Sphaerulariidae). He also discovered that, as in most sphaerulariids, only the free-living female nematode possessed a stylet and esophageal glands for penetrating a new host. The fate of the parasite in male hosts appeared to be a dead end.

The gall midges or gall gnats (family Cecidomyiidae) are normally associated with plants. The female fly usually oviposits on the plant, and in response to this, the plant produces a gall inside which the larvae usually develop and feed. Leuckart (1887) reported *Tripius gibbosus* (Leuck.) from the host, *Cecidomyia pini* DeGeer. It is surprising that more research has not been directed at finding nematode parasites of this group, since some members of this family are of considerable economic importance [i.e., the hessian fly, *Phytophaga destructor* (Say)].

B. Suborder Brachycera (Horse and Deer Flies)

1. Superfamily Tabanoidea

Seven families comprise this group, of which only the Tabanidae have been reported to have nematode parasites. In fact, this is really the only family of economic importance in this superfamily.

It is because of the blood feeding habit of the adult female that tabanids are such ideal vectors of parasites of humans and other animals (Krinsky, 1976). At the same time, many species are not vectors, but are nuisance species to people trying to enjoy various types of recreation. Interest in nematodes of this group has been spurred by individuals interested in using nematodes as biocontrol agents against the larvae and also by individuals interested in the adult fly as a vector of nematode parasites.

Poinar (1975) listed 10 examples of mermithids parasitizing tabanids. One of these was *Eurymermis chrysopidis* Mull., four were *Hexamermis* sp., two were *Bathymermis* sp., and three were undetermined at the generic level. Magnarelli and Anderson (1978) dissected 334 *Tabanus nigrovittatus* Macquart larvae and found 10 of them parasitized by a *Gastromermis* sp. (determined by Nickle). Figure 5 shows the tabanid larvae of *T. nigrovittatus* with the mermithid occupying a major portion of the hemocoel. Poinar and Lane (1978) reported the mermithid *Pheromermis myopis* Poinar and Lane parasitizing larvae of the horse fly, *Tabanus punctifer* Osten Sacken. Jenkins (1964b) and Anthony (1977) both reported on the literature of nematode parasites of tabanid larvae and pupae. The aquatic or semiaquatic habitat of most tabanid larvae should make them suitable hosts to nematode parasites. Lack of information on nematode parasites of this group of flies is believed to be due to inadequate sampling, which results from difficulty in finding the larval stages and failure to examine large enough samples for internal parasites.

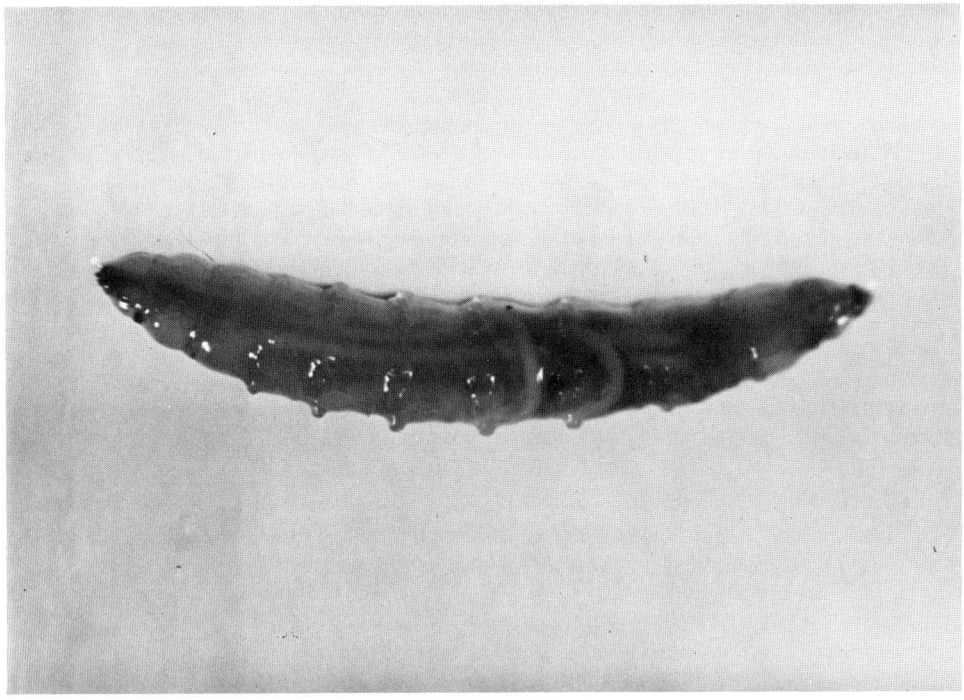

Fig. 5 *Tabanus nigrovittatus* larva parasitized by *Gastromermis* (Courtesy of L. A. Magnarelli and J. F. Anderson.)

In the most recent edition of their book, Harwood and James (1979) listed three filaroid nematode parasites of mammals, *Loa loa* (Guyot), *Dirofilaria roemeri* (Linstow), and *Elaeophora schneideri* W. and D., that are vectored by a tabanid intermediate host. Poinar (1975) listed three other filarial nematodes associated with tabanids: *Dipetalonema perstans*, *Seteria equina* (Abildy), and *Agamofilaria tabanicola* Johnston and Bancroft. Spratt (1970) reported on the synonymy of *Agamofilaria tabanicola* and *Dirofilaria roemeri*. Other than the research of Spratt (1974) on *D. roemeri*, no research has been conducted on either *D. perstans* or *S. equina* since the original report.

L. loa, the African eyeworm, causes human loiasis and is limited in distribution to the rain forest belt of West and Central Africa. Four *Chrysops* spp. have been shown to vector this parasite: *C. silacea* (Austen) and *C. dimidiatus* (van der Wulp) (vectors of the diurnal or human form, *L. loa*) and *C. centurionis* Austen and *C. langi* Macq. (vectors of the nocturnal or simian form, *L. loa papionis*). Since Duke (1954) speculated that all *Chrysops* species could serve as vectors of this parasite, Orihel and Lowrie (1975) attempted to infect *C. atlanticus* Pechuman, a salt marsh tabanid of the Atlantic coastal states, with *L. loa* being maintained in monkeys or baboons. When fed on microfilaremic hosts, *C. atlanticus* ingested parasite larvae, which successfully developed to third stage invasive larvae. These larvae were fully infective to experimentally injected monkeys, producing patent infections five months later.

Dirofilaria roemeri (Linstow), a parasite of the subcutaneous and intermuscular connective tissues of certain members of the kangaroo family, has been shown to complete its life cycle in and is vectored by *Dasybasis hebes* (Walker), a tabanid in southeastern Queensland, Australia (Spratt, 1974). The development and morphogenesis of this filarial nematode is similar to that reported for other species of the genus; however, all other reported filarial species develop in and are vectored by mosquitoes.

The tabanid, *Hybomitra laticornis* (Hine), was shown to be the most important vector of *Elaeophora schneideri* in the Gila National Forest of New Mexico (Clark and Hibler, 1973). Additional studies demonstrated that this filarial nematode is the causative agent of elaeophorosis in mule deer, domestic sheep, elk, and moose. The mule deer is considered the normal definitive host and rarely shows the signs of this disease, but domestic sheep, elk, and moose often show signs of blindness, deafness, and circling behavior (Hibner and Metzger, 1974). The role of tabanids as intermediate hosts of filarial nematodes has barely been explored, and it is believed that further studies involving filarial parasites of domestic and wild animals will yield a high return.

2. Superfamily Asiloidea

The life habits of the majority of species belonging to the eight families found in this group are such that the larvae feed on decaying debris in the soil, decaying wood, or fungi, or are internal parasites of spiders and insects. The adults are generally predaceous on other insects. The lack of reports on nematode associations with this group is either due to the life style of the insects being incompatible with a nematode group and/or failure of investigators to take a serious look at these associations. The latter is due mainly to the fact that this group is not economically important.

3. Superfamily Empidoidea

Members of this group are of no economic importance; consequently, failure to find nematodes associated with this group are mainly due to lack of interested investigators.

C. Suborder Cyclorrhapha, Division Aschiza (Phorids)

1. Superfamily Lonchopteroidea

Only one family, Lonchopteridae, is found in this group. These are yellowish or brownish flies, the larvae of which live in decaying vegetation. They are of no economic importance and thus have no reported nematode associations.

2. Superfamily Phoroidea

Of the two families (Phoridae and Platypezidae) making up this group, nematodes have only been reported from the former. The larval habitat is decaying animal or vegetable matter. One species in particular, *Megaselia halterata* Wood, is a major pest of mushrooms in England. As with the work on the sciarid pests of mushrooms, nematodes are being used as effective biocontrol agents in this particular type of agricultural system. The phorid fly can reduce the yield of mushrooms, is irritating to pickers, and can transmit various pathogens of the mushrooms. The problem of toxic residues in the mushroom crop is another reason leading to investigations designed to seek alternate methods of control of this pest. The nematode parasite, *Howardula husseyi* Richardson, Hesling, and Riding, was discovered by Hussey (1959) while dissecting flies. Since then considerable information has been gathered on this parasite.

Riding and Hague (1974) demonstrated that under laboratory conditions the fly population was eliminated in three to five generations by the nematode parasite. They concluded that the parasite reduces fecundity in the female, and when multiple infections occur, causes sterility in the female fly. They also noted that the parasite is host-density dependent and raised the question of the viability of the free-living stages over long periods of storage. Richardson and Chanter (1979) addressed this last question and proposed an alternative method of disseminating the parasite rather than using the larval stage of the nematode. They suggested rearing large numbers of parasitized adults and releasing these into the greenhouses so that they could naturally disseminate the parasites. Results showed that such a program would be most effective if the parasitized flies were released in two batches, with one release occurring in the middle of each week of the spawn-run. In an earlier study, Richardson and Hesling (1977) examined the effect of parasitism on the overwintering population of the mushroom phorid. Virtually nothing was known concerning the overwintering habits of the phorid, and since spring populations had a lower incidence of parasitism, Richardson and Hesling (1977) proposed that if winter decline in parasitism could be averted, the nematode would be more effective in controlling the fly population. Their study showed that winter decline in parasitism did not occur, but rather that parasitism enhanced fly survival through the winter.

Since sphaerulariid nematodes rarely kill their host, many investigators have questioned their potential as biocontrol agents. The research on *Howardula husseyi*, however, in controlling the phorid fly of mushrooms, clearly demonstrates that they can be effectively manipulated as biocontrol agents.

3. *Superfamily Syrphoidea*

No nematodes have been reported from the small group of flies known as the Pipunculidae. Lack of reported nematode associations may be due to the larvae being parasitic on other insects. The Syrphidae, although not economically important, is a group interesting to entomologists because syrphids are predators of aphids. Because of this, it is important to know the parasites of the predator in order to rear or efficiently utilize the biocontrol agent. Laumond and Lyon (1971) provided the only known work on the nematode parasite, *Syrphonema intestinalis* L. and L., that reproduces in the digestive tract of the adult fly.

The conopids (Conopidae) are medium-sized, brownish flies, and are usually found on flowers. The larvae are endoparasites of adult bumblebees and wasps. Their parasitic association and lack of economic importance is probably the main reason for the absence of reports on nematode associations.

D. Suborder Cyclorrhapha, Division Schizophora (Fruit Flies, Black Scavenger Flies, Marsh, Frit Flies, Leaf Miner and Galling Flies, House-flies, Onion and Cabbage Maggots

With the movement of the so-called higher Diptera into semiaquatic, and in general, more terrestrial habitats, there is seen an accompanying shift in the parasitization strategies of the associated nematodes. Due to the generally greater mobility of the adult stage of the insect and their frequent predilection for temporary and patchily distributed larval habitats, such as manure, decaying fruit, particular stages of annual plants, or carrion, mermithid-fly associations are relatively rare in this host group. Rather, one sees an in-

creasing prevalence of parasites whose life history and physiology are more
tightly bound to that of the various host stages. Thus, the tactics of host
castration and parasite-induced "mock oviposition," seen in some members of
the Sphaerulariidae, Tetradonematidae and Allantonematidae, are well represented in this insect group.

As is generally the case, more information is available on those families
that include members of agricultural or medical-veterinary importance. Of
the 59 families of North American Schizophora currently recognized (Stone et
al., 1965), only 11 have, to date, been reported parasitized by nematodes
(Poinar, 1975, 1979). Not surprisingly, these families include most of the
economically important pests of the group, such as the Tephritidae, Drosophilidae, Chloropidae, Agromyzidae, Anthomyiidae, Muscidae, Hippoboscidae,
Calliphoridae, and Sarcophagidae.

1. Superfamily Micropezoidea

The Micropezoidea is comprised of two small families of flies. The Micropezi-
dae, or stilt-legged flies, are saprovores which, while fairly abundant in the
tropics, are uncommon in North America. The cactus flies of the family Nerii-
dae consist of two North American species found in the southwestern United
States, which breed in decaying cacti. No nematodes have been reported
from either of these families.

2. Superfamily Nothyboidea

This superfamily consists of three families, none of which have been found
parasitized by nematodes. The diopsids, or stalk-eyed flies, breed in sphag-
num bogs and are taxonomically unique in the positioning of the eyes at the
ends of long stalks. Rust flies of the family Psilidae live in plant galls and
roots and occasionally cause damage to carrots and related crops. The Tany-
pezidae are uncommon flies, occasionally found in moist woods. Nothing is
known of their biology.

3. Superfamily Tephritoidea

The tephritoids consist of five families of flies characterized by the frequent
presence of colored markings on the wings. Richardiids are uncommon in
North America and breed in decaying plant material. Most of the picture-
winged flies of the families Otitidae and Platystomatidae appear to be plant
feeding as larvae; however, little is known about them. Pyrgotid flies parasitize adult June beetles as larvae. None of the above families have been
found parasitized by nematodes.

The Tephritidae, or fruit flies, is the largest family in this group. Larval fruit flies feed primarily on plants, and many are serious pests of fruit
crops, particularly apples, blueberries, and citrus. To date, only a single
member of this family has been reported parasitized by nematodes. The apple
maggot, Rhagoletis pomonella (Walsh), is a very destructive pest of apples
in many areas of North America due to the tunneling of fly larvae through
the fruit, frequently rendering the product unsuitable for retail sale. In a
survey of pathogens of apple maggot larvae and pupae collected from a Massachusetts orchard, Poinar et al. (1977) found two fly pupae parasitized by an
unidentified species of Neoaplectana. No specimens were recovered from
adults collected from the same area, however, the number of flies examined
was small (65). Poinar earlier (1968) had found N. carpocapsae in apple
maggot pupae (Poinar, 1979). Although no further work has been conducted

on this association, this nematode has been shown to possess considerable potential as a biocontrol agent of numerous other insect pests, as previously indicated.

4. Superfamily Sciomyzoidea

Of the six families included in the Sciomyzoidea, only two, the Sepsidae and Sciomyzidae, have been reported as hosts of nematodes. Larvae of the Helcomyzidae (seabeach flies) and Coelopidae (seaweed flies) live in decaying seaweed and are mostly restricted to the northern Pacific coastline. Ropalomerid flies are found, as adults, around palm trees, while dryomyzids are generally collected in moist woods. The larval habits of these latter families are unknown.

The Sepsidae, or black scavenger flies, breed in animal excrement and various decaying materials. Although not commonly regarded as serious pests, they often occur in large numbers in the vicinity of the larval habitat. Goodey (1941) studied a tetradonematid parasite of *Sepsis cynpsea* L., *Mermithonema entomophilum* Goodey. Since the description was based on material from a single host collected at St. Albans, much of the life cycle remains conjectural. The mode of initial entry is unclear. Either newly hatched nematode larvae penetrate the host's larval cuticle, as is the case with *Aproctonema entomophagum*, or embryonated eggs are ingested by the larva, as with *Tetradonema plicans*, both of which are closely related to this species. The parasites complete development free in the host hemocoel and persist into the adult stage of the fly where, at least for female hosts, castration results (Goodey, 1941). It appears likely that the cycle is continued via host deposition of parasites into the larva medium.

The Sciomyzidae, or marsh flies, are of interest not because of their own pest status but rather because they are natural enemies of aquatic snails, many of which serve as intermediate hosts of blood fluke parasites of the genus *Shistosoma*. The larvae are aquatic, primarily in standing water such as marshes and bogs, where they burrow into snails or snail egg masses to feed (Merrit and Cummins, 1978). Knutson (in Poinar, 1975) has reported finding unknown mermithids parasitizing *Dictya* sp. This potentially confounding factor of natural enemies of beneficial insects is one that must be carefully assessed prior to large-scale production and introduction of biological control agents.

5. Superfamily Lauxanoidea

The Lauxanoidea consists of three families, none of which include any reported nematode hosts. Lauxaniid flies breed, as larvae, in decaying plant material, and are quite common as adults in moist, shady areas. Larvae of aphid flies (Chamaemyiidae) are predators of mealy bugs, scale insects, and aphids. Periscelids are rare flies that are thought to breed in fermenting sap around tree wounds.

6. Superfamily Pallopteroidea

No nematodes have been reported from any of the five small families of the Pallopteroidea. Skipper flies of the Piophilidae are mostly scavengers as larvae, although *Piophila casei* (L.) is a serious pest of cheese and preserved meat. Lonchaeid flies invade, as larvae, tissues of wounded or diseased trees. Nearly nothing is known of the habits of the remaining members of this superfamily, the Thyreophoridae, Neottiophilidae, and Pallopteridae.

7. *Superfamily Milichioidea*

The small dung flies, or Sphaeroceridae, are the most abundant of the five families in this group. They breed in excrement and refuse and can cause considerable annoyance in enclosed poultry houses. The single species of the Braulidae, or bee lice, are wingless as adults and appear to feed on pollen and nectar from the mouthparts of bees. Tethinids are coastal flies found in beach grass, seaweed, and salt marshes. The larval habits of the Milichiidae are quite varied, ranging from blood feeding on birds to saprophagy in decaying plant and animal material. Larvae of beach flies (Canaceidae) feed on algae along the southeastern and California coast. No nematodes have been reported from any of these families.

8. *Superfamily Drosophiloidea*

Of the five families of the Drosophiloidea, only members of the Drosophilidae have been found parasitized by nematodes. The Ephydridae, or shore flies, is a large family of fairly common flies found near marshes, ponds, streams, and the seashore, Nearly nothing is known of the habits of the uncommon flies of the families Curtonotidae, Diastatidae, and Camillidae.

The small fruit flies, or pomace flies, of the family Drosophilidae constitute a large group of flies worldwide, nearly all of which feed as larvae on yeasts occurring in decaying fruit and vegetation. They are common pests in households and any place where fruit may be present. The family is parasitized principally by allantonematid nematodes; however, in Asia, *Phortica* (*Amiota*) *variegata* Fall. vectors the oriental eyeworm *Thelazia callipaeda* Railliet and Henry (Kozlov, in Skrjabin et al., 1967). Eyeworms will be discussed in further detail under the Muscidae.

Welch (1959) described in detail the taxonomy and life history of two drosophilophilic allantonematids, *Howardula aoronymphium* Welch and *Parasitylenchus diplogenus* Welch. In the case of the former, fertilized female nematodes pierce the larval cuticle of the host by means of a stylet, mature, and larviposit approximately 500 larvae in the host's hemocoel. These larvae penetrate the peritoneal sheath surrounding the ovary, resulting in nearly complete destruction of the ovary, and enter the oviducts, from which they either escape or are actively discharged by the female host into a new fruit. During the brief free-living stage that follows the nematodes mate, the males die, and fertilized females penetrate a new host to complete the cycle. An anomalous feature of the development of this parasite is that females mate in either the last or penultimate larval stage, rather than as adults. *H. aoronymphium* is a parasite of *Drosophila kuntzei* Duda and *D. phalerata* Mg.

Parasitylenchus diplogenus parasitizes *D. obscura* Fall., *D. silvestris* Basden *D. subobscura* Coll., and has a life cycle consisting of two sexual generations. An inseminated female enters the larval host's hemocoel via penetration and develops slowly into a large, saclike form that deposits eggs into the hemolymph. Larvae from these eggs develop rapidly to the adult stage, mate, and again oviposit. These larvae are deposited into the host larval habitat after passing through the ovaries, mature, mate, and resume the cycle. This two-generation strategy results in great proliferation of the parasites; a single female of the first generation produces 10-20 second-generation larvae, each of which can produce 100-200 larvae.

In addition to the strictly mechanical damage of the ovaries caused by the exiting larvae, Welch also noted that, prior to parasite exit, ovarian development was depressed and that the fat body of parasitized individuals was smaller in size than in uninfected flies, presumably due to host-parasite competition for nutrients.

9. Superfamily Chloropoidea

This superfamily is represented by a single family, the Chloropidae. Chloropid larvae tunnel through the stems of grasses and include several serious pests of cereal crops. One such pest, *Oscinella frit* (L.), which feeds on the stems of oat plants, has been found parasitized by the allantonematid *Howardula oscinellae* (Goodey) (Goodey, 1930). In England, the frit fly has two generations per season. Overwintering larvae develop into spring generation adults that oviposit on small oat seedlings. Larvae tunnel through the shoots and pupate on the plants. Emerging flies of this second generation oviposit on the older oat panicles; the larvae hatch, and soon begin to feed on the developing grains. The larvae pupate, emerge, and oviposit on wild grasses, where the larvae hatch, feed for a time, and overwinter in the plants.

The life history of *H. oscinellae* is tightly bound to that of its host. Sexually mature female parasites overwinter in the host and, in the spring, larviposit into the hemocoel. Emerging parasitized hosts mate and fly to oat seedlings where male and female nematode larvae escape, via the anus, into the plant tissues. Larvae develop to the adult stage, mate, males die, and the females penetrate a new host larva. This pattern continues for the remaining two generations of the host, the parasites each time being carried by adult stages of the fly to the new host plant, thus ensuring continuity of the cycle of parasitism.

As noted above, Goodey found that the mode of parasite exit from the host was via the alimentary tract and that therefore the mechanical injury to the ovaries seen in the above-mentioned cases does not occur. In female flies, however, ovarian development is arrested at an early stage, again probably due to nutrient competition. In addition, parasitized males were observed to have greatly reduced testes and "almost complete suppression" of the accessory reproductive glands (Goodey, 1930).

10. Unplaced Acalyptrate Families

None of the 13 unplaced acalyptrate families of the Schizophora, except for the Agromyzidae, have been found parasitized by nematodes. The Odiniidae, Acartophthalmidae, Rhinotoridae, Aulacigastridae, Trixoscelididae, and Asteiidae are small families of rare and poorly known flies. Clusiids are small flies whose larvae feed on decaying wood. Larvae of the Heleomyzidae feed on rotting animal and plant material and on fungi; immature opomyzids feed in stems of grasses. Chyromyid larvae have been found in rotting wood and birds' nests. The Cryptochetidae parasitize scale insects as larvae and have been used as biological control agents against certain scale pests.

The Agromyzidae is a family of leaf miner and gall-forming flies that includes a number of economically important pests of a variety of crops. To date, three members of the family, *Fergusonina carteri* Tonn. (Currie, 1937), *F. tillyardi* Tonn. (Fisher and Nickle, 1968), and *F. lockharti* Tonn. (Fisher and Nickle, 1968), have been reported parasitized by nematodes. The former two result in the production of flower galls on Australian *Eucalyptus* plants and are hosts of *Fergusobia curriei* (John.); the latter results in the formation of shoot tip galls on the same plant and is parasitized by an undetermined species of *Fergusobia*.

F. curriei exhibits alternation of sexual and parthenogenic generations. Sexually mature female parasites live free in the hemocoel of female flies, where they release eggs into the hemolymph. Newly hatched larvae migrate to the oviduct and are passed, along with healthy host eggs, into plant tissues. The larvae develop rapidly into parthenogenic females within the de-

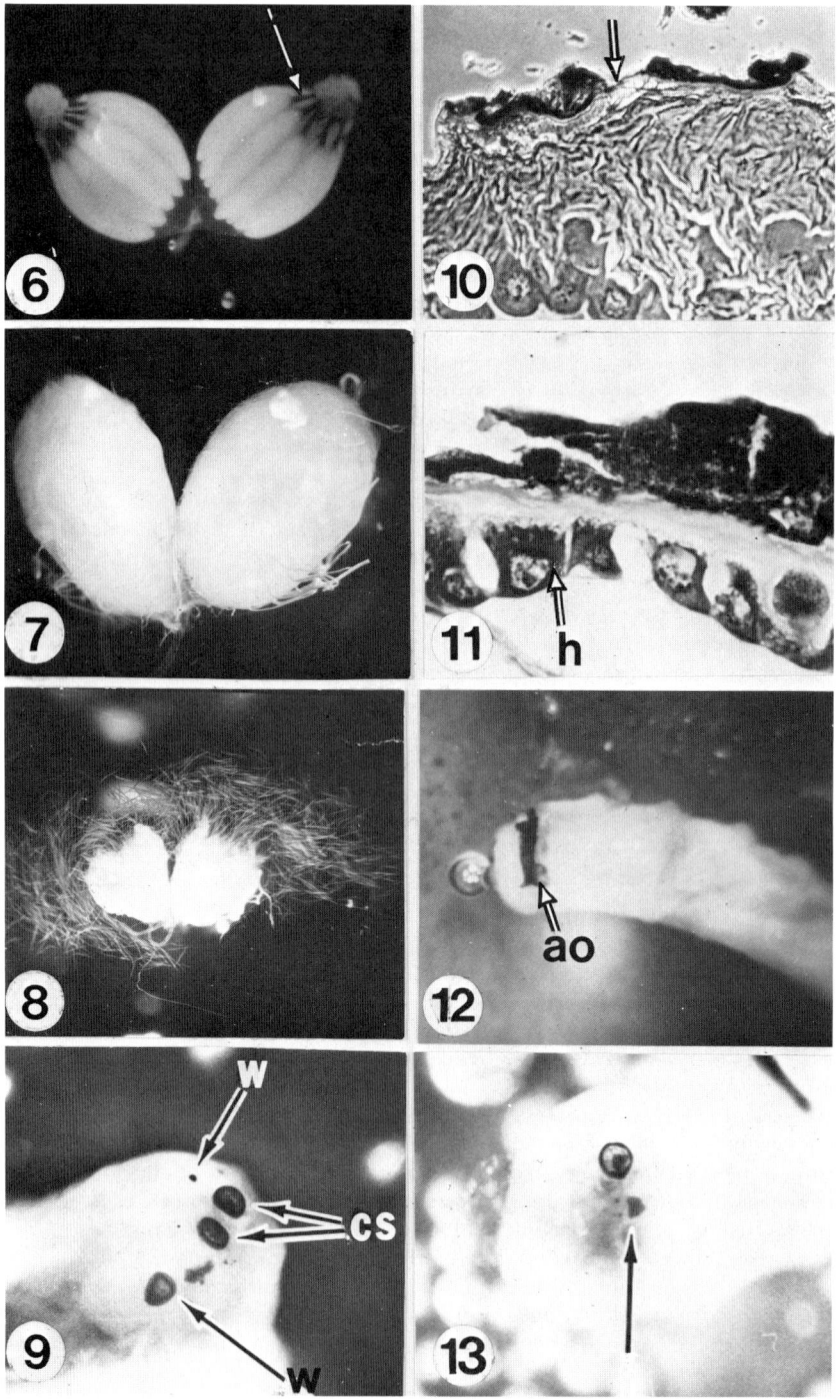

Fig. 6 Ovaries of noninfected face fly, *M. autumnalis*, containing several individual eggs each terminated by a dark respiratory mast (see arrow).

veloping galls and produce exclusively male larvae until host larvae reach the third instar. At this time both females and males are produced, which mate, followed by death of the males. Just prior to host pupation, fertilized female nematodes penetrate the fly larvae, develop to sexual maturity, and, following host development to the adult stage, resume the cycle. No parasitized male hosts have been found. However, it is not known whether this is the result of host discrimination by the parasites or of their inability to develop normally in male flies (Fisher and Nickle, 1968).

11. Superfamily Muscoidea

The calyptrate muscoids comprise a very large group of flies, many of which are serious livestock pests. For example, the genus *Musca* includes the house fly (*M. domestica* L.), face fly (*M. autumnalis* DeGeer), bazarre fly (*M. sorbens* Wiedemann), Australian bush fly (*M. vetustissima* Walker), and a host of other zoophilus, filth-breeding flies. Also among the muscids are the stable fly [*Stomoxys calcitrans* (L.)], the horn fly [*Haematobia irritans* (L.)], and the notorious tsetse fly (*Glossina* spp.), which vectors of African trypanosomes.

The anthomyiid genus *Hylemya* includes several pests of crops, including the onion maggot [*H. antiqua* (Meigen)], the seedcorn maggot [*H. platura* (Meigen)], which also attacks newly set tobacco plants, and the cabbage maggot [*H. brassicae* (Bouche)]. Filth-breeding anthomyiids include the lesser house fly [*Fannia canicularis* (L.)], some *Hylemya* spp., *Paregle* spp., and the yellow dung fly [*Scatophaga stercoraria* (L.)].

Because of the pest status of many members of these families, and the close proximity to humans and their animals displayed particularly by many muscids, a considerable amount of information is available concerning their nematode parasites. In addition to the nematodes for which the fly is the definitive host, several parasites of livestock utilize zoophilous flies as intermediate hosts in their development. These will be discussed in a separate section.

Fig. 7 Ovaries of an infected face fly containing thousands of male and female gametogenetic larvae instead of eggs.

Fig. 8 Same ovaries as in Fig. 7, but the ovary wall was ruptured to release the nematode larvae.

Fig. 9 House fly larva showing melanotic spots or cuticular wounds (W) on the caudal, dorsal end of the maggot just above and to the left of the caudal spiracles (CS). These areas represent spots where nematodes penetrated the cuticle of the host.

Fig. 10 Cross section through a melanotic spot or wound on the surface of a larval house fly showing the dark melanin in the epicuticle and the highly distorted endocuticle. Arrow shows the direction taken by nematode in penetrating the host cuticle.

Fig. 11 Longitudinal section through the anal organ of an infected fly larva with the anal organ darkened. Notice the dark hypodermal cells (h).

Fig. 12 Lateral view of a parasitized first-stage larva of face fly showing a host response taking the form of darkening of the anal organ (ao).

Fig. 13 Posterior end of an infected house fly larva showing the absence of the caudal spiracle on the right which was due to nematode penetration. All that is left of this spiracle is a melanotic spot (arrow).

E. Calyptrate Musoids as Definitive Hosts of Nematodes

1. Heterotylenchus

The genus *Heterotylenchus* Bovien is a member of the Sphaerulariidae, which includes a number of host-castrating species parasitizing flies and beetles. Although Slobodyanyuk (1975b) has proposed splitting the genus into *Heterotylenchus* and *Paraiotonchium* n. gen., this revision has not yet been generally accepted. For this reason, we shall retain the former generic name for all members of the group except for those that have been described by the above author as new species of *Paraiotonchium*.

Although Bovien (1937) described the type species, *Heterotylenchus aberrans* Bovien, from the onion maggot, *Hylemya antiqua*, considerably more biological information is available on *Heterotylenchus autumnalis*, a parasite of face fly in North America and Europe. The life cycle of this species described by Nickle (1967) is typical of the genus as a whole.

H. autumnalis displays an alternation of gametogenetic and parthenogenetic generations. Mated female nematodes in bovine manure penetrate the cuticle of developing face fly (*Musca autumnalis*) larvae and enter the hemocoel. Once within the hemocoel, the female worms develop to maturity and begin releasing eggs, which hatch into cigar-shaped parthenogenetic females. These females in turn deposit eggs into the hemolymph, which hatch into thousands of male and female gametogenetic larvae, and develop until nearly mature. By this time, the host has emerged as an adult fly. The larvae then enter the ovaries of female hosts, distending the ovaries to the point where the host perceives them to contain a full egg compliment (Figs. 6-8). Such parasitized hosts fly to newly deposited manure and undergo mock oviposition, depositing hundreds of larvae into the host's larval habitat. The parasites mate, the males die, and fertilized female nematodes penetrate a new host (Figs. 9 to 13), resuming the cycle (Stoffolano and Nickle, 1966). Male flies are parasitized as well; however, this appears to be a "dead end" for the parasites.

Stoffolano (1967) studied the synchronization between host and parasite life cycles in this association. Face flies overwinter as nonreproductive adults, with large stores of energy reserves stored in the fat body for general maintenance rather than ovarian development. At least in North America, it is known that *H. autumnalis* also halts reproductive processes during host diapause. This host-parasite synchrony probably results from the scarcity of available hemolymph protein in overwintering hosts. Kaya and Moon (1980) have shown that nematodes in protein-deprived face flies fail to produce gametogenetic males and females as rapidly as, or in equivalent numbers to, hosts provided with a normal diet.

It is as yet unclear what effects *H. autumnalis* has on the host other than castration of the females. Nappi (1973) reported that although infected male flies were able to mate, less than 50% of females inseminated by such males oviposited. In females, Kaya et al. (1979) found a shift in the feeding behavior such that parasitized flies, rather than alternating between feeding on lacrimal secretions of cattle and manure, became "terminal dung feeders." Robinson and Combs (1976) studied the effect of parasitism on face fly longevity and concluded that this parameter of fly fitness was unaffected by *Heterotylenchus* infection. Stoffolano and Streams (1971) examined the effects the parasite had on the host, and vice versa. If nematodes penetrated the anal organ of the fly larva, the entire organ would darken (Fig. 12); females penetrating the spiracular area often resulted in destruction of the undifferentiated cells producing the spiracle at the next molt with consequent loss of the spiracle (Fig. 13). Insects have evolved specific responses to attack by nematodes (Stoffolano and Streams, 1971). The major response is a blood cell or hemocytic response whereby the nematode is encapsulated and melanized (Figs. 14 to 19). It is extremely important in any biocontrol program to evaluate the potential of a host to respond to the parasite since this aspect of a host-parasite relationship may result in the parasite becoming ineffective.

Perhaps more important is the question of the extent to which this parasite regulates natural populations of face flies. Thomas and Puttler (1970) found low rates of fly infestation of cattle (2.9 flies per head) associated with high rates of parasitism (40%) in Missouri; however, in a later study, Thomas et al. (1972) concluded that parasitism was a host density-independent factor. Robinson and Combs (1976) found that fly pressure remained fairly constant, at about 10 flies per head, from mid-June through mid-December, but parasitism in the flies ranged from 4 to 37% during the same period. Similarly, following a study in northern California, Kaya and Moon (1978) suggested that *Heterotylenchus* ". . . may not be regulating face fly population density." Treece and Miller (1968) have pointed out that infected flies are capable of completing one gonotropic cycle prior to the invasion of the ovaries by gametogenic larvae. Thus, it appears that local climatic and other factors may play as important a role as parasitism on fly density. Nonetheless, as Nickle (1974) implied, the overall effect of *H. autumnalis* may be to retain fly populations at much lower levels than they would be otherwise.

Although in the United States *H. autumnalis* appears to be host specific to face fly (Jones and Perdue, 1967; Stoffolano, 1973), this may not be true elsewhere. Vilagiova (1968) found, in Czechoslovakia, this nematode parasitizing *Musca larvipara* Portchinsky, *M. tempestiva* Fallen, *Morellia simplex* Loew, and *Hydrotaea meterorica* Linne. It is possible that the forms found in the above hosts represent new species. For example, Nicholas and Hughes (1970) found a parasite of the Australian bush fly, *Musca vetustissima*, which closely resembled *H. autumnalis*, but that has since been described as a new species, *Paraiontonchium nicholasi* Slobodyanyuk (Slobodyanyuk, 1975b).

A new and as yet undescribed species of *Heterotylenchus* has been found in house fly (*M. domestica*) in Brazil (Coler et al., 1980). Unidentified *Heterotylenchus* spp. have also been recovered from *Morellia hortorum* (Fallen) (Stoffolano, 1969), *Musca conducens* Walk. (Nicholas, personal communication), and *M. sorbens* (Poinar, personal communcation). Slobodyanyuk (1975a, 1976) has described new species from *Morellia simplex* and *Musca osiris* Wied.

One of the strategies used in biocontrol programs is the introduction of natural enemies from one location to those where they are absent. A confounding factor for such introductions can be strain differences, resulting in host-parasite incompatibility. For example, *Heterotylenchus aberrans* is a parasite of the onion maggot native to Denmark; attempts to introduce this nematode into North American hosts were unsuccessful, possibly due to host strain differences between Danish and North American stock (Stoffolano, 1969).

2. *Neoaplectana*

Two pilot experiments have been conducted assessing the biocontrol potential of the steinernematid *Neoaplectana carpocapsae* (strain DD-136, see above) against *Hylemya* spp. Welch and Briand (1961) examined the effectiveness of this nematode against *H. brassicae*, the cabbage maggot, both in the laboratory and in the field in Ontario. In the laboratory, using thin slices of rutabagas with maggots, nematode treatment resulted in 60-70% larval mortality. In the field, where comparisons of cabbage damage in control and nematode-treated plots were made, a 30-50% reduction in damage in the treated plots

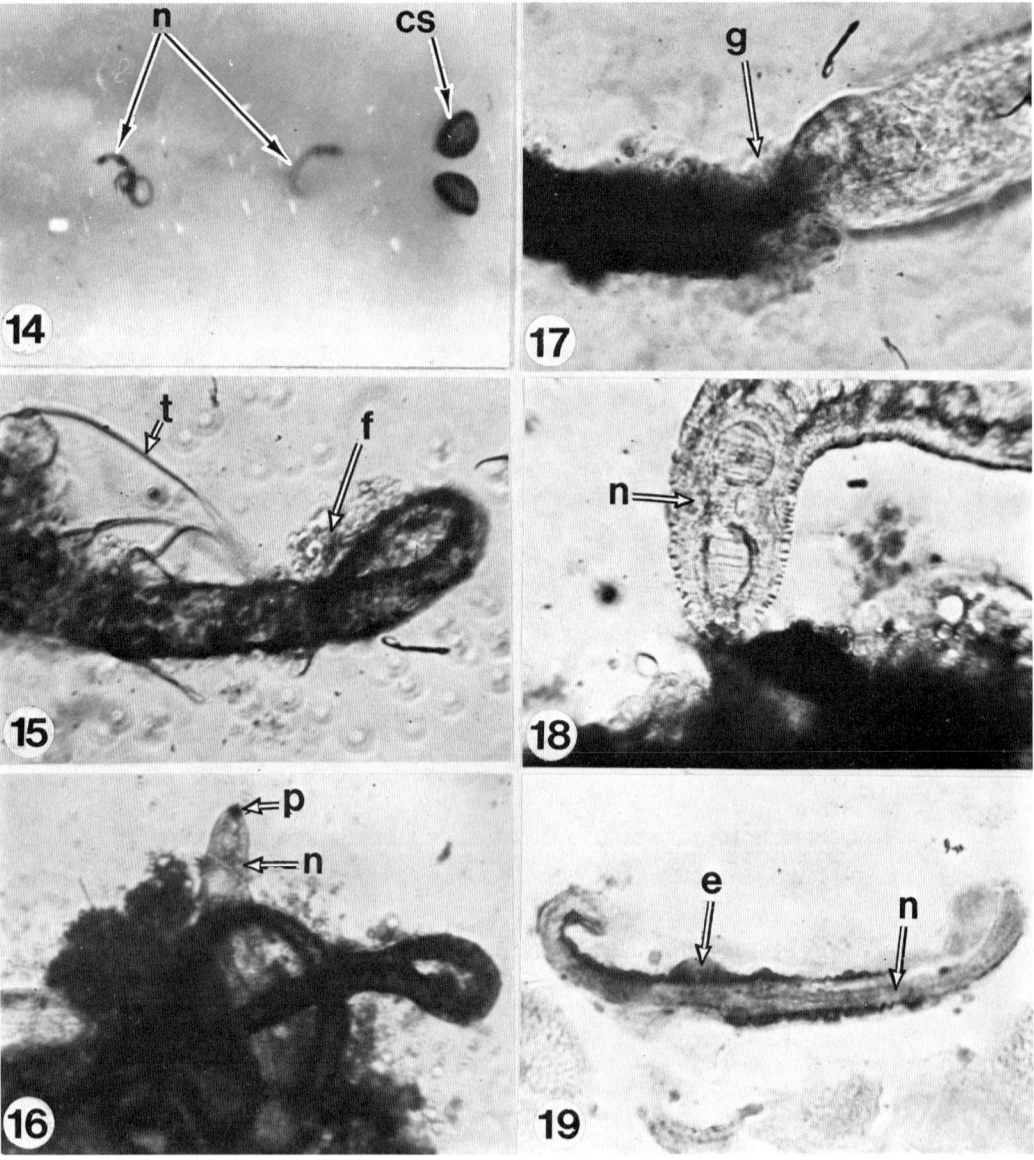

Fig. 14 Encapsulated, small gametogenetic female nematodes (n) just under the cuticle of an infected three-day-old house fly larva. The two black structures at the right are the caudal spiracles (cs).

Fig. 15 Posterior portion of an encapsulated, small gametogenetic female nematode removed from a house fly adult and showing the globules of fat (f) material stuck to the cuticle and the tracheae (t) going to the nematode surface.

Fig. 16 A mass of encapsulated, small gametogenetic female nematodes (n). The nematode at the top has a plug (p) of melanin covering the mouth.

Fig. 17 An encapsulated, girdled, small gametogenetic female nematode. Note the constriction (g) of the nematode at the junction of the encapsulated and unencapsulated portions.

was observed. No significant difference in effectiveness between serial and single-dose application regimes was observed. This fact, in addition to the observation of live infective-nematode larvae in the soil around the roots of the plants at two weeks postapplication, suggested that the nematodes are capable of persisting for some time in the field (Welch and Briand, 1961).

Cheng and Bucher (1972), also in Ontario, compared the effectiveness of this nematode with diazinon to suppress two tobacco pests, *Hylemya platura* and *H. florilega* (Zetterstedt). Both nematode and chemical treatments were ineffective in preventing stem surface damage relative to controls; however, both treatments were equally superior to controls at suppressing tunneling damage, which is more difficult for plants to recover from. The authors concluded that the nematode has ". . . potential value as a biological control agent against *Hylemya* spp. on tobacco when applied in the planting water." Although inexpensive to mass produce, the DD-136 nematode has the drawback of being rather inconvenient to apply as compared with pesticides (Cheng and Bucher, 1972).

In the laboratory, experimental infections with *N. carpocapsae* have been successful against the following: tsetse fly (*Glossina morsitans* Westwood), house fly (*Musca domestica*), and *Hylemya floralis* (Fallén) (in Poinar, 1979).

3. Mermithids and Others

As mentioned earlier, there is a relative dearth of reports of mermithids parasitizing the so-called higher Diptera. Rubtsov (1976) recently described three new mermithid species from zoophilous flies in Mongolia: *Onchiomermis haematobiae* Rubtsov from the horn fly *Haematobia irritans*, *O. hydrotaeae* Rubtzov from *Hydrotaea* sp., and *Dipteromermis lyperosiae* Rubtsov from *Lyperosia titillans* (Bezzi).

There have been several scattered reports of mermithids found in *Glossina* spp. in Africa. Foster (1963) has reviewed the earlier literature concerning findings of "Mermis-type" nematodes in *G. palpalis* (Robineau Desvoidy) and *G. morsitans* Westwood. He also reports having found 15 cases of parasitism of *G. palpalis* by mermithids out of 4001 flies examined, all but one of which were single-worm infections. None of the *G. fusca* Walker, *G. pallicera* Bigot, or *G. nigrofusca* Newstead that were also inspected were infected. In all cases no positive identification of the parasites could be made due to the absence of adult nematodes in any flies (Foster, 1963). Moloo (1972) found 1 of 5000 examined *G. brevipalpis* Newstead parasitized by unidentified mermithids in Uganda, but no *G. fuscipes* Newstead or *G. pallidipes* Austen were infected. He pointed out that all records indicated wet season infection of flies and that, given the very low parasitism rates observed in the field that such infections are ". . . almost certainly accidental."

El-Kifl et al. (1970), in a study of insect-nematode associations in Egypt, found *Musca domestica vicina* Macquart parasitized by *Alaimus primitivus* Rubtsov and the spirurid *Physocephalus sexalatus* Molin. Poinar and Bai (1979)

Fig. 18 Small, gametogenetic female nematode stuck by a cephalic cap to a mass of encapsulated material. Note nematode (n) is not encapsulated except at mouth.

Fig. 19 Longitudinal section of an encapsulated gametogenetic female nematode (n) removed from an infected house fly larva and showing the cellular capsule (e) surrounding the nematode.

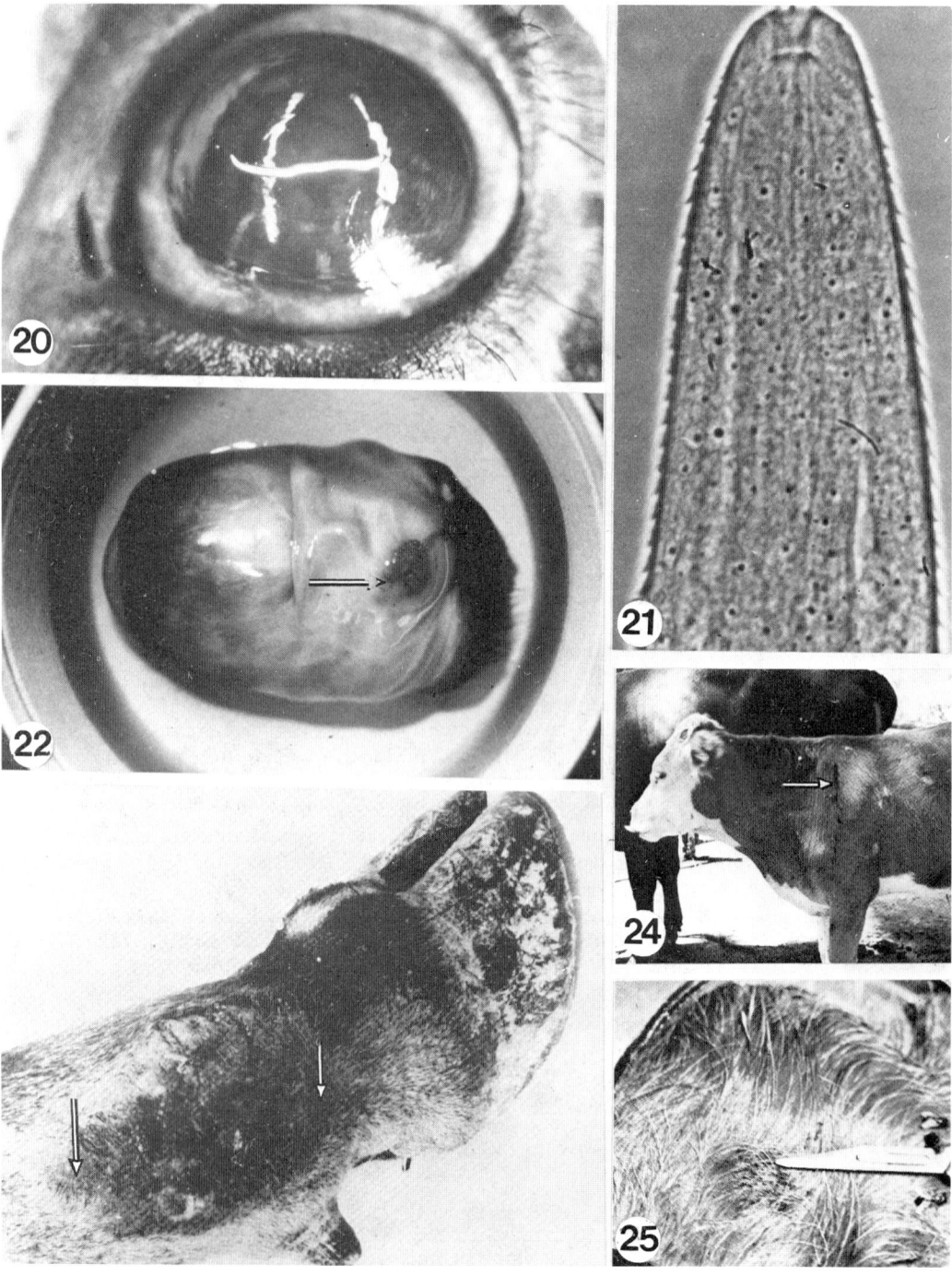

Fig. 20 Mature *Thelazia gulosa* parasite laid across bovine eye to give an indication of the size of these nematodes.
Fig. 21 Invasive third-stage larva of *T. gulosa* dissected from an infected
face fly. Note the serrate annulations surrounding the parasite.

have recently described a new species of Panagrolaimidae, *Panagrolaimus migo-phrlus* Poinar and Bai, from *M. domestica* larvae in India. This was the first report of a member of this otherwise freeliving genus, which occasionally is found in phoretic associations with insects, causing pathogenicity in flies.

F. Calyptrate Muscoids as Intermediate Hosts of Nematodes

1. *Thelazia*

The nematode genus *Thelazia* comprises a cosmopolitan group of spirurid parasites of the eyes of mammals, nearly all of which are vectored by filth flies of the genus *Musca* (Stoffolano, 1970). Adult parasites live under the nictitating membrane or in the conjunctival sac or lacrimal ducts of the host, where they deposit first-stage larvae into the lacrimal secretions. Adult nematodes may be relatively large in comparison to the eye of the host (Fig. 20). Flies that come to the host to feed on these secretions ingest nematode larvae that then penetrate the midgut (Geden and Stoffolano, 1982) and develop within capsules attached to either fat body or to the inner lining of the abdominal body wall. After reaching the third stage, invasive larvae rupture the capsules and migrate anteriorly to the proboscis of the vector, where they await the next visit of the fly to the eyes of a new host, at which time they exit, molt to the adult stage, mate, and resume the cycle. Invasive larvae are easy to identify because they have cuticular striations on their anterior cuticle (Fig. 21). For a review of *Thelazia* life cycles and a vector checklist, the reader should consult Skrjabin et al. (1967), Stoffolano (1970), or Geden (1979).

Pathogenicity to the definitive host varies along geographic lines within species, and between species worldwide. The most pathogenic eyeworm is *T. rhodesii* (Desmarest), an Old World species that produces a full spectrum of symptoms ranging from lacrimation, conjunctivitis, and photophobia (Okoshi and Kitano, 1967) to corneal ulcerations and permanent blindness (Divljanovic, 1958; Vorhadsky, 1970). In Senegal, Gretillat and Touré (1970) have reported *T. rhodesii* infection to result in 10-15% mortality among calves.

Eyeworm species in the United States include the bovine parasites *T. gulosa* Railliet and Henry and *T. skrjabini* Erschow, the equine parasite *T. lacrymalis* (Gurlt), and an indigenous species, *T. californiensis* Price, which parasitizes a wide range of mammals in the western United States. Bovine and equine eyeworms in North America tend to produce subclinical infections (Lyons and Drudge, 1975); however, conjunctival lesions associated with these parasites have been seen (Fig. 22) (Geden and Stoffolano, 1980).

As indicated above, most eyeworms are vectored by members of the genus *Musca*. The face fly (*M. autumnalis*) has been incriminated in transmission of *T. gulosa*, *T. skrjabini*, and *T. lacrymalis* in the United States (Chitwood and Stoffolano, 1971; Lyons et al., 1976; Geden and Stoffolano, 1981). Al-

---

Fig. 22 Bovine eye everted through lids to show conjunctival lesion found associated with *T. gulosa*.

Fig. 23 Skin lesion on cattle leg (region between arrows) caused by *Stephano-filaria* infection. (Courtesy of *J. Helminthol.*)

Fig. 24 "Summer bleeding" lesion on cattle infected with *Parafilaria bovicola*. (Courtesy of E. Nevill.)

Fig. 25 Skin lesion of cattle caused by tissue reaction to dead *P. bovicola*. (Courtesy of E. Nevill.)

though house fly (*M. domestica*) has been suggested as a vector in Czechoslovakia (Vilagiova, 1964) and India (Gupta, 1970), this fly appears to be an incompetent intermediate host in the United States (Geden and Stoffolano, 1981).

Two exceptions to the *Musca*-eyeworm association are *T. californiensis*, which is vectored by the anthomyiid *Fannia thelaziae* Turner (Weinmann et al., 1974; Turner, 1976, and the Oriental eyeworm of the dog, cat, and human, *T. callipaeda*, which is transmitted by a drosophilid fly, *Phortica variegata* Fall. (in Skrjabin et al., 1967).

2. *Habronema*

The genus *Habronema* consists of three species of spirurid parasites of equines: *H. muscae* (Carter), *H. megastoma* (Rudolphi), and *H. microstoma* (Schneider). The former generic name, *Draschia*, is occasionally encountered in the older literature.

Normally, the definitive host acquires the parasites by ingestion of third-stage invasive larvae deposited by flies feeding on various host secretions about the head of the animal. The house fly, *Musca domestica*, has been incriminated in the transmission of all three species. *H. muscae* and *H. megastoma* are also vectored by numerous other *Musca* species (Stoffolano, 1970) as well as by the sarcophagid *Sarcophaga misera* Wlk. (Johnston and Bancroft, 1920) and an unidentified species of *Pseudopyrellia* (*Orthellia*) (Hill, 1918). The principal vector of *H. microstoma* is the stable fly, *Stomoxys calcitrans* (Soulsby, 1968). One exception to the filth-fly-*Habronema* association comes from the early observations of Crawford (1926), who concluded that in the tropics drosophilids may play an important role in the life cycle of these parasites.

Upon reaching the stomach of the host, the larvae begin to feed and develop to the adult stage. *H. microstoma* and *H. muscae* feed on the surface of the gastric mucosa, with the head embedded in the gastric glands. Such infections may result in a chronic catarrhal gastritis, accompanied by the sloughing off of considerable amounts of mucus and ulcer formation (deJesus, 1963; Soulsby, 1968). *H. megastoma* is more pathogenic due to the burrowing of larvae deep into the stomach wall. Initially, individual larvae form small, craterlike nodules. As the parasites develop, these nodules expand and unite to form large tumors (Ashizawa et al., 1973). Macruz et al. (1973) have found ∝-hemolytic streptococci associated with such tumors in Brazil.

After developing for about two months in the host's stomach, adult female parasites deposit first-stage larvae that are passed out of the host along with excrement. These larvae enter developing fly larvae *per os*, penetrate the alimentary tract, and enter the hemocoel. They then migrate to the Malpighian tubules where, within a tissue capsule of unknown origin, they develop to invasive-stage larvae. By this time the vector has emerged as an adult. Nematode larvae escape from the capsules, migrate to the proboscis of the fly, and await the next feeding of the vector about the face of a horse.

Often, invasive larvae escape while the fly feeds on wounds or pools of host secretions on the skin of the animal. When this occurs, the larvae attempt to burrow through the skin, producing a form of dermatitis variously known as cutaneous habronemiasis, summer sores, granular dermatitis, Bursati, Esponje, and swamp cancer. This condition is particularly common during periods of peak vector abundance, and is most often seen on portions of the animal most likely to be injured, such as the legs and withers.

Because invasive larvae are unable to migrate from the hide to the stomach, such infections are dead ends for the parasites, with most larvae surviving less than four weeks. The lesions produced, however, ensure the con-

tinued presence of more larvae due to their attractiveness to flies. These lesions also render the host more susceptible to agents of secondary infection.

In some relatively rare instances, larvae have been found to migrate into
the lungs, where they produce fibrotic nodules around smaller brachioles
(Soulsby, 1968; Bain et al., 1969). In Brazil, Vasconcellos and Macruz (1973)
found three cases of *Habronema* parasites in the spleens of purebred English
racehorses. Adult *H. muscae* have also been found in the brain of horses
(Walker, in Bain et al., 1969).

3. *Stephanofilaria*

This genus of filariid parasites of large mammals is widespread in its distribution throughout Asia and North America. Groups of parasites are found under
the skin of the host where they produce a form of dermatitis known as stephanofilarial dermatitis, humpsore, earsore, Cascado, equine dhobie, filarial
sore, and Krian sore (Fig. 23).

Of the nine species of *Stephanofilaria*, the most common are *S. stilesi*
Chitwood, prevalent throughout the western United States and much of the
USSR and India, and *S. kaeli* Buckley and *S. assamensis* Pande, which are
restricted to Asia.

Vectors of these parasites are all muscids, with the particular vector
varying with parasite species and geographic area. *S. stilesi* is transmitted
in the United States by the horn fly, *Haematobia irritans* (Hibler, 1966), while
in the USSR, the European horn fly *Lyperosia titillans* (Ivashkin et al., 1972)
and the stable fly *Stomoxys calcitrans* (Golovanov et al., 1977) have been incriminated as vectors. *Stephanofilaria assamensis*, *S. kaeli*, and *S. zaheeri*
Singh are all transmitted by members of the genus *Musca* (Kono and Fukoyoshi, 1967; Dutt, 1971; Patnaik and Kumar, 1972; Fadzil, 1975).

Life history studies of these parasites have been rare. Patnaik (1973)
has studied the cycle of *S. assamensis* in the vector (*Musca conducens*) and
definitive host. Earlier, Hibler (1966) examined the development of *S. stilesi*
in horn fly and cattle.

Vectors feeding on exudate from lesions associated with adult parasites
ingest microfilariae, which for three days following are found in the crop of
the fly, three days in the foregut, and six days in the midgut, after which
they penetrate the gut wall and develop for 7-10 days in the thorax. The
invasive larvae then migrate to the head and proboscis of the fly, from which
they escape during vector feeding (Patnaik, 1973) on host wounds and sores.

Following inoculation with parasite larvae, the lesions first appear as
small hemorrhagic spots that later swell into larger (<15 cm diameter) areas
characterized by purulent discharge and encrustation with indented bleeding
cracks (Bhattacharjee and Dass, 1967; Loke and Ramachandran, 1967). On
reaching the adult stage, gravid females deposit microfilariae into the dermis, which are ingested by flies attracted to the lesions.

Fadzil (1977) and Dewan and Rahman (1970) have presented evidence
that pathogenic bacteria introduced by flies feeding about host wounds can
produce many of the symptoms of stephanofilarial dermatitis in the absence
of parasites. These workers suggest that initial lesion formation may be
accomplished by such pathogens, thus predisposing the region to successful
invasion by *Stephanofilaria* larvae.

4. *Parafilaria*

Two species, *Parafilaria multipapillosa* (Condamine and Drouilly) and *P. bovicola* deJesus, comprise this small genus of filariid parasites of horses and

cattle, respectively. *P. multipapillosa* is found in horses in the USSR, Eastern Europe, and the Middle East (Soulsby, 1968; Hedjazi and Mirzayans, 1978), and has been found in imported horses in England (Thomsett, 1968). *P. bovicola* has a similar range, which, in addition, includes portions of Asia and Africa (Nevill, 1975). It has also been reported in Sweden (Nilsson, 1978) and, on two occasions, in imported French cattle in Canada (Niilo, 1968; Webster and Wilkins, 1970).

Adult parasites live in dermal nodules of the host where they deposit microfilariae into the blood and other exudates which flow from the lesions. This condition is known as "summer bleeding" (Fig. 24). These larvae are ingested by flies attracted to the sores. Nevill (1975, 1979) has demonstrated *Musca lusoria* Wied. and *M. xanthomelas* Wied. to vector *P. bovicola* in South Africa. Development to the invasive-stage larva was found to require 14 days. Following deposition of invasive larvae, development to the adult stage required 7-10 months.

The precise mode of entry of invasive larvae is poorly understood. Larvae deposited on host skin migrate considerable distances before reaching adulthood, leading Nevill (1975) to suggest that inoculation may be via the orbital route, followed by larval migration to the neck and shoulder regions, where lesions are most commonly found (Fig. 25) (Viljoen, 1976). Chauhan et al. (1974) have reported finding an immature *P. bovicola* in the anterior chamber of the eye of an animal, lending support to this hypothesis.

Very little is known of the life cycle of *P. multipapillosa*. In the USSR, the vector is *Haematobia atripalpus* Bezzi, in which development to the invasive stage requires 10-15 days (Gnedina and Osipov, 1960).

5. *Others*

The filariid genus *Setaria* is a cosmopolitan group of parasites of the peritoneal cavity of ungulates. Microfilariae are released from adult females and enter the peripheral blood circulation where they are ingested by hematophagous flies. All species are vectored by mosquitoes, with the exception of *Setaria cervi* Maplestone, a parasite of deer and antelope, which is transmitted by *Haematobia stimulans* (Mg) in the USSR (Shol', 1971). The New World vector is uncertain. After being ingested, microfilariae molt to the second stage within 30-60 minutes in the alimentary tract, penetrate the gut wall, and migrate to the fat body, where they develop to invasive-stage larvae in 17-23 days. Larvae then migrate to the head of the vector and enter the blood of a new host when the fly takes a blood meal (Shol', 1971). Development to the adult stage requires five to six months and is preceded by a migration through the host's brain and spinal cord prior to reaching the peritoneum (Shol', 1969). The parasites are generally nonpathogenic, with occasional damage to host tissue resulting from the migration of preadult parasites (Pachauri, 1972).

Syngamus trachea (Montagu) is a syngamid parasite that occurs in the trachea of various domestic and wild birds. Adult parasites attach to the tracheal mucosa and feed on blood, often resulting in host anemia and emaciation. Eggs laid by females are coughed up, swallowed by the host, and passed out with excrement. Larvae hatch and may either infect the host directly by being swallowed, or first be ingested by any of a large number of invertebrate reservoir hosts, in which they encyst. When such a reservoir host is ingested by a bird, the larvae excyst, enter the host blood, and are carried to the lungs, from which they migrate to the trachea. Although the most important reservoir hosts of this parasite are earthworms (Clapham, 1935), fly larvae have also been found to play a role. Clapham (1939) found the house fly

(*Musca domestica*) and the blow fly (*Lucilia sericata* Meigen) to transmit *S. trachea*, and Baruš (1966) reported the lesser house fly (*Fannia canicularis*) and the anthomyiid [*Paregle cinerella* (Fallén)] to be effective reservoir hosts of this parasite.

The Hippoboscidae is a family of ectoparasitic flies known as "louse flies" that feed on the blood of mammals and birds. Nelson (1963) reported the only instance of nematode parasitism of this group in Kenya, where he found *Hippobosca longipennis* Fab. to be the vector of the filariid parasite *Dipetalonema dracunculoides* (Cobbold). This nematode is a parasite of the peritoneal membrane of dogs and hyenas in Africa. Microfilariae circulate in the blood and are ingested by *H. longipennis*, penetrate the gut, and migrate to the fat body, where development occurs. Invasive larvae migrate to the proboscis of the fly and presumably are introduced during vector feeding. Other members of this nematode genus are parasites of humans, primates, camels, and dogs and are transmitted by various mosquitoes, biting midges, fleas, and ticks.

III. CONCLUSION

Recent years have witnessed an increase in interest in dipteran-nematode relationships, bearing out Bovien's prediction in 1932 that such associations would ". . . offer a rich field for investigation." This renewed interest in nematode parasites of flies has grown out of the need for alternative control strategies for many dipterous pests of economic and medical-veterinary importance, due to the combined factors of insecticide resistance, increased cost of chemical control, and concern for the environment. In particular, biological control methods for such medical-veterinary pests as mosquitoes and blackflies are being sought as part of an international trend toward the application of integrated pest management (IPM) concepts and technologies to the control of insects affecting humans and domestic animals.

This emphasis on practical aspects of dipteran-nematode associations, however, has left many potential host groups unexplored. Thus, of the 105 families of true flies currently recognized, only 27% have been reported as hosts of nematodes, leaving many gaps in our understanding of the evolution of these associations.

Several evolutionary trends may be discerned nonetheless. Mermithid-fly associations abound in those families of flies that have relatively long-lived immature stages occurring in relatively permanent habitats, such as standing bodies of water and rivers, as is the case with many families of Nematocera. In the so-called higher dipteran families, which are generally characterized by the exploitation of more temporary, patchily distributed, terrestrial habitats, there is seen an accompanying shift toward nematode groups that utilize the strategies of persistence into the adult stage of the host, parasitic castration, and induced mock oviposition to ensure dispersal to the host's larval habitat. Also, alternation of sexual and parthenogenetic generations within the host is widely employed to guarantee sufficient production of nematode offspring to locate these more sparsely dispersed dipteran hosts.

A major exception to this scheme is the guild of blood- and tissue-feeding flies that serve as intermediate hosts for nematodes whose definitive hosts are the animals on which the flies feed. For these parasites, the common denominator of dipteran life history is the feeding habits of the adult fly, rather than larval habitat. Therefore, for example, nematode groups such as the filariids are vectored by such diverse dipteran families as the Simuliidae, Culicidae, Tabanidae, Muscidae, Anthomyiidae, and Hippoboscidae.

Clearly, a vast amount of work lies ahead for researchers interested in the area of dipteran-nematode associations. In the field of biological control of flies, major problems include the following: (1) development of mass-rearing techniques for the nematodes in question: (2) development of delivery systems that will ensure adequate parasite dispersal over the host habitat; (3) establishment of quality control standards for nematode colonies so that the parasites remain vigorous and aggressive in their host-seeking behavior; and (4) research on host-parasite genetics to determine the compatability of various strains of nematodes and their dipteran hosts.

In the area of nematodes that utilize flies as vectors of disease, major areas of future study might include: (1) development of a thorough understanding of the genetics and physiology of vector competence; (2) development of better vector survey techniques for epidemiological studies; (3) development of more sensitive diagnostic techniques for infections in the definitive host; and (4) development of superior prophylactic and curative pharmacological agents for such infections.

For many of the above areas that are open for investigation, the private sector should be urged to participate: biocontrol on a large scale and better treatment of nematode related medical-veterinary diseases will succeed only if it is practical and profitable for industry.

REFERENCES

Anthony, D. W. (1977). Pathogens of Tabanidae (horse flies). pp. 239-243. In *Pathogens of Medically Important Arthropods* D. W. Roberts and M. A. Strand, eds.). *Bull. WHO 55* (Suppl. No. 1): 419.

Ashizawa, H., Nosaka, D., Tateyama, S., and Usui, M. (1973). Stomach worm disease of horse caused by *Habronema megastoma*. III. Pathological findings on nodular foci caused by larvae. *Bull. Fac. Agric. Miyazaki Univ. 20*: 217-224.

Bacon, P. R. (1970). The natural enemies of the Ceratopogonidae – a review. *Tech. Bull. Commonwealth Inst. Biol. Control No. 13*, pp. 71-82.

Bain, A. M., Rofe, J. C., Hotson, I. K., and Murphy, S. (1969). *Habronema megastoma* larvae associated with pulmonary abscesses in a foal. *Aust. Vet. J. 45*: 101-102.

Baruš, V. (1966). Two new reservoir hosts of invasive larvae of the nematode *Syngamus trachea* (Montagu 1811) Chapin, 1925. *Helminthologica 7*: 323-327.

Bhattacharjee, M. L., and Dass, D. D. (1967). The study of ear sore of buffaloes due to stephanofilariasis in Assam. *Indian Vet. J. 44*: 400-402.

Borror, D. J., DeLong, D. M., and Triplehorn, C. A. (1981). *An Introduction to the Study of Insects*. Saunders, New York, p. 827.

Bovien, P. (1932). On a new nematode, *Scatonema wülkeri* gen. et sp. n. parasitic in the body-cavity of *Scatopse fuscipes* Meig. (Diptera Nematocera). *Vidensk. Medd. Dansk. Naturh. Foren. Kobenhavn. 94*: 13-32.

Bovien, P. (1937). Some types of association between nematodes and insects. *Vidensk. Medd. Dansk. Naturh. Foren. Kobenhavn. 101*: 1-257.

Bovien, P. (1944). *Proatractonema sciarae* n.g., n. sp., a parasitic nematode from the body cavity of a dipterous larva. *Vidensk. Medd. Dansk. Naturh. Foren. Kobenhavn. 108*: 1-14.

Buckley, J. J. C. (1938). On *Culicoides* as a vector of *Onchocerca gibsoni* (Cleland & Johnston, 1910). *J. Helminthol. 41*: 121-158.

Carter, J. B. (1976). A survey of microbial, insect and nematode parasites of Tipulidae (Diptera) larvae in north-east England. *J. Appl. Ecol. 13*: 103-122.

Chapman, H. C., Petersen, J. J., Woodard, D. B., and Clark, T. B. (1968). New records of parasites of Ceratopogonidae. *Mosq. News 28*: 122-123.

Chapman, H. C., Woodard, D. B., and Petersen, J. J. (1967). Nematode parasites of Culicidae and Chaoboridae in Louisiana. *Mosq. News 27*: 490-492.

Chapman, J., and Ecke, D. H. (1969). Study of a population of chironomid midges (*Tanytarsus*) parasitized by mermithid nematodes in Santa Clara County, California. *Calif. Vector Views 16*: 83-88.

Chauhan, P. P. S., Arora, G. S., and Ahluwalia, S. S. (1974). A note on the occurrence of an immature parafilariid worm in the anterior chamber of the eye of a buffalo (*Bubalis bubalis*). *J. Helminthol. 48*: 289-291.

Cheng, H. H., and Bucher, G. E. (1972). Field comparison of the neoaplectanid nematode DD-136 with Diazinon for control of *Hylemya* spp. on tobacco. *J. Econ. Entomol. 65*: 1761-1763.

Chitwood, M. B., and Stoffolano, Jr., J. G. (1971). First report of *Thelazia* sp. in the face fly, *Musca autumnalis*, in North America. *J. Parasitol. 57*: 1363-1364.

Clapham, P. A. (1935). The treatment of gapeworm diseases. *J. Helminthol. 13*: 3-8.

Clapham, P. A. (1939). On flies as intermediate hosts of *Syngamus trachea*. *J. Helminthol. 17*: 61-64.

Clark, G. C., and Hibler, C. P. (1973). Horse flies and *Elaeophora schneideri* in the Gila National Forest, New Mexico. *J. Wildlife Dis. 9*: 21-25.

Cobb. N. A. (1919). *Tetradonema plicans* nov. gen. et spec., representing a new family, Tetradonematidae as now found parasitic in larvae of the midge-insect *Sciara coprophila* Lintner. *J. Parasitol. 5*: 176-185.

Coler, R. R., Stoffolano, Jr., J. G. and Barreto, S. P. (1980). *Heterotylenchus* sp. (Nematoda: Sphaerulariidae), a nematode parasite of house fly, *Musca domestica* L. (Diptera: Muscidae). *Proc. Helm. Soc. Wash. 47*: 135-136.

Craig, D. A. (1963). The occurrence of nematodes in the family Blepharoceridae (Diptera). *N. Z. Entomol. 3*: 25.

Crawford, M. (1926). Development of *Habronema* larvae in drosophilid flies. *J. Comp. Pathol. Ther. 39*: 321-323.

Currie, G. A. (1937). Galls on eucalyptus trees. A new type of association between flies and nematodes. *Proc. Linn. Soc. N.S.W. 62*: 147-174.

Dewan, M. L., and Rhaman, M. M. (1970). Isolation of microorganisms from stephanofilariasis (humpsore) and their roles in the initiation of the disease. *Bangladesh Vet. J. 4*: 25-30.

Divljanovic, D. K. (1958). Eye lesions in cattle associated with *Thelazia rhodesii* infestation in Yugoslavia. *Vet. Glasn. 12*: 1017-1018.

Duke, B. O. L. (1954). The transmission of loiasis in the forest-fringe area of the British Cameroons. *Ann. Trop. Med. Parasitol. 48*: 349-355.

Dutky, S. R., Thompson, J. V., and Hough, W. S. (1962). A new nematode parasite of codling moth showing promise in insect control. Susceptibility of various insects to DD-136 nematode (mimeogr. abstr.). *8th Int. Congr. Microbiol. Montreal*, 1962.

Dutt, S. C. (1971). Preliminary studies on the life history of *Stephanofilaria zaheeri* Singh, 1958. *Indian J. Helminthol. 22*: 139-143.

El-Kifl, A. H., Abdel-Wahab, A. I., and Ali, M. A. M. (1970). Some insects parasitized by nematodes in Egypt. *Bull. Soc. Entomol. Egypt 54*: 87-89.

Fadzil, M. (1975). The development of *Stephanofilaria kaeli* Buckley, 1937 in *Musca conducens* Walker, 1859. *Kajian Vet. 7*: 1-7.

Fadzil, M. (1977). *Stephanofilaria kaeli* infection in cattle in Peninsular Malaysia: Prevalence and treatment. *Vet. Med. Rev. 1*: 44-52.

Fisher, J. M., and Nickle, W. R. (1968). On the classification and life history of *Fergusobia curriei* (Sphaerulariidae: Nematoda). *Proc. Helm. Soc. Wash. 35*: 40-46.

Foster, R. (1963). Infestation of *Glossina palpalis* R.D. 1830 (Diptera) by larval Mermithidae Braun 1883 (Nematoda) in West Africa, with some comments on the parasitization of man by the worms. *Ann. Trop. Med. Parasitol. 57*: 347-358.

Geden, C. J. (1979). Studies on the interactions of *Thelazia* sp., introduced eyeworm parasites of cattle, with their definitive and intermediate hosts in Massachusetts. M.S. thesis, University of Massachusetts, Amherst.

Geden, C. J., and Stoffolano, Jr., J. G. (1980). Bovine thelaziasis in Massachusetts. *Cornell Vet. 70*: 344-359.

Geden, C. J., and Stoffolano, Jr., J. G. (1981). Geographic range and temporal patterns of parasitization of *Musca autumnalis* by *Thelazia* sp. in Massachusetts, with observations on *Musca domestica* as an unsuitable intermediate host. *J. Med. Entomol. 18*: 449-456.

Geden, C. J., and Stoffolano, Jr., J. G. (1982). Development of the bovine eyeworm, *Thelazia gulosa*, in experimentally infected female *Musca autumnalis*. *J. Parasitol. 68*: 287-292.

Gnedina, M. P., and A. N. Osipov. (1960). The biology of the causative agent of parafilariasis in horses. *Veterinarya Moscow 37*: 49-50.

Golovanov, V. I., Zimin, Y. M., Dadaev, S., Isakova, D. T., Azimov, D. A., and Tukhmanyants, A. A. (1977). The role of some Diptera as intermediate hosts of nematodes of domestic animals in Uzbekistan. *Materialy Nauchnoi Konferentsii Vsesoyuznogo Obshchestva Gel 'mintologov 29*: 30-34.

Goodey, T. (1930). On a remarkable new nematode, *Tylenchinema oscinellae* gen. et sp. n., parasitic in the fritfly, *Oscinella frit* L., attacking oats. *Phil. Trans. R. Soc. Lond. 218*: 315-343.

Goodey, T. (1941). On the morphology of *Mermithonema entomophilum* n.g., n. sp., a nematode parasite of the fly *Sepsis cynipsea* L. *J. Helminthol. 19*: 105-114.

Gorden, R., and Webster, J. M. (1974). Biological control of insects by nematodes. *Helminthol. Abst. 43*: 327-349.

Gretillat, S., and Touré, S. (1970). First studies on the epidemiology of bovine thelaziasis and determination of the vector in West Africa. *C.R. Hebd. Seanc. Acad. Sci. Paris 270D*: 239-241.

Gupta, V. P. (1970). *Musca domestica* an intermediate host of two spiruroid parasites of livestock. *Orissa Vet. J. 5*: 147-150.

Harwood, R. F., and James, M. T. (1979). *Entomology in Human and Animal Health*. Macmillan, New York.

Hedjazi, M., and Mirzayans, A. (1978). Equine parafilariasis in Tehran district (Iran). Clinical aspects and treatment. *Rev. Med. Vet. 129*: 1685-1686; 1689-1691.

Heimpel, A. M. (1967). A critical review of *Bacillus thuringiensis* var. *thuringiensis* Berliner and other crystalliferous bacteria. *Ann. Rev. Entomol. 12*: 287-322.

Hibler, C. P. (1966). Development of *Stephanofilaria stilesi* in the horn fly. *J. Parasitol. 52:* 890-898.

Hibler, C. P., and Metzger, C. J. (1974). Morphology of the larval stages of *Elaeophora schneideri* in the intermediate and definitive hosts with some observations in their pathogenesis in abnormal definitive hosts. *J. Wildlife Dis. 10:* 361-369.

Hill, G. F. (1918). Relationship of insects to parasitic diseases in stock. *Proc. R. Soc. Vict. 31:* 11-107.

Hominick, W. M., and Welch, H. E. (1971). Synchronization of life cycles of three mermithids (Nematoda) with their chironomid (Diptera) hosts and some observations on the pathology of the infections. *Can. J. Zool. 49:* 975-982.

Hudson, K. E. (1974). Regulation of greenhouse sciarid fly populations using *Tetradonema plicans* (Nematoda: Mermithidae). *J. Invert. Pathol. 23:* 85-91.

Hungerford, H. B. (1919). Biological notes on *Tetradonema plicans* Cobb, a nematode parasite of *Sciara coprophila* Lintner. *J. Parasitol. 6:* 186-192.

Hussey, N. W. (1959). Biology of mushroom phorids. *Mushroom Sci. 4:* 260-269.

Hussey, N. W. (1965). Economics in pest control suggested by a study of phorid flight activity. *Mushroom Growers Assoc. Bull.* No. 182, pp. 71-83.

Ivashkin, V. M., Maklakova, C. P., and Veselkin, G. A. (1972). The study of flies associated with cattle and their role as intermediate hosts of helminths. In: Trudy VII. *Vsesoyuznoi Konferentsii Prirodnoy Ochagovosti Boleznei i Obshchim Voprosam Parasitologii Zhivotnykh*, 14-18 Oktyalorya 1969, Samarkand. *6:* 95-99.

Jenkins, D. W. (1964a). IV. Sandflies and Midges. In *Pathogens, Parasites and Predators of Medically Important Arthropods: Annotated List and Bibliography. World Health Organization*, Geneva, pp. 30-32.

Jenkins, D. W. (ed.) (1964b). Horseflies and deerflies-Tabanidae. In *Pathogens, Parasites and Predators of Medically Important Arthropods: Annotated List and Bibliography. Bull. WHO 30:* 35-39.

deJesus, Z. (1963). Observations on habronemiasis in horses. *Philipp. J. Vet. Med. 2:* 133-152.

Johnston, T. H., and Bancroft, M. J. (1920). The life history of *Habronema* in relation to *Musca domestica* and native flies of Queensland. *Proc. R. Soc. Queensland 32:* 61-88.

Jones, C. M., and Perdue, J. M. (1967). *Heterotylenchus autumnalis*, a parasite of the face fly. *J. Econ. Entomol. 60:* 1393-1395.

Kaya, H. K., and Moon, R. D. (1978). The nematode, *Heterotylenchus autumnalis* and face fly, *M. autumnalis*; a field study in northern California. *J. Nematol. 10:* 333-341.

Kaya, H. K., and Moon, R. D. (1980). Influence of protein in the diet of face fly (*Musca autumnalis*) on the development of its nematode parasite, *Heterotylenchus autumnalis. Ann. Entomol. Soc. Am. 73:* 547-552.

Kaya, H. K., Moon, R. D., and Witt, P. L. (1979). Influence of the nematode, *Heterotylenchus autumnalis* on the behavior of face fly, *Musca autumnalis. Environ. Entomol. 8:* 537-540.

Keilin, D., and Robinson, V. C. (1933). On the morphology and life history of *Aproctonema entomophagum* Keilin, a nematode parasite in the larva of *Sciara pullula* Winn. (Diptera-Nematocera). *Parasitology 25:* 285-295.

Kettle, D. S. (1965). Biting-ceratopogonids as vectors of human and animal diseases. *Acta Trop. 22*: 356-362.

Killick-Kendrick, R., Leaney, A. J., Molyneux, D. H., and Rioux, J. A. (1976). Parasites of *Phlebotomus ariasi*. *Trans. R. Soc. Trop. Med. Hyg. 70*: 22.

Kono, I., and Fukoyoshi, S. (1967). Leucoderma of the muzzle of cattle induced by a new species of *Stephanofilaria* II. *Jpn. J. Vet. Sci. 29*: 301-303.

Krinsky, W. L. (1976). Animal disease agents transmitted by horse flies and deer flies (Diptera: Tabanidae). *J. Med. Entomol. 13*: 225-275.

Lam, A. B. Q., and Webster, J. M. (1971). Morphology and biology of *Panagrolaimus tipulae* n.sp. (Panagrolaimidae) and *Rhabditis (Rhabditella) tipulae* n.sp. (Rhabditidae), from leather jacket larvae, *Tipula paludosa* (Diptera: Tipulidae). *Nematologica 17*: 201-212.

Lam, A. B. Q., and Webster, J. M. (1972). Effect of the DD-136 nematode and of a β-exotoxin preparation of *Bacillus thuringiensis* var. *thuringiensis* on leatherjackets, *Tipula paludosa* larvae. *J. Invert. Pathol. 20*: 141-149.

Laumond, C., and Lyon, J.-P. (1971). Le parasitisme de *Syrphonema intestinalis* n.g., n. sp., aux dépens des Syrphides (Insectes diptères) et la nouvelle famille des Syrphonematidae (Nematoda: Rhabditida). *C.R. Acad. Sci. Paris 272*: 1789-1792.

Leuckart, K. G. F. R. (1887). Neue Beitrage zur Kenntniss des Baues und der Lebensgeschichte der Nematoden. *Abh. Sachs. Akad. Wiss. Leipzig Math.-Naturwiss. Kl. 13*: 565-704.

Loke, Y. W., and Ramachandran, C. P. (1967). The pathology of lesions in cattle caused by *Stephanofilaria kaeli* Buckley, 1937. *J. Helminthol. 41*: 161-166.

Lyons, E. T., and Drudge, J. H. (1975). Two eyeworms, *Thelazia gulosa* and *Thelazia skrjabini*, in cattle in Kentucky. *J. Parasitol. 61*: 1119-1122.

Lyons, E. T., Drudge, J. H., and Tolliver, S. C. (1976). *Thelazia lacrymalis* in horses in Kentucky and observations on the face fly (*Musca autumnalis*) as a probable intermediate host. *J. Parasitol. 62*: 877-880.

Macruz, R., Giorgi, W., and Santos, M. R. S. (1973). Gastric habronemiasis in the horse. Bacteriological examination of the nodules. *Atualidades Vet. 9*: 54.

Magnarelli, L., and Anderson, J. (1978). Distribution and development of immature salt marsh Tabanidae (Diptera). *J. Med. Entomol. 14*: 573-578.

Mellor, P. S. (1973). Studies on *Onchocerca cervicalis* Railliet and Henry, 1910. II. Pathology in the horse. *J. Helminthol. 47*: 111-118.

Mellor, P. S. (1974). Studies on *Onchocerca cervicalis* Railliet and Henry, 1910. IV. Behaviour of the vector *Culicoides nubeculosus* in relation to the transmission of *Onchocerca cervicalis*. *J. Helminthol. 48*: 283-288.

Mellor, P. S. (1975). Studies on *Onchocerca cervicalis* Railliet and Henry 1910. V. The development of *Onchocerca cervicalis* larvae in the vectors. *J. Helminthol. 49*: 33-42.

Merrit, R. W., and Cummins, K. W. (eds.) (1978). *An Introduction to the Aquatic Insects of North America*. Kendall/Hunt Publ., So. Dubuque, Iowa.

Moloo, S. K. (1972). Mermithid parasite of *Glossina brevipalpis* Newstead. *Ann. Trop. Med. Parasitol. 66*: 159.

Nappi, A. J. (1973). Effects of parasitization by the nematode *Heterotylenchus autumnalis* on mating and oviposition in the host, *Musca autumnalis*. *J. Parasitol. 59*: 963-969.

Nelson, G. S. (1963). *Dipetalonema dracunculoides* (Cobbold, 1870), from the dog in Kenya; with a note on its development in the lousefly, *Hippobosca longipennis*. *J. Helminthol. 37*: 235-240.

Nelson, G. S. (1970). Onchocerciasis. In *Advances in Parasitology*, vol. 8 (B. Dawes, ed.). Academic, New York, pp. 173-224.

Nevill, E. M. (1975). Preliminary report on the transmission of *Parafilaria bovicola* in South Africa. *Onderstepoort J. Vet. Res. 42*: 41-48.

Nevill, E. M. (1979). The experimental transmission of *Parafilaria bovicola* to cattle in South Africa using *Musca* species (subgenus *Eumusca*) as intermediate hosts. *Onderstepoort J. Vet. Res. 46*: 51-57.

Nicholas, W. L., and Hughes, R. D. (1970). *Heterotylenchus* sp. (Nematoda: Sphaerulariidae), a parasite of the Australian bush fly, *Musca vetustissima*. *J. Parasitol. 56*: 116-122.

Nickle, W. R. (1967). *Heterotylenchus autumnalis* sp. n. (Nematoda: Sphaerulariidae), a parasite of the face fly, *Musca autumnalis* de Geer. *J. Parasitol. 53*: 398-401.

Nickle, W. R. (1969). *Corethrellonema grandispiculosum* n. gen., n. sp. and *Aproctonema chapmani* n. sp. (Nematoda: Tetradonematidae), parasites of the dipterous insect genera, *Corethrella* and *Culicoides* in Louisiana. *J. Nematol. 1*: 49-54.

Nickle, W. R. (1972). A contribution to our knowledge of the Mermithidae (Nematoda). *J. Nematol. 4*: 113-146.

Nickle, W. R. (1974). Nematode infections. *Insect Diseases*, Vol. II (G. E. Cantwell, ed.). Marcel Dekker, New York,

Niilo, L. (1968). Bovine hemorrhagic filariasis in cattle imported into Canada. *Can. Vet. J. 9*: 132-137.

Nilsson, N. G. (1978). *Parafilaria bovicola* – a working party report. *Svensk Vet. 30*: 785-787.

Okoshi, S., and Kitano, N. (1967). Studies on thelaziasis of cattle. III. *Thelazia rhodesii* in Japan. *Jpn. J. Vet. Sci. 29*: 1-10.

Orihel, T. C., and Lowrie, Jr., R. C. (1975). *Loa loa*: Development to the infective stage in an American deerfly, *Chrysops atlanticus*. *Am. J. Trop. Med. Hyg. 24*: 610-615.

Pachauri, S. P. (1972). Cerebrospinal nematodiasis in a buffalo. *Indian J. Anim. Res. 6*: 17-19.

Patnaik, B. (1973). Studies on stephanofilariasis in Orissa. III. Life cycle of *S. assamensis*. *Z. Tropenmed. Parasitol. 24*: 457-466.

Patnaik. B., and Kumar, V. (1972). A research note on the occurrence of juveniles of *Stephanofilaria*, the causative parasite of ear-sore in buffaloes, in *Musca autumnalis* de Geer (1776). *Indian J. Anim. Sci. 42*: 351-352.

Phelps, R. J., and Mokry, J. E. (1976). *Culicoides* species (Diptera: Ceratopogonidae) as potential vectors of filariasis in Rhodesia. *J. Entomol. Soc. South Africa. 39*: 201-206.

Poinar, G. O., Jr. (1964). A new nematode, *Orthomermis oedobranchus* gen. n., sp. n. (Mermithidae) parasitizing *Smittia* larvae (Chironomidae) in England. *Nematologica 10*: 501-506.

Poinar, G. O., Jr. (1965). The bionomics and parasite development of *Tripius sciarae* (Bovien) (Sphaerulariidae: Aphelenchoidea), a nematode parasite of sciarid flies (Sciaridae: Diptera). *Parasitology 55*: 559-569.

Poinar, G. O., Jr. (1966). The presence of *Achromobacter nematophilus* in the infective stage of a *Neoaplectana* sp. (Steinernematidae: Nematoda). *Nematologica 12*: 105-108.

Poinar, G. O., Jr. (1968). *Hydromermis conopophaga*, n.sp., parasitizing midges (Chironomidae) in California. *Ann. Entomol. Soc. Am. 61*: 593-598.

Poinar, G. O., Jr. (1975). *Entomogenous Nematodes – a Manual and Host List of Insect-Nematode Associations.* E. J. Brill, Leiden, Netherlands.

Poinar, G. O., Jr. (1979). *Nematodes for Biological Control of Insects.* CRC Press, Boca Raton, Florida.

Poinar, G. O., Jr., and Bai, G. (1979). *Panagrolaimus migophilus* sp.n. associated with *Musca domestica* (Diptera: Muscidae) in India. *Indian J. Nematol. 9*: 1-4.

Poinar, G. O., Jr., and Doncaster, C. C. (1965). The penetration of *Tripius sciarae* (Bovien) (Sphaerulariidae: Aphelenchoidea) into its insect host, *Bradysia paupera* Thom. (Mycetophilidae: Diptera). *Nematologica 11*: 73-78.

Poinar, G. O., Jr., and Lane, R. S. (1978). *Pheromermis myopis* sp.n. (Nematoda: Mermithidae), a parasite of *Tabanus punctifer* (Diptera: Tabanidae). *J. Parasitol. 64*: 440-444.

Poinar, G. O., Jr., and Lindhardt, K. (1971). The re-isolation of *Neoaplectana bibionis* Bovien (Nematodea) from Danish bibionids (Diptera) and their possible use as biological control agents. *Entomol. Scand. 2*: 301-303.

Poinar, G. O., Jr., Thomas, G., Prokopy, R. J. (1977). Microorganisms associated with *Rhagoletis pomonella* (Tephritidae; Diptera) in Massachusetts. *Proc. Entomol. Soc. Ontario 108*: 19-22.

Richardson, P. N., and Chanter, D. O. (1979). Phorid fly (Phoridae: *Megaselia halterata*) longevity and the dissemination of nematodes (Allantonematidae: *Howardula husseyi*) by parasitized females. *Ann. Appl. Biol. 93*: 1-11.

Richardson, P. N., and Hesling, J. J. (1977). Studies on an overwintering population of the mushroom phorid *Megaselia halterata* (Diptera: Phoridae) parasitized by *Howardula husseyi* (Nematoda: Allantonematidae). *Ann. Appl. Biol. 86*: 321-327.

Riding, I. L., and Hague, N. G. M. (1974). Some observations on a Tylenchid nematode *Howardula* sp. parasitizing the mushroom phorid *Megaselia halterata* (Phoridae, Diptera). *Ann. Appl. Biol. 78*: 205-211.

Robinson, E. J., Jr. (1971). *Culicoides crepuscularis* (Malloch) (Diptera: Ceratopogonidae) as a host for *Chandlerella quiscali* (Von Linstow, 1904) Comb. n. (Filarioidea: Onchocercidae). *J. Parasitol. 57*: 772-776.

Robinson, J. V., and Combs, Jr., R. L. (1976). Incidence and effect of *Heterotylenchus autumnalis* on the longevity of face flies in Mississippi. *J. Econ. Entomol. 69*: 722-724.

Rubstov, I. A. (1970). A new species and genus of mermithids from biting midges (in Russian). In *New and Little Known Species of the Fauna of Siberia. Novosibirsk Sect. 3*: 94-101.

Rubstov, I. A. (1976). Mermithids (Nematoda: Mermithidae) in stable flies (Diptera: Muscidae) in Mongolia. In *Nasekomye Mongolii, Akad. Nauk SSR, Zoologicheskiy Institut*, No. 4: 615-621.

Shol', V. A. (1969). Development of *Setaria altaica* in maral deer. *Isv. Akad. Nauk Kazakh. SSR, Ser. Biol. Nauk 7*: 45-50.

Shol', V. A. (1971). Development of *Setaria cervi* from maral deer in the fly *Haematobia stimulans. Dokl. Akad. Nauk SSSR 199*: 503-504.

Skrjabin, K. I., Sobolev, A. A., and Ivashkin, V. M. (1967). *Principles of Nematology*, vol. XVI. Spirurata of animals and man and the diseases caused by them. Part 4: Thelazioidea. Published by the Israel Program for Scientific Translation, pp. 1-54. (1971).

Slobodyanyuk, O. V. (1975a). *Heterotylenchus simplex* n. sp. (Nematoda, Sphaerulariidae), a parasite of the zoophilous fly *Morellia simplex*. *Parasitologiya 9*: 127-134.

Slobodyanyuk, O. V. (1975b). Erection of *Paraiotonchium* n.g. (Nematoda: Sphaerulariidae) and redescription of the type species, *P. autumnalis* (Nickle, 1967) n. comb. *Trudy Gel-mintol. Lab. 25*: 156-168.

Slobodyanyuk, O. V. (1976). *Paraiotonchium osiris* (Iotonchiinae: Tylenchida) – a new species of nematodes from *Musca osiris* Wd. *Parasitologiyae 10*: 30-39.

Soulsby, E. J. L. (1968). *Helminths, Arthropods and Protozoa of Domesticated Animals*. Williams and Wilkins, Baltimore.

Spratt, D. M. (1970). The synonymy of *Agamofilaria tabanicola* and *Dirofilaria roemeri*. *J. Parasitol. 56*: 622-623.

Spratt, D. M. (1974). Comparative epidemiology of *Dirofilaria roemeri* infection in two regions of Queensland. *Int. J. Parasitol. 4*: 481-488.

Stannard, A. A., and Cello, R. M. (1975). *Onchocerca cervicalis* infection in horses from the Western United States. *Am. J. Vet. Res. 36*: 1029-1031.

Steward, J. S. (1935). Fistulous withers and poll-evil. Equine and bovine onchocerciasis compared, with an account of the fly-histories of the parasites concerned. *Vet. Rec. 15*: 1563-1575.

Stoffolano, J. G., Jr. (1967). The synchronization of the life cycle of diapausing face flies, *Musca autumnalis*, and of the nematode, *Heterotylenchus autumnalis*. *J. Invert. Pathol. 9*: 395-397.

Stoffolano, J. G., Jr. (1969). Nematode parasites of the face fly and the onion maggot in France and Denmark. *J. Econ. Entomol. 62*: 792-795.

Stoffolano, J. G., Jr. (1970). Nematodes associated with the genus *Musca*. *Bull. Entomol. Soc. Am. 16*: 194-203.

Stoffolano, J. G., Jr. (1973). Host specificity of entomophilic nematodes – a review. *Exp. Parasitol. 33*: 263-284.

Stoffolano, J. G., Jr., and Nickle, W. R. (1966). Nematode parasite (*Heterotylenchus* sp.) of face fly in New York State. *J. Econ. Entomol. 59*: 221-222.

Stoffolano, J. G., Jr., and Streams, F. A. (1971). Host reactions of *Musca domestica*, *Orthellia caesarion*, and *Ravinia l'herminieri* to the nematode *Heterotylenchus autumnalis*. *Parasitology 63*: 195-211.

Stone, A., Sabrosky, C. W., Wirth, W. W., Foote, R. H., and Coulson, J. R. (1965). *A Catalog of the Diptera of America North of Mexico*. USDA, Agric. Res. Serv.

Thomas, G. D., and Puttler, B. (1970). Seasonal parasitism of the face fly by the nematode *Heterotylenchus autumnalis* in Central Missouri, 1968. *J. Econ. Entomol. 63*: 1922-1923.

Thomas, G. D., Puttler, B., and Morgan, C. E. (1972). Further studies of field parasitism of the face fly by the nematode *Heterotylenchus autumnalis* in central Missouri, with notes on the gonotrophic cycles of the face fly. *Environ. Entomol. 1*: 759-763.

Thomsett, L. R. (1968). *Parafilaria multipapillosa* in the horse. *Vet. Rec. 83*: 27-28.

Treece, R. E., and Miller, T. A. (1968). Observations on *Heterotylenchus autumnalis* in relation to the face fly. *J. Econ. Entomol. 61*: 454-456.

Turner, W. J. (1976). *Fannia thelaziae*, a new species of eye-frequenting fly of the *benjamini* group from California and description of *F. conspicua* female. *Pan-Pacific Entomol. 52*: 234-241.

Vasconcellos, S. A., and Macruz, R. (1973). Parasitological survey in equines. *Atualidades Venterinarias 9*: 48.

Vilagiova, I. (1964). K otázke sézonnej dynamiky invadovanosti medzihostitelov telázú hovädzieho dobytaka. *Biologia Bratislava 19*: 126-129.

Vilagiova, I. (1968). *Heterotylenchus autumnalis* Nickle (1967) – a parasite of pasture flies. *Biologia Bratislava 23*: 397-400.

Viljoen, J. H. (1976). Studies on *Parafilaria bovicola* (Tubangui 1934). I. Clinical observations and chemotherapy. *J. South Africa Vet. Assoc. 47*: 161-169.

Vorhadsky, F. (1970). Clinical course of *Thelazia rhodesii* infection in Accra plains of Ghana. *Bull. Epizoot. Dis. Afr. 18*: 159-170.

Votava, C. I., and Thompson, P. E. (1978). *Onchocerca lienalis* of cattle in Georgia. *J. Parasitol. 59*: 938-939.

Wachek, F. (1955). Die entoparasitischen Tylenchiden. *Parasitol. Schr. 3*: 1-119.

Webster, W. A., and Wilkins, D. B. (1970). The recovery of *Parafilaria bovicola* Tubangui, 1934 from an imported Charlais bull. *Can. Vet. J. 11*: 13-14.

Weinmann, C. J., Anderson, J. R., Rubtzoff, P., Connally, C., and Longhurst, W. M. (1974). Eyeworms and face flies in California. *Calif. Agric. 28*: 4-5.

Welch, H. E. (1959). Taxonomy, life cycle, development, and habits of two new species of Allantonematidae (Nematoda) parasitic in drosophilid flies. *Parasitology 49*: 83-103.

Welch, H. E. (1960). *Hydromermis churchillensis* n.sp. (Nematoda: Mermithidae) a parasite of *Aedes communis* (DeG.) from Churchill, Manitoba, with observations of its incidence and bionomics. *Can. J. Zool. 38*: 465-474.

Welch, H. E., and Briand, L. J. (1961). Field experiment on the use of a nematode for the control of vegetable crop insects. *Proc. Entomol. Soc. Ont. 91*: 197-202.

Wirth, W. W. (1977). Pathogens of Ceratopogonidae (midges). In *Pathogens of Medically Important Arthropods* (D. W. Roberts and M. A. Strand, eds.). World Health Org., Geneva. pp. 197-212.

Wülker, W. (1961). Untersuchungen über die Intersexualitat der Chironomiden nach Paramermis-Infektion. *Arch. Hydrobiol. Suppl. 25*: 127-181.

Yao, Y. T., Wu, C. C., and Sun, C. J. (1938). The development of microfilaria of *Wuchereria bancrofti* in the sandfly, *Phlebotomus sergenti* var. *mongolensis*. A preliminary report. *Chinese Med. J. Suppl. 2*: 401-410.

Young, D. G., and Lewis, D. J. (1977). I. Pathogens of Psychodidae (Phlebotomine sand flies). In *Pathogens of Medically Important Arthropods* (D. W. Roberts and M. A. Strand, eds.). World Health Organization, Geneva, pp. 9-24.

Index